T0201802

Introduction to Engineering Heat Transfer

This new text integrates fundamental theory with modern computational tools such as EES, MATLAB, and FEHT to equip students with the essential tools for designing and optimizing real-world systems and the skills needed to become effective practicing engineers. Real engineering problems are illustrated and solved in a clear step-by-step manner. Starting from first principles, derivations are tailored to be accessible to undergraduates by separating the formulation and analysis from the solution and exploration steps to encourage a deep and practical understanding. Numerous exercises are provided for homework and self-study and include standard hand calculations as well as more advanced project-focused problems for the practice and application of computational tools. Appendices include reference tables for thermophysical properties, and answers to selected homework problems from the book. Complete with an online package of guidance documents on EES, MATLAB, and FEHT software, sample code, lecture slides, video tutorials, and a test bank and full solutions manual for instructors, this is an ideal text for undergraduate heat transfer courses and a useful guide for practicing engineers.

G. F. Nellis is Professor of Mechanical Engineering at the University of Wisconsin, Madison. His teaching expertise has been recognized through awards including the Polygon Engineering Council Outstanding Professor of Mechanical Engineering Award (2013 and 2007), the Pi Tau Sigma Distinguished Professor of Mechanical Engineering Award (2016, 2012, 2009, and 2006), and the J. G. Woodburn award for Excellence in Teaching (2008). He is a Fellow of the American Society of Heating, Refrigeration, and Air-Conditioning Engineers.

S. A. Klein is Emeritus Professor of Mechanical Engineering at the University of Wisconsin, Madison. He is the recipient of the American Society for Mechanical Engineers (ASME) James Harry Potter Gold Medal (2013), the Pi Tau Sigma Distinguished Professor of Mechanical Engineering Award (1991, 1992) and the Polygon Engineering Council Outstanding Professor of Mechanical Engineering Award (1991, 1992). He is a Fellow of ASME, the American Society of Heating, Refrigeration, and Air-Conditioning Engineers (ASHRAE), the American Solar Energy Society (ASES), and the International Building Performance Simulation Association (IBPSA).

"This excellent text on heat transfer continues the tradition of the strong analytical treatment of conduction and convection heat transfer, buttressed by strong EES, FEHT, and MATLAB examples . . . The emphasis on examples is substantial, and the use of the software is tastefully introduced in ways that emphasize the solution instead of the software . . . This edition is well organized, succinctly written, and well supported by software aids. The book is also a valuable reference for those in a wide variety of disciplines desiring to self-learn heat transfer. All the essential elements of a heat transfer course are well represented in this volume."

Ernest W. Tollner, University of Georgia

"No other text spells out real-world problems with computer-based solutions as clearly as this one. This text will allow readers to translate quickly heat transfer lessons learned into interesting applied solutions."

Thomas Merrill, Rowan University

"I've practiced heat transfer for 30 years as an engineer in industry, a scientist at a national lab, and an academic. Midway through my career, I studied Nellis and Klein's pedagogically pioneering text. It was only then that I obtained a firm grasp of the subject matter. Feedback from students in my classes on their book has been remarkably terrific."

Marc Hodes, Tufts University

Introduction to Engineering Heat Transfer

G. F. Nellis
University of Wisconsin, Madison

S. A. Klein
University of Wisconsin, Madison

CAMBRIDGE
UNIVERSITY PRESS

University Printing House, Cambridge CB2 8BS, United Kingdom

One Liberty Plaza, 20th Floor, New York, NY 10006, USA

477 Williamstown Road, Port Melbourne, VIC 3207, Australia

314–321, 3rd Floor, Plot 3, Splendor Forum, Jasola District Centre, New Delhi – 110025, India

79 Anson Road, #06–04/06, Singapore 079906

Cambridge University Press is part of the University of Cambridge.

It furthers the University's mission by disseminating knowledge in the pursuit of education, learning, and research at the highest international levels of excellence.

www.cambridge.org
Information on this title: www.cambridge.org/9781107179530

First published 2021

Printed in Singapore by Markono Print Media Pte Ltd 2021

A catalogue record for this publication is available from the British Library.

Library of Congress Cataloging-in-Publication Data
Names: Nellis, Gregory, 1969, author. | Klein, Sanford A., 1950, author.
Title: Introduction to engineering heat transfer / G. F. Nellis (University of Wisconsin, Madison),
    S. A. Klein (University of Wisconsin, Madison).
Description: Cambridge, United Kingdom ; New York, NY : Cambridge University Press, 2020. |
    Includes bibliographical references and index.
Identifiers: LCCN 2019013363 | ISBN 9781107179530 (hardback ; alk. paper) |
    ISBN 110717953X (hardback ; alk. paper)
Subjects: LCSH: Heat–Transmission.
Classification: LCC TJ260 .N453 2019 | DDC 621.402/2–dc23
LC record available at https://lccn.loc.gov/2019013363

ISBN 978-1-107-17953-0 Hardback

Additional resources for this publication at www.cambridge.org/nellisklein

Contents

Preface

The objective of this book is to provide engineering students with the capability, tools, and confidence to solve real-world heat transfer problems. This objective has resulted in a textbook that differs from existing heat transfer textbooks in an important way. This textbook introduces fundamental heat transfer concepts at an introductory, undergraduate level that is appropriate for a practicing engineer and integrates these concepts with modern computational tools. The text provides extensive examples and problems that utilize these tools. The practicing engineer of today is expected to be proficient with computer tools; engineering education must evolve accordingly. Most real engineering problems cannot be solved using a sequential set of calculations that can be easily carried out with a pencil and a hand calculator. Engineers must have the ability and confidence to utilize the powerful computational tools that are available and essential for design and optimization of real-world systems.

The text reinforces good engineering problem solving technique by delineating the formulation and analysis steps from the solution and exploration steps. In the formulation step, the problem itself is defined and, through appropriate approximations, simplified to the point where it can be represented by a set of mathematical equations. These equations are derived from first principles in the analysis step. Many textbooks stop their presentation at this point. However, the solution step where the equations are solved is equally important. In some cases hand calculations are appropriate for solving the equations. More typically, the complexity of the problem dictates that some type of computational software must be used for the solution step. Each of these steps is essential. It is not possible to move to the solution step until the formulation and analysis steps are complete. Separating these steps forces the student to understand that the computational software cannot be used to "think" for them, but rather provide powerful tools for helping them solve the relevant equations. Computational software is essential for the exploration step in which the engineer carries out parametric, optimization, and design studies that allow a deeper understanding of the problem and provide more useful results. Exploration studies are a natural first step to becoming an effective practicing engineer.

This book integrates the computational software Engineering Equation Solver (EES), MATLAB, and Finite Element Heat Transfer (FEHT) directly with the heat transfer material so that students can see the relevance of these tools. The specific commands and output associated with these software packages are used in the solution and exploration steps of numerous examples so that the integration is seamless and does not detract from the presentation of the heat transfer concepts. The computational software tools used in this book are all common in industry and have existed for more than a decade; therefore, while this software will certainly continue to evolve, it is not likely to disappear. Educational versions of these software packages are available and therefore the use of these tools should not represent an economic hardship to any academic institution or student. These tools are easy to learn and use, allowing students to become proficient with all of them in a reasonable amount of time. Therefore, learning the computer tools will not detract from material coverage. In fact, providing the capability to easily solve the equations developed in the analysis is a motivator to many students. To facilitate this learning process, tutorials for each of the software packages are provided as appendices in this book.

Traditionally, tables and charts have been required to solve heat transfer problems in order to, for example, determine properties, view factors, shape factors, convection relations, and related information. Limited versions of these tables and graphs are provided in the textbook; however, much more extensive libraries have been made available as functions and procedures in the EES software so that they can be easily accessed and used to solve problems. The Heat Transfer Library that has been developed and integrated with EES as part of the preparation of this textbook and the more advanced textbook, *Heat Transfer*, enables a profound shift in the focus of the educational process. It is trivial to obtain, for example, the value of a shape factor or a view factor using the Heat Transfer Library. Therefore, it is possible to assign problems involving design and optimization studies that would be computationally impossible without these computer tools.

Integrating the study of heat transfer with computer tools does not diminish the depth of understanding of the underlying physics that students obtain. Conversely, our experience indicates that the innate understanding of the subject matter is enhanced by appropriate use of these tools for several reasons. First, the software allows the student to tackle practical and relevant problems as opposed to the comparatively simple problems that must otherwise be assigned. Real-world engineering problems are more satisfying to the student. Therefore, the marriage of computer tools with theory motivates students to understand the governing physics as well as to learn how to apply the computer tools. When a solution is obtained, students can carry out a more extensive investigation of its behavior and therefore a more intuitive and complete understanding of the subject of heat transfer. Along with the typical homework problems, each chapter includes several project type problems that allow a guided exploration of advanced topics using computer tools. Real-world problems often require a combination of English and SI units. The EES software provides unit checking that should prevent the student (and practicing) engineer from making unit conversion errors. Therefore, the examples and problems in this book use mixed units.

This book is unusual in its linking of classical theory and modern computing tools. It fills an obvious void that we have encountered in teaching undergraduate heat transfer. The text was developed over many years from our experiences teaching Introduction to Heat Transfer (an undergraduate course) at the University of Wisconsin. It is our hope that this text will not only be useful during the heat transfer course, but also a life-long resource for practicing engineers.

Sample Program of Study

A sample program of study is laid out below for a one-semester undergraduate course. The format assumes that there are 45 lectures within a 15-week semester.

Lecture	Sections in book	Topics
1	Chapter 1	Introduction
2	2.1–2.2.2	Fourier's Law, 1-D steady-state conduction
3	2.2.3–2.2.5	Resistance concepts and circuits
4	2.3	1-D steady-state with generation
5	2.4	Numerical solutions
6	3.1–3.2	Extended surface approximation and analytical solution
7	3.3	Fin behavior, fin efficiency, and finned surfaces
8	3.4	Numerical solution to extended surface problems
9	4.1–4.2	2-D steady-state conduction, shape factors
10	4.3.1–4.3.3	Finite difference solutions with EES
11	4.3.4–4.3.5	Finite difference solutions using matrix decomposition and Gauss–Seidel iteration
12	4.4	Finite element solutions
13	5.1–5.2	Lumped capacitance approximation and analytical solution
14	5.3	Numerical solution to lumped capacitance problems
15	6.1	1-D transient conduction concepts
16	6.2	Analytical solutions to 1-D transient problems
17	6.3	Numerical solutions to 1-D transient problems
18	6.4.3	Finite element solution to 2-D transient problems
19	7.1–7.2	Laminar and turbulent boundary layer concepts
20	7.3–7.4	The boundary layer equations and dimensional analysis
21	8.1–8.2	External flow correlations and flow over a flat plate
22	8.3–8.5	Flow over extrusions and spheres
23	9.1.1	Internal flow hydrodynamic concepts
24	9.1.2	Internal flow thermal concepts
25	9.2	Internal flow correlations
26	9.3	The energy balance for an internal flow
27	10.1–10.2	Free convection concepts and dimensionless parameters
28	10.3–10.4	Free convection correlations
29	10.5	Combined free and forced convection
30	11.1–11.2	Pool boiling
31	11.3–11.5	Boiling and condensation correlations
32	12.1–12.2	Heat exchanger configurations & concepts
33	12.3	Log-mean temperature difference method
34	12.4.1–12.4.4	Effectiveness–NTU method
35	12.4.5	Behavior of ε-NTU solutions and heat exchanger design
36	13.1–13.2	Introduction to mass transfer and mass diffusion
37	13.3	Diffusion in a stationary solid
38	13.4	Diffusion in a fluid
39	13.5–13.6	Mass transfer analogies and simultaneous heat and mass transfer

Nomenclature

A	area (m^2)
$\underline{\underline{A}}$	the coefficient matrix in a system of linear equations
A_c	cross-sectional area (m^2)
A_p	projected area (m^2)
A_s	surface area (m^2)
$A_{s,fin}$	surface area of a single fin exposed to fluid (m^2)
$A_{s,fins}$	surface area of all of the fins on a finned surface (m^2)
$A_{s,prime}$	surface area of the base of a finned surface that is exposed to fluid (m^2)
$A_{s,total}$	total surface area of fins and base exposed to fluid (m^2)
AR	aspect ratio of a rectangular duct, defined as the ratio of the minimum to the maximum dimensions of the cross-section
AR_{tip}	tip to perimeter surface area ratio for a fin (-)
$\underline{b}$	the constant vector in a system of linear equations
Bi	Biot number (-)
Bo	boiling number (-)
c	specific heat capacity (J/kg-K) speed of light (299,792,000 m/s)
c_v	specific heat capacity at constant volume (J/kg-K)
c_p	specific heat capacity at constant pressure (J/kg-K)
C	thermal capacitance (J/K)
$\dot{C}$	capacitance rate (W/K)
C_1, C_2	undetermined constant of integration (varies)
C_{crit}	critical heat flux constant (-)
C_D	drag coefficient (-)
C_f	local friction coefficient (-)
$\bar{C}_f$	average friction coefficient (-)
C_i	the ith constant in a separation of variables solution (-)
C_{ms}	heat capacity of microscale energy carrier (J/K)

C_N	correction factor for number of tubes in a tube bank (-)
C_{nb}	nucleate boiling constant (-)
C_R	capacitance ratio (-)
Co	convection number (-)
COP	coefficient of performance (-)
D	diameter (m)
D_h	hydraulic diameter (m)
dx	differential distance in the x-direction (m)
e	specific energy (J/kg) surface roughness (m)
E_b	blackbody emissive power (W/m^2)
$E_{b,0-\lambda_1}$	blackbody emissive power for $\lambda < \lambda_1$(W/m^2)
$E_{b,\lambda}$	blackbody spectral emissive power (W/m^2-μm)
ed	energy density (J/kg)
err	iteration error (varies)
f	Moody (or Darcy) friction factor (-)
$F_{0-\lambda_1}$	fraction of blackbody radiation emitted at $\lambda < \lambda_1$ (-)
$F_{i,j}$	view factor from surface i to surface j (-)
$F_{\lambda_1-\lambda_2}$	fraction of blackbody radiation emitted at $\lambda_1 < \lambda < \lambda_2$(-)
f_l	friction factor associated with the flow of liquid alone (-)
$\bar{f}$	average Moody friction factor (-)
$f_{Fanning}$	Fanning friction factor (-)
fpl	number of fins per length (1/m)
Ec	Eckert number (-)
F_D	drag force (N)
Fo	Fourier number (-)
Fr	Froude number (-)
Fr_{mod}	modified Froude number (-)
g	gravitational acceleration (m/s^2)
G	mass velocity, also known as mass flux (kg/m^2-s)
$\dot{g}$	rate of thermal energy generation (W)

$\dot{g}'''$	rate of thermal energy generation per unit volume (W/m^3)	mL	fin constant (-)
Ga	Galileo number (-)	MW	molecular weight (kg/kmol)
Gr	Grashof number (-)	N	total number of time steps
Gz	Graetz number (-)		used (in numerical problems)
h	local heat transfer coefficient (W/m^2-K)		intermediate dimensionless parameter for flow boiling correlation (-)
$\bar{h}$	average heat transfer coefficient (W/m^2-K)	n_{ms}	number density of microscale energy carriers (#/m^3)
$\tilde{h}$	dimensionless heat transfer coefficient for flow boiling (-)	N_L	number of rows of tubes in the longitudinal direction in
h_l	superficial heat transfer coefficient of the liquid phase (W/m^2-K)	Nu Nu_x	a tube bank local Nusselt number (-) local Nusselt number based
$\bar{h}_{eff}$	effective heat transfer coefficient (W/m^2-K)	$\overline{Nu}$	on the characteristic length x (-) average Nusselt number (-)
$\bar{h}_{rad}$	radiation heat transfer coefficient (W/m^2-K)	NTU OUT	number of transfer units (-) amount or rate of some
i	specific enthalpy (J/kg) integer index for spatial location (in numerical problems)		arbitrary quantity leaving a system
j	integer index for time (in numerical problems)	p P $\tilde{p}$	pressure (Pa) LMTD effectiveness (-) dimensionless pressure (-)
j_H	Colburn j_H factor (-)	p_{atm}	atmospheric pressure (Pa)
I_c	current (ampere)	p_∞	free stream pressure (Pa)
IN	amount or rate of some arbitrary quantity entering a system	per per_h	wetted perimeter (m) perimeter exposed to heating (m)
k	thermal conductivity (W/m-K)	Pr $\dot{q}$ $\dot{q}_{cond}$	Prandtl number (-) heat transfer rate (W) heat transfer rate due to
k_c	contraction loss coefficient (-)		conduction (W)
k_e	expansion loss coefficient (-)	$\dot{q}_{conv}$	heat transfer rate due to
Kn	Knudsen number (-)		convection (W)
L	length (m)	$\dot{q}_{fin}$	heat transfer rate to a fin (W)
L_c	corrected length for fin calculation (m)	$\dot{q}_{fin,k\to\infty}$	heat transfer rate to a fin with $k\to\infty$ (W)
L_{char}	characteristic length (m)	$\dot{q}_{no\,fin}$	heat transfer rate that would
L_{cond}	conduction length (m)		occur from a surface if fin
L_{flow}	length in the flow direction (m)		was removed (W)
L_{ms}	average distance between energy carrier interactions (m)	$\dot{q}_{rad}$ $\dot{q}_r$	heat transfer rate due to radiation (W) heat transfer rate in the
L_{nb}	nucleate boiling length scale (m)	$\dot{q}_x, \dot{q}_y, \dot{q}_z$	r-direction (W) heat transfer rate in the x-, y-, and z-directions (W)
$\dot{m}$	mass flow rate (kg/s)	$\dot{q}''$	heat transfer rate per unit
m	mass (kg)		area, heat flux (W/m^2)
M	total number of nodes used (in numerical problems)	$\dot{q}''_{conv}$	heat flux due to convection (W/m^2)

$\dot{q}''_{rad}$	heat flux due to radiation (W/m^2)	R_{SF}	shape factor thermal resistance (K/W)
$\dot{q}''_x$, $\dot{q}''_y$, $\dot{q}''_z$	heat flux in the x-, y-, and z-directions (W/m^2)	R_{sph}	thermal resistance associated with radial conduction through a spherical shell (K/W)
$\dot{q}''_s$	surface heat flux (W/m^2)		
$\dot{q}''_{s,crit}$	critical heat flux (W/m^2)	$R_{surface\text{-}to\text{-}surroundings}$	thermal resistance between the surface of an object and its surroundings (K/W)
$\dot{q}''_{s,nb}$	nucleate boiling heat flux (W/m^2)		
Q	total amount of heat transfer (J)	R_{total}	total resistance of a finned surface (K/W)
$\tilde{Q}$	dimensionless heat transfer (-)	R_{univ}	universal gas constant (8314 J/kmol-K)
r	radial coordinate, radius (m)		
$\tilde{r}$	dimensionless radial coordinate (-)	R''_c	area-specific contact resistance (K-m^2/W)
R	thermal resistance (K/W)	Ra	Rayleigh number (-)
	gas constant (J/kg-K)	Re	Reynolds number (-)
	$LMTD$ capacitance ratio (-)	Re_{δ_m}	Reynolds number based on the momentum boundary layer thickness (-)
R_c	contact thermal resistance (K/W)		
R_{cond}	thermal resistance to conduction (K/W)	Re_x	Reynolds number based on the characteristic length x (-)
$R_{cond,int}$	thermal resistance to internal conduction within an object (K/W)	RR	radius ratio, ratio of inner to outer radius of an annular duct (-)
$R_{cond,x}$	thermal resistance to conduction in the x-direction (K/W)	s	a coordinate direction (m)
		S	shape factor (m)
$R_{cond,y}$	thermal resistance to conduction in the y-direction (K/W)		spacing between plates (m)
		S_L	tube pitch in the longitudinal direction in a bank of tubes (m)
R_{conv}	thermal resistance to convection (K/W)		
R_{cyl}	thermal resistance associated with radial conduction through a cylindrical shell (K/W)	S_T	tube pitch in the transverse direction in a bank of tubes (m)
R_e	electrical resistance (ohm)	St	Stanton number (-)
R_f	fouling resistance (K/W)	$STORED$	amount or rate of some arbitrary quantity being stored in a system
R_{fin}	thermal resistance of a single fin (K/W)		
R_{fins}	thermal resistance of all of the fins on a finned surface (K/W)	t	time (s)
		t_j	time at the jth time in a numerical solution (s)
R''_f	fouling factor (K-m^2/W)		
$R_{i,j}$	space resistance between surfaces i and j in a radiation problem (1/m^2)	t_{sim}	simulation time (s)
		T	temperature (K)
		$\bar{T}$	average temperature (K)
R_{pw}	thermal resistance to conduction through a plane wall (K/W)	T_f	film temperature (K)
		T_h	solution to a homogeneous differential equation (K)
R_{rad}	thermal resistance associated with radiation (K/W)	T_i	temperature of the ith node in a numerical solution (K)
R_s	surface resistance in a radiation problem (1/m^2)	T_j	temperature at the jth time in a numerical solution (K)

$T_{i,j}$	temperature of the ith node and jth time in a numerical solution (K)	x	x-coordinate (m) direction parallel to a surface and in the flow direction for convection problems (m) thermodynamic quality (-)
$\hat{T}_i$	an intermediate estimate of the temperature of the ith node in a numerical solution (K)	$x_{fd,h}$	hydrodynamic entry length (m)
$\hat{T}_j$	an intermediate estimate of the temperature at the jth time in a numerical solution (K)	x_i	x-location of the ith node in a numerical solution (m)
		$\tilde{x}$	dimensionless x-coordinate (-)
		$\underline{X}$	the vector of unknown temperatures in a system of linear equations (K)
T_{ini}	initial temperature (K)		
T_p	solution to a particular differential equation (K)		
T_{ref}	reference temperature (K)	X_{tt}	Lockhart Martinelli parameter (-)
T_s	surface temperature (K)	y	y-coordinate (m) direction perpendicular to a surface for convection problems (m)
T_{sur}	surrounding temperature (K)		
T_∞	free stream temperature (K)		
th	thickness (m)		
$time$	time duration (s)	$\tilde{y}$	dimensionless y-coordinate (-)
tol	tolerance (K)	z	z-coordinate (m)
u	velocity in the x-direction (m/s) specific internal energy (J/kg)		

Greek Symbols

UA	conductance (W/K)	α	absorption coefficient (1/m) thermal diffusivity (m^2/s) ratio of gas side surface area to volume (1/m)
u_{char}	characteristic velocity (m/s)		
u_f	fluid approach velocity for an external flow (m/s)		
u_m	mean velocity (m/s)		
u_{max}	maximum velocity (m/s)	β	volumetric thermal expansion coefficient (1/K)
u_∞	free stream velocity (m/s)		
$\tilde{u}$	dimensionless velocity in the x-direction (-)	χ	correction factor for pressure drop in tube bank (-)
U	total internal energy (J)	δ	boundary layer thickness (m)
v	velocity in the y-direction (m/s) velocity in the r-direction (m/s)	δ_m	momentum boundary layer thickness (m)
v_{ms}	average velocity of microscale energy carriers (m/s)	δ_t	thermal penetration depth (m) thermal boundary layer thickness (m)
$\tilde{v}$	dimensionless velocity in the y-direction (-)	δ_{vs}	viscous sublayer thickness (m)
V	volume (m^3)	Δi_{vap}	latent heat of vaporization (J/kg)
$\dot{V}$	volumetric flow rate (m^3/s)	Δp	pressure drop (Pa)
$\dot{V}_{oc}$	open circuit flow rate produced by a pump with no resistance (m^3/s)	Δp_{dh}	dead head pressure rise produced by a pump with no flow (Pa)
$\dot{w}$	work transfer rate, power (W)	Δp_{pump}	pressure rise generated by a pump (Pa)
W	total amount of work (J) width (m)	Δt	duration of time step (s)
		Δt_{crit}	duration of critical time step (s)
		ΔT	temperature difference (K)

ΔT_{cond}	temperature difference due to conduction (K)	τ_{yx}	viscous stress on the y-face of a control volume in the x-direction (Pa)
$\Delta T_{cond,x}$	temperature difference due to conduction in x-direction (K)	τ_{yy}	viscous stress on the y-face of a control volume in the y-direction (Pa)
$\Delta T_{cond,y}$	temperature difference due to conduction in the y-direction (K)	τ_{diff}	diffusive time constant (s)
ΔT_{conv}	temperature difference due to convection (K)	τ_{lumped}	lumped capacitance time constant (s)
ΔT_e	excess temperature, surface minus saturation temperature (K)	υ	kinematic viscosity (m²/s) frequency (Hz)
ΔT_{lm}	log mean temperature difference (K)	ζ	angle relative to horizontal (radian)
Δx	distance between nodes in the x-direction (m)	ζ_1	the 1st eigenvalue in a separation of variables solution (-)
Δy	distance between nodes in the y-direction (m)	ζ_i	the ith eigenvalue in a separation of variables solution (-)
ε	emissivity (-) effectiveness (-)		
ε_{fin}	fin effectiveness (-)	**Subscripts**	
ϕ	viscous dissipation function (W/m³)	b	base
η	efficiency (-)	c	contact, corrected
η_{fin}	fin efficiency (-)	C	cold
η_o	overall efficiency of a finned surface (-)	$cond$	conduction
κ	Von Kármán constant, 0.41 (-)	$conv$	convection
λ	wavelength (μm)	$crit$	critical time step where simulation becomes unstable critical Reynolds number for laminar-to-turbulent transition
μ	dynamic viscosity (N-s/m²)		
θ	temperature difference (K)	cyl	cylinder
$\tilde{\theta}$	dimensionless temperature difference (-)	$diff$	diffusive
		fc	forced convection
ρ	density (kg/m³)	fd	fully developed
ρ_e	electrical resistivity (Ω-m)	fin	fin
σ	Stefan–Boltzmann constant (5.67 × 10⁻⁸ W/m²-K⁴) surface tension (N/m) ratio of free flow to frontal area (-)	h	homogeneous
		H	hot solution for constant heat flux boundary condition
τ	shear stress (Pa)	in	entering a system, inner (e.g., diameter or radius)
τ_s	shear stress at a surface (Pa)	ini	initial, at time $t = 0$
$\overline{\tau_s}$	average shear stress on surface (Pa)	int	internal, within an object
		is	inner surface
τ_{xx}	viscous stress on the x-face of a control volume in the x-direction (Pa)	l,sat	saturated liquid
		lam	laminar
		max	maximum possible amount
τ_{xy}	viscous stress on the x-face of a control volume in the y-direction (Pa)	nc	natural convection
		p	particular
		pw	plane wall

o	overall	$a_{i,j}$	the value of a at node i and time j
os	outer surface		
out	leaving a system, outer (e.g., diameter or radius)	a^k	the value of a for iteration k
		$a_{x=x_1}$	the value of a evaluated at the x-location x_1
rad	radiation		
s	at the surface	$a(x)$	a is a function only of x
$semi\text{-}\infty$	related to the semi-infinite body solution	$\dfrac{da}{dx}$	the ordinary derivative of a with respect to x (a is only a function of x)
SF	shape factor		
sph	sphere	$\left.\dfrac{da}{dx}\right|_{x=x_1}$	the ordinary derivative of a with respect to x evaluated at x-location x_1
$surface\text{-}to\text{-}surroundings$	from the surface of an object to the surroundings		
T	solution for constant temperature boundary condition	$\dfrac{\partial a}{\partial x}$	the partial derivative of a with respect to x (a is a function of variables other than x)
$total$	total resistance		
$turb$	turbulent	$\underline{a}$	a one-dimensional vector of values
uh	unheated		
v,sat	saturated vapor	$\underline{\underline{a}}$	a two-dimensional matrix of values
vs	viscous sublayer		
x	in the x-direction	$\underline{\underline{a}}^{-1}$	the inverse of $\underline{\underline{a}}$, a two-dimensional matrix of values
y	in the y-direction		

Superscripts and Abbreviations (Where *a* and *b* are Arbitrary Quantities)

		$\text{Max}\,(a_i)\,i=1\ldots M$	the maximum value of the elements of vector a with indices $i=1$ to M
a'	per unit length		
a''	per unit area	$\text{Min}\,(a_i)\,i=1\ldots M$	the minimum value of the elements of vector a with indices $i=1$ to M
a'''	per unit volume		
$\bar{a}$	average value of a	$\sum_{i=1}^{M} a_i$	the sum of the elements in vector a with indices $i=1$ to M
$\hat{a}$	prediction of a obtained during a predictor step		
$\tilde{a}$	dimensionless form of the variable a	$a\|b$	quantity a in parallel with quantity b, shorthand for $\left(\frac{1}{a}+\frac{1}{b}\right)^{-1}$
a_i	the value of a at node i	$O(a)$	order of magnitude of the quantity a
a_j	the value of a at time j		

1 | Introduction

1.1 Relevance of Heat Transfer

Heat transfer is the term used to describe the movement of thermal energy (heat) from one place to another. Heat transfer drives the world that we live in. Look around. Heat transfer is at work no matter where you currently are.

Do you see any buildings? Most modern building are heated and cooled by equipment such as furnaces and air conditioning units that rely on heat exchangers for their operation. In addition, there are many devices within the building that rely on heat transfer to operate. Examples include your computer, light bulbs, toasters, ovens, water heaters, and refrigerators; all of these devices must transfer thermal energy to operate. Many engineers will eventually find jobs in the Heating Ventilation Air-Conditioning and Refrigeration (HVAC&R) industry, which employs engineers to design and maintain building conditioning and related energy systems.

Do you see any vehicles? Most vehicles today employ internal combustion engines in which the chemical energy in fuel is converted into mechanical power to drive the vehicle. A large fraction of the energy in the fuel is converted to thermal energy which must be transferred to the environment in order to keep the engine and related parts from overheating. The "radiator" is a heat exchanger in the vehicle that transfers thermal energy from the engine coolant to the surroundings. Heat transfer governs the design of much of the equipment in a vehicle.

Are you using any electrical power right now? Much of electrical power used in the world is produced by burning a fuel such as coal or natural gas and transferring the thermal energy to water to make steam. The steam drives turbines, which in turn drive generators that produce the electrical power. Thermodynamics dictates the maximum efficiency at which the electrical generation process can occur, but heat transfer governs the actual design of the boiler in which the steam is produced and the condensers in which thermal energy from the cycle is transferred to the environment.

Perhaps you are outdoors, far away from any man-made technology. Even here, heat transfer (and the related science of mass transfer) is important. Heat transfer dictates what clothes you are wearing to maintain comfort. The concept of "wind-chill" is a means of expressing the heat transfer enhancement that results from air movement. On a grander scale, heat transfer regulates the temperature of our environment, balancing solar energy gains with thermal energy radiation from the planet to outer space. A detailed examination of the surroundings, e.g., green leaves on trees, animal fur, etc., shows the different adaptations that living things have made to control their heat transfer rates.

You are likely familiar with the terms global warming and climate change, which are used to describe the increase in the temperature of the environment as a result of human activities. The entire concept of global warming is rooted in heat transfer. A major contributor to global warming is the reduction in thermal energy transport from Earth's surface to outer space by radiation as a consequence of increased levels of carbon dioxide in the atmosphere that has resulted from the combustion of fossil fuels.

It has always been important for engineers to design efficient products. A more efficient system can be smaller and less expensive to manufacture as well as requiring less energy to operate, thereby reducing operating costs. We find ourselves entering a period of human history where efficient design is more important than it ever has been. Aside from limiting costs, we must also reduce the amount of fuels we consume to preserve our supply of nonrenewable fuels and to limit the climate change that they cause. We must develop alternative systems that rely on renewable energy sources such as solar energy and wind. The efficiency of both renewable and nonrenewable energy systems is primarily driven by the performance of the components within them that are used to transfer heat. The backbone of these systems will be highly effective heat exchangers and they will be designed by engineers who have a thorough understanding of heat transfer. Engineering students with a strong background in thermodynamics, fluids, and heat transfer are increasingly in demand as many of the problems that face the nation and the world center around the effective transfer of heat.

I.2 Relationship to Thermodynamics

The concept of heat is introduced in the study of **thermodynamics**. In thermodynamics, **heat** is defined as the energy that transfers across the boundary of a system as the result of a temperature gradient. Students of thermodynamics typically define **systems** and use these systems to carry out **energy balances**; heat plays a major role in these balances. However, thermodynamics is unconcerned with time and therefore the rates of energy transfer.

Energy balances form the backbone of the study of heat transfer as well. The difference between thermodynamics and heat transfer is that heat transfer problems couple these energy balances with **rate equations** in order to predict and understand the *rate* of heat transfer rather than just the amount.

Thermodynamics teaches us that energy is a conserved quantity (in the absence of nuclear reactions); that is, energy is neither generated nor destroyed. Therefore, an energy balance on a system enforces the idea that the energy entering the system must be equal to the sum of the energy leaving the system and the energy stored within the system:

$$IN = OUT + STORED. \tag{1.1}$$

For energy balances that are written for some finite time period, *IN* and *OUT* represent the amount of energy entering and leaving the system, respectively, and *STORED* represents the amount of energy stored in the system during that time (i.e., the change in the amount of energy that is contained within the system). In a heat transfer problem, these terms will typically represent rates of energy transfer and the rate of energy storage.

Energy can cross a system boundary in the form of heat, work, or with mass (for an open system). Figure 1.1 illustrates an energy balance on an open system (neglecting kinetic and potential energy terms) and results in the equation below:

$$\dot{m}_{in}\, i_{in} + \dot{q}_{in} + \dot{w}_{in} = \dot{m}_{out}\, i_{out} + \dot{q}_{out} + \dot{w}_{out} + \frac{dU}{dt}, \tag{1.2}$$

where the subscript *in* indicates quantities entering the control volume and *out* quantities leaving the control volume. The variable $\dot{m}$ is the mass flow rate crossing the system boundary and i is the specific enthalpy associated with that mass (note that the symbol h is *not* used for specific enthalpy here as it is used extensively in heat transfer literature to represent the heat transfer coefficient). The quantities $\dot{q}$ and $\dot{w}$ are the rates of heat transfer and work transfer, respectively, passing the control surface. Finally, the quantity U is the total amount of internal energy contained within the control volume and $\frac{dU}{dt}$ is its time derivative.

Thermodynamics by itself does not provide any way to compute the rate of heat transfer based on the physical situation (i.e., the geometry, materials, and other conditions associated with the situation). Thermodynamic problems must therefore either provide the heat transfer rate as an input or calculate the heat transfer rate by solving an energy balance like Eq. (1.2) in which $\dot{q}$ is the only unknown. Many problems in thermodynamics assume that a system is **adiabatic** (i.e., $\dot{q} = 0$). *The goal of heat transfer is to provide the tools needed to compute $\dot{q}$ based on the details of the situation.*

Thermodynamics problems apply energy balances almost exclusively to finite sized systems or control volumes. For example, in thermodynamics a system might include an entire tank or compressor or heat

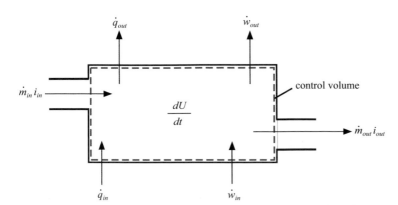

Figure 1.1 General energy balance on an open system.

exchanger. Heat transfer also utilizes energy balances, just like the one shown in Eq. (1.2), but in heat transfer the energy balances are often applied to a *differentially small* control volume. This is done in order to express the result of the energy balance as an ordinary or partial differential equation. Most thermodynamics problems do not require the solution of differential equations; however, many of the problems we encounter in heat transfer will involve differential equations.

Example 1.1

A small freezer uses a thermoelectric cooler to keep biological specimens frozen during shipping. The internal freezer space must be maintained at $T_C = -15°C$ and the ambient temperature is $T_H = 30°C$. The thermal resistance associated with the freezer walls is $R = 5.3$ K/W. This thermal resistance is used in the rate equation that computes the heat transfer rate from the ambient air to the contents of the freezer:

$$\dot{q}_{wall} = \frac{(T_H - T_C)}{R}. \tag{1}$$

Obviously, a well-designed freezer will have a large thermal resistance to reduce the need for the thermo-electric cooler to operate. The contents of this book will allow you to estimate the thermal resistance of a freezer given the details of its construction.

The thermoelectric cooler is powered by batteries and has a coefficient of performance, $COP = 0.5$; recall from thermodynamics that COP is defined as the ratio of the rate of cooling provided by the refrigerator ($\dot{q}_C$) to the power consumed by the cooler ($\dot{w}$). The freezer must operate for *time* = 2 days. Your research has indicated that the energy density of the lithium-ion batteries used to power the cooler is approximately $ed = 120$ W-hr/kg.

Determine:
- The electrical power consumed by the thermoelectric cooler and rate of heat transfer from the cooler to the ambient air.
- The mass of batteries required by the system.

Known Values

The known parameters are indicated on the sketch of the freezer shown in Figure 1; when necessary these parameters are converted to base SI units.

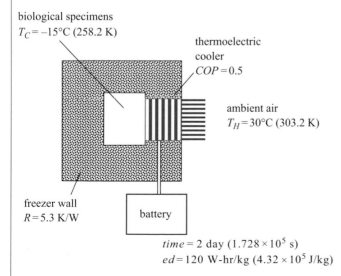

biological specimens
$T_C = -15°C$ (258.2 K)

thermoelectric
cooler
$COP = 0.5$

ambient air
$T_H = 30°C$ (303.2 K)

freezer wall
$R = 5.3$ K/W

battery

time = 2 day (1.728×10^5 s)
$ed = 120$ W-hr/kg (4.32×10^5 J/kg)

Figure 1 Freezer for biological specimens.

Continued

Example 1.1 (cont.)

Assumption

• Steady-state conditions exist.

Analysis

Equation (1) can be used to determine the rate of heat transfer from ambient to the contents of the freezer, $\dot{q}_{wall}$. An energy balance on the freezer compartment is shown in Figure 2 (left).

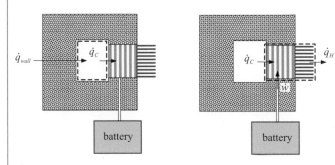

Figure 2 (Left) Energy balance on the freezer compartment and (right) energy balance on the thermoelectric cooler.

   Equation (1.2) can be applied to this situation by setting the power and mass flow rate terms to zero. The freezer is assumed to be at steady state so that the energy storage term is also set to zero:

$$\dot{q}_{wall} = \dot{q}_C. \tag{2}$$

The definition of COP can be used to determine the power consumed by the cooler:

$$\dot{w} = \frac{\dot{q}_C}{COP}. \tag{3}$$

Figure 2 (right) illustrates an energy balance on the cooler itself (again, assuming steady state):

$$\dot{q}_C + \dot{w} = \dot{q}_H. \tag{4}$$

The amount of energy consumed by the cooler from the battery during the shipping process is obtained by multiplying the steady-state power by the time:

$$W = \dot{w} \, time. \tag{5}$$

The mass of batteries required is obtained using the energy density:

$$m = \frac{W}{ed}. \tag{6}$$

Solution

Equations (1) through (6) can be solved sequentially and explicitly. Here we will solve them by hand:

$$\dot{q}_{wall} = \frac{(T_H - T_C)}{R} = \frac{(303.2 - 258.2)\text{K}}{} \left| \frac{\text{W}}{5.3 \text{ K}} \right. = 8.49 \text{ W}$$

$$\dot{q}_C = \dot{q}_{wall} = 8.49 \text{ W}$$

$$\dot{w} = \frac{\dot{q}_C}{COP} = \frac{8.49 \text{ W}}{} \left| \frac{}{0.5} \right. = \boxed{17.0 \text{ W}}$$

$$\dot{q}_H = \dot{q}_C + \dot{w} = 8.49 \text{ W} + 17.0 \text{ W} = \boxed{25.5 \text{ W}}$$

Example 1.1 (cont.)

$$W = \dot{w} \, time = \frac{17.0 \text{ W}}{} \left| \frac{1.728 \times 10^5 \text{s}}{} \right| \left| \frac{\text{J}}{\text{W-s}} \right| = 2.93 \times 10^6 \text{ J}$$

$$m = \frac{W}{ed} = \frac{2.93 \times 10^6 \text{ J}}{} \left| \frac{\text{kg}}{4.32 \times 10^5 \text{ J}} \right| = \boxed{6.79 \text{ kg} \, (15.0 \text{ lb}_\text{m})} .$$

Discussion

The mass of batteries required for this application is fairly large due to the low *COP* of the thermoelectric cooler. Further, the relatively large amount of heat that must be rejected to the ambient (also the result of the low *COP*) will lead to a large heat exchanger. The thermoelectric cooler may not be ideal for this application.

1.3 Problem Solving Methodology

Engineering is all about problem solving. As an engineer you should be one of the best problem solvers in any group of people. One objective of this text is to provide you with a lot of practice solving engineering problems. Heat transfer provides a great opportunity to tackle some very relevant and interesting problems. To this end, it is important to develop a general problem solving approach that can be systematically used to solve problems in this course as well as in any engineering discipline. The steps that we suggest are laid out below.

1. *Sketch the problem and list known values.* The problem statement will provide information about the problem and include important quantities that must be taken as inputs to the problem solution. It is best to draw a sketch that represents the problem and label these known quantities on the sketch. It is usually a good idea to convert these known values to a self-consistent unit system so that the solution is not complicated by unit conversions.
2. *List assumptions.* Engineering analysis almost always involves simplifying a complex problem so that it can be analyzed and understood. This process requires that you make assumptions and these assumptions should be carefully listed so that the limitations of the solution are clearly established. Some key assumptions may be included in the problem statement. However, it may be necessary for you to make additional assumptions and these often must be justified with appropriate calculations.
3. *Analysis.* Once the inputs and assumptions are clearly laid out, it is necessary to carry out an analysis. The analysis procedure will require that you apply generally useful tools such as energy balances, mass balances, rate equations, property information, etc., to the specific problem being examined. It is almost always advisable to draw additional sketches showing the system being analyzed (for energy balances), the geometric quantities of interest, the energy flows being calculated, etc. The analysis will lead to the systematic identification of a system of equations that can be used to solve the problem. The system of equations should be checked for completeness.
4. *Solution.* The system of equations derived in the analysis must be solved to provide useful numerical answers. First, examine the equations and identify a solution strategy. Can they be solved sequentially and explicitly? Is iteration required? If so, how will the iteration process be accomplished? The answer to these questions together with the intended purpose of the analysis will dictate the solution technique that is adopted. If your ultimate goal is to carry out a parametric study and generate a plot then you will eventually need to solve the equations using some computer software tool. In some cases it is sufficient to solve the equations using pencil, paper, and calculator. In either case, the analysis step that involves deriving a complete set of equations should be carried out *separately from and prior to* the solution step, in which these equations are solved. There is only one correct set of equations, but there are many equally correct methods that can be used to solve these equations.
5. *Discussion/Exploration.* This step is the most challenging and interesting one. Examine your solution and present any important conclusions that can be drawn. What is important in the problem? What is not so

important? Manipulate your solution to identify interesting trends and explain them. Carry out simple sanity checks or thought experiments to give you confidence in the solution. If you change an input then how should your solution change? Is this what actually happens? As you solve engineering problems in your career, you will become cognizant of how important it is to be right. Your solutions will guide decisions in design processes that are very costly and it is often expensive to be wrong. Obviously no one is right all the time, but it is usually possible to identify mistakes by careful and critical examination of your solution. The manipulation and exploration of a solution is easier if it is implemented using computer software.

1.4 Heat Transfer Mechanisms

The three mechanisms commonly used to understand and describe heat transfer are discussed in this section. The remainder of the book presents each of these mechanisms in much more detail, but it is useful to introduce the mechanisms and associated rate equations before proceeding.

1.4.1 Conduction

Conduction refers to heat transfer that is a result of the interactions of the micro-scale energy carriers that exist within a material. Conduction is a phenomenon that is conceptually easy to grasp, particularly in a gas or fluid where the energy carriers are typically molecules. Fast moving (i.e., higher temperature) molecules will strike slower moving (i.e., lower temperature) molecules causing an energy transfer from the hot to the cold molecules. This interaction occurs at a molecular level and the aggregated result of all of these individual collisions at the macroscale is the transfer of energy from hot to cold. Energy transfer by conduction is sometimes referred to as **thermal diffusion**. Conduction is the subject of Chapters 2 through 6 of this text.

In some materials the energy carriers may not be molecules, but the diffusion process is basically the same. The energy carriers in a solid may be electrons or phonons (i.e., vibrations in the structure of the solid). The transfer of energy by conduction is still related to the interactions of these microscale energy carriers. More energetic (i.e., higher temperature) energy carriers transfer energy to less energetic (i.e., lower temperature) ones, resulting in a net flow of energy from hot to cold (i.e., heat transfer).

No matter what material and energy carriers are involved, the rate equation that characterizes conduction heat transfer is **Fourier's Law**. Fourier's Law relates the **heat flux** (the heat transfer rate per unit area) in a particular direction to the temperature gradient in that same direction. For example, in the x-direction Fourier's Law can be written as:

$$\dot{q}''_x = -k \frac{\partial T}{\partial x}, \tag{1.3}$$

where $\dot{q}''_x$ is the heat flux in the x-direction. Note that in this text the ″ superscript indicates per unit area (just as the ′ superscript will indicate per unit length and the ‴ superscript per unit volume). Therefore, the heat flux $\dot{q}''_x$ has units of J/s-m^2 (or W/m^2). The parameter k in Eq. (1.3) is the **thermal conductivity** of the material. Thermal conductivity is a transport property that is discussed in more detail in Section 2.1. Evaluating the units of Eq. (1.3)

$$\underbrace{\left[\frac{W}{m^2}\right]}_{\dot{q}''_x} = \underbrace{\left[\frac{W}{m\text{-}K}\right]}_{k} \underbrace{\left[\frac{K}{m}\right]}_{\frac{\partial T}{\partial x}} \tag{1.4}$$

shows that thermal conductivity must have units of [W/m-K]. Transport properties, like thermodynamic properties (e.g., enthalpy or entropy), are fixed once you know the state of the material. In Section 1.5 methods for obtaining transport properties for specific materials are discussed. Table 1.1 provides the range of conductivity values that can be expected for some typical materials.

Fourier's Law makes physical sense in that we know heat must flow from hot to cold. For example, Figure 1.2 illustrates the temperature as a function of one dimension (x) in a stationary substance. This sketch might correspond to the temperature in the wall of a house on a cold winter day; the inner surface is adjacent to the

Table 1.1 Conductivity of typical materials.

Material	Conductivity (W/m-K)
Pure metals	50 to 500
Alloys	5 to 50
Polymers	0.1 to 2
Liquids	0.1 to 1
Gases	0.01 to 0.1

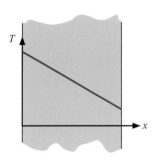

Figure 1.2 A sketch showing temperature as a function of position x.

heated interior of the house and is therefore warmer than the outer surface, which is adjacent to cold outdoor air. Physically, it seems obvious that there is heat transfer from left to right (i.e., from the warm inside of your house to the cold outdoors). Fourier's Law simply expresses this idea mathematically. The temperature is decreasing in the x-direction and therefore the temperature gradient, $\partial T/\partial x$, is negative. The negative sign appearing in Eq. (1.3) results in the rate of heat transfer in the x-direction being positive for this example (i.e., conduction is causing an energy flow from inside to outside).

Fourier's Law is an example of a **rate equation**. In heat transfer, rate equations relate the rate of energy transfer by heat to temperature gradients or temperature differences. Fourier's Law is the rate equation that we will always use for conduction heat transfer and it is also our primary tool for understanding conduction problems.

1.4.2 Convection

Convection refers to heat transfer between a surface and an adjacent flowing fluid as shown in Figure 1.3. Convection is not truly a unique mode of heat transfer. Within the fluid, conduction heat transfer occurs but the situation is complicated substantially by the motion of the fluid itself. The rate equation that characterizes convection heat transfer is **Newton's Law of Cooling**:

$$\dot{q}_{conv} = \bar{h}\, A_s (T_s - T_\infty), \tag{1.5}$$

where A_s is the surface area exposed to the fluid. The surface is at temperature T_s and the fluid has a free stream temperature T_∞. The **free stream temperature** refers to the temperature of fluid far away from the surface, where it has not been affected by the presence of the surface. The parameter $\bar{h}$ in Eq. (1.5) is the average **heat transfer coefficient** associated with the convection problem. Examination of the units in Eq. (1.5)

$$\underbrace{[\text{W}]}_{\dot{q}_{conv}} = \underbrace{\left[\frac{\text{W}}{\text{m}^2\text{-K}}\right]}_{\bar{h}} \underbrace{[\text{m}^2]}_{A_s} \underbrace{[\text{K}]}_{(T_s - T_\infty)} \tag{1.6}$$

Table 1.2 Typical values of heat transfer coefficient for some situations.

Situation	Heat transfer coefficient (W/m²-K)
Natural convection with gases	2 to 10
Natural convection with liquids	10 to 50
Forced convection with gases	10 to 70
Forced convection with liquids	50 to 1,000
Boiling and condensation	500 to 2,500

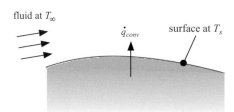

Figure 1.3 Convection.

shows that the average heat transfer coefficient must have units of W/m²-K. Note that the heat transfer coefficient is *not* a transport or a thermodynamic property like thermal conductivity or density. Rather, $\bar{h}$ is a complex function of the geometry, fluid properties, and flow conditions. Chapters 7 through 11 present some techniques that will allow us to understand and estimate the heat transfer coefficient for a variety of convection situations. Convection is classified as being either forced convection or natural convection. **Forced convection** refers to the situation where the fluid is being externally driven over the surface of interest by, for example, a fan or the wind. **Natural convection** refers to the situation where the fluid motion is driven by buoyancy forces. That is, the fluid that is being heated tends to rise as its density is reduced (or conversely, cooled fluid falls). In the absence of a temperature difference between the surface and the fluid, there would be no fluid motion in a natural convection situation. Typical ranges of values for convective heat transfer coefficients are provided in Table 1.2.

1.4.3 Radiation

Radiation refers to heat transfer between surfaces due to the emission and absorption of electromagnetic waves. Radiation heat transfer is complex when many surfaces at different temperatures are involved and Chapter 14 is dedicated to dealing with this type of problem. However, in the limit that a single surface at temperature T_s interacts only with surroundings at temperature T_{sur}, radiation heat transfer from the surface can be calculated according to:

$$\dot{q}_{rad} = A_s \, \sigma \, \varepsilon \left(T_s^4 - T_{sur}^4 \right), \tag{1.7}$$

where A_s is the area of the surface, σ is the Stefan–Boltzmann constant, and ε is the emissivity of the surface. The **Stefan–Boltzmann constant** is a universal constant, $\sigma = 5.67 \times 10^{-8}$ W/m²-K⁴. The **emissivity** is a parameter that ranges between near 0 (for highly reflective surfaces) to near 1 (for highly absorptive surfaces). Note that both T_s and T_{sur} must be expressed in terms of absolute temperature (i.e., in units K rather than °C) in Eq. (1.7). Near room temperature, the rate of radiation heat transfer is often quite small relative to forced convection heat transfer in situations where both phenomena occur simultaneously. However, in natural convection situations where the heat transfer coefficient is small or in a vacuum where convection is nonexistent, radiation can be quite important. Also, Eq. (1.7) shows that the rate of radiation heat transfer increases according to the fourth power of absolute temperature and therefore radiation becomes an important heat transfer mechanism at high temperature.

Example 1.2

A copper pipe carries hot water from a water heater to a shower through the basement of a house. The outer diameter of the pipe is $D = 0.5$ inch and the length of the pipe is $L = 20$ ft. The air in the basement is at $T_\infty = 68°F$ and the natural convection heat transfer coefficient between the pipe surface and surrounding air is $\bar{h} = 6.8$ W/m^2-K. The pipe radiates to surroundings at the same temperature as the air, $T_{sur} = T_\infty$, and the emissivity of the pipe surface is $\varepsilon = 0.62$. The water enters the pipe at $T_s = 130°F$. A typical shower requires $Q = 22{,}000$ Btu and takes $time = 10$ minutes.

Determine:
- The rate of heat loss from the pipe.
- The total amount of energy that is lost as the water flows through the pipe during a shower. Compare this value to the total energy required by the shower itself.

Known Values

The problem inputs are listed in the sketch shown in Figure 1.

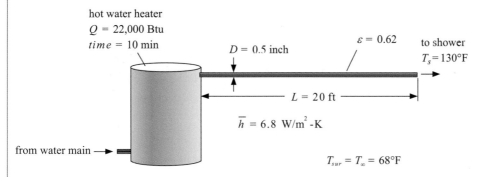

Figure 1 Pipe carrying water for a shower.

Assumptions

- Steady-state conditions exist.
- The water flowing through the pipe does not change temperature substantially as a result of the heat loss from the pipe.
- The surface temperature of the pipe is the same as the temperature of the water in the pipe.
- The energy associated with initially increasing the temperature of the pipe filled with water when the shower starts and then lost as the pipe returns to ambient temperature after the shower is stopped is neglected.

Analysis

The outside surface area of the pipe is:

Continued

Example 1.2 (cont.)

$$A_s = \pi D L. \tag{1}$$

The rate of heat loss by convection is obtained from Newton's Law of Cooling:

$$\dot{q}_{conv} = \bar{h} A_s (T_s - T_\infty). \tag{2}$$

The rate of heat loss by radiation is obtained from Eq. (1.7):

$$\dot{q}_{rad} = A_s \sigma \varepsilon (T_s^4 - T_{sur}^4). \tag{3}$$

The total rate of heat transfer from the pipe is the sum of the heat losses by radiation and convection:

$$\dot{q}_{loss} = \dot{q}_{conv} + \dot{q}_{rad}. \tag{4}$$

The total amount of energy lost from the pipe during the shower is the time integral of the rate of heat loss, which (assuming steady-state conditions) is:

$$Q_{loss} = \dot{q}_{loss} \, time. \tag{5}$$

Solution

Equations (1) through (5) are explicit and can easily be solved by hand, as was done in Example 1.1. However, in this problem we will use the computer software Engineering Equation Solver (EES) to implement the solution. A tutorial that will allow a new user to become familiar with the EES program can be found in Appendix E.

The inputs to the problem that are listed in Figure 1 are entered in EES.

D=0.5 [inch]***Convert**(inch,m)	"diameter"
L=20 [ft]***Convert**(ft,m)	"length"
T_s=**ConvertTemp**(F,K,130 [F])	"surface temperature"
T_infinity=**ConvertTemp**(F,K,68 [F])	"ambient air temperature"
T_sur=T_infinity	"external temperature for radiation"
h_bar=6.8 [W/m^2-K]	"heat transfer coefficient"
e=0.62 [-]	"emissivity"
Q=22000 [Btu]***Convert**(Btu,J)	"energy associated with shower"
time=10 [min]***Convert**(min,s)	"time associated with shower"

Notice that the Convert function in EES was used to convert the units of each variable to its base SI unit; the Convert function returns the unit conversion factor corresponding to the two units provided as arguments. The ConvertTemp function is used to convert the input temperatures from °F to K. Information about unit conversion factors available in EES can be obtained by selecting Unit Conversion Info from the Options menu.

Equations (1) through (5) are entered into the Equations Window. Notice that the constants π and σ are obtained using the built-in constants pi# and sigma# (the # at the end of their name indicates that they are constants in EES, although EES will also accept pi, without the # sign). Information about the constants in EES can be obtained by selecting Constants from the Options menu in EES.

A_s=pi#*D*L	"surface area of tube"
q_dot_conv=h_bar*A_s*(T_s-T_infinity)	"convection heat transfer rate"
q_dot_rad=sigma#*A_s*e*(T_s^4-T_sur^4)	"radiation heat transfer rate"
q_dot_loss=q_dot_conv+q_dot_rad	"total heat transfer rate"
Q_loss=q_dot_loss*time	"total energy lost by heat during shower"

Example 1.2 (cont.)

Solving provides $\dot{q}_{conv}$ = 56.97 W, $\dot{q}_{rad}$= 35.33 W, and $\boxed{\dot{q}_{loss} = 92.30 \text{ W}}$. The total energy lost from the pipe during a shower is $\boxed{Q_{loss} = 55,380 \text{ J}}$.

Discussion/Exploration

The rate of radiation heat transfer is comparable to the rate of convection heat transfer for this natural convection situation; therefore it would not be appropriate to neglect radiation for this problem. The total amount of energy lost from the pipe seems like a large number when expressed in units of J; however, a J is a very small unit of energy. It is easier to put the calculated value of Q_{loss} into context if we express it in terms of Btu. EES allows you to define not only a primary unit for each variable (the base SI unit that was used for calculation) but also a secondary unit. Right-click on the variable Q_loss in the Solution Window and enter the unit Btu into the secondary unit input box, as shown in Figure 2.

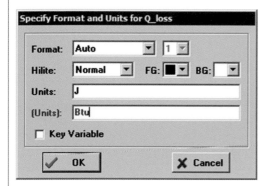

Figure 2 Specify Format and Units dialog.

The value of Q_loss will now be displayed in both J and Btu in the Solution Window, as shown in Figure 3.

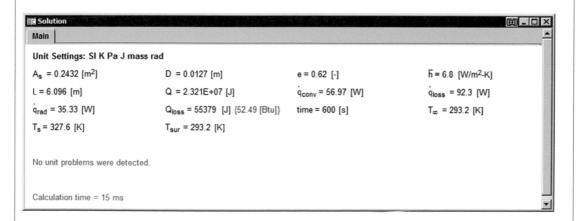

Figure 3 Solution Window.

Continued

Example 1.2 (cont.)

The calculated value of Q_{loss} is 52.5 Btu which is a small fraction (~0.2%) of the 22,000 Btu required by the shower. This result justifies the assumption that water flowing through the pipe does not change temperature substantially as a result of the heat loss from the pipe.

The convection coefficient was provided in the problem statement. In Chapters 7 through 11, we will see how the convection coefficient could be determined for this problem.

Example 1.3

A photovoltaic (PV) solar panel absorbs solar radiation and converts some fraction of this energy (defined as the efficiency of the panel) to electric power. The efficiency of the panel is a function of its temperature:

$$\eta = \eta_o - \beta(T - T_o),\qquad(1)$$

where $\eta_o = 0.18$ is the nominal efficiency obtained at the baseline temperature $T_o = 20°C$ and $\beta = 0.0015$ K^{-1} is the temperature coefficient. The size of the panel is $L = 4$ ft long and $W = 2$ ft wide and the solar flux is $\dot{q}''_s = 850$ W/m^2. The panel experiences convection with the surrounding air at $T_\infty = 20°C$ with heat transfer coefficient $\bar{h} = 12$ W/m^2-K. The panel also radiates to surroundings at $T_{sur} = 20°C$ and the emissivity of the panel surface is $\varepsilon = 1$.

Determine:
- The temperature of the panel.
- The rate at which the panel produces power.
- Prepare a plot of electrical power as a function of solar flux from $0 < \dot{q}''_s < 1000$ W/m^2.

Known Values

The problem inputs are listed in the sketch shown in Figure 1.

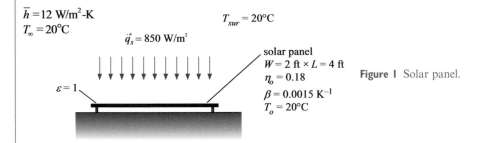

Figure 1 Solar panel.

Assumptions

- Steady-state conditions exist.
- The solar panel does not experience any heat loss from its back surface to the mounting structure.
- The solar panel absorbs all of the solar heat flux.
- The solar heat flux is normal to the panel surface.
- The solar panel is isothermal.

Example 1.3 (cont.)

Analysis

An energy balance on the solar panel is shown in Figure 2 and includes solar energy entering and radiation, convection, and electrical power leaving. There is no storage term for this (assumed) steady-state problem:

$$\dot{q}_s = \dot{q}_{conv} + \dot{q}_{rad} + \dot{w}. \tag{2}$$

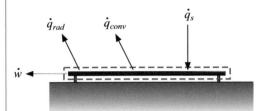

Figure 2 Energy balance on solar panel.

The area of the solar panel is:

$$A_s = W\,L. \tag{3}$$

The rate of solar energy absorbed is the product of the solar heat flux and the area:

$$\dot{q}_s = \dot{q}_s'' \, A_s. \tag{4}$$

The efficiency of the panel is calculated from its temperature using Eq. (1). The rate of electrical energy produced is:

$$\dot{w} = \dot{q}_s \eta. \tag{5}$$

The rate of convection heat transfer from the panel is:

$$\dot{q}_{conv} = \bar{h}\,A_s(T - T_\infty). \tag{6}$$

The rate of heat loss by radiation is:

$$\dot{q}_{rad} = A_s\,\sigma\varepsilon\left(T^4 - T_{sur}^4\right). \tag{7}$$

Solution

Equations (1) through (7) are seven equations in seven unknowns (η, T, A_s, $\dot{q}_s$, $\dot{w}$, $\dot{q}_{conv}$, and $\dot{q}_{rad}$). Unfortunately it is not easy to solve these equations by hand because it not possible to start with any of the seven equations and sequentially and explicitly work your way through the others to arrive at a solution. The equations are nonlinear and coupled and must be solved simultaneously.

We will use the EES program to carry out this solution. Notice that EES did not help with the analysis step – it was up to us to derive the appropriate equations and examine them to ensure that they were sufficient to obtain a solution. EES simply acts as a sophisticated calculator that is particularly useful for a problem like this where coupled, implicit equations must be solved. The use of a computer program such as EES is also useful when it is necessary to carry out a parametric study, such as examining the effect of solar flux as is necessary in this problem.

The inputs to the problem that are listed in Figure 1 are entered in EES and converted to base SI units.

Continued

Example 1.3 (cont.)

W=2 [ft]***Convert**(ft,m)   "width of panel"
L=4 [ft]***Convert**(ft,m)   "length of panel"
e=1 [-]   "emissivity"
T_sur=**ConvertTemp**(C,K,20 [C])   "external temperature"
T_infinity=**ConvertTemp**(C,K,20 [C])   "free stream temperature"
h_bar=12 [W/m^2-K]   "heat transfer coefficient"
eta_o=0.18 [-]   "efficiency at T_o"
T_o=**ConvertTemp**(C,K,20 [C])   "nominal temperature"
beta=0.0015 [1/K]   "temperature coefficient"
q``_dot_s=850 [W/m^2]   "solar flux"

Equations (1) through (7) are entered into the Equations Window.

eta=eta_o-beta*(T-T_o)   "efficiency of panel"

q_dot_s=w_dot+q_dot_rad+q_dot_conv   "energy balance on panel"
A_s=W*L   "surface area of panel"
q_dot_s=q``_dot_s*A_s   "rate of solar energy hitting panel"
w_dot=q_dot_s*eta   "rate of electrical energy produced by panel"
q_dot_rad=A_s*sigma#*e*(T^4-T_sur^4)   "rate of radiation from panel"
q_dot_conv=A_s*h_bar*(T-T_infinity)   "rate of convection from panel"

Solving provides $\boxed{T = 332.5 \text{ K } (59.38°\text{C})}$ and $\boxed{\dot{w} = 76.39 \text{ W}}$. A Parametric Table is created that includes the variables q``_dot_s, w_dot, T, and eta. The values of q``_dot_s are varied from 0 to 1000 W/m² and the line in the Equations Window that specifies the value of this variable is commented out. The Parametric Table is run and the results are used to generate a plot showing electrical power as a function of solar flux, Figure 3.

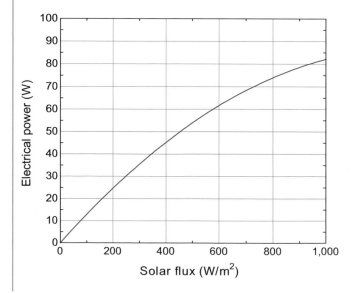

Figure 3 Electrical power generated by the panel as a function of the solar flux.

Example 1.3 (cont.)

Discussion/Exploration

The rate at which the panel produces electricity does not increase linearly with the rate of solar energy collected by the panel because the panel efficiency decreases as the solar flux (and thus the panel temperature) increases. Figure 4 shows the temperature and efficiency of the panel as a function of the solar flux.

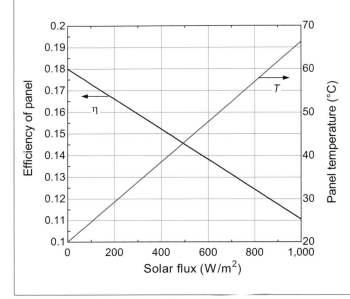

Figure 4 Efficiency and temperature of the panel as a function of the solar flux.

1.5 Thermophysical Properties

There are two types of properties commonly required to solve thermal-fluid problems. Thermodynamic properties refer to the equilibrium characteristics of a substance and include density and internal energy. Transport properties are not equilibrium properties as they describe the ability of a material to transfer some quantity like energy or momentum in a nonequilibrium process. These properties include conductivity and viscosity. In this section we will review some common models and references that can be used to estimate thermodynamic and transport properties.

1.5.1 Real Fluids

The **Phase Rule** is a basic theorem of equilibrium thermodynamics. According to the Phase Rule, the equilibrium state of a pure, single-phase substance is fixed by any two mass independent properties. That is, if two properties are known then all other properties are fixed. The Phase Rule also indicates that the state of a pure, two-phase substance is fixed by a single property. For example, the temperature at which water boils is fixed only by its pressure.

In general, the interrelationship between properties is complex and therefore properties are often presented in large tables that are two-dimensional in regions where the fluid is single-phase (i.e., subcooled liquid and superheated vapor) and one-dimensional in the region where the fluid is two-phase (i.e., the vapor dome). The Steam Tables that are used in most thermodynamics classes are examples of this type of information. Appendix D provides the saturated liquid and saturated vapor properties of a few fluids.

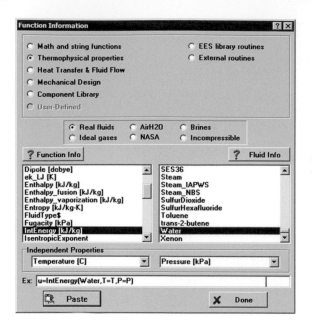

Figure 1.4 Function Information dialog showing properties available for real fluids in EES.

There are several computer tools available that can provide the properties of fluids. One example is the NIST Standard Reference Database 23, also called REFPROP (Lemmon *et al.*, 2018) and another is the database included in the Engineering Equation Solver (EES) software. EES allows easy access to the properties of many "real fluids", including not only water but also hundreds of other substances. The manner in which the properties of real fluids can be accessed using EES is discussed in the EES tutorial available in Appendix E. Select Function Info from the Options menu and then click the radio button Thermophysical properties, as shown in Figure 1.4. The various fluid property models are listed in the center box; select Real fluids to access a list of the real fluids programmed in EES (on the right) and the associated properties (on the left).

Select the property of interest from the list on the left and the fluid from the list on the right. Finally, select the two properties that fix the state using the drop down menus under the label Independent Properties. When you are satisfied with the selections, hit the Paste button to enter the code into the Equations Window. For example, the combination shown in Figure 1.4 will lead to the EES code below, which provides the specific internal energy of water given values of temperature and pressure.

```
u=IntEnergy(Water,T=T,P=P)
```

Specific internal energy is a thermodynamic property. The EES database of properties also includes transport properties that can be accessed in the same manner. For example, the thermal conductivity of water is obtained using the Conductivity function.

```
k=Conductivity(Water,T=T,P=P)
```

In order to use the properties programmed in EES it is necessary to specify the unit system; this choice will dictate the units that must be used for the inputs to the property functions (e.g., temperature T and pressure P in the code above) as well as the units of the properties that are returned (e.g., specific internal energy u and conductivity k). The unit system can be specified by selecting Unit System from the Options menu or, more conveniently, by using the $UnitSystem directive placed at the top of the Equations Window.

The EES code below computes the specific internal energy and conductivity of water at $T = 120°C$ and $P = 250$ kPa.

```
$UnitSystem SI Mass kJ kPa C Radian
T=120 [C]
P=250 [kPa]
u=IntEnergy(Water,T=T,P=P)
k=Conductivity(Water,T=T,P=P)
```

The result is $u = 503.6$ kJ/kg and $k = 0.6823$ W/m-K.

1.5.2 Ideal Gas Model

Under some conditions, it is appropriate to use the **ideal gas** model for thermophysical properties. The ideal gas model works well for very low-density gases where intermolecular forces are small. The density of an ideal gas is given by the well-known Ideal Gas Law, which is expressed on a mass basis as

$$\rho = \frac{P}{RT},$$ (1.8)

where P and T are the absolute pressure and temperature, respectively, and R is the **gas constant**. The value of R is related to the **universal gas constant** ($R_{univ} = 8314$ J/kmol-K) according to:

$$R = \frac{R_{univ}}{MW},$$ (1.9)

where MW is the **molar mass** of the gas.

A subtle outcome of the fact that a fluid obeys the Ideal Gas Law is that many of its thermodynamic and transport properties are only functions of temperature. For example, the specific internal energy (u) and specific enthalpy (i) of an ideal gas are only functions of temperature:

$$u = u(T) \text{ for an ideal gas}$$ (1.10)

$$i = i(T) \text{ for an ideal gas.}$$ (1.11)

As a result, the change in the specific internal energy and the change in specific enthalpy of an ideal gas between two temperatures (T_1 and T_2) can be evaluated by integrating the specific heat capacity at constant volume (c_v) and the specific heat capacity at constant pressure (c_p), respectively:

$$u_2 - u_1 = \int_{T_1}^{T_2} c_v(T)dT \text{ for an ideal gas}$$ (1.12)

$$i_2 - i_1 = \int_{T_1}^{T_2} c_p(T)dT \text{ for an ideal gas.}$$ (1.13)

We will often make the assumption that c_v and c_p are constant, allowing Eqs. (1.12) and (1.13) to be written as

$$u = c_v T \text{ for an ideal gas with constant } c_v$$ (1.14)

$$i = c_p T \text{ for an ideal gas with constant } c_p.$$ (1.15)

Note that Eqs. (1.14) and (1.15) assume a zero reference state for the property at 0 K.

The transport properties conductivity and viscosity are also functions only of temperature for an ideal gas:

$$k = k(T) \text{ for an ideal gas} \tag{1.16}$$

$$\mu = \mu(T) \text{ for an ideal gas.} \tag{1.17}$$

Appendix C contains a tabulation of some property data for ideal gases near room temperature. The EES software also includes many fluids that are modeled as ideal gases. Select Function Info from the Options menu and then select the Thermophysical properties radio button. Select Ideal gases from the selection of models that are available and you will again see a list of ideal gas fluids (on the right) and properties (on the left). Notice that most ideal gas fluids are indicated by their chemical structure as opposed to their full name; for example the fluid "CO2" in EES is carbon dioxide modeled as an ideal gas while "CarbonDioxide" is the real fluid model of the same substance. The only exception to this rule is "Air", which is modeled as an ideal gas whereas "Air_ha" is modeled as a real fluid. Also notice that many of the EES property functions will only allow one input (temperature, or a property or group of properties from which temperature can be determined) to be provided for an ideal gas.

Example 1.4

A closed cylindrical tank with diameter $D = 3$ inch and $L = 6$ inch is filled with air that is initially at $T = 70°F$ and $P = 1$ atm. The tank is placed in an oven causing its temperature to rise. The surface of the tank experiences convection with the air in the oven at $T_\infty = 450°F$ with average heat transfer coefficient $\bar{h} = 5.4$ W/m^2-K. The surface of the tank also experiences radiation with the oven walls which are at $T_{sur} = T_\infty$. The emissivity of the outer surface of the tank is $\varepsilon = 0.74$.

Determine:
- The rate at which the temperature of the air in the tank is initially changing.
- The rate at which the pressure of the air in the tank is initially changing.

Known Values

The problem inputs are converted to base SI units and shown in Figure 1 on a sketch of the problem.

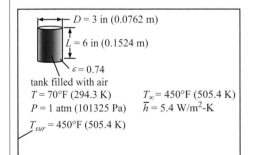

$D = 3$ in (0.0762 m)
$L = 6$ in (0.1524 m)
$\varepsilon = 0.74$
tank filled with air
$T = 70°F$ (294.3 K) $T_\infty = 450°F$ (505.4 K)
$P = 1$ atm (101325 Pa) $\bar{h} = 5.4$ W/m^2-K
$T_{sur} = 450°F$ (505.4 K)

Figure I Tank of air in an oven.

Assumptions

- The air in the tank and the tank wall are at a uniform temperature.
- The tank experiences radiation and convection from all of its surfaces, including the bottom (i.e., it is not on a shelf).

Example 1.4 (cont.)

- The air in the tank can be modeled as being an ideal gas with constant c_v and c_p.
- The tank walls do not store a significant amount of energy (this assumption may not be very good...).
- The tank volume does not change during heating.
- The tank does not leak.

Analysis

An energy balance on the tank is shown in Figure 2.

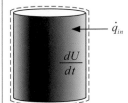

Figure 2 Energy balance on the tank.

There is no work if the tank is rigid and no mass flow across the system boundary if the tank does not leak. Therefore, Eq. (1.2) reduces to:

$$\dot{q}_{in} = \frac{dU}{dt}.$$ (1)

The rate of heat transfer to the tank is related to convection and radiation:

$$\dot{q}_{in} = \dot{q}_{conv} + \dot{q}_{rad}.$$ (2)

The rate of convection heat transfer can be computed using Eq. (1.5):

$$\dot{q}_{conv} = \bar{h} A_s (T_\infty - T).$$ (3)

Note that $\dot{q}_{conv}$ is the rate of convection heat transfer defined to conform with the direction of the arrow in Figure 2; that is, *from* the air *to* the tank and therefore this heat transfer is driven by $T_\infty - T$ in Eq. (3). Also note that Eq. (3) assumes that the temperature of the air in the tank is the same as the temperature of the surface of the tank. The rate of radiation heat transfer can be computed using Eq. (1.7):

$$\dot{q}_{rad} = A_s \sigma \varepsilon (T_{sur}^4 - T^4).$$ (4)

The surface area used in both Eqs. (3) and (4) is computed from:

$$A_s = 2\pi \frac{D^2}{4} + \pi D L.$$ (5)

The total internal energy in Eq. (1) is expressed as the product of the mass of air and its specific internal energy:

$$\dot{q}_{in} = \frac{d}{dt}(mu).$$ (6)

If the air is assumed to behave as an ideal gas with constant c_v then Eq. (1.14) can be used for u:

$$\dot{q}_{in} = \frac{d}{dt}(m c_v T).$$ (7)

Both m and c_v are constant for this process and therefore the energy balance becomes:

Continued

Example 1.4 (cont.)

$$\dot{q}_{in} = mc_v \frac{dT}{dt}, \tag{8}$$

which can be solved for the rate of temperature change:

$$\frac{dT}{dt} = \frac{\dot{q}_{in}}{mc_v}. \tag{9}$$

The mass of air in the tank is the product of the tank volume and the air density:

$$m = V\rho. \tag{10}$$

The density of air in the tank is computed using the Ideal Gas Law, Eq. (1.8):

$$\rho = \frac{P}{RT}. \tag{11}$$

Substituting Eq. (11) into Eq. (10) leads to:

$$m = V\frac{P}{RT}. \tag{12}$$

The volume of air is given by:

$$V = \pi \frac{D^2}{4} L. \tag{13}$$

The rate of change of the pressure in the tank is obtained using a mass balance on the air; because no mass is either entering or leaving the rate of storage must be zero:

$$0 = \frac{dm}{dt}. \tag{14}$$

Substituting Eq. (12) into Eq. (14) provides:

$$\frac{dm}{dt} = \frac{d}{dt}\left(V\frac{P}{RT}\right) = \frac{V}{R}\frac{d}{dt}\left(\frac{P}{T}\right) = \frac{V}{R}\left(\frac{1}{T}\frac{dP}{dt} - \frac{P}{T^2}\frac{dT}{dt}\right) = 0. \tag{15}$$

Solving Eq. (15) for the rate of change of pressure leads to:

$$\frac{dP}{dt} = \frac{P}{T}\frac{dT}{dt}. \tag{16}$$

Solution

The equations derived in the analysis section can be solved sequentially and by hand (albeit not in the order that they were derived). The properties required for air can be obtained using Table C.1 in Appendix C. Using the equations provided in Appendix C, the gas constant and constant volume specific heat capacity for air are $R = 287.0$ N-m/kg-K and $c_v = 720.0$ J/kg-K, respectively. Equations (5) and (13) are used to compute the surface area and volume of the tank:

$$A_s = 2\pi \frac{D^2}{4} + \pi D L = \frac{2\pi}{4}\left|\frac{(0.0762)^2 \text{ m}^2}{}\right| + \pi \left|\frac{0.0762 \text{ m}}{}\right|\frac{0.1524 \text{ m}}{} = 0.0456 \text{ m}^2$$

$$V = \pi \frac{D^2}{4} L = \frac{\pi}{4}\left|\frac{(0.0762)^2 \text{ m}^2}{}\right|\frac{0.1524 \text{ m}}{} = 6.950 \times 10^{-4} \text{ m}^3.$$

Equation (12) is used to compute the mass of air in the tank:

$$m = V\frac{P}{RT} = \frac{6.950 \times 10^{-4} \text{ m}^3}{}\left|\frac{101325 \text{ Pa}}{}\right|\frac{\text{kg K}}{287.0 \text{ N m}}\left|\frac{}{294.3 \text{ K}}\right|\left|\left|\frac{\text{N}}{\text{Pa m}^2}\right.\right. = 8.338 \times 10^{-4} \text{ kg}.$$

Example 1.4 (cont.)

Equations (3) and (4) are used to compute the rates of heat transfer by convection and radiation, respectively, and Eq. (2) is used to calculate the total rate of heat transfer.

$$\dot{q}_{conv} = \bar{h}\, A_s (T_\infty - T) = \frac{5.4\ \text{W}}{\text{m}^2\ \text{K}} \left| \frac{0.0456\ \text{m}^2}{} \right| \frac{(505.4 - 294.3)\ \text{K}}{} = 51.99\ \text{W}$$

$$\dot{q}_{rad} = A_s\, \sigma\, \varepsilon (T_{sur}^4 - T^4) = \frac{0.0456\ \text{m}^2}{} \left| \frac{5.67 \times 10^{-8}\ \text{W}}{\text{m}^2\text{-K}^4} \right| \frac{0.74}{} \left| \frac{(505.4^4 - 294.3^4)\,\text{K}^4}{} \right. = 110.5\ \text{W}$$

$$\dot{q}_{in} = \dot{q}_{conv} + \dot{q}_{rad} = 51.99\ \text{W} + 110.5\ \text{W} = 162.5\ \text{W}.$$

Equations (9) and (16) are used to compute the rate of temperature and pressure change, respectively:

$$\frac{dT}{dt} = \frac{\dot{q}_{in}}{mc_v} = \frac{162.5\ \text{W}}{} \left| \frac{}{8.338 \times 10^{-4}\ \text{kg}} \right| \frac{\text{kg K}}{720\ \text{J}} \left| \frac{\text{J}}{\text{W s}} \right. = \boxed{270.7\ \text{K/s}}$$

$$\frac{dP}{dt} = \frac{P}{T}\frac{dT}{dt} = \frac{101325\ \text{Pa}}{} \left| \frac{}{294.3\ \text{K}} \right| \frac{270.7\ \text{K}}{\text{s}} = \boxed{93{,}200\ \text{Pa/s}\ (13.52\ \text{psi/s})}.$$

Discussion/Exploration

The rate of change of temperature and pressure will not remain constant during the process. The initial rates of change are high due to the large rate of heat transfer that accompanies the large temperature difference between the tank and its surroundings. As the temperature of the air in the tank increases over time, the rates of change of temperature and pressure will decrease. Eventually the process will achieve **steady state** where the temperature and pressure stop changing. For this problem, steady state corresponds to the temperature of the tank reaching the same temperature as the oven, 450°F. At this point the rate of heat transfer is zero and, according to Eqs. (9) and (16), the rate of temperature and pressure change are also both zero. That is the definition of steady state: the properties of the system are unchanging with time.

By neglecting the amount of energy stored in the tank we likely have made a very poor assumption. Even a very thin wall tank will store more energy than the air that is contained within it due to the extremely high density of the tank wall material relative to air. We could have modeled the tank wall material as being an incompressible substance and included it in the analysis, as discussed in the subsequent section.

1.5.3 Incompressible Substance Model

It is often appropriate to model a substance as being **incompressible**. The density of an incompressible substance does not depend on pressure. In some cases, the density can also be assumed to be independent of temperature. Many, if not most, of the heat transfer problems that you will encounter involve substances that can, with little loss of accuracy, be approximated as being incompressible. One implication of the incompressible model is that the specific internal energy depends only on temperature:

$$u = u(T) \quad \text{for an incompressible substance.} \tag{1.18}$$

The change in the specific internal energy of an incompressible substance can be obtained by integrating the specific heat capacity (c):

$$u_2 - u_1 = \int_{T_1}^{T_2} c(T)\,dT \quad \text{for an incompressible substance.} \tag{1.19}$$

Notice that there is no need for a subscript v or p on the specific heat capacity of an incompressible substance because the two definitions lead to very nearly the same value and therefore only one specific heat is provided. We will often make the assumption that c is constant, allowing Eq. (1.19) to be written as:

$$u = cT \quad \text{for an incompressible substance with constant } c. \tag{1.20}$$

Equation (1.20) assumes a reference temperature of 0 K at which u is set to zero. Unlike specific internal energy, the specific enthalpy of an incompressible substance depends on both temperature and pressure. The change in the specific enthalpy of an incompressible substance from state 1 at T_1 and p_1 to state 2 at T_2 and p_2 is given by:

$$i_2 - i_1 = \int_{T_1}^{T_2} c(T)dT + \frac{(p_2 - p_1)}{\rho} \quad \text{for an incompressible substance.} \tag{1.21}$$

If the specific heat capacity is assumed constant then Eq. (1.21) can be written as:

$$i_2 - i_1 = c(T_2 - T_1) + \frac{(p_2 - p_1)}{\rho} \quad \text{for an incompressible substance with constant } c. \tag{1.22}$$

It is almost always the case that the enthalpy change caused by a change in temperature will be much larger than the enthalpy change caused by a change in pressure. In this case, the second term on the right side of Eq. (1.22) can be neglected as being small relative to the first term and specific enthalpy can be written approximately as:

$$i \approx cT \quad \text{for an incompressible substance with constant } c. \tag{1.23}$$

Appendices A and B contain a tabulation of data for solids and liquids near room temperature that can be treated approximately as incompressible substances. The EES software also includes many liquids and solids that are modeled as incompressible substances. Select Function Info from the Options menu and then select the Thermophysical properties radio button. Select Incompressible from the list of models that are available to obtain a list of incompressible substances (on the right) and their properties (on the left). The drop down menu enables subsets of these substances to be examined (e.g., Metals or Building Materials). Many of the properties for incompressible substances will only allow temperature to be provided as the one required input.

Example 1.5

A tube is wrapped with heater tape in order to elevate the temperature of the heat transfer fluid, Dowtherm A, that is flowing through it. The inlet temperature of the fluid is $T_{in} = 70°F$. The tube length is $L = 72$ ft and the outer diameter is $D_{out} = 0.5$ inch. The heat tape provides a uniform heat flux of $\dot{q}'' = 8.2$ kW/m² to the external surface of the tube. Insulation is wrapped around the heat tape. The volumetric flow rate of fluid flowing through the tube is $\dot{V} = 3.5$ gal/min.

Determine:
- The mean temperature at which the fluid leaves the tube.

Known Values

The problem inputs are shown in Figure 1 on a sketch of the problem.

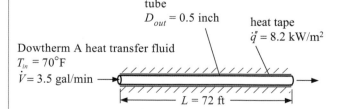

tube
$D_{out} = 0.5$ inch

heat tape
$\dot{q}'' = 8.2$ kW/m²

Dowtherm A heat transfer fluid
$T_{in} = 70°F$
$\dot{V} = 3.5$ gal/min

$L = 72$ ft

Figure 1 Fluid flowing through a tube wrapped with heat tape.

Example 1.5 (cont.)

Assumptions

- The energy provided by the heat tape goes entirely into the fluid and is not lost to the surroundings.
- The Dowtherm A fluid can be modeled as being incompressible with constant specific heat capacity.
- The effect of the pressure change on the enthalpy change can be neglected.
- The tube is at steady state.

Analysis

The surface area of the tube that is exposed to the heat flux is computed according to:

$$A_s = \pi D_{out} L. \tag{1}$$

The total heat transfer rate to the fluid is the product of the heat flux provided by the heat tape and the surface area to which it is applied:

$$\dot{q} = \dot{q}'' A_s. \tag{2}$$

The mass flow rate of Dowtherm A is the product of the volumetric flow rate and its density:

$$\dot{m} = \dot{V} \rho. \tag{3}$$

An energy balance on the pipe includes the enthalpy flow into the pipe, the enthalpy flow out of the pipe, and the heat transfer from the heat tape:

$$\dot{m} \, i_{in} + \dot{q} = \dot{m} \, i_{out}. \tag{4}$$

Substituting Eq. (1.23) for specific enthalpy and solving for the outlet temperature provides:

$$T_{out} = T_{in} + \frac{\dot{q}}{\dot{m} c}. \tag{5}$$

Solution

Equations (1) through (5) can be solved by hand or by using a computer program. The properties for Dowtherm A that are required include the density and specific heat capacity. EES is used because it has these properties included in its database. The unit system is specified using the $UnitSystem directive and the inputs to the problem are entered and converted to base SI units.

```
$UnitSystem SI Mass J K Pa
F$='Dowtherm_A'                                    "fluid"
D_out=0.5 [inch]*Convert(inch,m)                   "outer diameter of pipe"
L=72 [ft]*Convert(ft,m)                            "length of pipe"
V_dot=3.5 [gal/min]*Convert(gal/min,m^3/s)         "volumetric flow rate"
q``_dot=8.2 [kW/m^2]*Convert(kW/m^2,W/m^2)         "heat flux"
T_in=ConvertTemp(F,K,70 [F])                       "inlet temperature"
```

The Cp and Density functions are used to obtain c and ρ for Dowtherm A (which correspond to the substance "Dowtherm_A" in EES). Note that the Dowtherm_A fluid name is provided with string variable F$.

```
c=Cp(F$,T=T_in)                                    "specific heat capacity"
rho=Density(F$,T=T_in)                             "density"
```

Continued

Example 1.5 (cont.)

Equations (1) through (3) and Eq. (5) are entered in EES.

A_s=pi*D_out*L	"surface area"
q_dot=q``_dot*A_s	"heat transfer rate"
m_dot=rho*V_dot	"mass flow rate"
T_out=T_in+q_dot/(c*m_dot)	"exit temperature"

Solving provides $T_{out} = 313.8$ K ($105.1°$F).

Discussion/Exploration

We modeled the enthalpy change of the Dowtherm A assuming that the specific heat capacity is constant and also that the impact of the pressure change is small relative to the temperature change. It is possible to check that this assumption is reasonable. In Chapter 9, methods for predicting the pressure drop associated with this type of internal flow problem are presented; you may already have been exposed to these methods in your fluids class. The result of such an analysis would be a pressure drop of approximately $\Delta p = 430$ kPa (62 psi). Equation (1.22) provides the enthalpy change for an incompressible fluid with constant c including both the temperature- and pressure-driven components:

$$i_2 - i_1 = \underbrace{c(T_2 - T_1)}_{\substack{\text{temperature-}\\ \text{driven enthalpy}\\ \text{change}}} + \underbrace{\frac{(p_2 - p_1)}{\rho}}_{\substack{\text{pressure-driven}\\ \text{enthalpy change}}} . \tag{6}$$

For this problem, the magnitude of the first term in Eq. (6) is 30.7 kJ/kg while the magnitude of the second term in Eq. (6) is only 0.41 kJ/kg; the pressure-driven change in the enthalpy is on the order of 1% of the temperature-driven change.

1.6 Conclusions and Learning Objectives

This chapter presents a very brief overview of the relevance of heat transfer in today's world and an introduction to the three heat transfer mechanisms: conduction, convection, and radiation. Heat transfer relies on some of the same concepts that were introduced in thermodynamics, most importantly, the energy balance and models for the properties of substances. These concepts were reviewed briefly in this chapter.

A methodology for solving engineering problems was presented and used in the example problems. This methodology requires that a sufficient set of equations first be derived in the analysis step. These equations are independent of the subsequent solution step where they can be solved either by hand calculation or using computer software. Adhering to this approach requires that you consciously separate the analysis process from the solution process when working problems. This methodology will be adhered to for the remainder of the examples in this book.

Specific concepts and ideas that you should understand from this chapter are listed below.

- The difference between an energy balance and a rate equation (also called a mechanism equation).
- The three mechanisms for heat transfer and their basic description and rate equations.
- The significance of thermal conductivity, heat transfer coefficient, and emissivity.
- The fundamental steps for solving an engineering problem and in particular the difference between the analysis and solution steps in solving a problem.
- The units of heat transfer, heat transfer rate, and heat flux.

- How to estimate internal energy and enthalpy for an ideal gas.
- How to estimate internal energy and enthalpy for an incompressible substance.
- How to use the properties available in Appendices A through D of this text.
- How to enter inputs into EES and convert them to base SI units.
- How to enter equations into EES.
- How to carry out parametric studies and create plots in EES.
- How to access thermodynamic and transport properties in EES for real fluids, ideal gases, and incompressible substances.

Reference

Lemmon, E. W., M. L. Huber, and M. O. McLinden, NIST Standard Reference Database Number 23, NIST Reference Fluid Thermodynamics and Transport Properties Database (REFPROP): Version 10, June 2018.

Problems

1.1 An oven can be modeled as a cube that is 20 inches on a side. The front of the oven is a window made of glass that is 0.5 inch thick. The other five surfaces are made of brick with thickness 1 inch. Estimate the rate at which the electrical heaters within the oven must provide power in order to maintain the temperature of the inside surfaces at 400°F if the outside surface is at 70°F.

1.2 About one-third of the energy released with the combustion of fuel in an automobile engine is transferred to the surroundings through a heat exchanger called the radiator that is located in the front of the vehicle. About one-third of the energy is converted to mechanical power in the engine and the remaining one-third exits with the exhaust. A mixture of ethylene glycol and water causes the surface of the radiator to reach an average temperature of 200°F. Despite its name, heat transfer in the radiator occurs almost exclusively by convection to the surrounding air. In a particular case, the engine is producing 25 hp of mechanical power. The total surface area of the radiator experiencing convection is 6 m^2 and the temperature of the surrounding air is 70°F. Estimate the heat transfer coefficient between the radiator surface and the air.

1.3 The hot surface of a household iron has a surface area of 20 square inches. The iron is set for wool so that the surface temperature is 145°C. Thermal energy transfers from the iron surface to the 25°C surrounding air by free convection with an average heat transfer coefficient of 6.2 W/m^2-K. Estimate the electrical power needed to maintain the iron at this state.

1.4 Solar energy is absorbed on a metal plate with a flux of 375 W/m^2. The back side of the plate is insulated, but the top side is exposed to outdoor air at 25°C. A thermocouple mounted on the plate surface indicates a temperature of 42°C. Heat transfer takes place by convection due to the wind. Determine the average convective heat transfer coefficient between the plate and the surrounding air for these conditions.

1.5 A 60 W incandescent light bulb consists of an electrically heated filament surrounded by a glass enclosure having a surface area of 24.5 inch^2. Ten percent of the power is converted to light, which is transmitted through the glass. The remaining 90 percent is converted to heat, which must be transferred from the glass surface to the 70°F surroundings by radiation and free convection. The emissivity of glass is 0.90. The free convection coefficient is 12.5 W/m^2-K. Determine the temperature of the glass surface (in °F) under steady-state conditions. How important is the radiative heat transfer contribution in this situation?

1.6 An experiment to determine the convective heat transfer coefficient for air flow over a sphere consists of a 1.5 inch diameter copper sphere that is electrically heated to maintain its surface at a uniform temperature of 46°C as air at 25°C and 101.3 kPa is blown past the sphere with a velocity of 12 m/s. The sphere is polished to reduce radiation. The electrical power supplied to the heater that is embedded in the sphere is 7.2 W.

(a) Determine the average convection coefficient between the sphere and the air (neglect radiation).

(b) The polished sphere has an emissivity of 0.21. Determine the rate of radiative heat transfer from the sphere.

(c) Determine a corrected average convection coefficient that accounts for the radiation.

(d) Plot the corrected heat transfer coefficient as a function of the emissivity.

1.7 A microprocessor chip used in personal computers has a footprint of 45 mm × 42.5 mm and dissipates a peak power of 95 W. It is important to provide cooling to this chip as its performance deteriorates with increasing temperature.

(a) Determine the temperature that the chip would achieve under steady-state conditions in a 25°C environment if cooling were provided only by free convection with an average convection heat transfer coefficient of 6.0 W/m^2-K.

(b) A variety of techniques such as fans or heat-pipe coolers can be used to cool the chip. Determine the convection heat transfer coefficient that is required to maintain the chip at a maximum temperature of 42°C.

1.8 A single-glazed window is exposed to indoor air at 25°C on one side and cold outdoor air at –4°C on the other side. The average heat transfer coefficients are 6 W/m^2-K on the inner surface and 18 W/m^2-K on the outer surface. The glass is opaque to thermal radiation and has an emissivity of 0.9.

(a) Assume that the glass is sufficiently thin that the inside and outside temperatures are nearly the same. Determine the steady-state temperature of the glass and the rate of heat transfer per unit area from indoors to outdoors.

(b) The glass is 6 mm thick and its thermal conductivity is 0.95 W/m-K. Determine the temperature difference between the inside and outside surfaces of the glass. Is the assumption of uniform temperature made in part (a) justified?

1.9 The temperature distribution across a 0.5 m thick wall at a certain instant of time is given by the equation $T = 900 - 300x - 50x^2$, where T is the temperature in °C and x is the distance from the left side of the wall in m. The wall has conductivity $k = 0.72$ W/m-K.

(a) Calculate the rate of heat transfer per unit area in the positive x-direction at the left face of the wall ($x = 0$ m).

(b) Calculate the rate of heat transfer per unit area in the positive x-direction at the right face ($x = 0.5$ m).

(c) Indicate whether there is a net energy transfer rate into or out of the wall.

(d) The air temperature exposed to the left face of the wall is $T_\infty = 950$°C. Determine the heat transfer coefficient between the left surface of the wall and the air.

1.10 On a hot (86°F) summer day, solar radiation is absorbed on the asphalt at a rate of 625 W/m^2. The asphalt convects and radiates energy to the surroundings as well as conducting some energy into the ground. If there is no wind, the free convection coefficient is about 6 W/m^{2}K. The emissivity of the asphalt is 0.8. The thermal conductivity of the asphalt is 0.062 W/m-K and it is 2 inches thick. The temperature of the ground under the asphalt is 66°F. Calculate the steady-state temperature of the surface of the asphalt in °F? Will it be high enough to "fry an egg"?

1.11 A bike computer used for determining the speed and distance travelled by the biker also provides the outdoor temperature. However, it always reads higher than the actual temperature during the daytime. For example, on a sunny day when the actual temperature is 78°F, the computer shows a temperature of 86°F, while you are riding at 15 mph. Convection correlations indicate that the overall convection coefficient between the air and the computer surface is 42 W/m^2-K at these conditions. The emissivity of the computer surface is 0.9. Use this information to estimate the rate at which solar radiation is being absorbed per unit area on the computer surface in order to cause this discrepancy.

1.12 Water flows through a pipe with a volumetric flow rate of $\dot{V} = 20$ gal/min. The inner diameter of the pipe is $D = 2.25$ inch and the pipe is $L = 50$ ft long. The properties of water include specific heat capacity $c = 4200$ J/kg-K, density $\rho = 1000$ kg/m^3, and viscosity $\mu = 0.001$ kg/m-s. The heat transfer coefficient on the external surface of the pipe is $\bar{h} = 4$ Btu/hr-ft^2-R and the surrounding temperature is $T_\infty = 70°$F. The emissivity of the external surface of the pipe is $\varepsilon = 0.9$. The inlet temperature of the water is $T_{in} = 150°$F.

(a) Determine the pressure drop across the pipe (i.e., the amount of pressure rise that your pump will need to provide) using the formula

$$\Delta P = \frac{\rho u_m^2}{2} f \frac{L}{D},$$

where u_m is the average velocity of the flow in the pipe (the ratio of the volumetric flow rate to the pipe cross-sectional area) and $f = 0.016$ is the friction factor.

(b) Estimate the temperature drop experienced by the water as it flows through the pipe.

1.13 A pot is placed on an electric burner that consumes 1.2 kW of power and contains water that is boiling. The pot has diameter 8 inches and height 6 inches. The surfaces of the pan experience convection and radiation with surroundings at 20°C. The heat transfer coefficient is 8.0 W/m^2-K and the emissivity is 0.7. Estimate the efficiency of the burner, defined as the ratio of the rate of heat transfer that is provided to the water to the power provided to the burner.

1.14 Some hybrid and electric car manufacturers have begun to advertise that their cars are "green" because they can incorporate solar photovoltaic panels on their roof and use the power that is generated by the panel to power the wheels. In this problem we will assess the value of a solar panel installed on the roof of a car. Assume that the panel is 5 ft long and 4 ft wide. On a very sunny day (depending on your location and the time of year), the rate of solar energy per area hitting the roof of the car is 750 W/m^2. Assume that the panel's efficiency relative to converting solar energy to electrical energy that can be used to power the wheels is 12 percent. The cruising power required by the car is 22 hp.

(a) Estimate the rate of electrical power produced by the panel.

(b) Calculate the fraction of the *power* required by the car that is produced by the solar panel at a given instant of time.

(c) Estimate the temperature of the panel if the heat transfer coefficient is 45 W/m^2-K and the ambient temperature is 22°C.

(d) If the car (and therefore the panel) sits in the sun for 6 hours during a typical day then determine the total amount of electrical energy produced by the panel and stored in the car's battery.

(e) If the car is driven for 30 minutes during a typical day then determine the total amount of energy required by the car.

(f) What is the fraction of the *energy* required by the car that is produced by the solar panel during a typical day?

(g) Create a plot showing the fraction of energy required by the car that is produced by the solar panel as a function of the panel efficiency.

1.15 The quality of the construction of the walls of your house is typically characterized by an R-value. According to the Department of Energy, housing in Wisconsin should be constructed using R-values between 13 ft^2-F-hr/Btu to 23 ft^2-F-hr/Btu. Assume that you live in an average house constructed with R-18 walls. Your house has $A_{wall} = 600$ ft^2 of exposed walls. The total amount of heat transfer required to keep a house conditioned during the winter months is calculated using the formula

$$Q = \frac{A_{wall}}{R} HDD,$$

where *HDD* is the number of heating degree days at the location of the home. The number of heating degree days is the integral of the difference between the indoor and outdoor temperature with respect to time and has been tabulated for many locations in the USA (search online for heating degree days...). There are a number of places to go to find heating degree days for different locations; in Madison, WI, the number of heating degree days for a winter is approximately 7800°F-day.

(a) Calculate the amount of heat required to keep your house conditioned during the winter (in MJ).

(b) If the price of natural gas is 0.75 $/therm then calculate the cost of heating your home with natural gas (assuming your furnace is 100 percent efficient).

(c) If the price of electricity is 0.136 $/kW-hr then calculate the cost of heating your home with electricity.

1.16 An uninsulated, electrically heated water heater consists of a cylindrical tank made of metal with a 50 gallon capacity. The thin wall tank has a diameter of 18 inches. The outside surface of the tank is exposed to air in the basement at 20°C with a convection heat transfer coefficient of 4.5 W/m²-K. The bottom of the tank is sitting on a thick slab of Styrofoam and so the rate of heat loss from the bottom can be assumed to be negligible. A thermostat in the tank maintains the water at an average temperature of 52°C.

(a) Estimate the steady-state rate of heat loss from the tank when the water is at 52°C.

(b) An average of 42 gallons of hot water is used each day and must be heated from 10°C to 52°C. Determine the fraction of the daily total electrical energy supplied to the water heater that is required for the heat losses through the top and sides.

(c) Assuming that electricity costs 15 cents per kW-hr, determine the annual cost resulting from the heat loss from the tank.

1.17 A heater is to be operated in atmospheric air at 20°C. The emissivity of the heater surface is 0.8. Experiments have shown that the heat transfer coefficient can be estimated according to:

$$\bar{h} = 20[\text{m}^{-1}\ \text{K}^{-1/3}](T - T_\infty)^{1/3}k,$$

where T is the surface temperature, T_∞ is the surrounding temperature, and k is the conductivity of air evaluated at the average of T and T_∞.

(a) If the surface temperature of the heater is 300°C then determine the heat flux due to convection and radiation.

(b) Plot the heat flux due to convection and radiation as a function of the surface temperature.

1.18 The heat transfer rate equation associated with radiation was provided by Eq. (1.7):

$$\dot{q}_{rad} = A_s\,\sigma\,\varepsilon\left(T_s^4 - T_{sur}^4\right).$$

This equation can be expressed in the same form as Newton's Law of Cooling given in by Eq. (1.5):

$$\dot{q}_{conv} = \bar{h}\,A_s(T_s - T_\infty).$$

(a) Derive an equation that provides the "heat transfer coefficient for radiation."

(b) Plot the radiation heat transfer coefficient determined in part (a) for emissivity $\varepsilon = 1$, surroundings temperature $T_{sur} = 298.1$ K and surface temperatures ranging from 300 K to 1000 K.

1.19 A residential water heater consists of a well-insulated 40 gallon tank with two 2500 W immersion electrical heating elements. Each element resembles a U-shaped tube with a total (stretched) length of 620 mm and diameter of 15 mm. Water in the tank is heated by free convection from the surface of the heating element. Experiments have shown that the free convection heat transfer coefficient for elements of this design is given by:

$$\bar{h} = 285 + 10\,T_s,$$

where $\bar{h}$ is in W/m²-K and T_s is the surface temperature of the element in °C. Assume that the water in the tank is fully mixed (i.e., at a uniform temperature at any time).

(a) Determine the time required to heat the water from the supply temperature (10°C) to its final temperature of 55°C.

(b) Determine the surface temperature of the heating element shortly after the heating is initiated when the water temperature is at 10°C.

(c) Determine the surface temperature of the heating element at the end of the heating process when the water is 55°C.

(d) Plot the element surface temperature as a function of the water temperature.

1.20 A hot-wire anemometer is a device that is used to measure air velocity. The instrument is based on the fact that the heat transfer rate from an electrically heated wire in crossflow is related to the velocity of the air that is flowing perpendicular to the wire. In a particular experiment, a hot-wire anemometer is used to estimate the velocity of ambient pressure air at 24°C. A platinum wire having a diameter of 0.125 mm and length of 6.5 mm is maintained at 230°C. The power required to maintain the wire at this temperature is 0.36 W.

(a) Determine the average convection co-efficient between the wire and the free-stream air.

It is known that the average heat transfer coefficient ($\bar{h}$) is related to air velocity (u_∞) according to the following equation:

$$\frac{\bar{h}\,D}{k} = 0.82\,Re^{0.385},$$

where D is the diameter of the wire, k is the thermal conductivity of air, and Re is the Reynold's number, defined as:

$$Re = \frac{\rho\,u_\infty D}{\mu},$$

where ρ and μ are the density and viscosity of air, respectively. The air properties should be evaluated at the average of the surface temperature and the ambient temperature.

(b) Using this relation, what is the estimated air velocity?

(c) If the power is controlled in order to maintain the wire at a constant temperature, prepare a plot showing the power to the wire as a function of air velocity.

1.21 A double-glazed window (i.e., a window that includes two panes of glass) is exposed to indoor air at 25°C on one side and cold outdoor air at −4°C on the other side. The average heat transfer coefficients are 6 W/m²-K on the inside surface and 18 W/m²-K on the outside surface. The gap between the glazings is 1 cm wide and it contains air. Assume that the glass is opaque to thermal radiation and has emissivity $\varepsilon = 0.9$. Both radiation and convection occur in the gap. Radiation between the two glass surfaces (the inner one at $T_{g,i}$ and the outer one at $T_{g,o}$) is governed by the equation

$$\dot{q}''_{rad,gap} = \frac{\sigma\left(T_{g,i}^4 - T_{g,o}^4\right)}{\left(\dfrac{2}{\varepsilon} - 1\right)}.$$

Convection in the gap is governed by

$$\dot{q}''_{conv,gap} = \bar{h}_{gap}\left(T_{g,i} - T_{g,o}\right),$$

where the average convection coefficient for free convection in the gap is $\bar{h}_{gap} = 1.7$ W/m²-K.

Assume that the glass surfaces are at uniform temperatures (i.e., neglect temperature gradients in the glass due to conduction). Determine the steady-state temperatures of the inside and outside glazings and the rate of heat transfer per unit area from indoors to outdoors.

1.22 A sphere of copper with diameter 2 cm is at a uniform temperature of 200°C in an environment at 20°C. The copper surface has an emissivity of 0.2. The heat transfer coefficient is 25 W/m²-K. Estimate the rate of change of temperature with respect to time.

1.23 The duct connecting the outlet of a residential furnace to a room in a house is 16 ft in length and it carries air that has been heated to 105°F. The duct is made of thin sheet metal and it is rectangular, with dimensions 12 inch × 6 inch. Air is blown through the duct at an average velocity of 110 ft/min. The convection coefficient on the inside surface of the duct is 3.0 W/m²-K. The convection coefficient on the outside surface of the duct is 2.75 W/m²-K. The duct material itself offers negligible thermal resistance because it is thin and highly conductive.

(a) Estimate the rate of heat loss from the duct, neglecting radiation, to the 65°F surroundings.

(b) Estimate the change in temperature of the air as it flows through the duct.

1.24 A radiator in a car cools hot ethylene glycol coolant that is returning from the engine using atmospheric air in a heat exchanger. A fan blows air through the heat exchanger with volumetric flow rate 1600 cfm (ft³/min). The air enters the heat exchanger at 75°F and exits at 105°F. A pump causes ethylene glycol to flow through the heat exchanger with volumetric flow rate 5 gal/min. The ethylene glycol enters with temperature 150°F.

(a) Determine the rate of heat transfer from the ethylene glycol to the air.

(b) Determine the exit temperature of the ethylene glycol.

1.25 The convective heat transfer coefficient for crossflow of fluid past a long cylinder can be approximately calculated using the correlation:

$$\overline{Nu} = 0.3\, Re^{0.6}$$

where $\overline{Nu}$ is the average Nusselt number, a dimensionless quantity defined as $\overline{Nu} = \overline{h}\,D/k$ and Re is the dimensionless Reynold's number defined as $Re = \rho u_f D/\mu$. The Nusselt number is used to evaluate $\overline{h}$, the average convection coefficient. In these equations, D is the diameter of the cylinder. The variables k, ρ, and μ are respectively the thermal conductivity, density, and viscosity of the fluid evaluated at the average of the surface and free stream temperatures. The variable u_f is the velocity of the fluid flowing past the cylinder.

(a) A cylindrical heating element is 0.3 m in length and 0.025 m in diameter. The electrical power dissipated in the heater is 220 W. Air at 25°C, 1 bar is blown past the cylinder at 2 m/s. Estimate the surface temperature of the cylinder. Neglect radiation in this calculation.

(b) Repeat the calculation when water at 25°C, 1 bar pressure, instead of air, is the fluid.

Projects

1.26 The figure illustrates a Solar Electrical Generation System (a SEGS plant).

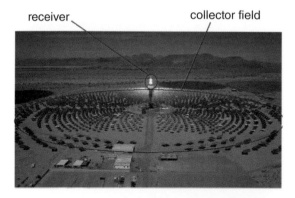

receiver collector field

A field of independently controlled mirrors focus the solar energy onto the central receiver in order to elevate the temperature of a molten salt which subsequently provides heat to a power cycle that produces electricity. We will model the central receiver as shown.

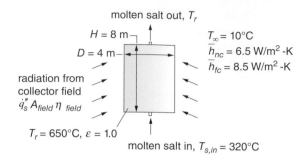

molten salt out, T_r

$H = 8$ m
$D = 4$ m

$T_\infty = 10°C$
$\overline{h}_{nc} = 6.5$ W/m²-K
$\overline{h}_{fc} = 8.5$ W/m²-K

radiation from collector field
$\dot{q}_s'' \, A_{field} \, \eta_{field}$

$T_r = 650°C$, $\varepsilon = 1.0$

molten salt in, $T_{s,in} = 320°C$

The receiver is a cylinder with diameter $D = 4$ m and height $H = 8$ m. The emissivity of the surface is $\varepsilon = 1$. The collector field has an area of $A_{field} = 20$ acre and an overall efficiency relative to transferring the incident solar flux to the receiver surface of $\eta_{field} = 0.76$. The solar flux that is incident on the field is $\dot{q}_s'' = 750$ W/m². The receiver absorbs all of the radiation that strikes its surface. The temperature of the receiver surface is $T_r = 650°C$. The molten salt is composed of 58 percent NaCl and 42 percent MgCl$_2$. The salt enters the receiver at $T_{s,in} = 320°C$ and leaves at the receiver surface temperature. Note that the properties of this molten salt are available in the Incompressible Substance Library in EES. The flow rate of the molten salt is controlled in order to maintain the required receiver temperature. The molten salt flows into the boiler of a power cycle. The efficiency of the power cycle is given by

$$\eta = \eta_2 \left(1 - \frac{T_\infty}{T_r}\right),$$

where $\eta_2 = 0.25$ is the second law efficiency of the power cycle.

The receiver experiences losses that are related to forced and natural convection to the surrounding fluid at $T_\infty = 10°C$. The natural convection heat transfer coefficient is $\overline{h}_{nc} = 6.5$ W/m²-K and the forced convection heat transfer coefficient is $\overline{h}_{fc} = 8.5$ W/m²-K. The heat transfer coefficient that characterizes the combined effect of natural and forced convection can be calculated approximately according to:

$$\overline{h} = (\overline{h}_{nc}^3 + \overline{h}_{fc}^3)^{1/3}.$$

The receiver also experiences radiation loss to surroundings at $T_\infty = 10°C$. Assume that the top and bottom surfaces are well insulated so that losses only occur from the sides.

(a) Determine the total rate of heat loss from the receiver.

(b) Determine the mass flow rate of molten salt.

(c) Determine the rate at which power is produced by the power cycle.

(d) Determine the overall efficiency of the SEGS plant; this is defined as the ratio of the power produced to the total solar energy incident on the field.

(e) Plot the overall efficiency as a function of the receiver temperature for various values of the solar heat flux. You should see that there is an optimal receiver temperature for each value of heat flux. Explain why this occurs.

(f) Plot the overall efficiency as a function of solar flux if the plant is operated with a constant receiver temperature of $T_r = 650°C$.

(g) Overlay on your plot from (f) the overall efficiency as a function of solar flux if the receiver temperature is optimized for the instantaneous value of the solar flux. You may want to use the Min/Max Table feature in EES to accomplish this.

1.27 Computer chips tend to work better if they are kept cold. You are examining the feasibility of maintaining the processor of a personal computer at the sub-ambient temperature of $T_{chip} = 0°F$. Assume that the operation of the computer chip itself generates $\dot{q}_{chip} = 10$ W of power. Model the processor unit as a box that is $a = 2$ inch $\times b = 6$ inch $\times c = 4$ inch. Assume that all six sides of the box are exposed to air at $T_{air} = 70°F$ with a convection heat transfer coefficient of $\bar{h} = 10$ W/m^2-K. The box experiences a radiation heat transfer with surroundings that are at $T_{sur} = 70°F$. The emissivity of the processor surface is $\varepsilon = 0.7$ and all six sides experience the radiation heat transfer. You are asked to size the refrigeration system required to maintain the temperature of the processor.

(a) What is the refrigeration load that your refrigeration system must be able to remove to maintain the processor at a steady-state temperature (W)?

(b) If the coefficient of performance (COP) of the refrigeration system is nominally 3.5, then how much heat must be rejected to the ambient air (W)? Recall that COP is the ratio of the amount of refrigeration provided to the amount of input power required.

(c) If electricity costs 12¢/kW-hr, how much does it cost to run the refrigeration system for a year, assuming that the computer is never shut off.

1.28 You are designing a cubical case that contains electronic components that drive remotely located instruments. You have been asked to estimate the maximum and minimum operating temperature limits that should be used to specify the components within the case. The case is $W = 8$ inch on a side. The emissivity of the paint used on the case is $\varepsilon = 0.85$. The operation of the electronic components within the case generates between $\dot{q} = 5$ and 10 W due to ohmic heating, depending on the intensity of the operation. The top surface of the case is exposed to a solar flux $\dot{q}''$. All of the surfaces of the case convect (with average heat transfer coefficient $\bar{h}$) and radiate to surroundings at T_∞. The case will be deployed in a variety of climates, ranging from very hot ($T_{\infty,max} = 110°F$) to very cold ($T_{\infty,min} = -40°F$), very sunny ($\dot{q}''_{max} = 850$ W/m^2) to night ($\dot{q}''_{min} = 0$ W/m^2), and very windy ($\bar{h}_{max} = 100$ W/m^2-K) to still ($\bar{h}_{min} = 5$ W/m^2-K). For the following questions, assume that the case is at a single, uniform temperature and at steady state.

(a) Come up with an estimate for the maximum operating temperature limit.

(b) Plot the maximum operating temperature as a function of the case size, W. Explain the shape of your plot (why does the temperature go up or down with W? if there is an asymptotic limit, explain why it exists).

(c) Come up with an estimate for the minimum operating temperature limit (with $W = 8$ inch).

(d) Do you feel that the emissivity of the case surface is very important for determining the minimum operating temperature? Justify your answer.

2 | One-Dimensional, Steady-State Conduction

In this chapter, we will begin the formal study of conduction heat transfer by introducing the fundamental rate equation, Fourier's Law. This rate equation will then be applied to situations that are both one-dimensional and steady state. This chapter also provides an initial exposure to the formal process of deriving a differential equation by applying the First Law of Thermodynamics to a differentially small control volume. The alternative process of solving a problem using numerical techniques in which the First Law of Thermodynamics is applied to small but finite control volumes is introduced. Both of these approaches are repeated throughout the text as the problems that are considered become increasingly complex. Finally, the concept of a thermal resistance is introduced in this chapter. Thermal resistances are a primary tool for understanding heat transfer problems in order to simplify and solve them.

2.1 Conduction Heat Transfer

2.1.1 Fourier's Law

Conduction heat transfer is a result of the interactions between the many, very small energy carriers that exist within any material. Conduction is easy to visualize in a gas where the energy carriers are the gas molecules. Faster (i.e., higher temperature) molecules strike slower (i.e., lower temperature) molecules causing an energy transfer from molecule to molecule. When agglomerated to the macroscale, this energy transfer is from hot regions to cold regions. The energy carriers responsible for conduction can also be electrons or lattice vibrations. Although conduction is inherently related to phenomena that occur at the microscale, our engineering analyses will not consider the motion of individual particles. Rather, we will model substances using the **continuum approach,** which is made possible by the fact that there are many energy carriers acting together. The continuum model for conduction heat transfer is captured by **Fourier's Law**.

Fourier's Law relates the **heat flux** (i.e., the heat transfer rate per unit area, W/m^2) in a particular direction to the temperature gradient in that direction. For example, if the temperature varies *only* in the x-direction, $T(x)$, then Fourier's Law should be written as:

$$\dot{q}''_x = -k\frac{dT}{dx},\tag{2.1}$$

where $\dot{q}''_x$ is the heat flux in the x-direction. Because temperature does not vary in the other directions (y and z) we know that there is no energy transfer by conduction in those directions. The parameter k in Eq. (2.1) is the **thermal conductivity** of the material, which characterizes the agglomerated behavior of the many individual energy carriers. Thermal conductivity is discussed in the next section.

In a situation in which temperature varies in multiple dimensions, $T(x, y, z)$, we can write Fourier's Law in each of the three coordinate directions:

$$\dot{q}''_x = -k\frac{\partial T}{\partial x}\tag{2.2}$$

$$\dot{q}''_y = -k\frac{\partial T}{\partial y}\tag{2.3}$$

$$\dot{q}''_z = -k\frac{\partial T}{\partial z}.\tag{2.4}$$

Equations (2.2) through (2.4) provide us with the instantaneous rate of heat transfer per unit area in the x-, y-, and z-directions ($\dot{q}''_x$, $\dot{q}''_y$, and $\dot{q}''_z$), respectively.

Fourier's Law applies for conduction in gases, liquids, and solids if the material is behaving as a continuum, which occurs when the distance between microscale energy carrier interactions is small relative to the dimension of the material that is being analyzed. The ratio of the average distance between energy carrier interactions (L_{ms}) to the length scale that characterizes the problem (L_{char}) is a dimensionless number that is referred to as the **Knudsen number**:

$$Kn = \frac{L_{ms}}{L_{char}}. \tag{2.5}$$

The Knudsen number can be estimated if there is any doubt that continuum concepts (like Fourier's Law) are applicable. If the Knudsen number is much less than unity then continuum theory holds. Otherwise, continuum concepts will begin to break down and solutions that are obtained using continuum models will be less accurate. This latter limit may be reached in microscale and nanoscale systems as well as in problems involving rarefied gas. In this text (and in almost any engineering problem you are likely encounter), materials are assumed to behave as a continuum.

Fourier's Law makes physical sense in that we know heat must flow *from* hot *to* cold. If we are moving in a coordinate direction and the temperature is decreasing as the position increases, then the temperature gradient, e.g. dT/dx in Eq. (2.1), is negative. Fourier's Law informs us that the rate of heat transfer in that direction is positive.

Fourier's Law is a **rate equation** in that it provides the linkage between the rate of conduction heat transfer and the temperature distribution. As such, it is our primary tool for understanding and analyzing conduction problems. We will often use Fourier's Law to prepare qualitative but useful sketches for conduction heat transfer problems that aid in our fundamental understanding. Examination of the temperature variation in the x-direction is sufficient to ascertain the qualitative features of the rate of heat flux in that direction using Fourier's Law. Alternatively, knowledge of the heat flux variation is sufficient to understand the relevant features of the temperature distribution.

Example 2.1

Figure 1 illustrates the heat flux in the x-direction (W/m^2) as a function of x for a particular problem.

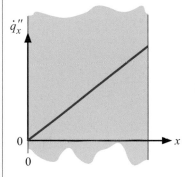

Figure 1 Heat flux in the x-direction as a function of x.

Sketch the temperature in the object as a function of x. Your sketch need not be quantitatively correct but it should have qualitative features that are consistent with Figure 1.

Continued

Example 2.1 (cont.)

Known Values

In this problem no specific numerical values are given. This conceptual problem requires you to apply your understanding of Fourier's Law in order to develop a qualitative solution as opposed to obtaining a more concrete, quantitative result.

Assumption

- No assumptions are required.

Analysis

According to Fourier's Law, repeated below,

$$\dot{q}''_x = -k\frac{\partial T}{\partial x},$$

the temperature gradient must be related to the magnitude of the heat flux:

$$\frac{\partial T}{\partial x} = -\frac{\dot{q}''_x}{k}.$$

Therefore, on the left side of the object (at $x = 0$) the temperature gradient must be zero; this requirement is typical of a location where there is no heat transfer rate, i.e., an adiabatic boundary. The magnitude of the heat flux increases as you move from the left to the right side of the object. To achieve this behavior, Fourier's Law requires that the temperature gradient (i.e., the slope of the temperature versus position x) becomes *more negative* as x increases. These characteristics are shown in Figure 2.

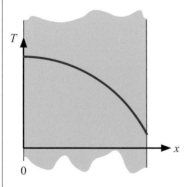

Figure 2 Temperature as a function of x.

Discussion/Exploration

In most heat transfer problems, we first understand the energy transfer rates using the First Law of Thermodynamics applied on a differential basis. Only then do we use Fourier's Law to infer the associated temperature distribution.

2.1.2 Thermal Conductivity

Conduction heat transfer is related to the underlying motion and interactions of many individual energy carriers within a material. We do not keep track of these particles separately but rather model the material as a **continuum** using Fourier's Law to evaluate the net contribution of all of the energy carriers:

$$\dot{q}''_x = -k\frac{\partial T}{\partial x}. \tag{2.6}$$

The parameter k in Eq. (2.6) is the **thermal conductivity** of the material and it provides the linkage between our macroscale model and the microscale reality of the energy transfer process. Thermal conductivity is therefore related to the ability of the underlying microscale energy carriers within a material to transport energy. It can be shown (Nellis and Klein, 2009) that thermal conductivity is related to the product of the number of energy carriers per unit volume (n_{ms}, their number density), their average velocity (v_{ms}), the mean distance between their interactions (L_{ms}), and their heat capacity (C_{ms}, the ratio of the amount of energy carried by each energy carrier to its temperature):

$$k \propto n_{ms} v_{ms} L_{ms} C_{ms}. \tag{2.7}$$

Equation (2.7) makes physical sense. Materials with many energy carriers that are moving at high velocity, each carrying a lot of energy for long distances, will be good conductors.

Physicists and material scientists might look at the underlying microstructure of a material in detail in order to predict or understand (or even modify) its thermal conductivity. As engineers, we will use thermal conductivity information that has been measured and tabulated in references. Appendices A through D contain a limited set of thermophysical properties for solids, liquids, gases, and saturated fluids that includes thermal conductivity. A much more comprehensive database for the thermal conductivity of many substances is available within the Engineering Equation Solver (EES) program.

Figure 2.1 illustrates the thermal conductivity of several common materials as a function of temperature. Notice that metals have the largest thermal conductivity, followed by other solids and liquids, while gases have the lowest conductivity. Gases are diffuse and thus the number density of the energy carriers, n_{ms} in Eq. (2.7), is substantially lower than for other forms of matter. Pure metals have the highest thermal conductivity because energy is carried primarily by electrons, which are numerous and fast moving. Alloys have lower thermal conductivity than the pure metals that they are composed of because the electron motion is substantially impeded by impurities within the structure of the material. In nonmetals, the energy is carried by phonons (or lattice vibrations), while in liquids the energy is carried by molecules.

2.2 Steady-State 1-D Conduction without Generation

2.2.1 Introduction

We will begin to study conduction by considering simple situations and gradually move to problems that are more complex. In this chapter one-dimensional (1-D), steady-state problems are considered. This class of problems can be solved both **analytically** and **numerically** and both solution techniques will be introduced in this chapter; however, this section focuses on analytical solutions.

An **analytical solution** is a **function** that is the solution to a **differential equation**. The governing differential equation is derived using essentially the same steps regardless of the complexity of the problem. In this section, we will practice applying these steps in a simple situation in order to become proficient with the process.

Analytical solutions have the advantages of being exact and computationally efficient. However, it is not always possible to obtain an analytical solution depending on the complexity of the problem. In subsequent sections, we will learn how to develop numerical solutions. **Numerical solutions** are approximate and can require significant computational time, but they are much more flexible and therefore more powerful. The best approach is to solve a problem using both techniques; the analytical solution verifies the numerical solution for a simple or limiting case and the numerical solution can then be extended and used with confidence to carry out design and optimization.

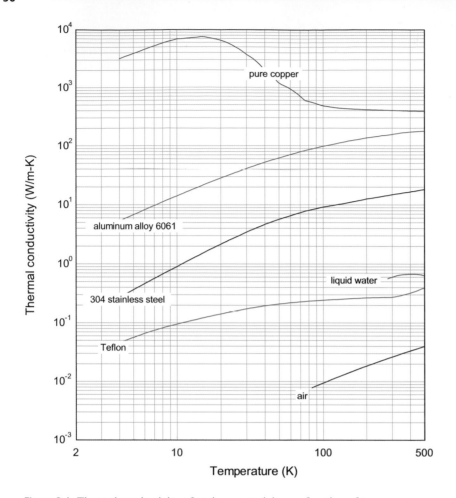

Figure 2.1 Thermal conductivity of various materials as a function of temperature.

2.2.2 The Plane Wall

In general, the temperature in any material is a function of position (x, y, and z, in Cartesian coordinates) and time (t):

$$T = T(x, y, z, t) \text{ in general.} \tag{2.8}$$

The definition of **steady state** is that temperature is unchanging with time:

$$T = T(x, y, z) \text{ at steady state.} \tag{2.9}$$

There are certain, idealized steady-state problems in which the temperature varies in only one direction (e.g., the x-direction). These are **one-dimensional** (1-D), steady-state problems:

$$T = T(x) \text{ for 1-D, steady state.} \tag{2.10}$$

One example of a 1-D steady-state problem is a plane wall (i.e., a wall with a constant cross-sectional area, A_c) that is insulated around its edges so heat flow only occurs in the x-direction, i.e., perpendicular to the

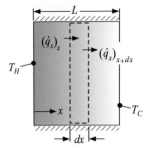

Figure 2.2 A plane wall with fixed temperature boundary conditions.

cross-sectional area. In order for the temperature distribution to be 1-D, each face of the wall must be subjected to a uniform boundary condition. For example, Figure 2.2 illustrates a plane wall in which the left face (at $x = 0$) is maintained at a uniform temperature, T_H, while the right face (at $x = L$) is held at a different uniform temperature, T_C.

The first step in the development of an analytical solution is the derivation of the governing differential equation. The steps required to derive the differential equation can be applied consistently to any problem and are discussed here in the context of the problem shown in Figure 2.2.

Define a Differential Control Volume

The first step involves the definition of a *differential* control volume, sometimes called a shell. The control volume should be drawn with a small thickness (dx) in the direction(s) in which the temperature will vary (in the x-direction for this case). This control volume, when taken to its infinitesimal limit, must encompass material that is at essentially the same temperature. For this problem, our control volume must be differentially small in the x-direction (i.e., it has width dx in the x-direction as shown in Figure 2.2) but can extend across the entire cross-sectional area of the wall as there are no temperature gradients in the y- or z-directions. In this example, we have assumed that insulation on the top, sides, and bottom surfaces prevent any heat transfer, and therefore there is no temperature variation in these directions.

Carry out an Energy Balance on the Control Volume

As discussed in Section 1.2, the First Law of Thermodynamics requires that energy cannot be generated or destroyed; that is, energy is a **conserved quantity**. Therefore, an energy balance in general leads to:

$$IN = OUT + STORED, \tag{2.11}$$

where *IN* and *OUT* represent energy transfers into and out of the control volume across the control surfaces that define it. The term *STORED* represents energy that is stored within the control volume. In heat transfer applications, Eq. (2.11) is usually applied on a rate basis, i.e., *IN* and *OUT* refer to the rates of energy transfer into and out of the control volume and *STORED* refers to the rate at which energy is stored within the control volume. This problem is a steady-state problem and therefore the *STORED* term will be zero, as the amount of energy within the control volume is not changing with time. In general, energy may be transferred across a boundary as heat, work, or with a mass transfer (for an open system). Here we have no work transfer across the system boundaries and the system is closed. Therefore, energy is only transferred by heat. Because there is a temperature gradient in the x-direction, there must be conduction heat transfer in the x-direction and we will therefore have a conduction heat transfer into the left side of the control volume (i.e., at position x) and out of the right side of the control volume (i.e., at position $x + dx$). The rate of heat transfer in the x-direction at location x is denoted $(\dot{q}_x)_x$ and the rate of heat transfer in the x-direction at location $x + dx$ is denoted $(\dot{q}_x)_{x+dx}$. A steady-state energy balance for the differential control volume is therefore:

$$(\dot{q}_x)_x = (\dot{q}_x)_{x+dx}. \tag{2.12}$$

Equation (2.12) is the general energy balance presented in Section 1.2, Eq. (1.2), simplified for this problem.

Take the Limit as $dx \to 0$

The energy balance on a differential control volume will lead to one or more terms that are evaluated at position $x + dx$. If the problem was multidimensional then you might also end up with terms evaluated at $y + dy$ or $z + dz$. In this step, we take advantage of the fact that the control volume is differentially small so that we can express these in terms of their derivatives.

Recall that the definition of a derivative for an arbitrary function $f(x)$ is given by:

$$\frac{df}{dx} = \lim_{dx \to 0} \frac{f_{x+dx} - f_x}{dx}. \tag{2.13}$$

Solving for f_{x+dx} provides:

$$f_{x+dx} = f_x + \frac{df}{dx} dx \text{ in the limit that } dx \to 0. \tag{2.14}$$

In this problem, we can write the definition of the derivative for the function $\dot{q}_x$, the rate of conduction heat transfer in the x-direction:

$$\frac{d\dot{q}_x}{dx} = \lim_{dx \to 0} \frac{(\dot{q}_x)_{x+dx} - (\dot{q}_x)_x}{dx}. \tag{2.15}$$

Solving Eq. (2.15) for $(\dot{q}_x)_{x+dx}$ provides:

$$(\dot{q}_x)_{x+dx} = (\dot{q}_x)_x + \frac{d\dot{q}_x}{dx} dx. \tag{2.16}$$

Substituting Eq. (2.16) into Eq. (2.12) leads to:

$$(\dot{q}_x)_x = \underbrace{(\dot{q}_x)_x + \frac{d\dot{q}_x}{dx} dx}_{(\dot{q}_x)_{x+dx}}. \tag{2.17}$$

Therefore, for our steady-state problem we obtain:

$$\frac{d\dot{q}_x}{dx} = 0. \tag{2.18}$$

Equation (2.18) is typical of the initial result that is obtained by considering a differential energy balance on a rate basis. Our analysis always starts out by considering an energy balance and therefore we obtain a differential equation that is expressed in terms of energy. This form of the differential equation should be checked against your intuition. Equation (2.18) indicates that the rate of conduction heat transfer in the x-direction is not a function of x. For the problem shown in Figure 2.2 there are no sources or sinks of energy and no energy storage within the material; therefore, there is no reason for the rate of heat transfer in the x-direction to vary with position. The rate of conduction heat transfer rate that is entering at the left face of the wall (at $x = 0$) must leave through the right face (at $x = L$) and the rate of conduction is the same at any x value between these two extremes.

Substitute Rate Equations into the Differential Equation

The final step in the derivation of the differential equation is the substitution of appropriate **rate equations** (sometimes called **mechanism equations**) into the differential equation. Rate equations in heat transfer will relate energy transfer rates to temperatures. The result of this substitution will therefore be a differential equation expressed in terms of temperature rather than in terms of energy. The rate equation for conduction is Fourier's Law:

$$\dot{q}''_x = -k\frac{dT}{dx}.$$ (2.19)

The rate of conduction in the x-direction, $\dot{q}_x$, is obtained by multiplying Eq. (2.19) by the area available for conduction in the x-direction, A_c:

$$\dot{q}_x = \dot{q}''_x A_c = -kA_c\frac{dT}{dx}.$$ (2.20)

Substituting Eq. (2.20) into Eq. (2.18) leads to:

$$\frac{d}{dx}\left[-kA_c\frac{dT}{dx}\right] = 0.$$ (2.21)

Equation (2.21) is equivalent to Eq. (2.18). It states that the rate of conduction heat transfer is constant with x. However, Eq. (2.21) is written in terms of the physical quantities that govern conduction heat transfer, specifically the area available for conduction, thermal conductivity, and the temperature gradient.

   If the thermal conductivity and the area for conduction are both constant, then Eq. (2.21) may be simplified to

$$\frac{d}{dx}\left[\frac{dT}{dx}\right] = 0.$$ (2.22)

The derivation of Eq. (2.22) is not difficult and yet it was valuable to go through the steps that are involved carefully as they are common to the derivation of the governing differential equation for more complex problems. To review, these steps include: (1) the definition of a differential control volume, (2) the development of an energy balance on that control volume, (3) the expansion of terms in the energy balance, and (4) the substitution of rate equations.

Define Boundary Conditions

In order to completely specify the mathematical problem it is necessary to define boundary conditions. **Boundary conditions** provide information about the solution at the extents of the computational domain (i.e., at the limits of the range of position and/or time over which the solution is valid). A second-order differential equation such as Eq. (2.22) requires two boundary conditions. For the problem shown in Figure 2.2, the boundary conditions are the specified temperatures at each end of the wall:

$$T_{x=0} = T_H$$ (2.23)

$$T_{x=L} = T_C.$$ (2.24)

Equations (2.22) through (2.24) represent a well-posed mathematical problem: a second-order differential equation with two boundary conditions. The technique used to solve the problem depends on the details of the governing differential equation. Equation (2.22) is very simple and can be solved by separation and direct integration:

$$d\left[\frac{dT}{dx}\right] = 0\,dx.$$ (2.25)

Equation (2.25) is integrated according to:

$$\int d\left[\frac{dT}{dx}\right] = \int 0\,dx.$$ (2.26)

The left side of Eq. (2.26) is the integral of an exact derivative and the right side integrates to zero. However, because Eq. (2.26) is an indefinite integral (i.e., there are no limits on the integrals), an undetermined constant (C_1) results from the integration:

$$\frac{dT}{dx} = C_1.$$ (2.27)

Equation (2.27) is separated and integrated again:

$$\int dT = \int C_1 \, dx$$ (2.28)

to yield

$$T = C_1 x + C_2.$$ (2.29)

Equation (2.29) shows that the temperature distribution must be linear. Any linear function will satisfy the differential equation; that is, any values of the constants C_1 and C_2 in Eq. (2.29) will result in a solution to Eq. (2.22). Equation (2.29) is therefore referred to as the **general solution** to the ODE. The constants of integration C_1 and C_2 are obtained by forcing the general solution to also satisfy the two boundary conditions. Equation (2.29) is substituted into Eqs. (2.23) and (2.24) in order to obtain:

$$T_H = C_1 0 + C_2$$ (2.30)

$$T_C = C_1 L + C_2.$$ (2.31)

Equations (2.30) and (2.31) are solved for C_1 and C_2 and the result is substituted into Eq. (2.29) to provide the solution:

$$T = \frac{(T_C - T_H)}{L} x + T_H.$$ (2.32)

The heat transfer rate at any location within the wall is obtained by substituting the temperature distribution, Eq. (2.32), into Fourier's Law, Eq. (2.20):

$$\dot{q}_x = -k A_c \frac{dT}{dx} = \frac{k A_c}{L} (T_H - T_C).$$ (2.33)

Notice a few things about our solution. First, Eq. (2.33) shows that the heat transfer rate is *not* a function of x; this behavior is consistent with the underlying differential equation written in terms of energy, Eq. (2.18), which specified mathematically that the rate of heat transfer should not change with position. Further, the linear temperature distribution that was obtained is also consistent with the underlying differential equation written in terms of temperature, Eq. (2.21). If cross-sectional area and thermal conductivity are both constant, then the temperature gradient must also be constant (i.e., the temperature varies linearly). It is always a good idea to critically examine your solution to ensure that it is consistent with the underlying differential equation and boundary conditions that it is supposed to be solving.

Example 2.2

The wall of an oven is composed of brick with conductivity $k = 0.72$ W/m-K and thickness $L = 10$ inch. The wall has width $W = 4$ ft and height $H = 5$ ft. The internal surface of the brick is held at $T_H = 450°$F and the external surface is at $T_C = 90°$F.

Determine:
- The total rate of heat transfer through the wall.
- The thickness of the brick wall that would be required to limit the rate of heat transfer to $\dot{q} = 900$ W.

Example 2.2 (cont.)

Known Values

The known values are indicated on the sketch of the wall shown in Figure 1; when necessary these parameters are converted to base SI units.

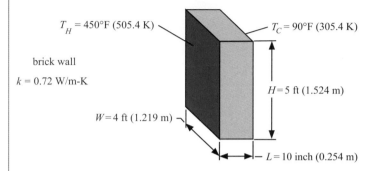

$T_H = 450°F$ (505.4 K)

$T_C = 90°F$ (305.4 K)

brick wall

$k = 0.72$ W/m-K

$H = 5$ ft (1.524 m)

$W = 4$ ft (1.219 m)

$L = 10$ inch (0.254 m)

Figure I Wall of an oven.

Assumptions

- Steady-state conditions exist.
- One-dimensional heat transfer occurs.
- The conductivity of the brick does not depend on temperature.

Analysis

The cross-sectional area of the wall is computed according to.

$$A_c = W H, \tag{1}$$

and the heat transfer rate is computed using Eq. (2.33):

$$\dot{q} = \frac{k A_c}{L}(T_H - T_C). \tag{2}$$

Equation (2) can be used to compute the rate of heat transfer for the situation where the thickness is given. In the second case the thickness must be calculated given a rate of heat transfer. Rearranging Eq. (2) to solve for L provides:

$$L = \frac{k A_c}{\dot{q}}(T_H - T_C). \tag{3}$$

Solution

Equations (1) and (2) can easily be solved by hand to compute the rate of heat transfer for the given value of wall thickness:

$$A_c = W H = \frac{1.219\,\text{m}}{} \left| \frac{1.524\,\text{m}}{} \right. = 1.858\,\text{m}^2$$

$$\dot{q} = \frac{k A_c}{L}(T_H - T_C) = \frac{0.72\,\text{W}}{\text{m K}} \left| \frac{1.858\,\text{m}^2}{} \right| \frac{(505.4 - 305.4)\,\text{K}}{} \left| \frac{}{0.254\,\text{m}} \right. = \boxed{1053\,\text{W}}.$$

Continued

Example 2.2 (cont.)

The thickness required to limit the heat transfer rate to $\dot{q} = 900$ W can be obtained using Eq. (3):

$$L = \frac{k A_c}{\dot{q}}(T_H - T_C) = \frac{0.72\,\text{W}}{\text{m K}} \left| \frac{1.858\,\text{m}^2}{} \right| \frac{(505.4 - 305.4)\text{K}}{} \left| \frac{}{900\,\text{W}} = \boxed{0.2973\,\text{m}\,(11.7\,\text{inch})}.$$

Example 2.3

A plane wall is made of a material that has temperature-dependent conductivity. The conductivity of the material is proportional to the absolute temperature according to:

$$k = bT, \tag{1}$$

where $b = 5$ W/m-K^2 and T is the temperature in K. The thickness of the wall is $L = 0.5$ m and its cross-sectional area is $A_c = 1$ m^2. The left side of the wall (at $x = 0$) is maintained at $T_H = 300$ K and the right side (at $x = L$) is held at $T_C = 20$ K.

- Without solving the problem mathematically, sketch the temperature distribution in the wall (i.e., sketch T as a function of x). Make sure that you get the qualitative features of your sketch right. Explain why your sketch looks the way it does.
- Derive the ordinary differential equation that governs this problem.
- Determine the boundary conditions.
- Solve the governing differential equation in order to obtain the general solution that involves two unknown constants of integration.
- Determine the values of the two constants of integration that cause the general solution to satisfy the boundary conditions.
- Prepare a plot of the temperature distribution and compare it to your initial sketch.
- Determine the rate of heat transfer through the wall.

Known Values

The known values are shown on the sketch of the problem in Figure 1.

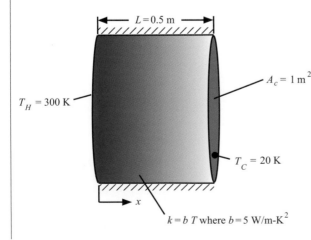

Figure 1 Plane wall with temperature-dependent conductivity.

Example 2.3 (cont.)

Assumptions

- Steady-state conditions exist.
- One-dimensional heat transfer occurs.

Analysis

The problem statement asks that we first sketch a temperature distribution. The rate of conduction heat transfer in the wall must be constant with position as there are no sources or sinks of energy and the problem is steady state. Therefore, whatever temperature distribution we sketch must be consistent with this understanding. Fourier's Law states that

$$\dot{q} = -k A_c \frac{dT}{dx}.$$

(2)

In this problem, the cross-sectional area is constant but the conductivity is not. As a result, in regions where there is high temperature corresponding to high conductivity (i.e., towards the left side of the wall), the magnitude of the temperature gradient must be small. In regions where there is low temperature (corresponding to low conductivity), the magnitude of the temperature gradient must increase in order to maintain a constant heat transfer rate. Figure 2 is qualitative, but reflects these characteristics while also satisfying the boundary conditions for the problem. Whatever exact solution we eventually derive should look something like the sketch shown in Figure 2.

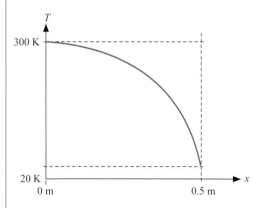

Figure 2 Qualitative sketch of the temperature distribution; note that there is a large temperature gradient when the conductivity is low and a small temperature gradient when conductivity is high in order to maintain a constant heat transfer rate.

   The steps in the derivation of the governing differential equation follow those presented in Section 2.2.2. A differential control volume (of width dx) is defined, as shown in Figure 2.2. An energy balance provides

$$(\dot{q}_x)_x = (\dot{q}_x)_{x+dx}.$$

(3)

Expanding the $x + dx$ term and taking the limit as dx approaches zero transforms Eq. (3) into

$$(\dot{q}_x)_x = (\dot{q}_x)_x + \frac{d\dot{q}_x}{dx} dx$$

(4)

Continued

Example 2.3 (cont.)

or

$$\frac{d\dot{q}_x}{dx} = 0. \tag{5}$$

Again we have found that the rate of conduction heat transfer in the x-direction must not change with position x. Substituting the rate equation for conduction heat transfer, Fourier's Law given by Eq. (2), into Eq. (5) leads to:

$$\frac{d}{dx}\left[-k A_c \frac{dT}{dx}\right] = 0. \tag{6}$$

In this case the thermal conductivity is *not* constant but rather varies with temperature according to Eq. (1). Therefore it is not possible to cancel k out of Eq. (6). Instead, we must substitute Eq. (1) into Eq. (6) in order to obtain the ordinary differential equation:

$$\frac{d}{dx}\left[-bT A_c \frac{dT}{dx}\right] = 0 \tag{7}$$

or, after dividing through by the constants b and A_c:

$$\boxed{\frac{d}{dx}\left[T \frac{dT}{dx}\right] = 0}. \tag{8}$$

The boundary conditions for this problem are specified temperatures:

$$\boxed{T_{x=0} = T_H} \tag{9}$$

$$\boxed{T_{x=L} = T_C}. \tag{10}$$

Equation (8) is a separable differential equation and can be solved by moving the dx to the right side and then integrating both sides:

$$\int d\left[T \frac{dT}{dx}\right] = \int 0 \, dx. \tag{11}$$

The left side of Eq. (11) is the integration of an exact differential and the right side integrates to zero. A constant of integration is required as there are no limits on the integrals:

$$T \frac{dT}{dx} = C_1. \tag{12}$$

Equation (12) can also be separated:

$$T \, dT = C_1 \, dx \tag{13}$$

and integrated:

$$\int T \, dT = \int C_1 \, dx \tag{14}$$

leading to:

$$\boxed{\frac{T^2}{2} = C_1 x + C_2}. \tag{15}$$

Equation (15) is the general solution to the problem; it satisfies the governing differential equation regardless of what values of C_1 and C_2 are used. These undetermined constants must be chosen so that the general solution satisfies the boundary conditions. Substituting Eqs. (9) and (10) into Eq. (15) provides two equations in the two unknown constants C_1 and C_2:

Example 2.3 (cont.)

$$\boxed{\frac{T_H^2}{2} = C_2} \tag{16}$$

$$\frac{T_C^2}{2} = C_1 L + C_2. \tag{17}$$

Substituting Eq. (16) into Eq. (17) and solving for C_1 provides:

$$\boxed{C_1 = -\frac{(T_H^2 - T_C^2)}{2L}.} \tag{18}$$

Substituting Eqs. (16) and (18) into Eq. (15) provides:

$$\boxed{T^2 = T_H^2 - \frac{x}{L}(T_H^2 - T_C^2).} \tag{19}$$

Fourier's Law, Eq. (2), relates the rate of heat transfer through the wall to the temperature gradient. Substituting Eqs. (1) and (12) into Eq. (2) provides:

$$\dot{q}_x = -k A_c \frac{dT}{dx} = -\underbrace{b T}_{k} A_c \underbrace{\frac{C_1}{T}}_{\frac{dT}{dx}} = -b A_c C_1. \tag{20}$$

Substituting Eq. (18) into Eq. (20) provides:

$$\boxed{\dot{q}_x = b A_c \frac{(T_H^2 - T_C^2)}{2L}.} \tag{21}$$

Solution

The problem statement asks for a plot and therefore the solution is implemented in EES. The inputs are entered into EES.

b = 5 [W/m-K^2]	"constant in conductivity function"
L = 0.5 [m]	"wall thickness"
T_H=300 [K]	"temperature at x=0"
T_C=20 [K]	"temperature at x=L"
A_c=1 [m^2]	"cross-sectional area"

Equation (21) is entered in order to solve for the heat transfer rate through the wall.

q_dot=b*A_c*(T_H^2-T_C^2)/(2*L)	"heat transfer rate"

Solving leads to $\boxed{\dot{q} = 448,000\,\text{W}}$. Equation (19) is entered in order to provide the temperature given a position, x:

T^2=T_H^2-x*(T_H^2-T_C^2)/L	"temperature"

Continued

Example 2.3 (cont.)

A Parametric Table is created that contains the variables x and T. The values in the column for x are varied from 0 to L. Solving the table provides the temperature at each value of position. Finally, the results in the table are used to generate the plot shown in Figure 3.

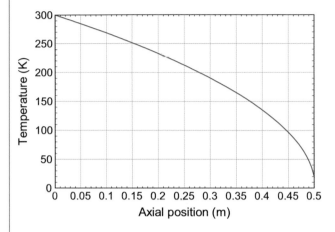

Figure 3 Temperature as a function of position.

Discussion/Exploration

Notice that Figure 3 has the qualitative characteristics of the sketch shown in Figure 2. It starts and ends at the right temperatures and has a slope that becomes increasingly negative as position increases.

It is a good idea to check for unit consistency as you solve a math problem such as this one. For example, the final solution for heat transfer rate, Eq. (21), can be checked for unit consistency:

$$\dot{q}_x = b \, A_c \frac{(T_H^2 - T_C^2)}{2L} \cdots \frac{\mathrm{W}}{\mathrm{m \, K^2}} \left| \frac{\mathrm{m^2}}{} \right| \frac{\mathrm{K^2}}{} \left| \frac{}{\mathrm{m}} \right| = \mathrm{W}. \tag{22}$$

2.2.3 The Resistance Concept

Equation (2.33), repeated below, provides the solution for conduction heat transfer through a plane wall with constant thermal conductivity at steady state:

$$\dot{q} = \frac{k \, A_c}{L} (T_H - T_C). \tag{2.34}$$

Equation (2.34) is an example of a **resistance equation** in that the rate of heat transfer passing through the wall ($\dot{q}$, in W) can be thought of as a flow that is driven by a temperature difference ($T_H - T_C$, in K) and resisted by a **thermal resistance** (R, in K/W):

$$\dot{q} = \frac{(T_H - T_C)}{R}. \tag{2.35}$$

This concept is illustrated symbolically in Figure 2.3(a). There is an analogy between heat transfer and electrical circuits. In your circuits class you were introduced to Ohm's Law, which stated that electrical current (I_c, in ampere) is driven by a voltage difference ($V_H - V_L$, in V) and resisted by an electrical resistance (R_e, in ohm):

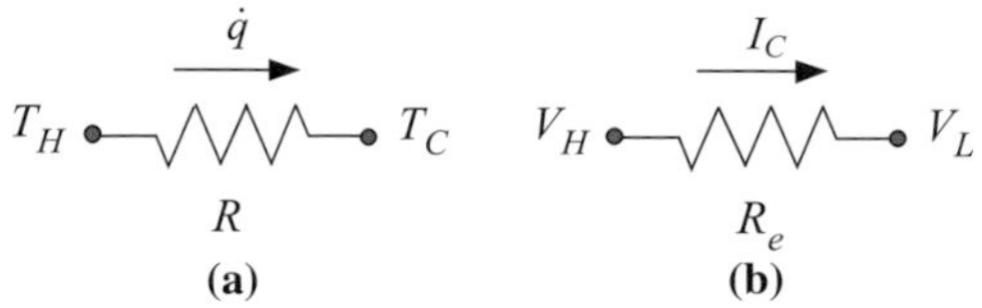

Figure 2.3 (a) Thermal resistance and (b) electrical resistance.

$$I_c = \frac{(V_H - V_L)}{R_e}.$$

(2.36)

This concept is illustrated symbolically in Figure 2.3(b).

Comparing Eq. (2.35) with Eq. (2.36) suggests that heat transfer rate and current are analogous flow quantities, and temperature and voltage are analogous potential quantities. Comparing Eq. (2.34) with Eq. (2.35) shows that the thermal resistance associated with conduction through a plane wall is given by:

$$R_{pw} = \frac{L}{k\,A_c}.$$

(2.37)

Notice that the resistance calculated using Eq. (2.37) has units of K/W; all thermal resistances described by Eq. (2.35) must also have units of K/W. We will encounter several other phenomena that can conveniently be described by a thermal resistance (e.g., convection and radiation) and each of these thermal resistances will have units of K/W.

The resistance formulation is useful for a variety of reasons and we will often return to this idea of a thermal resistance in order to help develop a conceptual understanding of various heat transfer problems. By using thermal resistances we can take a problem in which several heat transfer processes are occurring at once (e.g., conduction, convection, and radiation) and calculate the thermal resistance associated with each mechanism separately. This approach allows us to represent a complex problem more simply using a thermal resistance network. Further (and probably even more importantly) the relative values of the thermal resistances that are involved with the problem can be examined in order to understand which heat transfer mechanisms are important and which are not. We often calculate thermal resistances even for situations where they do not exactly apply in order to develop an understanding of the relative importance of various aspects of the problem. One of the most important skills that a good engineer possesses is the ability to simplify a complex, real-world problem to the point where it can be analyzed in a meaningful way. Thermal resistances are one of the primary tools that we use for this purpose in heat transfer.

Electrical resistances arranged in **series** indicate that the same current flows through each resistance. Thermal resistances arranged in series, as shown in Figure 2.4(a), indicate that the same rate of heat transfer flows through each resistance. For resistors in series, the most important thermal resistance will be the *largest* one because that is where the largest temperature change will occur.

Electrical resistances arranged in **parallel** indicate that the same voltage exists across each resistance but the current will split. Thermal resistances arranged in parallel, as shown in Figure 2.4(b), indicate that each resistance has the same temperature difference imposed across it. For resistors in parallel, the most important thermal resistance will be the *smallest* one because most of the energy transfer will pass through that resistance.

Resistance equations provide a method for succinctly summarizing a particular situation and we will derive resistance solutions for a variety of processes. By cataloging these resistance equations, it is possible to use the solution in the context of a particular problem without having to go through all of the steps that were required in the original derivation. For example, if we encounter a problem involving steady-state heat transfer through a plane wall then it is not necessary to rederive Eqs. (2.12) through (2.33); instead we can simply use Eq. (2.37) to compute the thermal resistance of the wall and apply the general resistance formula, Eq. (2.35), to obtain a solution.

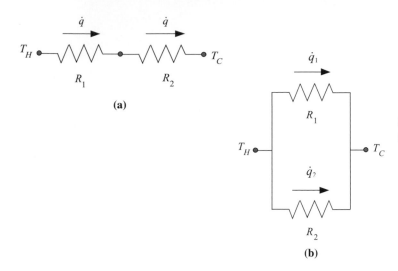

Figure 2.4 Thermal resistances arranged in (a) series and (b) parallel.

Example 2.4

A composite plane wall is made of two materials. Material A has conductivity $k_A = 10$ W/m-K and material B has conductivity $k_B = 1$ W/m-K. The thickness of both materials (in the x-direction) is $L = 0.1$ m and the cross-sectional area (perpendicular to the x-direction) is $A_c = 1$ m^2. The left side of material A (at $x = 0$) is maintained at $T_H = 300$ K and the right side of material B (at $x = 2L$) is kept at $T_C = 100$ K. The problem is steady state and 1-D.

Evaluate the rate of heat transfer through the composite structure and the temperature at the interface between materials A and B.

Known Values

Figure 1 illustrates the composite wall with the known values indicated.

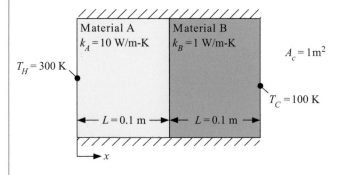

Figure 1 Plane wall composed of two materials, A and B.

Assumptions

- Steady-state conditions exist.
- The temperature distribution is one-dimensional.

Example 2.4 (cont.)

Analysis

The two materials are both plane walls and each can be represented using the conduction resistance given by Eq. (2.37). Energy moves from left to right in this problem, starting at $x = 0$ and passing first though material A and then through material B before finally leaving at $x = 2L$. Because the same energy transfer rate occurs within material A and material B, these two resistances should be placed in series, as shown in Figure 2.

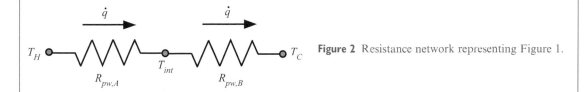

Figure 2 Resistance network representing Figure 1.

The resistances to conduction through materials A and B are both computed using Eq. (2.37):

$$R_{pw,A} = \frac{L}{k_A A_c} \tag{1}$$

$$R_{pw,B} = \frac{L}{k_B A_c}. \tag{2}$$

The heat transfer rate through the composite structure is obtained by evaluating the equivalent resistance of the network. The equivalent resistance for resistances in series is their sum:

$$R_{eq} = R_{pw,A} + R_{pw,B}. \tag{3}$$

Then the heat transfer rate is calculated using the resistance equation, Eq. (2.35):

$$\dot{q} = \frac{(T_H - T_C)}{R_{eq}}. \tag{4}$$

The temperature at the interface between the materials (T_{int} in Figure 2) is obtained by applying the resistance equation only to material A:

$$\dot{q} = \frac{(T_H - T_{int})}{R_{pw,A}}. \tag{5}$$

Equation (5) can be solved for T_{int}:

$$T_{int} = T_H - \underbrace{\dot{q} R_{pw,A}}_{\substack{\text{Temperature} \\ \text{drop across } R_{pw,A}.}} \tag{6}$$

Solution

Equations (1) through (4) and (6) are solved below:

$$R_{pw,A} = \frac{L}{k_A A_c} = \frac{0.1\,\text{m}}{10\,\text{W}} \left| \frac{\text{m K}}{1\,\text{m}^2} \right| = 0.01 \frac{\text{K}}{\text{W}}$$

$$R_{pw,B} = \frac{L}{k_B A_c} = \frac{0.1\,\text{m}}{1\,\text{W}} \left| \frac{\text{m K}}{1\,\text{m}^2} \right| = 0.1 \frac{\text{K}}{\text{W}}$$

Continued

Example 2.4 (cont.)

$$R_{eq} = R_{pw,A} + R_{pw,B} = 0.01 \frac{K}{W} + 0.1 \frac{K}{W} = 0.11 \frac{K}{W}$$

$$\dot{q} = \frac{(T_H - T_C)}{R_{eq}} = \frac{(300 - 100)K}{} \left| \frac{W}{0.11\,K} \right. = \boxed{1818\,W}$$

$$T_{int} = T_H - \dot{q}\,R_{pw,A} = 300K - \frac{1818\,W}{}\left|\frac{0.01\,K}{W}\right. = \boxed{281.8\,K}.$$

Discussion

Notice that the resistance to conduction through material B is 10× larger than the resistance to conduction through material A because of its smaller conductivity. T_{int} (281.8 K) is much closer to T_H (300 K) than it is to T_C (100 K) because $R_{pw,A}$ is so small compared to $R_{pw,B}$.

We can create a sketch of the temperature distribution in the wall (i.e., a sketch showing T as a function of x). The best way to generate a sketch like this is to first think about the heat transfer rate and then translate that into temperature. In this problem, the rate of conduction heat transfer is constant with x; that is, at any value of x, the rate of heat transfer is 1818 W. Fourier's Law states that:

$$\dot{q} = -k\,A_c \frac{dT}{dx}.$$

Within each material, both conductivity and area are constant and therefore the temperature gradient must also be constant:

$$\frac{dT}{dx} = \frac{-\dot{q}}{k\,A_c}. \tag{7}$$

In material A the temperature drops linearly from 300 K to 281.8 K and in material B the temperature drops linearly from 281.8 K to 100 K. Notice that the magnitude of the temperature gradient in material A is smaller than it is in material B because the value of k in the denominator of Eq. (7) is 10× larger. These characteristics are shown in Figure 3.

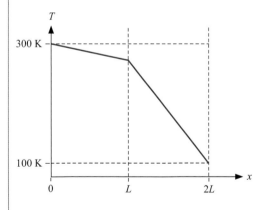

Figure 3 Sketch of the temperature distribution in the composite wall shown in Figure 1.

Notice in Figure 3 that the largest temperature drop occurs across the biggest resistance. This is generally true in a series resistance arrangement: *the largest resistance in a series network is the most important.* The rate of heat transfer through the wall is most affected by $R_{pw,B}$ for this problem and is not strongly affected by $R_{pw,A}$.

To see this more clearly, let's repeat our calculations by doubling the value of k_A (from 10 W/m-K to 20 W/m-K, which reduces the size of $R_{pw,A}$ by a factor of 2):

Example 2.4 (cont.)

$$R_{pw,A} = \frac{L}{k_A A_c} = \frac{0.1\,\text{m}}{20\,\text{W}}\left|\frac{\text{m K}}{1\,\text{m}^2}\right| = 0.005\,\frac{\text{K}}{\text{W}}$$

$$R_{pw,B} = \frac{L}{k_B A_c} = \frac{0.1\,\text{m}}{1\,\text{W}}\left|\frac{\text{m K}}{1\,\text{m}^2}\right| = 0.1\,\frac{\text{K}}{\text{W}}$$

$$R_{eq} = R_{pw,A} + R_{pw,B} = 0.005\,\frac{\text{K}}{\text{W}} + 0.1\,\frac{\text{K}}{\text{W}} = 0.105\,\frac{\text{K}}{\text{W}}$$

$$\dot{q} = \frac{(T_H - T_C)}{R_{eq}} = \frac{(300 - 100)\,\text{K}}{0.105\,\text{K}}\left|\frac{\text{W}}{}\right| = \boxed{1905\,\text{W}}.$$

The rate of heat transfer changes from 1818 W to 1905 W (an increase of only 4.8 percent). A 100 percent increase in k_A resulted in less than a 5 percent increase in $\dot{q}$! Clearly it is not very important that we know k_A (or any other parameters that affect $R_{pw,A}$) very accurately in order to solve this problem.

On the other hand, if we repeat our calculations by doubling the value of k_B (from 1 W/m-K to 2 W/m-K, which reduces the size of $R_{pw,B}$ by a factor of 2) then we will see that the rate of heat transfer changes substantially:

$$R_{pw,A} = \frac{L}{k_A A_c} = \frac{0.1\,\text{m}}{10\,\text{W}}\left|\frac{\text{m K}}{1\,\text{m}^2}\right| = 0.01\,\frac{\text{K}}{\text{W}}$$

$$R_{pw,B} = \frac{L}{k_B A_c} = \frac{0.1\,\text{m}}{2\,\text{W}}\left|\frac{\text{m K}}{1\,\text{m}^2}\right| = 0.05\,\frac{\text{K}}{\text{W}}$$

$$R_{eq} = R_{pw,A} + R_{pw,B} = 0.01\,\frac{\text{K}}{\text{W}} + 0.05\,\frac{\text{K}}{\text{W}} = 0.06\,\frac{\text{K}}{\text{W}}$$

$$\dot{q} = \frac{(T_H - T_C)}{R_{eq}} = \frac{(300 - 100)\,\text{K}}{0.06\,\text{K}}\left|\frac{\text{W}}{}\right| = \boxed{3333\,\text{W}}.$$

The rate of heat transfer increased from 1818 W to 3333 W (an increase of 83 percent). A 100 percent increase in k_B resulted in an 83 percent increase in $\dot{q}$. It is important that we know k_B accurately in order to solve this problem.

The relative magnitude of thermal resistances will be your best tool for understanding what aspects of a heat transfer problem are important and what aspects can be neglected.

Example 2.5

A composite plane wall is made of the same two materials examined in Example 2.4, but they are arranged in parallel rather than in series. Materials A and B have the same geometry and conductivities. However, in this arrangement the left sides (at $x = 0$) of both materials are maintained at $T_H = 300$ K while the right sides (at $x = L$) of both materials are maintained at $T_C = 100$ K.

Evaluate the rate of heat transfer through the composite structure.

Known Values

The known values are shown in the sketch of the problem in Figure 1.

Continued

Example 2.5 (cont.)

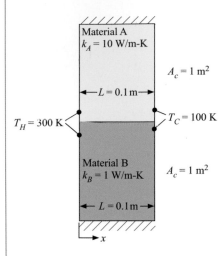

Figure 1 Plane wall composed of two materials, A and B, in parallel.

Assumptions

- Steady-state conditions exist.
- The temperature distribution within each of the materials is one-dimensional.

Analysis

The two materials experience the same temperatures but different heat transfer rates; therefore, they are in parallel as shown in Figure 2.

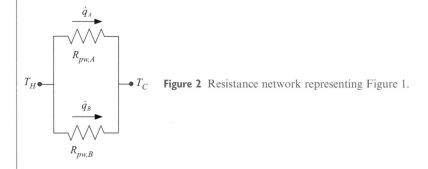

Figure 2 Resistance network representing Figure 1.

The values of the resistances are calculated in the same way as in Example 2.4:

$$R_{pw,A} = \frac{L}{k_A A_c} \tag{1}$$

$$R_{pw,B} = \frac{L}{k_B A_c}. \tag{2}$$

However, the resistors are arranged in parallel. Therefore, the equivalent resistance of the network shown in Figure 2 is given by:

$$R_{eq} = \left(\frac{1}{R_{pw,A}} + \frac{1}{R_{pw,B}} \right)^{-1}. \tag{3}$$

Example 2.5 (cont.)

The total rate of heat transfer is computed according to:

$$\dot{q} = \frac{(T_H - T_C)}{R_{eq}}. \tag{4}$$

Solution

Equations (1) through (4) can be solved by hand:

$$R_{pw,A} = \frac{L}{k_A A_c} = \frac{0.1\,\text{m}}{} \left| \frac{\text{m K}}{10\,\text{W}} \right| \frac{}{1\,\text{m}^2} = 0.01 \frac{\text{K}}{\text{W}}$$

$$R_{pw,B} = \frac{L}{k_B A_c} = \frac{0.1\,\text{m}}{} \left| \frac{\text{m K}}{1\,\text{W}} \right| \frac{}{1\,\text{m}^2} = 0.1 \frac{\text{K}}{\text{W}}$$

$$R_{eq} = \left(\frac{1}{R_{pw,A}} + \frac{1}{R_{pw,B}} \right)^{-1} = \left(\frac{\text{W}}{0.01\,\text{K}} + \frac{\text{W}}{0.1\,\text{K}} \right)^{-1} = 0.009091 \frac{\text{K}}{\text{W}}$$

$$\dot{q} = \frac{(T_H - T_C)}{R_{eq}} = \frac{(300 - 100)\text{K}}{} \left| \frac{\text{W}}{0.009091\,\text{K}} \right| = \boxed{22,000\,\text{W}}.$$

Discussion

The heat transfer has two paths available in Figure 2 and most of the heat will pass through the path of least resistance. Therefore, *the smallest resistance in a parallel network is the most important*. The rate of heat transfer through the wall is most affected by $R_{pw,A}$ for this problem and is not strongly affected by $R_{pw,B}$.

To see this more clearly, let's repeat our calculations but with double the value of k_A (from 10 W/m-K to 20 W/m-K, which reduces $R_{pw,A}$ by a factor of 2):

$$R_{pw,A} = \frac{L}{k_A A_c} = \frac{0.1\,\text{m}}{} \left| \frac{\text{m K}}{20\,\text{W}} \right| \frac{}{1\,\text{m}^2} = 0.005 \frac{\text{K}}{\text{W}}$$

$$R_{pw,B} = \frac{L}{k_B A_c} = \frac{0.1\,\text{m}}{} \left| \frac{\text{m K}}{1\,\text{W}} \right| \frac{}{1\,\text{m}^2} = 0.1 \frac{\text{K}}{\text{W}}$$

$$R_{eq} = \left(\frac{1}{R_{pw,A}} + \frac{1}{R_{pw,B}} \right)^{-1} = \left(\frac{\text{W}}{0.005\,\text{K}} + \frac{\text{W}}{0.1\,\text{K}} \right)^{-1} = 0.004762 \frac{\text{K}}{\text{W}}$$

$$\dot{q} = \frac{(T_H - T_C)}{R_{eq}} = \frac{(300 - 100)\,\text{K}}{} \left| \frac{\text{W}}{0.004762\,\text{K}} \right| = \boxed{42,000\,\text{W}}.$$

The rate of heat transfer changes from 22,000 W to 42,000 W (an increase of 91 percent). A 100 percent increase in k_A resulted in a 91 percent increase in $\dot{q}$; therefore it is important to know k_A (and the other parameters that affect $R_{pw,A}$) very accurately. On the other hand, if we double the value of k_B (from 1 W/m-K to 2 W/m-K, which reduces the value of $R_{pw,B}$ by a factor of 2) and repeat our calculations:

$$R_{pw,A} = \frac{L}{k_A A_c} = \frac{0.1\,\text{m}}{} \left| \frac{\text{m K}}{10\,\text{W}} \right| \frac{}{1\,\text{m}^2} = 0.01 \frac{\text{K}}{\text{W}}$$

$$R_{pw,B} = \frac{L}{k_B A_c} = \frac{0.1\,\text{m}}{} \left| \frac{\text{m K}}{2\,\text{W}} \right| \frac{}{1\,\text{m}^2} = 0.05 \frac{\text{K}}{\text{W}}$$

Continued

Example 2.5 (cont.)

$$R_{eq} = \left(\frac{1}{R_{pw,A}} + \frac{1}{R_{pw,B}}\right)^{-1} = \left(\frac{W}{0.01\,K} + \frac{W}{0.05\,K}\right)^{-1} = 0.008333\,\frac{K}{W}$$

$$\dot{q} = \frac{(T_H - T_C)}{R_{eq}} = \frac{(300 - 100)\,K}{0.008333\,K}\bigg|\frac{W}{} = \boxed{24{,}000\,W}.$$

The rate of heat transfer changes from 22,000 W to 24,000 W (an increase of 9 percent). A 100 percent increase in k_B resulted in less than a 10 percent increase in $\dot{q}$. Clearly it is not important that we know the value of k_B very accurately in order to solve the problem.

Again, the relative magnitude of thermal resistances is a great tool for developing an understanding of what aspects of a heat transfer problem are important and what aspects can be neglected. Generally, large resistances in a series configuration are important and small ones in a parallel configuration are important.

2.2.4 Radial Conduction

Radial conduction problems are slightly more complex than the plane wall problem considered in Section 2.2.2 because the area available for conduction varies in the radial direction.

Radial Conduction in a Cylinder

Figure 2.5 illustrates steady-state, radial conduction through a hollow cylinder (i.e., a tube) with length L, inner radius r_{in}, and outer radius r_{out}. The ends of the cylinder (at $x = 0$ and $x = L$) are insulated. The inner surface is held at a uniform temperature T_H and the outer surface is held at a uniform temperature T_C. These boundary conditions cause the temperature distribution to be one-dimensional; heat is only conducted in the r-direction and therefore temperature will only change in the r-direction.

The analytical solution to this problem is derived by employing the same steps described in Section 2.2.2. The control volume used for the process must be differentially small in the r-direction but extend circumferentially around the cylinder and axially along its length (imagine a tube with a very thin wall). The energy balance on the differential control volume is:

$$(\dot{q}_r)_r = (\dot{q}_r)_{r+dr}. \tag{2.38}$$

The $r + dr$ term in Eq. (2.38) can be written in terms of its derivative, leading to:

$$(\dot{q}_r)_{r+dr} = (\dot{q}_r)_r + \frac{d\dot{q}_r}{dr}dr. \tag{2.39}$$

Equation (2.39) is substituted into Eq. (2.38) in order to provide:

$$\frac{d\dot{q}_r}{dr} = 0. \tag{2.40}$$

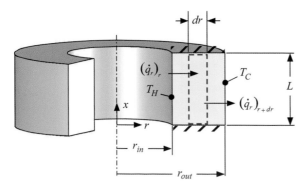

Figure 2.5 A cylinder with fixed temperature boundary conditions.

Equation (2.40) is the governing ordinary differential equation written in terms of energy. This ODE indicates that the rate of conduction heat transfer in the radial direction cannot change in the r direction. There is no sink or source of energy within the cylinder and therefore the rate of conduction must remain constant. The rate equation for conduction, Fourier's Law, is substituted into Eq. (2.40) in order to obtain:

$$\frac{d}{dr}\left[-k\,A_c\frac{dT}{dr}\right] = 0. \tag{2.41}$$

The difference between the plane wall geometry considered in Section 2.2.2 and the cylindrical geometry considered here is that the cross-sectional area for heat transfer, A_c in Eq. (2.41), is not constant but rather varies with radius according to

$$A_c = 2\pi r L. \tag{2.42}$$

At larger radii there is more area for conduction whereas the area becomes small towards the center of the cylinder. Substituting Eq. (2.42) into Eq. (2.41) provides:

$$\frac{d}{dr}\left[-k\,2\pi r L\frac{dT}{dr}\right] = 0 \tag{2.43}$$

where L is the length of the cylinder. Assuming that the thermal conductivity is constant, Eq. (2.43) can be simplified to:

$$\frac{d}{dr}\left[r\frac{dT}{dt}\right] = 0. \tag{2.44}$$

Notice that it is *not* possible to simply divide through by the r that exists within the derivative with respect to r in Eq. (2.43). Equation (2.44) is separable and can be integrated twice according to the following steps:

$$\int d\left[r\frac{dT}{dr}\right] = \int 0\,dr \tag{2.45}$$

$$r\frac{dT}{dr} = C_1 \tag{2.46}$$

$$\int dT = \int \frac{C_1}{r}\,dr \tag{2.47}$$

$$T = C_1 \ln(r) + C_2. \tag{2.48}$$

Equation (2.48) is the general solution to the ordinary differential equation given by Eq. (2.44), and C_1 and C_2 are constants of integration, evaluated by applying the boundary conditions:

$$T_{r=r_{in}} = T_H = C_1 \ln(r_{in}) + C_2 \tag{2.49}$$

$$T_{r=r_{out}} = T_C = C_1 \ln(r_{out}) + C_2. \tag{2.50}$$

Subtracting Eq. (2.50) from Eq. (2.49) eliminates C_2 so that

$$T_H - T_C = C_1 \ln(r_{in}) - C_1 \ln(r_{out}). \tag{2.51}$$

Solving for C_1 leads to:

$$C_1 = \frac{(T_H - T_C)}{\ln\left(\dfrac{r_{in}}{r_{out}}\right)}. \tag{2.52}$$

Equation (2.52) can be substituted into Eq. (2.50) to provide C_2:

$$C_2 = T_C - \frac{(T_H - T_C)}{\ln\left(\dfrac{r_{in}}{r_{out}}\right)} \ln(r_{out}). \tag{2.53}$$

Substituting Eqs. (2.52) and (2.53) into Eq. (2.48) leads to:

$$T = \frac{(T_H - T_C)}{\ln\left(\dfrac{r_{in}}{r_{out}}\right)} \ln(r) + T_C - \frac{(T_H - T_C)}{\ln\left(\dfrac{r_{in}}{r_{out}}\right)} \ln(r_{out}) \tag{2.54}$$

or

$$\frac{(T - T_C)}{(T_H - T_C)} = \frac{\ln\left(\dfrac{r}{r_{out}}\right)}{\ln\left(\dfrac{r_{in}}{r_{out}}\right)}. \tag{2.55}$$

The blue line in Figure 2.6 labeled "cylinder" illustrates the dimensionless temperature defined by Eq. (2.55) as a function of dimensionless radius (r/r_{out}) for the particular case where $r_{in}/r_{out} = 0.2$. Notice that, unlike a plane wall, the temperature distribution is *not* linear for radial heat transfer through a cylinder even with constant thermal conductivity because the cross-sectional area available for conduction changes with position. Fourier's Law states that the rate of conduction in the radial direction is given by

$$\dot{q}_r = -kA_c \frac{dT}{dr}. \tag{2.56}$$

In order for $\dot{q}_r$ to remain constant with radius it is necessary for the magnitude of the temperature gradient to increase at locations where the area is reduced (i.e., at small values of r) and decrease when the area is increased (at large values of r).

The rate of heat transfer at any radial location between r_{in} and r_{out} is given by

$$\dot{q}_r = -k2\pi rL\frac{dT}{dr}. \tag{2.57}$$

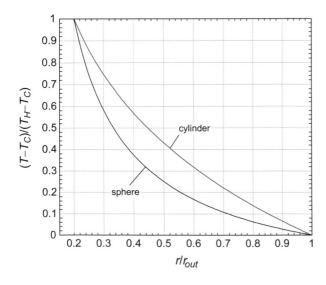

Figure 2.6 Dimensionless temperature within a cylinder, predicted by Eq. (2.55), and within a sphere, predicted by Eq. (2.73), as a function of r/r_{out} for the case where $r_{in}/r_{out} = 0.2$.

Substituting Eqs. (2.46) and (2.52) into Eq. (2.57) provides:

$$\dot{q}_r = -k\,2\pi r L \frac{C_1}{r} = -k\,2\pi L \frac{(T_H - T_C)}{\ln\left(\dfrac{r_{in}}{r_{out}}\right)} \tag{2.58}$$

or

$$\dot{q}_r = \underbrace{\frac{k\,2\pi L}{\ln\left(\dfrac{r_{out}}{r_{in}}\right)}}_{1/R_{cyl}}(T_H - T_C). \tag{2.59}$$

Equation (2.59) is another example of a resistance equation having the form indicated in Eq. (2.35). The thermal resistance associated with steady-state, radial conduction through a cylinder of material with constant conductivity is:

$$R_{cyl} = \frac{\ln\left(\dfrac{r_{out}}{r_{in}}\right)}{2\pi L k}. \tag{2.60}$$

It is worth noting that the thermal resistance to radial conduction through a cylinder must be computed using the ratio of the outer to the inner radii in the numerator, even if the heat transfer is into rather than out of the cylinder.

Radial Conduction in a Sphere

Figure 2.7 illustrates steady-state, radial conduction through a spherical shell where the inner and outer surfaces are both maintained at uniform temperatures, T_H and T_C, respectively. This is a one-dimensional problem and the appropriate differential energy balance (see Figure 2.7) leads to

$$(\dot{q}_r)_r = (\dot{q}_r)_{r+dr}, \tag{2.61}$$

which can be used with Fourier's Law to reach

$$\frac{d}{dr}\left[-k\,A_c \frac{dT}{dr}\right] = 0. \tag{2.62}$$

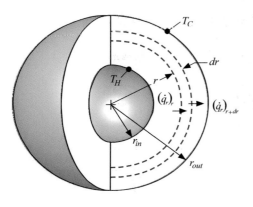

Figure 2.7 A sphere with fixed temperature boundary conditions.

The cross-sectional area for heat transfer is the surface area of a sphere at radius r:

$$\frac{d}{dr}\left[-k\underbrace{4\pi r^2}_{A_c}\frac{dT}{dr}\right] = 0. \tag{2.63}$$

Assuming that k is constant allows Eq. (2.63) to be simplified:

$$\frac{d}{dr}\left[r^2\frac{dT}{dr}\right] = 0. \tag{2.64}$$

Separating once and integrating leads to:

$$\int d\left[r^2\frac{dT}{dr}\right] = \int 0\,dr \tag{2.65}$$

$$r^2\frac{dT}{dr} = C_1. \tag{2.66}$$

Separating again and integrating leads to:

$$\int dT = \int \frac{C_1}{r^2}\,dr \tag{2.67}$$

$$T = -\frac{C_1}{r} + C_2. \tag{2.68}$$

The boundary conditions are:

$$T_{r=r_{in}} = T_H \tag{2.69}$$

$$T_{r=r_{out}} = T_C. \tag{2.70}$$

Substituting Eqs. (2.69) and (2.70) into Eq. (2.68) leads to two equations in the two unknown constants of integration:

$$T_H = -\frac{C_1}{r_{in}} + C_2 \tag{2.71}$$

$$T_C = -\frac{C_1}{r_{out}} + C_2. \tag{2.72}$$

Solving Eqs. (2.71) and (2.72) for C_1 and C_2 and substituting the result into Eq. (2.68) leads to:

$$\frac{(T - T_C)}{(T_H - T_C)} = \frac{\left(1 - \dfrac{r_{out}}{r}\right)}{\left(1 - \dfrac{r_{out}}{r_{in}}\right)}. \tag{2.73}$$

The black line in Figure 2.6 labeled "sphere" illustrates the dimensionless temperature distribution predicted by Eq. (2.73) for the particular case where $r_{in}/r_{out} = 0.2$. Notice that the curvature is even more pronounced for conduction through a sphere than it is for a cylinder because the area available for conduction varies more strongly with radial position.

The heat transfer at any radial location is given by Fourier's Law:

$$\dot{q}_r = -k\,4\pi r^2 \frac{dT}{dr}.$$ (2.74)

Substituting Eq. (2.73) into Eq. (2.74) provides:

$$\dot{q}_r = \frac{4\pi k}{\left(\dfrac{1}{r_{in}} - \dfrac{1}{r_{out}}\right)}(T_H - T_C).$$ (2.75)

Equation (2.75) provides the thermal resistance for steady-state, radial conduction through a sphere:

$$R_{sph} = \frac{\left(\dfrac{1}{r_{in}} - \dfrac{1}{r_{out}}\right)}{4\pi k}.$$ (2.76)

Resistance values must be positive and therefore the order of the terms in the numerator of Eq. (2.76) cannot change even if the rate of heat transfer is into instead of out of the sphere.

2.2.5 Other Resistance Formulae

Heat transfer rates are driven by temperature differences and therefore they may be cast in the form of a resistance formula. The advantage of the thermal resistance concept is that problems involving various types of heat transfer can be represented conveniently using thermal resistance networks. Resistance networks can be solved using techniques borrowed from electrical engineering. Also, by representing problems as thermal networks it is possible to obtain a physical feel for the situation. *Small resistances in series with large ones will tend to be unimportant. Large resistances in parallel with small ones can also be neglected.* This type of understanding is important and can be obtained quickly by computing the thermal resistances involved in a problem. In this section we will cast three additional heat transfer mechanisms in terms of resistances.

Convection Resistance

Convection was introduced in Section 1.4.2 and it is the subject of Chapters 7 through 11. Convection refers to heat transfer between a surface and an adjacent, flowing fluid as shown in Figure 2.8. The rate equation that characterizes the rate of convection heat transfer from a surface is **Newton's Law of Cooling**:

$$\dot{q}_{conv} = \underbrace{\bar{h}\,A_s}_{1/R_{conv}}(T_s - T_\infty),$$ (2.77)

where A_s is the surface area at temperature T_s that is exposed to fluid at temperature T_∞. The parameter $\bar{h}$ in Eq. (2.77) is the **average heat transfer coefficient** associated with the convection problem (which has units of W/m²-K). Methods of estimating the average heat transfer coefficient will be studied in Chapters 8 through 11 for different situations. By inspection of Eq. (2.77), the thermal resistance associated with convection (R_{conv}) is:

fluid at T_∞

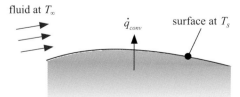

$\dot{q}_{conv}$ surface at T_s

Figure 2.8 Convection.

$$R_{conv} = \frac{1}{\bar{h}\,A_s}. \tag{2.78}$$

Contact Resistance

Contact resistance characterizes the localized impedance to heat transfer that occurs when two solid surfaces are brought into contact with one another. In Example 2.4 Figure 3 we sketched the temperature distribution that is expected when a high-conductivity material (material A) is placed in contact with a low-conductivity material (material B). The result is shown again in Figure 2.9(a).

Regardless of how well prepared the surfaces of materials A and B are, they are not flat at the microscale as shown in Figure 2.9(b). Therefore, the energy being transferred in the x-direction is constricted at the interface and must pass through an effectively smaller area than is available everywhere else in the material. Even if the surfaces were perfectly flat, the energy carriers in two dissimilar materials may not be the same and there will be an inefficiency associated with the transfer of energy from one type of energy carrier to another. This latter process is referred to as the Kapitza resistance and it becomes important at very low (i.e., cryogenic) temperatures (Barron and Nellis, 2016). The result is a temperature "jump" that, at the macroscale, appears to occur over an infinitesimally small spatial extent, as shown in Figure 2.9(b). This temperature change across the interface grows in proportion to the rate of heat transfer through the interface and can therefore be characterized by a contact resistance, R_c.

Contact resistance is usually characterized in terms of an **area-specific contact resistance**, R_c''. The contact resistance is calculated by dividing the area-specific contact resistance by the contact area:

$$R_c = \frac{R_c''}{A_c}. \tag{2.79}$$

The contact area, A_c, is the projected area of the surfaces ignoring their microstructure. Note that the area-specific contact resistance is *divided* (rather than multiplied) by the contact area; this makes sense, as the resistance increases if the contact area becomes smaller. The area-specific contact resistance, like the heat transfer

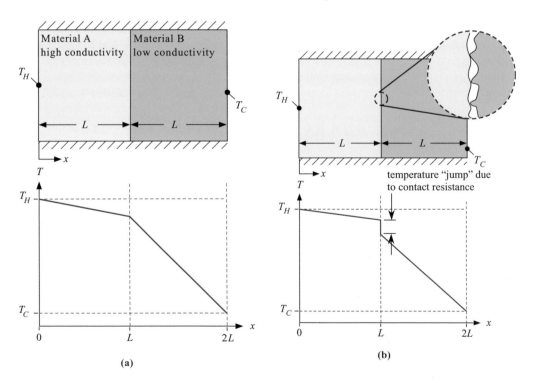

Figure 2.9 Temperature distribution expected for Example 2.4 (a) without contact resistance and (b) when contact resistance is considered.

coefficient, is not a material property. Instead, it is a complex function of the microstructure, the properties of the two materials involved, the contact pressure, any interstitial material placed in the gap, and other factors. The values of R_c'' for some interface conditions that are commonly encountered have been measured and tabulated in various references, for example Schneider (1985). Some representative values for R_c'' are listed in Table 2.1.

Notice in Table 2.1 that the area-specific contact resistance tends to be reduced with increasing clamping pressure and smaller surface roughness. One method for reducing contact resistance is to insert a soft metal (e.g., indium) or some type of grease into the interface in order to improve the heat transfer across the interstitial gap. The values listed in Table 2.1 can be used to determine whether contact resistance is likely to play an important role in a specific application by comparing its value to other resistances in the problem. However, if contact

Table 2.1 Area-specific contact resistance for some interfaces, from Schneider (1985) and Fried (1969).

Materials	Clamping pressure	Surface roughness	Interstitial material	Temperature	Area-specific contact resistance
copper-to-copper	100 kPa	0.2 μm	vacuum	46°C	1.5×10^{-4} K-m^2/W
copper-to-copper	1000 kPa	0.2 μm	vacuum	46°C	1.3×10^{-4} K-m^2/W
aluminum-to-aluminum	100 kPa	0.3 μm	vacuum	46°C	2.5×10^{-3} K-m^2/W
aluminum-to-aluminum	100 kPa	1.5 μm	vacuum	46°C	3.3×10^{-3} K-m^2/W
stainless-to-stainless	100 kPa	1.3 μm	vacuum	30°C	4.5×10^{-3} K-m^2/W
stainless-to-stainless	1000 kPa	1.3 μm	vacuum	30°C	2.4×10^{-3} K-m^2/W
stainless-to-stainless	100 kPa	0.3 μm	vacuum	30°C	2.9×10^{-3} K-m^2/W
stainless-to-stainless	1000 kPa	0.3 μm	vacuum	30°C	7.7×10^{-4} K-m^2/W
stainless-to-aluminum	100 kPa	1.2 μm	air	93°C	3.3×10^{-4} K-m^2/W
aluminum-to-aluminum	1000 kPa	0.3 μm	air	93°C	6.7×10^{-5} K-m^2/W
aluminum-to-aluminum	100 kPa	10 μm	air	20°C	2.8×10^{-4} K-m^2/W
aluminum-to-aluminum	100 kPa	10 μm	helium	20°C	1.1×10^{-4} K-m^2/W
aluminum-to-aluminum	100 kPa	10 μm	hydrogen	20°C	0.72×10^{-4} K-m^2/W
aluminum-to-aluminum	100 kPa	10 μm	silicone oil	20°C	0.53×10^{-4} K-m^2/W

resistance is important, then more precise data for the interface of interest should be obtained or measurements should be carried out.

Radiation Resistance

Radiation was introduced in Section 1.4.3. Radiation heat transfer occurs between surfaces due to the emission and absorption of electromagnetic waves. Chapter 14 is dedicated to dealing with complex radiation problems that involve many surfaces, but in the limit that a single surface at temperature T_s interacts with surroundings at temperature T_{sur} the rate equation for radiation is given by

$$\dot{q}_{rad} = A_s \sigma \varepsilon \left(T_s^4 - T_{sur}^4 \right) \tag{2.80}$$

where A_s is the area of the surface, σ is the Stefan–Boltzmann constant (5.67×10^{-8} W/m^2·K^4), and ε is the emissivity of the surface. The emissivity is a parameter that ranges between near 0 (for highly reflective surfaces) to near 1 (for highly absorptive surfaces). Note that both T_s and T_{sur} must be expressed in terms of absolute temperature (i.e., in units K rather than °C) in Eq. (2.80).

Equation (2.80) at first glance does not seem to be a resistance equation in the form of Eq. (2.35) because the heat transfer is not driven by a difference in temperatures but rather by a difference in the fourth power of the absolute temperatures. However, the fourth power difference in temperatures can be factored so that Eq. (2.80) can be equivalently written as:

$$\dot{q}_{rad} = A_s \underbrace{\underbrace{\sigma \varepsilon \left(T_s^2 + T_{sur}^2 \right)\left(T_s + T_{sur} \right)}_{\bar{h}_{rad}} \left(T_s - T_{sur} \right)}_{1/R_{rad}}. \tag{2.81}$$

Equation (2.81) suggests that an appropriate thermal resistance for radiation heat transfer (R_{rad}) is:

$$R_{rad} = \frac{1}{A_s \sigma \varepsilon \underbrace{\left(T_s^2 + T_{sur}^2 \right)\left(T_s + T_{sur} \right)}_{\approx 4 \bar{T}^3}}. \tag{2.82}$$

Because the absolute surface and surrounding temperatures are both typically large (i.e., >300 K) and also usually not too different from each other, Eq. (2.82) can be approximated by:

$$R_{rad} \approx \frac{1}{A_s \sigma \varepsilon 4 \bar{T}^3} \tag{2.83}$$

where $\bar{T}$ is the average of the surface and surrounding temperatures in K:

$$\bar{T} = \frac{T_s + T_{sur}}{2}. \tag{2.84}$$

Comparing Eq. (2.81) for radiation with Newton's Law of Cooling for convection, Eq. (2.77), shows that a **radiation heat transfer coefficient**, $\bar{h}_{rad}$, can be defined as:

$$\bar{h}_{rad} = \sigma \varepsilon \left(T_s^2 + T_{sur}^2 \right)\left(T_s + T_{sur} \right). \tag{2.85}$$

The radiation heat transfer coefficient is a useful quantity for many problems because it provides another way for convection and radiation to be compared directly in order to determine their relative magnitudes. Assuming that T_s and T_{sur} are not too different from one another allows the "radiation heat transfer coefficient" to be computed approximately according to:

$$\bar{h}_{rad} \approx \sigma \varepsilon 4 \bar{T}^3. \tag{2.86}$$

The resistance associated with radiation and the radiation heat transfer coefficient are both clearly temperature-dependent quantities. However, the conductivity, contact resistance, and average heat transfer coefficient that

Table 2.2 A summary of common resistance formulae.

Situation	Resistance formula	Nomenclature
Plane wall	$R_{pw} = \dfrac{L}{k\,A_c}$	L = wall thickness ($\|\|$ to heat flow) k = conductivity A_c = cross-sectional area ($\perp$ to heat flow)
Cylindrical shell (radial heat transfer)	$R_{cyl} = \dfrac{\ln\left(\dfrac{r_{out}}{r_{in}}\right)}{2\pi L k}$	L = cylinder length k = conductivity r_{in} and r_{out} = inner and outer radii
Spherical shell (radial heat transfer)	$R_{sph} = \dfrac{1}{4\pi k}\left[\dfrac{1}{r_{in}} - \dfrac{1}{r_{out}}\right]$	k = conductivity r_{in} and r_{out} = inner and outer radii
Convection	$R_{conv} = \dfrac{1}{\bar{h}\,A_s}$	$\bar{h}$ = average heat transfer coefficient A_s = surface area exposed to convection
Contact between surfaces	$R_c = \dfrac{R_c''}{A_c}$	R_c'' = area-specific contact resistance A_c = surface area in contact
Radiation (exact)	$R_{rad} = \dfrac{1}{A_s\,\sigma\varepsilon\left(T_s^2 + T_{sur}^2\right)\left(T_s + T_{sur}\right)}$	A_s = radiating surface area σ = Stefan–Boltzmann constant $\quad$(5.67×10^{-8} W/m²-K⁴) ε = emissivity T_s = absolute surface temperature T_{sur} = absolute surroundings temperature
Radiation (approximate)	$R_{rad} \approx \dfrac{1}{A_s\,\sigma\varepsilon 4\bar{T}^3}$	A_s = radiating surface area σ = Stefan–Boltzmann constant $\quad$(5.67×10^{-8} W/m²-K⁴) ε = emissivity $\bar{T}$ = average absolute temperature

are required to compute the other resistances we have discussed are also temperature dependent and therefore the resistance concept can only be approximate in any case. A summary of the thermal resistances that have been derived thus far is presented in Table 2.2.

Example 2.6

Figure 1 illustrates the composite plane wall that was considered in Example 2.4. The wall is made of two materials. Material A has conductivity $k_A = 10$ W/m-K and material B has conductivity $k_B = 1$ W/m-K. The thickness of both materials (in the x-direction) is $L = 0.1$ m and the cross-sectional area (perpendicular to the x-direction) is $A_c = 1$ m². The interface between the materials is characterized by an area-specific contact resistance of $R_c'' = 1 \times 10^{-3}$ K-m²/W. The left side of material A (at $x = 0$) is exposed to a hot fluid at $T_{\infty,H} = 300$ K with heat transfer coefficient $\bar{h}_H = 10$ W/m²-K. The right side of material B (at $x = 2L$) is exposed to a cold fluid at $T_{\infty,C} = 100$ K with heat transfer coefficient $\bar{h}_C = 100$ W/m²-K.

Determine the steady-state rate of heat transfer from the hot fluid to the cold fluid.

Continued

Example 2.6 (cont.)

Known Values

Figure 1 shows a sketch of the problem with the values from the problem statement indicated.

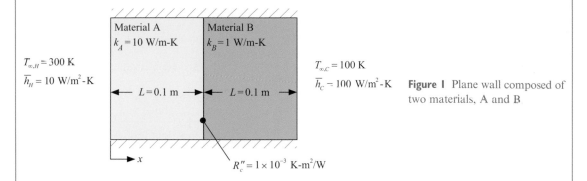

$T_{\infty,H} = 300$ K

$\overline{h}_H = 10$ W/m²-K

$T_{\infty,C} = 100$ K

$\overline{h}_C = 100$ W/m²-K

Figure 1 Plane wall composed of two materials, A and B

Assumptions

- Steady-state conditions exit.
- This is a 1-D heat transfer problem.
- Radiation is neglected.

Analysis

The resistance network that represents this situation is shown in Figure 2. Moving from the hot fluid to the cold fluid, the heat transfer must pass through a convection resistance on the hot side ($R_{conv,H}$), a conduction resistance associated with material A ($R_{pw,A}$), a contact resistance (R_c), a conduction resistance associated with material B ($R_{pw,B}$), and a convection resistance on the cold side ($R_{conv,C}$). These resistances are in series because the rate of heat transfer through each of them is the same.

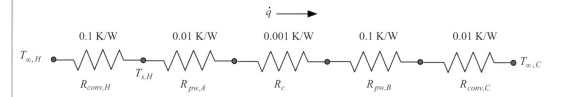

Figure 2 Resistance network that represents Figure 1.

The values of the resistances shown in Figure 2 are computed using the formulae summarized in Table 2.2. The convection resistances on the hot and code sides are:

$$R_{conv,H} = \frac{1}{\overline{h}_H A_c} \tag{1}$$

$$R_{conv,C} = \frac{1}{\overline{h}_C A_c}. \tag{2}$$

The conduction resistances associated with materials A and B are:

$$R_{pw,A} = \frac{L}{k_A A_c} \tag{3}$$

Example 2.6 (cont.)

$$R_{pw,B} = \frac{L}{k_B A_c}. \tag{4}$$

The contact resistance is:

$$R_c = \frac{R_c''}{A_c}. \tag{5}$$

The equivalent resistance of the network is given by:

$$R_{eq} = R_{conv,H} + R_{pw,A} + R_c + R_{pw,B} + R_{conv,C} \tag{6}$$

and the rate of heat transfer is given by:

$$\dot{q} = \frac{(T_{\infty,H} - T_{\infty,C})}{R_{eq}}. \tag{7}$$

Solution

Equations (1) through (7) are solved below:

$$R_{conv,H} = \frac{1}{\bar{h}_H A_c} = \frac{m^2 K}{10\,W}\left|\frac{}{1m^2}\right. = 0.1\,\frac{K}{W}$$

$$R_{conv,C} = \frac{1}{\bar{h}_C A_c} = \frac{m^2 K}{100\,W}\left|\frac{}{1m^2}\right. = 0.01\,\frac{K}{W}$$

$$R_{pw,A} = \frac{L}{k_A A_c} = \frac{0.1\,m}{}\left|\frac{m\,K}{10\,W}\right|\frac{}{1\,m^2} = 0.01\,\frac{K}{W}$$

$$R_{pw,B} = \frac{L}{k_B A_c} = \frac{0.1\,m}{}\left|\frac{m\,K}{1\,W}\right|\frac{}{1\,m^2} = 0.1\,\frac{K}{W}$$

$$R_c = \frac{R_c''}{A_c} = \frac{0.001\,K\,m^2}{W}\left|\frac{}{1m^2}\right. = 0.001\,\frac{K}{W}$$

$$R_{eq} = R_{conv,H} + R_{pw,A} + R_c + R_{pw,B} + R_{conv,C}$$

$$= 0.1\,\frac{K}{W} + 0.01\,\frac{K}{W} + 0.001\,\frac{K}{W} + 0.1\,\frac{K}{W} + 0.01\,\frac{K}{W} = 0.221\,\frac{K}{W}$$

$$\dot{q} = \frac{(T_{\infty,H} - T_{\infty,C})}{R_{eq}} = \frac{(300 - 100)\,K}{}\left|\frac{W}{0.221\,K}\right. = \boxed{905\,W}.$$

Discussion

The resistance values are superposed onto the resistance network shown in Figure 2. Figure 3 illustrates a qualitative sketch of the temperature distribution in the system. Notice that the largest temperature changes occur due to convection on the hot surface (i.e., between $T_{\infty,H}$ and the temperature at $x = 0$) and due to conduction across material B because these two resistances are the largest at 0.1 K/W. Relatively small temperature changes are associated with convection on the cold side (i.e., between the

Continued

Example 2.6 (cont.)

temperature at $x = 2L$ and $T_{\infty,C}$) and due to conduction across material A as these resistances are 10× smaller at 0.01 K/W. The temperature jump across the interface will be almost imperceptibly small as the value of R_c is extremely small relative to the others, at 0.001 K/W. Figure 3 reflects these characteristics.

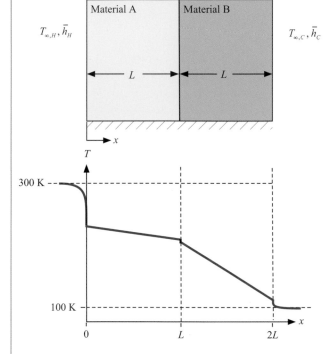

Figure 3 Sketch of the temperature distribution across the wall shown in Figure 1.

The most important resistances in this problem are the largest ones since they are all in series. Therefore, hot-side convection and conduction in material B are the most important heat transfer phenomena and these processes dominate the problem. Note that in Figure 3, the temperature drops due to the convective resistances occur over a small distance referred to as the boundary layer thickness. This subject is considered in Chapter 7.

We ignored radiation from the hot surface during our analysis. We can check to see if this is a reasonable assumption. The best way to determine whether a heat transfer process is important is to estimate the resistance associated with that process and compare its magnitude to other resistances involved in the problem. In this case we need to estimate the magnitude of the resistance to radiation from the hot surface to surroundings at $T_{\infty,H}$. Using Eq. (2.82), this resistance is computed according to

$$R_{rad,H} = \frac{1}{A_c \sigma \varepsilon_H \left(T_{s,H}^2 + T_{\infty,H}^2 \right) \left(T_{s,H} + T_{\infty,H} \right)}, \tag{8}$$

where $T_{s,H}$ is the surface temperature on the hot side of material A (shown in Figure 2); this is the temperature at $x = 0$. The value of $T_{s,H}$ can be estimated using our current solution:

$$T_{s,H} = T_{\infty,H} - \dot{q} R_{conv,H}. \tag{9}$$

Example 2.6 (cont.)

Equations (8) and (9) are used below to compute $R_{rad,H}$ with an assumed emissivity of $\varepsilon_H = 0.2$:

$$T_{s,H} = T_{\infty,H} - \dot{q} R_{conv,H} = 300\,\text{K} - \left.\frac{905\,\text{W}}{}\right|\frac{0.1\,\text{K}}{\text{W}} = 209.5\,\text{K}$$

$$R_{rad,H} = \frac{1}{A_c \sigma \varepsilon_H \left(T_{s,H}^2 + T_{\infty,H}^2\right)(T_{s,H} + T_{\infty,H})}$$

$$= \frac{1}{1\,\text{m}^2}\left|\frac{\text{m}^2\text{K}^4}{5.67 \times 10^{-8}\,\text{W}}\right|\frac{1}{0.2}\left|\frac{1}{(209.5^2 + 300^2)\text{K}^2}\right|\frac{1}{(209.5 + 300)\,\text{K}} = \boxed{1.293\,\frac{\text{K}}{\text{W}}}.$$

The hot surface *both* convects and radiates to the surroundings; therefore, the radiation resistance exists in *parallel* with the convection resistance as shown in Figure 4.

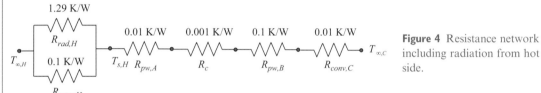

Figure 4 Resistance network including radiation from hot side.

Notice that $R_{rad,H}$ is a relatively large resistance compared to $R_{conv,H}$ and they are in parallel; therefore most of the energy transfer between $T_{\infty,H}$ and $T_{s,H}$ will occur through $R_{conv,H}$ and almost no energy will flow through $R_{rad,H}$. Large resistances in parallel with small ones are not important and radiation is therefore *not* a significant effect in this problem.

Example 2.7

A spherical Dewar contains saturated liquid oxygen that is kept at a pressure $p_{LOx} = 25$ psia; the saturation temperature of oxygen at this pressure is $T_{LOx} = 95.6$ K. The Dewar consists of an inner metal liner constructed of 304 stainless steel surrounded by an outer layer of polystyrene foam insulation. The inner metal liner has inner radius $r_{in} = 15.0$ cm and thickness $th_m = 3.5$ mm. The heat transfer coefficient between the oxygen within the Dewar and the inner surface of the Dewar is $\bar{h}_{in} = 150$ W/m²-K. The outer surface of the Dewar is surrounded by air at $T_\infty = 20°$C and radiates to surroundings that are also at T_∞. The emissivity of the outer surface of the Dewar is $\varepsilon = 0.82$. The heat transfer coefficient between the outer surface of the Dewar and the surrounding air is $\bar{h}_{out} = 6$ W/m²-K. The thickness of the insulation layer is $th_{ins} = 8.5$ mm.

Determine the rate of heat transfer to the liquid oxygen in the Dewar.

Known Values

Figure 1 illustrates a sketch of the problem with the known values included.

Continued

Example 2.7 (cont.)

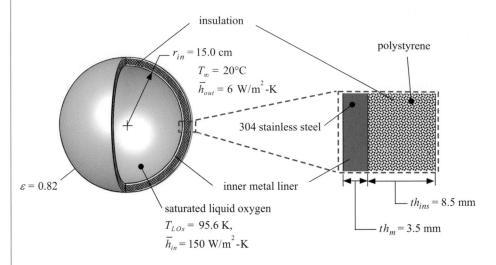

Figure 1 Spherical Dewar containing saturated liquid oxygen.

Assumptions

- Steady-state conditions exist.
- This is a 1-D conduction problem.
- Constant conductivity can be assumed in the stainless steel and polystyrene.
- There is no contact resistance between the metal liner and the insulation.

Analysis

The resistance network that represents this problem is illustrated in Figure 2.

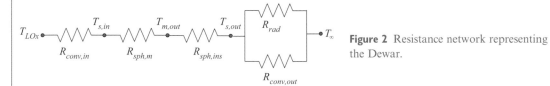

Figure 2 Resistance network representing the Dewar.

The resistance network includes interactions with the surrounding air and surroundings (at T_∞) and the saturated liquid oxygen (at T_{LOx}). The resistance to convection between the inner surface of the Dewar and the oxygen is

$$R_{conv,in} = \frac{1}{\bar{h}_{in} 4\pi r_{in}^2}. \tag{1}$$

The resistance to conduction through the metal liner is:

$$R_{sph,m} = \frac{1}{4\pi k_m (T_{m,avg})} \left[\frac{1}{r_{in}} - \frac{1}{(r_{in} + th_m)} \right], \tag{2}$$

where k_m is the conductivity of the 304 stainless steel. The conductivity of the 304 stainless steel is a function of temperature and should be evaluated at $T_{m,avg}$, the average temperature within the metal (i.e., the average of $T_{s,in}$ and $T_{m,out}$ in Figure 2):

Example 2.7 (cont.)

$$T_{m,avg} = \frac{T_{s,in} + T_{m,out}}{2}. \tag{3}$$

The resistance to conduction through the insulation layer is:

$$R_{sph,ins} = \frac{1}{4\pi k_{ins}(T_{ins,avg})} \left[\frac{1}{(r_{in} + th_m)} - \frac{1}{(r_{in} + th_m + th_{ins})} \right], \tag{4}$$

where k_{ins} is the conductivity of polystyrene. Again, the conductivity of polystyrene is a function of temperature and should be evaluated at $T_{ins,avg}$, the average of $T_{m,out}$ and $T_{s,out}$ in Figure 2:

$$T_{ins,avg} = \frac{T_{m,out} + T_{s,out}}{2}. \tag{5}$$

The convection resistance between the outer surface of the Dewar and the surrounding air is:

$$R_{conv,out} = \frac{1}{\bar{h}_{out} \, 4\pi(r_{in} + th_m + th_{ins})^2}. \tag{6}$$

The radiation resistance between the outer surface and the surroundings is computed according to:

$$R_{rad} = \frac{1}{4\pi(r_{in} + th_m + th_{ins})^2 \sigma \varepsilon \left(T_{s,out}^2 + T_\infty^2 \right)(T_{s,out} + T_\infty)}. \tag{7}$$

The total resistance separating the liquid oxygen from the surroundings is:

$$R_{eq} = R_{conv,in} + R_{sph,m} + R_{sph,ins} + \left(\frac{1}{R_{conv,out}} + \frac{1}{R_{rad}} \right)^{-1} \tag{8}$$

and the heat transfer rate from the surroundings to the liquid oxygen can be estimated:

$$\dot{q} = \frac{(T_\infty - T_{LOx})}{R_{total}}. \tag{9}$$

The intermediate temperatures required to compute the resistances ($T_{s,in}$, $T_{m,out}$, and $T_{s,out}$) can be computed according to:

$$T_{s,in} = T_{LOx} + \dot{q} R_{conv,in} \tag{10}$$

$$T_{m,out} = T_{s,in} + \dot{q} R_{sph,m} \tag{11}$$

$$T_{s,out} = T_{m,out} + \dot{q} R_{sph,ins}. \tag{12}$$

Equations (1) through (12) are a complete equation set when coupled with the problem inputs and information about the temperature-dependent conductivity of 304 stainless steel and polystyrene insulation. There are 12 equations in the 12 unknowns: $R_{conv,in}$, $R_{sph,m}$, $T_{m,avg}$, $T_{s,in}$, $T_{m,out}$, $R_{sph,ins}$, $T_{ins,avg}$, $T_{s,out}$, $R_{conv,out}$, R_{rad}, R_{eq}, and $\dot{q}$. They are not linear or explicit equations and therefore would be difficult to solve by hand. Instead, we will solve them using the EES software.

Solution

The inputs are entered in EES and converted to base SI units. The built-in property database will be used to determine the conductivity of 304 stainless steel and polyurethane. Therefore, it is necessary to specify the unit system that will be used using the $UnitSystem directive at the top of the EES code. Also, the names of the two substances, 'Stainless_AISI304' and 'Polystyrene', are assigned to the string variables S_m$ and S_ins$, respectively.

Continued

Example 2.7 (cont.)

```
$UnitSystem SI Mass J K Pa
T_LOx=95.6 [K]                          "temperature of liquid oxygen"
h_bar_in=150 [W/m^2-K]                  "heat transfer coef. between LOx and inner surface"
r_in=15 [cm]*Convert(cm,m)              "inner radius of Dewar"
S_m$='Stainless_AISI304'               "metal liner material"
th_m=3.5 [mm]*Convert(mm,m)             "thickness of metal liner"
S_ins$='Polystyrene'                    "insulation material"
th_ins=8.5 [mm]*Convert(mm,m)           "insulation thickness"
h_bar_out=6 [W/m^2-K]                   "heat transfer coef. from outer surface to surroundings"
e=0.82 [-]                              "emissivity of outer surface"
T_infinity=ConvertTemp(C,K,20 [C])     "surrounding temperature"
```

This problem is much like a calculation that you might encounter in actual engineering practice because only very fundamental quantities are provided to you. For example, the materials used to construct the Dewar and their dimensions are known, but the conductivity values to use in the calculations are not explicitly provided. In an even more realistic problem, the heat transfer coefficients would not be provided and you would need to calculate them from more fundamental quantities such as the geometry and fluid velocities. (This will have to wait until we cover Chapters 7 through 11.)

The Engineering Equation Solver (EES) software can deal with a problem like this in which a set of nonlinear and implicit equations must be solved. It is this capability that simultaneously makes EES so powerful and yet sometimes, ironically, difficult to use. EES should be able to solve equations regardless of the order in which they are entered. However, you should enter equations in a sequence that allows you to solve them as you enter them; this is exactly what you would be forced to do if you were to solve the problem using a typical programming language (e.g., MATLAB, FORTRAN, etc.). This technique of entering your equations in a systematic order provides you with the opportunity to debug each subset of equations as you move along, rather than waiting until all of the equations have been entered before you try to solve them. Another, more subtle, benefit of approaching a problem in this sequential manner is that you can consistently update the guess values associated with the variables in your problem. EES solves your equations using a nonlinear relaxation technique and therefore the closer the guess values of the variables are to "reasonable" values, the more likely it is that EES will find the correct solution. This problem provides an opportunity to demonstrate the best way to solve a problem that requires iteration using EES.

The internal convection resistance, Eq. (1), is entered in EES.

```
"Calculate resistances"
R_conv_in=1/(h_bar_in*4*pi*r_in^2)     "internal resistance to convection"
```

The resistance to conduction through the metal, Eq. (2), must be entered next; however, the temperature $T_{m,avg}$ that must be used to evaluate the conductivity is not known at this point in the solution process. Therefore, it is best to assume a reasonable value of the average metal temperature, $T_{m,avg,g}$, and use this temperature to evaluate conductivity in order to proceed with the solution process in a sequential manner.

```
T_m_avg_g=100 [K]                       "guess for average metal temperature"
k_m=Conductivity(S_m$,T=T_m_avg_g)     "conductivity of metal"
R_sph_m=(1/r_in-1/(r_in+th_m))/(4*pi*k_m)  "resistance to conduction through metal"
```

Example 2.7 (cont.)

Solving at this point allows you to verify that there are no errors in the equations, the units are consistent, and the results are reasonable. Eventually, we will remove the guessed value of $T_{m,avg}$ and instead calculate the average metal temperature precisely.

The resistance to conduction through the insulation, Eq. (4), is entered in EES and again we will assume a reasonable value of $T_{ins,avg,g}$ in order to calculate the conductivity of polystyrene and proceed.

```
T_ins_avg_g=150 [K]                          "guess for average insulation temperature"
k_ins=Conductivity(S_ins$,T=T_ins_avg_g)     "conductivity of insulation"
R_sph_ins=(1/(r_in+th_m)-1/(r_in+th_m+th_ins))/(4*pi*k_ins)
                                             "resistance to conduction through insulation"
```

The outer convection resistance, Eq. (6), is entered in EES.

```
R_conv_out=1/(h_bar_out*4*pi*(r_in+th_m+th_ins)^2)    "external resistance to convection"
```

The radiation resistance between the outer surface and the surroundings is computed according to Eq. (7). Again we will proceed by guessing the value of the outer surface temperature, $T_{s,out,g}$, in order to allow us to compute the resistance.

```
T_s_out_g=200 [K]                        "guess for external surface temperature"
R_rad=1/(e*sigma#*4*pi*(r_in+th_m+th_ins)^2*(T_s_out_g^2+T_infinity^2)*(T_s_out_g+T_infinity))
                                         "resistance to radiation"
```

The total resistance separating the liquid oxygen from the surroundings and the heat transfer rate from the surroundings to the liquid oxygen are computed using Eqs. (8) and (9).

```
"Calculate heat transfer rate"
R_eq=R_conv_in+R_sph_m+R_sph_ins+(1/R_rad+1/R_conv_out)^(-1)   "total equivalent resistance"
q_dot=(T_infinity-T_LOx)/R_eq                                 "heat transfer rate"
```

At this point we have a solution, but it is not correct because several of the resistance values were computed using guessed values of temperature. The solution to this point consists of a set of explicit equations that are easy to solve; you could have solved them as easily using a piece of paper and calculator since the right side of each equation referred only to variables that were either known, guessed, or previously calculated. In order to obtain a solution that is correct we must iterate. If you were not using EES, this iteration process would require that you compute the temperatures that were guessed and then carry out the solution again with these new temperatures. With luck, repeating this process a few times will lead to a self-consistent solution. EES makes this process substantially easier, but it is still important that you understand the iteration process.

Equations (10) through (12) are used to compute the intermediate temperatures and Eqs. (3) and (5) are used to compute the average temperatures required to obtain the conductivity values.

```
"Calculate temperatures"
T_s_in=T_LOx+R_conv_in*q_dot         "inner surface temperature"
T_m_out=T_s_in+R_sph_m*q_dot         "outer temperature of metal"
T_s_out=T_m_out+R_sph_ins*q_dot      "calculated outer surface temperature"
T_m_avg=(T_s_in+T_m_out)/2           "average metal temperature"
T_ins_avg=(T_m_out+T_s_out)/2        "average insulation temperature"
```

Continued

Example 2.7 (cont.)

Solving shows that $T_{m,avg}$ = 97.30 K, $T_{ins,avg}$ = 183.3 K, and $T_{s,out}$ = 269.2 K. These values are different from those that we guessed during the development of the solution: $T_{m,avg,g}$ = 100 K, $T_{ins,avg,g}$ = 150 K, and $T_{s,out,g}$ = 200 K. Therefore iteration is required. EES will accomplish this iteration automatically but it is good to have it start from a reasonable set of guess values. Our next step will be to comment out or delete the equations that provided the assumed temperatures and instead use the temperatures that were correctly calculated to compute these resistances. This step creates an implicit set of nonlinear equations. Before you ask EES to solve the set of equations, it is a good idea to update the guess values for each variable; this can be done by selecting Update Guesses from the Calculate menu. Note that the **$UpdateGuesses** directive placed at the top of the Equations Window will also cause EES to update guess values each time the code is successfully run. The resulting EES code is shown below:

```
$UnitSystem SI Mass J K Pa
$UpdateGuesses
T_LOx=95.6 [K]                          "temperature of liquid oxygen"
h_bar_in=150 [W/m^2-K]                  "heat transfer coef. between LOx and inner surface"
r_in=15 [cm]*Convert(cm,m)              "inner radius of Dewar"
S_m$='Stainless_AISI304'               "metal liner material"
th_m=3.5 [mm]*Convert(mm,m)             "thickness of metal liner"
S_ins$='Polystyrene'                    "insulation material"
th_ins=8.5 [mm]*Convert(mm,m)           "insulation thickness"
h_bar_out=6 [W/m^2-K]                   "heat transfer coef. from outer surface to surroundings"
e=0.82 [-]                              "emissivity of outer surface"
T_infinity=ConvertTemp(C,K,25 [C])      "surrounding temperature"

"Calculate resistances"
R_conv_in=1/(h_bar_in*4*pi*r_in^2)      "internal resistance to convection"
{T_m_avg_g=100 [K]}                      "guess for average metal temperature"
k_m=Conductivity(S_m$,T=T_m_avg_g)      "conductivity of metal"
R_sph_m=(1/r_in-1/(r_in+th_m))/(4*pi*k_m)   "resistance to conduction through metal"
{T_ins_avg_g=150 [K]}                    "guess for average insulation temperature"
k_ins=Conductivity(S_ins$,T=T_ins_avg_g) "conductivity of insulation"
R_sph_ins=(1/(r_in+th_m)-1/(r_in+th_m+th_ins))/(4*pi*k_ins)
                                        "resistance to conduction through insulation"
R_conv_out=1/(h_bar_out*4*pi*(r_in+th_m+th_ins)^2)
                                        "external resistance to convection"
{T_s_out_g=200 [K]}                      "guess for external surface temperature"
R_rad=1/(e*sigma#*4*pi*(r_in+th_m+th_ins)^2*(T_s_out_g^2+T_infinity^2)*(T_s_out_g+T_infinity))
                                        "resistance to radiation"

"Calculate heat transfer rate"
R_eq=R_conv_in+R_sph_m+R_sph_ins+(1/R_rad+1/R_conv_out)^(-1)
                                        "total equivalent resistance"
q_dot=(T_infinity-T_LOx)/R_eq           "heat transfer rate"

"Calculate temperatures"
T_s_in=T_LOx+R_conv_in*q_dot            "inner surface temperature"
T_m_out=T_s_in+R_sph_m*q_dot            "outer temperature of metal"
```

Example 2.7 (cont.)

```
T_s_out=T_m_out+R_sph_ins*q_dot          "calculated outer surface temperature"
T_m_avg=(T_s_in+T_m_out)/2               "average metal temperature"
T_ins_avg=(T_m_out+T_s_out)/2            "average insulation temperature"


"Use calculated temperatures to compute resistances"
T_m_avg_g=T_m_avg
T_ins_avg_g=T_ins_avg
T_s_out_g=T_s_out
```

The rate of heat transfer to the liquid oxygen that is predicted by the EES code is $\boxed{\dot{q} = 84.28\,\text{W}}$.

Discussion

The values of the resistances that are considered in the problem are overlaid onto the network shown in Figure 3.

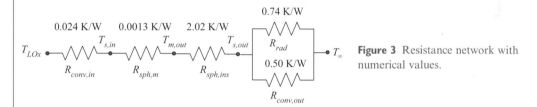

Figure 3 Resistance network with numerical values.

Examination of Figure 3 shows that the insulation conduction resistance is the most important (because it is the largest resistance in the series arrangement), followed by convection and radiation from the external surface. Convection from the internal surface and conduction through the metal liner are both unimportant for this problem.

We neglected contact resistance between the metal and insulation for this problem; it would be useful to know if this is a reasonable assumption. The contact resistance can be estimated using the values listed in Table 2.1. The values listed vary by as much as an order of magnitude; however, an upper bound on the area-specific contact resistance appears to be 4.5×10^{-3} K-m^2/W. Therefore, we will use this upper bound to estimate how large contact resistance might be:

$$R_c = \frac{R_c''}{4\pi(r_{in} + th_m)^2} = \frac{4.5 \times 10^{-3}\,\text{K m}^2}{\text{W}}\Bigg|_{4\pi(0.15 + 0.0035)^2\,\text{m}^2} = 0.0152\,\frac{\text{K}}{\text{W}}.$$

This value is 2 orders of magnitude less than the value of $R_{sph,ins}$ and therefore contact resistance is not likely to be important for this problem.

A solution implemented using a computer program can be used to carry out parametric studies. For example, we can investigate how the heat transfer rate is affected by the insulation thickness using a plot. In order to generate this plot, it is necessary to parametrically vary the insulation thickness. The specified value of the insulation thickness is commented out.

```
{th_ins=8.5 [mm]*Convert(mm,m)}          "insulation thickness"
```

A Parametric Table is generated (select New Parametric Table from the Tables menu) that includes the variables th_ins and q_dot (Figure 4).

Continued

Example 2.7 (cont.)

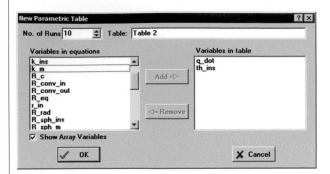

Figure 4 New Parametric Table Window.

Right-click on the th_ins column in the resulting Parametric Table and select Alter Values; vary the thickness from 5 mm (0.005 m) to 15 mm (0.015 m). Solve the table by selecting Solve Table from the Calculate menu. It is possible for each variable in EES to be assigned both a primary and secondary set of units; this is useful for situations where you would like to report a variable's value in a different set of units than you used for calculations. In this example it might be nice to plot the heat loss as a function of insulation thickness in mm rather than m as this is more physically meaningful to most system designers. In the Professional version of EES this can be accomplished by right-clicking on the th_ins column and selecting Properties. In the Format Parametric Table Column dialog that pops up you can select a secondary unit system and specify that the values will be shown in both primary and secondary units (Figure 5).

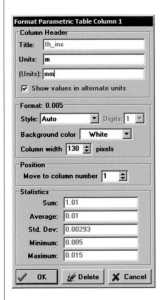

Figure 5 Format Parametric Table Column dialog; notice that the secondary units for the variable th_ins are specified to be mm.

Prepare a plot of the results by selecting New Plot Window from the Plots menu and then selecting X-Y Plot.[1] Selecting the variable th_ins [mm] for the X-Axis (corresponding to the thickness values in the secondary unit system, mm) and q_dot for the Y-Axis (Figure 6).

---

[1] Note that the ability to plot a variable in its alternate units as shown in this example requires a Professional version of EES. Alternatively, another variable can be defined in the Equations Window, e.g.,

th_ins_mm=th_ins***Convert**(m,mm).   "convert th_ins to mm"

Adding this variable to the Parametric table allows the plot to be constructed using units of mm instead of m.

Example 2.7 (cont.)

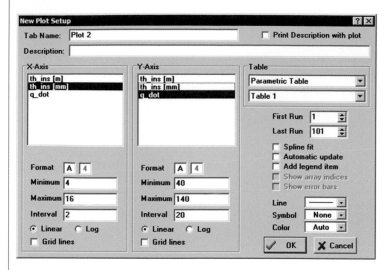

Figure 6 New Plot Setup Window.

Figure 7 illustrates the rate of heat transfer as a function of the insulation thickness.

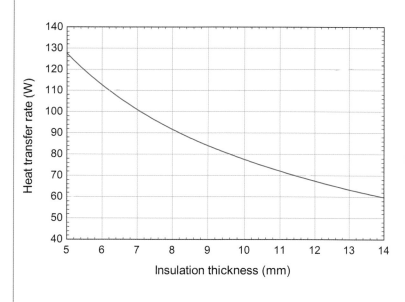

Figure 7 Heat transfer rate as a function of insulation thickness.

2.3 Steady-State 1-D Conduction with Generation

2.3.1 Introduction

The concept of an "energy generation" term in an energy balance should be troubling to anyone who has taken Thermodynamics. According to the First Law of Thermodynamics, energy is a *conserved* quantity; it is neither generated nor destroyed. In Heat Transfer, however, we are often doing a "thermal energy" balance and therefore we will encounter problems where thermal energy is "generated" because energy in a different form is being transformed to thermal energy. For example, ohmic dissipation occurs when electrical current is passed through a resistive material causing its temperature to rise because electrical energy is being transformed into thermal energy. Thermal energy generation can also occur due to chemical or nuclear reactions or by the absorption of radiation.

Thermal energy generation tends to occur *volumetrically*. That is, the material experiences a thermal energy generation that is distributed throughout its volume. For example, an electrical resistor may experience thermal energy generation that occurs equally or uniformly throughout the entire volume of the resistor. These phenomena will therefore be described by a **volumetric thermal energy generation rate**, $\dot{g}'''$, which specifies the rate of thermal energy generation per unit volume (W/m³).

The energy balance that we use to solve problems involving thermal energy generation is strictly a thermal energy conservation equation. The addition of thermal energy generation to the 1-D steady-state problems considered in Section 2.2 is relatively straightforward and the steps required to obtain an analytical solution are essentially the same. However, when carrying out the thermal energy balance on a control volume, there will be a new term related to the rate of thermal energy generation that occurs within that control volume. This term must be related to the product of $\dot{g}'''$ and the volume of material that is enclosed in the control volume.

Example 2.8

A composite plane wall is composed of two materials, A and B. Each section of the wall has the same thickness, L. There is a spatially uniform, volumetric rate of thermal energy generation present in material A ($\dot{g}'''_A = \dot{g}'''$) but no generation in material B ($\dot{g}'''_B = 0$). The conductivity of material B is twice that of material A ($k_B = 2k_A$). The left side of material A (at $x = 0$) is adiabatic and the right side of material B (at $x = 2L$) is maintained at temperature T_o.

Prepare a sketch of conduction heat transfer in the x-direction and another sketch of the temperature as a function of position within the wall. Each sketch should go from $x = 0$ (the left face of material A) to $x = 2L$ (the right face of material B). Note that the sketch should be qualitatively correct, but it cannot be quantitative as you have not been given any numbers for the problem.

Known Values

The problem setup is shown in Figure 1.

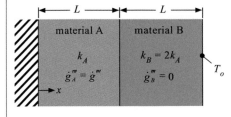

Figure 1 Plane wall composed of materials A and B.

Assumptions

- Steady-state conditions exist.
- Both materials have constant conductivity.
- There is no contact resistance at the interface separating materials A and B.

Analysis and Solution

Typically, the best way to understand a heat transfer problem is to first understand the energy flows using the First Law of Thermodynamics. Following this, you can determine how those energy flows will

Example 2.8 (cont.)

translate into a particular temperature distribution using rate equations. Within material A, the heat transfer rate must increase linearly from $\dot{q} = 0$ at the left face to $\dot{q} = \dot{g}''' A_c L$ at the right face, where A_c is the cross-sectional area of the wall. To see this, consider an energy balance on the control volume shown in Figure 2 (top).

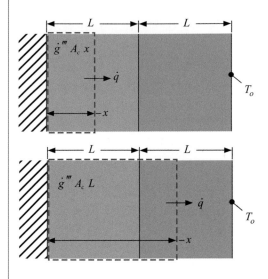

Figure 2 Energy balance on control volume (top) entirely within material A and (bottom) extending into material B.

The control volume shown in Figure 2 (top) is entirely within material A and has one control surface at the adiabatic wall and the other at position x where $x < L$. The First Law of Thermodynamics requires that:

$$IN = OUT + STORED, \tag{1}$$

where *IN*, *OUT*, and *STORED* represent the rates of energy entering, leaving, and being stored within the control volume, respectively. The storage term is zero as this is a steady-state problem. The thermal energy generated inside the control volume constitutes an *inflow* of thermal energy that is equal to the product of the volume of material contained in the control volume, $A_c\, x$, and the volumetric rate of thermal energy generation, $\dot{g}'''$. The outflow of energy is related to conduction out of the right-hand control surface. There is no heat transfer through the left-hand control surface as it is adiabatic. The result is:

$$\dot{g}''' A_c x = \dot{q}. \tag{2}$$

Clearly then the rate of conduction heat transfer increases linearly from 0 to $\dot{g}''' A_c L$ as x increases from 0 to L.

For x greater than L the rate of conduction heat transfer remains unchanged. To see this, consider an energy balance on the control volume shown in Figure 2 (bottom) which moves the right-hand control surface into material B. Again, the First Law of Thermodynamics requires that Eq. (1) be satisfied. In this case, the rate at which thermal energy generation is entering the system is the product of the entire volume of material A, $A_c\, L$, and the rate of thermal energy generation within that material, $\dot{g}'''$. There is no thermal energy generation within material B and so the volume of material B enclosed within the control volume does not contribute to the inflow of thermal energy. The outflow is again the rate of conduction through the right-hand control surface located at position x:

$$\dot{g}''' A_c L = \dot{q}. \tag{3}$$

Continued

Example 2.8 (cont.)

Figure 3 illustrates the rate of conduction as a function of position within the wall.

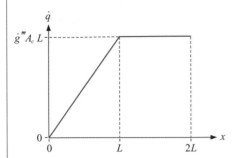

Figure 3 Sketch of the rate of conduction heat transfer as a function of position.

The temperature distribution can be inferred from the rate of conduction heat transfer using Fourier's Law:

$$\dot{q} = -kA_c \frac{dT}{dx}. \tag{4}$$

Rearranging Eq. (4) leads to:

$$\frac{dT}{dx} = -\frac{\dot{q}}{kA_c}. \tag{5}$$

Examination of Eq. (5) together with Figure 3 indicates that the temperature distribution must have a *slope* that starts at 0 at $x = 0$ and becomes more negative as x increases to L. Because the conductivity of material B is not equal to that of material A, the slope will change discontinuously at $x = L$. The heat transfer rate on both sides of the interface (i.e., in material A at $x = L^-$ and in material B at $x = L^+$) is the same. However, the conductivity in material B is twice that of material A and therefore, according to Eq. (5), the magnitude of the temperature gradient must be half as large. The temperature gradient does not change in material B as none of the quantities on the right side of Eq. (5) change. The temperature at $x = 2L$ must be T_o. These characteristics are all reflected in Figure 4.

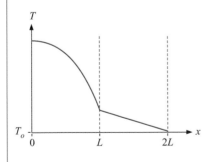

Figure 4 Sketch of the temperature as a function of position.

2.3.2 Uniform Thermal Energy Generation in a Plane Wall

Consider a plane wall with temperatures fixed at either edge that experiences a uniform volumetric rate of generation of thermal energy, $\dot{g}'''$, as shown in Figure 2.10. The problem is 1-D with temperature only varying in the x-direction and therefore an appropriate differential control volume has width dx (see Figure 2.10); notice

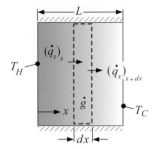

Figure 2.10 Plane wall with thermal energy generation and fixed temperature boundary conditions.

the additional energy term in the control volume that is related to the generation of thermal energy. A steady-state energy balance includes conduction into the left side of the control volume $(\dot{q}_x)_x$, generation within the control volume $\dot{g}$, and conduction out of the right side of the control volume $(\dot{q}_x)_{x+dx}$:

$$(\dot{q}_x)_x + \dot{g} = (\dot{q}_x)_{x+dx}. \tag{2.87}$$

After expanding the right-hand side and taking the limit as dx approaches zero we obtain:

$$(\dot{q}_x)_x + \dot{g} = (\dot{q}_x)_x + \frac{d\dot{q}_x}{dx}dx \tag{2.88}$$

or

$$\dot{g} = \frac{d\dot{q}_x}{dx}dx. \tag{2.89}$$

Equation (2.89) is the governing ordinary differential equation (ODE) for this problem written in terms of energy and it should be checked against our intuition. In Section 2.2 we considered problems *without* generation and found that the rate of conduction did not change with position. Here the rate of conduction must increase with position due to the presence of thermal energy generation. This makes sense and as a result we can expect a temperature gradient that is not constant even if the cross-sectional area and conductivity are constant.

In order to obtain a differential equation written in terms of temperature it is necessary to substitute rate equations into Eq. (2.89). The rate of thermal energy generation within the control volume can be expressed as the product of the volume of material within the control volume and the rate of thermal energy generation per unit volume, $\dot{g}'''$ (which in general may be a function of position or temperature):

$$\dot{g} = \dot{g}''' A_c\, dx, \tag{2.90}$$

where A_c is the cross-sectional area of the wall. The conduction term is expressed using Fourier's Law:

$$\dot{q}_x = -kA_c\frac{dT}{dx}. \tag{2.91}$$

Substituting Eqs. (2.91) and (2.90) into Eq. (2.89) results in

$$\dot{g}''' A_c\, dx = \frac{d}{dx}\left(-kA_c\frac{dT}{dx}\right)dx, \tag{2.92}$$

which can be simplified (assuming that conductivity is constant) to:

$$\frac{d}{dx}\left(\frac{dT}{dx}\right) = -\frac{\dot{g}'''}{k}. \tag{2.93}$$

Equation (2.93) is a separable second-order linear differential equation. It is separated and integrated:

$$\int d\left(\frac{dT}{dx}\right) = \int -\frac{\dot{g}'''}{k}\,dx.$$ (2.94)

If the volumetric rate of thermal energy generation is spatially uniform, then the integration leads to:

$$\frac{dT}{dx} = -\frac{\dot{g}'''}{k}x + C_1$$ (2.95)

where C_1 is a constant of integration, required because Eq. (2.94) is an indefinite integral. Equation (2.95) is integrated again:

$$\int dT = \int \left(-\frac{\dot{g}'''}{k}x + C_1\right)dx,$$ (2.96)

which leads to:

$$T = -\frac{\dot{g}'''}{2k}x^2 + C_1 x + C_2,$$ (2.97)

where C_2 is another constant of integration. Equation (2.97) is the general solution to the ODE provided by Eq. (2.93) with constant volumetric energy generation; the general solution solves the ODE regardless of what values of C_1 and C_2 are selected. All that remains is to force the general solution, Eq. (2.97), to satisfy the boundary conditions by adjusting the constants C_1 and C_2. The fixed temperature boundary conditions shown in Figure 2.10 correspond to:

$$T_{x=0} = T_H$$ (2.98)

$$T_{x=L} = T_C.$$ (2.99)

Substituting Eq. (2.97) into Eqs. (2.98) and (2.99) leads to:

$$C_2 = T_H$$ (2.100)

$$-\frac{\dot{g}'''}{2k}L^2 + C_1 L + T_H = T_C.$$ (2.101)

Solving Eq. (2.101) for C_1 results in

$$C_1 = \frac{\dot{g}'''}{2k}L - \frac{(T_H - T_C)}{L}.$$ (2.102)

Substituting Eqs. (2.100) and (2.102) into Eq. (2.97) provides the final result:

$$T = \frac{\dot{g}'''L^2}{2k}\left[\frac{x}{L} - \left(\frac{x}{L}\right)^2\right] - \frac{(T_H - T_C)}{L}x + T_H.$$ (2.103)

It is good practice to examine any solution you develop and verify that it makes sense. Eventually you will present your work to someone else (for example, your boss) and it's always better that you catch your errors before some other person points out some obvious inconsistency in your results. By inspection, it is clear that Eq. (2.103) limits to T_H at $x = 0$ and T_C at $x = L$, therefore the boundary conditions were implemented correctly. Furthermore, in the absence of any generation, Eq. (2.103) becomes

$$T = -\frac{(T_H - T_C)}{L}x + T_H,$$ (2.104)

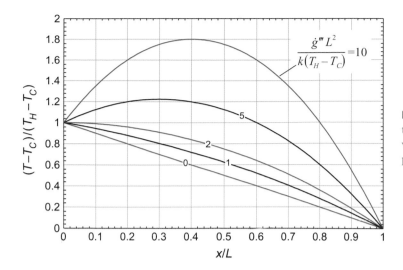

Figure 2.11 Dimensionless temperature distribution for various values of the dimensionless parameter $\dot{g}''' L^2/(k(T_H - T_C))$.

which is equivalent to Eq. (2.32), the solution that was derived for steady-state conduction through a plane wall without generation.

Equation (2.103) can be rearranged in order to provide a dimensionless temperature:

$$\frac{(T - T_C)}{(T_H - T_C)} = \frac{\dot{g}''' L^2}{2k(T_H - T_C)} \left[\frac{x}{L} - \left(\frac{x}{L} \right)^2 \right] + 1 - \frac{x}{L}. \tag{2.105}$$

Figure 2.11 illustrates the dimensionless temperature distribution predicted by Eq. (2.105) for various values of the dimensionless parameter $\dot{g}''' L^2/(k(T_H - T_C))$. Notice that the value of the slope of the temperature (i.e., dT/dx) decreases as x increases in Figure 2.11. According to Fourier's Law, this is consistent with an increasing rate of conduction heat transfer in the positive x-direction and therefore also consistent with our ODE, Eq. (2.89). The rate at which the temperature gradient changes becomes larger as the rate of volumetric thermal energy generation becomes larger, which is also consistent with Eq. (2.89). For very high values of $\dot{g}'''$ the rate of conduction heat transfer is actually in the negative x-direction at small values of x and gradually changes direction to the positive x-direction.

Example 2.9

Freshly cut hay is not really dead; chemical reactions continue in the plant cells and therefore a small amount of thermal energy is released within the hay bale. This is an example of the conversion of chemical to thermal energy. The amount of thermal energy generation within a hay bale depends on the moisture content of the hay when it is baled. Baled hay can become a fire hazard if the hay is baled when it is too wet causing the rate of volumetric generation to be high. If the interior temperature of the bale reaches $T_{fire} = 170°F$ then self-ignition can occur.

Consider a hay bale that can be modeled as a plane wall. The thickness of the bale is $L = 3.5$ ft. One edge (at $x = 0$) is adiabatic while the other edge (at $x = L$) is exposed to air in the barn at $T_\infty = 20°C$. The heat transfer coefficient between the air and the bale surface is $\bar{h} = 5$ W/m^2-K. Hay is a composite structure consisting of air and plant matter. The effective conductivity of the hay is $k = 0.04$ W/m-K. The volumetric rate of thermal energy generation within the hay bale is $\dot{g}''' = 2.5$ W/m^3.

- Develop an analytical solution for the temperature distribution within the bale.
- Prepare a plot showing the temperature as a function of position.

Continued

Example 2.9 (cont.)

Known Values

Figure 1 illustrates a sketch of the problem with the known values listed.

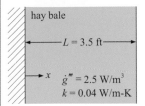

$\bar{h} = 5\,\text{W/m}^2\text{-K}$
$T_\infty = 20°\text{C}$

Figure 1 Hay bale stored in a barn.

Assumptions

- Steady-state conditions exist.
- The temperature distribution is one-dimensional.
- The conductivity and volumetric energy generation are not functions of temperature.
- Radiation from the outer surface is neglected.

Analysis

The derivation of the governing differential equation proceeds in the same way as in Section 2.3.2 and the general solution for the temperature gradient and temperature, Eqs. (2.95) and (2.97) respectively, are the same. However, the boundary conditions for this problem are different than those considered in Section 2.3.2. The temperatures at the boundaries are not given and therefore it becomes necessary to carry out an **interface energy balance** at each edge of the computational domain. Figure 2 illustrates an interface energy balance at $x = 0$. One control surface is located at $x = 0^-$ (i.e., just outside of the hay) while the other surface is located at $x = 0^+$ (just inside the hay).

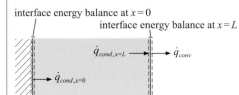

Figure 2 Interface energy balances at $x = 0$ and $x = L$.

The boundary at $x = 0$ is adiabatic and therefore no energy passes through the control surface located at $x = 0^-$. The conduction heat transfer passing through the boundary at $x = 0^+$ must be equal to zero:

$$\dot{q}_{cond,\,x=0} = 0. \tag{1}$$

Substituting Fourier's Law into Eq. (1) provides:

$$-k\,A_c \frac{dT}{dx}\bigg|_{x=0} = 0. \tag{2}$$

Substituting Eq. (2.95) evaluated at $x = 0$ into Eq. (2) provides:

$$-k\,A_c\left(-\frac{\dot{g}'''}{k}0 + C_1\right) = 0 \tag{3}$$

Example 2.9 (cont.)

or

$$C_1 = 0. \tag{4}$$

This leaves a single undetermined constant, C_2, in the general solution:

$$\frac{dT}{dx} = -\frac{\dot{g}'''}{k}x \tag{5}$$

$$T = -\frac{\dot{g}'''}{2k}x^2 + C_2. \tag{6}$$

Figure 2 also illustrates an interface energy balance carried out at $x = L$. In this case, the rate of conduction into the control surface at $x = L^-$ must equal the rate of convection leaving through the control surface at $x = L^+$:

$$\dot{q}_{cond,x=L} = \dot{q}_{conv}. \tag{7}$$

Substituting Fourier's Law and Newton's Law of Cooling into Eq. (7) provides:

$$-k\,A_c\frac{dT}{dx}\bigg|_{x=L} = \bar{h}\,A_c(T_{x=L} - T_\infty). \tag{8}$$

Substituting Eqs. (5) and (6) into Eq. (8) leads to an equation that defines the correct value of C_2:

$$-k\,A_c\left(-\frac{\dot{g}'''}{k}L\right) = \bar{h}\,A_c\left(-\frac{\dot{g}'''}{2k}L^2 + C_2 - T_\infty\right). \tag{9}$$

Solving for C_2 provides:

$$C_2 = T_\infty + \frac{\dot{g}'''\,L}{\bar{h}} + \frac{\dot{g}'''}{2k}L^2. \tag{10}$$

Substituting Eq. (10) into Eq. (6) leads to the analytical solution for this problem:

$$\boxed{T = -\frac{\dot{g}'''}{2k}x^2 + T_\infty + \frac{\dot{g}'''\,L}{\bar{h}} + \frac{\dot{g}'''}{2k}L^2}. \tag{11}$$

Solution

The analytical solution, Eq. (11), is implemented in EES in order to prepare the requested plot. The inputs are entered in EES and converted to base SI units.

```
$UnitSystem SI Mass J K Pa
T_fire = ConvertTemp(F,K,170 [F])        "self-ignition temperature"
T_infinity = ConvertTemp(C,K,20 [C])     "air temperature"
h_bar=5 [W/m^2-K]                        "heat transfer coefficient"
L = 3.5 [ft]*Convert(ft,m)               "thickness"
k = 0.04 [W/m-K]                         "effective conductivity of hay"
g_dot```=2.5 [W/m^3]                     "volumetric rate of thermal energy generation"
```

Equation (11) is programmed in EES. A Parametric Table is created and used to generate the plot shown in Figure 3.

```
x_bar=0 [-]                                                "dimensionless position, x/L"
x=L*x_bar                                                  "position"
T= -g_dot```*x^2/(2*k)+g_dot```*L/h_bar+g_dot```*L^2/(2*k)+T_infinity   "analytical solution"
```

Continued

Example 2.9 (cont.)

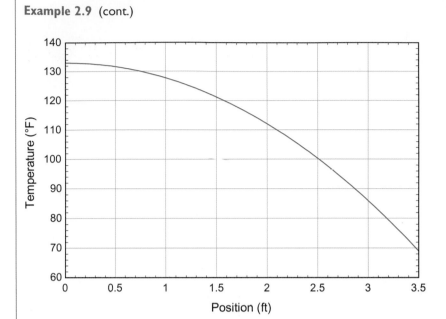

Figure 3 Temperature in hay bale as a function of position.

Discussion

The solution was developed by specifying a set of parameters including L in order to develop a model that can predict the maximum temperature in the bale. The parameters T_∞, $\bar{h}$, L, k, and $\dot{g}'''$ were all *inputs* and the *outputs* are the temperature distribution and also the maximum bale temperature. Once an engineering model is developed, we will often find ourselves specifying the values of one or more outputs in order to determine the corresponding value of an input. This is the crux of the design process – what input values give you a desired output? For this problem, we may be interested in knowing how large the bale can be before we are in danger of self-ignition; that is, what value of L causes the maximum temperature in the bale to reach T_{fire}?

If we set $x = 0$ in Eq. (11) we obtain an expression for the maximum temperature in the bale, which occurs next to the adiabatic edge:

$$T_{max} = T_\infty + \frac{\dot{g}''' L}{\bar{h}} + \frac{\dot{g}'''}{2k} L^2. \tag{12}$$

```
T_max=g_dot```*L/h_bar+g_dot```*L^2/(2*k)+T_infinity        "maximum temperature"
```

For the given value of $L = 3.5$ ft (1.067 m) we find that the maximum temperature in the bale is $T_{max} = 329.2$ K (133.0°F). However, we can specify the value of the maximum temperature:

```
T_max=T_fire                                "specify the maximum temperature"
```

and comment out the value of the bale thickness.

Example 2.9 (cont.)

{L = 3.5 [ft]*Convert(ft,m)}   "thickness"

Solving provides $L = 1.339$ m (4.392 ft); the thickest bale that should be stored in the barn is 4.392 ft in order to avoid a potential fire hazard.

Example 2.10

A shunt resistance is used to measure current. The current passes through the resistive material causing a voltage drop that can be measured and used to infer the current. The ohmic dissipation will cause the shunt resistor to become hot if it is not properly cooled; this is particularly true if the resistor is used to measure large currents. Because the electrical resistivity of the material may be a function of temperature, it is desirable to limit the temperature rise in a shunt resistance as much as possible so that the temperature change does not affect the accuracy of the current measurement.

A shunt resistor is mounted between two electrodes that provide the current. The resistor is a cylinder through which the current flows axially. The cylinder has length $L = 5$ cm and diameter $D = 4.5$ mm. The current is $I_c = 53$ amp and the electrical resistivity of the material is $\rho_e = 3.2 \times 10^{-8}$ Ω-m. The resistor material has thermal conductivity $k = 58$ W/m-K. The external surface (i.e., the outer diameter) of the resistor can be considered to be adiabatic. As a result, thermal energy generated in the resistor may only flow axially (not radially). The electrodes are both at $T_m = 20°C$. The mounting interface between the resistor and the electrode is not perfect. Rather there is an area-specific thermal contact resistance of $R_c'' = 1.4 \times 10^{-3}$ K-m^2/W associated with these joints. The result of the thermal contact resistance is that the temperatures at the two ends of the shunt resistor (at $x = 0$ and $x = L$) will *not* be equal to the electrode temperature, T_m.

- Determine the volumetric rate of thermal energy generation that is occurring within the resistor.
- Develop an analytical solution to the problem. That is, determine a function describing the temperature in the resistor as a function of position, x.
- Prepare a plot showing the temperature in the resistor as a function of position.

Known Values

Figure 1 illustrates the shunt resistor with the known values indicated.

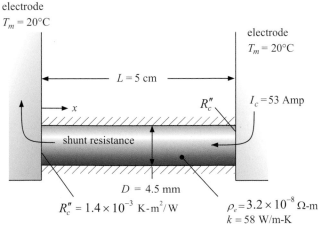

electrode
$T_m = 20°C$

electrode
$T_m = 20°C$

$L = 5$ cm

x

R_c''

$I_c = 53$ Amp

Figure 1 Shunt resistor.

shunt resistance

$D = 4.5$ mm

$R_c'' = 1.4 \times 10^{-3}$ K-m^2/W

$\rho_e = 3.2 \times 10^{-8}$ Ω-m
$k = 58$ W/m-K

Continued

Example 2.10 (cont.)

Assumptions

- Steady-state conditions exist.
- The temperature distribution is 1-D.
- The properties can be considered constant.

Analysis

The electrical resistance of the shunt resistor is calculated according to:

$$R_e = \frac{L\rho_e}{A_c} \tag{1}$$

where A_c is the cross-sectional area available for current flow:

$$A_c = \pi \frac{D^2}{4}. \tag{2}$$

The rate of ohmic dissipation is calculated according to:

$$\dot{g} = I_c^2 R_e. \tag{3}$$

The volumetric rate of thermal energy generation is the ratio of the rate of ohmic dissipation to the volume of the resistor:

$$\dot{g}''' = \frac{\dot{g}}{A_c L}. \tag{4}$$

Combining Eqs. (1), (3) and (4) provides:

$$\dot{g}''' = \frac{I_c^2 \rho_e}{A_c^2}. \tag{5}$$

The governing differential equation for the problem is the same as the one derived in Section 2.3.2:

$$\frac{d}{dx}\left(\frac{dT}{dx}\right) = -\frac{\dot{g}'''}{k}. \tag{6}$$

Integrating once leads to

$$\frac{dT}{dx} = -\frac{\dot{g}'''}{k}x + C_1, \tag{7}$$

and integrating again leads to the general solution

$$T = -\frac{\dot{g}'''}{2k}x^2 + C_1 x + C_2. \tag{8}$$

The constants C_1 and C_2 must be obtained from the boundary conditions for the problem. These boundary conditions are slightly more difficult than specified temperature boundary conditions. When the temperature at the edge of your computational domain is not specified then it will be necessary to carry out an interface energy balance. An interface energy balance simply enforces that the rate of energy entering at the edge of your domain must be balanced by the rate of conduction at that location.

Figure 2 illustrates an interface energy balance at $x = 0$. The control volume used for the energy balance is infinitesimally thin; one control surface is located at $x = 0^-$ (i.e., just outside of the material) while the other is located at $x = 0^+$ (i.e., just inside the material).

Example 2.10 (cont.)

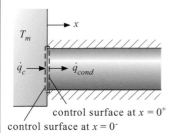

Figure 2 Interface energy balance at $x = 0$.

control surface at $x = 0^+$
control surface at $x = 0^-$

Heat transfer across the contact resistance passes through the control surface at $x = 0^-$, $\dot{q}_c$. Heat transfer due to conduction at $x = 0$, $\dot{q}_{cond}$, passes through the control surface at $x = 0^+$. The First Law of Thermodynamics requires that:

$$IN = OUT + STORED.$$

Energy storage does not occur in a steady-state problem in any case. However, there can *never* be any stored energy associated with an interface energy balance because there is no volume of material contained within the infinitesimally small control volume. In this case, the heat transfer rate across the contact resistance is *into* the control volume and the heat transfer rate due to conduction is out:

$$\dot{q}_c = \dot{q}_{cond}. \tag{9}$$

Substituting rate equations provides a boundary condition in terms of temperatures. The heat transfer rate across the contact resistance is:

$$\dot{q}_c = \frac{(T_m - T_{x=0})}{R_c}, \tag{10}$$

where R_c is the contact resistance, evaluated according to:

$$R_c = \frac{R_c''}{A_c}. \tag{11}$$

Notice that Eq. (10) is consistent with the direction of the arrow used in Figure 2. The heat transfer is drawn *from* T_m *to* $T_{x=0}$ which dictates the sign of the temperature difference. If T_m is greater than $T_{x=0}$ then we would expect $\dot{q}_c$ to be positive.

The heat transfer rate due to conduction at $x = 0$ is obtained from Fourier's Law:

$$\dot{q}_{cond} = -k A_c \frac{dT}{dx}\bigg|_{x=0}. \tag{12}$$

Equations (10) and (12) are substituted into Eq. (9):

$$\frac{(T_m - T_{x=0})}{R_c} = -k A_c \frac{dT}{dx}\bigg|_{x=0}. \tag{13}$$

Equation (13) provides guidance as to the characteristics of the temperature distribution that must exist in order for energy to be conserved at the boundary of our problem located at $x = 0$. The constants C_1 and C_2 must be selected so that Eq. (13) is satisfied. Substituting Eqs. (7) and (8), evaluated at $x = 0$, into Eq. (13) leads to:

$$\frac{\left(T_m - \overbrace{C_2}^{T_{x=0}}\right)}{R_c} = -k A_c \underbrace{C_1}_{\frac{dT}{dx}\big|_{x=0}}. \tag{14}$$

Continued

Example 2.10 (cont.)

Equation (14) is one algebraic equation in two unknowns, C_1 and C_2, and provides one boundary condition. The second boundary condition is obtained using an interface energy balance at $x = L$, as shown in Figure 3.

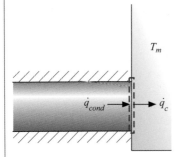

T_m

Figure 3 Interface energy balance at $x = L$.

$\dot{q}_{cond}$ $\dot{q}_c$

Heat transfer by conduction at $x = L$ enters the control surface at $x = L^-$ while heat transfer leaves across the contact resistance at $x = L^+$:

$$\dot{q}_{cond} = \dot{q}_c. \tag{15}$$

Again, we are being consistent with the arrows shown in Figure 3; the conduction heat transfer is defined as being *into* the interface while the contact heat transfer is defined as being *out*. The two rate equations are

$$\dot{q}_{cond} = -k A_c \frac{dT}{dx}\bigg|_{x=L} \tag{16}$$

and

$$\dot{q}_c = \frac{(T_{x=L} - T_m)}{R_c}. \tag{17}$$

Substituting Eqs. (16) and (17) into Eq. (15) leads to:

$$-k A_c \frac{dT}{dx}\bigg|_{x=L} = \frac{(T_{x=L} - T_m)}{R_c}. \tag{18}$$

Substituting Eqs. (7) and (8) evaluated at $x = L$ into Eq. (18) provides a second equation in terms of the unknown constants C_1 and C_2:

$$-k A_c \left(-\frac{\dot{g}'''}{k} L + C_1\right) = \frac{\left(-\dfrac{\dot{g}'''}{2k} L^2 + C_1 L + C_2 - T_m\right)}{R_c}. \tag{19}$$

Solution

Equations (14) and (19) together can be solved to obtain the unknown constants, C_1 and C_2, which can subsequently be substituted into the general solution, Eq. (8) in order to complete the problem. In order to conveniently solve the two equations in two unknowns and plot the solution we will implement our solution in EES. The problem inputs are entered in EES.

```
$UnitSystem SI Mass J K Pa Radian
T_m=Converttemp(C,K,20 [C])              "electrode temperature"
R``_c=1.4e-3 [K-m^2/W]                    "area specific contact resistance"
```

Example 2.10 (cont.)

```
L=5 [cm]*Convert(cm,m)                    "length"
D=4.5 [mm]*Convert(mm,m)                  "diameter"
I_c=53 [amp]                              "current"
rho_e=3.2e-8 [ohm-m]                      "electrical resistivity"
k=58 [W/m-K]                              "thermal conductivity"
```

Equations (2) and (5) are used to determine the cross-sectional area and the volumetric rate of thermal energy generation, respectively.

```
A_c=pi*D^2/4                             "area"
g```_dot=I_c^2*rho_e/A_c^2               "volumetric rate of thermal energy generation"
```

Solving provides $\dot{g}''' = 3.554 \times 10^5$ W/m^3. Equations (11), (14), and (19) are entered in order to solve for C_1 and C_2.

```
R_c=R``_c/A_c                           "thermal contact resistance"
(T_m-C_2)/R_c=-k*A_c*C_1                 "interface energy balance at x=0"
-k*A_c*(-g```_dot*L/k+C_1)=(-g```_dot*L^2/(2*k)+C_1*L+C_2-T_m)/R_c
                                        "interface energy balance at x=L"
```

The general solution, Eq. (8), is programmed in EES.

```
x=0 [m]                                  "position"
T=-g```_dot*x^2/(2*k)+C_1*x+C_2          "temperature"
```

A Parametric Table is used with x varying from 0 to 50 mm to generate the plot of temperature as a function of position shown in Figure 4.

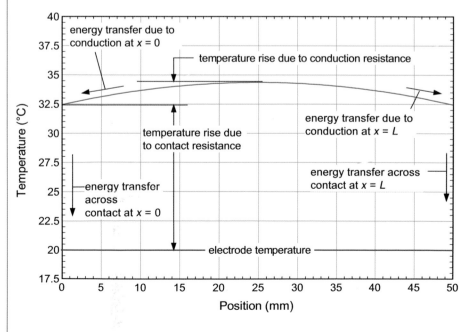

Figure 4 Temperature as a function of position.

Continued

Example 2.10 (cont.)

Discussion

There are a number of important points that can be emphasized by this problem.

Point 1: You should be able to critically examine your solution and see the characteristics that indicate that the boundary conditions have been implemented correctly.

We can examine the temperature distribution shown in Figure 4 and immediately see that the boundary conditions were imposed successfully. For example, at $x = L$ notice that the temperature is decreasing, which is consistent with a positive conduction heat transfer rate (i.e., energy is moving from left to right by conduction at $x = L$). Therefore conduction heat transfer is entering the interface at $x = L$. The temperature at $x = L$ is greater than the temperature of the electrodes, indicating that energy is leaving the right side of the shunt resistance across the thermal contact resistance. This is consistent with the interface energy balance. We can visually see the path of the energy by looking at the temperature distribution.

At $x = 0$ it should be clear that conduction is actually moving in the negative x-direction (i.e., energy is entering the interface at $x = 0$ by conduction). An equal amount of energy is leaving across the contact resistance. Again, it's clear that the energy is entering and then leaving the interface.

It is not difficult to make a sign error when you are carrying out your analysis. Typically this type of error can be detected by carefully examining your solution. For example, a sign error in Eq. (14) would lead to the temperature distribution shown in Figure 5.

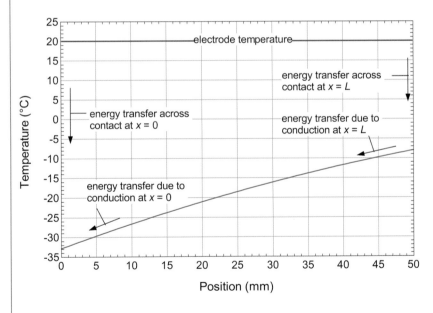

Figure 5 Incorrect temperature as a function of position due to a sign error in Eq. (14).

Examination of Figure 5 shows that energy is *not* being balanced at the interface at $x = 0$. Energy is entering the interface across the contact *and* entering by conduction. This behavior does not obey the First Law of Thermodynamics. Energy cannot be destroyed and there is no way it can be carried away at the interface at $x = 0$. The interface at $x = L$ is behaving itself in Figure 5; energy is entering across the contact and then leaving by conduction. If you were to plot your solution and see the behavior shown in Figure 5 you should be able to understand that it cannot be correct. A further clue is that the temperatures in the

Example 2.10 (cont.)

shunt are all lower than the temperature of the electrodes. This is clearly not possible as energy generation is occurring within the shunt. Further, by observation you know exactly which of your boundary conditions was not dealt with appropriately.

<u>Point 2:</u> When you are carrying out your interface energy balance it does not matter which direction you assume for the energy flows as long as you remain consistent with your assumption.

Figure 6 illustrates the interface energy balance at $x = 0$, revisited. This time, the direction of each of the heat flows is assumed to be in the *opposite* direction than was originally assumed in Figure 2.

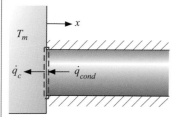

Figure 6 Interface energy balance at $x = 0$ with assumed direction of each heat flow reversed.

The First Law of Thermodynamics requires that:

$$IN = OUT + STORED.$$

In this case, the heat transfer across the contact resistance is *out of* the control volume and the heat transfer due to conduction is *into* the control volume:

$$\dot{q}_{cond} = \dot{q}_c. \tag{20}$$

The rate equations have to be written in a manner that is consistent with the assumed direction in Figure 6. The heat transfer rate across the contact is

$$\dot{q}_c = \frac{(T_{x=0} - T_m)}{R_c}. \tag{21}$$

Notice that Eq. (21) is consistent with the direction of the arrow used in Figure 6. The heat transfer is drawn *from* $T_{x=0}$ to T_m which dictates the form of the temperature difference in the numerator.

Fourier's Law states that the rate of conduction heat transfer in the *positive* x-direction is given by

$$\dot{q} = -kA_c\frac{dT}{dx}. \tag{22}$$

Figure 6 shows that the rate of conduction heat transfer into the interface at $x = 0$ is in the *negative* x-direction and therefore the appropriate rate equation is

$$\dot{q}_{cond} = kA_c\frac{dT}{dx}\bigg|_{x=0}. \tag{23}$$

Substituting Eqs. (21) and (23) into Eq. (20) provides:

$$kA_c\frac{dT}{dx}\bigg|_{x=0} = \frac{(T_{x=0} - T_m)}{R_c}. \tag{24}$$

Equation (24) is algebraically identical to Eq. (14), the boundary condition that was derived using the interface energy balance shown in Figure 2. The point of this exercise is that it makes no difference in what

Continued

Example 2.10 (cont.)

direction you assume for the energy flows when you do the interface energy balance. However, once you assume a direction you have to stick to that assumption when you do your energy balance and also when you write your rate equations.

<u>Point 3:</u> This problem cannot be solved using thermal resistances, because of the energy generation. However, we could have anticipated what the solution would look like by calculating the appropriate thermal resistances.

In this problem, thermal energy is generated within the resistor and exits by being first conducted to either end of the resistor and then transferred across the contact resistance. We can get a good feel for the problem by calculating the thermal resistances associated with the heat transfer processes. The contact resistance can be computed exactly and is equal to $R_c = 88.0$ K/W. The conduction resistance, on the other hand, cannot be computed exactly because there is no single conduction length that all of the energy generated within the resistor takes. Some is generated near the center and must travel a distance $L/2$ to get out, while some is generated very near the edge. In this case, the best that we can do is use some representative conduction length, like $L/4$, to compute an approximate resistance:

$$R_{cond} \approx \frac{L/4}{k \, A_c}. \qquad (25)$$

R_cond=(L/4)/(k*A_c)   "approximate resistance to conduction"

Solving provides $R_{cond} = 13.6$ K/W. Because the contact resistance is several times larger than the conduction resistance, it follows that the temperature rise associated with heat transfer across the interface will be much larger than the temperature rise associated with conduction within the resistor. Figure 4 shows that this is true.

It is hard to overstate the importance of being able to use thermal resistances to understand heat transfer problems. Without solving any ordinary differential equations it would have been possible to anticipate most of the important aspects of the solution. The total thermal energy generation within half of the resistance is $\dot{g}/2$. The temperature rise *within* the shunt resistor due to conduction will be approximately equal to the product of this thermal energy generation and the thermal resistance due to conduction:

$$\Delta T_{cond} \approx \frac{\dot{g}''' A_c L}{2} R_{cond} \qquad (26)$$

and the temperature rise between the edge of the shunt resistance and the electrode (across the contact resistance) will be

$$\Delta T_c \approx \frac{\dot{g}''' A_c L}{2} R_c.$$

DT_cond=(g```_dot*A_c*L/2)*R_cond   "approximate temperature rise due to conduction"
DT_c=(g```_dot*A_c*L/2)*R_c   "approximate temperature rise due to contact resistance"

Solving provides $\Delta T_{cond} = 1.92$ K and $\Delta T_c = 12.4$ K; while these are not exactly correctly, they are close to the values that can be observed in Figure 4.

<u>Point 4:</u> This problem has a **line of symmetry** that could have been exploited to reduce the size of the computational domain.

Figure 1 illustrates that the two halves of the resistor ($0 < x < L/2$ and $L/2 < x < L$) experience identical boundary conditions. That is, you could flip the problem over the midpoint and you would have the same

Example 2.10 (cont.)

problem. The problem is therefore **symmetric** around the line $x = L/2$. When a problem has a line of symmetry it is often possible to exploit this line in order to simplify the problem. In this case, we could have simulated only one side of the line of symmetry, $0 < x < L/2$, and obtained the same solution. While this is not a particular advantage for an analytical solution, reducing the size of the computational domain may be important for a numerical solution. Examination of Figure 4 shows that there is *no energy transfer* across the line of symmetry. This is evident by the fact that the slope of the temperature gradient is 0 at $x = L/2$ corresponding to a conduction heat transfer rate of zero according to Fourier's Law.

2.3.3 Uniform Thermal Energy Generation in Radial Geometries

The area for conduction through the plane wall discussed in Section 2.3.2 is constant in the coordinate direction (x). If the conduction area is a function of position, then the problem becomes slightly more complicated. Figure 2.12 illustrates the differential control volumes that should be defined to analyze radial heat transfer in (a) a cylinder and (b) a sphere with thermal energy generation.

Cylindrical Geometry
The differential energy balance suggested by Figure 2.12(a) is:

$$(\dot{q}_r)_r + \dot{g} = (\dot{q}_r)_{r+dr}, \tag{2.106}$$

which is expanded and simplified:

$$\dot{g} = \frac{d\dot{q}_r}{dr}dr. \tag{2.107}$$

The rate equations for $\dot{q}_r$ and $\dot{g}$ in a cylindrical geometry are:

$$\dot{q}_r = -k\,2\pi r L\frac{dT}{dr} \tag{2.108}$$

$$\dot{g} = 2\pi r L\,dr\,\dot{g}''' \tag{2.109}$$

where L is the length of the cylinder and $\dot{g}'''$ is the rate of thermal energy generation per unit volume. Substituting Eqs. (2.108) and (2.109) into Eq. (2.107) leads to:

$$2\pi r L\,dr\,\dot{g}''' = \frac{d}{dr}\left(-k\,2\pi r L\frac{dT}{dr}\right)dr, \tag{2.110}$$

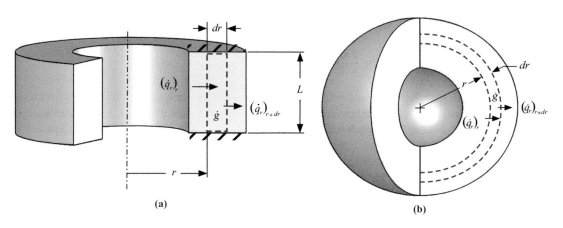

(a)

(b)

Figure 2.12 Differential control volume for (a) a cylinder and (b) a sphere with volumetric thermal energy generation.

which, assuming k is constant, can be simplified to:

$$\frac{d}{dr}\left(r\frac{dT}{dr}\right) = -\frac{r\dot{g}'''}{k}.$$

(2.111)

Notice that r cannot be cancelled from both sides of Eq. (2.110) as it appears within the derivative with respect to r. Equation (2.111) is separated and integrated:

$$\int d\left(r\frac{dT}{dr}\right) = -\int \frac{r\dot{g}'''}{k}\,dr$$

(2.112)

to obtain:

$$r\frac{dT}{dr} = -\frac{r^2\dot{g}'''}{2k} + C_1.$$

(2.113)

Equation (2.113) is again separated and integrated:

$$\int dT = -\int \frac{r\dot{g}'''}{2k}\,dr + \int \frac{C_1}{r}\,dr,$$

(2.114)

which provides the general 1-D relation for temperature as a function of radius in a cylinder with uniform internal generation:

$$T = -\frac{\dot{g}'''r^2}{4k} + C_1\ln(r) + C_2,$$

(2.115)

where C_1 and C_2 are constants of integration that depend on the boundary conditions.

Spherical Geometry

The differential energy balance suggested by Figure 2.12(b) is:

$$(\dot{q}_r)_r + \dot{g} = (\dot{q}_r)_{r+dr},$$

(2.116)

which is expanded and simplified:

$$\dot{g} = \frac{d\dot{q}_r}{dr}\,dr.$$

(2.117)

The rate equations for $\dot{q}$ and $\dot{g}$ in a spherical geometry are

$$\dot{q}_r = -k\,4\pi r^2 \frac{dT}{dr}$$

(2.118)

$$\dot{g} = 4\pi r^2\,dr\,\dot{g}'''.$$

(2.119)

Substituting Eqs. (2.118) and (2.119) into Eq. (2.117) leads to:

$$4\pi r^2\,dr\,\dot{g}''' = \frac{d}{dr}\left(-k\,4\pi r^2 \frac{dT}{dr}\right)dr,$$

(2.120)

which, assuming k is constant, can be simplified to:

$$\frac{d}{dr}\left(r^2\frac{dT}{dr}\right) = -\frac{r^2\dot{g}'''}{k}.$$

(2.121)

Equation (2.121) is separated and integrated once:

$$\int d\left(r^2\frac{dT}{dr}\right) = -\int \frac{r^2\dot{g}'''}{k}\,dr$$

(2.122)

Table 2.3 Summary of formulae for 1-D uniform thermal energy generation cases.

	Plane wall	Cylinder	Sphere
Governing differential equation	$\dfrac{d}{dx}\left(k\dfrac{dT}{dx}\right) = -\dot{g}'''$	$\dot{g}'''\,r = \dfrac{d}{dr}\left(-kr\dfrac{dT}{dr}\right)$	$\dot{g}'''\,r^2 = \dfrac{d}{dr}\left(-kr^2\dfrac{dT}{dr}\right)$
Temperature gradient	$\dfrac{dT}{dx} = -\dfrac{\dot{g}'''}{k}x + C_1$	$\dfrac{dT}{dr} = -\dfrac{\dot{g}'''\,r}{2k} + \dfrac{C_1}{r}$	$\dfrac{dT}{dr} = -\dfrac{\dot{g}'''\,r}{3k} + \dfrac{C_1}{r^2}$
General solution	$T = -\dfrac{\dot{g}'''}{2k}x^2 + C_1 x + C_2$	$T = -\dfrac{\dot{g}'''\,r^2}{4k} + C_1\ln(r) + C_2$	$T = -\dfrac{\dot{g}'''}{6k}r^2 - \dfrac{C_1}{r} + C_2$

to obtain:

$$r^2\frac{dT}{dr} = -\frac{r^3\dot{g}'''}{3k} + C_1. \tag{2.123}$$

Equation (2.123) can be separated and integrated again:

$$\int dT = -\int \frac{r\dot{g}'''}{3k}\,dr + \int \frac{C_1}{r^2}\,dr \tag{2.124}$$

to obtain the general 1-D relation for temperature in a sphere with internal generation:

$$T = -\frac{\dot{g}'''}{6k}r^2 - \frac{C_1}{r} + C_2. \tag{2.125}$$

The governing differential equation and general solutions for these 1-D geometries with a uniform rate of thermal energy generation are summarized in Table 2.3.

Example 2.11

Some magnets are constructed of superconducting material and therefore can handle extremely large amounts of current with no electrical resistance. However, superconductivity is a phenomenon that is only exhibited by materials at very low temperatures. Therefore, it is necessary to carry the current to and from the superconducting magnet using current leads (i.e., electrical conductors) that are not superconducting and have finite electrical resistance. In order to prevent these current leads from melting they are often cooled by convection to a flowing fluid.

One method of designing a gas-cooled current lead is in the form of a hollow cylinder. Current flows axially along the cylinder causing a uniform volumetric thermal energy generation and fluid flows through the center of the lead. The current carried by the lead is $I_c = 15{,}000$ amp. The outer radius of the lead is $R_{out} = 2$ cm and the inner radius is $R_{in} = 1$ cm. The outer radius is insulated and the inner radius is exposed to fluid at $T_\infty = 20°C$ with heat transfer coefficient $\bar{h} = 225$ W/m^2-K. The electrical resistivity of the current lead material is $\rho_e = 1.3 \times 10^{-8}$ ohm-m and the thermal conductivity is $k = 30$ W/m-K.

- Determine the volumetric rate of thermal energy generation in the lead.
- Develop an analytical solution for the temperature as a function of radial position within the current lead material.
- Plot the temperature in the conductor as a function of radial position.

Continued

Example 2.11 (cont.)

- Prepare a plot showing the maximum conductor temperature as a function of R_{in}, the inner radius of the conductor (with all other parameters held constant).
- Determine the optimal inner radius for this situation (i.e., the inner radius that keeps the conductor as cool as possible).

Known Values

Figure 1 shows a sketch of the problem with the known values indicated.

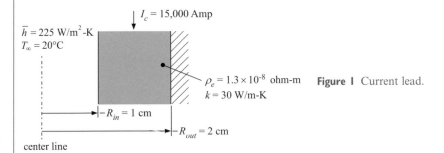

Figure 1 Current lead.

Assumptions

- Steady-state conditions exist.
- The temperature distribution is 1-D.
- The material has constant properties.
- The heat transfer coefficient does not change as the inner radius is varied.

Analysis

The cross-sectional area available for the current flow is:

$$A_c = \pi \left(R_{out}^2 - R_{in}^2 \right). \tag{1}$$

The electrical resistance of the current lead is:

$$R_e = \frac{\rho_e L}{A_c}, \tag{2}$$

where L is the length of the lead. The ohmic dissipation in the lead is:

$$\dot{g} = R_e I_c^2. \tag{3}$$

The rate of thermal energy generation per unit volume is:

$$\dot{g}''' = \frac{\dot{g}}{A_c L}. \tag{4}$$

Substituting Eqs. (2) and (3) into Eq. (4) provides:

$$\dot{g}''' = \frac{I_c^2 \rho_e}{A_c^2}. \tag{5}$$

The governing differential equation for this situation is Eq. (2.111), which leads to the general solution provided by Eqs. (2.113) and (2.115), repeated below:

$$\frac{dT}{dr} = -\frac{r \dot{g}'''}{2k} + \frac{C_1}{r} \tag{6}$$

Example 2.11 (cont.)

$$T = -\frac{\dot{g}'''}{4k}r^2 + C_1 \ln(r) + C_2. \tag{7}$$

The boundary condition at $r = R_{out}$ requires that the rate of conduction heat transfer be equal to zero:

$$\dot{q}_{cond, r = R_{out}} = 0. \tag{8}$$

Substituting Fourier's Law into Eq. (8) provides:

$$-k \, 2\pi \, R_{out} \, L \frac{dT}{dr}\bigg|_{r = R_{out}} = 0 \tag{9}$$

or

$$\frac{dT}{dr}\bigg|_{r = R_{out}} = 0. \tag{10}$$

Substituting Eq. (6) into Eq. (10) provides one equation for the unknown constants:

$$-\frac{R_{out}\dot{g}'''}{2k} + \frac{C_1}{R_{out}} = 0. \tag{11}$$

The boundary condition at $r = R_{in}$ requires that the rate of convection into the boundary from the fluid be equal to the rate of conduction away from the boundary, as shown by the interface energy balance in Figure 2:

$$\dot{q}_{conv} = \dot{q}_{cond, r = R_{in}}. \tag{12}$$

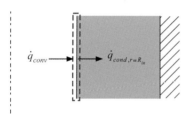

Figure 2 Interface energy balance at $r = R_{in}$.

center line

Substituting rate equations into Eq. (12) provides:

$$2\pi R_{in} L \bar{h}(T_\infty - T_{r = R_{in}}) = -2\pi R_{in} L k \frac{dT}{dr}\bigg|_{r = R_{in}} \tag{13}$$

or

$$\bar{h}(T_\infty - T_{r = R_{in}}) = -k \frac{dT}{dr}\bigg|_{r = R_{in}}. \tag{14}$$

Substituting Eqs. (6) and (7) into Eq. (14) provides another equation for the unknown constants:

$$h\left[T_\infty - \left(-\frac{\dot{g}'''}{4k}R_{in}^2 - C_1 \ln(R_{in}) + C_2\right)\right] = -k\left(-\frac{R_{in}\,\dot{g}'''}{2k} + \frac{C_1}{R_{in}}\right). \tag{15}$$

Equations (11) and (15) can solved simultaneously in order to solve for C_1 and C_2; substituting the result into Eq. (7) provides the temperature distribution.

Solution

The equations derived above provide the solution to the problem. Because parametric studies are required it is necessary that the solution be implemented using a computer. Here we will use EES. The inputs are entered in EES.

Continued

Example 2.11 (cont.)

```
$UnitSystem SI Mass J K Pa
R_out=2 [cm]*Convert(cm,m)              "outer radius"
R_in=1 [cm]*Convert(cm,m)              "inner radius"
rho_e=1.3e-8 [ohm-m]                   "electrical resistivity"
k=30 [W/m-K]                           "conductivity"
h_bar=225 [W/m^2-K]                    "heat transfer coefficient"
I_c=15000 [Amp]                        "current"
T_infinity=ConvertTemp(C,K,20 [C])     "coolant temperature"
```

The cross-sectional area available for the current flow and the rate of thermal energy generation per unit volume are computed using Eqs. (1) and (5).

```
A_c=pi*(R_out^2-R_in^2)                "cross sectional area"
g_dot```=I_c^2*rho_e/A_c^2             "volumetric thermal energy generation"
```

Solving provides $\dot{g}''' = 3.293 \times 10^6$ W/m^3. Equations (11) and (15) are entered in EES in order to solve for C_1 and C_2.

```
-g_dot```*R_out/(2*k)+C_1/R_out=0          "Boundary condition at r=R_out"
h_bar*(T_infinity-(-g_dot```*R_in^2/(4*k)+C_1*Ln(R_in)+C_2))=-k*(-g_dot```*R_in/(2*k)+C_1/R_in)
                                           "Boundary condition at r=R_in"
```

The general solution, Eq. (2), is entered in EES and used to generate the plot of temperature as a function of radius shown in Figure 3.

```
T= -g_dot```*r^2/(4*k)+C_1*Ln(r)+C_2        "temperature"
```

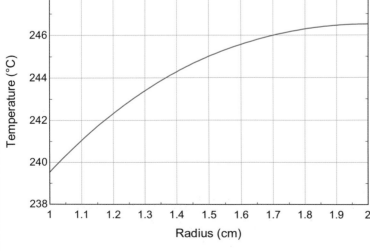

Figure 3 Temperature as a function of radius.

Example 2.11 (cont.)

The maximum temperature must occur at $r = R_{out}$; therefore, according to Eq. (7):

$$T_{max} = -\frac{\dot{g}'''}{4k} R_{out}^2 + C_1 \ln(R_{out}) + C_2.$$

> T_max= -g_dot```*R_out^2/(4*k)+C_1***Ln**(R_out)+C_2   "maximum temperature"

The given value of R_{in} that was entered is commented out

> {R_in=1 [cm]*Convert(cm,m)}   "inner radius"

and a Parametric Table is created that includes columns for both R_{in} and T_{max}. Figure 4 illustrates the maximum conductor temperature as a function of R_{in}.

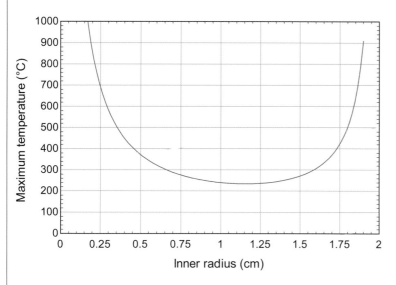

Figure 4 Maximum temperature as a function of inner radius.

Notice that there is an optimal inner radius that minimizes the maximum temperature. The optimal inner radius appears to be approximately $R_{in} = 1.1$ cm according to Figure 4.

Discussion

Whenever you see an optimal value of a parameter there must be at least two competing effects that are being balanced. When the inner radius is very small there is not much area for convection and a long conduction path; therefore, the temperature rise between the fluid and the material at $r = R_{out}$ becomes large. When the inner radius is very large there is not much cross-sectional area for the current and the electrical resistance becomes large, causing a large rate of thermal energy generation. The optimal radius balances these two effects.

A more exact value of the optimal inner radius can be obtained by employing the optimization algorithms programmed in EES. With the value of R_{in} commented out, select Min/Max from the Calculate menu. The Find Minimum or Maximum dialog will appear, as shown in Figure 5.

Continued

Example 2.11 (cont.)

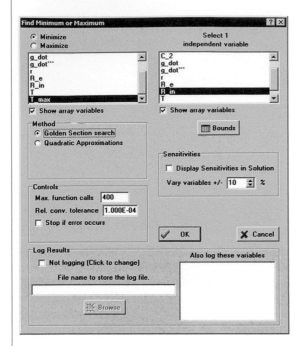

Figure 5 Find Minimum or Maximum dialog.

The radio buttons at the upper left of the window allow the user to select whether the objective is to be maximized or minimized. In this case, select Minimize. The left-hand window allows the user to select the parameter to be optimized; select the variable T_max. The right window allows the user to select the parameter(s) to be varied in order to carry out the optimization. For this problem we have only one degree of freedom and therefore only one variable, R_in, can be selected.

Before an optimization can be accomplished, it is necessary to set bounds for the independent variables (i.e., specify the extent over which it can be varied). Select the Bounds button to access the dialog shown in Figure 6.

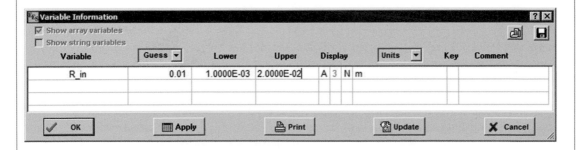

Figure 6 Variable information window allowing bounds to be set for variable R_in.

Select reasonable bounds for R_{in}, in this case a value arbitrarily close to zero for the lower limit and R_{out} for the upper limit. Select OK and then OK again in order to activate the optimization. EES will adjust the value of R_{in} between the bounds you selected in order to minimize T_{max} using the optimization technique selected in the Find Minimum or Maximum dialog (i.e., using the Golden Section search selected in

Example 2.11 (cont.)

Figure 5). The criteria that specify convergence to an optimal value can also be adjusted in this window. Once the optimization is complete, the result will be found in the Solution Window. In this case, EES has identified that the optimal value of R_{in} is 1.165 cm, which results in a maximum conductor temperature of 239.9°C.

2.3.4 Spatially Nonuniform Generation

The first few steps in solving conduction heat transfer problems include setting up an energy balance on a differential control volume and substituting in the appropriate rate equations. This process results in a differential equation that must be solved in order to determine the temperature distribution and heat transfer rates. The governing equations resulting from 1-D conduction with constant properties and constant volumetric generation are provided in Table 2.3 for Cartesian, cylindrical, and spherical geometries. These equations are relatively simple and we have demonstrated how to solve them analytically subject to various boundary conditions.

The complexity of the governing differential equation itself may increase if the thermal conductivity or volumetric generation depends on position or temperature. In these cases, an analytical solution to the governing equation may not be possible and the numerical solution techniques presented in Section 2.4 must be used to obtain a solution. In some situations, however, an analytical solution is still possible. Analytical solutions are concise and elegant as well as being accurate and therefore are preferable in many ways to numerical solutions. It is often best to have both an analytical and numerical solution to a problem as their agreement constitutes the best possible double-check of a solution.

Example 2.12

One side of a window is exposed to radiant energy. The window can be modeled as a plane wall with thickness $L = 1.0$ cm and thermal conductivity $k = 1.5$ W/m-K. The window is not perfectly transparent, but rather absorbs some of the illumination energy. The absorption coefficient of the material that is used to construct the window is $\alpha = 0.1$ mm^{-1}; the absorption coefficient is a material property that indicates how transparent the material is. The flux of radiant energy that is incident on the window surface is $\dot{q}''_{rad} = 0.1$ W/cm^2. Both surfaces of the window are exposed to air at $T_\infty = 20°C$ and the average heat transfer coefficient on these surfaces is $\bar{h} = 20$ W/m^2-K.

The volumetric rate at which radiation is absorbed and converted to thermal energy in the window ($\dot{g}'''$) is proportional to the local intensity of the radiant energy flux, which is itself reduced by absorption. The result is a spatially nonuniform volumetric generation of thermal energy given by

$$\dot{g}''' = \dot{q}''_{rad}\, \alpha \exp(-\alpha x), \qquad (1)$$

where x is the distance from the surface upon which the radiant flux is incident. Notice from Eq. (1) that the volumetric thermal energy generation is highest at $x = 0$ (i.e., at the surface being exposed to the radiant flux) and decays in the x-direction. As x increases, the amount of radiant energy that remains is reduced and thus the rate of absorption is reduced.

• Determine an analytical expression for the temperature distribution in the window and prepare a plot of temperature as a function of position.

Known Values

Figure 1 illustrates the problem and includes the values provided in the problem statement.

Continued

Example 2.12 (cont.)

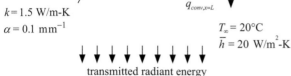

incident radiant energy, $\dot{q}''_{rad} = 0.1$ W/cm^2

$T_\infty = 20°C$

$\bar{h} = 20$ W/m^2-K

$\dot{q}_{conv,x=0}$

$\dot{q}_{cond,x=0}$ $(\dot{q}_x)_x$

x

dx

$\dot{g}$ $L = 1.0$ cm

$(\dot{q}_x)_{x+dx}$ $\dot{q}_{cond,x=L}$

$\dot{q}_{conv,x=L}$

$k = 1.5$ W/m-K

$\alpha = 0.1$ mm^{-1}

$T_\infty = 20°C$

$\bar{h} = 20$ W/m^2-K

transmitted radiant energy

Figure 1 Window absorbing radiant energy.

Assumptions

- Steady-state conditions exist.
- The temperature distribution is 1-D.
- The material has constant properties.

Analysis

An energy balance on an appropriate, differential control volume (see Figure 1) provides:

$$(\dot{q}_x)_x + \dot{g} = (\dot{q}_x)_{x+dx}. \tag{2}$$

The term at $x + dx$ can be written as:

$$(\dot{q}_x)_{x+dx} = (\dot{q}_x)_x + \frac{d(\dot{q}_x)}{dx} dx \tag{3}$$

and substituted into Eq. (2) to reach:

$$\dot{g} = \frac{d\dot{q}_x}{dx} dx. \tag{4}$$

Substituting the rate equations for $\dot{q}_x$ and $\dot{g}$ leads to:

$$\dot{g}''' A_c = \frac{d}{dx}\left(-k A_c \frac{dT}{dx}\right), \tag{5}$$

where A_c is the cross-sectional area of the window. Substituting Eq. (1) into Eq. (5) and simplifying leads to the governing differential equation for this problem.

$$\frac{d}{dx}\left(\frac{dT}{dx}\right) = -\frac{\dot{q}''_{rad}\,\alpha}{k} \exp(-\alpha x). \tag{6}$$

Equation (6) is a separable differential equation. Integrating once:

$$\int d\left(\frac{dT}{dx}\right) = -\int \frac{\dot{q}''_{rad}\,\alpha}{k} \exp(-\alpha x)\,dx \tag{7}$$

Example 2.12 (cont.)

leads to

$$\frac{dT}{dx} = \frac{\dot{q}''_{rad}}{k} \exp(-\alpha x) + C_1. \tag{8}$$

Equation (8) is separated and integrated again:

$$\int dT = \int \left[\frac{\dot{q}''_{rad}}{k} \exp(-\alpha x) + C_1 \right] dx, \tag{9}$$

which provides the general solution to this problem:

$$T = -\frac{\dot{q}''_{rad}}{k\alpha} \exp(-\alpha x) + C_1 x + C_2. \tag{10}$$

The constants of integration, C_1 and C_2 are obtained by enforcing the boundary conditions. The boundary conditions for this problem are derived from interface energy balances at the two edges of the computational domain ($x = 0$ and $x = L$, as shown in Figure 1). These interface balances include terms for convection and conduction:

$$\dot{q}_{conv, x=0} = \dot{q}_{cond, x=0} \tag{11}$$

$$\dot{q}_{cond, x=L} = \dot{q}_{conv, x=L}. \tag{12}$$

Substituting the rate equations for convection and conduction into the interface energy balances leads to the boundary conditions:

$$\bar{h} A_c (T_\infty - T_{x=0}) = -k A_c \frac{dT}{dx}\bigg|_{x=0} \tag{13}$$

$$-k A_c \frac{dT}{dx}\bigg|_{x=L} = \bar{h} A_c (T_{x=L} - T_\infty). \tag{14}$$

Note that it was important to consider the direction of the energy transfers during the substitution of the rate equations. For example, $\dot{q}_{conv, x=0}$ was defined in Figure 1 as being *into* the top surface of the window and therefore it is driven by $(T_\infty - T_{x=0})$ while $\dot{q}_{conv, x=L}$ is defined as being *out of* the bottom surface of the window and therefore it is driven by $(T_{x=L} - T_\infty)$. The general solution for the temperature distribution, Eqs. (8) and (10), must be substituted into the boundary conditions, Eqs. (13) and (14), and solved algebraically to determine the constants C_1 and C_2:

$$\bar{h} \left[T_\infty - \left(-\frac{\dot{q}''_{rad}}{k\alpha} + C_2 \right) \right] = -k \left(\frac{\dot{q}''_{rad}}{k} + C_1 \right) \tag{15}$$

$$-k \left(\frac{\dot{q}''_{rad}}{k} \exp(-\alpha L) + C_1 \right) = \bar{h} \left(-\frac{\dot{q}''_{rad}}{k\alpha} \exp(-\alpha L) + C_1 L + C_2 - T_\infty \right). \tag{16}$$

Solution

Equations (15) and (16) together can be solved for the two unknowns C_1 and C_2 which must be substituted into the general solution, Eq. (10) in order to complete the problem. In order to avoid the algebra associated with simultaneously solving Eqs. (15) and (16) and also to facilitate making the plot, the solution is accomplished using EES.

Continued

Example 2.12 (cont.)

The inputs are entered into EES.

```
$UnitSystem SI MASS RAD PA K J
k=1.5 [W/m-K]                                          "conductivity"
L=1 [cm]*Convert(cm,m)                                 "lens thickness"
alpha=0.1 [1/mm]*Convert(1/mm,1/m)                     "absorption coefficient"
q``_dot_rad=0.1 [W/cm^2]*Convert(W/cm^2,W/m^2)         "incident energy flux"
h_bar=20 [W/m^2-K]                                     "average heat transfer coefficient"
T_infinity=ConvertTemp(C,K,20 [C])                     "ambient air temperature"
```

Equations (15) and (16) are entered into EES and solved in order to obtain C_1 and C_2.

```
"Boundary conditions"
h_bar*(T_infinity-(-q``_dot_rad/(k*alpha)+C_2))=-k*(q``_dot_rad/k+C_1)
                                               "at x=0"
-k*(q``_dot_rad/k*Exp(-alpha*L)+C_1)=&
    h_bar*(-q``_dot_rad*Exp(-alpha*L)/(k*alpha)+C_1*L+C_2-T_infinity)
                                               "at x=L"
```

(Note that the ampersand, &, character provides a line break, but does not otherwise affect the equation.) The general solution, Eq. (10), is programmed

```
T=-q``_dot_rad*Exp(-alpha*x)/(k*alpha)+C_1*x+C_2           "temperature"
```

and used with a Parametric Table to generate the plot of temperature as a function of position that is shown in Figure 2.

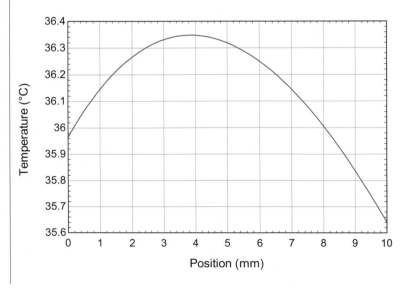

Figure 2 Temperature distribution.

> **Example 2.12** (cont.)
>
> **Discussion**
>
> The solution steps used to solve this problem involving spatially nonuniform generation are the same as those used for the other analytical problems that we have encountered. The differential equation that resulted was somewhat more difficult to solve.
>
> Notice in Figure 2 the asymmetry that is produced by the nonuniform volumetric generation (i.e., more thermal energy is generated towards the top of the window than the bottom and so the temperature is higher near the top surface). Also notice that the boundary conditions at both edges of the window are satisfied (i.e., conduction is into each of the interfaces and convection is out of those interfaces to the surroundings).

2.4 Numerical Solutions

2.4.1 Introduction

Thus far we have solved problems using analytical techniques. The governing differential equation was derived using a differentially small control volume and solved subject to boundary conditions. The result of this process is a *function* that satisfies the governing differential equation and boundary conditions *exactly* within the computational domain.

A **numerical solution** is very different than an analytical solution. A numerical solution will predict the values of temperatures at discrete locations (referred to as **nodes**) within the computational domain. The solution is obtained by solving a set of algebraic equations corresponding to energy balances around each node. No differential equations are solved using a numerical approach; however, the numerical solution will limit to the analytical solution as more nodes are used, causing each control volume to become smaller.

An analytical solution is convenient since it provides accurate results for arbitrary inputs with minimal computational effort. However, many problems of practical interest are too complicated to allow an analytical solution. In such cases, numerical solutions are required. Even when this is the case, analytical solutions remain useful as a way to test the validity of numerical solutions under limiting conditions. Numerical solutions generally require more computational resources (i.e., more equations must be solved) and are only approximations to the actual solution, albeit approximations that can be extremely accurate when done correctly. It is relatively straightforward to solve a problem using numerical techniques and, with care, it is possible to deal with problems numerically that are too complex to be solved analytically.

The steps required to set up a numerical solution using the **finite difference** approach will be the same even as the problems become more complex. Again, the result of a numerical model is not a functional relationship between temperature and position, but rather a prediction of the temperatures at many discrete positions (or nodes) within the computational domain. Therefore, the first step in preparing a numerical solution is to specify the locations where the numerical model will compute the temperatures (i.e., the nodes). The control volumes used in the numerical model are small but finite, as opposed to the infinitesimally small (differential) control volume that is defined in order to derive an analytical solution. It is necessary to perform an energy balance on each control volume. This requirement may seem daunting, given that many control volumes will be required to provide an accurate solution. However, most of the control volumes utilize the same balance equation and computers are very good at repetitive calculations. If your numerical code is designed in a systematic manner, then these operations can be done quickly for many thousands of control volumes.

Once the energy balance equations for each control volume have been set up, it is necessary to include rate equations that approximate each term using the nodal temperatures or other input parameters. The result of this step will be a set of algebraic equations (one for each control volume) in an equal number of unknown temperatures (one for each node). This set of equations must be solved to provide the numerical prediction of the temperature at each node and there are a number of ways to implement this solution.

It is tempting to declare victory after successfully solving the finite difference equations and obtaining a set of temperatures that looks reasonable. In reality, your work is only half done as there are several important steps remaining. First, it is necessary to verify that you have chosen a large enough number of nodes that your numerical solution has sufficiently converged to the actual solution. This verification can be accomplished by examining some aspect of your solution (e.g., a maximum temperature or an energy transfer rate that is particularly important) as the number of nodes increases. You should observe that your solution converges towards some value (the exact solution) as the number of control volumes (and therefore the computational effort) increases. Some engineering judgment is required for this step. You have to decide what aspect of the solution is most important and also how accurately it must be predicted in order to determine the number of nodes that are required.

Next, you should make sure that the solution makes sense. There are a number of "sanity checks" that can be applied to verify that the numerical model is behaving according to physical expectations. You can change an input parameter (e.g., thermal conductivity) and verify that the predicted temperatures respond in a manner that makes sense. Also, you could estimate the size of the temperature differences that you expect using the thermal resistance concepts that were discussed in Section 2.2 and verify that the observed temperature variation agrees at least approximately with the estimated one.

Finally, it is important that you verify the numerical solution against an analytical solution in some appropriate limit. This step may be the most difficult one, but it also provides the strongest possible verification. If the numerical model is to be used to make decisions that are important (e.g., to public safety or to your company's bottom line or to your career) then you probably will want to be completely sure that you have implemented the numerical model correctly. In this case, you should strive to find a limit where an analytical solution can be derived (e.g., using constant properties or simplified boundary conditions) and show that your numerical model matches the analytical solution to within numerical error for these conditions.

2.4.2 Developing the Finite Difference Equations

The development of a numerical model is best discussed in the context of a problem. Figure 2.13 illustrates the window that was considered in Example 2.12. The window is experiencing a nonuniform volumetric thermal energy generation due to the absorption of radiation:

$$\dot{g}''' = \dot{q}''_{rad}\, \alpha \exp(-\alpha x), \qquad (2.126)$$

where $\alpha = 0.1$ mm^{-1} is the absorption coefficient of the material and $\dot{q}''_{rad} = 0.1$ W/cm^2 is the incident radiation heat flux on the top surface of the window. The top and bottom surfaces (at $x = 0$ and $x = L$, respectively) are cooled by convection to fluid at $T_\infty = 20°C$ with heat transfer coefficient $\bar{h} = 20$ W/m^2-K. The thermal conductivity of the material is $k = 1.5$ W/m-K and the thickness is $L = 1$ cm. In Example 2.12 we developed an analytical solution for the problem, which is possible provided that the properties k and α are constant. Here we will develop a numerical model for the same situation, starting with the case where the properties are constant. The primary advantage of numerical techniques over analytical ones is that it is relatively straightforward to simulate more complex situations, for example where the conductivity or absorption coefficient are temperature dependent.

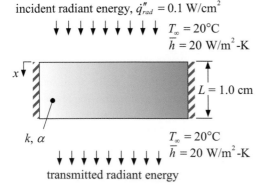

incident radiant energy, $\dot{q}''_{rad} = 0.1$ W/cm^2

$T_\infty = 20°C$
$\bar{h} = 20$ W/m^2-K

x

$L = 1.0$ cm

$k,\ \alpha$

$T_\infty = 20°C$
$\bar{h} = 20$ W/m^2-K

transmitted radiant energy

Figure 2.13 Window absorbing radiation from Example 2.12.

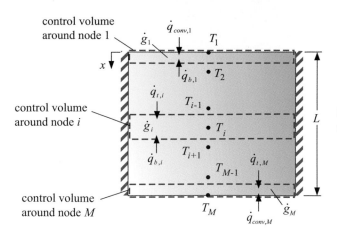

Figure 2.14 Nodes and control volumes for the numerical model.

The numerical solution technique divides the continuous medium into many small control volumes that are each treated using simple approximations. The computational domain in this problem lies between the top and bottom surfaces of the window ($0 < x < L$). The first step in the solution process is to position nodes (i.e., the positions where the temperature will be predicted) throughout the computational domain. The easiest way to distribute the nodes is uniformly, as shown in Figure 2.14 (note that only nodes 1, 2, $i-1$, i, $i+1$, $M-1$, and M are shown). The extreme nodes (i.e., nodes 1 and M) are placed on the surfaces of the window.

In some problems, it may not be computationally efficient to distribute the nodes uniformly. For example, if there are very sharp temperature gradients at some location then it may be wise to concentrate nodes in that region. Placing closely spaced nodes throughout the entire domain may be prohibitive from a computational resource viewpoint. For the uniform distribution shown in Figure 2.14, the location of each node (x_i) is

$$x_i = L\frac{(i-1)}{(M-1)} \quad \text{for } i = 1\ldots M, \tag{2.127}$$

where M is the number of nodes. The distance between adjacent nodes (Δx) is

$$\Delta x = \frac{L}{(M-1)}. \tag{2.128}$$

A control volume is defined around each node by bisecting the distance between the node and its adjacent nodes, as shown in Figure 2.14. The next step in the numerical solution is to write an energy balance for the control volume associated with every node. The internal nodes (i.e., nodes 2 through $M-1$) must be considered separately from the nodes on the edges of the computational domain (i.e., nodes 1 and M). The control volume for an arbitrary, internal node (node i) is shown in Figure 2.14. For this problem there are three energy terms associated with each of these internal control volumes: conduction heat transfer passing through the surface at the top of the control volume ($\dot{q}_{t,i}$, coming from node $i-1$), conduction heat transfer passing through the surface at the bottom of the control volume ($\dot{q}_{b,i}$, coming from node $i+1$), and generation of thermal energy within the control volume ($\dot{g}_i$). A steady-state energy balance for the internal control volume is

$$\dot{q}_{t,i} + \dot{q}_{b,i} + \dot{g}_i = 0 \quad \text{for } i = 2\ldots(M-1). \tag{2.129}$$

Note that Eq. (2.129) is *exactly* correct since no approximations have been used in its development. In the next step, however, each of the terms in the energy balance are modeled using a rate equation that is only *approximately* valid; it is this step that makes the numerical solution only an approximation of the actual solution. Conduction through the top surface is approximately driven by the temperature difference between nodes $i-1$ and i and resisted by the conduction thermal resistance associated with the material that lies between these nodes:

$$\dot{q}_{t,i} = \frac{(T_{i-1} - T_i)}{R_{cond}}.$$

(2.130)

The resistance in the denominator of Eq. (2.130) must represent the thermal resistance associated with the material separating these two nodes. In this case, the resistance of a plane wall with thickness Δx, conductivity k, and cross-sectional area A_c can be used:

$$R_{cond} = \frac{\Delta x}{k A_c}.$$

(2.131)

Substituting Eq. (2.131) into Eq. (2.130) provides:

$$\dot{q}_{t,i} = \frac{k A_c (T_{i-1} - T_i)}{\Delta x}.$$

(2.132)

A similar process for the conduction heat transfer through the bottom surface provides:

$$\dot{q}_{b,i} = \frac{k A_c (T_{i+1} - T_i)}{\Delta x}.$$

(2.133)

Note that it does not matter which direction the heat flow arrows associated with $\dot{q}_{t,i}$ and $\dot{q}_{b,i}$ are drawn in Figure 2.14; that is, the heat transfers could have been defined either as being into or out of the control volume. However, once the direction is defined, it is absolutely necessary to remain consistent with these arrows when you write your energy balance and define your rate equations. For the energy balance shown in Figure 2.14, the conduction heat transfer rates in the energy balance were written as flowing into the control volume in Eq. (2.129). These heat transfers are driven by $(T_{i-1} - T_i)$ and $(T_{i+1} - T_i)$ in Eqs. (2.132) and (2.133), respectively.

The rate of generation of thermal energy is the product of the volume of the control volume and the rate of thermal energy generation per unit volume:

$$\dot{g}_i = \dot{g}'''_{x=x_i} A_c \Delta x.$$

(2.134)

The rate of volumetric thermal energy generation depends on position in this problem. The best single value to use in Eq. (2.134) is $\dot{g}'''_{x=x_i}$, which is the value evaluated at the location of the node (i.e., the center of the control volume). Obviously this approximation improves as Δx is reduced. Substituting Eq. (2.126) into Eq. (2.134) provides:

$$\dot{g}_i = \dot{q}''_{rad} \alpha \exp(-\alpha x_i) A_c \Delta x.$$

(2.135)

Substituting Eqs. (2.132), (2.133), and (2.135) into Eq. (2.129) provides:

$$\frac{k A_c (T_{i-1} - T_i)}{\Delta x} + \frac{k A_c (T_{i+1} - T_i)}{\Delta x} + \dot{q}''_{rad} \alpha \exp(-\alpha x_i) A_c \Delta x = 0 \text{ for } i = 2 \ldots (M-1).$$

(2.136)

Equation (2.136) provides $M - 2$ algebraic equations (i.e., it can be written for every value of i from 2 to $M - 1$) for the M unknown temperatures. This equation set is not complete because we have not yet considered the nodes placed on the two boundaries.

Figure 2.14 illustrates the control volume associated with the node that is placed on the upper surface of the window (i.e., node 1). The energy balance for the control volume associated with node 1 includes a conduction term ($\dot{q}_{b,1}$), a generation term ($\dot{g}_1$), and a convection term ($\dot{q}_{conv,1}$). For steady-state conditions and energy terms directed as indicated in Figure 2.14, the energy balance on control volume 1 is:

$$\dot{q}_{conv,1} + \dot{q}_{b,1} + \dot{g}_1 = 0.$$

(2.137)

The conduction term rate equation is driven by $(T_2 - T_1)$ and resisted by the same conduction resistance experienced between the internal nodes, given by Eq. (2.131):

$$\dot{q}_{b,1} = \frac{k A_c (T_2 - T_1)}{\Delta x}.$$

(2.138)

Even though the control volume for node 1 is half as wide as the others, the distance between nodes 1 and 2 is still Δx and therefore the resistance to conduction between nodes 1 and 2 does not change. The generation term *is* different because the control volume is half as wide as the internal control volumes:

$$\dot{g}_1 = \dot{g}'''_{x=x_1} A_c \frac{\Delta x}{2} \tag{2.139}$$

or

$$\dot{g}_1 = \dot{q}''_{rad} \alpha \exp(-ax_1) A_c \frac{\Delta x}{2}. \tag{2.140}$$

Convection from the surrounding air to the surface is given by

$$\dot{q}_{conv,1} = \bar{h} A_c (T_\infty - T_1). \tag{2.141}$$

Substituting Eqs. (2.138), (2.140), and (2.141) into Eq. (2.137) leads to an additional algebraic equation:

$$\bar{h} A_c (T_\infty - T_1) + \frac{k A_c (T_2 - T_1)}{\Delta x} + \dot{q}''_{rad} \alpha \exp(-ax_1) A_c \frac{\Delta x}{2} = 0. \tag{2.142}$$

A similar procedure applied to the control volume associated with node M (see Figure 2.14) leads to:

$$\dot{q}_{conv,M} + \dot{q}_{t,M} + \dot{g}_M = 0, \tag{2.143}$$

where

$$\dot{q}_{conv,M} = \bar{h} A_c (T_\infty - T_M) \tag{2.144}$$

$$\dot{q}_{t,M} = \frac{k A_c (T_{M-1} - T_M)}{\Delta x} \tag{2.145}$$

$$\dot{g}_M = \dot{q}''_{rad} \alpha \exp(-ax_M) A_c \frac{\Delta x}{2}. \tag{2.146}$$

Substituting Eqs. (2.144) through (2.146) into Eq. (2.143) provides the final algebraic equation required to complete our equation set:

$$\bar{h} A_c (T_\infty - T_M) + \frac{k A_c (T_{M-1} - T_M)}{\Delta x} + \dot{q}''_{rad} \alpha \exp(-ax_M) A_c \frac{\Delta x}{2} = 0. \tag{2.147}$$

At this point we have a complete set of algebraic equations, repeated below for clarity:

node 1. $\quad \bar{h} A_c (T_\infty - T_1) + \dfrac{k A_c (T_2 - T_1)}{\Delta x} + \dot{q}''_{rad} \alpha \exp(-ax_1) A_c \dfrac{\Delta x}{2} = 0$

node 2. $\quad \dfrac{k A_c (T_1 - T_2)}{\Delta x} + \dfrac{k A_c (T_3 - T_2)}{\Delta x} + \dot{q}''_{rad} \alpha \exp(-ax_2) A_c \Delta x = 0$

node 3. $\quad \dfrac{k A_c (T_2 - T_3)}{\Delta x} + \dfrac{k A_c (T_4 - T_3)}{\Delta x} + \dot{q}''_{rad} \alpha \exp(-ax_3) A_c \Delta x = 0$

$\cdots$

node i. $\quad \dfrac{k A_c (T_{i-1} - T_i)}{\Delta x} + \dfrac{k A_c (T_{i+1} - T_i)}{\Delta x} + \dot{q}''_{rad} \alpha \exp(-ax_i) A_c \Delta x = 0$

$\cdots$

node $M-1$. $\quad \dfrac{k A_c (T_{M-2} - T_{M-1})}{\Delta x} + \dfrac{k A_c (T_M - T_{M-1})}{\Delta x} + \dot{q}''_{rad} \alpha \exp(-ax_{M-1}) A_c \Delta x = 0$

node M. $\quad \bar{h} A_c (T_\infty - T_M) + \dfrac{k A_c (T_{M-1} - T_M)}{\Delta x} + \dot{q}''_{rad} \alpha \exp(-ax_M) A_c \dfrac{\Delta x}{2} = 0. \tag{2.148}$

Depending on the number of nodes that you use, M, this might be a very large system of equations. Computers are good at solving large systems of equations, particularly linear equations such as these. Regardless of the computer tool that you use, the system of equations remains the same. The "heat transfer" part of the problem is complete at this point and what remains is related to understanding the details of the software that you will use to solve these equations. With most of software packages, you must take control of the solution process. You might take the system of equations, carefully put them into a matrix format, and then decompose or otherwise solve the matrix in order to obtain the solution. Alternatively, iterative techniques can be used to arrive at the solution without ever having to set up a matrix. Neither of these steps is necessary with EES, which saves considerable effort for the user (although EES must do this internally). We will go through the process of implementing the solution using EES and also using MATLAB, which is representative of a formal programming language.

2.4.3 Solving the Equations with EES

The inputs to the problem are entered in EES.

```
$UnitSystem SI MASS RAD PA K J
k=1.5 [W/m-K]                                    "conductivity"
L=1 [cm]*Convert(cm,m)                           "lens thickness"
q``_dot_rad=0.1 [W/cm^2]*Convert(W/cm^2,W/m^2)   "incident energy flux"
h_bar=20 [W/m^2-K]                               "average heat transfer coefficient"
T_infinity=ConvertTemp(C,K,20 [C])               "ambient air temperature"
alpha=0.1 [1/mm]*Convert(1/mm,1/m)               "absorption coefficient"
A_c=1 [m^2]                                      "do problem on a per unit area basis"
```

It is necessary to specify the number of nodes used in the numerical solution. We will start with a small number of nodes, $M = 6$, and increase the number of nodes once the solution is complete.

```
M=6 [-]                                          "number of nodes"
DELTAx=L/(M-1)                                   "distance between adjacent nodes"
```

Variables that are specific to each node should be placed in an array. For example, each node has an axial position and this value will be placed in an array. An **array** is a variable that contains more than one element (rather than a scalar, as we've used previously). EES recognizes a variable name to be an element of an array if it ends with square brackets surrounding an array index, e.g., x[4]. Array variables are just like any other variable in EES and they can be assigned to values, e.g., x[4]=0.16. Therefore, one way of setting up the position array would be to individually assign each value; x[1]=0, x[2]=0.002, x[3]=0.004, etc. However, this process is tedious. It is much more convenient to use the Duplicate command for this process. The Duplicate command literally duplicates the equations that lie within its domain, allowing for varying array index values. The Duplicate command must be followed by an integer index, in this case i, that passes through a range of values, in this case 1 to M. The EES code shown below will copy (duplicate) the statement(s) that are located between the key words Duplicate and End M times; each time, the value of i is incremented by 1, starting with 1 and ending with M.

```
Duplicate i=1,M
    x[i]=L*(i-1)/(M-1)                           "axial position of each node"
End
```

The EES code above is exactly like writing:

```
x[1]= L*(1-1)/(M-1)
x[2]= L*(2-1)/(M-1)
x[3]= L*(3-1)/(M-1),
etc.
```

Be careful not to put statements that you do not want to be duplicated between the Duplicate and End statements. For example, if you accidentally placed the statement M=6 inside the Duplicate loop it would be like writing M=6 six times (each time the duplicate loop iterates) which corresponds to six equations in a single unknown and is not solvable.

By default, arrays are not displayed in the Solution Window but rather in the Arrays Window. Solve the problem and then select Arrays from the Windows menu (or use the shortcut key Ctrl-Y) to access the Arrays Window shown in Figure 2.15(a). Units should be assigned to array variables, just as they are to any other EES

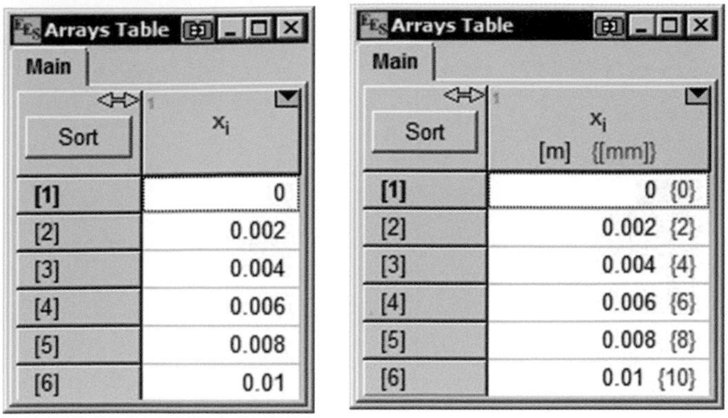

Figure 2.15 Arrays Window (a) with no units set and (b) with units set.

variables. It is possible to set the units by right-clicking on the column in the Arrays Table and specifying the units (and, if desired, the secondary units) in the resulting Format Arrays Table Column dialog. The result is shown in Figure 2.15(b). Another way to enter the units for all members of the array is to use the $VarInfo directive.

```
$VarInfo x[]  Units='m'  AltUnits='mm'
```

The equation set that must be solved is programmed next. The equation for node 1, Eq. (2.142), is entered.

```
"node 1"
h_bar*A_c*(T_infinity-T[1])+k*A_c*(T[2]-T[1])/DELTAx+&
      q``_dot_rad*alpha*Exp(-alpha*x[1])*A_c*DELTAx/2=0
```

Note that the temperatures of each node are also placed in an array, therefore T[1] and T[2] refer to the first and second elements of the array T, respectively. (The ampersand at the end of the first line indicates that the remainder of the equation will continue on the following line.) The equations for nodes 2 through $M - 1$, Eq. (2.136), are programmed using a Duplicate loop that begins with $i = 2$ and ends with $i = M - 1$.

```
"nodes 2 through M-1"
Duplicate i=2,(M-1)
    k*A_c*(T[i-1]-T[i])/DELTAx+k*A_c*(T[i+1]-T[i])/DELTAx+&
        q``_dot_rad*alpha*Exp(-alpha*x[i])*A_c*DELTAx=0
End
```

Finally, the equation for node M, Eq. (2.147), is programmed.

```
"node M"
h_bar*A_c*(T_infinity-T[M])+k*A_c*(T[M-1]-T[M])/DELTAx+&
      q``_dot_rad*alpha*Exp(-alpha*x[M])*A_c*DELTAx/2=0
```

It is good practice to assign the units for all variables including those in the arrays before attempting to solve the equations. The unit consistency of each equation is checked when the equations are solved. The solution is provided in the Solution Window for the scalar quantities and in the Arrays Window for the array of predicted temperatures. A plot showing the predicted temperature as a function of radius is provided in Figure 2.16 for $M = 6$ and $M = 20$. Also shown in Figure 2.16 is the analytical function that was derived in Example 2.12 to solve this problem. Clearly the numerical solution approaches the exact analytical solution as the number of nodes used in the simulation increases.

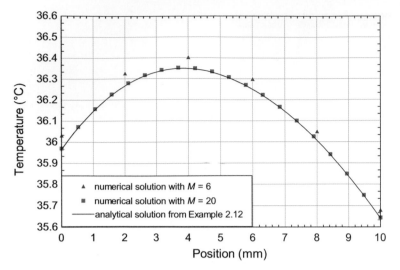

Figure 2.16 Predicted temperature distribution for $M = 6$ and $M = 20$ and analytical solution.

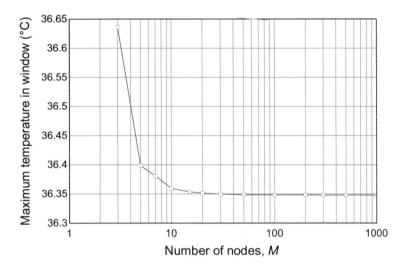

Figure 2.17 Predicted maximum temperature as a function of the number of nodes.

Before the numerical solution can be used with confidence, it is necessary to verify its accuracy. The first step in this process is to ensure that the mesh is adequately refined. Figure 2.16 shows that the solution becomes smoother and represents the actual temperature distribution better as the number of nodes is increased. The general approach to choosing a mesh is to pick an important characteristic of the solution and examine how this characteristic changes as the number of nodes in the computational domain is increased. In this case, an appropriate characteristic might be the maximum predicted temperature within the window. The following EES code extracts this value from the solution using the Max function, which returns the maximum of the arguments provided to it.

```
T_max=Max(T[1..M])                    "maximum  temperature"
```

The maximum temperature as a function of the number of nodes is shown in Figure 2.17, which was created by making a Parametric Table that includes the variables M and T_max. (Note that the values of M in the table must be set to *integers* that approximately *logarithmically* span from 3 to 1000 nodes.) The solution appears to have converged after approximately 20 nodes and further refinement is not likely to be necessary.

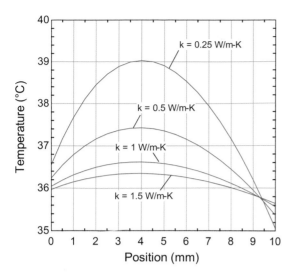

Figure 2.18 Temperature as a function of position for various values of conductivity.

The next step is to check that the solution agrees with physical intuition. For example, if the conductivity of the material is decreased, then the temperatures within the window should increase. Figure 2.18 illustrates the temperature as a function of position for various values of k (with $M = 20$) and shows that reducing the conductivity does tend to increase the temperature in the window.

There are many additional "sanity checks" that could be tested. For example, decrease the heat transfer coefficient or increase the incident radiation and make sure that the temperatures in the computational domain increase as they should.

2.4.4 Solving the Equations with Matrix Decomposition

The EES software internally provides all of the numerical manipulations that are needed to solve the system of algebraic equations that constitutes a numerical model. This capability greatly reduces the complexity of obtaining a solution. However, there are disadvantages to using EES. For example, EES will generally require significantly more time to solve the equations than is required by a compiled computer language. There is an upper limit to the number of variables that EES can handle, which places an upper bound on the number of nodes that can be used in the model. The structure of EES requires that every variable in the main program be retained in the final solution. Therefore, you cannot define and then erase intermediate variables in the course of obtaining a solution and as a result numerical solutions in EES often require a lot of memory. Finally, some models require logic statements (e.g., if-then-else statements), which require the use of functions or procedures in EES. For these reasons, as well as to achieve a better understanding of how a solution is obtained, it is useful to learn how to implement numerical models in a formal programming language, e.g., FORTRAN, C++, or MATLAB. The steps required to solve the algebraic equations associated with a numerical model by **matrix decomposition** are demonstrated in this section using MATLAB.

The equations that must be solved were derived in Section 2.4.2 and are repeated below:

$$\text{node 1.} \quad \bar{h}\,A_c(T_\infty - T_1) + \frac{k\,A_c(T_2 - T_1)}{\Delta x} + \dot{q}''_{rad}\,\alpha\,\exp\left(-\alpha\,x_1\right)A_c\,\frac{\Delta x}{2} = 0$$

$$\text{node 2.} \quad \frac{k\,A_c(T_1 - T_2)}{\Delta x} + \frac{k\,A_c(T_3 - T_2)}{\Delta x} + \dot{q}''_{rad}\,\alpha\exp\left(-\alpha\,x_2\right)A_c\,\Delta x = 0$$

$$\text{node 3.} \quad \frac{k\,A_c(T_2 - T_3)}{\Delta x} + \frac{k\,A_c(T_4 - T_3)}{\Delta x} + \dot{q}''_{rad}\,\alpha\exp\left(-\alpha\,x_3\right)A_c\,\Delta x = 0$$

$$\ldots$$

$$\text{node } i. \quad \frac{k\,A_c(T_{i-1} - T_i)}{\Delta x} + \frac{k\,A_c(T_{i+1} - T_i)}{\Delta x} + \dot{q}''_{rad}\,\alpha\exp\left(-\alpha\,x_i\right)A_c\,\Delta x = 0$$

$\ldots$

$$\text{node } M-1. \quad \frac{k A_c (T_{M-2} - T_{M-1})}{\Delta x} + \frac{k A_c (T_M - T_{M-1})}{\Delta x} + \dot{q}''_{rad} \, \alpha \exp(-\alpha x_{M-1}) A_c \Delta x = 0$$

$$\text{node } M. \quad \bar{h} A_c (T_\infty - T_M) + \frac{k A_c (T_{M-1} - T_M)}{\Delta x} + \dot{q}''_{rad} \, \alpha \exp(-\alpha x_M) A_c \frac{\Delta x}{2} = 0. \tag{2.149}$$

Equation (2.149) shows that there are M linear algebraic equations in an equal number of unknown temperatures. The most direct way to solve this system of equations using a formal programming language is as a **matrix equation**. Recall from linear algebra that a linear system of equations, such as

$$\begin{aligned} 2\,x_1 + 3\,x_2 + 1\,x_3 &= 1 \\ 1\,x_1 + 5\,x_2 + 1\,x_3 &= 2 \\ 7\,x_1 + 1\,x_2 + 2\,x_3 &= 5 \end{aligned} \tag{2.150}$$

can be written as a matrix equation

$$\underbrace{\begin{bmatrix} 2 & 3 & 1 \\ 1 & 5 & 1 \\ 7 & 1 & 2 \end{bmatrix}}_{\underline{\underline{A}}} \underbrace{\begin{bmatrix} x_1 \\ x_2 \\ x_3 \end{bmatrix}}_{\underline{X}} = \underbrace{\begin{bmatrix} 1 \\ 2 \\ 5 \end{bmatrix}}_{\underline{b}} \tag{2.151}$$

or

$$\underline{\underline{A}}\, \underline{X} = \underline{b}, \tag{2.152}$$

where $\underline{\underline{A}}$ is the **coefficient matrix**, $\underline{X}$ is a vector containing the unknowns, and $\underline{b}$ is the **constant vector**:

$$\underline{\underline{A}} = \begin{bmatrix} 2 & 3 & 1 \\ 1 & 5 & 1 \\ 7 & 1 & 2 \end{bmatrix}, \quad \underline{X} = \begin{bmatrix} x_1 \\ x_2 \\ x_3 \end{bmatrix}, \quad \text{and } \underline{b} = \begin{bmatrix} 1 \\ 2 \\ 5 \end{bmatrix}. \tag{2.153}$$

Most programming languages, including MATLAB, have built-in routines for decomposing the system of equations and solving for the vector of unknowns, $\underline{X}$. This is a mature area of numerical analysis and advanced methods exist for quickly solving matrix equations, particularly when most of the entries in $\underline{\underline{A}}$ are 0 (i.e., $\underline{\underline{A}}$ is a **sparse matrix**).

MATLAB is specifically designed to handle large matrix equations. In this section, we will use MATLAB to solve the conduction problem considered in Section 2.4.2 which requires that we understand how to place large systems of equations, corresponding to the energy balances, into a matrix format. Notice by comparing Eq. (2.150) and Eq. (2.153) that each *row* of the coefficient matrix $\underline{\underline{A}}$ and the constant vector $\underline{b}$ correspond to an equation whereas each *column* of $\underline{\underline{A}}$ is the coefficient that multiplies the corresponding unknown in that equation. To set up equations in matrix format, it is only necessary to carefully define how the equations will be ordered in the rows of $\underline{\underline{A}}$ and $\underline{b}$ and how the unknowns will be ordered in $\underline{X}$. In a problem as simple as the one considered here, this is fairly trivial; however, in more complex problems this will be less obvious.

The first step is to define the vector of unknowns, the vector $\underline{X}$ in Eq. (2.152). It does not really matter what order the unknowns are placed in $\underline{X}$, but the implementation of the solution is much easier if a logical order is used. In this problem, the unknowns are the nodal temperatures. Therefore, the most logical technique for ordering the unknown temperatures in the vector $\underline{X}$ is:

$$\underline{X} = \begin{bmatrix} X_1 = T_1 \\ X_2 = T_2 \\ \ldots \\ X_M = T_M \end{bmatrix}. \tag{2.154}$$

Equation (2.154) shows that the unknown temperature at node i (i.e., T_i) corresponds to element i of vector $\underline{X}$ (i.e., X_i).

The next step is to define how the rows in the matrix $\underline{\underline{A}}$ and the vector $\underline{b}$ correspond to the M control volume energy balances that must be solved. Again the solution is easiest if a logical order is used:

$$\underline{\underline{A}}, \underline{b} = \begin{bmatrix} \text{row } 1 = \text{control volume 1 equation} \\ \text{row } 2 = \text{control volume 2 equation} \\ \dots \\ \text{row } M = \text{control volume } M \text{ equation} \end{bmatrix}. \tag{2.155}$$

Equation (2.155) shows that the equation for control volume i is placed into row i of matrix $\underline{\underline{A}}$ and vector $\underline{b}$.

Appendix G includes a brief introduction to the MATLAB software that will be useful if you've never used the program before. Using the MATLAB editor, open a new script. Enter the inputs to the problem at the top of the script and save it. Note that the % symbol indicates that anything that follows on that line will be a comment. MATLAB will not assign units to any of the variables; they are all dimensionless as far as the software is concerned. This limitation puts the burden squarely on the user to understand the units of each variable and ensure that they are consistent. The use of a semicolon after each assignment statement prevents the variables from being echoed in the working environment. The clear command at the top of the script clears all variables from the workspace.

```
clear;                          %clear the workspace

%Inputs
k=1.5;                          %[W/m-K] conductivity
L=0.01;                         %[m] lens thickness
qf_dot_rad=1000;                %[W/m^2] incident energy flux
h_bar=20;                       %[W/m^2-K] average heat transfer coefficient
T_infinity=293.2;               %[K] ambient air temperature
alpha=100;                      %[1/m] absorption coefficient
A_c=1;                          %[m^2] do problem on a per unit area basis
```

In order to run your script from the MATLAB working environment, save the file as window.m and then select the green run button in the Script Editor. Depending on what folder you saved the file in, you may be presented with a dialog that asks you to either change to the appropriate folder (Change Folder) or add the folder to the path that MATLAB searches (Add to Path). Nothing appears to have happened when you run the script. However, all of the variables that are defined in the script are now available in the work space. For example, if the name of any variable is entered at the command prompt then the value is displayed.

```
>> k
k =
    1.5
```

For a complete list of variables in the workspace, use the command who.

```
>> who
Your variables are:
A_c    L    T_infinity  alpha    h_bar    k    qf_dot_rad
```

The number and location of the nodes for the solution must be specified. A vector of locations (x) is setup using a for loop. Each of the statements within the loop is executed for values of the loop index i ranging from 1 to M. Enter the following lines into the window M-file script.

```
%Setup nodes
M=11;                                   %[-] number of nodes
DELTAx=L/(M-1);                         %[m] distance between adjacent nodes
for i=1:M
    x(i)=L*(i-1)/(M-1);                 %[m] axial position of each node
end
```

The coefficient matrix is assigned to the variable A and the constant vector to the variable b. Recall that our matrix $\underline{\underline{A}}$ needs to have as many rows as there are equations (the M control volume energy balances) and as many columns as there are unknowns (the M unknown temperatures) and $\underline{b}$ is a vector with as many elements as there are equations (M). We will start with $\underline{\underline{A}}$ and $\underline{b}$ composed entirely of zeros and subsequently add nonzero elements as dictated by the equation set. As we do this you should notice that the $\underline{\underline{A}}$ matrix ends up being composed almost entirely of zeros and thus it is referred to as a sparse matrix. While we will not take advantage of this sparse characteristic of $\underline{\underline{A}}$ in this example, the solution can be accelerated considerably by using specialized matrix solution techniques that are designed for sparse matrices. MATLAB makes it extremely easy to use these sparse matrix solution techniques.

The zeros function used below simply returns a matrix filled with zeros with its size determined by the input arguments; the variable A will begin as an $M \times M$ matrix filled with zeros and the variable b will begin as an $M \times 1$ vector filled with zeros.

```
%Setup A and b
A=zeros(M,M);
b=zeros(M,1);
```

According to Eq. (2.155), the first row in $\underline{\underline{A}}$ must correspond to the energy balance for control volume 1, which is given by:

$$\bar{h} A_c (T_\infty - T_1) + \frac{k A_c (T_2 - T_1)}{\Delta x} + \dot{q}''_{rad} \alpha \exp(-\alpha x_1) A_c \frac{\Delta x}{2} = 0. \tag{2.156}$$

It is a good idea to algebraically manipulate Eq. (2.156) so that the coefficients that multiply each of the unknowns in this equation (i.e., T_1 and T_2) and the constant term in the equation (i.e., terms that are known) can be easily identified:

$$T_1 \underbrace{\left[-\bar{h} A_c - \frac{k A_c}{\Delta x} \right]}_{A_{1,1}} + T_2 \underbrace{\left[\frac{k A_c}{\Delta x} \right]}_{A_{1,2}} = \underbrace{-\dot{q}''_{rad} \alpha \exp(-\alpha x_1) A_c \frac{\Delta x}{2} - \bar{h} A_c T_\infty}_{b_1}. \tag{2.157}$$

Equation (2.157) corresponds to the first row of $\underline{\underline{A}}$ and $\underline{b}$. The coefficient in the first equation that multiplies the first unknown in $\underline{X}$, T_1 according to Eq. (2.154), must be $A_{1,1}$:

$$A_{1,1} = -\bar{h} A_c - \frac{k A_c}{\Delta x}. \tag{2.158}$$

The coefficient in the first equation that multiplies the second unknown in $\underline{X}$, T_2 according to Eq. (2.154), must be $A_{1,2}$:

$$A_{1,2} = \frac{k A_c}{\Delta x}. \tag{2.159}$$

Finally, the constant terms associated with the first equation must be b_1:

$$b_1 = -\dot{q}''_{rad} \alpha \exp(-\alpha x_1) A_c \frac{\Delta x}{2} - \bar{h} A_c T_\infty. \tag{2.160}$$

These assignments are accomplished in MATLAB with the following code.

```
%Energy balance for control volume 1
A(1,1)=-h_bar*A_c-k*A_c/DELTAx;
A(1,2)=k*A_c/DELTAx;
b(1)=-qf_dot_rad*alpha*exp(-alpha*x(1))*A_c*DELTAx/2-h_bar*A_c*T_infinity;
```

According to Eq. (2.155), rows 2 through $M - 1$ of matrix $\underline{\underline{A}}$ correspond to the energy balances for the corresponding control volumes; these equations are given by Eq. (2.136), which is repeated below:

$$\frac{k A_c (T_{i-1} - T_i)}{\Delta x} + \frac{k A_c (T_{i+1} - T_i)}{\Delta x} + \dot{q}''_{rad} \alpha \exp(-\alpha x_i) A_c \Delta x = 0 \qquad \text{for } i = 2 \dots (M-1). \tag{2.161}$$

Equation (2.161) is also rearranged to identify coefficients and constants.

$$T_i \underbrace{\left[-\frac{2 k A_c}{\Delta x} \right]}_{A_{i,i}} + T_{i-1} \underbrace{\left[\frac{k A_c}{\Delta x} \right]}_{A_{i,i-1}} + T_{i+1} \underbrace{\left[\frac{k A_c}{\Delta x} \right]}_{A_{i,i+1}} = \underbrace{-\dot{q}''_{rad} \alpha \exp[-\alpha x_i] A_c \Delta x}_{b_i} \qquad \text{for } i = 2 \dots (M-1) \tag{2.162}$$

All of the coefficients for control volume i must go into row i of $\underline{\underline{A}}$, the column depends on which unknown they multiply. Therefore:

$$A_{i,i} = -\frac{2 k A_c}{\Delta x} \text{ for } i = 2 \dots (M-1) \tag{2.163}$$

$$A_{i,i-1} = \frac{k A_c}{\Delta x} \text{ for } i = 2 \dots (M-1) \tag{2.164}$$

$$A_{i,i+1} = \frac{k A_c}{\Delta x} \text{ for } i = 2 \dots (M-1). \tag{2.165}$$

The constant for control volume i must go into row i of $\underline{b}$:

$$b_i = -\dot{q}''_{rad} \alpha \exp(-\alpha x_i) A_c \Delta x \text{ for } i = 2 \dots (M-1). \tag{2.166}$$

These equations are programmed most conveniently in MATLAB using a for loop.

```
%Energy balances for internal control volumes
for i=2:(M-1)
    A(i,i)=-2*k*A_c/DELTAx;
    A(i,i-1)=k*A_c/DELTAx;
    A(i,i+1)=k*A_c/DELTAx;
    b(i)=-qf_dot_rad*alpha*exp(-alpha*x(i))*A_c*DELTAx;
end
```

Finally, the last row of $\underline{\underline{A}}$ (i.e., row M) corresponds to the energy balance for the last control volume (node M), which is given by Eq. (2.147), repeated below:

$$\bar{h} A_c (T_\infty - T_M) + \frac{k A_c (T_{M-1} - T_M)}{\Delta x} + \dot{q}''_{rad} \alpha \exp(-\alpha x_M) A_c \frac{\Delta x}{2} = 0. \tag{2.167}$$

Equation (2.167) is rearranged:

$$T_M \underbrace{\left[-\bar{h} A_c - \frac{k A_c}{\Delta x} \right]}_{A_{M,M}} + T_{M-1} \underbrace{\left[\frac{k A_c}{\Delta x} \right]}_{A_{M,M-1}} = \underbrace{-\bar{h} A_c T_\infty - \dot{q}''_{rad} \alpha \exp(-\alpha x_M) A_c \frac{\Delta x}{2}}_{b_M}. \tag{2.168}$$

The coefficients in the last row of $\underline{\underline{A}}$ and $\underline{b}$ are

$$A_{M,M} = -\bar{h} A_c - \frac{k A_c}{\Delta x} \tag{2.169}$$

$$A_{M,M-1} = \frac{k A_c}{\Delta x} \qquad (2.170)$$

and

$$b_M = -\bar{h} A_c T_\infty - \ddot{q}''_{rad}\, \alpha \exp\left(-\alpha x_M\right) A_c \frac{\Delta x}{2}. \qquad (2.171)$$

```
%Energy balance for control volume M
A(M,M)=-h_bar*A_c-k*A_c/DELTAx;
A(M,M-1)=k*A_c/DELTAx;
b(M)=-h_bar*A_c*T_infinity-qf_dot_rad*alpha*exp(-alpha*x(M))*A_c*DELTAx/2;
```

At this point, the matrix $\underline{\underline{A}}$ and vector $\underline{b}$ are completely set up and can be used to determine the unknown temperatures. The solution to the matrix equation, Eq. (2.152), is:

$$\underline{X} = \underline{\underline{A}}^{-1}\underline{b}, \qquad (2.172)$$

where $\underline{\underline{A}}^{-1}$ is the inverse of matrix $\underline{\underline{A}}$. The solution is obtained using the backslash operator in MATLAB. (Note that this is much more efficient than explicitly solving for the inverse of $\underline{\underline{A}}$ using the inv command.)

```
%Solve for unknowns
X=A\b;                              %solve for unknown vector, X
```

The vector of unknowns, $\underline{X}$, contains the temperatures for this problem and these can be easily converted from K to °C.

```
T=X;                                %assign temperatures from X
T_C=T-273.15;                       %in C
```

The script window can be executed from the working environment by typing the name of the script (window) at the command prompt. After execution, the variables that were defined in the M-file and the solution will reside in the workspace; for example, you can view the solution vector (T_C) by typing T_C and you can plot temperature as a function of radius using the plot command.

```
>> window
>> T_C

T_C =

   35.9815
   36.1613
   36.2807
   36.3456
   36.3610
   36.3318
   36.2621
   36.1559
   36.0165
   35.8472
   35.6508

>> plot(x,T_C);
```

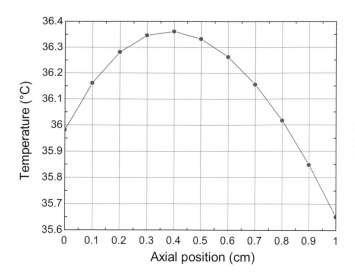

Figure 2.19 Temperature as a function of position predicted by the numerical model that uses matrix decomposition.

Figure 2.19 illustrates the temperature predicted by the MATLAB numerical model as a function of position.

2.4.5 Solving the Equations with Gauss–Seidel Iteration

In Section 2.4.3 we learned how to solve a system of equations using EES, which is an equation solver program. In Section 2.4.4 we placed the equations into matrix format so that the solution could be obtained using a matrix decomposition process using MATLAB. In this section we will solve the equations iteratively using the **Gauss–Seidel** technique. The finite difference equations that must be solved are repeated below:

node 1. $\quad \bar{h} A_c (T_\infty - T_1) + \dfrac{k A_c (T_2 - T_1)}{\Delta x} + \dot{q}''_{rad}\, \alpha \exp\left(-\alpha x_1\right) A_c \dfrac{\Delta x}{2} = 0$

node 2. $\quad \dfrac{k A_c (T_1 - T_2)}{\Delta x} + \dfrac{k A_c (T_3 - T_2)}{\Delta x} + \dot{q}''_{rad}\, \alpha \exp\left(-\alpha x_2\right) A_c \Delta x = 0$

node 3. $\quad \dfrac{k A_c (T_2 - T_3)}{\Delta x} + \dfrac{k A_c (T_4 - T_3)}{\Delta x} + \dot{q}''_{rad}\, \alpha \exp\left(-\alpha x_3\right) A_c \Delta x = 0$

. . .

node i. $\quad \dfrac{k A_c (T_{i-1} - T_i)}{\Delta x} + \dfrac{k A_c (T_{i+1} - T_i)}{\Delta x} + \dot{q}''_{rad}\, \alpha \exp\left(-\alpha x_i\right) A_c \Delta x = 0$

. . .

node $M-1$. $\quad \dfrac{k A_c (T_{M-2} - T_{M-1})}{\Delta x} + \dfrac{k A_c (T_M - T_{M-1})}{\Delta x} + \dot{q}''_{rad}\, \alpha \exp\left(-\alpha x_{M-1}\right) A_c \Delta x = 0$

node M. $\quad \bar{h} A_c (T_\infty - T_M) + \dfrac{k A_c (T_{M-1} - T_M)}{\Delta x} + \dot{q}''_{rad}\, \alpha \exp\left(-\alpha x_M\right) A_c \dfrac{\Delta x}{2} = 0.$ (2.173)

One advantage of the Gauss–Seidel technique is that the computer only needs to solve explicit equations one at a time, rather than requiring it to solve a simultaneous set of equations that depend upon one another. We start from a set of guess temperatures and then use the equations to improve these guesses over and over again until, hopefully, the temperatures stop changing as they approach the correct result.

To set up the equations, we will solve each of the nodal energy balances for the temperature at that node. For example, the energy balance for node 1:

$$\bar{h} A_c (T_\infty - T_1) + \dfrac{k A_c (T_2 - T_1)}{\Delta x} + \dot{q}''_{rad}\, \alpha \exp\left(-\alpha x_1\right) A_c \dfrac{\Delta x}{2} = 0 \qquad (2.174)$$

should be solved explicitly for T_1:

$$T_1 = \frac{\frac{k A_c}{\Delta x} T_2 + \dot{q}''_{rad} \alpha \exp\left(-\alpha x_1\right) A_c \frac{\Delta x}{2} + \bar{h} A_c T_\infty}{\left(\bar{h} A_c + \frac{k A_c}{\Delta x}\right)}. \tag{2.175}$$

Similarly, the energy balance equations for the internal nodes:

$$\frac{k A_c (T_{i-1} - T_i)}{\Delta x} + \frac{k A_c (T_{i+1} - T_i)}{\Delta x} + \dot{q}''_{rad} \alpha \exp\left(-\alpha x_i\right) A_c \Delta x = 0 \qquad \text{for } i = 2 \ldots (M-1) \tag{2.176}$$

should be solved for T_i:

$$T_i = \frac{\frac{k A_c}{\Delta x} T_{i-1} + \frac{k A_c}{\Delta x} T_{i+1} + \dot{q}''_{rad} \alpha \exp\left(-\alpha x_i\right) A_c \Delta x}{\frac{2 k A_c}{\Delta x}} \qquad \text{for } i = 2 \ldots (M-1). \tag{2.177}$$

Finally, the energy balance for node M:

$$\bar{h} A_c (T_\infty - T_M) + \frac{k A_c (T_{M-1} - T_M)}{\Delta x} + \dot{q}''_{rad} \alpha \exp\left(-\alpha x_M\right) A_c \frac{\Delta x}{2} = 0 \tag{2.178}$$

is solved for T_M:

$$T_M = \frac{\frac{k A_c}{\Delta x} T_{M-1} + \bar{h} A_c T_\infty + \dot{q}''_{rad} \alpha \exp\left(-\alpha x_M\right) A_c \frac{\Delta x}{2}}{\left(\bar{h} A_c + \frac{k A_c}{\Delta x}\right)}. \tag{2.179}$$

The resulting equations are then used as the engine for the iteration process. We start with a set of guess values for each of the nodal temperatures; let's call them T_i^k where the subscript i indicates the node and the superscript k indicates the iteration number ($k = 1$ will be our initial set of guess temperatures). Next we employ Eqs. (2.175), (2.177), and (2.179) to solve for each of the temperatures in the next iteration, T_i^{k+1} where $i = 1 \ldots M$. The temperature on the left side of the equal sign is the new temperature at node i while the temperatures used on the right side are the most recent value of the other temperatures. Note that the most recent value of temperature may either come from the prior iteration or the current iteration. The easiest way to accomplish this is to use one array for the unknown temperatures and overwrite the elements as you move through the iteration process. That way, whatever temperature is present in the array is the most recent one.

The values of the temperatures obtained for successive iterations must be compared in order to determine an error (*err*) that indicates whether the iteration process has sufficiently converged. There are different metrics that can be used for this purpose. The simplest is to take the maximum absolute value of the change of temperature between iterations for any node:

$$err = \text{Max}\left(\left|T_i^{k+1} - T_i^k\right|\right) \text{ for } i = 1 \ldots M. \tag{2.180}$$

Alternatively, we could calculate the rms value of the error associated with all of the nodes:

$$err = \sqrt{\frac{1}{M} \sum_{i=1}^{M} \left(T_i^{k+1} - T_i^k\right)^2} \tag{2.181}$$

or the average absolute change between iterations:

$$err = \frac{1}{M} \sum_{i=1}^{M} \left|T_i^{k+1} - T_i^k\right|. \tag{2.182}$$

Regardless of how the convergence error is calculated, it must be compared to some tolerance, *tol*. If the error remains above the tolerance then the iteration must be repeated, this time using the most recent temperature values.

We can program this process using any formal programming language as it depends only on the solution of explicit equations. Here we will implement the solution using MATLAB. A script is created, called window_GS. The inputs are entered at the top of the script.

```
clear;                              %clear the workspace

%Inputs
k=1.5;                              %[W/m-K] conductivity
L=0.01;                             %[m] lens thickness
qf_dot_rad=1000;                    %[W/m^2] incident energy flux
h_bar=20;                           %[W/m^2-K] average heat transfer coefficient
T_infinity=293.2;                   %[K] ambient air temperature
alpha=100;                          %[1/m] absorption coefficient
A_c=1;                              %[m^2] do problem on a per unit area basis
```

The positions of the nodes and distance between the nodes are setup using Eqs. (2.127) and (2.128).

```
%Setup nodes
M=11;                               %[-] number of nodes
DELTAx=L/(M-1);                     %[m] distance between adjacent nodes
for i=1:M
    x(i)=L*(i-1)/(M-1);            %[m] axial position of each node
end
```

The initial guess values of the temperatures (i.e., our initial iteration $\underline{T}^k$) are stored in vector T_k; we will start the iteration by assuming that the temperature of each node is T_∞. The temperatures for the next iteration, $\underline{T}^{k+1}$, are stored in vector T_kp. Note that the temperatures in the vector T_kp start with the values in T_k and are overwritten during the iteration process, ensuring that the most recently calculated value of temperature is always being used.

```
for i=1:M
    T_k(i)=T_infinity;             %[K] initial guess temperature
end
T_kp=T_k;                          %start with old temperatures and overwrite during iteration
```

We will put the iteration process within a while loop in MATLAB. A while loop continues until some termination criteria is satisfied; in this case the termination criteria is that the convergence error (*err*) is less than our specified tolerance (*tol*). We need to specify both the tolerance (the variable tol) and some initial value of the error (the variable err). The initial value of the error is not important provided that it is larger than the tolerance so that the while loop executes at least once.

```
tol=1e-6;                 %[K] convergence tolerance
err=tol+1;                %initial value of convergence error
while(err>tol)
    %put your iteration commands here
end
```

Within the while loop we will carry out the iteration process by programming Eqs. (2.175), (2.177), and (2.179) and storing the results in the vector T_kp. The convergence error is computed using Eq. (2.180). Finally, the values stored in T_kp are transferred to T_k in order to repeat the iteration if that is required (i.e., if the while loop does not terminate). The final script is shown below.

```
clear;                      %clear the workspace

%Inputs
k=1.5;                      %[W/m-K] conductivity
L=0.01;                     %[m] lens thickness
qf_dot_rad=1000;            %[W/m^2] incident energy flux
h_bar=20;                   %[W/m^2-K] average heat transfer coefficient
T_infinity=293.2;           %[K] ambient air temperature
alpha=100;                  %[1/m] absorption coefficient
A_c=1;                      %[m^2] do problem on a per unit area basis

%Setup nodes
M=11;                       %[-] number of nodes
DELTAx=L/(M-1);             %[m] distance between adjacent nodes
for i=1:M
    x(i)=L*(i-1)/(M-1);     %[m] axial position of each node
end

for i=1:M
    T_k(i)=T_infinity;      %[K] initial guess temperature
end
T_kp=T_k;                   %start with old temperatures and overwrite during iteration

tol=1e-6;                   %[K] convergence tolerance
err=tol+1;                  %initial value of convergence error
while(err>tol)
    %compute new temperatures
    T_kp(1)=(k*A_c*T_kp(2)/DELTAx+qf_dot_rad*alpha*exp(-alpha*x(1))*A_c*...
        DELTAx/2+h_bar*A_c*T_infinity)/(h_bar*A_c+k*A_c/DELTAx);
    for i=2:(M-1)
        T_kp(i)=(k*A_c*T_kp(i-1)/DELTAx+k*A_c*T_kp(i+1)/DELTAx+...
            qf_dot_rad*alpha*exp(-alpha*x(i))*A_c*DELTAx)/(2*k*A_c/DELTAx);
    end
    T_kp(M)=(k*A_c*T_kp(M-1)/DELTAx+qf_dot_rad*alpha*exp(-alpha*x(M))*...
        A_c*DELTAx/2+h_bar*A_c*T_infinity)/(h_bar*A_c+k*A_c/DELTAx);
    err=max(abs(T_kp-T_k))   %determine the convergence error
    T_k=T_kp;                %use calculated temperatures for next iteration
end
```

Because the line that computes the convergence error was not terminated in a semicolon, the value of the convergence error will be written in the command space during each iteration. Running the script produces the following output.

```
>> window_GS
err =
    0.0506
err =
    0.0491
```

```
err =
    0.0484
err =
    0.0480...
```

The iteration process continues in this fashion until finally the convergence error is reduced below the tolerance (1×10^{-6} K).

```
...
err =
    1.0068e-06
err =
    1.0042e-06
err =
    1.0016e-06
err =
    9.9897e-07
```

The predicted temperature as function of position is identical to the result shown in Figure 2.19.

2.4.6 Temperature-Dependent Properties

The properties of most materials are a function of temperature, although they can sometimes be approximated as being constant within a small temperature range. It is usually necessary to assume a constant value of thermal conductivity in order to solve a problem analytically, since to assume otherwise generally causes the differential equation to be much more complicated or even intractable. A major advantage of a numerical solution is that the temperature dependence of physical properties can be considered with little additional effort.

The problem considered in Sections 2.4.2 through 2.4.5 assumed both a constant thermal conductivity and a constant absorption coefficient in order to develop the algebraic set of finite difference equations that are provided below:

$$\text{node 1.} \qquad \bar{h} A_c (T_\infty - T_1) + \frac{k A_c (T_2 - T_1)}{\Delta x} + \dot{q}''_{rad} \alpha \exp(-\alpha x_1) A_c \frac{\Delta x}{2} = 0$$

$$\text{node 2.} \qquad \frac{k A_c (T_1 - T_2)}{\Delta x} + \frac{k A_c (T_3 - T_2)}{\Delta x} + \dot{q}''_{rad} \alpha \exp(-\alpha x_2) A_c \Delta x = 0$$

$$\text{node 3.} \qquad \frac{k A_c (T_2 - T_3)}{\Delta x} + \frac{k A_c (T_4 - T_3)}{\Delta x} + \dot{q}''_{rad} \alpha \exp(-\alpha x_3) A_c \Delta x = 0$$

...

$$\text{node } i. \qquad \frac{k A_c (T_{i-1} - T_i)}{\Delta x} + \frac{k A_c (T_{i+1} - T_i)}{\Delta x} + \dot{q}''_{rad} \alpha \exp(-\alpha x_i) A_c \Delta x = 0$$

...

$$\text{node } M-1. \qquad \frac{k A_c (T_{M-2} - T_{M-1})}{\Delta x} + \frac{k A_c (T_M - T_{M-1})}{\Delta x} + \dot{q}''_{rad} \alpha \exp(-\alpha x_{M-1}) A_c \Delta x = 0$$

$$\text{node } M. \qquad \bar{h} A_c (T_\infty - T_M) + \frac{k A_c (T_{M-1} - T_M)}{\Delta x} + \dot{q}''_{rad} \alpha \exp(-\alpha x_M) A_c \frac{\Delta x}{2} = 0. \qquad (2.183)$$

If the conductivity and absorption coefficient both depend on temperature then these equations are no longer a linear set of equations for the unknown temperatures. Everywhere k appears in Eq. (2.183) it cannot be considered to be a constant, but rather it must be thought of as a function that must be evaluated at some value of temperature. The same thing is true of α. There are two questions that must be answered then. First, what value of temperature should be used in the evaluation of k and α and, second, how can we solve the resulting set of nonlinear equations?

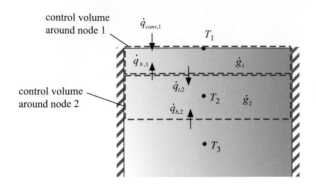

Figure 2.20 Control volumes used to develop the energy balances for nodes 1 and 2.

It is appropriate to evaluate the absorption coefficient using the associated nodal temperature, but it is important to be more careful with the temperature-dependent thermal conductivity. It is tempting to also evaluate the conduction terms associated with each energy balance using the thermal conductivity evaluated at the temperature of the node. However, doing so will result in an error in the energy balance. To see this, we can examine the energy balances for nodes 1 and 2, shown below:

$$\text{node 1.} \quad \bar{h} A_c (T_\infty - T_1) + \underbrace{\frac{k A_c (T_2 - T_1)}{\Delta x}}_{\dot{q}_{b,1}} + \dot{q}''_{rad}\, \alpha \exp(-\alpha x_1) A_c \frac{\Delta x}{2} = 0$$

$$\text{node 2.} \quad \underbrace{\frac{k A_c (T_1 - T_2)}{\Delta x}}_{\dot{q}_{t,2}} + \frac{k A_c (T_3 - T_2)}{\Delta x} + \dot{q}''_{rad}\, \alpha \exp(-\alpha x_2) A_c \Delta x = 0. \tag{2.184}$$

The second term in the energy balance for node 1, labeled $\dot{q}_{b,1}$ in Eq. (2.184), represents the conduction heat transfer crossing the control surface that separates nodes 1 and 2 and entering node 1, shown in Figure 2.20. The first term in the energy balance for node 2, labeled $\dot{q}_{t,2}$ in Eq. (2.184), represents the conduction heat transfer crossing the same control surface but entering node 2, also shown in Figure 2.20.

It should be clear that $\dot{q}_{b,1}$ and $\dot{q}_{t,2}$ must be equal and opposite, otherwise energy will be artificially generated or destroyed at the boundary between control volumes 1 and 2. If $\dot{q}_{b,1}$ is evaluated using the conductivity calculated at T_1 ($k_{T=T_1}$):

$$\dot{q}_{b,1} = \frac{k_{T=T_1} A_c (T_2 - T_1)}{\Delta x} \tag{2.185}$$

and $\dot{q}_{t,2}$ is evaluated using the conductivity calculated at T_2 ($k_{T=T_2}$):

$$\dot{q}_{t,2} = \frac{k_{T=T_2} A_c (T_1 - T_2)}{\Delta x} \tag{2.186}$$

then the magnitude of these two heat transfer rates will *not* necessarily be the same. To avoid this problem, the thermal conductivity should be evaluated at the average temperature of the two nodes that are involved in the conduction heat transfer (i.e., the temperature at the boundary). Equations (2.185) and (2.186) become

$$\dot{q}_{b,1} = \frac{k_{T=(T_1+T_2)/2} A_c (T_2 - T_1)}{\Delta x} \tag{2.187}$$

$$\dot{q}_{t,2} = \frac{k_{T=(T_1+T_2)/2} A_c (T_1 - T_2)}{\Delta x}, \tag{2.188}$$

which must be of equal magnitude and have opposite sign.

The resulting algebraic equations that must be solved if both absorption coefficient and thermal conductivity are functions of temperature are:

$$\bar{h}A_c(T_\infty - T_1) + \frac{k_{T=(T_1+T_2)/2}A_c(T_2 - T_1)}{\Delta x} + \dot{q}''_{rad}\,\alpha_{T=T_1}\exp\left(-\alpha_{T=T_1}x_1\right)A_c\frac{\Delta x}{2} = 0$$

$$\frac{k_{T=(T_{i-1}+T_i)/2}A_c(T_{i-1} - T_i)}{\Delta x} + \frac{k_{T=(T_{i+1}+T_i)/2}A_c(T_{i+1} - T_i)}{\Delta x} + \dot{q}''_{rad}\,\alpha_{T=Ti}\exp\left(-\alpha_{T=T_i}x_i\right)A_c\Delta x = 0$$

$$\text{for } i = 2\ldots(M-1)$$

$$\bar{h}A_c(T_\infty - T_M) + \frac{k_{T=(T_{M-1}+T_M)/2}A_c(T_{M-1} - T_M)}{\Delta x} + \dot{q}''_{rad}\,\alpha_{T=T_M}\exp\left(-\alpha_{T=T_M}x_M\right)A_c\frac{\Delta x}{2} = 0. \tag{2.189}$$

The equations are still algebraic but they are no longer linear.

Implementation in EES

Because EES can solve sets of nonlinear implicit equations, it is not much more difficult to solve the set of equations represented by Eq. (2.189) than it was to enter the linear equations represented by Eq. (2.183). To illustrate this process, we will assume that the conductivity and absorption coefficient for the glass used in the window are linear functions of temperature:

$$k(T) = k_o + \beta_k(T - T_o) \tag{2.190}$$

$$\alpha(T) = \alpha_o + \beta_\alpha(T - T_o). \tag{2.191}$$

The values k_o = 1.5 W/m-K and α_o = 0.1 mm^{-1} are the conductivity and absorption coefficient, respectively, at room temperature, T_o = 20°C. The quantities β_k = 0.02 W/m-K^2 and β_α = 0.002 mm^{-1} K^{-1} are the temperature coefficients of conductivity and absorption coefficient, respectively. Two functions, k_g and alpha_g, are written in EES that return the conductivity and absorption coefficient, respectively, given a value of temperature. An EES **function** is a self-contained code segment that is provided with one or more input parameters and returns a single value that is calculated based on these parameters. Functions in EES must be placed at the top of the Equations Window, before the main body of equations.

```
$UnitSystem SI MASS RAD PA K J
Function K_g(T)
   "This function returns the conductivity given the temperature"
   k_o=1.5 [W/m-K]                              "conductivity at room temperature"
   T_o=ConvertTemp(C,K,20 [C])                  "room temperature"
   beta_k=0.02 [W/m-K^2]                        "temperature coefficient"
   K_g=k_o+beta_k*(T-T_o)                       "conductivity"
End


Function Alpha_g(T)
   "This function returns the absorption coefficient given the temperature"
   alpha_o=0.1 [1/mm]*Convert(1/mm,1/m)         "absorption coefficient at room temperature"
   T_o=ConvertTemp(C,K,20 [C])                  "room temperature"
   beta_alpha=0.002 [1/mm-K]*Convert(1/mm-K,1/m-K)   "temperature coefficient"
   Alpha_g=alpha_o+beta_alpha*(T-T_o)           "absorption coefficient"
End
```

The functions begin with the key word **Function** followed by the name of the function (e.g., K_g) and the input arguments (T) and are terminated by the statement End. None of the variables in the main body of the EES program are accessible within the function other than those explicitly passed to the function as inputs. Unlike statements in the main program, the expressions within a function are executed in the order that they are entered and all variables on the right-hand side of each expression must be defined (i.e., the statements within a function are *assignments* rather than *equations*). The statements within the function are used to define the value of the function (K_g, for example). Units for the variables in the function should be set using the Variable Information dialog or the appropriate tab in the Solutions Window, in the same way that units are set for variables in the main program.

The EES code that solves the algebraic equations associated with Eq. (2.189) follows the function definition in the same Equations Window.

```
L=1 [cm]*Convert(cm,m)                           "lens thickness"
q``_dot_rad=0.1 [W/cm^2]*Convert(W/cm^2,W/m^2)   "incident energy flux"
h_bar=20 [W/m^2-K]                               "average heat transfer coefficient"
T_infinity=ConvertTemp(C,K,20 [C])               "ambient air temperature"
A_c=1 [m^2]                                      "do problem on a per unit area basis"

M=21 [-]                                          "number of nodes"
DELTAx=L/(M-1)                                    "distance between adjacent nodes"
Duplicate i=1,M
    x[i]=L*(i-1)/(M-1)                            "axial position of each node"
End

"node 1"
h_bar*A_c*(T_infinity-T[1])+K_g((T[1]+T[2])/2)*A_c*(T[2]-T[1])/DELTAx+&
    q``_dot_rad*Alpha_g(T[1])*Exp(-Alpha_g(T[1])*x[1])*A_c*DELTAx/2=0

"nodes 2 through M-1"
Duplicate i=2,(M-1)
    K_g((T[i-1]+T[i])/2)*A_c*(T[i-1]-T[i])/DELTAx+K_g((T[i+1]+T[i])/2)*A_c*(T[i+1]-T[i])/DELTAx+&
        q``_dot_rad*Alpha_g(T[i])*Exp(-Alpha_g(T[i])*x[i])*A_c*DELTAx=0
End

"node M"
h_bar*A_c*(T_infinity-T[M])+K_g((T[M-1]+T[M])/2)*A_c*(T[M-1]-T[M])/DELTAx+&
    q``_dot_rad*Alpha_g(T[M])*Exp(-Alpha_g(T[M])*x[M])*A_c*DELTAx/2=0

T_max=Max(T[1..M])                               "maximum temperature"
```

Select Solve from the Calculate menu and you are likely to find that the problem either fails to converge or converges to some ridiculous and nonphysical set of temperatures. This is not surprising since the temperature-dependent properties have transformed the algebraic equations from a set of equations that are linear in the unknown temperatures to a set of equations that are nonlinear in the unknown temperatures. Therefore, EES must start from some guess values for the unknown temperatures and attempt to iterate until a solution is obtained. The success of this process is highly dependent on the guess values that are used. It may be possible to simply set better guess values for the unknown temperatures. Select Variable Info from the Options menu and deselect the Show array variables button at the upper left. Set the guess value for the array T[ ] to something more reasonable than its default value of 1 K (e.g., 300 K), as shown in Figure 2.21. This process of setting guess values and units of the entries in an array can also be carried out using the $VarInfo directive, as shown below.

```
$VarInfo T[] Guess=300 Units = 'K'
```

The temperature distribution predicted using the temperature-dependent properties is shown in Figure 2.22.

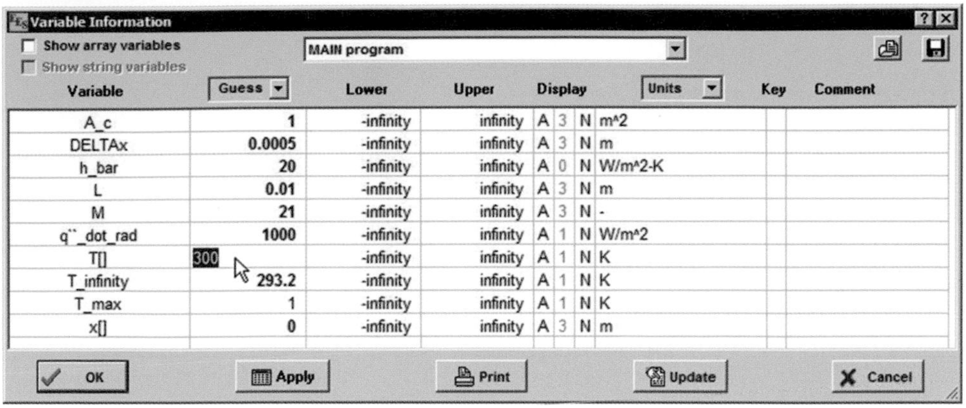

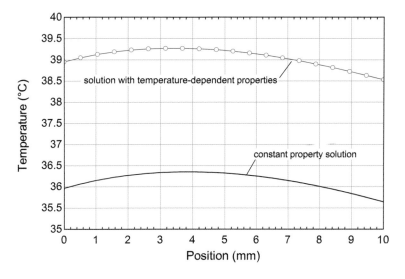

Figure 2.21 Setting the guess values for the array T[] in the Variable Information Window.

Figure 2.22 Temperature as a function of position for the case where the properties are constant and for the case with temperature-dependent properties.

Implementation using Matrix Decomposition

It is possible to solve the nonlinear finite difference equations, Eq. (2.189), using the direct matrix decomposition technique described in Section 2.4.4 with some iteration. The original set of equations derived in Section 2.4.2 is an example of a linear set of equations that can be represented in the form $\underline{\underline{A}}\,\underline{X} = \underline{b}$ where $\underline{\underline{A}}$ is a matrix and $\underline{b}$ is a vector, as discussed in Section 2.4.4, and solved directly and without iteration by matrix decomposition. The inclusion of temperature-dependent properties causes the problem to be nonlinear which complicates the solution using matrix decomposition because the equations can no longer be placed directly into matrix format. The coefficients multiplying the unknown temperatures themselves depend on the unknown temperatures. It is therefore necessary to use some type of iterative relaxation process in order to solve the problem.

The **successive substitution** process begins by assuming a temperature distribution throughout the computational domain (i.e., assume a value of temperature for each node, $\hat{T}_i$ for $i = 1...M$). The assumed values of temperature are used to compute the coefficients that are required to set up the matrix equation (e.g., the temperature-dependent conductivity). The matrix equation is subsequently solved, as discussed in Section 2.4.4, which results in a prediction for the temperature distribution throughout the computational domain (i.e., a predicted value of the temperature for each node, T_i for $i = 1...M$). The assumed and predicted temperatures at

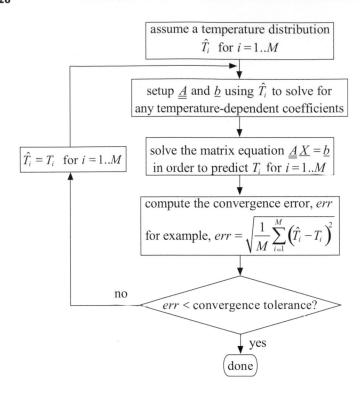

Figure 2.23 Successive substitution technique for solving problems with temperature-dependent properties using matrix decomposition.

each node are compared and used to compute an error; for example, the sum of the square of differences between the values $\hat{T}_i$ and T_i at every node. If the error is greater than some threshold or tolerance value then the process is repeated, this time using the solution T_i as the assumed temperature distribution $\hat{T}_i$ in order to calculate the coefficients of the matrix equation. The process continues until the error becomes sufficiently small, as shown schematically in Figure 2.23.

The successive substitution solution is implemented in MATLAB. The functions for thermal conductivity and absorption coefficient that implement Eqs. (2.190) and (2.191) are written as two separate m-files.

```
function [k] = k_g(T)
      k_o=1.5;                              %[W/m-K] conductivity at room temperature
      T_o=293.15;                           %[K] room temperature
      beta_k=0.02;                          %[W/m-K^2] temperature coefficient
      k=k_o+beta_k*(T-T_o);                 %[W/m-K] conductivity
end
```

```
function [alpha] = alpha_g(T)
      alpha_o=100;                          %[1/m] absorption coefficient at room temperature
      T_o=293.15;                           %[K] room temperature
      beta_alpha=2;                         %[1/m-K] temperature coefficient
      alpha=alpha_o+beta_alpha*(T-T_o);     %[1/m] absorption coefficient
end
```

The nonlinear finite difference equations, Eq. (2.189), are rearranged to make it clear what the coefficients and constants are. Notice that the temperature-dependent properties are evaluated using the known guess temperatures, $\hat{T}_i$, rather than the unknown nodal temperatures, T_i:

$$T_1 \underbrace{\left[-\bar{h} A_c - \frac{k_{T=(\hat{T}_1+\hat{T}_2)/2} A_c}{\Delta x} \right]}_{A_{1,1}} + T_2 \underbrace{\left[\frac{k_{T=(\hat{T}_1+\hat{T}_2)/2} A_c}{\Delta x} \right]}_{A_{1,2}} = \underbrace{-\dot{q}''_{rad} \alpha_{T=\hat{T}_1} \exp\left(-\alpha_{T=\hat{T}_1} x_1 \right) A_c \frac{\Delta x}{2} - \bar{h} A_c T_\infty}_{b_1}$$

$$\text{(2.192)}$$

$$T_i \underbrace{\left[-\frac{k_{T=(\hat{T}_{i-1}+\hat{T}_i)/2} A_c}{\Delta x} - \frac{k_{T=(\hat{T}_{i+1}+\hat{T}_i)/2} A_c}{\Delta x} \right]}_{A_{i,i}} + T_{i-1} \underbrace{\left[\frac{k_{T=(\hat{T}_{i-1}+\hat{T}_i)/2} A_c}{\Delta x} \right]}_{A_{i,i-1}} + T_{i+1} \underbrace{\left[\frac{k_{T=(\hat{T}_{i+1}+\hat{T}_i)/2} A_c}{\Delta x} \right]}_{A_{i,i+1}}$$

$$= \underbrace{-\dot{q}''_{rad} \alpha_{T=\hat{T}_i} \exp\left(-\alpha_{T=\hat{T}_i} x_i \right) A_c \Delta x}_{b_i} \qquad \text{for } i = 2 \ldots (M-1) \qquad \text{(2.193)}$$

$$T_M \underbrace{\left[-\bar{h} A_c - \frac{k_{T=(\hat{T}_M+\hat{T}_{M-1})/2} A_c}{\Delta x} \right]}_{A_{M,M}} + T_{M-1} \underbrace{\left[\frac{k_{T=(\hat{T}_M+\hat{T}_{M-1})/2} A_c}{\Delta x} \right]}_{A_{M,M-1}}$$

$$= \underbrace{-\bar{h} A_c T_\infty - \dot{q}''_{rad} \alpha_{T=\hat{T}_M} \exp\left(-\alpha_{T=\hat{T}_M} x_M \right) A_c \frac{\Delta x}{2}}_{b_M}. \qquad \text{(2.194)}$$

The MATLAB code below is saved as the script window_TD.m and carries out the iteration required to solve Eqs. (2.192) through (2.194).

```
clear;                      %clear the workspace

%Inputs
L=0.01;                     %[m] lens thickness
qf_dot_rad=1000;            %[W/m^2] incident energy flux
h_bar=20;                   %[W/m^2-K] average heat transfer coefficient
T_infinity=293.2;           %[K] ambient air temperature
A_c=1;                      %[m^2] do problem on a per unit area basis

%Setup nodes
M=11;                       %[-] number of nodes
DELTAx=L/(M-1);             %[m] distance between adjacent nodes
for i=1:M
    x(i)=L*(i-1)/(M-1);     %[m] axial position of each node
end

%Setup A and b
A=zeros(M,M);
b=zeros(M,1);

That=T_infinity*ones(M,1);  %initial guess for temperatures
err=1;                      %[K] initial value of error
tol=1e-6;                   %[K] convergence criteria

while(err>tol)
    %Energy balance for control volume 1
    A(1,1)=-h_bar*A_c-k_g((That(1)+That(2))/2)*A_c/DELTAx;
```

```
      A(1,2)=k_g((That(1)+That(2))/2)*A_c/DELTAx;
      b(1)=-qf_dot_rad*alpha_g(That(1))*exp(-alpha_g(That(1))*x(1))...
        *A_c*DELTAx/2-h_bar*A_c*T_infinity;

      %Energy balances for internal control volumes
      for i=2:(M-1)
          A(i,i)=-k_g((That(i-1)+That(i))/2)*A_c/DELTAx-k_g((That(i+1)+That(i))/2)*A_c/DELTAx;
          A(i,i-1)=k_g((That(i-1)+That(i))/2)*A_c/DELTAx;
          A(i,i+1)=k_g((That(i+1)+That(i))/2)*A_c/DELTAx;
          b(i)=-qf_dot_rad*alpha_g(That(i))*exp(-alpha_g(That(i))*x(i))*A_c*DELTAx;
      end

      %Energy balance for control volume M
      A(M,M)=-h_bar*A_c-k_g((That(M-1)+That(M))/2)*A_c/DELTAx;
      A(M,M-1)=k_g((That(M-1)+That(M))/2)*A_c/DELTAx;
      b(M)=-h_bar*A_c*T_infinity-qf_dot_rad*alpha_g(That(M))*...
        exp(-alpha_g(That(M))*x(M))*A_c*DELTAx/2;

      %Solve for unknowns
      X=A\b;            %solve for unknown vector, X
      T=X;              %assign temperatures from X
      err=sqrt(sum((That-T).^2)/M)
      That=T;
  end
```

Executing the code causes the value of variable err, which is convergence error, to be displayed after each iteration because the equation for err is not terminated in a semicolon.

```
>> window_TD
err =
      16.1375
err =
      2.5451
err =
      0.3335
err =
      0.0425
...
err =
      8.7114e-05
err =
      1.1062e-05
err =
      1.4046e-06
err =
      1.7837e-07
```

The temperatures predicted using this script are identical to those shown in Figure 2.22.

Implementation using Gauss–Seidel Iteration

Modification of the Gauss–Seidel iteration technique discussed in Section 2.4.5 to deal with temperature-dependent properties is straightforward. The iteration equations are shown below:

$$
T_1 = \frac{\dfrac{k_{T=(T_1+T_2)/2}\,A_c}{\Delta x}\,T_2 + \dot{q}''_{rad}\,\alpha_{T=T_1}\exp\left(-\alpha_{T=T_1}x_1\right)A_c\dfrac{\Delta x}{2} + \bar{h}A_c T_\infty}{\left(\bar{h}A_c + \dfrac{k_{T=(T_1+T_2)/2}\,A_c}{\Delta x}\right)}
\tag{2.195}
$$

$$
T_i = \frac{\dfrac{k_{T=(T_{i-1}+T_i)/2}\,A_c}{\Delta x}\,T_{i-1}^k + \dfrac{k_{T=(T_{i+1}+T_i)/2}\,A_c}{\Delta x}\,T_{i+1}^k + \dot{q}''_{rad}\,\alpha_{T=T_i}\exp\left(-\alpha_{T=T_i}x_i\right)A_c\Delta x}{\dfrac{k_{T=(T_{i-1}+T_i)/2}\,A_c}{\Delta x} + \dfrac{k_{T=(T_{i+1}+T_i)/2}\,A_c}{\Delta x}}
\tag{2.196}
$$

$$
\text{for } i = 2 \ldots (M-1)
$$

$$
T_M = \frac{\dfrac{k_{T=(T_M+T_{M+1})/2}\,A_c}{\Delta x}\,T_{M-1} + \bar{h}A_c T_\infty + \dot{q}''_{rad}\,\alpha_{T=T_M}\exp\left(-\alpha_{T=T_M}x_M\right)A_c\dfrac{\Delta x}{2}}{\left(\bar{h}A_c + \dfrac{k_{T=(T_M+T_{M+1})/2}\,A_c}{\Delta x}\right)}.
\tag{2.197}
$$

The script window_GS_TD.m, shown below, implements the iteration equations provided by Eqs. (2.195) through (2.197) in MATLAB. It requires use of functions k_g and alpha_g.

```
clear;                          %clear the workspace

%Inputs
L=0.01;                         %[m] lens thickness
qf_dot_rad=1000;                %[W/m^2] incident energy flux
h_bar=20;                       %[W/m^2-K] average heat transfer coefficient
T_infinity=293.2;               %[K] ambient air temperature
A_c=1;                          %[m^2] do problem on a per unit area basis

%Setup nodes
M=11;                           %[-] number of nodes
DELTAx=L/(M-1);                 %[m] distance between adjacent nodes
for i=1:M
    x(i)=L*(i-1)/(M-1);         %[m] axial position of each node
end
for i=1:M
    T_k(i)=T_infinity;          %[K] initial guess temperature
end
T_kp=T_k;                       %start with the last iteration temperatures
tol=1e-6;                       %[K] convergence tolerance
err=tol+1;                      %initial value of convergence error
while(err>tol)
    %compute new temperatures
    T_kp(1)=(k_g((T_kp(2)+T_kp(1))/2)*A_c*T_kp(2)/DELTAx+qf_dot_rad*...
        alpha_g(T_kp(1))*exp(-alpha_g(T_kp(1))*x(1))*A_c*DELTAx/2+...
        h_bar*A_c*T_infinity)/(h_bar*A_c+k_g((T_kp(2)+T_kp(1))/2)*A_c/DELTAx);
```

```
for  i=2:(M-1)
    T_kp(i)=(k_g((T_kp(i-1)+T_kp(i))/2)*A_c*T_kp(i-1)/DELTAx+...
        k_g((T_kp(i+1)+T_kp(i))/2)*A_c*T_kp(i+1)/DELTAx+qf_dot_rad*...
        alpha_g(T_kp(i))*exp(-alpha_g(T_kp(i))*x(i))*A_c*DELTAx)/...
        (k_g((T_kp(i-1)+T_kp(i))/2)*A_c/DELTAx+...
        k_g((T_kp(i+1)+T_kp(i))/2)*A_c/DELTAx);
end
T_kp(M)=(k_g((T_kp(M-1)+T_kp(M))/2)*A_c*T_kp(M-1)/DELTAx+qf_dot_rad*...
    alpha_g(T_kp(M))*exp(-alpha_g(T_kp(M))*x(M))*A_c*DELTAx/2+...
    h_bar*A_c*T_infinity)/(h_bar*A_c+...
    k_g((T_kp(M-1)+T_kp(M))/2)*A_c/DELTAx);
err=max(abs(T_kp-T_k))   %determine  the  convergence  error
T_k=T_kp;                %use  calculated  temperatures  for  next  iteration
end
```

Executing the script produces the same result shown in Figure 2.22. For any of the solution techniques discussed here with temperature-dependent properties, iteration is required and the iteration process must start from an initial set of guess temperatures. The selection of appropriate guess temperatures is critical to the success and speed of the iteration process. One approach for selecting appropriate guess values is to solve the problem using constant properties and use the resulting solution to start the iteration process.

2.5 Conclusions and Learning Objectives

This chapter presented both analytical and numerical methods for solving one-dimensional, steady-state conduction heat transfer problems. The solutions for a few useful situations were represented as resistance formulae that can be used to quickly solve and understand complex problems with multiple heat transfer mechanisms. This chapter also introduced the concept of thermal energy generation and showed how to deal with this effect in both analytical and numerical solutions.

Some specific concepts that you should understand are listed below.
- The definition of Fourier's Law and how it relates the conduction heat transfer rate to the temperature gradient.
- The fact that thermal conductivity is a material property.
- The relationship between thermal conductivity and the underlying energy carriers within a material that are collectively responsible for conduction heat transfer.
- The process of defining a differentially small control volume in order to derive a differential equation in a conduction problem.
- The importance of boundary conditions and the ability to derive boundary conditions using interface energy balances.
- The steps required to solve a separable, ordinary differential equation.
- The definition of a thermal resistance and the limitations of the concept.
- The thermal resistance due to conduction through various shapes including a plane wall, a cylindrical shell, and a spherical shell.
- The thermal resistances associated with convection, radiation, and contact.
- The ability to combine thermal resistances in order to make networks that represent a physical situation.
- The ability to use thermal resistances to estimate the relative importance of various effects in a heat transfer problem.
- The idea that thermal energy generation occurs volumetrically within a material and some mechanisms that give rise to this effect.
- The ability to incorporate volumetric thermal energy generation into an energy balance on a differential control volume in order to derive an appropriate ordinary differential equation.
- The ability to define differential control volumes using cylindrical and spherical shells.

- The ability to carry out energy balances on small but finite control volumes that are then coupled with rate equations in order to generate a system of equations for numerical models.
- The idea that numerical solutions are approximate because the rate equations used to develop them are only approximate.
- An understanding of several methods for solving the system of equations that result from a numerical model.
- The ability to assess the convergence of a numerical model.
- The ability to deal with temperature-dependent properties in a numerical model.

References

Bejan, A., *Heat Transfer*, Wiley, New York (1993).

Barron, R. and G. Nellis, *Cryogenic Heat Transfer*, Second Edition, Taylor and Francis, New York (2016).

Fried, E., *Thermal Conduction Contribution to Heat Transfer at Contacts*, in *Thermal Conductivity*, Volume 2, R. P. Tye, ed., Academic Press, London (1969).

Nellis, G. F. and S. A. Klein, *Heat Transfer*, Cambridge University Press, New York (2009).

Schneider, P. J., *Conduction*, in *Handbook of Heat Transfer*, Second Edition, W. M. Rohsenow *et al.*, eds., McGraw-Hill, New York (1985).

Problems

Conduction without Generation: Concepts and Analytical Solutions

2.1 The steady-state temperature distribution within a plane wall with a constant thermal conductivity is shown.

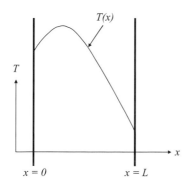

Circle the letter of the correct statement.
(a) The left-hand side of the wall is adiabatic.
(b) There is no internal generation of energy within the wall.
(c) At $x = 0$ (the left side of the wall) heat is being transferred out of the wall.
(d) At $x = L$ (the right side of the wall) heat is being transferred into the wall.

2.2 The thermal conductivity of air is 2–3 orders of magnitude less than that of water. Explain briefly why this is so.

2.3 A plane wall is a composite of a low-conductivity material (with thickness L_1 and conductivity k_1) and a high-conductivity material (with thickness $L_2 = L_1$ and conductivity $k_2 > k_1$). The edge of the wall at $x = 0$ is at a high temperature T_1 and the edge at $x = L_1 + L_2$ is at a low temperature T_2, as shown. The wall is at steady state and the temperature distribution in the wall is one-dimensional in x.

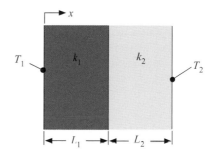

Sketch the heat flux (i.e., the heat transfer rate per unit area) as a function of x and the temperature as a function of x.

2.4 The surface temperature of a heat shield on a spacecraft that is re-entering the Earth's atmosphere is, at some instant in time, $T_s = 1000$ K. The surrounding air temperature can be approximated as being at $T_g = 1100$ K due to the kinetic energy of the spacecraft. The heat transfer coefficient between the surface of the spacecraft and the surroundings is $h = 10$ W/m²-K. The conductivity of the heat shield is $k = 1.0$ W/m-K and you may neglect radiation. The back surface of the heat shield is adiabatic and you may assume that the temperature distribution in the shield is 1-D in x.

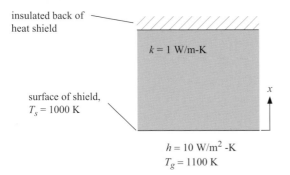

insulated back of heat shield

$k = 1$ W/m-K

surface of shield, $T_s = 1000$ K

x

$h = 10$ W/m² -K
$T_g = 1100$ K

Determine the conductive heat transfer per unit area in the positive x-direction at the surface of the heat shield and the temperature gradient in the heat shield material at the surface of the heat shield.

2.5 A composite wall is composed of two layers. Layer A has a thermal conductivity that is higher than layer B. The thermal conductivity in both layers is constant. For steady-state conduction of heat through the wall from left to right, with no internal generation of energy, which temperature distribution shown below is possible?

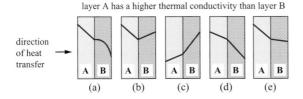

layer A has a higher thermal conductivity than layer B

direction of heat transfer

| A | B | | A | B | | A | B | | A | B | | A | B |

(a)   (b)   (c)   (d)   (e)

2.6 The temperature distribution for a shape can be assumed to be 1-D in the direction defined by the coordinate s. The problem is at steady state and the area available for conduction changes with s according to the function:

$$A(s) = \frac{0.01}{s}.$$

The temperatures of the two ends of the shape are specified; $T_H = 300$ K at $s_1 = 0.1$ m and $T_C = 200$ K at $s_2 = 0.2$ m.

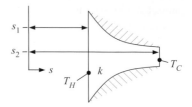

s_1
s_2
s
k
T_H
T_C

(a) Derive the governing differential equation for the problem.
(b) What are the boundary conditions for this problem?
(c) Solve the governing differential equation from (a); you should end up with a solution that involves two unknown constants of integration.
(d) Use the boundary conditions from (b) with the solution from (c) in order to obtain two equations in the two unknown constants.
(e) Enter the inputs for the problem and the equations from (d) into EES in order to evaluate the undetermined constants.
(f) Prepare a plot of the temperature as a function of position using EES.
(g) If the conductivity is $k = 10$ W/m-K then what is the rate of heat transfer through the shape?

2.7 The temperature distribution for the shape shown can be assumed to be 1-D in the direction defined by the coordinate s. The problem is at steady state and the area available for conduction changes with s according to the function:

$$A(s) = 0.01s^2.$$

The temperatures of the two ends of the shape are specified; $T_H = 300$ K at $s_1 = 0.02$ m and $T_C = 100$ K at $s_2 = 0.05$ m.

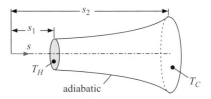

s_2
s_1
s
T_H
T_C
adiabatic

(a) Derive the governing differential equation for the problem.
(b) What are the boundary conditions for this problem?
(c) Solve the governing differential equation from (a) – you should end up with a solution that

involves two unknown constants of integration.

(d) Use the boundary conditions from (b) with the solution from (c) in order to obtain two equations in the two unknown constants.

(e) Enter the inputs for the problem and the equations from (d) into EES in order to evaluate the undetermined constants.

(f) Prepare a plot of the temperature as a function of position using EES.

(g) If the conductivity is $k = 1$ W/m-K then what is the rate of heat transfer through the shape?

2.8 The figure illustrates a plane wall made of a material with a temperature-dependent conductivity. The conductivity of the material is given by

$$k = bT,$$

where $b = 5$ W/m-K^2 and T is the temperature in K.

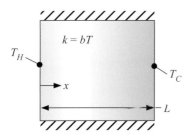

The thickness of the wall is $L = 0.5$ m. The left side of the wall (at $x = 0$) is maintained at $T_H = 300$ K and the right side (at $x = L$) is kept at $T_C = 20$ K. The problem is steady state and 1-D.

(a) Sketch the temperature distribution in the wall (i.e., sketch T as a function of x). Make sure that you get the qualitative features of your sketch right. Explain why your sketch looks the way it does.

(b) Derive the ordinary differential equation that governs this problem.

(c) What are the boundary conditions for this problem?

(d) Solve the governing differential equation from (b) – you should end up with a solution that involves two unknown constants of integration.

(e) Use the boundary conditions from (c) with the solution from (d) in order to obtain two equations in the two unknown constants.

(f) Enter the inputs for the problem and the equations from (e) into EES in order to evaluate the undetermined constants.

(g) Prepare a plot of the temperature as a function of position in the wall using EES. Compare it with your sketch from (a).

2.9 A plane wall is made of a material with a temperature-dependent conductivity. The conductivity of the material is given by:

$$k = bT^2,$$

where $b = 0.1$ W/m^2-K^3 and T is the temperature in K. The thickness of the wall is $L = 0.5$ m. The left side of the wall (at $x = 0$) is maintained at $T_H = 300$ K and the right side (at $x = L$) is kept at $T_C = 30$ K. The problem is steady state and 1-D.

(a) Sketch the temperature distribution in the wall (i.e., sketch T as a function of x). Make sure that you get the qualitative features of your sketch right.

(b) Derive the ordinary differential equation that governs this problem. Clearly show your steps. Please do not neglect the fact that conductivity is temperature dependent.

(c) What are the boundary conditions for this problem?

(d) Solve the governing differential equation from (b) – you should end up with a solution that involves two unknown constants of integration.

(e) Use the boundary conditions from (c) with the solution from (d) in order to obtain two equations in the two unknown constants.

(f) Enter the inputs for the problem and the equations from (e) into EES in order to evaluate the undetermined constants.

(g) Prepare a plot of the temperature as a function of position in the wall using EES.

2.10 The wall of a furnace is shown.

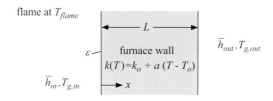

You may model the plane wall as being 1-D and at steady state. The conductivity is a function of temperature: $k(T) = k_o + a (T - T_o)$ where k_o is the conductivity at room temperature (T_o) and a is the temperature coefficient of conductivity.

The outer surface of the wall experiences only convection with gas at $T_{g,out}$ and with heat transfer coefficient $\bar{h}_{out}$. The inner surface experiences both radiation and convection. The radiation is with a flame at T_{flame} while the convection is with gas at $T_{g,in}$ and $\bar{h}_{in}$. The emissivity of the inner surface is ε and the thickness of the wall is L.

(a) Derive the differential equation for the temperature in the wall.
(b) Derive the boundary conditions for the problem.
(c) Solve the differential equation from (a). You should end up with a relationship (maybe an implicit one) for the temperature and temperature gradient in terms of position (x), symbols that are inputs to the problem, and two unknown constants. Don't forget that conductivity is a function of temperature.
(d) Use EES to determine the two constants of integration for the case where $T_{flame} = 1900$ K, $\varepsilon = 0.5$, $T_{g,in} = 800$ K, $\bar{h}_{in} = 65$ W/m²-K, $L = 8$ cm, $k_o = 0.62$ W/m-K, $T_o = 300$ K, $a = 0.0012$ W/m-K², $\bar{h}_{out} = 25$ W/m²-K, and $T_{g,out} = 300$ K.
(e) What is the heat transfer per unit area through the wall for the conditions in (d)?
(f) Plot the temperature as a function of position in the wall for the conditions in (d).
(g) Explain the shape of your plot from (f).

2.11 The figure shows a hollow sphere made of a material with a temperature-dependent conductivity. The conductivity of the material is given by:

$$k = bT,$$

where $b = 0.01$ W/m-K² and T is the temperature in K.

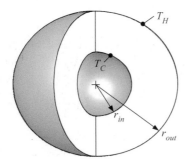

The sphere is used to contain liquid neon; therefore, the inner surface of the sphere at $r_{in} = 2$ cm

is held at $T_C = 30$ K. The outer surface of the sphere is exposed to ambient temperature; therefore, the temperature at $r_{out} = 5$ cm is 300 K.

(a) Sketch the temperature distribution in the sphere (i.e., sketch T as a function of r). Make sure that you get the qualitative features of your sketch right. Explain why your sketch looks the way it does.
(b) Derive the ordinary differential equation that governs this problem.
(c) What are the boundary conditions for this problem?
(d) Solve the governing differential equation from (b) - you should end up with a solution that involves two unknown constants of integration.
(e) Use the boundary conditions from (c) with the solution from (d) in order to obtain two equations in the two unknown constants.
(f) Type the inputs for the problem and the equations from (e) into EES in order to evaluate the undetermined constants.
(g) Prepare a plot of the temperature as a function of radius using EES. Compare it with your sketch from (a).
(h) Determine the rate of heat transfer to the liquid neon contained in the sphere.

Now that you have a model constructed you would like to optimize the outer radius of your liquid neon container. A larger outer radius should reduce the rate of heat transfer to the neon, thereby reducing the cost associated with buying liquid neon that is boiling away. On the other hand, a larger outer radius requires more material which is expensive. The material costs $C_m = \$250$/kg and has density $\rho_m = 7600$ kg/m³.

(i) Determine the cost of the material and plot the material cost as a function of outer radius (all other inputs held constant).

The system must operate 25 days per year and you are using a 5 year time frame for your analysis. The latent heat of fusion for neon is $i_{fg} = 81.73$ kJ/kg. The cost of liquid neon is $C_n = \$1.5$/liter and the density of liquid neon is $\rho_n = 1152$ kg/m³.

(j) Determine the cost of the liquid neon that boils off during the 5 years of operation and plot the neon cost as a function of outer radius (all other inputs held constant).

(k) Plot the total cost of the system (your answers from (i) and (j) added together) as a function of the outer radius. You should see an optimal outer radius exists that minimizes the cost.

2.12 The figure shows a washer with a triangular cross-section.

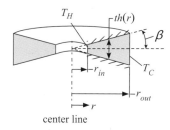

center line

The washer becomes thicker as it extends radially outwards. The angle associated with the taper is β. The inner surface of the washer is at r_{in} and is maintained at temperature T_H. The outer surface is at r_{out} and is maintained at temperature T_C. Assume that the temperature distribution is 1-D (i.e., temperature is only a function of radius) and the problem is steady state. The conductivity of the washer, k, is constant. The upper and lower surfaces of the washer are adiabatic; therefore, energy is conducted radially through the washer from r_{in} to r_{out}.

(a) Develop an expression for the thickness of the washer as a function of radius.

(b) Develop an expression for the conduction area in the radial direction as a function of radius.

(c) Derive the ordinary differential equation that governs the temperature distribution within the washer. Clearly show your steps.

(d) Solve the ODE from (c) in order to obtain a general solution that involves two constants of integration.

(e) Apply boundary conditions in order to obtain an expression for the two constants.

(f) Determine an expression for the rate of heat transfer through the washer.

(g) Determine an expression for the thermal resistance of the washer to radial conduction in terms of the parameters given in the problem statement.

2.13 A bracket is used to hold a cryogenic tank in place within a vacuum space. One end of the bracket is held at a cold temperature, T_C, while the other end is held at near room temperature, T_H. The length of the bracket is L. The thickness of the bracket

increases linearly with x so that the cross-sectional area for conduction is given by:

$$A_c = W(th_C + \beta x),$$

where th_C is the thickness at the cold end ($x = 0$), W is the width of the bracket (into the page), and β is the rate at which the thickness increases.

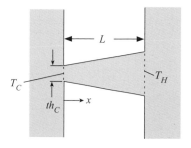

Assume that the temperature in the bracket is one-dimensional and the bracket is at steady state so that $T(x)$. Further, assume that there is no heat transfer from the outer surfaces of the bracket exposed to the vacuum and that the conductivity of the bracket material, k, is constant.

(a) Sketch the temperature distribution that you expect in the bracket (i.e., sketch T as a function of x). Make sure that you get the qualitative features of your sketch right.

(b) Derive the ordinary differential equation that governs this problem.

(c) What are the boundary conditions for this problem?

(d) Solve the governing differential equation from (b) – you should end up with a solution that involves two unknown constants of integration.

(e) Use the boundary conditions from (c) with the solution from (d) in order to obtain two equations in the two unknown constants.

Implement your solution for the case where $T_C = 70$ K, $T_H = 300$ K, $L = 12$ cm, $th_C = 0.6$ cm, $W = 4$ cm, $\beta = 0.25$, and $k = 15.2$ W/m-K.

(f) Enter the inputs for the problem and the equations from (e) into EES in order to evaluate the undetermined constants.

(g) Prepare a plot of the temperature as a function of position using EES.

(h) Determine the rate of heat transfer through the bracket.

(i) Determine the thermal resistance associated with the bracket. Thermal resistance is defined as the ratio of the temperature difference applied across the bracket to the rate of heat transfer through the bracket.

(j) Plot the thermal resistance of the bracket as a function of β for several values of L. You should produce a single plot showing R as a function of β and there should be several, labeled lines on the plot corresponding to different values of L.

2.14 An insulating material has a thickness of 13 cm. When a temperature difference of 85°C is imposed on the material, a steady-state heat flux of 23 W/m² is measured. From this information, determine the thermal conductivity of the material.

2.15 A material has a temperature-dependent thermal conductivity that can be described by the equation $k = 36 + 0.047T$, where k is the thermal conductivity (in W/m-K) and T is temperature (in K). Determine the steady-state rate of heat transfer per unit area across a 0.1 m thick section of material in the case where the temperature is 165°C on one face and 38°C on the opposite face.

2.16 A furnace wall design requires that the heat transfer rate per unit area be no larger than 1950 W/m². The maximum temperature within the furnace is 1125°C. To avoid discomfort, the outside surface temperature can be no higher than 32°C. A high-temperature insulating material having a thermal conductivity of 0.10 W/m-K is used for the wall. Determine the thickness of the insulation required to meet these specifications.

Thermal Resistance Problems

2.17 During the cleaning cycle, the inside surface temperature of a conventional household oven can reach $T_{in} = 280°C$. The convective heat transfer coefficient between the outside surface of the oven and the air in the kitchen at $T_\infty = 20°C$ is expected to be $\bar{h} = 12$ W/m²-K. The insulating material used to make the oven walls has an average thermal conductivity of $k = 0.030$ W/m-K and thickness $L = 2$ cm. The emissivity of the outer surface of the wall is $\varepsilon = 0.42$. The oven wall outer surface radiates to surroundings at $T_{sur} = T_\infty$.

(a) Draw a resistance network that represents this problem.

(b) Determine the rate of heat transfer per unit area from the oven.

(c) Rank the resistances in terms of their overall importance to your result from (b).

(d) Determine the temperature of the outer surface of the oven.

(e) Prepare a plot showing the outer surface temperature as a function of the insulation thickness.

(f) Determine the minimum insulation thickness that will keep the outer surface of the oven below $T_{out} = 35°C$.

2.18 Refrigerant flows in a 1.9 inch outer diameter copper tube with a 0.281 inch wall thickness. The inside surface temperature of the tube is held at 5°F and the room temperature is 70°F. Determine the thickness of an insulating pipe covering (with $k_{ins} = 0.428$ Btu/hr-ft-F) that must be added in order to reduce the heat gain to the pipe by 25 percent compared to its uninsulated value assuming that the heat transfer coefficient on the outer surface is $\bar{h} = 25$ W/m²-K. Neglect radiation.

2.19 Two rods are made of the same material with the same thermal conductivity. They are insulated on the sides and have the same temperatures on either end, as shown. The diameter of rod B is twice that of rod A and the length of rod B is half that of rod A.

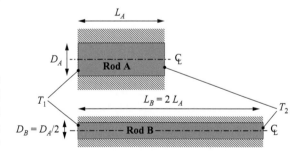

Circle the letter of the correct statement about the thermal resistance to conduction associated with the flow of heat through these rods.

(a) The thermal resistances of the two rods are equal ($R_B = R_A$).

(b) The thermal resistance of rod B is 2 times that of rod A ($R_B = 2R_A$).

(c) The thermal resistance of rod B is 4 times that of rod A ($R_B = 4R_A$).

(d) The thermal resistance of rod B is 8 times that of rod A ($R_B = 8R_A$).

(e) The thermal resistance of rod B is equal to the thermal resistance of rod A squared ($R_B = R_A^2$).

2.20 A cylindrical heater element is encased in a layer of ceramic. The outer surface of the ceramic both radiates and convects to its surroundings.

suface of ceramic experiences convection and radiation with surroundings
$R_{conv} = 10$ K/W
$R_{rad} = 100$ K/W

heat is generated by cylindrical heater

interface between the heater and the ceramic

heater is surrounded by ceramic
$R_{cond,cer} = 1.0$ K/W

(a) The thermal resistances have been calculated for you; the conduction resistance of the ceramic is $R_{cond,cer} = 1.0$ K/W, the resistance due to convection is $R_{conv} = 10$ K/W, and the resistance due to radiation is $R_{rad} = 100$ K/W. Which quantity listed below has the largest effect on the temperature at the interface between the cylindrical heater and the ceramic?

• The emissivity of ceramic surface.
• The heat transfer coefficient at the surface of the ceramic.
• The conductivity of the ceramic material.
• The conductivity of the heater material.

(b) Assume that the radius of the heater remains constant but the radius of the insulation is increased. Sketch how you expect $R_{cond,cer}$ and R_{conv} to change with the insulation radius.

2.21 A plane wall is composed of two materials, A and B. The interface between the materials is characterized by a contact resistance. The left surface of material A is held at $T_H = 400$ K and the right surface of material B radiates to surroundings at $T_C = 200$ K and is also exposed to convection to a fluid at T_C.

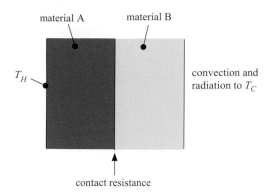

material A   material B

T_H

convection and radiation to T_C

contact resistance

The resistance network that represents the situation should include five thermal resistors; their values are provided below:

$R_{cond,A} = 0.1$ K/W, resistance to conduction through material A

$R_{contact} = 0.5$ K/W, contact resistance

$R_{cond,B} = 0.05$ K/W, resistance to conduction through material B

$R_{conv} = 1$ K/W, resistance to convection

$R_{rad} = 10.0$ K/W, resistance to radiation.

(a) Draw a resistance network that represents this situation. Each resistance in the network should be labeled according to $R_{cond,A}$, $R_{contact}$, $R_{cond,B}$, R_{conv}, and R_{rad}. Show where the temperatures T_H and T_C appear on your network.

(b) What is the most important resistor in the network? That is, the heat transfer from T_H to T_C is most sensitive to which of the five resistances?

(c) What is the least important resistor in the network?

(d) Determine the rate of heat transfer through the wall.

(e) Determine the temperature of the right surface of the wall.

(f) Sketch the temperature as a function of position in the wall. Make sure that your sketch is qualitatively consistent with the magnitude of the resistance values listed above.

2.22 Saturated steam at 110 kPa travels through an uninsulated pipe having an inner diameter of 12 cm and a length of 2.5 m. The heat transfer coefficient between the steam and the inner pipe wall is estimated to be 80 W/m²-K. The pipe is made of cast iron and the pipe wall is 0.0125 cm thick. The outside surface of the pipe has an emissivity of 0.7 and a heat transfer coefficient of 7.5 W/m²-K with the surrounding air. The air and surroundings are at 24°C.

(a) Draw a resistance network for this problem.

(b) Determine the surface temperature of the pipe.

(c) Determine the rate of steam condensation over the 2.5 m length in kg/s.

2.23 A flat polished copper plate having dimensions 0.2 m by 0.4 m is maintained at a uniform temperature of 42°C with electric heaters as air is blown across the top and bottom surfaces at 12 m/s. The electrical energy provided to the heaters is 55 W. The emissivity of copper is 0.025. The plate is in a laboratory that is maintained at 25°C. Determine the average convection coefficient at these conditions. If you are trying to use this setup to measure the heat transfer coefficient indicate why it is a good idea to polish the copper plate for this experiment.

2.24 A wall consists of two parallel sheets of 0.5 inch plywood (with $k = 0.12$ W/m-K) spaced 2.5 inches apart. The spacing is maintained with metal brackets that are very thin and occupy negligible volume. The space between the plywood sheets is filled with atmospheric air, which participates in heat transfer through free convection. Tests have shown that the air gap provides an effective conductivity of 5.4 W/m²-K. The heat transfer coefficient (including radiation and convection) on both outside surfaces of the plywood is 7.8 W/m²-K.

(a) What is the area-specific thermal resistance of this wall in units of m²-K/W?

(b) What is the thermal resistance of this wall after the air gap is filled with a foam ($k = 0.022$ W/m-K)?

(c) After curing, foam shrinks 5 percent resulting in a small air gap (5% of the 2.5 inch). Assuming that the resulting air gap has the same characteristics as the original air gap, determine the R-value of the wall after the foam is cured? What is the percentage reduction in R-value resulting from the curing?

2.25 An iron steam pipe carrying saturated steam at 2.5 bar has an outer diameter of 6 cm and is insulated with a magnesia insulation that has a thermal conductivity of 0.054 W/m-K. The heat transfer coefficient between the steam and the inner pipe surface is expected to be 2650 W/m²-K. The pipe

wall thickness is 0.5 cm. The area-specific contact resistance between the outer pipe surface and the insulation is estimated to be 2.8×10^{-4} K-m²/W. The insulation is exposed to air at 25°C and the heat transfer coefficient is 12 W/m²-K. Prepare plots of:

(a) the interface temperature between the outer pipe surface and the insulation, and

(b) the rate of heat loss from the pipe per unit length as a function of the insulation thickness, varying from 0 to 10 cm.

2.26 A composite wall is made of three layers. The first layer is an 8 inch thick fire-clay brick ($k = 0.6$ Btu/hr-ft-F), followed by a 6 inch thick layer of diatomaceous earth brick ($k = 0.16$ Btu/hr-ft-F) and finally a 4 inch thick outer layer of common brick ($k = 0.4$ Btu/hr-ft-F). The inside wall surface temperature is 1900°F and the outside wall surface is held at 300°F.

(a) Calculate the steady-state heat loss rate per square foot of wall area.

(b) Calculate the temperatures at the junctions between the different layers of brick.

(c) Carefully sketch the temperature distribution in the wall.

2.27 Spherical tanks are often used to store cryogenic fluids such as liquid nitrogen because this geometry provides the largest volume per unit surface area. One such tank consists of a stainless steel ($k = 11.9$ W/m-K) shell with an inside diameter of 0.675 m and a wall thickness of 6 mm. The shell contains liquid nitrogen at atmospheric pressure so the inside surface may be assumed to be at 77.4 K. The outer surface is insulated with 3.8 cm of insulation having a thermal conductivity of 0.032 W/m-K. The outside of the insulation is covered with a 2 mm thick shell of polished stainless steel. The convection coefficient on the outside surface is estimated to be 8.4 W/m²-K and the emissivity of the outside surface is 0.12.

(a) Determine the rate of heat transfer to the liquid nitrogen from a 25°C surrounding.

(b) Determine the temperature on the outside surface of the outer shell.

(c) Determine the rate of mass loss through the gas vent on the tank.

2.28 A 1-inch schedule 40 iron pipe carries 45°F water in surroundings that are at 78°F. The top half of the pipe is insulated with 1 inch of a foam insulation having a thermal conductivity of 0.045 W/m-K. The bottom half is inaccessible and consequently not insulated at all. The convection coefficient between the water and inside surface of the pipe is 18 Btu/hr-ft²-R. The outside surface convection coefficient (on both the top and bottom) is estimated to be 0.70 Btu/hr-ft²-R. Note that a schedule 40, 1 inch pipe has an inside diameter of 1.049 inches and an outside diameter of 1.315 in. The thermal conductivity of the iron is 47 Btu/hr-ft-R. Neglect radiation.

(a) Determine the rate of heat loss from the pipe per foot of length in Btu/hr-ft.

(b) Compare the result from part (a) to a totally uninsulated pipe.

(c) Compare the result from part (a) to a totally insulated pipe.

(d) Plot the heat transfer rate from the pipe (with insulation on top) as a function of insulation thickness for values of 0.01 to 4 inches.

2.29 A smooth (surface roughness = 10 µm) aluminum rod (k = 240 W/m-K) insulated on its outer surface is butted end-to-end with an identical, insulated aluminum rod with a clamping pressure of 100 kPa. Both rods are 1 cm in diameter and 0.25 m in length. The ends of the rods opposite of the butt joint are maintained at 118°C and 32°C, respectively, so that heat flows in the direction of the rod centerline. Determine the rate of heat transfer from the hot end to the cold end of the assembly. How significant is the resistance of the contact relative to the other thermal resistances?

2.30 A wall is composed of four materials as shown in the figure. The thermal conductivities of the materials are: k_A = 175 W/m-K, k_B = 36 W/m-K, k_C = 60 W/m-K and k_D = 80 W/m-K. The left face of the wall is maintained at 124°C while the right face is maintained at 36°C. Determine the rate of heat transfer per meter of width.

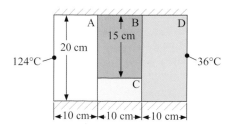

2.31 A spherical vessel contains a nuclear material that is steadily releasing thermal energy. The inside surface of the vessel has radius 2.5 cm and temperature 238°C. The vessel is thin and conductive and is surrounded by insulation with outer radius 8.5 cm and thermal conductivity 0.065 W/m-K. The insulation is surrounded by a metal jacket with a thermal conductivity of 18 W/m-K and outer radius 9 cm. The jacket is surrounded by 25°C air and the convection heat transfer coefficient between the surface and the air is 9.4 W/m²-K. Determine the rate at which thermal energy is released and the temperature on the outside surface of the sphere.

2.32 You have decided to install a strip heater under the linoleum in your bathroom in order to keep your feet warm on cold winter mornings. The bathroom is located on the first story of your house and is W = 2.5 m wide × L = 2.5 m long. The linoleum thickness is th_L = 5.0 mm and has conductivity k_L = 0.05 W/m-K. The strip heater under the linoleum is negligibly thin. Beneath the heater is a piece of plywood with thickness th_P = 0.25 inch and conductivity k_P = 0.4 W/m-K. The plywood is supported by th_s = 3.5 inch thick studs that are W_s = 1.75 inch wide with thermal conductivity k_s = 0.4 W/m-K. The center-to-center distance between studs is p_s = 18 inch. Between each stud are pockets of air that can be considered to be stagnant with conductivity k_a = 0.025 W/m-K. A sheet of drywall is nailed to the bottom of the studs. The thickness of the drywall is th_d = 0.63 inch and the conductivity of drywall is k_d = 0.1 W/m-K. The air above the linoleum in the bathroom is at $T_{air,1}$ = 15°C while the air in the basement below the drywall is at $T_{air,2}$ = 5°C. The heat transfer coefficient on both sides of the floor is $\bar{h}$ = 15 W/m²-K. You may neglect radiation and contact resistance for this problem.

(a) Draw a thermal resistance network that can be used to represent this situation. Be sure to label the temperatures of the air above and below the floor ($T_{air,1}$ and $T_{air,2}$), the temperature at the surface of the linoleum (T_L), the temperature of the strip heater (T_h), and the heat input to the strip heater ($\dot{q}_h$) on your diagram.

(b) Compute the value of each of the resistances from part (a).

(c) Determine the rate of heat transfer that must be supplied by the heater to raise the temperature of the surface of the floor to a comfortable 20°C?

(d) What physical quantities are most important to your analysis? What physical quantities are unimportant to your analysis?

(e) Discuss at least one technique that could be used to substantially reduce the amount of heater power required while still maintaining the floor at 20°C. Note that you have no control over $T_{air,1}$ or $\bar{h}$.

2.33 The figure illustrates the cross-section of the wall of a water tank.

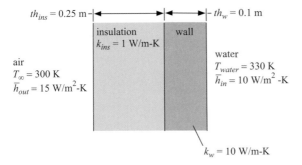

$th_{ins} = 0.25$ m | $th_w = 0.1$ m

insulation
$k_{ins} = 1$ W/m-K

wall

air
$T_\infty = 300$ K
$\bar{h}_{out} = 15$ W/m²-K

water
$T_{water} = 330$ K
$\bar{h}_{in} = 10$ W/m² -K

$k_w = 10$ W/m-K

You may treat the wall as a plane wall with cross-sectional area $A_c = 1$ m². The wall is $th_w = 0.1$ m thick and has conductivity $k_w = 10$ W/m-K. The wall is surrounded by a layer of insulation that is $th_{ins} = 0.25$ m thick and has conductivity $k_{ins} = 1$ W/m-K. The temperature of the water in the tank is $T_{water} = 330$ K and the heat transfer coefficient between the water and the wall is $\bar{h}_{in} = 10$ W/m²-K. The outer surface of the insulation is exposed to air at $T_\infty = 300$ K. The heat transfer coefficient between the air and the outer surface of the insulation is $\bar{h}_{out} = 15$ W/m²-K. You may neglect radiation from both surfaces and contact resistance between the wall and the insulation.

(a) Draw a resistance network that represents this problem. Label and calculate the value of each resistance.

(b) Determine the rate of heat transfer from the water to the air.

(c) Sketch the temperature distribution in the wall and the insulation. You do not have to be exact but your sketch should reflect the qualitative features that you expect based

on your calculations in parts (a) and (b). You may want to explain some of the features you are trying to draw.

(d) You neglected radiation from the surface exposed to the air – if the emissivity of this surface is $\varepsilon = 0.2$ determine whether this is a good assumption.

(e) You neglected contact resistance – if the area-specific contact resistance is $R_c'' = 0.001$ K-m²/W determine whether this is a good assumption.

In order to heat the water in the tank, a thin heater is installed between the wall and the insulation. The heater will provide a uniform heat flux at this interface. The heater power is adjusted until the temperature of the heater (i.e., the temperature at the interface between the insulation and the wall) is equal to $T_{htr} = 400$ K. The temperature of the water in the tank is unchanged.

(f) Determine the rate at which heat is added to the water.

(g) Determine the heater power.

(h) Sketch the temperature distribution in the wall and the insulation with the heater on. Again, you do not have to be exact but your sketch should reflect the qualitative features that you expect. You may want to explain some of the features you are trying to draw.

2.34 A plane wall is made of a thin ($th_w = 0.001$ m) and conductive ($k = 100$ W/m-K) material that separates two fluids, A and B. Fluid A is at $T_A = 100$°C and the heat transfer coefficient between the fluid and the wall is $\bar{h}_A = 10$ W/m²-K while fluid B is at $T_B = 0$°C with $\bar{h}_B = 100$ W/m²-K.

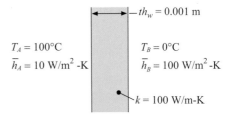

$th_w = 0.001$ m

$T_A = 100$°C
$\bar{h}_A = 10$ W/m² -K

$T_B = 0$°C
$\bar{h}_B = 100$ W/m² -K

$k = 100$ W/m-K

(a) Draw a resistance network that represents this situation and calculate the value of each resistor (assuming a unit area for the wall, $A = 1$ m²).

(b) If you wanted to predict the heat transfer rate from fluid A to fluid B very accurately, then which parameters (e.g., th_w, k, etc.)

would you try to understand/measure very carefully and which parameters are not very important? Justify your answer.

2.35 An electric burner for a stove is formed by taking a cylindrical piece of metal that is $D = 0.32$ inch in diameter and $L = 36$ inch long and winding it into a spiral shape. The burner consumes electrical power at a rate of $\dot{q} = 900$ W. The burner surface has an emissivity of $\varepsilon = 0.80$. The heat transfer coefficient between the burner and the surrounding air ($\bar{h}$) depends on the surface temperature of the burner (T_s) according to:

$$\bar{h} = 10.7\left[\frac{W}{m^2\text{-}k}\right] + 0.0048\left[\frac{W}{m^2\text{-}K^2}\right]T_s,$$

where $\bar{h}$ is in W/m²-K and T_s is in K. The surroundings and the surrounding air temperature are at $T_{sur} = 20°C$.
(a) Determine the steady-state surface temperature of the burner.
(b) Prepare a plot showing the surface temperature as a function of the burner input power.

2.36 A copper wire, 0.5 cm in diameter is insulated with a material having a thermal conductivity of 0.15 W/m-K. Because of electrical dissipation, the wire experiences a thermal generation of 20,000 W/m³. The wire is strung in 25°C ambient air and the effective heat transfer coefficient (including radiation and convection) between the air and insulation is 10.5 W/m²-K. The thermal conductivity of the copper is sufficiently high to assume that the copper is at uniform temperature.
(a) What is the rate of heat transfer per meter if the insulation thickness is 2 mm?
(b) What is the temperature of the copper for the conditions of part (a)?
(c) Is there an insulation thickness will cause the copper to be at a minimum temperature? If so, what is this thickness?

2.37 The figure shows the cross-section of the wall of a freezer.

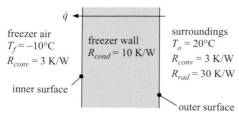

The inner surface of the wall experiences convection with the air in the freezer at $T_f = -10°C$. The outer surface of the wall experiences both convection and radiation with the surroundings at $T_o = 20°C$. The resistance to convection on the inside and outside of the wall is $R_{conv} = 3$ K/W. The resistance to radiation on the outer surface is $R_{rad} = 30$ K/W, you may neglect radiation from the inner surface. The resistance to conduction through the wall is $R_{cond} = 10$ K/W.
(a) Draw a resistance diagram that represents the freezer wall.
(b) Rank the following effects in terms of their importance (most important to least important): convection, radiation, conduction. Justify your answer.
(c) What is the rate of heat transfer to the freezer through the wall, $\dot{q}$?
(d) What is the temperature of the outer surface of the wall?

If the temperature of the outer surface of the wall is too low then condensation will form on the freezer. This is undesirable and therefore a thin heater is installed on the outer surface in order to maintain the temperature of the outer surface above the dew point temperature.

(e) What is the rate of heat transfer that must be provided by the heater in order to keep the temperature of the outer surface of the wall at $T_s = 18°C$?

2.38 In a residential heat pump, unit refrigerant at 0°F flows through 0.5 inch diameter thin-wall copper tubes. The heat transfer coefficient between the refrigerant and tube wall is high, due to the phase change. Air at 45°F is blown across the outside of the tubes. A layer of frost having thermal conductivity $k_f = 0.25$ Btu/hr-ft-F forms on the outside of the tubes with a thickness of 0.125 inch. Assuming that the convection coefficient between the outer surface and the air remains at 2.0 Btu/hr-ft²-F with or without the presence of the frost, determine the effect (expressed as a percentage up or down) that the frost will have on the heat transfer rate between the air and the refrigerant.

2.39 Double-glazed windows are used to decrease the rate of heat transfer between the inside of a house and outdoors. On a day in which the temperature is –10°F outdoors, the air inside of the building is maintained at 70°F. The average heat transfer coefficient on the inside and outside of the window surfaces are 3.0 and 7.5 Btu/hr-ft²-R, respectively. The effective conductance associated with the gap between the two window surfaces is 2.5 Btu/hr-ft²-R. (Radiative effects are included in these values.) The thermal conductivity of the glass is 0.45 Btu/hr-ft-F and each pane is 1/8 inch in thickness. Determine the rate of heat loss through a large window measuring 6 ft in height and 12 ft in width under these conditions.

2.40 The cross-section of a wall is shown in the figure. The wall separates the freezer air at $T_f = -10°C$ from air within the room at $T_r = 20°C$. The heat transfer coefficient between the freezer air and the inner wall of the freezer is $\bar{h}_f = 10$ W/m²-K and the heat transfer coefficient between the room air and the outer wall of the freezer is $\bar{h}_r = 10$ W/m²-K. The wall is composed of a $th_b = 1.0$ cm thick layer of fiberglass blanket sandwiched between two $th_w = 5.0$ mm sheets of stainless steel. The thermal conductivity of fiberglass and stainless steel are $k_b = 0.06$ W/m-K and $k_w = 15$ W/m-K, respectively. Assume that the cross-sectional area of the wall is $A_c = 1$ m². Neglect radiation from either the inner or outer walls.

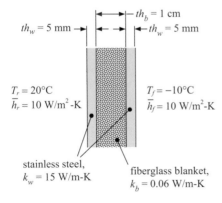

$th_b = 1$ cm

$th_w = 5$ mm → | ← | ← $th_w = 5$ mm

$T_r = 20°C$
$\bar{h}_r = 10$ W/m²-K

$T_f = -10°C$
$\bar{h}_f = 10$ W/m²-K

stainless steel,
$k_w = 15$ W/m-K

fiberglass blanket,
$k_b = 0.06$ W/m-K

(a) Draw a resistance network to illustrate this problem. Be sure to label the resistances in your network so that it is clear what each resistance is meant to represent.

(b) Enter all of the inputs in the problem into an EES program. Convert each input into the corresponding base SI units (i.e., m, kg, K, W, N, etc.) and set the units for each variable. Using comments, indicate what each variable means. Make sure that you set and

check the units of each variable that you use in the remainder of the solution process.

(c) Calculate the net rate of heat transfer to the freezer (W).

(d) Your boss wants to make a more energy-efficient freezer by reducing the rate of heat transfer to the freezer. He suggests that you increase the thickness of the stainless steel wall panels in order to accomplish this. Is this a good idea? Justify your answer briefly.

(e) Prepare a plot showing the rate of heat transfer to the freezer as a function of the thickness of the stainless steel walls. Prepare a second plot showing the rate of heat transfer to the freezer as a function of the thickness of the fiberglass. Make sure that your plots are clear (axes are labeled, etc.)

(f) What design change would you suggest in order to improve the energy efficiency of the freezer.

(g) One of your design requirements is that no condensation must form on the external surface of the freezer wall, even if the relative humidity in the room reaches 75 percent. This implies that the temperature of the external surface of the freezer wall must be greater than 15°C. Does your freezer wall satisfy this requirement? Calculate the external surface temperature (°C).

(h) In order to prevent condensation, you suggest placing a heater between the outer stainless steel wall and the fiberglass. What rate of heat transfer would be required to keep condensation from forming? Assume that the heater is very thin and conductive.

(i) Prepare a plot showing the transfer rate of heat required by the heater as a function of the freezer air temperature.

2.41 A temperature sensor is mounted in a pipe and used to measure the temperature of a flow of air.

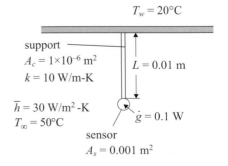

$T_w = 20°C$

support
$A_c = 1 \times 10^{-6}$ m²
$k = 10$ W/m-K

$L = 0.01$ m

$\bar{h} = 30$ W/m²-K
$T_\infty = 50°C$

sensor
$A_s = 0.001$ m²

$\dot{g} = 0.1$ W

The operation of the sensor leads to the dissipation of $\dot{g} = 0.1$ W of electrical power. This power is either convected to the air at $T_\infty = 50°C$ or conducted along the support to the wall at $T_w = 20°C$. Treat the support as conduction through a plane wall (i.e., neglect convection from the edges of the support). The heat transfer coefficient between the air and the sensor is $\bar{h} = 30$ W/m²-K. The surface area of the sensor exposed to air is $A_s = 0.001$ m². The support has cross-sectional area $A_c = 1 \times 10^{-6}$ m², length $L = 0.01$ m, and conductivity $k = 10$ W/m-K.

(a) What is the temperature of the temperature sensor?

(b) The sensor error is defined as the difference between the sensor and the air temperature. Is the error primarily due to self-heating of the sensor associated with $\dot{g}$ or due to the thermal communication between the sensor and the wall? Justify your answer.

(c) Radiation has been neglected for this problem. If the emissivity of the sensor surface is $\varepsilon = 0.02$, then assess whether radiation is truly negligible.

2.42 You have a problem with your house. Every spring at some time, the snow immediately adjacent to your roof melts and runs along the roof line until it reaches the gutter. The water in the gutter is exposed to air at temperature less than 0°C and therefore freezes, blocking the gutter and causing water to run into your attic.

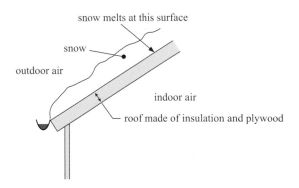

The air in the attic is at $T_{in} = 16°C$ and the heat transfer coefficient between the inside air and the inner surface of the roof is $\bar{h}_{in} = 7$ W/m²-K. The roof is composed of a $L_{ins} = 3.0$ inch thick piece of insulation with conductivity $k_{ins} = 0.035$ W/m-K that is sandwiched between two $L_p = 0.625$ inch thick pieces of plywood with conductivity

$k_p = 0.21$ W/m-K. There is an $L_s = 3.5$ inch thick layer of snow on the roof with conductivity $k_s = 0.08$ W/m-K. The heat transfer coefficient between the outside air at temperature $T_{out} = -15°C$ and the surface of the snow is $\bar{h}_{out} = 15$ W/m²-K. You may do this problem on a per unit area of roof basis, $A_c = 1$ m². Neglect radiation and contact resistances for this problem.

(a) Make a sketch of the situation.

(b) Draw a resistance network that represents the situation.

(c) Calculate the values of the resistances in your network.

(d) Calculate the temperature of the interface between the snow and the house. Will the snow melt?

(e) Plot the temperature of the interface between the snow and the house as a function of outdoor air temperature. At what outdoor air temperature should you start to worry about your gutters plugging up with ice?

2.43 The figure shows a cross-section of a thermal protection suit that is being designed for an astronaut.

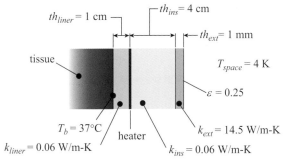

The suit consists of a liner that is immediately adjacent to the skin. The skin temperature is maintained at $T_b = 37°C$ by the flow of blood in the tissue. The liner is $th_{liner} = 1$ cm thick and has conductivity $k_{liner} = 0.06$ W/m-K. A thin heater is installed at the outer surface of the liner. Outside of the heater is a layer of insulation that is $th_{ins} = 4$ cm with conductivity $k_{ins} = 0.06$ W/m-K. Finally, the outer layer of the suit is $th_{ext} = 1$ mm thick with conductivity $k_{ext} = 14.5$ W/m-K. The outer surface of the external layer has emissivity $\varepsilon = 0.25$ and is exposed by radiation only to outer space at $T_{space} = 4$ K.

(a) You want to design the heater so that it completely eliminates any heat loss from the skin. What is the heat transfer to the heater per unit area required?

(b) In order, rate the importance of the following design parameters to your result from (a): k_{ins}, k_{ext}, and ε.

(c) Plot the heat transfer rate per unit area required to eliminate heat loss as a function of the emissivity, ε.

While the average emissivity of the suit's external surface is $\varepsilon = 0.25$, you have found that this value can change substantially based on how dirty or polished the suit is. You are worried about these local variations causing the astronaut discomfort due to local hot and cold spots.

(d) Assume that the heater power is kept at the value calculated in (a). Plot the rate of heat transfer rate from the skin as a function of the fractional change in the emissivity of the suit surface.

2.44 The figure shows a composite wall. The wall is composed of two materials (A with $k_A = 1$ W/m-K and B with $k_B = 5$ W/m-K), each has thickness $L = 1.0$ cm. The surface of the wall at $x = 0$ is perfectly insulated. A very thin heater is placed between the insulation and material A; the heating element provides $\dot{q}'' = 5000\,\text{W/m}^2$ of heat. The surface of the wall at $x = 2L$ is exposed to fluid at $T_{f,in} = 300$ K with heat transfer coefficient $\bar{h}_{in} = 100$ W/m²-K.

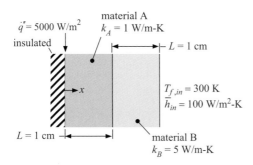

You may neglect radiation and contact resistance for parts (a) through (c) of this problem.

(a) Draw a resistance network to represent this problem; clearly indicate what each resistance represents and calculate the value of each resistance.

(b) Use your resistance network from (a) to determine the temperature of the heating element.

(c) Sketch the temperature distribution through the wall. Make sure that the sketch is consistent with your solution from (b).

The same composite wall is shown but with an additional layer added to the wall, material C with $k_C = 2.0$ W/m-K and $L = 1.0$ cm.

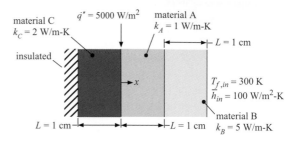

Neglect radiation and contact resistance for parts (d) through (f) of this problem.

(d) Draw a resistance network to represent the problem shown above; clearly indicate what each resistance represents and calculate the value of each resistance.

(e) Use your resistance network from (d) to determine the temperature of the heating element.

(f) Sketch the temperature distribution through the wall. Make sure that the sketch is consistent with your solution from (e).

The same composite wall is shown below but now there is a contact resistance between materials A and B, $R''_c = 0.01\,\text{K-m}^2/\text{W}$, and the surface of the wall at $x = -L$ is exposed to fluid at $T_{f,out} = 400$ K with a heat transfer coefficient $\bar{h}_{out} = 10$ W/m²-K.

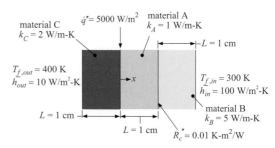

Neglect radiation for parts (g) through (i) of this problem.

(g) Draw a resistance network to represent the problem; clearly indicate what each resistance represents and calculate the value of each resistance.

(h) Use your resistance network from (g) to determine the temperature of the heating element.

(i) Sketch the temperature distribution through the wall.

2.45 The figure illustrates a cross-sectional view of a water heater.

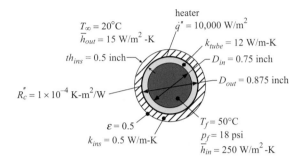

The water flows through a tube with inner diameter D_{in} = 0.75 inch and outer diameter D_{out} = 0.875 inch. The conductivity of the tube material is k_{tube} = 12 W/m-K. The water in the tube is at mean temperature T_f = 50°C and pressure p_f = 18 psi. The heat transfer coefficient between the water and the internal surface of the tube is $\bar{h}_{in}$ = 250 W/m²-K. A very thin heater is wrapped around the outer surface of the tube. The heater provides a heat transfer rate of $\dot{q}''$ = 10,000 W/m². Insulation is wrapped around the heater. The thickness of the insulation is th_{ins} = 0.5 inch and the conductivity is k_{ins}= 0.5 W/m-K. There is a contact resistance between the heater and the tube and the heater and the insulation. The area-specific contact resistance for both interfaces is R_c'' = 1 × 10⁻⁴ K-m²/W. The outer surface of the insulation radiates and convects to surroundings at T_∞ = 20°C. The heat transfer coefficient between the surface of the insulation and the air is $\bar{h}_{out}$ = 15 W/m²-K. The emissivity of the outer surface of the insulation is ε = 0.5.

(a) Draw a resistance network that represents this problem. Label each resistance and clearly indicate what it represents. Show where the heater power enters your network.

(b) Using EES, determine the temperature of the heater and the rate of heat transfer to the water per unit length of tube.

(c) What is the efficiency of the heater (the ratio of the power provided to the water to the power provided to the heater)?

(d) The efficiency of the heater is less than 100 percent due to heat loss to the atmosphere. Rank the following parameters in terms of their relative importance with respect to limiting heat loss to the atmosphere: ε, R_c'', k_{ins}, $\bar{h}_{out}$. Justify your answers using your resistance network and a discussion of the magnitude of the relevant resistances.

(e) Plot the efficiency as a function of the insulation thickness for 0 inch < th_{ins} < 1.5 inch. Explain the shape of your plot.

(f) The temperature on the internal surface of the tube must remain below the saturation temperature of the water in order to prevent any local boiling of the water. Based on this criterion, determine the maximum possible heat flux that can be applied to the heater (for th_{ins} = 0.5 inch).

2.46 You have designed a wall for a walk-in cooler. The wall separates the freezer air (at T_f = –10°C) from air within the room (at T_r = 20°C). The heat transfer coefficient between the freezer air and the inner wall of the freezer (h_f) is 10 W/m²-K and the heat transfer coefficient between the room air and the outer wall of the freezer (h_r) is also 10 W/m²-K. The wall is composed of a 1.0 cm thick layer of fiberglass blanket sandwiched between two 5.0 mm sheets of stainless steel. The thermal conductivity of fiberglass (k_b) and stainless steel (k_w) are 0.06 W/m-K and 15 W/m-K, respectively. Assume that the area of the wall is 1 m².

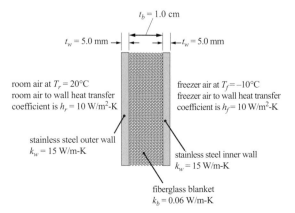

(a) Draw a resistance network to illustrate this problem. Be sure to label the resistances in your network so that it is clear what each is meant to represent. Calculate the value of each of the resistances in your network (K/W).

(b) Determine the rate of heat transfer from the room air to the freezer (W).

(c) Your boss wants to make a more energy-efficient freezer by reducing the rate of heat transfer to the freezer. He suggests that you double the thickness of the stainless steel wall panels from 5.0 mm to 10.0 mm. Is this a good idea? Justify your answer briefly.

(d) What design change to your wall would you suggest in order to improve the energy efficiency of the freezer.

(c) One of your design requirements is that no condensation must form on the external surface of your freezer wall, even if the relative humidity in the room reaches 75 percent. This implies that the external surface of the freezer wall must be greater than 15°C, as shown. Does your freezer wall satisfy this requirement? Calculate the external surface temperature (°C).

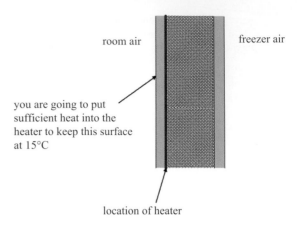

room air   freezer air

you are going to put sufficient heat into the heater to keep this surface at 15°C

location of heater

(e) An alternative strategy (to the heater) for preventing condensation is to use a fan to force the room air to flow along the outer wall, as shown in the figure. The impact of this is to increase the heat transfer coefficient between the outer wall and the room air (h_r). How high must the heat transfer coefficient be in order to prevent condensation (W/m²-K)? What is the impact of this approach on the power consumption of the freezer?

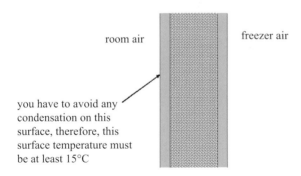

room air   freezer air

you have to avoid any condensation on this surface, therefore, this surface temperature must be at least 15°C

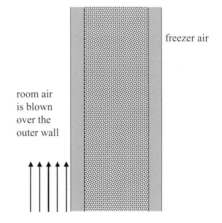

freezer air

room air is blown over the outer wall

(d) In order to prevent condensation, you decide to install a thin heater (i.e., neglect its thermal resistance) on the outer wall in order to keep the wall surface temperature at 15°C, as shown. How much heater power is required to prevent condensation (W)? By approximately how much (in percent) does the power consumed by your freezer increase or decrease due to the addition of the heater?

(f) For structural reasons, you must put some cross-members through the fiberglass blanket, as shown. The structural engineers inform you that 1 percent of the area of the wall must be occupied by these structural members and they would like some guidance on material selection.

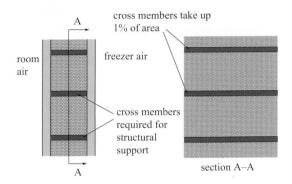

cross members take up 1% of area

A

room air

freezer air

cross members required for structural support

A

section A–A

Determine approximately a thermal conductivity value *above which* the power consumption of the freezer will be significantly negatively affected. Explain clearly the basis of your analysis.

2.47 An outside wall is constructed as shown in the figure. Wood studs are located 16 inches apart on center. Average thermal conductivities for the materials used in the wall are provided in the table.

Material	Thermal conductivity (W/m-K)
brick	0.72
plywood	0.12
insulation	0.034
sheet rock	0.22
wood stud	0.11

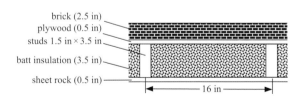

brick (2.5 in)
plywood (0.5 in)
studs 1.5 in × 3.5 in
batt insulation (3.5 in)
sheet rock (0.5 in)

16 in

The air in contact with the inside sheet rock surface is maintained at 70°F and the heat transfer coefficient (including radiation and convection) between the indoor air and wall surface is estimated to be 7.8 W/m²-K. The heat transfer coefficient between the brick and outdoor air is affected by wind speed; a conservative value is 12 W/m²-K.

(a) Draw a resistance network for this wall construction.

(b) Determine the rate of heat transfer for a wall that is 8 ft tall and 12 ft wide when the outdoor temperature is 0°F.

(c) In the United States, it is common to refer to the R-value of a wall to quantify its resistance to heat loss. The R-value is the resistance multiplied by the area and has units of hr-ft²-F/Btu. What is the R-value of this wall?

(d) What percent error would result if the effect of the studs was not considered in the estimate?

(e) Since the outdoor heat transfer coefficient is not well known, prepare a plot of the heat transfer rate versus the outdoor heat transfer coefficient for values ranging between 8 and 25 W/m²-K.

2.48 A composite material is formed from laminations of high-conductivity material (k_{high} = 100 W/m-K) and low-conductivity material (k_{low} = 1 W/m-K). Both laminations have the same thickness, *th*. A small section of the composite is shown in the figure.

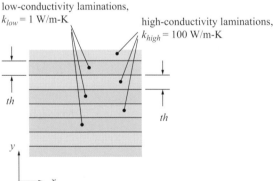

low-conductivity laminations, k_{low} = 1 W/m-K

high-conductivity laminations, k_{high} = 100 W/m-K

th

th

y

x

(a) Do you expect the equivalent conductivity of the composite to be higher in the *x*- or *y*-directions? Note that the *x*-direction is parallel to the laminations while the *y*-direction is perpendicular to the laminations.

(b) Estimate the equivalent conductivity of the composite in the *x*-direction. You should not need any calculations to come up with a good estimate for this quantity.

2.49 The ceiling of a residence consists of a 5/8 inch thick sheet of plaster board (k_p = 0.028 Btu/hr-ft-R) supported by wooden (k_w = 0.14 Btu/hr-ft-R) ceiling joists having dimensions of 1.625 inch by 5.5 inch spaced 24 inches apart, center to center. The space in between the joists is filled with R-24 insulation, i.e., its thermal resistance is 24 hr-R-ft²/Btu. Calculate the equivalent thermal resistance of the ceiling and compare it to the R-value of the insulation.

2.50 A pipe carries a flow of water at T_f = 6°C. The pipe is installed in a cold environment and therefore a layer of ice forms on the inner surface of the pipe. You need to estimate the layer of ice that will form at steady state. The surroundings are at T_{amb} = –20°C and the external surface of the pipe experiences both convection and radiation with the surroundings. The heat transfer coefficient between the surrounding air and the outer surface of the pipe is $\bar{h}_{out}$ = 150 W/m²-K and the emissivity of the pipe surface is ε = 0.82. The outer diameter of the pipe is D_p = 10 cm. The thickness of the pipe wall is th_p = 0.6 cm and the conductivity of the pipe material is k_p = 14 W/m-K. Initially, assume that the ice layer thickness is th_{ice} = 0.25 cm; we will eventually vary this value to determine the actual ice layer thickness that forms. The conductivity of the ice is k_{ice} = 2.1 W/m-K. The heat transfer coefficient between the inner surface of the ice layer and the water is $\bar{h}_{in}$ = 400 W/m²-K.

(a) Sketch the physical situation.
(b) Draw a resistance network that represents the situation. Label on your network the fluid temperature, the ambient air temperature, and the temperature of the ice/water interface. Note that the temperature of the ice/water interface must be equal to T_{freeze} = 0°C.
(c) Determine the rate of heat transfer per length of pipe from the water to the ice/water interface.
(d) Determine the rate of heat transfer per length of pipe from the ice/water interface to the surroundings.
(e) Based on your answers from (c) and (d), is the ice layer growing or shrinking when th_{ice} = 0.25 cm?
(f) Plot the answers from (c) and (d) as a function of the ice layer thickness. What is the steady-state ice layer thickness?

2.51 The cornea of an eye can be modeled as one-third of a spherical shell with inner radius r_i = 1.2 cm and thickness th_c = 3.2 mm. The conductivity of the cornea is k_c = 0.25 W/m-K. The inside of the cornea is filled with fluid at $T_{\infty,i}$ = 37°C. The heat transfer between the inner surface of the cornea and the fluid is $\bar{h}_i$ = 15 W/m²-K. On a cold day, the outside of the cornea is exposed to air at $T_{\infty,o}$ = 5°C. The heat transfer between the outer surface of the cornea and the air is $\bar{h}_o$ = 12 W/m²-K. Neglect radiation between the cornea and the surroundings.

(a) Determine the outer surface temperature of the cornea.

A contact lens can be modeled as an additional layer on the outside of the cornea that is th_l = 2.5 mm thick and has conductivity k_l = 0.85 W/m-K. The heat transfer coefficient between the surface of the lens and the surrounding air is unchanged. Neglect contact resistance between the cornea and the contact lens.

(b) Determine the outer surface temperature of the cornea with a contact lens.
(c) Your analysis from (a) and (b) should indicate that the temperature of the cornea actually decreases with a contact lens installed on the eye. Explain why this happens.
(d) We neglected contact resistance and radiation in our analysis; assess the validity of these assumptions.

2.52 A wall in a residence consists of an outer layer of common brick (k = 0.40 Btu/hr-ft-R) that is 4 inches thick, followed by 0.5 inch of sheathing (k = 0.028 Btu/hr-ft-R), a 3.75 inch air gap, and 0.5 inch of plaster (k = 0.43 Btu/hr-ft-R). The conductance of the air gap is 1.1 Btu/hr-ft²-R. The air inside of the house is maintained at 70°F. The convective heat transfer coefficients on the inside and outside surfaces of the wall are 1.5 and 4.5 Btu/hr-ft²-R, respectively.

(a) Calculate the overall thermal resistance of this wall for an area of 1 ft².
(b) Calculate the rate of heat transfer through a wall that is 8 ft tall and 14 ft wide if the outdoor temperature is 5°F.

2.53 A plane wall is shown in the figure. The left side of the wall (at x = 0) is insulated. There are two thin heaters installed in the wall. Heater 1 is at x = L = 30 cm and this heater provides a uniform heat flux $\dot{q}_1'' = 100$ W/m². Heater 2 is at $x = 2L$ and provides a uniform heat flux $\dot{q}_2'' = 1000$ W/m². The right side of the wall, at $x = 3L$, experiences both convection and radiation to $T_{sur} = 20°C$. The heat transfer coefficient is $\bar{h} = 10$ W/m²-K and the emissivity of the surface is $\varepsilon = 0.25$. The conductivity of the wall material is $k = 1.8$ W/m-K. There is no contact resistance in this problem and the wall is at steady state. You cannot neglect radiation for this problem.

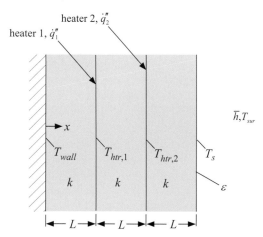

(a) Draw a resistance network that represents this problem. Clearly indicate on your network the locations where the heat transfers from heaters 1 and 2 enter as well as the location of the temperatures of heater 1 ($T_{htr,1}$ at $x = L$) and the temperature of heater 2 ($T_{htr,2}$ at $x = 2L$).

(b) Determine the temperatures of the two heaters.

(c) Sketch the temperature as a function of position.

2.54 A plane wall is shown.

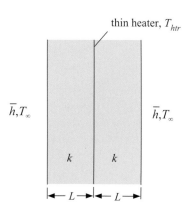

A thin heater is installed between two sections of material, each with thickness $L = 0.1$ m and conductivity $k = 0.5$ W/m-K. The cross-sectional area of the wall is $A = 1$ m². The two edges are cooled by convection to $T_\infty = 300$ K with heat transfer coefficient $\bar{h} = 10$ W/m²-K. There is no contact resistance in this problem and the wall is at steady state. You may neglect radiation for the initial part of this problem.

(a) Draw a resistance network that represents this problem. Clearly indicate on your network the locations where the heat transfer from the heater enters.

(b) Determine the rate of heat transfer required by the heater in order to reach a heater temperature of $T_{htr} = 400$ K.

(c) Determine the temperature of the surface of the wall.

(d) In parts (a), (b), and (c) you ignored radiation. If the emissivity of the surfaces is $\varepsilon = 0.1$ and both surfaces radiate to surroundings at $T_{sur} = 300$ K then evaluate whether this assumption is correct.

(e) Thus far you have also ignored contact resistance. If the area-specific contact resistance is $R_c'' = 1 \times 10^{-4}$ K-m²/W then evaluate whether this assumption is correct.

2.55 A photovoltaic solar panel is used to convert sunlight into electricity.

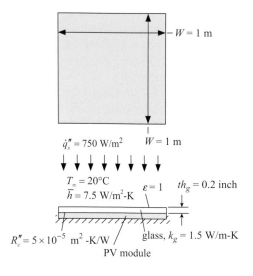

The panel is square with dimension $W = 1$ m. The incident solar flux is $\dot{q}_s'' = 750$ W/m². The

PV module is protected by a pane of glass that is $th_g = 0.2$ inch thick and has conductivity $k_g = 1.5$ W/m-K. The glass is transparent to the incident solar flux and therefore all of the incident solar flux is absorbed by the PV module. There is an area-specific contact resistance between the glass and the panel, $R_c'' = 5 \times 10^{-5}$ m²-K/W. The surface of the glass experiences convection and radiation with the surroundings at $T_\infty = 20°C$. The heat transfer coefficient is $\bar{h} = 7.5$ W/m²-K and the emissivity of the surface is $\varepsilon = 1.0$. The efficiency of the module is a decreasing function of its temperature:

$$\eta = \eta_o - \beta(T - T_o)$$

where $\eta_o = 0.15$ is the nominal efficiency of the module at the nominal temperature $T_o = 293.2$ K and $\beta = 0.002$ K^{-1} is the degradation coefficient. The backside of the module is well insulated.

(a) Draw a resistance network that represents the problem.

(b) Use your resistance network from (a) to estimate the cell temperature.

(c) Rank the following in terms of their importance to this problem (most important first, least important last): contact resistance, conduction through the glass, radiation, and convection. Justify your ranking.

(d) Plot the power produced by the solar cell as a function of the incident solar flux for a range of 0 W/m² (night) to 1000 W/m² (on the sunniest clear day of the year). You should see that there is a solar flux that maximizes the power produced; explain why this occurs.

(e) You can install a fan to help cool the solar panel. The effect of the fan is to increase the heat transfer coefficient to $\bar{h} = 35$ W/m²-K. However, the fan requires $\dot{w}_{fan} = 20$ W of electrical power to operate. Overlay on your plot from (d) the power produced as a function of solar flux with the fan installed. At what solar flux should you activate the fan?

2.56 The figure shows a very simple model of your skin on a very cold day in Wisconsin. The skin is composed of two layers. The inner layer is the dermis which has a thickness of $th_d = 3$ mm and conductivity $k_d = 0.32$ W/m-K.

The epidermis is the outer layer which has a thickness of $th_{ed} = 0.12$ mm and conductivity $k_{ed} = 0.14$ W/m-K. The inner surface of the dermis is exposed to blood flow and therefore remains at $T_{blood} = 34°C$. The outer surface of the epidermis is exposed to cold air at $T_\infty = -24°C$. The skin both radiates and convects to the surroundings at T_∞. The emissivity of the skin is $\varepsilon = 1$ and the heat transfer coefficient is $\bar{h} = 50$ W/m²-K.

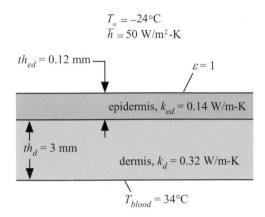

(a) Determine the temperature of the skin exposed to the air (in °C). Sketch the resistance network that you use for this calculation.

(b) Prepare a plot showing your skin temperature as a function of the heat transfer coefficient.

(c) You begin to feel cold-related discomfort when your skin temperature reaches approximately $T_{skin} = 25°C$. Determine the heat transfer coefficient (which is related to the wind speed) at which you experience cold-related discomfort.

(d) You are in danger of frost-bite when your skin temperature reaches approximately $T_{skin} = 10°C$. Determine the heat transfer coefficient at which you are in danger of frost-bite.

The figure illustrates a simple model of your skin on a very cold day in Wisconsin *after* you put gloves on. The thickness of the glove is $th_g = 5$ mm and the effective conductivity of the filler within the glove is $k_g = 0.14$ W/m-K. The outer surface of the glove has an emissivity of $\varepsilon = 1$.

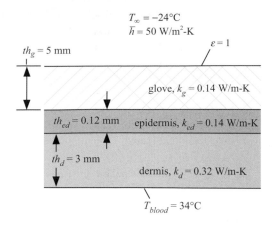

(e) Determine the temperature of the skin within the glove (in °C). Sketch the resistance network that you use for this calculation.

(f) Rank the thermal resistances in your resistance network in terms of their importance. Explain.

(g) Overlay on your plot from (b) the skin temperature with gloves as a function of heat transfer coefficient. At what heat transfer coefficient will you experience discomfort? Will you experience frostbite?

Finally, the figure below illustrates a simple model of your skin on a very cold day in Wisconsin after you put gloves on *and* insert a hand warmer. The hand warmer is a thin (assume negligibly thin) package filled with a chemical that reacts exothermically with air. The energy given off per unit area by the hand warmer is $\dot{g}'' = 250$ W/m².

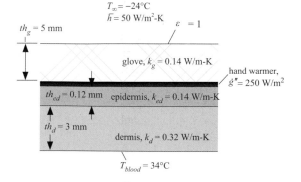

(h) Determine the temperature of the skin within the glove with the hand warmer (in °C). Sketch the resistance network that you use for this calculation.

(i) Overlay on your plot from (g) the skin temperature with gloves and hand warmer as a function of heat transfer coefficient. At what heat transfer coefficient will you experience discomfort? Will you ever experience frostbite?

2.57 You are a fan of ice fishing but don't enjoy the process of augering out your fishing hole in the ice. Therefore, you want to build a device, the "super ice-auger", that melts a hole in the ice.

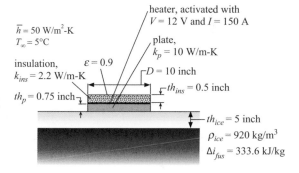

A heater is attached to the back of a $D = 10$ inch plate and electrically activated by your truck battery, which is capable of providing $V = 12$ V and $I = 150$ A. The plate is $th_p = 0.75$ inch thick and has conductivity $k_p = 10$ W/m-K. The back of the heater is insulated; the thickness of the insulation is $th_{ins} = 0.5$ inch and the insulation has conductivity $k_{ins} = 2.2$ W/m-K. The surface of the insulation experiences convection with surrounding air at $T_\infty = 5$°C and radiation with surroundings also at $T_\infty = 5$°C. The emissivity of the surface of the insulation is $\varepsilon = 0.9$ and the heat transfer coefficient between the surface and the air is $\bar{h} = 50$ W/m²-K. The super ice-auger is placed on the ice and activated, resulting in a heat transfer to the plate–ice interface that melts the ice. Assume that the water under the ice is at $T_{ice} = 0$°C so that no heat is conducted away from the plate–ice interface; all of the energy transferred to the plate–ice interface goes into melting the ice. The thickness of the ice is $th_{ice} = 5$ inch and the ice has density $\rho_{ice} = 920$ kg/m³.

The latent heat of fusion for the ice is $\Delta i_{fus} =$ 333.6 kJ/kg.

(a) Determine the heat transfer rate to the plate–ice interface.
(b) How long will it take to melt a hole in the ice?
(c) What is the efficiency of the melting process?
(d) If your battery is rated at 100 amp-hr at 12 V then what fraction of the battery's charge is depleted by running the super ice-auger.

2.58 Ice for an ice skating rink is formed by running refrigerant at $T_r = -30°C$ through a series of cast iron pipes that are embedded in concrete. The cast iron pipes have an outer diameter of $D_{o,p} = 4$ cm and an inner diameter of $D_{i,p} = 3$ cm. The pipes are spaced $L_{ptp} = 8.0$ cm apart. The heat transfer coefficient between the refrigerant and the pipe surface is $\bar{h}_r = 100$ W/m²-K. The concrete slab is $L_c = 8$ cm thick and the pipes are in the center of the slab. The bottom of the slab is insulated (assume perfectly). The thermal conductivity of concrete and iron are $k_c = 4.5$ W/m-K and $k_{iron} = 51$ W/m-K, respectively.

An $L_{fill} = 1$ cm thick layer of water is placed on the top of the concrete slab. The refrigerant cools the top of the slab and the water turns to ice slowly. Assume that the water is stagnant. The conductivity of ice and water are $k_{ice} = 2.2$ W/m-K and $k_w = 0.6$ W/m-K, respectively. The heat transfer coefficient between the top of the water layer and the surrounding air at $T_a = 15°C$ is $\bar{h}_a = 10.0$ W/m²-K. The top of the water surface has an emissivity of $\varepsilon = 0.90$ and radiates to surroundings at $T_{sur} = 15°C$.

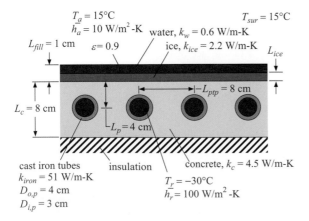

$T_a = 15°C$
$\bar{h}_a = 10$ W/m²-K   water, $k_w = 0.6$ W/m-K
$T_{sur} = 15°C$
$L_{fill} = 1$ cm $\varepsilon = 0.9$ ice, $k_{ice} = 2.2$ W/m-K L_{ice}
$L_c = 8$ cm
$|\!-L_{ptp} = 8$ cm
$L_p = 4$ cm
cast iron tubes   insulation   concrete, $k_c = 4.5$ W/m-K
$k_{iron} = 51$ W/m-K
$D_{o,p} = 4$ cm
$D_{i,p} = 3$ cm
$T_r = -30°C$
$\bar{h}_r = 100$ W/m² -K

(a) Draw a network that represents this situation using 1-D resistances. (Some of the resistances must be approximate since it is not possible to exactly calculate a 1-D resistance to the conduction heat flow in the concrete). Include an energy term related to the energy that is added to the system by the generation of ice. Clearly label the resistors.
(b) *Estimate* the magnitude of each of the resistances in your network when the ice is 0.5 cm thick (i.e., $L_{ice} = 0.5$ cm).
(c) Calculate the rate of change in the thickness of the ice when the ice thickness is 0.5 cm.
(d) Plot the rate of change in the thickness of the ice (in cm/hr) as a function of the ice thickness (in cm). Based on your plot – about how long do you expect it to take to freeze the ice?

Conduction with Generation: Concepts and Analytical Solutions

2.59 A plane wall is shown. Assume that the wall can be treated as being a one-dimensional, steady-state problem in which heat is transferred in the x-direction but not in the y- or z-directions.

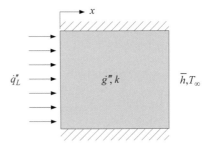

x
$\dot{q}''_L$ $\dot{g}''', k$ $\bar{h}, T_\infty$

The wall thickness is L and has conductivity k. The material experiences a uniform rate of volumetric thermal energy generation, $\dot{g}'''$. The edge of the wall at $x = 0$ experiences a heat flux to the wall of magnitude $\dot{q}''_L$ and the edge at $x = L$ experiences convection with surroundings at T_∞ and heat transfer coefficient $\bar{h}$. The differential equation and general solution for this problem can be found in Section 2.3.2.

(a) Write two equations that can be used to determine the constants C_1 and C_2 in the general solution for this situation. You do not need to solve your equations.
(b) Sketch the rate of conduction heat transfer and the temperature as a function of position. The sketch need not be exact, but you may want to justify some of the

characteristics of your sketch so that it is clear what you are thinking.

2.60 One of the engineers that you supervise has been asked to simulate the heat transfer problem shown.

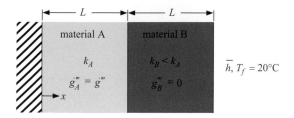

This is a 1-D, plane wall problem (i.e., the temperature varies only in the x-direction and the area for conduction is constant with x). Material A (from $0 < x < L$) has conductivity k_A and experiences a uniform rate of volumetric thermal energy generation, $\dot{g}'''$. The left side of material A (at $x = 0$) is completely insulated. Material B (from $L < x < 2L$) has *lower* conductivity, $k_B < k_A$. The right side of material B (at $x = 2L$) experiences convection with fluid at room temperature (20°C). Based on the facts above, critically examine the solution that has been provided to you by the engineer and is shown. There should be a few characteristics of the solution that do not agree with your knowledge of heat transfer; list as many of these characteristics as you can identify and provide a clear reason why you think the engineer's solution must be wrong.

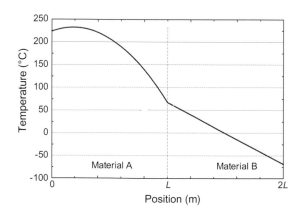

2.61 A composite plane wall is composed of two materials, A and B.

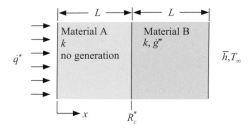

Materials A and B have the same thermal conductivity, k, and the same thickness, L. There is no generation of thermal energy in material A. Material B experiences a uniform generation of thermal energy, $\dot{g}'''$. The left side of material A is exposed to a heat flux, $\dot{q}''$. The right side of material B is cooled by convection with heat transfer coefficient $\bar{h}$ to fluid at T_∞. There is a contact resistance, R_c'', at the interface between materials A and B. This is a 1-D, steady-state problem.

(a) Sketch the temperature as a function of position. Your sketch should be qualitatively correct – you may want to justify some of the features in your sketch.

The general solution to the ordinary differential equation that applies within material B (i.e., in the computational domain $L < x < 2L$) can be found in Section 2.3.2.

(b) Provide two algebraic equations that can be used to solve for C_1 and C_2.

2.62 A composite plane wall is composed of two materials, A and B.

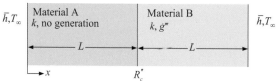

Materials A and B have the same thermal conductivity, k, and the same thickness, L. There is no generation of thermal energy in material A. Material B experiences a uniform generation of thermal energy, $\dot{g}'''$. The left side of material A and the right side of material B are both cooled by convection with heat transfer coefficient $\bar{h}$ to fluid at T_∞. There is a contact resistance, R_c'', at the interface between materials A and B. Sketch the steady-state temperature distribution from $x = 0$ to $x = 2\,L$. Be sure to get the qualitative features of your sketch correct. You can add some justification for your sketch.

2.63 A plane wall is composed of two materials, A and B, with the same thickness, L. The same, spatially uniform, volumetric rate of generation is present in both materials ($\dot{g}''' = \dot{g}_A''' = \dot{g}_B'''$) and the wall is at steady state. The conductivity of material B is twice that of material A ($k_B = 2\,k_A$). The left side of the wall is adiabatic and the right side is maintained at a temperature $T_{x=2L} = T_o$.

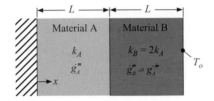

(a) Sketch the rate of heat transfer as a function of position within the wall from $x = 0$ (the left face of material A) to $x = 2L$ (the right face of material B). Note that the sketch should be qualitatively correct, but cannot be quantitative as you have not been given any numbers for the problem.

(b) Sketch the temperature as a function of position within the wall. Again, be sure that your sketch has the correct qualitative features.

2.64 A cylinder with conductivity k experiences a uniform rate of volumetric generation $\dot{g}'''$. The cylinder experiences 1-D, steady-state conduction heat transfer in the radial direction and therefore the general solution to the ordinary differential equation is given in Section 2.3.3. At the inner radius of the cylinder ($r = r_{in}$), a heater applies a uniform rate of heat transfer, $\dot{q}_{in}$. At the outer radius of the cylinder ($r = r_{out}$), the temperature is fixed at T_{out}. The length of the cylinder is L. Write the two algebraic equations that can be solved in order to obtain the constants C_1 and C_2. Your equations must contain only the following symbols from the problem statement: $\dot{q}_{in}$, T_{out}, k, r_{in}, r_{out}, L, $\dot{g}'''$, C_1, and C_2. Do not solve these equations.

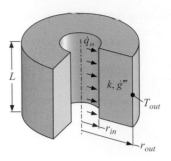

2.65 Coal powder is held in a large container. The thickness of the coal powder layer is $L = 2$ m. The coal powder interacts chemically with the interstitial air causing a uniform volumetric rate of energy generation $\dot{g}''' = 75$ W/m³. The effective thermal conductivity of the coal powder is $k = 1.5$ W/m-K. The heat transfer coefficient between the upper surface of the powder (at $x = L$) and the surrounding air is $\bar{h} = 5$ W/m²-K. The surrounding air is at $T_{amb} = 20°$C. The temperature of the lower surface of the coal powder (at $x = 0$) is equal to the ground temperature, $T_g = 22°$C.

(a) Sketch the physical situation.

(b) Develop an analytical solution for the temperature as a function of position in the coal powder.

(c) Implement your solution in EES and plot T vs. x.

(d) What is the maximum temperature in the coal?

(e) Plot the maximum temperature in the coal as a function of the coal powder layer thickness, L.

2.66 The rear window defroster in an automobile consists of uniformly spaced high-resistance wires molded into the glass. The glass is $L = 3/8$ inch thick with a thermal conductivity of $k = 0.94$ W/m-K. The heated portion of this window has an area of $H = 0.25$ m $\times W = 0.8$ m. When power is applied to the wires, thermal energy generation due to ohmic dissipation occurs. We are going to model this as occurring with a uniform volumetric rate in the glass, $\dot{g}'''$. This thermal energy is convected from both the interior and exterior surfaces of the window. One purpose of this problem is to estimate the total power required by the defroster to ensure that the window is defrosted. To ensure defrosting, the exterior surface of the window should be maintained at a minimum temperature of $T_{s,out} = 4°C$. The convection coefficients are expected to be $\bar{h}_{out} = 50$ W/m²K on the exterior and $\bar{h}_{in} = 7.0$ W/m²-K on the interior of the glass. Assume that the cabin air temperature is T_{cabin} 15°C and the outdoor temperature is $T_{outdoor} = -20°C$.

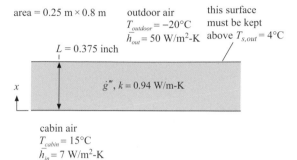

area = 0.25 m × 0.8 m

outdoor air
$T_{outdoor} = -20°C$
$\bar{h}_{out} = 50$ W/m²-K

this surface must be kept above $T_{s,out} = 4°C$

$L = 0.375$ inch

x

$\dot{g}''',\, k = 0.94$ W/m-K

cabin air
$T_{cabin} = 15°C$
$\bar{h}_{in} = 7$ W/m²-K

(a) Calculate and plot the steady-state temperature distribution in the glass.

(b) Determine the steady-state power required by the defroster.

2.67 A voltage difference of 50 volts is imposed on a 4.5 m long, 4.0 mm diameter stainless steel wire ($k = 15.2$ W/m-K). The electrical resistivity of stainless steel at these conditions is 0.72×10^{-6} ohm-m. The wire is uninsulated and submerged in water at 25°C with a convection coefficient of 750 W/m²-K.

(a) What is the rate of heat transfer from the wire?

(b) What is the surface temperature of the wire?

(c) What is the center temperature of the wire?

2.68 A current of 30 amp is passed through a bare stainless steel wire of 0.25 cm diameter. The thermal conductivity and electrical resistance per unit length of the wire are 15 W/m-K and 0.14 ohm/m, respectively. The wire is submerged in an oil that is maintained at 30°C. The steady-state temperature at the center of the wire is measured to be 180°C.

(a) Determine the temperature at the outer surface of the wire.

(b) Determine the convection coefficient between the wire surface and the oil.

(c) A plastic material having a thermal conductivity of 0.05 W/m-K can be used as insulation to the wire. Plot the center temperature of the wire as a function of the insulation thickness for thicknesses between 0 and

1 cm. Assume that the convection coefficient is the value from part (b).

2.69 A plane wall having a thickness of 5 cm generates thermal energy at a rate of 0.25 MW/m³. The outside surface of this wall is well insulated. The inside surface is exposed to a fluid that has a free stream temperature of 25°C. The heat transfer coefficient between the inside surface of the wall and the fluid is 470 W/m²-K. Prepare a plot of the temperature as a function of position for wall thermal conductivity values of 10, 20, and 200 W/m-K.

2.70 An uninsulated 1 inch schedule 40 iron pipe, 12 ft in length, is used in a residence to bring in water from the supply source. This pipe has an inner diameter of 1.049 inch, an outer diameter of 1.315 inch, and a thermal conductivity of 84 W/m-K. Owing to a power outage, the pipe was exposed to below freezing conditions and it now contains a mixture of ice and water at 0°C. A plumber plans to melt the ice by passing electric current through the pipe at a power level of 250 W. The convection coefficient between the inside surface of the pipe and the ice/water mixture is 24 W/m²-K. The convection coefficient between the outside surface of the pipe and the 18°C surroundings is 8.5 W/m²-K.

(a) Determine the temperatures of the inside and outside surface of the pipe.

(b) Determine the fraction of the electrical power that is transferred to the ice/water mixture.

2.71 An electrical current passes through a long copper wire having a 2 mm diameter. The wire both radiates and convects to its surrounding, which are at 25°C. The emissivity of the bare copper is 0.15 and the convection coefficient is 12 W/m²-K.

(a) Plot the surface temperature of the wire as a function of the current for values ranging from 0 to 40 amp.

(b) The wire is insulated with a plastic material having a thermal conductivity of 0.018 W/m-K. The emissivity of the outer surface of the insulation is 0.88 and the convection heat transfer coefficient on the outside of the insulation is the same value as for the bare wire. Plot the surface temperature of the copper and the temperature of the outer surface of the insulation as a function of the insulation thickness for values between 0.1 mm and 1 mm when the current is 20 amp. Assume perfect contact between the wire and the insulation and steady conditions.

2.72 A square extrusion of silicon with thermal conductivity 148 W/m-K and electrical resistivity of 2.73×10^{-8} m²-K/W is 4 inches in length and 0.8 inch on each side. A voltage of 200 volts is applied over the length of the bar, resulting in a uniform volumetric rate of thermal energy generation. The bar is well insulated except on its end surfaces, which are held by a 75°F clamp that has a contact resistance of 2.5×10^{-4} m²-K/W. Prepare a plot of the steady-state temperature of the silicon as a function of axial position.

2.73 A current of 100 amp passes through a bare stainless steel wire of $D = 1.0$ mm diameter. The thermal conductivity and electrical resistance per unit length of the wire are $k = 15$ W/m-K and $R'_e = 0.14$ ohm/m, respectively. The wire is submerged in an oil that is maintained at $T_\infty = 30°C$. The steady-state temperature at the center of the wire is measured to be 180°C, independent of axial position within the oil bath.
(a) What is the temperature at the outer surface of the wire?
(b) Estimate the convection coefficient between the submerged wire and the oil.

2.74 A cylindrical container is filled with hot coffee. In this problem we will assume that the container can be treated as being a 1-D steady-state problem in which heat is transferred radially through the sides of the container, but not from the top or bottom.

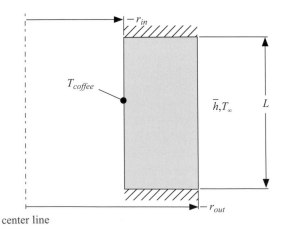

center line

The container is L long and it has inner and outer radii, r_{in} and r_{out}. The conductivity of the container material is k. The internal surface of the container can be assumed to be at the temperature of the coffee, T_{coffee}. The outer surface of the container experiences convection with surroundings at T_∞. The heat transfer coefficient at the external surface is $\bar{h}$. The container experiences a uniform rate of volumetric thermal energy generation in order to heat the coffee, $\dot{g}'''$. The ordinary differential equation and general solution that describes the temperature in the container material can be found in Section 2.3.3 of the text.
(a) Write two equations that can be used to determine the constants C_1 and C_2 in the general solution for this situation. You may neglect radiation from the external surface and you do not need to solve your equations.
(b) If the value of $\dot{g}'''$ is adjusted until the rate of heat loss from the coffee is zero (i.e., the coffee is neither being cooled nor heated) sketch the resulting temperature distribution that you expect. The sketch need not be exact, but you may want to justify some of the characteristics of your sketch so that it is clear what you are thinking.

2.75 The temperature distribution in a plane wall is 1-D and the problem is steady state.

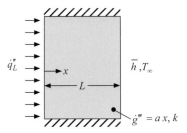

There is generation of thermal energy in the wall. The generation per unit volume is not uniform but rather depends on position according to: $\dot{g}''' = ax$ where a is a constant and x is position. The left side of the wall experiences a specified heat flux, $\dot{q}''_L$. The right side of the wall experiences convection with heat transfer coefficient $\bar{h}$ to fluid at temperature T_∞. The thickness of the wall is L and the conductivity of the wall material, k, is constant.
(a) Derive the ordinary differential equation that governs this problem. Clearly show your steps.

(b) Solve the differential equation that you obtained in (a). Your solution should include two undetermined constants.

(c) Specify the boundary conditions for the differential equation that you derived in (a).

(d) Use the results of (b) and (c) to obtain two equations that can be solved for the two undetermined constants.

2.76 A motor is constructed using windings that surround laminated iron poles. You have been asked to estimate the maximum temperature that will occur within the windings. The windings and poles are both approximated as being cylindrical, as shown.

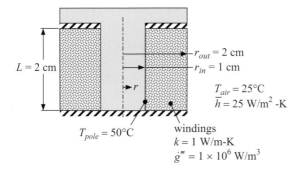

$r_{out} = 2$ cm
$r_{in} = 1$ cm
$L = 2$ cm

$T_{air} = 25°C$
$\bar{h} = 25$ W/m^2-K

$T_{pole} = 50°C$

windings
$k = 1$ W/m-K
$\dot{g}''' = 1 \times 10^6$ W/m^3

The windings are a complicated composite formed from copper conductor, insulation and air that fills the gaps between adjacent wires. However, the windings can be represented by a solid with equivalent properties that account for this underlying structure. The electrical current in the windings causes an ohmic dissipation that can be modeled as a uniform volumetric generation rate of $\dot{g}''' = 1 \times 10^6$ W/m^3. The conductivity of the windings is $k = 1.0$ W/m-K. The inner radius of the windings is $r_{in} = 1.0$ cm and the outer radius is $r_{out} = 2.0$ cm. The windings are $L = 2.0$ cm long and the upper and lower surfaces may be assumed to be insulated so that the temperature in the windings varies only in the radial direction. The stator pole is conductive and cooled externally; therefore, you can assume that the stator tooth has a uniform temperature of $T_{pole} = 50°C$. Neglect any contact resistance between the inner radius of the winding and the pole; therefore, the temperature of the windings at $r = r_{in}$ is T_{pole}. The outer radius of the windings is exposed to air at $T_{air} = 25°C$ with a heat transfer coefficient of $\bar{h} = 25$ W/m^2-K.

(a) Derive the governing differential equation for the temperature within the windings (i.e., the differential equation that is valid from $r_{in} < r < r_{out}$).

(b) Specify the boundary conditions for your differential equation.

(c) Solve the governing differential equation that you derived in part (a) by integrating twice. You should end up with a solution that involves two constants of integration, C_1 and C_2.

(d) Substitute your answer from part (c) into the boundary conditions specified in part (b) to obtain two equations for your two unknown constants of integration, C_1 and C_2.

(e) Implement your results from (c) and (d) in EES and prepare a plot of the temperature in the stator as a function of radius.

2.77 You need to transport water through a pipe from one building to another in an arctic environment. The water leaves the building very close to freezing, at $T_w = 5°C$, and is exposed to a high-velocity, very cold wind. The temperature of the surrounding air is $T_a = -35°C$ and the heat transfer coefficient between the outer surface of the pipe and the air is $\bar{h}_a = 50$ W/m^2-K. The pipe has an inner radius of $r_{h,in} = 2$ inch and an outer radius of $r_{h,out} = 4$ inch and is made of a material with a conductivity $k_h = 5$ W/m-K. The heat transfer coefficient between the water and the inside surface of the pipe is very large and therefore the inside surface of the pipe can be assumed to be at the water temperature. Neglect radiation from the external surface of the pipe.

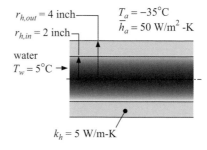

$r_{h,out} = 4$ inch
$r_{h,in} = 2$ inch

$T_a = -35°C$
$\bar{h}_a = 50$ W/m^2-K

water
$T_w = 5°C$

$k_h = 5$ W/m-K

(a) Determine the rate of heat lost from the water to the air for a unit length, $L = 1$ m, of pipe.

(b) Plot the rate of heat loss from the water as a function of the pipe outer radius from 0.06 m to 0.3 m (keep the same inner radius for this study). Explain the shape of your plot.

In order to reduce the heat loss from the water and therefore prevent freezing, you run current through the pipe material so that it generates thermal energy with a uniform volumetric rate, $\dot{g}''' = 5 \times 10^5$ W/m^3.

(c) Develop an analytical model capable of predicting the temperature distribution within the pipe. Implement your model in EES.

(d) Prepare a plot showing the temperature as a function of position within the pipe.

(e) Calculate the heat transfer rate from the water when you are heating the pipe.

(f) Determine the volumetric generation rate that is required so that there is no heat transferred from the water.

2.78 The heating element for the trim heater in a chemical reactor is shown in the figure.

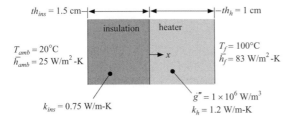

$th_{ins} = 1.5$ cm$\quad\quad\quad th_h = 1$ cm

insulation $\quad$ heater

$T_{amb} = 20°C$
$\bar{h}_{amb} = 25$ W/m^2-K

$T_f = 100°C$
$\bar{h}_f = 83$ W/m^2-K

$k_{ins} = 0.75$ W/m-K

$\dot{g}'' = 1 \times 10^6$ W/m^3
$k_h = 1.2$ W/m-K

The heater can be modeled as a plane wall with volumetric thermal energy generation (due to ohmic heating) of $\dot{g}''' = 1 \times 10^6$ W/m^3. The thickness of the heater is $th_h = 1$ cm and the conductivity of the heater material is $k_h = 1.2$ W/m-K. The edge of the heater at $x = th_h$ are exposed to the chemicals being heated at $T_f = 100°C$ with heat transfer coefficient $\bar{h}_f = 83$ W/m^2-K. The other edge of the heater (at $x = 0$) is insulated. The thickness of the insulation is $th_{ins} = 1.5$ cm and the insulation has conductivity $k_{ins} = 0.75$ W/m-K. There is no generation of thermal energy in the insulation. The insulation is exposed to ambient air at $T_{amb} = 20°C$ with heat transfer coefficient $\bar{h}_{amb} = 25$ W/m^2-K. The problem is steady state and 1-D. Ignore contact resistance and radiation.

(a) Develop an analytical model of the heating element. Implement your model in EES and plot the temperature as a function of position, x, within the heater (i.e., for $0 < x < th_h$).

(b) The efficiency of the heater is defined as the ratio of the heat transfer to the chemicals (i.e., the rate of heat transfer leaving the heater at $x = th_h$) to the total power provided to the heater. Determine the efficiency of the heater.

(c) Determine the maximum temperature in the heater.

(d) Overlay on one plot the maximum temperature in the heater and the efficiency of the heater as a function of the insulation thickness.

(e) The heater must provide 20 kW of heat transfer to the chemicals. Determine the required area of the heater (for $th_{ins} = 1.5$ cm).

(f) The chemical reactor operates 365 days a year 24 hours a day and the electrical energy required to run the heater costs $ec = 0.18$ \$/kW-hr. Determine the yearly operating cost (for $th_{ins} = 1.5$ cm).

(g) The insulation required for the heater is expensive: \$500 per cm of insulation thickness per m^2 of insulation area ($ic = 500 \$/cm-m^2). Determine the cost of insulation for the heater and the total cost associated with owning and operating the system for 1 year (for $th_{ins} = 1.5$ cm).

(h) Plot the total cost as a function of the insulation thickness. You should see that an economically optimal insulation thickness exists; explain why.

2.79 The inside wall of a nuclear reactor is exposed to radiation having a thermal conductivity of 25 W/m-K that is absorbed within the wall at rate $\dot{g}''' = \dot{g}_o'' \alpha \exp(-\alpha x)$ where $\dot{g}_o'' = $ 1e5 W/m^2, $\alpha = 0.25$ m^{-1} and x is the distance from the inside surface of the wall. The wall is 1.2 m thick. The inside surface of the wall (at $x = 0$) is exposed to a fluid at 250°C with a heat transfer coefficient of 140 W/m^2-K. The outside surface is maintained at 35°C.

(a) Derive a solution for the temperature of the wall as a function of position x.

(b) Prepare a plot of the temperature.

2.80 A plane wall experiences a volumetric generation of thermal energy.

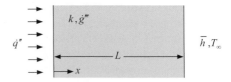

$k, \dot{g}'''$

q''

$\bar{h}, T_\infty$

L

x

One edge of the wall (at $x = 0$) is exposed to a heat flux, $\dot{q}''$ and the other edge (at $x = L$) is cooled by convection with fluid at T_∞ and heat transfer coefficient $\bar{h}$. The thickness of the wall is L and the wall material has thermal conductivity k. Assume that the temperature in the wall is a function only of x and the problem is steady-state.

(a) Without solving the problem, sketch the temperature as a function of position assuming that the volumetric rate of thermal energy generation is uniform (i.e., $\dot{g}'''$ is a constant). Justify any characteristics of your sketch that you think might help with grading.

For the remainder of this problem assume that the rate of volumetric generation is temperature dependent and increases linearly with temperature from its nominal value of $\dot{g}_o'''$ at T_o according to the equation

$$\dot{g}''' = \dot{g}_o''' + a(T - T_o).$$

(b) Derive the ordinary differential equation that governs the steady-state temperature distribution within the wall.

(c) Solve the ODE you derived in (b). Your solution should include two unknown constants, C_1 and C_2.

(d) Develop two equations for the two undetermined constants C_1 and C_2. You do not need to solve these equations.

2.81 A plane wall experiences a volumetric rate of thermal energy generation that depends on position according to:

$$\dot{g}'''(x) = G \sin\left(\frac{x\pi}{L}\right)$$

where L is the width of the wall and G is the amplitude of the generation (in W/m^3). The problem is one-dimensional and energy only moves in the x-direction. The edge of the wall at $x = 0$ experiences convection with heat transfer coefficient $\bar{h}$ and fluid temperature T_∞. The edge of the wall at $x = L$ is at a specified temperature, T_L.

(a) Derive the governing differential equation for the problem. Clearly show your steps.

(b) Derive the two boundary conditions for the problem.

(c) Solve the differential equation to obtain a general solution that includes two undetermined constants.

(d) Use your boundary conditions to develop two equations for the two undetermined constants.

2.82 An electrical conductor carrying a large amount of current must be cooled in order to avoid overheating. The figure illustrates a cylindrical conductor that is cooled on its inner and outer surfaces.

current carried by conductor, $I_c = 5000$ amp

conductor material
$k = 25$ W/m-K
$\rho_e = 1 \times 10^{-6}$ ohm-m

$T_f = 20°C$
$\bar{h}_{in}$

$T_f = 20°C$
$\bar{h}_{out} = 350$ W/m^2-K

$-r_{in} = 1$ cm

$-r_{out} = 3$ cm

center line

The current passing through the conductor (traveling in the axial direction) is $I_c = 5000$ amp. The electrical resistivity of the material is $\rho_e = 1 \times 10^{-6}$ ohm-m and the thermal conductivity of the material is $k = 25$ W/m-K. The outer radius of the conductor is $r_{out} = 3$ cm. The outer surface is cooled by convection to fluid at $T_f = 20°C$ with heat transfer coefficient $\bar{h}_{out} = 350$ W/m^2-K. The inner radius is $r_{in} = 1$ cm. The inner surface is also cooled by convection to fluid at T_f. The heat transfer coefficient on the inner surface depends on the inner radius according to

$$\bar{h}_{in} = \frac{B}{r_{in}}$$

where $B = 19.5$ W/m-K.

(a) Determine the volumetric rate of thermal energy generation in the conductor due to the ohmic dissipation.

(b) Develop an analytical model capable of predicting the temperature distribution in the conductor. Assume that the problem is one-dimensional (in r) and steady state. You may use EES to determine the constants of integration in the general solution. Plot the temperature as a function of radial position within the conductor.

(c) The current that can be carried by the conductor is limited by boiling; neither of the cooled surfaces (the inner or outer surface) can exceed $T_{boil} = 100°C$ or the coolant will begin to evaporate. Determine the maximum possible current that can be carried by the conductor.

(d) You would like to optimize the inner radius of the conductor in order to maximize its current carrying capability. Plot the maximum current that can be carried by the conductor (i.e., the current that leads to $T_{r=r_{out}} = T_{boil}$) as a function of r_{in}. You should see that there is an optimal value of the inner radius that maximizes the current carrying capacity.

2.83 You are designing a device to measure the thermal conductivity of a material. The device holds a cylindrical sample between two electrodes within a vacuum chamber. A voltage is applied between the ends of the sample causing current to flow through the sample which induces ohmic dissipation and causes the temperature of the sample to rise. The temperature difference between the center of the sample (at $x = L$) and the sample mount can be related to the conductivity of the sample.

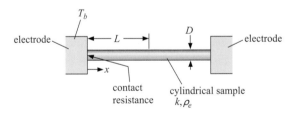

You are going to build a model of the measurement device so that we can understand how precisely it can be used to measure the thermal conductivity of a sample. Assume that the sample has diameter $D = 1.2$ mm and half-length (i.e., the distance from the electrode mount to the midpoint) $L = 1.67$ cm. The electrical resistivity (a material property) of the sample is $\rho_e = 1.43 \times 10^{-6}$ Ω-m. The voltage across the sample is $V = 50$ mV. The electrode mounts are maintained at $T_b = 300$ K. There is a thermal contact resistance between the sample and the electrode

characterized by an area-specific contact resistance of $R_c'' = 8.3 \times 10^{-5}$ K-m^2/W. The sample has conductivity $k = 14$ W/m-K.

This is a 1-D steady-state problem; the temperature in the sample varies only in the x-direction. We are neglecting radiation from the outer surface of the sample and the vacuum eliminates all convection.

(a) Determine the electrical resistance of the sample (in Ω).

(b) Determine the total ohmic dissipation induced in the sample (in W).

(c) Determine the volumetric generation of thermal energy induced in the sample (in W/m^3)

(d) Derive the ODE that governs the temperature distribution in the sample.

(e) Solve the ODE from (d) in order to obtain a general solution that includes two constants of integration.

(f) Derive the two boundary conditions required for the problem. Both should be based on interface energy balances (one at $x = 0$ and the other at $x = L$). Note that the center of the sample ($x = L$) is a line of symmetry. Also note that the temperature at $x = 0$ is not equal to the temperature of the mount (T_b) due to the presence of the contact resistance.

(g) Substitute the general solution from (e) into the boundary conditions from (f).

(h) Prepare a plot of temperature vs. position.

(i) For this sample, what is the measured value of temperature rise in the sample $\Delta T = T_{x=L} - T_b$?

Now that your model is working we can use it to assess the device as a measurement tool. In a real situation, you would measure the temperature rise induced in the sample and, using your model, determine the conductivity.

(j) Comment out the specified value of conductivity and instead specify a value of $\Delta T = 10$ K. What is the associated conductivity of the sample?

(k) Prepare a plot of thermal conductivity as a function of the measured temperature rise. One could use this plot to employ the instrument – measure the value of ΔT and then read the associated value of k off of the plot. (An alternative way of using the instrument is adjusting the voltage until a specified value of ΔT is achieved.)

Next let's see how well your device works in terms of the uncertainty of the measured conductivity. In order to measure thermal conductivity you have to measure or somehow know the following quantities: D, L, ρ_e, V, ΔT, and R_c''. You can't actually know or measure any of these quantities exactly; each has some uncertainty associated with it. The uncertainty in each of these quantities leads to some uncertainty in your measured value of k. It's important to keep track of this; fortunately EES does this automatically. The uncertainty in the measured diameter is $\delta D = 50$ μm, in the length is $\delta L = 100$ μm, in the value of the area-specific contact resistance is $\delta R_c'' / R_c'' = 35\%$, in the value of electrical resistivity is $\delta \rho_e / \rho_e = 5\%$, and in the voltage that is applied is $\delta V = 0.5$ mV. Your ability to measure the temperature difference is $\delta \Delta T = 0.25$ K. Let's see how these uncertainties affect the uncertainty in your measured value of k when $\Delta T = 10$ K.

(l) Select Uncertainty Propagation from the Calculate Window. Your calculated variable is k and your measured variables are D, L, ρ_e, V, ΔT, and R_c''. Select the Set Uncertainties button and set the uncertainties specified above. Select OK twice and EES will carry out an uncertainty analysis on your problem. What is the uncertainty in your measured value of k? If you wanted to improve your measurement (i.e., reduce δk) what would be the best way to go about it?

2.84 Freshly cut hay is not really dead; chemical reactions continue in the plant cells and therefore a small amount of heat is released within the hay bale. This is an example of the conversion of chemical to thermal energy. The amount of thermal energy generation within a hay bale depends on the moisture content of the hay when it is baled. Baled hay can become a fire hazard if the rate of volumetric generation is sufficiently high and the hay bale sufficiently large so that the interior temperature of the bale reaches 170°F, the temperature at which self-ignition can occur. Here, we will model a round hay bale that is wrapped in plastic to protect it from the rain. You may assume that the bale is at steady state and is sufficiently long that it can be treated as a 1-D radial conduction problem. The radius of the hay bale is $R_{bale} = 5$ ft and the bale is wrapped in plastic that is $t_p = 0.045$ inch

thick with conductivity $k_p = 0.15$ W/m-K. The bale is surrounded by air at $T_\infty = 20°C$ with $\bar{h} = 10$ W/m²-K. You may neglect radiation. The conductivity of the hay is $k = 0.04$ W/m-K.

(a) If the volumetric rate of thermal energy generation is constant and equal to $\dot{g}''' = 2$ W/m³ then determine the maximum temperature in the hay bale.

(b) Prepare a plot showing the maximum temperature in the hay bale as a function of the hay bale radius. How large can the hay bale be before there is a problem with self-ignition?

2.85 A spherical, nuclear fuel element consists of a sphere of fissionable material (fuel) with radius $r_{fuel} = 5$ cm and $k_{fuel} = 2$ W/m-K that is surrounded by a spherical shell of metal cladding with outer radius $r_{clad} = 7$ cm and $k_{clad} = 0.25$ W/m-K. The outer surface of the cladding is exposed to fluid that is being heated by the reactor. The convection coefficient between the fluid and the cladding surface is $\bar{h} = 50$ W/m²-K and the temperature of the fluid is $T_\infty = 500°C$. Neglect radiation heat transfer from the surface.

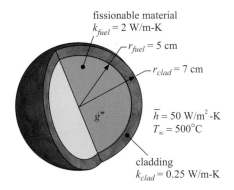

fissionable material
$k_{fuel} = 2$ W/m-K

$r_{fuel} = 5$ cm

$r_{clad} = 7$ cm

$\bar{h} = 50$ W/m²-K
$T_\infty = 500°C$

$\dot{g}'''$

cladding
$k_{clad} = 0.25$ W/m-K

Inside the fuel element, thermal energy is being generated for the reactor. This process can be modeled as a volumetric heat generation in the material that is not uniform throughout the fuel. The volumetric generation ($\dot{g}'''$) can be approximated by the function

$$\dot{g}''' = \frac{\beta}{r},$$

where $\beta = 5 \times 10^3$ W/m².

(a) Determine an analytical solution for the temperature distribution within the fuel element. Implement your solution in EES and plot the temperature as a function of radius for $0 < r < r_{fuel}$.

(b) The maximum allowable temperature in the fuel element is $T_{max} = 1100°C$. What is the maximum value of β that can be used? What is the associated total rate that heat is transferred to the fluid?

(c) You are designing the fuel elements. You can vary r_{fuel} and β. The cladding must always be 2 cm thick (that is $r_{clad} = r_{fuel} + 2$ cm). The constraint is that the fuel temperature cannot exceed $T_{max} = 1100°C$ and the design target (the figure of merit to be maximized) is the rate of heat transfer per unit volume of material (fuel and cladding). What values of r_{fuel} and β are optimal?

2.86 Lighting represents one of the largest uses of electrical energy in residential and commercial buildings; lighting loads are highest during on-peak hours when electrical energy is most costly. Also, the thermal energy associated with the lighting that is deposited into the conditioned space will add significantly to the air-conditioning load, which in turn, adds to the electrical energy required to run the air conditioning system. A novel lighting system that is being considered gathers the visible portion of the incident solar radiation while eliminating the other portions. The visible radiation is delivered to the building via a fiber optic bundle (FOB) that transmits the light from the collector to the luminaire. The fiber optic bundle is composed of many small diameter PMMA fibers that are each coated with a thin layer of polymer cladding and packed in approximately a hexagonal close-packed array.

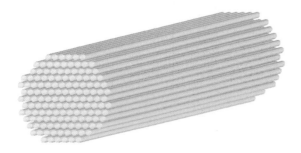

The porosity of the FOB is the ratio of the open area of the FOB to the total area of the FOB. The porosity of the FOB face is an important characteristic because any radiation that does not fall directly upon the PMMA fibers will not be transmitted to the luminaires and instead contributes to a thermal load on the FOB. The PMMA fibers are designed so that any radiation that strikes the face of a fiber within the design range of incident angles is "trapped" by total internal reflection. However, radiation that strikes the interstitial areas between the fibers will instead be absorbed in the cladding very close to the FOB face. Therefore, the FOB thermal loads can be divided into two components. Radiation that is incident on the face of the PMMA fibers contributes a low level of volumetric generation within the FOB that is related to transmission loss and can be neglected for the purposes of this problem. More importantly, radiation that is incident on the air gaps between PMMA fibers is absorbed by the cladding very quickly and contributes a high level of volumetric generation that is concentrated near the face of the FOB.

You have been asked to prepare an approximate model of the FOB; here you can assume that the outer edge of the FOB is completely insulated (i.e., no heat transfer from the outer radius of the bundle) and the forward face is exposed to air (at T_{air}) with a heat transfer coefficient (h). The individual fibers need not be considered; rather the bundle is a composite that can be represented by an effective conductivity in the axial direction ($k_{eff,x}$). The volumetric generation associated with the thermal energy related to the radiation incident on the interstitial regions ($\dot{g}'''$) can be represented by

$$\dot{g}''' = \frac{\phi \dot{q}''}{L_{ch}} \exp\left(-\frac{x}{L_{ch}}\right),$$

where ϕ is the porosity of the bundle, $\dot{q}''$ is the energy flux on the face of the FOB, x is the axial distance from the face, and L_{ch} is a characteristic length that depends on the details of the spectral and angular distribution of the incident flux and the structure of the FOB.

(a) Derive an analytical solution for the temperature as a function of axial position (x) within the FOB using only the symbols provided in the problem statement. Assume that the FOB is infinitely long.

(b) For the parameters provided in the table below prepare a plot showing the temperature as a function of position within the FOB for the first 20 cm of the FOB.

Parameter	Symbol	Value
Porosity	ϕ	0.05
Characteristic length for absorption	L_{ch}	0.025 m
Incident energy flux on FOB	$\dot{q}''$	100,000 W/m^2
Effective axial conductivity	$k_{eff,x}$	2.0 W/m-K
Ambient air temperature	T_{air}	20°C
Heat transfer coefficient	h	20 W/m^2-K

(c) Prepare a plot showing the maximum temperature within the FOB as a function of the heat transfer coefficient for various values of the effective axial conductivity. Explain any trends that you notice in your plot.

2.87 A PV panel is used on a satellite in order to produce electrical power.

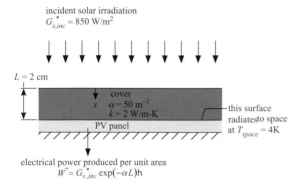

incident solar irradiation
$G''_{s,inc} = 850$ W/m^2

$L = 2$ cm

cover
x $\quad \alpha = 50$ m^{-1}
$k = 2$ W/m-K
PV panel

this surface radiates to space at $T_{space} = 4$K

electrical power produced per unit area
$\dot{W}'' = G''_{s,inc} \exp(-\alpha L) h$

The PV panel must be protected from debris and therefore a thick cover is required. The purpose of this problem is to evaluate the influence of absorption in the cover on the performance of the device. The cover is $L = 2$ cm thick with conductivity $k = 2$ W/m-K. The top surface of the cover is exposed to solar irradiation with a flux $\dot{G}''_{s,inc} = 850$ W/m^2.

The cover is not completely transparent to the solar irradiation; the absorption coefficient of the cover material is $\alpha = 50$ m^{-1}. Therefore, the

solar irradiation as a function of position decays exponentially according to:

$$\dot{G}''_s = \dot{G}''_{s,inc} \exp(-\alpha x).$$

The absorption of solar irradiation causes a volumetric generation within the cover that varies with position according to:

$$\dot{g}''' = \dot{G}''_{s,inc} \alpha \exp(-\alpha x).$$

The solar irradiation at the surface of the panel $(x = L)$ is converted to electricity with an efficiency that depends on the temperature of the PV panel:

$$\eta = \eta_{max} - \beta(T - T_{nom}),$$

where T is the temperature of the PV panel (which is identical to the temperature of the cover at $x = L$), $\eta_{max} = 0.15$, $\beta = 0.0015$ k^{-1}, and $T_{nom} = 20°$C. The cover is completely transparent to radiation emitted at high wavelengths. Therefore, the surface of the PV panel radiates directly to space (assume the emissivity of this surface is unity). The effective temperature of space is $T_{space} = 4$ K. The surface of the cover that is exposed to space (at $x = 0$) is adiabatic.

(a) Develop an analytical model for the temperature distribution within the cover. Plot the temperature as a function of x.
(b) Determine the rate of electrical power produced per area of the PV panel.
(c) Plot the rate of electrical power produced per area of the PV panel as a function of the cover thickness.

Numerical Solution Concepts

2.88 A cylindrical rod with length L and diameter D is made of a material that has a thermal conductivity described by the equation $k = a + bT$, where k is the thermal conductivity (in W/m-K) and T is the temperature (in K). One end of the rod is maintained at T_1, which is known. The other end convects to air at T_∞ with a convection coefficient of $\bar{h}$. The sides of the rod are well insulated. Derive the general equation for an interior node in the rod and the equation for a node located at the end of the rod where convection occurs (node M).

2.89 A tapered rod of length L has diameter of D_1 at one end and diameter D_2 at the other end. The rod is made of a material having a thermal conductivity k. The smaller end of the rod is maintained at T_1 whereas the larger end is exposed to a fluid at temperature T_∞ with a heat transfer coefficient of $\bar{h}$. The sides of the rod are well insulated. Derive the general equation for an interior node in the rod (node i) and the equation for a node that is located on the end of the rod where convection occurs (node M).

2.90 A plane wall with thickness L and cross-sectional area A has a specified temperature T_H on the left side (at $x = 0$) and a specified temperature T_C on the right side (at $x = L$). There is no volumetric generation in the wall. However, the conductivity of the wall material is a function of temperature such that $k = b + cT$, where b and c are constants. You would like to model the wall using a finite difference solution; a model with only three nodes is shown (right).

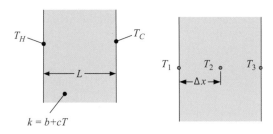

$k = b + cT$

The distance between adjacent nodes for the three-node solution is: $\Delta x = L/2$.

Write down the system of equations that could be solved in order to obtain the temperatures at the three nodes. Your equations should include the temperature of the nodes (T_1, T_2, and T_3) and the other parameters listed in the problem: T_H, T_C, Δx, A, b, and c.

2.91 The temperature distribution in the plane wall shown in the figure is 1-D and the problem is steady state.

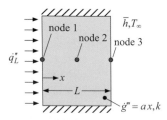

There is a generation of thermal energy in the wall. The generation per unit volume is not uniform but rather depends on position according to:

$$\dot{g}''' = a x,$$

where a is a constant and x is position. The left side of the wall experiences a specified heat flux, $\dot{q}''_L$. The right side of the wall experiences convection with heat transfer coefficient $\bar{h}$ to fluid at temperature T_∞. The thickness of the wall is L and the conductivity of the wall material, k, is constant. You are going to develop a numerical model with three nodes, as shown in the figure. The nodes are distributed uniformly throughout the domain. Derive the three equations that must be solved in order to implement the numerical model. Do not solve these equations.

Numerical Solutions

2.92 Obtain a numerical model using EES to the situation considered in Problem 2.66.
   (a) Calculate and plot the steady-state temperature distribution in the glass using a numerical solution. Overlay on this plot the analytical solution you obtained from Problem 2.66.
   (b) Determine the steady-state power required by the defroster.

2.93 Let's revisit Problem 2.65 but this time use a numerical model that will allow us to examine the fact that the rate of thermal energy generation actually increases strongly with temperature because the reaction is accelerated. Assume that all of the problem inputs are the same as in Problem 2.65 except that the volumetric rate of energy generation increases with temperature according to:

$$\dot{g}''' = \dot{g}'''_o + \beta(T - T_o),$$

where $\dot{g}'''_o = 75$ W/m^3, $\beta = 1$ W/m^3-K, and $T_o = 20°$C.
   (a) Develop a numerical solution for the temperature as a function of position in the coal powder.
   (b) Implement your solution in EES and plot T vs. x.
   (c) Set the value of $\beta = 0$ W/m^3-K (corresponding to a temperature-independent volumetric rate of thermal energy generation) and show that your numerical model limits to your analytical solution from Problem 2.65.

(d) With $\beta = 1$ W/m³-K, plot the maximum temperature in the coal as a function of the number of nodes in the solution.

(e) With $\beta = 1$ W/m³-K the coal pile can hit a thermal runaway situation if it gets too thick. As the coal powder layer thickness gets larger, the maximum temperature increases, causing the volumetric rate of thermal energy generation to increase and setting up an unstable situation. Plot the maximum temperature in the coal as a function of L, the thickness of the layer. You should see that as the length approaches a critical value, the temperature spikes upwards dramatically and there is no sensible solution above that critical value. Use your plot to estimate the critical pile thickness.

2.94 A plane wall is $L = 10$ cm thick with conductivity $k = 2.0$ W/m-K. The wall is adiabatic on one side ($x = 0$) with the other side ($x = L$) exposed to a fluid ($T_f = 20°C$) with heat transfer coefficient $h = 100$ W/m²-K. There is a volumetric thermal generation within the wall that varies with position according to:

$$\dot{g}''' = \frac{a}{1 + bx},$$

where $a = 1 \times 10^5$ W/m³ and $b = 10$ m⁻¹.

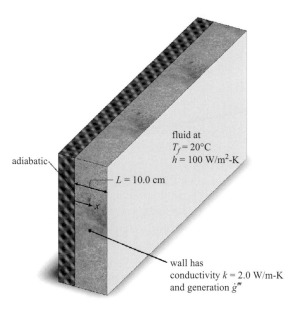

adiabatic

fluid at
$T_f = 20°C$
$h = 100$ W/m²-K

$L = 10.0$ cm

x

wall has
conductivity $k = 2.0$ W/m-K
and generation $\dot{g}'''$

(a) Develop a numerical model of the situation and plot the temperature as a function of position.

(b) Verify that your model has a sufficient number of nodes.

(c) Verify that your model matches an analytical solution in the limit that generation is a constant (i.e., $b = 0$).

2.95 An aluminum oxide cylinder is exposed to fluid with different temperatures on its internal and external surfaces ($T_{f,in} = 20°C$ and $T_{f,out} = 100°C$) and different heat transfer coefficients ($h_{in} = 100$ W/m²-K and $h_{out} = 200$ W/m²-K). The thermal conductivity of the aluminum oxide is approximately 10.0 W/m-K in the temperature range of interest. There is a volumetric generation within the cylinder that varies with radius according to:

$$\dot{g}''' = a + br + cr^2,$$

where $a = 1 \times 10^4$ W/m³, $b = 2 \times 10^5$ W/m⁴, and $c = 5 \times 10^7$ W/m⁵. The inner radius of the cylinder is $r_{in} = 10$ cm and the outer radius is $r_{out} = 20$ cm.

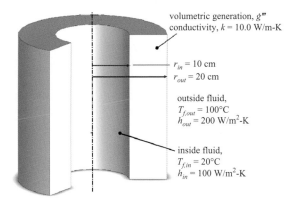

volumetric generation, $\dot{g}'''$
conductivity, $k = 10.0$ W/m-K

$r_{in} = 10$ cm
$r_{out} = 20$ cm

outside fluid,
$T_{f,out} = 100°C$
$h_{out} = 200$ W/m²-K

inside fluid,
$T_{f,in} = 20°C$
$h_{in} = 100$ W/m²-K

(a) Develop a numerical model and use it to plot the temperature as a function of radius.

(b) Verify that the model has numerically converged.

(c) Compare the numerical result with an analytical solution in the limit that generation is constant ($b = c = 0$).

2.96 Repeat Problem 2.95 but do not make the assumption that the conductivity of aluminum oxide is constant. Rather include the temperature-dependence of the conductivity. The conductivity for this material (in W/m-K) can be expressed as a function of temperature with the following equation where T_C is the temperature in °C:

$$k = 18.306 - 0.04246 T_C + 5.9733$$
$$\times 10^{-5} T_C^2 - 3.814 \times 10^{-8} T_C^3.$$

2.97 Reconsider Problem 2.84 but obtain a solution numerically using EES. Plot the temperature as a function of position and compare your answer to the analytical solution from Problem 2.84.

2.98 Solve Problem 2.67 using a 1-D steady-state numerical solution using EES and compare the result with the analytical solution. Vary the number of nodes used in your analysis to ensure that your solution has converged.

2.99 A cylindrical rod with length 0.1 m and diameter 0.05 m is made of a material that has a thermal conductivity described by the equation $k = 24 + 0.056T$, where k is the thermal conductivity in W/m-K and T is the temperature in °C. One end of the rod is maintained at 282°C. The other end convects to air at 25°C with a convection coefficient of 22 W/m²-K. The sides of the rod are well insulated. Determine the temperature at the end of the rod where convection occurs and the steady-state rate of heat transfer using a numerical method implemented in EES.

2.100 A tapered rod with length 0.06 m has a diameter that is 2.5 cm at one end and 7.0 cm at the other. The rod is made of steel ($k = 21$ W/m-K) and it is well insulated everywhere except on its ends. The smaller end is maintained at 250°C. The larger end exposed to a fluid at 25°C with a convection coefficient of 118 W/m²-K. The sides of the rod are well insulated. Determine the temperature at the end of the rod where convection occurs and the steady-state rate of heat transfer using a numerical method implemented in EES.

2.101 If you bale hay without allowing it to dry sufficiently then the hay bales will contain a lot of water. Besides making the bales heavy and therefore difficult to put in the barn, the water in the hay bales causes an exothermic chemical reaction to occur within the bale (i.e., the hay is rotting). The chemical reaction proceeds at a rate that is related to temperature and the bales may be thermally isolated (they are placed in a barn and surrounded by other hay bales); as a result, the hay can become very hot and even start a barn fire. The figure illustrates a cross-section of a barn wall with hay stacked against it.

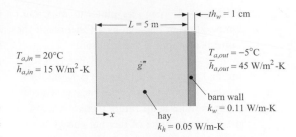

The air within the barn is maintained at $T_{a,in} = 20$°C and the heat transfer coefficient between the air and the inner surface of the hay is $\bar{h}_{a,in} = 15$ W/m²-K. The outside air is at $T_{a,out} = -5$°C with $\bar{h}_{a,out} = 45$ W/m²-K. Neglect radiation from the surfaces in this problem. The barn wall is composed of wood ($k_w = 0.11$ W/m-K) and is $th_w = 1$ cm thick. The hay has been stacked $L = 5$ m thick against the wall. Hay is a composite structure composed of plant fiber and air. However, hay can be modeled as a single material with an effective conductivity $k_h = 0.05$ W/m-K. The volumetric generation of the hay due to the chemical reaction is given by:

$$\dot{g}''' = 1.5 \left[\frac{\text{W}}{\text{m}^3}\right] \left[\exp\left(\frac{T}{320[K]}\right)\right]^{0.5},$$

where T is temperature in K.
(a) Develop a numerical model using EES that can predict the temperature within the hay.
(b) Prepare a plot that shows the temperature distribution as a function of position in the hay.
(c) Prepare a plot that shows that you are using a sufficient number of nodes in your numerical solution.
(d) Verify that your solution is correct by comparing it with an analytical solution in an appropriate limit. Prepare a plot that overlays your numerical solution and the analytical solution in this limit.
(e) What is the maximum allowable thickness of hay (L_{max}) based on keeping the maximum temperature below T_{fire} (200°F)?
(f) If $L = L_{max}$ from (e) then how much of the hay will remain usable (what percent of the hay is lost to heat degradation because it reaches a temperature that is in excess of 140°F)?

2.102 A current lead must be designed to carry current to a cryogenic superconducting magnet.

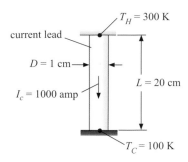

$T_H = 300$ K

current lead

$D = 1$ cm

$L = 20$ cm

$I_c = 1000$ amp

$T_C = 100$ K

The current lead carries $I_c = 1000$ amp and therefore experiences substantial generation of thermal energy due to ohmic dissipation. The electrical resistivity of the lead material (ohm-m) depends on temperature (K) according to:

$$\rho_e = 17 \times 10^{-9} + (T - 300)5 \times 10^{-11}.$$

The length of the current lead is $L = 20$ cm and the diameter is $D = 1$ cm. The hot end of the lead (at $x = 0$) is maintained at $T_{x=0} = T_H = 300$ K and the cold end (at $x = L$) is maintained at $T_{x=L} = T_C = 100$ K. The conductivity of the lead material is $k = 400$ W/m-K. The lead is installed in a vacuum chamber and therefore you may assume that the external surfaces of the lead (the outer surface of the cylinder) are adiabatic.

(a) Develop a numerical model in EES that can predict the temperature distribution within the current lead. Plot the temperature as a function of position.

(b) Determine the rate of energy transfer into the superconducting magnet at the cold end of the current lead. This parasitic heat transfer must be removed in order to keep the magnet cold and therefore must be minimized in the design of the current lead.

(c) Prepare a plot showing the rate of energy transfer into the magnet as a function of the number of nodes used in your model.

(d) Plot the rate of heat transfer to the cold end as a function of the diameter of the lead. You should see a minimum value and therefore an optimal diameter – explain why this occurs.

(e) Prepare a plot showing the optimal diameter and minimized rate of heat transfer to the cold end as a function of the current that must be carried by the lead. You may want to use the Min/Max Table selection from the Calculate menu to accomplish this.

2.103 A plane wall is 8.5 cm thick and uniformly generates heat at a rate of 0.4 MW/m³. One side of the wall is insulated and the other side is exposed to fluid at 93°C with a heat transfer coefficient of 500 W/m²-K. The thermal conductivity of the wall varies with position according to:

$$k = 20(1 + 2.3x),$$

where k is in W/m-K and x is the distance from the insulated side in meters.

(a) Develop a numerical model using EES to solve for temperature as a function of position x.

(b) Confirm that your numerical model is accurate by comparing it to an analytical solution to this problem for a constant thermal conductivity.

(c) Determine the maximum temperature of the wall and the location of the maximum temperature using the numerical model. Be sure that your solution does not change more than +/-0.1°C as you add more nodes.

2.104 Prepare a numerical model of the situation described in Problem 2.86 using the finite difference approach in EES. Compare your answer with the analytical solution from Problem 2.86.

2.105 Prepare a numerical model of the situation described in Problem 2.86 using matrix decomposition.

(a) Compare your answer with the analytical solution from Problem 2.86.

(b) One method for controlling the temperature within the FOB is to insert a high-conductivity material in the interstitial regions near the FOB face. Use your numerical model to evaluate the situation where the effective axial thermal conductivity is high at the face and drops off to the composite value:

$$k_{eff,x} = k_\infty + \Delta k \exp\left(-\frac{x}{L_k}\right)$$

where $k_\infty = 2.0$ W/m-K, $\Delta k = 28.0$ W/m-K, and $L_k = 0.05$ m. Prepare a plot of temperature vs. axial position for this situation and compare it to the original FOB bundle (i.e., one where the axial conductivity is constant and equal to 2.0 W/m-K).

2.106 A spherical reactor vessel has an outer diameter of 0.35 m and an inner diameter of 0.12 m. The vessel contains a chemical that is reacting to produce 465 W at steady state. The heat transfer coefficient between the chemical and the inside surface of the sphere is 280 W/m²-K. The outside surface is exposed to air at 25°C with a heat transfer coefficient of 10 W/m²-K. The material for the spherical shell has a thermal conductivity described by:

$$k = 9.918 + 0.0165T,$$

where k is in W/m-K and T is in K. Use a numerical solution developed in EES to determine the inside and outside temperatures of the sphere.

2.107 A bearing has the shape of a truncated cone. The height of the cone is 5 cm and the diameters of the two ends are 1.5 cm and 1.0 cm. The bearing is made of a ceramic material that has a thermal conductivity that depends on temperature according to:

$$k = 1.32 + 0.00135T,$$

where T is the temperature in K and k is in units of W/m-K. The larger face of the bearing is exposed to a fluid at 374 K with a convection coefficient of 84 W/m²-K. The smaller face is exposed to a fluid at 298 K with a convection coefficient of 124 W/m²-K. The outer edges of the bearing are insulated. Develop a numerical model using EES and use it to determine the steady-state heat transfer rate through the bearing and the temperature distribution.

2.108 A brass cylinder has diameter 0.375 inch and length 5 ft. The cylinder is exposed to air at 75°F on its outer edge with a heat transfer coefficient of 3.2 Btu/hr-ft²-F. An electrical current is passed through the cylinder with a voltage drop of 120 V across the 5 ft length. The thermal conductivity of brass can be represented as:

$$k = -0.0333 + 0.03T - 0.000030T^2,$$

where T is temperature in units °F and k is conductivity in units Btu/hr-ft-F. The electrical resistivity of brass is 8.495×10^{-8} ohm-m.

(a) Develop a numerical model using EES and plot the temperature of the rod as a function of radial position.
(b) Determine the maximum temperature in the brass rod and its radial location.
(c) Repeat part (b) assuming the thermal conductivity is constant at its average value. Comment on the importance of including the temperature dependence.

2.109 Reconsider Problem 2.84, but obtain a solution numerically using matrix decomposition in MATLAB. The description of the hay bale is provided in Problem 2.84. Prepare a model that can consider the effect of temperature on the volumetric generation. Increasing temperature tends to increase the rate of reaction and therefore increase the rate of generation of thermal energy; the volumetric rate of generation can be approximated by: $\dot{g}''' = a + bT$ where $a = -1$ W/m³ and $b = 0.01$ W/m³-K. Note that at $T = 300$ K, the generation is 2 W/m³ but that the generation increases with temperature.

(a) Prepare a numerical model of the hay bale using MATLAB. Plot the temperature as a function of position within the hay bale.
(b) Show that your model has numerically converged; that is, show some aspect of your solution as a function of the number of nodes and discuss an appropriate number of nodes to use.
(c) Verify your numerical model by comparing your answer to an analytical solution in some, appropriate limit. The result of this step should be a plot that shows the temperature as a function of radius predicted by both your numerical solution and the analytical solution and demonstrates that they agree.

2.110 Reconsider Problem 2.101 parts (a) and (b) using matrix decomposition in MATLAB.

2.111 Reconsider Problem 2.101 parts (a) and (b) using Gauss–Seidel Iteration in MATLAB.

2.112 Reconsider Problem 2.102 part (a) using matrix decomposition in MATLAB.

2.113 Reconsider Problem 2.102 part (a) using Gauss–Seidel Iteration in MATLAB.

2.114 Reconsider Problem 2.103 part (a) using matrix decomposition in MATLAB.

2.115 Reconsider Problem 2.103 part (a) using Gauss–Seidel Iteration in MATLAB.

Projects

2.116 You must design a current lead to carry 100 kamp (100,000 amp) to a superconducting magnet used for energy storage. The current lead will be a cylinder of high-purity copper.

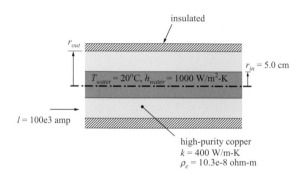

Water flows through the inside of the cylinder; the radius of the water passage is 5.0 cm. The temperature of the water is 20°C and the heat transfer coefficient between the cylinder surface and the water is 1000 W/m²-K. The external surface of the cylinder is insulated. The electrical resistivity of copper is 10.3×10^{-8} ohm-m. The thermal conductivity of copper is 400 W/m-K.

(a) Develop a numerical model for this 1-D, steady-state conduction problem using EES. Write a short (1–2 page) report that describes your model. The report must include the following things:
   - an energy balance for a typical internal node that shows the important energy terms,
   - a description of how each of the energy terms are approximated (don't forget that you are in a cylindrical coordinate system so the area associated with each conduction term changes), and
   - energy balances that show how the boundary nodes are treated.

The report should be clearly written using a word processor. All equations should be numbered and entered using an equation editor. Symbols used in equations should be clearly defined. Print out the Equation Window of your EES program.

(b) Verify your numerical model against an analytical solution. Write a short paragraph that describes the appropriate analytical solution and include a figure (with captions, units, labels, etc.) that shows that the numerical and analytical solutions agree.

(c) Design the current lead by preparing a plot of the temperature at the interface between the copper and the external insulation as a function of the outer radius. If the insulation is rated for 100°C then what is the minimum value of the outer radius?

2.117 Some conduction problems can be solved by using a 1-D model in which the temperature and area are a function of position (x) and the conductivity is a function of temperature; this is shown generally.

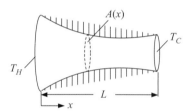

The boundary conditions for the situation are specified temperatures at the two ends of the computational domain: $T_{x=0} = T_H$ and $T_{x=L} = T_C$ where L is the length of the computational domain.

(a) Derive the ODE that describes this situation. Remember that conductivity is a function of temperature and area is a function of position.

(b) Integrate your ODE once to obtain a first-order ODE that includes an unknown constant of integration – examination of your resulting equation should show that this unknown constant of integration is equal to the rate of heat transfer through the solid, $\dot{q}$.

(c) Separate your first-order ODE; on one side you should have an integral with respect to temperature and on the other side an integral with respect to position.

The integral with respect to position can be expressed in terms of the shape factor, S, which is defined as:

$$\frac{1}{S} = \int_0^L \frac{dx}{A(x)}.$$

You will find conduction shape factors for many situations tabulated in various handbooks and also contained in the Conduction Shape Factors Library in EES which can be accessed by selecting Function Information from the Options menu and then clicking the Heat Transfer & Fluid Flow radio button and navigating to the Conduction Shape Factors menu.

(d) Express the integral with respect to x from part (c) in terms of the shape factor S. Derive the shape factor for conduction through:
  - a plane wall with area A and thickness L,
  - radial conduction through a cylindrical shell with height H and inner and outer radii r_{in} and r_{out}, respectively, and
  - radial conduction through a spherical shell with inner and outer radii r_{in} and r_{out}, respectively.

The integral with respect to temperature can be expressed in terms of the integrated thermal conductivity, K, which is defined as:

$$K(T) = \int_{T_{ref}}^T k(T)dT,$$

where T_{ref} is some reference temperature (which won't matter, just as the reference state used to define enthalpy or entropy does not matter for thermodynamic calculations). The integrated thermal conductivity is a property that is particularly useful for conduction problems in which thermal conductivity is a strong function of temperature. It is tabulated in many handbooks (particularly for cryogenic materials where k varies a lot with temperature) and can be accessed using the property function IntK in EES. The IntK function is accessed just as any other property function in EES is used. Select Function Information from the Options menu and then click the Thermophysical properties radio button. Navigate to the type of material of interest (e.g., incompressible) and then the property and fluid in the drop down lists.

(e) Express the integral with respect to T from part (c) in terms of the integrated thermal conductivity K. Prepare a single plot showing the thermal conductivity (left axis) and integrated thermal conductivity (right axis) of aluminum alloy 6061 from 4 K to 300 K.

Your result after part (e) should be:
$$\dot{q} = S[K(T_H) - K(T_C)],$$
which is convenient as the variation in the area is captured by the shape factor while the variation in the conductivity is captured by using K.

A spacer used in small Dewar flasks is fabricated in the form of the frustum of a cone. The small end of the spacer has a radius of $r_1 = 6$ mm and is maintained at a temperature of $T_H = 300$ K. The larger end of the spacer has a radius of $r_2 = 18$ mm and is maintained at $T_C = 300$ K. The height of the spacer is $L = 48$ mm. The material used to fabricate the spacer is polyurethane.

(f) Derive the equation for the shape factor for a truncated cone.
(g) Determine the heat transfer rate through the spacer.
(h) Plot the heat transfer rate as a function of r_1.

2.118 A tube with inner radius $r_{in} = 2.1$ mm carries fluid at $T_i = 100°C$. The heat transfer coefficient between the fluid and the tube inner surface is $\bar{h}_i = 850$ W/m²-K. The thickness of the tube is $th_t = 1.0$ mm and the conductivity of the tube material is $k_t = 25$ W/m-K. Insulation is applied to the outer surface of the tube to reduce heat loss from the fluid to the surroundings. There is an area-specific contact resistance between the tube and the insulation, $R_c'' = 1 \times 10^{-4}$ K-m²/W. The insulation thickness is $th_{ins} = 2.0$ mm and the conductivity of the insulation is $k_{ins} = 0.12$ W/m-K. The outer surface of the insulation experiences both convection and radiation with the surroundings at $T_o = 20°C$. The heat transfer coefficient at the outer surface of the insulation is $\bar{h}_o = 22$ W/m²-K and the emissivity of the surface of the insulation is $\varepsilon = 0.3$.

(a) Determine the rate of heat transfer per unit length from the fluid in the tube to the surroundings. Use EES to implement the solution to the problem.

(b) Examine the values of the various resistances in your solution and rank them in order of importance.

(c) Create a plot showing the rate of heat transfer per unit length as a function of the insulation thickness, th_{ins}.

(d) In your plot from (c) you should see that there is a maximum value of heat transfer. That is, below some insulation thickness the rate of heat transfer actually increases as you add insulation thickness. Explain why this is so.

(e) Your examination in part (b) should have revealed that the two most important resistances in the problem are conduction through the insulation and convection from the outer surface of the insulation. In the limit that these are the only two resistances considered, derive an expression for the critical insulation radius (i.e., the radius of insulation at which point the rate of heat transfer actually increases as more insulation is added).

2.119 Most conductors have electrical resistivity that increases with temperature. If you run current through such a conductor you run the risk of becoming thermally unstable because higher temperatures lead to higher resistivity which leads to higher rates of thermal energy generation which, in turn, leads to even higher temperatures – this is an unstable system.

At every point inside a conductor the volumetric rate of thermal energy generation is related to the current density (J in amp/m^2) and the electrical resistivity (ρ_e in ohm-m) according to:

$$\dot{g}''' = J^2 \rho_e.$$

In many conductors above room temperature, the conductivity is relatively constant but the electrical resistivity increases linearly according to:

$$\rho_e = \rho_{e,o} + \beta(T - T_o)$$

where $\rho_{e,o}$ is the electrical resistivity at room temperature (T_o) and β is the temperature coefficient of electrical resistivity.

(a) Plot the conductivity and electrical resistivity of copper from 300 K to 1200 K and estimate the values of k, $\rho_{e,o}$, and β that should be used to represent copper.

In this problem we will use two different ways to estimate the critical current density for a $r_o = 1$ cm radius, cylindrical copper conductor. Assume that the heat transfer coefficient between the external surface of the conductor is high so that its outer temperature is $T_e = 300$ K.

(b) Prepare a numerical model of the temperature distribution within the copper conductor capable of predicting temperature as a function of radial position, r.

(c) Plot the temperature as a function of radial position for the case where $J = 3 \times 10^8$ A/m^2 when the electrical resistivity is constant, $\rho_e = \rho_{e,o}$.

(d) Overlay on your plot from (c) the temperature as a function of radial position for constant electrical resistivity predicted using an analytical model.

The first method that we will use to estimate the critical current density is by slowly increasing the value of J in our numerical model (with the temperature-dependent resistivity on, of course).

(e) Start at a low value (e.g., $J = 1 \times 10^7$ A/m^2) and plot the maximum temperature as a function of J (updating your guess values each time). You should see that the maximum temperature increases extremely rapidly with J at some point and much above that point your model is unable to converge to a physical solution. Use your plot to estimate the critical current density.

The second method that we will use to estimate the critical current density is more approximate. There are two things going on that must balance: (1) energy dissipated within the conductor must be balanced by (2) energy conducted out of the conductor. Both of these phenomena tend to increase with ΔT, the temperature rise experienced within the conductor. At some point, the rate of change of the first thing overtakes the rate of change of the second thing and thermal runaway occurs. We can (approximately) write an equation for both of these phenomena and then take the derivative of both sides of the equation with respect to ΔT – the critical current occurs when these derivatives are equal.

(f) Write an *approximate* algebraic equation for the rate at which energy is conducted out of the conductor as a function of ΔT. Write another *approximate* algebraic equation for the rate at which energy is dissipated within the conductor as a function of ΔT. Take the derivative of each of these equations and use this approach to estimate the critical current density. Compare this value with the one calculated in (e).

2.120 A valve must be installed on a cryogenic experiment that is maintained at a cold temperature, T_C, by a container of liquid helium. The actuator for the valve is a thin wall stainless tube with cross-sectional area A_c and length L. The tube must extend from a platform at T_C to the walls of the vacuum vessel that are at room temperature, T_H.

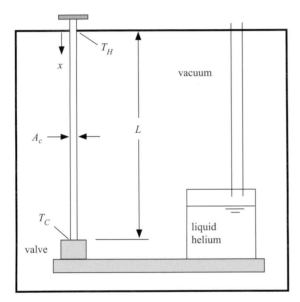

Because the experiment is installed in a vacuum space there is no convection to the surface of the actuator. We will also neglect any radiation to the actuator. As a result, this is a steady-state, 1-D conduction problem where the temperature of the actuator varies only in the x-direction. The temperature of the actuator at $x = 0$ is T_H and the temperature of the actuator at $x = L$ is T_C. The conductivity of the actuator material is a strong, increasing function of temperature between T_C and T_H. The relationship between conductivity and temperature is given approximately by the equation:

$$k = aT^b,$$

where a and b are constants.

(a) Derive the governing differential equation that must be solved in order to obtain temperature as a function of position along the actuator.

(b) Integrate the differential equation twice to obtain a solution for temperature as a function of position in terms of two unknown constants. Note that it is not necessary to explicitly solve for temperature, an implicit equation is fine.

(c) Write down the two equations that must be solved in order to determine the unknown constants of integration.

(d) Sketch the rate of heat transfer in the x-direction as a function of x. Justify your sketch.

(e) Sketch the temperature as a function of x. Justify your sketch.

The temperature of the experiment is $T_C = 4$ K and the temperature of the vessel walls is $T_H = 300$ K. The actuator cross-sectional area is $A_c = 0.0001$ m^2 and length is $L = 1$ m. The constants in the equation for thermal conductivity are $a = 0.171$ W/m-K$^{1.8}$ and $b = 0.8$.

(f) Substitute the two equations that you derived in part (c) for the constants of integration into EES in order to solve for them.

(g) Prepare a plot of temperature vs. position; comment on whether your plot matches your sketch from part (e).

(h) Determine the rate of heat transfer from the actuator into the experiment, $\dot{q}_e$ (i.e., conduction in the x-direction evaluated at $x = L$).

(i) Prepare a plot showing $\dot{q}_e$ as a function of L.

Your answer to part (h) should have been $\dot{q}_e = 0.27$ W of heat load on the experiment. This has been found to be unacceptably large and therefore you have been asked to analyze the idea of adding a thermal intercept to the system as shown in the figure below. A cryocooler (i.e., a refrigerator that provides cooling at low temperature) is added and a thermal strap extends from the cold head of the cryocooler to the actuator at some location, fL where $0 < f < 1$. The thermal intercept connects the cryocooler to the actuator so that the temperature of the actuator at that location is equal to the load temperature of the cryocooler, T_{mid}.

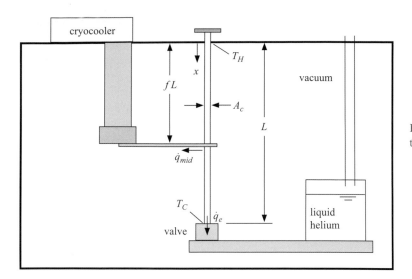

Figure for Problem 2.120 with a thermal intercept.

Assume that the cryocooler load temperature is set to T_{mid} = 50 K and f = 0.5.

(j) Determine the rate of heat transfer to the experiment, $\dot{q}_e$, and the rate of heat transfer to the cryocooler, $\dot{q}_{mid}$.

(k) Keeping f = 0.5 (i.e., assuming the thermal intercept is kept at the mid-point of the actuator), plot the rate of heat transfer to the cryocooler as a function of T_{mid}. Explain any interesting characteristics in your plot.

The cryocooler is characterized by a load curve. The manufacturer has measured the cooling capacity of the cryocooler as a function of the cold head temperature and found it to be:

$$\dot{q}_{cc} = 0.00013 \left[\frac{\text{W}}{\text{K}^2}\right] (T_{mid} - 10[\text{K}])^2.$$

(l) Overlay on your plot from (k) the cryocooler load curve. You should see that there is an intersection between the cooling capacity of the cryocooler and the cooling required by the actuator – this intersection corresponds to the actual operating temperature of the cryocooler and therefore the actual value of T_{mid}.

(m) Use EES to determine the operating temperature by commenting out the specified value of T_{mid} and instead requiring that $\dot{q}_{cc}$ must be equal to $\dot{q}_{mid}$. What is the actual operating temperature of the cryocooler and the corresponding actual value of $\dot{q}_e$ for the case where f = 0.5?

(n) Plot the rate of heat transfer to the experiment as a function of f, the location of the thermal intercept. You should see that there is an optimal value of f; explain why this is so.

2.121 The figure illustrates a cross-section of a nuclear fuel rod.

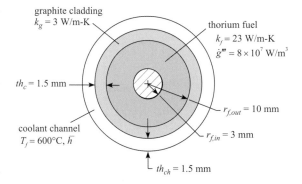

The thorium fuel material has inner radius $r_{f,in}$ = 3 mm and an outer radius of $r_{f,out}$ = 10 mm. The nuclear reaction leads to a volumetric generation of thermal energy within the thorium of $g''' = 8 \times 10^7$ W/m^3. The thorium has conductivity k_f = 23 W/m-K. The center of the thorium is empty and can be considered adiabatic. There is a graphite cladding surrounding the fuel with thickness th_c = 1.5 mm and conductivity k_g = 3 W/m-K. The area-specific contact resistance between the graphite and the thorium is R_c'' = 8.2 × 10^{-4} K-m^2/W. The outer surface of the cladding is exposed to the coolant that transfers energy to the nuclear power cycle. The coolant flows through a channel with thickness $th_{ch} = 1.5$ mm. The fluid temperature is $T_f = 600°C$ and the heat transfer coefficient between the graphite and the coolant depends on the thickness of the coolant channel according to:

$$\bar{h} = 5.4 \times 10^6 \left[\frac{W}{m^{3.5}K} \right] th_{ch}^{1.5}.$$

This problem concerns modeling and designing the fuel rod. Do the problem on a per unit length basis (i.e., $L = 1$ m).

(a) Determine the contact resistance between the thorium and graphite, the convection resistance between the graphite surface and coolant, and the conduction resistance associated with the graphite. Comment on which resistance is most important and which is less important. Are any negligible in your opinion?

It is not possible to represent the thorium using a thermal resistance because it experiences thermal energy generation. Instead we will need to derive and solve the ODE associated with the temperature in the fuel itself.

(b) Derive the governing ODE within the thorium.
(c) Determine the boundary conditions associated with the ODE.
(d) Implement your solution in EES in order to obtain the constants of integration. Prepare a plot showing the temperature (in °C) as a function of radius (in mm) within the thorium.

The rest of the assignment involves using your model to design the fuel rod. The fuel rod should be made as compact as possible while doing the most in order to make a cost-effective nuclear power plant. The thorium can't operate above a temperature of about $T_{melt} = 1600°C$.

(e) Comment out the value of $r_{f,out}$ that you set in your Equations Window and plot the maximum temperature in the fuel rod as a function of $r_{f,out}$. Use your graph to determine the maximum value of $r_{f,out}$ that you can select for the fuel rod under these conditions?
(f) Use EES to directly determine the answer you graphically found in (e). That is, set the

maximum temperature to the melting temperature and let EES calculate the value of $r_{f,out}$.

You now have a model that automatically adjusts $r_{f,out}$ so that the maximum temperature in the fuel rod is equal to the melting temperature. You can start to adjust other parameters to look at their effect on the design.

(g) One metric for the cost-effectiveness of the design is the ratio of the heat transferred to the coolant to the total volume of the fuel rod $\dot{g}/V$, the volume including the center material, thorium, graphite, and the coolant channel. Plot $\dot{g}/V$ as a function of the coolant channel thickness, th_{ch}. Note that your value of $r_{f,out}$ will always be adjusted so that the maximum fuel rod temperature is T_{melt} if you have done the problem correctly. You should see that an optimal coolant channel thickness exists – explain why.

Now develop a numerical model of the fuel rod. Your numerical model need only position nodes within the thorium itself – the contact resistance, cladding, and convection can all be modeled using an appropriate boundary condition for the outer node of your numerical model.

(h) Develop a numerical model of the thorium fuel rod for the conditions listed above. Plot the temperature as a function of radius.
(i) Overlay on your plot from (h) the analytical solution that you derived previously – they should lay nearly exactly on one another.
(j) Plot some aspect of your solution that you deem most important as a function of the number of nodes used in your numerical model.

One of the advantages of using a numerical model is that complications such as temperature-dependent properties can be relatively easily taken into account. Assume that the volumetric rate of thermal energy generation depends on temperature according to:

$$g'''(T) = 8 \times 10^7 \left[\frac{W}{m^3} \right] \left\{ 1 + \frac{(T - 2400[K])}{500[K]} \right\}.$$

(k) Modify your numerical model so that it considers the temperature-dependent volumetric rate of thermal energy generation. Prepare a single plot showing the temperature distribution within the thorium assuming constant generation and also temperature-dependent generation.

The conductivity of thorium is also temperature dependent and is given approximately by the equation:

$$k_f(T) = 23\left[\frac{W}{m\text{-}K}\right] - 0.025\left[\frac{W}{m\text{-}K^2}\right](T - 2000[K]).$$

(l) Modify your numerical model so that it considers the temperature-dependent conductivity as well as the temperature-dependent volumetric rate of thermal energy generation. Overlay on your plot from (k) the temperature distribution including both of these effects.

2.122 Ablation is a technique for treating cancerous tissue that occurs by heating it to a lethal temperature; the dead tissue can subsequently be left to be absorbed by the body or excised surgically. A number of techniques have been suggested in order to apply heat locally to the cancerous tissue and therefore spare surrounding healthy tissue. One interesting technique would embed small metallic spheres at precise locations surrounding the cancer tumor and then expose the region to an oscillating magnetic field. The magnetic field does not generate energy in the tissue; however, the conducting spheres experience a volumetric generation of thermal energy which thereby increases their temperature and results in the conduction of heat to the surrounding tissue. Precise placement of the spheres can be used to control the application of thermal energy.

sphere, k_{sp} = 10 W/m-K

tissue, k_t = 0.5 W/m-K

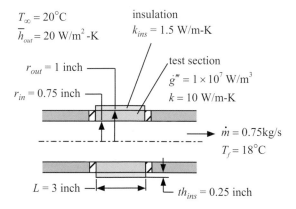

R_{sp} = 1.0 mm

$\dot{g}_{sp}$ = 1.0 W T_b = 37°C

It is necessary to determine the temperature field associated with a single sphere placed in an infinite medium of tissue. The sphere has a radius, R_{sp} = 1.0 mm and conductivity, k_{sp} = 10 W/m-K. A total of $\dot{g}_{sp}$ = 1.0 W of generation is uniformly distributed throughout the sphere. The temperature far from the sphere (the body temperature, T_b) is 37°C. The tissue has thermal conductivity k_t = 0.5 W/m-K; the effects of metabolic heat generation (i.e., volumetric generation in the tissue) and blood perfusion (i.e., the heat removed by blood flow in the tissue) are not considered.

(a) Prepare a plot showing the temperature as a function of radius from $r = 0$ to $r \to \infty$.

The maximum tissue temperature ($T_{max,t}$) is the temperature at the interface between the sphere and the tissue. The extent of the lesion (R_{lesion}) is defined as the radius where the tissue temperature reaches the lethal temperature for tissue (T_{lethal}) which is assumed to be 50°C.

(b) Plot the maximum temperature in the tissue and the extent of the lesion as a function of the heat generated in the sphere.

2.123 The figure illustrates a simple mass flow meter for use in an industrial refinery.

T_∞ = 20°C

$\overline{h}_{out}$ = 20 W/m²-K

insulation

k_{ins} = 1.5 W/m-K

r_{out} = 1 inch

r_{in} = 0.75 inch

test section

$\dot{g}'''$ = 1 × 10⁷ W/m³

k = 10 W/m-K

$\dot{m}$ = 0.75 kg/s

T_f = 18°C

L = 3 inch

th_{ins} = 0.25 inch

A flow of liquid passes through a test section consisting of an L = 3 inch section of pipe with inner and outer radii, r_{in} = 0.75 inch and r_{out} = 1.0 inch, respectively. The test section is uniformly heated by electrical dissipation at a rate $\dot{g}'''$ = 1 × 10⁷ W/m³ and has conductivity k = 10 W/m-K. The pipe is surrounded with insulation that is $th_{ins} = 0.25$ inch thick and has conductivity $k_{ins} = 1.5$ W/m-K. The external surface of the insulation experiences convection with air at $T_{\infty} = 20°C$. The heat transfer coefficient on the external surface is $\bar{h}_{out} = 20$ W/m²-K. A thermocouple is embedded at the center of the pipe wall. By measuring the temperature of the thermocouple, it is possible to infer the mass flow rate of fluid because the heat transfer coefficient on the inner surface of the pipe ($\bar{h}_{in}$) is strongly related to mass flow rate ($\dot{m}$). Testing has shown that the heat transfer coefficient and mass flow rate are related according to:

$$\bar{h}_{in} = C\left(\frac{\dot{m}}{1\,[kg/s]}\right)^{0.8},$$

where $C = 2500$ W/m²-K. Under nominal conditions, the mass flow rate through the meter is $\dot{m} = 0.75$ kg/s and the fluid temperature is $T_f = 18°C$. Assume that the ends of the test section are insulated so that the problem is 1-D. Neglect radiation and assume that the problem is steady state.

(a) Develop an analytical model in EES that can predict the temperature distribution in the test section. Plot the temperature as a function of radial position for the nominal conditions.

(b) Using your model, develop a calibration curve for the meter; that is, prepare a plot of the mass flow rate as a function of the measured temperature at the midpoint of the pipe. The range of the instrument is 0.2 kg/s to 2.0 kg/s.

The meter must be robust to changes in the fluid temperature. That is, the calibration curve developed in (b) must continue to be valid even as the fluid temperature changes by as much as 10°C.

(c) Overlay on your plot from (b) the mass flow rate as a function of the measured temperature for $T_f = 8°C$ and $T_f = 28°C$. Is your meter robust to changes in T_f?

In order to improve the meter's ability to operate over a range of fluid temperature, a temperature sensor is installed in the fluid in order to measure T_f during operation.

(d) Using your model, develop a calibration curve for the meter in terms of the mass flow rate as a function of ΔT, the difference between the measured temperatures at the midpoint of the pipe wall and the fluid.

(e) Overlay on your plot from (d) the mass flow rate as a function of the difference between the measured temperatures at the midpoint of the pipe wall and the fluid if the fluid temperature is $T_f = 8°C$ and $T_f = 28°C$. Is the meter robust to changes in T_f?

(f) If you can measure the temperature difference to within $\delta\Delta T = 1$ K then what is the uncertainty in the mass flow rate measurement? (Use your plot from part (d) to answer this question.)

You can use the built-in uncertainty propagation feature in EES to assess uncertainty automatically.

(g) Set the temperature difference to the value you calculated at the nominal conditions and allow EES to calculate the associated mass flow rate. Now, select Uncertainty Propagation from the Calculate menu and specify that the mass flow rate is the calculated variable while the temperature difference is the measured variable. Set the uncertainty in the temperature difference to 1 K and verify that EES obtains an answer that is approximately consistent with part (f).

(h) The nice thing about using EES to determine the uncertainty is that it becomes easy to assess the impact of multiple sources of uncertainty. In addition to the uncertainty $\delta\Delta T$, the constant C has an uncertainty of 5 percent and the conductivity of the material is only known to within 3 percent. Use EES' built-in uncertainty propagation to assess the resulting uncertainty in the mass flow rate measurement. Which source of uncertainty is the most important?

(i) The meter must be used in areas where the ambient temperature and heat transfer coefficient may vary substantially. Prepare a plot showing the mass flow rate predicted by your model for $\Delta T = 50$ K as a function of T_∞ for various values of $\bar{h}_{out}$. If the operating range of your meter must include $-5°C < T_\infty < 35°C$ then use your plot to determine the range of $\bar{h}_{out}$ that can be tolerated without substantial loss of accuracy.

2.124 Reconsider the mass flow meter that is investigated in Problem 2.123. The conductivity of the material that is used to make the test section is not actually constant, as was assumed in Problem 2.123, but rather depends on temperature according to:

$$k = 10 \frac{W}{\text{m-K}} + 0.035 \left[\frac{W}{\text{m-K}^2} \right] (T - 300[\text{K}]).$$

(a) Develop a numerical model of the mass flow meter using EES. Plot the temperature as a function of radial position for the conditions given in Problem 2.123 with the temperature-dependent conductivity.

(b) Verify that your numerical solution limits to the analytical solution from Problem 2.123 in the limit that the conductivity is constant.

(c) What effect does the temperature-dependent conductivity have on the calibration curve

that you generated in part (d) of Problem 2.123?

2.125 You are interested in optimizing the design of a low-temperature, low-current lead that spans the temperature range $T_H = 10$ K to $T_C = 2$ K for a cryogenic experiment. The current lead is $L = 0.1$ m long and must carry $I = 3.5$ amp. You want to select the size of the wire and also examine the impact of copper purity on the design.

The figure illustrates the thermal conductivity of copper for several values of the residual-resistivity ratio (RRR). The residual-resistivity ratio is defined as the electrical resistivity of the material at room temperature to its (extrapolated) value at 0 K. The RRR for a material varies strongly depending on the amount of impurities or other defects that are present and therefore RRR is often used as an indicator of the purity or overall quality of a sample of material. Materials with a high value of RRR are generally more pure than those with a lower value. Note that the figure was generated using the **Conductivity** function in EES with the materials "Copper_RRR50","'Copper_RRR100", etc.

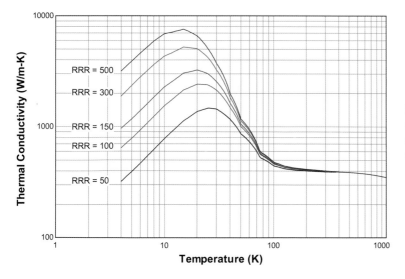

Figure for Problem 2.125 showing conductivity as a function of temperature for various values of copper purity (RRR).

Notice a few things about the figure. First, more pure material (i.e., with a higher value of RRR) tends to have higher conductivity. Second, there is a temperature (T_{max}) at which the thermal conductivity is highest for each value of RRR and the value of T_{max} tends to decrease as the material is more pure.

The thermal conductivity of metals is related almost entirely to the motion of electrons. Electron motion is impeded by interaction of the electrons with impurities as well as the interaction of electrons with lattice vibrations. One model for the conductivity of metals considers the two effects separately (Bejan, 1993). The thermal resistance (i.e., the inverse of the thermal conductivity) that is related to phonon scattering ($1/k_p$) tends to increase with temperature as the lattice vibrations become more significant. This component of the conductivity can be written as:

$$\frac{1}{k_p} = a_p T^2,$$

where a_p is a characteristic constant of the metal. The thermal resistance that is related to impurities ($1/k_i$) tends to decrease with temperature and can be written as:

$$\frac{1}{k_i} = \frac{a_i}{T},$$

where a_i is another characteristic constant of the metal. The total thermal resistance (i.e., the inverse of the total conductivity, k) can be written as the sum of these two effects:

$$\frac{1}{k} = \frac{1}{k_p} + \frac{1}{k_i}.$$

Notice that the two components of the thermal conductivity scale in opposite ways with temperature, thus explaining the peak that is observed in the conductivity curves above.

(a) Derive expressions that relate a_i and a_p to the observed value of the maximum conductivity (k_{max}) and the temperature at which the conductivity reaches a maximum (T_{max}).

(b) Use the Min/Max function in EES to identify k_{max} and T_{max} for the substances "Copper_RRR50", "Copper_RRR_150", and "Copper_RRR500". Report your results in a table.

(c) Use your relations from (a) to determine a_i and a_p for the substances "Copper_RRR50", "Copper_RRR_150", and "Copper_RRR500".

You should see that a_i decreases dramatically as the purity of the copper increases. Add these results to your table.

(d) Overlay the conductivity predicted by your model as a function of temperature onto the conductivity obtained from EES for the substances "Copper_RRR50", "Copper_RRR_150", and "Copper_RRR500". Based on your plot comment on the region where your model works the best.

In the low-temperature ($T \ll T_{max}$) region you should see that thermal conductivity is dominated by the thermal resistance related to impurities (i.e., $k \sim k_i$). Use this model for conductivity to derive the differential equation for 1-D, steady-state conduction through the current lead ignoring thermal energy generation, as shown.

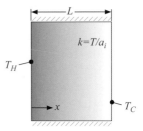

(e) Derive the governing differential equation, clearly show your steps.

(f) Obtain the general solution to the governing differential equation.

(g) Use the boundary conditions $T_{x=0} = T_H$ and $T_{x=L} = T_C$ to obtain the solution for the temperature distribution.

(h) Plot the temperature distribution for copper with RRR = 50 using the value of a_i that you found in part (c) with $T_H = 10$ K, $T_C = 2$ K, and $L = 0.1$ m.

(i) Develop an expression for the heat transfer rate through the lead due to conduction (again, neglect the thermal energy generation due to ohmic dissipation for now).

In the low-temperature ($T \ll T_{max}$) region, the electrical resistivity of metals approaches a constant, referred to as the residual resistivity, $\rho_{e,o}$. When designing a low-current lead for a cryogenic experiment you want to minimize the sum of ohmic dissipation in the lead and conduction heat leak through the lead. At low current we can treat these two effects separately and then add them together to get the total loss associated with the current lead.

(j) Assume that electrical resistivity is constant and equal to $\rho_{e,o} = 0.31$ nΩ-m for RRR = 50 copper, 0.1 nΩ-m for RRR = 150 copper, and 0.021 nΩ-m for RRR = 500 copper. Determine the diameter of the current lead that is optimal for each of these values of RRR and the resulting values of ohmic dissipation, conduction heat leak, and their sum – the total loss. Comment on what purity copper and what size lead would be best for your application.

2.126 The figure shows a cylindrical crucible that you are designing to hold molten material.

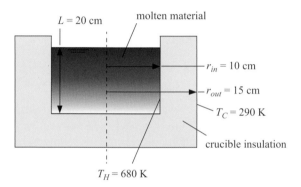

The crucible insulation has inner radius of r_{in} = 10 cm and outer radius r_{out} = 15 cm. The length is L = 20 cm. You are interested in determining the rate of heat transfer radially through the insulation (neglect heat loss from the top or bottom). The outer surface of the insulation is held at T_C = 290 K and the inner surface is at T_H = 680 K. The insulation is made of a metal-reinforced fiber insulation material that has a strongly temperature-dependent conductivity. The conductivity at a few values of temperature is listed in the table.

Temperature (K)	Conductivity (W/m-K)
310	0.038
365	0.046
420	0.056
530	0.078
700	0.140

(a) Plot the conductivity of the insulation as a function of temperature using EES. Use EES to determine the best second-order polynomial curve fit of the form:

$$k = a + bT + cT^2$$

where a, b, and c are the best-fit constants identified by EES, k is the conductivity in W/m-K, and T is the temperature in K. You will need to enter the data in a Lookup table and then use the Curve Fit option from the Plots menu.

(b) Estimate the rate of heat transfer from the crucible by using the solution for steady-state conduction through a cylinder with constant conductivity. Evaluate the conductivity at the average temperature using your best-fit curve from part (a).

You would like to take into account the temperature-dependent conductivity in your solution. The remainder of this problem asks you to develop an analytical solution for 1-D, steady conduction through a cylinder with conductivity that is given by the equation above.

(c) Derive the governing differential equation for the problem.

(d) Obtain the general solution to the differential equation that you derived in (c). Your general solution should include two undetermined constants.

(e) Write the two equations, related to the boundary conditions, that should be used in order to set the undetermined constants in the general solution derived in part (d).

(f) Enter the two equations derived in (e) into EES in order to determine the two undetermined constants.

(g) Determine the rate of heat transfer from the crucible based on your analytical solution.

(h) Plot the error (in percent) between the constant conductivity solution from part (b) and the varying conductivity solution from part (g) as a function of T_H. You should see that as T_H approaches T_C your varying conductivity solution approaches the constant conductivity solution and therefore the error approaches zero.

2.127 A large infrared detector array must be cooled to 100 K within a vacuum enclosure in order to be tested.

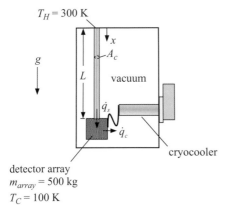

detector array
m_{array} = 500 kg
T_C = 100 K

The array mass is m_{array} = 500 kg and the array is supported by a structural support. One end of the support is at the array temperature, T_C = 100 K, while the other end is at room temperature T_H = 300 K. The majority of the heat leak to the array is related to conduction through the structure, $\dot{q}_s$. Eventually, you will be asked to optimize the design of the support. For now, assume that the support is made of 304 stainless steel (the solid "ss304_cryogenic" in EES) with a cross-sectional area of A_c = 0.001 m² and a length of L = 0.3 m.

(a) Develop a numerical model of the structural support that includes the temperature-dependent conductivity of the support material. Neglect radiation to the support (convection is negligible because the support is in a vacuum). What is the rate of conduction heat transfer to the array from the support?

(b) Show that your model has numerically converged by plotting the rate of conduction heat transfer to the array from the support as a function of the number of nodes used in your numerical model.

Cryogenic design engineers typically use the concept of an integrated average thermal conductivity in order to deal with conduction through materials with temperature-dependent properties without resorting to the numerical models that you generated in (a). The integrated average thermal conductivity is defined as:

$$\bar{k} = \frac{1}{(300\,\text{K} - T)} \int_{T}^{300\,\text{K}} k\, dT.$$

The integrated average conductivity is an appropriately defined average of the conductivity from room temperature down to some temperature, T, which allows you to calculate the rate of conduction heat transfer according to:

$$\dot{q} = \frac{\bar{k}\, A_c}{L} (300[\text{K}] - T).$$

Handbooks for cryogenic design have charts and formulae for the integrated average thermal conductivity of various materials as a function of temperature. For example, the following figure and table summarize the integrated average thermal conductivity of several common structural materials used in cryogenic applications as a function of temperature.

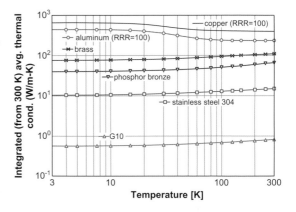

Material	Integrated average thermal conductivity			
	300 K to 150 K	**300 K to 80 K**	**300 K to 20 K**	**300 K to 4 K**
Copper (RRR 100)	404.7 W/m-K	418.6 W/m-K	596.4 W/m-K	655.4 W/m-K
304 Stainless steel	13.5 W/m-K	12.4 W/m-K	10.9 W/m-K	10.4 W/m-K
G-10	0.73 W/m-K	0.67 W/m-K	0.59 W/m-K	0.57 W/m-K
Brass	100.0 W/m-K	90.9 W/m-K	79.6 W/m-K	76.0 W/m-K
Phosphor bronze	54.9 W/m-K	48.2 W/m-K	42.1 W/m-K	39.9 W/m-K

(c) Show that you get approximately the same answer using the integrated average thermal conductivity that you obtained in (a) using a numerical model of the strut.

The array is cooled by a cryocooler. The refrigeration provided by the cryocooler at 100 K is $\dot{q}_c = 1$ W. Your calculated conduction heat transfer through the support from (a) or (c) should be much higher than this. Therefore, the detector array will never reach the desired test temperature.

(d) Allow EES to adjust the cross-sectional area of the support until the rate of heat transfer through the support is equal to the cooling power of the cryocooler.

(e) The allowable stress in the stainless steel support is $\sigma_{all} = 50$ MPa. Determine whether the stress in the support due to the mass of the array is below the allowable stress.

(f) The cost of stainless steel and manufacturing for this application is $C = 1750$ $/kg. The density of stainless steel is $\rho = 7900$ kg/m^3. Determine the cost of the support.

(g) Plot the stress induced in the support as a function of length for lengths between 0.05 m and 1 m. Be sure that you continue to require that the rate of heat transfer through the support match the cryocooler cooling power. Explain the shape of the plot.

(h) Overlay on your plot from (g) the cost of the support as a function of length (you may want to use a secondary y-axis). What is the optimal support geometry and the lowest possible cost?

(i) Use your model to evaluate the composite material G10 (the material "g10_cryogenic" in EES) and brass ('brass' in EES). Would you make the support out of stainless steel, G10, or brass? The allowable stress, density, and mass specific cost of G10 and brass are summarized in the table below.

Material	Allowable stress	Density	Cost
304 stainless steel	50 MPa	7900 kg/m^3	1750 $/kg
G10	2 MPa	1240 kg/m^3	3000 $/kg
brass	30 MPa	8530 kg/m^3	1000 $/kg

(j) Would you expect aluminum to be a good material choice for a strut? Why or why not?

(k) Would you expect polystyrene (Styrofoam) to be a good material choice for a strut? Why or why not?

2.128 The temperature dependence of many materials can be modeled using a simple power law equation:

$$k(T) = k_{300K} \left(\frac{T}{300\,K} \right)^{\alpha},$$

where k_{300K} and α depend on the material. In this problem we will be dealing with silicon ($k_{300K,Si} = 148$ W/m-K, $\alpha_{Si} = -1.3$) and germanium ($k_{300K,Ge} = 60$ W/m-K, $\alpha_{Ge} = -1.25$).

(a) Compare the conductivity predicted using the equation above with the thermal conductivity obtained using the database in EES for both silicon and germanium over a temperature range from 300 K to 800 K.

A cylindrical container has an inner radius of $r_{in} = 0.05$ m that is held at $T_H = 800$ K and an outer radius of $r_{out} = 0.15$ m that is held at $T_c = 300$ K. Assume that the container is made of a material with conductivity given by the equation above.

(b) Derive the differential equation and two boundary conditions for the problem.
(c) Solve the differential equation and determine the two algebraic equations that must be solved in order to obtain the two constants of integration in the general solution.
(d) Compute the heat transfer per unit length for the cylinder assuming its made of silicon and repeat the calculation assuming its made of germanium.

The conductivity of a germanium/silicon alloy, Si_xGe_{1-x}, is given approximately by the same equation where k_{300K} is computed according to (Bejan, 1993):

$$k_{300K} = \cfrac{1}{\cfrac{(1-x)}{k_{300K,\,Si}} + \cfrac{x}{k_{300K,\,Ge}} + \cfrac{(1-x)x}{C_n}},$$

and $C_n = 2.8$ W/m-K is referred to as the bowing factor. The exponent, a, can be computed according to:

$$a = (1 - x)a_{Si} + xa_{Ge}.$$

(e) Plot the conductivity of the alloy as a function of x, the concentration of silicon, at $T = 500$ K. Explain the shape of the plot.
(f) Plot the heat transfer per unit length as a function of x. If your goal was to minimize the heat transfer what alloy would you select?

3 | Extended Surface Problems

Chapter 2 considered problems that were truly one-dimensional. Energy transfer occurred only in one coordinate direction and therefore temperature varied only in that direction. In this chapter, we will examine problems that are only *approximately* one-dimensional, referred to generally as **extended surfaces**. Extended surfaces are typically thin pieces of conductive material that can be approximated as being isothermal in two dimensions and having temperature variations in only one direction. In an extended surface, energy *is* transferred laterally (i.e., across the thickness) but the temperature change induced by the energy transfer is sufficiently small that it can be neglected. The **extended surface approximation** greatly reduces the complexity of the problem and can often be applied with little loss in accuracy.

In this chapter, we will learn how to justify the extended surface approximation and how to solve extended surface problems both analytically and numerically. In addition, we will learn about extended surface solutions that apply specifically to **fins**, which are structures meant to enhance heat transfer in most heat exchangers.

3.1 Extended Surfaces

3.1.1 The Extended Surface Approximation

An extended surface problem is not truly one-dimensional. For example, Figure 3.1 shows a simple **fin**, which is a piece of material attached to a heat transfer surface in order to increase the surface area that is available for convection. You will see fins installed on almost every heat exchanger (e.g., a car radiator or the evaporator in an air conditioner) in order to increase the heat transfer rate.

The fin length (in the x-direction) is L and its thickness (in the y-direction) is th. The fin is assumed to be very long in the z-direction; here we will assume a width, W. The conductivity of the fin material is k. The base of the

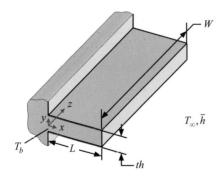

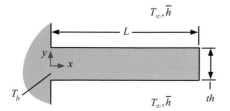

Figure 3.1 A constant cross-section fin.

fin (at $x = 0$) is maintained at temperature T_b and the surface of the fin experiences convection to fluid at temperature T_∞ with heat transfer coefficient $\bar{h}$.

This is *not* a one-dimensional problem because energy is moving in both the x- and y-directions. Energy is conducted axially from the base of the fin along its length in the x-direction. Energy is also conducted laterally (i.e., in the y-direction) to the fin surface where it is finally transferred by convection to the surrounding fluid. Temperature gradients always accompany heat transfer; thus, the temperature must vary in both the x- and y-directions and the temperature distribution in the fin must be two-dimensional (2-D). However, there are many situations where the temperature gradient in one coordinate direction within the material is small relative to the temperature gradient in the other direction so that it can be neglected in the solution without significant loss of accuracy. This is the **extended surface approximation**.

3.1.2 The Biot Number

Figure 3.2 illustrates the temperature in the fin shown in Figure 3.1 as a function of x/L, where x is varied from 0 (at the base) to L (at the tip). The temperature is shown at two values of y, corresponding to $y = 0$ (at the center) and $y = th/2$ (at the surface). Notice that the temperature decreases in both the x- and y-directions because conduction heat transfer occurs in both of these directions. The temperature decrease in the x-direction is evident in both of the two lines shown in Figure 3.2. The temperature decrease in the y-direction is less obvious but manifests itself as the temperature difference between the $y = 0$ and the $y = th/2$ lines. At every value of axial location, x, there is a temperature drop within the material in the y-direction due to conduction. At $x/L = 0.2$, this temperature drop is labeled $\Delta T_{cond,y}$ in Figure 3.2. There is another temperature drop that occurs from the surface of the fin to the surrounding fluid due to convection; this temperature difference at $x/L = 0.2$ is labeled ΔT_{conv} in Figure 3.2.

The extended surface approximation makes the assumption that the temperature in the material is approximately a function of x only, and not of y. This approximation turns a 2-D problem into a 1-D problem, which is much easier to solve. For the situation shown in Figure 3.2, the extended surface approximation would be valid in the situation when the temperature drop due to conduction in the y-direction is much less than the temperature drop due to convection, that is $\Delta T_{cond,y} << \Delta T_{conv}$. In this case, the two lines shown in Figure 3.2 will approximately collapse into one.

The best way to compare the magnitude of these two temperature drops is to think in terms of the heat transfer resistances that are involved in these processes. While this problem cannot be solved exactly using thermal resistances, it can certainly be examined and understood using them. In fact, thermal resistances are our primary tool for understanding what is important in almost every heat transfer problem. There are two thermal resistances that oppose heat transfer in the y-direction, conduction ($R_{cond,y}$) and convection (R_{conv}). The resistance due to conduction in the y-direction cannot be calculated exactly, but it can be estimated according to:

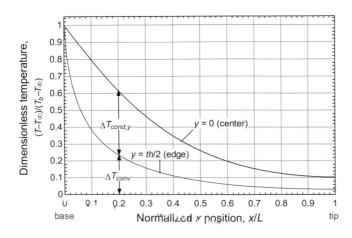

Figure 3.2 Dimensionless temperature in the fin shown in Figure 3.1 as a function of x/L for two values of y/th.

$$R_{cond,y} = \frac{(\text{distance to conduct})}{k(\text{area for conduction})}. \tag{3.1}$$

The distance to conduct in the y-direction must be related to the thickness. Here we will use $th/2$. The area for conduction in the y-direction is $W\,L$. With these substitutions, Eq. (3.1) becomes:

$$R_{cond,y} = \frac{th}{2\,k\,W\,L}. \tag{3.2}$$

The resistance due to convection (R_{conv}) is:

$$R_{conv} = \frac{1}{\bar{h}\,W\,L}. \tag{3.3}$$

The ratio of the two temperature differences shown in Figure 3.2 is related to the ratio of the resistances that we just calculated:

$$\frac{\Delta T_{cond,y}}{\Delta T_{conv}} \approx \frac{R_{cond,y}}{R_{conv}}. \tag{3.4}$$

The validity of the extended surface approximation increases as the ratio of these two resistances becomes much less than one. A ratio of resistances used for this purpose results in a dimensionless number that is referred to as the **Biot number** (Bi). In this case:

$$Bi = \frac{R_{cond,y}}{R_{conv}}. \tag{3.5}$$

Substituting Eqs. (3.2) and (3.3) into Eq. (3.5) leads to

$$Bi = \frac{th}{2\,k\,W\,L}\,\frac{\bar{h}\,W\,L}{1} = \frac{th\,\bar{h}}{2\,k}. \tag{3.6}$$

As the Biot number becomes smaller, there is less error introduced by the extended surface approximation. A typical threshold used to justify the extended surface approximation is that the Biot number should be less than 0.1. However, this is clearly a matter of engineering judgment and the threshold for an allowable Biot number cannot be stated without some knowledge of the application and the required accuracy of the solution. In the limit that the Biot number calculated using Eq. (3.6) is equal to 0.1, the temperature distribution is shown in Figure 3.3. Notice that the difference between the temperature at the center of the fin ($y = 0$) and the edge of the fin ($y = th/2$) is small and therefore the temperature in the fin is approximately only a function of x. The extended surface approximation is justified: $T(x, y) \approx T(x)$.

The Biot number calculated using Eq. (3.6) is specific to the common situation in which there is a conductive and a convective resistance in series. It would not be appropriate if, for example, there were also an important

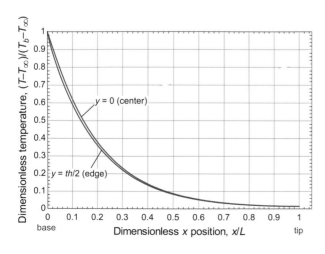

Figure 3.3 Dimensionless temperature in the fin shown in Figure 3.1 as a function of x/L for two values of y/th when the Biot number is $Bi = 0.1$.

radiation resistance from the surface. In general, the Biot number is the ratio of two thermal resistances; one resistance is related to conduction within a material in some direction and the other is related to heat transfer from the surface of the material to the surroundings:

$$Bi = \frac{\text{resistance to conduction in some direction}}{\text{resistance from the surface to the surroundings}}. \tag{3.7}$$

If the Biot number is much less than unity, then it is possible to neglect temperature gradients due to conduction in the corresponding direction, allowing a simpler model to be used to examine the situation. The extended surface approximation inherently neglects a conduction resistance in one or more directions. The resulting extended surface model is simpler and can be justified because the conduction resistance that you are neglecting is suitably small. However, "small" is always a relative term and the size of the conduction resistance can only be judged in relation to the resistance from the surface of the object to the surroundings.

3.2 Analytical Solutions to Extended Surface Problems

Extended surface problems can be solved both analytically and numerically. In this section, we will go through the process of developing an analytical solution to extended surface problems.

3.2.1 Deriving the ODE and Boundary Conditions

The analytical solution of an extended surface problem begins with the derivation of the governing differential equation. The steps involved in this process are the same as those discussed in Chapter 2 for the one-dimensional problems that were considered there. We will illustrate the process in the context of the fin that is discussed in Section 3.1 and shown in Figure 3.1. The first step is to define a differential control volume. The differential control volume should include material that is at a uniform temperature. Therefore, according to the extended surface approximation, the control volume must be differential in x but not in y or z, as shown in Figure 3.4. The energy balance suggested by Figure 3.4 is:

$$(\dot{q}_x)_x = \dot{q}_{conv} + (\dot{q}_x)_{x+dx}. \tag{3.8}$$

Expanding the higher-order term provides:

$$(\dot{q}_x)_{x+dx} = (\dot{q}_x)_x + \frac{d\dot{q}_x}{dx}dx. \tag{3.9}$$

Substituting Eq. (3.9) into Eq. (3.8) and simplifying leads to:

$$0 = \dot{q}_{conv} + \frac{d\dot{q}_x}{dx}dx. \tag{3.10}$$

Equation (3.10) is an ODE written in terms of energy, and indicates that the rate of conduction heat transfer is not constant but rather changes with position due to the convection from the surface. In order to recast our ODE in terms of temperature it is necessary to substitute rate equations. The rate equation for convection is given by Newton's Law of Cooling:

$$\dot{q}_{conv} = \underbrace{per\,dx}_{\substack{A_s\ \text{within} \\ \text{control volume}}} \bar{h}(T - T_\infty), \tag{3.11}$$

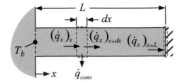

Figure 3.4 Differential control volume used to derive the governing differential equation for an extended surface.

where *per* is the perimeter of the fin; for the rectangular cross-section shown in Figure 3.1, $per = 2\,(W + th)$. However, Eq. (3.11) remains valid for other fin cross-sections. The rate equation for the conduction term is Fourier's Law:

$$\dot{q}_x = -k\,A_c\,\frac{dT}{dx},\tag{3.12}$$

where A_c is the cross-sectional area of the fin. For the fin shown in Figure 3.1, $A_c = W\,th$ but, again, other fin shapes are possible. Substituting Eqs. (3.11) and (3.12) into Eq. (3.10) leads to:

$$0 = per\,dx\,\bar{h}(T - T_\infty) + \frac{d}{dx}\left(-k\,A_c\,\frac{dT}{dx}\right)dx.\tag{3.13}$$

The cross-sectional area and conductivity are assumed to be constant, allowing Eq. (3.13) to be simplified:

$$\frac{d^2 T}{dx^2} - \frac{per\,\bar{h}}{k\,A_c}T = -\frac{per\,\bar{h}}{k\,A_c}T_\infty.\tag{3.14}$$

The boundary conditions for the problem shown in Figure 3.4 are a specified base temperature

$$T_{x=0} = T_b\tag{3.15}$$

and an adiabatic tip

$$(\dot{q}_x)_{x=L} = 0.\tag{3.16}$$

Substituting Fourier's Law into Eq. (3.16) provides:

$$\left.\frac{dT}{dx}\right|_{x=L} = 0.\tag{3.17}$$

3.2.2 Solving the ODE

Equation (3.14) is a second-order, nonhomogeneous, linear ODE. It is worth understanding what each of these terms mean before proceeding. The **order** of the equation refers to order of the highest-order derivative that is present; in Eq. (3.14), the highest order derivative is second order. A **homogeneous** equation is one where any multiple of a solution ($C\,T$ where C is some arbitrary constant and T is a solution) is itself a solution. Substituting $C\,T$ into Eq. (3.14) for T leads to:

$$C\left(\frac{d^2 T}{dx^2} - \frac{per\,\bar{h}}{k\,A_c}T\right) = -\frac{per\,\bar{h}}{k\,A_c}T_\infty.\tag{3.18}$$

Substituting Eq. (3.14) into Eq. (3.18) leads to:

$$C\left(-\frac{per\,\bar{h}}{k\,A_c}T_\infty\right) = -\frac{per\,\bar{h}}{k\,A_c}T_\infty,\tag{3.19}$$

which can only be true if $C = 1$. Therefore, Eq. (3.14) is **nonhomogeneous**. A **linear** ODE does not contain any products of the dependent variable (T) or its derivative; therefore, Eq. (3.14) is linear.

Equation (3.14) is *not* separable. The ODE therefore cannot be solved by direct integration (as was possible for the problems encountered in Chapter 2) because it is not possible to separate the x and T portions of the differential equation. A differential equation like Eq. (3.14) is typically solved by dividing it into a **homogeneous differential equation** and a **nonhomogeneous** (or **particular**) **differential equation**. We express the solution T as the sum of a homogeneous solution (T_h) and particular (nonhomogeneous) solution (T_p):

$$T = T_h + T_p.\tag{3.20}$$

Substituting Eq. (3.20) into Eq. (3.14) leads to:

$$\frac{d^2\left(T_h + T_p\right)}{dx^2} - \frac{per\,\bar{h}}{k\,A_c}\left(T_h + T_p\right) = -\frac{per\,\bar{h}}{k\,A_c}T_\infty\tag{3.21}$$

or

$$\underbrace{\frac{d^2 T_h}{dx^2} - \frac{per\,\bar{h}}{k\,A_c}\,T_h}_{\substack{=0 \\ \text{for homogeneous} \\ \text{differential equation}}} + \underbrace{\frac{d^2 T_p}{dx^2} - \frac{per\,\bar{h}}{k\,A_c}\,T_p = -\frac{per\,\bar{h}}{k\,A_c}\,T_\infty}_{\substack{\text{whatever is left over must be the} \\ \text{particular differential equation}}}. \tag{3.22}$$

Extract from Eq. (3.22) the homogeneous differential equation for T_h:

$$\frac{d^2 T_h}{dx^2} - \frac{per\,\bar{h}}{k\,A_c}\,T_h = 0 \tag{3.23}$$

and whatever is left over must be the particular differential equation for T_p:

$$\frac{d^2 T_p}{dx^2} - \frac{per\,\bar{h}}{k\,A_c}\,T_p = -\frac{per\,\bar{h}}{k\,A_c}\,T_\infty. \tag{3.24}$$

We can start by "solving" the homogeneous differential equation, Eq. (3.23). How do we solve this equation? Actually, functions have been defined specifically to solve various types of homogeneous equations. One function that solves Eq. (3.23) is the **exponential**. To see that this is true, assume a solution with an exponential form:

$$T_h = C \exp(m\,x), \tag{3.25}$$

where m and C are both arbitrary constants. Substitute Eq. (3.25) into Eq. (3.23):

$$C\,m^2 \exp(m\,x) - \frac{per\,\bar{h}}{k\,A_c}\,C \exp(m\,x) = 0. \tag{3.26}$$

Equation (3.26) is satisfied provided that

$$m^2 = \frac{per\,\bar{h}}{k\,A_c}. \tag{3.27}$$

Equation (3.27) is a quadratic equation for m and there are two roots to this equation. Therefore, we have actually identified two solutions ($T_{h,1}$ and $T_{h,2}$) to Eq. (3.23), corresponding to the positive and negative roots of Eq. (3.27):

$$T_{h,1} = C_1 \exp(m\,x) \tag{3.28}$$

and

$$T_{h,2} = C_2 \exp(-m\,x), \tag{3.29}$$

where

$$m = \sqrt{\frac{per\,\bar{h}}{k\,A_c}}. \tag{3.30}$$

Because Eq. (3.23) is a linear, homogeneous ODE, the sum of the two solutions is also a solution:

$$T_h = C_1 \exp(m\,x) + C_2 \exp(-m\,x). \tag{3.31}$$

Equation (3.31) is the homogeneous solution and it will solve the homogeneous differential equation, Eq. (3.23), regardless of the choice of C_1 and C_2.

Next, the nonhomogeneous (particular) differential equation, Eq. (3.24), must be solved. Substituting Eq. (3.30) into Eq. (3.24) leads to:

$$\frac{d^2 T_p}{dx^2} - m^2\,T_p = -m^2\,T_\infty. \tag{3.32}$$

The particular solution is solved by the **method of undetermined coefficients**. The method of undetermined coefficients proceeds by examining the form of the right side of the differential equation and then formulating a solution that includes that functional form as well as its derivatives, each multiplied by an unknown coefficient. In Eq. (3.32), the right side of the differential equation is a constant ($-m^2 T_\infty$) and therefore our assumed solution should include a constant (1) and its derivative (zero):

$$T_p = C_3(1). \tag{3.33}$$

The parameter C_3 in Eq. (3.33) is the *undetermined coefficient* that multiplies the constant. The method of undetermined coefficients substitutes the assumed form of the solution, with its undetermined coefficients, into the particular differential equation. Then the undetermined coefficients are selected so that the particular differential equation is satisfied. Substituting Eq. (3.33) into Eq. (3.24) leads to:

$$-m^2 C_3 = -m^2 T_\infty \tag{3.34}$$

or

$$C_3 = T_\infty. \tag{3.35}$$

Substituting Eq. (3.35) into Eq. (3.33) leads to the particular solution:

$$T_p = T_\infty. \tag{3.36}$$

Substituting the homogeneous and particular solutions, Eqs. (3.31) and (3.36), into Eq. (3.20) provides the general solution to the ODE:

$$T = C_1 \exp(m\,x) + C_2 \exp(-m\,x) + T_\infty. \tag{3.37}$$

3.2.3 Applying the Boundary Conditions

Equation (3.37) represents the solution to Eq. (3.14) to within two undetermined constants, C_1 and C_2. We must select C_1 and C_2 in such a way that the solution also satisfies the boundary conditions. In our example, we know the temperature at the base of the fin and also that its tip is insulated, which provides the boundary conditions given by Eqs. (3.15) and (3.17). Substituting Eq. (3.37) into Eq. (3.15) provides:

$$C_1 + C_2 + T_\infty = T_b. \tag{3.38}$$

Substituting Eq. (3.37) into Eq. (3.17) provides:

$$C_1 m \exp(m\,L) - C_2 m \exp(-m\,L) = 0. \tag{3.39}$$

Equations (3.38) and (3.39) are two equations in the two unknowns C_1 and C_2. Equation (3.38) is multiplied by $m \exp(m\,L)$ and rearranged:

$$C_1 m \exp(m\,L) + C_2 m \exp(m\,L) = (T_b - T_\infty)m \exp(m\,L). \tag{3.40}$$

Equation (3.40) is subtracted from Eq. (3.39):

$$\begin{aligned} &C_1 m \exp(m\,L) - C_2 m \exp(-m\,L) = 0 \\ &\underline{-[C_1 m \exp(m\,L) + C_2 m \exp(m\,L) = (T_b - T_\infty)\,m \exp(m\,L)]} \\ &-C_2 m \exp(-m\,L) - C_2 m \exp(m\,L) = -(T_b - T_\infty)\,m \exp(m\,L) \end{aligned}. \tag{3.41}$$

Equation (3.41) can be solved for C_2:

$$C_2 = \frac{(T_b - T_\infty)\exp(m\,L)}{\exp(-m\,L) + \exp(m\,L)}. \tag{3.42}$$

A similar sequence of operations leads to:

$$C_1 = \frac{(T_b - T_\infty)\exp(-m\,L)}{\exp(-m\,L) + \exp(m\,L)}. \tag{3.43}$$

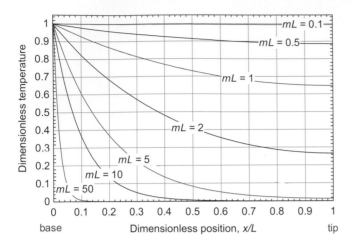

Figure 3.5 Dimensionless temperature of a fin with an adiabatic tip as a function of x/L for various values of the parameter mL.

Substituting Eqs. (3.42) and (3.43) into Eq. (3.37) leads to:

$$T = \underbrace{\frac{(T_b - T_\infty)\exp(-mL)}{\exp(-mL) + \exp(mL)}}_{C_1}\exp(mx) + \underbrace{\frac{(T_b - T_\infty)\exp(mL)}{\exp(-mL) + \exp(mL)}}_{C_2}\exp(-mx) + T_\infty, \tag{3.44}$$

which can be simplified:

$$\frac{(T - T_\infty)}{(T_b - T_\infty)} = \frac{\exp\left[-mL\left(1 - \frac{x}{L}\right)\right] + \exp\left[mL\left(1 - \frac{x}{L}\right)\right]}{\exp(-mL) + \exp(mL)}. \tag{3.45}$$

Figure 3.5 illustrates the dimensionless temperature predicted by Eq. (3.45) as a function of x/L for various values of mL.

Example 3.1

A frying pan has a handle that is in the form of a tube made of brass. The purpose of the handle is to allow the cook to grab the pan without burning his or her hand even when the pan itself is quite hot. Examine a situation where the outer diameter of the handle is $D_{out} = 1$ inch and the handle is $L = 12$ inch long. The base of the handle, where it connects to the pan, has temperature $T_b = 250°F$. The outer surface of the handle is surrounded by air at $T_\infty = 70°F$ with heat transfer coefficient $\bar{h} = 1.32$ Btu/ft²-hr-R.

Determine:
- The temperature at $x = 6$ inch from the base if the inner diameter of the tube is $D_{in} = 0.5$ inch.
- The inner diameter of the tube that is required to keep the temperature at $x = 6$ inch from the base at a safe temperature to touch, $T = 120°F$.

Known Values

Figure 1 illustrates a sketch of the situation with all known values indicated and converted to base SI units. The conductivity of brass is obtained from Appendix A, $k = 79.3$ W/m-K.

Example 3.1 (cont.)

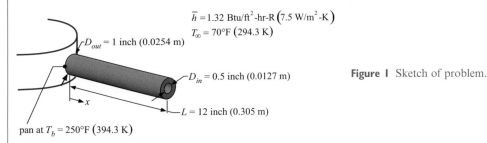

$\bar{h} = 1.32 \text{ Btu/ft}^2\text{-hr-R} \left(7.5 \text{ W/m}^2\text{-K}\right)$

$T_\infty = 70°\text{F} \left(294.3 \text{ K}\right)$

$D_{out} = 1 \text{ inch } (0.0254 \text{ m})$

$D_{in} = 0.5 \text{ inch } (0.0127 \text{ m})$

Figure 1 Sketch of problem.

$L = 12 \text{ inch } (0.305 \text{ m})$

pan at $T_b = 250°\text{F} \left(394.3 \text{ K}\right)$

Assumptions

- There is no convection to the air inside the handle, only from the external surface. The heat transfer coefficient on the outer surface is likely to be higher than on the inner surface and the temperature of the air trapped within the handle will be higher than T_∞, justifying this assumption. Also, the assumption makes the solution conservative in that the actual temperature at $x = 6$ inch will be lower than what is calculated.
- The end of the handle at $x = L$ is adiabatic. This allows the use of the extended surface solution derived in Section 3.2 and shown in Figure 3.5. The cross-sectional area at the tip is small relative to the area of the handle that is exposed to air and the temperature of the tip surface is low; therefore, ignoring convection from the tip will not significantly affect our solution.
- Radiation is neglected.
- Steady-state conditions are assumed.
- The extended surface approximation is valid. This assumption should be justified with a Biot number, which is the ratio of the resistance to conduction in the radial direction to convection from the outer surface:

$$Bi = \frac{R_{cond,r}}{R_{conv}} = \frac{(D_{out} - D_{in})\,\bar{h}\,\pi\,D_{out}\,L}{2\,k\,\pi\,D_{out}\,L} = \frac{(D_{out} - D_{in})\bar{h}}{2\,k}$$

$$= \frac{(0.0254 - 0.0127)\text{m}}{2} \left| \frac{7.5 \text{ W}}{\text{m}^2 \text{ K}} \right| \frac{\text{m K}}{79.3 \text{ W}} = 0.0006. \tag{1}$$

The Biot number is much less than one, justifying the extended surface approximation.

Analysis and Solution

The solution derived in Section 3.2 will be used for this problem. When the inner diameter is given, it becomes possible to compute mL and then use either Eq. (3.45) or Figure 3.5 to determine the dimensionless temperature at $x = 6$ inch which corresponds to $x/L = 0.5$. The cross-sectional area for conduction is:

$$A_c = \frac{\pi}{4}\left(D_{out}^2 - D_{in}^2\right) = \frac{\pi}{4}\left|\frac{(0.0254^2 - 0.0127^2)\text{m}^2}{}\right| = 0.00038 \text{ m}^2. \tag{2}$$

The perimeter for convection is:

$$per = \pi\,D_{out} = \pi\,(0.0254 \text{ m}) = 0.0798 \text{ m}. \tag{3}$$

The fin parameter is:

$$mL = \sqrt{\frac{\bar{h}\,per}{k\,A_c}}\,L = \sqrt{\frac{7.5 \text{ W}}{\text{m}^2 \text{ K}}\left|\frac{0.0798 \text{ m}}{}\right|\frac{\text{m K}}{79.3 \text{ W}}\left|\frac{}{0.00038 \text{ m}^2}\right|}\,(0.3048 \text{ m}) = 1.358. \tag{4}$$

Continued

Example 3.1 (cont.)

Examination of Figure 3.5 shows that the dimensionless temperature for $mL = 1.358$ at $x/L = 0.5$ is approximately 0.65. Therefore, the handle temperature 6 inches from the pot is:

$$T = T_\infty + 0.65(T_b - T_\infty) = 294.3\text{K} + 0.65\,(394.3 - 294.3)\text{K} = \boxed{359.3\text{ K}(187°\text{F})}. \tag{5}$$

In order to reduce the temperature at $x = 6$ inch from 187°F to 120°F (322.0 K), the dimensionless temperature at $x/L = 0.5$ must be reduced to:

$$\frac{(T - T_\infty)}{(T_b - T_\infty)} = \frac{(322.0 - 294.3)\text{ K}}{(394.3 - 294.3)\text{ K}} = 0.278. \tag{6}$$

Examination of Figure 3.5 shows that the fin constant must be increased to somewhere between $mL = 2$ and $mL = 5$ in order for this to occur. Here let's estimate $mL = 3.5$ as the appropriate value.

$$m\,L = \sqrt{\frac{\bar{h}\,per}{k\,A_c}}L = 3.5. \tag{7}$$

Solving Eq. (7) for the required cross-sectional area provides:

$$A_c = \frac{\bar{h}\,per\,L^2}{k(3.5)^2} = \frac{7.5\text{ W}}{\text{m}^2\text{ K}}\left|\frac{0.0798\text{ m}}{}\right|\frac{(0.3048)^2\text{ m}^2}{}\left|\frac{\text{m K}}{79.3\text{ W}}\right|\frac{1}{(3.5)^2} = 5.72 \times 10^{-5}\text{ m}^2. \tag{8}$$

The inner diameter of the handle must be selected to provide the required cross-sectional area:

$$A_c = \frac{\pi}{4}\left(D_{out}^2 - D_{in}^2\right). \tag{9}$$

Solving Eq. (9) for the inner diameter provides:

$$D_{in} = \sqrt{D_{out}^2 - \frac{4A_c}{\pi}} = \sqrt{(0.0254)^2\text{ m}^2 - \frac{4}{\pi}\left|\frac{5.72 \times 10^{-5}\text{ m}^2}{}\right|} = \boxed{0.0239\text{ m }(0.942\text{ inch})}. \tag{10}$$

Discussion

In order for the handle temperature to be reduced to an acceptable value, it was necessary to decrease the cross-sectional area, which increased the resistance to conduction in the x-direction causing a larger temperature drop along the fin. The result is a handle with a very thin wall that may not be structurally sound. An alternative technique might be to select a material with a lower thermal conductivity, such as stainless steel.

3.2.4 Hyperbolic Trigonometric Functions

The solution for extended surface problems can be derived and expressed more easily using hyperbolic functions as opposed to exponentials. The **hyperbolic cosine** (cosh, pronounced "kosh") and the **hyperbolic sine** (sinh, pronounced "cinch") are defined according to:

$$\cosh(A) = \frac{1}{2}[\exp(A) + \exp(-A)] \tag{3.46}$$

$$\sinh(A) = \frac{1}{2}[\exp(A) - \exp(-A)]. \tag{3.47}$$

These hyperbolic functions behave in much the same way that the trigonometric functions, cosine and sine, do. For example, you can show that

$$\cosh^2(A) - \sinh^2(A) = 1, \tag{3.48}$$

which is analogous to the trigonometric identity

$$\cos^2(A) + \sin^2(A) = 1. \tag{3.49}$$

Furthermore, the derivative of cosh is sinh and vice versa, which is analogous to the derivatives of sine and cosine (albeit, without the sign change):

$$\frac{d}{dx}[\sinh(A)] = \cosh(A)\frac{dA}{dx} \tag{3.50}$$

$$\frac{d}{dx}[\cosh(A)] = \sinh(A)\frac{dA}{dx}. \tag{3.51}$$

The solution for the fin shown in Figure 3.4, Eq. (3.45), can be expressed in terms of hyperbolic cosines:

$$\frac{(T - T_\infty)}{(T_b - T_\infty)} = \frac{\exp\left[-mL\left(1 - \frac{x}{L}\right)\right] + \exp\left[mL\left(1 - \frac{x}{L}\right)\right]}{2} \frac{2}{\exp(-mL) + \exp(mL)} \tag{3.52}$$

$$\frac{(T - T_\infty)}{(T_b - T_\infty)} = \frac{\cosh\left[mL\left(1 - \frac{x}{L}\right)\right]}{\cosh(mL)}. \tag{3.53}$$

Equation (3.53) is more concise but functionally identical to Eq. (3.45).

3.2.5 Solutions to Linear Homogeneous ODEs

The solution to the homogeneous ODE that was derived for this extended surface problem, Eq. (3.23), is repeated below:

$$\frac{d^2 T_h}{dx^2} - m^2 T_h = 0. \tag{3.54}$$

The solution was expressed in terms of exponentials in Section 3.2.2:

$$T_h = C_1 \exp(mx) + C_2 \exp(-mx). \tag{3.55}$$

We could just as easily and equivalently have expressed the solution in terms of hyperbolic cosine and hyperbolic sine:

$$T_h = C_1 \cosh(mx) + C_2 \sinh(mx), \tag{3.56}$$

where, in Eqs. (3.55) and (3.56), C_1 and C_2 are undetermined constants that must be obtained by satisfying the boundary conditions; note that the undetermined constants in Eq. (3.55) will be different from those in Eq. (3.56). Typically it will be more convenient to use the hyperbolic form of the solution, Eq. (3.56), when solving extended surface problems. Table 3.1 lists solutions to some common, linear homogeneous ODEs that arise when solving extended surface problems with different geometries. For example, Entry 1 results from the analysis of an extended surface that has a constant cross-sectional area and perimeter, as presented in Section 3.2.

Table 3.1 Solutions to some common linear homogeneous ODEs.

Entry	Linear homogeneous ODE	Solution
1	$\dfrac{d^2 T_h}{dx^2} - m^2 T_h = 0$	$T_h = C_1 \exp(m\,x) + C_2 \exp(-m\,x)$ $T_h = C_1 \cosh(m\,x) + C_2 \sinh(m\,x)$
2	$\dfrac{d^2 T_h}{dx^2} + m^2 T_h = 0$	$T_h = C_1 \cos(m\,x) + C_2 \sin(m\,x)$
3	$\dfrac{dT_h}{dt} + \dfrac{T_h}{\tau} = 0$	$T_h = C_1 \exp\left(-\dfrac{t}{\tau}\right)$
4	$\dfrac{d}{dx}\left(x\dfrac{dT_h}{dx}\right) - m^2 T_h = 0$	$T_h = C_1\,\mathrm{BesselI}(0, 2m\sqrt{x}) + C_2\,\mathrm{BesselK}(0, 2m\sqrt{x})$
5	$\dfrac{d}{dx}\left(x\dfrac{dT_h}{dx}\right) - m^2 x T_h = 0$	$T_h = C_1\mathrm{BesselI}(0, m\,x) + C_2\mathrm{BesselK}(0, m\,x)$
6	$\dfrac{d}{dx}\left(x\dfrac{dT_h}{dx}\right) - m^2 x^2 T_h = 0$	$T_h = C_1\,\mathrm{BesselI}\left(0, \tfrac{2}{3}m\,x^{3/2}\right) + C_2\,\mathrm{BesselK}\left(0, \tfrac{2}{3}m\,x^{3/2}\right)$
7	$\dfrac{d}{dx}\left(x\dfrac{dT_h}{dx}\right) + m^2 T_h = 0$	$T_h = C_1\,\mathrm{BesselJ}(0, 2m\sqrt{x}) + C_2\,\mathrm{BesselY}(0, 2m\sqrt{x})$
8	$\dfrac{d}{dx}\left(x\dfrac{dT_h}{dx}\right) + m^2 x T_h = 0$	$T_h = C_1\,\mathrm{BesselJ}(0, m\,x) + C_2\,\mathrm{BesselY}(0, m\,x)$
9	$\dfrac{d}{dx}\left(x\dfrac{dT_h}{dx}\right) + m^2 x^2 T_h = 0$	$T_h = C_1\,\mathrm{BesselJ}\left(0, \tfrac{2}{3}m\,x^{3/2}\right) + C_2\,\mathrm{BesselY}\left(0, \tfrac{2}{3}m\,x^{3/2}\right)$

The functions involved are summarized below:
- exp is the exponential function
- cos is the trigonometric cosine function
- sin is the trigonometric sine function
- cosh is the hyperbolic cosine function
- sinh is the hyperbolic sine function
- BesselI(n, x) is the nth-order Modified Bessel function of the first kind
- BesselK(n, x) is the nth-order Modified Bessel function of the second kind
- BesselJ(n, x) is the nth-order Bessel function of the first kind
- BesselY(n, x) is the nth-order Bessel function of the second kind.

Example 3.2

A cylindrical mounting bracket extends from the wall of a furnace and holds an object that is being heat-treated. The radiation from the flame in the furnace can be modeled as a heat flux, $\dot{q}'' = 4.5$ kW/m^2, that is uniformly incident on the outer surface of the bracket. The outer surface of the cylinder is also exposed to convection to air at $T_\infty = 50°$C with heat transfer coefficient $\bar{h} = 25$ W/m^2-K. The diameter of the bracket is $D = 1.1$ cm and its length is $L = 10.2$ cm. The conductivity of the bracket material is $k = 24$ W/m-K. The end of the bracket mounted to the wall ($x = 0$) is held at temperature $T_w = 30°$C while the end of the bracket attached to the object ($x = L$) is at temperature $T_o = 400°$C.

Determine:
- The temperature as a function of position (x) within the bracket.
- The rate of heat transfer from the object to the bracket.

Example 3.2 (cont.)

Known Values

The known parameters are indicated on the sketch of the bracket shown in Figure 1.

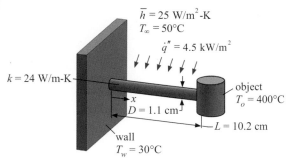

$\bar{h} = 25$ W/m²-K
$T_\infty = 50°C$
$\dot{q}'' = 4.5$ kW/m²

$k = 24$ W/m-K

x
$D = 1.1$ cm

object
$T_o = 400°C$

$L = 10.2$ cm

wall
$T_w = 30°C$

Figure 1 Bracket holding object in a furnace.

Assumptions

- Steady-state conditions exist.
- Thermal conductivity is constant.
- All of the incident heat flux is absorbed
- The temperature in the bracket varies approximately only in the x-direction. This extended surface approximation can be justified by calculating a Biot number, which is the ratio of the resistance to conduction in the radial direction to the resistance to convection from the surface:

$$Bi = \frac{R_{cond,r}}{R_{conv}} = \frac{(D/2)}{k \pi D L} \bar{h} \pi D L = \frac{D\bar{h}}{2k} = \frac{0.011 \text{ m}}{2} \left| \frac{25 \text{ W}}{\text{m}^2\text{K}} \right| \frac{\text{m K}}{24 \text{ W}} = 0.0057. \tag{1}$$

The Biot number is sufficiently less than 1 to justify the extended surface approximation.

Analysis

A differential energy balance on the bracket is shown in Figure 2. The energy balance must be differential in the x-direction but can extend over the entire cross-section of the bracket according to the extended surface approximation.

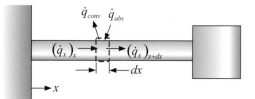

$\dot{q}_{conv}$ $\dot{q}_{abs}$

$(\dot{q}_x)_x$ → ← $(\dot{q}_x)_{x+dx}$

dx

x

Figure 2 Differential control volume.

The differential energy balance shown in Figure 2 leads to:

$$(\dot{q}_x)_x + \dot{q}_{abs} = \dot{q}_{conv} + (\dot{q}_x)_{x+dx}. \tag{2}$$

The final term can be expanded:

$$(\dot{q}_x)_x + \dot{q}_{abs} = \dot{q}_{conv} + (\dot{q}_x)_x + \frac{d\dot{q}_x}{dx} dx. \tag{3}$$

The rate of incident heat transfer absorbed within the control volume is the product of the heat flux and the surface area, *per dx*, where *per* is the perimeter of the cylinder:

$$\dot{q}_{abs} = \dot{q}'' \, per \, dx. \tag{4}$$

Continued

Example 3.2 (cont.)

The rate of convection leaving the control volume is

$$\dot{q}_{conv} = \bar{h}\, per\, dx(T - T_\infty). \tag{5}$$

The rate of conduction in the x-direction is

$$\dot{q}_x = -k\, A_c \frac{dT}{dx}, \tag{6}$$

where A_c is the cross-sectional area of the cylinder. Substituting Eqs. (4) through (6) into Eq. (3) provides:

$$\dot{q}''\, per\, dx = \bar{h}\, per\, dx(T - T_\infty) + \frac{d}{dx}\left(-k\, A_c \frac{dT}{dx}\right) dx. \tag{7}$$

Equation (7) can be simplified by dividing through by $k\, A_c\, dx$ in order to provide the governing differential equation for this situation:

$$\frac{d^2 T}{dx^2} - \frac{\bar{h}\, per}{k\, A_c} T = -\frac{\bar{h}\, per}{k\, A_c} T_\infty - \frac{\dot{q}''\, per}{k\, A_c}. \tag{8}$$

The boundary conditions are:

$$T_{x=0} = T_w \tag{9}$$

$$T_{x=L} = T_o. \tag{10}$$

Equation (8) is a nonhomogeneous, linear, second-order ODE; the solution is assumed to be the sum of a homogeneous and particular solution:

$$T = T_h + T_p. \tag{11}$$

Equation (11) is substituted into Eq. (8):

$$\underbrace{\frac{d^2 T_h}{dx^2} - \frac{\bar{h}\, per}{k\, A_c} T_h}_{\substack{=0 \\ \text{for homogeneous} \\ \text{differential question}}} + \underbrace{\frac{d^2 T_p}{dx^2} - \frac{\bar{h}\, per}{k\, A_c} T_p = -\frac{\bar{h}\, per}{k\, A_c} T_\infty - \frac{\dot{q}''\, per}{k\, A_c}}_{\text{whatever is left over must be the particular differential equation}}. \tag{12}$$

The homogeneous differential equation is:

$$\frac{d^2 T_h}{dx^2} - \frac{\bar{h}\, per}{k\, A_c} T_h = 0, \tag{13}$$

which according to Table 3.1 is solved by

$$T_h = C_1 \cosh(m\,x) + C_2 \sinh(m\,x), \tag{14}$$

where

$$m = \sqrt{\frac{per\, \bar{h}}{k\, A_c}}. \tag{15}$$

The particular differential equation is:

$$\frac{d^2 T_p}{dx^2} - \frac{\bar{h}\, per}{k\, A_c} T_p = -\frac{\bar{h}\, per}{k\, A_c} T_\infty - \frac{\dot{q}''\, per}{k\, A_c}. \tag{16}$$

Examination of the right side of Eq. (16) shows that it is a constant. Therefore, the particular solution is a constant:

$$T_p = C_3. \tag{17}$$

Example 3.2 (cont.)

Substituting Eq. (17) into Eq. (16) leads to:

$$-\frac{\bar{h}\,per}{k\,A_c}C_3 = -\frac{\bar{h}\,per}{k\,A_c}T_\infty - \frac{\dot{q}''\,per}{k\,A_c}.$$ (18)

Solving for C_3:

$$C_3 = T_\infty + \frac{\dot{q}''}{\bar{h}}.$$ (19)

Substituting Eq. (19) into Eq. (17) leads to:

$$T_p = T_\infty + \frac{\dot{q}''}{\bar{h}}.$$ (20)

Substituting Eqs. (14) and (20) into Eq. (11) leads to the general solution:

$$\boxed{T = C_1 \cosh\,(m\,x) + C_2 \sinh\,(m\,x) + T_\infty + \frac{\dot{q}''}{\bar{h}}}.$$ (21)

The boundary conditions must be used to evaluate C_1 and C_2 for a specific situation. Substituting Eq. (21) into Eq. (9) provides:

$$C_1 \cosh\,(m\,0) + C_2 \sinh\,(m\,0) + T_\infty + \frac{\dot{q}''}{\bar{h}} = T_w,$$ (22)

which can be solved for C_1:

$$\boxed{C_1 = T_w - T_\infty - \frac{\dot{q}''}{\bar{h}}}.$$ (23)

Substituting Eq. (21) into Eq. (10) provides:

$$C_1 \cosh\,(m\,L) + C_2 \sinh\,(m\,L) + T_\infty + \frac{\dot{q}''}{\bar{h}} = T_o,$$ (24)

which can be solved for C_2:

$$\boxed{C_2 = \frac{\left(T_o - T_\infty - \dfrac{\dot{q}''}{\bar{h}} - C_1 \cosh\,(m\,L)\right)}{\sinh\,(m\,L)}}.$$ (25)

Equations (21), (23), and (25) together provide the temperature distribution in the bracket. The rate of heat transfer from the object to the bracket ($\dot{q}_{obj}$) can be evaluated using the interface energy balance shown in Figure 3.

Figure 3 Rate of heat transfer from object to bracket.

The rate of conduction evaluated at $x = L$ and $\dot{q}_{obj}$ both enter the interface and sum to zero:

$$(\dot{q}_x)_{x=L} + \dot{q}_{obj} = 0.$$ (26)

Solving for $\dot{q}_{obj}$ and substituting Fourier's Law provides:

$$\dot{q}_{obj} = k\,A_c\frac{dT}{dx}\bigg|_{x=L}.$$ (27)

Continued

Example 3.2 (cont.)

Substituting Eq. (21) into Eq. (27) provides:

$$\dot{q}_{obj} = k\,A_c[C_1 m \sinh\,(m\,L) + C_2 m \cosh\,(m\,L)]. \tag{28}$$

Solution

The solution derived in the analysis section is implemented using EES in order to facilitate the development of the requested plot. The inputs are entered in EES.

```
$UnitSystem SI Mass J K Radian
q``_dot=4.5 [kW/m^2]*Convert(kW/m^2,W/m^2)        "incident heat flux"
D=1.1 [cm]*Convert(cm,m)                          "diameter of bracket"
L=10.2 [cm]*Convert(cm,m)                         "length of bracket"
k=24 [W/m-K]                                      "conductivity"
T_w=ConvertTemp(C,K,30 [C])                       "wall temperature"
T_o=ConvertTemp(C,K,400 [C])                      "object temperature"
T_infinity=ConvertTemp(C,K,50 [C])               "surrounding air temperature"
h_bar=25 [W/m^2-K]                                "heat transfer coefficient"
```

The perimeter (*per*), cross-sectional area (A_c), and solution parameter m, given by Eq. (15) are computed.

```
per=pi*D                            "perimeter"
A_c=pi*D^2/4                        "area"
m=Sqrt(per*h_bar/(k*A_c))          "solution parameter"
```

The solution provided by Eqs. (21), (23), (25), and (28) is entered.

```
C_1=T_w-T_infinity-q``_dot/h_bar                        "solution constant 1"
C_2=(T_o-T_infinity-q``_dot/h_bar-C_1*cosh(m*L))/sinh(m*L)   "solution constant 2"
T=C_1*cosh(m*x)+C_2*sinh(m*x)+T_infinity+q``_dot/h_bar  "general solution"
q_dot_obj=k*A_c*(C_1*m*sinh(m*L)+C_2*m*cosh(m*L))       "heat transfer rate from object"
```

A Parametric Table is created that includes the variables x and T in order to prepare the temperature distribution plot shown in Figure 4. The rate of heat transfer from the object to the bracket is found to be $\boxed{\dot{q}_{obj} = 10.32\ \text{W}}$.

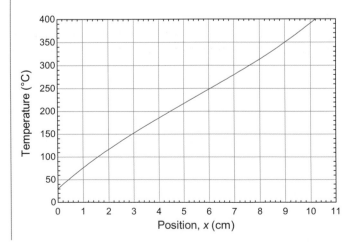

Figure 4 Temperature as a function of position.

Example 3.2 (cont.)

Discussion/Exploration

We can explore the effect of varying the parameters using our model. For example, we can overlay the temperature distributions that will result as the heat transfer coefficient is varied. The result is shown in Figure 5. Notice as $\bar{h}$ increases, the bracket temperature tends to be drawn towards T_∞ (50°C) except near the edges where the temperatures are set. As $\bar{h}$ decreases the bracket temperature increases due to the combined heating effect of the incident radiation and the heating from the object.

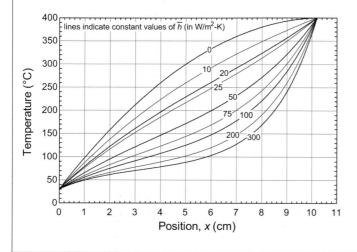

Figure 5 Temperature as a function of position for various values of the heat transfer coefficient.

3.3 Fins

The most common extended surface application is a fin. A fin is a piece of material that is attached to a heat transfer surface in order to increase the surface area that is available for convection. Fins are common structures in heat transfer devices like heat sinks or heat exchangers. In Sections 3.1 and 3.2 we considered a fin that has a constant cross-sectional area and perimeter and an adiabatic tip. The solution for the temperature distribution was obtained analytically and is given by Eq. (3.53), which is repeated below:

$$\frac{(T - T_\infty)}{(T_b - T_\infty)} = \frac{\cosh\left[m\,L\left(1 - \dfrac{x}{L}\right)\right]}{\cosh\,(m\,L)}. \tag{3.57}$$

As an engineer designing a heat transfer device it is most useful to know the rate of heat transfer to or from the base of the fin ($\dot{q}_{fin}$), which can be obtained from Fourier's Law evaluated at the base of the fin ($x = 0$ in Figure 3.1):

$$\dot{q}_{fin} = -k\,A_c\frac{dT}{dx}\bigg|_{x=0}. \tag{3.58}$$

Equation (3.58) provides the heat transfer rate *to* the fin; if the actual rate of heat transfer is from the fin to the wall then the result will be negative. Substituting Eq. (3.57) into Eq. (3.58) leads to:

$$\dot{q}_{fin} = -k\,A_c\frac{d}{dx}\left[(T_b - T_\infty)\frac{\cosh\,(m(L - x))}{\cosh\,(m\,L)} + T_\infty\right]_{x=0}$$

$$= -\frac{k\,A_c(T_b - T_\infty)}{\cosh\,(m\,L)}\frac{d}{dx}\left[\cosh\,(m(L - x))\right]_{x=0}. \tag{3.59}$$

Recalling that the derivative of cosh is sinh, according to Eq. (3.51):

$$\dot{q}_{fin} = \frac{k A_c (T_b - T_\infty)}{\cosh (m L)} m [\sinh (m(L - x))]_{x=0} \tag{3.60}$$

or

$$\dot{q}_{fin} = (T_b - T_\infty) k A_c m \frac{\sinh (m L)}{\cosh (m L)}. \tag{3.61}$$

The ratio of sinh to cosh is defined as the **hyperbolic tangent**, tanh (just as the ratio of sine to cosine is tangent); therefore, Eq. (3.61) may be written as

$$\dot{q}_{fin} = (T_b - T_\infty) \sqrt{\bar{h}\, per\, k\, A_c} \tanh (m L). \tag{3.62}$$

3.3.1 Fin Efficiency

There are many different fin geometries that can be used in heat transfer devices. Each of these fin types is accompanied by its own solution that provides the heat transfer rate to the base. These solutions are typically recorded in terms of the **fin efficiency**. The fin efficiency is defined as the ratio of the rate of heat transfer to the fin ($\dot{q}_{fin}$) to the rate of heat transfer to an *ideal* fin. An ideal fin is defined as one made of an infinitely conductive material, $k \rightarrow \infty$:

$$\eta_{fin} = \frac{\text{heat transfer to fin}}{\text{heat transfer to fin as } k \rightarrow \infty} = \frac{\dot{q}_{fin}}{\dot{q}_{fin, k \rightarrow \infty}}. \tag{3.63}$$

The ideal limit corresponds to a fin in which the resistance to conduction along the fin is zero and therefore the fin surface temperature is uniform and equal to the base temperature, T_b. Note that an ideal fin with infinite conductivity corresponds to the limit of mL approaching 0 in Figure 3.5. Because the surface temperature of an ideal fin is uniform, the rate of heat transfer is given by

$$\dot{q}_{fin, k \rightarrow \infty} = \bar{h} A_{s,fin} (T_b - T_\infty), \tag{3.64}$$

where $A_{s,fin}$ is the surface area of the fin that is exposed to fluid. Substituting Eq. (3.64) into (3.63) provides:

$$\eta_{fin} = \frac{\dot{q}_{fin}}{\bar{h} A_{s,fin} (T_b - T_\infty)}. \tag{3.65}$$

The fin efficiency represents the degree to which the temperature drop along the fin due to conduction has reduced the temperature difference driving convection from the fin surface.

The fin efficiency solution depends on the geometry of the fin. For the constant cross-sectional area fin with an adiabatic tip, Eq. (3.62) can be substituted into Eq. (3.65) to provide:

$$\eta_{fin} = \frac{(T_b - T_\infty) \sqrt{\bar{h}\, per\, k\, A_c} \tanh (m L)}{\bar{h} A_{s,fin} (T_b - T_\infty)}. \tag{3.66}$$

The surface area that must be used in the denominator of Eq. (3.66) for this fin is the product of the perimeter and length (note that the tip is not included in the surface area as the fin tip is assumed to be adiabatic):

$$\eta_{fin} = \frac{(T_b - T_\infty) \sqrt{\bar{h}\, per\, k\, A_c} \tanh (m L)}{\bar{h}\, per\, L (T_b - T_\infty)} \tag{3.67}$$

or

$$\eta_{fin} = \frac{\tanh (m L)}{\sqrt{\frac{\bar{h}\, per}{k\, A_c}} L}. \tag{3.68}$$

Equation (3.68) can be simplified by substituting in the definition of m, Eq. (3.30):

$$\eta_{fin} = \frac{\tanh\,(m\,L)}{m\,L}.$$ (3.69)

Figure 3.6 illustrates the fin efficiency predicted by Eq. (3.69) as a function of mL. Notice that the fin efficiency decreases as mL increases, which is consistent with the temperature distribution shown in Figure 3.5; a large value of mL corresponds to a large temperature drop due to conduction along the fin.

Fin efficiency solutions for many common fin geometries have been determined and several are listed in Table 3.2. (Note that all of the fins with a finite area at the tip have been assumed to be adiabatic at their tip in the derivation of the solution.) Fins having variable cross-section result in more complex differential equations, but these are solved using essentially the same steps that were carried out for the constant cross-section example. A more comprehensive set of fin efficiency solutions has been programmed in the Heat Transfer Library in EES. To access these solutions, select Function Info from the Options menu and then select the radio button labeled Heat Transfer & Fluid Flow and the Fin Efficiency category from the drop down menu (Figure 3.7).

It is possible to scroll through the various functions that are available or see more detailed information about any of these functions by pressing the Info button. Note that the fin efficiency can be accessed either in

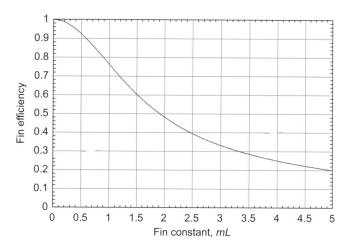

Figure 3.6 Fin efficiency for a constant cross-section, adiabatic tipped fin as a function of the parameter mL.

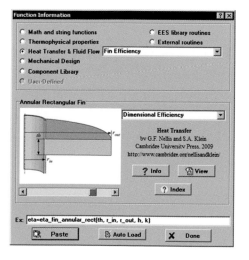

Figure 3.7 Fin efficiency function information in EES.

Table 3.2 Fin efficiency solutions (all assume adiabatic tip).

	Shape	Efficiency
Constant cross-section	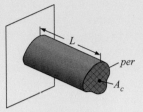	$\eta_{fin} = \dfrac{\tanh\ (mL)}{mL}$ $A_{s,fin} = per\ L$ $mL = \sqrt{\dfrac{\bar{h}\ per}{k\ A_c}}L$
Straight triangular, infinitely wide ($W \gg th$)	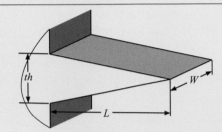	$\eta_{fin} = \dfrac{\text{BesselI}(1, 2mL)}{mL\ \text{BesselI}(0, 2\ mL)}$ $A_{s,fin} = 2W\sqrt{L^2 + \left(\dfrac{th}{2}\right)^2}$ $mL = \sqrt{\dfrac{2\bar{h}}{k\ th}}L$
Straight parabolic, infinitely wide ($W \gg th$)	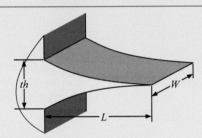	$\eta_{fin} = \dfrac{2}{\left[\sqrt{4(mL)^2 + 1} + 1\right]}$ $A_{s,fin} = W\left[C_1 L + \dfrac{L^2}{th}\ln\left(\dfrac{th}{L} + C_1\right)\right]$ $mL = \sqrt{\dfrac{2\bar{h}}{k\ th}}L,\ C_1 = \sqrt{1 + \left(\dfrac{th}{L}\right)^2}$
Spine triangular	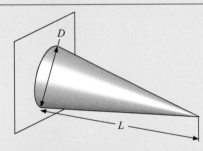	$\eta_{fin} = \dfrac{2\ \text{BesselI}(2, 2mL)}{mL\ \text{BesselI}(1, 2mL)}$ $A_{s,fin} = \dfrac{\pi D}{2}\sqrt{L^2 + \left(\dfrac{D}{2}\right)^2}$ $mL = \sqrt{\dfrac{4\bar{h}}{k\ D}}L$
Rectangular annular	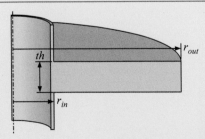	$A_{s,fin} = 2\pi\left(r_{out}^2 - r_{in}^2\right)$ $mr_{out} = \sqrt{\dfrac{2\bar{h}}{k\ th}}\,r_{out}$ $mr_{in} = \sqrt{\dfrac{2\bar{h}}{k\ th}}\,r_{in}$

$$\eta_{fin} = \frac{2mr_{in}}{\left[(mr_{out})^2 - (mr_{in})^2\right]}\frac{\left[\text{BesselK}(1, mr_{in})\text{BesselI}(1, mr_{out}) - \text{BesselI}(1, mr_{in})\text{BesselK}(1, mr_{out})\right]}{\left[\text{BesselI}(0, mr_{in})\text{BesselK}(1, mr_{out}) + \text{BesselK}(0, mr_{in})\text{BesselI}(1, mr_{out})\right]}$$

where	$\bar{h}$ = heat transfer coefficient	k = thermal conductivity

dimensional form (in which case the geometric parameters such as conductivity and heat transfer coefficient must be supplied) or nondimensional form (in which case the nondimensional parameters such as mL must be supplied).

Example 3.3

A heat exchange surface that employs a spine triangular fin (i.e., a cone-shaped fin) is to be made using additive manufacturing. The filled polymer used in the process has a conductivity of $k = 8.3$ W/m-K. The base of the cone has diameter $D = 5$ mm and the length of the cone is $L = 6$ cm. The base and ambient temperatures are $T_b = 60°C$ and $T_\infty = 30°C$, respectively. The heat transfer coefficient is $\bar{h} = 35$ W/m²-K.

Determine the rate of heat transfer to the fin and plot the rate of heat transfer as a function of the fin length.

Knowns

Figure 1 illustrates a sketch of the problem with the known information included.

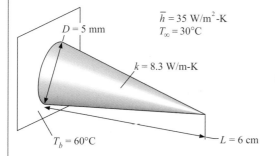

$\bar{h} = 35$ W/m²-K
$T_\infty = 30°C$

$D = 5$ mm

$k = 8.3$ W/m-K

$T_b = 60°C$

$L = 6$ cm

Figure I Sketch of problem.

Assumptions

- Steady-state conditions exist.
- Radiation is neglected.
- The extended surface approximation can be used to analyze the fin. This should be justified with the Biot number:

$$Bi = \frac{R_{cond,r}}{R_{conv}} = \frac{D\bar{h}}{2k} = \frac{0.005 \text{ m}}{2} \left| \frac{35 \text{ W}}{\text{m}^2\text{K}} \right| \frac{\text{m K}}{8.3 \text{ W}} = 0.011. \tag{1}$$

The Biot number is sufficiently less than 1 to justify the extended surface approximation.

Analysis

The fin efficiency (η_{fin}) will be computed using the Heat Transfer Library in EES. The surface area of the conical fin is given by:

$$A_{s,fin} = \frac{\pi D}{2} \sqrt{L^2 + \left(\frac{D}{2}\right)^2}. \tag{2}$$

The rate of heat transfer to the fin is obtained from Eq. (3.65):

$$\dot{q}_{fin} = \eta_{fin} \bar{h} A_{s,fin}(T_b - T_\infty). \tag{3}$$

Continued

Example 3.3 (cont.)

Solution

The solution is implemented in EES both to facilitate making the requested plot as well as to allow the use of the built-in Fin Efficiency Library. The inputs are entered as follows.

```
$UnitSystem SI Mass J K Pa
D=5 [mm]*Convert(mm,m)                    "diameter"
L=6 [cm]*Convert(cm,m)                    "length"
k=8.3 [W/m-K]                             "conductivity"
h_bar=35 [W/m^2-K]                        "heat transfer coefficient"
T_b=ConvertTemp(C,K,60 [C])              "base temperature"
T_infinity=ConvertTemp(C,K,30 [C])       "air temperature"
```

The fin efficiency is obtained by selecting Function Info from the Options menu and then selecting the Heat Transfer & Fluid Flow radio button. Navigate to the Fin Efficiency item in the pull-down menu and then select Dimensional Efficiency in the lower window and scroll to the correct entry in the library. Select Paste to insert the function call in the Equations Window.

```
eta_fin=eta_fin_spine_triangular(D, L, h_bar, k)      "fin efficiency"
```

Equations (2) and (3) are entered.

```
A_s_fin=(pi*D/2)*Sqrt(L^2+(D/2)^2)        "surface area"
q_dot_fin=h_bar*A_s_fin*eta_fin*(T_b-T_infinity)   "rate of heat transfer to the fin"
```

Solving provides $\eta_{fin} = 0.4556$ (45.56%) and $\boxed{\dot{q}_{fin} = 0.226 \text{ W}}$. A Parametric Table is generated that includes columns for the variables L and q_dot_fin. The value of L is varied from near zero to 10 cm and the results are used to prepare Figure 2.

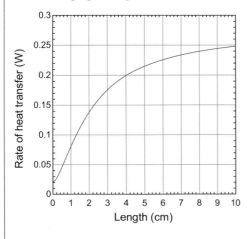

Figure 2 Rate of heat transfer to the fin as a function of the fin length.

Discussion

One metric that can be used to examine the performance of a fin is the **fin effectiveness**, ε_{fin}, which is defined as the ratio of the rate of heat transfer to the fin to the rate of heat transfer that would have occurred from the base material in the absence of a fin ($\dot{q}_{no \, fin}$):

Example 3.3 (cont.)

$$\varepsilon_{fin} = \frac{\dot{q}_{fin}}{\dot{q}_{no\,fin}}.$$ (4)

The rate of heat transfer that would occur in the absence of a fin is given by:

$$\dot{q}_{no\,fin} = A_{c,b}\,\bar{h}(T_b - T_\infty),$$ (5)

where $A_{c,b}$ is the cross-sectional area of the base of the fin, which corresponds to the convection area that would be present if the fin were removed. For the fin shown in Figure 2, the cross-sectional area at the base is given by:

$$A_{c,b} = \pi\frac{D^2}{4}.$$ (6)

Implementing Eqs. (4) through (6) in EES for $L = 6$ cm gives the following.

```
A_c_b=pi*D^2/4                          "cross-sectional area at the base of the fin"
q_dot_nofin=h_bar*A_c_b*(T_b-T_infinity) "rate of heat transfer in the absence of the fin"
eff_fin=q_dot_fin/q_dot_nofin            "effectiveness of the fin"
```

This provides $\varepsilon_{fin} = 10.94$; the rate of heat transfer from the 5 mm diameter area of the base has increased by an order of magnitude by adding the 6 cm long fin to the surface. Note that the *fin effectiveness* (ε_{fin}) is very different from the *fin efficiency* (η_{fin}) that was discussed in this section.

3.3.2 Convection from the Fin Tip

For most fins, the surface area that is available for convection at the tip is not significant relative to the total area available for convection from the surface and therefore the adiabatic tip solution for fin efficiency is sufficient. However, it is possible to derive the solution for a fin that is experiencing convection from its tip, as shown in Figure 3.8.

In Section 3.2.1 we derived the governing ordinary differential equation for this situation, Eq. (3.14):

$$\frac{d^2 T}{dx^2} - \frac{per\,\bar{h}}{k\,A_c}T = -\frac{per\,\bar{h}}{k\,A_c}T_\infty.$$ (3.70)

The boundary conditions for the problem shown in Figure 3.8 are a set base temperature:

$$T_{x=0} = T_b.$$ (3.71)

At the tip, conduction must balance convection:

$$-k\frac{dT}{dx}\bigg|_{x=L} = \bar{h}(T_{x=L} - T_\infty).$$ (3.72)

The general solution to Eq. (3.70) is:

$$T = C_1 \cosh(m\,x) + C_2 \sinh(m\,x) + T_\infty,$$ (3.73)

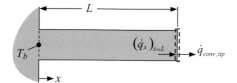

Figure 3.8 Constant cross-section fin experiencing convection from its tip.

where

$$m = \sqrt{\frac{\bar{h}\, per}{k\, A_c}}. \tag{3.74}$$

Substituting Eq. (3.73) into Eqs. (3.71) and (3.72) provides:

$$C_1 + T_\infty = T_b \tag{3.75}$$

$$C_1 \sinh(m\,L) + C_2 \cosh(m\,L) = -\frac{\bar{h}}{k\,m}[C_1 \cosh(m\,L) + C_2 \sinh(m\,L)]. \tag{3.76}$$

Solving Eqs. (3.75) and (3.76) for C_1 and C_2 provides:

$$C_1 = (T_b - T_\infty) \tag{3.77}$$

$$C_2 = -(T_b - T_\infty)\frac{\left[\sinh(m\,L) + \dfrac{\bar{h}}{k\,m}\cosh(m\,L)\right]}{\left[\cosh(m\,L) + \dfrac{\bar{h}}{k\,m}\sinh(m\,L)\right]}. \tag{3.78}$$

The rate of heat transfer to the fin is computed by determining the rate of heat transfer to the base of the fin according to:

$$\dot{q}_{fin} = -k\,A_c \frac{dT}{dx}\bigg|_{x=0}. \tag{3.79}$$

Substituting Eq. (3.73) into Eq. (3.79) provides:

$$\dot{q}_{fin} = -k\,A_c\,m\,C_2. \tag{3.80}$$

Substituting Eq. (3.78) into Eq. (3.80) leads to:

$$\dot{q}_{fin} = (T_b - T_\infty)\sqrt{\bar{h}\, per\, k\, A_c}\,\frac{\left[\sinh(m\,L) + \dfrac{\bar{h}}{m\,k}\cosh(m\,L)\right]}{\left[\cosh(m\,L) + \dfrac{\bar{h}}{m\,k}\sinh(m\,L)\right]}. \tag{3.81}$$

So the fin efficiency is:

$$\eta_{fin} = \frac{\dot{q}_{fin}}{\bar{h}\,(per\,L + A_c)(T_b - T_\infty)} = \frac{\sqrt{\bar{h}\, per\, k\, A_c}}{\bar{h}\,(per\,L + A_c)}\frac{\left[\sinh(m\,L) + \dfrac{\bar{h}}{m\,k}\cosh(m\,L)\right]}{\left[\cosh(m\,L) + \dfrac{\bar{h}}{m\,k}\sinh(m\,L)\right]}, \tag{3.82}$$

which can be simplified somewhat to

$$\eta_{fin} = \frac{\left[\tanh(m\,L) + m\,L\,AR_{tip}\right]}{m\,L\left[1 + m\,L\,AR_{tip}\tanh(m\,L)\right]\left(1 + AR_{tip}\right)}, \tag{3.83}$$

where AR_{tip} is the ratio of the area for convection from the tip to the surface area along the fin length:

$$AR_{tip} = \frac{A_c}{per\,L}. \tag{3.84}$$

The solution for a fin with convection from the tip is included in the Fin Efficiency Library in EES as eta_fin_constantCS_convtip and eta_fin_constantCS_convtip_nd. Notice that if AR_{tip} approaches zero then Eq. (3.83) reduces to:

$$\eta_{fin} = \frac{\tanh(m\,L)}{m\,L}, \tag{3.85}$$

which is equivalent to the result obtained for an adiabatic tip, Eq. (3.69). Figure 3.9 illustrates the fin efficiency associated with a fin with a convective tip as a function of the fin parameter (mL) for various values of the tip

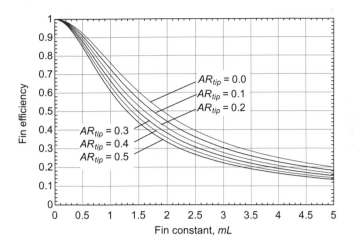

Figure 3.9 Fin efficiency for a constant cross-section fin with convection from the tip as a function of the parameter mL and various values of the tip area ratio, AR_{tip}.

area ratio (AR_{tip}). Note that the fin efficiency is reduced as the tip area ratio increases. This counterintuitive result is related to the fact that the tip area is included in the surface area available for convection from an ideal fin in the definition of the fin efficiency. The heat transfer rate from the fin will *increase* as the tip area is increased. However, the rate of heat transfer from an ideal (i.e., isothermal) fin *increases by a larger amount* since the tip is the least efficient portion of the fin.

It is also possible to use the simpler adiabatic tip fin efficiency equation and *approximately* correct for convection from the tip by modifying the length of the fin slightly to define a **corrected length**, L_c, that is used in place of the actual length:

$$L_c = L + \frac{A_c}{per}. \tag{3.86}$$

The corrected length should be used both to compute the fin constant, mL_c, as well as the surface area, $L_c\, per$.

The term added to the length in Eq. (3.86) is typically very small. For a fin with a circular cross-section, for example, the correction would be equal to one fourth the diameter ($D/4$). In most cases the correction associated with the tip convection is so small that it is not worth considering. In any case, neglecting convection from the tip is slightly conservative and other uncertainties in the problem (e.g., in the heat transfer coefficient) are likely to be more important.

Example 3.4

A fin of length L = 6 cm has a constant, square cross-section with side W = 4 mm. The fin material has conductivity k = 54 W/m-K. The base of the fin is at T_b = –5°C and the fin is surrounded by air at T_∞ = 10°C with heat transfer coefficient $\bar{h}$ = 68 W/m²-K.

Estimate the heat transfer to the fin using three different techniques.
- Ignore convection from the tip by using the adiabatic tip solution without correcting for the length.
- Include convection from the tip using the exact solution, Eq. (3.83), which is derived in Section 3.3.2 and shown in Figure 3.9.
- Approximately include convection from the tip by using the adiabatic tip solution but correcting for the length according to Eq. (3.86).

Continued

Example 3.4 (cont.)

Known Values

The known information is listed on the sketch in Figure 1.

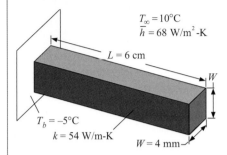

$T_\infty = 10°C$
$\bar{h} = 68 \text{ W/m}^2\text{-K}$

$L = 6$ cm

W

$T_b = -5°C$

$k = 54$ W/m-K

$W = 4$ mm

Figure 1 Sketch of the fin showing the known information.

Assumptions

- Steady-state conditions exist.
- Radiation is neglected.
- The extended surface approximation is valid. This should be justified by calculating a Biot number:

$$Bi = \frac{R_{cond,y}}{R_{conv}} = \frac{W\,\bar{h}}{2k} = \frac{0.004 \text{ m}}{2} \left| \frac{68 \text{ W}}{\text{m}^2\text{K}} \right| \frac{\text{m K}}{54 \text{ W}} = 0.003. \tag{1}$$

The Biot number is sufficiently less than 1 to justify the extended surface approximation.

Analysis

The cross-sectional area for conduction is computed according to:

$$A_c = W^2 \tag{2}$$

and the perimeter is:

$$per = 4\,W. \tag{3}$$

The fin constant is:

$$m = \sqrt{\frac{\bar{h}\,per}{k\,A_c}}. \tag{4}$$

Method 1 Ignore convection from the tip

The fin efficiency for a fin with an adiabatic tip using an uncorrected length should be calculated according to:

$$\eta_{fin,1} = \frac{\tanh\,(m\,L)}{m\,L}. \tag{5}$$

The heat transfer rate is given by:

$$\dot{q}_{fin,1} = \eta_{fin,1}\,\bar{h}\,per\,L(T_\infty - T_b). \tag{6}$$

Notice that the surface area of the fin used in Eq. (6) is the product of the perimeter and uncorrected length.

Example 3.4 (cont.)

Method 2 Include convection from the tip exactly

The fin efficiency for a fin with a convecting tip should be calculated according to Eq. (3.83):

$$\eta_{fin,2} = \frac{\left[\tanh{(mL)} + mL\,AR_{tip}\right]}{mL\left[1 + mL\,AR_{tip}\,\tanh{(mL)}\right]\left(1 + AR_{tip}\right)}, \tag{7}$$

where AR_{tip} is the ratio of the area for convection from the tip to the surface area along the fin length:

$$AR_{tip} = \frac{A_c}{per\,L}. \tag{8}$$

The heat transfer rate is given by:

$$\dot{q}_{fin,2} = \eta_{fin,2}\,\bar{h}(per\,L + A_c)(T_\infty - T_b). \tag{9}$$

Notice that the surface area of the fin used in Eq. (9) includes the tip.

Method 3 Include convection from the tip approximately with a corrected length

A corrected length is computed that approximately accounts for convection from the tip:

$$L_c = L + \frac{A_c}{per}. \tag{10}$$

The fin efficiency is computed using the adiabatic tip solution but with a corrected length:

$$\eta_{fin,3} = \frac{\tanh{(mL_c)}}{mL_c}. \tag{11}$$

The heat transfer rate is given by:

$$\dot{q}_{fin,3} = \eta_{fin,3}\,\bar{h}\,per\,L_c(T_\infty - T_b). \tag{12}$$

Notice that the surface area of the fin used in Eq. (12) is the product of the perimeter and corrected length.

Solution

Equations (2) through (4) can be solved in order to establish the values of m, A_c, and per for the analysis; these values do not change regardless of the method used to calculate the performance of the fin:

$$A_c = W^2 = (0.004)^2\,\text{m}^2 = 1.60 \times 10^{-5}\,\text{m}^2$$

$$per = 4W = 4(0.004\,\text{m}) = 0.016\,\text{m}$$

$$m = \sqrt{\frac{\bar{h}\,per}{k\,A_c}} = \sqrt{\frac{68\,\text{W}}{\text{m}^2\text{K}}\left|\frac{0.016\,\text{m}}{}\right|\frac{\text{m K}}{54\,\text{W}}\left|\frac{}{1.60 \times 10^{-5}\,\text{m}^2}\right|} = 35.5\,\text{m}^{-1}.$$

Equations (5) and (6) can be used to estimate the fin performance using Method 1:

$$mL = (35.5\,\text{m}^{-1})(0.06\,\text{m}) = 2.129$$

$$\eta_{fin,1} = \frac{\tanh{(mL)}}{mL} = \frac{\tanh{(2.129)}}{2.129} = 0.4566$$

$$\dot{q}_{fin,1} = \eta_{fin,1}\,\bar{h}\,per\,L(T_\infty - T_b)$$

$$= \frac{0.4566}{}\left|\frac{68\,\text{W}}{\text{m}^2\,\text{K}}\right|\frac{0.0160\,\text{m}}{}\left|\frac{0.06\,\text{m}}{}\right|\frac{15\,\text{K}}{} = \boxed{0.4471\,\text{W}}.$$

Continued

Example 3.4 (cont.)

Equations (7) through (9) can be used to estimate the fin performance using Method 2:

$$AR_{tip} = \frac{A_c}{per\,L} = \frac{1.6 \times 10^{-5}\,\text{m}^2}{\left|0.0160\,\text{m}\right|0.06\,\text{m}} = 0.0167$$

$$\eta_{fin,2} = \frac{\left[\tanh\,(m\,L) + m\,L\,AR_{tip}\right]}{m\,L\left[1 + m\,L\,AR_{tip}\,\tanh\,(m\,L)\right](1 + AR_{tip})}$$

$$= \frac{\left[\tanh\,(2.129) + (2.129)(0.0167)\right]}{(2.129)[1 + (2.129)(0.0167)\tanh\,(2.129)](1 + 0.0167)} = 0.4500$$

$$\dot{q}_{fin,2} = \eta_{fin,2}\,\bar{h}(per\,L + A_c)(T_\infty - T_b)$$

$$= \frac{0.4500}{\left|\frac{68\,\text{W}}{\text{m}^2\text{K}}\right|}\frac{\left[(0.0160)(0.06) + 1.6 \times 10^{-5}\right]\text{m}^2}{15\,\text{K}} = \boxed{0.4479\,\text{W}}.$$

Finally, Equations (10) through (12) can be used to estimate the fin performance using Method 3:

$$L_c = L + \frac{A_c}{per} = 0.06\,\text{m} + \frac{1.6 \times 10^{-5}\,\text{m}^2}{0.016\,\text{m}} = 0.061\,\text{m}$$

$$m\,L_c = (35.5\,\text{m}^{-1})(0.061\,\text{m}) = 2.165$$

$$\eta_{fin,3} = \frac{\tanh\,(2.165)}{2.165} = 0.4500$$

$$\dot{q}_{fin,3} = \eta_{fin,3}\,\bar{h}\,per\,L_c(T_\infty - T_b)$$

$$= \frac{0.4500}{\left|\frac{68\,\text{W}}{\text{m}^2\,\text{K}}\right|0.0160\,\text{m}\left|0.061\,\text{m}\right|}15\,\text{K} = \boxed{0.4479\,\text{W}}.$$

Discussion

The solution shows that ignoring convection from the tip altogether (Method 1) results in an answer that is only 0.2% in error relative to the correct answer (obtained from Method 2). Interestingly, the approximate method of correcting the length (Method 3) results in very nearly the correct answer in this problem; you have to go to the seventh decimal place to find a difference between $\dot{q}_{fin,2}$ and $\dot{q}_{fin,3}$.

Exploration

The equations used to solve the problem are entered in EES in order to facilitate a parametric study.

```
$UnitSystem SI Mass J K Pa
L=6 [cm]*Convert(cm,m)                              "length"
W=4 [mm]*Convert(mm,m)                              "width"
k=54 [W/m-K]                                        "conductivity"
h_bar=68 [W/m^2-K]                                  "heat transfer coefficient"
T_infinity=ConvertTemp(C,K, 10 [C])                 "ambient air temperature"
T_b=ConvertTemp(C,K, -5 [C])                        "base temperature"

A_c=W^2                                             "cross-sectional area"
per=4*W                                             "perimeter"
m=Sqrt(h_bar*per/(k*A_c))                           "fin constant"

"Approach 1: adiabatic tip, uncorrected length"
mL=m*L                                              "fin parameter"
eta_1=Eta_Fin_Constantcs_ND(mL)                     "fin efficiency"
```

Example 3.4 (cont.)

q_dot_fin_1=eta_1*h_bar*per*L*(T_infinity-T_b)   "rate of heat transfer to fin"

"Approach 2: convecting tip"
AR_tip=A_c/(per*L)   "fin tip area ratio"
eta_2=**Eta_Fin_Constantcs_Convtip_ND**(mL,AR_tip)   "fin efficiency"
q_dot_fin_2=eta_2*h_bar*(per*L+A_c)*(T_infinity-T_b)   "rate of heat transfer to fin"

"Approach 3: adiabatic tip, corrected length"
L_c=L+A_c/per   "corrected length"
mL_c=m*L_c   "fin parameter"
eta_3=**Eta_Fin_Constantcs_ND**(mL_c)   "fin efficiency"
q_dot_fin_3=eta_3*h_bar*per*L_c*(T_infinity-T_b)   "rate of heat transfer to fin"

A Parametric Table is generated in which the length is varied. Figure 2 illustrates the rate of heat transfer predicted by the three methods as a function of fin length. As the fin length is reduced, the adiabatic tip solution begins to deviate from the convecting tip solution. It is interesting to note that the technique of correcting the length to account for the tip convection provides nearly exactly the right answer for all lengths.

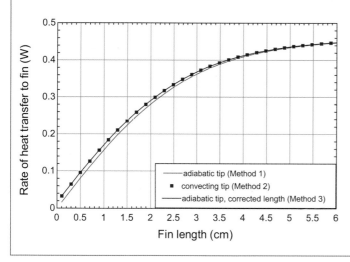

Figure 2 Rate of heat transfer to the fin as a function of the fin length predicted using the three methods discussed in Example 3.4.

3.3.3 Fin Resistance

The fin efficiency is the most useful format for presenting the results of a fin solution because it allows the calculation of a **fin resistance**, R_{fin}. The fin resistance is the thermal resistance that opposes heat transfer from the base of the fin to the surrounding fluid. The definition of the fin efficiency, Eq. (3.65), can be rearranged:

$$\dot{q}_{fin} = \eta_{fin}\,\bar{h}\,A_{s,fin}(T_b - T_\infty) = \frac{(T_b - T_\infty)}{R_{fin}}, \tag{3.87}$$

where $A_{s,fin}$ is the surface area of the fin exposed to the fluid. Note that *without* the fin efficiency, Eq. (3.87) is equivalent to Newton's Law of Cooling and therefore the thermal resistance is defined in basically the same way as a convection resistance but with the addition of the fin efficiency in the denominator:

$$R_{fin} = \frac{1}{\eta_{fin}\,\bar{h}\,A_{s,fin}}. \tag{3.88}$$

The fin efficiency is always less than one and therefore the fin resistance will be larger than the corresponding convection resistance; this increase in resistance is related to the conduction resistance within the fin. The fin resistance is convenient since it allows the effect of fins to be incorporated into a more complex problem (for example, one in which fins are attached to other structures) as additional resistances in a network.

Example 3.5

Two large pipes must be soldered together using a propane torch. Each of the two pipes is $L = 2.5$ ft long with inner diameter $D_{in} = 4.0$ inch and a thickness $th = 0.375$ inch. The pipe material has conductivity $k = 150$ W/m-K. The surrounding air is at $T_\infty = 20°C$ and the heat transfer coefficient between the external surface of the pipe and the air is $\bar{h} = 20$ W/m²-K. The temperature at the interface between the two pipes must be elevated to $T_m = 230°C$ in order to melt the solder.

Determine the heat transfer rate, $\dot{q}$, that must be applied by the propane torch in order to accomplish the soldering process.

Known Values

Figure 1 illustrates a sketch of the problem with the known information indicated, converted to base SI units.

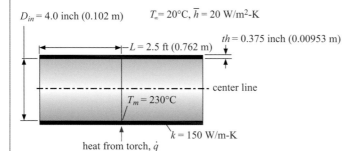

$D_{in} = 4.0$ inch (0.102 m) $T_\infty = 20°C$, $\bar{h} = 20$ W/m²-K

$th = 0.375$ inch (0.00953 m)

$-L = 2.5$ ft (0.762 m)

center line

$T_m = 230°C$

$k = 150$ W/m-K

heat from torch, $\dot{q}$

Figure 1 Two pipes being soldered together.

Assumptions

- Steady-state conditions exist.
- Radiation is neglected.
- Convection to the air on the inside of the pipe is neglected.
- The extended surface approximation is made, which allows the temperature of the pipe to be expressed as a function of x only. This can be justified by an appropriately defined Biot number:

$$Bi = \frac{R_{cond,r}}{R_{conv}} = \frac{th\,\bar{h}}{k} = \frac{0.00953\,\text{m}}{} \left|\frac{20\,\text{W}}{\text{m}^2\text{K}}\right|\frac{\text{m K}}{150\,\text{W}} = 0.0013. \tag{1}$$

The Biot number is sufficiently less than 1 to justify the extended surface approximation.
- The adiabatic tip solution is used to analyze the pipes.

Analysis

The cross-sectional area for conduction (A_c) is:

$$A_c = \frac{\pi}{4}\left[(D_{in} + 2\,th)^2 - D_{in}^2\right] \tag{2}$$

Example 3.5 (cont.)

and the perimeter exposed to air (*per*) is:

$$per = \pi(D_{in} + 2th). \tag{3}$$

Notice that the internal surface of the pipe (which is assumed to be adiabatic) is not included in the perimeter. According to Table 3.2 the fin constant (*mL*) is

$$m L = \sqrt{\frac{per\, \bar{h}}{k\, A_c}} L \tag{4}$$

and the fin efficiency (η_{fin}) is

$$\eta_{fin} = \frac{\tanh\,(m\,L)}{m\,L}. \tag{5}$$

The resistance of each fin (R_{fin}) is calculated according to Eq. (3.88), where the surface area of the fin is the product of the perimeter and length:

$$R_{fin} = \frac{1}{\eta_{fin}\, \bar{h}\, per\, L}. \tag{6}$$

The problem can be represented by the resistance network shown in Figure 2; the two pipes are represented by two resistors in parallel connecting the interface to the air and the heat input from the propane torch enters at the interface. In order for the solder to melt, the interface temperature must reach T_m.

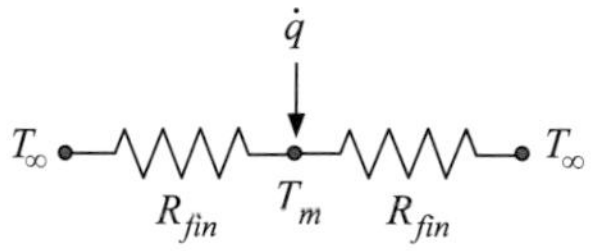

Figure 2 Resistance network associated with soldering two pipes.

The heat transfer rate required from the torch is therefore:

$$\dot{q} = \frac{(T_m - T_\infty)}{(R_{fin}/2)}, \tag{7}$$

where $R_{fin}/2$ is the equivalent resistance of the two fins in parallel.

Solution

Equations (2) through (7) are solved below:

$$A_c = \frac{\pi}{4}\left[(D_{in} + 2th)^2 - D_{in}^2\right] = \frac{\pi}{4}\left[(0.102 + 2(0.00953))^2 - (0.102)^2\right]\mathrm{m}^2 = 0.00333\,\mathrm{m}^2$$

$$per = \pi(D_{in} + 2th) = \pi(0.102 + (2)0.00953)\mathrm{m} = 0.379\,\mathrm{m}$$

$$m L = \sqrt{\frac{per\, \bar{h}}{k\, A_c}} L = \sqrt{\frac{0.379\,\mathrm{m}}{}\left|\frac{20\,\mathrm{W}}{\mathrm{m}^2\mathrm{K}}\right|\frac{\mathrm{m\,K}}{150\,\mathrm{W}}\left|\frac{}{0.00333\,\mathrm{m}^2}\right.}(0.762)\mathrm{m} = 2.971$$

$$\eta_{fin} = \frac{\tanh\,(m\,L)}{m\,L} = \frac{\tanh\,(2.971)}{2.971} = 0.335$$

$$R_{fin} = \frac{1}{\eta_{fin}\, \bar{h}\, per\, L} = \frac{1}{0.335}\left|\frac{\mathrm{m}^2\mathrm{K}}{20\,\mathrm{W}}\right|\frac{}{0.379\,\mathrm{m}}\left|\frac{}{0.762\,\mathrm{m}}\right. = 0.517\,\frac{\mathrm{K}}{\mathrm{W}}$$

$$\dot{q} = \frac{(T_m - T_\infty)}{(R_{fin}/2)} = \frac{2\,\mathrm{W}}{0.517\,\mathrm{K}}\left|210\,\mathrm{K}\right. = \boxed{812.4\,\mathrm{W}}.$$

Continued

Example 3.5 (cont.)

Discussion

The solution assumed that the internal surface of the pipe is adiabatic. In reality, there is some heat transfer to the air within the pipe. However, if the air is stagnant then the heat transfer coefficient is small (and thus the thermal resistance is large). In addition, the temperature of the air trapped in the pipe will rise as the heating process continues, further reducing convective loss.

The propane torch may not be able to provide the required heat transfer rate. One method of achieving the melting temperature at the interface between the pipes without requiring a more powerful torch is to place insulating sleeves of length L_{ins} = 8 inch over the pipes adjacent to the soldering torch, as shown in Figure 3.

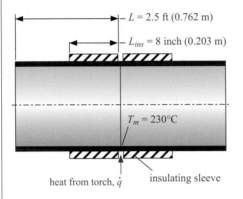

Figure 3 Tubes with insulating sleeves placed over them to reduce the heat transfer required.

If the insulation is perfect (i.e., convection is completely eliminated from the section of the pipe that is covered by the insulating sleeves), then we can compute the required heat transfer rate from the torch. The tubes with insulating sleeves can be represented by a resistance network similar to the one shown in Figure 2, but with additional resistances inserted between the interface and the base of the fins. These additional resistances correspond to the insulated sections of pipes, as shown in Figure 4.

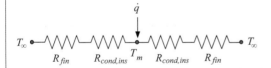

Figure 4 Resistance network with additional resistors associated with insulated tubes.

The resistance of the insulated sections of the pipe is calculated using the equation for the resistance associated with conduction through a plane wall:

$$R_{cond,ins} = \frac{L_{ins}}{k\,A_c}. \tag{8}$$

The length of the un-insulated section of pipe is reduced and therefore the fin efficiency and fin resistance must be re-calculated. The fin constant becomes

Example 3.5 (cont.)

$$m(L - L_{ins}) = \sqrt{\frac{per\,\bar{h}}{k\,A_c}}(L - L_{ins}) \tag{9}$$

and the fin efficiency (η_{fin}) becomes

$$\eta_{fin} = \frac{\tanh\left[m(L - L_{ins})\right]}{m(L - L_{ins})}. \tag{10}$$

The resistance of each fin (R_{fin}) is

$$R_{fin} = \frac{1}{\eta_{fin}\bar{h}\,per(L - L_{ins})}. \tag{11}$$

Using the resistance network shown in Figure 4, the required heat transfer rate can be expressed as:

$$\dot{q} = 2\frac{(T_m - T_\infty)}{\left(R_{cond,\,ins} + R_{fin}\right)}. \tag{12}$$

Solving Eqs. (8) through (12) provides:

$$R_{cond,\,ins} = \frac{L_{ins}}{k\,A_c} = \frac{0.203\,\text{m}}{\left|\frac{\text{m K}}{150\,\text{W}}\right|\frac{1}{0.00333\,\text{m}^2}} = 0.407\frac{\text{K}}{\text{W}}$$

$$m(L - L_{ins}) = \sqrt{\frac{per\,\bar{h}}{k\,A_c}}(L - L_{ins})$$

$$= \sqrt{\frac{0.379\,\text{m}}{\left|\frac{20\,\text{W}}{\text{m}^2\,\text{K}}\right|\frac{\text{m K}}{150\,\text{W}}\left|\frac{1}{0.00333\,\text{m}^2}\right.}}(0.762 - 0.203)\text{m} = 2.178$$

$$\eta_{fin} = \frac{\tanh\left[m(L - L_{ins})\right]}{m(L - L_{ins})} = \frac{\tanh(2.178)}{2.178} = 0.447$$

$$R_{fin} = \frac{1}{\eta_{fin}\,\bar{h}\,per\,(L - L_{ins})} = \frac{1}{0.447}\left|\frac{\text{m}^2\,\text{K}}{20\,\text{W}}\right|\frac{1}{0.379\,\text{m}}\left|(0.762 - 0.203\text{m})\right. = 0.528\frac{\text{K}}{\text{W}}$$

$$\dot{q} = 2\frac{(T_m - T_\infty)}{\left(R_{cond,\,ins} + R_{fin}\right)} = \frac{2\,\text{W}}{(0.407 + 0.528)\,\text{K}}\left|210\,\text{K}\right. = \boxed{449.2\,\text{W}}.$$

The addition of the insulating sleeves has reduced the required heat transfer rate by 45 percent.

3.3.4 Finned Surfaces

Fins are often placed on surfaces in order to improve their heat transfer capability. Examples of finned surfaces can be found within nearly every appliance in your house, from the evaporator and condenser in your refrigerator and air conditioner to the processor in your personal computer. Fins are essential to the design of economical, high-performance thermal devices.

The concept of a fin resistance makes it easier to consider the thermal performance of an array of fins that are placed on a surface. For example, Figure 3.10 illustrates a finned surface. An array of fins is placed onto a base. The **prime surface area** ($A_{s,prime}$) is the area of the base that is *directly exposed to the fluid*. If the dimension of the square base is $W \times W$ and there are N_{fin} fins, each with cross-sectional area A_c installed on the base then the prime surface area is:

$$A_{s,prime} = W^2 - N_{fin}A_c. \tag{3.89}$$

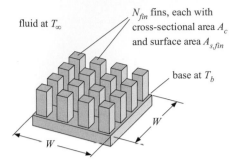

fluid at T_∞

N_{fin} fins, each with cross-sectional area A_c and surface area $A_{s,fin}$

base at T_b

W

W

Figure 3.10 A finned surface.

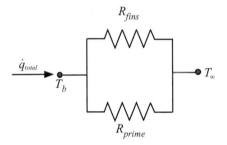

R_{fins}

$\dot{q}_{total}$

T_b

T_∞

R_{prime}

Figure 3.11 Resistance network representing the finned surface.

The **finned surface area** ($A_{s,fins}$) is the total surface area of *all of the fins*. If each of the fins in Figure 3.10 has surface area $A_{s,fin}$ then the finned surface area is:

$$A_{s,fins} = N_{fin} A_{s,fin}.$$
(3.90)

The **total surface area** ($A_{s,total}$) is the total area that is exposed to fluid; it is the sum of the prime and the finned surface area:

$$A_{s,total} = A_{s,fins} + A_{s,prime}.$$
(3.91)

The heat transfer from the base can either pass through the fins or be convected from the un-finned surface area on the base. Each of these parallel paths is associated with a thermal resistance, as shown in Figure 3.11. This analysis assumes that the heat transfer coefficient for the fins is the same as for the prime surface area.

The resistance of a single fin is given by Eq. (3.88):

$$R_{fin} = \frac{1}{\eta_{fin} \, \overline{h} \, A_{s,fin}},$$
(3.92)

where η_{fin} is the fin efficiency, computed using the formula or function that is appropriate given the geometry of the fin. The resistance of all of the fins in parallel, R_{fins} in Figure 3.11, is given by

$$R_{fins} = \frac{1}{\eta_{fin} \, \overline{h} \, A_{s,fins}}.$$
(3.93)

The resistance of the prime surface (the un-finned surface of the base) is

$$R_{prime} = \frac{1}{\overline{h} \, A_{s,prime}}.$$
(3.94)

These paths are in parallel and therefore the equivalent (or total) thermal resistance of the finned surface is

$$R_{total} = \left[\frac{1}{R_{prime}} + \frac{1}{R_{fins}}\right]^{-1} \tag{3.95}$$

or, substituting Eqs. (3.93) and (3.94) into Eq. (3.95):

$$R_{total} = \left[\bar{h} A_{s,prime} + \eta_{fin} \bar{h} A_{s,fins}\right]^{-1}. \tag{3.96}$$

The total rate of heat transfer from the surface is

$$\dot{q}_{total} = \frac{(T_b - T_\infty)}{R_{total}} = (T_b - T_\infty)\left[\bar{h} A_{s,prime} + \eta_{fin} \bar{h} A_{s,fins}\right]. \tag{3.97}$$

The **overall surface efficiency** (η_o) is defined as the ratio of the total heat transfer rate from the surface to the heat transfer rate that would result if the entire surface (the prime surface and the fins) were all at the base temperature; as with the fin efficiency, this limit corresponds to using a material with an infinite conductivity:

$$\eta_o = \frac{\dot{q}_{total}}{\bar{h}\underbrace{\left[A_{s,prime} + A_{s,fins}\right]}_{\text{total surface area, } A_{s,total}} (T_b - T_\infty)}. \tag{3.98}$$

Substituting Eq. (3.97) into Eq. (3.98) leads to:

$$\eta_o = \frac{\left[A_{s,prime} + \eta_{fin} A_{s,fins}\right]}{\left[A_{s,prime} + A_{s,fins}\right]}. \tag{3.99}$$

Equation (3.99) can be rearranged:

$$\eta_o = \frac{\left[A_{s,prime} + \eta_{fin} A_{s,fins} + A_{s,fins} - A_{s,fins}\right]}{\left[A_{s,prime} + A_{s,fins}\right]} \tag{3.100}$$

or

$$\eta_o = 1 - \frac{A_{s,fins}}{A_{s,total}}\left(1 - \eta_{fin}\right). \tag{3.101}$$

Rearranging Eq. (3.97), the total resistance to heat transfer from a finned surface can be expressed in terms of the overall surface efficiency and the total surface area:

$$R_{total} = \frac{1}{\eta_o \bar{h} A_{s,total}}. \tag{3.102}$$

Example 3.6

Heat rejection from a power electronics module is accomplished using a 10 × 10 array of cone-shaped fins. Each fin has a base diameter $D_{fin} = 1.5$ mm and is $L_{fin} = 15$ mm long. The fins are attached to a square base plate that is $W_b = 3$ cm on a side. The conductivity of the fin material is $k = 70$ W/m-K. The base temperature is $T_b = 30°C$ and the surrounding air temperature is $T_\infty = 20°C$. The average heat transfer coefficient between the air and the surface of the heat sink is $\bar{h} = 50$ W/m²-K.

Determine:
- The efficiency of the fins.
- The overall efficiency of the finned surface.
- The rate of heat transfer from the finned surface.

Continued

Example 3.6 (cont.)

Knowns

Figure 1 illustrates a sketch of the problem with the known information indicated.

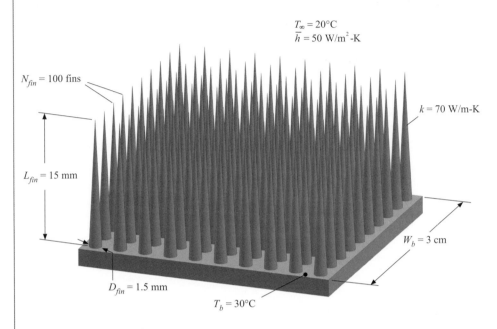

$T_\infty = 20°C$
$\bar{h} = 50 \text{ W/m}^2\text{-K}$

$N_{fin} = 100$ fins

$k = 70$ W/m-K

$L_{fin} = 15$ mm

$W_b = 3$ cm

$D_{fin} = 1.5$ mm

$T_b = 30°C$

Figure 1 Heat sink for a power electronics module.

Assumptions

- Steady-state conditions exist.
- The heat transfer coefficient is the same on the base and the fin surface.
- The fins can be treated as extended surfaces.

Analysis

According to Table 3.2, the fin efficiency for a cone-shaped fin (η_{fin}) is a function of the fin constant, defined as:

$$m\,L = \sqrt{\frac{4\bar{h}}{k\,D_{fin}}}L_{fin} \tag{1}$$

and the surface area of the fins is given by:

$$A_{s,fins} = N_{fin}\frac{\pi\,D_{fin}}{2}\sqrt{L_{fin}^2 + \left(\frac{D_{fin}}{2}\right)^2}. \tag{2}$$

The prime surface area is given by:

$$A_{s,prime} = W_b^2 - N_{fin}\frac{\pi D_{fin}^2}{4}. \tag{3}$$

Example 3.6 (cont.)

The total surface area is the sum of the finned surface area and the prime surface area:

$$A_{s,total} = A_{s,fins} + A_{s,prime}. \tag{4}$$

The overall surface efficiency is given by Eq. (3.101), repeated below:

$$\eta_o = 1 - \frac{A_{s,fins}}{A_{s,total}}\left(1 - \eta_{fin}\right). \tag{5}$$

The thermal resistance of the surface is given by:

$$R_{total} = \frac{1}{\eta_o \, \bar{h} \, A_{s,total}}. \tag{6}$$

The rate of heat transfer from the surface is:

$$\dot{q} = \frac{(T_b - T_\infty)}{R_{total}}. \tag{7}$$

Solution

The solution is implemented in EES in order to use the fin efficiency function for the conical fin and also to facilitate the parametric studies discussed in the subsequent *Exploration* section. The inputs are entered in EES.

```
$UnitSystem SI MASS RAD PA K J

"Inputs"
T_infinity=ConvertTemp(C,K,20 [C])        "air temperature"
T_b=ConvertTemp(C,K,30 [C])               "base temperature"
D_fin=1.5 [mm]*Convert(mm,m)              "fin diameter"
L_fin=15 [mm]*Convert(mm,m)               "fin length"
N_fin=100                                 "number of fins"
W_b=3 [cm]*Convert(cm,m)                  "width of base (square)"
k=70 [W/m-K]                              "conductivity of fin"
h_bar=50 [W/m^2-K]                        "heat transfer coefficient"
```

Equation (1) is used to compute the fin constant, mL, and the function eta_fin_spine_triangular_nd is used to determine the fin efficiency.

```
mL=Sqrt(4*h_bar/(k*D_fin))*L_fin          "fin constant"
eta_fin=Eta_Fin_Spine_Triangular_ND(mL)   "fin efficiency"
```

Solving provides $\boxed{\eta_{fin} = 0.9354(93.54 \text{ percent})}$. Equations (2) through (5) together are entered to provide the overall surface efficiency.

```
A_s_fins=pi*D_fin/2*Sqrt(L_fin^2+(D_fin/2)^2)*N_fin    "surface area of all fins"
A_s_prime=W_b^2-N_fin*pi*D_fin^2/4                     "prime surface area"
A_s_total=A_s_fins+A_s_prime                           "total surface area"
eta_o=1-A_s_fins*(1-eta_fin)/A_s_total                 "overall efficiency"
```

Solving provides $\boxed{\eta_o = 0.9464(94.64 \text{ percent})}$. Finally, Eqs. (6) and (7) are used to determine the rate of heat transfer.

Continued

Example 3.6 (cont.)

R_total=1/(eta_o*h_bar*A_s_total)   "total resistance"
q_dot=(T_b-T_infinity)/R_total   "heat transfer"

Solving provides $\boxed{\dot{q} = 2.017\,\text{W}}$.

Discussion/Exploration

Through material selection and manipulation of the air flow across the heat sink, it is possible to affect design changes to k and $\bar{h}$. One of the nice things about solving problems using a computer program as opposed to pencil and paper is that parametric studies and optimization are relatively straightforward once the problem is solved. In order to prepare a contour plot with EES, it is necessary to setup a Parametric Table in which both of the parameters of interest vary over a specified range. Open a new Parametric Table and include the two independent variables, k and h_bar, as well as the dependent variable of interest (the variable q_dot). In order to run the simulation for 20 values of k and 20 values of $\bar{h}$ it is necessary to include $20 \times 20 = 400$ runs in the table. (Add runs using the Insert/Delete Runs option from the Tables menu.)

It is necessary to set the values of k and h_bar in the table. It is possible to vary k from 20 to 150 W/m-K, 20 times by using the "Repeat pattern every" option in the Alter Values dialog that appears when you right-click on the k column, as shown in Figure 2.

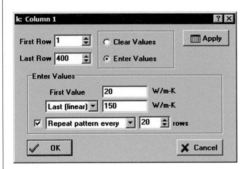

Figure 2 Vary k from 20 to 150 W/m-K 20 times.

In order to completely cover the parameter space, it is necessary to evaluate a different value of h_bar at each unique value of k; this can be accomplished using the "Apply pattern every" option in the Alter Values dialog for the h_bar column of the table, see Figure 3.

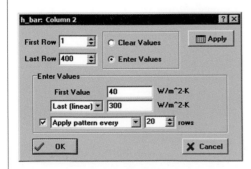

Figure 3 Vary $\bar{h}$ from 40 to 300 W/m-K with 20 runs for each of 20 values.

Example 3.6 (cont.)

When the specified values of the variables k and h_bar are commented out in the Equations Window, it is possible to run the Parametric Table using the Solve Table command in the Calculate menu (F3); 400 values of $\dot{q}$ are determined, one for each combination of k and h_bar set in the Parametric Table. To generate a contour plot, select X-Y-Z plot from the New Plot Window option in the Plots menu. Select k as the variable on the x-axis, h_bar as the y-axis variable and q_dot as the contour variable. You can adjust the appearance of the resulting contour plot by altering the resolution, smoothing, color options, and the type of function used for interpolation. The contour plot using isometric lines is shown in Figure 4.

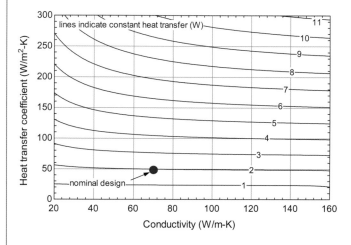

Figure 4 Contours of constant heat transfer rate in the parameter space of fin material conductivity and heat transfer coefficient.

The nominal design point shown in Figure 1 is also indicated in Figure 4. Contour plots are useful in that they can clarify the impact of design changes. For example, Figure 4 shows that it would be more beneficial to explore methods to increase the heat transfer coefficient (i.e., move the design point vertically) than the fin conductivity (which would move it horizontally) at the nominal design conditions.

3.4 Numerical Solutions to Extended Surface Problems

Section 3.2 discusses analytical solutions to extended surface problems. Only simple problems with constant properties can be solved analytically. There will be situations where these simplifications are not justified and it will be necessary to use a numerical model. Numerical modeling of extended surface problems is carried out using the same techniques that are discussed in Section 2.4.

If the extended surface approximation discussed in Section 3.1 is valid, then it is possible to obtain a numerical solution by dividing the computational domain into many small (but finite) 1-D control volumes. Energy balances are written for each control volume; the energy balances can include convective and/or radiative terms in addition to the conductive and generation terms that were considered in Section 2.4. Each term in the energy balance is represented by a rate equation that reflects the governing heat transfer mechanism. The result is a system of algebraic equations that can be solved using any of the techniques presented in Section 2.4. The solution should be checked for convergence, checked against your physical intuition, and compared with an analytical solution in the limit where one is valid.

Example 3.7

An expensive power electronics module normally receives only a moderate current. However, under certain conditions it might experience currents in excess of 100 amp. The module cannot survive such a high current and therefore you have been asked to design a fuse that will protect the module by limiting the current that it can experience.

The space available for the fuse allows a wire that is $L = 2.5$ cm long to be placed between the module and the surrounding structure. The surface of the fuse wire convects to air at $T_\infty = 20°C$ and radiates to surroundings at the same temperature. The heat transfer coefficient between the surface of the fuse and the air is $\bar{h} = 5$ W/m²-K. The fuse wire surface has an emissivity of $\varepsilon = 0.90$. The fuse is made of an aluminum alloy with conductivity $k = 150$ W/m-K. The electrical resistivity of the aluminum alloy is $\rho_e = 1 \times 10^{-7}$ ohm-m and the alloy melts at approximately $T_m = 500°C$. The ends of the fuse are maintained at $T_{end} = 20°C$ by contact with the surrounding structure and the module. The current passing through the fuse, $I_c = 100$ amp, results in a uniform volumetric generation within the fuse material.

- Develop a numerical model that can predict the temperature as a function of position within the fuse if the diameter of the fuse is $D = 0.9$ mm.
- Determine the diameter that will cause the fuse to melt (i.e., at some location within the fuse, the temperature will reach 500°C) when the current is 100 amp.

Knowns

A sketch of the problem with the known values is shown in Figure 1.

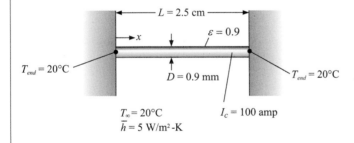

Figure 1 A fuse that protects a power electronics module from high current.

Assumptions

- Steady-state conditions exist.
- The properties do not depend on temperature.
- The volumetric rate of thermal energy generation due to ohmic dissipation is uniform throughout the fuse volume.
- The fuse can be treated as an extended surface; that is, the temperature gradient in the fuse in the radial direction is small so that the problem becomes approximately one-dimensional, $T(x)$. This assumption should be justified by an appropriately defined Biot number. The Biot number for this problem should be the ratio of the resistance to conduction in the radial direction ($R_{cond,r}$) to the parallel combination of the resistances to convection and radiation from the surface ($R_{conv}\|R_{rad}$). The resistance to conduction in the radial direction is approximately:

$$R_{cond,r} = \frac{(D/2)}{\pi D L k} = \frac{1}{2\pi L k} = \frac{1}{2\pi}\left|\frac{1}{0.025\text{ m}}\right|\frac{\text{m K}}{150\text{ W}} = 0.042 \frac{\text{K}}{\text{W}}. \tag{1}$$

Example 3.7 (cont.)

The resistance to convection is:

$$R_{conv} = \frac{1}{\pi \, D \, L \, \bar{h}} = \frac{1}{\pi} \left| \frac{1}{0.0009 \text{ m}} \right| \left| \frac{1}{0.025 \text{ m}} \right| \left| \frac{\text{m}^2 \text{ K}}{5 \text{ W}} \right| = 2830 \frac{\text{K}}{\text{W}}. \tag{2}$$

The resistance to radiation can be computed approximately according to:

$$R_{rad} = \frac{1}{\pi \, D \, L \, \sigma \, \varepsilon \, 4 \, \overline{T}^3}, \tag{3}$$

where $\overline{T}$ is the average absolute temperature of the surface and the surroundings. Of course, the temperature of the fuse surface varies between T_{end} to a temperature that might be as high as T_{melt}. We can estimate the value of $\overline{T}$ well enough to calculate a Biot number according to

$$\overline{T} = \frac{T_{end} + T_{melt}}{2} = \frac{293.2 \text{ K} + 773.2 \text{ K}}{2} = 533.2 \text{ K}, \tag{4}$$

which allows the calculation of R_{rad} according to Eq. (3):

$$R_{rad} = \frac{1}{\pi \, D \, L \, \sigma \, \varepsilon \, 4 \, \overline{T}^3} = \frac{1}{4\pi} \left| \frac{1}{0.0009 \text{ m}} \right| \left| \frac{1}{0.025 \text{ m}} \right| \left| \frac{\text{m}^2 \text{ K}^4}{5.67 \times 10^{-8} \text{ W}} \right| \left| \frac{1}{0.9} \right| \left| \frac{1}{(533.2)^3 \text{ K}^3} \right| = 457 \frac{\text{K}}{\text{W}}. \tag{5}$$

The Biot number is then

$$Bi = \frac{R_{cond,r}}{R_{conv} \| R_{rad}} = R_{cond,r} \left(\frac{1}{R_{conv}} + \frac{1}{R_{rad}} \right) = 0.042 \frac{\text{K}}{\text{W}} \left(\frac{\text{W}}{2829 \text{ K}} + \frac{\text{W}}{457 \text{ K}} \right) = 0.00011. \tag{6}$$

This process shows that the extended solution is valid. Also, because the resistance to radiation is less than the resistance to convection we have learned that radiation is a more important heat transfer mechanism than convection for this problem.

Analysis

The development of the numerical model follows the same steps that were previously discussed in the context of numerical models for 1-D geometries in Section 2.4. Nodes (i.e., locations where the temperature will be determined) are positioned uniformly along the length of the fuse. The location of each node (x_i) is:

$$x_i = \frac{(i-1)}{(M-1)} L \quad \text{for } i = 1 \dots M, \tag{7}$$

where M is the number of nodes used for the simulation. The distance between adjacent nodes (Δx) is:

$$\Delta x = \frac{L}{(M-1)}. \tag{8}$$

A control volume is defined around each node; the control surface bisects the distance between the nodes. Figure 2 illustrates an energy balance on a control volume that is defined around one of the internal nodes (i.e., a node where $2 < i < M - 1$).

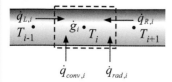

Figure 2 Energy balance on an internal node.

Continued

Example 3.7 (cont.)

The control volume shown in Figure 2 is subject to conduction heat transfer at each edge ($\dot{q}_{L,i}$ and $\dot{q}_{R,i}$), convection ($\dot{q}_{conv,i}$), radiation ($\dot{q}_{rad,i}$), and generation ($\dot{g}_i$). The energy balance is:

$$\dot{q}_{L,i} + \dot{q}_{R,i} + \dot{q}_{conv,i} + \dot{q}_{rad,i} + \dot{g}_i = 0 \quad \text{for } i = 2 \dots (M-1). \tag{9}$$

The conduction terms are approximated according to:

$$\dot{q}_{L,i} = \frac{k \pi D^2}{4 \Delta x}(T_{i-1} - T_i) \tag{10}$$

$$\dot{q}_{R,i} = \frac{k \pi D^2}{4 \Delta x}(T_{i+1} - T_i). \tag{11}$$

The convection term is modeled according to:

$$\dot{q}_{conv} = \bar{h} \pi D \Delta x(T_\infty - T_i). \tag{12}$$

The radiation term is given by:

$$\dot{q}_{rad,i} = \varepsilon \sigma \pi D \Delta x(T_\infty^4 - T_i^4). \tag{13}$$

The generation term is:

$$\dot{g}_i = \frac{4 \rho_e \Delta x}{\pi D^2} I_c^2. \tag{14}$$

Substituting Eqs. (10) through (14) into Eq. (9) leads to:

$$\frac{k \pi D^2}{4 \Delta x}(T_{i-1} - T_i) + \frac{k \pi D^2}{4 \Delta x}(T_{i+1} - T_i) + \bar{h} \pi D \Delta x(T_\infty - T_i)$$
$$+ \varepsilon \sigma \pi D \Delta x(T_\infty^4 - T_i^4) + \frac{4 \rho_e \Delta x}{\pi D^2} I_c^2 = 0 \quad \text{for } i = 2 \dots (M-1). \tag{15}$$

The nodes at the edges of the domain must be treated separately; the temperature at both edges of the fuse are specified:

$$T_1 = T_{end} \tag{16}$$

$$T_M = T_{end}. \tag{17}$$

Solution

Equations (15) through (17) are a system of M equations in an equal number of unknown temperatures. These equations can be solved in any of the ways discussed in Section 2.4. Here we will use EES to solve the algebraic equations. The inputs are entered in EES.

```
"Inputs"
L=2.5 [cm]*Convert(cm,m)                    "length"
D=0.9 [mm]*Convert(mm,m)                    "diameter"
T_infinity=ConvertTemp(C,K,20)             "air temperature"
T_end=ConvertTemp(C,K,20)                  "end temperature"
h_bar=5.0 [W/m^2-K]                        "heat transfer coefficient"
e=0.9 [-]                                  "emissivity"
k=150 [W/m-K]                              "conductivity"
rho_e=1e-7 [ohm-m]                         "electrical resistivity"
T_melt=ConvertTemp(C,K,500)                "melting temperature"
I_c=100 [amp]                              "current"
```

Example 3.7 (cont.)

The nodes are distributed according to Eqs. (7) and (8).

```
"Distribute nodes"
M=11 [-]                                      "number of nodes"
Duplicate i=1,M
    x[i]=(i-1)*L/(M-1)                        "location of each node"
End
DELTAx=L/(M-1)                                "distance between adjacent nodes"
```

Equations (15) through (17) are entered. The internal energy balances are entered using a **Duplicate** loop.

```
"System of algebraic equations"
T[1]=T_end                                    "specified temperature at x=0"
Duplicate i=2,(M-1)
    k*pi*D^2*(T[i-1]-T[i])/(4*DELTAx)+k*pi*D^2*(T[i+1]-T[i])/(4*DELTAx)&
        +h_bar*pi*D*DELTAx*(T_infinity-T[i])&
        +e*sigma#*pi*D*DELTAx*(T_infinity^4-T[i]^4)+4*rho_e*DELTAx*I_c^2/(pi*D^2)=0
                                              "energy balance on each node"
End
T[M]=T_end                                    "specified temperature at x=L"
```

Solving leads to the temperature distribution shown in Figure 3 for $M = 11$ nodes.

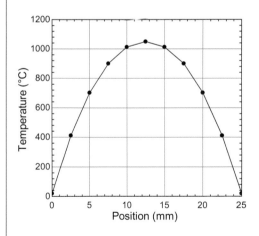

Figure 3 Temperature distribution for $M = 11$ nodes with $D = 0.9$ mm.

Figure 3 shows that the temperature in the fuse reaches approximately 1050°C, which is much higher than the melting point. The maximum temperature within the fuse is obtained using the **Max** command.

```
T_max=Max(T[1..M])                            "maximum temperature in the fuse"
```

The problems asks that the appropriate fuse diameter be determined which leads to the fuse just melting at $I_c = 100$ amp. Computer models make it easy to answer questions like this. EES in particular is well-suited to this kind of analysis. Before proceeding, it is important that we update the guess values in EES in order to

Continued

Example 3.7 (cont.)

start our iterative solution procedure from a good starting point. Select Update Guesses from the Calculate menu and select Yes from the resulting confirmation window. Now we can set the value of the variable T_max to be equal to T_melt.

T_max=T_melt	"force T_max to be equal to T_melt"

This over-constrains the problem and therefore we have to remove one of the inputs. In this case, comment out the specified diameter:

{D=0.9 [mm]*convert(mm,m)}	"diameter"

Solve the problem and you will find that $D = 0.001136$ m (1.136 mm) is the correct diameter for the fuse.

Discussion

With any numerical simulation it is important to verify that a sufficient number of nodes have been used so that the numerical solution has converged. There are different ways of accomplishing this. One method of exploring numerical convergence is to set the diameter to the value we found in the solution, 1.136 mm, and prepare a Parametric Table that includes both the variables M and T_max. The variable M is varied from 3 to 500 using an approximately logarithmic spacing (but making sure that the values are all integers) and the value of M in the Equations Window is commented out. Figure 4 illustrates the maximum temperature in the fuse as a function of the number of nodes used in the solution. It appears that the solution has converged to within about 0.1 K of the actual solution when more than 40 nodes are used.

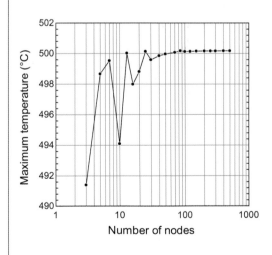

Figure 4 Maximum temperature as a function of the number of nodes for $D = 1.136$ mm.

3.5 Conclusions and Learning Objectives

This chapter presented the extended surface approximation that can be used to reduce the dimensionality of a problem and therefore simplify it by neglecting temperature variations due to conduction in one or more directions. The extended surface approximation is justified by the Biot number. The analytical and numerical solutions to extended surface problems were presented. The extended surface solutions to a class of problems referred to as fins was presented in terms of the fin efficiency. The utility of the fin efficiency to analyze heat transfer equipment was studied and the proper way to consider finned surfaces was discussed.

Some specific concepts that you should understand are listed below.
- The extended surface approximation and its justification in terms of the Biot number.
- How to define a differentially small control volume for an extended surface problem in order to derive a differential equation.
- How to solve a nonseparable, ordinary differential equation by dividing it into a homogeneous and particular solution.
- The method of undetermined coefficients for solving nonhomogeneous ordinary differential equations.
- The ability to identify the appropriate functional solutions to a homogeneous differential equation.
- The function of a fin for improving heat transfer and the definition of fin efficiency for various geometries.
- The ability to calculate a fin resistance and the use of this concept to incorporate fins into heat transfer problems.
- How to correct an adiabatic tip fin solution for convection from the tip.
- The calculation of an overall surface efficiency that can be used to analyze a finned surface.
- The ability to solve extended surface problems using numerical techniques.

Problems

The Extended Surface Approximation and the Biot Number

3.1 A wire is subjected to ohmic heating (i.e., a current runs through it) while it is convectively cooled. The ends of the wire have fixed temperatures. Describe how you would determine whether the extended surface approximation (i.e., the approximation in which you treat the temperature of the wire as being one-dimensional varying only along its length) is appropriate when solving this problem.

3.2 A 0.75 m long rod with 2.5 cm diameter is made of stainless steel (k = 10 W/m-K) and projects from the side of a space capsule. The emissivity of the rod surface is 0.35. One end of the rod is connected to the outside surface of the capsule which, due to the heat source onboard, is maintained at a temperature of 112 K. The capsule is in orbit where there is no air and the effective temperature for radiation from the rod to space is 4 K. Determine if this rod could be treated as a 1-D extended surface heat transfer problem.

3.3 A wire hanger is unbent and straightened and then used to roast marshmallows over a camp fire. The hanger has a diameter of 0.125 inch and is 2 ft long with thermal conductivity 34 Btu/hr-ft-R. The

convection coefficient is estimated to be 3.25 Btu/hr-ft²-R. Indicate whether you can treat the hanger as a 1-D extended surface.

3.4 Your family is planning to bake a turkey for the holidays. The turkey is approximately spherical with a mass of 14 lb_m and its properties (density, specific heat, thermal conductivity) are about the same as those for liquid water at standard temperature and pressure. The convection coefficient on the surface of the turkey is estimated to be 36 W/m²-K due to the use of a convection oven. Use an appropriate dimensionless parameter to determine whether the internal conduction resistance is likely to be an important factor related to the time to cook the turkey.

3.5 The figure illustrates a triangular fin with a circular cross-section.

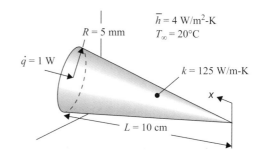

The fin is surrounded by fluid at $T_\infty = 20°C$ with heat transfer coefficient $\bar{h} = 4$ W/m²-K. The base of the fin is subjected to heat transfer at a rate of $\dot{q} = 1$ W. The length of the fin is $L = 10$ cm and the radius of the fin at its base is $R = 5$ mm. The fin is made of material with conductivity $k = 125$ W/m-K.

(a) Is an extended surface model of the fin appropriate? That is, can the temperature be modeled as being only a function of x and not of r. Justify your answer.

(b) Assume that your answer to (a) showed that an extended surface model is appropriate. Derive the governing differential equation for the problem. Note that the x-coordinate is defined as *starting* from the tip of the fin and moving toward the base.

3.6 The electrical connection between two conductors is imperfect and therefore the joint has a large electrical resistance that results in an ohmic generation of $\dot{g}$, as shown below. Conductor 1 has conductivity k_1 and conductor 2 has conductivity k_2. Both conductors have the same cross-sectional area (A_c) and perimeter (*per*). Both conductors experience convection with heat transfer coefficient $\bar{h}$ to an ambient fluid at T_∞. You may assume that the Biot number for each conductor is small and the problem is steady state so that the temperature is only a function of axial position. Both conductors extend infinitely away from the joint.

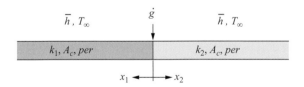

Derive the ordinary differential equations for T_1 (the temperature of material 1) as a function of x_1 and T_2 (the temperature of material 2) as a function of x_2. Also list the *four* boundary conditions needed for this problem.

3.7 The figure illustrates a thin disk used in a brake system. The disk is rotating with angular velocity ω. The inner and outer radii are r_i and r_o and the

thickness of the disk is b. Assume that the temperature distribution in the disk is 1-D and steady state, $T(r)$.

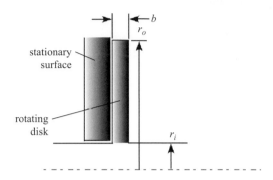

When the brake is engaged, a stationary surface is moved into contact with the disk on its left side. The pressure applied at the interface is p and the friction coefficient is μ. The relative motion of the rotating disk and the stationary surface causes a frictional heating that you can assume is entirely transferred as heat to the rotating disk. The back (right) side of the disk is cooled by fluid at T_∞. The heat transfer coefficient is proportional to the linear velocity of the disk surface and therefore is proportional to radius: $h = c\,r$. The end of the disk at $r = r_o$ is insulated and the inner radius of the disk is maintained at T_∞. Derive the ordinary differential and boundary conditions for this problem.

3.8 A thin, disk-shaped window is used to provide optical access to a combustion chamber. The thickness of the window is b and the outer radius of the window is R_o. The window is composed of material with conductivity k and absorption coefficient α. The combustion chamber side of the window is exposed to convection with hot gas at T_f and heat transfer coefficient h. Convection with the air outside of the chamber can be neglected. There is a radiation heat flux, $\dot{q}''_{rad}$, that is incident on the combustion chamber side of the glass. The amount of this radiation that is absorbed by the glass is, approximately, $\dot{q}''_{rad}\,\alpha\,b$. The remainder of this radiation, $\dot{q}''_{rad}(1 - \alpha b)$, exits the opposite surface of the glass. The outer edge (at $r = R_o$) of the glass is held at temperature T_{edge}.

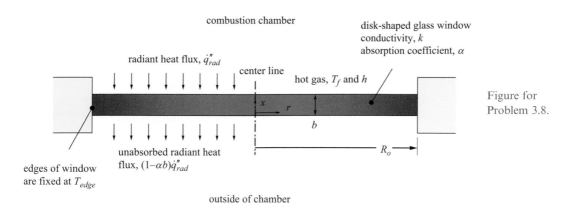

combustion chamber

disk-shaped glass window
conductivity, k
absorption coefficient, α

radiant heat flux, $\dot{q}''_{rad}$

center line

hot gas, T_f and h

x

r

b

Figure for
Problem 3.8.

unabsorbed radiant heat
flux, $(1-\alpha b)\dot{q}''_{rad}$

R_o

edges of window
are fixed at T_{edge}

outside of chamber

(a) How would you justify using a 1-D model of the glass? That is, what dimensionless number would you calculate in order to verify that the temperature does not vary substantially in the x-direction?

(b) Derive the ordinary differential equation in r that must be solved. Make sure that your differential equation includes the effect of conduction, convection with the gas within the chamber, and generation of thermal energy due to absorption.

(c) What are the boundary conditions for the ordinary differential equation that you derived in part (b)?

3.9 A support rod in a cryogenic experiment is made of the material G10 ($k = 0.24$ W/m-K). The outer surface of the G10 material has emissivity $\varepsilon = 0.65$. The rod is 12 cm in length and 1 cm in diameter. One end of the rod is fastened to a cold plate maintained at 4 K, whereas the other end is connected to an 80 K plate. The experiment is located in a Dewar that is evacuated so there is no convection. However, the rod radiates to the container walls, which are at 280 K. It is necessary to calculate the heat transfer through the rod. Can the rod be accurately approximated as a 1-D extended surface?

3.10 A long stack of square laminations has a thermal conductivity of 15 W/m-K in the direction parallel to the laminations and 1.8 W/m-K in the direction perpendicular to the laminations. The side length of the cross-section is 5 cm. One edge perpendicular to the laminations is heated while

all of the other surfaces convect to air at 25°C with a heat transfer coefficient of 9.5 W/m²-K. Can this material be considered to be a 1-D extended surface?

Analytical Solutions to Extended Surface Problems

3.11 A bucket contains water at 90°C and atmospheric pressure. The handle on the bucket is a metal wire with a 6.5 mm diameter that is bent into a semicircle with radius 15 cm. You are about to pick up the bucket at the center of the handle, but are concerned that you will experience a painful burn. The handle transfers energy to the surrounding air at 25°C with an estimated average convection heat transfer coefficient of 11.5 W/m²-K. The handle is made of metal with thermal conductivity 43 W/m-K. Your hand will burn if it contacts the metal handle at a point in which the temperature is greater than 40°C. Should you be concerned?

(a) Is the extended surface approximation appropriate?

(b) Plot the temperature distribution in the handle and indicate whether it is safe to grab.

3.12 A circular pin fin is used in a situation where the base temperature T_b is greater than the surrounding fluid temperature T_∞. Create a table that includes each of the following parameters: length (L), diameter (D) heat transfer coefficient ($\bar{h}$), fluid temperature (T_∞), and conductivity (k). If each of these parameters are increased, indicate whether:

(a) the tip temperature $(T_{x=L})$ will increase, remain the same, or decrease;

(b) the heat transfer rate $(\dot{q}_{fin})$ will increase, remain the same, or decrease.

3.13 A copper rod is 0.3 m in length and 0.05 m in diameter and located in a duct that carries air. One end of the rod is maintained at 200°C while the other end is held at 95°C. Air at 1 atm and 25°C flows across the rod at a rate of 0.025 kg/s. The heat transfer coefficient between the rod and the air is 12 W/m²-K.

(a) Justify that this problem can be treated as one-dimensional.

(b) Plot the temperature of the rod as a function of position along its length.

(c) Determine the total rate of convective heat transfer from the rod.

(d) Determine the temperature of the air downstream of the rod.

3.14 The heated end of a soldering gun consists of a 3 inch long copper rod with the last 0.25 inch being the tip that has been sharpened to a point. Heat transfer from the soldering gun occurs by free convection to air at 25°C with an estimated heat transfer coefficient of 16 W/m²-K. It is necessary to determine the temperature at the base of the copper rod and the required power needed to maintain the sharpened tip at 200°C, which is the temperature required to melt the solder. The engineer working on this problem is unsure how to treat the sharpened tip. You suggest that solution can be bracketed by considering two alternatives.

(a) Assume that the rod is 2.75 inches in length and the tip is insulated.

(b) Neglect the geometric details of the tip. Consider the soldering rod to be a 3 inch long copper rod with a convection occurring at the tip as well as along the length of the rod.

3.15 A bracket is in the shape of a solid rod that is L = 4 cm long with diameter D = 5 mm. The bracket extends between a wall at T_H = 100°C (at x = 0) and a wall at T_C = 20°C (at x = L). The conductivity of the bracket is k = 25 W/m-K. The bracket is surrounded by gas at

T_∞ = 200°C and the heat transfer coefficient is $\bar{h}$ = 250 W/m²-K.

(a) Is an extended surface approximation appropriate for this problem? Justify your answer.

(b) Assume that your answer to (a) was yes. Develop an analytical model. Plot the temperature as a function of position within the bracket.

(c) Overlay on your plot from (b) the temperature as a function of position with $\bar{h}$ = 2.5, 25 and $\bar{h}$ = 2500 W/m²-K. Explain the shape of your plots.

(d) Plot the rate of heat transfer from the wall at T_H into the bracket (i.e., the rate of heat transfer into the bracket at x = 0) as a function of $\bar{h}$. Explain the shape of your plot.

3.16 A small-diameter metal rod has a heater installed on its tip.

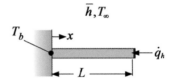

The rate of heat transfer provided by the heater, $\dot{q}_h$, is transferred by conduction *into* the tip of the rod (at x = L), as shown. The base of the rod (at x = 0) is maintained at temperature, T_b. The cross-sectional area of the rod is A_c and the perimeter of the rod is *per*. The rod is surrounded by fluid at temperature T_∞ with heat transfer coefficient $\bar{h}$. The conductivity of the rod material is k. Assume that the extended surface approximation is appropriate (i.e., the temperature within the rod is a function only of x). Neglect radiation for this problem.

(a) Sketch the temperature distribution in the rod. Assume that $T_b > T_\infty$. Justify your sketch by listing some of the characteristics that you expect to see.

The ordinary differential equation that governs this problem is the same as the ODE that was derived for a constant cross-sectional area fin:

$$\frac{d^2 T}{dx^2} - m^2 T = -m^2 T_\infty \quad \text{where } m = \sqrt{\frac{per\, \bar{h}}{k\, A_c}}.$$

The general solution to this ODE is:

$$T = C_1 \exp(mx) + C_2 \exp(-mx) + T_\infty.$$

(b) Use the correct boundary conditions in order to obtain two equations that can be solved simultaneously in order to provide C_1 and C_2.

3.17 A long metal rod with a diameter of 5 cm and thermal conductivity $k = 15$ W/m-K has one end placed in a hot furnace, while the remainder of the rod is exposed to 25°C air. It is not possible to measure the temperature at the hot end, but measurements made at positions 0.3 m and 0.36 m from the hot end show temperatures of 125°C and 92°C, respectively. Based on these measurements, estimate:

(a) the average convection coefficient between the rod to the air, and

(b) the temperature at the hot end of the rod ($x = 0$).

3.18 A brass rod that is 1 m in length with a diameter of 2.5 cm has one end maintained at 295°C and the other at 50°C. The temperature in the middle of the rod is 102°C. Determine:

(a) the average convection coefficient between the rod and the air, and

(b) the temperature of the surrounding air.

3.19 A wire of length L and radius r_o is attached at each end to a wall; one wall is at temperature T_1 and the other is at temperature T_2. A steady electrical current is passed through the wire so that thermal energy is uniformly generated volumetrically at rate $\dot{g}'''$. Convective heat transfer occurs between the surface of the wire and the air at temperature T_∞ with an average heat transfer coefficient of $\bar{h}$. Derive the differential equation that will provide the temperature of the wire as a function of axial position x. Indicate the boundary conditions required to solve this equation.

3.20 Two long pieces of copper wire ($k = 380$ W/m-K) that are each 0.16 cm in diameter are to be soldered end to end. The solder melting temperature is 225°C. The heat transfer coefficient between the 25°C air and the wire surface is 15 W/m²-K.

(a) Derive the governing differential equation for the temperature distribution along one of wires and specify the boundary conditions.

(b) Solve the differential equation and plot the temperature as a function of distance from the joint.

(c) Calculate the rate of heat transfer at the joint.

3.21 A disk-shaped bracket connects a cylindrical heater to an outer shell.

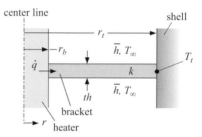

The thickness of the bracket is th and it is made of material with conductivity k. The bracket extends radially from $r = r_b$ at the heater to $r = r_t$ at the outer shell. The temperature of the bracket location where it meets the shell (at $r = r_t$) is T_t. The heater provides a known rate of heat transfer to the bracket, $\dot{q}$, at $r = r_b$. Both the upper and lower surfaces of the bracket are exposed to fluid at T_∞ with average heat transfer coefficient $\bar{h}$.

(a) What calculation would you do in order to justify treating the bracket as an extended surface (i.e., justify the assumption that temperature is only a function of r); provide an expression in terms of the symbols in the problem statement.

For the remainder of the problem, assume that the bracket can be treated as an extended surface.

(b) Sketch the temperature distribution that you would expect if the heat transfer coefficient $\bar{h}$ is very low ($\bar{h} \rightarrow 0$). Assume that $T_t < T_\infty$ and indicate the values of these temperatures on your plot; your sketch should be consistent with these values.

(c) Overlay onto your sketch from (b), the temperature distribution that you would expect if the heat transfer coefficient $\bar{h}$ is very high ($\bar{h} \rightarrow \infty$).

(d) Derive the governing ordinary differential equation for the bracket.

(e) What are the boundary conditions for the ordinary differential equation?

3.22 A current lead must be designed to carry current to a cryogenic superconducting magnet.

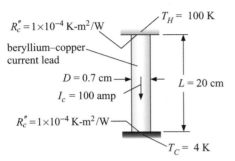

The current lead carries I_c = 100 amp and therefore experiences substantial generation of thermal energy due to ohmic dissipation. The lead is made of beryllium–copper, which is one of the substances whose properties are available in EES (it is an Incompressible Substance). The electrical resistivity and thermal conductivity of beryllium–copper can be obtained from the ElectricalResistivity and Conductivity functions in EES, respectively. The length of the current lead is L = 20 cm and the diameter is D = 0.7 cm. The hot end of the lead (at x = 0) is bolted to a flange that is maintained at T_H = 100 K and the cold end (at x = L) is bolted to a flange that is maintained at T_C = 4 K. The contact resistance between the ends of the leads and these flanges is $R''_c = 1 \times 10^{-4}$ K-m²/W. The lead is installed in a vacuum chamber and therefore you may assume that the external surfaces of the lead (the outer surface of the cylinder) are adiabatic.

(a) Develop an analytical model of the lead that assumes that the electrical resistivity and thermal conductivity of the material are constant; use the values calculated at the average temperature in the lead $(T_H + T_C)/2$. Implement your model in EES and plot the temperature as a function of position.

(b) Predict the rate of heat transfer from the current lead to the flange at T_C. This parasitic heat transfer rate must be removed in order to keep the magnet cold and therefore must be minimized in the design of the current lead.

(c) Plot the rate of heat transfer from the current lead to the flange as a function of the diameter of the current lead. You should see a minimum value and therefore an optimal diameter – explain why this occurs.

3.23 A fuse consists of a wire with diameter D = 0.006 inch and total length L = 4 inch that is coiled up and placed in an evacuated glass enclosure. There is no convection experienced by the wire but the surface of the wire does experience radiation with surroundings at T_{sur} = 20°C. The emissivity of the surface of the wire is ε = 0.65. The wire material has thermal conductivity k = 14.2 W/m-K and electrical resistivity ρ_e = 2.4 × 10^{-8} ohm-m. The current carried by the wire is 1.5 amp. The fuse "blows" when the material becomes hot enough to melt, which occurs at T_m = 650°C. The two ends of the fuse (at x = 0 and x = L) are maintained at the temperature of the surroundings by the end caps. This is a slow blow fuse which means that we are interested in the steady-state temperature distribution within the fuse.

Radiation heat flux is given by:

$$\dot{q}''_{rad} = \varepsilon \sigma (T^4 - T^4_{sur}),$$

where σ is the Stefan–Boltzmann constant and T is the surface temperature of the wire. In order to develop an analytical solution to this problem, it is necessary to linearize the radiation rate equation:

$$\dot{q}''_{rad} = \underbrace{\varepsilon \sigma (T^2 + T^2_{sur})(T + T_{sur})}_{\bar{h}_{rad}} (T - T_{sur}).$$

The coefficient in the above equation that multiplies the temperature difference is sometimes referred to as the radiation heat transfer coefficient:

$$\bar{h}_{rad} = \varepsilon \sigma (T^2 + T^2_{sur})(T + T_{sur}).$$

We will assume for this problem that $\bar{h}_{rad}$ is constant regardless of the local temperature of the fuse and therefore the radiation heat flux can be written as:

$$\dot{q}''_{rad} = \bar{h}_{rad}(T - T_{sur}).$$

(a) Estimate the value of $\bar{h}_{rad}$ that should be used for the analysis.

(b) Is it appropriate to treat the fuse as an extended surface (i.e., can we assume that the temperature in the fuse is a function only of x)? Justify your answer.

(c) What is the volumetric rate of thermal energy generation that is occurring within the fuse material?

(d) Develop an analytical model of the fuse that will predict the temperature distribution in the fuse and therefore the maximum temperature achieved by the fuse material. Clearly show how you derived the governing differential equation, solved it to find the general solution, and used the boundary conditions to determine the constants of integration. Plot the temperature of the fuse as a function of x.

(e) You are trying to design the fuse so that it "blows" when the current reaches 1.5 amp. What diameter should you choose?

(f) There is some manufacturing variability in all of the parameters associated with your fuse; you are interested in determining which parameters are most critical in order to ensure that your fuse actually blows at 1.5 amp. Use your model to rank the following parameters in order of importance (from most important to least important): diameter, length, emissivity, conductivity, electrical resistivity, and melting temperature. Clearly explain how you used your model to answer this question.

3.24 A 0.01 m diameter rod made of a material with thermal conductivity $k = 36$ W/m-K extends through fluid A, through a separating wall of 0.1 m thickness and through fluid B, as shown in the figure. The length of rod in both fluids exposed to convection is 1.0 m. The cross-hatched areas in the figure represent an insulating material with very low thermal conductivity. The convection coefficient between the fluid and the rod surface is 100 W/m²-K for both fluids A and B. Fluid A is maintained at 120°C while fluid B is maintained at 20°C. Calculate the rate of heat transfer from fluid A to fluid B.

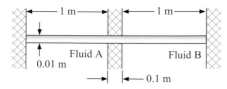

3.25 The figure illustrates a brake pad used in a large industrial machine.

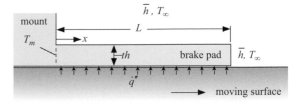

The brake pad can be considered to be infinitely wide (into the page). The thickness of the pad is th and its length is L. The pad material has conductivity k. The top surface of the pad is exposed to coolant at T_∞ with heat transfer coefficient $\bar{h}$. When the brake is engaged, it is pressed down upon a moving surface. The friction between the brake and the moving surface causes a uniform heat flux, $\dot{q}''$, to the bottom of the brake pad. The end of the brake pad (at $x = L$) is not insulated; rather, the end is exposed to convection with $\bar{h}$ and T_∞. The base of the brake pad (at $x = 0$) is attached to a mount with temperature T_m. The system is at steady state.

(a) What calculation would you carry out in order to justify treating the brake pad as an extended surface? Specifically show the quantity that you would calculate in terms of the symbols provided in the problem statement.

For the remainder of the problem, assume that you can treat the brake pad as an extended surface.

(b) Derive the ordinary differential equation and boundary conditions that govern this problem. Clearly show your steps.

(c) Solve the ODE from part (b) in order to obtain a general solution that includes two undetermined constants, C_1 and C_2.

(d) Develop two equations that can be used to solve for the two constants of integration, C_1 and C_2, in the general solution.

(e) Sketch the temperature distribution that you expect if the brake is engaged (i.e., $\dot{q}'' > 0$) but the cooling is turned off (i.e., $\bar{h} = 0$). Assume that $T_m > T_\infty$ and indicate the value of these temperatures in your sketch. Focus on getting the qualitative features of your temperature distribution correct.

You may want to briefly describe some characteristics of your temperature distribution sketch.

(f) Overlay on your sketch from (e) the temperature distribution that you expect if the brake is disengaged (i.e., $\dot{q}'' = 0$) but the cooling is turned on (i.e., $\bar{h} > 0$). Clearly label the two sketches from (e) and (f). Again, focus on getting the qualitative features of your temperature distribution correct. You may want to briefly describe some characteristics of your temperature distribution sketch.

3.26 In a cryogenic experiment, a low-temperature platform at T_C is suspended from an intermediate-temperature platform at T_H by a thermal standoff.

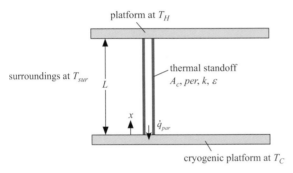

platform at T_H

surroundings at T_{sur}

L

x

thermal standoff A_c, per, k, ε

$\dot{q}_{par}$

cryogenic platform at T_C

The thermal standoff used to separate the platforms is a thin-walled metal tube. The purpose of the thermal standoff is to limit the parasitic heat transfer to the low-temperature platform, $\dot{q}_{par}$ in the figure. The outer surface of the tube is exposed to radiation from surroundings at T_{sur}. The conductivity of the thermal standoff is k, the cross-sectional area is A_c, and the perimeter exposed to radiation is per. The length of the thermal standoff is L. The radiation heat flux experienced by the surface of the standoff is

$$\dot{q}''_{rad} = \sigma\varepsilon\left(T_{sur}^4 - T^4\right)$$

where σ is the Stefan–Boltzmann constant, ε is the emissivity of the surface of the tube, and T is the local temperature of the tube surface. The radiation rate equation can be expanded:

$$\dot{q}''_{rad} = \sigma\varepsilon\left(T_{sur}^2 + T^2\right)(T_{sur} + T)(T_{sur} - T).$$

Because the surrounding temperature is much higher than the temperature of either of the platforms ($T_{sur} \gg T_H$ and $T_{sur} \gg T_C$), the equation can be written approximately as:

$$\dot{q}''_{rad} \approx \sigma\varepsilon T_{sur}^3(T_{sur} - T).$$

Assume that the temperature in the thermal is only a function of x.

(a) Derive the governing differential equation for the temperature in the thermal standoff. Clearly show your steps.

(b) What are the boundary conditions for the differential equation?

(c) Solve the differential equation derived in (a) subject to the boundary conditions developed in (b).

(d) Develop an expression for the parasitic heat transfer to the cryogenic plate, $\dot{q}_{par}$.

3.27 You are designing a manipulator for use within a furnace. The arm must penetrate the side of the furnace, as shown in the figure. The arm has a diameter of $D = 0.8$ cm and protrudes $L_i = 0.5$ m into the furnace, terminating in the actuator that can be assumed to be adiabatic. The portion of the arm in the furnace is exposed to flame and hot gas; these effects can be represented by a heat flux of $\dot{q}''' = 1 \times 10^4$ W/m^2 and convection to gas at $T_f = 500°$C with heat transfer coefficient $\bar{h}_f = 50$ W/m^2-K. The conductivity of the arm material is $k = 150$ W/m-K. The portion of the arm outside of the furnace has the same diameter and conductivity, but is exposed to air at $T_a = 20°$C with heat transfer coefficient $\bar{h}_a = 30$ W/m^2-K. The length of the arm outside of the furnace is $L_o = 0.75$ m and this portion of the arm terminates in the motor system which can also be considered to be adiabatic.

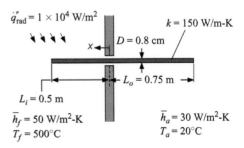

$\dot{q}''_{rad} = 1 \times 10^4$ W/m^2

$k = 150$ W/m-K

x $D = 0.8$ cm

$L_o = 0.75$ m

$L_i = 0.5$ m

$\bar{h}_f = 50$ W/m^2-K
$T_f = 500°$C

$\bar{h}_a = 30$ W/m^2-K
$T_a = 20°$C

(a) Is an extended surface model appropriate for this problem? Justify your answer.

(b) Develop an analytical model of the manipulator arm; implement your model in EES. Plot the temperature as a function of axial position x (see the figure) for $-L_o < x < L_i$.

(c) Prepare a plot showing the maximum temperature at the end of the arm (within the furnace) as a function of the internal length of the arm (L_i) for various values of the diameter (D).

3.28 A resistance temperature detector (RTD) utilizes a material that has a resistivity that is a strong function of temperature. The temperature of the RTD is inferred by measuring its electrical resistance. The figure shows an RTD that is mounted at the end of a metal rod and inserted into a pipe in order to measure the temperature of a flowing liquid. The RTD is monitored by passing a known current through it and measuring the voltage across it. This process results in ohmic heating that may tend to cause the RTD temperature to rise relative to the temperature of the surrounding liquid; this effect is referred to as a self-heating error. Also, conduction from the wall of the pipe to the temperature sensor through the metal rod can result in a temperature difference between the RTD and the liquid; this effect is referred to as a mounting error.

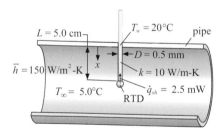

The thermal energy generation associated with ohmic heating is $\dot{q}_{sh} = 2.5$ mW. All of this ohmic heating is assumed to be transferred from the RTD into the end of the rod at $x = L$. The rod has a thermal conductivity $k = 10$ W/m-K, diameter $D = 0.5$ mm, and length $L = 5$ cm. The end of the rod that is connected to the pipe wall (at $x = 0$) is maintained at a temperature of $T_w = 20°C$. The liquid is at a uniform temperature, $T_\infty = 5°C$ and the heat transfer coefficient between the liquid and the rod is $\bar{h} = 150$ W/m²-K.

(a) Is it appropriate to treat the rod as an extended surface (i.e., can we assume that the temperature in the rod is a function only of x)? Justify your answer.

(b) Develop an analytical model of the rod that will predict the temperature distribution in the rod and therefore the error in the temperature measurement; this error is the difference between the temperature at the tip of the rod and the liquid.

(c) Prepare a plot of the temperature as a function of position and compute the temperature error.

(d) Investigate the effect of thermal conductivity on the temperature measurement error. Identify the optimal thermal conductivity and explain why an optimal thermal conductivity exists.

3.29 A flux meter is illustrated in the figure.

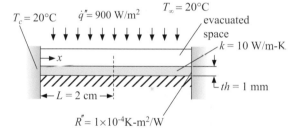

A thin plate is clamped on either end to a casing that is maintained at $T_c = 20°C$. There is a contact resistance between the plate and the casing, $R'' = 1 \times 10^{-4}$ K-m²/W. The thickness of the plate is $th = 1$ mm and the half-width of the plate is $L = 2$ cm. The plate conductivity is $k = 10$ W/m-K. The back of the plate is insulated and the front of the plate is mounted within an evacuated enclosure in order to eliminate convection. You may model radiation loss from the plate surface using an effective "radiation" heat transfer coefficient, calculated according to: $\bar{h}_{rad} \approx 4\varepsilon\sigma\overline{T}^3$, where $\overline{T}$ is the average absolute temperature of the plate and the surroundings (use $\overline{T} = 300$ K for this problem). The plate radiates to surroundings at $T_\infty = 20°C$. The nominal flux on the plate is $\dot{q}'' = 900$ W/m². You may assume that the plate temperature distribution is 1-D, varying only in the x-direction. You may also assume that the emissivity of the plate, ε, is unity and therefore all of the flux on the plate is absorbed. The flux meter operates by correlating the difference between the temperature at the center of the plate and the casing with the applied heat flux.

(a) Derive the differential equation that governs the temperature within the plate. Clearly show your steps.

(b) What are the boundary conditions for the differential equation?

(c) Determine the solution to the differential equation from (a) subject to the boundary conditions from (b).

(d) Plot the temperature in the plate as a function of position.

(e) Prepare a calibration curve for the flux meter – plot the heat flux as a function of the difference between the temperature at the center of the plate and the casing.

(f) If the uncertainty in the measurement of the temperature difference is $\delta\Delta T = 0.5$ K then what is the uncertainty in the measurement of the heat flux?

Fins and Finned Surfaces

3.30 A cylindrical rod with a diameter of 12 mm and a length of 5 cm is used as a fin. The rod is made of AISI 304 stainless steel. The base of the rod is maintained at 154°C. The outer surface of the rod is exposed to air at 25°C and convection occurs with a convection coefficient of 12 W/m²-K. The rod is polished so radiation is not important. Assume that the tip is insulated.

(a) Plot the temperature of the rod as a function of axial position.

(b) Determine the rate of heat transfer from the rod.

3.31 The efficiency of a fin is defined as the ratio of the actual heat transfer rate from the fin to the *maximum possible heat transfer rate*. Circle the letter of the statement below that best describes the physical meaning of how the *maximum possible heat transfer rate* could be achieved.

(a) The fin has a uniform temperature that is equal to the temperature of the surrounding fluid.

(b) The fin is infinitely long.

(c) The fin is made of a material with zero thermal conductivity.

(d) The convection heat transfer coefficient between the fin and the surrounding fluid is infinite.

(e) The fin has a uniform temperature that is equal to that of its base.

3.32 An annular fin with a rectangular cross-section having a thickness of $th = 1.6$ mm and outer radius $r_{out} = 12.5$ cm is attached to a circular thin-walled tube. The inner radius of the fin is $r_{in} = 2.5$ cm. The fin is made of aluminum. Boiling water at 100°C flows inside the tube. Because the boiling heat transfer coefficient is high, the outside tube wall where the fin is attached can be assumed to be at 100°C. Air at $T_\infty = 25$°C flows over the tube and fin. Prepare a plot of the heat transfer rate from the fin as a function of the average convection coefficient $\bar{h}$ for $20 < \bar{h} < 80$ W/m²-K.

3.33 A cylindrical rod with a diameter of 12 mm and a length of 5 cm is used as a fin. The rod is made of AISI 304 stainless steel. The base of the rod is maintained at 154°C. The outer surface of the rod and its tip are exposed to air at 25°C and convection occurs with a convection coefficient of 12 W/m²-K. The rod is polished so radiation is not important. Assume that the tip convects to the surrounding air.

(a) Plot the temperature of the rod as a function of axial position.

(b) Determine the rate of heat transfer from the rod.

(c) Determine the fin efficiency for this rod.

(d) Compare the values for parts (b) and (c) with a rod that has an insulated tip (Problem 3.30).

3.34 Rectangular fins are used on the outside wall of a computer heat sink. The fins are made of aluminum. Each fin is 5 mm thick, 3 cm in length and 3 cm in depth. The base temperature of the fins is 38°C and the environment is 20°C. However, the heat transfer coefficient is not precisely known. Prepare a plot of the heat transfer rate for each fin and the fin efficiency as a function of convection coefficient for values ranging from 10 to 35 W/m²-K.

3.35 A heat exchanger that transfers energy between steam and air consists of a 16.4 mm outer diameter tube made of copper. Annular fins made of brass are bonded to the tube surface. The fins are 0.25 mm thick with an outer diameter of 30.7 mm and are spaced every 3.63 mm. Steam at 100°C is flowing within the tube. Because of the high heat transfer coefficient and the conductive tube wall, the tube wall can be assumed to be at 100°C. Air at 16°C flows perpendicular to the tube with a heat transfer coefficient of 18 W/m²-K. Assume that there is no contact resistance between the fins and the tube wall.

(a) Determine the efficiency of the aluminum fins.
(b) Determine the overall surface efficiency.
(c) Determine the total rate of heat transfer per meter of length.

3.36 Reconsider the brass fins in Problem 3.35. If the fins are press-fit rather than soldered to the wall then the contact resistance between the fins and the tube is 1.2×10^{-4} m²-K/W. Determine the rate of heat transfer per meter of length.

3.37 The purpose of this problem is to compare the heat transfer rate from a triangular fin with that of a rectangular fin. The length of the fin is 2.25 cm and the thickness at its base is 8 mm where it connects to a surface maintained at 365°C. The fin is made of stainless steel 304. The convection coefficient between the 25°C air surrounding the fin and the fin surfaces is 18 W/m²-K.

(a) Calculate the fin efficiency and heat transfer rate per unit width for the triangular fin.
(b) Calculate the fin efficiency and heat transfer rate per unit width for the rectangular fin. Assume that the heat transfer coefficient for the tip is the same value as for the rest of the fin.
(c) Indicate the benefit, if any, of the triangular fin.

3.38 In a heat exchanger, thermal energy is transferred from water to air through a 3 mm brass wall with conductivity $k = 79.5$ W/m-K as shown in the figure on the left. The convection coefficient between fluid and the metal is 170 W/m²-K on the water side and 20 W/m²-K on the air side.

The manufacturer is considering the addition of brass fins that are 0.75 mm in width, 2.5 cm long spaced 1.175 cm apart, as shown in the figure on the right for the water side. Determine the percentage increase in the heat transfer rates compared to having no fins for the following situations:

(a) fins only on the water side,
(b) fins only on the air side, and
(c) fins on both sides.

3.39 A water heater consists of a copper tube that carries water through hot gas in a furnace, as shown in the figure. The copper tube has an outer radius, $r_{o,tube} = 0.25$ inch and a tube wall thickness of $th = 0.033$ inch. The conductivity of the copper is $k_{tube} = 300$ W/m-K. Water flows through the pipe at a temperature of $T_w = 30°C$. The heat transfer coefficient between the water and the internal surface of the pipe is $\bar{h}_w = 500$ W/m²-K. The external surface of the tube is exposed to hot gas at $T_g = 500°C$. The heat transfer coefficient between the gas and the outer surface of the pipe is $\bar{h}_g = 25$ W/m²-K. Neglect radiation from the tube surface.

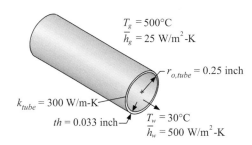

(a) At what rate is heat is added to the water for a unit length of tube, $L = 1$ m (W/m)?
(b) What is the dominant resistance to heat transfer in your water heater?

In order to increase the capacity of the water heater, you decide to slide washers over the tube, as shown in the figure. The washers are $w = 0.06$ inch thick with an outer radius of $r_{o,washer} = 0.625$ inch and have a thermal conductivity of $k_{washer} = 45$ W/m-K. The contact resistance between the washer and the tube is $R''_c = 5 \times 10^{-4}$ m²-K/W. The distance between two adjacent washers is $b = 0.25$ inch.

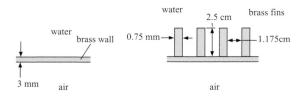

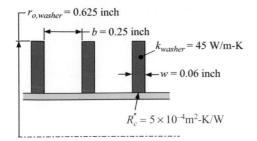

$r_{o,washer} = 0.625$ inch

$b = 0.25$ inch

$k_{washer} = 45$ W/m-K

$w = 0.06$ inch

$R''_c = 5 \times 10^{-4}$ m²-K/W

(c) Can the brass washers be treated as extended surfaces (i.e. can the temperature in the washers be considered to be only a function of radius)? Justify your answer with a calculation.

Assume that your answer to (c) showed that the washers can be treated as extended surfaces and therefore modeled as a fin with the appropriate fin resistance.

(d) Draw a thermal resistance network that can be used to represent this situation. Be sure to draw and label resistances associated with convection with the water ($R_{conv,w}$), conduction through the copper tube (R_{cond}), heat transfer through contact resistance ($R_{contact}$), heat transfer through the washers ($R_{washers}$), and convection to gas from unfinned outer surface ($R_{conv,g,unfinned}$).

(e) Determine the rate of heat transfer to the water with the washers installed on the tube for a 1 m length of tube?

3.40 A heat exchanger consists of 100 aluminum tubes each 3 ft in length. Each tube has a 1.0 inch inner diameter and a 1.125 in outer diameter. Saturated steam at 4.5 psia is flowing in the tubes and it interacts with the inner surface with an average convection coefficient of 200 Btu/hr-ft²-F. The outer surface of the tube is exposed to air at 75°F with a convection coefficient of 8.5 Btu/hr-ft²-F. To increase the heat transfer rate, it is proposed to place annular, rectangular aluminum fins on the outside of the tubes. The fins each have a thickness of 0.02 inch and the radial distance between the tube surface and the tip of the fin is 0.5 inch.

(a) Determine the rate of heat transfer without the fins.

(b) Determine the rate of heat transfer with the fins for fin spacings of 5, 10, and 20 fins per inch.

3.41 Your company has developed a micro-end milling process that allows you to fabricate an array of very small fins in order to make heat sinks for various types of electrical equipment. The end milling process removes material in order to generate the array of fins. Your initial design is the array of pin fins shown in the figure. You have been asked to optimize the design of the fin array for a particular application where the base temperature is $T_{base} = 120°C$ and the air temperature is $T_{air} = 20°C$. The heat sink is square; the size of the heat sink is $W = 10$ cm. The conductivity of the material is $k = 70$ W/m-K. The distance between the edges of two adjacent fins is a, the diameter of a fin is D, and the length of each fin is L.

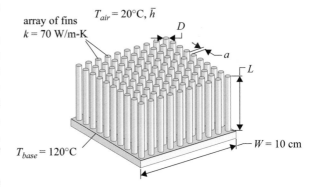

array of fins
$k = 70$ W/m-K

$T_{air} = 20°C, \bar{h}$

D

a

L

$T_{base} = 120°C$

$W = 10$ cm

Air is forced to flow through the heat sink by a fan. The heat transfer coefficient between the air and the surface of the fins as well as the unfinned region of the base, $\bar{h}$, has been measured for the particular fan that you plan to use and can be calculated according to:

$$\bar{h} = 40 \left[\frac{W}{m^2 \, K} \right] \left(\frac{a}{0.005 \, [m]} \right)^{0.4} \left(\frac{D}{0.01 \, [m]} \right)^{-0.3}.$$

Mass is not a concern for this heat sink; you are only interested in maximizing the heat transfer rate from the heat sink to the air given the operating temperatures. Therefore, you will want to make the fins as long as possible. However, in order to use the micro-end milling process you cannot allow the fins to be longer than $10\times$ the distance between two adjacent fins. That is, the length of the fins may be computed according to $L = 10a$. You must choose the optimal value of a and D for this application.

(a) Prepare a model using EES that can predict the heat transfer rate for a given value of a

and D. Use this model to predict the heat transfer rate from the heat sink for $a =$ 0.5 cm and $D = 0.75$ cm. Note that the number of fins should be computed based on how many can fit onto the heat sink base (rather than counting the number of fins in the figure).

(b) Prepare a plot that shows the heat transfer rate from the heat sink as a function of the distance between adjacent fins, a, for a fixed value of $D = 0.75$ cm. Be sure that the fin length is calculated using $L = 10a$. Also be sure that the number of fins installed on the heat sink varies as a is varied. Your plot should exhibit a maximum value, indicating that there is an optimal value of a.

(c) Prepare a plot that shows the heat transfer rate from the heat sink as a function of the diameter of the fins, D, for a fixed value of $a = 0.5$ cm. Be sure that the fin length is calculated using $L = 10a$ and the number of fins is computed. Your plot should exhibit a maximum value, indicating that there is an optimal value of D.

(d) Determine the optimal value of a and D using EES' built-in optimization capability.

3.42 A heat sink is composed of an array of rectangular fins.

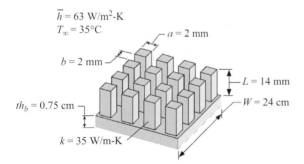

$\bar{h} = 63$ W/m²-K
$T_\infty = 35°C$
$a = 2$ mm
$b = 2$ mm
$L = 14$ mm
$W = 24$ cm
$th_b = 0.75$ cm
$k = 35$ W/m-K

The heat sink is square with base side dimension $W = 24$ cm and base thickness $th_b = 0.75$ cm. The fins are square and have side dimension $a = 2$ mm and length $L = 14$ mm. Fins are separated by a distance $b = 2$ mm. (Note that there are *more* than the 16 fins shown in the figure on the heat sink.) The fin and base material have conductivity $k = 35$ W/m-K. The heat transfer coefficient between the surface and the air is $\bar{h} = 63$ W/m²-K. The ambient air is at $T_\infty = 35°C$.

(a) Determine the efficiency of the fins.

(b) Draw a resistance network that extends from the underside of the heat sink to the ambient air.

(c) Calculate the values of each of the resistances in your network from (b) and the total resistance of the heat sink.

(d) Plot the temperature at the underside of the heat sink as a function of the rate of heat transfer between 0 and 2 kW to the heat sink.

3.43 Your company has developed a technique for forming very small fins on a plastic substrate. The diameter of the fins at their base is $D = 1$ mm. The ratio of the length of the fin to the base diameter is the aspect ratio, $AR = 10$. The fins are arranged in a hexagonal close-packed pattern. The ratio of the distance between fin centers to the base diameter is called the pitch ratio, $PR = 2$. The conductivity of the filled plastic material is $k = 2.8$ W/m-K. The heat transfer coefficient between the surface of the plastic and the surrounding gas is $\bar{h} = 35$ W/m²-K. The base temperature is $T_b = 60°C$ and the gas temperature is $T_\infty = 35°C$.

(a) Determine the number of fins per unit area and the thermal resistance of the unfinned region of the base per unit of base area.

(b) You have been asked to evaluate whether triangular, parabolic concave, or parabolic convex pin fins will provide the best performance. Plot the heat transfer per unit area of base surface for each of these fin shapes as a function of aspect ratio. Note that the performance of these fins can be obtained from an appropriate EES function. Explain the shape of your plot.

(c) Assume that part (b) indicated that parabolic convex fins are the best. You have been asked whether it is most useful to spend time working on techniques to improve (increase) the aspect ratio or improve (reduce) the pitch ratio. Answer this question using a contour plot that shows contours of the heat transfer per unit area in the parameter space of AR and PR.

Numerical Solutions to Extended Surface Problems

3.44 A fuse is a long, thin piece of metal that will heat up when current is passed through it. If a large amount of current is passed through the fuse, then the material will melt and this protects the electrical components downstream of the fuse. The figure illustrates a fuse that is composed of a piece of metal with a square cross-section ($a \times a$) that has length L. The conductivity of the fuse material is k. The fuse surface experiences convection with air at temperature T_a with heat transfer coefficient $\bar{h}$. Radiation from the surface can be neglected for this problem. The ohmic heating associated with the current passing through the fuse results in a uniform rate of volumetric thermal energy generation, $\dot{g}'''$. The two ends of the fuse (at $x = 0$ and $x = L$) are held at temperature T_b. You have been asked to generate a numerical model of the fuse. The figure below shows a simple numerical model that includes only four nodes, which are positioned uniformly along the length of the fuse.

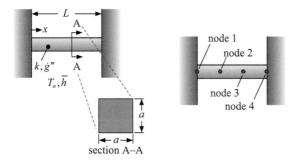

section A–A

(a) How would you determine if the extended surface approximation was appropriate for this problem?

For the remainder of this problem, assume that you can use the extended surface approximation.
(b) Derive a system of algebraic equations that can be solved in order to predict the temperatures at each of the four nodes in the figure (T_1, T_2, T_3, and T_4). Your equations should include only those symbols defined in the problem statement. Do not solve these equations.
(c) How would you determine the amount of heat transferred from the fuse to the wall at $x = 0$ using your solution from (b)?
(d) Derive the differential equation and boundary conditions that you would need in order to solve this problem analytically. Show your steps clearly.

3.45 You have decided to model the rod considered in Problem 3.16 using a three-node numerical solution, as shown in the figure.

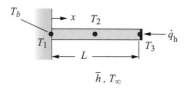

Derive the three equations that must be solved simultaneously in order to provide the temperatures of the three nodes (T_1, T_2, and T_3).

3.46 Prepare a numerical solution to a pin-fin composed of a 1 cm diameter copper ($k = 400$ W/m-K) rod that is 20 cm long with a base temperature of 500 K and an ambient temperature of 300 K. Assume that the heat transfer coefficient is 100 W/m²-K.
(a) Prepare a plot of the temperature as a function of axial position.
(b) Calculate the fin efficiency and compare it with the analytical solution.
(c) Prepare a plot of predicted fin efficiency as a function of number of nodes.

3.47 Reconsider the disk-shaped bracket that was discussed in Problem 3.21. You have decided to generate a numerical model of the bracket that has three nodes, positioned as shown in the figure.

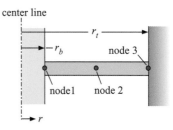

Derive a system of algebraic equations that can be solved in order to predict the temperatures at each of the three nodes (T_1, T_2, and T_3). Your equations should include only those symbols defined in the problem statement for Problem 3.21 as well as the radial locations of the three nodes (r_1, r_2, and r_3). Do not solve these equations.

3.48 The figure illustrates a cylindrical fin composed of two materials. From $0 < x < L$ the fin is made of material 1 with k_1 while from $L < x < 2L$ the fin is composed of material 2 with $k_2 < k_1$. The fin is surrounded with fluid at T_∞ and heat transfer coefficient $\bar{h}$. Neglect contact resistance. The diameter of the cylinder is D and there is a heat transfer $\dot{q}$ into the tip of the fin, as shown. The base of the fin is kept at T_∞. The problem is steady state.

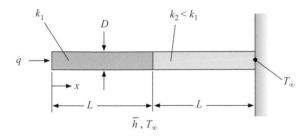

(a) How would you justify making the extended surface approximation for this problem?
(b) Derive the governing differential equations and the boundary conditions (all four of them) required to get an analytical solution to the problem.
(c) Derive the algebraic equations that must be solved in order to obtain a numerical solution to the problem that uses only three nodes, placed at $x = 0$, $x = L$, and $x = 2L$.

3.49 You have decided to solve Problem 3.25 numerically. The figure illustrates a three-node numerical model of the brake pad. Derive the three algebraic equations that must be solved in order to predict the temperature at each of the nodes. Your equations should include the variables defined in the problem statement for Problem 3.25 and the distance between each node, Δx.

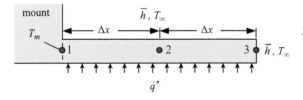

3.50 A triangular fin is shown in the figure.

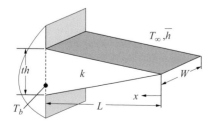

The fin can be treated as an extended surface. The thickness of the fin base is $th = 1$ cm and the length is $L = 10$ cm. The conductivity of the material is $k = 24$ W/m-K. The base temperature is $T_b = 140°C$ and the ambient temperature is $T_\infty = 25°C$. The heat transfer coefficient is $\bar{h} = 15$ W/m^2-K. The width of the fin, W, is much larger than its length.

(a) Develop a numerical model of the fin.
(b) Plot the temperature distribution within the fin.
(c) Determine the fin efficiency. Compare your answer with the fin efficiency obtained from the EES function eta_fin_straight_triangular.
(d) Plot the fin efficiency as a function of the number of nodes used in the solution.

3.51 A cylindrical rod with a diameter of 12 mm and a length of 5 cm is used as a fin. The rod is made of AISI 304 stainless steel. The base of the rod is maintained at 154°C. The outer surface of the rod and its tip are exposed to air at 25°C and convection occurs with a heat transfer coefficient of 12 W/m^2-K. Use a 1-D steady-state numerical method to plot the temperature distribution in the rod as a function of position. Be sure to determine the number of nodes needed to provide a reasonably accurate result. Determine the total rate of convective heat transfer from the rod (including the tip) for the following two situations.

(a) Assume that the conductivity of the rod is constant and equal to the value evaluated at the average of the base and air temperatures.
(b) Include in your model the conductivity of the rod material as a function of temperature.

3.52 A camper has found an old coat hanger that he plans to straighten and use to toast marshmallows over an open campfire. He is concerned about how hot the hanger will get at the end where it is touching his hand. The maximum temperature of the wood fire is 1000°F and the last 4 inches of the hanger are exposed to this temperature with a convection coefficient of 8.5 Btu/hr-ft^2-F. The remainder of the hanger is exposed to an ambient temperature of 60°F with a heat transfer coefficient of 5 Btu/hr-ft^2-F. The hanger has a total length of 3 ft with a 1/8 inch diameter. The thermal conductivity is 25 Btu/hr-ft-F. Determine the temperature distribution in the hanger. Is the camper in danger of burning his hand? Make sure that your solution shows clearly a derivation of the finite difference equations and a plot of the temperature distribution for the number of nodes you recommend for this problem.

3.53 Reconsider Problem 3.22.

(a) Develop a numerical model in EES that can predict the temperature distribution within the current lead. Do not assume that the conductivity and electrical resistivity are constant but rather include their temperature dependence. Plot the temperature as a function of position.

(b) Prepare a plot showing the rate of energy transfer into the magnet (i.e., into the flange at T_C) as a function of the number of nodes used in your model.

3.54 The figure illustrates a flat plate solar collector.

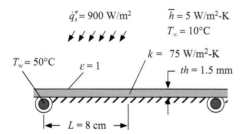

$$\dot{q}_s'' = 900 \text{ W/m}^2$$
$$\bar{h} = 5 \text{ W/m}^2\text{-K}$$
$$T_\infty = 10°C$$
$$k = 75 \text{ W/m}^2\text{-K}$$
$$th = 1.5 \text{ mm}$$
$$T_w = 50°C$$
$$\varepsilon = 1$$
$$L = 8 \text{ cm}$$

The collector consists of a flat plate that is $th = 1.5$ mm thick with conductivity $k = 75$ W/m-K. The plate is insulated on its back side and experiences a solar flux of $\dot{q}_s'' = 900$ W/m² which is all absorbed. The surface is exposed to convection and radiation to the surroundings. The emissivity of the surface is $\varepsilon = 1$. The heat transfer coefficient is $\bar{h} = 5$ W/m²-K and the surrounding temperature is $T_\infty = 10°C$. The temperature of the water flowing in the tubes is $T_w = 50°C$. The center-to-center distance between adjacent tubes is $2L$ where $L = 8$ cm.

(a) Develop a numerical model that can predict the rate of energy transfer to the water per unit length of collector. List any assumptions that you make.

(b) Determine the efficiency of the collector; efficiency is defined as the ratio of the energy

delivered to the water to the solar energy incident on the collector.

(c) Plot the efficiency as a function of the number of nodes used in the solution.

(d) Plot the efficiency as a function of $T_w - T_\infty$. Explain your plot.

3.55 An expensive power electronics module normally receives only a moderate current. However, under certain conditions it might experience currents in excess of 100 amps. The module cannot survive such a high current and therefore, you have been asked to design a fuse that will protect the module by limiting the current that it can experience.

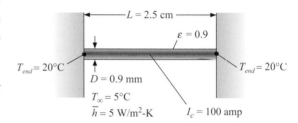

$$L = 2.5 \text{ cm}$$
$$\varepsilon = 0.9$$
$$T_{end} = 20°C$$
$$T_{end} = 20°C$$
$$D = 0.9 \text{ mm}$$
$$T_\infty = 5°C$$
$$\bar{h} = 5 \text{ W/m}^2\text{-K}$$
$$I_c = 100 \text{ amp}$$

The space available for the fuse allows a wire that is $L = 2.5$ cm long to be placed between the module and the surrounding structure. The surface of the fuse wire is exposed to air at $T_\infty = 20°C$ and the heat transfer coefficient between the surface of the fuse and the air is $\bar{h} = 5.0$ W/m²-K. The fuse surface has an emissivity of $\varepsilon = 0.90$. The fuse is made of aluminum. Your model should account for the fact that the electrical resistivity and thermal conductivity of aluminum are both functions of temperature. The ends of the fuse (i.e., at $x = 0$ and $x = L$) are maintained at $T_{end} = 20°C$ by contact with the surrounding structure and the module. The current passing through the fuse, I_c, results in a volumetric generation within the fuse material. If the fuse operates properly, then it will melt (i.e., at some location within the fuse, the temperature will exceed the melting temperature of aluminum) when the current reaches 100 amp. Your job will be to select the fuse diameter; to get your model started you may assume a diameter of $D = 0.9$ mm.

(a) Prepare a numerical model of the fuse that can predict the steady-state temperature distribution within the fuse material. Plot the temperature as a function of position within the wire when the current is 100 amp and the diameter is 0.9 mm.

(b) Verify that your model has numerically converged by plotting the maximum temperature in the wire as a function of the number of nodes in your model.

(c) Prepare a plot of the maximum temperature in the wire as a function of the diameter of the wire for $I_c = 100$ amp. Use your plot to select an appropriate fuse diameter.

3.56 Revisit Problem 3.23. Develop a numerical model of the situation. Because your model is numerical rather than analytical it is possible to include radiation from the surface without the use of the linearizing radiation coefficient. Also, the temperature dependence of the material properties can be taken into account. The table provides the thermal conductivity and electrical resistivity of the fuse material at several values of temperature. In Problem 3.23, constant values of k and ρ_e corresponding to the average temperature were used; however, the table shows that these material properties are actually strong functions of temperature over the temperature range experienced by the fuse.

Thermal conductivity and electrical resistivity vs. temperature.

Temperature (K)	Thermal conductivity (W/m-K)	Electrical resistivity (ohm-m)
300	18.0	3.80×10^{-8}
450	15.5	3.00×10^{-8}
600	14.2	2.40×10^{-8}
750	13.5	2.15×10^{-8}
900	13.1	2.10×10^{-8}

(a) Enter the thermal conductivity data from the table into a lookup table in EES and create a plot showing the thermal conductivity data as a function of temperature.

(b) Use the curve fit option from the plot menu to obtain a curve fit to the data. Overlay your curve fit onto the plot made in (a). Use the curve fit to develop a function in

EES that returns thermal conductivity given temperature.

(c) Repeat parts (a) and (b) for electrical resistivity.

(d) Develop a numerical model of the fuse. Implement your model for the same conditions used in Problem 3.23: $D = 0.006$ inch, $L = 4$ inch, $T_{sur} = 20°C$, $\varepsilon = 0.65$, $I_c = 1.5$ amp. Overlay the predicted temperature distribution onto the plot generated in part (d) of Problem 3.23.

(e) Verify that your model has numerically converged by plotting the maximum temperature in the fuse as a function of the number of nodes in your model.

(f) Determine the wire diameter that will cause your fuse to blow for a current of 1.5 amp. Compare your answer to the wire diameter obtained in part (e) of Problem 3.23.

3.57 A very simple solar collector is shown in the figure.

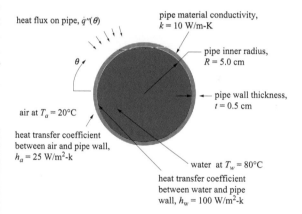

heat flux on pipe, $\dot{q}''(\theta)$

pipe material conductivity, $k = 10$ W/m-K

pipe inner radius, $R = 5.0$ cm

pipe wall thickness, $t = 0.5$ cm

air at $T_a = 20°C$

heat transfer coefficient between air and pipe wall, $h_a = 25$ W/m²-k

water at $T_w = 80°C$

heat transfer coefficient between water and pipe wall, $h_w = 100$ W/m²-k

A pipe is exposed to solar radiation over its top half (from 0 to π rad); the variation of the solar radiation depends on the position of the Sun and the geometry of the concentrating mirrors. The functional form of the flux variation is:

$$\dot{q}''(\theta) = \begin{cases} \dot{q}''_{max} \sin(\theta) & 0 < \theta < \pi \\ 0 & \pi \leq \theta \leq 2\pi \end{cases},$$

where $\dot{q}''_{max}$ is the maximum heat flux, 5000 W/m². The pipe inner radius is 5.0 cm and its thickness is 0.5 cm. The thermal conductivity of the pipe material is 10 W/m-K. The solar collector is used to heat water which is at 80°C and flowing through the pipe. The heat transfer coefficient between the water and the internal surface of the pipe is 100 W/m²-K. The external surface of the pipe is exposed to air at 20°C. The heat transfer coefficient between the air and the external surface of the pipe is 25 W/m²-K.

(a) Can the pipe be treated as an extended surface for this problem (i.e., can we neglect temperature gradients in the radial direction in the pipe material)? Justify your answer clearly.

(b) Develop a numerical model and plot the temperature in the pipe wall as a function of angular position.

(c) Prepare a plot of the predicted maximum pipe wall temperature as a function of the number of nodes used in the numerical model. Use this plot to determine the number of nodes necessary to yield good results.

(d) Overlay on one plot the temperature distributions for the cases where k = 1.0 W/m-K, 10 W/m-K, and 100 W/m-K (all other inputs are as shown in the figure). Explain the results.

3.58 Your company manufactures heater wire. Heater wire is applied to surfaces that need to be heated and then current is passed through the wire in order to develop ohmic dissipation. A key issue with your product is failures that occur when a length of the wire becomes detached from the surface and therefore the wire is not well connected thermally to the surface. The wire in the detached region tends to get very hot and it can melt.

The wire diameter is D = 0.4 mm and the current passing through the wire is I_c = 10 amp. The detached wire is exposed to surroundings at T_∞ = 20°C through convection and radiation. The average convection heat transfer coefficient is $\bar{h}$ = 30 W/m²-K and the emissivity of the wire surface is ε = 0.5. The length of the wire that is detached from the surface is L = 1 cm. The ends of the wire at x = 0 and x = L are held at T_{end} = 50°C. The wire conductivity is k = 10 W/m-K and the electrical resistivity is ρ_e = 1 × 10⁻⁷ ohm-m.

(a) Develop a numerical model of the wire and use the model to plot the temperature distribution within the wire.

(b) Plot the maximum temperature in the wire as a function of the number of nodes in the numerical model.

(c) Plot the maximum temperature in the wire as a function of the length of the detached region. If the maximum temperature of the wire before failure is T_{max} = 400°C, then what is the maximum allowable length of detached wire?

You are looking at methods to alleviate this problem and have identified an alternative material, material D, in which the electrical resistivity is ρ_e = 1 × 10⁻⁷ ohm-m but the conductivity increases with temperature according to:

$$k = 10 \left[\frac{W}{m\,K}\right] + 0.05 \left[\frac{W}{m\,K^2}\right](T - 300[K]).$$

(d) Modify your numerical model in order to simulate material D. Overlay on your plot from (c) the maximum temperature as a function of the length of the detached wire for material D.

You have identified another alternative material, material E, in which the thermal conductivity is k = 10 W/m-K but the electrical resistivity decreases with temperature according to:

$$\rho = 1 \times 10^{-7}[ohm\,m] - 1 \times 10^{-10}\left[\frac{ohm\,m}{K}\right](T - 300[K]).$$

(e) Modify your numerical model in order to simulate material E. Overlay on your plot from (c) the maximum temperature as a function of the length of the detached wire for material E.

Projects

3.59 The figure illustrates a fin that has the shape of a truncated cone.

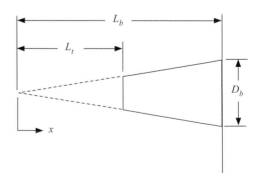

The fin diameter at the point where it meets the wall ($x = L_b$) is $D_b = 0.01$ m, where $L_b = 0.1$ m is the distance from the position where the fin diameter would be zero. The fin terminates at $x = L_t = 0.05$. Using this coordinate system, the diameter of the fin as a function of x can be written as

$$D = \frac{x}{L_b} D_b,$$

and the computational domain for the problem is $L_t < x < L_b$. The fin base is at $T_b = 373$ K and the fin is surrounded by fluid at $T_\infty = 293$ K with heat transfer coefficient $\bar{h} = 20$ W/m²-K. The conductivity of the fin material is $k = 25$ W/m-K. Assume the end of the fin (at $x = L_t$) is adiabatic.

(a) Is an extended surface approximation for this fin valid? That is, can we treat the temperature distribution as being approximately only a function of x? Justify your answer with a calculation.

(b) Derive the ODE and boundary conditions that govern this equation.

(c) Solve the ODE and enforce the boundary conditions to compute the two undetermined constants.

(d) Plot the temperature as a function of position.

(e) Determine the rate of heat transfer to the fin.

(f) Determine the fin efficiency.

(g) Plot the fin efficiency as a function of L_t. As L_t approaches zero your solution should limit to the solution programmed in EES in the function eta_fin_spine_triangular – show that is does.

An array of these fins are mounted on a base in order to create a heat sink. Each fin occupies a $D_b \times D_b$ square area of the base and the base is $th_b = 0.01$ m thick.

(h) Determine the total heat transfer rate from the $D_b \times D_b$ square area of the base – this should include the heat transfer to the fin and the heat transfer from the unfinned area of the base.

(i) Determine the total volume of material within the $D_b \times D_b$ square area of the base – this should include the volume of the base material and the volume of the fin material.

(j) Determine the ratio of the rate of heat transfer (h) to the total volume of material (i). For cost

or mass considerations it is desirable to design the fin so that this quantity is maximized.

(k) Plot heat transfer per volume (j) as a function of L_t – you should see that your plot exhibits a maximum value.

(l) Simultaneously optimize the values of L_b and L_t/L_b in order to optimize the fin (keeping $D_b = 0.01$ m). What is the optimal fin design and what is the associated value of the heat transfer rate per unit volume?

3.60 Dismounted soldiers and emergency response personnel are routinely exposed to high-temperature/humidity environments as well as external energy sources such as flames, motor heat, or solar radiation. The protective apparel required by chemical, laser, biological, and other threats tend to have limited heat removal capability. These and other factors can lead to severe heat stress. One solution is a portable, cooling system integrated with an encapsulating garment to provide metabolic heat removal. A portable metabolic heat removal system that is acceptable for use by a dismounted soldier or emergency response personnel must satisfy a unique set of criteria. The key requirements for such a system is that it be extremely low mass and very compact in order to ensure that any gain in performance due to active cooling is not offset by fatigue related to an increase in pack load. In order to allow operation for an extended period of time, a system must either be passive (require no consumable energy source), very efficient (require very little consumable energy), or draw energy from a high-energy density source. One alternative for providing portable metabolic heat removal is using a thermoelectric cooling system.

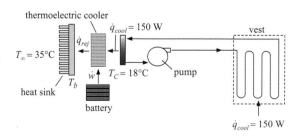

A pump forces a liquid solution to flow through plastic tubes in the vest in order to transfer the cooling from the thermoelectric cooler to the person. The cooler consumes power from a battery and rejects heat to ambient air using a heat sink. The thermoelectric cooler operates between the liquid temperature (T_C = 18°C) on its cold end and the base temperature of the heat sink, T_b, on its hot end. Initially, we will assume that the base temperature of the heat sink is T_b = 40°C; although this value will be adjusted as we do the problem.

The soldier requires $\dot{q}_{cool}$=150 W. The maximum possible coefficient of performance (COP, defined as the ratio of the cooling provided to the power required) of a thermoelectric cooler can be computed according to:

$$COP = \eta\, COP_{max},$$

where η = 0.5 is the efficiency of the cooler and COP_{max} is the maximum possible value, calculated according to:

$$COP_{max} = \frac{\overline{T}}{\Delta T}\frac{\left(\sqrt{1 + Z\overline{T}} - 1\right)}{\sqrt{1 + Z\overline{T}} + 1},$$

where Z = 0.003 [1/K] is the figure of merit of the cooler, $\overline{T}$ is the absolute average temperature seen by the cooler:

$$\overline{T} = \frac{T_C + T_b}{2},$$

and ΔT is the temperature difference across the cooler:

$$\Delta T = T_b - T_C.$$

(a) For the assumed base temperature, T_b = 40°C, determine the COP of the cooler and the rate that it consumes power and rejects heat to the heat sink.

The heat sink is composed of an array of rectangular fins, as shown in the figure (with only 16 fins drawn – the actual number of fins would be much larger).

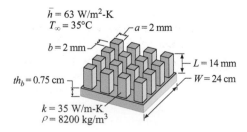

$\bar{h}$ = 63 W/m²-K
T_∞ = 35°C
a = 2 mm
b = 2 mm
L = 14 mm
W = 24 cm
th_b = 0.75 cm
k = 35 W/m-K
ρ = 8200 kg/m³

The heat sink is square with base side dimension W = 24 cm and base thickness th_b = 0.75 cm. The fins are square and have side dimension a = 2 mm and length L = 14 mm. Fins are separated by a distance b = 2 mm. The fin and base material have conductivity k = 35 W/m-K and density ρ = 8200 kg/m³. The heat transfer coefficient between the surface and the air is $\bar{h}$ = 63 W/m²-K. The ambient air is at T_∞ = 35°C.

(b) For the assumed base temperature, T_b = 40°C, determine the rate of heat rejection from the heat sink.

(c) Adjust the heat sink base temperature, T_b, so that the heat rejection calculated in (a) matches the heat rejection calculated in (b). What is the base temperature of the heat sink and the power consumed by the cooler?

(d) Determine the mass of the heat sink.

(e) A mission requires time – 4 hours of cooling at 150 W. The energy density of an advanced lithium ion battery is ed = 0.22 kW-hr/kg. Determine the mass of batteries required.

(f) Plot the total system mass (the sum of the battery and the heat sink – assume the rest of the system has a negligible mass) as a function of the size of the heat sink (W). Explain the shape of your plot.

3.61 A heat sink for a high-performance processor uses a fan that directs air through the passages formed between adjacent fins that extend in the radial direction. Close inspection of the heat sink shows that it has a split fin design. A schematic of a single split-fin is shown in the figure.

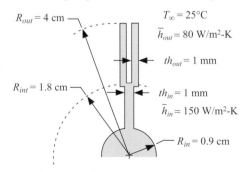

R_{out} = 4 cm
T_∞ = 25°C
$\bar{h}_{out}$ = 80 W/m²-K
th_{out} = 1 mm
R_{int} = 1.8 cm
th_{in} = 1 mm
$\bar{h}_{in}$ = 150 W/m²-K
R_{in} = 0.9 cm

The fin extends from the hub at radius R_{in} = 0.9 cm to the tip at radius R_{out} = 4 cm. The fin splits into two fins of equal size at an intermediate radius, R_{int} = 1.8 cm. The width of the fin (into the page) is W = 0.85 cm. The thickness of the fin in the inner region (i.e., from $R_{in} < r < R_{int}$) is th_{in} = 1 mm. The thickness of each of the two fins in the outer region (i.e., from $R_{int} < r < R_{out}$) is th_{out} = 1 mm. The base of the inner fin is maintained at T_b = 85°C. The fin surface is exposed to air at T_∞ = 25°C. The heat transfer coefficient between the surface of the fin in the inner region and the air is $\bar{h}_{in}$ = 150 W/m²-K and the heat transfer coefficient in the outer region is $\bar{h}_{out}$ = 80 W/m²-K. The conductivity of the fin material is k = 115 W/m-K.

(a) Is an extended surface model of the split-fin appropriate? Justify your answer.

(b) Develop an analytical model for the temperature distribution in the split fin. You will need to have one equation for the temperature in the inner region and a different equation for the temperature in the outer region. Plot the temperature as a function of radius.

(c) Determine the rate of heat transfer to a single split fin and the fin efficiency.

(d) The gap between adjacent fins at the base, at $r = R_{in}$, is equal to the thickness of the fin, th_{in}. Determine the number of fins and the total rate of heat transfer to the heat sink.

The next step in the problem is to vary the intermediate radius, R_{int}. As the intermediate radius is adjusted, the gap between adjacent fins at the interface radius, R_{int}, must be kept equal to the thickness of the fins in the outer region, th_{out}. Use this information to set the thickness of the fins in the outer region.

(e) Prepare a plot showing the total rate of heat transfer to the heat sink as a function of the intermediate radius. You should see that an optimal intermediate radius exists. Use the Min/Max function in EES to identify the optimal intermediate radius.

(f) Plot the optimized heat transfer rate and optimal intermediate radius as a function of the thickness of the inner fin, th_{in}. Use the Min/Max Table option in EES to accomplish this.

3.62 Optical cooling has been proposed as a method for providing very low-temperature cooling without any physical contact. The figure illustrates a method for implementing optical cooling as the lowest temperature stage of a cryocooler. With the optical cooling turned off, a sample is cooled by connecting it to the cold head of a conventional

cryocooler that is maintained at T_{ch} = 4.2 K. When optical cooling is turned on, the sample must be thermally disconnected from the cold head and exposed to the optical cooling effect; this allows its temperature to be reduced to a value that is below T_{ch} – lower than can be achieved using conventional cooling technology. This process requires a "thermal switch" of some type.

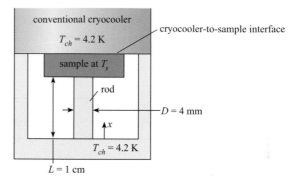

There are many designs for a thermal switch; you have been asked to evaluate the design shown in the figure. The sample is mounted on a rod of material with diameter D = 4 mm and length L = 1.0 cm. The material has conductivity k = 2.8 W/m-K and an electrical resistivity that depends on temperature according to:

$$\rho_e = \rho_{e,o} - \alpha_e T,$$

where $\rho_{e,o}$ = 7.2 × 10⁻⁴ Ω-m and α_e = 5.5 × 10⁻⁵ Ω-m/K. In order to activate the switch, a current I = 2.5 amp is passed through the rod, causing its temperature to rise. The material will tend to expand as it is heated, which will cause the sample to be pushed against the cold head along the interface labeled cryocooler-to-sample contact in the figure. The thermal resistance that characterizes the cryocooler-to-sample contact decreases as the force exerted by the thermal expansion increases. In order to deactivate the switch, the current is removed causing the rod to contract until the sample moves away from the cold head so that the cryocooler-to-sample contact resistance becomes infinitely large. In the off state, the sample is only connected to the cold head via conduction through the low-conductivity rod.

(a) Neglect radiation from the edges of the rod so that the rod temperature can be assumed to be 1-D in x; x = 0 corresponds to the end of the rod connected to the cold head and $x = L$ corresponds to the end of the rod connected to the sample. Derive the governing differential equation for the problem.

(b) Determine the general solution to the differential equation that you obtained in part (a).

(c) The temperature of the rod at $x = 0$ is equal to the cold head temperature. For now, we will assume that the sample temperature with the switch on is equal to $T_{s,on} = 4.4$ K (i.e., $T_{x=L} = T_{s,on}$). Develop two equations using these boundary conditions that can be used to determine the constants in your general solution from part (b).

(d) Implement your solution in EES in order to determine the two constants. Plot the temperature as a function of position in the rod.

Next we need to compute how much force the rod is exerting on the sample and therefore how much force is exerted at the cryocooler-to-sample interface. If the rod were allowed to expand freely, its length would increase by an amount given by

$$\Delta L = \int_0^L CTE(T - T_{ch}) \, dx,$$

where $CTE = 1 \times 10^{-5}$ 1/K is the coefficient of thermal expansion.

(e) Substitute your solution for the temperature distribution in the rod into the equation above and determine the amount of expansion that the rod would undergo if it were allowed to expand freely.

Because the rod is constrained, it cannot expand and therefore instead exerts a force on the sample. The force generated by the rod is equal to the force that would be required to compress the rod back to its original length:

$$F = \frac{A_c \, E \, \Delta L}{L},$$

where $E = 100$ GPa is Young's modulus for the material. The thermal resistance between the sample and the cryocooler (at the sample-to-cryocooler interface) depends on force according to

$$R = \frac{B}{F},$$

where $B = 38$ K-N/W. Note that if the force approaches zero (i.e., the current is turned off) then the resistance approaches infinity. Larger values of the force result in smaller values of the thermal resistance.

(f) Determine the force exerted by the rod and the resistance of the sample-to-crycooler interface.

In part (c) we assumed a value of the sample temperature, $T_{s,on}$, in order to proceed with developing the model. In fact, the sample temperature must be determined by an energy balance on the sample. The sample experiences radiation heat transfer from the surroundings equal to $\dot{q}_{rad} = 0.042$ W.

(g) If the sample is at steady state with the switch on, determine the actual temperature of the sample.

(h) Plot the sample temperature with the switch on as a function of the rod diameter.

When the optical cooling is activated there is an energy extraction from the sample (optically) that occurs at a rate that is proportional to the sample temperature:

$$\dot{q}_c = 0.025 \left[\frac{W}{K}\right] T_s.$$

The switch is deactivated (i.e., the current is turned off) but the sample experiences conduction through the rod from the cold head. The sample also continues to experience radiation heat transfer from the surroundings at a rate of $\dot{q}_{rad} = 0.042$ W.

(i) Determine the sample temperature with the switch off and optical cooling activated.

(j) Overlay on your plot from part (h) the temperature of the sample when the switch is off and optical cooling activated as a function of the rod diameter.

3.63 The figure illustrates a temperature sensor used to measure the dry bulb temperature of an air flow.

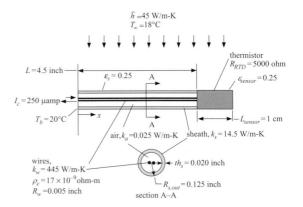

The temperature sensor is a thermistor installed at the end of a sheath – a stainless steel tube. In order to measure the resistance of the thermistor (which is related to its temperature) a current of $I_c = 250$ μamp passes through the sensor and the resulting voltage is measured. The nominal resistance of the thermistor is $R_{RTD} = 5000$ ohm. The two wires used to measure the resistance of the thermistor run along the center of the sheath, as shown in the figure. The inner surface of the sheath and the wires communicate thermally by conduction through the air space surrounding the wires. Assume that the air is stagnant and has conductivity $k_a = 0.025$ W/m-K.

The wires and sheath have temperature $T_b = 20°C$ at $x = 0$ where they are attached to the wall of the duct. The air flow being measured flows across the sheath and the sensor with temperature $T_\infty = 18°C$ and $\bar{h} = 45$ W/m²-K. The radius of the wires is $R_w = 0.005$ inch. The outer radius of the sheath is $R_{s,out} = 0.125$ inch and the thickness of the sheath is $th_s = 0.020$ inch. The conductivity of the sheath material is $k_s = 14.5$ W/m-K. The conductivity of the wire material is $k_w = 445$ W/m-K and the electrical resistivity of the wire material is $\rho_e = 17 \times 10^{-9}$ ohm-m. The length of the sheath and wires is $L = 4.5$ inch.

The sheath and sensor both radiate to surroundings at T_b (not T_∞). The emissivity of the sheath's outer surface is $\varepsilon_s = 0.25$ and the emissivity of the sensor is $\varepsilon_{sensor} = 0.25$. The sensor is a cylinder with radius $R_{sensor} = R_{s,out}$ and $L_{sensor} = 1$ cm. The sensor also experiences convection with the air flow.

(a) Can the sheath material be modeled as an extended surface? That is, can the temperature gradients within the sheath material be neglected? Justify your answer with an appropriately defined Biot number.

(b) Is it appropriate to assume that the wires and the sheath are at the same temperature at every axial position? Justify your answer with an appropriately defined Biot number.

Assume that your answer to (b) is no – therefore, you must develop a model that predicts the temperature distribution in the wire and the temperature distribution in the sheath.

(c) Develop a numerical model using EES. Assume that the sensor is at a uniform temperature. On a single plot, show the temperature

distribution in the wire and the temperature distribution in the sheath. What is the measurement error associated with the sensor?

(d) Prepare a plot of the measurement error as a function of the number of nodes used in the simulation. How many nodes are required for accurate results?

(e) Compare your numerical solution with an analytical solution in some limiting condition where the analytical solution is valid (clearly describe this limiting condition). Show that they agree in this limit by preparing a single plot showing both the temperature distribution predicted by the numerical model and the temperature distribution predicted by the analytical model.

(f) One of the nice things about a numerical model is that it allows you to run numerical experiments in order to determine the important factors. For example, in this problem you can at least approximately examine the importance of various effects by "turning them off" in your model and seeing how much the temperature sensor error changes. Make a table of the various phenomena that lead to the temperature measurement error and use your model to delineate the contribution of that effect. The first row of the table is shown for you (with numbers left off) in order to provide an idea of what is expected. Based on your table, discuss at least two strategies that could be used to improve the sensor design.

Examination of the effects that lead to the sensor measurement error.

Effect	Method of examining the effect	New error	Percent change in the measurement error
Radiation from sheath	Set $\varepsilon_s = 0$		

3.64 Heat switches are useful for a variety of cryogenic applications. A heat switch is an active device that can be turned on in order to establish a thermal link between two objects and turned off in order to thermally isolate the two objects. The figure illustrates one concept for a heat switch.

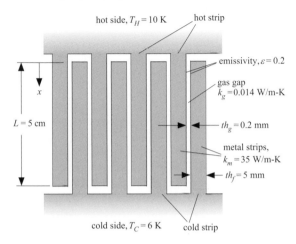

The heat switch is formed from interleaved metal strips; alternate strips extend from the hot and cold sides. The hot side is at $T_H = 10$ K and the cold side is at $T_C = 6$ K. The strips transfer heat across the small gas gap that separates them. The thickness of the gas gap is $th_g = 0.2$ mm and the active length of the strips is $L = 5$ cm. The strips are $th_f = 5$ mm wide and composed of material with conductivity $k_m = 35$ W/m-K. When the heat switch is deactivated, the gas gap is evacuated so that only radiation heat transfer occurs from the hot strips to the cold strips. The emissivity of the surfaces of each strip is $\varepsilon = 0.2$. In this problem we will assume that the surface of a strip at location x radiates only to the surface of the strip that is directly across from it. In this limit, the radiation heat flux between the hot and cold strips at any location x is given by:

$$\dot{q}''_{rad} = \frac{\sigma \varepsilon}{(1 - \varepsilon)} \left(T_h^4 - T_c^4 \right)$$

where T_h is the temperature of the surface of the hot strip at position x and T_c is the temperature of the surface of the cold strip at the same position x. The equation above can be rewritten as

$$\dot{q}''_{rad} = h_{rad}(T_h - T_c),$$

where h_{rad} is a radiation heat transfer coefficient that depends on temperature according to:

$$h_{rad} = \frac{\sigma \varepsilon}{(1 - \varepsilon)} \left(T_h^2 + T_c^2 \right)(T_h + T_c).$$

When the heat switch is activated, the gas gap is filled with helium gas with conductivity $k_g = 0.014$ W/m-K and heat transfer between the strips also occurs by conduction. Assume that the conduction heat flux between the hot and cold strips at any location x is given by:

$$\dot{q}''_{cond} = \frac{k_g}{th_g}(T_h - T_c).$$

(a) Is an extended surface model of a metal strip appropriate?

(b) Develop a numerical model of the heat switch using matrix decomposition in MATLAB assuming that the extended surface approximation is valid. Use a successive substitution technique in order to account for the nonlinear nature of the radiation heat transfer. Prepare a plot showing the temperature distribution in the hot and cold strips (on the same plot).

(c) Plot the resistance ratio of the switch as a function of the thickness of the gas gap. Use a log–log plot and extend the plot from 0.001 mm $< th_g < 1$ mm. The resistance ratio of the heat switch is defined as the ratio of the thermal resistance of the switch when it is deactivated to the thermal resistance when it is activated.

3.65 A fiber optic bundle (FOB) is shown in the figure and used to transmit light for a building application.

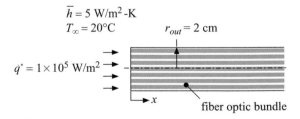

The FOB is composed of several, small-diameter fibers that are each coated with a thin layer of polymer cladding and packed in approximately a hexagonal close-packed array. The porosity of the FOB is the ratio of the open area of the FOB face to its total area. The porosity of the FOB face is an important characteristic because any radiation that does not fall directly upon the fibers will not be transmitted and instead contributes to a thermal load on the FOB. The fibers are designed so that any radiation that strikes the face of a fiber is "trapped" by total internal reflection. However, radiation that strikes the interstitial areas between the fibers will instead be absorbed in the cladding very close to the FOB face. The volumetric generation of thermal energy associated with this radiation can be represented by:

$$\dot{g}''' = \frac{\phi \dot{q}''}{L_{ch}} \exp\left(-\frac{x}{L_{ch}}\right),$$

where $\dot{q}'' = 1 \times 10^5$ W/m^2 is the energy flux incident on the face, $\phi = 0.05$ is the porosity of the FOB, x is the distance from the face, and $L_{ch} = 0.025$ m is the characteristic length for absorption of the energy. The outer radius of the FOB is $r_{out} = 2$ cm. The face of the FOB as well as its outer surface are exposed to air at $T_\infty = 20°$C with heat transfer coefficient $\bar{h} = 5$ W/m^2-K. The FOB is a composite structure and therefore conduction through the FOB is a complicated problem involving conduction through several different media. The effective thermal conductivity of the FOB in the radial direction is $k_{eff,r} = 2.7$ W/m-K. In order to control the temperature of the FOB near the face, where the volumetric generation of thermal energy is largest, it has been suggested that high-conductivity filler material be inserted in the interstitial regions between the fibers. The result of the filler material is that the effective conductivity of the FOB in the axial direction varies with position according to:

$$k_{eff,x} = k_{eff,x,\infty} + \Delta k_{eff,x} \exp\left(-\frac{x}{L_k}\right)$$

where $k_{eff,x,\infty} = 2.0$ W/m-K is the effective conductivity of the FOB in the x-direction without filler material, $\Delta k_{eff,x} = 28$ W/m-K is the augmentation of the conductivity near the face, and $L_k = 0.05$ m is the characteristic length over which the effect of the filler material decays. The length of the FOB is effectively infinite.

(a) Is it appropriate to use a 1-D model of the FOB?

(b) Assume that your answer to (a) was yes. Develop a numerical model of the FOB.

(c) Overlay on a single plot the temperature distribution within the FOB for the case where the filler material is present ($\Delta k_{eff,x} = 28$ W/m-K) and the case where no filler material is present ($\Delta k_{eff,x} = 0$).

3.66 The receiver tube of a concentrating solar collector is shown in the figure.

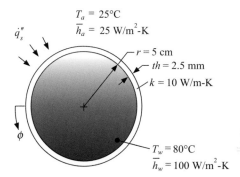

The receiver tube is exposed to solar radiation that has been reflected from a concentrating mirror. The heat flux received by the tube is related to the position of the Sun and the geometry and efficiency of the concentrating mirrors. For this problem, you may assume that all of the radiation heat flux is absorbed by the collector and neglect the radiation emitted by the collector to its surroundings. The flux received at the collector surface ($\dot{q}_s''$) is not circumferentially uniform but rather varies with angular position; the flux is uniform along the top of the collector, $\pi < \phi < 2\pi$ rad, and varies sinusoidally along the bottom, $0 < \phi < \pi$ rad, with a peak at $\phi = \pi/2$ rad:

$$\dot{q}_s''(\phi) = \begin{cases} \dot{q}_t'' + \left(\dot{q}_p'' - \dot{q}_t''\right) \sin(\phi) \text{ for } 0 < \phi < \pi \\ \dot{q}_t'' \text{ for } \pi < \phi < 2\pi \end{cases},$$

where $\dot{q}_t'' = 1000$ W/m^2 is the uniform heat flux along the top of the collector tube and $\dot{q}_p'' = 5000$ W/m^2 is the peak heat flux along the bottom. The receiver tube has an inner radius of $r = 5.0$ cm and thickness of $th = 2.5$ mm (because $th/r \ll 1$ it is possible to ignore the small difference in convection area on the inner and outer surfaces of the tube). The thermal conductivity of the tube material is $k = 10$ W/m-K. The solar collector is used to heat water, which is at $T_w = 80°$C at the axial position of interest. The average heat transfer coefficient between the water and the internal surface of the collector is $\bar{h}_w = 100$ W/m²-K. The external surface of the collector is exposed to air at $T_a = 25°C$. The average heat transfer coefficient between the air and the external surface of the collector is $\bar{h}_a = 25$ W/m²-K.

(a) Can the collector be treated as an extended surface for this problem (i.e., can the temperature gradients in the radial direction in the collector material be neglected)?

(b) Develop an analytical model that will allow the temperature distribution in the collector wall to be determined as a function of circumferential position. You may find that a symbolic software package such as Maple is helpful.

4 | Two-Dimensional, Steady-State Conduction

Chapters 2 and 3 discussed the analytical and numerical solution of one-dimensional (1-D), steady-state problems. These are problems in which the temperature within the material is independent of time and varies in only one spatial dimension (e.g., x). Examples of such problems are the plane wall studied in Section 2.2, which is truly a 1-D problem, and the extended surface problems in Chapter 3 that are only approximately 1-D. The governing differential equation for these problems is an ordinary differential equation (ODE) and the mathematics required to solve the problem are straightforward.

In this chapter, we will examine conduction problems in which the temperature varies significantly in two spatial coordinates, making them two-dimensional (2-D). Two-dimensional problems are necessarily more complicated both conceptually and mathematically than 1-D problems. In this chapter we will learn how to derive the partial differential equation that governs 2-D steady-state problems. Shape factors are introduced as one way of conveniently recording solutions to commonly encountered 2-D and 3-D steady-state problems. The numerical solution to 2-D problems will be presented using both the finite difference and finite element approaches. The finite difference approach for 2-D problems is an extension of the numerical solution techniques encountered for 1-D problems in Chapters 2 and 3. The finite element approach is widely used by commercially available numerical software. The underlying theory of the finite element technique is not discussed, but the basic steps for using software that is based on the finite element approach are illustrated using the FEHT software.

4.1 The Governing Differential Equation and Boundary Conditions

Figure 4.1 illustrates an example of a 2-D problem. An object is being heated at its top surface (at $y = H$) by a uniform heat flux, $\dot{q}''$, while it is being cooled from its bottom ($y = 0$) and left ($x = 0$) surfaces by convection with fluid temperature T_∞ and heat transfer coefficient $\bar{h}$. The right surface ($x = W$) is adiabatic. We will assume that the length of the object into the page (L) is sufficiently long that the problem can be considered to be 2-D rather than 3-D.

The governing differential equation required to solve this problem is derived using the same steps that were used in Chapters 2 and 3 for 1-D problems. A differential control volume is defined (see Figure 4.1) and used to develop a steady-state energy balance. Note that the control volume for this situation must be differential in both the x- and y-directions as there are temperature gradients in both of these directions:

$$(\dot{q}_x)_x + (\dot{q}_y)_y = (\dot{q}_x)_{x+dx} + (\dot{q}_y)_{y+dy}. \tag{4.1}$$

The $x + dx$ and $y + dy$ terms are expanded as usual:

$$(\dot{q}_x)_x + (\dot{q}_y)_y = (\dot{q}_x)_x + \frac{\partial \dot{q}_x}{\partial x}dx + (\dot{q}_y)_y + \frac{\partial \dot{q}_y}{\partial y}dy. \tag{4.2}$$

Note that the ordinary derivatives that were used in Chapters 2 and 3 have become *partial derivatives* in Eq. (4.2), indicating that the heat transfer rates are functions of more than one independent variable. Equation (4.2) can be simplified to:

$$\frac{\partial \dot{q}_x}{\partial x}dx + \frac{\partial \dot{q}_y}{\partial y}dy = 0. \tag{4.3}$$

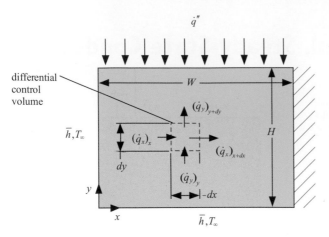

Figure 4.1 Example of a 2-D steady-state problem.

Fourier's Law is used to determine the conduction heat transfer rates in the x- and y-directions. In the x-direction, the area for conduction is $L\ dy$ and conduction is driven by the temperature gradient in the x-direction, leading to:

$$\dot{q}_x = - k\,L\,dy\,\frac{\partial T}{\partial x}. \tag{4.4}$$

In the y-direction, the area for conduction becomes $L\ dx$ and conduction is driven by the temperature gradient in the y-direction:

$$\dot{q}_y = - k\,L\,dx\,\frac{\partial T}{\partial y}. \tag{4.5}$$

Equations (4.4) and (4.5) are substituted into Eq. (4.3):

$$\frac{\partial}{\partial x}\left(-k\,L\,dy\,\frac{\partial T}{\partial x}\right)dx + \frac{\partial}{\partial y}\left(-k\,L\,dx\,\frac{\partial T}{\partial y}\right)dy = 0. \tag{4.6}$$

If the thermal conductivity is constant, then Eq. (4.6) can be simplified to

$$\frac{\partial^2 T}{\partial x^2} + \frac{\partial^2 T}{\partial y^2} = 0, \tag{4.7}$$

which is the governing *partial* differential equation for this problem. Equation (4.7) is sometimes referred to as **Laplace's equation** for the French mathematician Pierre-Simon Laplace. Laplace's equation shows up not only in heat transfer but also in other fields such as fluid dynamics and electrostatics. Equation (4.7) is second order in both the x- and y-directions and therefore two boundary conditions are required in each of these directions. For a 2-D problem, boundary conditions are defined along the lines that define the edges of the computational domain. For the problem illustrated in Figure 4.1, none of the boundaries have a specified temperature. Therefore we will need to carry out an interface energy balance on each of the edges to obtain the appropriate boundary conditions.

Figure 4.2 illustrates interface energy balances along two of the four boundaries that together define the computational domain. The interface energy balance on the boundary at $x = 0$ has area $dy\ L$. Heat transfer due to convection enters the left side while energy leaves the right side by conduction. An energy balance is always of the form

$$IN = OUT + STORED. \tag{4.8}$$

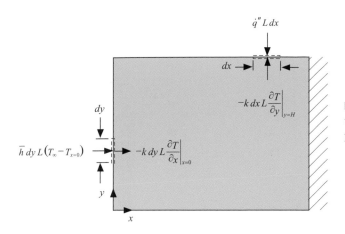

Figure 4.2 Interface energy balances that are used to obtain the boundary conditions for the problem shown in Figure 4.1.

For the interface energy balance, convection is defined as entering and conduction as leaving. There is no storage for an interface because the control surfaces do not enclose any finite volume. With these substitutions, Eq. (4.8) becomes:

$$\bar{h}\,dy\,L(T_\infty - T_{x=0}) = -k\,dy\,L\frac{\partial T}{\partial x}\bigg|_{x=0},$$

(4.9)

which can be simplified to:

$$\bar{h}(T_\infty - T_{x=0}) = -k\frac{\partial T}{\partial x}\bigg|_{x=0}.$$

(4.10)

The interface energy balance on the boundary at $y = H$ is also shown in Figure 4.2. In this case, energy is conducted into the interface through the bottom and the heat flux is incident on the top. These are both defined as inflows of energy to the interface and therefore should be placed on the same side of Eq. (4.8):

$$-k\frac{\partial T}{\partial y}\bigg|_{y=0} + \dot{q}'' = 0.$$

(4.11)

A similar process carried out for the boundaries at $x = W$ and $y = 0$ leads to:

$$\frac{\partial T}{\partial x}\bigg|_{x=W} = 0$$

(4.12)

$$\bar{h}(T_\infty - T_{x=0}) = -k\frac{\partial T}{\partial y}\bigg|_{y=0}.$$

(4.13)

The governing differential equation, Eq. (4.7), coupled with the boundary conditions, Eqs. (4.10) to (4.13), is a well-specified mathematical problem. However, the techniques required to obtain an analytical solution to a partial differential equation are beyond the scope of this text. Readers are referred to Nellis and Klein (2009) or Myers (1998) for a discussion of separation of variables and other relevant analytical solution techniques for this type of conduction problem.

The remaining sections in this chapter will discuss methods of dealing with multi-dimensional steady-state conduction problems. Many 2-D and 3-D conduction problems that are commonly encountered have been solved and the solutions have been tabulated in the form of a shape factor, as discussed in Section 4.2. The numerical techniques that were discussed in Section 2.4 for 1-D steady-state conduction problems can be extended to 2-D problems, as presented in Section 4.3. Finally, many engineers utilize commercially available numerical analysis software packages (e.g., ANSYS) that can be used to simulate multidimensional conduction

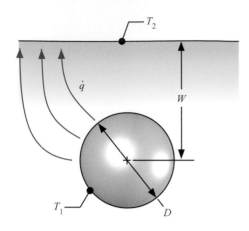

Figure 4.3 Sphere buried in a semi-infinite medium.

problems. In Section 4.4 we will use the software FEHT to explore best practices for this type of numerical simulation. A student version of this software is available at no cost.

4.2 Shape Factors

There are many 2-D and 3-D conduction problems involving heat transfer between two well-defined surfaces (referred to as surface 1 and surface 2) that commonly appear in heat transfer applications and have previously been solved analytically and/or numerically. These solutions are often tabulated in the form of a **shape factor**.

4.2.1 Definition of Shape Factor

Figure 4.3 illustrates a sphere that is buried in a **semi-infinite medium**. The term semi-infinite shows up again in Chapter 6 in the context of transient conduction problems and refers to a situation where a material extends without end from a surface. Of course there is no such thing as a truly semi-infinite body, but in many situations the extent of the material is so vast that this approximation is a good one.

In Figure 4.3 there are two distinct surfaces at different temperatures; surface 1 is the surface of the sphere at T_1 while surface 2 is the surface of the semi-infinite medium at T_2. Provided that T_1 and T_2 are different, there will be a conduction heat transfer ($\dot{q}$) between these two surfaces through the material that separates them. The rate of heat transfer is written in terms of the shape factor (S) according to

$$\dot{q} = Sk(T_1 - T_2),\tag{4.14}$$

where k is the thermal conductivity of the material through which the conduction is occurring. In order for Eq. (4.14) to have consistent units, the unit of a shape factor solution must be consistent with a length (e.g., m in the SI unit system). If k is constant then the shape factor, S, in Eq. (4.14) is a function only of the geometry of the problem. For example, the shape factor solution for the completely buried sphere in a semi-infinite medium shown in Figure 4.3 is given by

$$S = \frac{2\pi D}{1 - \dfrac{D}{4W}},\tag{4.15}$$

where D is the diameter of the sphere and W is the distance between the center of the sphere and the surface. Shape factors for other situations are tabulated in various references, for example, Rohsenow *et al.* (1998). Table 4.1 summarizes a few shape factors solutions.

A library of shape factors, including those shown in Table 4.1 as well as others, is provided in EES. To access the Shape Factor Library, select Function Information from the Options menu and then select Heat Transfer &

Table 4.1 Shape factor solutions.

Buried sphere

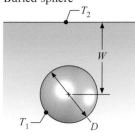

$$S = \frac{2\pi D}{1 - \frac{D}{4W}}$$

Buried beam ($L \gg a, b$)

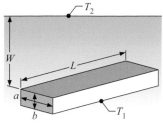

$$S = 2.756L \left[\ln\left(1 + \frac{W}{a}\right) \right]^{-0.59} \left(\frac{W}{b}\right)^{-0.078}$$

Circular extrusion with off-center hole ($L \gg D_{out}$)

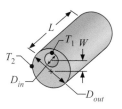

$$S = \frac{2\pi L}{\cosh^{-1}\left(\frac{D_{out}^2 + D_{in}^2 - 4W^2}{2 D_{out} D_{in}}\right)}$$

Parallel cylinders ($L \gg D_1, D_2$)

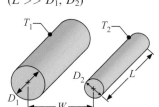

$$S = 2\pi L \cosh^{-1}\left(\frac{4W^2 - D_1^2 - D_2^2}{2 D_1 D_2}\right)$$

Buried cylinder ($L \gg D$)

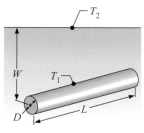

$$S = \begin{cases} \dfrac{2\pi L}{\cosh^{-1}\left(\dfrac{2W}{D}\right)} & \text{if } W \le \dfrac{3D}{2} \\[3ex] \dfrac{2\pi L}{\ln\left(\dfrac{4W}{D}\right)} & \text{if } W > \dfrac{3D}{2} \end{cases}$$

Cylinder half-way between parallel plates ($W \gg D$ and $L \gg W$)

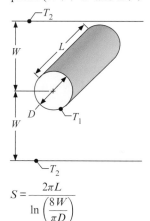

$$S = \frac{2\pi L}{\ln\left(\frac{8W}{\pi D}\right)}$$

Square extrusion with a centered circular hole ($L \gg W$)

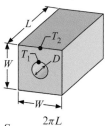

$$S = \frac{2\pi L}{\ln\left(\frac{1.08 W}{D}\right)}$$

Disk on surface of semi-infinite body

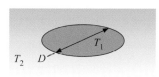

$$S = 2D$$

Square extrusion ($L \gg a$)

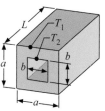

$$S = \begin{cases} \dfrac{2\pi L}{0.785 \ln\left(\dfrac{a}{b}\right)} & \text{if } \dfrac{a}{b} \le 1.41 \\[3ex] \dfrac{2\pi L}{\left[0.93 \ln\left(\dfrac{a}{b}\right) - 0.0502\right]} & \text{if } \dfrac{a}{b} > 1.41 \end{cases}$$

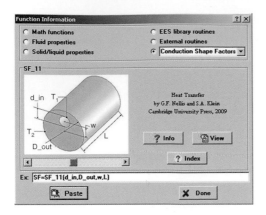

Figure 4.4 Accessing the Shape Factor Function Library from EES.

Fluid Flow. Select Conduction Shape Factors from the drop down menu and then select from the shape factor functions that are available by moving the scroll bar below the picture, as shown in Figure 4.4.

Example 4.1

Two tubes are buried side by side in a trench and subsequently covered with soil. One tube carries cool water and has a surface temperature of $T_c = 25°C$. The other tube carries hot water and has a surface temperature of $T_h = 40°C$. The tubes each have outer diameter $D = 0.5$ inch and length $L = 15$ ft. The center-to-center spacing between the tubes is $W = 3$ inch. The conductivity of the soil is $k = 0.52$ W/m-k.

Determine the rate of heat transfer from the hot tube to the cold tube.

Known Values

Figure 1 illustrates a sketch of the problem with all known values indicated and converted to SI units.

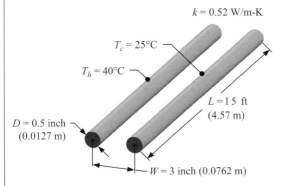

Figure 1 Sketch of problem.

Assumptions

. Steady-state conditions exit.
. The soil extends infinitely in all directions.
. The tube surfaces are isothermal; this assumption implies that the water in the tubes does not change temperature significantly as it flows.

Example 4.1 (cont.)

Analysis and Solution

The solution can be obtained using the shape factor for parallel tubes shown in Table 4.1 simplified for the case where $D_1 = D_2 = D$:

$$S = \frac{2\pi L}{\cosh^{-1}\left(\dfrac{4W^2 - 2D^2}{2D^2}\right)} = \frac{2\pi(4.572)\,\text{m}}{\cosh^{-1}\left[\dfrac{4(0.0762)^2 - 2(0.0127)^2}{2(0.0127)^2}\right]} = 5.797\,\text{m}. \tag{1}$$

The shape factor can be used to compute the heat transfer rate according to Eq. (4.14):

$$\dot{q} = S\,k(T_1 - T_2) = \frac{5.797\,\text{m}}{}\left|\frac{0.52\,\text{W}}{\text{m K}}\right|15\,\text{K} = \boxed{45.2\,\text{W}}. \tag{2}$$

Discussion

The change in water temperature induced by a 45.2 W heat transfer rate should be computed in order to evaluate the validity of the third assumption. An energy balance on one of the tubes requires that

$$\dot{q} = \dot{m}c\,\Delta T, \tag{3}$$

where $\dot{m}$ is the mass flow rate of water, c is the specific heat capacity of the water (4180 J/kg-K), ΔT is the change in temperature that the water experiences between entering and leaving the tube. The temperature change becomes larger as the flow rate is reduced. If the volumetric flow rate of water is $\dot{V} = 1$ gpm (6.3×10^{-5} m³/s) then the mass flow rate is:

$$\dot{m} = \rho\dot{V} = \frac{1000\,\text{kg}}{\text{m}^3}\left|\frac{6.3\times10^{-5}\,\text{m}^3}{\text{s}}\right. = 0.063\,\frac{\text{kg}}{\text{s}} \tag{4}$$

and the temperature change is:

$$\Delta T = \frac{\dot{q}}{\dot{m}c} = \frac{45.2\,\text{W}}{}\left|\frac{\text{s}}{0.063\,\text{kg}}\right|\frac{\text{kg K}}{4180\,\text{J}} = 0.17\,\text{K}, \tag{5}$$

which is small relative to the tube-to-tube temperature difference. This calculation provides some justification for the assumption of isothermal tubes.

In many cases we may want to consider other mechanisms of heat transfer in addition to the conduction heat transfer through the soil separating the tubes. For example, in this problem it would be useful to consider the convection resistance between the water and the internal surface of the tube. This becomes possible using the concept of a shape factor resistance, discussed in Section 4.2.2.

4.2.2 Shape Factor Resistance

The heat transfer rate expressed in terms of a shape factor, Eq. (4.14), is another example of a resistance equation:

$$\dot{q} = Sk(T_1 - T_2) = \frac{(T_1 - T_2)}{R_{SF}}, \tag{4.16}$$

where R_{SF} is the thermal resistance to conduction between surfaces 1 and 2 calculated using the shape factor. Comparing the two sides of Eq. (4.16) suggests that the **shape factor resistance** is

$$R_{SF} = \frac{1}{kS}. \tag{4.17}$$

Note that the shape factor resistance computed using Eq. (4.17) has units K/W and can be used in conjunction with other thermal resistances (e.g., radiation or convection) in a larger resistance network in order to approximately deal with several heat transfer mechanisms at once.

Example 4.2

Let's reconsider Example 4.1 but include the effect of the convection resistance that exists between the fluid flowing in the tubes and the tube inner surface. Again, the two tubes are buried side by side in a trench. One tube carries cool water and the other carries hot water. The temperatures of the fluids within the tube (not the tube surface temperature) are $T_c = 25°C$ and $T_h = 40°C$. The heat transfer coefficient between the water and the internal tube surface in both tubes is $\bar{h} = 43$ W/m²-K. The thin wall tubes each have outer diameter $D = 0.5$ inch and length $L = 15$ ft. The center-to-center spacing between the tubes is $W = 3$ inch. The conductivity of the soil is $k = 0.52$ W/m-k.

Determine the rate of heat transfer from the hot tube to the cold tube.

Known Values

Figure 1 illustrates a sketch of the problem with all known values indicated and converted to SI units.

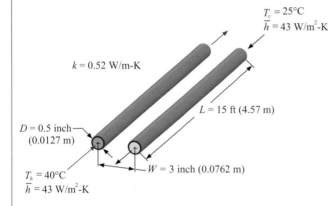

$T_c = 25°C$
$\bar{h} = 43$ W/m²-K

$k = 0.52$ W/m-K

$L = 15$ ft (4.57 m)

$D = 0.5$ inch (0.0127 m)

$T_h = 40°C$
$\bar{h} = 43$ W/m²-K

$W = 3$ inch (0.0762 m)

Figure 1 Sketch of problem.

Assumptions

- Steady-state conditions exist.
- The soil extends infinitely in all directions.
- The resistance to conduction through the tube wall is negligible.
- The tube surfaces are isothermal.

Analysis

The solution to the conduction problem is obtained using the shape factor for parallel tubes that was calculated in Example 4.1, $S = 5.797$ m. The resistance to conduction between the tube outer surfaces is given by Eq. (4.17):

$$R_{SF} = \frac{1}{k\,S} = \frac{\text{m K}}{0.52 \text{ W}}\bigg|\frac{}{5.797 \text{ m}} = 0.332 \frac{\text{K}}{\text{W}}. \tag{1}$$

The convection resistance can be included in the analysis using the network shown in Figure 2. The energy transfer goes from the hot fluid via a convection resistance to the tube surface. The energy must pass through the conduction resistance associated with the soil separating the tubes and finally through a second convection resistance to the cold fluid.

Example 4.2 (cont.)

Figure 2 Resistance network.

The resistance to convection within each tube is computed according to:

$$R_{conv} = \frac{1}{\bar{h}\,\pi\,D\,L} = \frac{m^2 K}{43\ W}\left|\frac{1}{\pi}\right|\frac{1}{0.0127\ m}\left|\frac{1}{4.57\ m}\right| = 0.128\,\frac{K}{W}. \tag{2}$$

The rate of heat transfer is given by:

$$\dot{q} = \frac{(T_h - T_c)}{(R_{conv} + R_{SF} + R_{conv})} = \frac{15\ K}{}\left|\frac{W}{(0.128 + 0.332 + 0.128)K}\right| = \boxed{25.6\ W}. \tag{3}$$

Discussion

The resistance to convection (R_{conv}) is on the same order as the resistance to conduction (R_{SF}) which indicates that both are important. Considering the impact of convection on the heat transfer rate caused the predicted value to decrease by almost 45 percent relative to the value calculated in Example 4.1, where convection was neglected.

4.2.3 The Meaning of a Shape Factor

The shape factor solution obtained from a library of such solutions does not immediately appear to be physically intuitive. Yet it is important to recognize that the shape factor does have some physical meaning so that you can at least carry out a sanity check to verify that the calculated result is in the right order of magnitude. Recall that the resistance of a plane wall, derived in Section 2.2.3, is given by:

$$R_{pw} = \frac{L}{kA_c}, \tag{4.18}$$

where L is the length of conduction path and A_c is the area for conduction. Comparing Eq. (4.18) with the shape factor resistance, Eq. (4.17), repeated below,

$$R_{SF} = \frac{1}{kS} \tag{4.19}$$

suggests that the shape factor can be interpreted as the ratio of the effective area for conduction to the effective length for conduction:

$$S \approx \frac{A_c}{L}. \tag{4.20}$$

Any shape factor solution should be checked against your intuition using Eq. (4.20). Given a specific situation involving conduction it should be possible to approximately identify the area and length that characterize the conduction process; the ratio of these quantities should have the same order of magnitude as the shape factor solution.

Example 4.3

A fluid heater is made by inserting a cylindrical electric cartridge heater into a hole drilled in the center of a metal bar with square cross-section. The metal heater block is made of 316 stainless steel and is $W = 1$ inch on each side. The heater block and the heater cartridge are both $L = 4$ inch long and the diameter of the heater is $D = 0.5$ inch. The heater power is $\dot{q} = 100$ W. There is an area-specific contact resistance of $R_c'' = 1 \times 10^{-4}$ K-m^2/W between the heater surface and the heater block inner surface. The outer surface of the heater block is exposed to fluid at $T_\infty = 40°C$ with heat transfer coefficient $\bar{h} = 725$ W/m^2-K.

Determine the surface temperature of the heater for the given conditions. Prepare a plot showing the temperature of the heater as a function of the heater power. Overlay on this plot the results that would be obtained if the heater block were composed of (1) nickel steel and (2) of copper.

Known Information

Figure 1 illustrates a sketch of the problem with the known information shown.

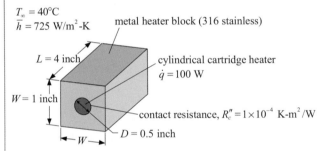

$T_\infty = 40°C$
$\bar{h} = 725$ W/m^2-K   metal heater block (316 stainless)

$L = 4$ inch   cylindrical cartridge heater
$\dot{q} = 100$ W

Figure I Sketch of the problem.

$W = 1$ inch

contact resistance, $R_c'' = 1 \times 10^{-4}$ K-m^2/W

$D = 0.5$ inch

W

Assumptions

- The conductivity of the metal is not a function of temperature.
- Steady-state conditions exist.
- The surfaces of the heater block exposed to the fluid and the heater are isothermal.
- Heat transfer from the ends of the heater block and the heater is neglected.

Analysis

The resistance network that represents this problem is shown in Figure 2.

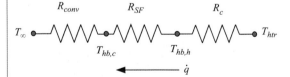

R_{conv} R_{SF} R_c

T_∞ $T_{hb,c}$ $T_{hb,h}$ T_{htr}

$\dot{q}$

Figure 2 Resistance network representation of the problem.

Example 4.3 (cont.)

The heater power leaves the heater (at T_{htr}) and must pass through a contact resistance (R_c), a conduction resistance computed using a shape factor (R_{SF}), and a convection resistance (R_{conv}) in order to reach the fluid at T_∞. The contact resistance is computed according to:

$$R_c = \frac{R_c''}{\pi\,D\,L}. \tag{1}$$

The convection resistance is computed according to:

$$R_{conv} = \frac{1}{4W\,L\,\bar{h}}. \tag{2}$$

The conduction resistance is computed according to Eq. (4.17):

$$R_{SF} = \frac{1}{k\,S}, \tag{3}$$

where S is the shape factor for a square extrusion with a round hole and k is the thermal conductivity of the metal. Note that the thermal conductivity of the material has some temperature dependence while the shape factor solution assumes a constant value of conductivity. The most appropriate thermal conductivity to use in Eq. (3) is the value computed at the average temperature within the metal (i.e., at the average of $T_{hb,c}$ and $T_{hb,h}$ in Figure 2). The heater temperature is computed according to:

$$T_{htr} = T_\infty + \dot{q}(R_{conv} + R_{SF} + R_c). \tag{4}$$

Solution

The solution to this problem will be implemented in EES in order to allow the requested plot to be generated. The use of EES also allows us to access the internal property database for thermal conductivity as well as the Shape Factor Library. The known information is entered in EES and converted to base SI units. Notice that the $UnitSystem directive is used to specify the units; this is necessary because the program will access the property database. Also notice that the name of the material used to construct the heater block ('Stainless_AISI316') is assigned to the string variable S$; this is done so that it is easy to switch between materials.

```
$UnitSystem SI Mass J K Pa
T_infinity=ConvertTemp(C,K,40 [C])        "fluid temperature"
h_bar=725 [W/m^2-K]                       "heat transfer coefficient"
R``_c=1e-4 [K-m^2/W]                      "area-specific contact resistance"
L=4 [inch]*Convert(inch,m)                "length of heater block"
W=1 [inch]*Convert(inch,m)                "width of heater block"
D=0.5 [inch]*Convert(inch,m)              "diameter of heater block"
q_dot=100 [W]                             "heater power"
S$='Stainless_AISI316'                    "material"
```

The contact and convection resistances are computed according to Eqs. (1) and (2).

```
R_c=R``_c/(pi*D*L)                        "contact resistance"
R_conv=1/(4*W*L*h_bar)                    "convection resistance"
```

The Shape Factor Library in EES is accessed by selecting Function Info from the Options menu and then selecting the Heat Transfer & Fluid Flow radio button. Select Conduction Shape Factors from the drop

Continued

Example 4.3 (cont.)

down menu and navigate to the function corresponding to the correct geometry (SF_12) using the scroll bar. Select Paste to paste the function call into the Equation Window.

> S=**SF_12**(D,W,L)   "shape factor"

Solving provides $S = 0.829$ m. It is probably a good idea to check that this shape factor makes physical sense before we use it to solve the problem. Equation (4.20) suggests that the shape factor can be thought of approximately as the ratio of the area to the length for conduction. Figure 3 illustrates the heater block and shows an *approximately* defined conduction area and length that can be used to check the value of the shape factor.

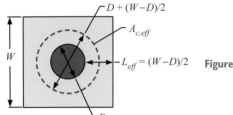

Figure 3 Approximate conduction area and length.

The approximate conduction length is defined by

$$L_{eff} = \frac{(W - D)}{2} = \frac{(0.0254 - 0.0127)\text{m}}{2} = 0.00635\text{ m},\tag{5}$$

and the approximate conduction area is based on the perimeter of a circle that is midway between the diameter of the heater and the outer edge of the block, given by

$$A_{c,eff} = \pi\left[D + \frac{(W - D)}{2}\right]L = \pi\left[0.0127\text{ m} + \frac{(0.0254 - 0.0127)\text{ m}}{2}\right](0.102\text{ m}) = 0.00608\text{ m}^2.\tag{6}$$

According to Eq. (4.20), the shape factor should be approximately given by:

$$SF \approx \frac{A_{c,eff}}{L_{eff}} = \frac{0.00608\text{ m}^2}{0.00635\text{ m}} = 0.958\text{ m}.\tag{7}$$

This value is within 15 percent of the value of S that was obtained from the Shape Factor Library.

The resistance to conduction can be computed using Eq. (3) given the conductivity of the metal. The EES function Conductivity can be used to determine the conductivity of the metal; this function requires the name of the substance (contained in the string S$) as well as the temperature at which to evaluate the thermal conductivity. The average temperature in the material ($\overline{T}$, the average of the temperatures $T_{hb,c}$ and $T_{hb,h}$) is not known at this point. The best way to proceed is to assume a reasonable value for $\overline{T}$ and use that value to determine the conductivity. Before completing the problem we will calculate the actual value of $\overline{T}$ and use it for this purpose.

> T_bar=T_infinity   "initial guess for the average temperature of the metal"
> k=**Conductivity**(S$,*T*=T_bar)   "conductivity"

The conduction resistance is determined according to Eq. (3) and the heater temperature is computed using Eq. (4).

Example 4.3 (cont.)

```
R_SF=1/(S*k)                           "resistance to conduction"
T_htr=T_infinity+q_dot*(R_conv+R_SF+R_c)   "heater temperature"
```

Solving provides T_{htr} = 337.8 K (64.65°C). At this point we can compute the minimum and maximum heater block temperatures:

$$T_{hb,c} = T_\infty + \dot{q} R_{conv} \tag{8}$$

$$T_{hb,h} = T_{hb,c} + \dot{q}\, R_{SF}. \tag{9}$$

```
T_hb_c=T_infinity+q_dot*R_conv     "cold, external temperature of heater block"
T_hb_h=T_hb_c+q_dot*R_SF           "hot, internal temperature of heater block"
```

Before proceeding we should update the guess values that EES uses to solve an iterative set of equations by selecting Update Guesses from the Calculate menu. Then remove (highlight and comment out) the assumed value of $\overline{T}$ and instead compute the actual value:

$$\overline{T} = \frac{(T_{hb,c} + T_{hb,h})}{2}. \tag{10}$$

```
{T_bar=T_infinity}          "initial guess for the average temperature of the metal"
T_bar=(T_hb_c+T_hb_h)/2     "recalculate average temperature of metal"
```

Solving leads to T_{htr} = 337.6 K (64.46°C), which is almost exactly the same value that we obtained using the assumed value of $\overline{T}$ to compute k; this result occurs because the conductivity of 316 stainless steel is not a strong function of temperature at the conditions associated with this problem.

It is possible to generate the figure requested in the problem statement by creating a Parametric Table that contains $\dot{q}$ and T_{htr} and using the result to create the plot shown in Figure 4. Changing substances is accomplished by changing the name assigned to the string variable S$ from "Stainless_AISI316" to "Nickel_Steel 9 Ni" and then to "Copper".

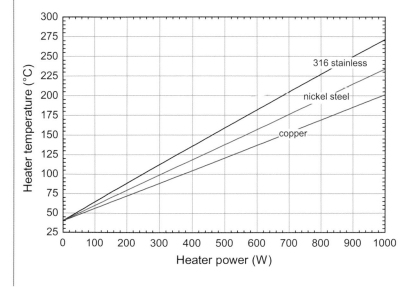

Figure 4 Heater temperature as a function of the heater power for 316 stainless, nickel steel, and copper.

Continued

Example 4.3 (cont.)

Discussion

The resistances that govern the heat transfer problem are $R_{conv} = 0.1336$ K/W, $R_c = 0.02467$ K/W, and $R_{SF} = 0.0863$ K/W (for stainless steel). These values suggest that convection is the dominant resistance but conduction is also important. The contact resistance is not very significant. Changing from stainless steel to copper essentially removes the conduction resistance as the conductivity of copper is very high. Figure 4 shows that the heater temperature rise (relative to T_∞) for a copper heater block is approximately 60 percent that of 316 stainless steel, which is consistent with setting R_{SF} to zero.

4.3 Finite Difference Solution

4.3.1 Introduction

The shape factors discussed in Section 4.2 can be used only in a limited set of situations where a shape factor solution is already available. A numerical solution is much more flexible, allowing the consideration of relatively arbitrary geometries and boundary conditions. There are two numerical techniques that are used to solve multi-dimensional conduction problems: finite difference solutions and finite element solutions. Finite difference solutions were presented in Section 2.4 in the context of 1-D problems and they are extended to 2-D problems in this section. Finite difference solutions are intuitive and powerful, allowing you to develop your own solution to a relatively complex problem. However, finite difference solutions can be difficult to apply to complex geometries. Finite element solutions can be applied more easily to complex geometries and are used more often in industry. A complete description of the finite element technique is beyond the scope of this book. However, finite element solutions to heat transfer problems are extremely powerful and many commercial software packages are available that use this technique. Therefore you should be aware of how best to use this tool, as discussed in Section 4.4.

The finite difference solution breaks a large computational domain into many smaller ones that are referred to as control volumes. The control volumes are modeled approximately in order to generate a system of equations that can be efficiently solved using a computer. The approximate, numerical solution will approach the actual solution as the number of control volumes or elements is increased. It is important to remember that it is not sufficient to obtain a solution. You should also:

(1) verify that your solution has an adequately large number of control volumes or elements,
(2) verify that your solution makes physical sense and obeys your intuition, and
(3) compare your numerical solution to an analytical solution in an appropriate limit.

These steps are widely accepted as being "best practice" when working with numerical solutions of any type.

4.3.2 Developing the Finite Difference Equations

Finite difference solutions to 1-D steady-state problems were presented in Sections 2.4 and 3.4. The steps required to set up a numerical solution to a 2-D problem are essentially the same; however, the book-keeping process (i.e., the process of entering the algebraic equations into the computer) may be somewhat more cumbersome.

The first step is to define small control volumes that are distributed through the computational domain and to precisely define the locations at which the numerical model will compute the temperatures (i.e., the locations of the nodes). The control volumes are small but finite; for the 1-D problems that were investigated in Chapters 2 and 3, the control volumes were small in a single dimension whereas they must be small in two dimensions for a 2-D problem. It is necessary to perform an energy balance on each differential control volume and provide rate equations that approximate each term in the energy balance based upon the nodal temperatures or other input parameters. The result of this step will be a set of equations (one for each control volume) in an equal number of unknown temperatures (one for each node). This set of equations can be solved using any of the techniques

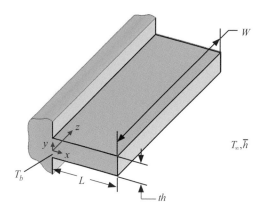

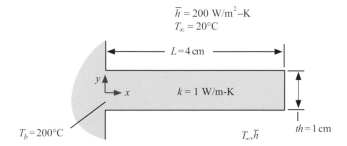

Figure 4.5 Constant cross-sectional area fin.

presented in Sections 2.4.3 through 2.4.5 in order to provide a numerical prediction of the temperature at each node.

In Chapter 3, the constant cross-section fin shown in Figure 4.5 was analyzed under the assumption that it could be treated as an extended surface (i.e., temperature gradients in the y-direction could be neglected). In order to introduce the finite difference solution technique, the fin is revisited here but this time under conditions where the extended surface approximation is not appropriate so that it becomes a 2-D steady-state problem.

The tip of the fin is insulated and the width (W) is much larger than its thickness (th) so that convection from the edges of the fin can be neglected. The length of the fin is $L = 4.0$ cm and its thickness is $th = 1.0$ cm. The fin base temperature is $T_b = 200°C$ and it transfers heat to the surrounding fluid at $T_\infty = 20°C$ with average heat transfer coefficient, $\bar{h} = 200$ W/m²-K. The conductivity of the fin material is $k = 1$ W/m-K. Under these conditions, the Biot number is:

$$Bi = \frac{\bar{h}\, th}{2\, k} = \frac{200\ \text{W}}{\text{m}^2\ \text{K}} \left| \frac{0.01\text{m}}{2} \right| \frac{\text{m k}}{1\ \text{W}} = 1, \tag{4.21}$$

which is not much less than one and therefore the extended surface approximation is *not* appropriate. The temperature gradients in the x- and y-directions must both be important and the problem is 2-D. The computational domain associated with a half-symmetry model of the fin is shown in Figure 4.6, assuming that the tip of the fin is adiabatic.

The first step in obtaining a numerical solution is to position the nodes throughout the computational domain. A regularly spaced grid of nodes is used with the first and last nodes in each dimension placed on the boundaries of the domain, as shown in Figure 4.6. Each node is identified with two subscripts, i and j. The first subscript indicates the position of the node in the x-dimension. Therefore, a node that has its first subscript $i = 1$ corresponds to a node located at $x = 0$. The number of nodes in the x-direction is defined by the value of the variable M which must be an integer. Therefore, a node with first subscript $i = M$ corresponds to a node

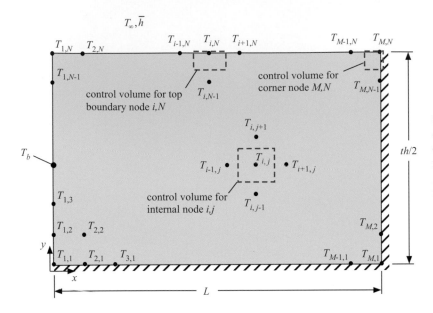

Figure 4.6 The computational domain associated with the constant cross-sectional area fin and the regularly spaced grid that is used to develop the finite difference equations.

located at $x = L$. Similarly, the second subscript indicates the position of the node in the y-dimension; $j = 1$ corresponds to a node located at $y = 0$ (the center of the fin) and $j = N$ corresponds to a node located at $y = th/2$. The total number of nodes used in the simulation is $M \times N$.

The x- and y-positions of any node designated with subscripts i and j are given by:

$$x_i = \frac{(i-1)L}{(M-1)} \quad \text{for} \quad i = 1 \ldots M \tag{4.22}$$

$$y_j = \frac{(j-1)th}{2(N-1)} \quad \text{for} \quad j = 1 \ldots N. \tag{4.23}$$

The x- and y-distance between adjacent nodes (Δx and Δy, respectively) are:

$$\Delta x = \frac{L}{(M-1)} \tag{4.24}$$

$$\Delta y = \frac{th}{2(N-1)}. \tag{4.25}$$

The next step in the derivation of the finite difference equations is to write an energy balance for each node. Figure 4.7 illustrates a control volume and the associated energy transfers for an internal node (see Figure 4.6); these include conduction from each side ($\dot{q}_{RHS}$ and $\dot{q}_{LHS}$), the top ($\dot{q}_{top}$), and the bottom ($\dot{q}_{bottom}$). Note that the direction associated with these energy transfers is arbitrary (e.g., they could have been taken as being positive if energy leaves the control volume). However, once the direction is chosen you must be careful to be consistent when writing your energy balance and rate equations.

The steady-state energy balance suggested by Figure 4.7 is

$$\dot{q}_{RHS} + \dot{q}_{LHS} + \dot{q}_{top} + \dot{q}_{bottom} = 0. \tag{4.26}$$

The next step is to approximate each of the terms in the energy balance with an appropriate rate equation. The material separating the nodes is assumed to behave as a plane wall thermal resistance. Therefore, $\Delta y\, W$ (where W is the width of the fin into the page) is the area for conduction between nodes i, j and $i+1, j$ and Δx is the distance over which the conduction heat transfer occurs:

$$\dot{q}_{RHS} = \frac{k\Delta y W}{\Delta x}\left(T_{i+1,j} - T_{i,j}\right). \tag{4.27}$$

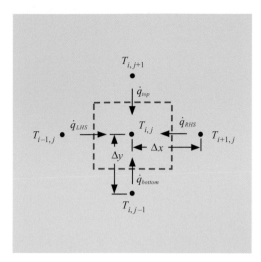

Figure 4.7 Energy balance for an internal node.

Note that the temperature difference in Eq. (4.27) is consistent with the direction of the arrow in Figure 4.7. The other conductive heat transfers are approximated using a similar model:

$$\dot{q}_{LHS} = \frac{k\Delta y W}{\Delta x}\left(T_{i-1,j} - T_{i,j}\right)$$ (4.28)

$$\dot{q}_{top} = \frac{k\Delta x W}{\Delta y}\left(T_{i,j+1} - T_{i,j}\right)$$ (4.29)

$$\dot{q}_{bottom} = \frac{k\Delta x W}{\Delta y}\left(T_{i,j-1} - T_{i,j}\right).$$ (4.30)

Substituting Eqs. (4.27) through (4.30) into the energy balance, Eq. (4.26), written for all of the internal nodes leads to a total of $(M-1) \times (N-1)$ equations:

$$\frac{k\Delta y W}{\Delta x}\left(T_{i+1,j} - T_{i,j}\right) + \frac{k\Delta y W}{\Delta x}\left(T_{i-1,j} - T_{i,j}\right) + \frac{k\Delta x W}{\Delta y}\left(T_{i,j+1} - T_{i,j}\right) + \frac{k\Delta x W}{\Delta y}\left(T_{i,j-1} - T_{i,j}\right) = 0$$

$$\text{for } i = 2\ldots(M-1) \text{ and } j = 2\ldots(N-1).$$ (4.31)

Equation (4.31) can be simplified to:

$$\frac{\Delta y}{\Delta x}\left(T_{i+1,j} - T_{i,j}\right) + \frac{\Delta y}{\Delta x}\left(T_{i-1,j} - T_{i,j}\right) + \frac{\Delta x}{\Delta y}\left(T_{i,j+1} - T_{i,j}\right) + \frac{\Delta x}{\Delta y}\left(T_{i,j-1} - T_{i,j}\right) = 0$$

$$\text{for } i = 2\ldots(M-1) \text{ and } j = 2\ldots(N-1).$$ (4.32)

Boundary nodes must be treated separately from internal nodes. This was also true for the 1-D problems discussed in Chapter 2 and 3. However, 2-D problems have many more boundary nodes than 1-D problems and so dealing with the boundaries is more tedious. The left boundary ($x = 0$) is easy because the temperature is specified. Therefore, every node in which $i = 1$ must be set to temperature T_b:

$$T_{1,j} = T_b \quad \text{for} \quad j = 1\ldots N.$$ (4.33)

The remaining boundary nodes do not have specified temperatures and therefore the finite difference equations must be determined using energy balances. Figure 4.8 illustrates an energy balance associated with a node that is located on the top boundary (at $y = th/2$, see Figure 4.6).

The steady-state energy balance suggested by Figure 4.8 is

$$\dot{q}_{RHS} + \dot{q}_{LHS} + \dot{q}_{bottom} + \dot{q}_{conv} = 0.$$ (4.34)

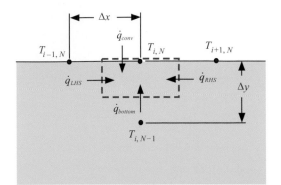

Figure 4.8 Energy balance for a node on the top boundary.

The conduction terms in the x-direction differ from those for the internal nodes:

$$\dot{q}_{RHS} = \frac{k\Delta y W}{2\Delta x}(T_{i+1,N} - T_{i,N}) \tag{4.35}$$

$$\dot{q}_{LHS} = \frac{k\Delta y W}{2\Delta x}(T_{i-1,N} - T_{i,N}). \tag{4.36}$$

Notice the factor of 2 that appears in the denominator of Eqs. (4.35) and (4.36) but did not appear in any of the rate equations for the internal node, Eqs. (4.27) through (4.30). This factor of 2 appears because the area available for conduction through the sides of the control volumes located on the top boundary is $(\Delta y/2) \times W$. The remaining conduction term is approximated as before:

$$\dot{q}_{bottom} = \frac{k\Delta x W}{\Delta y}(T_{i,N-1} - T_{i,N}). \tag{4.37}$$

The convection term is

$$\dot{q}_{conv} = \bar{h}\Delta x W(T_\infty - T_{i,N}). \tag{4.38}$$

Substituting Eqs. (4.35) through (4.38) into Eq. (4.34) for all of the nodes on the upper boundary (but not the corners) leads to:

$$\frac{k\Delta y W}{2\Delta x}(T_{i+1,N} - T_{i,N}) + \frac{k\Delta y W}{2\Delta x}(T_{i-1,N} - T_{i,N}) + \frac{k\Delta x W}{\Delta y}(T_{i,N-1} - T_{i,N}) + W\Delta x\bar{h}(T_\infty - T_{i,N}) = 0$$

$$\text{for } i = 2\ldots(M-1), \tag{4.39}$$

which can be simplified, assuming k is constant, to:

$$\frac{\Delta y}{2\Delta x}(T_{i+1,N} - T_{i,N}) + \frac{\Delta y}{2\Delta x}(T_{i-1,N} - T_{i,N}) + \frac{\Delta x}{\Delta y}(T_{i,N-1} - T_{i,N}) + \frac{\Delta x\bar{h}}{k}(T_\infty - T_{i,N}) = 0$$

$$i = 2\ldots(M-1). \tag{4.40}$$

Notice that the control volume at the top left corner, node 1, N, has already been specified by the equations for the left boundary, Eq. (4.33). The upper right corner must be considered separately. A control volume and energy balance for node M, N (see Figure 4.6) is shown in Figure 4.9.

The energy balance suggested by Figure 4.9 is

$$\frac{k\Delta x W}{2\Delta y}(T_{M,N-1} - T_{M,N}) + \frac{k\Delta y W}{2\Delta x}(T_{M-1,N} - T_{M,N}) + \bar{h}\frac{\Delta x W}{2}(T_\infty - T_{M,N}) = 0, \tag{4.41}$$

$T_\infty, \overline{h}$

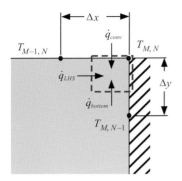

Figure 4.9 Energy balance for a node on the top right corner.

which can be simplified to

$$\frac{\Delta x}{2\Delta y}\left(T_{M,N-1} - T_{M,N}\right) + \frac{\Delta y}{2\Delta x}\left(T_{M-1,N} - T_{M,N}\right) + \frac{\overline{h}\Delta x}{2k}\left(T_\infty - T_{M,N}\right) = 0. \tag{4.42}$$

The remaining boundary nodes are treated similarly. Energy balances for the nodes on the right-hand boundary $(x = L)$ lead to:

$$\frac{\Delta x}{2\Delta y}\left(T_{M,j+1} - T_{M,j}\right) + \frac{\Delta x}{2\Delta y}\left(T_{M,j-1} - T_{M,j}\right) + \frac{\Delta y}{\Delta x}\left(T_{M-1,j} - T_{M,j}\right) = 0 \ \text{for} \ j = 2\ldots(N-1). \tag{4.43}$$

The energy balance for the right lower corner, node $M, 1$, leads to:

$$\frac{\Delta x}{2\Delta y}\left(T_{M,2} - T_{M,1}\right) + \frac{\Delta y}{2\Delta x}\left(T_{M-1,1} - T_{M,1}\right) = 0. \tag{4.44}$$

A similar procedure for the nodes on the lower boundary leads to:

$$\frac{\Delta y}{2\Delta x}\left(T_{i+1,1} - T_{i,1}\right) + \frac{\Delta y}{2\Delta x}\left(T_{i-1,1} - T_{i,1}\right) + \frac{\Delta x}{\Delta y}\left(T_{i,2} - T_{i,1}\right) = 0 \ \text{for} \ i = 2\ldots(M-1). \tag{4.45}$$

Notice that there is no convection term in Eq. (4.45) because the lower boundary is adiabatic.

We have derived a total of $M \times N$ equations in the $M \times N$ unknown temperatures; these finite difference equations completely specify the problem and they are summarized in Table 4.2. These coupled equations must be solved simultaneously in order to obtain the solution. This can be accomplished by entering them in EES (as discussed in Section 4.3.3) or by carrying out a matrix decomposition or Gauss–Seidel elimination process using a formal programming language like MATLAB (as discussed in Sections 4.3.4 and 4.3.5).

4.3.3 Solving the Equations with EES

The equations can be entered directly in EES and the software will internally carry out the numerical technique required to obtain a solution. The inputs are entered.

```
$VarInfo T[] Units='C'
$VarInfo x[] Units='m'
$VarInfo y[] Units='m'
T_b=200 [C]                          "base temperature"
T_infinity=20 [C]                    "fluid temperature"
h_bar=200 [W/m^2-K]                  "heat transfer coefficient"
th=1 [cm]*Convert(cm,m)              "thickness"
L=4 [cm]*Convert(cm,m)              "length"
k=1 [W/m-K]                          "thermal conductivity"
```

Table 4.2 Finite difference equations for the problem shown in Figure 4.6.

Equation source	Number	Number of equations	Equation	Subscript range
internal	Eq. (4.32)	$(M-2) \times (N-2)$	$\frac{\Delta y}{\Delta x}\left(T_{i+1,j} + T_{i-1,j} - 2T_{i,j}\right) + \frac{\Delta x}{\Delta y}\left(T_{i,j+1} + T_{i,j-1} - 2T_{i,j}\right) = 0$	$i = 2 \ldots (M-1)$ $j = 2 \ldots (N-1)$
left boundary	Eq. (4.33)	N	$T_{1,j} = T_b$	$i = 1$ $j = 1 \ldots N$
top boundary	Eq. (4.40)	$(M-2)$	$\frac{\Delta y}{2\Delta x}\left(T_{i+1,N} + T_{i-1,N} - 2T_{i,N}\right) + \frac{\Delta x}{\Delta y}\left(T_{i,N-1} - T_{i,N}\right) + \frac{\Delta x \bar{h}}{k}\left(T_\infty - T_{i,N}\right) = 0$	$i = 2 \ldots (M-1)$ $j = N$
upper right corner	Eq. (4.42)	1	$\frac{\Delta x}{2\Delta y}\left(T_{M,N-1} - T_{M,N}\right) + \frac{\Delta y}{2\Delta x}\left(T_{M-1,N} - T_{M,N}\right) + \frac{\bar{h}\Delta x}{2k}\left(T_\infty - T_{M,N}\right) = 0$	$i = M$ $j = N$
right boundary	Eq. (4.43)	$(N-2)$	$\frac{\Delta x}{2\Delta y}\left(T_{M,j+1} + T_{M,j-1} - 2T_{M,j}\right) + \frac{\Delta y}{\Delta x}\left(T_{M-1,j} - T_{M,j}\right) = 0$	$i = M$ $j = 2 \ldots (N-1)$
lower right corner	Eq. (4.44)	1	$\frac{\Delta x}{2\Delta y}\left(T_{M,2} - T_{M,1}\right) + \frac{\Delta y}{2\Delta x}\left(T_{M-i,1} - T_{M,1}\right) = 0$	$i = M$ $j = 1$
bottom boundary	Eq. (4.45)	$(M-2)$	$\frac{\Delta y}{2\Delta x}\left(T_{i+1,1} + T_{i-1,1} - 2T_{i,1}\right) + \frac{\Delta x}{\Delta y}\left(T_{i,2} - T_{i,1}\right) = 0$	$i = 2 \ldots (M-1)$ $j = 1$

Note that the $VarInfo T[] directive sets the units of all of the elements of the array T[] to be °C regardless of how many nodes are used. This directive is also used to set the units for arrays x and y. The grid is setup using Eqs. (4.22) through (4.25).

```
"Setup grid"
M=51 [-]                                    "number of x-nodes"
N=21 [-]                                    "number of y-nodes"
Duplicate i=1,M
    x[i]=(i-1)*L/(M-1)                      "x-position of each node"
    x_bar[i]=x[i]/L                         "dimensionless x-position of each node"
End
DELTAx=L/(M-1)                             "x-distance between adjacent nodes"
Duplicate j=1,N
    y[j]=(j-1)*th/(2*(N-1))                 "y-position of each node"
    y_bar[j]=y[j]/th                        "dimensionless y-position of each node"
End
DELTAy=th/(2*(N-1))                        "y-distance between adjacent nodes"
```

The finite difference equations are entered next. The equations for the internal nodes, Eq. (4.32), are entered using nested Duplicate loops.

```
"Internal node energy balances"
Duplicate i=2,(M-1)
    Duplicate j=2,(N-1)
        DELTAy*(T[i+1,j]-T[i,j])/DELTAx+DELTAy*(T[i-1,j]-T[i,j])/DELTAx&
            +DELTAx*(T[i,j+1]-T[i,j])/DELTAy+DELTAx*(T[i,j-1]-T[i,j])/DELTAy=0
    End
End
```

The equations on the boundaries, Eqs. (4.33), (4.40), (4.43), and (4.45), are entered using single Duplicate loops.

```
"left boundary"
Duplicate j=1,N
    T[1,j]=T_b
End

"top boundary"
Duplicate i=2,(M-1)
    DELTAy*(T[i+1,N]-T[i,N])/(2*DELTAx)+DELTAy*(T[i-1,N]-T[i,N])/(2*DELTAx)&
        +DELTAx*(T[i,N-1]-T[i,N])/DELTAy+DELTAx*h_bar*(T_infinity-T[i,N])/k=0
End

"right boundary"
Duplicate j=2,(N-1)
    DELTAx*(T[M,j+1]-T[M,j])/(2*DELTAy)+DELTAx*(T[M,j-1]-T[M,j])/(2*DELTAy)&
        +DELTAy*(T[M-1,j]-T[M,j])/DELTAx=0
End

"bottom boundary"
Duplicate i=2,(M-1)
    DELTAy*(T[i+1,1]-T[i,1])/(2*DELTAx)+DELTAy*(T[i-1,1]-T[i,1])/(2*DELTAx)&
        +DELTAx*(T[i,2]-T[i,1])/DELTAy=0
End
```

Main									
Sort	$T_{i,1}$ [C]	$T_{i,2}$ [C]	$T_{i,3}$ [C]	$T_{i,4}$ [C]	$T_{i,5}$ [C]	$T_{i,6}$ [C]	$T_{i,7}$ [C]	$T_{i,8}$ [C]	T_i [C
[1]	200	200	200	200	200	200	200	200	
[2]	182.1	182.1	182	181.9	181.7	181.4	181.1	180.7	
[3]	164.7	164.7	164.5	164.3	163.9	163.5	162.9	162.2	
[4]	148.3	148.3	148.1	147.8	147.3	146.8	146	145.2	
[5]	133.2	133.2	132.9	132.6	132.1	131.5	130.7	129.8	
[6]	119.5	119.4	119.2	118.9	118.4	117.7	117	116	
[7]	107.2	107.2	107	106.6	106.1	105.5	104.8	103.9	
[8]	96.31	96.25	96.06	95.74	95.31	94.74	94.05	93.23	
[9]	86.68	86.62	86.45	86.16	85.77	85.25	84.63	83.89	
[10]	78.21	78.16	78.01	77.75	77.4	76.94	76.38	75.71	
[11]	70.79	70.75	70.61	70.38	70.07	69.66	69.16	68.58	

Figure 4.10 Array containing the numerical solution in EES.

Finally the two corner equations, Eqs. (4.42) and (4.44), are entered.

```
"upper right corner"
DELTAx*(T[M,N-1]-T[M,N])/(2*DELTAy)+DELTAy*(T[M-1,N]-T[M,N])/(2*DELTAx)+&
    h_bar*DELTAx*(T_infinity-T[M,N])/(2*k)=0

"lower right corner"
DELTAx*(T[M,2]-T[M,1])/(2*DELTAy)+DELTAy*(T[M-1,1]-T[M,1])/(2*DELTAx)=0
```

Select Solve from the Calculate menu. The solution is displayed in the Arrays Window as shown in Figure 4.10.

Each column of the array corresponds to the temperatures associated with one value of the subscript j and all of the values of the subscript i (i.e., the temperatures in a column are at a constant value of y and varying values of x). The solution is obtained for $M = 51$ and $N = 21$. Therefore, the column $T_{i,1}$ corresponds to $y = 0$, $T_{i,5}$ corresponds to $y = 0.20$ $th/2$, etc., to $T_{i,21}$ which corresponds to $y = th/2$. These columns are plotted in Figure 4.11 as a function of the dimensionless x-position. The solution corresponds to our physical intuition as it exhibits temperature gradients in the x- and y-directions that result from the conduction heat transfer in both of these directions. Notice that the temperature distribution is not 1-D, as there is a significant variation in temperature across the thickness of the fin. This is consistent with the fact that the Biot number, calculated in Eq. (4.21), is not small.

It is important to verify that the grid used in the solution is adequately refined (i.e., that you are using a sufficiently large number of nodes). One way to evaluate the mesh refinement is to examine an important parameter that can be predicted by the simulation as the number of nodes is increased; at some point we should find that the value of the parameter is insensitive to the grid. For the problem we are studying here, the fin efficiency is a good choice for our grid sensitivity study. As discussed in Section 3.3.1, the fin efficiency is defined as the ratio of the actual ($\dot{q}_{fin}$) to the maximum possible ($\dot{q}_{max}$) heat transfer rates:

$$\eta_{fin} = \frac{\dot{q}_{fin}}{\dot{q}_{max}}. \tag{4.46}$$

The actual heat transfer rate is computed by evaluating the convective heat transfers from each of the nodes that are located on the top boundary (i.e., all of the nodes where $j = N$):

$$\dot{q}_{fin} = 2\bar{h}\Delta x W \left[\frac{(T_{1,N} - T_\infty)}{2} + \sum_{i=2}^{M-1} (T_{i,N} - T_\infty) + \frac{(T_{M,N} - T_\infty)}{2} \right]. \tag{4.47}$$

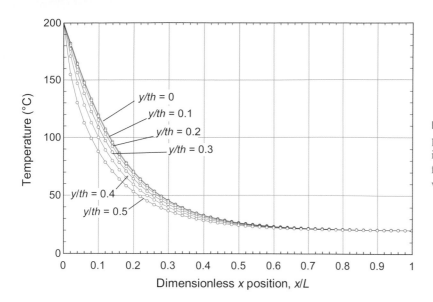

Figure 4.11 Temperature predicted by numerical model implemented in EES as a function of x/L for various values of y/th.

Note that the corner nodes must be considered outside of the summation in Eq. (4.47) as they have one-half the surface area available for convection. The maximum heat transfer rate is associated with an isothermal fin where the entire surface is at T_b:

$$\dot{q}_{max} = 2LW\bar{h}(T_b - T_\infty). \qquad (4.48)$$

The factor of 2 that appears in both Eqs. (4.47) and (4.48) is related to the fact that the numerical model only considers one-half of the fin. Substituting Eqs. (4.47) and (4.48) into Eq. (4.46) leads to:

$$\eta_{fin} = \frac{\Delta x}{L(T_b - T_\infty)} \left[\frac{(T_{1,N} - T_\infty)}{2} + \sum_{i=2}^{M-1} (T_{i,N} - T_\infty) + \frac{(T_{M,N} - T_\infty)}{2} \right]. \qquad (4.49)$$

```
"calculate fin efficiency"
eta_fin=DELTAx*((T[1,N]-T_infinity)/2+sum((T[i,N]-T_infinity),i=2,(M-1))&
        +(T[M,N]-T_infinity)/2)/(L*(T_b-T_infinity))
```

Figure 4.12 illustrates the predicted fin efficiency as a function of the number of nodes where an equal number of nodes are used in the x- and y-directions ($M = N$). Note that the fin efficiency calculated for this problem will not agree with the value obtained from Eq. (3.69) for a fin that is modeled as an extended surface because of the two-dimensional heat transfer that occurs in this fin.

EES can solve thousands of simultaneous equations, so a reasonably large problem can be considered using EES. However, EES is not really the best tool for dealing with very large sets of equations. In the next section, we will look at how these equations can be solved using a more powerful, formal programming language.

4.3.4 Solving the Equations with Matrix Decomposition

Section 4.3.3 describes how 2-D, steady-state problems can be solved using a finite difference solution implemented in EES. This process is intuitive and easy because EES will automatically solve a set of implicit equations. However, EES is not well-suited for problems that involve very large numbers of equations.

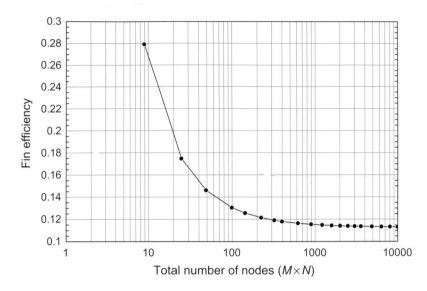

The finite difference method results in a system of algebraic equations that can be solved in a number of ways using different computer tools. Large problems will normally be implemented in a formal programming language such as C++, FORTRAN, or MATLAB. This section describes the method of solving the finite difference equations using matrix decomposition and illustrates the process using the programming language MATLAB.

The system of equations that results from applying the finite difference technique to a steady-state problem can be solved by placing these equations into a matrix format, as discussed previously in Section 2.4.4 in the context of a 1-D problem:

$$\underline{\underline{A}}\ \underline{X} = \underline{b}. \tag{4.50}$$

The vector $\underline{X}$ contains the unknown temperatures. Each *row* of the $\underline{\underline{A}}$ matrix and $\underline{b}$ vector corresponds to an equation (for one of the control volumes in the computational domain) whereas each *column* of the $\underline{\underline{A}}$ matrix holds the coefficients that multiply the corresponding unknown (the nodal temperature) in that equation. To place a system of equations in matrix format, it is necessary to carefully define how each of the rows in the $\underline{\underline{A}}$ matrix and $\underline{b}$ vector map to an equation and also to define the order in which the unknown temperatures are placed into the vector $\underline{X}$. This process was easy for the 1-D steady-state problems considered in Section 2.4.4, but it becomes somewhat more difficult for 2-D problems.

For a 2-D problem with M nodes in one dimension and N in the other, a logical technique for ordering the unknown temperatures in the vector $\underline{X}$ is:

$$\underline{X} = \begin{bmatrix} X_1 = T_{1,1} \\ X_2 = T_{2,1} \\ X_3 = T_{3,1} \\ \cdots \\ X_M = T_{M,1} \\ X_{M+1} = T_{1,2} \\ X_{M+2} = T_{2,2} \\ \cdots \\ X_{MN} = T_{M,N} \end{bmatrix}. \tag{4.51}$$

Equation (4.51) indicates that temperature $T_{i,j}$ corresponds to element $X_{M(j-1)+i}$ of the vector $\underline{X}$. A logical technique for mapping the equations to the rows of $\underline{\underline{A}}$ and $\underline{b}$ is:

$$\underline{A} \text{ and } \underline{b} = \begin{bmatrix} \text{row } 1 = \text{control volume equation for node } (1,1) \\ \text{row } 2 = \text{control volume equation for node } (2,1) \\ \text{row } 3 = \text{control volume equation for node } (3,1) \\ \cdots \\ \text{row } M = \text{control volume equation for node } (M,1) \\ \text{row } M+1 = \text{control volume equation for node } (1,2) \\ \text{row } M+2 = \text{control volume equation for node } (2,2) \\ \cdots \\ \text{row } MN = \text{control volume equation for node } (M,N) \end{bmatrix}. \qquad (4.52)$$

Equation (4.52) indicates that the equation for the control volume defined around node i,j should be placed into row $M(j-1)+i$. The mappings represented by Eqs. (4.51) and (4.52) should be kept in mind as we build the matrix $\underline{A}$ and the vector $\underline{b}$.

The process of implementing a numerical solution in MATLAB is illustrated in the context of the problem shown in Figure 4.6 which resulted in the finite difference equations summarized in Table 4.2. The input parameters are entered in the MATLAB script Fin2D.m.

```
clear all;

T_b=200;              %[C] base temperature
T_infinity=20;        %[C] fluid temperature
h_bar=200;            %[W/m^2-K] heat transfer coefficient
th=0.01;              %[m] thickness
L=0.04;               %[m] length
k=1;                  %[W/m-K] thermal conductivity
```

The x- and y-coordinates of each node are provided by Eqs. (4.22) and (4.23) and the distance between adjacent nodes is given by Eqs. (4.24) and (4.25).

```
%Setup grid
M=51;                 %number of x-nodes
N=21;                 %number of y-nodes

for i=1:M
    x(i)=(i-1)*L/(M-1);       %x-position of each node
    x_bar(i)=x(i)/L;          %dimensionless x-position of each node
end
DELTAx=L/(M-1);               %x-distance between adjacent nodes

for j=1:N
    y(j)=(j-1)*th/(2*(N-1));  %y-position of each node
    y_bar(j)=y(j)/th;         %dimensionless y-position of each node
end
DELTAy=th/(2*(N-1));          %y-distance between adjacent nodes
```

Each of the $M \times N$ equations that are summarized in Table 4.2 must be placed into one of the rows of the matrix $\underline{A}$ and the vector $\underline{b}$. We can start with the equations that correspond to energy balances on the internal nodes, Eq. (4.32), repeated below:

$$\frac{\Delta y}{\Delta x}\left(T_{i+1,j} - T_{i,j}\right) + \frac{\Delta y}{\Delta x}\left(T_{i-1,j} - T_{i,j}\right) + \frac{\Delta x}{\Delta y}\left(T_{i,j+1} - T_{i,j}\right) + \frac{\Delta x}{\Delta y}\left(T_{i,j-1} - T_{i,j}\right) = 0$$

$$\text{for } i = 2 \ldots (M-1) \text{ and } j = 2 \ldots (N-1). \qquad (4.53)$$

Equation (4.53) is rearranged to identify the coefficients that multiply each unknown temperature and the constant:

$$T_{i,j}\left[-2\frac{\Delta y}{\Delta x}-2\frac{\Delta x}{\Delta y}\right]+T_{i+1,j}\left[\frac{\Delta y}{\Delta x}\right]+T_{i-1,j}\left[\frac{\Delta y}{\Delta x}\right]+T_{i,j+1}\left[\frac{\Delta x}{\Delta y}\right]+T_{i,j-1}\left[\frac{\Delta x}{\Delta y}\right]=0$$

$$\text{for } i=2\ldots(M-1) \text{ and } j=2\ldots(N-1).$$
(4.54)

The control volume equations must be placed into the matrix equation

$$\underline{\underline{A}}\,\underline{X}=\underline{b},$$
(4.55)

where the equation for the control volume around node i,j is placed into row $M(j-1)+i$ of $\underline{\underline{A}}$ and $\underline{b}$ and $T_{i,j}$ corresponds to element $X_{M(j-1)+i}$ in the vector $\underline{X}$, as required by Eqs. (4.52) and (4.51), respectively. Each coefficient in Eq. (4.54) (i.e., each term multiplying an unknown temperature on the left side of the equation) must be placed in row $M(j-1)+i$ of $\underline{\underline{A}}$ and in the column corresponding to the position of the unknown temperature in $\underline{X}$. Therefore, the matrix assignments consistent with Eq. (4.54) are

$$\text{for } i=2\ldots(M-1) \text{ and } j=2\ldots(N-1):$$

$$A_{M(j-1)+i,M(j-1)+i}=\left[-2\frac{\Delta y}{\Delta x}-2\frac{\Delta x}{\Delta y}\right]$$

$$A_{M(j-1)+i,M(j-1)+i+1}=\left[\frac{\Delta y}{\Delta x}\right]$$

$$A_{M(j-1)+i,M(j-1)+i-1}=\left[\frac{\Delta y}{\Delta x}\right]$$
(4.56)

$$A_{M(j-1)+i,M(j+1-1)+i}=\left[\frac{\Delta x}{\Delta y}\right]$$

$$A_{M(j-1)+i,M(j-1-1)+i}=\left[\frac{\Delta x}{\Delta y}\right].$$

A matrix is allocated in MATLAB for $\underline{\underline{A}}$ and a vector is allocated for $\underline{b}$. The assignments summarized in Eq. (4.56) are made.

```
A=zeros(M*N,M*N);                    %initialize A matrix
b=zeros(M*N,1);                      %initialize b vector

%internal nodes
for i=2:(M-1)
  for j=2:(N-1)
      A(M*(j-1)+i, M*(j-1)+i)=-2*DELTAy/DELTAx-2*DELTAx/DELTAy;
      A(M*(j-1)+i, M*(j-1)+i+1)=DELTAy/DELTAx;
      A(M*(j-1)+i, M*(j-1)+i-1)=DELTAy/DELTAx;
      A(M*(j-1)+i, M*(j+1-1)+i)=DELTAx/DELTAy;
      A(M*(j-1)+i, M*(j-1-1)+i)=DELTAx/DELTAy;
  end
end
```

We can move on to the equations corresponding to the left boundary, Eq. (4.33), repeated below:

$$T_{1,j}[1]=T_b \text{ for } j=1\ldots N.$$
(4.57)

The matrix assignments consistent with Eq. (4.57) are

$$\text{for } j = 1 \ldots N:$$

$$A_{M(j-1)+1, M(j-1)+1} = [1]$$

(4.58)

$$b_{M(j-1)+1} = [T_b].$$

```
%left boundary
for  j=1:N
   A(M*(j-1)+1,  M*(j-1)+1)=1;
   b(M*(j-1)+1)=T_b;
end
```

The energy balances on the top boundary nodes are given by Eq. (4.40), repeated below:

$$\frac{\Delta y}{2\Delta x}(T_{i+1,N} - T_{i,N}) + \frac{\Delta y}{2\Delta x}(T_{i-1,N} - T_{i,N}) + \frac{\Delta x}{\Delta y}(T_{i,N-1} - T_{i,N}) + \frac{\Delta x \bar{h}}{k}(T_\infty - T_{i,N}) = 0,$$

$$i = 2 \ldots (M-1)$$

(4.59)

which is rearranged:

$$T_{i,N}\left[-\frac{\Delta y}{\Delta x} - \frac{\Delta x}{\Delta y} - \frac{\Delta x \bar{h}}{k}\right] + T_{i+1,N}\left[\frac{\Delta y}{2\Delta x}\right] + T_{i-1,N}\left[\frac{\Delta y}{2\Delta x}\right] + T_{i,N-1}\left[\frac{\Delta x}{\Delta y}\right] = -\frac{\Delta x \bar{h}}{k}T_\infty$$

$$i = 2 \ldots (M-1).$$

(4.60)

The matrix assignments consistent with Eq. (4.60) are

$$\text{for } i = 2 \ldots (M-1):$$

$$A_{M(N-1)+i, M(N-1)+i} = \left[-\frac{\Delta y}{\Delta x} - \frac{\Delta x}{\Delta y} - \frac{\Delta x \bar{h}}{k}\right]$$

$$A_{M(N-1)+i, M(N-1)+i+1} = \left[\frac{\Delta y}{2\Delta x}\right]$$

$$A_{M(N-1)+i, M(N-1)+i-1} = \left[\frac{\Delta y}{2\Delta x}\right]$$

(4.61)

$$A_{M(N-1)+i, M(N-1-1)+i} = \left[\frac{\Delta x}{\Delta y}\right]$$

$$b_{M(N-1)+i} = \left[-\frac{\Delta x \bar{h}}{k}T_\infty\right].$$

```
%top  boundary
for  i=2:(M-1)
   A(M*(N-1)+i,  M*(N-1)+i)=-DELTAy/DELTAx-DELTAx/DELTAy-DELTAx*h_bar/k;
   A(M*(N-1)+i,  M*(N-1)+i+1)=DELTAy/(2*DELTAx);
   A(M*(N-1)+i,  M*(N-1)+i-1)=DELTAy/(2*DELTAx);
   A(M*(N-1)+i,  M*(N-1-1)+i)=DELTAx/DELTAy;
   b(M*(N-1)+i)=-DELTAx*h_bar*T_infinity/k;
end
```

The energy balance on the node in the upper right corner is given by Eq. (4.42), repeated below:

$$\frac{\Delta x}{2\Delta y}(T_{M,N-1} - T_{M,N}) + \frac{\Delta y}{2\Delta x}(T_{M-1,N} - T_{M,N}) + \frac{\bar{h}\Delta x}{2k}(T_\infty - T_{M,N}) = 0,$$

(4.62)

which is rearranged:

$$T_{M,N}\left[-\frac{\Delta x}{2\Delta y}-\frac{\Delta y}{2\Delta x}-\frac{\bar{h}\Delta x}{2k}\right]+T_{M,N-1}\left[\frac{\Delta x}{2\Delta y}\right]+T_{M-1,N}\left[\frac{\Delta y}{2\Delta x}\right]=-\frac{\bar{h}\Delta x}{2k}T_\infty. \tag{4.63}$$

The matrix assignments consistent with Eq. (4.60) are:

$$A_{M(N-1)+M,M(N-1)+M}=\left[-\frac{\Delta x}{2\Delta y}-\frac{\Delta y}{2\Delta x}-\frac{\bar{h}\Delta x}{2k}\right]$$

$$A_{M(N-1)+M,M(N-1-1)+M}=\left[\frac{\Delta x}{2\Delta y}\right]$$

$$A_{M(N-1)+M,M(N-1)+M-1}=\left[\frac{\Delta y}{2\Delta x}\right] \tag{4.64}$$

$$b_{M(N-1)+M}=\left[-\frac{\bar{h}\Delta x}{2k}T_\infty\right].$$

```
%upper right corner
A(M*(N-1)+M, M*(N-1)+M)=-DELTAx/(2*DELTAy)-DELTAy/(2*DELTAx)-h_bar*DELTAx/(2*k);
A(M*(N-1)+M, M*(N-1-1)+M)=DELTAx/(2*DELTAy);
A(M*(N-1)+M, M*(N-1)+M-1)=DELTAy/(2*DELTAx);
b(M*(N-1)+M)=-h_bar*DELTAx*T_infinity/(2*k);
```

This process is repeated with the right boundary, Eq. (4.43):

$$\text{for } j=2\ldots(N-1):$$

$$A_{M(j-1)+M,M(j-1)+M}=\left[-\frac{\Delta y}{\Delta x}-\frac{\Delta x}{\Delta y}\right]$$

$$A_{M(j-1)+M,M(j-1+1)+M}=\left[\frac{\Delta x}{2\Delta y}\right]$$

$$A_{M(j-1)+M,M(j-1-1)+M}=\left[\frac{\Delta x}{2\Delta y}\right] \tag{4.65}$$

$$A_{M(j-1)+M,M(j-1)+M-1}=\left[\frac{\Delta y}{\Delta x}\right].$$

```
%right boundary
for j=2:(N-1)
    A(M*(j-1)+M, M*(j-1)+M)=-DELTAx/DELTAy-DELTAy/DELTAx;
    A(M*(j-1)+M, M*(j+1-1)+M)=DELTAx/(2*DELTAy);
    A(M*(j-1)+M, M*(j-1-1)+M)=DELTAx/(2*DELTAy);
    A(M*(j-1)+M, M*(j-1)+M-1)=DELTAy/DELTAx;
end
```

The lower right corner energy balance, Eq. (4.44), becomes:

$$A_{M(1-1)+M,M(1-1)+M}=\left[-\frac{\Delta x}{2\Delta y}-\frac{\Delta y}{2\Delta x}\right]$$

$$A_{M(1-1)+M,M(2-1)+M}=\left[\frac{\Delta x}{2\Delta y}\right] \tag{4.66}$$

$$A_{M(1-1)+M,M(1-1)+M-1}=\left[\frac{\Delta y}{2\Delta x}\right].$$

```
%lower right corner
A(M*(1-1)+M, M*(1-1)+M)=-DELTAx/(2*DELTAy)-DELTAy/(2*DELTAx);
A(M*(1-1)+M, M*(2-1)+M)=DELTAx/(2*DELTAy);
A(M*(1-1)+M, M*(1-1)+M-1)=DELTAy/(2*DELTAx);
```

Finally, the bottom boundary, Eq. (4.45), becomes

$$\text{for } i = 2 \ldots (M-1):$$

$$A_{M(1-1)+i, M(1-1)+i} = \left[-\frac{\Delta y}{\Delta x} - \frac{\Delta x}{\Delta y} \right]$$

$$A_{M(1-1)+i, M(1-1)+i+1} = \left[\frac{\Delta y}{2\Delta x} \right]$$

$$A_{M(1-1)+i, M(1-1)+i-1} = \left[\frac{\Delta y}{2\Delta x} \right]$$

$$A_{M(1-1)+i, M(2-1)+i} = \left[\frac{\Delta x}{\Delta y} \right].$$

(4.67)

```
%bottom boundary
for i=2:(M-1)
    A(M*(1-1)+i, M*(1-1)+i)=-DELTAy/DELTAx-DELTAx/DELTAy;
    A(M*(1-1)+i, M*(1-1)+i+1)=DELTAy/(2*DELTAx);
    A(M*(1-1)+i, M*(1-1)+i-1)=DELTAy/(2*DELTAx);
    A(M*(1-1)+i, M*(2-1)+i)=DELTAx/DELTAy;
end
```

The vector $\underline{X}$ is obtained using MATLAB's backslash command and the temperature of each node is placed in the matrix T.

```
X=A\b;
for i=1:M
    for j=1:N
        T(i,j)=X(M*(j-1)+i);
    end
end
```

The solution is plotted as follows.

```
plot(x_bar,T);
```

The result is shown in Figure 4.13. There are a variety of 3-D plotting functions in MATLAB; these can be investigated by typing help graph3d at the command window. For example, the mesh command can be used to generate the plot shown in Figure 4.14.

4.3.5 Solving the Equations with Gauss–Seidel Iteration

Section 2.4.5 introduced the method of Gauss–Seidel iteration for solving the sets of equations that arise in 1-D conduction problems. The method can be extended to the larger set of equations associated with 2-D problems. Each of the nodal energy balances must be solved to obtain an explicit equation providing the temperature at that node in terms of the other nodal temperatures. The balances derived in Section 4.3.1 are written in the format required for Gauss–Seidel iteration and summarized in Table 4.3.

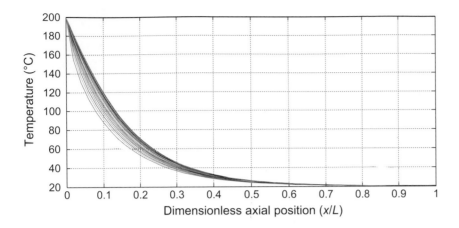

Figure 4.13 Plot of temperature as a function of dimensionless x position (x/L) at every value of y position.

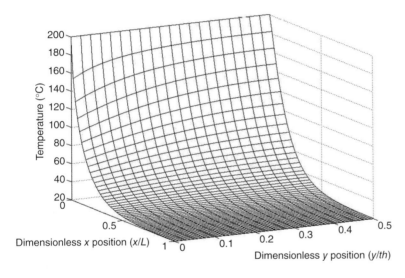

Figure 4.14 Mesh plot of temperature distribution.

We start with a set of guess values for each of the nodal temperatures, $T_{i,j}^k$, where the superscript k indicates the iteration number. Next we employ the explicit equations in Table 4.3 to solve for each of the nodal temperatures in the next iteration, $T_{i,j}^{k+1}$. The temperature on the left side of the equal sign is the new temperature at node i, j while the temperatures on the right side are the most recent values of the other nodal temperatures. A single matrix of the unknown temperatures is used and overwritten as the iteration process progresses so that whatever temperature is currently present in the matrix must be the most recent one.

The values of the temperatures obtained for successive iterations must be compared in order to determine an error (err) that can be used to terminate the iteration process. The maximum absolute value of the change of temperature between iterations for any node can be used for this purpose:

$$err = \text{Max} \left(\left| T_{i,j}^{k+1} - T_{i,j}^k \right| \right) \text{ for } i = 1 \ldots M \text{ and } j = 1 \ldots N. \tag{4.68}$$

Alternatively, we could calculate the rms value of the error associated with all of the nodes:

$$err = \sqrt{\frac{1}{M\,N} \sum_{i=1}^{M} \sum_{j=1}^{N} \left(T_{i,j}^{k+1} - T_{i,j}^k \right)^2} \tag{4.69}$$

Table 4.3 Finite difference equations in Table 4.2 corresponding to the problem shown in Figure 4.6 solved to provide the explicit equation for each nodal temperature that is necessary for Gauss–Seidel iteration.

Equation source	Number	Number of equations	Equation	Subscript range
internal	Eq. (4.32)	$(M-2)\times(N-2)$	$T_{i,j} = \dfrac{\frac{\Delta y}{\Delta x}\left(T_{i+1,j}+T_{i-1,j}\right)+\frac{\Delta x}{\Delta y}\left(T_{i,j+1}+T_{i,j-1}\right)}{2\left(\frac{\Delta y}{\Delta x}+\frac{\Delta x}{\Delta y}\right)}$	$i = 2\ldots(M-1)$ $j = 2\ldots(N-1)$
left boundary	Eq. (4.33)	N	$T_{1,j} = T_b$	$i = 1$ $j = 1\ldots N$
top boundary	Eq. (4.40)	$(M-2)$	$T_{i,N} = \dfrac{\frac{\Delta y}{2\Delta x}\left(T_{i+1,N}+T_{i-1,N}\right)+\frac{\Delta x}{\Delta y}T_{i,N-1}+\frac{\Delta x\bar h}{k}T_\infty}{\left(\frac{\Delta y}{\Delta x}+\frac{\Delta x}{\Delta y}+\frac{\Delta x\bar h}{k}\right)}$	$i = 2\ldots(M-1)$ $j = N$
upper right corner	Eq. (4.42)	1	$T_{M,N} = \dfrac{\frac{\Delta x}{2\Delta y}T_{M,N-1}+\frac{\Delta y}{2\Delta x}T_{M-1,N}+\frac{\bar h\Delta x}{2k}T_\infty}{\left(\frac{\Delta x}{2\Delta y}+\frac{\Delta y}{2\Delta x}+\frac{\bar h\Delta x}{2k}\right)}$	$i = M$ $j = N$
right boundary	Eq. (4.43)	$(N-2)$	$T_{M,j} = \dfrac{\frac{\Delta x}{2\Delta y}\left(T_{M,j+1}+T_{M,j-1}\right)+\frac{\Delta y}{\Delta x}T_{M-1,j}}{\left(\frac{\Delta x}{\Delta y}+\frac{\Delta y}{\Delta x}\right)}$	$i = M$ $j = 2\ldots(N-1)$
lower right corner	Eq. (4.44)	1	$T_{M,1} = \dfrac{\frac{\Delta x}{\Delta y}T_{M,2}+\frac{\Delta y}{\Delta x}T_{M-1,1}}{\left(\frac{\Delta x}{\Delta y}+\frac{\Delta y}{\Delta x}\right)}$	$i = M$ $j = 1$
bottom boundary	Eq. (4.45)	$(M-2)$	$T_{i,1} = \dfrac{\frac{\Delta y}{2\Delta x}\left(T_{i+1,1}+T_{i-1,1}\right)+\frac{\Delta x}{\Delta y}T_{i,2}}{\left(\frac{\Delta y}{\Delta x}+\frac{\Delta x}{\Delta y}\right)}$	$i = 2\ldots(M-1)$ $j = 1$

or the average absolute change between iterations:

$$err = \frac{1}{M\,N} \sum_{i=1}^{M} \sum_{j=1}^{N} \left| T_{i,j}^{k+1} - T_{i,j}^{k} \right|. \tag{4.70}$$

We can program this process using any formal programming language as it depends only on the solution of explicit equations. Here we will implement the solution using MATLAB. A script is created, called Fin2D_GS.m. The inputs are entered at the top of the script.

```
clear all;

T_b=200;                        %[C] base temperature
T_infinity=20;                  %[C] fluid temperature
h_bar=200;                      %[W/m^2-K] heat transfer coefficient
th=0.01;                        %[m] thickness
L=0.04;                         %[m] length
k=1;                            %[W/m-K] thermal conductivity
```

The positions of the nodes and distance between the nodes are set up.

```
%Setup grid
M=21;                           %number of x-nodes
N=11;                           %number of y-nodes

for i=1:M
    x(i)=(i-1)*L/(M-1);         %x-position of each node
    x_bar(i)=x(i)/L;            %dimensionless x-position of each node
end
Dx=L/(M-1);                     %x-distance between adjacent nodes

for j=1:N
    y(j)=(j-1)*th/(2*(N-1));    %y-position of each node
    y_bar(j)=y(j)/th;           %dimensionless y-position of each node
end
Dy=th/(2*(N-1));                %y-distance between adjacent nodes
```

The initial guess values of the temperatures (i.e., our initial iteration for the temperature matrix, $\underline{T}^k$) are stored in matrix T_k. We will start the iteration by assuming that the temperature of each node is equal to the base temperature, T_b. Note that the temperatures of the nodes on the left side of the boundary are specified to be equal to T_b according to Eq. (4.33) and therefore these nodal temperatures do not need to be adjusted during the iteration. The matrix of temperatures for the next iteration, $\underline{T}^{k+1}$, is stored in matrix T_kp. The temperatures in the matrix T_kp start with the values in T_k and are overwritten during the iteration process, ensuring that the most recently calculated value of temperature is always being used.

```
for i=1:M
    for j=1:N
        T_k(i,j)=T_b;           %[K] initial guess temperature
    end
end
T_kp=T_k;                       %start with old temps and overwrite as we go
```

We will put the iteration process within a while loop in MATLAB that terminates when the convergence error (*err*) is less than a specified tolerance (*tol*). Within the while loop each of the equations in Table 4.3 except Eq. (4.33) is programmed and used to compute the matrix T_kp. The convergence error is computed using Eq. (4.68). Finally, the values stored in T_kp are transferred to T_k in order to repeat the iteration.

```
tol=1e-6;                              %[K] convergence tolerance
err=tol+1;                             %initial value of convergence error
while(err>tol)
    %compute new temperatures
    %internal nodes
    for  i=2:(M-1)
        for  j=2:(N-1)
            T_kp(i,j)=(Dy*(T_kp(i+1,j)+T_kp(i-1,j))/Dx+...
                (Dx*(T_kp(i,j+1)+T_kp(i,j-1))/Dy))/(2*(Dy/Dx+Dx/Dy));
        end
    end

    %top boundary
    for  i=2:(M-1)
        T_kp(i,N)=(Dy*(T_kp(i+1,N)+T_kp(i-1,N))/(2*Dx)+...
            Dx*T_kp(i,N-1)/Dy+Dx*h_bar*T_infinity/k)/(Dy/Dx+Dx/Dy+Dx*h_bar/k);
    end

    %upper right corner
    T_kp(M,N)=(Dx*T_kp(M,N-1)/(2*Dy)+Dy*T_kp(M-1,N)/(2*Dx)+...
        h_bar*Dx*T_infinity/(2*k))/(Dx/(2*Dy)+Dy/(2*Dx)+h_bar*Dx/(2*k));

    %right boundary
    for  j=2:(N-1)
        T_kp(M,j)=(Dx*(T_kp(M,j+1)+T_kp(M,j-1))/(2*Dy)+Dy*T_kp(M-1,j)/Dx)/(Dx/Dy+Dy/Dx);
    end

    %lower right corner
    T_kp(M,1)=(Dx*T_kp(M,2)/Dy+Dy*T_kp(M-1,1)/Dx)/(Dx/Dy+Dy/Dx);

    %bottom boundary
    for  i=2:(M-1)
        T_kp(i,1)=(Dy*(T_kp(i+1,1)+T_kp(i-1,1))/(2*Dx)+Dx*T_kp(i,2)/Dy)/(Dy/Dx+Dx/Dy);
    end
    err=max(max(abs(T_kp-T_k)))        %determine the convergence error
    T_k=T_kp;                          %use calculated temperatures for next iteration
end
```

Running the script shows the progress towards convergence. The result is the same as was shown in Figure 4.14.

4.4 Finite Element Solution

4.4.1 Introduction

Section 4.3 presents the finite difference method for solving 2-D steady-state conduction problems. The finite difference method is not convenient for systems with irregular geometry. Many of the advanced heat transfer simulation tools that are commercially available do not use finite difference techniques but rather employ the finite element method because it can be more easily applied to complex shapes. The finite element technique divides the solution domain into simply shaped regions (or elements) that are typically triangular or quadrilateral for 2-D problems. Rather than relying directly on energy balances, as in the finite difference method, the finite element method assumes a computationally simple functional form to describe the temperature distribution within each element. Typically the temperature is assumed to be a function of position within the element and the coefficients of the function used within each element are selected so that the temperature distribution most closely satisfies the underlying governing differential equation. The total solution links (or assembles) the individual solutions over each element in a manner that ensures continuity at the boundaries. The finite element technique can be applied to many different problems in engineering. The details of applying the finite element approach to conduction heat transfer are discussed in Nellis and Klein (2009) and Myers (1998).

The program FEHT solves 2-D steady-state conduction problems using the finite element technique (Klein, 2019). It is not necessary to understand the finite element in order to use FEHT. A student version of FEHT can be downloaded from http://fchartsoftware.com/feht/demo.php and Appendix F provides an introduction to using the software. There are many commercially available finite element packages that are more powerful than FEHT. Such software can handle 3-D problems and provide the user with more control of the specification of the geometry (usually allowing the user to import geometry from solid modeling programs). Most commercial programs automatically generate and refine the mesh, and they provide methods for examining and manipulating the solution. Some widely used programs include ANSYS$^{©}$, COSMOS$^{©}$, and COMSOL$^{©}$. However, the steps needed to use any of these commercial software packages are similar and include specifying the geometry, the material properties, the boundary conditions, and the mesh. The solution should always be verified for grid convergence, compared to your physical intuition, and if possible checked against an analytical solution.

This section demonstrates the proper use of a finite element software package to solve conduction heat transfer problems in the context of the problem shown in Figure 4.15 (left). A heat sink is used to transfer heat

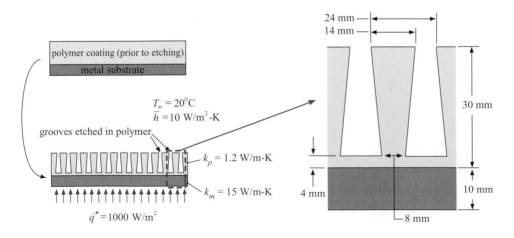

Figure 4.15 A heat sink coated with polymer that has been etched away to form fins (left) and the geometry of a single fin (right).

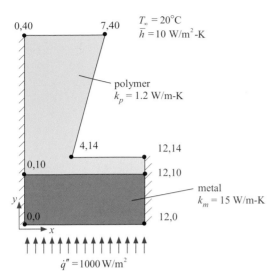

Figure 4.16 Unit cell used for the simulation (points shown in mm).

to air. The heat sink is formed by coating a metal substrate with a polymer that can be etched away to form fin-like structures. The etching process is not perfect and therefore the edges of the fins are not vertical but rather slope inwards as shown. The fins increase the surface area that is exposed to the air but the low thermal conductivity of the polymer adds a conduction resistance.

The geometry of a single fin and the adjacent metal substrate is shown in Figure 4.15 (right). The fin extends sufficiently far (into the page) that this can be considered a 2-D conduction problem. The conductivity of the metal is k_m = 15 W/m-K and the conductivity of the polymer is k_p = 1.2 W/m-K. A uniform heat flux of $\dot{q}''$ = 1000 W/m² is applied to the bottom surface. The temperature of the surrounding air is T_∞ = 20°C and the heat transfer coefficient is $\bar{h}$ = 10 W/m²-K.

4.4.2 Specifying the Problem

When using any numerical software the most critical part of the process is specifying the problem. What geometry, material properties, and boundary conditions are appropriate to simulate the actual situation? Is there symmetry that can be taken advantage of? What effects are important and must be included and which ones can be neglected? These questions require engineering judgment and dictate how efficient and effective the simulation is.

Figure 4.16 illustrates the geometry of the unit cell that is used for the simulation. Notice that the geometry takes advantage of the two natural lines of symmetry in the problem (down the middle of each fin and in the middle of the material separating each fin). There is no heat transfer across a line of symmetry and therefore they can be modeled as being adiabatic. The bottom surface has a specified heat flux and the remaining surfaces experience a convection boundary condition. The material properties required for this steady-state problem are the conductivities of the polymer and metal, both provided in the problem statement. The problem specification is implemented in FEHT using the procedure discussed in Appendix F. The result is shown in Figure 4.17. The points defining each of the lines have been exactly positioned, the material properties specified, and the boundary conditions on each of the external surfaces set.

4.4.3 Specifying the Mesh and Solving

The next step in the solution process is the specification of a mesh. Figure 4.18 (left) shows a coarse mesh that is manually created by drawing element lines; the mesh is then refined (right). A plot showing temperature contours and gradient vectors is given in Figure 4.19.

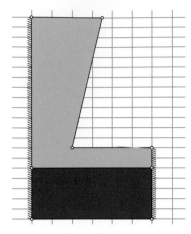

Figure 4.17 Problem specification in FEHT.

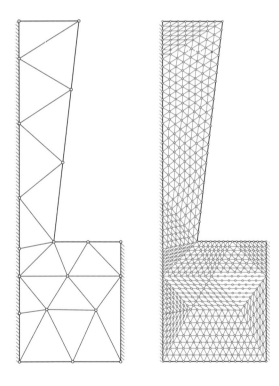

Figure 4.18 Coarse mesh manually created (left) and refined mesh (right).

4.4.4 Examination of the Solution

The final step in using a finite element software package is the critical examination of your solution in order to determine that a sufficient number of elements are being used and to assure yourself that it behaves according to your engineering judgment.

Mesh Convergence
It is important to check for mesh convergence of your solution. The typical method of doing this is to monitor some important aspect of the solution as the mesh is progressively refined. For this problem, the parameter of interest might be the maximum temperature in the fin, which occurs at the lower right corner of the domain.

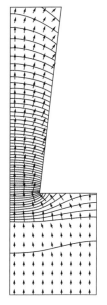

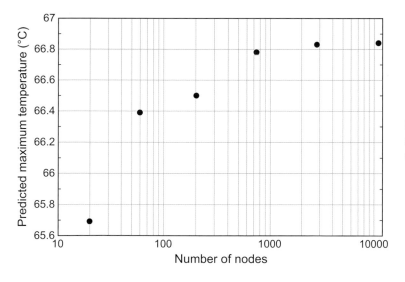

Figure 4.19 Temperature contours and gradient vectors.

Figure 4.20 Maximum temperature as a function of the number of nodes.

Figure 4.20 illustrates the predicted maximum temperature as a function of the number of nodes. The degree of convergence required depends on the accuracy that is necessary for the problem. Figure 4.20 suggests that approximately 700 nodes provides convergence to within about 0.1°C of the actual solution. (The student version of FEHT provides a maximum of 1000 nodes.)

Engineering Judgment
It is important that you critically examine your solution using engineering judgment. Figure 4.21 illustrates the temperature distribution with a few of the contours labeled. We can check some of the observed

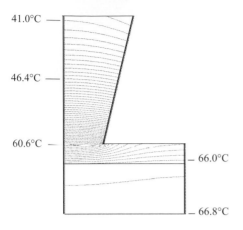

41.0°C

46.4°C

60.6°C

— 66.0°C

— 66.8°C

Figure 4.21 Temperature distribution.

characteristics against our engineering judgment. First, there is a relatively small (~0.8°C) temperature change across the metal substrate. The rate of heat transfer through the metal is

$$\dot{q} = \frac{1000\ \text{W}}{\text{m}^2} \left| \frac{1\ \text{m}}{} \right| \frac{0.012\ \text{m}}{} = 12\ \text{W} \tag{4.71}$$

and the resistance of the metal is, approximately, the thickness divided by the conductivity and the area for conduction:

$$R_m = \frac{0.01\ \text{m}}{} \left| \frac{\text{m K}}{15\ \text{W}} \right| \frac{1\ \text{m}}{} \left| \frac{}{0.012\ \text{m}} \right| = 0.056\ \frac{\text{K}}{\text{W}}. \tag{4.72}$$

Therefore, an estimate of the temperature difference across the metal substrate is

$$\Delta T_m \approx \dot{q}\ R_m = \frac{12\ \text{W}}{} \left| \frac{0.056\ \text{K}}{\text{W}} \right| = 0.7\ \text{K}, \tag{4.73}$$

which is consistent with the FEHT model result. Notice also that the fin structure in Figure 4.21 is approximately isothermal in the x-direction. This is consistent with a small Biot number, defined according to:

$$R_{cond,x} \approx \frac{0.0055\ \text{m}}{} \left| \frac{\text{m K}}{1.2\ \text{W}} \right| \frac{1\ \text{m}}{} \left| \frac{}{0.03\ \text{m}} \right| = 0.15\ \frac{\text{K}}{\text{W}} \tag{4.74}$$

$$R_{conv} \approx \frac{\text{m}^2\ \text{K}}{10\ \text{W}} \left| \frac{}{1\ \text{m}} \right| \frac{}{0.03\ \text{m}} = 3.3\ \frac{\text{K}}{\text{W}} \tag{4.75}$$

$$Bi = \frac{R_{cond,x}}{R_{conv}} = 0.05. \tag{4.76}$$

The fin itself appears to have moderate efficiency; its tip temperature is about 41°C with a base temperature of approximately 60.6°C relative to a fluid temperature of 20°C. Using the extended surface solution for a constant cross-sectional area fin discussed in Section 3.3.1 we obtain:

$$mL \approx \sqrt{\frac{10\ \text{W}}{\text{m}^2\ \text{K}} \left| \frac{\text{m K}}{1.2\ \text{W}} \right| \frac{}{0.0055\ \text{m}}} (0.03\ \text{m}) = 1.2 \tag{4.77}$$

$$\eta_{fin} \approx \frac{\tanh{(1.2)}}{1.2} = 0.71. \tag{4.78}$$

There are other checks of this type that can be carried out, but the point is that we should apply our basic knowledge of heat transfer to check that the solution makes sense. If we found some inconsistency between our intuition and the results then it is likely that some mistake was made in setting up the finite element solution. Alternatively, it is possible that the inconsistency is real and there is an opportunity to learn something from the unexpected behavior.

4.5 Conclusions and Learning Objectives

This chapter presented the steps required to derive the governing differential equation that must be solved for a 2-D steady-state conduction problem. Where a 1-D problem resulted in an ordinary differential equation coupled to boundary conditions that are defined at points, a 2-D problem leads to a partial differential equation coupled to a set of boundary conditions defined along lines. The methods required to solve such a problem analytically are beyond the scope of this text. However, this chapter showed how to use shape factors to solve commonly encountered problems of this type. The solution to 2-D problems can be accomplished using numerical techniques based on either the finite difference or finite element approach. The development of the finite difference equations was extended from 1-D to 2-D problems and the various techniques used to solve these equations were illustrated. Finally, the proper use of finite element software was demonstrated although the theory associated with this technique is beyond the scope of this text.

Some specific concepts that you should understand are listed below.
- The steps required for the derivation of the partial differential equation for a 2-D problem and the specification of boundary conditions using interface energy balances.
- The definition of a shape factor.
- The physical meaning of a shape factor.
- The use of a shape factor solution to define a thermal resistance,
- The development of a set of finite difference equations that can be used to numerically solve 2-D problems.
- The various techniques that can be used to solve the large set of algebraic equations resulting from a finite difference solution.
- The proper steps involved in employing numerical software to develop solutions to conduction problems.

References

Klein, S. A., FEHT – Finite Element Heat Transfer, version 8.025, http://fchartsoftware.com/feht/, 2019.
Myers, G. E., *Analytical Methods in Conduction Heat Transfer*, Second Edition, AMCHT Publications, Madison, WI (1998).
Nellis, G. F. and S. A. Klein, *Heat Transfer*, Cambridge University Press, New York (2009).
Rohsenow, W. M., J. P. Hartnett, and Y. I. Cho, *Handbook of Heat Transfer*, Third Edition, McGraw-Hill, New York (1998).

Problems

The Governing Differential Equation and Boundary Conditions

4.1 The figure illustrates a thin plate that is exposed to air on its top and bottom surfaces. The heat transfer coefficient between the top and bottom surfaces is $\bar{h}$ and the air temperature is T_f. The thickness of the plate is th and its width and height are a and b, respectively. The conductivity of the plate is k. The top edge is fixed at a uniform temperature, T_1. The right edge is fixed at a different, uniform temperature, T_2. The left edge of the plate is insulated. The bottom edge of the plate is exposed to a heat flux, $\dot{q}''$.

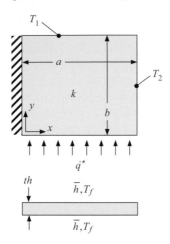

(a) The temperature distribution within the plate can be considered 2-D (i.e., temperature variations in the z-direction can be neglected) if the plate is thin and conductive. How would you determine if this approximation is valid?

(b) Derive the partial differential equation and boundary conditions that would need to be solved in order to obtain an analytical solution to this problem.

4.2 A laminated composite structure is shown in the figure.

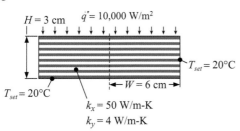

The structure is anisotropic. The effective conductivity of the composite in the x-direction is $k_x = 50$ W/m-K and in the y-direction it is $k_y = 4$ W/m-K. The top of the structure is exposed to a heat flux of $\dot{q}'' = 10,000$ W/m^2. The other edges are maintained at $T_{set} = 20°C$. The height of the structure is $H = 3$ cm and the half-width is $W = 6$ cm. Derive the governing partial differential equation and boundary conditions for this problem.

4.3 The figure illustrates a cylinder that is exposed to a concentrated heat flux at one end.

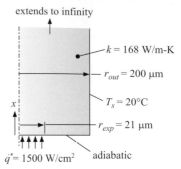

The cylinder extends infinitely in the x-direction. The surface at $x = 0$ experiences a uniform heat flux of $\dot{q}'' = 1500$ W/cm^2 for $r < r_{exp} = 21$ μm and is adiabatic for $r_{exp} < r < r_{out}$ where $r_{out} = 200$ μm is the outer radius of the cylinder. The outer surface of the cylinder is maintained at a uniform temperature of $T_s = 20°C$. The conductivity of the cylinder material is $k = 168$ W/m-K. Derive the governing partial differential equation and the boundary conditions for this problem.

4.4 A disk-shaped window in an experiment is shown in the figure.

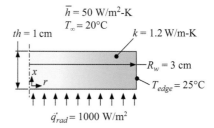

The inside of the window (the surface at $x = 0$) is exposed to vacuum and therefore does not experience any convection. However, this surface is exposed to a radiation heat flux $\dot{q}''_{rad} = 1000$ W/m^2. The window is assumed to be completely opaque to this radiation and therefore it is absorbed at $x = 0$. The edge of the window at $R_w = 3$ cm is maintained at a constant temperature $T_{edge} = 25°C$. The outside of the window (the surface at $x = th$) is cooled by air at $T_\infty = 20°C$ with heat transfer coefficient $\bar{h} = 50$ W/m^2-K. The conductivity of the window material is $k = 1.2$ W/m-K.

(a) Is the extended surface approximation appropriate for this problem? That is, can the temperature in the window be approximated as being 1-D in the radial direction? Justify your answer.

(b) Assume that your answer to (a) is no. Derive the governing partial differential equation and boundary conditions that must be solved.

4.5 The plate shown in the figure is exposed to a uniform heat flux $\dot{q}'' = 1 \times 10^5$ W/m^2 along its top surface and is adiabatic at its bottom surface. The left side of the plate is kept at $T_L = 300$ K and the right side is at $T_R = 500$ K. The height and width of the plate are $H = 1$ cm and $W = 5$ cm, respectively. The conductivity of the plate is $k = 10$ W/m-K.

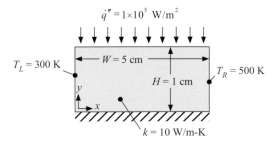

Derive the governing partial differential equation and boundary conditions for this problem.

4.6 The figure illustrates a 2-D, steady-state heat transfer problem.

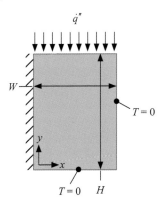

Derive the governing partial differential equation and boundary conditions for this problem.

4.7 The figure shows a thin plate that has dimension W in the x-direction and dimension H in the y-direction.

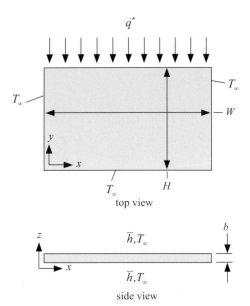

top view

side view

The thickness of the plate is $b = 0.01$ m. The top and bottom surfaces (at $z = 0$ and $z = b$) are exposed to convection with $\bar{h} = 100$ W/m^2-K to fluid at temperature T_∞. The conductivity of the plate material is $k = 100$ W/m-K. The upper edge of the plate (at $y = H$) is exposed to a heat flux, $\dot{q}''$. The other edges (at $x = 0$, $x = W$, and $y = 0$) are maintained at temperature T_∞. The plate is at steady state.

(a) The temperature distribution in the plate is three-dimensional, $T(x, y, z)$. Can the temperature distribution be modeled as being 2-D, $T(x, y)$, with little loss of accuracy? Justify your answer.

(b) Assume that the temperature distribution in the plate can be modeled as 2-D, $T(x, y)$. Derive the partial differential equation and boundary conditions for the problem.

4.8 The figure illustrates a rectangular bar of metal that extends between two walls. The bar is 0.5 mm thick, 5.0 cm wide, and 25.0 cm long. The thermal conductivity of the bar material is 10 W/m-K. The walls are cooled and maintained at 40°C. The top edge of the bar is exposed to a heat flux of 5×10^4 W/m^2 and the bottom edge is insulated. The front and back surfaces are insulated. Derive the governing differential equation and boundary conditions for this 2-D problem.

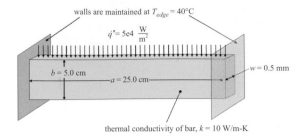

walls are maintained at $T_{edge} = 40°C$

$\dot{q}'' = 5e4 \ \frac{W}{m^2}$

$b = 5.0$ cm

$a = 25.0$ cm

$w = 0.5$ mm

thermal conductivity of bar, $k = 10$ W/m-K

Shape Factors

4.9 Currently, the low-pressure steam exhausted from a steam turbine at a power plant is condensed by heat transfer to cooling water. An alternative that has been proposed is to transport the steam via an underground pipe to a large building complex and use the steam for space heating. You have been asked to evaluate the feasibility of this proposal. The building complex is located 0.2 miles from the power plant. The pipe is made of uninsulated PVC (thermal conductivity of 0.19 W/m-K) with an inner diameter of 8.33 inch and wall thickness of 0.148 inch. The pipe will be buried underground at a depth of 4 ft in soil that has an estimated thermal conductivity of 0.5 W/m-K. The steam leaves the power plant at 6.5 lb$_m$/min, 8 psia with a 95 percent quality. The outdoor temperature is 5°F. Condensate is returned to the power plant in a separate pipe as, approximately, saturated liquid at 8 psia.

(a) Neglecting the inevitable pressure loss, estimate the state of the steam that is provided to the building complex.

(b) Are the thermal losses experienced in the underground pipe transport process significant in your opinion? Do you recommend insulating this pipe?

(c) Provide a sanity check on the shape factor that you used to solve this problem.

4.10 Two tubes are buried in the ground behind your house that transfer water to and from a wood burner (see the figure). The left-hand tube carries hot water from the burner back to your house at $T_{w,h} = 135°F$ while the right-hand tube carries cold water from your house back to the burner at $T_{w,c} = 70°F$. Both tubes have outer diameter $D_o = 0.75$ inch and thickness $th = 0.065$ inch. The conductivity of the tubing material is

$k_t = 0.22$ W/m-K. The heat transfer coefficient between the water and the tube internal surface (in both tubes) is $\bar{h}_w = 250$ W/m²-K. The center-to-center distance between the tubes is $w = 1.25$ inch and the length of the tubes is $L = 20$ ft (into the page). The tubes are buried in soil that has conductivity $k_s = 0.30$ W/m-K.

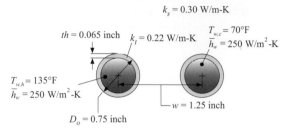

$k_s = 0.30$ W/m-K

$th = 0.065$ inch $k_t = 0.22$ W/m-K

$T_{w,c} = 70°F$
$\bar{h}_w = 250$ W/m²-K

$T_{w,h} = 135°F$
$\bar{h}_w = 250$ W/m²-K

$w = 1.25$ inch

$D_o = 0.75$ inch

(a) Estimate the heat transfer from the hot water to the cold water due to their proximity to one another.

(b) To do part (a) you should have needed to determine a shape factor; calculate an approximate value of the shape factor and compare it to the accepted value.

(c) Plot the rate of heat transfer from the hot water to the cold water as a function of the center-to-center distance between the tubes.

4.11 A pipe carrying water for a ground source heat pump is buried horizontally in soil with conductivity $k = 0.4$ W/m-K. The center of the pipe is $W = 6$ ft below the surface of the ground. The pipe has inner diameter $D_i = 1.5$ inch and outer diameter $D_o = 2$ inch. The pipe is made of material with conductivity $k_p = 1.5$ W/m-K. The water flowing through the pipe has temperature $T_w = 35°F$ with heat transfer coefficient $\bar{h}_w = 200$ W/m-K. The temperature of the surface of the soil is $T_s = 0°F$.

(a) Determine the rate of heat transfer between the water and the air per unit length of pipe.

(b) Plot the heat transfer as a function of the depth of the pipe.

(c) Carry out a sanity check on the value of the shape factor that you used in (a).

4.12 A solar electric generation system (SEGS) employs molten salt as both the energy transport and storage fluid. The molten salt is heated to

500°C and stored in a buried semi-spherical tank. The top (flat) surface of the tank is at ground level. The diameter of the tank before insulation is applied is 14 m. The outside surfaces of the tank are insulated with 0.30 m thick fiberglass having a thermal conductivity of 0.035 W/m-K. Sand having a thermal conductivity of 0.27 W/m-K surrounds the tank, except on its top surface. Estimate the rate of heat loss from this storage unit to the 25°C surroundings.

4.13 An 8 inch schedule 40 pipe has an inner diameter of 7.98 inch, an outer diameter of 8.62 inch, and is made of steel with conductivity 37 Btu/hr-ft-F. The pipe carries saturated steam at 24 psia and is buried 4 ft below the ground. The thermal conductivity of the ground is approximately 0.30 Btu/hr-ft-F. The surface of the ground exposed to air is at 65°F. Determine the heat loss per 100 ft of pipe in Btu/hr if:

(a) the pipe is not insulated, and

(b) the pipe is insulated with 0.5 inch thick layer of magnesia with conductivity 0.030 Btu/hr-ft-F.

4.14 Ablation is a technique for treating cancerous tissue that occurs by heating it to a lethal temperature; the dead tissue can subsequently be left to be absorbed by the body or excised surgically. A number of techniques have been suggested in order to apply heat locally to the cancerous tissue and therefore spare surrounding healthy tissue. One interesting technique would embed small metallic spheres at precise locations surrounding the cancer tumor and then expose the region to an oscillating magnetic field. The magnetic field does not generate energy in the tissue; however, the conducting spheres experience a volumetric generation of thermal energy which thereby increases their temperature and results in the conduction of heat to the surrounding tissue. Precise placement of the spheres can be used to control the application of thermal energy.

You want to measure the power generated by the sphere used for the ablation process. The radius of the sphere is $R_{sp} = 1.0$ mm. You place the sphere $w = 5$ cm below the surface of a solution of agar; agar is a material with well-known thermal properties that resembles jello

and is sometimes used as a surrogate for tissue in biological experiments. The agar is allowed to solidify around the sphere and you may assume that the container of agar is large enough to be considered semi-infinite. The surface of the agar is exposed to an ice-water bath in order to keep it at a constant temperature, $T_{ice} = 0°C$. The sphere is heated by an oscillating magnetic field and its temperature is measured using a thermocouple. The conductivity of agar is $k = 0.35$ W/m-K.

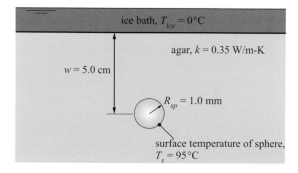

(a) If the surface temperature of the sphere is determined to be $T_s = 95°C$, then how much energy is transferred from the sphere?

(b) Estimate the uncertainty in your measurement of the power. Assume that your temperature measurements (ice and sphere surface temperature) are accurate to $\delta T = 1.0°C$, the conductivity of agar is known to within 10 percent ($\delta k = 0.035$ W/m-K), the depth measurement is known to $\delta w = 2.0$ mm, and the sphere radius is known to within $\delta R_{sp} = 0.1$ mm. You may want to use the uncertainty propagation feature in EES for this.

4.15 Saturated steam flows within a square channel extruded from AISI304 stainless steel. The extrusion has outer dimension 10 cm × 10 cm and the inner channel has dimension 7.5 cm × 7.5 cm. The heat transfer coefficient between the steam and inside channel surface is estimated to be 540 W/m²-K. Air at 25°C surrounds the outside of the channel with a heat transfer coefficient of 12.5 W/m²-K ± 2.5 W/m²-K. Prepare a plot of the rate of heat loss from the steam per meter of channel as a function of pressure for pressures ranging from 1 to 10 bar. Indicate the effect of the uncertainty in the air-side convection coefficients in the plot with dotted lines.

4.16 An electrical cable made of aluminum with diameter 1.25 cm is buried 1 m below the surface in sand. The sand can be assumed to be a semi-infinite medium with a surface temperature of 25°C. Electricity flows through the cable at 100 amp.
(a) What is rate of heat loss from the cable to the sand per meter of length?
(b) What is the temperature of the surface of the cable?
(c) What is the center temperature of the cable?

4.17 A large university campus produces saturated steam at 170 psig at a central location and distributes it underground to campus buildings. The steam is used to power chillers, pumps, and autoclaves, as well as to provide building heat. The steam pipes have conductivity 37 Btu/hr-ft-F with inner diameter of 7.625 in and a 0.5 in wall thickness. The mass flow rate of steam is 250 lb$_m$/hr. The pipes are buried 8 ft below grade in a mixture of clay and soil with conductivity 0.35 Btu/hr-ft-F.
(a) Calculate the energy flow rate provided by the steam assuming that water is returned to the physical plant at 125°F and atmospheric pressure.
(b) Assuming that the ground can be treated as a semi-infinite solid with a surface temperature equal to annual average temperature of 47°F, calculate the rate that condensate accumulates due to heat loss in a 100 ft pipe. Compare the rate of heat loss to the energy flow rate determined in part (a).
(c) It has been suggested that the pipe should be insulated with a 1 inch thick layer of diatomaceous pipe insulation with conductivity 0.05 Btu/hr-ft-F. Redo the calculation in part (b) for this situation and compare the results. Do you recommend applying the insulation?

4.18 A copper pipe having a 5 cm inner diameter and 5 mm wall thickness carries liquid nitrogen at atmospheric pressure. The pipe is insulated with a low-temperature insulation having a thermal conductivity of 0.046 W/m-K. The insulation thickness is 3.0 cm and the air temperature outside of the pipe is 25°C. The convection coefficient between the air and insulation surface is 12 W/m²-K. Owing to sloppy workmanship, the pipe and insulation center points do not coincide. Prepare a plot of the heat transfer rate to the liquid nitrogen per meter of pipe as a function of the distance between the center of the pipe and the center of the insulation for values ranging from 0 to 2.95 cm.

4.19 Derive the shape factor formula that should be used for conduction through:
(a) a cylindrical shell, and
(b) a spherical shell.

4.20 A tall, square chimney is made of brick with conductivity 0.72 Btu/hr-ft-F. The chimney has outside dimensions of 16 inch × 16 inch and inside dimensions of 12 inch × 12 inch. Combustion gas flows in the chimney with an average temperature of 130°F and a convection coefficient of 12 W/m²-K. The outside surface of the chimney is exposed to outdoor air at 10°F with a convection coefficient of 18 W/m²-K.
(a) Estimate the rate of heat loss per ft of chimney height.
(b) Estimate the temperature on the outside surface of the chimney.

Finite Difference Solutions

4.21 A heater extends from a wall into fluid that is to be heated.

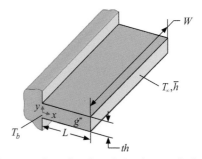

Assume that the tip of the heater is insulated and that the width (W) is much larger than the thickness (th) so that convection from the edges can be neglected. The length of the heater is

$L = 5.0$ cm. The base temperature is $T_b = 20°C$ and the heater experiences convection with fluid at $T_\infty = 100°C$ with average heat transfer coefficient, $\bar{h} = 100$ W/m²-K. The heater is $th = 3.0$ cm thick and has conductivity $k = 1.5$ W/m-K. The heater experiences a uniform volumetric generation of $\dot{g}''' = 5 \times 10^5$ W/m³.

(a) Develop a numerical solution for the temperature distribution in the heater using a finite difference technique and implement the solution in EES.

(b) Use the numerical solution to predict and plot the temperature distribution in the heater.

(c) Use the numerical solution to predict the heater efficiency; the heater efficiency is defined as the ratio of the rate of heat transfer to the fluid to the total rate of thermal energy generation in the fin.

(d) Plot the heater efficiency as a function of the length for various values of the thickness. Explain your plot.

4.22 The figure illustrates a device used to measure solar radiation.

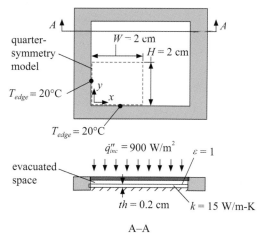

A–A

A thin membrane is located beneath a transparent window. The space between the membrane and the window is evacuated, eliminating convection from the surface of the membrane. The solar flux, $\dot{q}''_{inc} = 900$ W/m², passes through the window and is completely absorbed on the surface of the membrane. The membrane is $th = 0.2$ mm thick and has conductivity $k = 15$ W/m-K. The half-width of the membrane is $W = 2$ cm and the half-height of the membrane is $H = 2$ cm. The emissivity of

the membrane surface is $\varepsilon = 1.0$. The back-side of the membrane is completely insulated. The top of the membrane (where the solar flux is absorbed) experiences a radiation heat transfer with the surroundings at $T_\infty = 20°C$. The edges of the membrane are held at $T_{edge} = 20°C$. The difference between the temperature at the center of the membrane ($x = W$ and $y = H$) and the edge can be correlated to the magnitude of the incident flux.

(a) Is it appropriate to model the membrane as having a temperature that varies only in the x- and y-directions but not the z-direction? Justify your answer.

(b) Develop a quarter-symmetry numerical model of the membrane (see the figure) and implement the solution in EES.

(c) Use the numerical model to prepare a plot of the temperature in the membrane as a function of x at several values of y.

(d) Use the numerical solution to prepare a calibration curve for the instrument: the incident flux as a function of the measured center-to-edge temperature difference.

4.23 The figure illustrates a composite material that is being processed using a laser.

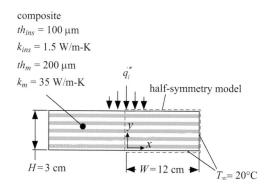

The composite is composed of alternating layers of insulating material and metal. The insulating layers have thickness $th_{ins} = 100$ μm and conductivity $k_{ins} = 1.5$ W/m-K. The metal layers have thickness $th_m = 200$ μm and conductivity $k_m = 35$ W/m-K. The work piece is very long in the z-direction and therefore the temperature distribution can be considered

2-D in x and y. The half-width of the work piece is $W = 12$ cm and the thickness is $H = 3$ cm. We will develop a half-symmetry model using the coordinates shown in the figure. The bottom surface ($y = 0$) and edges ($x = W$) are clamped into a fixture and therefore these surfaces are maintained at $T_\infty = 20°C$. The top surface ($y = H$) is exposed to a heat flux that is concentrated at the center of the piece. The heat flux varies with position according to

$$\dot{q}_l'' = a \exp\left[-\left(\frac{x}{pw}\right)^2\right],$$

where $a = 320{,}000$ W/m^2 and $pw = 1$ cm.
(a) What is the effective thermal conductivity of the composite in the x- and y-directions?
(b) Develop a 2-D numerical model of the work piece in EES. Plot the temperature as a function of x at various values of y, including at least $y = 0$, $H/2$, and H.

4.24 A disk-shaped window in an experiment is shown in the figure.

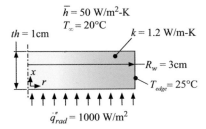

$\bar{h} = 50$ W/m^2-K
$T_\infty = 20°C$
$th = 1$cm
$k = 1.2$ W/m-K
$R_w = 3$cm
$T_{edge} = 25°C$
$\dot{q}_{rad}'' = 1000$ W/m^2

The inside of the window (the surface at $x = 0$) is exposed to vacuum and therefore does not experience any convection. However, this surface is exposed to a radiation heat flux $\dot{q}_{rad}'' = 1000$ W/m^2. The window is assumed to be completely opaque to this radiation and therefore it is absorbed at $x = 0$. The edge of the window at $R_w = 3$ cm is maintained at a constant temperature $T_{edge} = 25°C$. The outside of the window (the surface at $x = th$) is cooled by air at $T_\infty = 20°C$ with heat transfer coefficient $\bar{h} = 50$ W/m^2-K. The conductivity of the window material is $k = 1.2$ W/m-K.
(a) Is the extended surface approximation appropriate for this problem? That is, can the temperature in the window be approximated as being 1-D in the radial direction? Justify your answer.

(b) Assume that your answer to (a) is no. Develop a 2-D numerical solution to this problem using the finite difference technique and implement your solution in MATLAB using matrix decomposition.
(c) Plot the temperature as a function of r for various values of x.
(d) Prepare a contour plot of the temperature in the window.

4.25 Reconsider Problem 4.8. Prepare a numerical model of the problem using matrix decomposition and plot the temperature as a function of x for $y = 2.5$ cm (i.e., at the center of the bar).

4.26 Reconsider Problem 4.24. Assume that the window is not opaque to the radiation but rather is semi-transparent so that only some of the radiation is absorbed. The radiation that is absorbed is transformed to thermal energy. The volumetric rate of thermal energy generation is given by: $\dot{g}''' = \dot{q}_{rad}'' \alpha \exp(-\alpha x)$ where $\alpha = 100$ m^{-1} is the absorption coefficient. The radiation that is not absorbed is transmitted. Otherwise the problem remains the same.
(a) Develop a finite difference numerical solution and implement the solution in MATLAB using matrix decomposition.
(b) Plot the temperature as a function of r for various values of x.

4.27 A plate with thickness $w = 5.0$ cm is subjected to a uniform volumetric generation of 1×10^5 W/m^3. The width of the plate is $a = 1.0$ m and its height is $b = 0.5$ m. The conductivity of the plate material is $k = 100$ W/m-K. The top and bottom surfaces of the plate are exposed to air at $T_{air} = 300$ K with a convection coefficient, $\bar{h} = 10$ W/m^2-K. Three sides of the plate are maintained at $T_{edge} = 300$ K while the fourth edge is insulated.

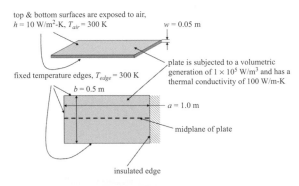

top & bottom surfaces are exposed to air, $h = 10$ W/m^2-K, $T_{air} = 300$ K
$w = 0.05$ m
plate is subjected to a volumetric generation of 1×10^5 W/m^3 and has a thermal conductivity of 100 W/m-K
fixed temperature edges, $T_{edge} = 300$ K
$b = 0.5$ m
$a = 1.0$ m
midplane of plate
insulated edge

(a) Can the temperature distribution in the plate be considered as 2-D? In other words, can you neglect the temperature gradient across the thickness (w) of the plate? Justify your answer.

(b) Prepare a numerical model of the plate using EES. Using your model, prepare a plot of the temperature as a function of position along the mid-plane of the plate indicated in the figure.

4.28 A long square bar, 0.35 m on each side is made of stainless steel ($k = 15$ W/m-K). The top side is painted flat black and exposed to direct sunlight so that it absorbs thermal energy at a rate of 1000 W/m². The left, top and right sides experience free convection to air at 25°C with a convection coefficient of 15 W/m²-K. The bottom side is insulated. Determine the temperature distribution in the bar using a finite difference numerical method implemented in EES. Check to ensure that sufficient nodes are used in the solution. Use the solution to prepare a contour plot showing the temperature of the bar as a function of x and y position.

4.29 The figure illustrates a composite material that is being machined on a lathe.

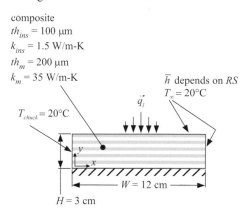

composite
$th_{ins} = 100$ μm
$k_{ins} = 1.5$ W/m-K
$th_m = 200$ μm
$k_m = 35$ W/m-K
$T_{chuck} = 20°C$
$\dot{q}_l''$
$\bar{h}$ depends on RS
$T_\infty = 20°C$
$W = 12$ cm
$H = 3$ cm

The composite is composed of alternating layers of insulating material and metal. The insulating layers have thickness $th_{ins}= 100$ μm and conductivity $k_{ins} = 1.5$ W/m-K. The metal layers have thickness $th_m = 200$ μm and conductivity $k_m = 35$ W/m-K. The work piece is actually cylindrical and rotating. However, because the radius is large relative to it thickness and there are no circumferential variations we can model the work piece as a 2-D problem in Cartesian coordinates, x and y, as shown in the figure. The width of the work piece is

$W = 12$ cm and the thickness is $H = 3$ cm. The left surface of the work piece at $x = 0$ is attached to the chuck and therefore maintained at $T_{chuck} = 20°C$. The inner surface at $y = 0$ is insulated. The outer surface (at $y = H$) and right surface (at $x = W$) are exposed to air at $T_\infty = 20°C$ with heat transfer coefficient $\bar{h}$ that depends on the rotational speed of the chuck, RS in rev/min, according to

$$\bar{h} = 2 \left[\frac{\text{W min}^2}{\text{m}^2 \text{ K rev}^2} \right] RS^2.$$

In order to extend the life of the tool used for the machining process, the work piece is preheated by applying laser power to the outer surface. The heat flux applied by the laser depends on the rotational speed and position according to

$$\dot{q}_l'' = a\, RS \exp\left[-\left(\frac{x - x_c}{pw} \right)^2 \right],$$

where $a = 5000$ W-min/m²-rev, $x_c = 8$ cm, and $pw = 1$ cm.

(a) What is the effective thermal conductivity of the composite in the x- and y-directions?

(b) Develop a 2-D numerical model of the work piece in EES. Plot the temperature as a function of x at various values of y, including at least $y = 0$, $H/2$, and H.

(c) Plot the maximum temperature in the work piece as a function of the rotational speed, RS. If the objective is to preheat the material to its maximum possible temperature, then what is the optimal rotational speed?

4.30 The steady-state temperature distribution in a cylindrical section is to be numerically determined using the seven nodes shown in the figure. The radial and angular spacing of the nodes is uniform. Surface 1-4-5 is adiabatic. Surface 1-2-3 is maintained at temperature T_s. Surface 5-7-3 is exposed to air at temperature T_∞ with heat transfer coefficient $\bar{h}$. Derive a finite difference equation for node 7.

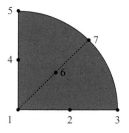

4.31 A long square bar has a cut-out section as shown in the figure. Surface 12-13-14-15 is maintained at temperature T_h and has width L. Surface 3-7-11-15 has height L and convects to air at T_∞ with convection coefficient $\bar{h}$. Surfaces 1-2-3 and 4-8-12 are insulated. Surface 4-5-1 also convects to air at T_∞ with convection coefficient $\bar{h}$. Derive the steady-state finite difference equation for node 5.

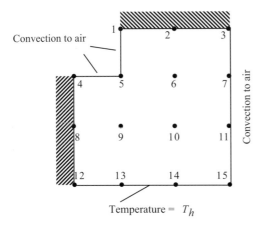

4.32 A large plate is insulated on its left and bottom surfaces, as shown in the figure. The top surface receives a solar flux of $\dot{q}''$. The top and right surfaces convect to air at T_∞ with convection coefficient $\bar{h}$. Derive the steady-state finite difference equation for node 1, located at the upper right corner. Express your answer in terms of Δx and Δy.

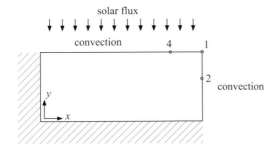

4.33 The figure illustrates a cut-away view of two plates that are being welded together. Both edges of the plate are clamped and effectively held at temperatures $T_s = 25°C$. The top of the plate is exposed to a heat flux that varies with position x, measured from joint, according to $\dot{q}''_m(x) = \dot{q}''_j \exp(-x/L_j)$, where

$\dot{q}''_j = 1 \times 10^6$ W/m² is the maximum heat flux (at the joint, $x = 0$) and $L_j = 2.0$ cm is a measure of the extent of the heat flux. The back side of the plates are exposed to liquid cooling by a jet of fluid at $T_f = -35°C$ with $\bar{h} = 5000$ W/m²-K. A half-symmetry model of the problem is shown in the figure. The thickness of the plate is $b = 3.5$ cm and the width of a single plate is $W = 8.5$ cm. Assume that the welding process is steady state and 2-D. You may neglect convection from the top of the plate. The conductivity of the plate material is $k = 38$ W/m-K.

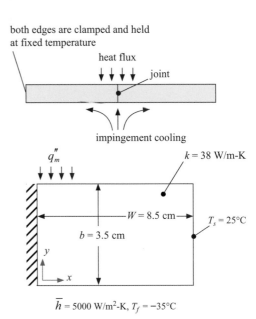

(a) Develop a numerical model of the problem. Implement the solution in MATLAB and prepare a contour or surface plot of the temperature in the plate.
(b) Plot the temperature as a function of x at $y = 0$, $b/2$, and b.

Finite Element Method using FEHT

4.34 Gas turbine power cycles are used for the generation of power; the size of these systems can range from tens of kilowatts for the microturbines that are being installed on-site at some commercial and industrial locations to hundreds of megawatts for natural gas fired power plants. The efficiency of a gas turbine power plant increases with the temperature of the gas entering from the combustion chamber; this temperature is constrained by the material limitations of the turbine blades which tend to creep (i.e., slowly grow over time) in the high-temperature environment due to their high centrifugal stress state. One technique for achieving high gas temperatures is to cool the blades internally; often the air is bled through the blade surface using a technique called transpiration. A simplified version of a turbine blade that will be analyzed in this problem is shown in the figure.

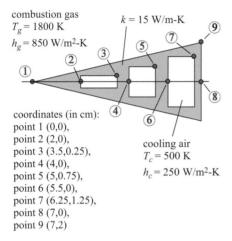

combustion gas
$T_g = 1800$ K
$h_g = 850$ W/m²-K
$k = 15$ W/m-K

coordinates (in cm):
point 1 (0,0),
point 2 (2,0),
point 3 (3.5,0.25),
point 4 (4,0),
point 5 (5,0.75),
point 6 (5.5,0),
point 7 (6.25,1.25),
point 8 (7,0),
point 9 (7,2)

cooling air
$T_c = 500$ K
$h_c = 250$ W/m²-K

The high-temperature combustion gas is at $T_g = 1800$ K and the heat transfer coefficient between the gas and blade external surface is $h_g = 850$ W/m²-K. The blades are cooled by three internal air passages. The cooling air in the passages is at $T_c = 500$ K and the air-to-blade heat transfer coefficient is $h_c = 250$ W/m²-K. The blade material has conductivity $k = 15$ W/m-K. The coordinates of the points required to define the geometry are indicated in the figure.

(a) Generate a half-symmetry model of the blade in FEHT. Generate a figure showing the temperature distribution in the blade predicted using a very crude mesh.

(b) Refine your mesh and keep track of the temperature experienced at the trailing edge of the blade (i.e., at position 9 in the figure) as a function of the number of nodes in your mesh. Prepare a plot of this data that can be used to establish that your model has converged to the correct solution.

(c) Do your results make sense? Use a very simple, order-of-magnitude analysis based on thermal resistances to decide whether your predicted blade surface temperature is reasonable (hint – there are three thermal resistances that govern the behavior of the blade, estimate each one and show that your results are approximately correct given these thermal resistances).

4.35 The figure illustrates a power electronics chip that is used to control the current to a winding of a motor.

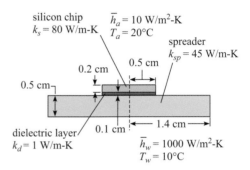

silicon chip
$k_s = 80$ W/m-K
$\bar{h}_a = 10$ W/m²-K
$T_a = 20°C$
spreader
$k_{sp} = 45$ W/m-K
0.2 cm
0.5 cm
0.5 cm
dielectric layer
$k_d = 1$ W/m-K
0.1 cm
1.4 cm
$\bar{h}_w = 1000$ W/m²-K
$T_w = 10°C$

The silicon chip has dimensions 0.2 cm × 0.5 cm and conductivity $k_s = 80$ W/m-K. A generation of thermal energy occurs due to losses in the chip; the thermal energy generation can be modeled as being uniformly distributed with a value of $\dot{g}''' = 1 \times 10^8$ W/m³ in the upper 50 percent of the silicon. The chip is thermally isolated from the spreader by a dielectric layer with thickness 0.1 cm and conductivity $k_d = 1$ W/m-K. The spreader has dimension 0.5 cm × 1.4 cm and conductivity $k_{sp} = 45$ W/m-K. The external surfaces are all air cooled with $\bar{h}_a = 10$ W/m²-K and $T_a = 20°C$ except for the bottom surface of the spreader which is water cooled with $\bar{h}_w = 1000$ W/m²-K and $T_w = 10°C$.

(a) Develop a numerical model of the system using FEHT.

(b) Plot the maximum temperature in the system as a function of the number of nodes.

(c) Develop a simple sanity check of your results using a resistance network.

4.36 The figure illustrates a heat exchanger in which hot fluid and cold fluid flow through alternating rows of square channels that are installed in a piece of material.

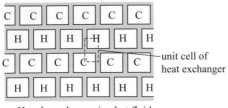

H = channels carrying hot fluid
C = channels carrying hot fluid

You are analyzing this heat exchanger and will develop a model of the unit cell that is shown in the figure.

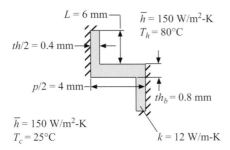

The metal struts separating the square channels form fins. The length of the fin (the half-width of the channel) is $L = 6$ mm and the fin thickness is $th = 0.8$ mm. The thickness of the material separating the channels is $th_b = 0.8$ mm. The distance between adjacent fins is $p = 8$ mm. The channel structure for both sides (hot and cold) is identical. The conductivity of the metal is $k = 12$ W/m-K. The hot fluid has temperature $T_h = 80°C$ and heat transfer coefficient $\bar{h} = 150$ W/m²-K. The cold fluid has temperature $T_c = 25°C$ and the same heat transfer coefficient.
(a) Prepare a numerical model of the unit cell shown using FEHT.
(b) Plot the rate of heat transfer from the hot fluid to the cold fluid within the unit cell as a function of the number of nodes.
(c) Develop a simple model of the unit cell using a resistance network and show that your result from (b) makes sense.

4.37 A 1.4 cm thick stainless steel plate, 20 cm in width and 10 cm in height, is insulated on its

left and bottom sides. The top side is in contact with boiling water at 100°C. The convection coefficient between the plate and the water is estimated to be 1840 W/m²-K. The right side is exposed to 25°C fluid with a convection coefficient of 278 W/m²-K. The plate has a thermal conductivity of 16 W/m-K. It is insulated on its front and back faces so that the heat transfer can be assumed to occur in two dimensions.
(a) Using FEHT, prepare a contour plot of the steady-state temperatures in the plate.
(b) Determine the steady-state rate of heat transfer to the 25°C fluid.

4.38 The figure illustrates a double-paned window. The window consists of two panes of glass each of which is $tg = 0.375$ inch thick and $W = 4$ ft wide by $H = 5$ ft high. The glass panes are separated by an air gap of $g = 0.75$ inch. You may assume that the air is stagnant with $k_a = 0.025$ W/m-K. The glass has conductivity $k_g = 1.4$ W/m-K. The heat transfer coefficient between the inner surface of the inner pane and the indoor air is $\bar{h}_{in} = 10$ W/m²-K and the heat transfer coefficient between the outer surface of the outer pane and the outdoor air is $\bar{h}_{out} = 25$ W/m²-K. You keep your house heated to $T_{in} = 70°F$.

width of window, $W = 4$ ft

$T_{in} = 70°F$
$\bar{h}_{in} = 10$ W/m²-K

$H = 5$ ft

$tg = 0.375$ inch
$tg = 0.375$ inch
$g = 0.75$ inch

$T_{out} = 23°F$
$\bar{h}_{out} = 25$ W/m²-K

$k_a = 0.025$ W/m-K
$k_g = 1.4$ W/m-K

casing

The average heating season where you live lasts about *time* = 130 days and the average outdoor temperature during this time is $T_{out} = 23°F$. You heat with natural gas and pay, on average, $ec = 1.415$ $/therm (a therm is an energy unit $=1.055 \times 10^8$ J).

(a) Calculate the average rate of heat transfer through the double-paned window during the heating season.

(b) How much does the energy lost through the window cost during a single heating season?

There is a metal casing that holds the panes of glass and connects them to the surrounding wall, as shown in the figure. Because the metal casing has a high conductivity, it seems likely that you could lose a substantial amount of heat by conduction through the casing (potentially negating the advantage of using a double-paned window). The geometry of the casing is shown in the figure; note that the casing is symmetric about the center of the window.

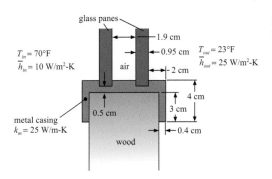

All surfaces of the casing that are adjacent to glass, wood, or the air between the glass panes can be assumed to be adiabatic. The other surfaces are exposed to either the indoor or outdoor air.

(c) Prepare a 2-D thermal analysis of the casing using FEHT. Turn in a print out of your geometry as well as a contour plot of the temperature distribution. What is the rate of energy lost via conduction through the casing per unit length (W/m)?

(d) Show that your numerical model has converged by recording the rate of heat transfer per length for several values of the number of nodes.

(e) How much does the casing add to the cost of heating your house?

4.39 Consider the 2-D problem shown in the figure. Thermal energy is generated throughout

the region at a rate of $\dot{g}''' = 2000$ W/m^3. The material has conductivity $k = 54$ W/m-K. There are six boundaries. Boundary 1 is exposed to convection with $h = 50$ W/m^2-K and $T_\infty = 300$ K. Boundaries 2 and 5 are adiabatic. Boundary 3 is exposed to a heat flux $q'' = 50,000$ W/m^2. Boundary 4 is exposed to a heat flux $q'' = 1000$ W/m^2. Boundary 6 is exposed to convection with $h = 50$ W/m^2-K and $T_\infty = 300$ K.

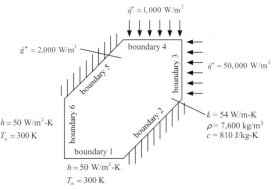

Use FEHT to develop finite element solution to the problem; start with the mesh shown in the figure.

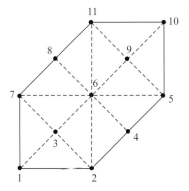

(a) Plot the temperature distribution.

(b) Plot the temperature of node 6 as a function of the total number of nodes.

4.40 A wall in a residential building is 4 m wide and 2.6 m high. The wall is typical residential construction with 4 cm wood studs ($k = 0.16$ W/m-K) that are spaced 40 cm on center. The inside surface is 1.5 cm gypsum board ($k = 0.22$ W/m-K). The outside surface is 2 cm exterior sheathing

($k = 0.055$ W/m-K). The space between the studs is filled with fiberglass insulation ($k = 0.038$). Determine the rate of heat loss from the wall using FEHT. If the studs were not needed, how much would the heat transfer rate be reduced?

4.41 A triangular fin on the cylinder head of a motorcycle engine has a base width of 1 cm and is 5 cm in length. The fin thickness is 2 cm. The cylinder head is made of aluminum with a thermal conductivity of 168 W/m-K, density of 2790 kg/m³ and specific heat of 883 J/kg-K. Using FEHT, estimate the heat transfer rate and fin efficiency for this fin when the base temperature is 180°C and the fin convects to 20°C air with a convection coefficient of 100 W/m²-K.

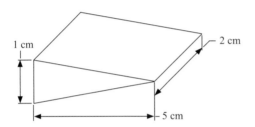

4.42 A long cylindrical pipe having an outer diameter of 10 cm and an inner diameter of 5 cm is made of a material having a thermal conductivity of 2.5 W/m-K, as shown in the figure. An oil is flowing in the pipe at 64°C and the convection coefficient between the oil and the pipe wall is 88 W/m²-K. The bottom outside surface of the pipe is well-insulated and the top convects to air at 25°C with a convection coefficient of 12 W/m²-K. Using FEHT, determine the heat loss per unit length of pipe.

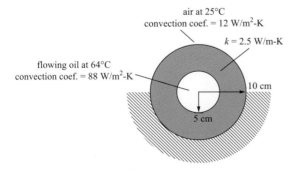

4.43 A wall is made from bricks having a thermal conductivity of 0.416 Btu/hr-ft-R. Each brick is 8 inches long and 3 inches high with two 1 inch holes placed as shown in the figure.

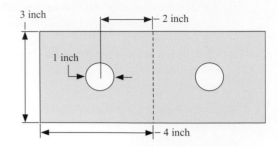

(a) Using FEHT, determine the equivalent thermal resistance of this wall in hr-ft²-R/Btu and K-m²/W. Note that the circulation of the air causes the air in the holes to be at nearly a uniform temperature. Therefore, the holes should be modeled as a fluid/lumped component and the convection coefficient between the air in the hole and the brick surface is 1.4 Btu/hr-ft²-R.

(b) Compare the equivalent resistance determined in part (a) with the resistance that the brick wall would have if there were no holes.

4.44 A long fin made of stainless steel ($k = 14.9$ W/m-K) is 0.25 m in length and 0.04 m thick. The surface of the fin convects to a fluid at 25°C with a convection coefficient of 148 W/m²-K. The base of the fin is maintained at 72°C. The tip is not insulated.

(a) Can this fin be treated as an extended surface problem?

(b) Estimate the rate of heat transfer using an appropriate solution method.

4.45 A radiator panel extends from a spacecraft; both surfaces of the radiator are exposed to space (for the purposes of this problem it is acceptable to assume that space is at 0 K); the emissivity of the surface is $\varepsilon = 1.0$. The plate is made of aluminum ($k = 200$ W/m-K) and has a fluid line attached to it, as shown in the figure. The half-width of the plate is $a = 0.5$ m wide while the height of the plate is $b = 0.75$ m. The thickness of the plate is a design variable and will be varied in this analysis; begin by assuming that the thickness is $th = 1.0$ cm. The fluid lines carry coolant at $T_c = 320$ K. Assume that the fluid temperature is constant although the fluid temperature will actually decrease as it transfers heat to the radiator. The combination of convection and conduction through the panel-to-fluid line mounting leads to an effective heat transfer coefficient of $h = 1000$ W/m^2-K over the 3.0 cm strip occupied by the fluid line.

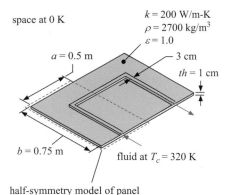

space at 0 K

$k = 200$ W/m-K
$\rho = 2700$ kg/m^3
$\varepsilon = 1.0$

$a = 0.5$ m

3 cm

$th = 1$ cm

$b = 0.75$ m

fluid at $T_c = 320$ K

half-symmetry model of panel

The radiator panel is symmetric about its half-width and the critical dimensions that are required to develop a half-symmetry model of the radiator are shown in the figure. There are three regions associated with the problem that must be defined separately so that the surface conditions can be set differently. Regions 1 and 3 are exposed to space on both sides while Region 2 is exposed to the coolant fluid one side and space on the other; for the purposes of this problem, the effect of radiation to space on the back side of Region 2 is neglected.

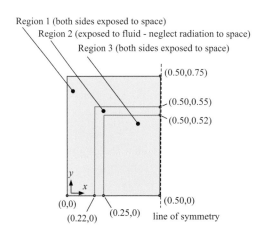

Region 1 (both sides exposed to space)
Region 2 (exposed to fluid - neglect radiation to space)
Region 3 (both sides exposed to space)

(0.50,0.75)

(0.50,0.55)

(0.50,0.52)

y
x
(0,0)

(0.22,0)   (0.25,0)

(0.50,0)   line of symmetry

(a) Prepare a FEHT model that can predict the temperature distribution over the radiator panel.
(b) Export the solution to EES and calculate the total heat transferred from the radiator and the radiator efficiency (defined as the ratio of the radiator heat transfer to the heat transfer

from the radiator if it were isothermal and at the coolant temperature).
(c) Explore the effect of thickness on the radiator efficiency and mass.

4.46 A circular pipe having a 5 cm outer diameter is insulated with a 2.5 cm thick layer of glass wool ($k = 0.035$ W/m-K). However, the insulation is positioned in an eccentric manner such that the top of the pipe only has 1.25 cm of insulation whereas the bottom has 3.75 cm of insulation. The pipe wall is at 220°C and the outside of the insulation convects to air at 25°C with a convection coefficient of 12 W/m^2-K.

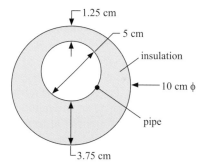

1.25 cm

5 cm

insulation

10 cm ϕ

pipe

3.75 cm

Use FEHT to determine the rate of heat transfer per meter of pipe. Compare the result with a calculation that uses shape factors.

4.47 A tall, square chimney made of brick ($k = 0.72$ W/m-K) has a square hole through which combustion gas at an average temperature of 96°C is flowing. The hole is offset within the chimney, as shown in the figure. The heat transfer coefficient between the gas flowing in the chimney and the walls of the hole is 18 W/m^2-K. The convection coefficient on the outside surfaces of the chimney is 24 W/m^2-K and the outdoor temperature is –6°C.

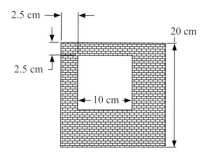

2.5 cm

20 cm

2.5 cm

10 cm

Use FEHT to calculate the rate of heat transfer from the gas per meter of height.

Projects

4.48 A 10 cm × 10 cm square block of clay ($k = 1.2$ W/m²-K) has a 2.5 cm × 2.5 cm square hole in its center. The inner surface is maintained at 150°C while the outer surface is maintained at 30°C. The block is very long, so 2-D heat transfer can be assumed.

(a) Determine the rate of heat transfer from the inner to the outer surface using a shape factor solution.

(b) Set up a numerical finite-difference model for this problem. Using the model, compare the rate of heat transfer from the inner to the outer surface with the result obtained in part (a). Vary the number of nodes to ensure that a sufficient number are being employed.

4.49 Power plants operating on the Rankine cycle need to transfer low-temperature heat from the cycle to the environment in order to operate continuously. This heat transfer process is most commonly accomplished by evaporating water in a cooling tower. However, increasingly the resulting water usage is a concern, particularly for solar power plants that operate in the desert where water is scarce. A heat rejection system has been proposed using water storage coupled to night-sky radiators. This problem focuses on the analysis of a night-sky radiator, shown in the figure. The radiator is an aluminum plate with thickness 2 mm that has water tubes connected to it spaced 30 cm apart. It is 10 m long in the flow direction. Using symmetry, the centerline between tubes is adiabatic and therefore only a 15 cm width needs to be analyzed. However, note that each tube is attached to a plate on its left and right sides so the heat transfer rate from the water to the plate calculated using one plate, as shown in the figure, should be multiplied by a factor of 2. The bottom of the plate and all of its edges except the edge connected to the tube can be considered to be adiabatic. Assume that the temperature difference between the top and bottom of the plate is negligible so that steady-state heat transfer occurs in the plate only parallel and perpendicular to the direction of the water flow. The top of the plate experiences both convection to ambient air at 25°C with a convection coefficient of

15 W/m²-K and also radiates to the night sky at a rate per unit area given by

$$\dot{q}'' = \sigma\left(T^4 - T_{sky}^4\right)$$

where T is the local temperature of the plate and T_{sky} is 281 K (5°C). In your analysis, assume that the temperature of the plate at the point it touches the tube is the same as the water temperature at that location due to a high heat transfer coefficient between the water and the tube and the high-conductivity solder joint. Using a finite difference numerical method, estimate the heat transfer rate per unit area of radiator to the environment and the outlet temperature of the water. Check to ensure that you have sufficiently resolved the mesh so that your answer is not affected by the mesh size.

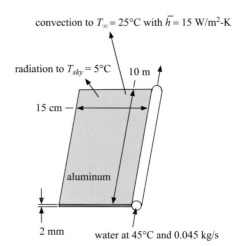

convection to $T_\infty = 25°C$ with $\bar{h} = 15$ W/m²-K

radiation to $T_{sky} = 5°C$

10 m

15 cm

aluminum

2 mm   water at 45°C and 0.045 kg/s

4.50 A small heater block provides heat to the backside of a fin array that must be tested.

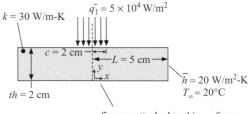

$k = 30$ W/m-K $\dot{q}_1'' = 5 \times 10^4$ W/m²

$c = 2$ cm $L = 5$ cm

y
x

$th = 2$ cm

$\bar{h} = 20$ W/m²-K
$T_\infty = 20°C$

fins are attached to this surface

The fins are installed on a base plate that has half-width of $L = 5$ cm, thickness of $th = 2$ cm, and width $W = 10$ cm (into the page). The base plate material has conductivity $k = 30$ W/m-K. The edge of the base plate (at $x = L$) is exposed to air at $T_\infty = 20°C$ with heat transfer coefficient $\bar{h} = 20$ W/m^2-K. The middle of the plate (at $x = 0$) is a line of symmetry and can be modeled as being adiabatic. The bottom of the plate (at $y = 0$) has an array of circular fins installed on it. The fins have the same conductivity as the base plate. Each fin has radius $r_f = 1$ mm and length $L_f = 1.4$ cm. The fins are installed in a hexagonal close packed array with center-to-center distance $S_f = 4$ mm between fins. The fin surface and the unfinned base material on the bottom of the plate are exposed to fluid at $T_f = 35°C$ with heat transfer coefficient $\bar{h}_f = 20$ W/m^2-K. The top of the plate (at $y = th$) is exposed to the heat flux from the heater block. The heat flux is distributed according to:

$$\dot{q}''_{y=th} = \begin{cases} \dot{q}''_1 & \text{if } x < c \\ 0 & \text{if } c < x < L \end{cases},$$

where $\dot{q}''_1 = 5 \times 10^4$ W/m^2 and $c = 2$ cm.

(a) Determine an effective heat transfer coefficient that can be applied to the surface at $y = th$ in your model in order to capture the combined effect of the fins and the un-finned base area.

(b) Develop a numerical solution to the problem.

(c) Prepare a plot showing the temperature as a function of x at various values of y.

(d) The goal of the base plate is to provide a uniform heat flow to each fin. Assess the performance of the base plate by plotting the rate of heat flux transferred to the fluid as a function of x at $y = 0$.

5 | Lumped Transient Problems

Chapters 1 through 4 discuss steady-state problems, i.e., problems in which temperature depends on position (e.g., x and y) but does not change with time (t). Steady-state problems become progressively more difficult as the dimensionality of the problem increases from 1-D to 2-D (and even to 3-D, although this was not covered). This chapter begins the consideration of *transient* conduction problems, i.e., problems where temperature depends on time. This chapter specifically considers the simplest transient problem, one in which the temperature approximately depends *only* on time and not on position.

5.1 The Lumped Capacitance Assumption

Conduction problems in which temperature depends on time are referred to as *transient* (vs. steady state). In general, temperature can be a function of time as well as position; the most complex situation being one that is both transient and 3-D: $T(x, y, z, t)$. The simplest transient problem is one in which the temperature depends on time but *not* on position: $T(t)$. This type of problem is *zero*-dimensional (0-D); that is, the temperature does not vary in any spatial coordinate direction. This approximation is often referred to as the **lumped capacitance assumption** and it is appropriate for an object that is thin and conductive.

The use of the lumped capacitance assumption allows a control volume to be defined that includes the entire object (as it is all assumed to be at the same temperature at any given time). An energy balance will include all of the relevant energy transfers (e.g., convection and radiation) as well as the rate of energy storage, which is related to the rate at which the temperature is changing. The result will be a first-order differential equation that governs the temperature of the object as a function of time. A single boundary condition, typically the initial temperature of the object, is required to obtain a solution using either analytical or numerical techniques.

5.1.1 The Biot Number

The lumped capacitance approximation is similar to the extended surface approximation that was discussed in Section 3.1 and it is justified in a similar way. The resistance to conduction within the object is neglected as being small relative to the resistance to heat transfer from the surface of the object to the surroundings. Therefore, the lumped capacitance approximation is justified using an appropriate **Biot number**, defined as the ratio of the resistance to conduction within the object to the resistance to heat transfer from the surface:

$$Bi = \frac{\text{resistance to internal conduction}}{\text{resistance from surface to surroundings}} = \frac{R_{cond,int}}{R_{surface-to-surroundings}}. \tag{5.1}$$

It is important to understand that the resistance to internal conduction can only be approximated, but not calculated exactly. The approximate value of the thermal resistance to conduction within an object is given by:

$$R_{cond,int} = \frac{(\text{distance to conduct})}{k(\text{area for conduction})}. \tag{5.2}$$

The conduction length and area for conduction in Eq. (5.2) are not always obvious. In some cases it is relatively easy to make a good estimate by keeping in mind that thermal energy will be conducted along the easiest (i.e., shortest) path. For example, consider a regularly shaped object like a thin square plate of thickness th and side width W. In this case, the conduction length could reasonably be estimated as the half-thickness of the plate ($th/2$) and the area as $W \times W$. For irregularly shaped objects, the value of the conduction length and area may be less obvious. In this case, it is common practice to estimate the conduction length as the ratio of the volume of the object (V) to its surface area (A_s):

$$\text{distance to conduct} = L_{cond} \approx \frac{V}{A_s} \tag{5.3}$$

and the conduction area as the surface area:

$$\text{area for conduction} \approx A_s. \tag{5.4}$$

Setting the conduction area equal to the surface area suggests that energy is everywhere seeking the shortest path to the surface and the distance of this path is, on average, L_{cond} from Eq. (5.3).

For the case where only convection occurs from the surface of the object, the resistance to heat transfer from the surface of the object to the surroundings is given by:

$$R_{surface-to-surroundings} = R_{conv} = \frac{1}{\bar{h} A_s} \tag{5.5}$$

and the Biot number that should be calculated is:

$$Bi = \frac{R_{cond,int}}{R_{conv}} = \frac{L_{cond}\,\bar{h}}{k}. \tag{5.6}$$

Note that it is often the case that heat transfer from the surface of the object will be resisted by mechanisms other than or in addition to convection (e.g., radiation or conduction through a thin insulating layer). These additional resistances should be included in the estimate of $R_{surface-to-surroundings}$ that appears in the denominator of an appropriately defined Biot number.

If the Biot number is much less than unity, then the lumped capacitance assumption is justified. A common criterion is that the Bi must be less than 0.1. However, this is certainly a case where engineering judgment is required based on the level of accuracy that is required for the model.

5.1.2 The Lumped Capacitance Time Constant

Before we begin developing detailed solutions to lumped capacitance problems using either analytical or numerical techniques, it is worth discussing the concept of a lumped capacitance time concept. The **lumped capacitance time constant** (τ_{lumped}) is the product of the thermal resistance to heat transfer from the surface of the object to the surroundings ($R_{surface-to-surroundings}$) and the **thermal capacitance** of the object (C):

$$\tau_{lumped} = R_{surface-to-surroundings}\,C. \tag{5.7}$$

The thermal capacitance is equal to the product of the object's mass (m) and its specific heat capacity (c):

$$C = m\,c. \tag{5.8}$$

The units of C is J/K and indeed the thermal capacitance is a measure of how much energy is required in order to change the temperature of the object. In the simple situation where the object experiences only convection we may substitute Eqs. (5.5) and (5.8) into Eq. (5.7) to obtain:

$$\tau_{lumped} = \frac{m\,c}{\bar{h} A_s}. \tag{5.9}$$

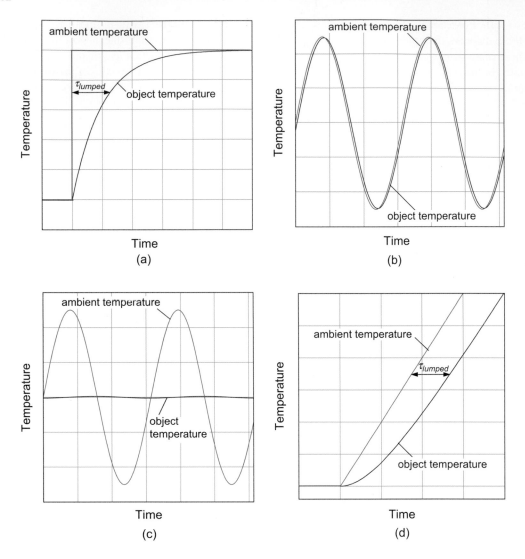

Figure 5.1 Approximate temperature response for an object subjected to (a) a step change in its ambient temperature, (b) and (c) an oscillatory ambient temperature where (b) the frequency of the oscillation is much less than the inverse of the time constant and (c) the frequency of the oscillation is much greater than the inverse of the time constant, and (d) a ramped ambient temperature.

The lumped capacitance time constant for more complex situations can be calculated by modifying the resistance term in Eq. (5.7), $R_{surface\text{-}to\text{-}surroundings}$. For example, if the object experienced both convection and radiation from its surface, then an appropriate thermal resistance would be formulated as the parallel combination of a convective and radiative resistance:

$$R_{surface-to-surroundings} = \left[\frac{1}{R_{conv}} + \frac{1}{R_{rad}} \right]^{-1}. \tag{5.10}$$

The lumped capacitance time constant is analogous to the electrical time constant associated with an R-C circuit and many of the concepts that may be familiar from electrical circuits can be applied to lumped capacitance transient heat transfer problems.

A quick estimate of the lumped capacitance time constant for a problem can provide substantial insight into the behavior of the system. The lumped capacitance time constant is, approximately, the amount of time that it will take the object to respond to any change in its thermal conditions. For example, if the object is subjected to a step change in the ambient temperature, T_∞, then its temperature will exponentially approach the new temperature with a time constant of τ_{lumped}, as shown in Figure 5.1(a). If the object is subjected to an oscillatory ambient temperature (e.g., within an engine cylinder or some other cyclic device) then the temperature of the object will follow the ambient temperature nearly exactly if the period of oscillation is much greater than the time constant (i.e., if the frequency is much less than the inverse of the time constant). In the opposite extreme, the object's temperature will be essentially constant if the period of oscillation is much less than the time constant (i.e., if the frequency is much greater than the inverse of the time constant). These extremes in behavior are shown in Figure 5.1(b) and (c). If the object is subjected to a ramped (i.e., linearly increasing) ambient temperature, then its temperature will tend to increase linearly as well but its response will be delayed by approximately one time constant as shown in Figure 5.1(d).

In steady-state problems we often obtain a physical understanding of the situation by estimating and comparing thermal resistances. In transient problems it is usually wise to begin by calculating time constants in order to develop a similar, physical understanding of the situation.

Example 5.1

A platinum resistance thermometer (PRT) is placed in an oven. The air temperature in the oven increases linearly from 20°C to 200°C over the course of 1 minute and then decreases linearly back to 20°C over another minute, as shown in Figure 1. The initial temperature of the PRT is $T_{ini} = 20$°C.

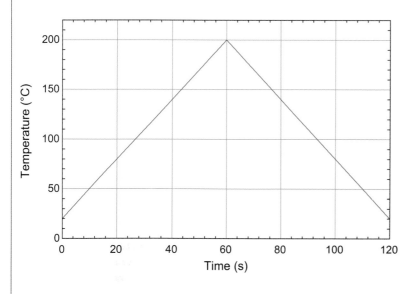

Figure 1 Ambient air temperature in oven.

The PRT is a cylinder with diameter $D = 5$ mm and length $L = 20$ mm. The heat transfer coefficient between the surface of the PRT and the ambient air temperature is $\bar{h} = 30$ W/m²-K.

Determine:
- whether the lumped capacitance assumption is appropriate, and
- the lumped capacitance time constant.

Continued

Example 5.1 (cont.)

Known Values

The known values are indicated on the sketch of the PRT shown in Figure 2.

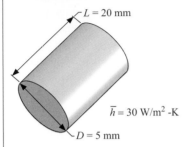

$L = 20$ mm

$\bar{h} = 30$ W/m^2 -K

$D = 5$ mm

Figure 2 PRT in an oven.

The properties of platinum are obtained from Appendix A: $k = 73$ W/m-K, $\rho = 21{,}447$ kg/m^3, and $c = 132.9$ J/kg-K.

Assumptions

- The properties of platinum do not depend on temperature.
- Radiation is neglected.

Analysis and Solution

The surface area of the PRT is computed according to:

$$A_s = 2\frac{\pi D^2}{4} + \pi D L = \frac{\pi}{2}\left|\frac{(0.005)^2 \text{ m}^2}{}\right| + \frac{\pi}{}\left|0.005 \text{ m}\right|\frac{0.02 \text{ m}}{} = 3.53 \times 10^{-4}\text{m}^2. \tag{1}$$

The volume of the PRT is computed according to:

$$V = \frac{\pi D^2}{4}L = \frac{\pi}{4}\left|\frac{(0.005)^2 \text{ m}^2}{}\right|\frac{0.02 \text{ m}}{} = 3.93 \times 10^{-7}\text{m}^3. \tag{2}$$

In order to compute an internal conduction resistance we need to determine both a conduction length and area. The conduction length is computed using Eq. (5.3):

$$L_{cond} = \frac{V}{A_s} = \frac{3.93 \times 10^{-7}\text{m}^3}{}\left|\frac{}{3.53 \times 10^{-4} \text{ m}^2}\right| = 0.0011 \text{ m } (1.1 \text{ mm}). \tag{3}$$

Note that the conduction length is approximately half the radius of the PRT which makes physical sense in terms of the average distance that energy must be conducted in order to move into or out of the object. The area for conduction is computed according to Eq. (5.4):

Example 5.1 (cont.)

$$A_c = A_s = 3.53 \times 10^{-4} \text{ m}^2. \tag{4}$$

The resistance to internal conduction is then:

$$R_{cond, int} = \frac{L_{cond}}{k \; A_c} = \frac{0.0011 \text{ m}}{} \left| \frac{\text{m K}}{73 \text{ W}} \right| \frac{}{3.53 \times 10^{-4} \text{ m}^2} = 0.043 \frac{\text{K}}{\text{W}}. \tag{5}$$

The resistance to convection from the surface of the PRT to the surroundings is:

$$R_{conv} = \frac{1}{\overline{h} A_s} = \frac{\text{m}^2 \text{ K}}{30 \text{ W}} \left| \frac{}{3.53 \times 10^{-4} \text{ m}^2} \right. = 94.3 \frac{\text{K}}{\text{W}}. \tag{6}$$

Clearly the resistance to conduction within the PRT is extremely small compared to the resistance to convection from its surface and therefore any temperature gradients that occur within the PRT will be much smaller than the temperature difference from the surface to the surroundings. The Biot number is:

$$Bi = \frac{R_{cond, int}}{R_{conv}} = \frac{0.043 \text{ K}}{\text{W}} \left| \frac{\text{W}}{94.3 \text{ K}} \right. = \boxed{0.00046}. \tag{7}$$

Because the Biot number is so much less than unity, the lumped capacitance approximation is valid. The total heat capacity of the PRT is:

$$C = mc = \rho V c = \frac{21,447 \text{ kg}}{\text{m}^3} \left| \frac{3.93 \times 10^{-7} \text{ m}^3}{} \right| \frac{132.9 \text{ J}}{\text{kg K}} = 1.12 \frac{\text{J}}{\text{K}}. \tag{8}$$

The lumped capacitance time constant is:

$$\tau_{lumped} = R_{conv} \; C = \frac{94.3 \text{ K}}{\text{W}} \left| \frac{1.12 \text{ J}}{\text{K}} \right| \frac{\text{W s}}{\text{J}} = \boxed{106 \text{ s}}. \tag{9}$$

Discussion

The lumped capacitance time constant computed in Eq. (9) is approximately equal to the time required for the oven temperature to increase and decrease, as shown in Figure 1. This suggests that the PRT will not be able to respond to the changes in the oven temperature with any degree of accuracy. The sensor is not a good choice for this application. Without resorting to solving a differential equation or carrying out a numerical analysis, it is possible to anticipate the sensor response approximately based only on the knowledge of the time constant. This approximate response is sketched in Figure 3. Notice that the sensor lags the oven temperature by more than a minute. Also, the temperature of the sensor is always increasing when the temperature of the air is above the sensor temperature and decreasing otherwise. The rate of increase (or decrease) is proportional to the temperature difference. These qualitative characteristics are all evident in Figure 3.

Continued

Example 5.1 (cont.)

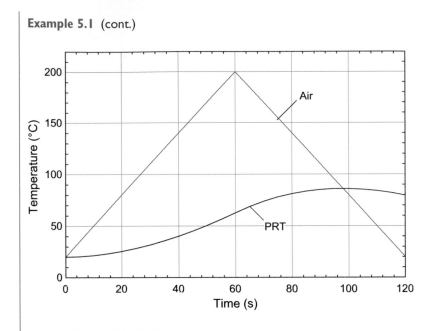

Figure 3 Sketch of the PRT response.

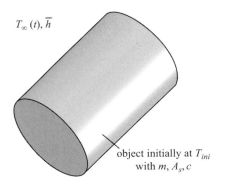

Figure 5.2 An object exposed to a time varying fluid temperature.

5.2 Analytical Solutions

The development of an analytical solution to a lumped capacitance problem will be discussed in the context of the object shown in Figure 5.2. We will assume that the object has a small Biot number that can therefore be modeled using the lumped capacitance assumption. The object has mass m, surface area A_s, and is composed of a material with specific heat capacity c. The object is initially at a uniform temperature T_{ini} and is exposed to a time varying fluid temperature, $T_\infty(t)$, with heat transfer coefficient, $\bar{h}$. We are interested in predicting the temperature of the object as a function of time, $T(t)$.

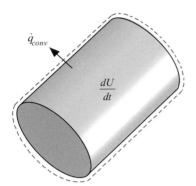

Figure 5.3 Control volume used to derive the differential equation.

5.2.1 Deriving the Differential Equation

In order to derive the differential equation, it is necessary to do an energy balance on a control volume that includes material that is all at the same temperature at an instant of time. For the problems considered in Chapters 2 through 4 these control volumes were necessarily differentially small in one or more coordinate directions. However, because the lumped capacitance assumption requires that the entire object be modeled as being at a single temperature, the appropriate control volume encompasses the *entire object*, as shown in Figure 5.3.

An energy balance on the control volume requires that

$$IN = OUT + STORED. \tag{5.11}$$

For the first time in this book we need to include the *STORED* term because the amount of energy in the system is changing with time for this transient problem. Energy can cross the control surface as convection heat transfer and energy can be stored within the control volume as the temperature of the object changes:

$$0 = \dot{q}_{conv} + \frac{dU}{dt}. \tag{5.12}$$

In Eq. (5.12) $\dot{q}_{conv}$ is the rate of convective heat transfer from the surface and U is the total energy associated with the object. Equation (5.12) is the differential equation written in terms of energy. In order to obtain a differential equation in terms of temperature we must substitute a rate equation for the convection term and relate the energy of the object to its temperature. The rate equation for convective heat transfer is Newton's Law of Cooling:

$$\dot{q}_{conv} = \bar{h} A_s [T - T_\infty(t)]. \tag{5.13}$$

Note that the convection heat transfer was defined as being from the object to the surroundings in Figure 5.3 and therefore it took the role of an *outflow* of energy in Eq. (5.12) and is written in terms of $(T - T_\infty)$ in Eq. (5.13). The total energy is the product of the mass and the specific internal energy (u):

$$U = mu. \tag{5.14}$$

Substituting Eq. (5.14) into the rate of energy storage provides:

$$\frac{dU}{dt} = \frac{d(mu)}{dt}. \tag{5.15}$$

Recognizing that the mass of the object is constant:

$$\frac{dU}{dt} = m\frac{du}{dt}.$$ (5.16)

For solids or incompressible fluids, Eq. (5.16) can be accurately represented as:

$$\frac{dU}{dt} = m\underbrace{\frac{du}{dT}}_{c}\frac{dT}{dt} = mc\frac{dT}{dt},$$ (5.17)

where c is the specific heat capacity of the object material. Substituting Eqs. (5.17) and (5.13) into Eq. (5.12) leads to:

$$0 = \bar{h}A_s[T - T_\infty(t)] + mc\frac{dT}{dt}.$$ (5.18)

If c is assumed to be constant, Eq. (5.18) is a first-order linear differential equation. Equation (5.18) can be rearranged to put it into a standard form where all of the temperature terms are on the left side of the equation and the highest-order derivative is isolated:

$$\frac{dT}{dt} + \underbrace{\frac{\bar{h}A_s}{mc}}_{\underset{\tau_{lumped}}{1}}T = \frac{\bar{h}A_s}{mc}T_\infty(t).$$ (5.19)

Note that the group of variables that multiply the temperatures on the left and right sides of Eq. (5.19) must have units of inverse time; this group is equal to the inverse of the lumped time constant, τ_{lumped}, that is discussed in Section 5.1.2:

$$\tau_{lumped} = \frac{mc}{\bar{h}A_s}.$$ (5.20)

Substituting Eq. (5.20) into Eq. (5.19) leads to:

$$\frac{dT}{dt} + \frac{T}{\tau_{lumped}} = \frac{T_\infty(t)}{\tau_{lumped}}.$$ (5.21)

Equation (5.21) is first order and therefore a single boundary condition is required. In a transient problem, the **initial condition** is typically known. The initial condition is the temperature at time $t = 0$:

$$T_{t=0} = T_{ini}.$$ (5.22)

If additional heat transfer mechanisms or thermal loads are included in the problem (e.g., radiation or volumetric generation of thermal energy) then the governing differential equation will be different than Eq. (5.21). However, the steps associated with the derivation will remain the same.

5.2.2 Solving the Differential Equation

Equation (5.21) is a first-order linear (assuming τ_{lumped} is constant) differential equation. Unless $T_\infty(t)$ is a constant, Eq. (5.21) is not separable. There are several techniques available for solving such an ODE; here we will use the same method that was discussed in Section 3.2 for solving extended surface problems.

The first step is to split the solution into the sum of a homogeneous solution (T_h) and a particular (or nonhomogeneous) solution (T_p):

$$T = T_h + T_p.$$ (5.23)

Substituting Eq. (5.23) into Eq. (5.21) provides:

$$\frac{d(T_h + T_p)}{dt} + \frac{(T_h + T_p)}{\tau_{lumped}} = \frac{T_\infty(t)}{\tau_{lumped}} \tag{5.24}$$

or

$$\underbrace{\frac{dT_h}{dt} + \frac{T_h}{\tau_{lumped}}}_{\substack{= 0 \\ \text{for homogeneous} \\ \text{differential equation}}} + \underbrace{\frac{dT_p}{dt} + \frac{T_p}{\tau_{lumped}} = \frac{T_\infty(t)}{\tau_{lumped}}}_{\substack{\text{whatever is left over must be the} \\ \text{particular differential equation}}} \tag{5.25}$$

Remove the homogeneous differential equation for T_h from Eq. (5.25):

$$\frac{dT_h}{dt} + \frac{T_h}{\tau_{lumped}} = 0 \tag{5.26}$$

leaving behind the particular differential equation for T_p:

$$\frac{dT_p}{dt} + \frac{T_p}{\tau_{lumped}} = \frac{T_\infty(t)}{\tau_{lumped}}. \tag{5.27}$$

The homogeneous differential equation, Eq. (5.26), is separable and therefore can be solved relatively easily. Separating the temperature and time components leads to:

$$\frac{dT_h}{T_h} = -\frac{dt}{\tau_{lumped}}. \tag{5.28}$$

Integrating once leads to:

$$\int \frac{dT_h}{T_h} = -\int \frac{dt}{\tau_{lumped}} \tag{5.29}$$

$$\ln(T_h) = -\frac{t}{\tau_{lumped}} + C_1, \tag{5.30}$$

where C_1 is a constant of integration. Solving Eq. (5.30) for T_h leads to:

$$T_h = \exp\left(-\frac{t}{\tau_{lumped}} + C_1\right) \tag{5.31}$$

which can be rearranged to provide:

$$T_h = \exp\left(-\frac{t}{\tau_{lumped}}\right) \underbrace{\exp(C_1)}_{C_2}. \tag{5.32}$$

Recognizing that the exponential of the undetermined constant C_1 is just another undetermined constant (C_2), Eq. (5.32) can be written as:

$$T_h = C_2 \exp\left(-\frac{t}{\tau_{lumped}}\right). \tag{5.33}$$

Equation (5.33) is the solution to the homogeneous differential equation.

The solution to the particular differential equation, Eq. (5.27), can often be obtained using the method of undetermined coefficients. A functional form of the particular solution should be assumed, where each of the functions involved is multiplied by an undetermined coefficient. Substitution of the assumed solution into the particular differential equation allows the selection of the coefficients. The assumed form of the particular solution should be based on the form of the right side of the particular differential equation and should include whatever functions are included there as well as their derivatives.

Step Change in Ambient Temperature

For example, consider the particular solution that would result if the ambient temperature were a constant:

$$T_\infty(t) = T_\infty. \tag{5.34}$$

Substituting Eq. (5.34) into Eq. (5.27) leads to the particular differential equation:

$$\frac{dT_p}{dt} + \frac{T_p}{\tau_{lumped}} = \frac{T_\infty}{\tau_{lumped}}. \tag{5.35}$$

Applying the method of undetermined coefficients to this particular differential equation requires examination of the right side (the nonhomogeneous part) of the particular differential equation. In Eq. (5.35), the right side is a constant. Therefore, our assumed form of the particular solution should include a constant and its derivative (which is zero), multiplied by an undetermined coefficient:

$$T_p = C_3. \tag{5.36}$$

Substituting Eq. (5.36) into Eq. (5.35) leads to:

$$0 + \frac{C_3}{\tau_{lumped}} = \frac{T_\infty}{\tau_{lumped}}. \tag{5.37}$$

Solving Eq. (5.37) for the undetermined coefficient:

$$C_3 = T_\infty. \tag{5.38}$$

So the particular solution to Eq. (5.35) is:

$$T_p = T_\infty. \tag{5.39}$$

Substituting Eqs. (5.39) and (5.33) into Eq. (5.23) provides the general solution:

$$T = C_2 \exp\left(-\frac{t}{\tau_{lumped}}\right) + T_\infty. \tag{5.40}$$

Equation (5.40) is a solution to the differential equation regardless of the value of C_2; this coefficient must be selected in order to satisfy the boundary condition. The boundary condition is the initial condition, Eq. (5.22):

$$T_{t=0} = C_2 \exp\left(-\frac{0}{\tau_{lumped}}\right) + T_\infty = T_{ini}. \tag{5.41}$$

Solving Eq. (5.41) for C_2 provides:

$$C_2 = T_{ini} - T_\infty, \tag{5.42}$$

which leads to the solution:

$$T = (T_{ini} - T_\infty) \exp\left(-\frac{t}{\tau_{lumped}}\right) + T_\infty. \tag{5.43}$$

Equation (5.43) can be rearranged as:

$$\frac{(T - T_\infty)}{(T_{ini} - T_\infty)} = \exp\left(-\frac{t}{\tau_{lumped}}\right). \tag{5.44}$$

Ramped Ambient Temperature

A more complex particular differential equation is obtained if T_∞ is a function of time. For example, a ramped ambient temperature would start at T_{ini} and then increase linearly with time according to:

$$T_\infty(t) = T_{ini} + \beta t. \tag{5.45}$$

The homogeneous solution remains the same, but substituting Eq. (5.45) into Eq. (5.27) leads to a different particular differential equation:

$$\frac{dT_p}{dt} + \frac{T_p}{\tau_{lumped}} = \frac{(T_{ini} + \beta t)}{\tau_{lumped}}. \tag{5.46}$$

Examination of the right side of Eq. (5.46) shows that the nonhomogeneous terms now include both a linear term and a constant. Therefore, the assumed form of the particular solution should include a linear term and a constant as well as their derivatives (a constant and zero, respectively) each multiplied by undetermined coefficients.

$$T_p = C_3 + C_4 t. \tag{5.47}$$

Substituting Eq. (5.47) into Eq. (5.46) provides:

$$C_4 + \frac{C_3}{\tau_{lumped}} + \frac{C_4}{\tau_{lumped}} t = \frac{T_{ini}}{\tau_{lumped}} + \frac{\beta}{\tau_{lumped}} t. \tag{5.48}$$

The undetermined constants C_3 and C_4 must be selected such that Eq. (5.48) is satisfied at all times. This can only be true if the constant terms and the linear terms on the two sides of the equation separately balance each other correctly. Therefore, we obtain one equation for the constant terms:

$$C_4 + \frac{C_3}{\tau_{lumped}} = \frac{T_{ini}}{\tau_{lumped}} \tag{5.49}$$

and a second equation for the linear terms:

$$\frac{C_4}{\tau_{lumped}} t = \frac{\beta}{\tau_{lumped}} t. \tag{5.50}$$

Solving Eq. (5.50) provides:

$$C_4 = \beta. \tag{5.51}$$

Substituting this result into Eq. (5.49):

$$C_3 = T_{ini} - \beta \tau_{lumped}. \tag{5.52}$$

Substituting Eqs. (5.51) and (5.52) into Eq. (5.47):

$$T_p = T_{ini} - \beta \tau_{lumped} + \beta t. \tag{5.53}$$

Substituting Eqs. (5.53) and (5.33) into Eq. (5.23) leads to the general solution:

$$T = C_2 \exp\left(-\frac{t}{\tau_{lumped}}\right) + T_{ini} - \beta \tau_{lumped} + \beta t. \tag{5.54}$$

The constant C_2 is selected in order to enforce the initial condition:

$$T_{t=0} = C_2 + T_{ini} - \beta \tau_{lumped} = T_{ini}. \tag{5.55}$$

Solving Eq. (5.55) for C_2 leads to:

$$C_2 = \beta \tau_{lumped}. \tag{5.56}$$

Substituting Eq. (5.56) into Eq. (5.54) provides the final solution:

$$T = T_{ini} + \beta t + \beta \tau_{lumped} \left[\exp\left(-\frac{t}{\tau_{lumped}}\right) - 1 \right]. \tag{5.57}$$

Example 5.2

Blueberries are being processed in a flash freezer. The purpose of the flash freezer is to quickly reduce the temperature of the fruit to below the freezing temperature. Each blueberry has diameter $D = 1$ cm and properties $k = 0.63$ W/m-K, $c = 4200$ J/kg-K, and $\rho = 1000$ kg/m^3. The blueberries are initially at $T_{ini} = 20°C$ when they are placed on a tray and the freezer is activated. The freezer produces a flow of cold air. The temperature of the air produced by the freezer (T_∞) begins at T_{ini} and exponentially approaches a

Continued

Example 5.2 (cont.)

temperature that is reduced by $\Delta T = 100°C$. The time variation of the air flowing through the freezer can be described as an exponential:

$$T_\infty(t) = T_{ini} - \Delta T\left[1 - \exp\left(-\frac{t}{\tau_{ff}}\right)\right],$$ (1)

where $\tau_{ff} = 3$ minutes is the time constant associated with the freezer activation process (*not* the time constant of the blueberries). The heat transfer coefficient between the surface of the blueberry and the air is $\bar{h} = 36$ W/m²-K.

Determine the temperature of a representative blueberry as a function of time.

Known Values

The known values are indicated on the sketch shown in Figure 1.

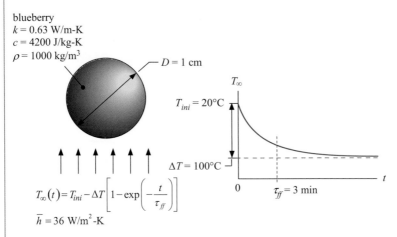

blueberry
$k = 0.63$ W/m-K
$c = 4200$ J/kg-K
$\rho = 1000$ kg/m³

$D = 1$ cm

$T_{ini} = 20°C$

$\Delta T = 100°C$

$T_\infty(t) = T_{ini} - \Delta T\left[1 - \exp\left(-\frac{t}{\tau_{ff}}\right)\right]$

$\bar{h} = 36$ W/m²-K

$\tau_{ff} = 3$ min

Figure I Blueberry in a flash freezer.

Assumptions

- The properties of fruit do not depend on temperature.
- The latent heat associated with the freezing is ignored.
- The surface area and average convection heat transfer coefficient associated with each blueberry are unaffected by the presence of other blueberries.
- The fruit is at a spatially uniform temperature. This is the lumped capacitance assumption which should be verified by computing a Biot number. The volume and surface area of a berry are computed:

$$V = \frac{4\pi}{3}\left(\frac{D}{2}\right)^3 = \frac{4\pi}{3}\left(\frac{0.01 \text{ m}}{2}\right)^3 = 5.24 \times 10^{-7} \text{ m}^3$$ (2)

$$A_s = 4\pi\left(\frac{D}{2}\right)^2 = 4\pi\left(\frac{0.01 \text{ m}}{2}\right)^2 = 3.14 \times 10^{-4} \text{ m}^2.$$ (3)

The conduction length and cross-sectional area are computed:

$$L_{cond} = \frac{V}{A_s} = \frac{5.24 \times 10^{-7} \text{ m}^3}{3.14 \times 10^{-4} \text{ m}^2} = 0.00167 \text{ m}$$ (4)

$$A_c = A_s = 3.14 \times 10^{-4} \text{ m}^2.$$ (5)

Example 5.2 (cont.)

The resistance to internal conduction is computed:

$$R_{cond, int} = \frac{L_{cond}}{k\, A_c} = \frac{0.00167 \text{ m}}{} \left| \frac{\text{m K}}{0.63 \text{ W}} \right| \frac{}{3.14 \times 10^{-4} \text{ m}^2} = 8.42 \frac{\text{K}}{\text{W}}. \tag{6}$$

The resistance to convection is:

$$R_{conv} = \frac{1}{\bar{h}\, A_s} = \frac{\text{m}^2 \text{ K}}{36 \text{ W}} \left| \frac{}{3.14 \times 10^{-4} \text{ m}^2} \right. = 88.4 \frac{\text{K}}{\text{W}}. \tag{7}$$

The Biot number is computed according to:

$$Bi = \frac{R_{cond, int}}{R_{conv}} = \frac{8.42 \text{ K/W}}{88.4 \text{ K/W}} = 0.095. \tag{8}$$

The Biot number is small relative to unity and so the lumped capacitance approximation is justified. The Biot number is approximately 0.1 which can be interpreted as indicating that the spatial temperature variation within the blueberry (e.g., from its center to its edge) will be approximately 10 percent of the temperature difference from the surface of the blueberry to the surroundings.

Analysis

The derivation of the governing ordinary differential equation follows the steps outlined in Section 5.2.1 and leads to the same result:

$$\frac{dT}{dt} + \frac{T}{\tau_{lumped}} = \frac{T_\infty(t)}{\tau_{lumped}}, \tag{9}$$

where τ_{lumped} is the lumped capacitance time constant:

$$\tau_{lumped} = \frac{V \rho c}{\bar{h}\, A_s}. \tag{10}$$

Equation (9) is divided into a homogeneous and particular differential equation, as discussed in Section 5.2.2. The homogeneous solution is:

$$T_h = C_1 \exp\left(-\frac{t}{\tau_{lumped}} \right) \tag{11}$$

where C_1 is an undetermined constant of integration. The particular differential equation is:

$$\frac{dT_p}{dt} + \frac{T_p}{\tau_{lumped}} = \frac{T_\infty(t)}{\tau_{lumped}}. \tag{12}$$

Substituting Eq. (1) into Eq. (12) provides:

$$\frac{dT_p}{dt} + \frac{T_p}{\tau_{lumped}} = \frac{T_{ini}}{\tau_{lumped}} - \frac{\Delta T}{\tau_{lumped}} \left[1 - \exp\left(-\frac{t}{\tau_{ff}} \right) \right]. \tag{13}$$

Equation (13) is solved using the method of undetermined coefficients. The right side of Eq. (13) is the nonhomogeneous component of the equation and it includes both a constant and an exponential function. Therefore our assumed functional form for the solution should include both a constant and its derivative (zero) as well as an exponential and its derivative (which is the same exponential). This leads to:

$$T_p = C_2 + C_3 \exp\left(-\frac{t}{\tau_{ff}} \right), \tag{14}$$

where C_2 and C_3 are undetermined coefficients. Equation (14) is substituted into Eq. (13):

$$-\frac{C_3}{\tau_{ff}} \exp\left(-\frac{t}{\tau_{ff}} \right) + \frac{C_2}{\tau_{lumped}} + \frac{C_3}{\tau_{lumped}} \exp\left(-\frac{t}{\tau_{ff}} \right) = \frac{T_{ini}}{\tau_{lumped}} - \frac{\Delta T}{\tau_{lumped}} \left[1 - \exp\left(-\frac{t}{\tau_{ff}} \right) \right]. \tag{15}$$

Continued

Example 5.2 (cont.)

Because Eq. (15) must be satisfied at all values of time, it is necessary that the constant terms on each side always balance:

$$\frac{C_2}{\tau_{lumped}} = \frac{T_{ini}}{\tau_{lumped}} - \frac{\Delta T}{\tau_{lumped}} \tag{16}$$

and also that the exponential terms on each side always balance:

$$-\frac{C_3}{\tau_{ff}} \exp\left(-\frac{t}{\tau_{ff}}\right) + \frac{C_3}{\tau_{lumped}} \exp\left(-\frac{t}{\tau_{ff}}\right) = \frac{\Delta T}{\tau_{lumped}} \exp\left(-\frac{t}{\tau_{ff}}\right) \tag{17}$$

or

$$-\frac{C_3}{\tau_{ff}} + \frac{C_3}{\tau_{lumped}} = \frac{\Delta T}{\tau_{lumped}}. \tag{18}$$

Solving Eqs. (17) and (18) provides:

$$C_2 = T_{ini} - \Delta T \tag{19}$$

$$C_3 = \frac{\Delta T}{\left(1 - \dfrac{\tau_{lumped}}{\tau_{ff}}\right)}. \tag{20}$$

Substituting Eqs. (19) and (20) into Eq. (14) leads to the particular solution:

$$T_p = T_{ini} - \Delta T + \frac{\Delta T}{\left(1 - \dfrac{\tau_{lumped}}{\tau_{ff}}\right)} \exp\left(-\frac{t}{\tau_{ff}}\right). \tag{21}$$

The sum of the homogeneous and particular solutions is the general solution:

$$T = C_1 \exp\left(-\frac{t}{\tau_{lumped}}\right) + T_{ini} - \Delta T + \frac{\Delta T}{\left(1 - \dfrac{\tau_{lumped}}{\tau_{ff}}\right)} \exp\left(-\frac{t}{\tau_{ff}}\right). \tag{22}$$

The constant C_1 is obtained by enforcing the initial condition:

$$T_{t=0} = T_{ini} = C_1 + T_{ini} - \Delta T + \frac{\Delta T}{\left(1 - \dfrac{\tau_{lumped}}{\tau_{ff}}\right)}. \tag{23}$$

Solving Eq. (23) provides:

$$C_1 = \Delta T - \frac{\Delta T}{\left(1 - \dfrac{\tau_{lumped}}{\tau_{ff}}\right)} = \frac{\Delta T}{\left(1 - \dfrac{\tau_{ff}}{\tau_{lumped}}\right)}. \tag{24}$$

Substituting Eq. (24) into Eq. (22) provides the solution to the problem:

$$T = \frac{\Delta T}{\left(1 - \dfrac{\tau_{ff}}{\tau_{lumped}}\right)} \exp\left(-\frac{t}{\tau_{lumped}}\right) + T_{ini} - \Delta T + \frac{\Delta T}{\left(1 - \dfrac{\tau_{lumped}}{\tau_{ff}}\right)} \exp\left(-\frac{t}{\tau_{ff}}\right). \tag{25}$$

Example 5.2 (cont.)

Solution

The solution is implemented in EES.

```
D=1 [cm]*Convert(cm,m)                          "diameter"
k=0.63 [W/m-K]                                  "conductivity"
c=4200 [J/kg-K]                                 "specific heat capacity"
rho=1000 [kg/m^3]                               "density"
h_bar=36 [W/m^2-K]                              "heat transfer coefficient"
T_ini=20 [C]                                    "initial temperature"
DT=100 [C]                                      "change in temperature"
tau_ff=3 [min]*Convert(min,s)                   "time constant for flash freezer activation"

Vol=4*pi*(D/2)^3/3                              "volume"
A_s=4*pi*(D/2)^2                                "surface area"
tau_lumped=Vol*c*rho/(h_bar*A_s)                "lumped capacitance time constant"

{time=0 [s]}                                    "time"
T_infinity=T_ini-DT*(1-Exp(-time/tau_ff ))      "air temperature"
T=DT*Exp(-time/tau_lumped)/(1-tau_ff/tau_lumped)+&
    T_ini-DT+DT*Exp(-time/tau_ff )/(1-tau_lumped/tau_ff )   "blueberry temperature"
```

A Parametric Table is created that includes columns for time, air temperature, and blueberry temperature. Figure 2 illustrates the temperature of the air and the temperature of the fruit as a function of time.

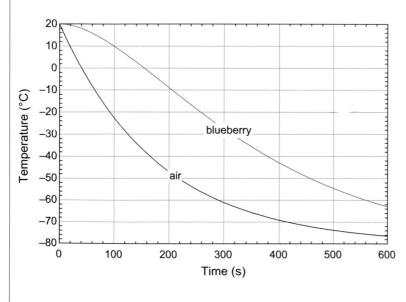

Figure 2 Blueberry and air temperature as a function of time.

Continued

Example 5.2 (cont.)

Discussion

Notice that the air temperature decays with a time constant of 3 minutes (180 s) and the blueberry temperature lags the air temperature by approximately 200 s. This lag is due to the lumped capacitance time constant of the blueberry, which is:

$$\tau_{lumped} = \frac{V \rho c}{\bar{h} A_s} = \frac{5.24 \times 10^{-7} \text{ m}^3}{} \left| \frac{1000 \text{ kg}}{\text{m}^3} \right| \frac{4200 \text{ J}}{\text{kg K}} \left| \frac{\text{m}^2 \text{ K}}{36 \text{ W}} \right| \frac{}{3.14 \times 10^{-4} \text{ m}^2} \left\| \frac{\text{W s}}{\text{J}} \right. = 194 \text{ s}. \qquad (26)$$

Example 5.3

A wire is placed in a flow of liquid. The wire is $L = 25$ mm long and $D = 0.7$ mm in diameter and is made of a metal with $k = 24.3$ W/m-K, $\rho = 8200$ kg/m^3, and $c = 387$ J/kg-K. The wire is initially in equilibrium with the liquid at $T_{ini} = T_\infty = 20°C$ when it is subjected to a time varying current flow that causes a volumetric rate of thermal energy generation in the wire that is approximately sinusoidal:

$$\dot{g}''' = \bar{\dot{g}}''' + \Delta \dot{g}''' \sin(2\pi f\, t), \qquad (1)$$

where $\bar{\dot{g}}''' = 4.8 \times 10^6$ W/m^3 and $\Delta \dot{g}''' = 1.2 \times 10^6$ W/m^3 are the average and the amplitude of the volumetric thermal energy generation, respectively, and $f = 0.2$ Hz is the frequency of the oscillation. The heat transfer coefficient between the liquid and the wire is $\bar{h} = 100$ W/m^2-K.

Determine the temperature of the wire as a function of time.

Known Values

The known values are indicated on the sketch shown in Figure 1.

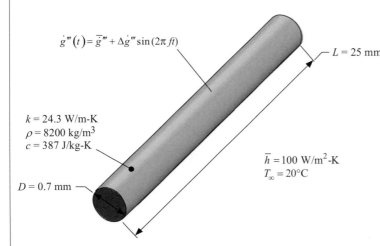

$$\dot{g}'''(t) = \bar{\dot{g}}''' + \Delta \dot{g}''' \sin(2\pi f t)$$

$L = 25$ mm

$k = 24.3$ W/m-K
$\rho = 8200$ kg/m^3
$c = 387$ J/kg-K

$\bar{h} = 100$ W/m^2-K
$T_\infty = 20°C$

$D = 0.7$ mm

Figure 1 Wire installed in a flow of liquid.

Assumptions

- The properties of the wire do not depend on temperature.
- The wire has a spatially uniform temperature. This is the lumped capacitance assumption which should be verified by computing a Biot number. The volume and surface area of the wire are computed:

$$V = \frac{\pi}{4} D^2 L = \frac{\pi}{4} (0.007 \text{ m})^2 (0.025 \text{ m}) = 9.62 \times 10^{-9} \text{ m}^3 \qquad (2)$$

Example 5.3 (cont.)

$$A_s = \pi D L = \pi (0.007 \ \text{m})(0.025 \ \text{m}) = 5.50 \times 10^{-5} \ \text{m}^2. \tag{3}$$

The conduction length and cross-sectional area are computed:

$$L_{cond} = \frac{V}{A_s} = \frac{9.62 \times 10^{-9} \ \text{m}^3}{5.50 \times 10^{-5} \ \text{m}^2} = 0.000175 \ \text{m} \tag{4}$$

$$A_c = A_s = 5.50 \times 10^{-5} \ \text{m}^2. \tag{5}$$

The resistance to internal conduction is computed:

$$R_{cond,int} = \frac{L_{cond}}{k \, A_c} = \frac{0.000175 \ \text{m}}{} \left| \frac{\text{m K}}{24.3 \ \text{W}} \right| \frac{}{5.50 \times 10^{-5} \ \text{m}^2} = 0.131 \ \frac{\text{K}}{\text{W}}. \tag{6}$$

The resistance to convection is:

$$R_{conv} = \frac{1}{\bar{h} \, A_s} = \frac{\text{m}^2 \ \text{K}}{100 \ \text{W}} \left| \frac{}{5.50 \times 10^{-5} \ \text{m}^2} \right. = 181.9 \ \frac{\text{K}}{\text{W}}. \tag{7}$$

The Biot number is computed according to:

$$Bi = \frac{R_{cond,int}}{R_{conv}} = \frac{0.131 \ \text{K/W}}{181.9 \ \text{K/W}} = 0.00072. \tag{8}$$

The Biot number is very small relative to unity and so the lumped capacitance approximation is justified.

Analysis

The derivation of the governing ordinary differential equation follows the steps outlined in Section 5.2.1 but will lead to a different result due to the additional energy term related to the ohmic generation within the wire. A control volume is defined that encompasses the entire wire, as shown in Figure 2. The energy balance suggested by the control volume is:

$$\dot{g} = \dot{q}_{conv} + \frac{dU}{dt}. \tag{9}$$

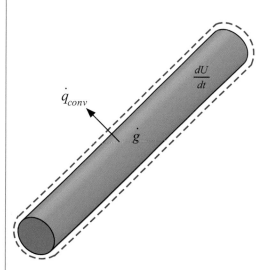

Figure 2 Energy balance used to derive the differential equation.

Substituting rate equations into the energy balance provides:

$$\left[\bar{g}''' + \Delta \dot{g}''' \sin\left(2\pi \, f \, t\right) \right] V = \bar{h} \, A_s (T - T_\infty) + \rho \, c \, V \frac{dT}{dt}. \tag{10}$$

Continued

Example 5.3 (cont.)

Equation (10) is put in standard form by placing the homogeneous portion on the left side and the nonhomogeneous portion on the right:

$$\frac{dT}{dt} + \frac{T}{\tau_{lumped}} = \frac{T_\infty}{\tau_{lumped}} + \frac{\left[\bar{g}''' + \Delta\dot{g}''' \sin(2\pi f t)\right]}{\rho c}, \tag{11}$$

where τ_{lumped} is the lumped capacitance time constant, computed according to:

$$\tau_{lumped} = R_{conv} C = \frac{\rho V c}{\bar{h} A_s}. \tag{12}$$

Equation (11) is divided into a homogeneous and a particular differential equation, as discussed in Section 5.2.2. The homogeneous differential equation is

$$\frac{dT_h}{dt} + \frac{T_h}{\tau_{lumped}} = 0 \tag{13}$$

and the homogeneous solution is

$$T_h = C_1 \exp\left(-\frac{t}{\tau_{lumped}}\right), \tag{14}$$

where C_1 is an undetermined constant of integration. The particular differential equation is:

$$\frac{dT_p}{dt} + \frac{T_p}{\tau_{lumped}} = \frac{T_\infty}{\tau_{lumped}} + \frac{\left[\bar{g}''' + \Delta\dot{g}''' \sin(2\pi f t)\right]}{\rho c}. \tag{15}$$

Equation (15) is solved using the method of undetermined coefficients. The right side of Eq. (15) is the nonhomogeneous component of the equation and it includes both a constant and a sinusoidal function. Therefore our functional form for the solution should include both a constant and its derivative (zero) as well as a sinusoid and its derivative (cosine). This leads to

$$T_p = C_2 + C_3 \sin(2\pi f t) + C_4 \cos(2\pi f t), \tag{16}$$

where C_2, C_3, and C_4 are undetermined coefficients. Equation (16) must be substituted into Eq. (15):

$$2\pi f \, C_3 \cos(2\pi f t) - 2\pi f \, C_4 \sin(2\pi f t) + \frac{C_2 + C_3 \sin(2\pi f t) + C_4 \cos(2\pi f t)}{\tau_{lumped}} =$$
$$\frac{T_\infty}{\tau_{lumped}} + \frac{\left[\bar{g}''' + \Delta\dot{g} \sin(2\pi f t)\right]}{\rho c}. \tag{17}$$

Because Eq. (17) must be satisfied at all values of time, it is necessary that the constant terms on each side always balance:

$$\frac{C_2}{\tau_{lumped}} = \frac{T_\infty}{\tau_{lumped}} + \frac{\bar{g}'''}{\rho c} \tag{18}$$

that the sine terms on each side always balance:

$$-2\pi f \, C_4 \sin(2\pi f t) + \frac{C_3 \sin(2\pi f t)}{\tau_{lumped}} = \frac{\Delta\dot{g}''' \sin(2\pi f t)}{\rho c} \tag{19}$$

and that the cosine terms always balance:

$$2\pi f \, C_3 \cos(2\pi f t) + \frac{C_4 \cos(2\pi f t)}{\tau_{lumped}} = 0. \tag{20}$$

Example 5.3 (cont.)

Solving Eq. (18) provides:

$$C_2 = T_\infty + \frac{\overline{g}'''}{\rho c} \tau_{lumped} = T_\infty + \frac{\overline{g}''' V}{\overline{h} A_s}.$$ (21)

Equations (19) and (20) are two equations in the two unknowns C_3 and C_4 that must be solved simultaneously. Solving Eq. (20) for C_3 provides:

$$C_3 = -\frac{C_4}{2\pi f \tau_{lumped}}.$$ (22)

Substituting Eq. (22) into Eq. (19) leads to:

$$-2\pi f C_4 - \frac{C_4}{2\pi f \tau_{lumped}^2} = \frac{\Delta \dot{g}'''}{\rho c},$$ (23)

which can be solved for C_4:

$$C_4 = -\frac{\Delta \dot{g}'''}{\rho c \left(2\pi f + \dfrac{1}{2\pi f \tau_{lumped}^2} \right)}.$$ (24)

Substituting Eq. (24) into Eq. (22) provides an expression for C_3:

$$C_3 = \frac{\Delta \dot{g}'''}{\rho c \left((2\pi f)^2 \tau_{lumped} + \dfrac{1}{\tau_{lumped}} \right)}.$$ (25)

Substituting Eqs. (21), (24), and (25) for the undetermined constants into the particular solution, Eq. (16):

$$T_p = T_\infty + \frac{\overline{g}''' V}{\overline{h} A_s} + \frac{\Delta \dot{g}'''}{\rho c \left((2\pi f)^2 \tau_{lumped} + \dfrac{1}{\tau_{lumped}} \right)} \sin(2\pi f t)$$

$$- \frac{\Delta \dot{g}'''}{\rho c \left(2\pi f + \dfrac{1}{2\pi f \tau_{lumped}^2} \right)} \cos(2\pi f t).$$ (26)

The sum of the homogeneous and particular solutions is the general solution:

$$T = C_1 \exp\left(-\frac{t}{\tau_{lumped}} \right) + T_\infty + \frac{\overline{g}''' V}{\overline{h} A_s} + \frac{\Delta \dot{g}'''}{\rho c \left((2\pi f)^2 \tau_{lumped} + \dfrac{1}{\tau_{lumped}} \right)} \sin(2\pi f t)$$

$$- \frac{\Delta \dot{g}'''}{\rho c \left(2\pi f + \dfrac{1}{2\pi f \tau_{lumped}^2} \right)} \cos(2\pi f t).$$ (27)

The constant C_1 is obtained by enforcing the initial condition:

$$T_{t=0} = T_{ini} = T_\infty = C_1 + T_\infty + \frac{\overline{g}''' V}{\overline{h} A_s} - \frac{\Delta \dot{g}'''}{\rho c \left(2\pi f + \dfrac{1}{2\pi f \tau_{lumped}^2} \right)}.$$ (28)

Solving Eq. (28) provides:

$$C_1 = -\frac{\overline{g}''' V}{\overline{h} A_s} + \frac{\Delta \dot{g}'''}{\rho c \left(2\pi f + \dfrac{1}{2\pi f \tau_{lumped}^2} \right)}.$$ (29)

Continued

Example 5.3 (cont.)

Substituting Eq. (29) into Eq. (27) provides the solution to the problem:

$$
T = \left[-\frac{\overline{\dot{g}}''' V}{\overline{h} A_s} + \frac{\Delta \dot{g}'''}{\rho c \left(2\pi f + \dfrac{1}{2\pi f \tau_{lumped}^2} \right)} \right] \exp\left(-\frac{t}{\tau_{lumped}} \right) + T_\infty + \frac{\overline{\dot{g}}''' V}{\overline{h} A_s}
$$
$$
+ \frac{\Delta \dot{g}'''}{\rho c \left((2\pi f)^2 \tau_{lumped} + \dfrac{1}{\tau_{lumped}} \right)} \sin\left(2\pi f\, t \right) - \frac{\Delta \dot{g}'''}{\rho c \left(2\pi f + \dfrac{1}{2\pi f \tau_{lumped}^2} \right)} \cos\left(2\pi f\, t \right)
$$

(30)

Solution

The solution is implemented in EES.

```
$UnitSystem Radian
k=24.3 [W/m-K]                        "conductivity"
rho=8200 [kg/m^3]                     "density"
c=387 [J/kg-K]                        "specific heat capacity"
D=0.7 [mm]*Convert(mm,m)              "diameter"
L=25 [mm]*Convert(mm,m)              "length"
h_bar=100 [W/m^2-K]                   "heat transfer coefficient"
g_dot```_bar=4.8e6 [W/m^3]           "average rate of vol. thermal energy generation"
Dg_dot```=1.2e6 [W/m^3]              "amplitude of vol. thermal energy generation"
f=0.2 [Hz]                            "frequency of oscillation"
T_infinity=20 [C]                     "ambient temperature"

A_s=pi*D*L                            "surface area"
Vol=pi*D^2/4*L                        "volume"
tau_lumped=rho*Vol*c/(h_bar*A_s)      "lumped time constant"

{time=0 [s]}                          "time"
T=(-g_dot```_bar*Vol/(h_bar*A_s)+&
   Dg_dot```/(rho*c*(2*pi*f+1/(2*pi*f*tau_lumped^2))))*Exp(-time/tau_lumped)+&
   T_infinity+g_dot```_bar*Vol/(h_bar*A_s)+&
   Dg_dot```/(rho*c*((2*pi*f)^2*tau_lumped+1/tau_lumped))*Sin(2*pi*f*time)-&
   Dg_dot```/(rho*c*(2*pi*f+1/(2*pi*f*tau_lumped^2)))*Cos(2*pi*f*time)
                                      "temperature"
```

A Parametric Table is created that includes columns for time and temperature. Figure 3 illustrates the temperature of the wire as a function of time.

Example 5.3 (cont.)

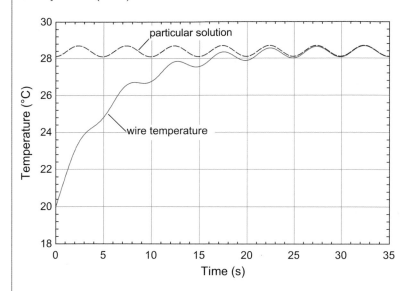

Figure 3 Temperature as a function of time.

Discussion

Notice that the wire temperature eventually (after about 15 s) reaches a periodic steady-state solution in which each oscillation looks the same. The periodic steady-state solution occurs after the effect of the initial condition has died away. The "long-time" solution that persists is simply the particular solution, as shown in Figure 3. The homogeneous solution, Eq. (14), will necessarily decay to zero after a few time constants, leaving only the particular solution. This is a general characteristic of this type of solution.

Also notice that we could have avoided much of the tedious algebra associated with the method of undetermined coefficients by programming the equations that lead to the undetermined coefficients, Eqs. (18), (19), and (20), into EES. Similarly, the equation that provides C_1, the undetermined coefficient in the homogeneous solution could be solved using EES. The result is the much simpler program shown below.

```
$UnitSystem Radian
k=24.3 [W/m-K]                                "conductivity"
rho=8200 [kg/m^3]                             "density"
c=387 [J/kg-K]                                "specific heat capacity"
D=0.7 [mm]*Convert(mm,m)                       "diameter"
L=25 [mm]*Convert(mm,m)                        "length"
h_bar=100 [W/m^2-K]                           "heat transfer coefficient"
g_dot```_bar=4.8e6 [W/m^3]                     "average rate of vol. thermal energy generation"
Dg_dot```=1.2e6 [W/m^3]                        "amplitude of vol. thermal energy generation"
f=0.2 [Hz]                                    "frequency of oscillation"
T_infinity=20 [C]                             "ambient temperature"

A_s=pi*D*L                                    "surface area"
Vol=pi*D^2/4*L                                "volume"
tau_lumped=rho*Vol*c/(h_bar*A_s)              "lumped time constant"

"Equations from the method of undetermined coefficients"
C_2/tau_lumped=T_infinity/tau_lumped+g_dot```_bar/(rho*c)
-2*pi*f*C_4+C_3/tau_lumped=Dg_dot```/(rho*c)
```

Continued

Example 5.3 (cont.)

2*pi*f*C_3+C_4/tau_lumped=0

"initial condition"
T_infinity=C_1+C_2+C_4

"temperature"
T=C_1***Exp**(-time/tau_lumped)+C_2+C_3***Sin**(2*pi*f*time)+C_4***Cos**(2*pi*f*time)

5.3 Numerical Solutions

5.3.1 Introduction

Section 5.1 discussed the lumped capacitance model that neglects any spatial temperature gradients within an object and therefore approximates the temperature in a transient problem as being only a function of time. The analytical solution to such 0-D (or lumped capacitance) problems was examined in Section 5.2. In this section, lumped capacitance problems will be solved numerically. The numerical techniques presented in this section are discussed in the context of the lumped capacitance problem from Section 5.2. An object is exposed to a time varying ambient temperature through convection. The resulting differential equation is:

$$\frac{dT}{dt} + \frac{T}{\tau_{lumped}} = \frac{T_\infty(t)}{\tau_{lumped}}, \tag{5.58}$$

where

$$\tau_{lumped} = \frac{m\,c}{h\,A_s} \tag{5.59}$$

and

$$T_{t=0} = T_{ini}. \tag{5.60}$$

This problem was solved analytically in Section 5.2.2 for the case where T_∞ changes in a ramped manner:

$$T_\infty(t) = T_{ini} + \beta t. \tag{5.61}$$

The resulting solution to this problem is:

$$T = T_{ini} + \beta\,t + \beta\,\tau_{lumped}\left[\exp\left(-\frac{t}{\tau_{lumped}}\right) - 1\right]. \tag{5.62}$$

The numerical solution technique can be applied to much more complex problems than the one examined here. However, it is convenient to begin our exploration of numerical solutions by considering this same problem because we have the exact solution and therefore can compare the numerical and analytical results.

The numerical solution relies on the identification of the **state equation**, which provides the time rate of change of a variable (or several variables) given its own value (or values). State equations characterize the transient behavior of many engineering problems in kinematics, fluids, electrical circuits, and other fields. Therefore, the numerical solution techniques provided in this section are relevant to a wide range of engineering problems. In a heat transfer problem, the state equation will provide the time rate of change of temperature given the current value of temperature and of time. Rearranging Eqs. (5.58) and (5.61) provides the state equation for the problem that we are considering here:

$$\frac{dT}{dt} = \frac{(T_{ini} + \beta\,t - T)}{\tau_{lumped}}. \tag{5.63}$$

Each problem will have its own state equation and the first step in any numerical solution will be its identification. Before you begin to apply any of the numerical techniques that are described in this section,

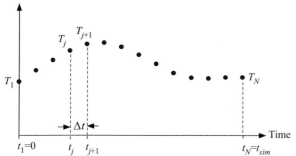

Figure 5.4 Dividing the simulation time into small time steps.

ask yourself if you can reliably calculate the time rate of temperature change given the current value of temperature and time through the application of an appropriate equation or set of equations. You *must* be able to do this *before* attempting a numerical solution.

The numerical technique will predict the value of temperature at many discrete values of time and therefore requires that the total simulation time (t_{sim}) be broken into small time steps, as shown in Figure 5.4. The simplest method for accomplishing this uses equal-sized steps, each with the same duration Δt:

$$\Delta t = \frac{t_{sim}}{(N-1)}, \tag{5.64}$$

where N is the number of times at which the temperature will be evaluated. The temperature at each time step is T_j, where j is an integer index that indicates the time. The temperature T_1 is the initial temperature of the object and T_N is the temperature at the end of a simulation. The time corresponding to each time step is therefore:

$$t_j = \frac{(j-1)}{(N-1)} t_{sim} \quad \text{for } j = 1 \ldots N. \tag{5.65}$$

Each of the numerical techniques that are discussed will provide an algorithm for estimating the value of temperature at the end of a times step (T_{j+1}) given the value of temperature at the beginning of the time step (T_j) and the state equation.

The same caveats that were discussed previously in the context of the numerical solution of steady-state conduction problems also apply to the numerical solution of transient conduction problems. Any numerical solution is only approximate and therefore it should be evaluated for numerical convergence (i.e., to ensure that you are using a sufficient number of time steps). Every solution should be examined against your physical intuition and, if possible, compared to an analytical solution in some limit.

5.3.2 The Euler Method

The simplest (and generally the least efficient) technique for numerical integration is the Euler Method. The **Euler Method** approximates the rate of temperature change within the time step as being constant and equal to its value calculated at the beginning of the time step. Therefore, for any time step j we can write:

$$T_{j+1} = T_j + \underbrace{\frac{dT}{dt}\Big|_{T=T_j, t=t_j}}_{\substack{\text{state equation} \\ \text{evaluated at the} \\ \text{beginning of} \\ \text{the time step}}} \Delta t \quad \text{for} \quad j = 1 \ldots (N-1). \tag{5.66}$$

In order to implement the Euler Method for a specific problem, it is necessary to insert the appropriate state equation into Eq. (5.66). For example, the Euler Method is implemented for the ramped ambient temperature problem by inserting Eq. (5.63) into Eq. (5.66):

$$T_{j+1} = T_j + \frac{(T_{ini} + \beta t_j - T_j)}{\tau_{lumped}} \Delta t \quad \text{for } j = 1 \ldots (N-1). \tag{5.67}$$

Using Eq. (5.67), the temperature at the end of each time step (T_{j+1}) can be calculated *explicitly* using known information at the beginning of the time step (T_j); for this reason, the Euler Method is referred to as an **explicit numerical technique**.

As an example, we can implement the solution for the case where $\tau_{lumped} = 20$ s, $T_{ini} = 300$ K, and $\beta = 1$ K/s. These inputs are entered in EES.

```
"Inputs"
tau_lumped=20 [s]                           "lumped capacitance time constant"
T_ini=300 [K]                               "initial temperature"
beta=1 [K/s]                                "ramp rate"
```

The time step duration and array of times for which the solution will be computed are specified using Eqs. (5.64) and (5.65) with $t_{sim} = 100$ s and $N = 101$ times (i.e., 100 time steps). This leads to a time step duration of $\Delta t = 1$ s.

```
t_sim=100 [s]                               "simulation time"
N=101 [-]                                    "number of times"
DELTAt=t_sim/(N-1)                          "duration of a time step"
Duplicate j=1,N
    time[j]=(j-1)*t_sim/(N-1)               "times"
End
```

The initial temperature (T_1) is known and Eq. (5.67) is used to compute the remaining N temperatures.

```
T[1]=T_ini                                  "initial temperature"
"The Euler method"
Duplicate j=1,(N-1)
    T[j+1]=T[j]+(T_ini+beta*time[j]-T[j])*DELTAt/tau_lumped
End
```

The numerical technique is not exact. Because we have the analytical solution to this problem, Eq. (5.62), it is possible to examine the deviation of the numerical solution from the exact solution. The analytical solution is computed at each value of time and the absolute value of the difference between the analytical and numerical solutions is computed. The global error is then defined as the maximum deviation that occurs during the simulation.

```
"The analytical solution"
Duplicate j=1,N
    T_an[j]=T_ini+beta*time[j]+beta*tau_lumped*(Exp(-time[j]/tau_lumped)-1)
    error[j]=Abs(T_an[j]-T[j])             "error between analytical and numerical solutions"
End
globalerror=Max(error[1..N])               "global error"
```

Running the EES program indicates that the global error is 0.19 K for the simulation. If the time step duration is increased by a factor of 10, to $\Delta t = 10$ s (by reducing N from 101 to 11), then the value of the global error increases to 2.4 K. Therefore, the error between the numerical and analytical solutions also increases by approximately a factor 10. This behavior is a characteristic of the Euler Method; the global error is proportional to the duration of the time step to the first power. This is referred to as a **first-order** numerical integration technique.

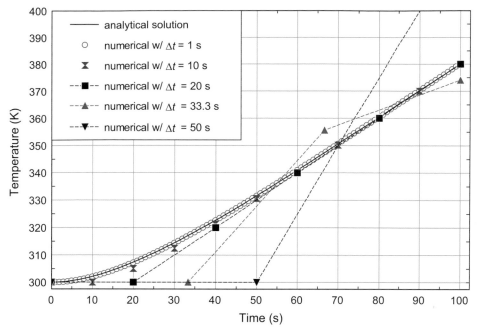

Figure 5.5 Numerical solution obtained using the Euler Method with various values of Δt. Also shown is the analytical solution to the same problem.

The order of the Euler technique can be inferred by examination of Eq. (5.66), which is essentially the first two terms of a Taylor series expansion of the temperature about time t_j:

$$T_{t=t+\Delta t} = \underbrace{T_j + \left.\frac{dT}{dt}\right|_{T=T_j,\,t=t_j}\Delta t}_{\text{Euler method approximation, } T_{j+1}} + \underbrace{\left.\frac{d^2 T}{dt^2}\right|_{T=T_j,\,t=t_j}\frac{\Delta t^2}{2!} + \left.\frac{d^3 T}{dt^3}\right|_{T=T_j,\,t=t_j}\frac{\Delta t^3}{3!} + \cdots}_{\text{local error} \approx \text{neglected terms}} \tag{5.68}$$

Examination of Eq. (5.68) shows that the **local error** which is associated with taking a single time step corresponds to the neglected terms in the Taylor series. The local error is therefore proportional to Δt^2 (assuming that the smaller, higher-order terms can be neglected). The Euler Method is referred to as a first-order method because it estimates only the first term in the Taylor series when taking each step. The *local error* associated with a first-order technique is proportional to the *second power* of the time step duration (and the second derivative of temperature). The *global error* associated with a first-order technique is proportional to the *first power* of the time step duration because the local errors tend to accumulate by an amount that is related to the number of time steps taken. Most numerical techniques that are commonly used are higher order and therefore achieve higher accuracy for a given time step duration.

A drawback of using the Euler Method (or any explicit numerical technique) is that it will become **unstable** if the duration of the time step used in the simulation becomes too large. For this example, if the time step duration is increased beyond the time constant of the object, $\tau_{lumped} = 20$ s, then the problem becomes unstable. Figure 5.5 illustrates the numerical solution for various values of the time step duration. The analytical solution provided by Eq. (5.62) is also shown. Notice that if the simulation is carried out with a time step duration that is less than $\Delta t = \tau_{lumped} = 20$ s, then the solution is stable and tends to follow the analytical solution more precisely as the time step is reduced. If the numerical solution is carried out with a time step duration that is between $\Delta t = \tau_{lumped} = 20$ s and $\Delta t = 2\,\tau_{lumped} = 40$ s then the prediction oscillates about the actual temperature but remains bounded. For time step durations greater than $\Delta t = 2\,\tau_{lumped} = 40$ s, the solution oscillates in an unbounded manner.

Figure 5.5 shows that the solutions for which $\Delta t > \tau_{lumped}$ are unstable. This threshold time step duration that governs the stability of the solution is called the **critical time step**, Δt_{crit}; the value of the critical time step can be

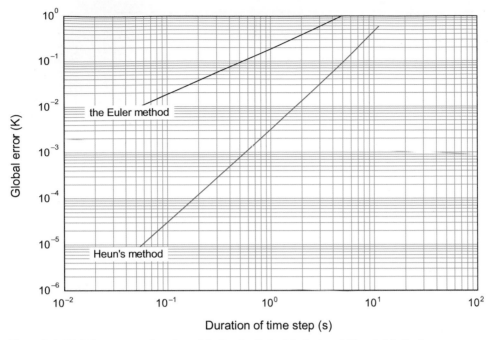

Figure 5.6 Global error as a function of Δt for the Euler Method and Heun's Method.

determined from the details of the problem. The implementation of the Euler Method for this problem was provided by Eq. (5.67), which is rearranged below:

$$T_{j+1} = T_j \underbrace{\left(1 - \frac{\Delta t}{\tau_{lumped}}\right)}_{\text{must be} >0 \text{ for stability}} + \frac{(T_{ini} + \beta\, t_j)}{\tau_{lumped}}\Delta t \quad \text{for } j = 1 \ldots (N-1). \tag{5.69}$$

The solution becomes unstable when the coefficient multiplying the temperature at the beginning of the time step (T_j) becomes negative; therefore for this problem:

$$\Delta t_{crit} = \tau_{lumped}. \tag{5.70}$$

Not surprisingly, the numerical simulation of objects that have small time constants will require the use of small time steps and thus increased computational effort.

The accuracy of the solution (i.e., the degree to which the numerical solution agrees with the analytical solution) improves as the duration of the time step is reduced. Figure 5.6 illustrates the global error as a function of the duration of the time step for the Euler Method. Also shown are the global error for Heun's Method, which is an alternative, higher-order numerical integration technique discussed in the subsequent section.

Any of the numerical integration techniques discussed in this section can be implemented in most software. The Euler solution is implemented in MATLAB using a script, as shown below. Because the Euler Method is explicit, the solution in MATLAB looks essentially identical to the solution in EES with some small differences related to the details of the language.

```
tau_lumped=20;          %lumped capacitance time constant (s)
T_ini=300;              %initial temperature (K)
beta=1;                 %ramp rate (K/s)

t_sim=100;              %simulation time (s)
N=51;                   %number of times
```

```
DELTAt=t_sim/(N−1);                      %duration of a time step
for j=1:N
    time(j)=(j−1)*t_sim/(N−1);           %times
end

T(1)=T_ini;                              %initial temperature
T_infinity(1)=T_ini;                     %initial ambient temperature
for j=1:(N−1)
    T(j+1)=T(j)+(T_ini+beta*time(j)−T(j))*DELTAt/tau_lumped;
    T_infinity(j+1)=T_ini+beta*time(j+1);  %ambient temperature
end

hold off;
plot(time,T_infinity,'b−');
hold on;
plot(time,T,'o');
grid on;
xlabel('Time (s)');
ylabel('Temperature (K)');
text(35,365,'ambient temperature');
text(40,315,'body temperature');
```

Running the script should result in the plot of temperature as a function of time that is shown in Figure 5.7.

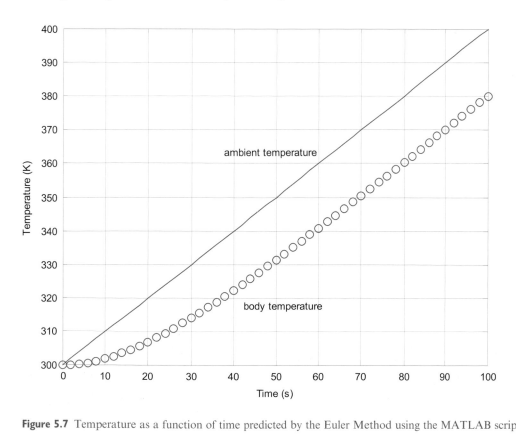

Figure 5.7 Temperature as a function of time predicted by the Euler Method using the MATLAB script.

5.3.3 Predictor–Corrector Methods

The Euler method is the simplest example of a numerical integration technique; it is an *explicit* technique with *first-order* accuracy. **Predictor–corrector methods** extend the Euler method in order to improve the accuracy for a specified time step duration. There are many different predictor–corrector methods. In this section, we will examine one of the simplest options which is referred to as **Heun's Method**.

The approach that is employed by any predictor–corrector technique is to take one or more *predictor steps*. Predictor steps are simple, Euler Method steps through or partially through the time step that result in some additional information. Finally, the *corrector step* applies all of the information that was obtained from the predictor step(s) in order to simulate the time step.

In order to simulate an arbitrary time step (i.e., the process of moving from time t_j with temperature T_j to time t_{j+1} by predicting T_{j+1}), Heun's Method begins with an Euler step to obtain an initial prediction for the temperature at the conclusion of the time step ($\hat{T}_{j+1}$). This first step in the solution is referred to as the predictor step (there is only one such step for Heun's Method) and the details are essentially identical to the Euler Method:

$$\hat{T}_{j+1} = T_j + \frac{dT}{dt}\bigg|_{T=T_j, t=t_j} \Delta t. \tag{5.71}$$

However, Heun's Method uses the result of the predictor step to carry out a subsequent corrector step. The temperature predicted at the end of the time step ($\hat{T}_{j+1}$) is used to predict the temperature rate of change at the end of the time step ($\frac{dT}{dt}\big|_{T=\hat{T}_{j+1}, t=t_{j+1}}$). The corrector step predicts the temperature at the end of the time step (T_{j+1}) based on the average of the time rates of change estimated at the beginning and end of the time step:

$$T_{j+1} = T_j + \left[\frac{dT}{dt}\bigg|_{T=T_j, t=t_j} + \frac{dT}{dt}\bigg|_{T=\hat{T}_{j+1}, t=t_{j+1}} \right] \frac{\Delta t}{2}. \tag{5.72}$$

This process is repeated for each step, from $j = 1$ to $(N - 1)$. Heun's Method is illustrated in the context of the simple problem considered in Section 5.3.2 which used the state equation:

$$\frac{dT}{dt} = \frac{(T_{ini} + \beta t - T)}{\tau_{lumped}}. \tag{5.73}$$

The inputs ($\tau_{lumped} = 20$ s, $T_{ini} = 300$ K, and $\beta = 1$ K/s) are entered in EES and the times and time step duration are setup as before. The temperature at time $t_1 = 0$ is the initial temperature.

```
"Inputs"
tau_lumped=20 [s]                       "lumped capacitance time constant"
T_ini=300 [K]                           "initial temperature"
beta=1 [K/s]                            "ramp rate"
t_sim=100 [s]                           "simulation time"
N=11 [-]                                "number of times"
DELTAt=t_sim/(N-1)                      "duration of a time step"
Duplicate j=1,N
    time[j]=(j-1)*t_sim/(N-1)           "times"
End
T[1]=T_ini                             "initial temperature"
```

The process of moving through each time step now requires two computations. It is often a good idea to move through the first time step (i.e., from time t_1 to time t_2) and debug your program in this simple limit before automating the process of moving through all of your time steps. The process of moving through the first time step begins with the predictor step:

$$\hat{T}_2 = T_1 + \frac{dT}{dt}\bigg|_{T=T_1, t=t_1} \Delta t \tag{5.74}$$

or, substituting Eq. (5.73) into Eq. (5.74):

$$\hat{T}_2 = T_1 + \frac{(T_{ini} + \beta\, t_1 - T_1)}{\tau_{lumped}} \Delta t. \tag{5.75}$$

The corrector step follows:

$$T_2 = T_1 + \left[\frac{dT}{dt}\bigg|_{T=T_1, t=t_1} + \frac{dT}{dt}\bigg|_{T=\hat{T}_2, t=t_2}\right]\frac{\Delta t}{2} \tag{5.76}$$

or, substituting Eq. (5.73) into Eq. (5.76):

$$T_2 = T_1 + \left[\frac{(T_{ini} + \beta\, t_1 - T_1)}{\tau_{lumped}} + \frac{(T_{ini} + \beta\, t_2 - \hat{T}_2)}{\tau_{lumped}}\right]\frac{\Delta t}{2}. \tag{5.77}$$

```
"Move through first time step"
T_hat[2]=T[1]+(T_ini+beta*time[1]-T[1])*DELTAt/tau_lumped            "predictor step"
T[2]=T[1]+((T_ini+beta*time[1]-T[1])/tau_lumped+(T_ini+beta*time[2]-&
   T_hat[2])/tau_lumped)*DELTAt/2                                    "corrector step"
```

Once the solution for taking a single step has been debugged it is easy to automate the process of taking all of the steps. Set up a Duplicate loop that has index j starting at 1 and ending at (N -1). Cut and paste your code into the Duplicate loop and change all of the 1 indices to j and all of the 2 indices to j+1.

```
"Automate moving through all of the time steps"
Duplicate j=1,(N-1)
   T_hat[j+1]=T[j]+(T_ini+beta*time[j]-T[j])*DELTAt/tau_lumped       "predictor step"
   T[j+1]=T[j]+((T_ini+beta*time[j]-T[j])/tau_lumped+&
      (T_ini+beta*time[j+1]-T_hat[j+1])/tau_lumped)*DELTAt/2         "corrector step"
End
```

The analytical solution is used to determine the global error, as in Section 5.3.2.

```
"The analytical solution"
Duplicate j=1,N
   T_an[j]=T_ini+beta*time[j]+beta*tau_lumped*(Exp(-time[j]/tau_lumped)-1)
   error[j]=Abs(T_an[j]-T[j])              "error between analytical and numerical solutions"
End
globalerror=Max(error[1..N])               "global error"
```

Figure 5.6 illustrates the global error as a function of the duration of the time step. Notice that the global error for Heun's technique is much smaller that what can be obtained using the Euler Method for a given time step duration (although two calculations are required for each time step vs. one for the Euler method). Also notice that the global error for Heun's Method is proportional to Δt^2 whereas the global error for the Euler Method is proportional to Δt^1. The two-step predictor–corrector method is a *second*-order method whereas the single step Euler Method is only a *first*-order method. In general, commonly used predictor–corrector methods are higher-order with more predictor steps.

Predictor–corrector methods are explicit. Examination of the predictor and corrector steps, Eqs. (5.71) and (5.72), respectively, show that both rely on explicit equations. Therefore, predictor–corrector methods will become unstable for time step durations that are greater than the critical time step.

Because Heun's Method is explicit, its implementation in a formal programming language like MATLAB is straightforward, as shown below.

```
tau_lumped=20;                    %lumped capacitance time constant (s)
T_ini=300;                        %initial temperature (K)
beta=1;                           %ramp rate (K/s)

t_sim=100;                        %simulation time (s)
N=51;                             %number of times
DELTAt=t_sim/(N−1);               %duration of a time step
for j=1:N
    time(j)=(j−1)*t_sim/(N−1);    %times
end

T(1)=T_ini;                       %initial temperature
for j=1:(N−1)
    T_hat=T(j)+(T_ini+beta*time(j)−T(j))*DELTAt/tau_lumped;
    T(j+1)=T(j)+((T_ini+beta*time(j)−T(j))/tau_lumped+...
        (T_ini+beta*time(j+1)−T_hat)/tau_lumped)*DELTAt/2;
end
```

Notice that because MATLAB utilizes assignment statements it is not necessary to store the intermediate variable $\hat{T}_{j+1}$ (i.e., the result of the predictor step) for each time step in an array. Instead, the value of the variable $\hat{T}_{j+1}$ (the variable T_hat) is over-written during each iteration of the for loop; this saves memory. EES uses equations rather than assignment statements and so it will not allow the value of a variable to be overwritten in the main body of the program, although it is possible to overwrite values with assignment statements in an EES function or procedure.

5.3.4 Implicit Methods

The numerical methods discussed thus far are *explicit*, meaning that the solution can be obtained directly by solving for the unknown temperature at each time step without having to iterate. Examination of Eq. (5.66) for the Euler Method or Eqs. (5.71) and (5.72) for Heun's Method shows that these are all explicit equations or assignments. The values of the terms on the right side of the equation are known at the time that the equation is being evaluated. Explicit methods share the characteristic of becoming unstable when the time step exceeds a critical value. An **implicit technique** avoids this problem.

Implicit techniques utilize implicit equations to move through each time step. Implicit equations are those in which the unknown variable (in this case, the temperature at the end of the time step), is not isolated on one side of the equation. There are several implicit methods that can be used; the simplest one is the **fully implicit method** that is discussed in this section. The fully implicit method is similar to Euler's method in that the time rate of change is assumed to be constant throughout the time step. However, the time rate of change is computed at the *end* of the time step rather than at the beginning. Therefore, for any time step *j*:

$$T_{j+1} = T_j + \underbrace{\left. \frac{dT}{dt} \right|_{T=T_{j+1},\, t=t_{j+1}}}_{\substack{\text{state equation} \\ \text{evaluated at the} \\ \text{end of the time step}}} \Delta t \quad \text{for } j = 1 \ldots (N-1). \tag{5.78}$$

The time rate of change at the end of the time step depends on the temperature at the end of the time step (T_{j+1}), which is yet to be determined. Therefore, T_{j+1} cannot be calculated explicitly using information at the beginning of the time step (T_j) and an implicit equation is obtained for T_{j+1}. For the example problem that has been considered in the previous sections, the implicit equation is obtained by substituting Eq. (5.63) into Eq. (5.78):

$$T_{j+1} = T_j + \frac{\left(T_{ini} + \beta\, t_{j+1} - T_{j+1} \right)}{\tau_{lumped}} \Delta t \quad \text{for } j = 1 \ldots (N-1). \tag{5.79}$$

Notice that T_{j+1} appears on both sides of Eq. (5.79). Because EES solves implicit equations, it is not necessary to re-arrange Eq. (5.79). The implicit solution is obtained by modifying the EES code developed in Section 5.3.2 to implement the Euler Method; the modifications are highlighted in bold/red below.

```
"Inputs"
tau_lumped=20 [s]                        "lumped capacitance time constant"
T_ini=300 [K]                            "initial temperature"
beta=1 [K/s]                             "ramp rate"
t_sim=100 [s]                            "simulation time"
N=101 [-]                                "number of times"
DELTAt=t_sim/(N-1)                       "duration of a time step"
Duplicate j=1,N
   time[j]=(j-1)*t_sim/(N-1)             "times"
End
T[1]=T_ini                               "initial temperature"

"The fully implicit method"
Duplicate j=1,(N-1)
   T[j+1]=T[j]+(T_ini+beta*time[j+1]-T[j+1])*DELTAt/tau_lumped
End
"The analytical solution"
Duplicate j=1,N
   T_an[j]=T_ini+beta*time[j]+beta*tau_lumped*(Exp(-time[j]/tau_lumped)-1)
   error[j]=Abs(T_an[j]-T[j])            "error between analytical and numerical solutions"
End

globalerror=Max(error[1..N])            "global error"
```

The fully implicit technique is no more accurate than the Euler technique for a given time step duration. However, Figure 5.8 shows that the fully implicit solution does not become unstable, even when the duration of the time step is greater than the critical time step.

The implementation of the fully implicit method cannot be accomplished directly using a formal programming language because Eq. (5.79) is not an explicit equation for T_{j+1}; while EES can solve this implicit equation, programming languages rely on assignment statements. Therefore, is necessary to solve Eq. (5.79) for T_{j+1} in order to carry out the integration step:

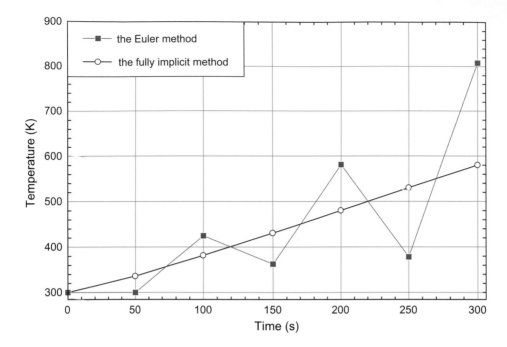

Figure 5.8 Temperature as a function of time predicted using the Euler Method and the fully implicit method with a time step that is larger than the critical time step ($\Delta t_{crit} = \tau_{lumped} = 20$ s).

$$T_{j+1} = \frac{T_j + \dfrac{\left(T_{ini} + \beta\, t_{j+1}\right)}{\tau_{lumped}}\Delta t}{\left(1 + \dfrac{\Delta t}{\tau_{lumped}}\right)} \quad \text{for } j = 1\ldots(N-1). \quad (5.80)$$

The resulting program in MATLAB is listed below.

```
tau_lumped=20;                          %lumped capacitance time constant (s)
T_ini=300;                              %initial temperature (K)
beta=1;                                 %ramp rate (K/s)

t_sim=100;                              %simulation time (s)
N=51;                                   %number of times
DELTAt=t_sim/(N−1);                     %duration of a time step
for j=1:N
    time(j)=(j−1)*t_sim/(N−1);          %times
end

T(1)=T_ini;                             %initial temperature
T_infinity(1)=T_ini;                    %initial ambient temperature
for j=1:(N−1)
    T(j+1)=(T(j)+(T_ini+beta*time(j+1))*DELTAt/tau_lumped)/(1+DELTAt/tau_lumped);
    T_infinity(j+1)=T_ini+beta*time(j+1);   %ambient temperature
end
```

5.3.5 Using ODE Solvers

Most engineering software includes **ODE solvers** that can be used to conveniently obtain the numerical solution to one or more ordinary differential equations. These solvers are typically more sophisticated (e.g., higher order) than any of the options that were reviewed in Sections 5.3.2 through 5.3.4, although the basic idea is the same. One or more state equations are integrated through time, one time step at a time.

The implementation of the techniques discussed in this section has been accomplished using a fixed duration time step for the entire simulation. This fixed time step implementation is often not efficient because there are periods of time during the simulation where the value of the state equation is not changing substantially and therefore large time steps can be taken with little loss of accuracy. One advantage of using ODE solvers is that they typically use **adaptive step size** techniques in which the size of the time step used by the simulation is automatically adjusted based on the local characteristics of the state equation in order to guarantee a certain level of accuracy.

In this section we will discuss the ODE solvers that are available in EES and MATLAB. Other software packages have their own versions of this capability, but the implementation will be similar: program and check the state equation and then provide the ODE solver access to it.

EES' Integral Command

The ODE solver included in the EES software is accessed using the Integral command. EES' Integral command requires four arguments.

```
F = Integral(Integrand, VarName, LowerLimit, UpperLimit)
```

Here Integrand is the EES variable or expression that must be integrated, VarName is the integration variable, and the parameters LowerLimit and UpperLimit define the limits of integration. A fixed step size can optionally be supplied as a fifth parameter; if one is not supplied then an adaptive step size technique is used.

When using the Integral command in EES (or indeed any numerical integration technique) it is important that you first program and check the state equation. For the lumped capacitance problems examined in this chapter you should verify that, given temperature and time, the EES code in the Equations Window is capable of computing the time rate of change of the temperature. For the problem considered here, the first step is to implement Eq. (5.63) for arbitrary (but reasonable) values of T and t:

```
"Inputs"
tau_lumped=20 [s]                     "lumped capacitance time constant"
T_ini=300 [K]                         "initial temperature"
beta=1 [K/s]                          "ramp rate"

"Program and check the state equation"
time=0 [s]                            "set a value of time"
T=300 [K]                            "and temperature"
dTdt=(T_ini+beta*time-T)/tau_lumped   "and check the state equation"
```

The next step is to comment out the arbitrary values of temperature and time that were used to test the computation of the state equation and instead let EES' Integral function control these variables for the numerical integration. The temperature of the object is given by

$$T = T_{ini} + \int_0^{t_{sim}} \frac{dT}{dt} dt, \tag{5.81}$$

where t_{sim} is the total amount of time to be simulated. Therefore, the solution to our example problem is obtained by calling the Integral function; the argument Integrand is replaced with the variable dTdt, VarName is replaced with time, LowerLimit is replaced with 0, and UpperLimit is replaced with the variable t_sim.

```
"Inputs"
tau_lumped=20 [s]                              "lumped capacitance time constant"
T_ini=300 [K]                                  "initial temperature"
beta=1 [K/s]                                   "ramp rate"

"Program and check the state equation"
{time=0 [s]                                    "set a value of time"
T=300 [K]                                      "and temperature"}
dTdt=(T_ini+beta*time-T)/tau_lumped            "and check the state equation"

"EES' Integral method"
t_sim=100 [s]                                  "simulation time"
T=T_ini+Integral(dTdt,time,0,t_sim)            "EES' ODE solver"
```

In order to accomplish the numerical integration, EES adjusts the value of variable time from 0 to t_sim in small time steps so as to maintain a specified accuracy criterion. At each value of time, EES will iteratively solve all of the equations in the Equations Window that depend on time. For the example above, this process will result in the variable T being evaluated at each value time. When the calculation is complete the Solution Window will provide the temperature at the end of this integration process (i.e., the temperature at time = t_sim); in this case $T_{t=t_{sim}} = 380.1$ K.

Often it is interesting to know how temperature varies with time *during* the process; this information can be obtained by using an Integral Table, which is created by including the $IntegralTable directive in the file. The format of the $IntegralTable directive is shown below.

```
$IntegralTable VarName, x, y, z
```

Here VarName is the integration variable (the first column in the Integral Table will hold values of this variable). The variables x, y, z ... must correspond to variables in the EES program. A separate column will be created in the Integral Table for each specified variable.

Solving the EES code will result in the generation of an Integral Table that is filled with intermediate values resulting from the numerical integration. The values in the Integral Table can be plotted, printed, saved, and copied in exactly the same manner as for other tables in EES. The Integral Table is saved when the EES file is saved and the table is restored when the EES file is loaded. If an Integral Table exists when calculations are initiated, it will be deleted and a new Integral Table will be created.

Use the EES code below to generate an Integral Table containing the results of the numerical simulation.

```
$IntegralTable time, T
```

After running the code, an Integral Table similar to the one shown in Figure 5.9 will be generated.

The results of the integration can be plotted by selecting the Integral Table as the source of the data to plot. The result is shown in Figure 5.10; notice that the time step used in the simulation varies. At small times, the ODE solver used small time steps (the points in Figure 5.10 are very close together) while at longer times the solver used larger time steps (the points are further apart).

Any ODE solver will include some method of controlling the integration process. For example, in EES it is possible to adjust the adaptive step size parameters by selecting Preferences from the Options menu and then selecting the Integration Tab, as shown in Figure 5.11. Another option for controlling these parameters is to use the $IntegralAutoStep directive.

Note that the increment between entries in the Integral Table can be controlled with an optional stepsize parameter following the integration variable in the $IntegralTable directive. An example is given below.

```
$IntegralTable VarName:stepsize, x, y, z
```

Integral Table

Main

	time [s]	T [K]
Row 1	0	300
Row 2	0.025	300
Row 3	0.05	300
Row 4	0.075	300
Row 5	0.1	300
Row 6	0.125	300
Row 7	0.15	300
Row 8	0.175	300
Row 9	0.2	300
Row 10	0.225	300
Row 11	0.25	300

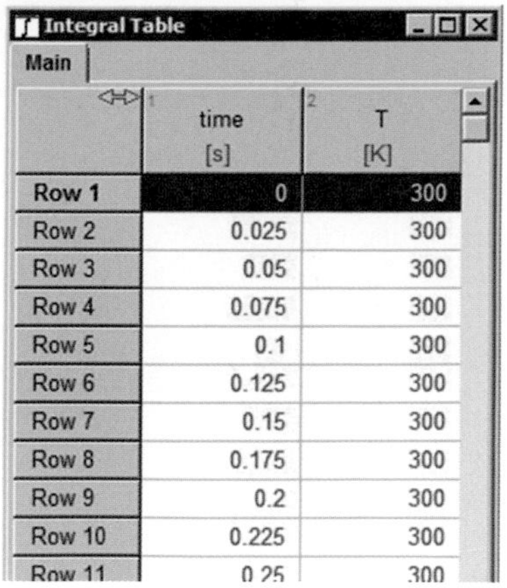

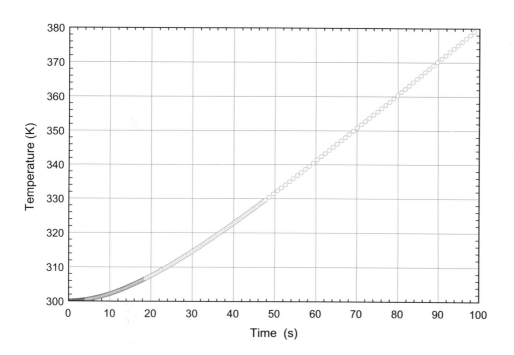

Figure 5.9 Integral Table generated by EES.

Figure 5.10 Temperature as a function of time predicted using EES' Integral command.

This will limit the output stored in the table to one value at each value of time that is a multiple of the parameter stepsize. The use of the optional stepsize parameter in the $IntegralTable directive will *not* affect the step size that is actually used to accomplish the integration.

MATLAB's ODE Solvers

MATLAB also includes a set of solvers for initial value problems that implement advanced numerical integration algorithms (e.g., the functions ode45, ode23, etc.). These ordinary differential equation solvers all have a similar calling protocol.

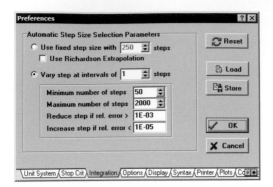

Figure 5.11 Integration Tab of the Preferences Window.

```
[time, T] = ode45('dTdt', tspan, T0)
```

The ODE solvers require three inputs. The first ('dTdt') is the name of the function that returns the derivative of temperature with respect to time given the current value of temperature and time (i.e., the state equation). The ODE solver therefore requires that you have created a function that implements the state equation. The solver assumes that this function will accept two inputs, corresponding to the current values of time and temperature, and return a single output, the rate at which temperature is changing with time (although it is possible to pass a function handle as this argument). The remaining inputs include the simulation time span (tspan) and the initial temperature (T0). The ODE solver returns two vectors containing the solution times (time) and the temperatures predicted at the solution times (T).

The function dTdt_function must be created to implement the state equation, Eq. (5.63).

```
function[dTdt]=dTdt_function(time,T)

% Inputs
% T - temperature (K)
% time - time (s)
%
% Outputs
% dTdt - time rate of change (K/s)

tau=20;                          % time constant (s)
T_ini=300;                       % initial temperature (K)
beta=1;                          % rate of temperature change (K/s)
dTdt=(T_ini+beta*time-T)/tau;    % state equation
end
```

Then a script can be written to solve the problem. The script calls the ODE solver ode45 and specifies that the state equation is computed using the function dTdt_function, the simulation time is from 0 to t_sim and the initial condition is T_ini.

```
clear all;

%Inputs
T_ini=300;                                % initial temperature (K)
t_sim=100;                                % simulation time (s)
[time,T]=ode45('dTdt_function',t_sim,T_ini);  % use ode45 to solve problem
```

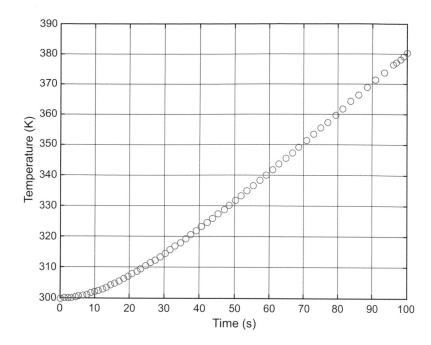

Figure 5.12 Predicted temperature as a function of time obtained using MATLAB's ode45 solver.

The vectors time and T are the times and temperatures used to carry out the simulation. The MATLAB solver uses an adaptive step size. In order to control the integration process it is possible to provide an optional fourth argument to the solver, which is a vector that sets the integration characteristics; in the absence of a fourth argument, MATLAB will use the default settings for the integration properties. The most convenient way to adjust the integration settings is with the odeset function, which has the following format.

```
options = odeset('property1', value1, 'property2', value2, ...)
```

Here 'property1' refers to one of the integration characteristics and value1 refers to its specified value, 'property2' refers to another integration characteristic and value2 refers to its specified value, etc. The properties that are not explicitly set in the odeset command remain at their default values. Type help odeset at the command line prompt in order to see a list and description of the integration properties as well as their default values.

The property RelTol refers to the relative error tolerance and has a default value of 0.001 (0.1% accuracy). The relative error tolerance of the integration process can be changed to 1×10^{-7} by changing the script according to the following.

```
clear all;

%Inputs
T_ini=300;                              % initial temperature (K)
t_sim=100;                              % simulation time (s)
options=odeset('RelTol',1e-7);          % change the relative error tolerance
[time,T]=ode45('dTdt_function',t_sim,T_ini,options);
                                        %call ode45

hold off;
plot(time,T,'o');                       %plot results
xlabel('Time (s)');
ylabel('Temperature (K)');
grid on;
```

The results of the integration are shown in Figure 5.12; notice the variation in the time step duration used by the ODE solver during the simulation.

Example 5.4

One technique for detecting chemical threats uses a laser to ablate (i.e., thermally destroy) small particles so that they can subsequently be analyzed using ion mobility spectroscopy. Consider a particle that is spherical with radius $r_p = 7$ μm and properties $c = 1500$ J/kg-K, $k = 3.5$ W/m-K, and $\rho = 80$ kg/m³. The particle is surrounded by air at $T_\infty = 20°C$. The heat transfer coefficient is $\bar{h} = 60,000$ W/m²-K. The particle is initially at T_∞. The laser power as a function of time is given by

$$\dot{q}_{laser}(t) = \dot{q}_{max} \exp\left[-\frac{(t - t_p)^2}{2t_d^2}\right], \tag{1}$$

where $\dot{q}_{max} = 0.2$ W is the maximum value of the laser power during the pulse, $t_p = 3$ μs is the time at which the maximum power occurs, and $t_d = 0.5$ μs is related to the duration of the pulse.

Determine the temperature of the particle as a function of time.

Known Values

The known values are indicated on the sketch shown in Figure 1.

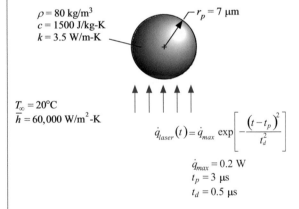

$\rho = 80$ kg/m³
$c = 1500$ J/kg-K
$k = 3.5$ W/m-K

$r_p = 7$ μm

$T_\infty = 20°C$
$\bar{h} = 60,000$ W/m²-K

$\dot{q}_{laser}(t) = \dot{q}_{max} \exp\left[-\frac{(t - t_p)^2}{t_d^2}\right]$

$\dot{q}_{max} = 0.2$ W
$t_p = 3$ μs
$t_d = 0.5$ μs

Figure 1 Particle exposed to a laser pulse.

Assumptions

- The properties of the particle do not depend on temperature.
- Radiation is neglected. This approximation should be justified by computing the resistance to radiation and comparing it to the resistance to convection. These resistances act in parallel from the surface of the object and therefore if the resistance to radiation is substantially *larger* than the resistance to convection, radiation can be neglected. The surface area of the particle is computed:

$$A_s = 4\pi r_p^2 = 4\pi (7 \times 10^{-6} \text{ m})^2 = 6.16 \times 10^{-10} \text{ m}^2. \tag{2}$$

The convection resistance is computed:

$$R_{conv} = \frac{1}{\bar{h} A_s} = \frac{\text{m}^2 \text{ K}}{60,000 \text{ W}} \left| \frac{1}{6.16 \times 10^{-10} \text{ m}^2} \right| = 27,000 \frac{\text{K}}{\text{W}}. \tag{3}$$

Example 5.4 (cont.)

As discussed in Section 2.2.5, the radiation resistance can be estimated according to:

$$R_{rad} \approx \frac{1}{A_s \, \sigma \, \varepsilon \, 4\bar{T}^3}, \tag{4}$$

where ε is the emissivity of the particle and $\bar{T}$ is the average of the surface and the surrounding temperature. Neither ε nor $\bar{T}$ are known and therefore it is only possible to estimate R_{rad}. In order to compute the *smallest* possible estimate of the radiation resistance (which corresponds to it being *most important*) we will use the relatively large values of $\varepsilon = 1$ and $\bar{T} = 600$ K:

$$R_{rad} \approx \frac{1}{A_s \, \sigma \, \varepsilon \, 4\bar{T}^3} = \frac{1}{6.16 \times 10^{-10} \text{ m}^2} \left| \frac{\text{m}^2 \text{ K}^4}{5.67 \times 10^{-8} \text{ W}} \right| \frac{1}{4} \left| \frac{1}{600^3 \text{ K}^3} \right| = 3.32 \times 10^7 \frac{\text{K}}{\text{W}}. \tag{5}$$

The estimated value of R_{rad} is approximately 1000× larger than the value of R_{conv} and therefore it is justified to ignore radiation in this problem.

- The particle is at a spatially uniform temperature. This is the lumped capacitance assumption which should be verified by computing a Biot number. The volume of the particle is computed:

$$V = \frac{4}{3}\pi r_p^3 = \frac{4}{3}\pi \left(7 \times 10^{-6} \text{ m}\right)^3 = 1.44 \times 10^{-15} \text{ m}^3. \tag{6}$$

The conduction length and cross-sectional area are computed:

$$L_{cond} = \frac{V}{A_s} = \frac{1.44 \times 10^{-15} \text{ m}^3}{6.16 \times 10^{-10} \text{ m}^2} = 2.33 \times 10^{-6} \text{ m} \tag{7}$$

$$A_c = A_s = 6.16 \times 10^{-10} \text{ m}^2. \tag{8}$$

The resistance to internal conduction is computed:

$$R_{cond,int} = \frac{L_{cond}}{k \, A_c} = \frac{2.33 \times 10^{-6} \text{ m}}{3.5 \text{ W}} \left| \frac{\text{m K}}{6.16 \times 10^{-10} \text{ m}^2} \right| = 1080 \frac{\text{K}}{\text{W}}. \tag{9}$$

The Biot number is computed according to:

$$Bi = \frac{R_{cond,int}}{R_{conv}} = \frac{1080 \text{ K/W}}{27,000 \text{ K/W}} = 0.04. \tag{10}$$

The Biot number is small relative to unity and so the lumped capacitance approximation is justified.

Analysis

A state equation must be derived in order to employ any numerical method. A control volume is defined that encompasses the entire particle, as shown in Figure 2. The energy balance suggested by the control volume is:

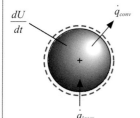

Figure 2 Control volume used to derive the state equation.

Continued

Example 5.4 (cont.)

$$\dot{q}_{laser} = \dot{q}_{conv} + \frac{dU}{dt}. \tag{11}$$

Substituting rate equations into the energy balance provides:

$$\dot{q}_{max} \exp\left[-\frac{(t-t_p)^2}{2t_d^2}\right] = \bar{h} A_s (T - T_\infty) + \rho c V \frac{dT}{dt}. \tag{12}$$

Equation (12) is rearranged to provide the state equation, an equation that allows the calculation of the time rate of temperature change given the value of temperature and time:

$$\frac{dT}{dt} = \frac{(T_\infty - T)}{\tau_{lumped}} + \frac{\dot{q}_{max}}{\rho c V} \exp\left[-\frac{(t-t_p)^2}{2t_d^2}\right], \tag{13}$$

where τ_{lumped} is the lumped capacitance time constant, computed according to:

$$\tau_{lumped} = R_{conv} C = \frac{\rho c V}{\bar{h} A_s}. \tag{14}$$

A numerical simulation based on the Euler Method will evaluate the state equation at the beginning of each time step and use that value of the temperature rate of change to move through the time steps one at a time according to:

$$T_{j+1} = T_j + \frac{dT}{dt}\bigg|_{T=T_j, t=t_j} \Delta t \quad \text{for} \quad j = 1 \ldots (N-1). \tag{15}$$

Substituting Eq. (13) into Eq. (15) provides:

$$T_{j+1} = T_j + \left\{\frac{(T_\infty - T_j)}{\tau_{lumped}} + \frac{\dot{q}_{max}}{\rho c V} \exp\left[-\frac{(t_j - t_p)^2}{2t_d^2}\right]\right\} \Delta t \quad \text{for} \quad j = 1 \ldots (N-1). \tag{16}$$

Solution

The solution is implemented in EES.

```
"Inputs"
rho=80 [kg/m^3]                      "density"
c=1500 [J/kg-K]                      "specific heat capacity"
k=3.5 [W/m-K]                        "thermal conductivity"
r_p=7 [micron]*Convert(micron,m)     "radius"
T_infinity=ConvertTemp(C,K,20 [C])   "ambient temperature"
h_bar=60000 [W/m^2-K]                "heat transfer coefficient"
q_dot_max=0.2 [W]                    "maximum laser power"
time_p=3e-6 [s]                      "time of laser pulse"
time_d=0.5e-6 [s]                    "duration of laser pulse"

A_s=4*pi*r_p^2                       "surface area"
Vol=4*pi*r_p^3/3                     "volume"

tau_lumped=rho*c*Vol/(h_bar*A_s)     "lumped capacitance time constant"

t_sim=20e-6 [s]                      "simulation time"
N=201 [-]                            "number of times"
```

Example 5.4 (cont.)

```
DELTAt=t_sim/(N-1)                          "duration of a time step"
Duplicate j=1,N
    time[j]=(j-1)*t_sim/(N-1)               "times"
End


T[1]=T_infinity                             "initial temperature"


Duplicate j=1,(N-1)
    T[j+1]=T[j]+((T_infinity-T[j+1])/tau_lumped+&
        q_dot_max*Exp(-(time[j+1]-time_p)^2/(2*time_d^2))/(rho*c*Vol))*Deltat
End
```

Figure 3 illustrates the temperature of the particle as a function of time.

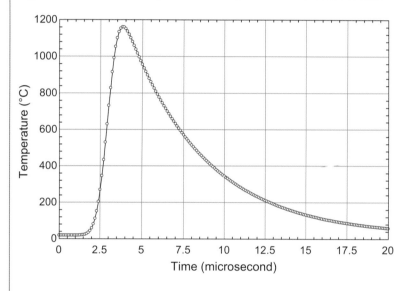

Figure 3 Temperature as a function of time predicted using the Euler Method.

Discussion

Notice that the particle temperature increases rapidly near $t = t_p = 3$ μs due to the energy directly deposited by the laser pulse and then decays slowly and exponentially towards the ambient temperature with time constant $\tau_{lumped} = 4.7$ μs. The EES program could be modified, as shown below, to make use of EES' internal ODE solver with the Integral command.

```
"Inputs"
rho=80 [kg/m^3]                             "density"
c=1500 [J/kg-K]                             "specific heat capacity"
k=3.5 [W/m-K]                               "thermal conductivity"
r_p=7 [micron]*Convert(micron,m)            "radius"
T_infinity=ConvertTemp(C,K,20 [C])          "ambient temperature"
h_bar=60000 [W/m^2-K]                       "heat transfer coefficient"
q_dot_max=0.2 [W]                           "maximum laser power"
time_p=3e-6 [s]                             "time of laser pulse"
time_d=0.5e-6 [s]                           "duration of laser pulse"
```

Continued

Example 5.4 (cont.)

```
A_s=4*pi*r_p^2                        "surface area"
Vol=4*pi*r_p^3/3                      "volume"
tau_lumped=rho*c*Vol/(h_bar*A_s)      "lumped capacitance time constant"

t_sim=20e-6 [s]                       "simulation time"

dTdt=(T_infinity-T)/tau_lumped+q_dot_max*Exp(-(time-time_p)^2/(2*time_d^2))/(rho*c*Vol)
T=T_infinity+Integral(dTdt, time, 0, t_sim)

$IntegralTable time,T
```

Figure 4 illustrates the results obtained using EES' internal ODE solver; notice the adaptive step size that is utilized to maintain accuracy.

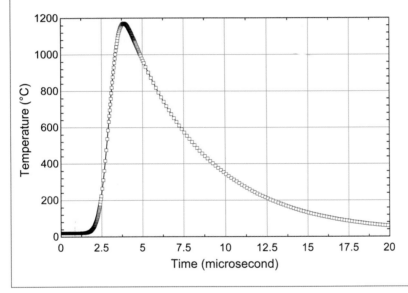

Figure 4 Temperature as a function of time predicted using the EES Integral command.

5.4 Conclusions and Learning Objectives

This chapter discusses lumped capacitance problems, which are the simplest transient problem that we will encounter because the spatial temperature gradients within a body are ignored. The lumped capacitance assumption should be justified with a Biot number that compares the internal conduction resistance to an external resistance between the object and its surroundings. The lumped capacitance time constant provides an easy way to gain some physical insight into the problem.

Analytical and numerical solutions to lumped capacitance problems are presented. The analytical solutions follow naturally from the solutions of the nonseparable ordinary differential equations that were encountered in Chapter 3 in the context of extended surface problems. The differential equation is split into its homogeneous and particular portions. The homogeneous differential equation is solved to within an undetermined constant and the particular differential equation is solved using the method of undetermined coefficients. The homogeneous and particular solutions are assembled to form the general solution. Finally, the undetermined constant in the homogeneous solution is selected in order to satisfy the initial condition.

Several numerical solution techniques are discussed; each use the state equation to move through many finite time steps in order to obtain an approximate solution to the problem. Low-order, explicit techniques like the

Euler Method suffer from poor accuracy and instability when the step size exceeds the critical time step. Higher-order, explicit methods like Heun's Method are more accurate but may still become unstable if the time step exceeds the critical time step. Implicit techniques like the fully implicit technique are unconditionally stable, but the accuracy is affected by the step size. The details of implementing each of these techniques were presented using both EES and MATLAB.

Some specific concepts that you should understand are listed below.
- The physical meaning of the lumped capacitance assumption and how it is justified by calculating a Biot number.
- The method of calculating a lumped capacitance time constant and an understanding of how to use a lumped capacitance time constant to develop some intuition into the expected behavior of a lumped capacitance problem.
- The ability to develop an analytical solution to a lumped capacitance problem by solving a nonseparable, first-order, linear differential equation subject to an initial condition.
- The meaning of a state equation and the ability to derive a state equation using an energy balance for a lumped capacitance problem.
- The meaning of the order of a numerical technique and implication of the order on the accuracy of the technique.
- How to calculate the critical time step for an explicit technique.
- The difference between an explicit and an implicit numerical technique.
- The ability to implement the Euler technique, Heun's Method, and the fully implicit method using an equation solver (e.g., EES) or a programming language (e.g., MATLAB).
- The ability to use native ODE solvers in software such as EES and MATLAB.

Problems

The Lumped Capacitance Approximation and the Biot Number

5.1 A 3 pound chicken, initially at 10°C, is placed in a 350°F convection oven. The convection coefficient between the chicken and the air in the oven is estimated to be 34 W/m²-K. The chicken is approximately spherical and it has about the same thermal properties as liquid water at 25°C. Can the chicken be treated as a lumped capacitance for the purpose of determining its center temperature as a function of time?

5.2 A spherical object experiences a cross-flow of fluid at T_∞. The diameter of the sphere is $D = 0.02$ m and the average heat transfer coefficient between the object surface and the fluid is $\bar{h} = 5{,}000$ W/m²-K. The object is composed of material with $k_s = 450$ W/m-K, $c_s = 420$ J/kg-K, and $\rho_s = 4000$ kg/m³. You may neglect radiation for this problem.
  (a) Is a lumped capacitance approximation appropriate for this problem? Justify your answer.
  (b) Determine the lumped capacitance time constant of the object.

The free stream temperature of the fluid, T_∞, varies with time as shown in the figure.

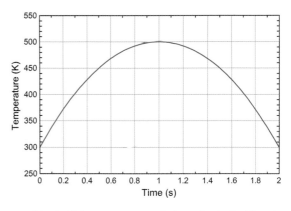

The initial temperature of the spherical object is $T_{ini} = 400$ K.
  (c) Sketch the temperature of the object as a function of time. Also sketch the temperature of the object with respect to time that you would expect if the lumped capacitance time constant was 50 s and 0.05 s. Clearly label each of the three sketches.

5.3 The door of a fire-resistant safe is made of steel ($k = 37$ W/m-K, $\rho = 7900$ kg/m³, $c = 530$ J/kg-K). The door is 0.25 m high, 0.25 m wide, and 1.8 cm thick. The inside surface of the door is coated with a thin layer of insulation. Assume that the outside surface of the door is suddenly exposed to flame at 1250 K with a convection coefficient of 34 W/m²-K.

(a) Can the safe door be accurately modeled as a lumped capacitance?

(b) Determine the value of the lumped capacitance time constant

(c) Sketch the temperature of the door as a function of time. Assume that the door is initially at room temperature (300 K).

5.4 Your company manufactures a product that consists of many small metal bars that run through a polymer matrix. The material can be used as a thermal path, allowing heat to transfer efficiently in the z-direction (the direction that the metal bars run) because the heat can travel without interruption through the metal bars. However, the material blocks heat flow in the y- and z-directions because the energy must be conducted through the low-conductivity polymer. Because the size of the metal bars is small relative to the size of the composite structure, it is appropriate to model the material as a composite with an effective conductivity that is different in the z-direction than in the x- and y-directions.

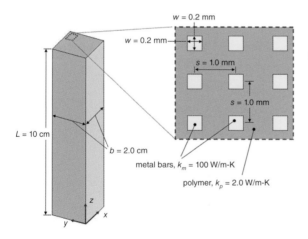

The metal bars are square with edge width $w = 0.2$ mm and are aligned with the z-direction. The bars are arrayed in a regularly spaced matrix with a center-to-center distance of $s = 1.0$ mm. The conductivity of the metal is $k_m = 100$ W/m-K. The length of the material in the z-direction is $L = 10$ cm. The polymer fills the space between the bars and has a thermal conductivity $k_p = 2.0$ W/m-K. The cross-section of the material in the x–y plane is square with edge width $b = 2.0$ cm.

(a) Determine the effective conductivity in the x- and y-directions and in the z-direction.

The outer edges of the material are insulated and the faces of the material (at $z = 0$ and $z = L$) are exposed to a convective boundary condition with $h = 10$ W/m²-K.

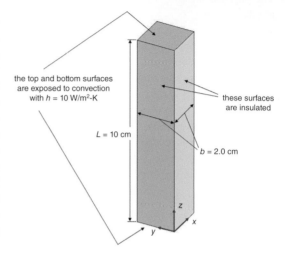

(b) Is it appropriate to treat this as a lumped capacitance problem? Justify your answer.

5.5 A thick aluminum baking sheet with dimensions 12 inch × 16 inch × 1/4 inch is initially at 70°F when it is placed in a 350°F oven. The convection coefficient between the air in the oven and the baking sheet is estimated to be 12 W/m²-K on both its top and bottom surfaces. The baking sheet also participates in radiative heat transfer with the oven walls. The emissivity of the sheet is 0.25 and oven walls are approximately black.

(a) Develop an appropriate expression for the Biot number for this situation

(b) Evaluate the Biot number to determine if the baking sheet can be modeled using the lumped capacitance approximation.

(c) Determine the lumped time capacitance constant for the baking sheet.

5.6 In a manufacturing process, a 1 mm thick rubber sheet with dimensions 0.5 m × 1 m is heated by contact on its top and bottom surfaces with a hot metal surface. The properties of the rubber are $k = 0.015$ W/m-K, $\rho = 1190$ kg/m³, and $c = 1680$ J/kg-K. Owing to the uneven surface, the area-specific contact resistance between the metal and rubber is 1.5×10^{-2} K-m²/W.

(a) Develop an appropriate dimensionless expression to judge whether the rubber can be treated as a lumped material

(b) Can the rubber be considered to be lumped?

5.7 A carbon steel rod having a diameter of 2.5 cm and a length of 0.4 m is heated to a uniform temperature of 532°C. The heating is then stopped and the rod cools by free convection to 25°C air and by radiation to 25°C surroundings. The free convection coefficient is approximately described by the following equation:

$$\bar{h} = 10.3 + 2190/T_s - 1.16e6/T_s^2$$

where the convection coefficient has units of W/m²-K and T_s is the surface temperature in K. The physical properties of the rod are $k = 37$ W/m-K, $\rho = 7900$ kg/m³, $c = 530$ J/kg-K. The emissivity of the surface is 0.82.

(a) Plot the convective and radiative thermal resistances as a function of the surface temperature for temperatures ranging between the initial and final rod temperatures.

(b) Plot the Biot number as a function of temperature.

(c) Indicate if the rod can be considered to be lumped throughout the cooling process.

Analytical Solutions

5.8 A household iron is made of 3 pounds of aluminum. It has a surface area of 0.5 ft² and it is equipped with a 50 W heater. The iron is initially at 70°F (the temperature of the surroundings) and the heat transfer coefficient between the iron surface and the surrounding air is 2.0 Btu/hr-ft²-F. How long will it take for the iron surface to reach 220°F after it is turned on?

5.9 The heat transfer coefficient associated with the flow of 25°C air over a 1.0 cm diameter copper sphere is measured by observing the time–temperature history of a thermocouple that is located at the center of the copper sphere. In one test, the initial temperature indicated by the thermocouple was 65°C. The temperature dropped by 11°C in 75 s. What is your estimate of the heat transfer coefficient?

5.10 A sphere (with conductivity $k = 50$ W/m-K, density $\rho = 5000$ kg/m³, specific heat capacity

$c = 500$ J/kg-K, and radius, $r = 0.002$ m), is removed from a heat treating oven at $T_{ini} = 500$°C and dropped into a large bucket of water at $T_\infty = 10$°C. The heat transfer between the sphere and the water is characterized by a convection coefficient, $\bar{h} = 400$ W/m²-K.

(a) Can you model the sphere as a lumped capacitance (that is, can you assume that the sphere material is all at the same temperature, T, throughout the process)? Justify your answer.

(b) Assuming that you can treat the sphere as a lumped capacitance, how long (in seconds) does it take before the sphere temperature reaches 40°C?

(c) If the sphere thermal conductivity were doubled (from 50 W/m-K to 100 W/m-K) how would this affect your answer to (b)? Your answer need not be exact – just qualitative (i.e., no effect, a large effect, a small effect, etc.) Justify your answer.

(d) If the sphere radius were doubled (from 0.002 m to 0.004 m) how would this affect your answer to (b)? Your answer need not be exact – just qualitative (i.e., no effect, a large effect, a small effect, etc.). Justify your answer.

5.11 An aluminum sphere having a diameter of 2.5 cm is at a uniform temperature of 260°C. It is suddenly subjected to a flowing stream of 40°C water with a heat transfer coefficient of 680 W/m²-K.

(a) Determine if the sphere can be considered to be lumped.

(b) What is the time constant for this system?

(c) Prepare a plot of the sphere temperature as a function of time for 100 s.

5.12 Stainless steel ($k = 14.9$ W/m-K, $\rho = 7900$ kg/m³, and $c = 477$ J/kg-K) ball bearings, which have been heated to a uniform temperature of 850°C, are hardened by quenching them in an oil bath that is maintained at 50°C. The heat transfer coefficient between the surface of the ball bearings and the oil is 1000 W/m²-K. The diameter of the ball bearings is 20 mm. The ball bearings have been coated with a thin coating of a dielectric material. This coating has a negligible effect on the dimensions and properties of the ball bearings, but it adds an additional heat transfer resistance of 33.1 K/W at the surface of the ball bearings. Determine the time required for the center of the coated ball bearings to reach a temperature of 110°C.

5.13 A 0.5 inch diameter copper sphere is uniformly heated to a temperature that is 100°F above the surrounding air temperature. The sphere is then allowed to cool. In 25 s it is found that the sphere temperature is 50°F above the surrounding air temperature. The sphere is polished to minimize radiative heat transfer. The properties of the sphere are $\rho = 558.5$ lb_m/ft^3, $c = 0.0911$ Btu/lb_m-R, and $k = 228.1$ Btu/hr-ft-R. Using this information, determine the average heat transfer coefficient between the sphere and the air.

5.14 An experimental method to determine the value of a convection coefficient for cylinders uses a long copper cylinder, 2 inches in diameter, that has been fitted with a thermocouple at its center. The cylinder is heated until it is uniformly at 90°F and then it is plunged into a water bath at 35°F. After 3 minutes, the temperature of the cylinder is measured to be 44°F and this measurement is used to determine the heat transfer coefficient. Using these data, estimate the heat transfer coefficient for this experiment.

5.15 Cylindrical steel rods, 50 mm in diameter and 5 m long, are heat treated by placing them in an oven in which air is maintained at 750°C. The overall heat transfer coefficient between the rods and the air is 125 W/m²-K. The rods enter the oven at a uniform temperature of 50°C. The heat treating is complete when the rods reach a temperature of 575°C. How long must the rods remain in the oven? The properties of the steel are: thermal diffusivity $\alpha = k/(\rho\, c) = 17.3 \times 10^{-6}$ m²/s and thermal conductivity $k = 62.6$ W/m-K.

5.16 A hailstone (ice) falls through the atmosphere at a constant velocity of 15 m/s. The hailstone has a diameter of 0.01 m, a mass of 0.5×10^{-3} kg and is initially at a uniform temperature of –20°C. The air temperature is 20°C. The convective heat transfer coefficient is estimated to be 150 W/m²-K at these conditions.
  (a) Assuming that the hailstone can be represented as a lumped system, determine the distance it falls until it reaches –5°C.

  (b) Indicate whether the lumped system assumption is warranted.

5.17 A coffee can is filled with 2 kg of water and heated to 80°C. The surface area of the can is 0.1 m² and the heat transfer coefficient between the outside surfaces of the can and the air is 15 W/m²-K. Because of the internal mixing you may assume that the water can be represented as a lumped system.

The heater power is cut and the water cools due to heat transfer to the surrounding air. Assuming the heat transfer coefficient remains unchanged, determine the time required to cool the water to 25°C.

5.18 A hot-wire anemometer is a device that is used to measure air velocity. Its operation is based on the fact that the heat flow from an electrically heated wire in crossflow is related to the velocity of the fluid flowing perpendicular to the wire. In a particular experiment, a hot-wire anemometer is used to estimate the velocity of 70°F, atmospheric air. A platinum wire having a diameter of 0.005 inch and a length of $\frac{1}{4}$ inch is maintained at 450°F. The power required to maintain the wire at this temperature is 0.36 watt.
  (a) Calculate the average heat transfer coefficient between the air and the wire at these conditions.
  (b) The power to the device is turned off. How long does it take the wire to reach 75°F?

5.19 Nuclear fuel pellets must be cooled in order to prevent their overheating. In this problem, we will consider a spherical nuclear fuel pellet with radius $R = 10$ mm that is stored in a retention pond. The pellet has thermal conductivity $k = 2$ W/m-K, density $\rho = 1000$ kg/m³, and specific heat $c = 450$ J/kg-K. The pellet is initially at a uniform temperature, $T_{ini} = 20°C$. The pellet is subjected to convection with a heat transfer coefficient $\bar{h} = 10$ W/m²-K with the pond water that is at temperature T_∞. The temperature of the pond varies sinusoidally with time according to:

$$T_\infty(t) = T_o + \Delta T_m \sin\left(2\pi\frac{t}{\tau_p}\right),$$

where $T_o = 20°C$ is the average pond temperature, $\Delta T_m = 50$ K is the amplitude of the temperature variation, and $\tau_p = 60$ s is the duration of one cycle. The nuclear reaction occurring in the pellet leads to the generation of thermal energy at the rate of $\dot{g} = 1.2$ W.

(a) Calculate the Biot number for this problem. Is using a lumped capacitance model of the pellet justified in order to find its transient temperature?

(b) Determine the time constant associated with the pellet coming to equilibrium via convection.

(c) Assume that the pond temperature is constant (i.e., assume $\Delta T_m = 0$ K). In this limit, develop a transient, analytical solution that provides the temperature response of the pellet as a function of time. Plot the pellet's temperature versus time between 0 s and 900 s. What steady-state temperature does the pellet reach?

5.20 A very long aluminum rod with 5.0 cm diameter is heat treated in a furnace by exposing it to hot gas at 1000°C. The heat transfer coefficient between the gas and the aluminum bar is 100 W/m²-K. The bar is initially at 20°C. You would like the bar to reach 400°C in order to anneal it, but cannot allow it to reach the melting point of aluminum, 660°C.

(a) Is it possible to use a lumped capacitance model to investigate the transient process associated with the aluminum bar? Justify your answer.

(b) What are the minimum and maximum lengths of time that you can allow the aluminum to be in the furnace?

For parts (c) and (d), assume that the manufacturing process has left behind a very thin (2.0 mm) layer of material on the aluminum bar. The material has low thermal conductivity (0.5 W/m-K) but a negligible thermal mass relative to the bar.

(c) Is it possible to use a lumped capacitance model that considers the combined resistance of both the convection and conduction resistance associated with the low-conductivity material (but lumps the aluminum at one temperature)? Justify your answer.

(d) With the resistance of the material included, what are the minimum and maximum lengths of time that you can allow the aluminum to be in the furnace?

5.21 You are building an instrument for measuring the heat transfer coefficient ($\bar{h}$) between a sphere and a flowing fluid. The instrument is a spherical temperature sensor with diameter $D = 3$ mm that is initially in equilibrium with the fluid at T_∞. The sensor has density $\rho = 7500$ kg/m³, specific heat capacity $c = 820$ J/kg-K, and conductivity $k = 75$ W/m-K. The sensor is heated with a constant rate of thermal energy generation of $\dot{g} = 0.1$ W and the time required for the sensor temperature to increase by $\Delta T = 10$ K is recorded. The range of the instrument is expected to be from $\bar{h} = 30$ W/m²-K to $\bar{h} = 300$ W/m²-K.

(a) Can the sensor be treated as a lumped capacitance? Justify your answer.

(b) Assume that your answer from (a) is yes. Develop an equation that relates the measured time to the heat transfer coefficient.

(c) Plot the heat transfer coefficient as a function of measured time.

(d) Assume that you can measure time with an uncertainty of $\delta t_{meas} = 0.1$ s. Use your plot from (c) to estimate the uncertainty of your measurement of the heat transfer coefficient over the range of the sensor.

5.22 During normal operation, an electrical component experiences a constant rate of ohmic dissipation, $\dot{g}_{ini} = 0.01$ W, that causes the conversion of electrical to thermal energy. The component has conductivity $k = 10$ W/m-K, density $\rho = 1000$ kg/m³, and specific heat capacity $c = 100$ J/kg-K. The surface of the component is cooled by convection with air at $T_\infty = 20$°C and heat transfer coefficient $\bar{h} = 10$ W/m²-K. The volume of the component is $V = 1 \times 10^{-7}$ m³ and its surface area is $A_s = 1 \times 10^{-5}$ m².

(a) Is a lumped capacitance model of the component appropriate? Justify your answer.

(b) Assume that your answer to (a) is yes. What is the steady-state temperature of the component?

(c) What is the lumped capacitance time constant of the component?

At time $t = 0$, the power to the circuit is shut off and the ohmic dissipation in the component decays to zero according to:

$$\dot{g} = \dot{g}_{ini} \exp\left(-\frac{t}{\tau_e}\right),$$

where $\tau_e = 25$ s is the electrical time constant of the circuit.

(d) Sketch the temperature of the component as a function of time after the circuit is shut off. Be sure to indicate where the temperature starts, where it ends up, and approximately how long it takes to get there.

(e) Derive the ordinary differential equation for this problem.

(f) What is the initial condition for this problem?

(g) Determine the homogeneous solution for the problem.

(h) Determine the particular solution for the problem.

(i) Provide an equation that can be solved for the undetermined constant.

5.23 You want to bake a batch of delicious chocolate chip cookies from scratch. You remember all of the ingredients and correct proportions, but can't recall the correct time to allow for baking. The cookies have thermal conductivity $k = 6$ W/m-K, density $\rho = 850$ kg/m³, and specific heat capacity $c = 1500$ J/kg-K. Each cookie may be considered to be a disk of radius $R = 2$ inch and thickness $th = 0.5$ inch. The cookies are initially at a uniform temperature, $T_{ini} = 70°$F. They are then placed in an oven where they convect from all surfaces to surrounding air. The heat transfer coefficient is $\bar{h} = 25$ W/m²-K and the oven's air temperature is $T_\infty = 350°$F. As the cookies bake, an endothermic chemical process *removes* thermal energy at a rate

$$\dot{g} = G(T - T_{ini})$$

where $G = 0.5$ W/K.

(a) Calculate the Biot number for this problem. Is using a lumped capacitance model of the cookie justified in order to find its transient temperature?

(b) Determine the time constant that characterizes the time required for the cookie to equilibrate with its surroundings via convection.

(c) Develop an analytical model of the temperature response of the cookie. Plot the cookie's

temperature as a function of time up to $t = 600$ s.

(d) If the cookies burn when their temperature reaches $T_{burn} = 200°$F, at what time should you take them out of the oven?

5.24 A thin disk of material is shown in the figure.

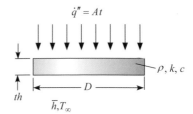

The disk is exposed to a flux of energy on its top side that increases in time linearly according to:

$$\dot{q}'' = At,$$

where A is a constant. Initially, the material is in equilibrium with the surroundings at T_∞. The thickness of the disk is th and the diameter is D. The properties of the material are ρ, k, and c. The disk is cooled on its bottom surface by convection to the surroundings with heat transfer coefficient $\bar{h}$. Assume that a lumped capacitance model is appropriate for the disk.

(a) Derive the governing differential equation that describes the problem.

(b) Determine the solution to the differential equation that you derived in part (a).

5.25 Your cabin is located close to a source of geothermal energy and therefore you have decided to heat it during the winter by lowering spheres of metal into the ground in the morning so that they are heated to a uniform temperature, $T_{gt} = 300°$C during the day. In the evenings you remove the spheres and carry them to your cabin; this trip requires about $\tau_{travel} = 30$ minutes. The spheres are placed in your cabin and give off heat during the night as they cool; the night is $\tau_{night} = 6$ hours long. The heat transfer coefficient between a sphere and the surrounding air (outdoor or cabin) is $\bar{h} = 20$ W/m²-K (neglect radiation) and the temperature of the surrounding air (outdoor or cabin) is $T_{amb} = 10°$C. You can carry about $m = 100$ lb$_m$ of metal and are trying to decide what radius of sphere would work best. You can carry a lot of little spheres (as small as $r_{min} = 5.0$ mm) or a single very large sphere. The properties of the metal are $k = 80$ W/m-K, $\rho = 9000$ kg/m^3, and $c = 1000$ J/kg-K.

(a) What is the largest sphere you could use, r_{max}? That is, what it is the size of a sphere with mass $m = 100$ lb$_m$?

(b) What is the Biot number associated with the maximum size sphere from (a)? Is a lumped capacitance model of the sphere appropriate for this problem?

(c) Prepare a plot showing the amount of energy released from the metal (all of the spheres) during τ_{travel}, the period of time that is required to transport the metal back to your cabin, as a function of sphere radius. Explain the shape of your plot (that is, explain why it increases or decreases).

(d) Prepare a plot showing the amount of energy released from the metal to your cabin during the night (i.e., from $t = \tau_{travel}$ to $t = \tau_{travel} + \tau_{night}$) as a function of sphere radius. Explain the shape of your plot (again, why does it look the way it does?).

(e) Prepare a plot showing the efficiency of the heating process, η, as a function of the radius of the metal. The efficiency is defined as the ratio of the amount of energy provided to your cabin to the maximum possible amount of energy you could get from your mass of metal. (Note that this limit occurs if the metal is delivered to the cabin at T_{gt} and removed at T_{amb}.)

5.26 An instrument on a spacecraft must be cooled to cryogenic temperatures in order to function. The instrument has mass $m = 0.05$ kg and specific heat capacity $c = 300$ J/kg-K. The surface area of the instrument is $A_s = 0.02$ m^2 and the emissivity of its surface is $\varepsilon = 0.35$. The instrument is exposed to a radiative heat transfer from surroundings at $T_{sur} = 300$ K. It is connected to a cryocooler that can provide $\dot{q}_{cooler} = 5$ W. The instrument is exposed to a solar flux that oscillates according to:

$$\dot{q}_s'' = \overline{\dot{q}_s''} + \Delta \dot{q}_s'' \sin(\omega t)$$

where $\overline{\dot{q}_s''} = 100$ W/m^2, $\Delta \dot{q}_s'' = 100$ W/m^2, and $\omega = 0.02094$ rad/s. The initial temperature of the

instrument is $T_{ini} = 300$ K. Assume that the instrument can be treated as a lumped capacitance.

(a) Develop an analytical model of the cool-down process and implement your model in EES.

(b) Plot the temperature as a function of time.

5.27 Small spherical metal pieces are coated with a thin protective layer that must be subjected to an elevated temperature in order to harden. The metallic spheres have diameter $D = 5$ mm and have properties $k = 58.8$ W/m-K, $\rho = 6500$ kg/m^3, and $c = 1200$ J/kg-K. The coating thickness is $th_c = 0.5$ mm and the coating has thermal conductivity $k_c = 0.42$ W/m-K. The processing line exposes the outer surface of the coating to high-temperature air with heat transfer coefficient $\bar{h} = 250$ W/m^2-K. The coating has negligible heat capacity but may act as a significant resistance to heat transfer from the metal to the air that you need to account for in your calculation of the Biot number, the time constant, as well as in your model. You may neglect radiation for this problem.

(a) Estimate the total thermal resistance to heat transfer between the metal and the surroundings (note that this must include both convection and conduction through the coating). Is a lumped capacitance model of the metal appropriate? Justify your answer with a calculation.

(b) Assume that your calculations from (a) justify a lumped capacitance model. Determine the lumped capacitance time constant of the spheres.

As the spheres enter and leave the oven they experience a time varying temperature of their surroundings that is given by:

$$T_\infty(t) = T_{ini} + \Delta T_{oven} \sin\left(\frac{\pi t}{a}\right),$$

where $T_{ini} = 20°$C is the initial temperature of the spheres, $\Delta T_{oven} = 350$ K is the maximum temperature elevation of the air, and $a = 200$ s is the processing time. Note that the argument of the sinusoid must be in radians not degrees.

(c) Derive the governing differential equation for this problem.

(d) Determine the solution to the homogeneous differential equation; your solution should include one undetermined constant.

(e) Determine the solution to the particular differential equation. Note that your particular ODE should include both a constant and a sinusoidal term on the right side. Therefore, your particular equation should include a constant, sine, and cosine term; you need to select the constants that multiply these terms using the method of undetermined coefficients.

(f) Determine the constant of integration by enforcing the initial condition.

(g) Plot the temperature of the sphere and the temperature of the surrounding air as a function of time for $0 < t < a$.

(h) Plot the rate of temperature change of the sphere as a function of time for $0 < t < a$.

(i) You should see in your plot from (h) that $dTdt$ is zero two times during the process: at $t = 0$ and again when the temperature of the spheres reaches its maximum value. Use this fact to determine the maximum temperature reached by the sphere. Plot the maximum temperature as a function of a, the time period associated with the processing. Explain the shape of your plot.

5.28 A temperature sensor is used to measure the temperature in a chemical reactor that operates in a cyclic fashion; the temperature of the fluid in the reactor varies in an approximately sinusoidal manner with a mean temperature $\bar{T}_f = 320°C$, an amplitude $\Delta T_f = 50°C$, and a frequency of $f = 0.5$ Hz. The sensor can be modeled as a sphere of diameter $D = 1.0$ mm. The sensor is made of a material with conductivity $k_s = 50$ W/m-K, specific heat capacity $c_s = 150$ J/kg-K, and density $\rho_s = 16000$ kg/m³. In order to provide corrosion resistance, the sensor has been coated with a thin layer of plastic; the coating is $t_c = 100$ µm thick with conductivity $k_c = 0.2$ W/m-K and has negligible heat capacity relative to the sensor itself. The heat transfer coefficient between the surface of the plastic and the fluid is $\bar{h} = 500$ W/m²-K.

(a) Is a lumped capacitance model of the temperature sensor appropriate? Your lumped capacitance model will account for the resistance due to conduction through the plastic and convection with the fluid but neglect any temperature gradients within the sensor itself.

(b) What is the time constant associated with the sensor? Do you expect there to be a substantial temperature measurement error related to the dynamic response of the sensor?

5.29 A resistance temperature detector (RTD) is placed in a flow of fluid in which the velocity oscillates.

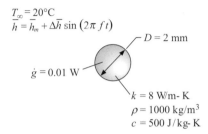

$T_\infty = 20°C$
$h = \bar{h}_m + \Delta\bar{h} \sin(2\pi f t)$
$D = 2$ mm
$\dot{g} = 0.01$ W
$k = 8$ W/m-K
$\rho = 1000$ kg/m³
$c = 500$ J/kg-K

The fluid always has the same temperature, $T_\infty = 20°C$, but the heat transfer coefficient between the fluid and the sensor varies with time due to the oscillating velocity. The heat transfer coefficient $(\bar{h})$ as a function of time (t) is given by:

$$\bar{h} = \bar{h}_m + \Delta\bar{h} \sin(2\pi f t),$$

where $\bar{h}_m = 100$ W/m²-K is the average heat transfer coefficient, $\Delta\bar{h} = 50$ W/m²-K is the amplitude of the variation in the heat transfer coefficient, and $f = 0.1$ Hz is the frequency of oscillation. The RTD is a sphere with diameter $D = 2.0$ mm. The material has properties $k = 8$ W/m-K, $\rho = 1000$ kg/m³, and $c = 500$ J/kg-K. The ohmic dissipation within the RTD that results from energizing it with current leads to the generation of thermal energy in the sensor at a rate of $\dot{g} = 0.01$ W. Neglect radiation for this problem.

(a) Is a lumped capacitance model of the temperature sensor appropriate? Justify your answer.

For the remainder of this problem, assume that your answer to (a) was yes – a lumped capacitance model of the sensor is appropriate.

(b) Estimate the lumped capacitance time constant of the sphere (note that the time constant varies with heat transfer coefficient – determine an approximate value).

The sensor has been operating in the oscillatory environment for a long time and you notice that the temperature measured by the sensor is not equal to $T_\infty = 20°C$. Instead, the temperature of the sensor is higher than T_∞ and oscillates.

(c) Sketch the temperature of the sensor as a function of time; on the same sketch show the time variation of the heat transfer coefficient on a secondary axis. Make sure that your sketch for temperature (the primary axis) is consistent with the variation of the heat transfer coefficient and also with the time constant that you calculated in (b). Justify various characteristics of your sketch.

(d) Derive the governing ordinary differential equation for this problem.

5.30 Fresh cherries with diameter $D = 18$ mm are initially at $T_{ini} = 30°C$ and must be frozen by suddenly placing them in cold nitrogen gas at $T_\infty = -180°C$. The freezing point for the cherries is $T_s = -1.8°C$. The thermal properties of fresh cherries are as follows.

Density	$\rho = 1050$ kg/m^3
Thermal conductivity	$k = 0.545$ W/m-K
Specific heat, unfrozen	$c = 3.52$ kJ/kg-K
Specific heat, frozen	$c_s = 1.85$ kJ/kg-K
Latent heat of fusion	$\Delta h_{sf} = 267$ kJ/kg

These thermal properties may be treated as constants. The convective heat transfer coefficient between the cherry and the gaseous nitrogen is $\bar{h} = 15$ W/m^2-K for natural convection. The frozen cherries can be packaged once their temperature reaches $T_{final} = -80°C$.

(a) Is the lumped capacitance model appropriate for this problem?

(b) Determine the time required to cool the cherry from its initial temperature to the freezing point.

(c) Once the cherry has reached its freezing temperature, determine the time required for it to completely freeze.

(d) Determine the time required to take the frozen cherry from T_s to T_{final}.

(e) Plot the total time required to process the cherry (i.e., the sum of your answers from parts (b), (c), and (d)) as a function of the cherry diameter. Indicate on your plot the diameter at which the lumped capacitance model is no longer valid.

5.31 One technique for detecting chemical threats uses a laser to ablate small particles so that they can subsequently be analyzed using ion mobility spectroscopy. The laser pulse provides energy to a particle according to:

$$\dot{q}_{laser} = \dot{q}_{max} \exp\left[-\frac{(t - t_p)^2}{2t_d^2}\right],$$

where $\dot{q}_{max} = 0.22$ W is the maximum value of the laser power, $t_p = 2$ μs is the time at which the peak laser power occurs, and $t_d = 0.5$ μs is a measure of the duration of the pulse. The particle has radius $r_p = 5$ μm and has properties $c = 1500$ J/kg-K, $k = 2.0$ W/m-K, and $\rho = 800$ kg/m^3. The particle is surrounded by air at $T_\infty = 20°C$. The heat transfer coefficient is $\bar{h} = 60000$ W/m^2-K. The particle is initially at T_∞.

(a) Is a lumped capacitance model of the particle justified?

(b) Assume that your answer to (a) is yes; develop an analytical model of the particle (you may find that a symbolic software package such as Maple is useful for this). Plot the temperature of the particle as a function of time. Overlay on your plot (on a secondary axis) the laser power.

5.32 A metallic sphere with conductivity, k, and radius, r, is implanted in tissue at T_{inf}. Initially, the sphere is in equilibrium with the tissue. At some time ($t = 0$), the tissue is exposed to an electric field which causes a constant rate of volumetric generation ($\dot{g}$) within the sphere. The heat transfer between the sphere and the tissue is characterized by a convection coefficient $\bar{h}$.

(a) Assume that you can treat the sphere as a lumped capacitance (that is, assume that the sphere material is all at the same temperature, T, throughout the process). Derive the governing differential equation for the sphere.

(b) At very long times, what will the temperature of the sphere be (i.e., if you keep the volumetric generation rate activated for a long time, what temperature will the sphere eventually reach)?

(c) Provide a solution for the temperature of the sphere as a function of time in terms of the symbols listed in the problem statement.

Numerical Solutions

5.33 You have decided to develop a numerical solution to Problem 5.22. You are going to use uniform time steps of duration $\Delta t = 10$ s.
  (a) Write down the equation required to take an Euler step from $t = 0$ to $t = \Delta t$. What is the predicted temperature at $t = \Delta t$?
  (b) Will your Euler solution be stable? Justify your answer.

5.34 A pan is placed above a camp fire flame. The pan can be modeled as a circular disk with thickness $th = 0.25$ inch and diameter $D = 8$ inch. The pan is made of material with density $\rho = 7870$ kg/m^3, specific heat capacity $c = 450$ J/kg-K, and conductivity $k = 70$ W/m-K. The bottom side of the pan is exposed to radiation from the flame at $T_{flame} = 650°C$. The upper side of the pan is exposed to radiation with the environment at $T_\infty = 15°C$. The emissivity of the pan surface is $\varepsilon = 0.8$. The upper side of the pan is also exposed to convection with the environment. The heat transfer coefficient depends on the surface temperature of the pan (T) according to:

$$\bar{h} = 2\left[\frac{W}{m^2\text{-K}}\right] + 0.01\left[\frac{W}{m^2\text{-K}^2}\right](T - T_\infty).$$

  (a) What is the steady-state temperature of the pan (T_{ss})?
  (b) Estimate the lumped capacitance time constant of the pan during the time that it is heated from T_∞ to T_{ss}.
  (c) Is a lumped capacitance model of the pan appropriate?
  (d) Develop a numerical model of the temperature of the pan using the Euler technique; assume that the pan is initially in thermal equilibrium with the surroundings. Plot the temperature of the pan as a function of time.

5.35 The water storage tank in a solar water heating system can be considered to be a lumped system because of the internal mixing that occurs when water flows into and out of the tank. The solar water tank has a volume of 100 liters and is initially filled with water at a uniform temperature of 34°C. Starting at 6:00 a.m., water is

constantly withdrawn from the tank for use at a rate of 20 liters per hour and replaced with water at 20°C. Also starting at 6:00 a.m., a pump is activated allowing heated water from the solar collector to enter the tank at 10 liters per hour and a temperature that depends on time according to:

$$T = 82[°C]\sin\left(\frac{t}{12[\text{hr}]}\frac{\pi}{2}\right),$$

where t is the military time (i.e., the number of hours that have passed since midnight). This process continues until 6:00 p.m. at which point the collector pump is stopped. The specific heat and density of water can be assumed to be constants evaluated at 50°C.
  (a) What is the maximum time step duration for which the solution will be stable using the Euler Method?

Use the Euler Method to do the remainder of the problem. Choose a time step duration that you consider to be small enough to guarantee accurate results.
  (b) Prepare a numerical solution for this problem that will calculate and plot the average temperature of the water in the tank as function of time for a 12 hour period.
  (c) Determine the total energy added to the tank from the solar collector over the 12 hour period.
  (d) Determine the total energy withdrawn from the tank during the 12 hour period.

5.36 Reconsider Problem 5.29. You have decided to model the temperature sensor using an Euler technique with time steps of duration $\Delta t = 0.5$ s. The temperature of the sensor at $t_1 = 0$ is $T_1 = 20°C$.
  (a) Predict the temperature of the sensor at the end of the first time step (i.e., what is your prediction for T_2, the temperature at $t_2 = \Delta t$)?
  (b) Determine the critical time step for your simulation; that is, what is the largest time step that you can take before the simulation becomes unstable?
  (c) We ignored radiation for this problem – was this a reasonable thing to do? Justify your answer.

5.37 Reconsider Problem 5.19 but allow the temperature of the pond to vary sinusoidally with time according to:

$$T_\infty = T_o + \Delta T_m \sin\left(2\pi \frac{t}{\tau_p}\right),$$

where $T_o = 20°C$ is the average pond temperature, $\Delta T_m = 50$ K is the amplitude of the temperature variation, and $\tau_p = 60$ s is the duration of one cycle.

Develop a transient, numerical model of the temperature response of the pellet. Use the Euler Method and implement your solution in EES. Overlay the pellet's temperature as a function of time between 0 s and 900 s onto a plot of the analytical solution from Problem 5.19 that neglected the time variation of the pond temperature.

5.38 A long steel bar is 0.04 m × 0.02 m. The back and top sides are well insulated. The front and bottom sides convect to 20°C water with a convection coefficient of 80 W/m²-K. Electrical power is dissipated within the bar at a rate per unit length of 400 W/m. The bar is initially in thermal equilibrium with the water when the power is turned on. The steel has a thermal conductivity of 60 W/m-K, a density of 8000 kg/m³, and a specific heat of 460 J/kg-K.

back and top sides insulated

front and bottom sides convect to 20°C water with convection coefficient of 80 W/m²-K

Assume that the bar can be lumped and validate this assumption.

(a) Derive an algebraic equation that will provide the value of the bar one time step in the future using the Euler Method.

(b) Determine the critical time step.

(c) Use the model to plot the temperature of the bar versus time for a 1 hour period. Indicate the time at which boiling will occur on the bar surface.

5.39 A disk-shaped piece of material is used as the target of a laser, as shown in the figure. The laser target diameter is D and its thickness is b. The target is made of a material with density ρ, conductivity k, and specific heat capacity c. The front side of the target is exposed to a time varying heat flux from a laser, $\dot{q}''_{laser}(t)$ which is given by the function

$$\dot{q}''_{laser}(t) = At^2 \exp\left(-\frac{t}{t_{pulse}}\right),$$

where A and t_{pulse} are constants and t is time. The back side of the target is cooled by exposure to fluid at T_f with heat transfer coefficient $\bar{h}$. You may neglect radiation from the target and convection from the front side of the target.

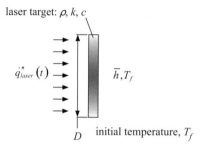

laser target: ρ, k, c

$\dot{q}''_{laser}(t)$

$\bar{h}, T_f$

D initial temperature, T_f

(a) How would you determine whether a lumped capacitance model of the laser target is appropriate? Write down the equation that you would need in terms of symbols given in the problem statement.

(b) Assume that a lumped capacitance model is appropriate for the laser target. Derive the ordinary differential equation that governs the temperature of the laser target (T). You should end up with an equation for the time rate of change of the target temperature in terms of the symbols listed in the problem statement.

(c) You want to simulate the problem using a numerical method based on the Euler technique. Write the equation that you would use to move through the first time step (i.e., from T[1] at $t = 0$ to T[2] at $t = \Delta t$). The duration of the time step is Δt.

(d) Will the Euler step in part (c) become unstable if Δt is very large? If so then what is the critical time step; that is, how large of a time step can you take before the solution becomes unstable?

(e) You want to simulate the problem using a numerical method based on the fully implicit technique. Write the equation that you would use to move through the first time step (i.e., from T[1] at $t = 0$ to T[2] at $t = \Delta t$).

(f) Will the implicit step in part (e) become unstable if Δt is very large? If so then what is the critical time step; that is, how large of a time step can you take before the solution becomes unstable?

5.40 A temperature sensor is used to measure the temperature in a chemical reactor that operates in a cyclic fashion; the temperature of the fluid in the reactor varies in an approximately sinusoidal manner with a mean temperature $\bar{T}_f = 320°C$, an amplitude $\Delta T_f = 50°C$, and a frequency of $f = 0.5$ Hz. The sensor can be modeled as a sphere of diameter $D = 1.0$ mm. The sensor is made of a material with conductivity $k_s = 50$ W/m-K, specific heat capacity $c_s = 150$ J/kg-K, and density $\rho_s = 16,000$ kg/m³. In order to provide corrosion resistance, the sensor has been coated with a thin layer of plastic; the coating is $t_c = 100$ μm thick with conductivity $k_c = 0.2$ W/m-K and has negligible heat capacity relative to the sensor itself. The heat transfer coefficient between the surface of the sensor and the fluid is $\bar{h} = 500$ W/m²-K. The initial temperature of the sensor is 25°C. Using a numerical method, determine the sensor temperature as a function of time. Plot the sensor and fluid temperatures as a function of time for 10 s and comment on the ability of this sensor to accurately measure the fluid temperature.

5.41 An instrument on a spacecraft must be cooled to cryogenic temperatures in order to function. The instrument has mass $m = 0.05$ kg and specific heat capacity $c = 300$ J/kg-K. The surface area of the instrument is $A_s = 0.02$ m² and the emissivity of its surface is $\varepsilon = 0.35$. The instrument is exposed to a radiative heat transfer from surroundings at $T_{sur} = 300$ K. It is connected to a cryocooler. The cooling power of the cryocooler (W) is not constant but rather is a function of temperature (K) given by:

$$\dot{q}_{cooler} = \begin{cases} -4.995 + 0.1013T - 0.0001974T^2 & \text{if } T > 55.26 \text{ K} \\ 0 & \text{if } T < 55.26 \text{ K} \end{cases}$$

The instrument is exposed to a solar flux that oscillates according to: $\dot{q}''_s = \bar{\dot{q}}''_s + \Delta \dot{q}''_s \sin(\omega t)$, where $\bar{\dot{q}}''_s = 100$ W/m², $\Delta \dot{q}''_s = 100$ W/m², and $\omega = 0.02094$ rad/s. The initial temperature of the instrument is $T_{ini} = 300$ K. Assume that the instrument can be treated as a lumped capacitance.

(a) Develop a numerical model using Heun's Method. Plot the temperature of the instrument as a function of time for 2000 s after the cryocooler is activated.

(b) Develop a numerical model in EES using the Integral command. Plot the temperature of the instrument as a function of time for 2000 s after the cryocooler is activated.

5.42 A spherical object experiences a cross-flow of fluid with velocity $u_\infty = 4$ m/s. The diameter of the sphere is $D = 0.015$ m. The heat transfer coefficient is $\bar{h} = 7,000$ W/m²-K. The object is composed of material with $k_s = 450$ W/m-K, $c_s = 420$ J/kg-K, and $\rho_s = 4,000$ kg/m³. You may neglect radiation for this problem.

(a) Is a lumped capacitance approximation appropriate for this problem? Justify your answer.

(b) Determine the lumped capacitance time constant of the object.

The free stream temperature of the fluid, T_∞, varies with time as shown in the figure. The initial temperature of the object is $T_{ini} = 300$ K. The variation of T_∞ with time that is shown in the figure can be described according to the following polynomial:

$$T_\infty = a_0 + a_1 t + a_2 t^2,$$

where $a_0 = 300$ K, $a_1 = 400$ K/s, and $a_2 = -200$ K/s², and t is time.

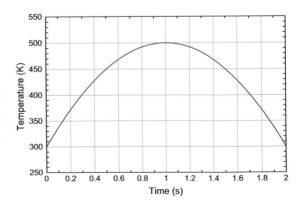

(c) Sketch the temperature of the object as a function of time on the figure. Also sketch the temperature of the object with respect to time that you would expect if the time constant was 50 s and 0.05 s. Clearly label each of the three sketches.

(d) Use the differential equation to take the first time step of a numerical solution using Heun's technique with $\Delta t = 0.25$ s. That is, predict T_2 at $t_2 = \Delta t$ given $T_1 = 300$ K at $t_1 = 0$ s.

(e) Develop an analytical solution to the problem.

5.43 The Integrated Power Electronics Module (IPEM) used to control the motor on one wheel of a hybrid electric vehicle experiences a large amount of ohmic heating, particularly during periods of acceleration. The IPEM is cooled by an array of fins, as shown in the figure.

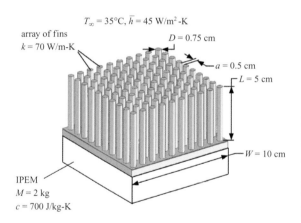

$T_\infty = 35°C, \bar{h} = 45$ W/m²-K

array of fins
$k = 70$ W/m-K

$D = 0.75$ cm

$a = 0.5$ cm

$L = 5$ cm

$W = 10$ cm

IPEM
$M = 2$ kg
$c = 700$ J/kg-K

The heat sink is square; the size of the heat sink is $W = 10$ cm. The conductivity of the material is $k = 70$ W/m-K. The distance between the edges of two adjacent fins is $a = 0.5$ cm, the diameter of a fin is $D = 0.75$ cm, and the length of each fin is $L = 5$ cm. Note that the number of fins on the heat sink should be computed based on the dimensions rather than by counting fins in the figure. The fins are cooled by a flow of air at $T_\infty = 35°C$ provided by a fan. The heat transfer coefficient is $\bar{h} = 45$ W/m²-K. The IPEM is initially at T_∞ and has mass $M = 2$ kg and specific heat capacity $c = 700$ J/kg-K. You may neglect the heat capacity of the fin material and treat the

IPEM as a lumped capacitance for this problem. The figure illustrates the ohmic dissipation in the IPEM as a function of time associated with a typical acceleration event.

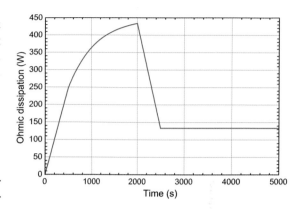

The EES code below will return $\dot{g}$ (the rate of thermal energy generation in the IPEM) given a value of time.

```
function g_dot(time)
  "returns thermal energy generation (W)
  given time (s)"
  if (time<500 [s]) then
    g_dot=250 [W]*time/500 [s]
  else
    if (time<2000 [s]) then
      g_dot=250 [W]+200[W]*(1-exp(-(time-&
        500 [s])/600 [s]))
    else
      if (time<2500 [s]) then
        g_dot=250 [W]+200[W]*(1-exp(-1500&
          [s]/600 [s]))-300[W]*(time-2000 [s])/&
          500 [s]
      else
        g_dot=250 [W]+200[W]*(1-exp(-1500&
          [s]/600 [s]))-300[W]
      endif
    endif
  endif
end
```

You have been asked to simulate the thermal response of the IPEM during the acceleration event shown in the figure.

(a) Determine the efficiency of the fins in the heat sink.

(b) Determine the total thermal resistance between the IPEM and the surroundings.

(c) Determine the lumped capacitance time constant of the IPEM.

(d) Develop a numerical model of the response of the IPEM using the Crank–Nicolson technique. The Crank–Nicolson technique is a slight modification to the fully implicit technique in which the time rate of temperature change used to take the step is the average of the value determined at the beginning and the end of each time step (vs. just the end). It is both implicit and also second order. Plot the temperature of the IPEM as a function of time for $t_{sim} = 5000$ s.

5.44 A block of cheese that is 5.0 cm on each side is placed in a refrigerator with a faulty temperature control unit. The refrigerator temperature (T_{ref}) varies sinusoidally with an average temperature of 2.0°C and an amplitude of 3.0°C over the course of each day:

$$T_{ref}(t) = 2.0 + 3.0 \sin(\omega t),$$

where ω is the angular frequency associated with a 24 hour period (note – make sure that you have EES' unit system set to radians rather than degrees to avoid problems with your sine function), and t is time. The properties of cheese can be considered to be equivalent to liquid water. The heat transfer coefficient between the cheese and the air in the refrigerator is constant and equal to 3.0 W/m²-K. The air in the refrigerator is assumed to be at the refrigerator temperature. Assume that the cheese is initially at 2.0°C.

(a) Create a program using EES that can predict the cheese's temperature (assuming that it behaves as a lumped capacitance) for two days. Your program should use Heun's technique.

(b) Does the cheese ever freeze during the two days?

(c) Is the use of a lumped capacitance model justified for this situation?

5.45 An aluminum sphere having a diameter of 2.5 cm is at a uniform temperature of 260°C. It is then subjected to a flowing stream of 40°C water with a convection coefficient of 680 W/m²-K. Prepare a plot of the sphere temperature as a function of time for 100 s using

each of the methods listed. For each case, use a time step duration that is one-half of the critical time step duration and comment on the accuracy of each method relative to the analytical solution to this problem which was determined in Problem 5.11 by calculating the maximum error.

(a) Euler's Method.

(b) Heun's Method.

(c) The Fully Implicit Method.

(d) EES' Integral function.

5.46 Repeat Problem 5.35 using the EES Integral function instead of the Euler Method.

5.47 A copper sphere having a diameter of 4 cm is equipped with a 2.5 W electric heating element. The sphere is placed in a Dewar container which is evacuated. Because of the vacuum, there is no convective heat transfer from the sphere to its surroundings. However, the sphere radiates to the wall of the Dewar, which is at 25°C. The emissivity of the surface of the sphere is 0.2. The sphere is initially at 25°C. Using a numerical solution with the EES' Integral function, prepare a plot of the sphere temperature as a function of time for a 3 hour (10,800 s) period. How long does it take for the sphere to reach 100°C?

5.48 The wall of a storage vessel for high-pressure carbon dioxide gas is made of AISI316 stainless steel with a thickness of 4 cm. The outside surface of the wall is well insulated. The inside surface has been coated with a thin layer of an anti-corrosion material that provides an effective, area-specific thermal resistance of 0.1 K-m²/W. The wall is initially at a uniform temperature when it is exposed to carbon dioxide at 220°C through a heat transfer coefficient of 14 W/m²-K.

(a) Determine if it is appropriate to consider the wall to be lumped by evaluating an appropriate Biot number for this wall.

(b) Using a numerical solution with the EES Integral function, prepare a plot of the average temperature of the wall as a function of time for a period of 12 hours after its initial exposure to the hot gas.

5.49 A 1 inch diameter sphere made of aluminum has an embedded electrical heater that provides heat at a rate that depends on time according to:

$$\dot{q}_{htr} = 0.5[W] + 2[W]\left|\sin\left(\frac{t}{45}\pi\right)\right|,$$

where t is time (in seconds). The sphere is initially at 25°C. The convection coefficient between the surrounding 25°C air and the sphere is 25 W/m²-K. Show that the sphere can be assumed to be lumped and then plot the sphere temperature for a period of 10 minutes. Use a numerical solution with the EES Integral function.

5.50 A glass bottle containing 1.1 liters of water at 75°C is placed in an ice batch. The water in the bottle is stirred so it can be assumed to be at a uniform temperature at any time. The walls of the bottle are 6 mm thick. You may neglect the heat capacity of the glass. The ratio of the height to the diameter of the bottle is 2.2. The heat transfer coefficient between the water inside the bottle and the glass wall is 84 W/m²-K. The heat transfer coefficient between the ice bath and the outside surface of the bottle is 124 W/m²-K. Prepare a plot of the average water temperature as a function of time for a period of 60 minutes. Use a numerical solution with the EES Integral function.

5.51 An experiment to determine the heat transfer coefficient consists of a cube of pure aluminum ($c = 900.9$ J/kg-K, $\rho = 2699$ kg/m³, $k = 234.9$ W/m-K) that is 2.5 cm in each dimension. The cube has an embedded heater that is repeatedly turned on and off at 1200 s intervals. The cube, which is initially at 25°C, convects from all six faces to the environment at 25°C with a heat transfer coefficient that is expected to be between 10 and 30 W/m²-K.
(a) Show that the cube can be treated as a lumped system
(b) Calculate and plot the temperature of the cube as a function of time for a 1 hour period for heat transfer coefficients of 10, 15, 20, 25, and 30 W/m²-K. Place all of the plots on one graph. Use a numerical solution with the EES Integral function.
(c) The maximum temperature observed during the experiment was 342 K. Using the data in the plots, what is your estimate of the heat transfer coefficient?

5.52 Reconsider Problem 5.26 using a numerical model. The cooling power of the cryocooler (W) is not constant but is a function of temperature (K):

$$\dot{q}_{cooler} = \begin{cases} -4.995 + 0.1013T - 0.0001974T^2 & \text{if } T > 55.26 \text{ K} \\ 0 & \text{if } T < 55.26 \text{ K} \end{cases},$$

where T is the temperature of the instrument.
Develop a numerical model in EES using the Integral command. Plot the temperature of the instrument as a function of time for 2000 s after the cryocooler is activated.

5.53 Reconsider Problem 5.52 using a numerical model developed in MATLAB using the ODE solver. Plot the temperature of the instrument as a function of time for 2000 s after the cryocooler is activated.

Projects

5.54 A calorimetric detector can be generated using MEMS techniques by suspending a tiny plate containing a temperature sensor from one or more cantilever beams. When energy is applied to the plate (for example, from a reaction or an energetic particle) then the plate temperature rise is measured. The magnitude of the temperature rise indicates the amount of energy and the time resolution of the temperature provides a signature of the time evolution of the process being monitored. One design of a calorimetric detector is shown in the figure.

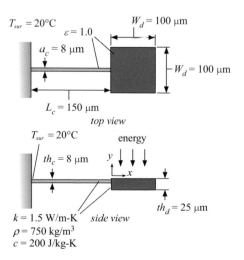

The cantilever that supports the detector is L_c = 150 μm long, a_c = 8 μm wide, and th_c = 8 μm thick. The cantilever is anchored at a structure that remains always at T_{sur} = 20°C. The detector is square with dimension W_d = 100 μm and thickness th_d = 25 μm. The detector and cantilever surfaces have emissivity ε = 1 and radiate to surroundings at T_{sur} = 20°C. The detector is placed in a vacuum and therefore there is no convection. The properties of the detector and the cantilever are k = 1.5 W/m-K, ρ = 750 kg/m³, and c = 200 J/kg-K. Assume that the sensor is detecting energy from some source. The total energy associated with an event is Q_p = 0.1 × 10⁻⁶ J. You may model the interaction with the detector as occurring over t_p = 0.02 s and being uniformly distributed in time. The detector is initially at T_{sur}.

(a) Can the cantilever be modeled as a conduction thermal resistance for a plane wall between the detector and the surrounding structure? Justify your answer.

(b) Can the detector be modeled as a lumped capacitance? Justify your answer.

(c) Assume that your answers to (a) and (b) are both yes. Prepare an analytical lumped capacitance model of the detector during the time that the energy is applied ($0 < t < t_p$) and after the time that the energy is applied ($t > t_p$).

(d) Prepare a plot showing the temperature of the detector as a function of time for $0 < t < 0.1$ s.

(e) If the resolution of the temperature detector is δT = 0.1 K then what is the smallest amount of energy that you can detect?

(f) For Q_p = 0.1 × 10⁻⁶ J, what is the time resolution of the detector (i.e., your detector will indicate that an event happens some time after the actual event happens – what is this time lag?).

5.55 A power electronics component is exposed to a flow of evaporating refrigerant in order to cool it. The component has a square cross-section with width and length W = 1 cm and thickness th = 1 mm. The top surface ($W \times W$) of the component is exposed to the refrigerant while the other surfaces are adiabatic. The density of the material is ρ = 2330 kg/m³, the specific heat

capacity is c = 712 J/kg-K, and the conductivity is k = 148 W/m-K. The heat transfer coefficient associated with convection from the top surface of the component to the refrigerant is strongly dependent on the heat flux, $\dot{q}''$:

$$\bar{h} = \bar{h}_0 + A\left(\frac{\dot{q}''}{\bar{q}''}\right)^b$$

where $\bar{h}_0$ = 250 W/m²-K is the heat transfer coefficient at zero flux, A = 2500 W/m²-K, b = 0.8, and $\bar{q}''$ = 2.5×10⁴ W/m² is the average heat flux experienced by the device. The temperature of the refrigerant is T_∞ = 22°C. The heat flux from the surface varies with time cyclically, with a period of t_p = 5 s. During the first half of the cycle ($0 < t < t_p/2$), the heat flux from the device is low and during the second half of the cycle ($t_p/2 < t < t_p$) it is high. The heat flux as a function of time is given by:

$$\dot{q}'' = \begin{cases} (1-f)\bar{q}'' & \text{for } 0 < t < t_p/2 \\ (1+f)\bar{q}'' & \text{for } t_p/2 < t < t_p \end{cases},$$

where f = 0.5 is the fractional variation of the heat flux from its average value. We are interested only in the cyclic steady-state variation in the temperature of the component. Therefore, we want the solution where the temperature at the beginning and end of the cycle is the same.

(a) Is a lumped capacitance model of the component appropriate? Justify your answer.

(b) Develop an analytical solution that provides the cyclic steady-state temperature variation of the component. Note that you will need a solution that is valid during the low heat flux period of time (the first half of the cycle) and another that is valid during the high heat flux period (the second half).

(c) Plot the temperature as a function of time.

(d) Plot the maximum temperature experienced by the device as a function of the parameter f for various values of t_p. Notice that f = 0 corresponds to the case where the heat flux is not varied and f = 1 corresponds to the case where the flux is switched completely off during the low heat flux periods.

5.56 A hot metal sphere at 100°C with a total heat capacity of 100 J/K is dropped into an insulated container of water initially at 10°C with total heat capacity 400 J/K. The heat transfer coefficient between the sphere and the water is 25 W/m²-K and the area of the sphere is 0.01 m². The sphere can be treated as a lumped capacitance because of its high thermal conductivity and the water can be treated as a lumped capacitance because of mixing related to natural convection. Develop an analytical solution for the temperature of the sphere and the temperature of the water as a function of time, implement your solution in EES and plot these temperature variations.

5.57 Repeat Problem 5.56 using a numerical technique.

5.58 The figure illustrates a simple model of an industrial soldering iron tip. The soldering iron can be approximated as a cylinder of metal with radius r_{out} = 5.0 mm and length L = 20 mm. The metal is carbon steel; assume that the steel has constant density ρ = 7854 kg/m³ and constant conductivity k = 50.5 W/m-K, but a specific heat capacity that varies with temperature according to:

$$c = 374.9 \left[\frac{J}{kg\text{-}K} \right] + 0.0992 \left[\frac{J}{kg\text{-}K^2} \right] T$$
$$+ 3.596 \times 10^{-4} \left[\frac{J}{kg\text{-}K^3} \right] T^2.$$

The surface of the iron radiates and convects to surroundings that have temperature T_{amb} = 20°C. The heat transfer coefficient is $\bar{h}$ = 10 W/m²-K and the surface of the iron has an emissivity ε = 1.0. The iron is heated electrically by ohmic dissipation; the rate at which electrical energy is added to the iron is $\dot{g}$. The soldering iron can be used once the tip temperature reaches its operating temperature, T_{op} = 520°C.

(a) What is the value of generation, $\dot{g}_{ss}$, that is required in order to sustain a steady-state tip temperature of T_{op}?

(b) Can the soldering iron be treated as a lumped capacitance? Justify your answer.

(c) What is the approximate time constant of the soldering iron? That is, about how long would someone have to wait for it to warm up if it were activated with a constant rate of generation?

(d) Develop a numerical model implemented in EES that can predict the temperature of the soldering iron as a function of time. Assume that it is activated at ambient temperature with the generation that was calculated in (b).

(e) Develop an analytical model that can predict the temperature of the soldering iron as a function of time in the limit that radiation is neglected and the heat capacity is constant (and equal to its value at the average of the ambient and the operating temperatures). Show that your numerical model from part (d) agrees with the analytical model in this limit (no radiation and constant heat capacity) by overlaying a plot of the analytical solution on top of the numerical solution (suitably modified so that it is consistent with your analytical solution).

(f) You want to evaluate methods for accelerating the soldering iron's heat up process. Assume that the maximum heater power that can be applied is 100 W and that you can sense the tip temperature and control the power based upon the instantaneous tip temperature. Modify your model from (d) so that the power applied to the tip obeys a simple proportional control algorithm:

$$\dot{g} = \dot{g}_{ss} + K_p \left(T_{op} - T \right)$$

with Max $(\dot{g})$ = 100 W and Min $(\dot{g})$ = 0 W

where K_p = 0.5 W/K is the proportional controller gain and $\dot{g}_{ss}$ is the generation required at steady state, computed in part (a). Overlay a plot of the temperature vs. time attained using this controlled soldering iron onto the original, uncontrolled response that was predicted in part (d).

5.59 One technique for detecting chemical threats uses a laser to ablate (i.e., thermally destroy)

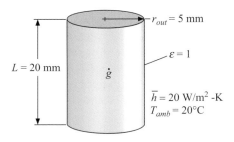

r_{out} = 5 mm

ε = 1

L = 20 mm

$\dot{g}$

$\bar{h}$ = 20 W/m² -K
T_{amb} = 20°C

small particles so that they can subsequently be analyzed using ion mobility spectroscopy. The laser pulse provides energy to a particle at a rate of $\dot{q}_{laser}$ = 0.025 W. The particle is spherical with radius r_p = 7 μm and has properties c = 1500 J/kg-K, k = 3.5 W/m-K, and ρ = 800 kg/m³. The particle is surrounded by air at T_∞ = 20°C. The heat transfer coefficient is $\bar{h}$ = 60,000 W/m²-K. The particle is initially at T_∞. Neglect radiation.

(a) Is a lumped capacitance model of the particle justified?

(b) Assume that your answer from (a) is yes. Determine a lumped capacitance time constant for the particle.

(c) Develop an analytical solution for the temperature of the particle as a function of time after the laser is activated. Clearly show the derivation of your solution.

(d) Plot the temperature of the particle as a function of time for $0 < t < 200$ μs.

(e) Prepare a numerical solution to the problem. Use a function that returns the laser power as a function of time (a constant for now, but later on we will explore time varying laser power) and use that function in your numerical solution.

(f) Overlay your numerical solution on your plot from (d).

(g) Determine the global error associated with the numerical solution (the maximum difference between the analytical solution from (c) and the numerical solution from (e)). Plot the global error as a function of the time step duration on a log–log plot.

(h) What is the order of the numerical technique that you used? What characteristic of your plot from (g) indicates the order?

You have been unable to find a laser that can continuously provide a sufficient amount of power to ablate the material. Therefore, you are exploring the use multiple pulses. The laser that you are looking at can provide four closely spaced pulses, each pulse has a Gaussian time distribution. The laser power at any time is approximately given by

$$\dot{q}_{laser} = \dot{q}_{max} \sum_{i=1}^{4} \exp\left[-\frac{(t - t_{p,i})^2}{2t_d^2}\right],$$

where $\dot{q}_{max}$ = 0.2 W is the maximum value of the laser power, t_d = 0.5 μs is a measure of the

duration of the pulse, and $t_{p,i}$ is the time at which the peak of each pulse occurs. Your laser has a 333 kHz repetition rate; that is, the pulses can occur every 3 μs. Therefore, the pulse peak times are $t_{p,1}$ = 3 μs, $t_{p,2}$ = 6 μs, $t_{p,3}$ = 9 μs, and $t_{p,4}$ = 12 μs.

(i) Modify your function for the laser power to reflect the 4-pulse train time variation. Plot laser power as a function of time for $0 < t < 20$ μs in order to verify that your function is operating correctly; you should see 4 pulses that reach 0.2 W separated by a time of 3 μs with a duration of approximately 0.5 μs.

(j) Use your numerical model to predict the temperature of the particle as a function of time when it is exposed to the time varying laser power discussed in (i). Plot the temperature as a function of time for $0 < t < 20$ μs.

5.60 The figure illustrates an apparatus that is used to cure a thin film on the surface of a target wafer.

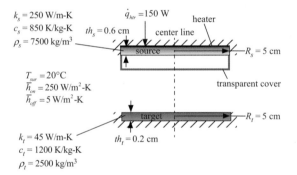

The apparatus consists of a source of radiation that is in close proximity to the target wafer. The source disk is heated electrically so that it emits the radiation required to cure the film on the target. The target wafer is cooled by a flow of air driven by a fan. The target remains in the position shown until the process is complete. At that point, the heater and fan are both deactivated. The operator has observed that the deactivation of the device actually causes the target wafers to fail; they seem to get too hot and crack due to thermal expansion. This is surprising – the heater is deactivated so you expect that everything would cool down. Therefore, you have been asked to carry out a thermal analysis on the system.

The source is a disk with radius R_s = 5 cm and thickness th_s = 0.6 cm. The top and sides are insulated and do not experience any heat transfer. The bottom surface experiences radiation heat transfer with the target ($\dot{q}_{rad,s,t}$) and the surroundings ($\dot{q}_{rad,s,sur}$). The source, target, and surroundings can all be modeled as being black surfaces and therefore these radiation heat transfers can be written as

$$\dot{q}_{rad,s,t} = A_s F_{s,t}\, \sigma\left(T_s^4 - T_t^4\right)$$

and

$$\dot{q}_{rad,s,sur} = A_s F_{s,sur}\, \sigma\left(T_s^4 - T_{sur}^4\right),$$

where $F_{s,t} = 0.8$ is the view factor between the source and the target and $F_{s,sur} = 0.2$ is the view factor between the source and the surroundings. The quantity A_s is the surface area of the bottom surface of the source, T_s and T_t are the absolute temperatures of the source and the target, respectively, and $T_{sur} = 20°C$ is the temperature of the surroundings. The quantity $\sigma = 5.67 \times 10^{-8}$ W/m²-K⁴ is the Stefan–Boltzmann constant. The source is protected by a transparent covering and therefore does not experience convection. The electrical heater embedded in the source provides $\dot{q}_{htr} = 150$ W when the apparatus is activated.

The target is a disk with radius $R_t = 5$ cm and thickness $th_t = 0.2$ cm. The bottom and sides of the target are insulated and experience neither convection nor radiation. The top surface experiences radiation heat transfer with the target as well as radiation heat transfer with the surroundings:

$$\dot{q}_{rad,t,sur} = A_t F_{t,sur}\, \sigma\left(T_t^4 - T_{sur}^4\right),$$

where T_t is the temperature of the top surface of the target and $F_{t,sur} = 0.2$ is the view factor between the target and the surroundings. The target is cooled by convection with the surrounding air at T_{sur}. The heat transfer coefficient that exists when the apparatus is activated and the fan is on is $\bar{h}_{on} = 250$ W/m²-K. When the apparatus is deactivated, the fan is shut off and the heat transfer coefficient drops to $\bar{h}_{off} = 5$ W/m²-K. The properties of the source material are $k_s = 250$ W/m-K, $c_s = 850$ J/kg-K, and $\rho_s = 7500$ kg/m³. The properties of the target material are $k_t = 45$ W/m-K, $c_t = 1200$ J/kg-K, and $\rho_t = 2500$ kg/m³.

(a) Determine the initial steady-state operating temperatures of the source and the target.

These are the temperatures that exist after the apparatus has been operating for a long time and are present immediately before the apparatus is deactivated.

(b) Is a lumped capacitance transient model of the system appropriate for the time after it is deactivated? That is, can the temperature of the target and the source both be assumed to be uniform? Justify your answer by computing (approximately) the resistances that characterize each of the heat transfers involved in the problem as well as the resistance to conduction within each of the objects.

(c) Calculate the four lumped capacitance time constants that characterize the equilibration process that occurs after the device is deactivated. These should include $\tau_{s,t}$, the time constant associated with the source equilibrating to the target; $\tau_{t,s}$, the time constant associated with the target equilibrating to the source; $\tau_{t,sur}$, the time constant associated with the target equilibrating to the surroundings; and $\tau_{s,sur}$, the time constant associated with the source equilibrating to the surroundings.

(d) Based on your time constant calculations from (c) sketch the anticipated temperature variation of the target and the source after the apparatus is deactivated.

(e) Develop a numerical model in EES using Euler's technique. Plot the temperature of the source and the target as a function of time for the period of time following the deactivation of the apparatus.

(f) Show that the Euler solution becomes unstable when the time step is increased to approximately the size of the smallest time constant identified in (c).

(g) Solve the problem using the fully implicit technique and show that it remains stable even at time steps that are larger than the smallest time constant that you identified in (c).

(h) Implement the solution using the Integral command in EES.

5.61 The figure shows a simple Joule–Thomson system that is used to cool a platform containing a cryogenic detector.

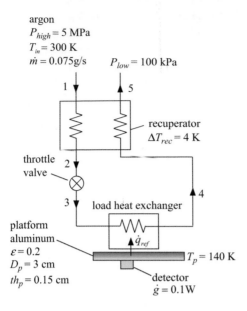

argon
$P_{high} = 5$ MPa
$T_{in} = 300$ K
$\dot{m} = 0.075$ g/s $P_{low} = 100$ kPa

recuperator
$\Delta T_{rec} = 4$ K

throttle valve

load heat exchanger

platform aluminum
$\varepsilon = 0.2$
$D_p = 3$ cm
$th_p = 0.15$ cm

$\dot{q}_{ref}$

$T_p = 140$ K

detector
$\dot{g} = 0.1$ W

$T_{sur} = 300$ K

High-pressure argon at $T_{in} = 300$ K and $P_{high} = 5$ MPa enters a recuperator at state 1 with a mass flow rate of $\dot{m} = 0.075$ g/s. The argon is pre-cooled in the recuperator and leaves with no pressure loss at state 2. The argon enters a throttle valve where its pressure is reduced to $P_{low} = 100$ kPa at state 3. The argon then passes through a load heat exchanger that is interfaced with the detector platform. The pressure at state 4 is equal to the pressure at state 3 and the argon leaves the load heat exchanger at the platform temperature, $T_p = 140$ K. The argon leaving the load heat exchanger passes through the recuperative heat exchanger where it precools the incoming high-pressure argon. The recuperator has an approach temperature difference of $\Delta T_{rec} = 4$ K, which implies that the temperature at state 5 is ΔT_{rec} less than the temperature at state 1. There is no pressure loss in the recuperator. The detector platform is made of aluminum and has diameter $D_p = 3$ cm and thickness $th_p = 0.15$ cm. The platform surface has emissivity $\varepsilon = 0.2$ and experiences radiation with surroundings at $T_{sur} = 300$ K. The detector dissipation is $\dot{g} = 0.1$ W.

(a) Determine the rate of refrigeration heat transfer provided by the Joule–Thomson system, $\dot{q}_{ref}$.

(b) Determine the time rate of change of the platform temperature. Assume that the platform can be treated as a lumped capacitance.

(c) Determine the mass flow rate that will let the detector platform reach a steady-state temperature of $T_p = 140$ K.

(d) Set the mass flow rate equal to the value calculated in part (c). Use the Integral command in EES to simulate the cooldown of the detector platform from its initial temperature (T_{sur}) to its steady-state temperature. Plot the temperature of the platform as a function of time.

5.62 A large number of spherical parts are dumped into a container of water in order to quickly cool them after a treatment. The objects are initially at $T_p = 800°$C and the water is initially at $T_a = 20°$C. The volume of water is $V_w = 2$ liter and the water can be assumed to have constant properties $\rho_w = 1000$ kg/m^3 and $c_w = 4200$ J/kg-K. There are $N_o = 200$ objects placed in the water, each with diameter $D_o = 1$ cm. The properties of the object can be assumed to be constant at $k_o = 42$ W/m-K, $c_o = 2500$ J/kg-K, and $\rho_o = 4500$ kg/m^3. The heat transfer coefficient between the surface of the object and the water is $\bar{h} = 350$ W/m^2-K. The outside of the container is insulated so no heat is lost from the water to the surroundings during the process. However, the volume of the water is sufficiently small that the water itself changes temperature substantially during the cooling process.

(a) Evaluate whether the objects can be treated as a lumped capacitance.

(b) Assume that both the water and the objects can be treated as lumped capacitances. Derive two ordinary differential equations that describe the temperature of the objects and the temperature of the water. You should end up with two first-order ODEs, one for T_o and one for T_w, each characterized by an initial condition. These ODEs will be coupled (i.e., T_o and T_w will appear in both).

(c) Subtract the ODEs you derived in (b) from one another. You should end up with a single ODE that can be expressed terms of $\theta = T_o - T_w$.

(d) Solve the ODE you derived in (c) in order to obtain an expression for θ as a function of t.

(e) Substitute the result from (d) back into either of the ODEs for T_o and T_p you derived in (b) and solve in order to end up with expressions for T_o and T_p as a function of time.

(f) Plot T_o and T_p as a function of time (on a single plot).

(g) Prepare a single plot that shows T_o and T_p as a function of time when various values of N_o are used (i.e., what happens to the process as you dump more and more spheres into the same amount of water).

(h) Prepare a numerical model of the process (with $N_o = 200$). Overlay your result on the plot from (f) to show that the answer is the same as the analytical solution.

5.63 You are interested in using a thermoelectric cooler to quickly reduce the temperature of a small detector from its original temperature of $T_{ini} = 300$ K to its operating temperature.

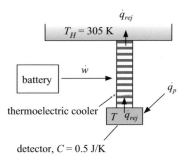

detector, $C = 0.5$ J/K

The thermoelectric cooler receives power at a rate of $\dot{w} = 5.0$ W from a small battery and rejects heat at a rate of $\dot{q}_{ref}$ to ambient temperature $T_H = 305$ K. The cooler removes energy at a rate of $\dot{q}_{ref}$ from the detector, which is at temperature T. (Note that the detector temperature T will change with time, t.) The detector can be modeled as a lumped capacitance and has a total heat capacity, $C = 0.5$ J/K. Despite your best efforts to isolate the detector from the ambient, the detector is subjected to a parasitic heat gain, $\dot{q}_p$, that can be modeled as occurring through a fixed resistance $R_p = 100$ K/W; this resistance represents the combined effect of radiation and conduction:

$$\dot{q}_p = \frac{T_H - T}{R_p}.$$

The thermoelectric cooler has a second-law efficiency $\eta_c = 10$ percent regardless of its operating temperatures. This means that the cooler provides 10 percent of the refrigeration that a reversible cooler could provide given the same operating temperatures and input power. If you review your thermodynamics, you will find that the rate of refrigeration provided to the detector can be related to the input power provided to the thermoelectric cooler and its operating temperatures according to:

$$\dot{q}_{ref} = \frac{\dot{w}\eta_c}{\left(\dfrac{T_H}{T} - 1\right)}.$$

(a) Derive the state equation for the detector. That is, provide a symbolic equation for the rate of temperature change of the detector as a function of the quantities given in the problem (i.e., T_H, R_p, C, $\dot{w}$, η_c) and the instantaneous value of the detector temperature (T).

(b) Develop an EES program that numerically solves this problem for the values given in the problem statement. Using your program, prepare a plot showing the temperature of the detector as a function of time for 120 s after the cooler is activated.

(c) Modify your program so that it accounts for the fact that your battery only has 100 J of energy storage capacity. That is, once the 100 J of energy in the battery is depleted, the power driving the thermoelectric cooler (and the refrigeration that it provides) goes to zero. Prepare a plot showing the temperature of the detector as a function of time for 120 s after the cooler is activated.

(d) Assume that the objective of your cooler is to keep the detector at a temperature below 240 K for as long as possible, given that your battery only has 100 J of energy. What power ($\dot{w}$) would you use to run the thermoelectric cooler? Justify your answer with plots and an explanation.

5.64 The figure shows the wall of a hypersonic aircraft. The outer wall is composed of steel that is 2.0 cm thick; assume that steel has thermal conductivity $k_s = 50$ W/m-K, density $\rho_s = 8200$ kg/m^3, and specific heat capacity $c_s = 468$ J/kg-K. During re-entry the aircraft wall is exposed to a high heat flux ($\dot{q}''$) related to the

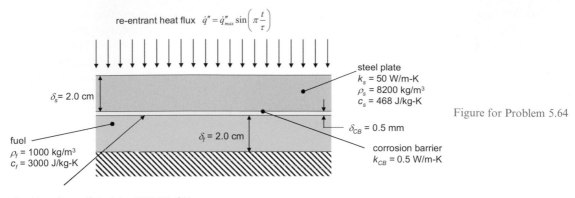

Figure for Problem 5.64

high velocity. The heat flux is approximately sinusoidal and lasts for the duration of the re-entry process (τ):

$$\dot{q}'' = \begin{cases} \dot{q}''_{max} \sin\left(\pi\dfrac{t}{\tau}\right) & t < \tau \\ 0 & t \geq \tau \end{cases}.$$

where $\dot{q}''_{max} = 50 \times 10^4 \; \frac{W}{m^2}$ and $\tau = 300$ s. The temperatures seen by the steel plate are too high and therefore engineers have decided to put a layer of fuel adjacent to the plate to help cool it. In order to protect the plate from the fuel, a corrosion barrier coating is placed on the plate. The thickness of the corrosion barrier (δ_{CB}) is 0.5 mm and its conductivity, k_{CB} is 0.5 W/m-K. The corrosion barrier has a negligible heat capacity when compared to the fuel or the steel.

The fuel layer has a thickness (δ_f) of 2.0 cm. The fuel has density $\rho_f = 1000$ kg/m³, and specific heat capacity $c_f = 3000$ J/kg-K. The fuel can be assumed to be well mixed and therefore at a uniform temperature. The heat transfer coefficient between the fuel and the corrosion barrier, h, is 1000 W/m²-K. The steel and the fuel are both initially at 300 K.

(a) Develop an analytical solution for the temperature of the steel as a function of time during the re-entry process (i.e., for $0 < t < \tau$) in the limit of an infinite fuel specific heat capacity ($c_f \to \infty$).

(b) Develop a numerical solution capable of predicting both the temperature of the steel and the fuel for a finite fuel heat capacity. Use one of the numerical techniques discussed in Section 5.3 and implement your solution in EES.

(c) Verify your numerical solution by comparing it with the analytical solution in the limit of an infinite fuel specific heat capacity ($c_f \to \infty$). Prepare a plot that shows the numerical and analytical temperature variation during the re-entry process (i.e., for $0 < t < \tau$) in this limit.

(d) Use your numerical solution to generate a plot showing the temperature of the wall without fuel (i.e., if it's inner surface is adiabatic) and with fuel (i.e., as shown in the figure). Also show the fuel temperature on the plot. Make sure that your plot extends beyond the re-entry process (i.e., for $0 < t < 2\tau$).

6 | Transient Conduction

Chapter 5 discussed transient problems in which the spatial temperature gradients within a solid object can be neglected and therefore the problem is approximately zero-dimensional (0-D). In these *lumped capacitance* transient problems the solution is a function only of time. Lumped capacitance problems essentially ignore the process of conduction as being unimportant. This chapter discusses transient problems where internal, spatial temperature gradients related to energy transfer by conduction are nonnegligible (i.e., the Biot number is not much less than unity). The first section provides some conceptual tools that are not exact, but can be used to develop an understanding of transient conduction problems. More sophisticated analytical and numerical solutions that provide more exact solutions are presented in the remaining sections.

6.1 Conceptual Tools

This section is meant to provide you with simple conceptual tools that can be used to understand transient conduction problems at a high level prior to going through the effort of developing more exact but more complicated analytical or numerical solutions. These tools should be viewed in the same manner as the thermal resistances and lumped capacitance time constant presented previously. These concepts allow an engineer to quickly determine what is important and what is not, in order to develop an engineering solution that includes only those mechanisms that are important.

6.1.1 Diffusive Energy Transport

Diffusive energy transport refers to the movement of energy due to conduction. In a transient situation, this energy transport causes the material to change temperature and so there is an interplay between the conductivity of the material and its heat capacity. A simple case that can be used to develop our understanding of transient diffusive energy transport occurs when the object itself can be considered **semi-infinite**. Semi-infinite means that the material is bounded at one edge (e.g., at $x = 0$) but extends infinitely otherwise. No object is truly semi-infinite, although many heat transfer problems can be accurately modeled using this concept in situations where the time period of interest is short or the object is very large. The Earth, for example, is semi-infinite relative to most surface phenomena.

Figure 6.1 shows a semi-infinite body that is initially at a uniform temperature, T_{ini}, when at time $t = 0$ the temperature at the surface (i.e., at $x = 0$) is raised to T_s.

temperature of surface is suddenly raised to T_s

T_{ini}

Figure 6.1 Semi-infinite body subjected to a sudden change in the surface temperature.

initially all material is at temperature T_{ini}

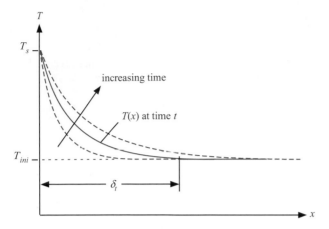

Figure 6.2 illustrates a sketch of the temperature as a function of position for increasing values of time. Notice that the conduction process resembles a "thermal wave" that penetrates or *diffuses* into the solid from the surface. The temperature of the material is, at first, affected by the surface temperature change only at positions that are very near the surface. As time increases, the thermal wave penetrates deeper into the solid; i.e., the depth of the penetration (δ_t) grows and therefore the amount of material affected by the surface change increases.

Before we begin to pursue exact solutions to problems like this one, it is worthwhile to pause and understand the behavior of this diffusive thermal wave that characterizes all transient conduction problems. Our objective is therefore to develop a simple, approximate equation that relates the distance that energy can be transferred by conduction (δ_t) to the elapsed time (t) and the properties of the material.

There are two phenomena that occur simultaneously in this transient conduction process: (1) thermal energy is being conducted from the surface into the body, and (2) the energy is stored in the material that is affected by the surface phenomena. By developing simple models for these two processes it is possible to understand, to a first approximation, how the thermal wave behaves. A control volume is drawn that includes material starting at the surface ($x = 0$) and extending very far from the surface (much further than the thermal wave has progressed, $x \gg \delta_t$), as shown in Figure 6.3.

An energy balance on the control volume shown in Figure 6.3 includes conduction into the control surface at $x = 0$ ($\dot{q}_{cond}$) and the rate of energy storage within the control volume (dU/dt):

$$\dot{q}_{cond} = \frac{dU}{dt}. \tag{6.1}$$

Neither of the terms in Eq. (6.1) can be computed exactly without a detailed model of the process. The rate of heat transfer by conduction is approximated using the resistance associated with a plane wall:

$$\dot{q}_{cond} = \frac{T_s - T_{ini}}{R_{pw}}, \tag{6.2}$$

where

$$R_{pw} = \frac{(\text{distance to conduct})}{k(\text{area for conduction})}. \tag{6.3}$$

The area for conduction is A_c, the cross-sectional area of the wall. The distance to conduct is approximately δ_t, the spatial extent of the material that has been affected by the process:

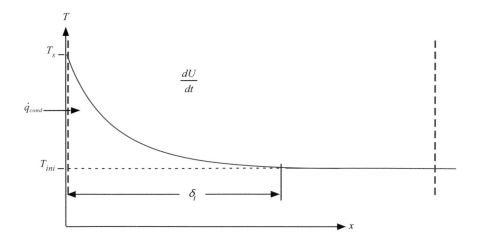

Figure 6.3 Control volume used to develop a simple model of the transient conduction process.

$$R_{pw} \approx \frac{\delta_t}{k\,A_c}. \tag{6.4}$$

Substituting Eq. (6.4) into Eq. (6.2) provides:

$$\dot{q}_{cond} \approx \frac{k\,A_c(T_s - T_{ini})}{\delta_t}. \tag{6.5}$$

Equation (6.5) is only approximate, but it shows that the rate of conduction into the wall will be inversely proportional to δ_t. At small times, when δ_t is small, the rate of conduction will be high and it will decrease as δ_t grows, requiring that the energy be conducted further.

The thermal energy stored in the material (U) is the product of the average temperature rise and the total heat capacity of the material that has been heated (i.e., the material that lies within the heated zone, $0 < x < \delta_t$). The average temperature rise of the material in the heated zone is, approximately:

$$\overline{\Delta T} \approx \frac{(T_s - T_{ini})}{2}. \tag{6.6}$$

The heat capacity of the material within the heated zone is, approximately:

$$C \approx \rho\,c\,\delta_t\,A_c, \tag{6.7}$$

where ρ and c are the density and specific heat capacity of the material, respectively. The thermal energy stored in the heated zone is then approximated by:

$$U \approx \underbrace{\frac{(T_s - T_{ini})}{2}}_{\overline{\Delta T}}\,\underbrace{\rho\,c\,\delta_t\,A_c}_{C}. \tag{6.8}$$

Substituting Eqs. (6.5) and (6.8) into Eq. (6.1) leads to:

$$\frac{k\,A_c(T_s - T_{ini})}{\delta_t} \approx \frac{d}{dt}\left[\frac{(T_s - T_{ini})}{2}\rho\,c\,\delta_t\,A_c\right]. \tag{6.9}$$

For the problem shown in Figure 6.1 only the penetration depth (δ_t) varies with time in Eq. (6.9); therefore, Eq. (6.9) can be rearranged to provide an ordinary differential equation for δ_t:

$$\frac{k}{\delta_t} \approx \frac{\rho c \, d\delta_t}{2 \; dt}.$$ (6.10)

Equation (6.10) can be rearranged:

$$\frac{2k}{\rho c} \approx \delta_t \frac{d\delta_t}{dt}.$$ (6.11)

Equation (6.11) can be simplified somewhat using the definition of **thermal diffusivity** (α):

$$\alpha = \frac{k}{\rho c}.$$ (6.12)

Note that because conductivity (k), density (ρ), and specific heat capacity (c) are all properties, thermal diffusivity is also a property. Substituting Eq. (6.12) into Eq. (6.11) leads to:

$$2\alpha \approx \delta_t \frac{d\delta_t}{dt}.$$ (6.13)

Assuming α to be constant, Eq. (6.13) can be separated and integrated:

$$\int_0^t 2\,\alpha\,dt \approx \int_0^{\delta_t} \delta_t \, d\delta_t$$ (6.14)

in order to obtain an approximate relationship between the penetration depth of the thermal wave and time:

$$2\alpha\, t \approx \frac{\delta_t^2}{2}$$ (6.15)

or

$$\delta_t \approx 2\sqrt{\alpha\,t}.$$ (6.16)

Equation (6.16) indicates that a diffusive thermal wave will grow in proportion to $\sqrt{\alpha\,t}$. Equation (6.16) is a very important and practical (but not an exact) result that governs transient conduction problems. The constant in Eq. (6.16) will change depending on the precise nature of the problem and the exact definition of the thermal penetration depth. However, Eq. (6.16) will be approximately correct for a large variety of transient conduction problems.

Example 6.1

A semi-infinite body is initially at steady state and has a uniform temperature T_{ini} when a heat flux of $\dot{q}_s''$ is applied to the surface at $x = 0$. The properties of the material are k, ρ, and c. Carry out an analysis that is similar to the one presented in Section 6.1.1 in order to establish an approximate relationship between the size of the thermally affected region (δ_t) and the elapsed time (t).

Known Values

Figure 1 provides a sketch of the situation.

Example 6.1 (cont.)

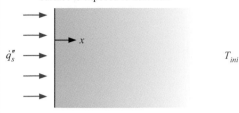

surface is exposed to heat flux

$\dot{q}''_s$

x

T_{ini}

initially all material is at temperature T_{ini}

T

increasing time

T_{ini}

x

Figure 1 Semi-infinite body exposed to a heat flux at the surface.

Assumptions

- The properties of the material are not a function of temperature.
- Radiation and convection from the surface are neglected.
- The problem is one dimensional (1-D).

Analysis and Solution

An energy balance is carried out using the control volume shown in Figure 6.3, which includes all of the material that lies within the thermally affected region. The rate at which energy is added to the material by the surface heat flux must be balanced by the rate at which energy is stored:

$$\dot{q}_{cond} = \frac{dU}{dt}. \tag{1}$$

The approximate rate equation for the rate of conduction is

$$\dot{q}_{cond} \approx \frac{k A_c (T_s - T_{ini})}{\delta_t} \tag{2}$$

and the amount of energy that has been stored in the material is

$$U \approx \frac{(T_s - T_{ini})}{2} \rho c \delta_t A_c. \tag{3}$$

Substituting Eqs. (2) and (3) into Eq. (1) provides:

$$\frac{k A_c (T_s - T_{ini})}{\delta_t} \approx \frac{d}{dt} \left[\frac{(T_s - T_{ini})}{2} \rho c \delta_t A_c \right] \tag{4}$$

Continued

Example 6.1 (cont.)

which is the same expression that was derived previously in Section 6.1.1 as Eq. (6.9). However, for this problem both the penetration depth (δ_t) and the surface temperature (T_s) are changing with time. The rate of heat flux into the surface is specified, however, and therefore:

$$\dot{q}_{cond} = \dot{q}''_s \, A_c. \tag{5}$$

Substituting Eq. (2) into Eq. (5) provides

$$\dot{q}''_s \, A_c \approx \frac{k \, A_c (T_s - T_{ini})}{\delta_t}. \tag{6}$$

Equation (6) can be solved for the surface temperature elevation:

$$(T_s - T_{ini}) \approx \frac{\dot{q}''_s \, \delta_t}{k}. \tag{7}$$

Substituting Eq. (7) into Eq. (4) leads to:

$$\frac{k \, A_c \, \dot{q}''_s \, \delta_t}{\delta_t \quad k} \approx \frac{d}{dt} \left[\frac{\dot{q}''_s \, \delta_t}{2 \, k} \rho \, c \, \delta_t \, A_c \right]. \tag{8}$$

Recognizing that the only time varying parameter remaining in Eq. (8) is δ_t and simplifying provides:

$$1 \approx \frac{\rho \, c}{2 \, k} \frac{d}{dt} \left[\delta_t^2 \right]. \tag{9}$$

Substituting the definition of thermal diffusivity, Eq. (6.12), into Eq. (9) and separating leads to:

$$2\alpha \, dt \approx d \left[\delta_t^2 \right]. \tag{10}$$

Integrating both sides of Eq. (10) provides:

$$2\alpha \, t \approx \delta_t^2. \tag{11}$$

Solving for the thermal penetration depth leads to:

$$\boxed{\delta_t \approx \sqrt{2} \sqrt{\alpha \, t}}, \tag{12}$$

which is similar to Eq. (6.16) except that the constant 2 is replaced by $\sqrt{2}$.

Discussion

The transport of energy by conduction requires time. Equation (6.16) and Eq. (12) from this example differ only in the constant that multiplies $\sqrt{\alpha \, t}$ and this exercise shows that this constant will not be the same for all processes and all definitions of penetration depth. However, the *scaling* $\delta_t \propto \sqrt{\alpha \, t}$ will persist and the constant should be not too different from unity. Therefore, Eq. (6.16) provides a powerful method of understanding most transient conduction processes.

6.1.2 The Diffusive Time Constant

Equation (6.16) allows us to define a **diffusive time constant** that characterizes how long it takes energy to pass through a certain region due to conduction. For example, if you heat one side of a plate of 304 stainless steel ($\alpha = 3.91 \times 10^{-6} \, \mathrm{m^2/s}$ according to Appendix A) that is $L = 1.0 \, \mathrm{cm}$ thick, then the temperature on the opposite side of the plate will begin to change in about $\tau_{diff} = 6.4 \, \mathrm{s}$; this result follows directly from Eq. (6.16), re-arranged:

$$\tau_{diff} \approx \frac{L^2}{4\alpha}. \tag{6.17}$$

The time required for the thermal wave to pass through a body is referred to as the diffusive time constant (τ_{diff}) and it is a broadly useful concept in the same way that the lumped capacitance time constant introduced in Chapter 5 is useful. The diffusive time constant characterizes, approximately, how long it takes for an object to equilibrate internally by conduction. The lumped capacitance time constant characterizes, approximately, how long it takes for an object to equilibrate with its environment. The first step in understanding any transient heat transfer problem involves the calculation of these two time constants.

Example 6.2

A metal wall is used to separate two liquids that are at different temperatures: $T_{hot} = 500$ K and $T_{cold} = 400$ K. The thickness of the wall is $L = 0.8$ cm. The properties of the wall material are $\rho = 8000$ kg/m^3, $c = 400$ J/kg-K, and $k = 20$ W/m-K. The average heat transfer coefficient between the wall and the liquid in both tanks is $\bar{h}_{liq} = 5000$ W/m^2-K. Initially, the system is in steady state and the wall temperature distribution is not changing with time. At time, $t = 0$, both tanks are drained after which both sides of the wall are exposed to cool gas at $T_{gas} = 300$ K. The average heat transfer coefficient between the walls and the gas is $\bar{h}_{gas} = 100$ W/m^2-K.

Sketch the temperature of the wall as a function of position at $t = 0$ and as $t \rightarrow \infty$. Also include the temperature as a function of position at several intermediate values of time during the equilibration process.

Known Values

The conditions that exist before the tank is emptied (i.e., at $t < 0$) are shown in Figure 1(a). The conditions following the draining of the tank (i.e., at $t > 0$) are shown in Figure 1(b).

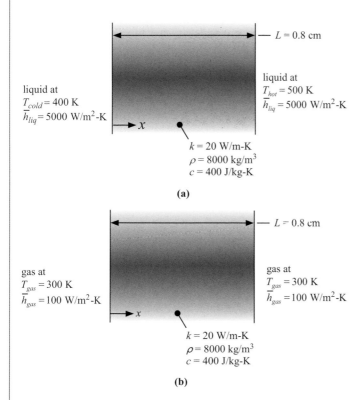

Figure I Tank wall (a) exposed to two liquids at different temperatures before it is emptied and (b) exposed to cool gas on both sides after draining the liquids.

Continued

Example 6.2 (cont.)

Assumptions

- The properties of the tank wall do not depend on temperature.
- Radiation is neglected.
- The process of draining the liquid and filling with gas occurs instantaneously.
- The problem is one dimensional (1-D).

Analysis and Solution

At time $t = 0$ the tank wall is at steady state and in thermal equilibrium with the liquid. This problem is a 1-D steady-state problem of the type encountered in Chapter 2 and it can be understood using the resistance network shown in Figure 2.

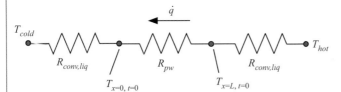

Figure 2 Resistance network used to model tank wall at steady state in equilibrium with the liquids, as shown in Figure 1(a).

The resistance to convection from the surface on either side of the wall is computed according to:

$$R_{conv,\ liq} = \frac{1}{\bar{h}_{liq}\ A_c} = \frac{m^2\ K}{5000\ W}\bigg|\frac{1}{1\ m^2} = 0.0002\ \frac{K}{W} \tag{1}$$

and the resistance to conduction through the metal wall is

$$R_{pw} = \frac{L}{k\ A_c} = \frac{0.008\ m}{20\ W}\bigg|\frac{m\ K}{1\ m^2}\bigg| = 0.0004\ \frac{K}{W}. \tag{2}$$

The total heat transfer rate through the wall before the liquid is drained is:

$$\dot{q} = \frac{(T_{hot} - T_{cold})}{(R_{conv,\ liq} + R_{pw} + R_{conv,\ liq})} = \frac{(500 - 400)K}{(0.0002 + 0.0004 + 0.0002)\ K/W} = 125,000\ W(125\ kW). \tag{3}$$

The temperature on the left side of the wall ($x = 0$) at time $t = 0$ is computed according to:

$$T_{x=0,\ t=0} = T_{cold} + \dot{q}R_{conv,\ liq} = 400\ K + (125,000\ W)(0.0002\ K/W) = 425\ K. \tag{4}$$

The temperature on the right side of the wall ($x = L$) at time $t = 0$ is computed according to:

$$T_{x=L,\ t=0} = T_{hot} - \dot{q}R_{conv,\ liq} = 500\ K - (125,000\ W)(0.0002\ K/W) = 475\ K. \tag{5}$$

The temperature as a function of position at time $t = 0$ is linear, starting at 425 K at $x = 0$ and ending at 475 K at $x = L$. This temperature distribution is shown in Figure 3 and labeled as $t = 0$.

After the tank is drained, both sides of the wall are exposed to gas at $T_{gas} = 300$ K and therefore the temperature everywhere in the wall will eventually (as $t \rightarrow \infty$) reach 300 K. This is also shown in Figure 3 and labeled as $t \rightarrow \infty$.

Example 6.2 (cont.)

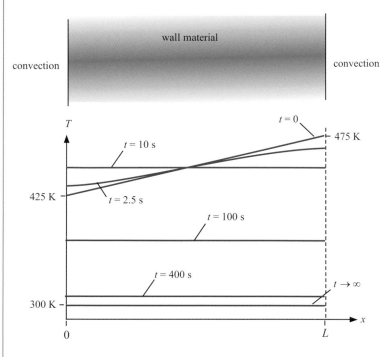

Figure 3 Sketch of the temperature distribution in the wall at $t = 0$, $t \to \infty$, and several intermediate times.

Now that we have established the initial and final temperature distributions we can examine, approximately, what happens *during* the transient equilibration process. There are two processes that occur after the tank is drained. The tank material is not in equilibrium with itself due to the temperature gradient that is present at $t = 0$. Therefore, there is a conduction process that occurs within the wall that tends to transfer energy from the hotter material on the right side of the wall to the colder material on the left side. This process alone would eventually cause the wall material to come to a uniform internal temperature. There is also an external equilibration process associated with the material transferring energy to the surrounding gas. This process will eventually cause the wall material to come to the gas temperature.

Both of these processes are characterized by a time constant. The internal equilibration process is governed by a diffusive time constant provided, approximately, by Eq. (6.17):

$$\tau_{diff} = \frac{L^2}{4\alpha} = \frac{(0.008)^2 \ \mathrm{m}^2}{4} \left| \frac{\mathrm{s}}{6.25 \times 10^{-6} \ \mathrm{m}^2} \right. = 2.56 \ \mathrm{s}, \tag{6}$$

where α is the thermal diffusivity of the wall material:

$$\alpha = \frac{k}{\rho c} = \frac{20 \ \mathrm{W}}{\mathrm{m} \ \mathrm{K}} \left| \frac{\mathrm{m}^3}{8000 \ \mathrm{kg}} \right| \frac{\mathrm{kg} \ \mathrm{K}}{400 \ \mathrm{J}} \left\| \frac{\mathrm{J}}{\mathrm{W} \ \mathrm{s}} \right. = 6.25 \times 10^{-6} \frac{\mathrm{m}^2}{\mathrm{s}}. \tag{7}$$

The diffusive time constant is about 2.6 s; this is, approximately, how long it will take for a thermal wave to pass from one side of the wall to the other and therefore this is also about the amount of time that is required for the wall to internally equilibrate. Equation (6) tells us that if the edges of the wall were adiabatic, then the wall will be at a nearly uniform temperature after a few (5–10) seconds.

The external equilibration process is governed by the lumped time constant, discussed previously in Chapter 5 and defined for this problem as

Continued

Example 6.2 (cont.)

$$\tau_{lumped} = R_{conv,\,gas} C = \frac{0.005 \text{ K}}{\text{W}} \left| \frac{25,600 \text{ J}}{\text{K}} \right| \frac{\text{W s}}{\text{J}} = 128 \text{ s},$$ (8)

where $R_{conv,gas}$ is the thermal resistance to convection between the wall and the gas:

$$R_{conv,\,gas} = \frac{1}{2\,\bar{h}_{gas}\,A_c} = \frac{1}{2} \left| \frac{\text{m}^2 \text{ K}}{100 \text{ W}} \right| \frac{1}{1 \text{ m}^2} = 0.005 \frac{\text{K}}{\text{W}}$$ (9)

and C is the thermal capacitance of the wall:

$$C = A_c\,L\rho\,c = \frac{1 \text{ m}^2}{} \left| \frac{0.008 \text{ m}}{} \right| \frac{8000 \text{ kg}}{\text{m}^3} \left| \frac{400 \text{ J}}{\text{kg K}} \right| = 25,600 \frac{\text{J}}{\text{K}}.$$ (10)

The lumped time constant is about 130 s. This tells us that it will take hundreds of seconds for the wall material to equilibrate with the surrounding gas.

Because the diffusive or internal time constant is two orders of magnitude smaller than the lumped or external time constant, we have a reasonably good idea of how the transient process will progress without doing any of the complex math that is required to solve the problem more exactly. First, the wall will internally equilibrate relatively rapidly and then the wall will externally equilibrate much more slowly. The initial internal equilibration process will be completed after about 5–10 s (the diffusive time constant is around 2.5 s, so after several times this value we can expect the associated process to be complete). The external equilibration process will require 200–400 s (the lumped capacitance time constant is around 100 s so after several times this value we can expect the associated process to be complete). The temperature distributions sketched in Figure 3 for intermediate times are consistent with these time constants. Figure 3 does not represent an exact solution, but it is consistent with the physical intuition that was gained through knowledge of the two time constants governing the process.

Discussion

A more exact solution to this problem can be obtained using the numerical techniques discussed in Section 6.3. However, the calculation of diffusive and lumped time constants provides important physical intuition about the problem and any solution that we eventually obtain should be checked against this intuition.

In this example, the results were presented in the form of sketches of temperature as a function of position at various values of time in Figure 3. We could also have shown sketches of temperature as a function of time at various values of position, as is done in Figure 4.

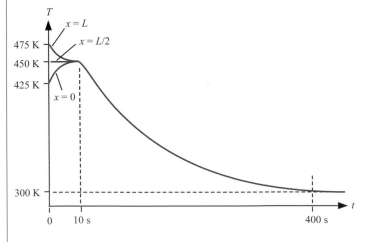

Figure 4 Sketch of the temperature as a function of time for various positions in the wall.

Example 6.2 (cont.)

Notice that the temperature at the right side of the wall, $x = L$, rapidly falls to 450 K and then slowly approaches 300 K. The first process is diffusive and therefore is done after about 10 s while the second is convective and takes about 400 s. The temperature at the left side of the wall actually rapidly *increases* initially and then slowly drops.

6.1.3 The Semi-Infinite Resistance

The approximate equation for diffusive energy transport, Eq. (6.16), is repeated below:

$$\delta_t \approx 2\sqrt{\alpha t}. \tag{6.18}$$

This equation was rearranged in Section 6.1.2 in order to define a diffusive time constant. Equation (6.18) can also be used to define a conduction resistance associated with moving energy into the thermally affected region of a semi-infinite body, as shown Figure 6.4.

The rate of heat transfer into the surface ($\dot{q}_s$) is driven by the temperature difference between the surface (T_s) and the material that has yet to be affected by the surface process (T_{ini}). The resistance to conduction is generally given by:

$$R_{cond} \approx \frac{(\text{distance to conduct})}{k(\text{area for conduction})}. \tag{6.19}$$

When dealing with the semi-infinite body shown in Figure 6.4 the distance to conduct is, approximately, the size of the thermally affected zone and therefore Eq. (6.19) can be written as:

$$R_{semi-\infty} \approx \frac{\delta_t}{k\,A_c}. \tag{6.20}$$

This result is both useful and intuitive; the material within the thermal wave in Figure 6.4 acts like a conduction thermal resistance to heat transfer from the surface ($R_{semi-\infty}$):

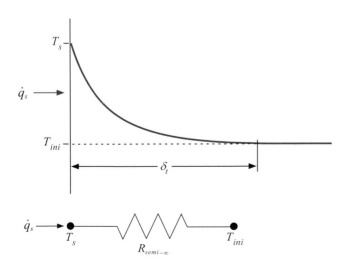

Figure 6.4 The concept of the semi-infinite resistance.

$$\dot{q}_s = \frac{(T_s - T_{ini})}{R_{semi-\infty}}. \tag{6.21}$$

The conduction resistance $R_{semi-\infty}$ increases with time as the thermal wave grows (and therefore the distance over which the conduction occurs increases):

$$R_{semi-\infty} \approx \frac{2\sqrt{\alpha t}}{k A_c}. \tag{6.22}$$

This concept is useful for understanding the physics associated with transient conduction problems. Of course, as soon as the thermal wave reaches a boundary the problem is no longer semi-infinite and therefore the concepts of the semi-infinite resistance and thermal penetration wave are no longer as relevant.

Example 6.3

A laser machining process uses a very short but high power heat flux to melt aluminum. Initially the aluminum is at a uniform temperature, $T_{ini} = 20°C$. The heat flux from the laser is applied at the surface and lasts only $t_{laser} = 0.1$ ms. You would like to increase the temperature at the surface to the melting temperature, $T_{melt} = 630°C$. Estimate the heat flux that is required for this purpose.

Known Values

The problem is illustrated in Figure 1 with the known values included. The properties of aluminum are obtained from Appendix A: $k = 235$ W/m-K and $\alpha = 96.6 \times 10^{-6}$ m^2/s.

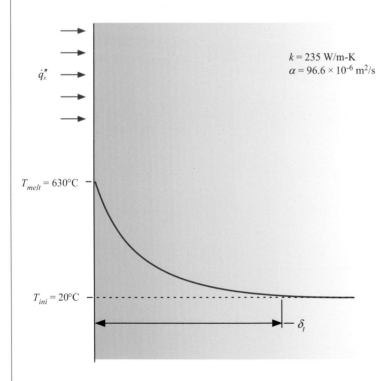

Figure 1 Aluminum slab subjected to a short-duration laser pulse.

Example 6.3 (cont.)

Assumptions

- The properties of aluminum do not depend on temperature.
- Convection and radiation from the surface to the surroundings are neglected.
- The aluminum can be treated as a semi-infinite body. This implies that the thickness of the aluminum slab (L) is larger than the penetration of the thermal wave:

$$L > \delta_t \approx 2\sqrt{\alpha\, t_{laser}} = 2\sqrt{\frac{96.6 \times 10^{-6}\ \text{m}^2}{\text{s}}\left|0.1 \times 10^{-3}\ \text{s}\right.} = 1.97 \times 10^{-4}\ \text{m}\ (0.197\ \text{mm}). \tag{1}$$

- The semi-infinite body can be treated using the approximate resistance formula given by Eq. (6.22).

Analysis

The thermally affected region of the aluminum is treated as a thermal resistance that grows with time according to Eq. (6.22). Substituting Eq. (6.22) into Eq. (6.21) provides:

$$\dot{q}_s \approx \frac{k\, A_c(T_s - T_{ini})}{2\sqrt{\alpha\, t}}. \tag{2}$$

The objective of the machining process is to cause T_s to reach T_{melt} at time $t = t_{laser}$. Therefore, Eq. (2) can be solved for the required heat flux:

$$\dot{q}_s'' \approx \frac{k(T_{melt} - T_{ini})}{2\sqrt{\alpha\, t_{laser}}}. \tag{3}$$

Solution

Substituting known values into Eq. (3) provides:

$$\dot{q}_s'' \approx \frac{k(T_{melt} - T_{ini})}{2\sqrt{\alpha\, t_{laser}}} = \frac{235\ \text{W}}{\text{m K}}\left|\frac{(630°\text{C} - 20°\text{C})}{2}\right|\sqrt{\frac{\text{s}}{96.6 \times 10^{-6}\ \text{m}^2}\left|0.1 \times 10^{-3}\ \text{s}\right.}$$

$$= \boxed{7.29 \times 10^8\ \frac{\text{W}}{\text{m}^2}\left(729\ \frac{\text{MW}}{\text{m}^2}\right)}. \tag{4}$$

Discussion

We can gain some understanding of the process by sketching the temperature distribution that would exist in the aluminum at various times. The sketch is qualitative but captures many of the expected features so that a more detailed model, for example using the numerical techniques discussed in Section 6.3, can be checked against our intuition. Figure 2 shows the temperature as a function of x (the distance from the surface) at a few times.

Continued

Example 6.3 (cont.)

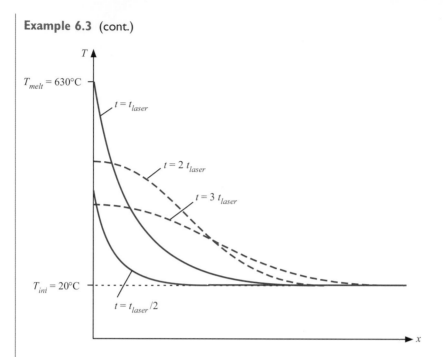

Figure 2 Temperature distributions (sketched) as a function of position at various times.

Notice that for times less than or equal to t_{laser} (the solid lines in Figure 2) the slopes of the curves at $x = 0$ are the same. The slope is directly related to the heat flux at the surface; since the heat flux is constant while the laser is on, the heat transfer rate due to conduction must be the same at $x = 0$. The thermal wave penetration continues to grow throughout the entire process, even at times greater than t_{laser} where the laser heat flux has been turned off. For times greater than t_{laser} (the dashed lines in Figure 2) the slope of the curves at $x = 0$ is zero because the heat flux applied to the surface is zero. The continued growth of the thermal wave after the laser is deactivated causes the total energy in the wall, which remains constant, to be distributed over a greater amount of material. As a result, the temperature near the surface of the material falls with time.

6.2 Analytical Solution

Section 6.1 discussed some useful conceptual tools for understanding transient conduction problems. The diffusive time constant discussed in Section 6.1.2 provides an understanding of how quickly energy can be transferred by conduction. The semi-infinite resistance discussed in Section 6.1.3 allows estimates of temperatures or heat transfers in situations where the semi-infinite body assumption is valid. These tools provide order of magnitude estimates only; more accurate solutions can be obtained either analytically or numerically. Numerical solutions to transient conduction problems are discussed in Sections 6.3 and 6.4. The development of analytical solutions to transient conduction problems is beyond the scope of this text; interested readers are directed to Nellis and Klein (2009). However, in Section 6.2.1 the partial differential equation and the initial and boundary conditions are derived for a typical situation; solving the problem then becomes a purely mathematical exercise. Some benchmark analytical solutions exist for common situations that arise in semi-infinite and bounded geometries and these are presented in Sections 6.2.2 and 6.2.3, respectively.

6.2.1 The Differential Equation

Figure 6.5 shows a semi-infinite body that is initially at a uniform temperature, T_{ini}, when at time $t = 0$ the temperature of the surface of the body (at $x = 0$) is raised to T_s. This problem was considered approximately in

temperature of surface is suddenly raised to T_s

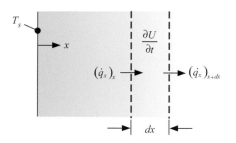

Figure 6.5 Differential control volume used to derive the differential equation.

initially all material is at temperature T_{ini}

Figure 6.3 using a control volume that included *all* of the material within the thermally affected region. Figure 6.5 shows the *differentially small* control volume that is defined in order to derive the partial differential equation that must be solved to obtain an exact solution.

The control volume is differentially small in the x-direction since the temperature is varying along that coordinate. The energy balance suggested by Figure 6.5 is:

$$(\dot{q}_x)_x = (\dot{q}_x)_{x+dx} + \frac{\partial U}{\partial t}.$$

(6.23)

Expanding the $x + dx$ term in Eq. (6.23) and taking the limit as dx approaches zero leads to:

$$(\dot{q}_x)_x = (\dot{q}_x)_x + \frac{\partial \dot{q}_x}{\partial x} dx + \frac{\partial U}{\partial t}.$$

(6.24)

The conduction term is evaluated using Fourier's Law:

$$\dot{q}_x = -k A_c \frac{\partial T}{\partial x},$$

(6.25)

where A_c is the area of the wall. The partial derivative of the total internal energy with respect to time can be written as:

$$\frac{\partial U}{\partial t} = \rho A_c dx \frac{\partial u}{\partial t}.$$

(6.26)

The time rate of change of specific internal energy is expressed in terms of the specific heat capacity and the rate of change of temperature:

$$\frac{\partial U}{\partial t} = \rho A_c dx \underbrace{\frac{du}{dT}}_{c} \frac{\partial T}{\partial t} = \rho A_c dx c \frac{\partial T}{\partial t}.$$

(6.27)

Substituting Eqs. (6.25) and (6.27) into Eq. (6.24) leads to

$$0 = \frac{\partial}{\partial x} \left[-k A_c \frac{\partial T}{\partial x} \right] dx + \rho A_c c dx \frac{\partial T}{\partial t},$$

(6.28)

which, for a constant thermal conductivity, can be simplified to

$$\alpha \frac{\partial^2 T}{\partial x^2} = \frac{\partial T}{\partial t},$$

(6.29)

where α is the thermal diffusivity. Equation (6.29) is the partial differential equation that must be solved to attain an analytical solution to this problem. The differential equation is second order in x and first order in t;

therefore, two spatial boundary conditions and an initial condition are required. For the problem shown in Figure 6.5 the temperature at the surface ($x = 0$) is equal to T_s for all values of time:

$$T_{x=0} = T_s. \tag{6.30}$$

Also, the material is assumed to be semi-infinite so that at large values of x we will always find material at T_{ini}:

$$T_{x \to 0} = T_{ini}. \tag{6.31}$$

The initial condition is:

$$T_{t=0} = T_{ini}. \tag{6.32}$$

The mathematical techniques that provide a solution to the partial differential equation, Eq. (6.29), subject to the boundary conditions, Eqs. (6.30) through (6.32), are beyond the scope of this text. The solution can be obtained using either self-similar techniques or the Laplace transform. Analytical solutions for this and several other semi-infinite body problems are presented in Section 6.2.2.

Example 6.4

A long, hollow cylinder is insulated on its outer surface ($r = r_{out}$). Initially, the cylinder material is at a uniform temperature, T_{ini}. At time $t = 0$ the inner surface ($r = r_{in}$) experiences convection with a fluid at T_∞. The heat transfer coefficient between the inner surface and the fluid is $\bar{h}$. The cylinder material has thermal conductivity k and thermal diffusivity α. Derive the differential equation and boundary conditions that must be solved in order to obtain an analytical solution to this problem.

Known Values

The problem is illustrated in Figure 1 with the known values included.

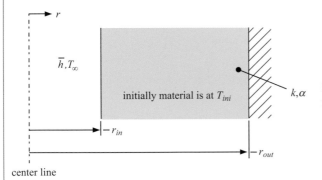

Figure 1 Cylinder experiencing convection on its inner surface.

Assumptions

- The properties do not depend on temperature.
- The temperature distribution is one-dimensional. That is, the temperature only varies in the r-direction and not the x-direction. This assumption is valid provided that the cylinder is long compared to its radius or if its top and bottom surfaces are insulated.

Example 6.4 (cont.)

Analysis and Solution

The control volume that is appropriate for this problem is differentially small in the *r*-direction, as shown in Figure 2.

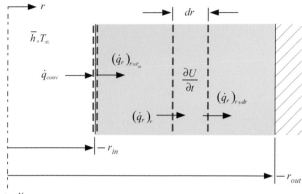

Figure 2 Differential control volume.

An energy balance on the differential control volume shown in Figure 2 provides:

$$(\dot{q}_r)_r = (\dot{q}_r)_{r+dr} + \frac{\partial U}{\partial t}. \tag{1}$$

Expanding the $r + dr$ term in Eq. (1) and taking the limit as dr approaches zero leads to:

$$(\dot{q}_r)_r = (\dot{q}_r)_r + \frac{\partial \dot{q}_r}{\partial r} dr + \frac{\partial U}{\partial t}. \tag{2}$$

The conduction term is evaluated using Fourier's Law:

$$\dot{q}_r = -k \, 2\pi r L \frac{\partial T}{\partial r}, \tag{3}$$

where L is the length of the cylinder and therefore $2\pi r L$ is the cross-sectional area for radial conduction heat transfer. The rate of change of internal energy contained in the differential control volume is:

$$\frac{\partial U}{\partial t} = \rho \, c \, 2\pi r L \, dr \frac{\partial T}{\partial t}, \tag{4}$$

where ρ is the material density and c is the specific heat capacity. Substituting Eqs. (3) and (4) into Eq. (2) provides:

$$0 = \frac{\partial}{\partial r}\left(-k \, 2\pi r L \frac{\partial T}{\partial r}\right) dr + \rho \, c \, 2\pi r L \, dr \frac{\partial T}{\partial t}, \tag{5}$$

which can be simplified to

$$\boxed{\frac{\alpha}{r} \frac{\partial}{\partial r}\left(r \frac{\partial T}{\partial r}\right) = \frac{\partial T}{\partial t}}. \tag{6}$$

Equation (6) is second order in r and first order in t. The boundary conditions in r must be derived using an interface energy balance, as shown in Figure 2 at $r = r_{in}$. The convection heat transfer rate to the inner surface must equal the rate of conduction away from the surface:

Continued

Example 6.4 (cont.)

$$\dot{q}_{conv} = (\dot{q}_r)_{r = r_{in}}. \tag{7}$$

Notice that Eq. (7) does *not* include a storage term even though this is a transient problem; this is because the interface is infinitesimally thin and therefore the interface energy balance includes no volume and no energy. Substituting Newton's Law of Cooling and Fourier's Law into Eq. (7) leads to:

$$\boxed{\bar{h}(T_\infty - T_{r = r_{in}}) = -k\left(\frac{\partial T}{\partial r}\right)_{r = r_{in}}.} \tag{8}$$

An interface energy balance at $r = r_{out}$ provides the second spatial boundary condition:

$$\boxed{\left(\frac{\partial T}{\partial r}\right)_{r = r_{out}} = 0.} \tag{9}$$

The initial condition is:

$$\boxed{T_{t = 0} = T_{ini}.} \tag{10}$$

Discussion

The partial differential equation and boundary conditions derived in this problem can be solved using a technique referred to as Separation of Variables, which is beyond the scope of this text. Separation of Variables is often presented in higher-level math courses. However, it is important that you are able to apply the principles of heat transfer in order to derive the math problem that must be solved, even if the math itself may be slightly out of reach at this point in your career.

6.2.2 Semi-Infinite Body Solutions

Analytical solutions to a semi-infinite body that is exposed to different boundary conditions at the surface have been developed and some of these are summarized in Table 6.1. Each of the entries in Table 6.1 is a solution to the partial differential equation derived in Section 6.2.1:

$$\alpha\frac{\partial^2 T}{\partial x^2} = \frac{\partial T}{\partial t} \tag{6.33}$$

with the spatial boundary condition at $x \to \infty$ corresponding to the semi-infinite body assumption:

$$T_{x \to \infty} = T_{ini} \tag{6.34}$$

and a uniform initial temperature distribution:

$$T_{t = 0} = T_{ini}. \tag{6.35}$$

Each solution corresponds to a different boundary condition at the surface, $x = 0$. These analytical solutions can be used to examine processes that occur over short time scales, specifically for times where the thermal penetration wave (δ_t) is small relative to the spatial extent of the object. Functions that return the temperature at a specified position and time for the semi-infinite body solutions in Table 6.1 are also available in EES. Select Function Info from the Options menu and select Transient Conduction from the pull-down menu at the lower right-hand corner of the upper box, toggle to the Semi-Infinite Body Library and scroll across to find the function of interest.

Table 6.1 Solutions to semi-infinite body problems.

Boundary condition	Analytical solution and function in EES

(a)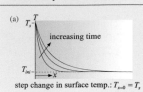

step change in surface temp.: $T_{x=0} = T_s$

$$\frac{T - T_{ini}}{T_s - T_{ini}} = 1 - \text{erf}\left(\frac{x}{2\sqrt{\alpha t}}\right) \quad \dot{q}''_{x=0} = \frac{k}{\sqrt{\pi \alpha t}}(T_s - T_{ini})$$

T = SemiInf1(T_ini, T_s, alpha, x, time)

(b)

surface heat flux: $\dot{q}''_s = -k\frac{\partial T}{\partial x}\Big|_{x=0}$

$$T - T_{ini} = \frac{\dot{q}''_s}{k}\left[2\sqrt{\frac{\alpha t}{\pi}}\exp\left(-\frac{x^2}{4\alpha t}\right) - x\,\text{erfc}\left(\frac{x}{2\sqrt{\alpha t}}\right)\right]$$

T = SemiInf2(T_ini, q``_dot_s, k, alpha, x, time)

(c)

convection to fluid: $\bar{h}\left(T_\infty - T_{x=0}\right) = -k\frac{\partial T}{\partial x}\Big|_{x=0}$

$$\frac{T - T_{ini}}{T_\infty - T_{ini}} = \text{erfc}\left(\frac{x}{2\sqrt{\alpha t}}\right)$$

$$- \exp\left(\frac{\bar{h}x}{k} + \frac{\bar{h}^2\alpha t}{k^2}\right)\text{erfc}\left(\frac{x}{2\sqrt{\alpha t}} + \frac{\bar{h}}{k}\sqrt{\alpha t}\right)$$

T = SemiInf3(T_ini, T_infinity, h_bar, k, alpha, x, time)

(d)

surface energy per unit area released at $t = 0$ wall adiabatic for $t > 0$

increasing time

surface energy pulse: $\lim_{t,\Delta t \to 0} \dot{q}_s \Delta t = E''$

$$T - T_{ini} = \frac{E''}{\rho c\sqrt{\pi \alpha t}}\exp\left(-\frac{x^2}{4\alpha t}\right)$$

T = SemiInf4(T_ini, E``, rho, c, alpha, x, time)

(e)

sinisoidal surface temp. at quasi-steady state
$T_{x=0} = T_{ini} + \Delta T\sin(\omega t)$

$$T - T_{ini} = \Delta T\exp\left(-x\sqrt{\frac{\omega}{2\alpha}}\right)\sin\left(\omega t - x\sqrt{\frac{\omega}{2\alpha}}\right)$$

$$\dot{q}''_{x=0} = k\Delta T\sqrt{\frac{\omega}{\alpha}}\sin\left(\omega t + \frac{\pi}{4}\right)$$

T = SemiInf5(T_ini, DeltaT, omega, alpha, x, time)

(f)

contact between two semi-infinite solids

$$T_{int} = \frac{\sqrt{k_A\rho_A c_A}\,T_{ini, A} + \sqrt{k_B\rho_B c_B}\,T_{ini, B}}{\sqrt{k_A\rho_A c_A} + \sqrt{k_B\rho_B c_B}}$$

$$\frac{T_A - T_{ini, A}}{T_{int} - T_{ini, A}} = 1 - \text{erf}\left(\frac{x_A}{2\sqrt{\alpha_A t}}\right),$$

$$\frac{T_B - T_{ini, B}}{T_{int} - T_{ini, B}} = 1 - \text{erf}\left(\frac{x_B}{2\sqrt{\alpha_B t}}\right)$$

T_{ini} = initial temperature t = time relative to surface disturbance k = conductivity
 x = position from surface ρ = density c = specific heat capacity
 α = thermal diffusivity erf() = Gaussian error function erfc() = complementary error function

Example 6.5

Revisit Example 6.3 using the exact solution that is available in Table 6.1. A laser machining process uses a very short but very high power heat flux to melt aluminum. Initially the material is at a uniform temperature, $T_{ini} = 20°C$. The heat flux from the laser is applied at the surface and lasts only $t_{laser} = 0.1$ ms. You would like to increase the temperature at the surface to the melting temperature, $T_{melt} = 630°C$. Determine the heat flux that is required for this process.

Known Values

The problem was previously illustrated in Figure 1 of Example 6.3 with the known values included. The properties of aluminum are obtained from Appendix A: $k = 235$ W/m-K and $\alpha = 96.6 \times 10^{-6}$ m^2/s.

Assumptions

- The properties of aluminum do not depend on temperature.
- The aluminum can be treated as a semi-infinite body.

Analysis

The solution in Table 6.1(b) corresponding to a constant heat flux can be applied to this problem:

$$T - T_{ini} = \frac{\dot{q}_s''}{k}\left[2\sqrt{\frac{\alpha t}{\pi}}\exp\left(-\frac{x^2}{4\alpha t}\right) - x\,\text{erfc}\left(\frac{x}{2\sqrt{\alpha t}}\right)\right]. \tag{1}$$

The surface temperature ($x = 0$) at the conclusion of the laser pulse ($t = t_{laser}$) must be equal to the melting temperature ($T = T_{melt}$). With these substitutions, Eq. (1) becomes:

$$T_{melt} - T_{ini} = \frac{\dot{q}_s''}{k}2\sqrt{\frac{\alpha t_{laser}}{\pi}}. \tag{2}$$

Solving Eq. (2) for the surface heat flux provides:

$$\dot{q}_s'' = \frac{\sqrt{\pi}}{2}\frac{k(T_{melt} - T_{ini})}{\sqrt{\alpha t_{laser}}}. \tag{3}$$

Solution

Substituting known values into Eq. (3) provides:

$$\dot{q}_s'' \approx \frac{\sqrt{\pi}}{2}\frac{k(T_{melt} - T_{ini})}{\sqrt{\alpha t_{laser}}} = \frac{\sqrt{\pi}}{2}\left|\frac{235}{m}\frac{W}{K}\right|\left|(630°C - 20°C)\right|\sqrt{\frac{s}{96.6 \times 10^{-6} \, m^2}\left|\frac{1}{0.1 \times 10^{-3} \, s}\right|}$$

$$\tag{4}$$

$$= \boxed{12.9 \times 10^8 \frac{W}{m^2}\left(1290\frac{MW}{m^2}\right)}.$$

Discussion

Example 6.3 used the approximate concept of a semi-infinite resistance to estimate that the required heat flux is 729 MW/m^2. In this example, the exact solution showed that a heat flux of 1290 MW/m^2 is required. The difference is on the order of 50 percent, which should be considered the level of accuracy that can be expected when using the concepts discussed in Section 6.1 to understand transient conduction problems.

Example 6.6

Liquid nitrogen is spilled onto a concrete floor that is initially at a uniform temperature of $T_{ini} = 20°C$. The liquid nitrogen boils at $T_{LN2} = 77$ K. The heat transfer coefficient between the surface of the floor and the boiling liquid nitrogen is $\bar{h} = 50$ W/m²-K. You are interested in the extent of concrete that might be damaged by this accident. According to the supplier, the concrete should not be exposed to temperatures below $T_{damage} = -40°C$. Prepare a plot showing the thickness of the layer of concrete that has achieved a temperature below T_{damage} as a function of time for 5 min $< t <$ 60 min.

Known Values

Figure 1 illustrates the problem with the known values indicated. The properties of concrete are obtained from Appendix A: $k = 1.4$ W/m-K and $\alpha = 0.692 \times 10^{-6}$ m²/s.

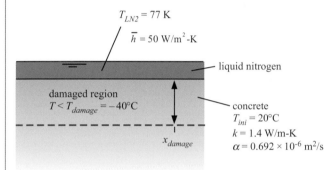

Figure 1 Concrete floor with liquid nitrogen spill.

Assumptions

- The properties of concrete do not depend on temperature.
- The concrete can be treated as a semi-infinite body for the duration of this process. This implies that the thickness of the concrete slab (L) is larger than the penetration of the thermal wave:

$$L > \delta_t \approx 2\sqrt{\alpha t} = 2\sqrt{\frac{0.692 \times 10^{-6} \text{ m}^2}{\text{s}} \left|\frac{1 \text{ hr}}{}\right|\left|\frac{3600 \text{ s}}{\text{hr}}\right|} = 0.0998 \text{ m (9.98 cm)}. \tag{1}$$

Analysis

The solution in Table 6.1(c) corresponding to a semi-infinite body with a surface experiencing convection can be applied to this problem:

$$\frac{T - T_{ini}}{T_\infty - T_{ini}} = \text{erfc}\left(\frac{x}{2\sqrt{\alpha t}}\right) - \exp\left(\frac{\bar{h}x}{k} + \frac{\bar{h}^2 \alpha t}{k^2}\right)\text{erfc}\left(\frac{x}{2\sqrt{\alpha t}} + \frac{\bar{h}}{k}\sqrt{\alpha t}\right). \tag{2}$$

We are interested in the x-position (x_{damage}) at which the temperature reaches $T = T_{damage}$:

$$\frac{T_{damage} - T_{ini}}{T_\infty - T_{ini}} = \text{erfc}\left(\frac{x_{damage}}{2\sqrt{\alpha t}}\right) - \exp\left(\frac{\bar{h}x_{damage}}{k} + \frac{\bar{h}^2 \alpha t}{k^2}\right)\text{erfc}\left(\frac{x_{damage}}{2\sqrt{\alpha t}} + \frac{\bar{h}}{k}\sqrt{\alpha t}\right). \tag{3}$$

Continued

Example 6.6 (cont.)

Equation (3) provides a relationship between x_{damage} and t. It is much easier to use the solution programmed in EES to develop a plot of x_{damage} as a function of t than it is to iteratively solve Eq. (3) manually.

Solution

The known information is entered in EES.

h_bar=50 [W/m^2-K]	"heat transfer coefficient"
T_LN2 = 77 [K]	"temperature of liquid nitrogen"
T_ini=**ConvertTemp**(C,K,20 [C])	"initial temperature"
T_damage = **ConvertTemp**(C,K,-40 [C])	"freezing temperature"
k=1.4 [W/m-K]	"conductivity of concrete"
alpha=0.692e-6 [m^2/s]	"thermal diffusivity of concrete"

Initially an arbitrary value of time is specified in order to ensure that we can compute an associated value of x_{damage}.

time=3000 [s]	"time"

The appropriate semi-infinite body solution is obtained by selecting Function Info from the Options menu and then selecting Transient Conduction from the drop down menu next to Heat Transfer & Fluid Flow. In the lower window select Semi-Infinite Body and scroll to the appropriate boundary condition, as shown in Figure 2.

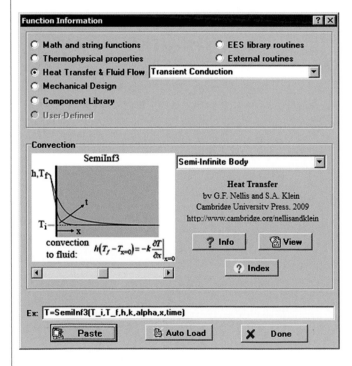

Figure 2 Function Information dialog for the Semilnf3 function.

Example 6.6 (cont.)

Select the Paste button to enter the function call into the Equations Window. Modify the call so that the temperature returned by the function must be T_{damage} which implies that EES will vary the value of x_{damage} until this equation is satisfied.

```
T_damage=SemiInf3(T_ini, T_LN2, h_bar,k ,alpha, x_damage, time)
                                        "find the freezing depth"
```

It is likely that you will receive an error upon trying to solve the program because EES will require both a reasonable guess value and limits in order to iterate to the correct solution. Select Variable Info from the Options menu and then make the guess value for the variable x_damage something more reasonable than 1 m (smaller, certainly, based on the fact that the thermal penetration depth is only 10 cm after an entire hour). Also specify that the lower limit on x_damage must be 0 (it makes no sense to have a negative x value). These changes are shown in Figure 3.

Variable Information
Show array variables
Show string variables

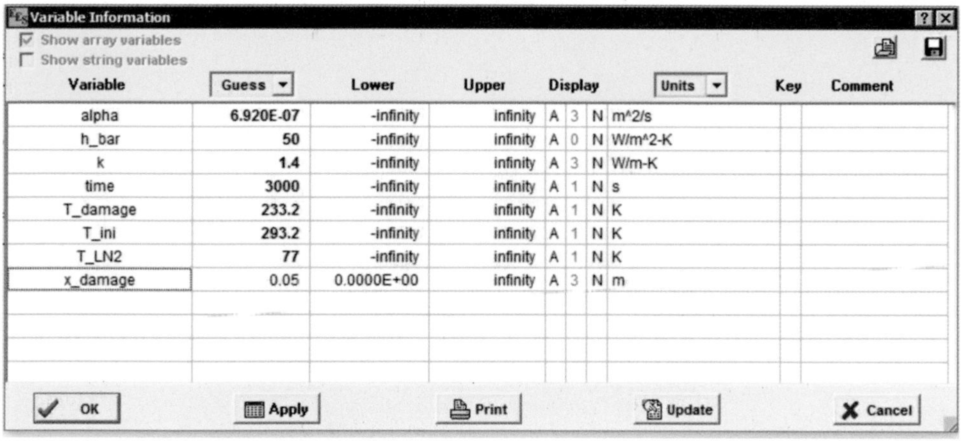

Variable	Guess	Lower	Upper	Display			Units	Key	Comment
alpha	6.920E-07	-infinity	infinity	A	3	N	m^2/s		
h_bar	50	-infinity	infinity	A	0	N	W/m^2-K		
k	1.4	-infinity	infinity	A	3	N	W/m-K		
time	3000	-infinity	infinity	A	1	N	s		
T_damage	233.2	-infinity	infinity	A	1	N	K		
T_ini	293.2	-infinity	infinity	A	1	N	K		
T_LN2	77	-infinity	infinity	A	1	N	K		
x_damage	0.05	0.0000E+00	infinity	A	3	N	m		

OK Apply Print Update Cancel

Figure 3 Variable Information Window with reasonable guess value and limits set for x_damage.

Generate a Parametric Table that contains the variables x_damage and time. Vary time from 5 minutes to 60 minutes and comment out the value of time set in the Equations Window. The data are used to generate the requested plot in Figure 4.

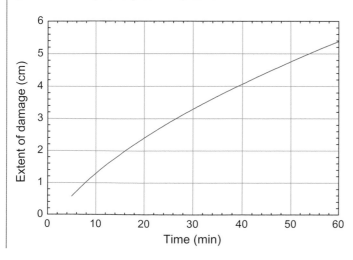

Figure 4 Extent of damaged concrete as a function of time.

Continued

Example 6.6 (cont.)

Discussion

In Figure 4 it is clear that the extent of damaged concrete (which corresponds to the position of the isotherm at T_{damage}) continues to grow with time but at a decreasing rate. This is consistent with our understanding of how thermal waves grow diffusively from Section 6.1:

$$\delta_t \approx 2\sqrt{\alpha t}. \tag{4}$$

Taking the derivative of Eq. (4) provides

$$\frac{d\delta_t}{dt} \approx \sqrt{\frac{\alpha}{t}}, \tag{5}$$

which shows that the rate of growth is proportional to $1/\sqrt{t}$. Also note that the properties of many substances are built into the EES database. Concrete, for example, is part of the Incompressible substance database. The thermal properties k and α could have been obtained using this database rather than by looking them up in Appendix A. First, set the unit system with the $UnitSystem directive. Then use the Conductivity and ThermalDiffusivity functions to obtain the values, as shown in the program below.

```
$UnitSystem SI Mass J K Pa
h_bar=50 [W/m^2-K]                              "heat transfer coefficient"
T_LN2 = 77 [K]                                  "temperature of liquid nitrogen"
T_ini=ConvertTemp(C,K,20 [C])                   "initial temperature"
T_damage = ConvertTemp(C,K,-40 [C])             "freezing temperature"
k=Conductivity('Concrete_ stone mix', T=T_ini)  "conductivity from EES database"
alpha=ThermalDiffusivity('Concrete_ stone mix', T=T_ini)
                                                "thermal diffusivity from EES database"

//time=3000 [s]                                 "time"
T_damage=SemiInf3(T_ini, T_LN2, h_bar,k ,alpha, x_damage, time)
                                                "find the freezing depth"
```

6.2.3 Bounded Problem Solutions

Section 6.1 discussed the behavior of a thermal wave within an unbounded (i.e., semi-infinite) solid and introduced the concept of a diffusion time constant. Some analytical solutions to this type of problem were presented in Section 6.2.2. This section presents the solution to 1-D transient problems that cannot be modeled as being semi-infinite. These *bounded* solutions remain valid even after the thermal wave has progressed to an adjacent boundary; that is, for times greater than the diffusive time constant discussed in Section 6.1.2. The method of separation of variables can be used to obtain analytical solutions for transient conduction problems in bounded systems. The solutions associated with the plane wall, solid cylinder, and solid sphere are available in most textbooks, as approximate formulae and in graphical format. These solutions are also available within EES both in dimensional and nondimensional form using the Transient Conduction Library. This section provides, without derivation, the solutions to these basic problems. A more thorough discussion of the techniques used to obtain these solutions can be found in Myers (1998) and in Nellis and Klein (2009).

The Plane Wall – Exact Solution

Consider the plane wall problem shown in Figure 6.6. Initially the material is at a uniform temperature, T_{ini}. The surface at $x = 0$ is adiabatic (or, equivalently, a line of symmetry). At time $t = 0$ the surface at $x = L$ begins to experience convection with fluid at temperature, T_∞, with heat transfer coefficient $\bar{h}$.

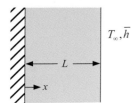

Figure 6.6 A uniform temperature (T_{ini}) subjected to a convective boundary condition ($\bar{h}$, T_∞).

The partial differential equation associated with this problem can be obtained using the steps discussed in Section 6.2.1:

$$\frac{\partial^2 T}{\partial x^2} = \frac{1}{\alpha}\frac{\partial T}{\partial t}. \tag{6.36}$$

The boundary conditions at $x = 0$ and $x = L$ for the problem are:

$$\left.\frac{\partial T}{\partial x}\right|_{x=0} = 0 \tag{6.37}$$

$$-k\left.\frac{\partial T}{\partial x}\right|_{x=L} = \bar{h}[T_{x=L} - T_\infty]. \tag{6.38}$$

The initial condition is:

$$T_{t=0} = T_{ini}. \tag{6.39}$$

The exact solution for this problem can be derived using the method of separation of variables and is expressed as an infinite series:

$$\tilde{\theta}(\tilde{x}, Fo) = \sum_{i=1}^{\infty} C_i \cos{(\zeta_i \tilde{x})} \exp\left[-\zeta_i^2 Fo\right], \tag{6.40}$$

where $\tilde{x}$, Fo, and $\tilde{\theta}$ are the dimensionless position, **Fourier number** (dimensionless time), and dimensionless temperature difference:

$$\tilde{x} = \frac{x}{L} \tag{6.41}$$

$$Fo = \frac{t\alpha}{L^2} \tag{6.42}$$

$$\tilde{\theta} = \frac{T - T_\infty}{T_{ini} - T_\infty}. \tag{6.43}$$

The parameters ζ_i in Eq. (6.40) are referred to as **eigenvalues** and correspond to each of the infinite number of roots of the **eigencondition**:

$$\tan{(\zeta_i)} = \frac{Bi}{\zeta_i}, \tag{6.44}$$

where Bi is the Biot number:

$$Bi = \frac{\bar{h} L}{k}. \tag{6.45}$$

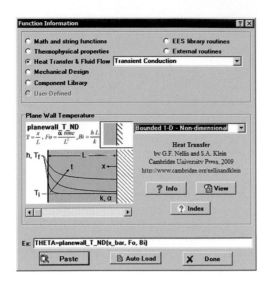

Figure 6.7 Function information for the EES function that provides the exact dimensionless solution to the problem shown in Figure 6.6.

The constants C_i in Eq. (6.40) are:

$$C_i = \frac{2 \sin (\zeta_i)}{\zeta_i + \cos (\zeta_i) \sin (\zeta_i)}. \tag{6.46}$$

The first 20 terms of Eq. (6.40) are programmed in EES as the function Planewall_T_ND, which can be accessed by selecting Function Information from the Options menu and then Transient Conduction from the Heat Transfer & Fluid Flow drop down menu. In the bottom window select Bounded 1-D – Non-dimensional and scroll to Planewall_T_ND, as shown in Figure 6.7.

A dimensional version of the function is also available by selecting Bounded 1-D – Dimensional. The dimensional solution, Planewall_T, returns the temperature at a given position and time as a function of the *dimensional* independent variables (i.e., L, α, k, $\bar{h}$, T_{ini}, T_∞).

The exact solutions for the dimensionless temperature are often presented graphically in a form that was initially published by Heisler (1947) and are now referred to as the Heisler charts; these charts were convenient when access to computer solutions was not readily available. Heisler charts typically include the dimensionless center temperature ($\tilde{\theta}_{\tilde{x}=0}$) as a function of the Fourier number for various values of the inverse of the Biot number, as shown in Figure 6.8(a). The information presented in this figure corresponds to the temperature at the position of the adiabatic boundary ($x = 0$) and is therefore referred to as the "center" temperature; the center-line of a plane wall would be a line of symmetry and therefore adiabatic if convection occurred on both sides of the wall. The temperature at other locations can be obtained using Figure 6.8(b), which shows the ratio of the dimensionless temperature to the dimensionless center temperature ($\tilde{\theta}/\tilde{\theta}_{\tilde{x}} = 0$) as a function of the inverse of the Biot number for various values of dimensionless position; note that the Fourier number does not affect this ratio. Also notice in Figure 6.8(b) that as the inverse of the Biot number becomes large (corresponding to a small Biot number), the x/L lines collapse indicating that the temperature of the plane wall becomes very nearly spatially uniform. This situation corresponds to the limit where the lumped capacitance approximation is valid, as discussed in Section 5.1.

It is sometimes important to calculate the total amount of energy transferred to the wall. An energy balance on the wall indicates that the total energy transfer (Q) is the difference between the energy stored in the wall (U) and the energy stored in the wall at its initial condition ($U_{t=0}$):

$$Q = U - U_{t=0}. \tag{6.47}$$

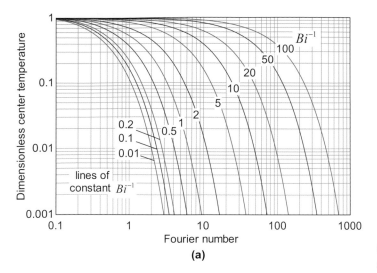

(a)

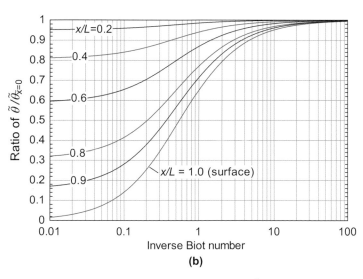

(b)

Figure 6.8 Heisler charts for a plane wall showing (a) $\tilde{\theta}_{\tilde{x}=0}$ as a function of Fo for various values of Bi^{-1} and (b) $\tilde{\theta}/\tilde{\theta}_{\tilde{x}=0}$ as a function of Bi^{-1} for various values of $\tilde{x}$.

The total energy transfer is made dimensionless ($\tilde{Q}$) by normalizing it against the maximum amount of energy that could be transferred to the wall (Q_{max}):

$$\tilde{Q} = \frac{Q}{Q_{max}}. \tag{6.48}$$

The maximum energy transfer would occur if the process continued to $t \to \infty$ and therefore the wall material had time to equilibrate completely with the fluid temperature at T_∞:

$$Q_{max} = \rho\, c\, L\, A_c (T_\infty - T_{ini}). \tag{6.49}$$

The dimensionless energy transfer is related to an integral of the dimensionless temperature according to:

$$\tilde{Q} = 1 - \int_0^1 \tilde{\theta}\, d\tilde{x}. \tag{6.50}$$

Substituting the exact solution, Eq. (6.40), into Eq. (6.50) and carrying out the integration leads to:

$$\tilde{Q} = 1 - \sum_{i=1}^{\infty} C_i \frac{\sin(\zeta_i)}{\zeta_i} \exp\left(-\zeta_i^2 Fo\right). \tag{6.51}$$

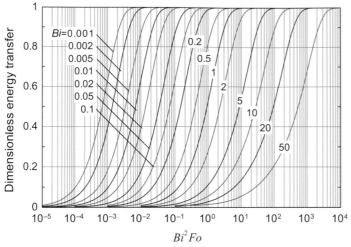

Figure 6.9 Dimensionless heat transfer, $\tilde{Q}$, as a function of $Bi^2\,Fo$ for various values of Bi.

The first 20 terms of the exact solution for the dimensionless energy transfer associated with a plane wall, Eq. (6.51), have been programmed in EES as the function **Planewall_Q_ND** which has a dimensional companion **Planewall_Q**. Figure 6.9 illustrates the associated Heisler chart; the dimensionless total energy transfer, $\tilde{Q}$, is provided as a function of $Bi^2\,Fo$ for various values of the Biot number.

Example 6.7

A slab of iron that is 20 cm thick is removed from an oven at a uniform temperature of $T_{ini} = 1000°C$ and cooled on both sides by exposure to convection with water at $T_\infty = 20°C$ with heat transfer coefficient $\bar{h} = 1500$ W/m²-K. How much time is required for the center temperature to reach $T_{center} = 300°C$? At this time, what is the surface temperature and how much energy has been removed per area of slab?

Known Values

Figure 1 illustrates the problem with the known values indicated. The properties of iron are obtained from Appendix A: $k = 80.5$ W/m-K and $\alpha = 23.2 \times 10^{-6}$ m²/s.

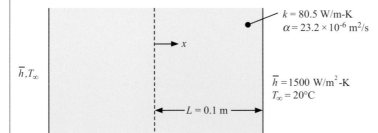

all material is initially at $T_{ini} = 1000°C$

Figure 1 Slab of hot iron exposed to water cooling.

Assumptions

- The properties of iron do not depend on temperature.
- The slab is sufficiently large in the y- and z-directions that the problem can be treated as being one dimensional (1-D).

Example 6.7 (cont.)

Analysis and Solution

Figure 6.8(a) presents the relationship between the dimensionless temperature difference at the center of the slab ($\tilde{\theta}_{x=0}$), Fourier number (Fo), and Biot number (Bi). Given any two of these quantities, the Heisler chart can be used to determine the third one. In this problem, the Biot number and required dimensionless center temperature can be computed:

$$Bi = \frac{\bar{h}\,L}{k} = \frac{1500 \text{ W}}{\text{m}^2 \text{ K}}\left|\frac{0.1 \text{ m}}{}\right|\frac{\text{m K}}{80.5 \text{ W}} = 1.863 \tag{1}$$

$$\tilde{\theta}_{x=0} = \frac{T_{center} - T_\infty}{T_{ini} - T_\infty} = \frac{300°\text{C} - 20°\text{C}}{1000°\text{C} - 20°\text{C}} = 0.286. \tag{2}$$

The Fourier number can then be read off of the chart at $Bi^{-1} = 1/1.863 = 0.54$ and $\tilde{\theta}_{x=0} = 0.29$, which leads to $Fo \approx 1.3$. The Fourier number is used to compute the time:

$$t = \frac{L^2}{\alpha} Fo = \frac{(0.1)^2 \text{ m}^2}{}\left|\frac{1.3}{}\right|\frac{\text{s}}{23.2 \times 10^{-6} \text{ m}^2} = \boxed{560 \text{ s } (9.3 \text{ min})}. \tag{3}$$

Figure 6.8(b) provides information about the dimensionless temperature difference at other values of dimensionless positions. At $Bi^{-1} = 0.54$ and $\tilde{x} = 1$ the value of $\tilde{\theta}/\tilde{\theta}_{x=0}$ is approximately 0.5. Therefore, the dimensionless temperature difference at the surface is

$$\tilde{\theta}_{\tilde{x}=1} = 0.5\,\tilde{\theta}_{\tilde{x}=0} = 0.5(0.286) = 0.14 \tag{4}$$

and the surface temperature is

$$T_s = T_\infty + \tilde{\theta}_{\tilde{x}=1}(T_{ini} - T_\infty) = 20°\text{C} + 0.14(1000°\text{C} - 20°\text{C}) = \boxed{158°\text{C}}. \tag{5}$$

The amount of heat transfer is obtained using Figure 6.9. At $Bi^2\,Fo = 4.4$ and $Bi = 1.86$ the dimensionless energy transfer is found to be $\tilde{Q} \approx 0.75$. The maximum possible amount of energy transfer is

$$Q_{max} = \rho\,c\,L\,A_c(T_\infty - T_{ini}). \tag{6}$$

The product of density and specific heat capacity is sometimes referred to as the volume specific heat capacity as it represents the heat capacity on a per unit volume basis. According to the definition of thermal diffusivity, $\alpha = k/(\rho\,c)$, the volume specific heat capacity is k/α. With this substitution, the maximum amount of energy transfer per area is

$$Q''_{max} = \frac{k\,L}{\alpha}(T_\infty - T_{ini}) = \frac{80.5 \text{ W}}{\text{m K}}\left|\frac{\text{s}}{23.2 \times 10^{-6} \text{m}^2}\right|\frac{0.1\text{m}}{}\left|(20°\text{C} - 1000°\text{C})\right|\frac{\text{J}}{\text{W s}} = -3.40 \times 10^8 \frac{\text{J}}{\text{m}^2}\left(-340\frac{\text{MJ}}{\text{m}^2}\right) \tag{7}$$

and the actual amount of energy transfer per area is

$$Q'' = \tilde{Q}\,Q''_{max} = 0.75\left(-3.40 \times 10^8\,\frac{\text{J}}{\text{m}^2}\right) = -2.55 \times 10^8 \frac{\text{J}}{\text{m}^2}\left(-255\frac{\text{MJ}}{\text{m}^2}\right). \tag{8}$$

Note that the answer is negative because energy is being removed rather than added to the material. The question asks for the amount of energy removed from the wall per unit area, which is therefore: $2.55 \times 10^8\,\frac{\text{J}}{\text{m}^2}\left(255\frac{\text{MJ}}{\text{m}^2}\right).$

Discussion

The solution to the problem could have been obtained more easily and accurately using the exact solution programmed in EES. The inputs are entered and the functions Planewall_T and Planewall_Q are used to determine the time and surface temperature and the energy transfer, respectively.

Continued

Example 6.7 (cont.)

```
$UnitSystem SI Mass C Pa J
k = 80.5 [W/m-K]                                           "conductivity of iron"
alpha=23.2e-6 [m^2/s]                                      "thermal diffusivity of iron"
T_ini=1000 [C]                                            "initial temperature"
T_infinity=20 [C]                                         "fluid temperature"
L=10 [cm]*Convert(cm,m)                                   "half-width of slab"
h_bar=1500 [W/m^2-K]                                      "heat transfer coefficient"
T_center = 300 [C]                                        "required center temperature"


T_center=Planewall_T(0 [m], time, T_ini, T_infinity, alpha, k, h_bar, L)    "get time"
T_s=Planewall_T(L, time, T_ini, T_infinity, alpha, k, h_bar, L)     "get surface temperature"
Q=-Planewall_Q(time, T_ini, T_infinity, alpha, k, h_bar, L)    "get energy transfer per unit area"
```

Solving provides time = 546.3 s (9.11 min), T_s = 158°C, and Q = 2.60 × 10^8 J/m^2 (260 MJ/m^2).

The Plane Wall – Approximate Solution

The exact solution to the plane wall problem is the infinite series given by Eq. (6.40) with the constants given by Eq. (6.46). Substitution of these equations leads to:

$$\tilde{\theta}(\tilde{x}, Fo) = \sum_{i=1}^{\infty} \underbrace{\frac{2\sin(\zeta_i)}{\zeta_i + \cos(\zeta_i)\sin(\zeta_i)}}_{i\text{th constant, } C_i} \cos(\zeta_i \tilde{x}) \underbrace{\exp\left[-\zeta_i^2 \, Fo\right]}_{i\text{th exponential term}},$$ (6.52)

where the dimensionless eigenvalues, ζ_i, are the repeated roots of Eq. (6.44). Examination of Eq. (6.52) shows that successive terms in the series become small relatively quickly because the magnitude of the eigenvalues are progressively larger. Therefore both the constant and the exponential parts of each successive term decreases. The exponential term in particular approaches zero quickly as the value of the Fourier number (*Fo*, dimensionless time) increases. As a result, the series solution can be adequately approximated using only the first term when *Fo* > 0.2. In this case, only the first eigenvalue (i.e., ζ_1, the first root of Eq. (6.44)) is required to implement this approximate solution. The value of the first dimensionless eigenvalue as a function Biot number is shown in Figure 6.10 and tabulated in Table 6.2 for the plane wall, as well as the cylinder and sphere solutions, discussed subsequently.

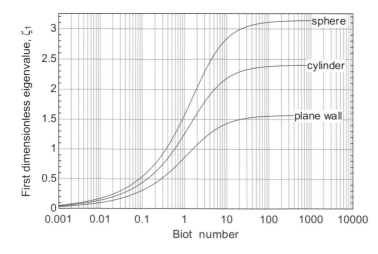

Figure 6.10 First dimensionless eigenvalue for use in the approximate solution to the plane wall, cylinder, and sphere bounded transient conduction problems.

Table 6.2 First dimensionless eigenvalue for use in the approximate solution to the plane wall, cylinder, and sphere bounded transient conduction problems.

Biot number	First eigenvalue		
Bi	Plane wall	Cylinder	Sphere
0.001	0.03162	0.04472	0.05477
0.002	0.04471	0.06323	0.07744
0.005	0.07065	0.09994	0.1224
0.01	0.09983	0.1412	0.1730
0.02	0.1410	0.1995	0.2445
0.05	0.2218	0.3143	0.3854
0.1	0.3111	0.4417	0.5423
0.2	0.4328	0.6170	0.7593
0.3	0.5218	0.7465	0.9208
0.4	0.5932	0.8516	1.053
0.5	0.6533	0.9408	1.166
0.6	0.7051	1.018	1.264
0.7	0.7506	1.087	1.353
0.8	0.7910	1.149	1.432
0.9	0.8274	1.205	1.504
1	0.8603	1.256	1.571
2	1.077	1.599	2.029
3	1.192	1.789	2.289
4	1.265	1.908	2.456
5	1.314	1.990	2.570
6	1.350	2.049	2.654
7	1.377	2.094	2.716
8	1.398	2.129	2.765
9	1.415	2.157	2.804
10	1.429	2.179	2.836
20	1.496	2.288	2.986
50	1.540	2.357	3.079
100	1.555	2.381	3.110
200	1.563	2.393	3.126
500	1.568	2.400	3.135
1000	1.569	2.402	3.138
∞	$\pi/2$	2.405	π

Using ζ_1, the approximate solutions for the dimensionless temperature difference and dimensionless energy transfer become:

$$\tilde{\theta}(\tilde{x}, Fo) \approx \left[\frac{2 \sin (\zeta_1)}{\zeta_1 + \cos (\zeta_1) \sin (\zeta_1)} \right] \exp \left(-\zeta_1^2 \, Fo \right) \cos \left(\zeta_1 \, \tilde{x} \right) \tag{6.53}$$

$$\tilde{Q}(Fo) \approx 1 - \left[\frac{2 \sin (\zeta_1)}{\zeta_1 + \cos (\zeta_1) \sin (\zeta_1)} \right] \frac{\sin (\zeta_1)}{\zeta_1} \exp \left(-\zeta_1^2 \, Fo \right). \tag{6.54}$$

Example 6.8

Reconsider Example 6.7 using the single term, approximate solution for the plane wall problem.

Known Values

Figure 1 in Example 6.7 illustrates the problem with the known values indicated.

Assumption

- The single term approximation to the exact solution is appropriate. This assumption is justified because the Fourier number calculated in Example 6.7 is $Fo = 1.3$, which is larger than 0.2.

Analysis and Solution

The Biot number and required dimensionless center temperature were computed in Example 6.7:

$$Bi = \frac{\bar{h}L}{k} = \frac{1500 \text{ W}}{\text{m}^2 \text{ K}} \left| 0.1 \text{ m} \right| \frac{\text{m K}}{80.5 \text{ W}} = 1.863 \tag{1}$$

$$\tilde{\theta}_{x=0} = \frac{T_{center} - T_\infty}{T_{ini} - T_\infty} = \frac{300°C - 20°C}{1000°C - 20°C} = 0.286. \tag{2}$$

Interpolating from Table 6.2, the first dimensionless eigenvalue is approximately $\zeta_1 = 1.047$. Equation (6.53) is used with this value of ζ_1 and $\tilde{x} = 0$ to approximate the relationship between the center temperature and the Fourier number:

$$\tilde{\theta}_{x=0} \approx \left[\frac{2 \sin(1.047)}{1.047 + \cos(1.047)\sin(1.047)} \right] \exp\left[-(1.047)^2 Fo \right] = 1.170 \exp(-1.096 \ Fo). \tag{3}$$

Solving Eq. (3) for the Fourier number when $\tilde{\theta}_{x=0} = 0.286$ provides:

$$Fo = \frac{-\ln\left(\dfrac{\tilde{\theta}_{x=0}}{1.170}\right)}{1.096} = \frac{-\ln\left(\dfrac{0.286}{1.170}\right)}{1.096} = 1.285. \tag{4}$$

The Fourier number is used to determine the time:

$$t = \frac{L^2}{\alpha} Fo = \frac{(0.1)^2 \text{ m}^2}{} \left| 1.285 \right| \frac{\text{s}}{23.2 \times 10^{-6} \text{ m}^2} = \boxed{554 \text{ s } (9.2 \text{ min})}. \tag{5}$$

The surface temperature is associated with $\tilde{\theta}_{x=L}$, which is found using Eq. (6.53) with $\tilde{x} = 1$:

$$\tilde{\theta}_{\tilde{x}=1} \approx \underbrace{\left[\frac{2\sin(\zeta_1)}{\zeta_1 + \cos(\zeta_1)\sin(\zeta_1)} \right] \exp\left(-\zeta_1^2 \ Fo\right)}_{\tilde{\theta}_{\tilde{x}=0}} \cos(\zeta_1) = 0.286\cos(1.047) = 0.143. \tag{6}$$

The associated surface temperature is:

$$T_s = T_\infty + \tilde{\theta}_{\tilde{x}=0}(T_{ini} - T_\infty) = 20°C + 0.143(1000°C - 20°C) = \boxed{160°C}. \tag{7}$$

Example 6.8 (cont.)

The amount of heat transfer is associated with $\tilde{Q}$, which is obtained using Eq. (6.54).

$$\tilde{Q} \approx 1 - \left[\frac{2\sin(\zeta_1)}{\zeta_1 + \cos(\zeta_1)\sin(\zeta_1)}\right] \frac{\sin(\zeta_1)}{\zeta_1} \exp\left(-\zeta_1^2 \, Fo\right)$$

$$\approx 1 - \left[\frac{2\sin(1.047)}{1.047 + \cos(1.047)\sin(1.047)}\right] \frac{\sin(1.047)}{1.047} \exp\left[-(1.047)^2 1.285\right] = 0.764. \tag{8}$$

The maximum possible amount of energy transfer per unit area was found in Example 6.7 to be $Q''_{max} = 3.4 \times 10^8$ J/m². Therefore, the actual amount of heat transfer per unit area removed is

$$Q'' = \tilde{Q} \, Q''_{max} = 0.764\left(3.40 \times 10^8 \, \frac{\text{J}}{\text{m}^2}\right) = \boxed{2.60 \times 10^8 \, \frac{\text{J}}{\text{m}^2} \left(260 \, \frac{\text{MJ}}{\text{m}^2}\right)}. \tag{9}$$

Discussion

The solutions obtained using the single term approximation are very close to those obtained in Example 6.7. The values obtained using the exact solution programmed in EES provide the best comparison as there may be errors associated with reading numbers from the Heisler charts. The comparison of the two solutions is: $t = 546.3$ s (exact) vs. 554.0 s (single term), $T_s = 158°C$ (exact) vs. 160°C (single term), and $Q'' = 260$ MJ/m² (exact) vs. 260 MJ/m² (single term).

The Cylinder – Exact Solution

Consider the problem associated with transient conduction in a solid cylinder, shown in Figure 6.11. Initially the material is at a uniform temperature T_{ini}. At time $t = 0$ the surface at $r = r_{out}$ begins to experience convection with fluid at temperature, T_∞, with heat transfer coefficient $\bar{h}$.

The partial differential equation associated with transient conduction within a cylinder is derived in Example 6.4:

$$\frac{\alpha}{r}\frac{\partial}{\partial r}\left[r\frac{\partial T}{\partial r}\right] = \frac{\partial T}{\partial t}. \tag{6.55}$$

The initial condition is

$$T_{t=0} = T_{ini} \tag{6.56}$$

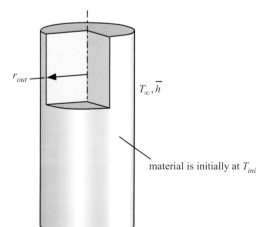

r_{out}

$T_\infty, \bar{h}$

material is initially at T_{ini}

Figure 6.11 A solid cylinder initially at a uniform temperature (T_{ini}) subjected a convective boundary condition ($\bar{h}$, T_∞).

and the spatial boundary conditions are

$$\frac{\partial T}{\partial r}\bigg|_{r=0} \quad \text{must be bounded} \tag{6.57}$$

$$-k\frac{\partial T}{\partial r}\bigg|_{r=r_{out}} = \bar{h}(T_{r=r_{out}} - T_\infty). \tag{6.58}$$

At the center of the cylinder the conduction heat transfer rate must be zero. This implies that the product of the area and the temperature gradient at $r = 0$ must be zero. However, the area for radial conduction at $r = 0$ is, itself, zero and therefore it is only necessary for the temperature gradient at $r = 0$ to remain finite in order for this boundary condition to be satisfied. The exact solution to the problem posed by Eqs. (6.55) through (6.58) can be derived using the method of separation of variables and is given by the infinite series:

$$\tilde{\theta}(\tilde{r}, Fo) = \sum_{i=1}^{\infty} C_i \, \text{BesselJ}\,(0, \zeta_i \, \tilde{r}) \exp\left[-\zeta_i^2 \, Fo\right]. \tag{6.59}$$

Note that the function BesselJ$(0, x)$ is the zeroth-order Bessel function of the first kind; this function is defined to solve a certain, linear homogeneous ordinary differential equation much like the functions exponential, sine, and cosine solve specific types of differential equations, as discussed in Section 3.2.5. The parameters $\tilde{r}$, Fo, and $\tilde{\theta}$ are the dimensionless radius, Fourier number, and dimensionless temperature difference, defined as:

$$\tilde{r} = \frac{r}{r_{out}} \tag{6.60}$$

$$Fo = \frac{t\alpha}{r_{out}^2} \tag{6.61}$$

$$\tilde{\theta} = \frac{T - T_\infty}{T_{ini} - T_\infty}. \tag{6.62}$$

The dimensionless eigenvalues, ζ_i in Eq. (6.59) are the roots of the eigencondition

$$\zeta_i \, \text{BesselJ}(1, \zeta_i) - Bi \, \text{BesselJ}(0, \zeta_i) = 0, \tag{6.63}$$

where BesselJ$(1, x)$ is the first-order Bessel function of the first kind and Bi is the Biot number, defined as

$$Bi = \frac{\bar{h} \, r_{out}}{k}. \tag{6.64}$$

The constants in Eq. (6.59) are given by:

$$C_i = \frac{2 \, \text{BesselJ}(1, \zeta_i)}{\zeta_i \left[\text{BesselJ}^2(0, \zeta_i) + \text{BesselJ}^2(1, \zeta_i)\right]}. \tag{6.65}$$

The dimensionless energy transfer is computed by integration of the dimensionless temperature difference according to:

$$\tilde{Q} = 1 - 2 \int_0^1 \tilde{\theta} \, \tilde{r} \, d\tilde{r}. \tag{6.66}$$

Substituting the exact solution, Eq. (6.59), into Eq. (6.66) and carrying out the integration leads to:

$$\tilde{Q} = 1 - \sum_{i=1}^{\infty} C_i \frac{2 \, \text{BesselJ}(1, \zeta_i)}{\zeta_i} \exp\left(-\zeta_i^2 \, Fo\right). \tag{6.67}$$

The equivalent to the Heisler chart solutions that are provided for a plane wall in Figure 6.8 and Figure 6.9 can be generated for the cylinder problem, but are not presented here. Rather, the reader is encouraged to use the functions programmed in EES. These are accessed from the Transient Conduction section of the Heat Transfer and Fluid Flow Library and include both nondimensional (Cylinder_T_ND and Cylinder_Q_ND) and dimensional (Cylinder_T and Cylinder_Q) versions.

The Cylinder – Approximate Solution

The series solution for the cylinder, like the plane wall, can be adequately approximated using only the first term when $Fo > 0.2$. The value of the first eigenvalue as a function Biot number is shown in Figure 6.10 and tabulated in Table 6.2. Using ζ_1, the approximate solutions for the dimensionless temperature distribution and energy transfer become:

$$\tilde{\theta}(\tilde{r}, Fo) = \frac{2 \, \text{BesselJ}(1, \zeta_1) \, \text{BesselJ}(0, \zeta_1 \tilde{r})}{\zeta_1 \left[\text{BesselJ}^2(0, \zeta_1) + \text{BesselJ}^2(1, \zeta_1) \right]} \exp\left[-\zeta_1^2 \, Fo \right] \tag{6.68}$$

$$\tilde{Q} = 1 - \frac{4 \, \text{BesselJ}^2(1, \zeta_1)}{\zeta_1^2 \left[\text{BesselJ}^2(0, \zeta_1) + \text{BesselJ}^2(1, \zeta_1) \right]} \exp\left(-\zeta_1^2 \, Fo \right). \tag{6.69}$$

The Sphere – Exact Solution

Consider the problem associated with transient conduction in a solid sphere, shown in Figure 6.12. Initially the material is at a uniform temperature T_{ini}. At time $t = 0$ the surface at $r = r_{out}$ begins to experience convection with fluid at temperature, T_∞, with heat transfer coefficient $\bar{h}$.

The partial differential equation associated with the sphere is:

$$\frac{\alpha}{r^2} \frac{\partial}{\partial r} \left[r^2 \frac{\partial T}{\partial r} \right] = \frac{\partial T}{\partial t}. \tag{6.70}$$

The initial condition is:

$$T_{t=0} = T_{ini} \tag{6.71}$$

and the spatial boundary conditions are:

$$\left. \frac{\partial T}{\partial r} \right|_{r=0} \quad \text{must be bounded} \tag{6.72}$$

$$\left. -k \frac{\partial T}{\partial r} \right|_{r=r_{out}} = \bar{h} \left(T_{r=r_{out}} - T_\infty \right). \tag{6.73}$$

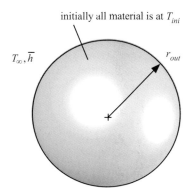

initially all material is at T_{ini}

$T_\infty, \bar{h}$

r_{out}

Figure 6.12 A solid sphere initially at a uniform temperature (T_{ini}) subjected to a convective boundary condition ($\bar{h}$, T_∞).

The exact solution for the spherical problem given by equations (6.70) through (6.73) is the infinite series

$$\tilde{\theta}(\tilde{r}, Fo) = \sum_{i=1}^{\infty} C_i \frac{\sin(\zeta_i \tilde{r})}{\zeta_i \tilde{r}} \exp\left[-\zeta_i^2 \ Fo\right], \tag{6.74}$$

where $\tilde{r}$, Fo, and $\tilde{\theta}$ are the dimensionless radius, Fourier number, and dimensionless temperature difference, defined as:

$$\tilde{r} = \frac{r}{r_{out}} \tag{6.75}$$

$$Fo = \frac{t\alpha}{r_{out}^2} \tag{6.76}$$

$$\tilde{\theta} = \frac{T - T_\infty}{T_{ini} - T_\infty}. \tag{6.77}$$

The dimensionless eigenvalues, ζ_i in Eq. (6.74) are the roots of the eigencondition:

$$\zeta_i \cos(\zeta_i) + (Bi - 1)\sin(\zeta_i) = 0, \tag{6.78}$$

where Bi is the Biot number, defined as

$$Bi = \frac{\bar{h}\, r_{out}}{k} \tag{6.79}$$

and the constants in Eq. (6.74) are

$$C_i \frac{2[\sin(\zeta_i) - \zeta_i \cos(\zeta_i)]}{\zeta_i - \sin(\zeta_i)\cos(\zeta_i)}. \tag{6.80}$$

The dimensionless energy transfer is related to the integral of the dimensionless temperature difference according to:

$$\tilde{Q} = 1 - 3\int_0^1 \tilde{\theta}\, \tilde{r}^2\, d\tilde{r}. \tag{6.81}$$

Substituting the exact solution, Eq. (6.74) into Eq. (6.81) leads to:

$$\tilde{Q} = 1 - \sum_{i=1}^{\infty} C_i \frac{3[\sin(\zeta_i) - \zeta_i \cos(\zeta_i)]}{\zeta_i^3} \exp\left(-\zeta_i^2 \ Fo\right). \tag{6.82}$$

The exact solution to this problem has been programmed in EES. These functions are accessed from the Transient Conduction section of the Heat Transfer and Fluid Flow Library and include both nondimensional (Sphere_T_ND and Sphere_Q_ND) and dimensional (Sphere_T and Sphere_Q) versions.

The Sphere – Approximate Solution

The series solution for the sphere can be adequately approximated using only the first term when $Fo > 0.2$. The value of the first eigenvalue as a function Biot number is shown in Figure 6.10 and tabulated in Table 6.2. Using ζ_1, the approximate solutions for the dimensionless temperature distribution and energy transfer become

$$\tilde{\theta}(\tilde{r}, Fo) = \frac{2[\sin(\zeta_1) - \zeta_1 \cos(\zeta_1)]}{[\zeta_1 - \sin(\zeta_1)\cos(\zeta_1)]} \frac{\sin(\zeta_1 \tilde{r})}{\zeta_1 \tilde{r}} \exp\left[-\zeta_1^2 \ Fo\right] \tag{6.83}$$

$$\tilde{Q} = 1 - \frac{6[\sin(\zeta_1) - \zeta_1 \cos(\zeta_1)]^2}{\zeta_1^3 [\zeta_1 - \sin(\zeta_1)\cos(\zeta_1)]} \exp\left(-\zeta_1^2 \ Fo\right). \tag{6.84}$$

Example 6.9

A long cylinder of food product must be cooled quickly after it is processed. The product is encased in a thin layer of plastic with thickness $th_c = 1$ mm and conductivity $k_c = 0.2$ W/m-K. The outer radius of the product is $r_{out} = 5$ cm and the product has properties $k_p = 0.8$ W/m-K, $\rho_p = 570$ kg/m³, and $c_p = 1800$ J/kg-K. Initially the product is at a uniform temperature $T_{ini} = 55°C$ when the outer surface of the plastic is exposed to very cold air at T_∞ (which is a design parameter) with heat transfer coefficient $\bar{h} = 100$ W/m²-K. The process must continue until all of the product has reached $T_f = -20°C$ or below.

You have been asked to help design the system, in particular to help examine the impact of the air temperature, T_∞, on the process. The two criteria to be considered include the processing time required and the efficiency of the process. The processing time required should be as short as possible for reasons related to throughput. This requirement suggests that a very low value of T_∞ be selected so that cooling occurs quickly. However, a very low value of T_∞ requires more energy-intense and costly refrigeration processes. Further, low values of T_∞ cause the product near the surface to be cooled well below the required temperature of T_f. This "over-cooling" effect make the process energy inefficient as more energy is removed from the product than necessary.

Prepare a plot that shows both the required processing time and the efficiency of the process as a function of the air temperature.

Known Values

Figure 1 illustrates the problem with the known values indicated.

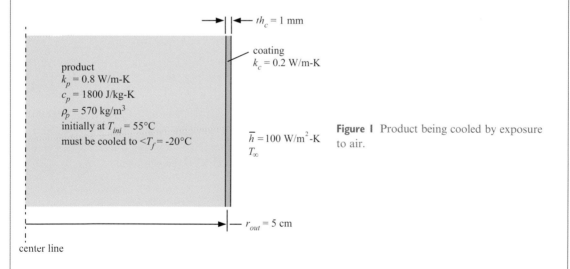

Figure I Product being cooled by exposure to air.

Assumptions

- The properties of the product do not depend on temperature.
- The product is sufficiently long that the process can be considered to be 1-D.
- Any latent heat associated with freezing is neglected.
- Radiation from the surface is neglected.
- The specific heat of the plastic coating is neglected.
- Contact resistance between the product and the plastic coating is neglected.

Continued

Example 6.9 (cont.)

Analysis

The solutions provided in this section only consider convection from the surface of the shape to the surroundings. However, there are two thermal resistances from the outer surface of the product to the surrounding air in this problem. The resistance to convection is

$$R_{conv} = \frac{1}{\bar{h}\,2\pi(r_{out} + th_c)L}, \tag{1}$$

where L is the length of the cylinder. The resistance to conduction through the plastic coating is:

$$R_{cond} = \frac{\ln\left(\dfrac{r_{out} + th_c}{r_{out}}\right)}{2\pi\,k_c\,L}. \tag{2}$$

Although the solutions considered in Section 6.2.3 and the associated functions in EES only consider the thermal resistance associated with convection, these functions can be used when other resistances are present. This is accomplished by defining an effective heat transfer coefficient ($\bar{h}_{eff}$) that provides the correct thermal resistance:

$$\frac{1}{\bar{h}_{eff}\,2\pi\,r_{out}\,L} = R_{conv} + R_{cond}. \tag{3}$$

Substituting Eqs. (1) and (2) into Eq. (3) and solving for $\bar{h}_{eff}$ provides:

$$\bar{h}_{eff} = \left[\frac{1}{\bar{h}\left(1 + \dfrac{th_c}{r_{out}}\right)} + \frac{\ln\left(1 + \dfrac{th_c}{r_{out}}\right)r_{out}}{k_c}\right]^{-1}. \tag{4}$$

The thermal diffusivity of the food product is obtained from:

$$\alpha_p = \frac{k_p}{\rho_p\,c_p}. \tag{5}$$

The exact solution for transient conduction in a cylinder can be used to identify the time (t) at which the center temperature reaches T_f and the associated energy transfer per unit length (Q'). The efficiency is defined based on the actual energy transfer compared to the minimum possible energy transfer (Q'_{min}); in this case, the minimum possible heat transfer occurs if the entire product is cooled uniformly to T_f:

$$Q'_{min} = \pi\,r_{out}^2\,\rho_p\,c_p\left(T_f - T_{ini}\right) \tag{6}$$

$$\eta = \frac{Q'_{min}}{Q'}. \tag{7}$$

Solution

The requirement that a plot be produced in which the air temperature is varied suggests that this problem should be solved using a computer in order to facilitate parametric studies. The inputs are entered in EES. Initially an air temperature of $T_\infty = -50°C$ is set. Eventually this input will be varied in a Parametric Table.

```
$UnitSystem SI Mass Radian J K Pa
T_infinity = ConvertTemp(C,K,-50 [C])      "air temperature"
T_ini=ConvertTemp(C,K,55 [C])              "initial product temperature"
h_bar=100 [W/m^2-K]                        "heat transfer coefficient"
```

Example 6.9 (cont.)

```
th_c=1 [mm]*Convert(mm,m)              "thickness of container"
k_c=0.2 [W/m-K]                        "conductivity of container"
k_p=0.8 [W/m-K]                        "conductivity of the product"
c_p=1800 [J/kg-K]                      "specific heat capacity of the product"
rho_p=570 [kg/m^3]                     "density of the product"
r_out=5 [cm]*Convert(cm,m)             "product radius"
T_f=ConvertTemp(C,K,-20[C])            "storage temperature"
```

The effective heat transfer coefficient and thermal diffusivity of the product are computed using Eqs. (4) and (5), respectively.

```
h_bar_eff=(1/(h_bar*(1+th_c/r_out))+Ln(1+th_c/r_out)*r_out/k_c)^(-1)
                                       "effective heat transfer coefficient"
alpha_p=k_p/(rho_p*c_p)                "thermal diffusivity"
```

The function Cylinder_T is used to determine the time required for processing, corresponding to the time at which the center temperature ($r = 0$) reaches T_f. Note that you will likely need to set a reasonable guess value for the variable time in order for EES to converge to the correct solution. The function Cylinder_Q is used to determine the energy transfer per length to the product at this value of time. (Note that this is a negative number because energy is removed during the cooling process.)

```
T_f=Cylinder_T(0 [m], time, T_ini, T_infinity, alpha_p, k_p, h_bar_eff, r_out)
                                       "determine process completion time"
Q`=Cylinder_Q(time, T_ini, T_infinity, alpha_p, k_p, h_bar_eff, r_out)
                                       "determine heat transfer per length"
```

Finally, the minimum possible energy transfer per length and efficiency are computed using Eqs. (6) and (7), respectively.

```
Q_min`=pi*r_out^2*rho_p*c_p*(T_f-T_ini)    "minimum heat transfer per length"
eta=Q_min`/Q`                              "efficiency of process"
```

A Parametric Table is created that contains T_∞, t, and η. Figure 2 illustrates the processing time and efficiency as a function of the air temperature.

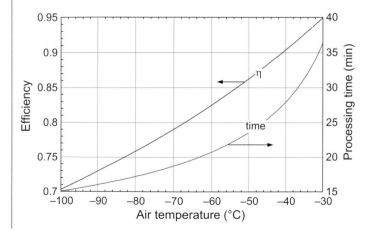

Figure 2 Efficiency and processing time as a function of the air temperature.

Continued

Example 6.9 (cont.)

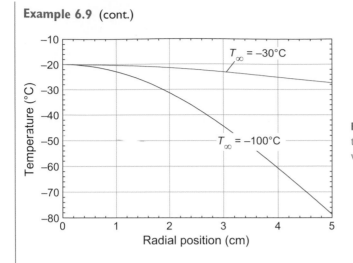

Figure 3 Final temperature distribution for the case where $T_\infty = -100°C$ and the case where $T_\infty = -30°C$.

Discussion

Figure 2 shows that very low values of air temperature result in a short processing time. However, low air temperatures also lead to a low efficiency as more energy is removed from the product than necessary because the edges are over-cooled. Figure 3 illustrates the temperature distribution that exists when the center has reached T_f for the case where $T_\infty = -100°C$ and clearly shows the over-cooling that occurs near the edges. Higher values of T_∞ result in a more uniform temperature distribution and therefore a higher efficiency, but at the cost of a longer processing time; the temperature distribution for the case where $T_\infty = -30°C$ is also shown in Figure 3. This is a typical result: high-efficiency processes tend to be slow.

6.3 1-D Numerical Solutions

6.3.1 Introduction

Sections 2.4 and 3.4 discuss the numerical solution to steady-state 1-D problems and Section 5.3 discusses the numerical solution to 0-D (lumped) transient problems. In this section, these concepts are combined and extended in order to numerically solve 1-D transient problems. The underlying methodology is the same; a set of equations is obtained from energy balances on small (but not differentially small) control volumes that are distributed throughout the computational domain. In addition to energy transfer terms related to conduction, convection, and/or radiation, these equations must include energy storage terms due to the transient nature of the problem. The energy storage terms involve the time rate of temperature change and therefore must be numerically integrated forward in time using one of the techniques presented in Section 5.3 (e.g., the Euler Method).

The process of obtaining a numerical solution to a 1-D, transient conduction problem will be illustrated in the context of a plane wall that is subjected to a convective boundary condition on one surface, as shown in Figure 6.13. The plane wall has thickness $L = 5.0$ cm and properties $k = 5.0$ W/m-K, $\rho = 2000$ kg/m^3, and $c = 200$ J/kg-K. The wall is initially at $T_{ini} = 20°C$ when at time $t = 0$, the surface (at $x = L$) experiences convection with fluid at $T_\infty = 200°C$ with average heat transfer coefficient $\bar{h} = 500$ W/m^2-K. The wall at $x = 0$ is adiabatic.

The plane wall problem that we are considering here corresponds exactly to the bounded plane wall problem considered in Section 6.2.3. This problem was selected so that the solution that we obtain numerically can be directly compared to the exact solution accessed by the Planewall_T function in EES. However, the major advantage of using a numerical technique is the capability to easily solve more complex problems that may not have an analytical solution.

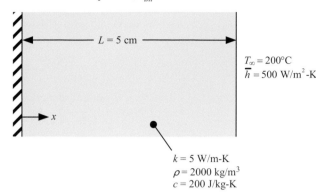

initial temperature, $T_{ini} = 20°C$

$L = 5$ cm

$T_\infty = 200°C$
$\overline{h} = 500$ W/m²-K

Figure 6.13 A plane wall exposed to a convective boundary condition at time $t = 0$.

$k = 5$ W/m-K
$\rho = 2000$ kg/m³
$c = 200$ J/kg-K

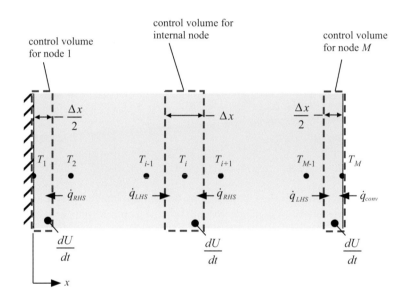

Figure 6.14 Nodes and control volumes distributed uniformly throughout computational domain.

6.3.2 The State Equations

The first step in developing a numerical solution is to partition the computational domain into a large number of small control volumes that are each analyzed using energy balances. These energy balances form the basis of a set of coupled ordinary differential equations that must be solved. In Figure 6.14, the nodes (i.e., the positions at which the temperature will be calculated) are distributed uniformly through the wall in the x-direction. The location of each node (x_i) is

$$x_i = \frac{(i-1)}{(M-1)} L \quad \text{for } i = 1 \ldots M \tag{6.85}$$

where M is the number of nodes used for the simulation. The distance between adjacent nodes (Δx) is

$$\Delta x = \frac{L}{(M-1)}. \tag{6.86}$$

A control volume is defined around each node, as shown in Figure 6.14; notice that the control surface separating each system bisects the distance between the nodes. An energy balance must be written for each control volume. The control volume for an arbitrary, internal node experiences conduction heat transfer with the adjacent nodes as well as energy storage (as shown in Figure 6.14 for node i):

$$\dot{q}_{LHS} + \dot{q}_{RHS} = \frac{dU}{dt}. \tag{6.87}$$

Each term in Eq. (6.87) must be approximated. The conduction terms from the adjacent nodes are modeled according to:

$$\dot{q}_{LHS} = \frac{k\, A_c (T_{i-1} - T_i)}{\Delta x} \tag{6.88}$$

$$\dot{q}_{RHS} = \frac{k\, A_c (T_{i+1} - T_i)}{\Delta x}, \tag{6.89}$$

where A_c is the cross-sectional area of the wall. The rate of energy storage is the product of the time rate of change of the nodal temperature and the thermal mass of the control volume:

$$\frac{dU}{dt} = A_c\, \Delta x\, \rho\, c\, \frac{dT_i}{dt}. \tag{6.90}$$

Substituting Eqs. (6.88) through (6.90) into Eq. (6.87) leads to:

$$\frac{k\, A_c (T_{i-1} - T_i)}{\Delta x} + \frac{k\, A_c (T_{i+1} - T_i)}{\Delta x} = A_c\, \Delta x\, \rho\, c\, \frac{dT_i}{dt} \quad \text{for } i = 2 \ldots (M-1). \tag{6.91}$$

Solving for the time rate of the temperature change provides:

$$\frac{dT_i}{dt} = \frac{\alpha}{\Delta x^2} (T_{i-1} + T_{i+1} - 2T_i) \quad \text{for } i = 2 \ldots (M-1). \tag{6.92}$$

The control volumes at the boundaries must be treated separately, both because they have a smaller volume and also because they experience different energy transfers. An energy balance on the control volume at the adiabatic wall (node 1 in Figure 6.14) leads to:

$$\dot{q}_{RHS} = \frac{dU}{dt} \tag{6.93}$$

or

$$\frac{k\, A_c (T_2 - T_1)}{\Delta x} = \frac{A_c\, \Delta x\, \rho\, c}{2} \frac{dT_1}{dt}. \tag{6.94}$$

The factor of two on the right-hand side of Eq. (6.94) results because the control volume around node 1 has half the width and thus half the heat capacity of the other nodes. Solving for the time rate of temperature change for node 1:

$$\frac{dT_1}{dt} = \frac{2\alpha}{\Delta x^2} (T_2 - T_1). \tag{6.95}$$

An energy balance on the control volume for the node located at the outer surface (node M in Figure 6.14) leads to:

$$\frac{dU}{dt} = \dot{q}_{LHS} + \dot{q}_{conv} \tag{6.96}$$

or

$$\frac{A_c\, \Delta x\, \rho\, c}{2} \frac{dT_M}{dt} = \frac{k\, A_c (T_{M-1} - T_M)}{\Delta x} + \bar{h}\, A_c (T_\infty - T_M). \tag{6.97}$$

Solving for the time rate of temperature change for node M:

$$\frac{dT_M}{dt} = \frac{2\alpha}{\Delta x^2}(T_{M-1} - T_M) + \frac{2\bar{h}}{\Delta x \rho c}(T_\infty - T_M). \tag{6.98}$$

Equations (6.92), (6.95), and (6.98) provide the time rate of change for the temperature of every node. This result is similar to the situation that we encountered in Section 5.3.1 when developing a numerical solution for lumped capacitance problems. In that case, the energy balance for the single control volume (defined around the object being studied) provided a single equation for the time rate of change of the object's temperature. Here, energy balances written for each of the control volumes (which are all modeled as being at a single, nodal temperature at any instant of time) has provided a set of equations for their time rates of change. In order to solve the problem, it is necessary to integrate each of these coupled ordinary differential equations forward in time. All of the numerical integration techniques that were discussed in Section 5.3 to solve lumped capacitance problems can be applied here to solve 1-D transient problems.

The temperature in the wall is a function both of position (x) and time (t). The index that specifies the node's position is i, where $i = 1$ corresponds to the adiabatic boundary ($x = 0$) and $i = M$ corresponds to the surface experiencing convection ($x = L$) as shown in Figure 6.14. A second index, j, must be added to each nodal temperature in order to indicate the time; $j = 1$ corresponds to the beginning of the simulation ($t = 0$) and $j = N$ corresponds to the end of the simulation ($t = t_{sim}$). The total simulation time, t_{sim}, is divided into N time steps. Most of the techniques discussed here will divide the simulation time into time steps of equal duration, Δt, although other approaches may also be used:

$$\Delta t = \frac{t_{sim}}{(N - 1)}. \tag{6.99}$$

The time associated with index j is:

$$t_j = (j - 1)\Delta t \text{ for } j = 1 \dots N \tag{6.100}$$

All of the nodal temperatures at $j = 1$ are known from the specified initial condition and should be assigned:

$$T_{i,1} = T_{ini} \text{ for } i = 1 \dots M. \tag{6.101}$$

Each of the numerical techniques discussed in this section can be used to determine the temperature of all of the nodes at the end of a time step (i.e., $T_{i,j+1}$ for all $i = 1 \dots M$) given the temperature of the nodes at the beginning of the time step (i.e., $T_{i,j}$ for all $i = 1 \dots M$) and the set of state equations associated with the problem.

In order to illustrate the implementation of the techniques presented in this section we will program the required equations in EES. The inputs are entered.

"Inputs"
L=5 [cm]*Convert(cm,m)   "wall thickness"
k=5.0 [W/m-K]   "conductivity"
rho=2000 [kg/m^3]   "density"
c=200 [J/kg-K]   "specific heat capacity"
T_ini=20 [C]   "initial temperature"
T_infinity=200 [C]   "fluid temperature"
h_bar=500 [W/m^2-K]   "heat transfer coefficient"
alpha=k/(rho*c)   "thermal diffusivity"

Arbitrarily, $M = 11$ nodes are distributed in the x-direction with 100 time steps (corresponding to $N = 101$ times). These choices can be changed to adjust the accuracy and computational time of the model. The nodal locations are entered into an array (x[]) according to Eq. (6.85) and the distance between adjacent nodes is defined using Eq. (6.86).

```
"Setup grid"
M=11 [-]                                        "number of nodes"
Duplicate i=1,M
    x[i]=(i-1)*L/(M-1)                          "position of each node"
End
Dx=L/(M-1)                                      "distance between adjacent nodes"
```

The time step duration is computed using Eq. (6.99) and an array containing each of the solution times (time[])
is setup using Eq. (6.100).

```
"Setup time steps"
N=101 [-]                                       "number of time steps"
t_sim=40 [s]                                    "simulation time"
Dt=t_sim/(N-1)                                  "time step duration"
Duplicate j=1,N
    time[j]=(j-1)* Dt                           "times"
End
```

The initial conditions for this problem are that all of the temperatures at $t = 0$ (i.e., $j = 1$) are equal to T_{ini}.
Therefore, the first column of the matrix containing temperatures (T[,]) is defined using Eq. (6.101).

```
Duplicate i=1,M
    T[i,1]=T_ini                                "initial condition"
End
```

6.3.3 The Euler Method

The simplest technique for numerical integration is the Euler Method, which approximates the time rate of
temperature change within each time step as being constant and equal to its value at the beginning of the time
step. Mathematically, this approach is stated as:

$$T_{i,j+1} = T_{i,j} + \underbrace{\frac{dT}{dt}\bigg|_{T=T_{i,j}, t=t_j}}_{\substack{\text{time rate of change} \\ \text{evaluated at the} \\ \text{beginning of the time step}}} \Delta t \quad \text{for } i = 1 \dots M \text{ and } j = 1 \dots (N-1). \tag{6.102}$$

Note that it is almost always a good idea to develop a transient numerical simulation by initially taking only a
single step and then, once that works, automating the process of simulating all of the time steps. For example,
using the Euler Method the temperatures of all M nodes at the end of the first time step (i.e., $j = 2$) can be
computed according to:

$$T_{i,2} = T_{i,1} + \frac{dT}{dt}\bigg|_{T=T_{i,1}, t=t_1} \Delta t \quad \text{for } i = 1 \dots M. \tag{6.103}$$

The state equations for the problem being considered are given by Eqs. (6.92), (6.95), and (6.98). These can be
substituted into Eq. (6.103) to provide:

$$T_{1,2} = T_{1,1} + \frac{2\alpha}{\Delta x^2}(T_{2,1} - T_{1,1})\Delta t \tag{6.104}$$

$$T_{i,2} = T_{i,1} + \frac{\alpha}{\Delta x^2}(T_{i-1,1} + T_{i+1,1} - 2T_{i,1})\Delta t \quad \text{for } i = 2 \dots (M-1) \tag{6.105}$$

$$T_{M,2} = T_{M,1} + \left[\frac{2\alpha}{\Delta x^2} (T_{M-1,1} - T_{M,1}) + \frac{2\bar{h}}{\Delta x \rho c} (T_\infty - T_{M,1}) \right] \Delta t. \tag{6.106}$$

These equations are entered in EES.

```
"Take a single Euler step"
T[1,2]=T[1,1]+2*alpha*(T[2,1]-T[1,1])*Dt/Dx^2              "node 1"
Duplicate i=2,(M-1)
    T[i,2]=T[i,1]+alpha*(T[i-1,1]+T[i+1,1]-2*T[i,1])*Dt/Dx^2  "internal nodes"
End
T[M,2]=T[M,1]+(2*alpha*(T[M-1,1]-T[M,1])/Dx^2+2*h_bar*(T_infinity-T[M,1])/(rho*c*Dx))*Dt
                                               "node M"
```

Solving the program will provide a solution for the temperature at each node at the end of the first time step (i.e., at $j = 2$). The solution can be examined by selecting Arrays from the Windows menu; notice that the second column in the matrix T is now filled in. The equations used to move through any arbitrary time step j follow logically from Eqs. (6.104) through (6.106):

$$T_{1,j+1} = T_{1,j} + \frac{2\alpha}{\Delta x^2} (T_{2,j} - T_{1,j}) \Delta t \tag{6.107}$$

$$T_{i,j+1} = T_{i,j} + \frac{\alpha}{\Delta x^2} (T_{i-1,j} + T_{i+1,j} - 2T_{i,j}) \Delta t \quad \text{for } i = 2 \ldots (M-1) \tag{6.108}$$

$$T_{M,j+1} = T_{M,j} + \left[\frac{2\alpha}{\Delta x^2} (T_{M-1,j} - T_{M,j}) + \frac{2\bar{h}}{\Delta x \rho c} (T_\infty - T_{M,j}) \right] \Delta t. \tag{6.109}$$

Equations (6.107) through (6.109) are solved for all of the time steps by placing the EES code within a second duplicate loop that steps from $j = 1$ to $j = (N-1)$. This is accomplished by nesting the entire EES code associated with taking the first Euler step within an outer duplicate loop. Wherever the *second* index was a 2 (i.e., referred to the end of time step 1) we replace it with $j + 1$ and wherever the second index was a 1 (i.e., referred to the beginning of time step 1) we replace it with j. The revised code is shown below; note that the code associated with taking a single step must be commented out.

```
Duplicate j=1,(N-1)
    T[1,j+1]=T[1,j]+2*alpha*(T[2,j]-T[1,j])*Dt/Dx^2              "node 1"
    Duplicate i=2,(M-1)
        T[i,j+1]=T[i,j]+alpha*(T[i-1,j]+T[i+1,j]-2*T[i,j])*Dt/Dx^2  "internal nodes"
    End
    T[M,j+1]=T[M,j]+(2*alpha*(T[M-1,j]-T[M,j])/Dx^2+2*h_bar*(T_infinity-T[M,j])/(rho*c*Dx))*Dt
                                                   "node M"
End
```

Figure 6.15 illustrates the temperature as a function of position at various values of time. The exact solution obtained using the Planewall_T function is also shown.

If the duration of the time step is increased from $\Delta t = 0.4$ s to 1.0 s (by reducing N from 101 to 41) then the solution becomes unstable (try it and see; the solution will oscillate between large positive and negative temperatures). The existence of a stability limit is one of the key disadvantages associated with the Euler Method, as we saw in Section 5.3.2 for lumped capacitance problems. The maximum time step that can be used before the solution becomes unstable (i.e., the critical time step, Δt_{crit}) can be determined by examining the algebraic equations that are used to step through time. Rearranging Eq. (6.107), which governs the behavior of the node at the adiabatic edge, leads to:

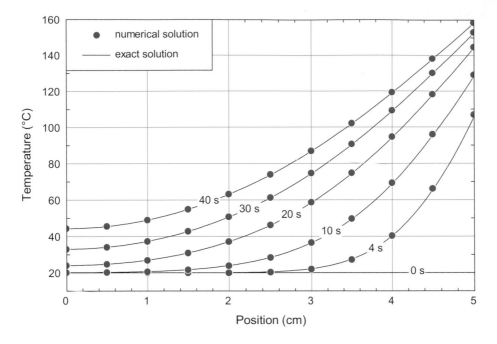

Figure 6.15 Temperature as a function of position predicted by the Euler Method at various times. Also shown is the exact solution accessed using the Planewall_T function in EES.

$$T_{1,j+1} = T_{1,j} \underbrace{\left[1 - \frac{2\alpha \, \Delta t}{\Delta x^2}\right]}_{\text{must be} > 0 \text{ for stability}} + \frac{2\alpha \, \Delta t}{\Delta x^2} T_{2,j}. \tag{6.110}$$

The solution will become unstable when the coefficient multiplying $T_{1,j}$ becomes negative. Therefore, Eq. (6.110) shows that the solution will tend to become unstable as Δt becomes larger or Δx becomes smaller. The critical time step associated with node 1 is

$$\Delta t_{crit,\,1} = \frac{\Delta x^2}{2\alpha}. \tag{6.111}$$

Applying the same process to Eq. (6.108), which governs the behavior of the internal nodes, leads to:

$$T_{i,j+1} = T_{i,j} \underbrace{\left[1 - \frac{2\alpha \, \Delta t}{\Delta x^2}\right]}_{\text{must be} > 0 \text{ for stability}} + \frac{\alpha \, \Delta t}{\Delta x^2} \left(T_{i-1,j} + T_{i+1,j}\right) \text{ for } i = 2 \ldots (M-1). \tag{6.112}$$

According to Eq. (6.112), the critical time step for the internal nodes is the same as for node 1:

$$\Delta t_{crit,\,i} = \frac{\Delta x^2}{2\alpha} \text{ for } i = 2 \ldots (M-1). \tag{6.113}$$

Equation (6.109), which governs the behavior of the node placed on the surface of the wall, is rearranged:

$$T_{M,j+1} = T_{M,j} \underbrace{\left[1 - \frac{2\alpha \, \Delta t}{\Delta x^2} - \frac{2\bar{h} \, \Delta t}{\Delta x \rho \, c}\right]}_{\text{must be} > 0 \text{ for stability}} + \frac{2\alpha \, \Delta t}{\Delta x^2} T_{M-1,j} + \frac{2\bar{h} \, \Delta t}{\Delta x \rho \, c} T_{\infty}. \tag{6.114}$$

Equation (6.114) leads to a different critical time step for node M:

$$\Delta t_{crit, M} = \frac{\Delta x}{2\left[\dfrac{\alpha}{\Delta x} + \dfrac{\bar{h}}{\rho c}\right]}.$$ (6.115)

Comparing Eq. (6.115) with Eqs. (6.111) and (6.113) indicates that the critical time step for node M will always be somewhat less than the critical time step of the other nodes for this problem and therefore Eq. (6.115) will govern the stability of the problem. For this problem with $\Delta x = 0.005$ m, the critical time step is $\Delta t_{crit} = 0.67$ s which is consistent with our observations; the numerical solution was stable with $\Delta t = 0.4$ s and it was unstable at $\Delta t = 1.0$ s.

The accuracy of a numerical solution is related to the size of the control volumes. Smaller values of Δx will provide more accurate solutions. However, according to Eq. (6.115), smaller values of Δx will also dramatically reduce the critical time step duration (which scales approximately as Δx^2). As a result, reducing Δx will greatly increase the number of time steps required to maintain stability when using an explicit numerical technique like the Euler Method. It is sometimes the case that the number of time steps required by stability is larger than the number required for sufficient accuracy and therefore the use of explicit techniques can be unnecessarily computationally intensive.

All of the numerical integration techniques presented here can be implemented in almost any software. The Euler Method solution discussed in this section is implemented in MATLAB using a script, as shown below. Because the Euler method is explicit, the solution in MATLAB looks essentially identical to the solution in EES with some small differences related to the details of the language.

```
%Inputs
L=0.05;                          %wall thickness (m)
k=5.0;                           %conductivity (W/m-K)
rho=2000;                        %density (kg/m^3)
c=200;                           %specific heat capacity (J/kg-K)
T_ini=20;                        %initial temperature (C)
T_infinity=200;                  %fluid temperature (C)
h_bar=500;                       %heat transfer coefficient (W/m^2-K)
alpha=k/(rho*c);                 %thermal diffusivity (m^2/s)

%Setup grid
M=11;                            %number of nodes
for i=1:M
    x(i)=(i-1)*L/(M-1);          %position of each node (m)
end
Dx=L/(M-1);                      %distance between adjacent nodes (m)

%Setup time steps
N=101;                           %number of time steps
t_sim=40;                        %simulation time (s)
Dt=t_sim/(N-1);                  %time step duration (s)
for j=1:N
    time(j)=(j-1)*Dt;            %times (s)
end

for i=1:M
    T(i,1)=T_ini;                %initial condition (C)
end
```

```
for j=1:(N-1)
    T(1,j+1)=T(1,j)+2*alpha*(T(2,j)-T(1,j))*Dt/Dx^2;                    %node 1
    for i=2:(M-1)
        T(i,j+1)=T(i,j)+alpha*(T(i-1,j)+T(i+1,j)-2*T(i,j))*Dt/Dx^2;   %internal nodes
    end
    T(M,j+1)=T(M,j)+(2*alpha*(T(M-1,j)-T(M,j))/Dx^2+2*h_bar*(T_infinity-T(M,j))/(rho*c*Dx))*Dt;
                                                                       %node M
end

hold off;
plot(time,T','b-');
hold on;
grid on;
xlabel('Time (s)');
ylabel('Temperature (C)');
text(10,132,'x=L','BackgroundColor',[1 1 1]);
text(32,30,'x=0','BackgroundColor',[1 1 1]);
```

Figure 6.16 illustrates the temperature as a function of time at the axial locations corresponding to each of the nodes.

   This example of the Euler Method uses a 2-D array to hold the values of all temperatures at every time step. EES has an upper limit on the number of variables it can hold in the main program and the use of 2-D arrays with many time steps can quickly exceed this limit. An alternative is to use 1-D arrays holding the temperatures at each position and vary the time in a Parametric Table. An example of this alternative method of implementing the Euler and Implicit integration methods is provided in the file Difeqn2.ees that can be found in the Examples folder with the EES program.

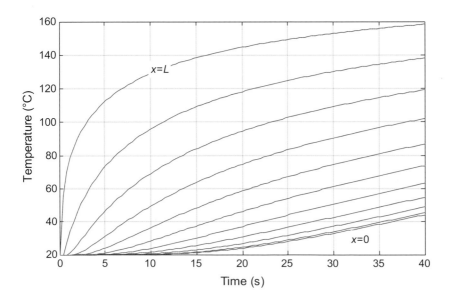

Figure 6.16 Temperature as a function of time at various positions.

6.3.4 Predictor–Corrector Methods

Predictor–corrector methods were discussed in Section 5.3.3 in the context of lumped capacitance problems as a way of extending the Euler Method in order to improve its accuracy. Predictor–corrector techniques take one or more predictor steps (based on the Euler Method) followed by a corrector step in which the knowledge obtained from the predictor step(s) is used to improve the integration process. Because the predictor–corrector techniques rely on explicit steps, they suffer from the same limitations related to stability that were discussed in the context of the Euler Method. However, because multiple evaluations of the state equations are used to move through each time step, the predictor–corrector methods are higher order and therefore higher accuracy than the Euler Method.

Heun's Method is a simple example of this class of numerical integration methods. In order to simulate a time step (moving from time j to time $j+1$) for any node i, Heun's Method begins with an Euler step to obtain an initial prediction for the temperature at the conclusion of the time step ($\hat{T}_i$):

$$\hat{T}_i = T_{i,j} + \left.\frac{dT}{dt}\right|_{T = T_{i,j},\, t = t_j} \Delta t \ \text{ for } i = 1 \ldots M. \tag{6.116}$$

Note that $\hat{T}_i$ is not the final prediction for the temperature of node i at time $j+1$; rather, it is an intermediate piece of information obtained from what is referred to as the predictor step. Heun's Method uses only one predictor step. More advanced methods use multiple predictor steps. The results of the predictor step(s) are used to carry out a final corrector step that leads to the final prediction of the temperature at the end of the time step, $T_{i,j+1}$. The corrector step used by Heun's Method is:

$$T_{i,j+1} = T_{i,j} + \left(\underbrace{\left.\frac{dT}{dt}\right|_{T = T_{i,j},\, t = t_j}}_{\substack{\text{time rate of change at} \\ \text{beginning of time step}}} + \underbrace{\left.\frac{dT}{dt}\right|_{T = \hat{T}_i,\, t = t_j + \Delta t}}_{\substack{\text{time rate of change at} \\ \text{end of time step}}} \right) \frac{\Delta t}{2} \ \text{ for } i = 1 \ldots M. \tag{6.117}$$

The average of the time rates of change computed at the beginning of the time step (using $T_{i,j}$ and t_j) and the end of the time step (using $\hat{T}_i$ and t_{j+1}) is used to traverse the time step.

Heun's Method is illustrated using the problem that was presented in Section 6.3.1, which results in the state equations developed in Section 6.3.2. Substituting the state equations given by Eqs. (6.92), (6.95), and (6.98) into Eq. (6.116) provides the following equations for the predictor step:

$$\hat{T}_1 = T_{1,j} + \frac{2\alpha}{\Delta x^2}\left(T_{2,j} - T_{1,j}\right)\Delta t \tag{6.118}$$

$$\hat{T}_i = T_{i,j} + \frac{\alpha}{\Delta x^2}\left(T_{i-1,j} + T_{i+1,j} - 2T_{i,j}\right)\Delta t \ \text{ for } i = 2 \ldots (M-1) \tag{6.119}$$

$$\hat{T}_M = T_{M,j} + \left[\frac{2\alpha}{\Delta x^2}\left(T_{M-1,j} - T_{M,j}\right) + \frac{2\bar{h}}{\Delta x \rho c}\left(T_\infty - T_{M,j}\right)\right]\Delta t. \tag{6.120}$$

Substituting the state equations into Eq. (6.117) provides the following equations for the corrector step:

$$T_{1,j+1} = T_{1,j} + \frac{2\alpha}{\Delta x^2}\left[\left(T_{2,j} - T_{1,j}\right) + \left(\hat{T}_2 - \hat{T}_1\right)\right]\frac{\Delta t}{2} \tag{6.121}$$

$$T_{i,j+1} = T_{i,j} + \frac{\alpha}{\Delta x^2}\left[\left(T_{i-1,j} + T_{i+1,j} - 2T_{i,j}\right) + \left(\hat{T}_{i-1} + \hat{T}_{i+1} - 2\hat{T}_i\right)\right]\frac{\Delta t}{2} \ \text{ for } i = 2 \ldots (M-1) \tag{6.122}$$

$$T_{M,j+1} = T_{M,j} + \left\{\frac{2\alpha}{\Delta x^2}\left[\left(T_{M-1,j} - T_{M,j}\right) + \left(\hat{T}_{M-1} - \hat{T}_M\right)\right] + \frac{2\bar{h}}{\Delta x \rho c}\left[\left(T_\infty - T_{M,j}\right) + \left(T_\infty - \hat{T}_M\right)\right]\right\}\frac{\Delta t}{2}. \tag{6.123}$$

Equations (6.118) through (6.123) are all explicit equations and therefore they can be programmed easily using any computer software. The MATLAB code from Section 6.3.3 is modified to accomplish the integration using Heun's Method rather than the Euler Method. The revised code is listed below.

```
%Inputs
L=0.05;                                           %wall thickness (m)
k=5.0;                                            %conductivity (W/m-K)
rho=2000;                                         %density (kg/m^3)
c=200;                                            %specific heat capacity (J/kg-K)
T_ini=20;                                         %initial temperature (C)
T_infinity=200;                                   %fluid temperature (C)
h_bar=500;                                        %heat transfer coefficient (W/m^2-K)
alpha=k/(rho*c);                                  %thermal diffusivity (m^2/s)

%Setup grid
M=11;                                             %number of nodes
for i=1:M
    x(i)=(i-1)*L/(M-1);                           %position of each node (m)
end
Dx=L/(M-1);                                       %distance between adjacent nodes (m)

%Setup time steps
N=101;                                            %number of time steps
t_sim=40;                                         %simulation time (s)
Dt=t_sim/(N-1);                                   %time step duration (s)
for j=1:N
    time(j)=(j-1)*Dt;                             %times (s)
end

for i=1:M
    T(i,1)=T_ini;                                 %initial condition (C)
end

%predictor step
for j=1:(N-1)
    T_hat(1)=T(1,j)+2*alpha*(T(2,j)-T(1,j))*Dt/Dx^2;       %node 1
    for i=2:(M-1)
        T_hat(i)=T(i,j)+alpha*(T(i-1,j)+T(i+1,j)-2*T(i,j))*Dt/Dx^2;   %internal nodes
    end
    T_hat(M)=T(M,j)+(2*alpha*(T(M-1,j)-T(M,j))/Dx^2+2*h_bar*(T_infinity-T(M,j))/(rho*c*Dx))*Dt;
                                                  %node M

    %corrector step
    T(1,j+1)=T(1,j)+2*alpha*(T(2,j)-T(1,j)+T_hat(2)-T_hat(1))*Dt/Dx^2/2;
                                                  %node 1
    for i=2:(M-1)
        T(i,j+1)=T(i,j)+alpha*(T(i-1,j)+T(i+1,j)-2*T(i,j)+T_hat(i-1)+T_hat(i+1)-2*T_hat(i))*Dt/Dx^2/2;
                                                  %internal nodes
    end
```

```
T(M,j+1)=T(M,j)+(2*alpha*(T(M-1,j)-T(M,j)+T_hat(M-1)-T_hat(M))/Dx^2+2*h_bar*(T_infinity-T(M,j)+...
    T_infinity-T_hat(M))/(rho*c*Dx))*Dt/2;
                                                    %node M
end
```

The numerical error associated with Heun's Method is smaller than the Euler Method and scales with the time step to the second power because Heun's Method is a second-order technique, as discussed in Section 5.3.3. Also note that if the time step is increased to a value above the critical timestep then Heun's Method becomes unstable, just as the Euler Method did.

6.3.5 Implicit Methods

The Euler Method and Heun's Method are explicit techniques. These methods both have the characteristic of becoming unstable when the time step exceeds a critical value. An implicit technique avoids this problem. There are several implicit methods that can be applied; all share the characteristic of using a set of *implicit* equations to navigate the time step. The **fully implicit method** is the simplest of these and was presented in Section 5.3.4 in the context of a lumped capacitance problem. The fully implicit method is similar to the Euler Method in that the time rate of change is assumed to be constant throughout the time step. The difference is that the time rate of change is computed at the *end* of the time step rather than at the beginning. Therefore, for any node i the process of moving from time j to $j+1$ is given by:

$$T_{i,j+1} = T_{i,j} + \underbrace{\left.\frac{dT}{dt}\right|_{T=T_{i,j+1},\, t=t_{j+1}}}_{\substack{\text{time rate of change computed} \\ \text{at the end of the time step}}} \Delta t \quad \text{for } i = 1 \ldots M. \tag{6.124}$$

The time rate of change computed at the end of the time step depends on the temperature of the nodes at the end of the time step ($T_{i,j+1}$), which are not yet determined. This leads to an implicit set of M coupled algebraic equations for the temperatures at each node at the end of the time step.

The fully implicit method is illustrated in the context of the problem that was presented in Section 6.3.1 with the state equations developed in Section 6.3.2. Substituting Eqs. (6.92), (6.95), and (6.98) into Eq. (6.124) provides the implicit set of equations required to take a step using the fully implicit method:

$$T_{1,j+1} = T_{1,j} + \frac{2\alpha}{\Delta x^2}\left(T_{2,j+1} - T_{1,j+1}\right)\Delta t \tag{6.125}$$

$$T_{i,j+1} = T_{i,j} + \frac{\alpha}{\Delta x^2}\left(T_{i-1,j+1} + T_{i+1,j+1} - 2T_{i,j+1}\right)\Delta t \quad \text{for } i = 2 \ldots (M-1) \tag{6.126}$$

$$T_{M,j+1} = T_{M,j} + \left[\frac{2\alpha}{\Delta x^2}\left(T_{M-1,j+1} - T_{M,j+1}\right) + \frac{2\bar{h}}{\Delta x \rho c}\left(T_\infty - T_{M,j+1}\right)\right]\Delta t. \tag{6.127}$$

Equations (6.125) through (6.127) represent M coupled, algebraic equations for the M unknown temperatures at the end of the time step ($T_{i,j+1}$, where $i = 1...M$). These equations can be solved directly using an equation solver like EES but must be solved using either matrix decomposition or Gauss–Seidel iteration when using a formal programming language.

Implementation with EES

Because EES solves implicit algebraic equations directly, it is not necessary to rearrange Eqs. (6.125) through (6.127) in order to solve them. The implicit solution is obtained using the following EES code; note that the changes from the EES code that was used to implement the Euler Method in Section 6.3.3 are indicated in bold/ red font.

```
"Inputs"
L=5 [cm]*Convert(cm,m)                                "wall thickness"
k=5.0 [W/m-K]                                         "conductivity
"rho=2000 [kg/m^3]                                    "density"
c=200 [J/kg-K]                                        "specific heat capacity"
T_ini=20 [C]                                          "initial temperature"
T_infinity=200 [C]                                    "fluid temperature"
h_bar=500 [W/m^2-K]                                   "heat transfer coefficient"
alpha=k/(rho*c)                                       "thermal diffusivity"


"Setup grid"
M=11 [-]                                              "number of nodes"
Duplicate i=1,M
    x[i]=(i-1)*L/(M-1)                                "position of each node"
End
Dx=L/(M-1)                                            "distance between adjacent nodes"


"Setup time steps"
N=41 [-]                                              "number of time steps"
t_sim=40 [s]                                          "simulation time"
Dt=t_sim/(N-1)                                        "time step duration"
Duplicate j=1,N
    time[j]=(j-1)* Dt                                 "times"
End


Duplicate i=1,M
   T[i,1]=T_ini                                       "initial condition"
End


Duplicate j=1,(N-1)
   T[1,j+1]=T[1,j]+2*alpha*(T[2,j+1]-T[1,j+1])*Dt/Dx^2    "node 1"
   Duplicate i=2,(M-1)
      T[i,j+1]=T[i,j]+alpha*(T[i-1,j+1]+T[i+1,j+1]-2*T[i,j+1])*Dt/Dx^2  "internal nodes"
   End
   T[M,j+1]=T[M,j]+(2*alpha*(T[M-1,j+1]-T[M,j+1])/Dx^2+2*h_bar*(T_infinity-T[M,j+1])/(rho*c*Dx))*Dt
                                                      "node M"
End
```

An attractive feature of the implicit technique is that it is possible to vary M and N independently in order to achieve sufficient accuracy without being constrained by stability considerations.

Implementation with Matrix Decomposition
The implementation of the implicit technique in a formal programming language is not a straightforward extension of the EES code. Programming languages utilize assignment statements rather than equations and therefore they cannot directly solve the set of implicit algebraic equations represented by Eqs. (6.125) through (6.127). One way to solve these equations is to place them in matrix format so that they can be solved by matrix decomposition, as discussed previously in Sections 2.4.4 and 4.3.4 in the context of 1-D and 2-D steady-state problems, respectively.

Equations (6.125) through (6.127) represent M linear equations for the M unknown nodal temperatures at the end of the time step ($T_{i,j+1}$ for $i = 1$ to M). In order to solve these implicit equations, they must be placed in matrix format:

$$\underline{\underline{A}}\,\underline{X} = \underline{b}. \tag{6.128}$$

It is important to clearly specify the order that the equations are placed into the coefficient matrix $\underline{\underline{A}}$ and the constant vector $\underline{b}$ as well as the order that the unknown temperatures are placed into the vector $\underline{X}$. The most logical method to set up the vector $\underline{X}$ is

$$\underline{X} = \begin{bmatrix} X_1 = T_{1,j+1} \\ X_2 = T_{2,j+1} \\ \dots \\ X_M = T_{M,j+1} \end{bmatrix} \tag{6.129}$$

so that $T_{i,j+1}$ corresponds to element i of the vector $\underline{X}$. The most logical method for placing the equations into $\underline{\underline{A}}$ and $\underline{b}$ is

$$\underline{\underline{A}} \text{ and } \underline{b} = \begin{bmatrix} \text{row } 1 = \text{control volume 1 equation} \\ \text{row } 2 = \text{control volume 2 equation} \\ \dots \\ \text{row } M = \text{control volume } M \text{ equation} \end{bmatrix}. \tag{6.130}$$

The coefficients associated with the equation that describes node i should be placed in row i of the matrix $\underline{\underline{A}}$ and the associated constant should be placed in row i of $\underline{b}$.

Equations (6.125) through (6.127) are rearranged so that the coefficients multiplying the unknowns and the constants for the linear equations are clear:

$$T_{1,j+1}\underbrace{\left[1 + \frac{2\alpha\,\Delta t}{\Delta x^2}\right]}_{A_{1,1}} + T_{2,j+1}\underbrace{\left[-\frac{2\alpha\,\Delta t}{\Delta x^2}\right]}_{A_{1,2}} = \underbrace{T_{1,j}}_{b_1} \tag{6.131}$$

$$T_{i,j+1}\underbrace{\left[1 + \frac{2\alpha\,\Delta t}{\Delta x^2}\right]}_{A_{i,i}} + T_{i-1,j+1}\underbrace{\left[-\frac{\alpha\,\Delta t}{\Delta x^2}\right]}_{A_{i,i-1}} + T_{i+1,j+1}\underbrace{\left[-\frac{\alpha\,\Delta t}{\Delta x^2}\right]}_{A_{i,j+1}} = \underbrace{T_{i,j}}_{b_i} \text{ for } i = 2\dots(M-1) \tag{6.132}$$

$$T_{M,j+1}\underbrace{\left[1 + \frac{2\alpha\,\Delta t}{\Delta x^2} + \frac{2\bar{h}\,\Delta t}{\Delta x \rho\, c}\right]}_{A_{M,M}} + T_{M-1,j+1}\underbrace{\left[-\frac{2\alpha\,\Delta t}{\Delta x^2}\right]}_{A_{M,M-1}} = T_{M,j} + \underbrace{\frac{2\bar{h}\,\Delta t}{\Delta x \rho\, c}T_\infty}_{b_M}. \tag{6.133}$$

For this problem with constant properties it is only necessary to construct the matrix $\underline{\underline{A}}$ once, although the vector $\underline{b}$ does change every step and therefore must be reconstructed. The MATLAB code that implements the fully implicit method using matrix decomposition is shown below.

```
%Inputs
L=0.05;                    %wall thickness (m)
k=5.0;                     %conductivity (W/m-K)
rho=2000;                  %density (kg/m^3)
c=200;                     %specific heat capacity (J/kg-K)
T_ini=20;                  %initial temperature (C)
T_infinity=200;            %fluid temperature (C)
h_bar=500;                 %heat transfer coefficient (W/m^2-K)
alpha=k/(rho*c);           %thermal diffusivity (m^2/s)
```

```
%Setup grid
M=11;                                              %number of nodes
for i=1:M
    x(i)=(i-1)*L/(M-1);                            %position of each node (m)
end
Dx=L/(M-1);                                        %distance between adjacent nodes (m)

%Setup time steps
N=101;                                             %number of time steps
t_sim=40;                                          %simulation time (s)
Dt=t_sim/(N-1);                                    %time step duration (s)
for j=1:N
    time(j)=(j-1)*Dt;                              %times (s)
end

for i=1:M
    T(i,1)=T_ini;                                  %initial condition (C)
end

%setup the coefficient matrix
A=spalloc(M,M,3*M);                                %initialize A as sparse w/3M nonzero elements

%equation for node 1
A(1,1)=1+2*alpha*Dt/Dx^2;
A(1,2)=-2*alpha*Dt/Dx^2;

%equations for nodes 2..(M-1)
for i=2:(M-1)
    A(i,i)=1+2*alpha*Dt/Dx^2;
    A(i,i-1)=-alpha*Dt/Dx^2;
    A(i,i+1)=-alpha*Dt/Dx^2;
end

%equation for node M
A(M,M)=1+2*alpha*Dt/Dx^2+2*h_bar*Dt/(Dx*rho*c);
A(M,M-1)=-2*alpha*Dt/Dx^2;

b=zeros(M,1);                                      %initialize constant vector
for j=1:(N-1)
    %setup constant vector
    for i=1:(M-1)
        b(i)=T(i,j);                               %equations for nodes 1..(M-1)
    end
    b(M)=T(M,j)+2*h_bar*Dt*T_infinity/(Dx*rho*c);  %equation for node M

    T(:,j+1)=A\b;                                  %matrix decomposition
end
```

Implementation with Gauss–Seidel Iteration

Sections 2.4.5 and 4.3.5 discussed the method of Gauss–Seidel iteration for solving a set of implicit equations. The technique can be used to solve the set of implicit equations that arise in an implicit numerical model for a transient conduction problem, for example Eqs. (6.125) through (6.127). Each of the equations must be solved to obtain an explicit equation providing the temperature at that node at the end of the time step in terms of other nodal temperatures:

$$T_{1,j+1} = \frac{T_{1,j} + \dfrac{2\alpha\,\Delta t}{\Delta x^2}\,T_{2,j+1}}{\left(1 + \dfrac{2\alpha\,\Delta t}{\Delta x^2}\right)} \tag{6.134}$$

$$T_{i,j+1} = \frac{T_{i,j} + \dfrac{\alpha\,\Delta t}{\Delta x^2}\left(T_{i-1,j+1} + T_{i+1,j+1}\right)}{\left(1 + \dfrac{2\alpha\,\Delta t}{\Delta x^2}\right)} \quad \text{for } i = 2 \ldots (M-1) \tag{6.135}$$

$$T_{M,j+1} = \frac{T_{M,j} + \dfrac{2\alpha\,\Delta t}{\Delta x^2}\,T_{M-1,j+1} + \dfrac{2\bar{h}\,\Delta t}{\Delta x \rho c}\,T_\infty}{\left(1 + \dfrac{2\alpha\,\Delta t}{\Delta x^2} + \dfrac{2\bar{h}\,\Delta t}{\Delta x \rho c}\right)}. \tag{6.136}$$

We start with a set of guess values for each of the unknown nodal temperatures, $T_{i,j+1}^k$, where the superscript k indicates the iteration number. Usually these guess temperatures are the values determined at the previous time step, or the initial conditions (for the first time step). Next we employ the explicit equations for the problem, in this case Eqs. (6.134) through (6.136) to solve for each of the nodal temperatures in the next iteration, $T_{i,j+1}^{k+1}$. The temperature on the left side of the equal sign is the new temperature at node i and time $j+1$ while the temperatures on the right side with subscript $j+1$ are the *most recent* values of the other nodal temperatures at the end of the time step. The values of the unknown temperatures are overwritten as the iteration process progresses so that whatever temperature is currently present in the vector must be the most recent one. The iteration process ends when the values of the unknown temperatures stop changing (to within some convergence) between iterations. The MATLAB code below implements the Gauss–Seidel iteration process.

```
%Inputs
L=0.05;                          %wall thickness (m)
k=5.0;                           %conductivity (W/m-K)
rho=2000;                        %density (kg/m^3)
c=200;                           %specific heat capacity (J/kg-K)
T_ini=20;                        %initial temperature (C)
T_infinity=200;                  %fluid temperature (C)
h_bar=500;                       %heat transfer coefficient (W/m^2-K)
alpha=k/(rho*c);                 %thermal diffusivity (m^2/s)

%Setup grid
M=11;                            %number of nodes
for i=1:M
    x(i)=(i-1)*L/(M-1);          %position of each node (m)
end
Dx=L/(M-1);                      %distance between adjacent nodes (m)

%Setup time steps
N=101;                           %number of time steps
t_sim=40;                        %simulation time (s)
Dt=t_sim/(N-1);                  %time step duration (s)
for j=1:N
    time(j)=(j-1)*Dt;            %times (s)
end
```

```
for i=1:M
    T(i,1)=T_ini;                          %initial condition (C)
end


tol=1e-6;                                  %convergence tolerance (C)
for j=1:(N-1)
    j
    T_k=T(:,j);                            %guess temperatures (C)
    T_kp=T_k;                              %start with guess temps and overwrite (C)
    err=tol+1;                             %initial value of convergence error
    while(err>tol)
        T_kp(1)=(T(1,j)+2*alpha*Dt*T_kp(2)/Dx^2)/(1+2*alpha*Dt/Dx^2);
        for i=2:(M-1)
            T_kp(i)=(T(i,j)+alpha*Dt*(T_kp(i-1)+T_kp(i+1))/Dx^2)/(1+2*alpha*Dt/Dx^2);
        end
        T_kp(M)=(T(M,j)+2*alpha*Dt*T_kp(M-1)/Dx^2+...
            2*h_bar*Dt*T_infinity/(Dx*rho*c))/(1+2*alpha*Dt/Dx^2+2*h_bar*Dt/(Dx*rho*c));
        err=max(max(abs(T_kp-T_k)));       %compute convergence error
        T_k=T_kp;                          %reset T_k
    end
    T(:,j+1)=T_kp;
end
```

6.3.6 Using ODE Solvers

In Section 5.3.5 the use of the ordinary differential equation (ODE) solvers that are provided with most engineering software is discussed and used to solve a lumped capacitance problem that requires a single state equation be integrated through time. These solvers can be extended to the 1-D transient conduction problems considered in this section. Instead of integrating a single state equation through time, it is necessary to integrate a set of coupled state equations through time. Otherwise, the techniques remain the same.

There are several advantages associated with using the ODE solvers that are available in most engineering software. They typically use adaptive step-size techniques and higher-order numerical methods in order to accomplish the simulation in a computationally efficient manner while maintaining a specified level of accuracy. In this section we will revisit the ODE solvers that are available in EES and MATLAB and demonstrate them in the context of the 1-D transient problem characterized by the state equations derived in Section 6.3.2.

EES' Integral Command

The Integral command in EES provides a powerful and computationally efficient method to solve 1-D transient conduction problems. The Integral command implements a third-order accurate integration scheme and can also use an adaptive time step size.

It is important that you first program and check the state equations when using an ODE solver. For the 1-D transient problems examined in this section you should verify that, given an array of temperatures and time, the EES code in the Equations Window is capable of computing the time rate of change of each of the temperatures using the set of state equations derived for the problem.

For the problem considered here, the first step is to implement Eqs. (6.92), (6.95), and (6.98) for an arbitrary (but reasonable) set of temperature values, T_i for $i = 1$ to M, and time, t, as shown below.

```
"Inputs"
L=5 [cm]*Convert(cm,m)                    "wall thickness"
k=5.0 [W/m-K]                             "conductivity"
rho=2000 [kg/m^3]                         "density"
c=200 [J/kg-K]                            "specific heat capacity"
T_ini=20 [C]                              "initial temperature"
T_infinity=200 [C]                        "fluid temperature"
h_bar=500 [W/m^2-K]                       "heat transfer coefficient"
alpha=k/(rho*c)                           "thermal diffusivity"


"Setup grid"
M=11 [-]                                  "number of nodes"
Duplicate i=1,M
   x[i]=(i-1)*L/(M-1)                     "position of each node"
End
Dx=L/(M-1)                                "distance between adjacent nodes"


"Program and check the state equations"
time=0 [s]                                "set a value of time (not needed for this problem)"
Duplicate i=1,M
   T[i]=T_ini                             "and temperature"
End


"and check the state equations"
dTdt[1]=2*alpha*(T[2]-T[1])/Dx^2          "node 1"
Duplicate i=2,(M-1)
   dTdt[i]=alpha*(T[i-1]+T[i+1]-2*T[i])/Dx^2    "internal nodes"
End
dTdt[M]=2*alpha*(T[M-1]-T[M])/Dx^2+2*h_bar*(T_infinity-T[M])/(Dx*rho*c)
                                          "node M"
```

The next step is to comment out the arbitrary values of the temperatures and time that were used to test the computation of the state equation and instead let EES' Integral function control these variables for the numerical integration. The EES Integral command requires four arguments, as shown below.

```
F=Integral(Integrand,VarName,LowerLimit,UpperLimit)
```

The variable F is the integrated quantity, which is the temperature at a node for our problem and Integrand is the EES variable or expression that is to be integrated (which, in this case, is each of the time derivative of the nodal temperature), VarName is the integration variable (time), and LowerLimit and UpperLimit define the limits of integration (0 and t_sim). The time step is automatically selected and varied as needed through the integration process. An optional fifth parameter can be provided to the Integral command if a fixed time step is to be used.

Comment out the arbitrary values of time and temperatures that were used to test the state equations and replace these equations with M calls to the Integral function, as shown below. The $IntegralTable directive is used to store the temperatures at each node and each time.

```
"Inputs"
L=5 [cm]*Convert(cm,m)                    "wall thickness"
k=5.0 [W/m-K]                             "conductivity"
rho=2000 [kg/m^3]                         "density"
c=200 [J/kg-K]                            "specific heat capacity"
```

```
T_ini=20 [C]                                        "initial temperature"
T_infinity=200 [C]                                  "fluid temperature"
h_bar=500 [W/m^2-K]                                 "heat transfer coefficient"
alpha=k/(rho*c)                                     "thermal diffusivity"


"Setup grid"
M=11 [-]                                            "number of nodes"
Duplicate i=1,M
    x[i]=(i-1)*L/(M-1)                              "position of each node"
End
Dx=L/(M-1)                                          "distance between adjacent nodes"


{"Program and check the state equations"
time=0 [s]                                          "set a value of time (not needed for this problem)"
Duplicate i=1,M
    T[i]=T_ini                                      "and temperature"
End}


"and check the state equations"
dTdt[1]=2*alpha*(T[2]-T[1])/Dx^2                    "node 1"
Duplicate i=2,(M-1)
    dTdt[i]=alpha*(T[i-1]+T[i+1]-2*T[i])/Dx^2       "internal nodes"
End
dTdt[M]=2*alpha*(T[M-1]-T[M])/Dx^2+2*h_bar*(T_infinity-T[M])/(Dx*rho*c)
                                                    "node M"


t_sim=40 [s]                                        "simulation time"
Duplicate i=1,M
    T[i]=T_ini+Integral(dTdt[i],time,0,t_sim)       "integrate the coupled set of state equations"
End


$IntegralTable time, T[1..M]                        "store variables in an Integral Table"
```

After running the code (select Solve from the Calculate menu), an Integral Table will be generated. The results in the Integral Table can be used to create plots in the same way as the results in a Parametric Table or an Array Table. Note that the integration options (e.g., the error tolerances) can be controlled by selecting Preferences from the Options menu and then selecting the Integration tab (or also by using the $IntegralAutoStep directive).

MATLAB's ODE Solver

MATLAB's suite of integration routines was discussed in Section 5.3.5 in the context of lumped capacitance problems that required that a single state equation be integrated through time. Each of the ODE solvers (e.g., ode45, ode23, etc.) can be extended to integrate systems of state equations through time. The calling protocol is as follows.

```
[time, T] = ode45( 'dTdt', tspan, T0)
```

In Section 5.3.5, the argument 'dTdt' specifies the name of the function that returns a *single* output, the time rate of change of temperature given the current temperature and time. For the 1-D transient problems considered in this section, an energy balance on each control volume leads to a system of M equations (one for each node) that must be solved in order to provide the time rate of change of each nodal temperature given the current values of all of the nodal temperatures and time. Therefore, the function 'dTdt' returns a *vector* that contains the time rate of change of each nodal temperature and it will require as inputs a *vector* of nodal temperatures and time.

The function below is saved as file dTdtS6p3p6.m and implements the state equations for the example problem considered in this section, Eqs. (6.92), (6.95), and (6.98).

```
function[dTdt]=dTdtS6p3p6(time,T)

    %Inputs:
    %time - time in simulation (s)
    %T - temperatures at each node (C)
    %
    %Outputs:
    %dTdt - temperature rate of change at each node (C/s)

    L=0.05;                                  %wall thickness (m)
    k=5.0;                                   %conductivity (W/m-K)
    rho=2000;                                %density (kg/m^3)
    c=200;                                   %specific heat capacity (J/kg-K)
    T_infinity=200;                          %fluid temperature (C)
    h_bar=500;                               %heat transfer coefficient (W/m^2-K)
    alpha=k/(rho*c);                         %thermal diffusivity (m^2/s)

    M=size(T,1);                             %determine number of nodes
    Dx=L/(M-1);                              %distance between adjacent nodes (m)
    dTdt=zeros(M,1);                         %initialize the dTdt vector

    dTdt(1)=2*alpha*(T(2)-T(1))/Dx^2;        %node 1 state eq.
    for i=2:(M-1)
        dTdt(i)=alpha*(T(i-1)+T(i+1)-2*T(i))/Dx^2;   %internal nodes state eq.
    end
    dTdt(M)=2*alpha*(T(M-1)-T(M))/Dx^2+2*h_bar*(T_infinity-T(M))/(Dx*rho*c);
                                             %node M state eq.

end
```

The ODE solver ode45 is called from a MATLAB script.

```
clear all;
T_ini=20;                            % initial temperature (C)
M=11;                                % number of nodes (-)
t_sim=40;                            % simulation time (s)
[time,T]=ode45('dTdtS6p3p6',[0,t_sim],T_ini*ones(M,1));
```

The solution is contained in the matrix T and includes temperatures at each node (corresponding to the columns) at each simulated time (corresponding to each row). The time values are returned in the vector time. (Note that the time values at which the solution is returned can be specified by providing a vector of specific time values as the second argument in place of [0, t_sim].)

It is not covenient that the function dTdtS6p3p6 has only two input arguments, time and T. The argument list can be expanded in order to include all of the parameters that are required to specify the problem. This modification allows these parameters to be set just once (e.g., in the script) and passed through to the function.

```
function[dTdt]=dTdtS6p3p6(time,T,L,k,rho,c,T_infinity,h_bar)

    %Inputs:
    %time - time in simulation (s)
    %T - temperatures at each node (C)
    %L - thickness of wall (m)
    %k - conductivity (W/m-K)
    %rho - density (kg/m^3)
    %c - specific heat capacity (J/kg-K)
    %T_infinity - fluid temperature (C)
    %h_bar - heat transfer coefficient (W/m^2-K)
    %
    %Outputs:
    %dTdt - temperature rate of change at each node (C/s)

    alpha=k/(rho*c);                                %thermal diffusivity (m^2/s)
    M=size(T,1);                                    %determine number of nodes
    Dx=L/(M-1);                                     %distance between adjacent nodes (m)
    dTdt=zeros(M,1);                                %initialize the dTdt vector

    dTdt(1)=2*alpha*(T(2)-T(1))/Dx^2;               %node 1 state eq.
    for i=2:(M-1)
        dTdt(i)=alpha*(T(i-1)+T(i+1)-2*T(i))/Dx^2;  %internal nodes state eq.
    end
    dTdt(M)=2*alpha*(T(M-1)-T(M))/Dx^2+2*h_bar*(T_infinity-T(M))/(Dx*rho*c);
                                                    %node M state eq.
end
```

The modified function is resaved as the file **dTdtS6p3p6.m**. The parameters are specified within the script and passed to the function. The inputs **time** and **T** are *mapped* onto the new function as shown below.

```
clear all;
L=0.05;                                             %wall thickness (m)
k=5.0;                                              %conductivity (W/m-K)
rho=2000;                                           %density (kg/m^3)
c=200;                                              %specific heat capacity (J/kg-K)
T_infinity=200;                                     %fluid temperature (C)
h_bar=500;                                          %heat transfer coefficient (W/m^2-K)
T_ini=20;                                           % initial temperature (C)
M=11;                                               % number of nodes (-)
t_sim=40;                                           % simulation time (s)
[time,T]=ode45(@(time,T) dTdtS6p3p6(time,T,L,k,rho,c,T_infinity,h_bar),[0,t_sim],T_ini*ones(M,1));
```

The integration options (tolerance, etc.) can be controlled using the **odeset** function and an optional fourth argument, OPTIONS, to ode45. For example, the relative error tolerance for the integration can be set according to the following.

```
OPTIONS=odeset('RelTol',1e-6);                      %set the relative tolerance
[time,T]=ode45(@(time,T) dTdtS6p3p6(time,T,L,k,rho,c,T_infinity,h_bar),...
        [0,t_sim],T_ini*ones(M,1),OPTIONS);
```

6.4 2-D Numerical Solutions

6.4.1 Introduction

Section 6.3 described some techniques for developing a numerical model for 1-D transient conduction problems. It is straightforward to extend each of these techniques to 2-D (or even 3-D) problems where the boundaries of the computational domain are simple (e.g., a rectangular region of space). As an example of this process, Section 6.4.2 illustrates the finite difference solution technique using the ODE solver that is built into EES. The finite element solution method presented in Section 4.4 for steady problems is an excellent tool for solving multidimensional transient problems and forms the basis of most commercial simulation software. The steps involved with using finite element software to solve a transient conduction problem are illustrated in Section 6.4.3 using the FEHT software.

6.4.2 The Finite Difference Solution

The first step in applying the finite difference technique to a 2-D transient conduction problem is the derivation of the state equations. A 2-D array of nodes is required to capture the temperature variation in two spatial coordinates; therefore, there will be many more state equations than there were for 1-D transient problems. The process of deriving the state equations will be illustrated in the context of the fin problem that was studied in Section 4.3.2. The problem is shown again in Figure 6.17. Because the Biot number of the fin is not small, the temperature gradients in the material in both the x- and y-directions are significant. In Section 4.3.2 the steady-state temperature distribution in the fin is predicted. Here, we will simulate the transient process associated with activating the fin.

The tip of the fin is insulated and the width (W) (into the page) is much larger than its thickness (th). The length of the fin is $L = 4.0$ cm and its thickness is $th = 1.0$ cm. The fin transfers heat to the surrounding fluid at $T_\infty = 20°C$ with average heat transfer coefficient, $\bar{h} = 200$ W/m²-K. The properties of the fin material are $k = 1$ W/m-K, $\rho = 2100$ kg/m³, and $c = 850$ J/kg-K. The fin material is initially in thermal equilibrium with the surrounding fluid at T_∞ when, at time $t = 0$, a heat flux of magnitude $\dot{q}'' = 32,000$ W/m² is applied to the base.

Deriving the State Equations

Figure 6.18 illustrates the distribution of nodes across the computational domain associated with the upper half of the fin, taking advantage of the line of symmetry associated with the fin centerline. In this problem, the position of each node is identified with two subscripts, i and j. The first subscript indicates the position of the node in the x-dimension and the second subscript indicates the position of the node in the y-dimension. The total number of nodes used in the simulation is $M \times N$. The x- and y-positions of any node designated with subscripts i and j are given by:

$$x_i = \frac{(i-1)L}{(M-1)} \quad \text{for } i = 1 \dots M \tag{6.137}$$

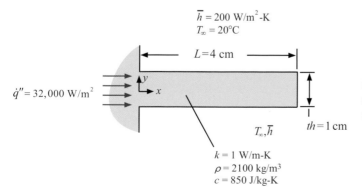

$\bar{h} = 200$ W/m²-K
$T_\infty = 20°C$

$L = 4$ cm

$\dot{q}'' = 32,000$ W/m²

$T_\infty, \bar{h}$

$th = 1$ cm

$k = 1$ W/m-K
$\rho = 2100$ kg/m³
$c = 850$ J/kg-K

Figure 6.17 Straight, constant cross-sectional area fin experiencing a transient activation process.

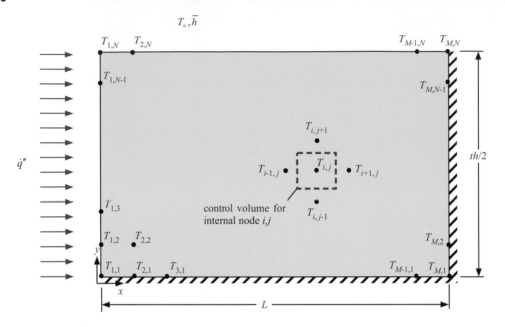

Figure 6.18 The distribution of nodes within the computational domain.

$$y_j = \frac{(j-1)th}{2(N-1)} \quad \text{for } j = 1 \dots N. \tag{6.138}$$

The x- and y-distance between adjacent nodes (Δx and Δy, respectively) are:

$$\Delta x = \frac{L}{(M-1)} \tag{6.139}$$

$$\Delta y = \frac{th}{2(N-1)}. \tag{6.140}$$

The state equations come from energy balances on control volumes that are defined around each of the nodes. The internal nodes, edge nodes, and corner nodes must all be considered separately. Figure 6.19 illustrates a control volume and the associated energy terms for an internal node. The energy balance suggested by Figure 6.19 is

$$\dot{q}_{RHS} + \dot{q}_{LHS} + \dot{q}_{top} + \dot{q}_{bottom} = \frac{dU}{dt}. \tag{6.141}$$

Substituting rate equations for the conduction terms and the storage term provides:

$$\frac{k\Delta y W}{\Delta x}\left(T_{i+1,j} - T_{i,j}\right) + \frac{k\Delta y W}{\Delta x}\left(T_{i-1,j} - T_{i,j}\right) + \frac{k\Delta x W}{\Delta y}\left(T_{i,j+1} - T_{i,j}\right)$$

$$+ \frac{k\Delta x W}{\Delta y}\left(T_{i,j-1} - T_{i,j}\right) = \Delta y \Delta x W \rho c \frac{dT_{i,j}}{dt}. \tag{6.142}$$

Rearranging Eq. (6.142) for the time rate of change of nodal temperature $T_{i,j}$ provides the state equation for the internal nodes:

$$\frac{dT_{i,j}}{dt} = \frac{\alpha}{\Delta x^2}\left(T_{i+1,j} + T_{i-1,j} - 2T_{i,j}\right) + \frac{\alpha}{\Delta y^2}\left(T_{i,j+1} + T_{i,j-1} - 2T_{i,j}\right)$$

$$\text{for } i = 2\dots(M-1) \text{ and } j = 2\dots(N-1). \tag{6.143}$$

Continuing this process for each of the boundary and corner nodes leads to the complete set of state equations summarized in Table 6.3.

Table 6.3 State equations for the problem shown in Figure 6.18.

Equation source	Number of equations	Equation	Subscript range
internal	$(M-1)\times$ $(N-1)$	$\dfrac{dT_{i,j}}{dt} = \dfrac{\alpha}{\Delta x^2}\left(T_{i+1,j} + T_{i-1,j} - 2T_{i,j}\right) + \dfrac{\alpha}{\Delta y^2}\left(T_{i,j+1} + T_{i,j-1} - 2T_{i,j}\right)$	$i = 2\ldots(M-1)$ $j = 2\ldots(N-1)$
left boundary	$(N-2)$	$\dfrac{dT_{1,j}}{dt} = \dfrac{2\dot{q}''}{\rho c\,\Delta x} + \dfrac{\alpha}{\Delta y^2}\left(T_{1,j+1} + T_{1,j-1} - 2T_{1,j}\right) + \dfrac{2\alpha}{\Delta x^2}\left(T_{2,j} - T_{1,j}\right)$	$i = 1$ $j = 2\ldots(N-1)$
top boundary	$(M-2)$	$\dfrac{dT_{i,N}}{dt} = \dfrac{\alpha}{\Delta x^2}\left(T_{i+1,N} + T_{i-1,N} - 2T_{i,N}\right) + \dfrac{2\alpha}{\Delta y^2}\left(T_{i,N-1} - T_{i,N}\right) + \dfrac{2\bar{h}}{\rho c\,\Delta y}\left(T_\infty - T_{i,N}\right)$	$i = 2\ldots(M-1)$ $j = N$
right boundary	$(N-2)$	$\dfrac{dT_{M,j}}{dt} = \dfrac{\alpha}{\Delta y^2}\left(T_{M,j+1} + T_{M,j-1} - 2T_{M,j}\right) + \dfrac{2\alpha}{\Delta x^2}\left(T_{M-1,j} - T_{M,j}\right)$	$i = M$ $j = 2\ldots(N-1)$
bottom boundary	$(M-2)$	$\dfrac{dT_{i,1}}{dt} = \dfrac{\alpha}{\Delta x^2}\left(T_{i+1,1} + T_{i-1,1} - 2T_{i,1}\right) + \dfrac{2\alpha}{\Delta y^2}\left(T_{i,2} - T_{i,1}\right)$	$i = 2\ldots(M-1)$ $j = 1$
upper right corner	1	$\dfrac{dT_{M,N}}{dt} = \dfrac{2\alpha}{\Delta y^2}\left(T_{M,N-1} - T_{M,N}\right) + \dfrac{2\alpha}{\Delta x^2}\left(T_{M-1,N} - T_{M,N}\right) + \dfrac{2\bar{h}}{\rho c\,\Delta y}\left(T_\infty - T_{M,N}\right)$	$i = M$ $j = N$
lower right corner	1	$\dfrac{dT_{M,1}}{dt} = \dfrac{2\alpha}{\Delta y^2}\left(T_{M,2} - T_{M,1}\right) + \dfrac{2\alpha}{\Delta x^2}\left(T_{M-1,1} - T_{M,1}\right)$	$i = M$ $j = 1$
lower left corner	1	$\dfrac{dT_{1,1}}{dt} = \dfrac{2\dot{q}''}{\rho c\,\Delta x} + \dfrac{2\alpha}{\Delta y^2}\left(T_{1,2} - T_{1,1}\right) + \dfrac{2\alpha}{\Delta x^2}\left(T_{2,1} - T_{1,1}\right)$	$i = 1$ $j = 1$
upper left corner	1	$\dfrac{dT_{1,N}}{dt} = \dfrac{2\dot{q}''}{\rho c\,\Delta x} + \dfrac{2\alpha}{\Delta y^2}\left(T_{1,N-1} - T_{1,N}\right) + \dfrac{2\alpha}{\Delta x^2}\left(T_{2,N} - T_{1,N}\right) + \dfrac{2\bar{h}}{\rho c\,\Delta y}\left(T_\infty - T_{1,N}\right)$	$i = 1$ $j = N$

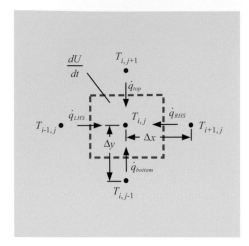

Figure 6.19 Energy balance for an internal node.

Integrating through Time

Any of the integration techniques provided in Section 6.3 can be extended to a 2-D problem. Table 6.3 shows that the number of equations that must be simultaneously integrated through time is much larger for a 2-D problem than it is for a 1-D problem, but the principle is the same. The built-in ODE solver in EES is used here to solve the problem. The EES program below sets arbitrary values of each of the nodal temperatures and, using those values, evaluates each of the state equations provided in Table 6.3. The $VarInfo directive specifies the units of all of the elements in the matrix T[].

```
$VarInfo T[] Units='C'

q``_dot=32000 [W/m^2]                  "heat flux to base"
T_infinity=20 [C]                      "fluid temperature"
h_bar=200 [W/m^2-K]                    "heat transfer coefficient"
th=1 [cm]*Convert(cm,m)                "thickness"
L=4 [cm]*Convert(cm,m)                 "length"
k=1 [W/m-K]                            "thermal conductivity"
rho=2100 [kg/m^3]                      "density"
c=850 [J/kg-K]                         "specific heat capacity"
alpha=k/(rho*c)                        "thermal diffusivity"

"Setup grid"
M=16 [-]                               "number of x-nodes"
N=11 [-]                               "number of y-nodes"
Duplicate i=1,M
   x[i]=(i-1)*L/(M-1)                  "x-position of each node"
   x_bar[i]=x[i]/L                     "dimensionless x-position of each node"
End
Dx=L/(M-1)                             "x-distance between adjacent nodes"
Duplicate j=1,N
   y[j]=(j-1)*th/(2*(N-1))             "y-position of each node"
   y_bar[j]=y[j]/th                    "dimensionless y-position of each node"
End
Dy=th/(2*(N-1))                        "y-distance between adjacent nodes"

"Set state variables"
Duplicate i=1,M
```

```
        Duplicate j=1,N
            T[i,j]=T_infinity
        End
    End

"Internal nodes"
Duplicate i=2,(M-1)
    Duplicate j=2,(N-1)
        dTdt[i,j]=alpha*(T[i+1,j]+T[i-1,j]-2*T[i,j])/Dx^2+alpha*(T[i,j+1]+T[i,j-1]-2*T[i,j])/Dy^2
    End
End

"left boundary"
Duplicate j=2,(N-1)
    dTdt[1,j]=2*q``_dot/(rho*c*Dx)+alpha*(T[1,j+1]+T[1,j-1]-2*T[1,j])/Dy^2+2*alpha*(T[2,j]-T[1,j])/Dx^2
End

"top boundary"
Duplicate i=2,(M-1)
    dTdt[i,N]=alpha*(T[i+1,N]+T[i-1,N]-2*T[i,N])/Dx^2+2*alpha*(T[i,N-1]-T[i,N])/Dy^2+&
        2*h_bar*(T_infinity-T[i,N])/(rho*c*Dy)
End

"right boundary"
Duplicate j=2,(N-1)
    dTdt[M,j]=alpha*(T[M,j+1]+T[M,j-1]-2*T[M,j])/Dy^2+2*alpha*(T[M-1,j]-T[M,j])/Dx^2
End

"bottom boundary"
Duplicate i=2,(M-1)
    dTdt[i,1]=alpha*(T[i+1,1]+T[i-1,1]-2*T[i,1])/Dx^2+2*alpha*(T[i,2]-T[i,1])/Dy^2
End

"upper right corner"
dTdt[M,N]=2*alpha*(T[M,N-1]-T[M,N])/Dy^2+2*alpha*(T[M-1,N]-T[M,N])/Dx^2+&
    2*h_bar*(T_infinity-T[M,N])/(rho*c*Dy)

"lower right corner"
dTdt[M,1]=2*alpha*(T[M,2]-T[M,1])/Dy^2+2*alpha*(T[M-1,1]-T[M,1])/Dx^2

"lower left corner"
dTdt[1,1]=2*q``_dot/(rho*c*Dx)+2*alpha*(T[1,2]-T[1,1])/Dy^2+2*alpha*(T[2,1]-T[1,1])/Dx^2

"upper left corner"
dTdt[1,N]=2*q``_dot/(rho*c*Dx)+2*alpha*(T[1,N-1]-T[1,N])/Dy^2+&
    2*alpha*(T[2,N]-T[1,N])/Dx^2+2*h_bar*(T_infinity-T[1,N])/(rho*c*Dy)
```

Once the state equations are checked, the code that sets the state variables is commented out and the Integral command is used to integrate the state equation associated with each node forward in time. This is accomplished using a nested set of Duplicate loops, as shown below. The $IntegralTable directive is used to create an

Integral Table that stores whatever nodal temperatures are of interest. In the code below, the nodal temperatures along the centerline of the fin (i.e., $j = 1$ and $i = 1..M$) are included in the Integral Table and their values are recorded every 1 s.

```
$VarInfo T[] Units='C'

q``_dot=32000 [W/m^2]                        "heat flux to base"
T_infinity=20 [C]                            "fluid temperature"
h_bar=200 [W/m^2-K]                          "heat transfer coefficient"
th=1 [cm]*convert(cm,m)                      "thickness"
L=4 [cm]*convert(cm,m)                       "length"
k=1 [W/m-K]                                  "thermal conductivity"
rho=2100 [kg/m^3]                            "density"
c=850 [J/kg-K]                               "specific heat capacity"
alpha=k/(rho*c)                              "thermal diffusivity"

"Setup grid"
M=16 [-]                                     "number of x-nodes"
N=11 [-]                                     "number of y-nodes"
Duplicate i=1,M
   x[i]=(i-1)*L/(M-1)                        "x-position of each node"
   x_bar[i]=x[i]/L                           "dimensionless x-position of each node"
End
Dx=L/(M-1)                                   "x-distance between adjacent nodes"
Duplicate j=1,N
   y[j]=(j-1)*th/(2*(N-1))                   "y-position of each node"
   y_bar[j]=y[j]/th                          "dimensionless y-position of each node"
End
Dy=th/(2*(N-1))                              "y-distance between adjacent nodes"

{"Set state variables"
Duplicate i=1,M
   Duplicate j=1,N
      T[i,j]=T_infinity
   End
End}

"Internal nodes"
Duplicate i=2,(M-1)
   Duplicate j=2,(N-1)
      dTdt[i,j]=alpha*(T[i+1,j]+T[i-1,j]-2*T[i,j])/Dx^2+alpha*(T[i,j+1]+T[i,j-1]-2*T[i,j])/Dy^2
   End
End

"left boundary"
Duplicate j=2,(N-1)
   dTdt[1,j]=2*q``_dot/(rho*c*Dx)+alpha*(T[1,j+1]+T[1,j-1]-2*T[1,j])/Dy^2+2*alpha*(T[2,j]-T[1,j])/Dx^2
End

"top boundary"
Duplicate i=2,(M-1)
   dTdt[i,N]=alpha*(T[i+1,N]+T[i-1,N]-2*T[i,N])/Dx^2+2*alpha*(T[i,N-1]-T[i,N])/Dy^2+&
```

```
        2*h_bar*(T_infinity-T[i,N])/(rho*c*Dy)
End

"right boundary"
Duplicate j=2,(N-1)
    dTdt[M,j]=alpha*(T[M,j+1]+T[M,j-1]-2*T[M,j])/Dy^2+2*alpha*(T[M-1,j]-T[M,j])/Dx^2
End

"bottom boundary"
Duplicate i=2,(M-1)
    dTdt[i,1]=alpha*(T[i+1,1]+T[i-1,1]-2*T[i,1])/Dx^2+2*alpha*(T[i,2]-T[i,1])/Dy^2
End

"upper right corner"
dTdt[M,N]=2*alpha*(T[M,N-1]-T[M,N])/Dy^2+2*alpha*(T[M-1,N]-T[M,N])/Dx^2+&
    2*h_bar*(T_infinity-T[M,N])/(rho*c*Dy)
"lower right corner"
dTdt[M,1]=2*alpha*(T[M,2]-T[M,1])/Dy^2+2*alpha*(T[M-1,1]-T[M,1])/Dx^2

"lower left corner"
dTdt[1,1]=2*q``_dot/(rho*c*Dx)+2*alpha*(T[1,2]-T[1,1])/Dy^2+2*alpha*(T[2,1]-T[1,1])/Dx^2

"upper left corner"
dTdt[1,N]=2*q``_dot/(rho*c*Dx)+2*alpha*(T[1,N-1]-T[1,N])/Dy^2+&
    2*alpha*(T[2,N]-T[1,N])/Dx^2+2*h_bar*(T_infinity-T[1,N])/(rho*c*Dy)

t_sim=1000
Duplicate i=1,M
    Duplicate j=1,N
        T[i,j]=T_infinity+Integral(dTdt[i,j],time,0,t_sim)
    End
End

$IntegralTable time:1, T[1..M,1]
```

Figure 6.20 illustrates the temperature as a function of time at $y = 0$ and the values of x that correspond to each node (i.e., temperatures along the center line of the fin). Note that the temperature at $x = 0$ responds instantly to the application of the heat flux at the base while nodes located at larger values of x require progressively more time to react due to the diffusion time constant.

6.4.3 The Finite Element Solution

The mathematical details behind the finite element technique are beyond the scope of this book; interested readers are referred to Nellis and Klein (2009) and Myers (1998) for a thorough treatment of this method. However, it is expected that a competent engineer will be capable of using commercial software based on the finite element technique in a professional manner. Therefore, the steps involved with using finite element software to solve a transient conduction problem are illustrated in this section using the FEHT software. FEHT can be downloaded from http://fchartsoftware.com/feht/ and an introduction to the use of FEHT can be found in Appendix F.

Figure 6.21 illustrates a disk brake that is used to bring a piece of rotating machinery to a smooth stop. The brake pad engages the machine when the brake is activated. Friction causes a heat flux at the interface during the braking process. All surfaces not exposed to the frictional heat flux experience convection with air at $T_\infty = 20°C$

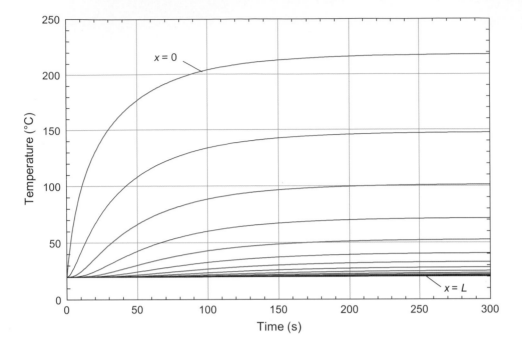

Figure 6.20 Temperature as a function of time for $y = 0$ and various values of x.

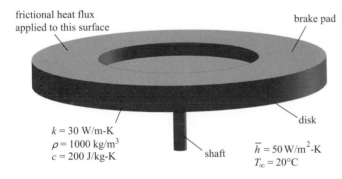

frictional heat flux
applied to this surface

brake pad

Figure 6.21 A disk brake on a rotating machine.

$k = 30$ W/m-K
$\rho = 1000$ kg/m^3
$c = 200$ J/kg-K

disk

shaft

$\bar{h} = 50$ W/m^2-K
$T_\infty = 20°$C

and $\bar{h} = 50$ W/m^2-K. The disk material properties are $k = 30$ W/m-K, $\rho = 1000$ kg/m^3, and $c = 200$ J/kg-K. The braking process lasts 50 s and during the process the heat flux continuously diminishes with time according to:

$$\dot{q}'' = 200,000 \left[1 - \left(\frac{t}{50} \right)^2 \right], \tag{6.144}$$

where $\dot{q}''$ is the heat flux (in W/m^2) and t is time (in s).

Specifying the Problem

Figure 6.22 illustrates the geometry that is used for the simulation. The problem is a 2-D transient conduction problem in radial coordinates. The top surface has a specified heat flux that is a function of time and the remaining surfaces experience a convection boundary condition. The material properties required for this transient problem are the thermal conductivity, density, and specific heat capacity.

The geometry is implemented in FEHT using the procedure discussed in Appendix F. In the Setup menu the problem is specified as being both Cylindrical and Transient. The points defining the lines have been exactly positioned, the material properties and initial temperature specified, and the boundary conditions on each of the external surfaces set. Note that the heat flux variation with time given by Eq. (6.144) can be implemented by

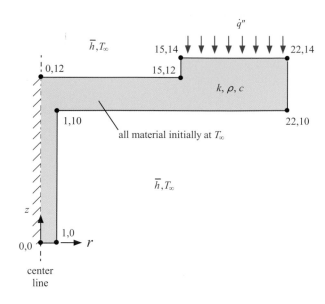

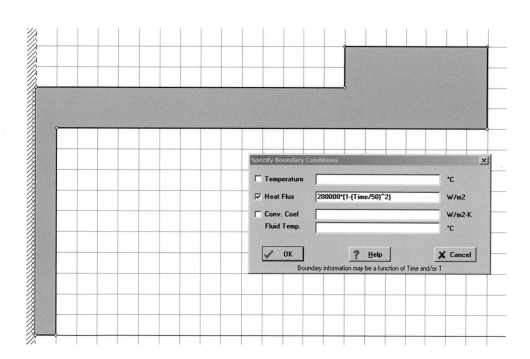

Figure 6.22 Specification of the problem (all dimensions are indicated in cm).

Figure 6.23 Problem specification in FEHT.

selecting the brake pad boundary and then typing the equation for the heat flux into the Boundary Conditions dialog accessed from the Specify menu. The result is shown in Figure 6.23.

Specifying the Mesh and Solving
Figure 6.24 illustrates a possible mesh that can be used; this mesh was obtained by manually drawing a coarse mesh and then reducing it several times. Alternatively, it is possible to select Automesh from the Draw menu and reduce the mesh.

Select Check from the Run menu and FEHT will check to make sure all boundary conditions, properties, etc., are specified. The critical time step will also be estimated; for the mesh shown in Figure 6.24 the critical time step is

$\Delta t_{crit} = 2.1 \times 10^{-3}$ s. Next it is necessary to specify the duration of the simulation (50 s) and the numerical integration technique by selecting Calculate from the Run menu. The duration of the simulation is more than 70,000 times larger than the critical time step and therefore an explicit technique like the Euler Method will require many more time steps than is necessary from an accuracy standpoint. Instead, specify that the Crank–Nicolson technique (which is an implicit method) be used for the 50 s simulation with 0.1 s time steps, as shown in Figure 6.25.

The simulation results can be viewed in various ways. The most direct way is by selecting Temperature Contours from the View menu and then selecting the From start to stop radio button in order to display the temperature contours as they evolve with time during the process.

Mesh Convergence

As with any numerical solution it is important to verify that you have achieved mesh convergence. This is done by examining how some important aspect of your solution changes as the mesh is refined. The temperature contours suggest that the node placed at $r = 22$ cm and $z = 14$ cm (i.e., the upper right corner of the pad) is close to the hottest point in the brake. Therefore, the temperature of this node as a function of time as the mesh is refined is shown in Figure 6.26, which suggests that a mesh of at least 384 elements is required to capture the maximum temperature in the brake material to within about 1°C.

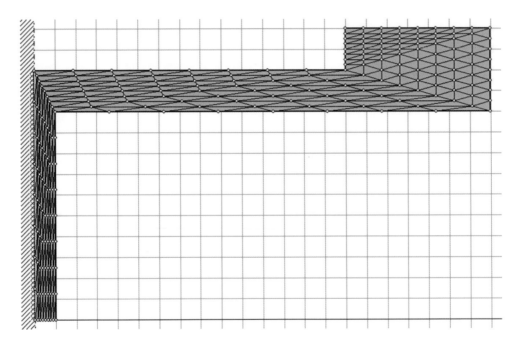

Figure 6.24 Mesh.

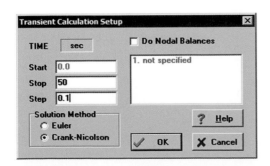

Figure 6.25 Transient Calculation Setup dialog.

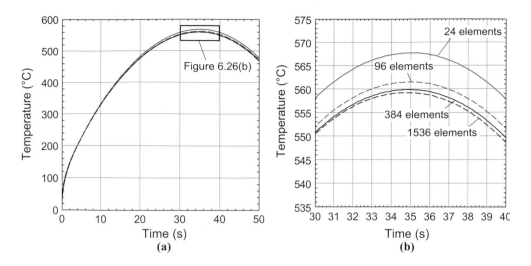

Figure 6.26 Temperature of the node placed at $r = 22$ cm and $z = 14$ cm as a function of time for several different mesh resolutions over (a) the entire simulation and (b) near the maximum temperature.

6.5 Conclusions and Learning Objectives

Chapter 5 discussed transient problems in which conduction is neglected as being unimportant relative to heat transfer processes that occur at the surface. These problems are called lumped capacitance problems and they ignore any spatial gradients of temperature within the object. This chapter discusses transient problems where conduction is important and therefore spatial gradients within the object cannot be ignored. The characteristics of transient energy transfer by conduction must be considered in order to understand and solve these problems.

Several conceptual tools were developed and applied in order to help understand transient conduction problems. The most important of these is the relationship $\delta_t \approx 2\sqrt{\alpha t}$ which approximately relates the distance that energy can be conducted to the thermal diffusivity (a material property) and time. This relation was used to develop both the concept of a diffusive time constant as well as the semi-infinite resistance. These conceptual tools are useful for developing a physical feel for the processes that are important in a transient conduction problem and to anticipate the behavior of these situations.

The analytical solution to transient conduction processes requires the solution of a partial differential equation subject to an initial condition and boundary conditions. The mathematical techniques required to do this are beyond the scope of this text but it is expected that you can derive the mathematical statement of the problem. The solution to semi-infinite body problems with various boundary conditions and bounded plane wall, cylinder, and spherical geometries are presented without derivation.

The numerical techniques used to solve 1-D and 2-D steady-state problems and lumped capacitance transient problems are extended in this chapter to allow the solution to 1-D and 2-D transient problems. These include explicit and implicit numerical techniques that can be developed using either EES or MATLAB as well as the native ODE solvers in these software packages. The procedure to develop a finite element solution is also presented in the context of the FEHT software.

Some specific concepts that you should understand are listed below.
- The relationship between the thermal penetration depth, thermal diffusivity, and time.
- The meaning of the property thermal diffusivity.
- The amount of time during which a conduction process can be considered to be semi-infinite.
- The diffusive time constant and how this concept can be used to anticipate the behavior of transient conduction problems.
- The idea of the semi-infinite resistance, including how it is derived and how it can be applied to understand the behavior of transient problems at short time scales.

- The ability to derive the partial differential equation, initial condition, and boundary conditions that together provide the mathematical statement of a transient conduction problem.
- How to use standard semi-infinite body solutions.
- How to use standard bounded geometry solutions.
- The conditions under which the single-term solutions are sufficient for a bounded geometry solution.
- The ability to develop the state equations that must be solved in order to solve a 1-D or 2-D transient conduction problem numerically.
- The ability to integrate these state equations through time using the Euler Method, a simple predictor–corrector method, and the fully implicit method both in EES and in MATLAB.
- The steps involved in using a commercial finite element heat transfer software package to solve a transient conduction problem.

References

Heisler, M. P., "Temperature charts for induction and constant temperature heating," *Trans. of the ASME*, Vol. 69, pp. 227–236 (1947).

Myers, G. E., *Analytical Methods in Conduction Heat Transfer*, Second Edition, AMCHT Publications, Madison, WI (1998).

Nellis, G. F. and S. A. Klein, *Heat Transfer*, Cambridge University Press, New York (2009).

Problems

Conceptual Tools

6.1 You are designing a wall that will be used to absorb solar radiation during the day and release it into the building space during the night. The plan is to choose a wall thickness so that the heat transfer from the back side of the wall (in the building space) will occur about 12 hours after solar energy is absorbed on the front face. The wall is made of reinforced concrete. What is your estimate of the required wall thickness?

6.2 A plane wall has an area $A_c = 1$ m^2 and is $L = 1$ m thick. The wall is composed of a material with thermal conductivity $k = 1$ W/m-K, density $\rho = 1$ kg/m^3, and $c = 1$ J/kg-K. The right side of the wall (at $x = L$) is exposed to fluid at $T_C = 0°$C with a heat transfer coefficient $\bar{h} = 1$ W/m^2-K. The wall is initially at a uniform temperature $T_{ini} = 0°$C when, at time $t = 0$, the left side of the wall is exposed to a hotter fluid, $T_H = 100°$C, with a heat transfer coefficient $\bar{h} = 1$ W/m^2-K.

$A_c = 1$ m^2, $k = 1$ W/m-K, $\rho = 1$ kg/m^3, $c = 1$ J/kg-K

$T_H = 100°$C
$\bar{h} = 1$ W/m^2-K

$T_C = 0°$C
$\bar{h} = 1$ W/m^2-K

x

$L = 1$ m

at time $t = 0$, the material is at $T_{ini} = 0°$C

(a) For some time, the wall can be treated as a semi-infinite body. How long will it take for the thermal wave associated with the disturbance at the left side of the wall to reach the right side of the wall? That is, estimate how long it takes for the temperature of the right side of the wall (at $x = L$) to begin to rise.

(b) Draw an approximate thermal resistance network that can be used to represent the problem for times that are less than what you calculated in (a). Label the resistors and show how they would be calculated using symbols provided in the problem statement and, if necessary, time (t).

(c) Sketch the temperature distribution at the time you calculated in (a) and at half of that time. That is, provide a qualitative sketch of the temperature as a function of position at the instant that the thermal wave hits the right side of the wall and some time before the thermal wave hits the right side of the wall.

(d) Draw a thermal resistance network that can be used to represent the problem when it reaches steady state (i.e., for times much larger than you calculated in (a)). Calculate the numerical value of each resistance in your network.

(e) What is the steady-state rate at which heat is conducted into the left-hand surface of the wall? That is, what is the numerical value (W) of the rate of heat conduction at $x = 0$ as time goes to infinity?

(f) What are the steady-state temperatures of the left-hand and right-hand surfaces of the wall?

That is, what are $T(x = 0, t \to \infty)$ and $T(x = L, t \to \infty)$?

(g) Sketch the steady-state temperature distribution (i.e., the temperature as a function position as $t \to \infty$).

(h) Sketch the rate of heat conduction at the left-hand side of the wall (i.e., at $x = 0$) as a function of time. Make sure that you indicate the numerical value of the heat transfer rate (in W) that you expect at $t = 0$ s, at the time you calculated in (a) (approximately), and at $t \to \infty$. You may want to refer to your resistance networks from parts (b) and (d) to help with this problem.

(i) Sketch the time variation of the temperatures at the left-hand and right-hand surfaces of the wall. That is, what are $T(x = 0, t)$ and $T(x = L, t)$? You may want to refer to your resistance networks from parts (b) and (d) to help with this problem.

6.3 A plane wall has the linear temperature distribution at time $t = 0$ as shown in the figure.

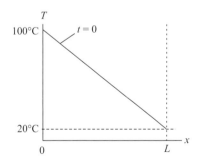

For times greater than $t = 0$, both edges of the wall are exposed to convection. The fluid temperature is $T_\infty = 20°C$ and the heat transfer coefficient is $\bar{h} = 1$ W/m²-K. The wall is $L = 0.1$ m thick and composed of a material with $k = 1$ W/m-K, $\rho = 1$ kg/m³ and $c = 1$ J/kg-K.

Sketch the temperature distribution that you expect as time approaches infinity. Also show temperature distributions at a few interesting intermediate times. Your sketch should be based on the time constants associated with the equilibration process, which you should calculate.

6.4 Two walls with different values of thermal diffusivity ($\alpha = k/\rho c$) are shown in the figure. Wall A has a high value of thermal diffusivity while Wall B has a low value of thermal diffusivity. Both walls are initially at a uniform, low temperature (T_C). At time $t = 0$, the temperature at the left-hand wall is increased to (T_H).

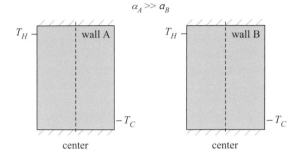

Sketch the center temperature of wall A and wall B as a function of time.

6.5 A large sphere of low-conductivity ceramic material with outer radius r_{cer} encloses a small, very high-conductivity copper sphere with radius r_{Cu}.

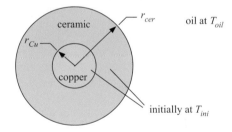

The ceramic and copper are both initially at a uniform temperature (T_{ini}) and then the sphere is dropped into a bath of hot oil at T_{oil}. Convection heat transfer occurs between the oil and the outer surface of the ceramic. Neglect any contact resistance between the two spheres and neglect radiation.

Sketch the temperature distribution from $r = 0$ to $r = r_{cer}$ at various times, including at least the beginning of the process, sometime during the heating process, and near the end of the heating process.

6.6 A wall is exposed to a heat flux for a long time, as shown the figure. The left side of the wall is exposed to liquid at $T_f = 20°C$ with a very high heat transfer coefficient; therefore, the left side of the wall ($T_{x=0}$) always has the temperature T_f. The right side of the wall is exposed to the heat flux and also convects to gas at $T_f = 20°C$ but with a heat transfer coefficient of $\bar{h} = 5000$ W/m²-K. The wall is $L = 0.5$ m thick and composed of a material with $k = 1.0$ W/m-K, $\rho = 4000$ kg/m³, and $c = 700$ J/kg-K. The wall is initially at steady state with the heat flux on when, at time $t = 0$, the heat flux is suddenly shut off. The wall subsequently equilibrates with the liquid and gas; eventually, it reaches a uniform temperature equal to T_f.

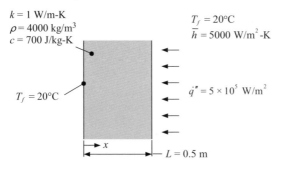

$k = 1$ W/m-K
$\rho = 4000$ kg/m³
$c = 700$ J/kg-K

$T_f = 20°C$
$\bar{h} = 5000$ W/m²-K

$q'' = 5 \times 10^5$ W/m²

$T_f = 20°C$

$L = 0.5$ m

(a) Calculate the temperature of the right-hand side of the wall at $t = 0$ (i.e., determine $T_{x=L, t=0}$).
(b) Sketch the temperature distribution at $t = 0$ and the temperature distribution as $t \rightarrow \infty$.
(c) Sketch the temperature distribution at $t = 10$ s, 100 s, 1×10^3 s, 1×10^4 s, and 1×10^5 s. Justify the shape of these sketches by calculating the characteristic time scales that govern the equilibration process.

6.7 A thin heater is sandwiched between two materials, A and B. Both materials are very thick and so they may be considered semi-infinite. Initially, both materials are at a uniform temperature of $T_{ini} = 0$. The heater is activated at $t = 0$ and delivers a uniform heat flux, $\dot{q}''_{heater} = 1$ W/m² to the interface between the materials; some of this energy will be conducted into material A ($\dot{q}''_A$) and some into material B ($\dot{q}''_B$). Materials A and B have the same thermal diffusivity, $\alpha = 0.001$ m²/s, but the conductivity of material A is twice that of material B, $k_A = 2$ W/m-K and $k_B = 1$ W/m-K. There is no contact resistance anywhere in this problem and it is a 1-D, transient conduction problem.

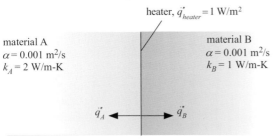

heater, $\dot{q}''_{heater} = 1$ W/m²

material A
$\alpha = 0.001$ m²/s
$k_A = 2$ W/m-K

material B
$\alpha = 0.001$ m²/s
$k_B = 1$ W/m-K

$\dot{q}''_A$ $\dot{q}''_B$

(a) Draw a thermal resistance network that you could use to model this problem approximately. Your resistances should be written in terms of time, t, and the symbols in the problem statement. Clearly indicate on your network where $\dot{q}''_{heater}$ is added and where the temperatures T_{in} and T_{heater} are located.
(b) Use your resistance network from (a) to predict the heater temperature, T_{heater}, at time $t = 1$ s.
(c) Sketch the temperature distribution at $t = 1$ s and $t = 2$ s. Label your plots clearly. Focus on getting the qualitative features of your plot correct; justify them in words if necessary.

6.8 The figure shows a slab of material that is $L = 5$ cm thick and is heated from one side ($x = 0$) by a radiant heat flux $\dot{q}''_s = 7500$ W/m². The material has conductivity $k = 2.4$ W/m-K and thermal diffusivity $\alpha = 2.2 \times 10^{-4}$ m²/s. Both sides of the slab are exposed air at $T_\infty = 20°C$ with heat transfer coefficient $\bar{h} = 15$ W/m²-K. The initial temperature of the material is $T_{ini} = 20°C$.

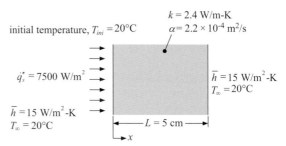

initial temperature, $T_{ini} = 20°C$

$k = 2.4$ W/m-K
$\alpha = 2.2 \times 10^{-4}$ m²/s

$\dot{q}''_s = 7500$ W/m²

$\bar{h} = 15$ W/m²-K
$T_\infty = 20°C$

$\bar{h} = 15$ W/m²-K
$T_\infty = 20°C$

$L = 5$ cm

(a) About how long do you expect it to take for the temperature of the material on the unheated side ($x = L$) to begin to rise?
(b) What do you expect the temperature of the material at the heated surface ($x = 0$) to be (approximately) at the time identified in (a)?
(c) Develop a simple and approximate model that can predict the temperature at the heated surface as a function of time for times that are less than the time calculated in (a). Plot the temperature at $x = 0$ as a function of time from $t = 0$ to the time identified in (a).
(d) Sketch the temperature as a function of position in the slab for several times less than the time identified in (a) and greater than the time identified in (a). Make sure that you get the qualitative features of the sketch correct. Also sketch the temperature as a function of position in the slab at steady state (make sure that you get the temperatures at either side correct).

6.9 The figure illustrates a very thin heater that is placed between two layers of material.

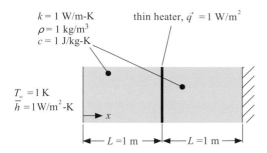

Each of the layers have thickness $L = 1$ m, conductivity $k = 1$ W/m-K, density $\rho = 1$ kg/m^3, and specific heat capacity $c = 1$ J/kg-K. Initially, the material in both layers is at $T_{ini} = 1$ K when the heater is activated, providing a heat flux of $\dot{q}'' = 1$ W/m^2 to the interface between the layers. The left surface (at $x = 0$) is exposed to fluid at $T_\infty = 1$ K with heat transfer coefficient $\bar{h} = 1$ W/m^2-K. The right surface (at $x = 2\,L$) is insulated. You may neglect contact resistance and radiation for this problem.

(a) About how long will it take for the temperature of the surface at $x = 0$ to begin to rise? Justify your answer.

(b) Draw a resistance network that can be used to approximately represent the problem for times that are less than t_a, where t_a is the time that you calculated in (a).

(c) Use your resistance network from (b) to estimate the heater temperature at $t = t_a$.

(d) Sketch the temperature as a function of position (for $0 < x < 2\,L$) at time $t = t_a$, Also sketch the temperature distribution at $t = t_a/2$.

(e) Determine the temperature of the heater when $t \rightarrow \infty$ (i.e., the steady-state temperature of the heater).

(f) Overlay on your sketch from (d) the temperature as a function of position (for $0 < x < 2\,L$) when $t \rightarrow \infty$ (i.e., the steady-state temperature distribution).

6.10 Two very long rods (A and B) are composed of the same material ($\alpha = $ 1e-5 m^2/s, $k = 10$ W/m-K) but have different initial temperatures; Rod A has an initial temperature of 0°C while Rod B has an initial temperature of 100° C. At time = 0, these two rods are brought into contact at their face, as shown. The contact between the surfaces of Rods A and B is not perfect, rather it is characterized by a contact resistance ($R''_c = 1 \times 10^{-4}$ K-m^2/W). The edges of rods are insulated and you may assume for this problem that the rods are infinitely long.

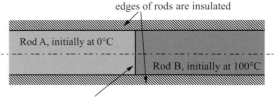

interface between rods A and B has contact resistance, $R_c'' = $ 1e-4 K-m^2/W

(a) Sketch (qualitatively) the temperature in the two rods as a function of axial position using axes like those shown below. Justify your answer.

Include at least the following four increasing times:

- time = 0, where the rods are just brought into contact (this one is done for you),
- time = t_1, where the effect of the contact resistance dominates,
- time = t_2, where the effect of contact resistance is still important, and
- time = t_3, where the effect of contact resistance is negligible.

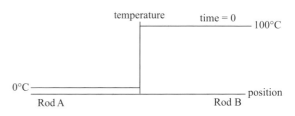

(b) Sketch (qualitatively), the surface temperature of Rod A and the surface temperature of Rod B as a function of time. Justify your answer.

(c) *About* how long will it take before the surface temperature of Rod A is nearly equal to the surface temperature of Rod B? Give a reasonable estimate but not an exact solution and justify your answer.

6.11 An annular fin is shown. The fin is one of many installed on an evaporator tube. During normal operation, the inner radius of the fin is at temperature T_{evap} and the surrounding air temperature is T_∞. The heat transfer coefficient between the fin and the surrounding air is $\bar{h}$. Heat is transferred from the surrounding air to the refrigerant in the tube and therefore the steady-state temperature distribution is as shown in the figure. The inner and outer radii of the fin are r_{in} = 1 cm and r_{out} = 4 cm, respectively. The thermal diffusivity of the fin material is $\alpha = 1 \times 10^{-5}$ m²/s. The thickness of the fin is th. The fin material has properties k, ρ, and c.

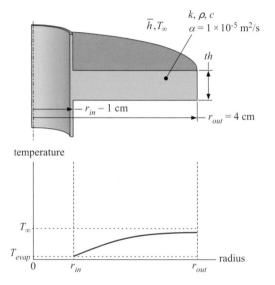

Every few hours, the fin has to be defrosted (i.e., the frost that forms on the fin surface must be melted away). This is accomplished by passing hot gas from the compressor discharge through the tube. At the initiation of the defrost process (i.e., at time $t = 0$), the temperature at the inner radius of the fin changes from T_{evap} to T_{comp}, where T_{comp} is greater than T_∞. The fan is turned off during the defrost process and therefore the heat transfer coefficient is reduced. The defrost process is maintained for sufficient time that the fin temperature stops changing (i.e., the fin achieves a new steady-state temperature distribution). Assume that the extended surface approximation is appropriate (i.e., temperature is a function only of radius and time).

(a) Sketch the new steady-state temperature distribution that you expect at the end of the defrost process.

(b) About how long do you think it will take for the temperature of the tip of the fin to begin to rise? Justify your answer with a calculation.

(c) Add to your sketch from (a) the temperature distribution that you expect for several times that are less than and greater than the time that you calculated in (b).

6.12 The figure illustrates a plane wall with thickness $L = 1$ m and area $A = 1$ m². The wall material has thermal diffusivity $\alpha = 0.01$ m²/s and conductivity $k = 2$ W/m-K. The left edge of the wall (at $x = 0$) is exposed to fluid at $T_\infty = 20°C$ with heat transfer coefficient $\bar{h} = 3$ W/m²-K. The right edge of the wall (at $x = L$) is insulated. Initially, the wall material is at a uniform temperature T_∞. At time $t = 0$, the left edge is exposed to a heat flux, $\dot{q}_s'' = 100$ W/m². Note that the left edge continues to be cooled by convection to the fluid at T_∞ during the process.

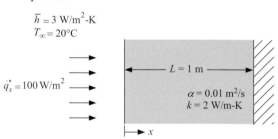

(a) Estimate the time required for the temperature of the material at $x = L$ to begin to rise.

(b) Develop a simple model of the heating process that is valid for times that are less than the time calculated in part (a). This model should be based on a resistance network – carefully label each resistance in your network and show how it would be calculated. Show where the energy associated with the heat flux enters your network.

(c) Use your model from part (b) to estimate the temperature at $x = 0$ at the time calculated in part (a).

(d) Sketch the temperature as a function of position in the wall at the time you calculated in part (a) as well as one-half that time and twice that time.

(e) Determine the temperature at $x = 0$ as time goes to infinity, $t \to \infty$.

(f) Overlay on your sketch from part (d) the temperature as a function of position as time goes to infinity.

6.13 A rod with uniform cross-sectional area, $A_c = 0.1$ m² and perimeter $per = 0.05$ m is placed in a vacuum environment. The length of the rod is $L = 0.09$ m and the external surfaces of the rod can be assumed to be adiabatic. For a long time, a heat transfer rate of $\dot{q}_h = 100$ W is provided to the end of the rod at $x = 0$. The end of the rod at $x = L$ is always maintained at $T_l = 20°C$. The rod material has density $\rho = 5000$ kg/m³, specific heat capacity $c = 500$ J/kg-K, and conductivity $k = 5$ W/m-K. The rod is at a steady-state operating condition when, at time $t = 0$, the heat transfer rate at $x = 0$ becomes zero.

(a) About how long does it take for the rod to respond to the change in heat transfer?

(b) Sketch the temperature distribution you expect at $t = 0$ and $t \to \infty$. Make sure that you get the temperatures at either end of the rod and the shape of the temperature distributions correct.

(c) Overlay on your sketch from (b) the temperature distributions that you expect at the time that you calculated in (a) as well as half that time and twice that time.

(d) Sketch the heat transfer from the rod at $x = L$ as a function of time. Make sure that your sketch clearly shows the behavior before and after the time identified in (a). Make sure that you get the rate of heat transfer at $t = 0$ and $t \to \infty$ correct.

6.14 A cylindrical heating element is used in an appliance to heat water. The heating element has radius $r_o = 0.25$ cm and properties $k = 15$ W/m-K, $\rho = 6000$ kg/m³, and $c = 750$ J/kg-K. The electrical resistivity of the material is $\rho_e = 6.2 \times 10^{-4}$ ohm-m. During normal operation, the element experiences a current of $I = 20$ amp and is exposed to a flow of water at $T_\infty = 20°C$. The heat transfer coefficient between the flowing water and the conductor is so large that the surface temperature of the conductor (at $r = r_o$) can be taken to be equal to T_∞. The conductor has been operating for a long time and therefore is at steady state.

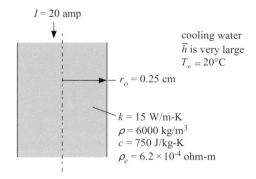

$I = 20$ amp

cooling water
$\bar{h}$ is very large
$T_\infty = 20°C$

$r_o = 0.25$ cm

$k = 15$ W/m-K
$\rho = 6000$ kg/m³
$c = 750$ J/kg-K
$\rho_e = 6.2 \times 10^{-4}$ ohm-m

(a) Determine the volumetric rate of thermal energy generation in the conductor.

(b) Determine the maximum temperature within the conductor during normal operation.

(c) If the current is deactivated and the surface of the conductor surface is completely insulated, instead of being cooled by water, determine the final uniform temperature that the material will reach.

(d) About how long does it take the conductor to reach the uniform temperature calculated in (c)?

At some point, the appliance is deactivated. The current is removed and the water is drained from the system and replaced with air at $T_\infty = 20°C$. The heat transfer coefficient between the air and the conductor is $\bar{h} = 20$ W/m²-K. We are interested in the transient process associated with the conductor going from the steady, normal operating condition examined in parts (a) and (b) to a new steady-state condition where it is in thermal equilibrium with the air.

(e) Qualitatively sketch the temperature of the surface of the conductor ($r = r_o$) and the temperature at the center of the conductor ($r = 0$) as a function of time during the equilibration process. Your sketch need not be accurate but it should be qualitatively correct and show the correct behavior. You should use the results of the calculations in parts (a) through (d) as well as any additional useful time constants to do this problem.

6.15 It is an interesting fact that different objects that are at the same temperature can feel as if they are at different temperatures. For example, a piece of metal that is initially at a uniform temperature of $T_{amb} = 20°C$ will feel colder to the touch than a piece of plastic that is initially at the same temperature. The figure illustrates a finger touching a large piece of material and, below that, a close up of the interface between the finger and the material.

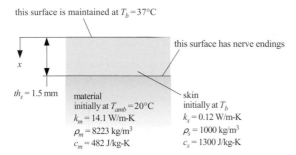

this surface is maintained at $T_b = 37°C$

this surface has nerve endings

$th_s = 1.5$ mm

material
initially at $T_{amb} = 20°C$
$k_m = 14.1$ W/m-K
$\rho_m = 8223$ kg/m³
$c_m = 482$ J/kg-K

skin
initially at T_b
$k_s = 0.12$ W/m-K
$\rho_s = 1000$ kg/m³
$c_s = 1300$ J/kg-K

We will model the skin as being a layer of material that is $th_s = 1.5$ mm thick with properties $k_s = 0.12$ W/m-K, $\rho_s = 1000$ kg/m³, and $c_s = 1300$ J/kg-K. The back side of the skin (at $x = 0$) is maintained at $T_b = 37°C$ by the flow of blood. The skin is initially at a uniform temperature that is equal to the blood temperature, T_b, when the surface at $x = th_s$ is pressed against the material at time $t = 0$. The material is initially at a uniform temperature $T_{amb} = 20°C$. The material is large enough that it can be assumed to be semi-infinite for this problem. The material has properties $k_m = 14.1$ W/m-K, $\rho_m = 8223$ kg/m³, and $c_m = 482$ J/kg-K. The nerve endings that actually sense the temperature are on the surface of the skin (at $x = th_s$).

(a) Estimate the time required for the thermal disturbance at the surface of the skin (at $x = th_s$) to travel to the back of the skin (at $x = 0$).

(b) Sketch the temperature as a function of x (in the skin and the material) for a single time that is less than the value calculated in part (a) but greater than 0. The sketch does not need to be quantitatively correct but it should be qualitatively correct and it should show some of the features that you expect to be present; you may want to briefly describe some of these features.

(c) Develop a simple model that is based on a thermal resistance network and is valid for times less than the value calculated in part (a). Use this model to derive an equation that predicts, approximately, the surface temperature of the skin (at $x = th_s$).

(d) Your model from part (c) should show that the skin surface temperature is constant with time for times less than the value calculated in part (a). Explain why this is so.

(e) Use your equation from part (c) to predict the temperature of the surface of the skin for times less than the value calculated in part (a). If the material was replaced with plastic, $k_p = 0.13$ W/m-K, $\rho_p = 1100$ kg/m³, and $c_p = 2010$, would it feel colder or hotter to the touch? Justify your answer with a calculation.

(f) Overlay on your sketch from part (b) the temperature as a function of x (in the skin and the material) for a single time that is greater than the value calculated in part (a). The sketch does not need to be quantitatively correct but it should be qualitatively correct and it should show some of the features that you expect to be present; you may want to briefly describe some of these features. Carefully label your sketch.

(g) Develop a simple model that is based on a thermal resistance network and is valid for times greater than the value calculated in part (a). Use this model to derive an equation that predicts, approximately, the surface temperature of the skin (at $x = th_s$).

(h) Your model from part (g) should show that the temperature of the surface of the skin does start to change once time increases beyond the value calculated in part (a). As time goes to infinity, what value will the temperature of the surface of the skin go to? Compute a characteristic time that will let you estimate how long this process takes.

6.16 A convection oven claims to reduce energy consumption by reducing the time required to bake foods. Unlike a conventional oven, the air within a convection oven is circulated by a fan so that the heat transfer to the food occurs by forced convection rather than by free convection in addition to radiation. Typical heat transfer coefficients are 15 W/m²-K for the conventional oven and 50 W/m²-K for the convection oven. Suppose that you are planning to bake a 7 kg turkey at 200°C. Do you expect that the baking time will be significantly reduced by using the convection oven? Assume that the turkey has a density of 96 g/cm³, a specific heat of 4.1 J/g-K and a thermal diffusivity of 0.14×10^{-6} m²/s.

6.17 A d_m = 5 mm thick piece of metal (with density ρ_m = 2700 kg/m³, specific heat capacity c_m = 900 J/kg-K, and thermal conductivity k_m = 240 W/m-K) is attached to a very thick piece of glass (with density ρ_g = 2600 kg/m³, specific heat capacity c_g = 800 J/kg-K, and thermal conductivity k_g = 1.1 W/m-K). The interface between the metal and the glass is characterized by a contact resistance $R''_{t,\,c}$ = 5.0 × 10⁻⁴ K-m²/W. The face of the metal is exposed to a constant heat flux $\dot{q}''$ = 50,000 W/m². Initially (i.e., at time t = 0), the metal and glass are at a uniform temperature T_{ini} = 20°C.

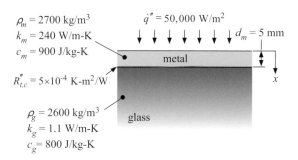

ρ_m = 2700 kg/m³
k_m = 240 W/m-K
c_m = 900 J/kg-K
$\dot{q}''$ = 50,000 W/m²
d_m = 5 mm
metal
$R''_{t,c}$ = 5×10⁻⁴ K-m²/W
x
ρ_g = 2600 kg/m³
k_g = 1.1 W/m-K
c_g = 800 J/kg-K
glass

(a) Calculate the time required for the thermal wave to penetrate to the back of the metal. That is, at what time after the application of the heat flux will the temperature at the interface between the metal and the glass begin to rise?

(b) Sketch the temperature as a function of position, x, a times t = 0, t = 0.020 s (20 ms), t = 0.050 s (50 ms), t = 0.2 s, and t = 1.0 s. Justify your sketches as much as possible. Do not

worry about the magnitude of the temperatures – concentrate on the shape of the temperature distributions and the relative temperature drop across the various portions of the system as well as the qualitative features that the temperature distribution must exhibit.

(c) For times greater than 0.2 s, is it possible to treat the metal as being lumped (i.e., is the metal all at essentially the same temperature)? Justify your answer.

(d) Is the contact resistance an important parameter in this problem? Are there some periods of time (e.g., short times or long times or both) where contact resistance does not play a role in the thermal behavior of the system? Quantify the time period where contact resistance is important.

(e) Assuming that you answered yes to question (c), derive the governing differential equation that describes the temperature of the metal as a function of time for times greater than 0.2 s. Your answer should be a symbolic expression for the rate of temperature change for the metal in terms of the quantities given in the problem (ρ_m, c_m, k_m, ρ_g, c_g, k_g, d_m, $R''_{t,\,c}$, $\dot{q}''$, and T_{ini}) as well as the time (t) and the instantaneous temperature of the aluminum (T). DO NOT attempt to explicitly solve the 1-D transient conduction problem associated with the glass – instead, your solution should be approximate, based on your knowledge of how semi-infinite bodies behave and it should use the concept of a penetration depth. *Hint*: a resistance network might be a good way to think about this problem.

The Differential Equation and Boundary Conditions

6.18 Your family is planning to bake a turkey for the holidays. The turkey is approximately spherical.

(a) Assuming constant properties, set up the partial differential equation that relates the temperature at radial position r in the turkey to time.

(b) The turkey has a mass of 14 lb$_m$ and its properties (density, specific heat, thermal conductivity) are about the same as those for liquid water. Using the concept of a diffusive time constant, estimate the time required for the center of the turkey to respond to step-change increase in the surface temperature.

6.19 Revisit Problem 6.11. Derive the partial differential equation that is required to mathematically specify the transient problem.

6.20 A plane wall experiences a volumetric generation of thermal energy.

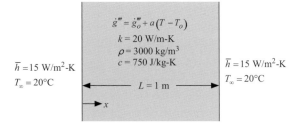

$$\dot{g}''' = \dot{g}'''_o + a(T - T_o)$$
$k = 20$ W/m-K
$\rho = 3000$ kg/m³
$c = 750$ J/kg-K

$\bar{h} = 15$ W/m²-K
$T_\infty = 20°C$

$\bar{h} = 15$ W/m²-K
$T_\infty = 20°C$

$L = 1$ m

x

The volumetric generation is temperature dependent. The rate of generation per unit volume increases linearly with temperature from its nominal value of $\dot{g}'''_o = 2000$ W/m³ at $T_o = 20°C$ according to:

$$\dot{g}''' = \dot{g}'''_o + a(T - T_o),$$

where $a = 1$ W/m³-K. Both edges of the wall are cooled by convection with $T_\infty = 20°C$ and $\bar{h} = 15$ W/m²-K. The thickness of the wall is $L = 1$ m. The wall material has properties $k = 20$ W/m-K, $\rho = 3000$ kg/m³, and $c = 750$ J/kg-K. Assume that the temperature in the wall is a function only of x and that the cross-sectional area is $A_c = 1$ m³.

(a) Without solving the problem, come up with a reasonable estimate (an approximation - not an exact solution) for the surface ($x = 0$) and middle ($x = L/2$) temperatures in the wall at steady state. Use these estimates to sketch qualitatively the steady-state temperature distribution. Clearly show how you are estimating these values.

(b) Derive the ordinary differential equation that governs the steady-state temperature distribution within the wall.

(c) Solve the ODE you derived in (b). Your solution should include two unknown constants, C_1 and C_2.

(d) Develop two equations for the two undetermined constants C_1 and C_2. You do not need to solve these equations.

After a long operating time the wall has achieved the steady-state temperature distribution that you sketched in part (a). Suddenly, the volumetric generation in the wall is switched off, causing the wall to thermally equilibrate with its surroundings.

(e) Calculate two time constants that will help you understand this equilibration process.

(f) Using the time constants that you calculated in (e), sketch the temperature distribution that you expect at a few interesting times using the plot from part (a). Clearly label the temperature distributions and the times – your sketch should clearly show how you expect the transient equilibration process to proceed and some of the characteristics that you expect. You may want to add a few sentences explaining your reasoning.

6.21 A semi-infinite body is exposed at its surface to a radiation heat flux with magnitude $\dot{q}''$. The radiation is **not** absorbed at the surface because the material is partially transparent. The absorption coefficient within the material is a. The absorption of the radiation results in a volumetric generation of thermal energy that depends on position according to:

$$\dot{g}''' = \dot{q}'' a \exp(-a x).$$

The surface is cooled convectively by a fluid at T_∞ with heat transfer coefficient $\bar{h}$. The thermal diffusivity of the material is α and the thermal conductivity is k. Initially the material is in equilibrium with the fluid. Determine the governing partial differential equation, boundary conditions, and initial condition for the problem.

6.22 A constant cross-sectional area fin is initially in equilibrium with its surroundings at T_∞ when, at time $t = 0$, the base temperature is changed to T_b. The cross-sectional area of the fin is A_c and its perimeter is per. The heat transfer coefficient between the fin and the surroundings is $\bar{h}$. The length of the fin is L. Derive the governing differential equation, boundary conditions, and initial conditions for the transient, 1-D problem.

6.23 A plane wall element in a nuclear reactor has thickness L. Both sides of the wall are maintained at a constant temperature T_∞ and the wall material is initially at T_∞. At time zero, a neutron flux strikes one face and causes thermal generation to begin within the wall. The volumetric rate of thermal energy generation decreases exponentially with distance into the material according to $\dot{g}''' \exp{(-\gamma x)}$, where γ is the inverse of a characteristic decay length (with units 1/m). Derive the governing differential equation, boundary conditions, and initial conditions for the transient, 1-D problem.

6.24 Regenerators are indirect heat exchangers used in many applications. During part of the cycle, hot fluid flows over a solid material causing its temperature to rise (storing energy). During another part of the cycle, cold fluid flows over the same material causing its temperature to fall (releasing energy). The solid material is often a packed bed of spherical particles. Let's look at the simple situation where the temperature at the surface of a sphere varies sinusoidally according to:

$$T_{r=R} = \bar{T} + \Delta T \sin{(\omega t)},$$

where R is the radius of the sphere, $\bar{T}$ is the average temperature of the sphere surface, ΔT is the amplitude of the temperature variation, and ω is the frequency of the oscillation. Assume that the material is initially at a uniform temperature that is equal to $\bar{T}$.

(a) Determine the governing partial differential equation, initial condition, and boundary conditions for this problem.

It is typical to assume that the particles are isothermal so that all of the solid material in the regenerator fully participates in the energy storage process. This is true provided that the diffusive time constant associated with transferring energy from the edge of the sphere to the center is much smaller than the period of the cycle.

(b) What dimensionless number should be calculated in order to determine whether the assumption that the material is isothermal is appropriate?

6.25 A large sphere of low-conductivity ceramic material (with density ρ_c, thermal conductivity

k_c, and specific heat capacity c_c) encloses a small, very high-conductivity copper sphere (with density ρ_{Cu} and specific heat capacity c_{Cu}). The ceramic sphere has a radius of 15 cm and the copper sphere has a radius of 5 cm.

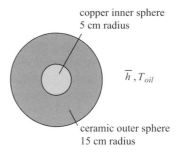

copper inner sphere
5 cm radius

$\bar{h}, T_{oil}$

ceramic outer sphere
15 cm radius

The ceramic and copper are both initially at a uniform temperature (T_{ini}) and then the sphere is dropped into a bath of hot oil at T_{oil}. The heat transfer coefficient between the oil and the outer surface of the ceramic is $\bar{h}$. Neglect contact resistance and radiation.

(a) *Sketch* the temperature distribution from $r = 0$ to the oil temperature at various times, including at least the beginning of the process, sometime during the heating process, and near the end of the heating process.

(b) Derive the differential equation that governs the temperature distribution in the outer ceramic shell (i.e., the governing equation that applies from 5 cm $< r <$ 15 cm). Clearly show the steps in the derivation.

(c) Assume that the copper center sphere can be considered to be at a uniform temperature. What is the boundary condition for the governing equation you derived in (b) that you would apply at $r_1 = 5$ cm?

(d) What is the boundary condition for the governing equation you derived in (b) that you would apply at $r_2 = 15$ cm?

(e) Describe how you would justify the assumption made in (c) that the copper is all at a uniform temperature.

6.26 The surface of a semi-infinite body ($x = 0$) is heated by a radiant heat flux $\dot{q}''_s = 7{,}500$ W/m². The material has conductivity $k = 2.4$ W/m-K and thermal diffusivity $\alpha = 2.2 \times 10^{-4}$ m²/s. The surface is also exposed to air at $T_\infty = 20°$C with heat transfer coefficient $\bar{h} = 15$ W/m²-K. The initial temperature of the material is spatially uniform and equal to $T_{ini} = 20°C$.

Develop the governing differential equation and boundary conditions for this problem.

Semi-Infinite Solutions

6.27 A semi-infinite body has conductivity $k = 1.2$ W/m-K and thermal diffusivity $\alpha = 5 \times 10^{-4}$ m²/s. At time $t = 0$, the surface is exposed to fluid at $T_{\infty} = 90°C$ with heat transfer coefficient $\bar{h} = 35$ W/m²-K. The initial temperature of the material is $T_{ini} = 20°C$.

(a) Develop an approximate model that can provide the temperature of the surface and the rate of heat transfer into the surface as a function of time.

(b) Based on your model, develop an expression that provides a characteristic time related to how long it will take for the surface of the solid to approach T_{∞}.

(c) Compare the results of your model from (a) with the exact solution programmed in EES and accessed using the SemiInf3 function.

6.28 A 7.5 cm thick concrete driveway has a uniform temperature of 38°C when it is sprayed with 12°C water. The convection coefficient is estimated to be 110 W/m²-K.

(a) Plot the time required for the thermal wave to penetrate the concrete as a function of its thickness.

(b) Plot the temperature at 5 cm from the surface as a function of time for a 1 hour period.

(c) Calculate the energy removed from the concrete per square meter after 1 hour of exposure to the water spray.

6.29 One technique that is being proposed for measuring the thermal diffusivity of a material is illustrated schematically in the figure.

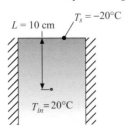

The material is placed in a long, insulated container and allowed to come to thermal equilibrium with its environment $T_{in} = 20°C$. A thermocouple is embedded in the material at a distance $L = 10$ cm below the surface. At time $t = 0$ the temperature of the surface is changed from T_{in} to $T_s = -20°C$ by applying a flow of chilled ethylene glycol to the surface. The time required for the thermocouple to change from T_{in} to $T_{target} = 0°C$ is found to be $t_{target} = 310.2$ s.

(a) What is the measured thermal diffusivity?

There is some error in your measurement from part (a) due to inaccuracies in your thermocouple and your measurement of time and position of the thermocouple. Assume that the following uncertainties characterize your experiment:

• the temperature measurements have an uncertainty of $\delta T_{in} = \delta T_{target} = \delta T_s = 0.2°C$
• the position measurement has an uncertainty of $\delta L = 0.1$ mm
• the time measurement has an uncertainty of $\delta t_{target} = 0.5$ s.

(b) What is the uncertainty in your measured value of thermal diffusivity from part (a)? You can answer this question in a number of ways including using the uncertainty propagation capability in EES.

6.30 The annual average air temperature variation in a northern location is described by $T = 8 + 16 \sin(\omega h)$ where T is the average temperature in °C, h is the hour of the year (0 to 8760) and $\omega = 0.041$ deg/hr. The properties of the soil at this location are $c = 1840$ J/kg-K, $k = 0.52$ W/m-K, and $\rho = 2050$ kg/m³. You would like to know how far below the surface one needs to go before freezing is no longer a concern.

(a) Plot the temperature at the surface as well as 0.5 m and 1 m below the ground as a function of day of the year (0 to 365) for a yearly period.

(b) At what depth do you estimate that freezing ceases to be a concern?

6.31 A thick plate of AISI 304 stainless steel is initially at a uniform temperature of 25°C when it receives a pulse of 1×10^6 J/m² from a laser.

(a) Using the exact solution, plot the temperature at 5, 10, 15, and 30 s as a function of distance from the surface.

(b) Calculate the diffusive time constant related to the energy reaching a position that is 2 cm from the surface. Is the value of the time constant consistent with the information in the plots?

6.32 A long carbon steel rod having a 2.5 cm diameter is insulated such that only its ends are exposed. The rod is at a uniform temperature of 25°C when it is butted against a similar rod made of AISI 304 that is initially at 136°C. Assume that there is no contact resistance.
(a) Determine the interface temperature.
(b) Determine the diffusive time constants for both rods at 1 cm from the interface.
(c) Plot the temperature of the rods as a function of position 1 s after the rods are brought together.

Plane Wall, Cylinder, and Sphere Solutions

6.33 A new concrete driveway has cured and it is recommended that it be cooled by spraying water on its surface. The concrete is 10 cm thick and initially it is at a uniform temperature of 38°C. Water at 20°C is then sprayed on the surface. The convection coefficient between the water and the concrete is estimated to be 80 W/m²-K. Assume the bottom edge of the concrete is insulated.
(a) Determine the time required to cool the concrete to 30°C at a position 5 cm below the surface. Neglect any energy generation that may result from additional curing
(b) Prepare plots of the temperature versus time for a 5 hour period at the following depths below the surface: 1 cm, 5 cm, and 10 cm.

6.34 You have been asked to provide an estimate of how long it will take snow to melt off of a car windshield after the defroster has been turned on. The windshield is 0.25 inch thick with properties similar to Pyrex glass and initially it is at a uniform temperature of 14°F. The defroster blows air at 95°F past the inner surface of the windshield with an average convection coefficient of 25 W/m²-K. State any assumptions you employ.
(a) Solve the problem using the appropriate graph or EES function that provides the exact solution.

(b) Compare the result obtained in part (a) with the approximate solution using the first term of the series solution for a plane wall.

6.35 The last step in a process used to make steel is to pour the molten steel into a crucible and let it solidify. Assume that the solidification process takes 30 minutes and during this time the molten steel maintains the internal surface of the crucible at a constant temperature of 1500°C. The crucible is made of a material with conductivity 4 W/m-K, density 2500 kg/m³, and specific heat capacity 800 J/kg-K. The crucible is initially at 100°C. The external surface of the crucible is insulated (assume it to be adiabatic). The temperature at the external surface cannot exceed 700°C during the solidification process. How thick must the crucible be?

6.36 An energy storage system that is proposed for a wind energy system converts excess electrical energy to thermal energy. One cell of the energy storage unit consists of a plane wall made of a masonry material with thickness 16 cm thick that is 1 m in height and 2 m in length. The material has a density of $\rho = 1920$ kg/m³, a specific heat of $c = 835$ J/kg-K and a thermal conductivity of $k = 0.72$ W/m-K. The energy storage unit is charged using electrical heaters that are embedded on both sides of the masonry material (i.e., at $x = 0$ and $x = 16$ cm). The heaters provide an energy flux of 7.5 kW/m² when energized. Air is passed over these same surfaces to reclaim the thermal energy after it is charged.
(a) The masonry material is initially at 20°C. The maximum safe temperature that the masonry can be heated to is 420°C. Assuming that the material can be considered a semi-infinite body during the charging process, estimate how long the charging process can occur before this temperature limit occurs?
(b) Estimate the penetration depth of the thermal wave for the time determined in part (a). Based on your answer, can the masonry material actually be considered a semi-infinite solid during charging?
(c) Plot the temperature distribution of the masonry material as a function of position

measured from the surface at the time determined in part (a).

(d) Determine the amount of thermal energy stored in the unit at the time found in part (a).

(e) An engineer has proposed a different control strategy that will allow more energy to be stored in the unit without exceeding the 420°C temperature limit. The proposal is to control the heater so that the surface temperature remains at a constant 420°C while it is being energized. Determine the time required for the center to reach 400°C, at which point the unit is considered to be fully charged.

6.37 A 1.5 inch thick slice of steak is cooked by convectively exposing both sides to air at 365°F air with a convection coefficient of 28 W/m^2-K. The intent is to continue the heating until the center temperature reaches 120°F. Assume that the steak has the same thermal properties as liquid water at 75°F.

(a) Estimate the cooking time required using a diffusive time constant.

(b) Estimate the cooking time using an analytical solution and compare the result with that obtained in part (a).

6.38 Ball bearings made from AISI 304 stainless steel are hardened by heating them to a uniform temperature of T_{ini} = 850°C and then quenching in an oil bath that is maintained at T_∞ = 50°C. The diameter of the ball bearings is D = 20 mm and the heat transfer coefficient between the oil and bearings is estimated to be $\bar{h}$ = 900 W/m^2-K.

(a) How much time must a ball bearing spend in the oil bath to ensure that the center is cooled to a temperature below T_{center} = 110°C?

(b) One type of ball bearing receives a thin coating of a dielectric material before it is hardened. This coating adds an additional resistance of 33.1 K/W at the surface of the ball bearing with negligible effect on its mass or heat capacity. Does this thin coating affect the time required for quenching? If so, estimate the time to cool the center of the coated bearing to 110°C from an initial temperature of 850°C.

6.39 A long solid oak log that is 15 cm in diameter is placed on a fire. The combustion gases surrounding the log are at 210°C and the convection coefficient is 38 W/m^2-K. The log is initially at 25°C. (Neglect the radiant flux from the flame, mass transfer within the log, and heat transfer from the ends of the log.)

(a) Determine the Biot number for this situation.

(b) Estimate the time required for the center to reach 100°C.

(c) Determine the surface temperature at this time.

(d) Determine the Fourier number at this time.

6.40 You are trying to model the thermal processing of a composite structure. A material is built by alternating thin (0.5 mm) copper laminations with thicker layers of plastic (1.0 mm), as shown.

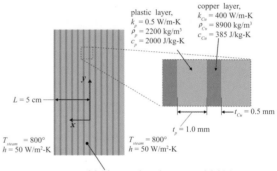

slab of composite to be processed, initially at 20°C

The copper and plastic properties are shown in the figure. In order to perform a simplified thermal analysis, it is appropriate to approximate the slab as having an effective thermal conductivity (k), effective density (ρ), and effective heat capacity (c).

(a) Determine an effective thermal conductivity for the composite material that could be used to characterize heat transfer in the x-direction (see the figure).

(b) Determine an effective thermal conductivity for the composite material that could be used to characterize heat transfer in the y-direction.

(c) Determine an effective density and heat capacity for the composite material.

The thermal process involves exposing both sides of a slab of the composite to steam at

800°C. The heat transfer coefficient between the steam and the slab surface is 50 W/m²-K and the slab is initially at 20°C.

(d) Determine how long it will take for the center of the slab to reach 400°C. What will the surface temperature be at this time?

6.41 During the early part of the twentieth century, many houses were heated with a central pot-belly stove. Bedrooms far from the stove were cold so people would heat a large rock and take the rock wrapped in a towel back to bed with them in order to stay warm. Assume that the rock is a perfect sphere with ρ = 2630 kg/m³, k = 2.79 W/m-K, and c = 775 J/kg-K. The towel and convection provide a combined thermal resistance of 0.15 K-m²/W between the surface of the rock and the surrounding air at 5°C. The rock is heated to a uniform temperature of 55°C.

(a) Determine the diameter of the rock that would be needed to provide a minimum heat transfer rate of 10 W for a 6 hour period.

(b) Determine the temperatures at the center and surface of the rock at this time.

(c) Determine the mass of the rock.

(d) Determine the total energy transferred from the rock during the 6 hour period.

(e) Determine the Biot and Fourier numbers for this transient heat transfer problem.

6.42 A regenerative heat exchanger is to be used in a super-critical carbon dioxide Brayton power cycle. The regenerator is packed with 25 mm diameter AISI304 stainless steel spheres. When the valve to the storage unit is opened, the spheres, which are initially at a uniform temperature of 320°C, are exposed to carbon dioxide gas at 560°C with a heat transfer coefficient of 180 W/m²-K for a period of 30 s. Neglect the interaction of the spheres where they touch. State all other assumptions that you employ.

(a) Determine the center temperature of the spheres at the end of the process.

(b) Determine the amount of energy stored in one sphere during the process.

(c) The size of the storage unit can be reduced if the spheres can store more energy. How much time is required for the center temperature to reach 550°C?

6.43 Ice cream containers are removed from a warehouse and loaded into a refrigerated truck. During this loading process, the ice cream may sit on the dock for a substantial amount of time. The dock temperature is substantially higher than the warehouse temperature, which can cause two problems. First, the temperature of the ice cream near the surface can become elevated, resulting in a loss of food quality. Second, the energy absorbed by the ice cream on the dock must subsequently be removed by the equipment on the refrigerated truck, causing a substantial load on this relatively under-sized and inefficient equipment. The ice cream is placed in cylindrical cardboard containers. Assume that the containers are very long and therefore, the temperature distribution of the ice cream is one dimensional, as shown in the figure. The inner radius of the cardboard ice cream containers is R_o = 10 cm and the thickness of the wall is th_{cb} = 2.0 mm. The conductivity of cardboard is k_{cb} = 0.08 W/m-K. The ice cream comes out of the warehouse at T_{ini} = 0°F and is exposed to the dock air at T_{dock} = 45°F with heat transfer coefficient, $\bar{h}$ = 20 W/m²-K. The ice cream has properties k_{ic} = 0.2 W/m-K, ρ_{ic} = 720 kg/m³, and c_{ic} = 3200 J/kg-K. (Assume that the ice cream does not melt.)

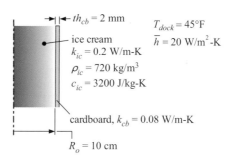

(a) Determine an effective heat transfer coefficient, $\bar{h}_{eff}$, that can be used in conjunction with the analytical solution for a cylinder subjected to a step change in fluid temperature but includes the conduction resistance associated with the cardboard as well as the convection to the air.

(b) If the ice cream remains on the dock for t_{load} = 5 min, what will the temperature of the surface of the ice cream be when it is loaded?

(c) How much energy must be removed from the ice cream (per unit length of container) after it is loaded in order to bring it back to a uniform temperature of $T_{ini} = 0°F$?

(d) What is the maximum amount of time that the ice cream can sit on the dock before the ice cream at the outer surface begins to melt?

6.44 Small spheres are injected into flows in order to act as "tracers" for particle-image velocimetry. The idea is to capture two images of the flow that are closely separated in time; by evaluating the distance and direction that each particle has traveled between the images it is possible to back out the velocity of the particle. Your company wants to introduce tracer particles that have optical properties that are strongly affected by temperature. The intensity of the particle images will therefore be related to temperature; in this way you can simultaneously measure the velocity and temperature distribution of a flow. The tracer particles are spherical with radius $r_{out} = 1.0$ mm. The material has properties $k = 7.0$ W/m-K, $\rho = 2300$ kg/m^3, and $c = 750$ J/kg-K. They are to be used in a liquid flow and the heat transfer coefficient between the fluid and the sphere is approximately $\bar{h} = 5000$ W/m^2-K.

(a) What is the Biot number associated with tracers? Can you treat them as a lumped capacitance?

(b) Estimate a diffusive and lumped capacitance time constant for the sphere.

(c) Assume that your answer from (a) shows that a lumped capacitance solution is NOT appropriate. Use the exact solution for a sphere exposed to a step change in the convective surface condition in order to prepare a plot of the surface temperature and center temperature of the sphere as a function of time for the case where the sphere is initially at a temperature of $T_{ini} = 20°C$ when at time $t = 0$ the fluid temperature is changed to $T_\infty = 40°C$.

(d) Explain how your plot from (c) is consistent with the time constants that you calculated in (b).

(e) Overlay on your plot from (c) the average temperature of the sphere material as a function of time. Use the Integral command in EES to compute the average temperature.

What is the time constant associated with the temperature measurement?

6.45 Ice cream is packaged in long cylindrical containers and stored in a refrigerated warehouse that is at $T_i = -20°C$. The ice cream is removed from the warehouse for shipping and may sit for as long as $\tau = 30$ min on the loading dock before it is loaded into a refrigerated truck. You have been asked to determine whether the ice cream at the outer surface of the container will begin to melt during this time. The properties of ice cream are $\rho = 920$ kg/m^3, $k = 1.88$ W/m-K, and $c = 2040$ J/kg-K. The inner radius of the ice cream container is $R_o = 12$ cm and the thickness of the cardboard container is $th_{cb} = 1.0$ mm. The conductivity of cardboard is $k_{cb} = 1.88$ W/m-K. The air on the loading dock is $T_{amb} = 20°C$ and $h = 15$ W/m^2-K.

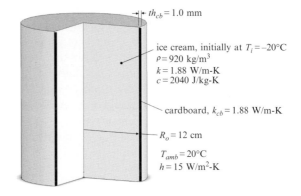

(a) What is the maximum temperature of the ice cream after it spends 30 min on the loading dock?

(b) Prepare a plot that shows the maximum time that the ice cream can spend on the loading dock as a function of the temperature on the loading dock. Assume that ice cream melts at approximately $T_{melt} = 0°C$.

1-D Transient Numerical Solutions

6.46 Revisit Problem 6.2. Use the three-node model shown ($\Delta x = L/2 = 0.5$ m) with time step $\Delta t = 0.1$ s and an Euler technique to predict the temperature of node 1 at the end of the first time step (at $t = \Delta t$). Provide a numerical answer for the temperature.

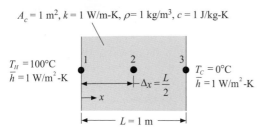

$A_c = 1 \text{ m}^2, k = 1 \text{ W/m-K}, \rho = 1 \text{ kg/m}^3, c = 1 \text{ J/kg-K}$

$T_H = 100°C$
$\bar{h} = 1 \text{ W/m}^2\text{-K}$

1   2   3

$\Delta x = \dfrac{L}{2}$

$T_C = 0°C$
$\bar{h} = 1 \text{ W/m}^2\text{-K}$

x

$L = 1 \text{ m}$

at time $t = 0$, the material is at $T_{ini} = 0°C$

6.47 Two fins protrude from each side of a very thin wall as shown in the figure. One fin is immersed in hot fluid at $T_H = 100°C$ while the other is immersed in cold fluid at $T_C = 20°C$. The diameter of both fins is $D = 0.005 \text{ m}$ and their length is $L = 0.05 \text{ m}$. The tips of the fins are adiabatic. The fin material has conductivity $k = 200 \text{ W/m-K}$, density $\rho = 1000 \text{ kg/m}^3$, and specific heat capacity $c = 430 \text{ J/kg-K}$. Initially (at time $t = 0$), the fins are at steady state and the temperature distribution along the fins is shown.

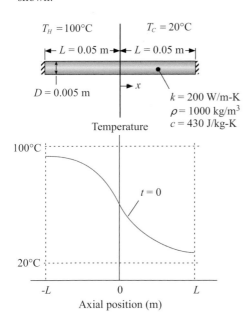

$T_H = 100°C$ $T_C = 20°C$

$\leftarrow L = 0.05 \text{ m} \rightarrow \leftarrow L = 0.05 \text{ m} \rightarrow$

$D = 0.005 \text{ m}$

x

$k = 200 \text{ W/m-K}$
$\rho = 1000 \text{ kg/m}^3$
$c = 430 \text{ J/kg-K}$

Temperature

$100°C$

$t = 0$

$20°C$

$-L$ 0 L

Axial position (m)

At $t = 0$, the fluid is removed from both sides of the wall and replaced with gas at $T_g = 20°C$. The heat transfer coefficient between the gas and the fin is $\bar{h} = 20 \text{ W/m}^2\text{-K}$. This problem is concerned with the equilibration process that occurs after $t = 0$ (the thermal equilibration of the fins with the surrounding gas at T_g). You may assume that the temperature distribution during this process is only a function of axial position (x) and time (t); that is, the Biot number, $\bar{h}D/k$, is much less than 1.0.

(a) Sketch the temperature distribution that you expect to see in the fins (i.e., the temperature as a function of position from $x = -0.05 \text{ m}$ to $x = +0.05 \text{ m}$) for $t = 0.2 \text{ s}$, $t = 2.0 \text{ s}$, $t = 20 \text{ s}$, $t = 200 \text{ s}$, and $t \to \infty$.

(b) You've decided to model the equilibration of the fins using an analytical model. Assume that the equilibration process can be modeled as a 1-D, transient process. Derive the governing partial differential equation that describes the process; your equation should include only symbols defined in the problem statement (not their numerical values) as well as temperature (T), position (x), and time (t).

(c) You've decided to model the equilibration process using the 5-node numerical model shown. Derive the state equation (i.e., the equation that provides the time rate of change of the temperature) for node 1. Your equation should include only symbols defined in the problem statement as well as the temperatures of the nodes (T_1, T_2, ...) and time (t).

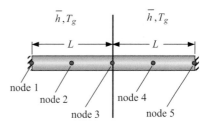

$\bar{h}, T_g$ $\bar{h}, T_g$

L L

node 1
       node 2
              node 3
                    node 4
                          node 5

(d) Use your equation from part (c) to take a single Euler step for node 1. That is, what equation would you use to predict the temperature of node 1 at time $j+1$ ($T_{1,j+1}$) given all of the nodal temperatures at time j? Assume that the time step duration for the step is Δt. Your equation should include only symbols defined in the problem statement as well as the temperatures of the nodes at time j ($T_{1,j}$, $T_{2,j}$, ...).

(e) Determine the critical time step for the Euler step in part (d). That is, what is the largest time step that you could take before the solution became unstable?

6.48 A metal plate of thickness L has a large width and height. One surface is insulated and the other is exposed to a fluid. Initially the plate and fluid are at the same temperature, but suddenly, the fluid temperature is changed. Derive the finite difference equations that will provide the temperature versus time history for:

(a) the node on the insulated surface,

(b) the node on the surface exposed to the fluid, and

(c) an interior node.

6.49 An orange has a diameter of 7 cm. Initially, the orange is at a uniform temperature. It is then placed in a refrigerator. The unsteady temperature distribution of the orange is to be determined numerically with eight nodes spaced 0.5 cm apart, as shown in the figure. The thermal properties of the peel, which is 0.5 cm in thickness, are markedly different from those of the orange itself. Set up the nodal equation for node 6, which is on the boundary between the orange and the peel.

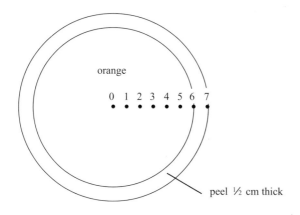

6.50 The figure illustrates an electrical fuse. The fuse material is enclosed in a glass tube that is evacuated (i.e., a vacuum exists in the space surrounding the fuse). The fuse radiates to surroundings at T_{sur}. The emissivity of the fuse surface is ε. The fuse is designed so that when current passes through it, a volumetric generation of thermal energy occurs, $\dot{g}'''$, which causes the temperature of the fuse material to increase. If the current is sufficiently high, then the material will reach its melting temperature, T_{melt}, and the fuse will "blow".

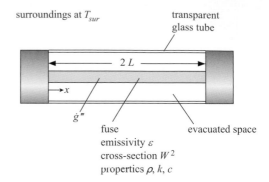

The fuse material is initially in equilibrium with the surroundings at T_{sur} when at time $t = 0$ it is exposed to an electrical current. During the ensuing transient process, the ends of the fuse (at $x = 0$ and $x = 2L$) may be assumed to remain at T_{sur}. The half-length of the fuse is L. The cross-section of the fuse is square with dimension W. The fuse material properties are k, ρ, and c.

(a) How would you decide whether you can model the fuse as an extended surface? Be specific; write the calculations that you would do in terms of symbols provided in the problem statement.

For the remainder of this problem, assume that the fuse can be modeled as an extended surface. You have decided to develop a three-node, half-symmetry numerical model of the fuse. Your three nodes should be placed at $x = 0$, $x = L/2$, and $x = L$ (i.e., the midpoint of the fuse).

(b) Derive the three state equations for your three nodes.

(c) Write the equations that must be solved in order to take a single Euler step from $t = 0$ to $t = \Delta t$.

6.51 The figure illustrates a flat plate solar collector.

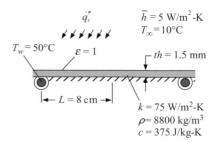

The collector consists of a flat plate that is $th = 1.5$ mm thick with conductivity $k = 75$ W/m-K, density $\rho = 8800$ kg/m^3, and specific heat capacity $c = 375$ J/kg-K. The plate is insulated on its back side and under normal conditions experiences a solar flux of $\dot{q}''_{s,max} = 900$ W/m^2 that is all absorbed. The surface is exposed to convection and radiation to the surroundings. The emissivity of the surface is $\varepsilon = 1$. The heat transfer coefficient is $\bar{h} = 5$ W/m^2-K and the surrounding temperature is $T_\infty = 10°$C. The temperature of the water in the tubes is $T_w = 50°$C. The center-to-center distance between adjacent tubes is $2L$ where $L = 8$ cm. This problem deals with the transient behavior of the collector. Assume that the plate is initially at the water temperature and that the heat flux from the sun varies due to clouds. Specifically, assume that the solar flux is given by

$$\dot{q}''_s = \begin{cases} \dot{q}''_{s,max} & 0 < t < 300 \text{ s} \\ 0 & t > 300 \text{ s} \end{cases}$$

so that the initial 300 s are related to the startup of the collector and the subsequent time is related to the shut down of the collector after the flux is removed.

(a) Develop a numerical model using the fully implicit technique implemented in EES that can predict the temperature distribution in the collector plate during the transient process associated with the sun going behind a cloud. Plot the temperature at $t = 0$ s, 300 s, and 600 s.

(b) Use your numerical model to compute the rate of heat transfer to the water at each timestep (per tube per unit length of collector). Plot the rate of heat transfer to the water as a function of time.

6.52 A plane wall experiences a volumetric generation of thermal energy.

$$\dot{g}''' = \dot{g}'''_o + a(T - T_o)$$
$k = 20$ W/m-K
$\rho = 3000$ kg/m^3
$c = 750$ J/kg-K

$\bar{h} = 15$ W/m^2-K
$T_\infty = 20°$C

$\bar{h} = 15$ W/m^2-K
$T_\infty = 20°$C

$L = 1$ m

x

The volumetric rate of generation is temperature dependent. The rate of generation per unit volume increases linearly with temperature from its nominal value of $\dot{g}'''_o = 2000$ W/m^3 at $T_o = 20°$C according to:

$$\dot{g}''' = \dot{g}'''_o + a(T - T_o),$$

where $a = 1$ W/m^3-K. Both edges of the wall are cooled by convection with $T_\infty = 20°$C and $\bar{h} = 15$ W/m^2-K. The thickness of the wall is $L = 1$ m. The wall material has $k = 20$ W/m-K, $\rho = 3000$ kg/m^3, and $c = 750$ J/kg-K. Assume that the temperature in the wall is a function only of x and that the cross-sectional area is $A_c = 1$ m^3.

(a) Derive the ordinary differential equation that governs the *steady-state* temperature distribution within the wall.

(b) Solve the ODE you derived in (b) and prepare a plot of the steady-state temperature distribution. Note that the homogeneous form of the ODE should have the form:

$$\frac{d^2 T_h}{dx^2} + m^2 T_h = 0,$$

which is solved by:

$$T_h = C_1 \sin(m\,x) + C_2 \cos(m\,x).$$

After a long operating time the wall has achieved the steady-state temperature distribution that you solved for in part (b). Suddenly, the volumetric generation in the wall is switched off, causing the wall to thermally equilibrate with its surroundings.

(c) Calculate two time constants that will help you understand this equilibration process.

(d) Develop a numerical model of the equilibration process (i.e., the process that occurs after the volumetric generation is shut off). Use the Crank–Nicolson (CN) method in EES. Note that the CN method is a slight modification of the fully implicit method. Rather than using only the time rate of change evaluated at the end of the time step, the CN method uses the average of the time rates of change evaluated at the beginning and end of the time step.

(e) Plot the temperature as a function of position at several values of time. Explain how your solution agrees with the time constants that you calculated in part (c).

6.53 A 3 inch thick brick ($k = 0.72$ W/m-K, $\rho = 1920$ kg/m^3, $c = 835$ J/kg-K) wall is initially at a uniform temperature of 25°C. Then, the front side of the wall is subjected to a heat flux of

600 W/m^2 for one hour. Thereafter, the heat flux is zero. Both sides experience convection to 25°C air with a heat transfer coefficient of 12 W/m^2-K. Develop a numerical model of the wall using the Integral function in EES. Use your program to plot the temperatures and the heat transfer rates associated with both the front and back sides of the wall as a function of time for a 4 hour period. Explain the behavior that you see.

6.54 A 3 inch diameter orange, originally at 80°F, is placed in a refrigerator where the air temperature is 40°F. The average heat transfer coefficient is 5 Btu/hr-ft^2-R. The properties of the orange are $k = 0.339$ Btu/hr-ft-F, $\rho = 62.4$ lb$_m$/ft^3, $c = 1.004$ Btu/lb$_m$-R. Using a numerical method based on the Integral function in EES, calculate and plot the center temperature as a function of time for a 4 hour period. How long does it take the orange to reach a center temperature of 50°F?

6.55 A 4 inch thick AISI 304 stainless steel plate that is insulated on one side is exposed to 180°C oil through a heat transfer coefficient of 82 W/m^2-K on the other side. Initially the plate is at a uniform temperature of 60°C. Using a numerical method based on the Integral function in EES, prepare plots of the temperature of the insulated surface, the convective surface, and a point at the half-thickness of the plate for a period of 1000 s.

6.56 A circular glass filter is used to protect a camera lens. The filter is made of Pyrex glass with a diameter of 72 mm and a thickness of 2 mm. The camera is used outdoors where the glass is exposed to an outdoor temperature of 0°C for an extended period of time. When the camera is brought back indoors where the temperature is 25°C, the filter becomes fogged with condensation. The photographer does some tests and notes that the fog disappears when the center of the filter glass reaches 10°C. She has proposed the novel idea of providing an electrical heater around the circumference of the filter to remove the condensate more quickly. Tests show that the convection coefficient (on both sides of the filter glass) is 15 W/m^2-K. Show justification that the temperature gradient across the thickness of the filter can be neglected. Neglect the thermal

effects of vaporizing the condensate. State any other assumptions you employ. Using a numerical method based on the Integral function in EES determine the following.

(a) The time for the center of the glass filter to reach 10°C if the edge is insulated.

(b) The time for the center of the glass filter to reach 10°C if a heater provides 1 W of thermal energy uniformly distributed to the circumference of the filter.

(c) Repeat part (b) assuming the heater provides 5 W.

6.57 A concrete wall is used to store solar energy so that it can be later used to provide thermal energy to a building. The wall is 2.5 m in height and 3 m in width. It is 25 cm thick and it is initially at a uniform temperature of 18°C. One side of the wall (at $x = 0$) is exposed to solar radiation, which is absorbed at rate

$$\dot{q}''_{solar} = 725 \cos\left[(time - 12 \ [hr])15\right],$$

where time is the clock time in hours, $\dot{q}''_{solar}$ is the solar absorption per area in W/m^2, and the argument of the cosine is in degree. Both sides of the wall (at $x = 0$ and $x = 25$ cm) convect to air at 18°C with a convection coefficient of 8 W/m^2-K. Using this information and a numerical model based on the Integral function in EES, prepare the following plots for a time period between 6 a.m. and 6 p.m.

(a) The temperature of the wall surface exposed to the Sun.

(b) The temperature of the wall surface exposed to the building (at $x = 25$ cm).

(c) The rate of energy transfer from the surface of the wall exposed to the building (at $x = 25$ cm).

6.58 Reconsider Problem 6.6. Prepare a numerical solution for the problem using an implicit technique. Implement your solution in MATLAB and prepare a plot of the temperature as a function of position at several values of times. Use the Crank–Nicolson (CN) method. Note that the CN method is a slight modification of the fully implicit method. Rather than using only the time rate of change evaluated at the end of the time step, the CN method uses the average of the time rates of change evaluated at the beginning and end of the time step.

6.59 You are designing a heating element for a washing machine. The heating element can be modeled as a plane wall with a 5.0 cm half-width (a) and is made of an alloy with properties $k = 4$, $\rho = 8930$ kg/m^3, and $c = 385$ J/kg-K. The heating element is energized by a current that causes a volumetric heating. The ohmic dissipation is transferred to water at 20°C that is flowing to the washing machine. The heat transfer coefficient between the surface of the heating element and the water is 200 W/m^2-K.

    (a) The maximum allowable temperature of the heating element during normal operation is 80°C. Determine the maximum allowable level of volumetric heating (W/m^3).

You need to design the relays that shut off the power to the heating element in the event of a dry-out event – an interruption of the water to the heating element. If the heating element is run in air then the heat transfer coefficient goes down drastically, from 200 W/m^2-K to 10 W/m^2-K.

    (b) Prepare an EES program that can predict the temperature in the heating element as a function of position and time after a dry-out event. The simulation should start from the temperature distribution associated with steady operation in water ($h = 200$ W/m^2-K) and, at time = 0, the heat transfer should switch to the lower value associated with air ($h = 10$ W/m^2-K).

    (c) Prepare a plot of temperature as a function of position for various times. Your plot should show at least time = 0 as well as the time where the temperature at the center reaches 100°C.

6.60 Small particles can be ablated (thermally destroyed) by applying a heat flux from a laser. This process allows the composition of the particles to be analyzed via ion mobility spectography; this is useful for identifying harmful substances at airports and other secure areas. Because ion mobility spectography occurs at near-ambient pressures it has a lower sensitivity than techniques that are carried out in a vacuum environment, such as traditional mass spectography. However, because no vacuum equipment is required the tool can be portable, low cost, and robust; ideal characteristics for a deployable technology. The low sensitivity of the technique makes it essential that the signal strength be maximized. Practically, this means

that near-complete ablation of all particles occurs in order to produce the maximum number of ions.

We will develop an approximate thermal analysis in which the surface of a spherical particle is exposed to a single laser pulse. Ablation (i.e., melting) is ignored here; rather we are going to understand how a single pulse causes the particle to heat up. The total energy per unit area delivered by the laser (*fluence*) characterizes the laser power. Assume that we are using a very-high-power laser with a *fluence* of 10,000 mJ/cm^2 (100 J/m^2). The laser pulse is assumed to be Gaussian in time with a fixed duration (τ) equal to 100 ns (100×10^{-9} s). The laser flux incident on the particle ($\dot{q}''$) is therefore given by the expression:

$$\dot{q}''(t) = fluence \frac{\exp\left[-\dfrac{(t - peaktime)^2}{2\tau^2}\right]}{\tau\sqrt{2\pi}},$$

where *peaktime* is some arbitrary time relative to the beginning of the simulation process (a good number to use is 1000e-9 s). The spherical particle (of radius $R = 5$ μm) is cooled by its communication with ambient air ($\bar{h} = 2 \times 10^5$ W/m^2-K, $T_a = 20$°C). The particle has conductivity, $k = 2.0$ W/m-K, density, $\rho = 800$ kg/m^3, and specific heat capacity, $c = 500$ J/kg-K. Initially, the particle is in thermal equilibrium with the air ($T_{ini} = 20$°C).

Develop a numerical model using the fully implicit method in MATLAB that is capable of predicting the temperature distribution in the sphere as it is exposed to the laser pulse. Use your model to generate a plot which shows the temperature distribution in the sphere as a function of radial position at various times including 0 ns, 250 ns, 500 ns, 750 ns, 1000 ns, 1250 ns, 1500 ns, 1750 ns, and 2000 ns (where 0 ns is 1000 ns prior to the peak of the laser pulse).

6.61 A wall is exposed to a heat flux for a long time, as shown in the figure. The left side of the wall is exposed to liquid at $T_f = 20$°C with a very high heat transfer coefficient; therefore, the left side of the wall ($T_{x=0}$) always has the temperature T_f. The right side of the wall is exposed to the heat flux and also convects to gas at $T_f = 20$°C but with a heat transfer coefficient of $\bar{h} = 5000$ W/m^2-K. The wall is $L = 0.5$ m thick and composed of a material with $k = 1.0$ W/m-K, $\rho = 4000$ kg/m^3, and $c = 700$ J/kg-K. The wall is initially at steady state with the heat flux when, at time $t = 0$, the heat flux is suddenly shut off. The wall subsequently equilibrates with the liquid and gas, eventually it reaches a uniform temperature equal to T_f.

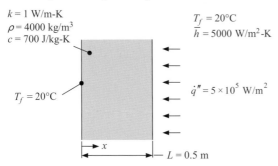

$k = 1$ W/m-K
$\rho = 4000$ kg/m^3
$c = 700$ J/kg-K

$T_f = 20°C$
$\bar{h} = 5000$ W/m^2-K

$T_f = 20°C$

$\dot{q}'' = 5 \times 10^5$ W/m^2

$L = 0.5$ m

Prepare a numerical solution for the equilibration process using the Crank–Nicolson (CN) technique. Note that the CN method is a slight modification of the fully implicit method. Rather than using only the time rate of change evaluated at the end of the time step, the CN method uses the average of the time rates of change evaluated at the beginning and end of the time step. Implement your solution in MATLAB and prepare a plot of the temperature as a function of position at $t = 10{,}000$ s.

6.62 Fusion reactors are fueled by frozen deuterium pellets that are shot into the plasma. The deuterium pellets are formed by solidifying the deuterium in the barrel of a cylindrical apparatus using a flow of liquid helium on its outer surface.

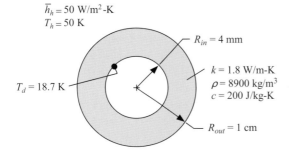

$\bar{h}_h = 50$ W/m^2-K
$T_h = 50$ K

$R_{in} = 4$ mm

$k = 1.8$ W/m-K
$\rho = 8900$ kg/m^3
$c = 200$ J/kg-K

$T_d = 18.7$ K

$R_{out} = 1$ cm

The inner radius of the device is exposed to freezing deuterium; you may assume that this surface is always maintained at the triple point of deuterium, $T_d = 18.7$ K. During the cooling mode, the external surface of the apparatus is cooled by a flow of liquid helium with $\bar{h}_c = 285$ W/m^2-K and $T_c = 4.2$ K. Once the deuterium is

completely frozen, it is injected using a high-pressure gas flow. However, in order to inject the pellet it must be released from the inner surface of the apparatus. One strategy for accomplishing this release is to quickly change the flow of fluid at the outer surface of the barrel from a coolant to a hotter fluid (e.g., neon or nitrogen). After some time, the deuterium at the inner surface of the apparatus is subjected to some heating (rather than cooling) causing it to release its hold and allowing it to be injected.

For this problem, assume that the apparatus has been operating in cooling mode for a long time so that the initial temperature distribution in the barrel is consistent with steady-state operation. Then, at time $t = 0$, the liquid helium is replaced with a heating fluid with $T_h = 50$ K and $\bar{h}_h = 50$ W/m^2-K. The inner and outer radii of the apparatus are $R_{in} = 4$ mm and $R_{out} = 1$ cm. The barrel is composed of a single material with $k = 1.8$ W/m-K, $\rho = 8900$ kg/m^3, and $c = 200$ J/kg-K.

(a) Develop a numerical simulation of the barrel material during the heating process. Use Heun's method and implement your solution in MATLAB.

(b) Plot the temperature in the barrel as a function of time for various radial locations.

(c) Plot the rate of heat transfer to the deuterium per length of bore as a function of time.

(d) Calculate the two time constants that govern this transient process. Point out how both time constants are evident in your plot from (c).

(e) Calculate the critical time step for your solution and demonstrate that it goes unstable if the time step is raised above the critical time step.

6.63 A pin fin is used as part of a thermal management system for a power electronics system.

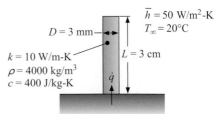

$D = 3$ mm

$\bar{h} = 50$ W/m^2-K
$T_\infty = 20°C$

$k = 10$ W/m-K
$\rho = 4000$ kg/m^3
$c = 400$ J/kg-K

$L = 3$ cm

$\dot{q}$

The diameter of the fin is $D = 3$ mm and the length is $L = 3$ cm. The fin material has conductivity $k = 10$ W/m-K, $\rho = 4000$ kg/m^3, and $c = 400$ J/kg-K. The surface of the fin is exposed to air at $T_\infty = 20°C$

with heat transfer coefficient $\bar{h} = 50$ W/m²-K. The tip of the fin can be assumed to be adiabatic. The power electronics system does not operate at steady state; rather, the load applied at the base of the fin cycles between a high and a low value with some frequency, ω (in rad/s). The average heat transfer rate is $\bar{q} = 0.5$ W and the amplitude of the fluctuation is $\Delta\dot{q} = 0.1$ W. The frequency of oscillation varies. The fin is initially in equilibrium with T_∞.

(a) Develop a 1-D transient model that can be used to analyze the startup and operating behavior of the pin fin. Use the ODE solver in MATLAB.

(b) Plot the temperature as a function of time at various values of axial position for the start up assuming a constant heat load ($\omega = 0$).

(c) Calculate a diffusive time constant and a lumped capacitance time constant for the equilibration process. Is the plot from (b) consistent with these values?

(d) Adjust the diameter of the fin so that the lumped time constant is much greater than the diffusive time constant. Plot the temperature as a function of time at various values of axial position for the start up assuming a constant heat load ($\omega = 0$). Explain your result.

(e) Return the diameter of the fin to $D = 3$ mm and set the oscillation frequency to $\omega = 1$ rad/s. Prepare a contour plot showing the temperature of the fin as a function of position and time. You should see that the oscillation of the heat load causes a disturbance that penetrates only part-way along the axis of the fin. Explain this result.

(f) Is the maximum temperature experienced by the fin under oscillating conditions at cyclic steady state (i.e., after the startup transient has decayed) greater than or less than the maximum temperature experienced under steady-state conditions (i.e., with $\omega = 0$)?

(g) Plot the ratio of the maximum temperature under oscillating conditions to the maximum temperature under steady-state conditions as a function of frequency.

(h) Define a meaningful dimensionless frequency and plot the maximum temperature under oscillating conditions to the maximum temperature under steady-state conditions as a function of this dimensionless frequency. Explain the shape of your plot.

Finite Element Method using FEHT

6.64 A composite material consists of three alternating layers of steel ($\rho = 7854$ kg/m³, $c = 434$ J/kg-K, $k = 60.50$ W/m-K) and wood ($\rho = 540$ kg/m³, $c = 2400$ J/kg-K, $k = 0.166$ W/m-K).

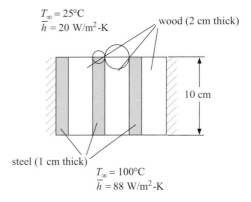

The left and right faces are insulated. The material is initially at 25°C when the bottom surface is exposed to 100°C with a convection coefficient of 88 W/m²-K. The top surface convects to air at 25°C with a convection coefficient of 20 W/m²-K. Use FEHT to solve this problem.

(a) Prepare a plot of the temperatures of the steel and the wood along the top surface (at the positions shown with a circle in the figure) as a function of time from the start of the process for 1 hour.

(b) Show a temperature contour plot of the temperatures at time 3600 s.

6.65 The figure illustrates a disk brake that is used to bring a piece of rotating machinery to a smooth stop.

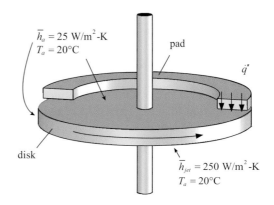

The brake pad engages the disk at its outer edge when the brake is activated. The outer edge and top surface of the disk (except under the pad) are exposed to air at $T_a = 20°C$ with $\bar{h}_a = 25$ W/m²-K. The bottom edge is exposed to air jets in order to control the disk temperature; the air jets have $T_a = 20°C$ and $\bar{h}_{jet} = 250$ W/m²-K. The problem can be modeled as a 2-D, radial problem as shown.

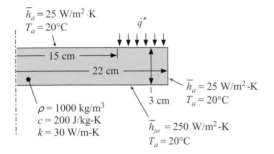

The dimensions of the brake and boundary conditions are shown in the figure; the friction between the disk and the brake causes a spatially uniform heat flux that varies with time according to:

$$\dot{q}'' = 200000 \ [\mathrm{W/m^2}] \left(1 - \left(\frac{t[\mathrm{s}]}{50[\mathrm{s}]}\right)^2\right).$$

You may neglect the contribution of the shaft (i.e., assume that the brake is just a disk). The disk is initially at a uniform temperature of 20°C. The density of the disk material is $\rho = 1000$ kg/m³, the conductivity is $k = 30$ W/m-K, and the specific heat capacity is $c = 200$ J/kg-K.

(a) Develop a FEHT model that can predict the temperature distribution in the disk as a function of time during the 50 s that is required for the rotating machine to stop.

(b) Plot the maximum temperature in the disk as a function of time for two values of the number of nodes in order to demonstrate that your mesh is sufficiently refined. You will need to generate two plots and the comparison will be qualitative.

(c) Prepare a contour plot showing the temperature distribution at $t = 10$ s, $t = 25$ s, and $t = 50$ s. You may also want to animate your temperature contours by selecting Temperature Contours from the View menu and selecting From start to stop.

(d) Plot the temperature on the lower surface (the surface exposed to the jets of air) at various locations as a function of time. Explain the shape of the plot – does the result make physical sense to you based on any time constants that you can compute?

6.66 A turkey has been sitting the refrigerator for a long time so that its temperature is uniform at 10°C. The turkey is approximately spherical with a diameter of 24 cm, a density of 1000 kg/m³, a conductivity of 0.7 W/m-K and a specific heat of 3800 J/kg-K. The cook has two cooking options. The first option is to put the turkey in a convection oven at 190°C. The convection coefficient between the air in the oven and the outside surface of the turkey is estimated to be 25 W/m²-K. The second option is to deep fry the turkey in 190°C oil. The convection coefficient between the oil and the turkey surface is estimated to be 750 W/m²-K. The turkey is considered to be done when the center temperature reaches 70°C. Using FEHT, determine the cooking time for both options.

6.67 The steel bracket shown is initially at a uniform temperature of 25°C. The left, bottom and right sides are insulated. The properties of steel are $k = 15$ W/m-K, $\rho = 7854$ kg/m³, and $c = 434$ J/kg-K. At time $t = 0$, the horizontal and vertical sections at the upper left of the bracket are exposed to 100°C fluid with a heat transfer coefficient of 140 W/m²-K. The top right surface convects to ambient 25°C fluid with a heat transfer coefficient of 84 W/m²-K. Use FEHT to calculate the temperatures within the steel as a function of time for a 1 hour period. Plot the temperature in the steel in the center of the bottom surface as a function of time and use the plot to determine the time required for the temperature to reach 80°C at this point.

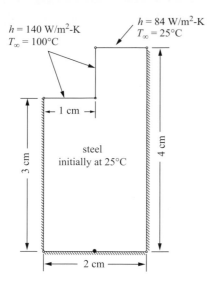

$h = 140$ W/m²-K
$T_\infty = 100°C$

$h = 84$ W/m²-K
$T_\infty = 25°C$

1 cm

steel
initially at 25°C

3 cm

4 cm

2 cm

6.68 The metal casting shown is initially at a uniform temperature of 184°C. All surfaces except the top are insulated. The top surface convects to 25°C fluid with a heat transfer coefficient of 148 W/m²-K. The properties of the casting are $k = 15$ W/m-K, $\rho = 7854$ kg/m³, and $c = 434$ J/kg-K.

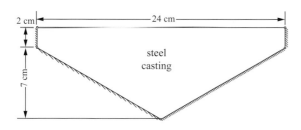

2 cm

24 cm

steel
casting

7 cm

Use FEHT to prepare the following plots for a period of 1800 s.

(a) The temperature at the center bottom of the casting as a function of time. What is the temperature at this point after 1800 s?

(b) The convective heat transfer rate from the top surface as a function of time.

6.69 A stainless steel billet is a long rod that is 10 cm in diameter. The rod is heated to a uniform temperature of 600°C in a furnace. When removed from the furnace, the rod convects to ambient air at 25°C with a convection coefficient of 14 W/m²-K. The metal can be formed

provided it is at a temperature above 500°C. The properties of the billet are $k = 15$ W/m-K, $\rho = 7854$ kg/m³, and $c = 434$ J/kg-K.

(a) Using FEHT, determine for how much time the rod can sit in the 25°C environment before the surface cools below the 500°C limit.

(b) Show a temperature contour within the rod at this time.

(c) Compare your results with the analytical solution for a cylinder.

6.70 A square concrete culvert with each side having a dimension of 30 cm encloses a 12 cm round hole. The concrete is initially at a uniform temperature of 25°C when saturated steam at 100°C starts to flow through it. The convection coefficient between the steam and the concrete is 380 W/m²-K. The bottom, left and right sides of the culvert are insulated. The top surface convects to water at 25°C with a convection coefficient of 86 W/m²-K. The properties of the concrete are $k = 1.4$ W/m-K, $\rho = 2300$ kg/m³, and $c = 880$ J/kg-K.

(a) Using FEHT, prepare a plot of the bottom and top corners of the culver as a function of time for a 2 hour period. Is the problem close to steady state at this time?

(b) Determine the total heat transfer from the steam during this 2 hour period.

(c) Show the steady-state temperature contours.

Projects

6.71 The figure illustrates a residence in a warm climate that utilizes a ground-coupled heat pump.

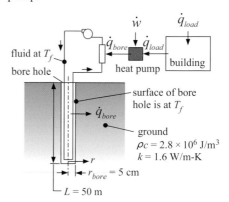

$\dot{w}$ $\dot{q}_{load}$

$\dot{q}_{bore}$ $\dot{q}_{load}$

fluid at T_f

bore hole

heat pump

building

surface of bore hole is at T_f

$\dot{q}_{bore}$

ground
$\rho c = 2.8 \times 10^6$ J/m³
$k = 1.6$ W/m-K

r

$r_{bore} = 5$ cm

$L = 50$ m

The load on the building ($\dot{q}_{load}$) is shown.

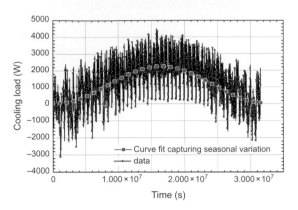

Notice the spikes that occur each day and the lower frequency, seasonal variation in the load. We are going to focus on the seasonal variation and ignore the daily fluctuations. The curve fit that captures the yearly variation is shown in the figure, and given by:

$$\dot{q}_{load} = 1148.78 [\text{W}]$$

$$-1088.8[\text{W}]\sin\left(1.992 \times 10^{-7} \left[\frac{\text{rad}}{\text{s}}\right] t - 425.81\right).$$

The load is removed by a heat pump that requires a work transfer ($\dot{w}$). The heat pump rejects heat ($\dot{q}_{bore}$) to fluid that is circulated through a bore hole. The fluid in the bore hole transfers energy to the ground. The coefficient of performance of the heat pump, COP, is defined as the ratio of $\dot{q}_{load}$ to $\dot{w}$. The coefficient of performance, COP, depends on the temperature of the fluid, T_f, according to:

$$COP = 4 - 0.05 [\text{K}^{-1}](T_f - 285[\text{K}]).$$

Note that the COP is reduced as the temperature of the fluid increases (it is harder to provide cooling when you have to reject heat at a higher temperature). You may neglect any change in the temperature of the fluid as it is circulated through the bore hole and also neglect the thermal resistance due to convection between the fluid and the internal surface of the bore hole. The ground around the bore hole can be treated using a 1-D transient model. The conductivity of the ground

is $k = 1.6$ W/m-K and the product of the density and specific heat capacity is $\rho c = 2.8 \times 10^6$ J/m³. The radius of the bore is $r_{bore} = 5$ cm. Develop a 1-D numerical model of the ground using an ode solver in MATLAB. Set the outer radius of the computational domain to a value that is large enough that it does not impact your results. The boundary condition at $r = r_{bore}$ should be the specified heat load, $\dot{q}_{bore}$, that is consistent with the fluid temperature and the building conditions. The length of the bore hole is $L = 50$ m and the temperature of the ground initially is $T_{ground} = 290$ K.

(a) Determine the state equations for the nodes in your numerical model. That is, derive a set of equations that will provide the derivatives for the temperatures at each of the nodes (T_i for $i = 1...N$) if you know t and the instantaneous values of the temperatures at each of the nodes.

(b) Implement the state equations in MATLAB and simulate 1 year. Plot the temperature as a function of time for various values of radial position.

(c) Plot the temperature at the bore surface as a function of time for 10 years. You should see both a seasonal variation related to load as well as a long-term buildup of energy in the ground (referred to as "annealing" the ground).

(d) The cost of the electricity required to run the heat pump is $ec = 0.1$ \$/kW-hr and the cost of installing a bore is $bfc = 20$ \$/m. Integrate the electrical cost in order to determine the operating cost for 10 years. Neglect the time value of money.

(e) Plot the operating cost, capital cost, and total cost (cost of the bore hole plus the operating cost) as a function of bore length, L, for 10 years of operation; again, neglect the time value of money. You should see an optimal bore length. Explain why this occurs.

6.72 A refrigerated warehouse is used to store cheese. A very simplified schematic of this situation is shown. The safe storage temperature for milk products ranges from 0.6 to 4.4°C.

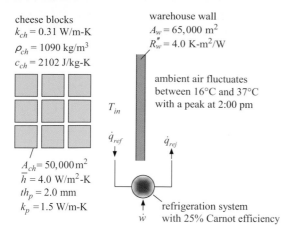

cheese blocks
$k_{ch} = 0.31$ W/m-K
$\rho_{ch} = 1090$ kg/m^3
$c_{ch} = 2102$ J/kg-K

$A_{ch} = 50,000$ m^2
$\bar{h} = 4.0$ W/m^2-K
$th_p = 2.0$ mm
$k_p = 1.5$ W/m-K

warehouse wall
$A_w = 65,000$ m^2
$R''_w = 4.0$ K-m^2/W

T_{in}

ambient air fluctuates
between 16°C and 37°C
with a peak at 2:00 pm

$\dot{q}_{ref}$

$\dot{q}_{rej}$

$\dot{w}$ refrigeration system
with 25% Carnot efficiency

The total resistance of the warehouse wall per unit area is $R''_w = 4.0$ K-m^2/W (this includes the effect of convection on the inside and outside surfaces) and there is 65,000 m^2 of warehouse wall separating the internal air temperature (T_{in}) from the ambient air temperature (T_∞). During a typical summer day, the temperature of the ambient air oscillates sinusoidally with a 24 hour period between 16°C and 37°C; the maximum temperature occurs at 2:00 p.m. The heat load through the wall ($\dot{q}_w$) is removed by an industrial refrigeration system with a Carnot efficiency of 25 percent (that is, the refrigerator requires four times as much power, $\dot{w}$, as a reversible refrigerator operating between T_{in} and T_∞). The cost of electrical energy to the warehouse during on-peak hours (9:00 a.m. to 9:00 p.m.) is 0.060$/kW-hr and drops to only 0.020$/kW-hr during off-peak hours (9:00 p.m. to 9:00 a.m.). You can assume that the refrigeration control system and refrigeration equipment is capable of exactly controlling the internal temperature of the air within the building. Currently, the warehouse operates the refrigeration system so that it maintains the internal air temperature at a constant value of $T_{in} = 4.4$°C.

(a) Prepare a plot of the outside air temperature as a function of time for a complete day and the refrigeration load and input power as a function of time for a complete day. What is total cost required to refrigerate the warehouse using this control scheme ($/day)?

You propose that the cheese warehouse take advantage of the thermal capacitance of the cheese within the factory by operating the

refrigeration system at a higher level during off-peak hours in order to sub-cool the cheese (i.e., cool it to near its lower safe storage temperature of 0.6°C). The sub-cooled cheese will warm during the on-peak hours and reduce the refrigeration load. To implement this strategy, you program the refrigeration system to cool the internal air to 0.6°C during off-peak hours and then switch to 4.4°C during on-peak hours. Again, assume that the refrigeration system is perfect so that the air temperature within the warehouse can be controlled exactly. This approach has two potential advantages: (1) you can "shift" your refrigeration load from on-peak to off-peak hours when electricity is cheaper, and (2) you can take advantage of the lower ambient air temperatures during the night that translate to higher refrigeration efficiency. There is a potential disadvantage in that you have to operate your refrigeration system at a lower temperature some of the time in order to sub-cool the cheese. The cheese has conductivity $k_{ch} = 0.31$ W/m-K, density $\rho_{ch} = 1090$ kg/m^3, and specific heat capacity $c_{ch} = 2102$ J/kg-K. You can assume that the cheese behaves like a semi-infinite body. That is, we can ignore the exact shape of the cheese blocks because the thermal wave associated with the periodically fluctuating temperature of the air within the warehouse never penetrates to the center of the cheese block. The total surface area of cheese blocks exposed to the internal air temperature is $A_{ch} = 50,000$ m^2. The heat transfer coefficient between the cheese surface and the internal air is $\bar{h} = 4.0$ W/m^2-K. Each block of cheese is wrapped with a thin layer of plastic with thickness $th_p = 2.0$ mm and conductivity $k_p = 1.5$ W/m-K.

(b) How big do the cheese blocks need to be (approximately) for the semi-infinite body assumption to be valid?

(c) Prepare a 1-D, transient numerical model of the cheese using an implicit technique. Implement your solution in MATLAB subject to the internal air temperature boundary condition discussed above (i.e., a stepwise change from 0.6°C to 4.4°C at 9:00 a.m. and back again at 9:00 p.m.). Note that your solution must start with the cheese all at some initial condition and then integrate forward in time until a cyclic steady-state condition is achieved (that is, the temperature of the cheese at the beginning and end of the day is the same). Prepare a plot of the cheese temperature at the surface and several positions within the cheese for this "typical" day.

(d) Prepare a plot that includes the refrigeration load associated with the heat transfer through the wall, the refrigeration load associated with heat transfer between the air and the cheese (note that this will be negative and positive depending on the time of the day), and the total refrigeration load required.

(e) Prepare a plot of the total refrigeration load and associated input power as a function of time for a complete day. What is total cost required to refrigerate the warehouse using this alternative control scheme ($/day)? Does it save money relative to your answer from (a)?

6.73 Friction stir welding is a method for joining dissimilar materials that cannot be joined using other techniques. The figure shows a friction stir welding process.

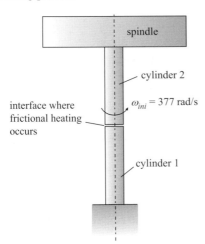

Two cylindrical samples are to be joined. Cylinder 1 is held stationary while cylinder 2 is held in a massive spindle that is brought to a high rotational speed, $\omega_{ini} = 377$ rad/s. The stir welding process is initiated by pressing the ends of the two cylinders together. The frictional heating associated with the relative motion between the two cylinders elevates the temperature at the interface between the cylinders,

accomplishing the welding process. The torque exerted by the friction causes cylinder 2 to slow down and eventually stop rotating.

Here we will focus on a thermal analysis of cylinder 1, as shown.

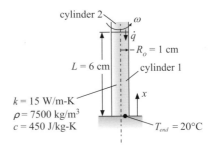

The radius of the cylinder is $R_o = 1$ cm and the length of the cylinder is $L = 6$ cm. The end of the cylinder at $x = 0$ is maintained at $T_{end} = 20°C$. The perimeter of the cylinder can be assumed to be adiabatic. The cylinder has properties $k = 15$ W/m-K, $\rho = 7500$ kg/m³, and $c = 450$ J/kg-K. The rotational speed of cylinder 2 as a function of time is

$$\omega = \omega_{ini} - \frac{2\pi \mu p R_o^3 t}{3 J_m},$$

where $\mu = 0.7$ is the coefficient of friction, $p = 1$ MPa is the pressure used to force the two cylinders together, t is time, and J_m is the mass moment of inertia of the spindle, $J_m = 0.25$ kg-m². The rate of frictional heating at the interface between the cylinders (i.e., at $x = L$) is given by

$$\dot{q} = \frac{2\pi \mu p R_o^3 \omega}{3},$$

where ω is the rotational speed. You may assume that all of the frictional heating is transferred into the end of cylinder 1. The cylinder material is initially at at $T_{edge} = 20°C$.

(a) Divide the cylinder into N nodes in the x-direction and derive the state equation (i.e., the equation that provides the time rate of change of that node) for the internal and boundary nodes.

(b) Develop a 1-D transient model of cylinder 1 using a numerical simulation that is based on the fully implicit integration technique. Simulate the entire spin down time (i.e., simulate the system from $t = 0$ to $t = t_{sd}$, where t_{sd} is the time at which the spindle comes to rest).

(c) Plot the temperature in the cylinder as a function of position for various values of time.

(d) Plot the temperature at the interface as a function of time. Overlay on this plot the interface temperature as a function of time for various values of the mass moment of inertia of the spindle (J_m).

6.74 Cryosurgery is a method for treating cancer in which malignant tissue is exposed to low temperature in order to destroy it. The cryosurgical probe is the cold end of a small refrigeration unit and therefore provides cooling to the tissue adjacent to its point of insertion. Depending on the geometry of the probe, the cryolesion that forms (i.e., the region of frozen tissue) is roughly spherical in size. The figure illustrates a spherical cryoprobe inserted in tissue.

tissue
$\rho = 1000$ kg/m^3
$k = 0.5$ W/m-K
$c = 3500$ J/kg-K
$\beta = 20{,}000$ W/m^3-K

$T_b = 37°C$

$R_p = 5$ mm

cryoprobe cooling

The tissue has conductivity $k = 0.5$ W/m-K, density $\rho = 1000$ kg/m^3, and specific heat capacity $c = 3500$ J/kg-K; we are going to ignore the latent heat of solidification for this problem and assume that frozen and unfrozen tissue have the same properties. If the temperature of tissue is changed relative to body temperature, $T_b = 37°C$, the blood flow tends to cause a volumetric heating or cooling that will attempt to restore the tissue to T_b. Therefore, when the tissue is cooled, the action of blood perfusion is to create a volumetric heating effect ($\dot{g}'''$) that is proportional to the difference between the body temperature and the local temperature:

$$\dot{g}''' = \beta(T_b - T),$$

where $\beta = 20000$ W/m^3-K is the blood perfusion constant. The radius of the cryoprobe is $R_p = 5$ mm and the cooling power provided at the surface of the cryoprobe depends on temperature. The cooling power provided at T_b is $\dot{q}_b = 100$ W and the cooling power decreases linearly to zero at $T_{nl} = 150$ K. Therefore, the cooling power is given by:

$$\dot{q} = \begin{cases} \dot{q}_b \dfrac{(T - T_{nl})}{(T_b - T_{nl})} & \text{for } T > T_{nl} \\ 0 & \text{for } T \le T_{nl} \end{cases}.$$

The tissue is initially at a uniform temperature of T_b. The temperature far from the probe will always remain at T_b. During a procedure, the probe is activated for $t_p = 10$ min and then deactivated.

(a) Develop a numerical simulation of the cryolesion formation process. Your model should simulate both the freezing process ($0 < t < t_p$) and the thawing process ($t > t_p$). Use Heun's Method and implement your solution in MATLAB.

(b) Plot the temperature in the tissue as a function of time for various radial locations.

(c) Plot the volume of the cryolesion as a function of time. The cryolesion volume is defined according to:

$$V_{lesion} = \int_{R_p}^{\infty} 4\pi r^2 p(T)\,dr,$$

where p is the probability of tissue destruction, which depends on temperature according to:

$$p = \begin{cases} 0 & T > 280 \text{ K} \\ \dfrac{T - 280 \text{ K}}{260 \text{ K} - 280 \text{ K}} & 260 \text{ K} < T < 280 \text{ K} \\ 1 & T < 260 \text{ K} \end{cases}.$$

(d) By re-engineering the cryoprobe it is possible to achieve a lower no-load temperature, T_{nl}, but at the expense of reduced refrigeration power at body temperature. The trade-off is given by:

$$\dot{q}_b = -50 \text{ W} + 1\frac{\text{W}}{\text{K}} T_{nl}.$$

Therefore, a no-load temperature of 150 K corresponds to a body temperature refrigeration power of 100 W. Plot the maximum cryolesion volume as a function of no-load temperature.

6.75 You are analyzing the barrel of a large calibre weapon. Each firing event causes a short ($t_f = 0.2$ s) but very large ($\dot{q}_f'' = 1.22 \times 10^5$ W/m^2) heat flux to be applied to the internal surface of the barrel. These firing events are repeated every $t_{bf} = 1$ s, as shown in the figure.

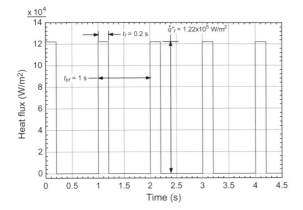

Because the barrel diameter is large, you can model it approximately as a plane wall with thickness $L = 1$ cm, as shown. The outer surface of the barrel ($x = L$) is cooled by fluid at $T_\infty = 20°C$ with heat transfer coefficient $\bar{h} = 400$ W/m²-K. The inner surface ($x = 0$) is not cooled by convection or radiation. The barrel material has properties $k = 15$ W/m-K, $\rho = 7500$ kg/m³, and $c = 450$ J/kg-K. Initially the barrel is at a uniform temperature, T_∞.

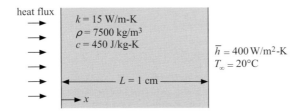

Before developing a detailed numerical model of the barrel, we are going to try to develop some understanding of the anticipated behavior.

(a) Estimate the thickness of the material adjacent to the inner surface of the barrel that is affected by the first firing event.

(b) Estimate the temperature of the material at the inner surface of the barrel at the conclusion of the first firing event.

(c) Sketch the temperature distribution in the barrel that you expect at various times. The sketch need not be exact, but should capture the qualitative features correctly; these include the approximate temperature and positions. Include sketches for the following times:

at $t = 0$ (i.e., the onset of the first firing event),

at $t = 0.1$ s (i.e., half-way through the first firing event),

at $t = 0.2$ s (i.e., the end of the first firing event),

at $t = 0.6$ s (i.e., 0.4 s after the first firing event has ended), and

at $t = 1$ s (i.e., immediately before the second firing event will start).

(d) After how many firing events will the outer surface of the barrel begin to get warm? Justify your answer with a calculation.

(e) Eventually, after a long time during which firing events are continuously repeated, the barrel will reach some kind of quasi-steady state where the temperature adjacent to the inner surface goes up during each firing event and then down between firing events but the average temperature of the barrel does not change. When this condition is reached, estimate the average temperature of the external surface of the barrel (at $x = L$). Explain your reasoning.

The next portion of the problem will develop a numerical model of the system that can be used to simulate the first few firing events.

(f) Write an EES function that takes in time as an input argument and returns heat flux according to the heat flux variation given in the figure. You may find the **Floor** function in EES useful for this.

(g) Develop a numerical model using the fully implicit technique in EES. Simulate the first firing event and the ensuing zero flux time (i.e., for $0 < t < t_{bf}$).

(h) Plot the temperature as a function of position at various values of time – compare your numerical results with the sketch that you made for part (c).

(i) Plot the surface temperatures (i.e., the temperature at $x = 0$ and $x = L$) as a function of time for the first five firing events.

6.76 Superconductors are materials with electrical resistivity that goes to zero below some critical temperature, T_c. This allows extremely small conductors to carry large amounts of current, enabling all sorts of applications from magnetically levitated trains, to power transmission systems, to electrical energy storage in magnets. If the temperature of the superconductor increases above T_c then the conductor returns to its normal (i.e., resistive) state instantaneously. When this happens while the superconductor is carrying large amounts of current, the result can be devastating. The ohmic heating that results causes the superconductor to melt, destroying itself.

Let's examine a niobium–tin superconductor. In its superconducting state, the material can carry current densities on the order of 2500 A/mm^2 (or 2.5×10^9 A/m^2); to put this into context, a conductor of diameter 0.7 mm could theoretically carry 1000 amp. However, small disturbances (mechanical or magnetic or thermal) can cause the temperature locally to increase above the critical temperature. The designer must therefore make sure that, given time, the local temperature will eventually drop back below the critical temperature. That is, the designer must design a superconductor that is locally stable so that a hypothetical normal zone will disappear, re-establishing the superconducting state, rather than expand, causing the entire superconductor to quench.

Local stability is achieved by axial conduction. The superconductor is created as a composite of pure superconducting strands within a high-purity copper matrix that has high conductivity. The effective properties of the superconductor are k = 1500 W/m-K, ρ = 9000 kg/m^3, and c = 0.5 J/kg-K. The electrical resistivity of the superconductor is given by the function

$$\rho_e(T) = \begin{cases} 0 & \text{if } T < T_c \\ \rho_{e,n} & \text{if } T \geq T_c \end{cases},$$

where T_c = 5.2 K is the critical temperature and $\rho_{e,n} = 1 \times 10^{-8}$ ohm-m is the electrical resistivity in the normal state.

To evaluate the local stability of a superconductor we will assume that a disturbance (i.e., hot spot) has a length $2L$ where L = 1 cm. Within the hot spot, the temperature is uniform and above the critical temperature, T_{ini} = 6 K. Outside of the hot spot, the temperature is uniform and equal to the steady-state operating temperature of the superconductor, T_{ss} = 4 K.

There is no convection or radiation from the surface of the superconductor and therefore this is a 1-D, transient problem, $T(x,t)$. In the absence of any current, the hot spot will cool back to T_{ss}. However, as the current density

(J – current per unit area, A/m^2) increases you will eventually reach a critical threshold (J_c) that distinguishes between stable disturbances (i.e., a normal zone that disappears) and unstable disturbances (i.e., a normal zone that expands).

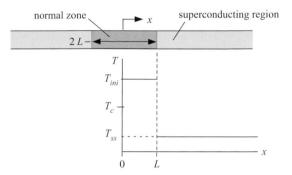

The behavior of the normal zone can be studied by starting the simulation with an initial temperature distribution shown in the figure and then simulating the temperature as a function of x and t. Prepare a numerical model of the superconductor in order to identify the critical current density.

(a) Determine the characteristic time for the problem – this is the time required for energy to travel by conduction from the center ($x = 0$) to the edge ($x = L$) of the normal zone. Your simulation needs to be 4–5× longer than this in order to capture the process.

(b) Determine the length of conductor that must be simulated. The conductor extends infinitely away from the normal zone but you, obviously, cannot extend your computational domain to infinity. Therefore, determine how large your computational domain must be (L_{sim}) such that the outer edge of computational domain (at $x = L_{sim}$) is not affected by the hot spot during the entire process.

(c) Create a numerical model in EES using an implicit technique that can simulate the evolution of the normal zone with time. Plot the center temperature (i.e., the temperature at $x = 0$) as a function of time for $J = 1 \times 10^7$ A/m^2. You should see the center temperature drop as this is below the critical current density threshold.

(d) Plot the temperature as a function of position for various values of time for $J = 1 \times 10^7$ A/m^2.

You should see the normal zone disappear as this is below the critical threshold.

(e) Increase the current density in your program until the center temperature increases rather than decreases indicating that the normal zone expands. To the best of your ability identify the critical threshold, J_c, that delineates the region between a stable and unstable conductor.

(f) Overlay on your plot from (c) the center temperature as a function time for twice the current density identified in (e); clearly label your different plots.

(g) Plot the temperature as a function of position for various values of time when the current density is twice the value identified in (e). You should see the normal zone expand rather than disappear.

7 Convection

Chapters 2 through 6 consider heat transfer in a stationary medium where energy transport occurs entirely by conduction and is governed by Fourier's Law. Thus far, convection has been considered primarily as a boundary condition for these conduction problems. **Convection** refers to the transfer of energy that occurs between a surface and a moving medium, most often a liquid or gas flowing through a duct or over an object. Convection processes include fluid motion and the related energy transfer. The additional terms in an energy balance related to the fluid flow often dominate the now familiar energy transport by conduction. The presence of fluid motion complicates heat transfer problems substantially and links the heat transfer problem with an underlying fluid dynamics problem. The complete solution to most convection problems therefore requires sophisticated computational fluid dynamic (CFD) tools that are beyond the scope of this book.

The presentation of convection heat transfer provided in this chapter introduces these processes at a conceptual level by considering the behavior of boundary layers. The differential equations that govern the problem are derived, simplified for application in a boundary layer, and finally made dimensionless. Some of these steps may be familiar to you from a prior class in fluid dynamics. The nondimensional numbers that result from this process are particularly important because most real engineering problems are solved through the application of appropriate correlations. **Convection correlations** are a particularly powerful example of nondimensionalization. A relatively small number of experiments or simulations can be used to solve problems over a tremendous range of conditions when the results are presented using the appropriate dimensionless numbers. We will examine specific convection correlations and understand their proper use and the limits of their applicability in Chapters 8 through 11 for various situations. The convection heat transfer correlations included in this book (as well as others) are also included as built-in functions and procedures in EES, which simplifies their application.

7.1 The Laminar Boundary Layer

As engineers, we are often interested in the interaction between a fluid and a surface; specifically in the transport of momentum and energy between the surface and the fluid. The transport of momentum is related to the force exerted on the surface or the loss of pressure in a flow that must be made up using a pump or fan. The transport of energy is related to the now-familiar heat transfer coefficient. These are the engineering quantities of interest and they are governed by the behavior of the **boundary layer**, the layer of fluid that is adjacent to the surface and is affected by its presence. In this section, we will attempt to obtain a conceptual understanding regarding the behavior of laminar boundary layers.

7.1.1 The Velocity Boundary Layer

Figure 7.1 illustrates, qualitatively, the laminar flow of a fluid over a surface. The x-direction is defined as being parallel to the surface in the direction of the flow, with $x = 0$ corresponding to the leading edge of the surface. The y-direction is defined as being normal to the surface with $y = 0$ corresponding to the interface between the surface and the fluid. Velocities in the x- and y-directions are denoted as u and v, respectively. The flow approaching the plate has a uniform velocity (u_∞) in the x-direction and no velocity in the y-direction. The quantity u_∞ is referred to as the **free stream velocity**. The difference between laminar and turbulent flow will be discussed in more detail later in this chapter. For now, it is sufficient to understand that the laminar flow is steady, provided that the free stream velocity is not changing in time; an instrument placed in the flow would report a constant value of velocity with no fluctuations.

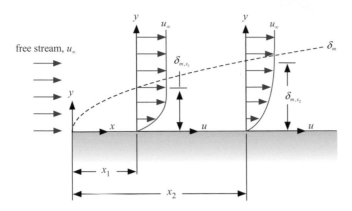

Figure 7.1 The velocity distribution associated with flow over a surface.

The presence of the surface affects the velocity in the fluid for $x > 0$. The plate is stationary and therefore the fluid particles immediately adjacent to the plate (i.e., at $y = 0$) will have zero velocity; this is referred to as the **no-slip condition**. These particles exert a shear stress on those that are slightly farther from the plate, causing them to slow down. The result is the velocity distribution shown in Figure 7.1, where the velocity gradually increases from zero at the surface to u_∞ far from the surface. The **velocity boundary layer** (sometimes called the **momentum boundary layer**) refers to the region of the flow where the velocity is affected by the presence of the plate (i.e., the region where $u < u_\infty$). The **velocity boundary layer thickness**, δ_m, is the thickness of this region at any x-position. There are different definitions of boundary layer thickness that can be used. A common definition of the velocity boundary layer thickness is the distance from the plate (i.e., the y-location) where the velocity has recovered to 99 percent of its free stream value, $u/u_\infty = 0.99$.

The thickness of the velocity boundary layer will grow as the fluid moves downstream (i.e., as x increases); notice in Figure 7.1 that $\delta_{m,\,x_2} > \delta_{m,\,x_1}$. One of the reasons that the velocity boundary layer is important is that its size is related to the gradient in the velocity that exists at the surface. Notice in Figure 7.1 that there is no velocity gradient at all outside of the velocity boundary layer. The velocity gradient is related to the **shear stress** (i.e., the diffusive transport of momentum) according to:

$$\tau = \mu \frac{\partial u}{\partial y},\tag{7.1}$$

where μ is the viscosity of the fluid. The similarity between Eq. (7.1) and Fourier's Law

$$\dot{q}''_y = -k\frac{\partial T}{\partial y}\tag{7.2}$$

is significant. Both energy and momentum are transported diffusively by virtue of a gradient in the potential that drives the transport process (i.e., by a gradient in temperature and velocity, respectively).

Within the boundary layer, the velocity gradient (and therefore the shear stress) is largest at the surface and diminishes to zero at the outer edge of the boundary layer. The shear stress at the surface (τ_s) is an engineering quantity of interest in a convection problem:

$$\tau_{y=0} = \tau_s = \mu \left(\frac{\partial u}{\partial y}\right)_{y=0}.\tag{7.3}$$

Example 7.1

Engine oil (10W) is flowing over a surface. The velocity of the oil has been measured as a function of the distance from the surface (y) at a fixed position relative to the leading edge (i.e., at a fixed value of x in Figure 7.1); these measurements are summarized in Table 1.

Example 7.1 (cont.)

Table 1 Velocity measurements as a function of distance from a surface.			
Position, y (mm)	Velocity, u (m/s)	Position, y (mm)	Velocity, u (m/s)
0	0	24.0	4.77
2.40	0.66	26.4	4.87
4.80	1.32	28.8	4.93
7.20	1.97	29.5	4.95
9.60	2.58	31.2	4.96
12.0	3.14	33.6	4.98
14.4	3.64	36.0	4.99
16.8	4.05	48.0	5.00
19.2	4.37	60.0	5.00
21.6	4.61	100	5.00

Based on these measurements, estimate the velocity boundary layer thickness as well as the shear stress on the surface.

Known Values

The viscosity of 10 W engine oil is obtained from Appendix B, $\mu = 0.133$ Pa-s.

Assumptions

- The viscosity of the engine oil is not very different from its value at 20°C (the basis for the value in Appendix B).
- The uncertainty in the measured velocity is not significant.

Analysis and Solution

The definition of the velocity boundary layer thickness used in this section is the y-position where $u/u_\infty = 0.99$. Table 1 suggests that the free stream velocity is $u_\infty = 5$ m/s (the value far from the plate surface). Therefore, the velocity boundary layer thickness is the y-position at which the velocity is 99 percent of 5 m/s, or $u = 4.96$ m/s. Examining Table 1 suggests that $\boxed{\delta_m = 31.2 \text{ mm}}$.

The shear stress at the surface is obtained by evaluating Eq. (7.1) at $y = 0$ (i.e., at the surface).

$$\tau_s = \mu \left(\frac{\partial u}{\partial y} \right)_{y=0}. \tag{1}$$

The velocity gradient at the surface is estimated using the measurements at $y = 0$ mm and $y = 2.40$ mm in Table 1:

$$\left(\frac{\partial u}{\partial y} \right)_{y=0} \approx \frac{(0.66 - 0) \text{ m/s}}{(2.40 - 0) \text{ mm}} = 0.275 \frac{\text{m/s}}{\text{mm}} \left(275 \frac{\text{m/s}}{\text{m}} \right). \tag{2}$$

Substituting Eq. (2) into Eq. (1) provides:

$$\tau_s = \mu \left(\frac{\partial u}{\partial y} \right)_{y=0} = \frac{0.133 \text{ Pa s}}{\text{m}} \left| \frac{275 \text{ m/s}}{\text{m}} \right. = \boxed{36.6 \text{ Pa}}. \tag{3}$$

Continued

Example 7.1 (cont.)

Discussion

The value of the shear stress that was calculated is the *local* shear stress which exists at the point on the surface that the measurements were taken. A more useful quantity for computing the force on the surface is *average* shear stress, which is defined as the average value of τ_s over the entire surface.

7.1.2 The Thermal Boundary Layer

Figure 7.2 illustrates the temperature distribution associated with the laminar flow of a cold fluid over a heated surface that is maintained at a uniform temperature (T_s). The flow approaching the surface has a uniform temperature (T_∞) that is referred to as the **free stream temperature**.

The presence of the hot surface affects the temperature of the fluid in a region of the flow that is referred to as the **thermal boundary layer**. Those fluid particles immediately adjacent to the hot plate will be at the surface temperature, T_s, and transfer energy to the cooler particles that are farther from the plate causing a temperature gradient. There are different definitions of the **thermal boundary layer thickness** (δ_t), but the one used here is the distance from the plate where the fluid-to-plate temperature difference has achieved 99 percent of its free stream value, $(T - T_s)/(T_\infty - T_s) = 0.99$.

The thickness of the thermal boundary layer grows in the downstream direction just as the velocity boundary layer thickness does; notice in Figure 7.2 that $\delta_{t,\,x_2} > \delta_{t,\,x_1}$. The growth of a laminar boundary layer is analogous to the penetration of a thermal wave by conduction into a semi-infinite body, discussed in Section 6.1. This similarity is the basis of our conceptual understanding of laminar boundary layers, as discussed in Section 7.1.3. One of the reasons that the thermal boundary layer is important is that its size is related to the gradient in temperature that drives the heat transfer from the surface. Consider a control surface that is defined at the surface, but on the fluid side (rather than the solid side), as shown in Figure 7.3.

The rate of energy transfer through the control surface shown in Figure 7.3 is equal to the convective heat transfer rate. There is no fluid flow through this control surface (the velocity v at $y = 0$ must be zero) and therefore the energy transfer is only related to conduction, evaluated using Fourier's Law:

$$\dot{q}''_{y=0} = \dot{q}''_{conv} = \dot{q}''_s = -k\left(\frac{\partial T}{\partial y}\right)_{y=0}. \tag{7.4}$$

Comparing Eqs. (7.3) and (7.4) is instructive as it reinforces the similarity between momentum transport (shear stress) and energy transport (heat flux).

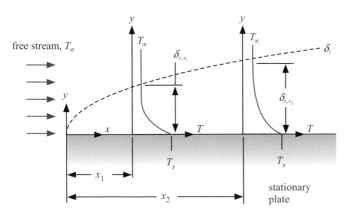

Figure 7.2 The temperature distribution associated with laminar flow over a heated surface.

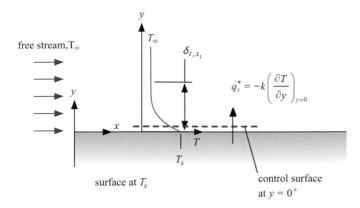

Figure 7.3 Control surface defined at $y = 0^+$ (i.e., on the fluid side of the surface).

Example 7.2

Reconsider the flow of engine oil (10 W) flowing over a surface. The temperature of the oil has also been measured as a function of the distance from the surface (y) at a fixed x-position relative to the leading edge; these measurements are summarized in Table 1.

Table 1 Temperature measurements as a function of distance from the surface.

Position, y (mm)	Temperature, T (°C)	Position, y (mm)	Temperature, T (°C)
0	20.00	0.95	49.13
0.095	24.17	1.04	49.60
0.19	28.31	1.14	49.84
0.28	32.35	1.23	49.95
0.38	36.17	1.33	49.99
0.47	39.65	1.42	50.00
0.57	42.67	1.52	50.00
0.66	45.13	1.61	50.00
0.76	47.00	1.71	50.00
0.85	48.30	1.80	50.00

Based on these measurements, estimate the thermal boundary layer thickness as well as the convective heat flux and the heat transfer coefficient.

Known Values

The conductivity of 10W engine oil is obtained from Appendix B, $k = 0.1442$ W/m-K.

Assumptions

- The thermal conductivity of the engine oil is not a strong function of temperature between 20°C (the basis for Appendix B) and 50°C (the highest value in Table 1).
- The uncertainty in the measured temperature is not significant.

Continued

Example 7.2 (cont.)

Analysis and Solution

The definition of the thermal boundary layer thickness is the y-position where $(T - T_s)/(T_\infty - T_s) = 0.99$. Table 1 suggests that the surface temperature is $T_s = 20°C$ and the free stream temperature is $T_\infty = 50°C$. Therefore, the thermal boundary layer thickness is the y-position at which the temperature is:

$$T = T_s + 0.99(T_\infty - T_s) = 20°C + 0.99(50°C - 20°C) = 49.70°C. \tag{1}$$

Examination of Table 1 suggests that $\boxed{\delta_t = 1.08 \text{ mm}}$.

The convective heat flux at the surface is obtained by evaluating Eq. (7.4):

$$\dot{q}''_s = -k\left(\frac{\partial T}{\partial y}\right)_{y=0}. \tag{2}$$

The temperature gradient at the surface is estimated using the measurements at $y = 0$ mm and $y = 0.095$ mm in Table 1:

$$\left(\frac{\partial T}{\partial y}\right)_{y=0} \approx \frac{(24.17 - 20.00) \text{ K}}{(0.095 - 0) \text{ mm}} = 43.89 \frac{\text{K}}{\text{mm}} \left(43,890 \frac{\text{K}}{\text{m}}\right). \tag{3}$$

Substituting Eq. (3) into Eq. (2) provides:

$$\dot{q}''_s = \dot{q}''_{conv} = -k\left(\frac{\partial T}{\partial y}\right)_{y=0} = -\frac{0.1442 \text{ W}}{\text{m K}}\left|\frac{43,890 \text{ K}}{\text{m}} = \boxed{-6330 \frac{\text{W}}{\text{m}^2}}\right.. \tag{4}$$

The heat flux is negative because the heat transfer is from the warmer fluid to the cooler surface. The heat transfer coefficient is defined from Newton's Law of Cooling as discussed in Section 1.4.3:

$$\dot{q}''_{conv} = \dot{q}''_s = h(T_s - T_\infty). \tag{5}$$

Therefore, the heat transfer coefficient is:

$$h = \frac{\dot{q}''_s}{(T_s - T_\infty)} = \frac{-6330 \text{ W}}{\text{m}^2}\left|\frac{}{(20 - 50) \text{ K}} = \boxed{211 \frac{\text{W}}{\text{m}^2 \text{ K}}}\right.. \tag{6}$$

Discussion

The thermal boundary layer thickness computed in this problem ($\delta_t = 1.08$ mm) is much smaller than the velocity boundary layer thickness computed in Example 7.1 for the same flow ($\delta_m = 31.2$ mm). We will see in Section 7.1.4 that this behavior is consistent with the laminar flow of a highly viscous fluid like engine oil because such a fluid can transport momentum more effectively than energy. The dimensionless parameter that describes this effect is called the Prandtl number (Pr).

The value of the heat transfer coefficient calculated in this example is the *local* heat transfer coefficient (h) which exists at the point on the surface that the measurements were taken. We have been using the *average* heat transfer coefficient ($\bar{h}$) in Chapters 2 through 6 because it is more useful for most problems. The average heat transfer coefficient is the area-weighted integration of the local heat transfer coefficient over the entire surface. The difference between the local and average heat transfer coefficient is discussed in Section 7.1.9.

7.1.3 A Conceptual Model of Laminar Boundary Layer Growth

The velocity distributions shown in Figure 7.1 and the temperature distributions shown in Figure 7.2 are both consistent with laminar flow. An important characteristic of laminar flow is that it is highly ordered with no velocity fluctuations in the boundary layer. There is also very little velocity in the y-direction associated with this type of flow. As a result, the transport of thermal energy in the y-direction must be primarily due to conduction, i.e., the diffusion of energy due to the interaction of the molecular scale energy carriers.

Conceptually, you can think of the free stream as being very similar to the semi-infinite body that was discussed in Section 6.1. The free stream experiences a step change in its surface temperature (i.e., the temperature at $y = 0$) at the instant that the fluid encounters the leading edge of the plate. This occurs at $x = 0$ which, from the standpoint of the moving fluid, corresponds to time $t = 0$. The disturbance associated with the change in the surface temperature diffuses as a thermal wave into the free stream (i.e., in the y-direction). This diffusion process takes time, allowing the fluid to move downstream from the leading edge (i.e., towards increasing x) even as the thermal wave propagates into the fluid (i.e., towards increasing y). Notice in Figure 7.2 that at $x = x_2$ the thermal wave has propagated farther into the free stream (i.e., in the y-direction) than it had at $x = x_1$.

In Section 6.1, we learned that the motion of a diffusive thermal wave can be approximately represented by

$$\delta_t \approx 2\sqrt{\alpha\, t}, \tag{7.5}$$

where α is the thermal diffusivity of the material and t is the time relative to the disturbance at the surface. In a convection problem, the transport time that is available for diffusion is approximately related to the distance from the leading edge and the free stream velocity according to:

$$t \approx \frac{x}{u_\infty}. \tag{7.6}$$

Equation (7.6) provides an estimate for the amount of time that the free stream has been in contact with the surface. Substituting Eq. (7.6) into Eq. (7.5) provides a simple, but conceptually useful, estimate for the thermal boundary layer thickness:

$$\delta_t \approx 2\sqrt{\frac{\alpha\, x}{u_\infty}}. \tag{7.7}$$

The transport of momentum by molecular diffusion is analogous to the transport of energy by conduction; a "momentum wave" will travel a distance δ_m according to:

$$\delta_m \approx 2\sqrt{\upsilon\, t}, \tag{7.8}$$

where υ is the **kinematic viscosity** (the ratio of the dynamic viscosity of the fluid to its density, μ/ρ). Substituting Eq. (7.6) into Eq. (7.8) provides a simple conceptual model for the velocity boundary layer thickness:

$$\delta_m \approx 2\sqrt{\frac{\upsilon\, x}{u_\infty}}. \tag{7.9}$$

Note that the thicknesses of both the thermal and velocity layers grow with the square root of the distance from the leading edge (x).

7.1.4 The Prandtl Number

The property kinematic viscosity (υ) in Eq. (7.8) has the same units as the property thermal diffusivity (α) in Eq. (7.5), m^2/s. Thermal diffusivity describes the ability of a fluid to transport energy by diffusion, whereas kinematic viscosity describes the ability of a fluid to transport momentum by diffusion. For many fluids, α and υ have similar values because the transport of energy and momentum both occur by the same mechanism. For example, the transport of energy and momentum in a gas both occur as a result of the collisions between individual gas molecules, as discussed in Section 1.4.1. Room temperature, ambient air, for example, has $\alpha = 22.3 \times 10^{-6}$ m^2/s and $\upsilon = 15.8 \times 10^{-6}$ m^2/s (Appendix C).

The ratio of the kinematic viscosity to the thermal diffusivity is an important dimensionless parameter in convection heat transfer, referred to as the **Prandtl number** (Pr):

$$Pr = \frac{\upsilon}{\alpha}. \tag{7.10}$$

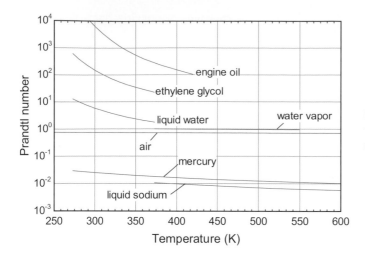

Figure 7.4 Prandtl number as a function of temperature for various fluids at atmospheric pressure.

The Prandtl number can be interpreted as the relative ability of a fluid to diffusively transport momentum compared to its ability to diffusively transport energy. Figure 7.4 illustrates the Prandtl number as a function of temperature for several fluids at atmospheric pressure.

A fluid with a large Prandtl number (e.g., engine oil) is viscous and nonconductive. Such a fluid will transport momentum much better than thermal energy. At the other extreme, a fluid with a small Prandtl number (e.g., a liquid metal such as liquid sodium) is very conductive but inviscid. Such a fluid will transport thermal energy much better than momentum.

The ratio of the velocity boundary layer thickness to the thermal boundary layer thickness in a laminar flow can be estimated using the conceptual models developed in Section 7.1.3, Eqs. (7.7) and (7.9):

$$\frac{\delta_m}{\delta_t} \approx \frac{2\sqrt{\dfrac{\upsilon x}{u_\infty}}}{2\sqrt{\dfrac{\alpha x}{u_\infty}}} = \sqrt{\frac{\upsilon}{\alpha}} = \sqrt{Pr}. \tag{7.11}$$

Equation (7.11) shows that the ratio of δ_m to δ_t in a laminar flow is only related to the Prandtl number. This is not surprising, given that the Prandtl number characterizes the ability of a fluid to transport momentum relative to its ability to transport thermal energy. Figure 7.5(a) illustrates the boundary layers for a fluid with a Prandtl number that is much higher than unity (e.g., engine oil). According to Eq. (7.11), δ_t will grow more slowly than δ_m because engine oil can transport momentum more efficiently than it can transfer energy. We saw this to be true in Examples 7.1 and 7.2. Figure 7.5(b) illustrates the converse situation, a fluid with a Prandtl number that is much less than unity (e.g., a liquid metal). According to Eq. (7.11), δ_t will be larger than δ_m because the liquid metal can transport energy more efficiently than momentum.

7.1.5 A Conceptual Model of Shear Stress and the Heat Transfer Coefficient

Why should we care about the thickness of the boundary layers? In this section, we will see that the engineering quantities that are of primary interest in a convection problem, shear stress and heat transfer coefficient, are actually directly related to δ_m and δ_t. In fact, you could say that understanding convection heat and momentum transfer depends on your understanding of these boundary layer thicknesses.

The shear stress at the surface is given by Eq. (7.3), which is repeated below:

$$\tau_s = \mu \frac{\partial u}{\partial y}\bigg|_{y=0}. \tag{7.12}$$

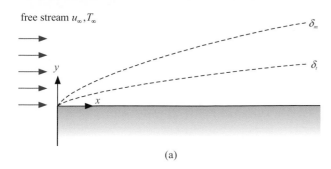

(a)

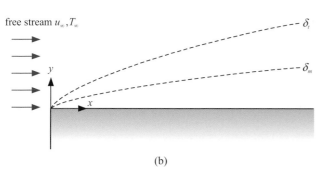

(b)

Figure 7.5 Sketch of the thermal and momentum boundary layers for (a) a liquid with $Pr \gg 1$ and (b) a liquid with $Pr \ll 1$.

The convective heat flux at the surface is given by Eq. (7.4), which is also repeated:

$$\dot{q}''_{conv} = -k \frac{\partial T}{\partial y}\bigg|_{y=0}. \tag{7.13}$$

The shear stress experienced by the surface is related to the velocity gradient at $y = 0$ and the heat flux at the surface is related to the temperature gradient at $y = 0$. The velocity and temperature gradients at the wall are directly related to the respective boundary layer thicknesses. These gradients can be approximately written as:

$$\frac{\partial u}{\partial y}\bigg|_{y=0} \approx \frac{u_\infty}{\delta_m} \tag{7.14}$$

$$\frac{\partial T}{\partial y}\bigg|_{y=0} \approx \frac{(T_\infty - T_s)}{\delta_t}. \tag{7.15}$$

Equations (7.14) and (7.15) are not exact because these gradients are not constant throughout the boundary layer. However, Eq. (7.14) does correctly reflect the fact that the velocity will change from zero to u_∞ between the surface and the outer edge of the velocity boundary layer (δ_m). Equation (7.15) indicates that the temperature of the fluid will change from T_s to T_∞ between the surface and the outer edge of the thermal boundary layer (δ_t).

Substituting Eq. (7.14) into Eq. (7.12) shows that the velocity boundary layer thickness governs the shear stress experienced at the plate surface:

$$\tau_s \approx \mu \frac{u_\infty}{\delta_m}. \tag{7.16}$$

Similarly, substituting Eq. (7.15) into Eq. (7.13) shows that the thermal boundary layer thickness governs the rate of heat transfer at the surface:

$$\dot{q}''_{conv} \approx k \frac{(T_s - T_\infty)}{\delta_t}. \tag{7.17}$$

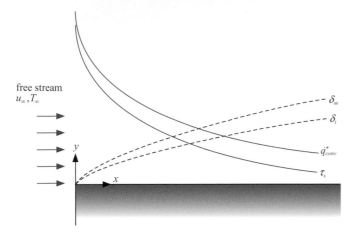

Figure 7.6 A sketch showing how the shear and heat transfer rate vary with position.

At the surface of the plate, the fluid motion is zero and so the shear and heat transfer rate represented by Eqs. (7.16) and (7.17), respectively, represent the total interaction between the surface and the fluid. Equations (7.16) and (7.17) suggest that the local shear and the rate of heat transfer are inversely proportional to the local thermal and momentum boundary layer thicknesses, respectively. Accordingly, we expect that the shear stress and the rate of heat transfer will be largest at the leading edge and decrease with x due to the thickening of the boundary layers and corresponding reduction in the gradients, as shown in Figure 7.6.

Figure 7.6 shows clearly one of the engineering challenges associated with the design of most energy conversion systems. It is often the case that you would like to maximize the convective heat flux, $\dot{q}''_{conv}$, in order to get the highest performance (for example, in a heat exchanger). However, you would also like to minimize the shear stress in order to reduce the pump or fan power required. Equations (7.16) and (7.17) show that it is usually not possible to increase the convective heat flux without simultaneously increasing the shear stress. Fluids with higher conductivity tend to also have higher viscosity and flow configurations that lead to lower values of δ_t also tend to have lower values of δ_m. It is a kind of Murphy's Law for the thermal engineer; you generally pay for improved heat transfer with increased shear stress.

Equation (7.17) has the same form as a thermal resistance equation:

$$\dot{q}''_{conv} \approx \frac{(T_s - T_\infty)}{R_{bl}},\tag{7.18}$$

where R_{bl} is the thermal resistance of the boundary layer:

$$R_{bl} \approx \frac{\delta_t}{k}.\tag{7.19}$$

Equations (7.18) and (7.19) show that laminar flow can be understood approximately as a conduction problem; the conduction length is the thermal boundary layer thickness. The laminar boundary layer can be thought of as a conduction resistance that exists between the surface and the free stream.

Comparing Eq. (7.17) with Newton's Law of Cooling (i.e., the definition of the local heat transfer coefficient, h) leads to:

$$\dot{q}''_{conv} = h(T_s - T_\infty) \approx k\frac{(T_s - T_\infty)}{\delta_t}.\tag{7.20}$$

Rearranging Eq. (7.20) leads to:

$$h \approx \frac{k}{\delta_t}.\tag{7.21}$$

Equation (7.21) is very important and provides a physical understanding of the heat transfer coefficient. Fluids with high conductivity will, in general, provide a high heat transfer coefficient. For example, liquid water will almost always provide a higher heat transfer coefficient than air due, in part, to its much higher conductivity (0.60 W/m-K for water vs. 0.026 W/m-K for air at room temperature and atmospheric pressure). Furthermore, flow situations where the thermal boundary layer is thin will, in general, provide a high heat transfer coefficient. For example, a higher velocity flow will almost always lead to a higher heat transfer coefficient than a lower velocity flow; this result follows directly from Eq. (7.7).

Example 7.3

Air at atmospheric pressure flows over a flat surface. The free stream velocity of the air is $u_\infty = 5$ m/s and the free stream air temperature is $T_\infty = 20°C$. The length of the surface is $L = 0.25$ m. There is a uniform heat flux from the surface that is equal to $\dot{q}_s'' = 500$ W/m².

(a) Estimate the shear stress experienced by the surface at the trailing edge (i.e., at $x = L$). Do you expect the average shear stress experienced by the surface to be larger or smaller than this value?

(b) Estimate the surface temperature at the trailing edge. Sketch the variation in surface temperature that you expect as a function of position, x.

Known Values

The properties of air are obtained from Appendix C: $\mu = 18.54 \times 10^{-6}$ Pa-s, $v = 15.75 \times 10^{-6}$ m²/s, $k = 0.02638$ W/m-K, and $\alpha = 22.27 \times 10^{-6}$ m²/s. Figure 1 illustrates a sketch with the known information indicated.

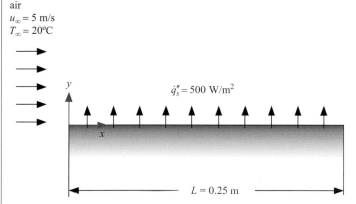

Figure 1 Sketch of the problem.

Assumptions

- The properties of air are constant and equal to the values obtained from Appendix C at 300 K.
- The flow remains laminar over the surface.

Analysis and Solution

The velocity boundary layer thickness at the trailing edge of the surface is estimated according to Eq. (7.9):

$$\delta_{m, x=L} \approx 2\sqrt{\frac{v L}{u_\infty}} = 2\sqrt{\frac{15.75 \times 10^{-6} \text{ m}^2}{\text{s}} \left|\frac{0.25 \text{ m}}{}\right| \frac{\text{s}}{5 \text{ m}}} = 0.00178 \text{ m}(1.78 \text{ mm}). \tag{1}$$

Continued

Example 7.3 (cont.)

The shear stress at the trailing edge is estimated from Eq. (7.16):

$$\tau_s \approx \mu \frac{u_\infty}{\delta_m} = \frac{18.54 \times 10^{-6} \text{ Pa s}}{\text{s}} \left| \frac{5 \text{ m}}{0.00178 \text{ m}} \right| = \boxed{0.0522 \text{ Pa}}. \tag{2}$$

The shear stress over the rest of the plate will be larger than the value at the trailing edge because the velocity boundary layer is thinner at smaller values of x. Therefore, the average shear stress experienced by the surface will be *larger* than the value calculated by Eq. (2).

The thermal boundary layer thickness at the trailing edge of the surface is estimated according to Eq. (7.7):

$$\delta_{t,\, x=L} \approx 2\sqrt{\frac{\alpha L}{u_\infty}} = 2\sqrt{\frac{22.27 \times 10^{-6} \text{ m}^2}{\text{s}} \left| \frac{0.25 \text{ m}}{} \right| \frac{\text{s}}{5 \text{ m}}} = 0.00211 \text{ m } (2.11 \text{ mm}). \tag{3}$$

The heat transfer coefficient is estimated using Eq. (7.21):

$$h \approx \frac{k}{\delta_t} = \frac{0.02638 \text{ W}}{\text{m K}} \left| \frac{1}{0.00211 \text{ m}} \right| = 12.5 \frac{\text{W}}{\text{m}^2 \text{ K}}. \tag{4}$$

The surface temperature is estimated using Eq. (7.20):

$$T_s = T_\infty + \frac{\dot{q}_s''}{h} = 20°\text{C} + \frac{500 \text{ W}}{\text{m}^2} \left| \frac{\text{m}^2 \text{ K}}{12.5 \text{ W}} \right| = \boxed{60°\text{C}}. \tag{5}$$

Figure 2 illustrates, qualitatively, the growth of the thermal boundary layer thickness with x. As the value of δ_t grows, the thermal resistance of the boundary layer increases according to Eq. (7.19), or equivalently the heat transfer coefficient is reduced according to Eq. (7.21). The result is that the surface temperature will rise in the x-direction, also shown in Figure 2.

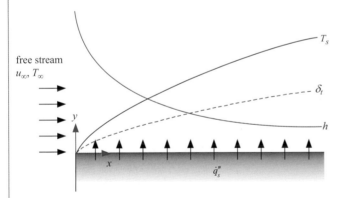

Figure 2 Sketch of the thermal boundary layer thickness, heat transfer coefficient, and surface temperature as a function of position.

Discussion

For a laminar flow, the thermal boundary layer thickness dictates the heat transfer coefficient. A thin boundary layer is desirable for heat transfer devices as it leads to a boundary layer with a small thermal resistance, which results in a high heat transfer coefficient. Understanding the thermal boundary layer thickness provides some qualitative understanding of how the heat transfer coefficient will vary along the surface.

The approximate relationships presented in this section should NOT be used to solve real engineering problems. More exact solutions for various types of flow configurations are presented in subsequent chapters as correlations between dimensionless variables. These correlations will provide results that are much more precise and useful.

7.1.6 The Reynolds Number

In Section 7.1.4 we encountered the Prandtl number, an important dimensionless number for convection heat transfer problems. A second dimensionless number that is important for these types of problems is the **Reynolds number** (Re). The Reynolds number is defined in a manner that depends on the flow situation and configuration. In general, the Reynolds number is defined according to:

$$Re = \frac{u_{char} \, L_{char} \, \rho}{\mu}, \tag{7.22}$$

where L_{char} is a characteristic length and u_{char} is a characteristic velocity associated with a problem, ρ is the density of the fluid, and μ is the viscosity of the fluid. The Reynolds number may be familiar to you from a fluid dynamics class. We will find in Section 7.3 that the Reynolds number is related to (but not directly equal to) the ratio of the inertial to the viscous forces that are present in the flow.

For flow over a flat surface, as shown in Figures 7.1 and 7.2, the characteristic length is the distance from the leading edge, x, and the characteristic velocity is the free stream velocity. Therefore, the local Reynolds number that is appropriate for this situation is

$$Re_x = \frac{u_\infty \, x \rho}{\mu}, \tag{7.23}$$

where the subscript x indicates that this definition of the Reynolds number uses x as the characteristic length. The simple models of the velocity and thermal boundary layer thicknesses that were developed in Section 7.1.3 can be expressed in terms of the Reynolds number and Prandtl number. Equation (7.9):

$$\delta_m \approx 2 \sqrt{\frac{\upsilon \, x}{u_\infty}} \tag{7.24}$$

can be rearranged to give

$$\frac{\delta_m}{x} \approx \frac{2}{x} \sqrt{\frac{\upsilon \, x}{u_\infty}} = 2 \sqrt{\frac{\upsilon}{u_\infty \, x}} = \frac{2}{\sqrt{\dfrac{u_\infty \, x}{\upsilon}}}. \tag{7.25}$$

Substituting the definition of kinematic viscosity into Eq. (7.25) leads to:

$$\frac{\delta_m}{x} \approx \frac{2}{\sqrt{\dfrac{u_\infty \, x \rho}{\mu}}}. \tag{7.26}$$

The argument of the square root in the denominator of Eq. (7.26) is the Reynolds number, Eq. (7.23):

$$\frac{\delta_m}{x} \approx \frac{2}{\sqrt{Re_x}} \tag{7.27}$$

The exact solution for the momentum boundary layer thickness can be obtained using a self-similar solution, as discussed by Schlichting (2000). The exact boundary layer thickness for laminar flow over a flat plate based on the definition that $u/u_\infty = 0.99$ is:

$$\frac{\delta_m}{x} \approx \frac{4.916}{\sqrt{Re_x}}. \tag{7.28}$$

Therefore, the conceptual model given by Eq. (7.27) is not perfect. However, it has certainly predicted the growth of the velocity boundary layer and its scale correctly, which should reinforce the idea that a laminar boundary layer is, to first order, a diffusive transport problem that can be understood using the concepts developed in Chapters 2 through 6 for conduction (the diffusive transport of energy).

The conceptual model of the thermal boundary layer thickness:

$$\delta_t \approx 2 \sqrt{\frac{\alpha \, x}{u_\infty}} \tag{7.29}$$

can also be re-arranged:

$$\frac{\delta_t}{x} \approx \frac{2}{x}\sqrt{\frac{\alpha x}{u_\infty}} = 2\sqrt{\frac{\alpha}{u_\infty x}} = 2\sqrt{\frac{\alpha}{u_\infty x}\frac{\upsilon}{\upsilon}} = \frac{2}{\sqrt{\frac{u_\infty x}{\upsilon}\frac{\upsilon}{\alpha}}} = \frac{2}{\sqrt{\frac{u_\infty x \rho}{\mu}}\sqrt{\frac{\upsilon}{\alpha}}}.$$ (7.30)

Introducing the definitions of the Reynolds number and Prandtl number provides:

$$\frac{\delta_t}{x} \approx \frac{2}{\sqrt{Re_x\, Pr}}.$$ (7.31)

The exact solution to this problem, discussed in Nellis and Klein (2009), shows that over a large range of Prandtl number the thermal boundary layer based on the definition $(T - T_s)/(T_\infty - T_s) = 0.99$ is given by:

$$\frac{\delta_t}{x} = \frac{4.916}{Re_x^{1/2}\, Pr^{1/3}}.$$ (7.32)

Again, the conceptual model has come close to the exact solution because the transport of thermal energy through the laminar boundary layer is similar to a conduction problem.

7.1.7 The Friction Coefficient and the Nusselt Number

Sections 7.1.4 and 7.1.6 introduced the Prandtl number and the Reynolds number, respectively. These dimensionless numbers are the independent variables used in the convective heat and momentum transfer correlations that are presented in Chapters 8 through 11 for a variety of situations. The engineering quantities of interest are typically the shear stress (or the drag force) and the heat transfer coefficient.

The correlations used to solve most convection heat transfer problems will provide the dimensionless heat transfer coefficient (the Nusselt number) in terms of the two independent dimensionless variables Pr and Re, for a specific flow situation (e.g., flow over a sphere). The major reason for using the dimensionless quantity (Nusselt number) as opposed to the dimensional quantity (heat transfer coefficient) is the dramatic reduction in the number of independent variables required to describe the problem. You may have seen this approach in a fluid dynamics class when, for example, a friction coefficient or drag coefficient for a specific flow situation was presented as a function of only a single variable, the Reynolds number. Section 7.3 will show how the governing differential equations associated with a convection problem *require* that the Nusselt number be a function only of the Reynolds and Prandtl numbers under most conditions. On the other hand, the heat transfer coefficient is a function of a larger number of dimensional parameters (e.g., velocity, viscosity, conductivity, characteristic length, etc.). It is therefore convenient to correlate the Nusselt number as a function of Reynolds number and Prandtl number based on either exact or approximate solutions or experimental data. Such a correlation can be used for a wide range of problems that span different operating conditions, fluid properties, and length scales. In this section we will define two dependent dimensionless parameters and examine them in the context of the simple, laminar boundary layer growth model that was the basis of Sections 7.1.3 and 7.1.5.

The shear stress is made dimensionless using the **friction coefficient** defined according to:

$$C_f = \frac{2\tau_s}{\rho\, u_\infty^2}.$$ (7.33)

The friction coefficient for a specific flow condition may be presented as a function only of the Reynolds number.

We can use our model of diffusive boundary layer growth to estimate the friction coefficient for the simple flow condition illustrated in Figure 7.1. Substituting Eq. (7.16) into Eq. (7.33) leads to:

$$C_f \approx \frac{2\mu}{\rho\, u_\infty\, \delta_m}.$$ (7.34)

Substituting the approximate equation for the momentum boundary layer thickness, Eq. (7.27), into Eq. (7.34) leads to:

$$C_f \approx \frac{2\mu}{\rho u_\infty} \frac{\sqrt{Re_x}}{2x}. \tag{7.35}$$

Equation (7.35) can be rearranged:

$$C_f \approx \frac{2\mu}{\rho u_\infty} \frac{\sqrt{Re_x}}{2x} = \frac{\mu}{\rho u_\infty x} \sqrt{Re_x} = \frac{1}{\sqrt{Re_x}}, \tag{7.36}$$

which should reinforce the idea that the friction coefficient can always be expressed in terms of an appropriately defined Reynolds number. The exact solution for laminar flow developing over a flat surface is presented by Schlichting (2000):

$$C_f = \frac{0.664}{\sqrt{Re_x}}. \tag{7.37}$$

Without solving the complicated partial differential equations that describe the boundary layer behavior, our simple conceptual model has shown that the friction coefficient varies with the inverse of the square root of Reynolds number.

The heat transfer coefficient is made dimensionless using the **Nusselt number** (Nu). The Nusselt number is defined in general according to:

$$Nu = \frac{h L_{char}}{k}, \tag{7.38}$$

where L_{char} is the characteristic dimension of the problem. Note that the characteristic length used to define the Nusselt number should be the same characteristic length that is used to define the Reynolds number for a specific problem. Therefore, for the flow situation shown in Figure 7.2, the characteristic length for the local Nusselt number is the distance from the leading edge of the surface, x :

$$Nu_x = \frac{h x}{k}. \tag{7.39}$$

Like the Reynolds number, a subscript on the Nusselt number can be used to indicate the specific characteristic length scale that is used in its definition.

We can again look at the Nusselt number using our simple model of diffusive boundary layer growth. Substituting Eq. (7.21) into Eq. (7.39) provides:

$$Nu_x \approx \frac{x}{\delta_t}. \tag{7.40}$$

Equation (7.40) is important as it suggests that the Nusselt number can be interpreted as the ratio of two lengths. The Nusselt number can be thought of as the ratio of the characteristic length that is appropriate for the problem (in this case x) to the distance that energy must be conducted in the fluid. For laminar flows over a flat surface, the conduction length scale is directly related to the thermal boundary layer thickness. Under most conditions, the laminar boundary layer thickness will be much smaller than the characteristic length of the surface. (This observation underlies the boundary layer simplifications that are discussed in Section 7.3.3.) Therefore the Nusselt number for this type of flow will always be much larger than unity. In other situations this will not be true. For example, internal laminar flow (e.g., flow through a pipe) is discussed in Chapter 9. In an internal flow, the thermal boundary layer is confined (by the other side of the pipe) and therefore cannot continue to grow. As a result, the thermal boundary layer thickness must eventually be approximately equal to the radius of the pipe ($\delta_t \approx R$). The characteristic length for an internal flow problem is defined as the pipe diameter ($L_{char} = 2R$). Therefore, if the Nusselt number is the interpreted as the ratio of the characteristic length ($2R$) to the thermal boundary layer thickness ($\approx R$) then the Nusselt number for fully developed laminar, internal flow through a round duct should be approximately 2.0. In fact, the Nusselt number for this situation ranges from 3.66 to 4.36, which demonstrates the validity of this approximation.

Substituting the conceptual model of the thermal boundary layer thickness, Eq. (7.31), into Eq. (7.40) leads to:

$$Nu \approx 0.5 \, Re_x^{0.5} \, Pr^{0.5}. \tag{7.41}$$

A more precise solution for the flow situation shown in Figure 7.2 is presented by Nellis and Klein (2009):

$$Nu_x = 0.332 \, Re_x^{0.5} \, Pr^{0.33}, \tag{7.42}$$

which is again impressively close to our approximate solution.

7.1.8 The Reynolds Analogy

The Reynolds analogy formalizes the similarity between the transport of energy and momentum. The Reynolds analogy is discussed formally in Section 7.4.4; however, it is worth introducing it here before diving into the details of the boundary layer equations in the next section. The Reynolds analogy simply states that for many fluids, the Prandtl number is near unity and therefore the momentum and thermal boundary layer thicknesses will be approximately the same ($\delta_m \approx \delta_t$). In this limit, it is possible to relate (approximately) the hydrodynamic characteristics of the problem (i.e., the friction coefficient) to the thermal characteristics of the problem (i.e., the Nusselt number).

Equation (7.40) expresses the Nusselt number in terms of the thermal boundary layer thickness:

$$Nu_x \approx \frac{x}{\delta_t}, \tag{7.43}$$

which can be solved for δ_t:

$$\delta_t \approx \frac{x}{Nu_x}. \tag{7.44}$$

Equation (7.34) expresses the friction coefficient as a function of the momentum boundary layer thickness:

$$C_f \approx \frac{2\mu}{\rho \, u_\infty \, \delta_m} \tag{7.45}$$

which can be solved for δ_m:

$$\delta_m \approx \frac{2\mu}{\rho \, u_\infty C_f}. \tag{7.46}$$

In the limit that the Prandtl number is near unity, Eq. (7.11) indicates that $\delta_t \approx \delta_m$ and therefore Eqs. (7.44) and (7.46) must be approximately equal:

$$\frac{x}{Nu_x} \approx \frac{2\mu}{\rho \, u_\infty C_f}. \tag{7.47}$$

Equation (7.47) can be written as a relationship between the local Nusselt number, friction coefficient, and Reynolds number:

$$Nu_x \approx \frac{C_f \, Re_x}{2}. \tag{7.48}$$

Equation (7.48) is the Reynolds analogy and expresses the thermal solution (Nu_x) in terms of the hydrodynamic solution (C_f). This is a useful result because it is often easier to measure or model the shear stress than the heat transfer coefficient.

The Reynolds analogy can be extended so that it is valid over a wider range of Prandtl number. According to Eq. (7.11), the ratio of the boundary layer thicknesses is, approximately:

$$\frac{\delta_t}{\delta_m} \approx \frac{1}{\sqrt{Pr}}. \tag{7.49}$$

Substituting Eqs. (7.44) and (7.46) into Eq. (7.49) leads to:

$$\frac{x}{Nu_x} \frac{\rho \, u_\infty C_f}{2\mu} \approx \frac{1}{\sqrt{Pr}} \tag{7.50}$$

or

$$Nu_x \approx \frac{Pr^{1/2} C_f Re_x}{2}.$$ (7.51)

Typically, the exponent on the Prandtl number is taken to be ⅓ rather than ½ in order to provide slightly better results:

$$Nu_x \approx \frac{Pr^{1/3} C_f Re_x}{2}.$$ (7.52)

The **modified Reynolds analogy** given by Eq. (7.52) is sometimes referred to as **the Chilton–Colburn analogy**. These analogies should be applied with some caution. There are physical phenomena that can cause the momentum and thermal boundary layers to be substantially different even when the Prandtl number is near unity. For example, large amounts of viscous dissipation or a strong pressure gradient will reduce the accuracy of the Reynolds or Chilton–Colburn analogy.

Example 7.4

Fluid with properties $\mu = 0.05$ Pa-s, $k = 0.62$ W/m-K, $\rho = 800$ kg/m^3, $c = 1200$ J/kg-K flows over a surface with free stream velocity $u_\infty = 7.2$ m/s. At a distance $x = 0.5$ m from the leading edge the local shear stress has been measured to be $\tau_s = 65$ Pa. Estimate the local heat transfer coefficient at that location.

Known Values

Figure 1 illustrates a sketch of the problem with the known information indicated.

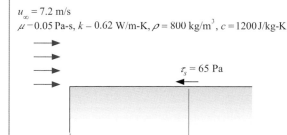

$u_\infty = 7.2$ m/s
$\mu - 0.05$ Pa-s, $k - 0.62$ W/m-K, $\rho = 800$ kg/m^3, $c = 1200$ J/kg-K

$\tau_s = 65$ Pa

$x = 0.5$ m

Figure 1 Sketch of the problem.

Assumptions

• The properties of the fluid are constant.
• The flow remains laminar over the surface.
• The Chilton–Colburn analogy applies.

Analysis and Solution

The Chilton–Colburn analogy relates the dimensionless heat transfer coefficient (the Nusselt number) to the Reynolds number, friction coefficient, and Prandtl number. The Reynolds number is computed according to:

Continued

Example 7.4 (cont.)

$$Re_x = \frac{\rho\, u_\infty\, x}{\mu} = \frac{800\ \text{kg}}{\text{m}^3}\left|\frac{7.2\ \text{m}}{\text{s}}\right|\frac{0.5\ \text{m}}{0.05\ \text{Pa s}}\left\|\frac{\text{Pa m}^2}{\text{N}}\right|\frac{\text{N s}^2}{\text{kg m}} = 5.76 \times 10^4. \tag{1}$$

The kinematic viscosity and thermal diffusivity are computed:

$$v = \frac{\mu}{\rho} = \frac{0.05\ \text{Pa s}}{}\left|\frac{\text{m}^3}{800\ \text{kg}}\right\|\frac{\text{N}}{\text{Pa m}^2}\left|\frac{\text{kg m}}{\text{N s}^2} = 6.25 \times 10^{-5}\ \frac{\text{m}^2}{\text{s}} \tag{2}$$

$$\alpha = \frac{k}{\rho c} = \frac{0.62\ \text{W}}{\text{m K}}\left|\frac{\text{m}^3}{800\ \text{kg}}\right|\frac{\text{kg K}}{1200\ \text{J}}\left\|\frac{\text{J}}{\text{W s}} = 6.46 \times 10^{-7}\ \frac{\text{m}^2}{\text{s}}. \tag{3}$$

The Prandtl number is:

$$Pr = \frac{v}{\alpha} = \frac{6.25 \times 10^{-5}\ \text{m}^2}{\text{s}}\left|\frac{\text{s}}{6.46 \times 10^{-7}\ \text{m}^2} = 96.8. \tag{4}$$

The shear stress is used to calculate the local friction coefficient:

$$C_f = \frac{2\tau_s}{\rho\, u_\infty^2} = \frac{2\left|65\ \text{Pa}\right|}{}\left|\frac{\text{m}^3}{800\ \text{kg}}\right|\frac{\text{s}^2}{(7.2)^2\ \text{m}^2}\left\|\frac{\text{N}}{\text{Pa m}^2}\right|\frac{\text{kg m}}{\text{N s}^2} = 0.00314. \tag{5}$$

The Chilton–Colburn analogy, Eq. (7.52), is used to calculate the local Nusselt number:

$$Nu_x \approx \frac{Pr^{1/3} C_f\, Re_x}{2} = \frac{(96.8)^{1/3}\left|0.00314\right|}{2}\left|\frac{5.7 \times 10^4}{} = 414.5. \tag{6}$$

The local Nusselt number is used to estimate the local heat transfer coefficient using Eq. (7.39):

$$h = \frac{Nu_x\, k}{x} = \frac{414.5\left|0.62\ \text{W}\right|}{}\left|\frac{}{\text{m K}}\right|\frac{}{0.5\ \text{m}} = \boxed{514\ \frac{\text{W}}{\text{m}^2\ \text{K}}}. \tag{7}$$

Discussion

The Chilton–Colburn analogy allowed a mechanical measurement (shear stress) to be used to infer a heat transfer coefficient. Because the Prandtl number is significantly different than unity ($Pr = 96.8$) it is necessary to use the Chilton–Colburn analogy rather than the Reynolds analogy for this problem.

7.1.9 Local vs. Average Quantities

The discussion about boundary layer behavior has been centered on the local value of the heat transfer coefficient (h) or shear stress (τ_s) that exists at some particular position on the surface. These quantities are used to define a local friction coefficient (C_f) and a local Nusselt number (Nu_x). It is almost always more useful to know the average value of these quantities when solving an engineering problem. You might have noticed that the Chapters 1 through 6 used an average heat transfer coefficient ($\bar{h}$) to specify boundary conditions rather than a local heat transfer coefficient (h).

The Average Friction Coefficient

The **average friction coefficient** ($\overline{C}_f$) is defined based on the average shear stress experienced by the plate ($\bar{\tau}_s$) over its entire surface:

$$\overline{C}_f = \frac{2\,\bar{\tau}_s}{\rho\, u_\infty^2}, \tag{7.53}$$

where the average shear stress is defined as

$$\bar{\tau}_s = \frac{1}{A_s} \int_{A_s} \tau_s \, dA_s \tag{7.54}$$

and A_s is the surface area of the surface. If you are interested in calculating the total force on a surface, then the average friction coefficient is much more useful than the local one. The local friction coefficient is correlated against a local Reynolds number defined based on local position x (Re_x). The average friction factor will be correlated against a Reynolds number defined based on a characteristic dimension of the surface. For a flat surface, this dimension is the length of the surface in the flow direction (Re_L):

$$Re_L = \frac{u_\infty L \rho}{\mu}. \tag{7.55}$$

The Drag Coefficient
Most objects experience both shear stress as well as substantial **form drag** (i.e., the force due to pressure gradients related to the flow changing direction). For example, flow over a sphere usually produces a low-pressure wake region on the downstream surface that is primarily responsible for the force on the sphere. In these situations the total **drag force** experienced by the object (F_D) is a combination of shear stress and pressure force. The total drag force is correlated using the **drag coefficient**, defined as

$$C_D = \frac{2 F_D}{\rho u_\infty^2 A_p}, \tag{7.56}$$

where A_p is the *projected area* of the object when viewed from the direction of the oncoming flow. The drag coefficient is inherently an average quantity as the drag force is related to the integration of the shear and pressure force over the entire surface. The drag coefficient is correlated against a Reynolds number defined based on a characteristic length, L_{char}. Care must be taken to be sure that the correct characteristic length is used when applying correlations for the drag coefficient.

The Average Nusselt Number
The total rate of heat transfer from an object is most conveniently calculated using the average rather than the local heat transfer coefficient. The average heat transfer coefficient ($\bar{h}$) is defined according to:

$$\bar{h} = \frac{1}{A_s} \int_{A_s} h \, dA_s \tag{7.57}$$

and the **average Nusselt number** ($\overline{Nu}$) is defined as:

$$\overline{Nu} = \frac{\bar{h} L_{char}}{k}. \tag{7.58}$$

The average Nusselt number is correlated against the Prandtl number and a Reynolds number defined based on a characteristic length, L_{char}. Again, care must be taken to ensure that the correct characteristic length is used when applying correlations for the average Nusselt number. Correlations for the average friction coefficient, drag coefficient, and the average Nusselt number for a variety of flow situations are presented in Chapter 8.

Example 7.5

The surface shown in Figure 1 is maintained at a constant temperature, $T_s = 100°C$, while fluid with free stream velocity $u_\infty = 2.3$ m/s and temperature $T_\infty = 20°C$ flows over it. The properties of the fluid are $\mu = 0.001$ Pa-s, $k = 0.2$ W/m-K, $\rho = 600$ kg/m^3, and $c = 3850$ J/kg-K.

Continued

Example 7.5 (cont.)

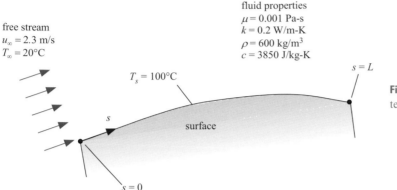

free stream
$u_\infty = 2.3$ m/s
$T_\infty = 20°C$

fluid properties
$\mu = 0.001$ Pa-s
$k = 0.2$ W/m-K
$\rho = 600$ kg/m³
$c = 3850$ J/kg-K

$T_s = 100°C$

$s = L$

surface

s

$s = 0$

Figure I Surface at a constant temperature.

Measurements of the local Nusselt number on the surface have provided the correlation:

$$Nu_s = 0.52 \, Re_s^{1/2} \, Pr^{1/3} \tag{1}$$

where s is a coordinate that follows the surface from the leading edge ($s = 0$) to the trailing edge ($s = L$, where $L = 0.14$ m). The width of the surface perpendicular to the flow is $W = 0.1$ m.

Determine:
(a) the local heat transfer coefficient at the trailing edge,
(b) the local heat flux at the trailing edge,
(c) the average heat transfer coefficient for the surface,
(d) the total rate of heat transfer from the surface.

Known Values

Figure 1 provided a sketch of the problem with the known information indicated.

Assumption

. The properties of the fluid are constant.

Analysis and Solution

The correlation provided by Eq. (1) can be used to determine the local Nusselt number and therefore the local heat transfer coefficient. The Reynolds number evaluated at the trailing edge of the surface is:

$$Re_{s=L} = \frac{\rho \, u_\infty L}{\mu} = \frac{600 \text{ kg}}{\text{m}^3} \left| \frac{2.3 \text{ m}}{\text{s}} \right| \frac{0.14 \text{ m}}{0.001 \text{ Pa s}} \left\| \frac{\text{Pa m}^2}{\text{N}} \right| \frac{\text{N s}^2}{\text{kg m}} = 1.93 \times 10^5. \tag{2}$$

The kinematic viscosity and thermal diffusivity are computed:

$$\upsilon = \frac{\mu}{\rho} = \frac{0.001 \text{ Pa s}}{} \left| \frac{\text{m}^3}{600 \text{ kg}} \right| \frac{\text{N}}{\text{Pa m}^2} \left| \frac{\text{kg m}}{\text{N s}^2} \right| = 1.67 \times 10^{-6} \frac{\text{m}^2}{\text{s}} \tag{3}$$

$$\alpha = \frac{k}{\rho c} = \frac{0.2 \text{ W}}{\text{m K}} \left| \frac{\text{m}^3}{600 \text{ kg}} \right| \frac{\text{kg K}}{3850 \text{ J}} \left\| \frac{\text{J}}{\text{W s}} \right| = 8.66 \times 10^{-8} \frac{\text{m}^2}{\text{s}}. \tag{4}$$

Example 7.5 (cont.)

The Prandtl number is:

$$Pr = \frac{\upsilon}{\alpha} = \frac{1.67 \times 10^{-6} \text{ m}^2}{\text{s}} \left| \frac{\text{s}}{8.66 \times 10^{-8} \text{ m}^2} = 19.3. \right. \tag{5}$$

Equation (1) is used to compute the local Nusselt number at the trailing edge:

$$Nu_{s=L} = 0.52 \, Re_{s=L}^{1/2} \, Pr^{1/3} = 0.52(1.93 \times 10^5)^{1/2}(19.3)^{1/3} = 612.6. \tag{6}$$

The local heat transfer coefficient at $s = L$ is therefore:

$$h_{s=L} = \frac{Nu_{s=L} k}{s} = \frac{612.6}{} \left| \frac{0.2 \text{ W}}{\text{m K}} \right| \frac{}{0.14 \text{ m}} = \boxed{875.1 \frac{\text{W}}{\text{m}^2 \text{ K}}}. \tag{7}$$

The local heat flux at the trailing edge is:

$$\dot{q}''_{s=L} = h_{s=L}(T_s - T_\infty) = \frac{875.1 \text{ W}}{\text{m}^2\text{K}} \left| \frac{(100 - 20)\text{K}}{} = \boxed{70,000 \frac{\text{W}}{\text{m}^2}}. \right. \tag{8}$$

The average heat transfer coefficient must be obtained by integrating the local heat transfer coefficient over the entire plate surface, as given by Eq. (7.57):

$$\bar{h} = \frac{1}{A_s} \int_{A_s} h \, dA_s. \tag{9}$$

The correlation provided by Eq. (1) is substituted into Eq. (7):

$$h = \frac{k}{s} 0.52 \, Re_s^{1/2} \, Pr^{1/3}. \tag{10}$$

Equation (2) (with local position s replacing L) is substituted into Eq. (10) in order to obtain a relationship between local heat transfer coefficient and position on the plate:

$$h = \frac{k}{s} 0.52 \sqrt{\frac{\rho u_\infty}{\mu}} \, Pr^{1/3} \, s^{1/2} = 0.52 \, k \sqrt{\frac{\rho u_\infty}{\mu}} \, Pr^{1/3} \, s^{-1/2}. \tag{11}$$

Equation (11) is substituted into Eq. (9):

$$\bar{h} = \frac{1}{A_s} \int_{A_s} 0.52 \, k \sqrt{\frac{\rho u_\infty}{\mu}} Pr^{1/3} \, s^{-1/2} \, dA_s. \tag{12}$$

The surface area and differential surface area are expressed as:

$$A_s = W \, L \tag{13}$$

$$dA_s = W \, ds. \tag{14}$$

Equations (13) and (14) are substituted into Eq. (12):

$$\bar{h} = \frac{1}{W \, L} \int_{s=0}^{s=L} 0.52 \, k \sqrt{\frac{\rho u_\infty}{\mu}} Pr^{1/3} \, s^{-1/2} \, W \, ds. \tag{15}$$

The integral indicated by Eq. (15) is carried out:

$$\bar{h} = \frac{1}{L} 0.52 \, k \sqrt{\frac{\rho u_\infty}{\mu}} Pr^{1/3} \int_{s=0}^{s=L} s^{-1/2} \, ds = \frac{2}{L} 0.52 \, k \sqrt{\frac{\rho u_\infty}{\mu}} Pr^{1/3} \left[s^{1/2} \right]_{s=0}^{s=L}$$

$$= 1.04 \frac{k}{L} \sqrt{\frac{\rho u_\infty}{\mu}} Pr^{1/3} L^{1/2}. \tag{16}$$

Continued

Example 7.5 (cont.)

The definition of the Reynolds number based on L is substituted into Eq. (16):

$$\bar{h} = 1.04 \frac{k}{L} Re_L^{1/2} Pr^{1/3}. \tag{17}$$

The average heat transfer coefficient is therefore:

$$\bar{h} = 1.04 \frac{k}{L} Re_L^{1/2} Pr^{1/3} = \frac{1.04}{\left|} \frac{0.2 \text{ W}}{\text{m K}} \right| \frac{\left(1.93 \times 10^5\right)^{1/2}}{0.14 \text{ m}} \left| (19.3)^{1/3} \right| = \boxed{1750 \frac{\text{W}}{\text{m}^2 \text{ K}}} \tag{18}$$

and the total rate of heat transfer from the surface is:

$$\dot{q} = \bar{h} \, W \, L (T_s - T_\infty) = \frac{1750 \text{ W}}{\text{m}^2 \text{ K}} \left| \frac{0.1 \text{ m}}{} \right| 0.14 \text{ m} \left| \frac{(100 - 20) \text{ K}}{} \right| = \boxed{1960 \text{ W}}. \tag{19}$$

Discussion

The average heat transfer coefficient is larger than the local heat transfer coefficient at the trailing edge because the heat transfer coefficient over the surface is smallest at the trailing edge where the thermal boundary layer is thickest.

Careful examination of Eq. (18) shows that the average Nusselt number is twice the value of the local Nusselt number evaluated at the trailing edge:

$$\frac{\bar{h} \, L}{k} = \overline{Nu}_L = 1.04 \, Re_L^{1/2} Pr^{1/3} = 2 \left(0.52 Re_{s=L}^{1/2} Pr^{1/3} \right) = 2 Nu_{s=L}. \tag{20}$$

This is a typical result for laminar flow.

7.2 Turbulent Boundary Layer Concepts

7.2.1 Introduction

In Section 7.1 we found that the existence of a *laminar* boundary layer is the result of the *diffusive* penetration of momentum and thermal energy into the free stream. As a result, laminar flow will persist only when the action of viscosity is strong relative to inertial forces. When inertial forces become sufficiently large, the flow "transitions" from laminar to turbulent. A laminar boundary layer is highly ordered. The trajectories of fluid particles (streamlines) are smooth and an instrument placed in a steady laminar flow will report a constant value of velocity, temperature, etc., as shown in Figure 7.7 (for location A). A **turbulent boundary layer** is characterized by vortices or eddies that randomly exist at many time and length scales. Streamlines cannot be easily identified because fluid particles, while always moving on average in the *x*-direction with the mean flow, will also move randomly up and down in the *y*-direction under the influence of turbulent eddies. The fluid may even occasionally move in the opposite direction of the mean flow. An instrument placed in a turbulent flow will report an *oscillating* value of velocity, temperature, etc., as shown in Figure 7.7 (for location B), even if overall behavior of the problem is otherwise at steady state.

The characteristics of a turbulent flow are discussed in this section so that the differences between laminar and turbulent flow are clear. A complete discussion of turbulent flow is beyond the scope of this book. There are several good references that provide a more thorough presentation of turbulent flows; for example, Tennekes and Lumley (1972), Hinze (1975), and Schlichting (2000).

7.2.2 The Critical Reynolds Number

In Section 7.1.6 the Reynolds number was introduced as

$$Re = \frac{u_{char} \, L_{char} \, \rho}{\mu}, \tag{7.59}$$

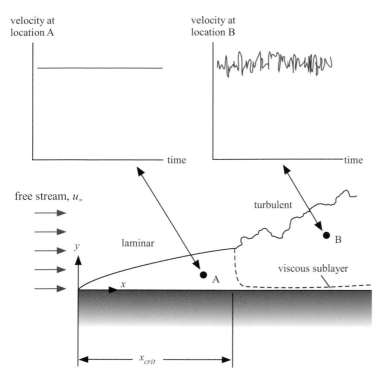

Figure 7.7 Flow over a surface transitioning from laminar to turbulent.

where L_{char} is a characteristic length associated with the flow situation. Further, it was stated that the Reynolds number is related to the ratio of inertial to viscous forces in the flow. As such, it is not surprising that the transition from laminar (viscous dominated) to turbulent (inertia dominated) flow corresponds to an increase in the Reynolds number. The local Reynolds number defined for a flat surface uses a characteristic length x, which is the distance from the leading edge:

$$Re_x = \frac{u_\infty x \rho}{\mu}. \tag{7.60}$$

This Reynolds number defined by Eq. (7.60) is not exactly *equal* to the ratio of inertial to viscous forces. The ratio of the flow inertia

$$\text{inertia} \approx \rho u_\infty^2 \tag{7.61}$$

to its viscous shear

$$\text{viscous shear} \approx \mu \frac{u_\infty}{\delta_m} \tag{7.62}$$

is given by

$$\frac{\text{inertia}}{\text{viscous}} \approx \frac{\rho u_\infty^2 \delta_m}{\mu u_\infty}. \tag{7.63}$$

Equation (7.63) can be rearranged to provide a Reynolds number that is defined using the velocity boundary layer thickness as the characteristic length:

$$Re_{\delta_m} = \frac{\rho u_\infty \delta_m}{\mu}. \tag{7.64}$$

The Reynolds number defined by Eq. (7.64) is more correctly equal to the ratio of inertial to viscous forces. We would expect laminar flow when Eq. (7.64) is less than unity and turbulent flow when it becomes much larger than unity, which corresponds physically to the boundary layer growing to the point where the viscous shear is small. However, experimental measurements show that there is no single value of Re_{δ_m} that characterizes the transition from laminar to turbulent flow. In fact, the transition to turbulent flow is neither sudden nor precise. The conditions under which the flow will transition to turbulence depend on the shape of the leading edge, the presence of de-stabilizing oscillations (e.g., structural vibrations), the roughness of the surface, the character of the free stream, etc. Also, the flow will not transition from completely laminar to completely turbulent flow at a particular location. Instead, there will be a region that is characterized by local, intermittent bursts of turbulence that progressively become more intense as the flow moves downstream; eventually the flow produces a fully turbulent boundary layer. Experimental measurements show that the critical Reynolds number based on velocity boundary layer thickness can range from $700 < Re_{\delta_m, crit} < 5000$ depending on flow conditions. A critical value of $Re_{\delta_m, crit} = 3500$ is typically assumed.

The Reynolds number based on the velocity boundary layer thickness, Re_{δ_m}, is not a convenient parameter for use in engineering calculations. Rather, the Reynolds number based on x-position, Re_x, given by Eq. (7.60) is typically used. These two definitions of Reynolds number, Re_{δ_m} and Re_x, can be related to one another using the solution for the velocity boundary layer thickness, Eq. (7.28), repeated below:

$$\frac{\delta_m}{x} = \frac{4.916}{\sqrt{Re_x}}. \tag{7.65}$$

Substituting Eq. (7.65) into Eq. (7.64) provides:

$$Re_{\delta_m} = \frac{\rho\, u_\infty \delta_m}{\mu} = \frac{\rho\, u_\infty}{\mu}\frac{4.916x}{\sqrt{Re_x}} = 4.916\sqrt{Re_x}. \tag{7.66}$$

Rearranging Eq. (7.66) provides:

$$Re_x = 0.041\, Re_{\delta_m}^2. \tag{7.67}$$

Equation (7.67) suggests that the critical Reynolds number based on x at which the transition to turbulence occurs may lie between $20{,}000 < Re_{x,crit} < 1 \times 10^6$ (corresponding to $700 < Re_{\delta_m, crit} < 5000$). A typical value of $Re_{x,crit} = 5 \times 10^5$ is assumed for most calculations.

Example 7.6

An $L = 6$ ft long kayak is being paddled at $u_\infty = 3$ knot (1.54 m/s) through still water at $T_\infty = 20°C$. Determine whether the velocity boundary layer that forms on the hull will be mostly laminar or turbulent.

Known Values

Figure 1 provided a sketch of the problem with the known information indicated (converted to base SI units). The properties of water are obtained from Appendix D: $\mu = 0.001$ Pa-s and $\rho = 998.2$ kg/m^3.

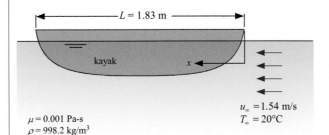

Figure 1 Sketch of problem with known information shown.

Example 7.6 (cont.)

Assumptions

- The properties of the fluid are constant.
- The critical Reynolds number based on x is $Re_{x,crit} = 5 \times 10^5$.
- The kayak hull can be treated as a flat plate.

Analysis and Solution

The Reynolds number based on x is defined as:

$$Re_x = \frac{u_\infty x \rho}{\mu}. \tag{1}$$

Solving for x provides:

$$x = \frac{\mu Re_x}{u_\infty \rho}. \tag{2}$$

We will introduce the critical Reynolds number, $Re_{x,crit} = 5 \times 10^5$, into Eq. (2) in order to estimate the position along the hull at which the flow transitions to turbulence:

$$x_{crit} = \frac{\mu Re_{x,\,crit}}{u_\infty \rho} = \frac{0.001 \text{ Pa s} \left|5 \times 10^5\right|}{\left|\frac{\text{s}}{1.54 \text{ m}}\right|\frac{\text{m}^3}{998.2 \text{ kg}}\left|\frac{\text{N}}{\text{Pa m}^2}\right|\frac{\text{kg m}}{\text{N s}^2}} = \boxed{0.32 \text{ m}(1.07 \text{ ft})}. \tag{3}$$

Therefore, approximately 80 percent of the hull will experience turbulent flow.

Discussion

The velocity boundary layer thickness at the point where the flow transitions to turbulence can be estimated using Eq. (7.65):

$$\delta_{m,\,crit} = \frac{4.916}{\sqrt{Re_{x,\,crit}}} x_{crit} = \frac{4.916}{\sqrt{5 \times 10^5}} 0.32 \text{ m} = 0.0022 \text{ m}(2.2 \text{ mm}). \tag{4}$$

When the velocity boundary layer thickness increases beyond this point, the viscous shear has diminished to the point where the flow transitions to turbulence.

7.2.3 A Conceptual Model of the Turbulent Boundary Layer

In laminar flow, the transport of momentum and energy in the y-direction (i.e., perpendicular to the free stream flow) occurs primarily due to diffusion because the y-directed velocity is quite small. Molecules of fluid with high energy collide with those at lower energy and this interaction leads to the transport of both energy and momentum. The fluid properties that characterize this microscale process are thermal conductivity and viscosity; fluid at a given state will always possess the same thermal conductivity and viscosity. When studying turbulence you will often see some properties identified as being **molecular**, for example, the *molecular* viscosity and *molecular* thermal conductivity, to emphasize that they are state properties.

In a turbulent flow, the transport of momentum and energy continues to occur due to microscale diffusive transport processes. However, the chaotic motion of macroscale packets of fluid under the influence of the turbulent eddies provides an additional (and very efficient mechanism) for transporting momentum and energy. Much more energy is transported when a large packet of fluid is moved away from the wall by a turbulent eddy than would be transported by diffusive conduction alone in the same amount of time. Therefore, the presence of the turbulent eddies increases the *effective* viscosity and conductivity of the fluid tremendously everywhere that these eddies exist. Turbulent eddies exist throughout the entire turbulent boundary layer except in a very thin layer near the wall that is referred to as the **viscous sublayer**. The turbulent eddies are suppressed in the viscous

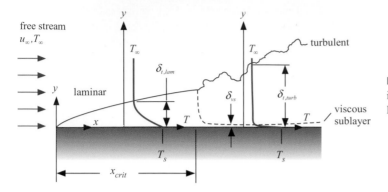

Figure 7.8 The temperature distribution in the laminar and turbulent boundary layers.

sublayer by the viscous action of the wall itself and therefore the flow in the viscous sublayer behaves as if it were laminar. The thickness of the viscous sublayer (δ_{vs}) is, however, much thinner than the thickness of a laminar boundary layer under comparable conditions.

This description of a turbulent boundary layer is sufficient to provide a conceptual model of turbulent flow that explains many of its characteristics. Figure 7.8 illustrates a sketch of the temperature distribution that exists within the laminar and turbulent boundary layers. In both the laminar and turbulent regions, the temperature of the fluid must vary from T_s at the plate surface ($y = 0$) to T_∞ at the edge of the thermal boundary layer ($y = \delta_t$). However, the temperature distribution in the laminar boundary layer is quite different than it is in the turbulent boundary layer. While the temperature distribution in the laminar boundary layer is not linear, it does change smoothly and gradually over the entire extent of the boundary layer as shown in Figure 7.8. This behavior is consistent with the discussion in Section 7.1.5, in which laminar convection heat transfer is treated, approximately, as conduction through a layer of fluid with thickness $\delta_{t,lam}$ (i.e., the laminar boundary layer thickness).

The heat flux through a laminar boundary layer ($\dot{q}''_{lam}$) can be expressed approximately as

$$\dot{q}''_{lam} \approx \frac{(T_s - T_\infty)}{\left(\dfrac{\delta_{t,\,lam}}{k}\right)} = h_{lam}(T_s - T_\infty), \tag{7.68}$$

where k is the conductivity of the fluid. The area-specific thermal resistance of a laminar boundary layer is, approximately, $\delta_{t,lam}/k$, and therefore the heat transfer coefficient for a laminar flow (h_{lam}) is, approximately:

$$h_{lam} \approx \frac{k}{\delta_{t,\,lam}}. \tag{7.69}$$

Energy that is transported through a turbulent boundary layer to the free stream must pass through two regions of fluid. The viscous sublayer is a very thin layer of fluid close to the wall where no turbulent eddies exist. Therefore, the area-specific thermal resistance of the viscous sublayer is, approximately, δ_{vs}/k, where k is the thermal conductivity of the fluid itself (i.e., the *molecular* conductivity). The bulk of the boundary layer (the turbulent core) is characterized by large turbulent eddies and therefore has a large, *effective* thermal conductivity (k_{turb}). The area-specific thermal resistance of the turbulent core is $\delta_{t,turb}/k_{turb}$, where $\delta_{t,turb}$ is the turbulent boundary layer thickness. It is important to recognize that k_{turb} is *not* a fluid property, but rather a characteristic of the flow itself; k_{turb} is sometimes referred to as the **eddy conductivity of the fluid**.

The heat flux through the turbulent boundary layer ($\dot{q}''_{turb}$) can be expressed, approximately, as:

$$\dot{q}''_{turb} \approx \frac{(T_s - T_\infty)}{\dfrac{\delta_{vs}}{k} + \dfrac{\delta_{t,\,turb}}{k_{turb}}} = h_{turb}(T_s - T_\infty). \tag{7.70}$$

A turbulent boundary layer therefore behaves somewhat like a composite wall consisting of a thin, low-conductivity material (the viscous sublayer) beneath a thicker, but extremely high-conductivity material (the turbulent core). The thermal resistance of the viscous sublayer is much larger than that of the turbulent core (i.e., the first term in the denominator of Eq. (7.70) is much larger than the second term). In such a series combination of thermal resistances, the largest resistance will dominate the problem. Therefore, the majority of the temperature change will occur across the viscous sublayer, as shown in the sketch of the temperature distribution in Figure 7.8.

The characteristics of the viscous sublayer will dictate the behavior of a turbulent boundary layer. According to Eq. (7.70), the heat transfer coefficient for a turbulent boundary layer can be written approximately as:

$$h_{turb} \approx \frac{1}{\dfrac{\delta_{vs}}{k} + \dfrac{\delta_{t,\,turb}}{k_{turb}}}. \tag{7.71}$$

Because the thermal resistance of the viscous sublayer is much larger than that of the turbulent core, Eq. (7.71) can be approximated by:

$$h_{turb} \approx \frac{k}{\delta_{vs}}. \tag{7.72}$$

Comparing Eqs. (7.69) and (7.72) and recalling that $\delta_{vs} << \delta_{t,turb}$ suggests that $h_{turb} >> h_{lam}$. Figure 7.9 illustrates, qualitatively, the variation of the heat transfer coefficient and shear stress with position along the plate. There is a transition region between a fully laminar and fully turbulent boundary layer where the variation of heat transfer coefficient is not clear. However, the heat transfer coefficient will increase substantially when the flow transitions from laminar to turbulent.

The shear stress at the surface of the plate in the laminar boundary layer ($\tau_{s,lam}$) is the product of the fluid viscosity and the velocity gradient at the plate surface and is given, approximately, by

$$\tau_{s,\,lam} = \mu \frac{\partial u}{\partial y}\bigg|_{y=0} \approx \mu \frac{u_{\infty}}{\delta_{m,\,lam}}, \tag{7.73}$$

where $\delta_{m,lam}$ is the velocity boundary layer thickness in a laminar flow and μ is the viscosity of the fluid. The shear stress at the surface of the plate in the turbulent boundary layer ($\tau_{s,turb}$) is also equal to the product of the viscosity of the fluid (i.e., the *molecular* viscosity) and the velocity gradient at the plate. However, in a turbulent flow the velocity gradient will be much steeper as the velocity change will occur primarily across the viscous sublayer. Therefore, the shear stress at the plate surface in a turbulent flow is given, approximately, by

$$\tau_{s,\,turb} = \mu \frac{\partial u}{\partial y}\bigg|_{y=0} \approx \mu \frac{u_{\infty}}{\delta_{vs}}. \tag{7.74}$$

Comparing Eqs. (7.73) and (7.74) suggests that the shear stress will increase substantially when the flow transitions from laminar to turbulent, as shown in Figure 7.9.

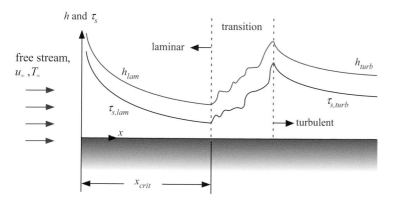

Figure 7.9 Qualitative variation of the heat transfer coefficient and shear stress along a surface.

The simple conceptual model described in this section provides some understanding of the differences between laminar and turbulent flow and the associated effect on the heat transfer coefficient and shear stress. While the correlations provided in subsequent chapters should be used to calculate these quantities, this discussion explains the behavior predicted by the correlations. Why does the heat transfer coefficient increase when flow becomes turbulent? What should the velocity distribution look like?

As an example, we can understand now why the surface roughness (e.g., imperfections on the surface) is often important for a turbulent flow. In order to affect the shear stress or heat transfer coefficient, the surface roughness must be sufficiently large that it disrupts the velocity and temperature gradient that exists at the surface ($y = 0$). Figure 7.8 suggests that the size of the roughness required to change the characteristics of a turbulent boundary layer is on the order of the viscous sublayer, which is much less than the size that is required to disrupt a laminar boundary layer. Therefore, the behavior of a turbulent flow is more sensitive to surface roughness than laminar flow. Indeed, correlations for the friction coefficient or Nusselt number in a laminar flow will almost never include any provision for specifying the surface roughness whereas turbulent flow correlations often do.

7.3 The Boundary Layer Equations

7.3.1 Introduction

In Sections 7.1 and 7.2, boundary layers are discussed at a conceptual level in order to obtain some physical feel for their behavior. In this section, the partial differential equations that govern the behavior of fluid flow are derived and then simplified for the special case of flow inside of a boundary layer. In Section 7.4, the boundary layer equations are made dimensionless in order to identify the minimum set of nondimensional parameters that govern the boundary layer problem and also identify their significance. These steps are accomplished in order to justify the use of the correlations that are presented in Chapters 8 through 11 and used extensively to solve convection heat transfer problems.

7.3.2 The Governing Equations for Viscous Fluid Flow

The governing differential equations for fluid flow are derived by enforcing the conservation of mass, momentum (in each direction), and thermal energy at every position within the fluid. In Chapters 2 through 6, differential equations are derived for a variety of conduction problems by applying conservation of thermal energy to a differential control volume. The sequence of steps used to derive the governing differential equations for a fluid flow in this chapter is essentially the same. A differentially small control volume is defined and used to write a conservation equation. The terms in the conservation equation are expanded and the first term in the Taylor series is retained. Finally, appropriate rate equations (e.g., Fourier's Law) are substituted into the equation. This process is illustrated for a 2-D flow with constant properties in Cartesian coordinates in order to demonstrate the steps. The resulting partial differential equations are provided in Table 7.1. Table 7.2 lists the governing equations for 2-D flow with constant properties in radial coordinates.

The Continuity Equation

The **continuity equation** enforces mass conservation for a differential control volume. A differential control volume is shown in Figure 7.10 for a 2-D flow using Cartesian coordinates. A mass balance on the differential control volume leads to:

$$(\rho u)_x \, dy \, W + (\rho v)_y \, dx \, W = (\rho u)_{x+dx} \, dy \, W + (\rho v)_{y+dy} \, dx \, W + dx \, dy \, W \frac{\partial \rho}{\partial t}, \tag{7.75}$$

where W is the width of the control volume in the z-direction (into the page) and u and v are the velocities in the x- and y-directions, respectively. We will assume that the fluid is **incompressible** so that the density is constant. Therefore, Eq. (7.75) can be simplified to:

$$u_x \, dy + v_y \, dx = u_{x+dx} \, dy + v_{y+dy} \, dx. \tag{7.76}$$

Table 7.1 The governing equations for 2-D constant property flow in Cartesian coordinates (u and v are the velocities in the x- and y-directions, respectively).

continuity	$\dfrac{\partial u}{\partial x} + \dfrac{\partial v}{\partial y} = 0$
x-momentum	$\rho\left(\dfrac{\partial u}{\partial t} + u\dfrac{\partial u}{\partial x} + v\dfrac{\partial u}{\partial y}\right) = -\dfrac{\partial p}{\partial x} + \mu\left(\dfrac{\partial^2 u}{\partial x^2} + \dfrac{\partial^2 u}{\partial y^2}\right) + \rho g_x$
y-momentum	$\rho\left(\dfrac{\partial v}{\partial t} + u\dfrac{\partial v}{\partial x} + v\dfrac{\partial v}{\partial y}\right) = -\dfrac{\partial p}{\partial y} + \mu\left(\dfrac{\partial^2 v}{\partial y^2} + \dfrac{\partial^2 v}{\partial x^2}\right) + \rho g_y$
energy	$\rho c\left(\dfrac{\partial T}{\partial t} + u\dfrac{\partial T}{\partial x} + v\dfrac{\partial T}{\partial y}\right) = k\left(\dfrac{\partial^2 T}{\partial x^2} + \dfrac{\partial^2 T}{\partial y^2}\right) + \Phi$
viscous dissipation function	$\Phi = 2\mu\left[\left(\dfrac{\partial u}{\partial x}\right)^2 + \left(\dfrac{\partial v}{\partial y}\right)^2\right] + \mu\left(\dfrac{\partial u}{\partial y} + \dfrac{\partial v}{\partial x}\right)^2$

Table 7.2 The governing equations for 2-D (x–r) constant property flow in cylindrical coordinates (u and v are the velocities in the x- and r-directions, respectively).

continuity	$\dfrac{\partial u}{\partial x} + \dfrac{1}{r}\dfrac{\partial (rv)}{\partial r} = 0$
x-momentum	$\rho\left(\dfrac{\partial u}{\partial t} + u\dfrac{\partial u}{\partial x} + v\dfrac{\partial u}{\partial r}\right) = -\dfrac{\partial p}{\partial x} + \mu\left(\dfrac{\partial^2 u}{\partial r^2} + \dfrac{1}{r}\dfrac{\partial u}{\partial r} + \dfrac{\partial^2 u}{\partial x^2}\right) + \rho g_x$
r-momentum	$\rho\left(\dfrac{\partial v}{\partial t} + u\dfrac{\partial v}{\partial x} + v\dfrac{\partial v}{\partial r}\right) = -\dfrac{\partial p}{\partial r} + \mu\left(\dfrac{\partial^2 v}{\partial r^2} + \dfrac{1}{r}\dfrac{\partial v}{\partial r} - \dfrac{v}{r^2} + \dfrac{\partial^2 v}{\partial x^2}\right)$
energy	$\rho c\left(\dfrac{\partial T}{\partial t} + u\dfrac{\partial T}{\partial x} + v\dfrac{\partial T}{\partial r}\right) = k\left[\dfrac{\partial^2 T}{\partial x^2} + \dfrac{1}{r}\dfrac{\partial}{\partial r}\left(r\dfrac{\partial T}{\partial r}\right)\right] + \Phi$
viscous dissipation	$\Phi = 2\mu\left[\left(\dfrac{\partial u}{\partial x}\right)^2 + \left(\dfrac{v}{r}\right)^2 + \left(\dfrac{\partial v}{\partial r}\right)^2\right] + \mu\left(\dfrac{\partial u}{\partial r} + \dfrac{\partial v}{\partial x}\right)^2$

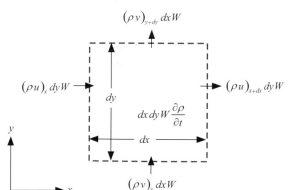

Figure 7.10 A mass balance on a differential control volume.

The terms at $x + dx$ and $y + dy$ are expanded:

$$u_x\, dy + v_y\, dx = \left(u_x + \frac{\partial u}{\partial x} dx \right) dy + \left(v_y + \frac{\partial v}{\partial y} dy \right) dx. \tag{7.77}$$

Equation (7.77) can be simplified to:

$$\frac{\partial u}{\partial x} + \frac{\partial v}{\partial y} = 0. \tag{7.78}$$

Equations (7.78) is referred to as the continuity equation for an incompressible flow. The equivalent equation in cylindrical coordinates is listed in Table 7.2.

The Momentum Equations

The concept of momentum equations may be familiar from fluid dynamics. Momentum can be defined relative to each of the coordinate directions. Therefore, there will be as many momentum equations as there are dimensions to the problem. For the 2-D problem in Cartesian coordinates considered here, momentum equations must be derived in both the x- and y-directions. Momentum is not a conserved quantity in the same way that mass or energy is. Rather, the momentum principle states that the sum of the forces in a given direction will add to the momentum in that direction. In the x-direction this leads to the equation:

$$\text{rate of } x\text{-momentum in} + \sum \text{forces in } x\text{-direction} =$$
$$\text{rate of } x\text{-momentum out} + \frac{\partial}{\partial t}(x\text{-momentum stored}). \tag{7.79}$$

Figure 7.11 illustrates the x-directed momentum terms and forces for a differential control volume.

It is worth discussing the source of the various terms that appear in Figure 7.11, starting with the momentum terms. The x-directed momentum per unit mass of fluid entering the control volume is simply the x-velocity, u. Therefore, the momentum transfer terms are the product of the x-directed momentum per unit mass and the appropriate mass flow rates (which are shown in Figure 7.10). The inflow of x-momentum passing through the control surface on the left side of the control volume is:

$$\text{momentum flow rate entering left side} = \underbrace{(\rho u)_x dy\, W}_{\text{mass flow rate}}\, u \tag{7.80}$$

and the inflow of x-momentum through the bottom control surface is:

$$\text{momentum flow rate entering bottom} = \underbrace{(\rho v)_y dx\, W}_{\text{mass flow rate}}\, u. \tag{7.81}$$

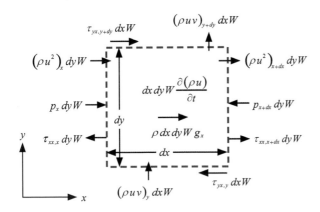

Figure 7.11 The x-directed momentum and forces on a differential control volume.

There are associated outflows of momentum through the control surfaces on the right and top control surfaces at $x + dx$ and $y + dy$, respectively.

The amount of momentum stored in the control volume is equal to the product of the momentum per unit mass (u) and the mass of fluid in the control volume. The rate of momentum storage is the time derivative of this quantity:

$$\text{rate of change of momentum} = \frac{\partial}{\partial t} \left(\underbrace{\rho \, dx \, dy \, W}_{\text{mass}} u \right). \tag{7.82}$$

The x-directed forces shown in Figure 7.11 include pressure forces, shear forces, and a gravitational force. The pressure forces in the x-direction are exerted on the left and right faces of the control volume and are the product of the area of these faces and the pressure (p) acting on these faces. The pressure force acting on the left side control surface is:

$$\text{pressure force on left side} = \underbrace{p_x}_{\substack{\text{pressure at} \\ \text{position } x}} \underbrace{dy \, W}_{\text{area}}. \tag{7.83}$$

There is an associated pressure force acting on the right side (at $x + dx$) in the opposite direction.

The gravitational force is the product of the mass of fluid that is contained in the control volume and the component of gravity acting in the x-direction, g_x. If the x-direction is perpendicular to the gravity vector, then this force is zero:

$$\text{gravitational force} = \underbrace{g_x}_{\substack{x\text{-component} \\ \text{of gravity}}} \underbrace{\rho \, dx \, dy \, W}_{\text{mass}}. \tag{7.84}$$

There are two additional forces considered in Figure 7.11 that are related to the viscous stresses caused by the fluid motion, τ_{yx} and τ_{xx}; note that the first subscript indicates the face that the stress acts *on* and the second subscript indicates the direction that the stress acts *in*. The tangential stress (or shear, τ_{yx}) acts on the bottom and top (i.e., the y-directed) faces while the normal stress (τ_{xx}) acts on the left and right (i.e., the x-directed) faces. The x-directed viscous shear force acting on the bottom control surface is

$$\text{viscous force on bottom surface} = \underbrace{\tau_{yx, \, y}}_{\substack{\text{viscous stress on} \\ \text{bottom surface at} \\ \text{position } y}} \underbrace{dx \, W}_{\text{area}} \tag{7.85}$$

and the x-directed viscous normal force acting on the left control surface is

$$\text{viscous force on left surface} = \underbrace{\tau_{xx, \, x}}_{\substack{\text{viscous stress on} \\ \text{left surface at} \\ \text{position } x}} \underbrace{dy \, W}_{\text{area}}. \tag{7.86}$$

There are corresponding forces acting on the top and right control surfaces at $y + dy$ and $x + dx$, respectively.

The development of the momentum equation in the x-direction on the differential control volume shown in Figure 7.11 is a matter of placing each of these terms into Eq. (7.79). Notice from Figure 7.11 that the forces acting in the positive x-direction are placed on the left side of Eq. (7.87), while those acting in the negative x-direction are placed on the right:

$$\rho \, dx \, dy \, W g_x + \left(\rho u^2 \right)_x dy \, W + \left(\rho u v \right)_y dx \, W + p_x dy \, W + \tau_{yx, \, y+dy} \, dx \, W + \tau_{xx, \, x+dx} \, dy \, W =$$
$$\left(\rho u^2 \right)_{x+dx} dy \, W + \left(\rho u v \right)_{y+dy} dx \, W + p_{x+dx} dy \, W + \tau_{yx, \, y} dx \, W + \tau_{xx, \, x} dy \, W + dx \, dy \, W \frac{\partial (\rho u)}{\partial t}. \tag{7.87}$$

The terms in Eq. (7.87) are expanded and taken to the limit of dx and dy approaching zero in order to achieve:

$$\underbrace{\rho\, g_x}_{\text{gravity force}} + \underbrace{\frac{\partial \tau_{yx}}{\partial y} + \frac{\partial \tau_{xx}}{\partial x}}_{\text{viscous stresses}} = \underbrace{\rho\frac{\partial\left(u^2\right)}{\partial x} + \rho\frac{\partial(uv)}{\partial y}}_{\text{momentum transfer}} + \underbrace{\frac{\partial p}{\partial x}}_{\text{pressure force}} + \underbrace{\rho\frac{\partial u}{\partial t}}_{\substack{\text{rate of}\\ \text{momentum storage}}} . \tag{7.88}$$

The rate equations that govern the viscous stress in an incompressible Newtonian fluid are given by Newton's Law of Viscosity (Alexandrou, 2001):

$$\tau_{xx} = 2\mu\frac{\partial u}{\partial x} \tag{7.89}$$

$$\tau_{yx} = \mu\left(\frac{\partial u}{\partial y} + \frac{\partial v}{\partial x}\right). \tag{7.90}$$

When Eqs. (7.89) and (7.90) are substituted into Eq. (7.88) the result is:

$$\rho g_x + \underbrace{\mu\frac{\partial^2 u}{\partial y^2} + \mu\frac{\partial^2 v}{\partial y\partial x}}_{\dfrac{\partial \tau_{yx}}{\partial y}} + \underbrace{2\mu\frac{\partial^2 u}{\partial x^2}}_{\dfrac{\partial \tau_{xx}}{\partial x}} = \rho\frac{\partial\left(u^2\right)}{\partial x} + \rho\frac{\partial(uv)}{\partial y} + \frac{\partial p}{\partial x} + \rho\frac{\partial u}{\partial t}. \tag{7.91}$$

In order to simplify Eq. (7.91) to the form of the x-momentum equation that is familiar from fluid dynamics it is necessary to use the continuity equation, Eq. (7.78), repeated below:

$$\frac{\partial v}{\partial y} = -\frac{\partial u}{\partial x}. \tag{7.92}$$

Differentiating Eq. (7.92) with respect to x provides:

$$\frac{\partial^2 v}{\partial x\,\partial y} = -\frac{\partial^2 u}{\partial x^2}. \tag{7.93}$$

Recognizing that the order of differentiation does not matter provided v is a continuous function provides:

$$\frac{\partial^2 v}{\partial y\,\partial x} = -\frac{\partial^2 u}{\partial x^2}. \tag{7.94}$$

Finally, applying the chain rule to the two momentum terms on the right side of Eq. (7.91) leads to:

$$\frac{\partial\left(u^2\right)}{\partial x} + \frac{\partial(uv)}{\partial y} = 2u\frac{\partial u}{\partial x} + u\frac{\partial v}{\partial y} + v\frac{\partial u}{\partial y} = u\frac{\partial u}{\partial x} + v\frac{\partial u}{\partial y} + u\underbrace{\left(\frac{\partial u}{\partial x} + \frac{\partial v}{\partial y}\right)}_{\substack{=\,0\text{ according to the}\\ \text{continuity equation}}} = u\frac{\partial u}{\partial x} + v\frac{\partial v}{\partial y}. \tag{7.95}$$

Substituting Eqs. (7.94) and (7.95) into Eq. (7.91) leads to our final form of the x-momentum equation:

$$\rho\left(\frac{\partial u}{\partial t} + u\frac{\partial u}{\partial x} + v\frac{\partial u}{\partial y}\right) = -\frac{\partial p}{\partial x} + \mu\left(\frac{\partial^2 u}{\partial x^2} + \frac{\partial^2 u}{\partial y^2}\right) + \rho g_x. \tag{7.96}$$

Carrying out the same steps for the y-directed momentum and force terms leads to a similar equation:

$$\rho\left(\frac{\partial v}{\partial t} + u\frac{\partial v}{\partial x} + v\frac{\partial v}{\partial y}\right) = -\frac{\partial p}{\partial y} + \mu\left(\frac{\partial^2 v}{\partial y^2} + \frac{\partial^2 v}{\partial x^2}\right) + \rho g_y. \tag{7.97}$$

Equations (7.96) and (7.97) together are referred to as the **Navier–Stokes Equations** for 2-D incompressible flow. The equivalent equations in cylindrical coordinates are listed in Table 7.2.

The Energy Conservation Equation

The energy conservation equation enforces the First Law of Thermodynamics on the differential control volume. The rate of energy flowing into the control volume must be balanced by the rate at which it flows out and the rate of storage. In general, energy can cross a control surface as heat transfer, work transfer, and with mass; in our differential control volume we have the possibility of all three forms of energy transfer. Again, we will take the time to examine the steps associated with deriving the governing differential equation rather than simply stating it without justification.

Figure 7.12 illustrates each of the energy terms that act on the differential control volume. The terms are represented on three different figures in order to make the presentation clear; however, it should be understood that all of the terms act simultaneously. Figure 7.12(a) illustrates the heat transfer terms crossing the control surfaces that define the differential control volume as well as the energy storage term; these have been the only terms considered in Chapters 2 through 6. The energy stored within the control volume may include kinetic and gravitational energy as well as internal energy. Therefore, the specific energy e in Figure 7.12(a) is defined as:

$$e = \hat{u} + \frac{u^2}{2} + \frac{v^2}{2} + g\,z \tag{7.98}$$

where $\hat{u}$ is the specific internal energy and gz represents the potential energy. Typically the kinetic energy and potential energy changes of the fluid are negligible relative to the internal energy change and therefore:

$$e = \hat{u}. \tag{7.99}$$

Figure 7.12(b) illustrates the energy terms that are associated with the mass flow rates through the control surfaces of the differential control volume. The parameter i represents the specific enthalpy of the fluid (the kinetic and potential energy of the flow are again neglected). Figure 7.12(c) illustrates the energy terms associated with work transfer crossing the control surfaces; each of the forces that act on the control surface must be multiplied by the velocity where the force is applied. When the force and velocity are in the same direction, the work transfer is into the control volume and when they are in opposite directions the work transfer must be out of the control volume. Note that the work transfers associated with pressure forces do not appear in Figure 7.12(c) because these terms are included in the specific enthalpy associated with the mass flow rates.

Combining the terms shown in Figure 7.12, expanding the $x + dx$ and $y + dy$ terms and taking the limit as dx and dy both approach zero provides an energy balance on the differential control volume:

$$\frac{\partial(\tau_{xx}u)}{\partial x} + \frac{\partial(\tau_{xy}v)}{\partial x} + \frac{\partial(\tau_{yx}u)}{\partial y} + \frac{\partial(\tau_{yy}v)}{\partial y} = \frac{\partial \dot{q}_x}{\partial x} + \frac{\partial \dot{q}_y}{\partial y} + \frac{\partial(\rho \hat{u})}{\partial t} + \frac{\partial(\rho u i)}{\partial x} + \frac{\partial(\rho v i)}{\partial y}. \tag{7.100}$$

Substituting the rate equations for conduction (Fourier's Law) and the viscous stress (Newton's Law of Viscosity) into Eq. (7.100) and taking advantage of the assumption that the fluid is incompressible allows Eq. (7.100) to be simplified:

$$\rho\frac{\partial \hat{u}}{\partial t} + \rho u\frac{\partial i}{\partial x} + \rho v\frac{\partial i}{\partial y} = k\frac{\partial^2 T}{\partial x^2} + k\frac{\partial^2 T}{\partial y^2} + \Phi, \tag{7.101}$$

where Φ is the viscous dissipation function

$$\Phi = 2\mu\left[\left(\frac{\partial u}{\partial x}\right)^2 + \left(\frac{\partial v}{\partial y}\right)^2\right] + \mu\left(\frac{\partial u}{\partial y} + \frac{\partial v}{\partial x}\right)^2. \tag{7.102}$$

The pressure-related contribution to changes in the specific enthalpy can usually be neglected. Therefore, Eq. (7.101) can be written as

$$\rho c\left(\frac{\partial T}{\partial t} + u\frac{\partial T}{\partial x} + v\frac{\partial T}{\partial y}\right) = k\frac{\partial^2 T}{\partial x^2} + k\frac{\partial^2 T}{\partial y^2} + \Phi, \tag{7.103}$$

where c is the specific heat capacity of the incompressible fluid.

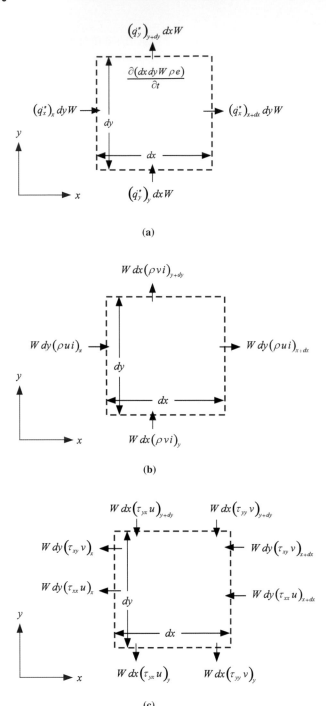

Figure 7.12 Energy terms on the differential control volume, including (a) heat transfer and storage terms, (b) mass transfer terms, and (c) work transfer terms.

7.3.3 The Boundary Layer Simplifications

A boundary layer is very thin in almost every flow situation and this fact can be used to simplify the mass, momentum and energy equations that were developed in Section 7.3.2. The conceptual model discussed in Section 7.1 for a laminar boundary layer provides an estimate of the ratio of the velocity boundary layer thickness to the length of the surface. (This quantity is sometimes referred to as the **slenderness ratio**.):

$$\frac{\delta_m}{L} \approx \frac{1}{\sqrt{Re_L}}. \tag{7.104}$$

For most practical flow situations, the Reynolds number in Eq. (7.104) will be very large. For example, the flow of air at room temperature and atmospheric pressure across a surface of size $L = 1$ m with velocity $u_\infty = 10$ cm/s is characterized by a Reynolds number of $Re_L = 6000$. A flow of water at the same conditions and velocity is characterized by $Re_L = 1 \times 10^5$. The boundary layer will therefore be between $100\times$ and $1000\times$ smaller than the dimension of the surface under these conditions. This observation leads to the **boundary layer simplifications**, which are based fundamentally on the assumption that

$$\delta_m \text{ and } \delta_t \ll L. \tag{7.105}$$

Unless the Prandtl number is very large or very small, δ_m will have the same order of magnitude as δ_t. Therefore, for the purposes of the scaling arguments discussed in this section we will refer only to a boundary layer thickness, δ, without specifying whether it is the thermal or momentum boundary layer thickness. Equation (7.104) can be used to provide an estimate of the scale of each of the terms in the governing equations that were developed in Section 7.1.2 when they are applied within a boundary layer. It is important to understand that we are not obtaining a solution to the governing differential equations; instead we are trying to understand the *order of magnitude* of each of the terms in order to (hopefully) eliminate some that are not important in a convection situation, thereby simplifying the problem.

The Continuity Equation

The continuity equation, Eq. (7.78), can be rearranged:

$$\frac{\partial u}{\partial x} = -\frac{\partial v}{\partial y} \tag{7.106}$$

Again, we are not interested in the exact value of the terms in Eq. (7.106), only their order of magnitude. The order of magnitude of a term is indicated by the notation $O(\text{term})$.

$$O\left(\frac{\partial u}{\partial x}\right) = O\left(\frac{\partial v}{\partial y}\right). \tag{7.107}$$

Equation (7.107) implies that the order of magnitude of the partial derivative of u with respect to x must be equal to the order of magnitude of the partial derivative of v with respect to y. It is relatively easy to estimate the order of magnitude of the term on the left side of Eq. (7.107). If you move along the surface in the x-direction from the leading edge ($x = 0$) to the trailing edge ($x = L$), then the largest change in u that is possible is u_∞. Therefore, the partial derivative of u with respect to x has order of magnitude:

$$O\left(\frac{\partial u}{\partial x}\right) = \frac{u_\infty}{L}. \tag{7.108}$$

There is no equally obvious scaling for the velocity in the y-direction. However, if you move in the y-direction from the surface ($y = 0$) to the outer edge of the boundary layer ($y = \delta$) then v will change from 0 to v_δ, where v_δ is an estimate of the y-directed velocity at the outer edge of the boundary layer. Note that v_δ can also be interpreted as the rate at which the boundary layer is growing, as seen from the perspective of the fluid. The partial derivative of v with respect to y therefore has order of magnitude v_δ / δ:

$$O\left(\frac{\partial v}{\partial y}\right) = \frac{v_\delta}{\delta}. \tag{7.109}$$

Equations (7.108) and (7.109) are substituted into Eq. (7.107):

$$\frac{u_\infty}{L} \approx \frac{v_\delta}{\delta} \tag{7.110}$$

or

$$v_\delta \approx \frac{\delta}{L} u_\infty. \tag{7.111}$$

Equation (7.111) indicates that when the boundary layer assumption, Eq. (7.105), is satisfied then the y-directed velocity in the boundary layer will be much less than the x-directed velocity. This result is intuitive. The x-directed velocity is driven by the free stream whereas the y-directed velocity originates because the boundary layer growth pushes against the free stream. The boundary layer is small and therefore the rate at which it grows is also small. Substituting Eq. (7.104) into Eq. (7.111) provides:

$$v_\delta \approx \frac{u_\infty}{\sqrt{Re_L}}. \tag{7.112}$$

The x-Momentum Equation

Equation (7.111) provides a convenient scaling for the velocity in the y-direction in a boundary layer and therefore allows an order of magnitude analysis of the x- and y-momentum equations. The x-directed momentum equation, Eq. (7.96), is simplified by assuming steady state and neglecting the gravitational force:

$$\rho\left(u\frac{\partial u}{\partial x} + v\frac{\partial u}{\partial y}\right) = -\frac{\partial p}{\partial x} + \mu\left(\frac{\partial^2 u}{\partial y^2} + \frac{\partial^2 u}{\partial x^2}\right). \tag{7.113}$$

The order of magnitude of each of the terms in Eq. (7.113) is estimated using the same type of scaling argument that was applied to the continuity equation:

$$\underbrace{O\left(\rho u\frac{\partial u}{\partial x}\right)}_{\rho\frac{u_\infty^2}{L}} + \underbrace{O\left(\rho v\frac{\partial u}{\partial y}\right)}_{\rho v_\delta \frac{u_\infty}{\delta} = \rho\frac{u_\infty^2}{L}} = \underbrace{O\left(-\frac{\partial p}{\partial x}\right)}_{\rho\frac{u_\infty^2}{L}} + \underbrace{O\left(\mu\frac{\partial^2 u}{\partial y^2}\right)}_{\mu\frac{u_\infty}{\delta^2}} + \underbrace{O\left(\mu\frac{\partial^2 u}{\partial x^2}\right)}_{\mu\frac{u_\infty}{L^2}}, \tag{7.114}$$

where the scale of the pressure gradient is taken to be the ratio of the inertial pressure drop based on the free stream velocity to the length scale and Eq. (7.112) is substituted for v_δ. The order of magnitude of the first three terms in Eq. (7.114) is the same and equal to $\rho u_\infty^2/L$. The last two terms are therefore divided by $\rho u_\infty^2/L$ in order to evaluate their relative importance:

$$\rho\frac{u_\infty^2}{L}\left[\underbrace{\frac{O\left(\rho u\frac{\partial u}{\partial x}\right)}{\rho\frac{u_\infty^2}{L}}}_{1} + \underbrace{\frac{O\left(\rho v\frac{\partial u}{\partial y}\right)}{\rho\frac{u_\infty^2}{L}}}_{1} = \underbrace{\frac{O\left(-\frac{\partial p}{\partial x}\right)}{\rho\frac{u_\infty^2}{L}}}_{1} + \underbrace{\frac{O\left(\mu\frac{\partial^2 u}{\partial y^2}\right)}{\rho\frac{u_\infty^2}{L}}}_{\mu\frac{u_\infty}{\delta^2}\frac{L}{\rho u_\infty^2}} + \underbrace{\frac{O\left(\mu\frac{\partial^2 u}{\partial x^2}\right)}{\rho\frac{u_\infty^2}{L}}}_{\mu\frac{u_\infty}{L^2}\frac{L}{\rho u_\infty^2}}\right]. \tag{7.115}$$

The relative order of the second viscous stress term, the last term in Eq. (7.115), is:

$$\frac{O\left(\mu\frac{\partial^2 u}{\partial x^2}\right)}{\rho\frac{u_\infty^2}{L}} = \mu\frac{u_\infty}{L^2}\frac{L}{\rho u_\infty^2} = \frac{1}{Re_L}. \tag{7.116}$$

Our basis for the boundary layer simplifications is that the Reynolds number is much larger than unity and therefore the last term will be small and can be neglected relative to the others.

The relative order of the first viscous stress term, the fourth term in Eq. (7.115), is:

$$\frac{O\left(\mu\frac{\partial^2 u}{\partial y^2}\right)}{\rho\frac{u_\infty^2}{L}} = \mu\frac{u_\infty}{\delta^2}\frac{L}{\rho u_\infty^2} = \frac{L^2}{\delta^2}\frac{1}{Re_L} = 1. \tag{7.117}$$

At first glance, the order of this term is not as clear; the Reynolds number is large but L^2/δ^2 is also large. However, substituting Eq. (7.104) into Eq. (7.117) suggests that the relative order of magnitude of the fourth term will be unity and therefore this term must be retained.

Two important conclusions have resulted from this analysis. First, the final term in Eq. (7.113) can be neglected relative to the others in the boundary layer:

$$\rho\left(u\frac{\partial u}{\partial x} + v\frac{\partial u}{\partial y}\right) = -\frac{\partial p}{\partial x} + \mu\frac{\partial^2 u}{\partial y^2}. \tag{7.118}$$

Second, the order of magnitude of the terms that have been retained in the x-directed momentum equation are all $\rho\, u_\infty^2/L$.

The y-Momentum Equation

The y-directed momentum equation, Eq. (7.97), is rewritten assuming steady state and neglecting the gravitational force:

$$\rho\left(u\frac{\partial v}{\partial x} + v\frac{\partial v}{\partial y}\right) = -\frac{\partial p}{\partial y} + \mu\left(\frac{\partial^2 v}{\partial y^2} + \frac{\partial^2 v}{\partial x^2}\right). \tag{7.119}$$

The order of magnitude of each of the terms in Eq. (7.119) in the boundary layer is estimated:

$$\underbrace{O\left(\rho u\frac{\partial v}{\partial x}\right)}_{\rho u_\infty \frac{v_\delta}{L} = \left(\rho\frac{u_\infty^2}{L}\right)\frac{\delta}{L}} + \underbrace{O\left(\rho v\frac{\partial v}{\partial y}\right)}_{\rho v_\delta \frac{v_\delta}{\delta} = \left(\rho\frac{u_\infty^2}{L}\right)\frac{\delta}{L}} = \underbrace{O\left(-\frac{\partial p}{\partial y}\right)}_{} + \underbrace{O\left(\mu\frac{\partial^2 v}{\partial y^2}\right)}_{\mu\frac{v_\delta}{\delta^2} = \left(\rho\frac{u_\infty^2}{L}\right)\frac{\delta}{L}} + \underbrace{O\left(\mu\frac{\partial^2 v}{\partial x^2}\right)}_{\mu\frac{v_\delta}{L^2} = \left(\rho\frac{u_\infty^2}{L}\right)\frac{\delta}{L}\frac{1}{Re_L}}. \tag{7.120}$$

The order of every term in Eq. (7.120) is $(\rho u_\infty^2/L)(\delta/L)$ or smaller; said differently, the order of every term in the y-momentum equation is much smaller than the order of all of the terms in the x-momentum equation in a boundary layer, Eq. (7.118). As a result, the entire y-momentum equation (i.e., the momentum in the direction perpendicular rather than parallel to the flow) can be neglected as being small. This scaling analysis suggests that the pressure gradient in the y-direction must be small. As a result, the free stream pressure variation that exists outside of boundary layer along the surface, $p_\infty(x)$, imposes itself through the boundary layer and the pressure in the boundary layer will be approximately only a function of x. The partial derivative of pressure with respect to x in Eq. (7.118) can be replaced by the ordinary derivative of the free stream pressure:

$$\rho\left(u\frac{\partial u}{\partial x} + v\frac{\partial u}{\partial y}\right) = -\frac{dp_\infty}{dx} + \mu\frac{\partial^2 u}{\partial y^2}. \tag{7.121}$$

Equation (7.121) is the x-directed momentum equation inside the boundary layer. The term related to the free stream pressure gradient represents the solution to an inviscid flow problem that exists beyond the boundary layer and therefore it is related to the shape of the object being considered. The function $\frac{dp_\infty}{dx}$ for flow over a flat plate or a sphere or a wedge-shaped object will all be somewhat different.

The Energy Conservation Equation

The energy conservation equation, Eq. (7.103), is rewritten assuming steady state:

$$\rho c\left(u\frac{\partial T}{\partial x} + v\frac{\partial T}{\partial y}\right) = k\left(\frac{\partial^2 T}{\partial x^2} + \frac{\partial^2 T}{\partial y^2}\right) + \Phi, \tag{7.122}$$

where

$$\Phi = 2\mu\left[\left(\frac{\partial u}{\partial x}\right)^2 + \left(\frac{\partial v}{\partial y}\right)^2\right] + \mu\left(\frac{\partial u}{\partial y} + \frac{\partial v}{\partial x}\right)^2. \tag{7.123}$$

Keeping in mind that changes in u are much larger than changes in v and also that changes in x are much larger than changes in y, the term in the viscous dissipation function, Eq. (7.123), with the largest order of magnitude is related to $\frac{\partial u}{\partial y}$. Substituting this result into Eq. (7.122) and evaluating the order of each of the terms leads to:

$$O\left(\rho c u \frac{\partial T}{\partial x}\right) + \quad O\left(\rho c v \frac{\partial T}{\partial y}\right) \quad = O\left(k \frac{\partial^2 T}{\partial x^2}\right) + O\left(k \frac{\partial^2 T}{\partial y^2}\right) + O\left(\mu\left(\frac{\partial u}{\partial y}\right)^2\right), \quad (7.124)$$

$$\rho c u_\infty \frac{\Delta T}{L} \qquad \rho c v_\delta \frac{\Delta T}{\delta} = \rho c u_\infty \frac{\Delta T}{L} \qquad k\frac{\Delta T}{L^2} \qquad k\frac{\Delta T}{\delta^2} \qquad \mu\frac{u_\infty^2}{\delta^2}$$

where ΔT is a reference temperature difference that can be used for scaling purposes; this temperature difference is usually chosen to be the surface to the free stream temperature difference. Equation (7.124) is divided through by the scale of the first two terms in order to ascertain the relative order of the remaining terms:

$$\rho c u_\infty \frac{\Delta T}{L}\left[\underbrace{\frac{O\left(\rho c u \frac{\partial T}{\partial x}\right)}{\rho c u_\infty \frac{\Delta T}{L}}}_{1} + \underbrace{\frac{O\left(\rho c v \frac{\partial T}{\partial y}\right)}{\rho c u_\infty \frac{\Delta T}{L}}}_{1} = \underbrace{\frac{O\left(k \frac{\partial^2 T}{\partial x^2}\right)}{\rho c u_\infty \frac{\Delta T}{L}}}_{k\frac{\Delta T}{L^2}\frac{L}{\rho c u_\infty \Delta T}} + \underbrace{\frac{O\left(k \frac{\partial^2 T}{\partial y^2}\right)}{\rho c u_\infty \frac{\Delta T}{L}}}_{k\frac{\Delta T}{\delta^2}\frac{L}{\rho c u_\infty \Delta T}} + \underbrace{\frac{O\left(\mu\left(\frac{\partial u}{\partial y}\right)^2\right)}{\rho c u_\infty \frac{\Delta T}{L}}}_{\mu\frac{u_\infty^2}{\delta^2}\frac{L}{\rho c u_\infty \Delta T}}\right]. \quad (7.125)$$

Introducing the definition of the Reynolds number and the Prandtl number into Eq. (7.125) provides:

$$\rho c u_\infty \frac{\Delta T}{L}\left[\underbrace{\frac{O\left(\rho c u \frac{\partial T}{\partial x}\right)}{\rho c u_\infty \frac{\Delta T}{L}}}_{1} + \underbrace{\frac{O\left(\rho c v \frac{\partial T}{\partial y}\right)}{\rho c u_\infty \frac{\Delta T}{L}}}_{1} = \underbrace{\frac{O\left(k \frac{\partial^2 T}{\partial x^2}\right)}{\rho c u_\infty \frac{\Delta T}{L}}}_{\frac{L}{Re_L Pr}} + \underbrace{\frac{O\left(k \frac{\partial^2 T}{\partial y^2}\right)}{\rho c u_\infty \frac{\Delta T}{L}}}_{\frac{L^2}{\delta^2}\frac{L}{Re_L Pr}=1} + \underbrace{\frac{O\left(\mu\left(\frac{\partial u}{\partial y}\right)^2\right)}{\rho c u_\infty \frac{\Delta T}{L}}}_{\frac{L^2}{\delta^2}\frac{u_\infty^2}{c\Delta T}\frac{1}{Re_L}}\right]. \quad (7.126)$$

The third term in Eq. (7.126) is clearly negligible in the boundary layer (recall that the Reynolds number will be large) provided that the Prandtl number is not too different from unity. Equation (7.104) suggests that the order of the fourth term in Eq. (7.126) will be near unity. The order of the last term is less clear and so it will be retained.

There are two important conclusions from this analysis. First, the axial conduction term (i.e., the term related to the second derivative of temperature with respect to x) can be neglected in the boundary layer. Second, the only significant term in the viscous dissipation function is related to the gradient of the x-velocity in the y-direction. The thermal energy equation in the boundary layer is therefore:

$$\rho c\left(u \frac{\partial T}{\partial x} + v \frac{\partial T}{\partial y}\right) = k\frac{\partial^2 T}{\partial y^2} + \mu\left(\frac{\partial u}{\partial y}\right)^2. \quad (7.127)$$

The boundary layer equations in Cartesian coordinates are summarized in Table 7.3. Carrying out the equivalent analysis of the fluid flow equations in radial coordinates provided in Table 7.2 will lead to the boundary layer equations summarized in Table 7.4, assuming that x is the flow direction.

The boundary layer equations are simplified, but still far from simple. Solutions to these equations remain quite challenging and experiments are often used to understand convection problems. Therefore, dimensional analysis is extremely important to convection heat transfer as discussed in the following section.

7.4 Dimensional Analysis in Convection

7.4.1 Introduction

In Section 7.3.3, it is shown that the governing equations within the boundary layer can be simplified relative to the general governing equations for an incompressible, viscous fluid derived in Section 7.3.2. The steady-state continuity, x-directed momentum, and thermal energy equations in a boundary layer are summarized in Table 7.3 (in Cartesian coordinates) and Table 7.4 (in radial coordinates). Although these equations have been simplified for steady flow in the boundary layer, they are still pretty imposing and difficult to solve analytically in most practical cases. Consequently, researchers have been forced to carry out experiments or develop computational

Table 7.3 The boundary layer equations for 2-D, steady constant property flow in Cartesian coordinates. The velocities u and v are in the x- and y-directions, respectively, where x is in the direction of the surface and y is perpendicular to the surface (gravity forces are not considered).

continuity	$\dfrac{\partial u}{\partial x} + \dfrac{\partial v}{\partial y} = 0$
x-momentum	$\rho\left(u\dfrac{\partial u}{\partial x} + v\dfrac{\partial u}{\partial y}\right) = -\dfrac{dp_\infty}{dx} + \mu\dfrac{\partial^2 u}{\partial y^2}$, all terms are of order $\rho\dfrac{u_\infty^2}{L}$
y-momentum	$\dfrac{\partial p}{\partial y} \approx 0$, all terms are of order $\left(\rho\dfrac{u_\infty^2}{L}\right)\dfrac{\delta}{L}$
energy	$\rho c\left(u\dfrac{\partial T}{\partial x} + v\dfrac{\partial T}{\partial y}\right) = k\dfrac{\partial^2 T}{\partial y^2} + \mu\left(\dfrac{\partial u}{\partial y}\right)^2$

Table 7.4 The boundary layer equations for 2-D, steady constant property flow in cylindrical coordinates. The velocities u and v are in the x- and r-directions, respectively.

continuity	$\dfrac{\partial u}{\partial x} + \dfrac{1}{r}\dfrac{\partial(rv)}{\partial r} = 0$
x-momentum	$\rho\left(u\dfrac{\partial u}{\partial x} + v\dfrac{\partial u}{\partial r}\right) = -\dfrac{dp_\infty}{dx} + \mu\left(\dfrac{\partial^2 u}{\partial r^2} + \dfrac{1}{r}\dfrac{\partial u}{\partial r}\right)$, all terms are of order $\rho\dfrac{u_\infty^2}{L}$
r-momentum	$\dfrac{\partial p}{\partial r} \approx 0$, all terms are of order $\left(\rho\dfrac{u_\infty^2}{L}\right)\dfrac{\delta}{L}$
energy	$\rho c\left(u\dfrac{\partial T}{\partial x} + v\dfrac{\partial T}{\partial r}\right) = k\dfrac{1}{r}\dfrac{\partial}{\partial r}\left(r\dfrac{\partial T}{\partial r}\right) + \mu\left(\dfrac{\partial u}{\partial r}\right)^2$

fluid dynamic (CFD) models of the physical situation. Experiments and CFD models are relatively expensive and time consuming. Often it is not possible to build an experiment that has the same length scale or operates under the same conditions as the physical device or situation of interest. The proper execution of a CFD model requires expertise and considerable computational resources, particularly as the Reynolds number becomes large. In order to maximize the utility of a set of experimental results or CFD simulations, it is necessary to identify the minimum set of nondimensional parameters that can be used to correlate the data.

Dimensional analysis provides a technique that can be used to reduce a complicated problem to its simplest form in order to get the maximum use from the information that is available. Dimensional analysis may also provide physical insight into the relative magnitude of the various physical processes occurring in a situation. The process of dimensional analysis has been the backbone of many scientific and engineering disciplines, as discussed by Bridgman (1922). Looking ahead, dimensional analysis provides the justification for correlating heat transfer data in the form that is encountered in most textbooks and handbooks: Nusselt number as a function of Reynolds number and Prandtl number. We tend to take this presentation for granted and use these correlations without giving any real thought to the remarkable simplification that they represent. The same correlation for a sphere can be used to estimate the heat transfer coefficient for a large cannonball traveling through air or a tiny spherical thermocouple mounted in a flowing liquid. The physical underpinnings and practical application of dimensional analysis is eloquently discussed by Sonin (1992) and others.

Given a physical problem of interest, there are at least two methods that can be used to identify the nondimensional parameters that govern the problem. The classic technique is Buckingham's Pi Theorem (Buckingham (1914)) in which all of the independent physical quantities that are involved in the problem are listed. A complete and dimensionally independent subset of these quantities is selected and then used to nondimensionalize all of the remaining quantities. The result is one set of nondimensional quantities that can be used to describe the problem.

An alternative approach is possible when additional information about the problem is available, for example the governing differential equation(s). In this situation, it is often more instructive to define physically meaningful, dimensionless quantities and substitute these into the governing equations which, through algebraic manipulation, are themselves made dimensionless. The nondimensional groups that result from this algebraic manipulation often represent a more physically meaningful set of parameters than those that are arrived at by using Buckingham's Pi Theorem. In this section, the governing equations for a boundary layer are made dimensionless and used to identify the important nondimensional quantities that govern a convective heat transfer problem.

7.4.2 The Dimensionless Boundary Layer Equations

The process begins by defining a set of meaningful, nondimensional quantities that can be identified by inspection. The coordinates x and y are made dimensionless by normalizing them against the length of the surface (L), which is the only characteristic length that is directly available in the problem:

$$\tilde{x} = \frac{x}{L} \tag{7.128}$$

$$\tilde{y} = \frac{y}{L}. \tag{7.129}$$

The x- and y-components of velocity are normalized against the free stream velocity (u_∞):

$$\tilde{u} = \frac{u}{u_\infty} \tag{7.130}$$

$$\tilde{v} = \frac{v}{u_\infty}. \tag{7.131}$$

A dimensionless temperature (difference) is defined by normalizing the temperature elevation relative to the surface by the free stream to surface temperature difference:

$$\tilde{\theta} = \frac{T - T_s}{T_\infty - T_s}. \tag{7.132}$$

A dimensionless pressure is defined by normalizing against the fluid inertia:

$$\tilde{p} = \frac{v}{\rho u_\infty^2}. \tag{7.133}$$

These dimensionless parameters are substituted into the governing equations in the following sections.

The Dimensionless Continuity Equation

Substituting the definitions for the dimensionless velocities and positions, Eqs. (7.128) through (7.131), into the continuity equation, Eq. (7.78), leads to:

$$\frac{\partial(\tilde{u}\, u_\infty)}{\partial(\tilde{x}\, L)} + \frac{\partial(\tilde{v}\, u_\infty)}{\partial(\tilde{y}\, L)} = 0 \tag{7.134}$$

or

$$\frac{u_\infty}{L}\frac{\partial \tilde{u}}{\partial \tilde{x}} + \frac{u_\infty}{L}\frac{\partial \tilde{v}}{\partial \tilde{y}} = 0. \tag{7.135}$$

Equation (7.135) is multiplied by L/u_∞ in order to obtain the dimensionless form of the continuity equation:

$$\frac{\partial \tilde{u}}{\partial \tilde{x}} + \frac{\partial \tilde{v}}{\partial \tilde{y}} = 0. \tag{7.136}$$

The dimensionless continuity equation has the same form as the original continuity equation and has not resulted in the identification of any new nondimensional groups.

The Dimensionless Momentum Equation

Substituting the definitions of the dimensionless quantities into the x-momentum equation for the boundary layer, Eq. (7.121), leads to:

$$\rho\left(\tilde{u}\,u_\infty\frac{\partial(\tilde{u}\,u_\infty)}{\partial(\tilde{x}\,L)} + \tilde{v}\,u_\infty\frac{\partial(\tilde{u}\,u_\infty)}{\partial(\tilde{y}\,L)}\right) = -\frac{d\left(\tilde{p}_\infty\,\rho u_\infty^2\right)}{d(\tilde{x}\,L)} + \mu\frac{\partial^2(\tilde{u}\,u_\infty)}{\partial(\tilde{y}\,L)^2} \tag{7.137}$$

or

$$\frac{\rho u_\infty^2}{L}\left(\tilde{u}\frac{\partial\tilde{u}}{\partial\tilde{x}} + \tilde{v}\frac{\partial\tilde{u}}{\partial\tilde{y}}\right) = -\frac{\rho u_\infty^2}{L}\frac{d\tilde{p}_\infty}{d\tilde{x}} + \mu\frac{u_\infty}{L^2}\frac{\partial^2\tilde{u}}{\partial\tilde{y}^2}. \tag{7.138}$$

Equation (7.138) is divided by $\rho u_\infty^2/L$ (the scale of the inertial terms on the left-hand side of the momentum equation) in order to obtain the dimensionless form of the x-momentum equation:

$$\tilde{u}\frac{\partial\tilde{u}}{\partial\hat{x}} + \tilde{v}\frac{\partial\tilde{u}}{\partial\hat{y}} = -\frac{d\tilde{p}_\infty}{d\hat{x}} + \underbrace{\mu\frac{u_\infty}{L^2}\frac{L}{\rho u_\infty^2}}\frac{\partial^2\tilde{u}}{\partial\tilde{y}^2}. \tag{7.139}$$

$$\frac{\mu}{\rho u_\infty L} = \frac{1}{Re_L}$$

The group that multiplies the last term in Eq. (7.139) (the viscous shear term) must be dimensionless and also must be an additional dimensionless group that governs the solution. Simplifying the last term leads to:

$$\tilde{u}\frac{\partial\tilde{u}}{\partial\tilde{x}} + \tilde{v}\frac{\partial\tilde{u}}{\partial\tilde{y}} = -\frac{d\tilde{p}_\infty}{d\tilde{x}} + \frac{1}{Re_L}\frac{\partial^2\tilde{u}}{\partial\tilde{y}^2}. \tag{7.140}$$

The Reynolds number is related to the ratio of the inertial to the viscous forces, as discussed in Section 7.2.2, and it appears in the dimensionless momentum equation when the viscous force is divided by the inertial force.

The Dimensionless Energy Equation

Substituting the definitions of the dimensionless quantities into the thermal energy equation for the boundary layer, Eq. (7.127), leads to:

$$\rho c\left(\tilde{u}\,u_\infty\frac{\partial\left(\tilde{\theta}(T_\infty - T_s)\right)}{\partial(\tilde{x}L)} + \tilde{v}\,u_\infty\frac{\partial\left(\tilde{\theta}(T_\infty - T_s)\right)}{\partial(\tilde{y}L)}\right) = k\frac{\partial^2\left(\tilde{\theta}(T_\infty - T_s)\right)}{\partial(\tilde{y}L)^2} + \mu\left(\frac{\partial(\tilde{u}\,u_\infty)}{\partial(\tilde{y}L)}\right)^2 \tag{7.141}$$

or

$$\frac{\rho c\,u_\infty(T_\infty - T_s)}{L}\left(\tilde{u}\frac{\partial\tilde{\theta}}{\partial\tilde{x}} + \tilde{v}\frac{\partial\tilde{\theta}}{\partial\tilde{y}}\right) = \frac{k(T_\infty - T_s)}{L^2}\frac{\partial^2\tilde{\theta}}{\partial\tilde{y}^2} + \frac{\mu u_\infty^2}{L^2}\left(\frac{\partial\tilde{u}}{\partial\tilde{y}}\right)^2. \tag{7.142}$$

Equation (7.142) is made dimensionless by dividing through by the scale of the convective terms on the left side, $\rho c\,u_\infty(T_\infty - T_s)/L$:

$$\tilde{u}\frac{\partial\tilde{\theta}}{\partial\tilde{x}} + \tilde{v}\frac{\partial\tilde{\theta}}{\partial\tilde{y}} = \underbrace{\frac{k(T_\infty - T_s)}{L^2}\frac{L}{\rho c\,u_\infty(T_\infty - T_s)}}\frac{\partial^2\tilde{\theta}}{\partial\tilde{y}^2} + \underbrace{\frac{\mu u_\infty^2}{L^2}\frac{L}{\rho c\,u_\infty(T_\infty - T_s)}}\left(\frac{\partial\tilde{u}}{\partial\tilde{y}}\right)^2. \tag{7.143}$$

$$\frac{k}{\rho c\,u_\infty L} = \frac{\alpha}{\upsilon}\frac{\mu}{\rho u_\infty L} = \frac{1}{Pr\,Re_L} \qquad \frac{u_\infty^2}{c(T_\infty - T_s)}\frac{\mu}{\rho u_\infty L} = \frac{Ec}{Re_L}$$

The dimensionless group that appears in the first term on the right side of Eq. (7.143) can be rearranged to provide the now familiar Reynolds number and Prandtl number. Because $(Pr\,Re_L)^{-1}$ appears when the conduction term is divided by the convection term, it is reasonable to assume that the product of the Reynolds number and Prandtl number is related to the relative importance of convection to conduction.

Similarly, the dimensionless group that appears in the second term on the right side of Eq. (7.143) can be rearranged to provide the ratio of the **Eckert number** to the Reynolds number, where the Eckert number is defined as:

$$Ec = \frac{u_\infty^2}{c(T_\infty - T_s)}. \tag{7.144}$$

Table 7.5 The dimensionless forms of the boundary layer equations for 2-D, steady constant property flow in Cartesian coordinates.

continuity	$\dfrac{\partial \tilde{u}}{\partial \tilde{x}} + \dfrac{\partial \tilde{v}}{\partial \tilde{y}} = 0$
x-momentum	$\tilde{u}\dfrac{\partial \tilde{u}}{\partial \tilde{x}} + \tilde{v}\dfrac{\partial \tilde{u}}{\partial \tilde{y}} = -\dfrac{d\tilde{p}_\infty}{d\tilde{x}} + \dfrac{1}{Re_L}\dfrac{\partial^2 \tilde{u}}{\partial \tilde{y}^2}$
energy	$\tilde{u}\dfrac{\partial \tilde{\theta}}{\partial \tilde{x}} + \tilde{v}\dfrac{\partial \tilde{\theta}}{\partial \tilde{y}} = \dfrac{1}{Re_L Pr}\dfrac{\partial^2 \tilde{\theta}}{\partial \tilde{y}^2} + \dfrac{Ec}{Re_L}\left(\dfrac{\partial \tilde{u}}{\partial \tilde{y}}\right)^2$

A common simplification in convection problems is to neglect viscous dissipation. The majority of the convection correlations that are used to solve these problems have been developed in this limit. Equation (7.143) shows that when the ratio of the Eckert number to the Reynolds number is much less than unity then the term related to viscous dissipation will disappear and very little error is introduced by neglecting viscous dissipation. However, as the value of Ec/Re_L increases, the effect of viscous dissipation becomes more important and as a result, conventional correlations may no longer be valid. The final form of the dimensionless form of the energy equation is:

$$\tilde{u}\frac{\partial \tilde{\theta}}{\partial \tilde{x}} + \tilde{v}\frac{\partial \tilde{\theta}}{\partial \tilde{y}} = \frac{1}{Re_L Pr}\frac{\partial^2 \tilde{\theta}}{\partial \tilde{y}^2} + \frac{Ec}{Re_L}\left(\frac{\partial \tilde{u}}{\partial \tilde{y}}\right)^2. \tag{7.145}$$

The dimensionless forms of the boundary layer equations in Cartesian coordinates are summarized in Table 7.5.

7.4.3 Correlations

The power of the nondimensionalization presented in Section 7.4.2 is the identification of the minimal set of parameters that govern the solution to the problem. In this section, the dimensionless equations are examined in order to understand how the important engineering quantities associated with boundary layer flows can be correlated.

The Friction and Drag Coefficients

The simultaneous solution of the dimensionless continuity and *x*-momentum equations, Eqs. (7.136) and (7.140), would provide the dimensionless *x*- and *y*-directed velocities. The functional form of these solutions must include the dimensionless coordinates ($\tilde{x}$ and $\tilde{y}$) as well as the Reynolds number and free stream pressure gradient function that appear in Eq. (7.140). The free stream pressure gradient function is an *input* in that it represents the solution to a separate, inviscid problem outside the boundary layer that depends on the shape of the surface being considered:

$$\tilde{u} = \tilde{u}\left(\tilde{x}, \tilde{y}, Re_L, \frac{d\tilde{p}_\infty}{d\tilde{x}}\right) \tag{7.146}$$

$$\tilde{v} = \tilde{v}\left(\tilde{x}, \tilde{y}, Re_L, \frac{d\tilde{p}_\infty}{d\tilde{x}}\right). \tag{7.147}$$

One engineering quantity of interest is the shear stress at the surface of the plate:

$$\tau_s = \mu\frac{\partial u}{\partial y}\bigg|_{y=0}. \tag{7.148}$$

Equation (7.148) can be expressed in terms of the dimensionless quantities

$$\tau_s = \mu\frac{\partial(\tilde{u}\, u_\infty)}{\partial(\tilde{y}L)}\bigg|_{\tilde{y}=0} \tag{7.149}$$

or

$$\tau_s = \frac{\mu\, u_\infty}{L} \frac{\partial \tilde{u}}{\partial \tilde{y}}\bigg|_{\tilde{y}=0}. \tag{7.150}$$

As discussed in Section 7.1.7, the shear stress is typically made dimensionless by normalizing it with the fluid inertia in order to obtain the friction coefficient, C_f:

$$C_f = \frac{2\tau_s}{\rho u_\infty^2}. \tag{7.151}$$

Substituting Eq. (7.150) into Eq. (7.151) leads to:

$$C_f = \frac{2}{\rho u_\infty^2} \frac{\mu\, u_\infty}{L} \frac{\partial \tilde{u}}{\partial \tilde{y}}\bigg|_{\tilde{y}=0} = \frac{2}{Re_L} \frac{\partial \tilde{u}}{\partial \tilde{y}}\bigg|_{\tilde{y}=0}. \tag{7.152}$$

Equation (7.152) shows that the solution for the friction factor depends on the solution for $\tilde{u}$. However, only the partial derivative of $\tilde{u}$ with respect to $\tilde{y}$ at $\tilde{y}=0$ is required. Therefore, while $\tilde{u}$ depends on $\tilde{y}$, C_f does not:

$$C_f = C_f\left(\tilde{x}, Re_L, \frac{d\tilde{p}_\infty}{d\tilde{x}}\right). \tag{7.153}$$

The friction coefficient defined by Eq. (7.152) is the local friction coefficient that characterizes the shear stress at any location on the surface. As discussed in Section 7.1.9, the average friction coefficient ($\bar{C}_f$) is a more useful quantity. The average friction coefficient is defined as

$$\bar{C}_f = \frac{2\, \bar{\tau}_s}{\rho\, u_\infty^2}, \tag{7.154}$$

where the average shear stress for the 2-D problem considered here is defined as:

$$\bar{\tau}_s = \frac{1}{L}\int_0^L \tau_s\, dx. \tag{7.155}$$

Substituting Eq. (7.150) into Eq. (7.155) and rearranging leads to:

$$\bar{\tau}_s = \frac{\mu\, u_\infty}{L}\int_0^1 \frac{\partial \tilde{u}}{\partial \tilde{y}}\bigg|_{\tilde{y}=0} d\tilde{x}. \tag{7.156}$$

Substituting Eq. (7.156) into Eq. (7.154) leads to:

$$\bar{C}_f = \frac{2}{\rho\, u_\infty^2} \frac{\mu\, u_\infty}{L}\int_0^1 \frac{\partial \tilde{u}}{\partial \tilde{y}}\bigg|_{\tilde{y}=0} d\tilde{x} = \frac{2}{Re_L}\int_0^1 \frac{\partial \tilde{u}}{\partial \tilde{y}}\bigg|_{\tilde{y}=0} d\tilde{x}. \tag{7.157}$$

Equation (7.157) shows that the solution for the average friction factor depends on the solution for $\tilde{u}$. However, only the partial derivative of $\tilde{u}$ with respect to $\tilde{y}$ at $\tilde{y}=0$ is required and this quantity is integrated over the entire range of $0 < \tilde{x} < 1$. Therefore, while $\tilde{u}$ depends on $\tilde{x}$ and $\tilde{y}$, $\bar{C}_f$ does not depend on either of these parameters:

$$\bar{C}_f = \bar{C}_f\left(Re_L, \frac{d\tilde{p}_\infty}{d\tilde{x}}\right). \tag{7.158}$$

For a given shape (e.g., for a flat plate), the dimensionless pressure gradient function is not an independent parameter with regard to the boundary layer problem, but rather an input related to the behavior of the inviscid free stream as it flows over the surface. Therefore, the average friction coefficient for a particular shape will depend only on the Reynolds number:

$$\bar{C}_f = \bar{C}_f(Re_L) \text{ for a given shape.} \tag{7.159}$$

Equation (7.159) is remarkable in that it indicates that a single dimensionless parameter (Re_L) can be used to correlate the functional behavior of the average friction coefficient for a particular shape. Equation (7.159) allows us to use small scale and therefore affordable wind tunnel tests to design large aircraft or, alternatively, to use large scale and therefore manageable and observable flow tests to characterize the flow around microscopic objects.

For most shapes, form drag is also an important consideration and therefore the drag coefficient, C_D, is more relevant. As discussed in Section 7.1.9, the drag coefficient is defined based on the drag force (F_D) exerted on the object by the flow:

$$C_D = \frac{2 F_D}{\rho u_\infty^2 A_p}, \tag{7.160}$$

where A_p is the projected area of the object when viewed from the direction of the flow. The drag coefficient can also be correlated against the Reynolds number defined based on a characteristic dimension of the problem, L_{char}:

$$C_D = C_D(Re_{L_{char}}) \text{ for a given shape.} \tag{7.161}$$

The Nusselt Number

The solution of the dimensionless thermal energy equation, Eq. (7.145), must provide the dimensionless temperature, $\tilde{\theta}$, as a function of $\tilde{x}$ and $\tilde{y}$. Notice that the dimensionless velocity field is required to solve Eq. (7.145) and therefore all of the functional dependence exhibited by $\tilde{u}$ and $\tilde{v}$ in Eqs. (7.146) and (7.147) must also be present in the solution for $\tilde{\theta}$. Consequently, $\tilde{\theta}$ must also depend on the Reynolds number and dimensionless pressure gradient function. In addition, $\tilde{\theta}$ must depend on the Prandtl number and Eckert number as these parameters both appear in Eq. (7.145):

$$\tilde{\theta} = \tilde{\theta}\left(\tilde{x}, \tilde{y}, Re_L, \frac{d\tilde{p}_\infty}{d\tilde{x}}, Pr, Ec\right). \tag{7.162}$$

The engineering quantity of interest is the heat transfer coefficient:

$$h = \frac{\dot{q}''_s}{(T_s - T_\infty)}, \tag{7.163}$$

where $\dot{q}''_s$ is the heat flux at the surface:

$$\dot{q}''_s = -k\frac{\partial T}{\partial y}\bigg|_{y=0}. \tag{7.164}$$

The heat transfer coefficient can be expressed in terms of dimensionless quantities:

$$h_s = -\frac{k}{(T_s - T_\infty)}\frac{\partial(\tilde{\theta}(T_\infty - T_s))}{\partial(\tilde{y}L)}\bigg|_{\tilde{y}=0} \tag{7.165}$$

or

$$h = \frac{k}{L}\frac{\partial\tilde{\theta}}{\partial\tilde{y}}\bigg|_{\tilde{y}=0}. \tag{7.166}$$

The dimensionless heat transfer coefficient is the Nusselt number, defined in Section 7.1.7:

$$Nu = \frac{hL}{k}. \tag{7.167}$$

Substituting Eq. (7.166) into Eq. (7.167) provides:

$$Nu = \frac{\partial \tilde{\theta}}{\partial \tilde{y}}\bigg|_{\tilde{y}=0}. \tag{7.168}$$

Therefore, the solution for the Nusselt number depends on the solution for $\tilde{\theta}$. However, the Nusselt number is only a function of the gradient in the dimensionless temperature difference at $\tilde{y} = 0$ and therefore the solution for the Nusselt number does not depend on $\tilde{y}$:

$$Nu = Nu\left(\tilde{x}, Re_L, \frac{d\tilde{p}_\infty}{d\tilde{x}}, Pr, Ec\right). \tag{7.169}$$

Also, if the average Nusselt number $(\overline{Nu})$ is required, then the local Nusselt number must be integrated from $\tilde{x} = 0$ to $\tilde{x} = 1$. As a result, the solution for the average Nusselt number does not depend on $\tilde{x}$ or $\tilde{y}$:

$$\overline{Nu} = \overline{Nu}\left(Re_L, \frac{d\tilde{p}_\infty}{d\tilde{x}}, Pr, Ec\right). \tag{7.170}$$

The dimensionless pressure gradient function depends on the shape being considered and therefore the average Nusselt number will depend on the Reynolds number, Prandtl number, and Eckert number for a specific shape:

$$\overline{Nu} = \overline{Nu}(Re_L, Pr, Ec) \text{ for a given shape.} \tag{7.171}$$

In most situations of general engineering interest, viscous dissipation can be neglected and therefore the effect of the Eckert number can be neglected:

$$\overline{Nu} = \overline{Nu}(Re_L, Pr) \text{ for a given shape, if viscous dissipation is negligible.} \tag{7.172}$$

Equation (7.172) is as remarkable as Eq. (7.159). It shows that only two parameters are required to correlate the thermal behavior of a given shape under most conditions. The experimental and theoretical results obtained by researchers for different flow situations are ultimately presented in terms of the Nusselt number as a function of Reynolds number and Prandtl number. Indeed handbooks of heat transfer are filled with figures and correlations cast in terms of these dimensionless parameters.

Example 7.7

The receiving element in a solar electrical generation system is exposed to an external flow of wind at $u_\infty = 5$ m/s. The characteristic length of the receiving element is $L_{char} = 1$ m. The element is an irregularly shaped object and no correlation exists that would allow you to compute its Nusselt number. Therefore, you have decided to measure the heat transfer coefficient that occurs when a $1/10^{th}$ scale model of the receiving element is placed in a wind tunnel; the characteristic length of the test element is $L_{char,t} = 0.1$ m. You want to use the average heat transfer coefficient measured during the test $(\bar{h}_t)$ to determine the average heat transfer coefficient associated with the full-scale element $(\bar{h})$.

(a) At what velocity should the test be carried out?
(b) If the test is run at the velocity calculated in (a) and the measured value of the heat transfer coefficient is $\bar{h}_t = 100$ W/m^2-K, then what is $\bar{h}$?

Known Values

Figure 1 provides a sketch of the problem with the known information indicated.

Continued

Example 7.7 (cont.)

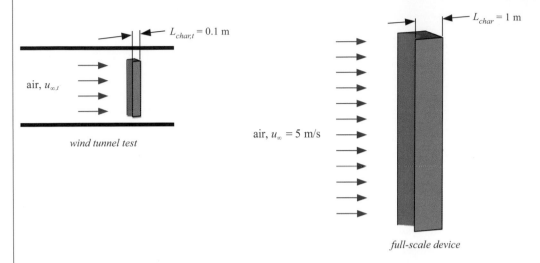

$L_{char,t} = 0.1$ m

air, $u_{\infty,t}$

wind tunnel test

$L_{char} = 1$ m

air, $u_\infty = 5$ m/s

full-scale device

Figure 1 Sketch of problem with known information shown.

Assumptions

- The properties of the air used in the wind tunnel are the same as those associated with the wind.
- The effect of viscous dissipation is negligible.

Analysis and Solution

According to Eq. (7.172), the average Nusselt number for a given shape is a function of Reynolds number and Prandtl number. Therefore, in order for the test measurements taken in the wind tunnel to be applicable to the full-scale system, it is necessary that both the Reynolds number and Prandtl associated with the test match the Reynolds number and Prandtl number associated with the full-scale system. Because the same fluid (air) with the same properties is used, the Prandtl numbers will be the same. Therefore, the wind tunnel velocity, $u_{\infty,t}$, should be selected so that the Reynolds number associated with the test is equal to the Reynolds number associated with the full-scale system. The Reynolds number for the test is:

$$Re_t = \frac{\rho_{air} L_{char,t} u_{\infty,t}}{\mu_{air}}. \tag{1}$$

While the Reynolds number for the full-scale system is:

$$Re = \frac{\rho_{air} L_{char} u_\infty}{\mu_{air}}. \tag{2}$$

Setting Eq. (1) equal to Eq. (2) and recognizing that the viscosity and density of the air used in the test and flowing over the full-scale device are assumed to be the same provides:

$$u_{\infty, t} = u_\infty \frac{L_{char}}{L_{char,t}} = \frac{5 \text{ m}}{\text{s}} \left| \frac{1 \text{ m}}{0.1 \text{ m}} \right| = \boxed{50 \text{ m/s}}. \tag{3}$$

Example 7.7 (cont.)

In order for the sub-scale test to be *similar* to the full-scale device, the testing must be carried out with a wind velocity of 50 m/s. If the Prandtl number and Reynolds number associated with the test are matched to the full scale device then the average Nusselt number measured during the test will be equal to the average Nusselt number associated with the full-scale system. The average Nusselt number associated with the test is defined as:

$$\overline{Nu_t} = \frac{\overline{h_t}\, L_{char,t}}{k_{air}} \tag{4}$$

and the average Nusselt number associated with the full-scale device is:

$$\overline{Nu} = \frac{\overline{h}\, L_{char}}{k_{air}}. \tag{5}$$

Setting Eq. (4) equal to Eq. (5) and recognizing that the conductivity of the air in the test is the same as the air flowing over the full-scale device provides:

$$\overline{h} = \frac{L_{char,\,t}\,\overline{h_t}}{L_{char}} = \frac{0.1\ \text{m}}{} \left|\frac{100\ \text{W}}{\text{m}^2\ \text{K}}\right|\frac{}{1\ \text{m}} = \boxed{10\ \text{W/m}^2\text{-K}}. \tag{6}$$

Discussion

The use of dimensional analysis to carry out experiments and scale results from models to full-scale devices is a powerful tool for engineers in a variety of disciplines. Testing the full-scale collector unit under the correct conditions is likely prohibitively expensive if not impossible. Testing a sub-scale model under conditions that provide matching of the appropriate dimensionless numbers is a much more practical alternative.

The assumption that viscous dissipation is not important can be justified by examining the value of Ec/Re_L, as discussed in Section 7.4.2. The Reynolds number is computed according to:

$$Re = \frac{\rho_{air}\,L_{char}\,u_\infty}{\mu_{air}} = \frac{1.177\ \text{kg}}{\text{m}^3}\left|\frac{1\ \text{m}}{}\right|\frac{5\ \text{m}}{\text{s}}\left|\frac{}{18.54 \times 10^{-6}\ \text{Pa s}}\right|\left|\frac{\text{Pa m}^2}{\text{N}}\right|\frac{\text{N s}^2}{\text{kg m}} = 3.2 \times 10^5 \tag{7}$$

where the density and viscosity of air are obtained from Appendix C. The Eckert number requires a temperature difference to compute, which was not provided in the problem statement. We can therefore determine how large the Eckert number must be in order for viscous dissipation to be important:

$$\frac{Ec}{Re_L} << 1. \tag{8}$$

Let's use as a threshold here 0.1, which leads to:

$$Ec = 0.1\,Re_L = 0.1\left(3.2 \times 10^5\right) = 3.2 \times 10^4. \tag{9}$$

If the Eckert number is above approximately 3.2×10^4 then we should start to question whether viscous dissipation is also important; in this case, achieving similarity would require that we match the Eckert number (as well at the Reynolds and Prandtl numbers). Fortunately, an exceptionally small temperature difference would be required achieve an Eckert number this large under the stated conditions. Rearranging the definition of the Eckert number, Eq. (7.144), in order to solve for the temperature difference leads to:

$$(T_\infty - T_s) = \frac{u_\infty^2}{c\,Ec} = \frac{(5)^2\ \text{m}^2}{\text{s}^2}\left|\frac{\text{kg K}}{1007\ \text{J}}\right|\frac{}{3.2 \times 10^4}\left|\frac{\text{J}}{\text{N m}}\right|\frac{\text{N s}^2}{\text{kg m}} = 7.7 \times 10^{-7}\ \text{K}. \tag{10}$$

Clearly in any realistic situation the temperature difference will be orders of magnitude larger than this and so the value of Ec/Re_L will be much less than one.

7.4.4 The Reynolds Analogy (Revisited)

The Reynolds analogy is discussed in Section 7.1.8 using a simple conceptual model of the boundary layer that is based on diffusive penetration into the free stream. It is informative to revisit the Reynolds analogy using the dimensionless governing equations that were derived in Section 7.4.2 in order to understand its limitations. The dimensionless momentum and energy equations for the boundary layer, Eqs. (7.140) and (7.145) are repeated below:

$$\tilde{u}\frac{\partial \tilde{u}}{\partial \tilde{x}} + \tilde{v}\frac{\partial \tilde{u}}{\partial \tilde{y}} = -\frac{\partial \tilde{p}_\infty}{\partial \tilde{x}} + \frac{1}{Re_L}\frac{\partial^2 \tilde{u}}{\partial \tilde{y}^2} \tag{7.173}$$

$$\tilde{u}\frac{\partial \tilde{\theta}}{\partial \tilde{x}} + \tilde{v}\frac{\partial \tilde{\theta}}{\partial \tilde{y}} = \frac{1}{Re_L Pr}\frac{\partial^2 \tilde{\theta}}{\partial \tilde{y}^2} + \frac{Ec}{Re_L}\left(\frac{\partial \tilde{u}}{\partial \tilde{y}}\right)^2. \tag{7.174}$$

In the limit that (1) $Pr = 1$, (2) there is a negligible free stream pressure gradient, and (3) viscous dissipation is not important, Eqs. (7.173) and (7.174) reduce to:

$$\tilde{u}\frac{\partial \tilde{u}}{\partial \tilde{x}} + \tilde{v}\frac{\partial \tilde{u}}{\partial \tilde{y}} = \frac{1}{Re_L}\frac{\partial^2 \tilde{u}}{\partial \tilde{y}^2} \tag{7.175}$$

$$\tilde{u}\frac{\partial \tilde{\theta}}{\partial \tilde{x}} + \tilde{v}\frac{\partial \tilde{\theta}}{\partial \tilde{y}} = \frac{1}{Re_L}\frac{\partial^2 \tilde{\theta}}{\partial \tilde{y}^2}, \tag{7.176}$$

which are partial differential equations of identical form defining the dimensionless velocity and dimensionless temperature. The boundary conditions for the dimensionless velocity and temperature are also the same. At the surface the velocity must be zero and the temperature must be T_s; therefore, both dimensionless quantities must be zero:

$$\tilde{u}_{\tilde{y}=0} = 0 \text{ and } \tilde{\theta}_{\tilde{y}=0} = 0. \tag{7.177}$$

As $\tilde{y}$ approaches infinity the free stream velocity and temperature must be recovered; therefore, both quantities must approach unity:

$$\tilde{u}_{\tilde{y}\rightarrow\infty} = 1 \text{ and } \tilde{\theta}_{\tilde{y}\rightarrow\infty} = 1. \tag{7.178}$$

Finally, at $\tilde{x} = 0$ the free stream velocity and temperature must be imposed; therefore, both quantities will again approach unity:

$$\tilde{u}_{\tilde{x}=0} = 1 \text{ and } \tilde{\theta}_{\tilde{x}=0} = 1. \tag{7.179}$$

Therefore, under the limiting conditions discussed above (i.e., $Pr \approx 1$, $\frac{d\tilde{p}_\infty}{d\tilde{x}} \approx 0$, and $Ec \approx 0$) the dimensionless temperature and velocity are governed by the same partial differential equation with the same boundary conditions. They must have the same solution. This is a more formal statement of the Reynolds analogy and it provides a clearer picture of its limitations.

Equations (7.152) and (7.168) express the friction factor and Nusselt number in terms of the dimensionless velocity and temperature difference, respectively:

$$C_f = \frac{2}{Re_L}\frac{\partial \tilde{u}}{\partial \tilde{y}}\bigg|_{\tilde{y}=0} \tag{7.180}$$

$$Nu = \frac{\partial \tilde{\theta}}{\partial \tilde{y}}\bigg|_{\tilde{y}=0}. \tag{7.181}$$

If the Reynolds analogy holds, then the gradients of $\tilde{\theta}$ and $\tilde{u}$ must be identical and therefore

$$\frac{\partial \tilde{u}}{\partial \tilde{y}}\bigg|_{\tilde{y}=0} = \frac{\partial \tilde{\theta}}{\partial \tilde{y}}\bigg|_{\tilde{y}=0} = \frac{C_f Re_L}{2} = Nu, \tag{7.182}$$

which is the same statement of the Reynolds analogy obtained in Section 7.1.8.

7.5 Conclusions and Learning Objectives

Convection is fundamentally the problem of momentum and energy transport from an object to a surrounding fluid flow. These transport processes are governed by gradients in the velocity (for momentum) and temperature (for energy), respectively, at the object's surface. These gradients are related to the behavior of a relatively thin layer of fluid that is directly adjacent to the object and is affected by its presence, referred to as the boundary layer. This chapter has provided a conceptual description of the boundary layer in order to gain some basic understanding of laminar and turbulent convection processes.

This chapter has also formally derived the set of coupled partial differential equations that govern a convection problem. This process should reinforce the practice of carrying out a balance (of mass, momentum, and energy) on a differentially small control volume. The result suggests that the solution to a convection problem is not easy, which motivates the simplification of the equations for application in a boundary layer. The resulting boundary layer equations remain complex and this provides an incentive to nondimensionalize them in order to identify the minimum set of dimensionless parameters that can be used to represent a solution. The result is the important conclusion that in many cases the solution to a convection problem associated with a particular shape can be represented as a correlation for Nusselt number (the dimensionless heat transfer coefficient) as a function of Reynolds number and Prandtl number. Specific examples of these correlations can be found in the following chapters for different geometric configurations as well as different flow situations.

Some specific concepts that you should understand are listed below.
- The definition of a velocity boundary layer as the region of flow where the velocity has been reduced, relative to the free stream, due to the presence of a surface.
- The definition of a temperature boundary layer as the region of flow where temperature varies from the surface to the free stream temperature.
- The concept that laminar boundary layers grow by diffusion processes; the diffusion of momentum (for a velocity boundary layer) or energy (for a thermal boundary layer).
- The idea that kinematic viscosity and thermal diffusivity indicate the ability of a fluid to transport momentum and thermal energy, respectively, using diffusion.
- The definition of the Prandtl number as the ratio of kinematic viscosity to thermal diffusivity and the idea that this important dimensionless number represents the relative ability of a fluid to transport momentum to energy.
- An understanding of the physical mechanisms that could lead to a Prandtl number that is very different from unity.
- The conceptual model of boundary layer growth that arises from our previous study of diffusion energy transport from Chapter 6 and the associated understanding of the heat transfer coefficient and shear stress.
- The definition of the friction coefficient, the drag coefficient, and the Nusselt number.
- The underpinnings of the Reynolds analogy and the modified Reynolds analogy.
- The difference between the local and average heat transfer coefficient and shear stress.
- The conceptual model of a turbulent boundary layer based on the motion of turbulent eddies.
- The concept of a viscous sublayer and the idea that the viscous sublayer rather than the boundary layer thickness governs the gradients of velocity and temperature in a turbulent flow.
- The derivation of the continuity, momentum, and energy equations that govern a convection process.
- The basis for the boundary layer simplifications.
- The ability to estimate the order of magnitude of terms within a partial differential equation in order to ascertain their relative importance.
- The ability to nondimensionalize a system of equations through the systematic identification and substitution of a few dimensionless quantities.
- The concept of dimensional similarity and what this means specifically for convection problems.
- The ability to use the idea of dimensional similarity in order to develop correlations.
- The ability to use the idea of dimensional similarity to translate test data or other results from one situation (size, fluid, velocity, etc.) to another situation.

References

Alexandrou, A., *Principles of Fluid Mechanics*, Prentice Hall, Upper Saddle River (2001).

Bejan, A., *Heat Transfer*, Wiley, New York (1993).

Bridgman, P. W., *Dimensional Analysis*, Yale University Press, New Haven (1922).

Buckingham, E., "On physically similar systems; illustrations of the use of dimensional analysis," *Phys. Rev.*, Vol. 4, pp. 345–376 (1914).

Hinze, J. O, *Turbulence*, Second Edition, McGraw-Hill, New York (1975).

Nellis, G. F. and S. A. Klein, *Heat Transfer*, Cambridge University Press, New York (2009).

Schlichting, H., *Boundary Layer Theory*, Springer Publishing, New York (2000).

Sonin, A. A., *The Physical Basis of Dimensional Analysis*, Class notes for Advanced Fluid Mechanics in the Department of Mechanical Engineering at the Massachusetts Institute of Technology, Cambridge, MA (1992).

Tennekes, H. and J. L. Lumley, *A First Course in Turbulence*, The MIT Press, Cambridge (1972).

Problems

Laminar Boundary Layer Concepts

7.1 The figure illustrates a fluid flowing over a flat plate subjected to a constant heat flux, $\dot{q}_s''$. The free stream temperature is T_∞.

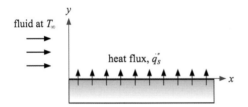

(a) Sketch the thermal boundary layer thickness (δ_t) as a function of position, x, from the leading edge of the plate. Assume that the flow is laminar.

(b) Sketch the surface temperature of the plate (T_s) as a function of position, x, from the leading edge of the plate. Assume that the flow is laminar.

7.2 Air flows over two, thin flat plate arrangements (A and B in the figure). The single plate in arrangement A is twice as long as the two plates in arrangement B. The plates have the same width (into the page), the same uniform surface temperature, and are exposed to the same free stream flow. The flow over the plates is laminar in both arrangements.

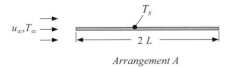

Arrangement A

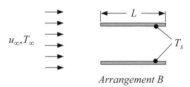

Arrangement B

Do you expect the total rate of heat transfer to the air to be higher for arrangement A or B? Justify your answer.

7.3 Two identical plates are oriented parallel to a flow, as shown.

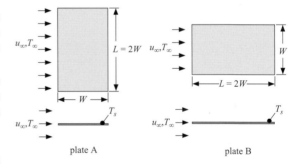

plate A   plate B

Plate B is oriented with its long dimension, L, oriented with the flow and its short dimension, W, oriented perpendicular to the flow. Plate A is oriented with its long dimension perpendicular to the flow. In both cases, $L = 2W$. Assume that the flow remains laminar in both cases. The free stream velocity and temperature, u_∞ and T_∞, and the plate temperature, T_s, are the same for both cases. Estimate the ratio of the total rate of heat transfer from plate A to the total rate of heat transfer from plate B. Do not use a correlation to answer this question.

7.4 A liquid at 25°C, 1 atm is flowing at 30 m/s over the top side of a metal plate that is 4 m wide and 15 cm long in the flow direction. The plate is maintained at 42°C by electrical heaters. The heat transfer coefficient between the plate and the liquid is 1500 W/m²-K. The average properties of the liquid are: $\rho = 1264$ kg/m³, $\mu = 1.49$ Pa-s, $k = 0.286$ W/m-K, and $c = 2.386$ kJ/kg-K. Determine the values of:

(a) the average Reynolds number,
(b) the Prandtl number,
(c) the average Nusselt number, and
(d) the required rate of electrical power to maintain the plate at constant temperature.
(e) Use the Chilton–Colburn analogy to provide an estimate of the force required to keep the plate from moving.

7.5 Cold fluid flows over a hot plate that has a uniform surface temperature. Indicate the letter of the curve that most closely describes the variation of the local heat flux from the surface of the plate with position.

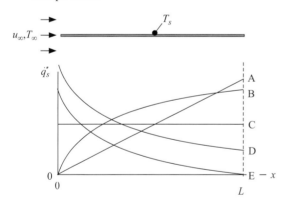

7.6 Air at atmospheric pressure flows over a flat plate that is maintained at 150°C. The free stream velocity is 10 m/s and the temperature far from the plate is 54°C. At a certain distance from the leading edge of the plate, the thermal boundary layer thickness (δ_t), is 1.25 mm and the temperature distribution in the direction normal to the plate is given by

$$T = 150 - 45\frac{y}{\delta_t} + 50\left(\frac{y}{\delta_t}\right)^3 - 100\left(\frac{y}{\delta_t}\right)^4,$$

where T is the temperature in °C and y is the vertical position above the plate.

(a) Estimate the value of the local convection coefficient.
(b) Estimate the local value of Reynolds number.

7.7 Water at 1 atm over a flat plate that is maintained at 80°C. At a certain distance from the leading edge of the plate, the temperature distribution in the direction normal to the plate is given by

$$T = 80 - 40(1 - \exp(-350y)),$$

where T is the temperature in °C and y is the vertical position above the plate in m.

(a) Determine the value of the local convection heat transfer coefficient at this axial position.
(b) Determine the approximate thickness of the boundary layer at this axial position.

7.8 A thin square metal plate, 15 cm on each side, is placed in a wind tunnel. Air at 25°C and 101.3 kPa flows over the top and bottom surfaces at 35 m/s. The force required to keep the plate stationary is measured to be 0.057 N.

(a) Calculate the friction coefficient.
(b) Estimate the average heat transfer coefficient using the Chilton–Colburn analogy.

7.9 Consider a flat surface oriented parallel to a flow that has velocity (u_∞) and temperature (T_∞). The properties of the flow are ρ, c, k, and μ. The length of the plate in the flow direction is L. Also, the flow is steady in time and laminar over the entire plate. The Prandtl number of the fluid is nominally equal to unity. The heat flux at the surface is a function of the axial location (x, measured relative to the leading edge of the plate). Here we will consider two cases; in case A, the heat flux is constant in x whereas in case B, the heat flux varies linearly with x. In both cases, the average heat flux integrated over the entire plate ($\overline{\dot{q}''}$) is the same:

$$\dot{q}''(x) = \begin{cases} \overline{\dot{q}''} & \text{for case A} \\ 2\dfrac{x}{L}\overline{\dot{q}''} & \text{for case B.} \end{cases}$$

Sketch the temperature at the surface of the plate, $T_s(x)$, for both cases.

7.10 A fluid with a constant velocity and a high temperature is flowing along a smooth, flat plate. The surface of the plate is cooled by some internal means to a uniform cold temperature over its initial length. After this initial length, the surface of the plate is insulated. Assume that the flow is laminar. Estimate and sketch how the surface temperature of the plate varies after the initial length.

7.11 The figure illustrates the flow of a fluid over a flat plate. The free stream temperature is T_∞ and the free stream velocity is u_∞. The temperature difference between the surface of the plate and the fluid temperature varies linearly with distance from the leading edge, according to:

$$T_s - T_\infty = Ax,$$

where A is a constant. The flow transitions from laminar to turbulent at position $x = L/2$, where L is the length of the plate.

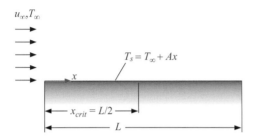

(a) Determine an approximate relationship for the heat flux at the plate surface ($\dot{q}_s''$) as a function of position that is valid for $0 < x < L/2$ (i.e., in the laminar region). Your answer should not rely on any correlation, but rather on your understanding of how laminar boundary layers develop.

(b) Use your answer from (a) to sketch the heat flux as a function of position. Your sketch should be qualitatively correct in that it should have the correct functional form from $0 < x < L/2$ and the change at $x = L/2$ should be clear.

7.12 A thin plate separates two fluids flowing in opposite directions. The fluid flowing along the top of the plate from left to right is hot, with free stream temperature T_H. The fluid flowing along the bottom of the plate is cold, with free stream temperature T_C.

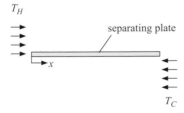

Assume that the separating plate is infinitely thin; therefore, it offers no resistance to heat transfer across it (in the stream-to-stream direction) and infinite resistance to heat transfer along it (in the direction parallel to the flow). Also, assume that the flow remains laminar on both sides of the plate and that the hot and cold fluids have approximately the same properties. You should not use any correlations in the development of your solution – only your conceptual understanding of how boundary layers grow. All of the required sketches can be qualitative with some written justification as you feel is appropriate.

(a) Sketch the thermal boundary layer thickness on the hot and cold sides of the plate as a function of x.

(b) Sketch the heat transfer coefficient on the hot and cold sides of the plate as a function of x.

(c) Sketch the temperature of the plate as a function of position.

7.13 A sun roof is installed in the roof of a car.

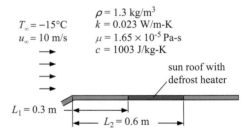

A defrost heater is integrated with the sun roof glass in order to prevent ice from forming. The sun roof begins at $L_1 = 0.3$ m from the front of the roof and extends to $L_2 = 0.6$ m. The roof is $W = 1.2$ m wide. The car is driving at $u_\infty = 10$ m/s through air at $T_\infty = -15°C$. Assume that the properties of the air are conductivity $k = 0.023$ W/m-K, density $\rho = 1.3$ kg/m³, viscosity $\mu = 1.65 \times 10^{-5}$ Pa-s, and specific heat capacity $c = 1003$ J/kg-K. The defrost heater is a thin sheet of electrically resistive material that covers the entire sun roof and transfers a uniform heat flux to the glass. The convection coefficient between the interior surface of the roof and the cabin air is sufficiently low that convection heat transfer with the interior can be neglected. Assume that the flow over the roof remains laminar and that the thermal boundary layer develops from the beginning of the roof (rather than the beginning of the sun roof). Do not use a correlation to solve this problem; instead, use your conceptual knowledge of boundary layer behavior to estimate the answers.

(a) What is the total power required to keep the sun roof above $T_{ice} = 0°C$?

(b) Where is the hottest temperature on the sun roof surface? Estimate this temperature.

7.14 The figure illustrates the flow of a fluid with $T_\infty = 0°C$, $u_\infty = 1$ m/s over a flat plate.

fluid at $T_\infty = 0°C$, $u_\infty = 1$ m/s
$Pr = 1$, $k = 1$ W/m-K, $\alpha = 1 \times 10^{-3}$ m²/s

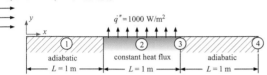

The flat plate is made up of three sections, each with length $L = 1$ m. The first and last sections are insulated and the middle section is exposed to a constant heat flux, $\dot{q}'' = 1000$ W/m². The properties of the fluid are Prandtl number $Pr = 1$, conductivity $k = 1$ W/m-K, and thermal diffusivity $\alpha = 1 \times 10^{-3}$ m²/s. Assume that the flow is laminar over the entire surface.

(a) Sketch the momentum and thermal boundary layers as a function of position, x. Do not worry about the qualitative characteristics of your sketch – get the quantitative characteristics correct.

(b) Sketch the temperature distribution (the temperature as a function of distance from the plate y) at the four locations indicated in the figure. Location 1 is half-way through the first adiabatic region, Location 2 is half-way through the heated region, Location 3 is at the trailing edge of the heated region (in the heated region), and Location 4 is at the trailing edge of the final adiabatic region. Again, focus on getting as many of the qualitative characteristics of your sketch correct as you can.

(c) Predict, approximately, the temperature of the surface at locations 1, 2, and 3 in the figure. Do not use a correlation. Instead, use your conceptual understanding of how boundary layers behave to come up with very approximate estimates of these temperatures.

Turbulent Boundary Layer Concepts

7.15 Air at 40°C and 1 atm flows over the top side a flat plate at a free stream velocity of 33 m/s. The plate is 40 cm wide and 25 cm in length.

(a) Plot the boundary layer thickness as a function of distance from the leading edge of the plate.

(b) Plot the local friction factor as a function of distance from the leading edge of the plate.

(c) At what point would you expect the flow to transition to turbulent flow?

7.16 A fluid with a low Prandtl number ($Pr \ll 1$) flows over a heated flat plate ($T_s > T_\infty$). Assume that the flow velocity, u_∞, is low enough that the flow remains laminar at a particular axial location x.

(a) Sketch the velocity and temperature distribution that you expect at location x. Indicate on your sketches the approximate extent of the momentum and thermal boundary layer thicknesses (δ_m and δ_t); make sure that the relative size of these boundary layer thicknesses is appropriate given the low Prandtl number of the fluid.

(b) Assume that the flow velocity, u_∞, is increased sufficiently that the flow transitions to turbulence. Sketch the temperature distribution that you would expect. Make sure that the qualitative differences between your answers to (a) and (b) are clear.

7.17 In an experiment, fluid flows over a flat plate that is 3.5 m long, in which the shear force and heat transfer coefficient are measured. A transition from laminar to turbulent flow is observed to occur at 2.4 m. At this point, the Reynold's number is 1.2×10^6. If the transition were to occur at a Reynold's number of 5×10^5 then:

(a) how far from the leading edge of the plate would the transition occur, and

(b) would the calculated shear force on the plate be less than or greater than the measured value?

7.18 Water at atmospheric pressure, free stream velocity $u_\infty = 1.0$ m/s and temperature $T_\infty = 25°C$ flows over a flat plate with a surface temperature $T_s = 90°C$. The plate is $L = 0.15$ m long.

Assume that the flow is laminar over the entire length of the plate.

(a) Estimate, using your knowledge of how boundary layers grow, the size of the momentum and thermal boundary layers at the trailing edge of the plate (i.e., at $x = L$). Do not use a correlation for this; instead use the approximate model for boundary layer growth.

(b) Use your answer from (a) to estimate the shear stress at the trailing edge of the plate and the heat transfer coefficient at the trailing edge of the plate.

(c) You measure a shear stress of $\tau_{s,meas} = 1.0$ Pa at the trailing edge of the plate; use the modified Reynolds analogy to predict the heat transfer coefficient at this location.

7.19 A flat plate is exposed to a flow of water ($k = 0.61$ W/m-K). At the trailing edge of the plate ($x = L = 0.2$ m) you have found that the thermal boundary layer thickness is $\delta_t = 0.48$ mm and the temperature distribution is given by:

$$\frac{(T_s - T)}{(T_s - T_\infty)} = \frac{y}{2\delta_t}\left(3 - \frac{y^2}{\delta_t^2}\right),$$

where $T_s = 340$ K and $T_\infty = 300$ K are the surface and free stream temperatures, respectively.

(a) Estimate the local heat flux at $x = L$ and the local heat transfer coefficient.

(b) Determine the local Nusselt number at $x = L$.

(c) If the heat transfer coefficient is proportional to $x^{-1/2}$ then determine the average heat transfer coefficient on the plate.

7.20 A flat plate with length $L = 0.1$ m is exposed to a flow of oil at $u_\infty = 0.05$ m/s. The surface of the plate is maintained at a constant temperature that is greater than the free stream temperature of the oil. Here we will consider the two different types of oil. The properties of oil A are $\rho_A = 1000$ kg/m^3, $\mu_A = 0.001$ Pa-s, $c_A = 1000$ J/kg-K, $k_A = 1$ W/m-K. Oil B has the same properties as oil A except that it has a much higher thermal conductivity, $k_B = 100$ W/m-K.

(a) Sketch the momentum boundary layer and shear stress as a function of x for $0 < x < L$ for both of the two oils (A and B).

(b) Sketch the thermal boundary layer and heat flux as a function of x for $0 < x < L$ for both of the two oils (A and B).

This chapter deals mostly with flow over "aerodynamically smooth" surfaces; that is, the surface roughness is small enough so that it does not affect the shear stress or heat flux at the surface.

(c) Based on your knowledge of the important length scales in the boundary layer, estimate the size of the roughness element at the trailing edge of the plate that would significantly affect the friction coefficient for both of the two oils (A and B).

(d) Based on your knowledge of the important length scales in the boundary layer, estimate the size of the roughness element at the trailing edge of the plate that would significantly affect the heat transfer coefficient for both of the two oils (A and B).

Now assume that the velocity of the flow is increased to 25 m/s.

(e) Sketch the momentum boundary layer and shear stress as a function of x for $0 < x < L$ for both of the two oils (A and B).

(f) Sketch the thermal boundary layer and heat flux as a function of x for $0 < x < L$ for both of the two oils (A and B). Be sure to address the following questions using your sketch and justify your answers – how different do you expect the size of the thermal boundary layer to be at the trailing edge? How different do you expect the heat flux to be at the trailing edge?

The Boundary Layer Equations and Dimensional Analysis for Convection

7.21 The rate equations that govern the shear stress in a Newtonian fluid are given by Bejan (1993) and elsewhere. The tangential stress and excess normal stress are given by:

$$\tau_{yx} = \mu\left(\frac{\partial u}{\partial y} + \frac{\partial v}{\partial x}\right),$$

$$\sigma_{xx} = 2\mu\frac{\partial u}{\partial x}.$$

Show that these rate equations can be substituted into the x-momentum balance and lead to the x-momentum equation.

7.22 Derive the continuity equation for a 2-D flow in Cartesian coordinates but do not make the assumption that the fluid is incompressible or that the flow is steady.

7.23 Dowtherm A is flowing over a surface with free stream velocity 6 m/s. At a particular location 0.25 m from the leading edge the local shear is found to be 18.8 Pa.
(a) Determine the local friction coefficient and the Reynolds number.
(b) Use the modified Reynolds analogy to estimate the local Nusselt number and local heat transfer coefficient.

7.24 Use the information from Example 7.1 to estimate the friction coefficient.

7.25 In Chapter 9 we will examine internal flow situations. In fully developed, steady laminar flow through a channel the pressure is only a function of x, the flow direction and the x-velocity (u) only depends on y, the direction perpendicular to the flow direction. Simplify the x-momentum equation in Cartesian coordinates for this situation. Neglect the effect of gravity.

7.26 Consider fully developed, steady laminar flow through a channel made by two parallel separated by a distance H. In this case, the pressure is only a function of x, the flow direction and the x-velocity (u) only depends on y, the direction perpendicular to the flow direction. The x-momentum equation for this situation, neglecting gravity, is

$$\frac{dp}{dx} = \mu \frac{d^2u}{dy^2},$$

where dp/dx is the pressure gradient driving the flow. The velocity u must be 0 at $y = 0$ (the bottom surface of the duct) and $y = H$ (the top surface of the duct). Derive the velocity distribution in the duct. Plot the velocity distribution for the case where $\mu = 0.01$ Pa-s, $H = 1$ cm, and $dpdx = -100$ Pa/m.

7.27 You have fabricated a 1/10th scale model of a new type of automobile and placed it in a wind tunnel. You want to test it at an air velocity that will allow you to compute the drag force on the car when it travels at 55 miles per hour. What air velocity should you choose and how is the drag force that you measure on the model related to the drag force on the automobile?

7.28 A microscale feature is to be fabricated on a micro-electro-mechanical system (MEMS) in order to sense the characteristics of the flow of a fluid within a microchannel.

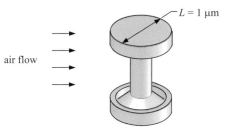

In order to design the device it is necessary to understand the heat transfer coefficient associated with the flow of air across the feature. The device is not a commonly encountered shape (e.g., a cylinder or sphere) and therefore you are unable to locate appropriate correlations in the literature. Further, the device is only $L = 1$ μm in size and is therefore too small to test accurately. You have decided to build a scaled-up model of the feature that can be more easily tested in a wind tunnel. The scaled up model is 10,000× larger than the MEMS feature (i.e., $L_{test} = 1$ cm and all of the object dimensions scale proportionally). The model is mounted in a wind tunnel and electrical heaters are embedded in the model and apply a measured, constant heating of $\dot{q}_{test} = 0.5$ W. The model is composed of a conductive material so that it has approximately a uniform surface temperature ($T_{s,test}$). The wind tunnel is used to provide a uniform flow of air at $T_{\infty,test} = 20°C$ and $P_{test} = 1$ atm over a range of velocities ($u_{\infty,test}$).

Air velocity, $u_{\infty,test}$ (m/s)	Surface temperature, $T_{s,test}$ (°C)
0.15	84.39
0.30	69.82
0.45	62.88
0.60	58.55
0.75	55.50
1.13	50.55
1.50	47.47
2.25	43.64
3.00	41.25
4.50	38.29
6.00	36.45
7.50	35.14
11.25	33.03

Note that the total surface area of the model is $A_{s,test} = 6.28 \times 10^{-4}$ m^2 and the surface area of the microscale device is $(10,000)^2 \times$ smaller.

(a) Use the experimental measurements to prepare a figure that shows the Nusselt number of the microscale device as a function of the Reynolds number for a Prandtl number that is consistent with air.

(b) Prepare a figure that shows the heat transfer coefficient for the microscale device as a function of the gas velocity across the device for air with $T_\infty = 80°C$ and $p = 10$ atm.

7.29 This problem is related to dimensionless parameters and testing of sub-scale models. The mast of a ship is to be changed to an extended pentagon shape; all sides are equal with dimension W and the length of the mast is L. The behavior of this shape when exposed to a high cross-wind during operation is important and not well known. In particular, you have been asked to determine the force per length imposed on the mast for a full-size ($W = 25$ cm) mast when it is exposed to a cross-wind at $u = 11.2$ m/s and $p = 1$ atm. It is not feasible to test a full-scale mast at prototypical conditions; the full-size device will not fit into any wind tunnels that you have access to. Therefore, you have tested a sub-scale replica of the mast in a small wind tunnel that can be operated using high pressure air ($p_{test} = 4$ atm). The sub-scale model is 1/25th scale (that is, $W_{test} = 1.0$ cm). The data taken in the wind tunnel at various air velocities are provided in the table. Assume (1) that air obeys the Ideal Gas Law, (2) that the air viscosity is not a function of pressure, and (3) that the test and full scale air temperatures are the same.

Data taken with sub-scale model.

Wind tunnel air velocity, u_{test} (m/s)	Measured force per unit length, F_{test}/L (N/m)
31.2	36.4
40.1	59.2
49.2	88.2
61.5	135
70.2	172.4
79.9	219.6
89.2	270.6
100.0	334.8

Use these data to estimate the drag force experienced by the actual mast per unit length. Explain the steps you are taking as well as any additional assumption that you make.

7.30 You have fabricated a 1000× scale model of a microscale feature that is to be used in a microchip. The device itself is only 1 μm in size and is therefore too small to test accurately. However, you'd like to know the heat transfer coefficient between the device and an air flow that has a velocity of 10 m/s. What velocity should you use for the test and how will the measured heat transfer coefficient be related to the actual one?

7.31 Your company makes an extrusion that can be used as a lightweight structural member; the extrusion is long and thin and has an odd cross-sectional shape that is optimized for structural performance. This product has been used primarily in the aircraft industry; however, your company wants to use the extrusion in an application where it will experience crossflow of water rather than air. There is some concern that the drag force experienced by the extrusion will be larger than it can handle. Because the cross-section of the extrusion is not simple (e.g., circular or square) you cannot go look up a correlation for the drag coefficient in the same way that you could for a cylinder. However, because of the use of the extrusion in the aircraft industry you have an extensive amount of data relating the drag force on the extrusion to velocity when it is exposed to a crossflow of air. These data have been collated and are shown graphically in the figure.

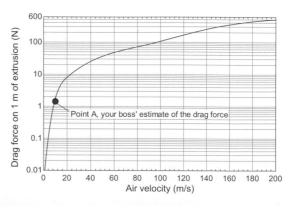

Your boss insists that the drag force for the extrusion exposed to water can be obtained by looking at the figure and picking off the data at the point where $V = 10$ m/s (Point A); this corresponds to a drag force of about 1.7 N/m of extrusion.

(a) Is your boss correct? Explain why or why not.

(b) If you think that your boss is not correct, then explain how you could use the data shown in the figure to estimate the drag force that will be experienced by the extrusion for a water crossflow velocity of 10 m/s.

(c) Use the data in the figure to estimate the drag force that will be experienced by the extrusion for a water crossflow velocity of 10 m/s.

Projects

7.32 You need to know the thermal-fluid characteristics of electronics modules, shaped like a cube, when they are placed in a flow stream. The modules that you are interested in are quite large (0.5 m per side) and will be towed behind a ship (in water that is at $20°$ C and with a speed of 1 m/s). The electronics dissipate power at a rate of 10 kW.

You are restricted to doing experiments on a much smaller cube (2 cm per side) that is placed in a pipe in a laboratory. The cube is suspended in the pipe in the same orientation relative to the flow that is expected for the electronics module. The pipe is quite large and some flow straighteners are placed ahead of the test section in order to try and replicate free stream conditions. A heater is embedded in the cube which is made of a high-conductivity material. The heater power is adjusted so that the temperature difference between the cube surface and the water is approximately 10°C. The heater power, cube temperature, free stream temperature, and the force exerted on the cube are all measured as a function of the free stream velocity. The results are summarized (with no attempt at quantifying experimental error – clearly a very important part of any real test of this type) in the table.

T_water (°C)	T_cube (°C)	v_water (m/s)	q_heater (W)	F_cube (N)
20.0	30.1	4.2	284	3.5
20.0	30.1	8.2	419	13.4
20.0	30.2	12.5	545	31.2
20.0	30.2	16.1	629	51.7
20.0	30.1	20.4	720	83.1
20.0	30.1	24.8	807	122.8
20.0	30.1	28.5	872	162.2
20.0	30.2	32.2	950	207.0
20.0	30.3	36.2	1023	261.6
20.0	30.1	40.1	1071	321.0

Use the results in the table to estimate the force exerted on the full-size electronics module as well as the operating temperature of the full-size electronics module (assume that the module is isothermal).

7.33 The figure illustrates a concept for a very high heat flux heat exchanger operating between two fluids.

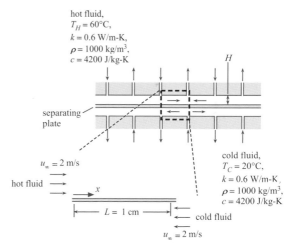

The heat exchanger consists of a separating plate and two fluid manifolds. A series of slots inject and remove the hot fluid so that it flows along the top of the plate. Every other slot is an injection slot where fluid is introduced and the fluid is removed from the remaining slots. The distance between adjacent slots is $L = 1$ cm. The hot fluid enters at temperature $T_H = 60°C$ and has conductivity $k = 0.6$ W/m-K, density $\rho = 1000$ kg/m^3, and specific heat capacity $c = 4200$ J/kg-K.

The cold fluid is injected and removed below the separating plate. The cold fluid enters at $T_C = 20°C$ and has the same conductivity, density, and specific heat capacity as the hot fluid. The initial design has the cold fluid flowing in the opposite direction as the hot fluid (i.e., in a counter-flow arrangement).

You can't find an exact correlation that corresponds to this impinging jet situation so you've decided to model the flow approximately. The flow rates are sufficiently high and the channels sufficiently large (i.e., H is sufficiently large) that you have decided to approximate the flow through the channel as an external flow over a plate. The viscosity of the fluid is large enough that the flow remains laminar. The free stream velocity of both the hot and cold fluids is $u_\infty = 2$ m/s. You should assume that a thermal boundary layer develops on the hot side according to:

$$\delta_{t,H} \approx 2\sqrt{\alpha t} = 2\sqrt{\alpha \frac{x}{u_\infty}},$$

where x is measured from the injection slot to the removal slot. The local heat transfer coefficient is given (approximately) by:

$$h \approx \frac{k}{\delta_{t,H}}.$$

The heat transfer coefficient on the cold-fluid side should be modeled in an analogous manner (although the thermal boundary layer on the cold side is growing in the opposite direction). Assume that the separating plate is infinitely thin; therefore, it offers no resistance to heat transfer across it (in the stream-to-stream direction) and infinite resistance to heat transfer along it (in the direction parallel to the flow).

(a) Plot the temperature of the separating plate as a function of position, x.
(b) Plot the heat flux from the hot fluid to the cold fluid as a function of position, x.
(c) Determine the average heat flux from the hot fluid to the cold fluid. Plot the average heat flux as a function of the slot-to-slot distance, L.

You have been asked to evaluate the possibility of re-plumbing the heat exchanger so that the hot and cold fluids flow in the same direction (i.e., in parallel-flow), as shown.

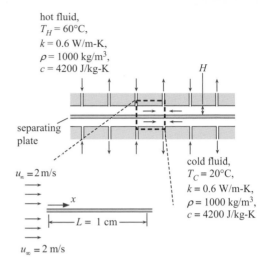

hot fluid,
$T_H = 60°C$,
$k = 0.6$ W/m-K,
$\rho = 1000$ kg/m^3,
$c = 4200$ J/kg-K

H

separating plate

$u_\infty = 2$ m/s

x

$L = 1$ cm

$u_\infty = 2$ m/s

cold fluid,
$T_C = 20°C$,
$k = 0.6$ W/m-K,
$\rho = 1000$ kg/m^3,
$c = 4200$ J/kg-K

(d) Overlay on your plot from (a) the temperature of the separating plate as a function of position, x.
(e) Overlay on your plot from (b) the heat flux from the hot fluid to the cold fluid as a function of position, x.
(f) Determine the average heat flux from the hot fluid to the cold fluid. Overlay on your plot from (c) the average heat flux as a function of the slot-to-slot distance, L. Explain why one configuration works better (i.e., provides a larger average heat flux) than the other.

7.34 Your company has come up with a randomly packed fibrous material that could be used as a regenerator packing. Currently there are no correlations available that would allow the prediction of the heat transfer coefficient for the packing. Therefore, you have carried out a series of tests to measure the heat transfer coefficient. A $D_{bed} = 2$ cm diameter bed is filled with these fibers with diameter $d_{fiber} = 200$ μm. The nominal temperature and pressure of the testing is $T_{nom} = 20°C$ and $p_{nom} = 1$ atm, respectively. The mass flow rate of the test fluid, $\dot{m}$, is varied and the heat transfer coefficient is measured. Several fluids, including air, water, and ethanol, are used for testing. The data are shown in the table.

Air		Water		Ethanol	
Mass flow rate (kg/s)	Heat transfer coefficient (W/m²-K)	Mass flow rate (kg/s)	Heat transfer coefficient (W/m²-K)	Mass flow rate (kg/s)	Heat transfer coefficient (W/m²-K)
0.0001454	170.7	0.00787	8464	0.009124	4162
0.0004073	311.9	0.02204	15470	0.02555	7607
0.0006691	413.7	0.0362	20515	0.04197	10088
0.0009309	491.8	0.05037	24391	0.05839	11993
0.001193	572.7	0.06454	28399	0.07481	13964
0.001454	631.1	0.0787	31296	0.09124	15388

(a) Plot the heat transfer coefficient as a function of mass flow rate for the three different test fluids.

(b) Plot the Nusselt number as a function of the Reynolds number for the three different test fluids. Use the fiber diameter as the characteristic length and the free flow velocity (i.e., the velocity in the bed if it were empty) as the characteristic velocity.

(c) Correlate the data for all of the fluids using a function of the form: $Nu = a\,Re^b\,Pr^c$. Note

that you may want to transform the results using a natural logarithm and use the Linear Regression option from the Tables menu in EES to determine a, b, and c.

(d) Use your correlation to estimate the heat transfer coefficient for 20 kg/s of oil passing through a 50 cm diameter bed composed of fibers with 2 mm diameter. The oil has density 875 kg/m³, viscosity 0.018 Pa-s, conductivity 0.14 W/m-K, and Prandtl number 20.

8 | External Forced Convection

Chapter 7 provides a discussion of the behavior of laminar and turbulent boundary layers at a conceptual level without presenting any specific correlations that can be used to solve an external flow problem. In Section 7.3 the boundary layer equations are derived and Section 7.4 shows how, with some limitations, their solution can be expressed in terms of a limited set of nondimensional parameters: the Reynolds number, Prandtl number, and Nusselt number. These relationships are referred to as *correlations*.

This chapter presents some useful correlations for external flow that can be used to solve a wide range of engineering problems. These results are based on careful experimental and theoretical work accomplished by many researchers. The correlations are examined as they are presented in order to verify that they agree with the physical understanding of boundary layer behavior that is developed in Chapter 7. A relatively complete and useful set of these correlations is provided as libraries of functions that are available in EES and the use of these functions to solve practical engineering problems is illustrated by example.

8.1 Methodology for using a Convection Correlation

The methodology for correctly using a convection correlation to solve an engineering problem is outlined in this section as a sequence of five steps.

1. *Understand the flow situation and geometry*

   The purpose of this step is to understand the type of flow being considered. It involves answering several questions.

   - Is the flow forced convection or natural convection?

     Forced convection flows are those in which the velocity is driven *externally* (e.g., by a pump or fan) and correlations for these situations are presented in this chapter and Chapter 9. **Natural convection** flows are driven by density differences within the fluid itself that are induced by the heating (or cooling) process. In the absence of any heating or cooling, the fluid in a natural convection problem would be quiescent. Correlations for this type of situation are presented in Chapter 10.

   - Is the flow an external flow or an internal flow?

     External flows occur in situations where the boundary layers can continue to grow without bound as the flow progresses along the surface. External flows correspond to flows over an object that is exposed to a large flow field (e.g., a sphere falling through the air or a building exposed to the wind). **Internal flows** occur in situations where the boundary layers become confined due to the presence of other surfaces. Internal flows are encountered in pipes and ducts. Correlations for external and internal flows are presented in this chapter and Chapter 9, respectively.

   - Is the fluid sufficiently far from the vapor dome that it remains single phase or is evaporation (boiling) or condensation occurring?

     Correlations for situations where phase change, either boiling or condensation, is occurring are presented in Chapter 11.

   - What are the important characteristics of the geometry associated with the flow condition?

     Specify the shape of the object or the cross-section of the duct as well as the orientation relative to the flow. In some cases it may be necessary to specify details of the boundary conditions as well (e.g., the surface experiences a uniform temperature or a uniform heat flux).

- Is an average or a local heat transfer coefficient required?

Most engineering problems rely on an average heat transfer coefficient, but there are some situations where you may be interested in the local conditions at some specific location on the surface.

The answers to all of these questions will allow you to identify the most appropriate correlation to use for the problem.

2. *Determine the properties of the fluid*

In order to compute the dimensionless quantities required by the correlations it is necessary to know the properties of the fluid (e.g., density, viscosity and conductivity). These properties can be obtained from tables or using software such as EES. The properties must be calculated using a single temperature even though the temperature of the fluid actually varies from the surface temperature (T_s) to the free stream temperature (T_∞). There are different ways to deal with this issue. The simplest way is to use the **film temperature** (T_f), defined as:

$$T_f = \frac{T_s + T_\infty}{2}. \tag{8.1}$$

The correlations presented in this book will typically assume that the properties are computed using the film temperature. An alternative technique that is used in some correlations computes all of the properties at the free stream temperature and then a property ratio is included in the correlation itself to account for property variations. For example, the Nusselt number may be multiplied by the ratio $(\mu/\mu_s)^{0.25}$ or $(Pr/Pr_s)^{0.25}$ where μ and Pr are the viscosity and Prandtl number, respectively, evaluated at T_∞, and μ_s and Pr_s are the same quantities evaluated at T_s.

It is worth noting that the outcome of many convection problems will be a prediction of the surface temperature. Therefore, it is often the case that T_s is not known a priori and an iterative approach will be required. A reasonable value of T_s must be assumed in order to compute the properties and then the calculated value of T_s can be used in a subsequent iteration. The iterative solution capability provided in EES is helpful for these situations.

3. *Calculate the dimensionless quantities required by the correlation*

As discussed in Section 7.4.3, most correlations for single phase, forced convection flow will be provided in the form of Nusselt number as a function of Reynolds number and Prandtl number. Therefore, this step will require that the Reynolds number be computed:

$$Re_{L_{char}} = \frac{\rho\, u_\infty\, L_{char}}{\mu}. \tag{8.2}$$

This seemingly simple step can lead to some relatively common mistakes. The characteristic length that should be used to define the Reynolds number is not always obvious. However, the researchers who formulated the correlation have chosen a characteristic length and you have to use the same one. If you are unsure of what characteristic length to use, you will need to look it up in an appropriate reference.

Also, the properties in Eq. (8.2) are the *fluid* properties determined in Step 2. This is obvious but it is remarkably easy to inadvertently use the density of the object (for example) or some other property value that happens to be relevant to the problem. This is one of those rare mistakes that will not be caught by checking for unit consistency.

4. *Access the correlations in order to compute the Nusselt number*

Once the Reynolds number and Prandtl number are known it should be possible to determine the Nusselt number. This could involve using a chart or an equation. Alternatively, this may involve calling a function or procedure in EES or some other software.

It is important to verify that the correlation being used remains valid for the Reynolds number and Prandtl number of interest. All correlations come with some range in which they are valid. They cannot be used outside of this range with any confidence.

5. *Determine the heat transfer coefficient*

The Nusselt number can be used to compute the heat transfer coefficient. The definition of the average Nusselt number provides:

$$\bar{h} = \overline{Nu}\,\frac{k}{L_{char}}.$$ (8.3)

The same cautions mentioned in Step 3 should be repeated here. The characteristic length used in Eq. (8.3) must be the same one that is used to compute the Reynolds number in Eq. (8.2). The conductivity must be the fluid conductivity, not the conductivity of some other material.

These steps are largely common sense but it helps to have a process that can be followed. The methodology provided here can be used for any convection problem.

8.2 Flow over a Flat Plate

Flow over a flat surface provides the basis for much of the discussion of convection in Chapter 7. This section presents accepted correlations for the friction coefficient and Nusselt number for flow over a flat plate, as shown in Figure 8.1.

8.2.1 The Friction Coefficient

This section provides both local and average friction coefficient correlations for a smooth plate under both laminar and turbulent conditions. The effect of roughness on the local friction coefficient and associated correlations are also discussed.

Local Friction Coefficient for a Smooth Plate

The local shear stress experienced by the plate (τ_s) is expressed in terms of the local friction coefficient:

$$C_f = \frac{2\,\tau_s}{\rho\,u_\infty^2}.$$ (8.4)

The local friction coefficient is correlated using the Reynolds number defined based on the position with respect to the leading edge, x:

$$Re_x = \frac{\rho\,u_\infty\,x}{\mu}.$$ (8.5)

As discussed in Section 7.2.2, the flow will remain laminar when the Reynolds number is less than a critical Reynolds, $Re_x < Re_{crit}$. The critical Reynolds number may vary based on the flow situation, but a typical value is $Re_{crit} = 5 \times 10^5$. If the Reynolds number is less than Re_{crit}, then the self-similar solution presented by Schlichting (2000) provides the local friction coefficient:

$$C_f = \frac{0.664}{\sqrt{Re_x}} \quad \text{for } Re_x < Re_{crit}.$$ (8.6)

free stream
u_∞, T_∞

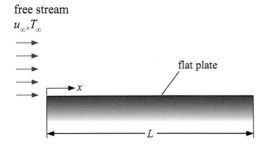

flat plate

Figure 8.1 Flow over a flat plate.

Correlations for turbulent flow are based on experimental data. Several alternative correlations are available in the literature. For a smooth plate, the local friction coefficient for a turbulent flow has been correlated using the simple power-law expression:

$$C_f = \frac{0.027}{Re_x^{1/7}} \quad \text{for} \quad Re_{crit} < Re_x < 1 \times 10^{10}. \tag{8.7}$$

A slightly more accurate expression is given by:

$$C_f = \frac{0.455}{\ln^2(0.06\,Re_x)} \quad \text{for} \quad Re_{crit} < Re_x < 1 \times 10^{10}. \tag{8.8}$$

Both of these expressions are found in White (1991).

Local Friction Coefficient for a Rough Plate

The thickness of the velocity boundary layer for a laminar flow will typically be substantially larger than the scale of the roughness (e) that is encountered on most engineering surfaces. Therefore, laminar flows tend to be insensitive to the roughness of the surface and Eq. (8.6) can be applied regardless of the plate roughness.

Shear stress is governed by the velocity gradient at the wall and in a turbulent flow this gradient occurs largely over the extent of the viscous sublayer. As discussed in Section 7.2, the viscous sublayer is extremely thin and can easily be on the same scale as roughness elements that might exist on an engineering surface. Therefore, the behavior of a turbulent flow can be substantially affected by the surface roughness. The presence of roughness will tend to disrupt the viscous sublayer and therefore increase both the shear and the heat transfer coefficient. The heat transfer surfaces in heat exchangers are sometimes intentionally roughened specifically to improve the heat transfer. This improvement will come at the expense of increased shear and therefore additional pump or fan power.

The local friction coefficient for a rough plate is a function of the **relative roughness** (e/x) and the Reynolds number. An implicit relationship between these three parameters is given by White (1991):

$$Re_x = 1.73\left(1 + 0.3\,e^+\right)\exp\left(Z\right)\left[Z^2 - 4Z + 6 - \frac{0.3\,e^+}{1 + 0.3\,e^+}(Z-1)\right]$$

$$\text{where} \quad Z = \kappa\sqrt{\frac{2}{C_f}} \quad \text{and} \quad e^+ = \frac{Re_x\left(\dfrac{e}{x}\right)}{\sqrt{\dfrac{2}{C_f}}} \quad \text{for} \quad Re_{crit} < Re_x < 1 \times 10^{10} \tag{8.9}$$

where $\kappa = 0.41$ is the Von Kármán constant. Equation (8.9) is not convenient for hand calculations but it can be solved using an equation solver like EES. Figure 8.2 illustrates the friction coefficient predicted by Eq. (8.6) in the laminar regime and Eq. (8.9) in the turbulent regime for various values of the relative roughness. The interaction between turbulent flows and roughness elements is the subject of substantial study and the results shown in Figure 8.2 are for a particular roughness pattern and geometry (specifically, sand grain roughness). Therefore, the results should be used with some caution for other situations; however, Figure 8.2 is useful in that it indicates the approximate magnitude of the effect of roughness on the friction coefficient.

The sharp discontinuity that occurs when the critical Reynolds number is reached obviously corresponds to the large increase in shear stress that is associated with turbulent vs. laminar flow. The effect of roughness is to increase the friction coefficient even further. It is interesting to note that there is a region in Figure 8.2 where the lines corresponding to nonzero values of relative roughness become horizontal, suggesting that the Reynolds number no longer affects the friction coefficient. This behavior corresponds to the **fully rough region** where the friction coefficient only depends on the relative roughness. Physically, the fully rough region occurs when the roughness is much larger than the viscous sublayer. The criterion for this behavior is when the parameter e^+ in Eq. (8.9) is greater than about 60:

$$e^+ = \frac{Re_x\left(\dfrac{e}{x}\right)}{\sqrt{\dfrac{2}{C_f}}} \geq 60. \tag{8.10}$$

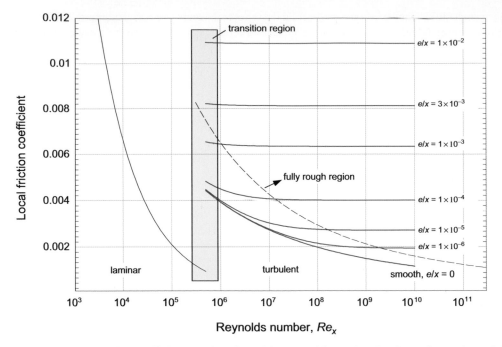

Figure 8.2 Local friction coefficient as a function of the Reynolds number Re_x for various values of the relative roughness, e/x.

Equation (8.10) is shown in Figure 8.2 as the dashed line. The local friction factor in the fully rough regime has been correlated in terms of the relative roughness (White, 1991):

$$C_f = \frac{1}{\left[2.87 + 1.58 \log_{10}\left(\frac{x}{e}\right)\right]^{2.5}}. \tag{8.11}$$

Average Friction Coefficient

The average friction coefficient is discussed in Section 7.1.9 and defined according to:

$$\bar{C}_f = \frac{2\,\bar{\tau}_s}{\rho\,u_\infty^2}, \tag{8.12}$$

where $\bar{\tau}_s$ is the average shear stress:

$$\bar{\tau}_s = \frac{1}{L}\int_0^L \tau_s\, dx. \tag{8.13}$$

Substituting the definition of the local friction coefficient, Eq. (8.4), into Eq. (8.13) leads to:

$$\bar{\tau}_s = \frac{\rho\,u_\infty^2}{2L}\int_0^L C_f\, dx. \tag{8.14}$$

Substituting Eq. (8.14) into Eq. (8.12) leads to:

$$\bar{C}_f = \frac{2}{\rho\,u_\infty^2}\frac{\rho\,u_\infty^2}{2L}\int_0^L C_f\, dx \tag{8.15}$$

or

$$\bar{C}_f = \frac{1}{L}\int_0^L C_f\, dx. \tag{8.16}$$

Therefore, the average friction factor is the average of the local friction factor over the plate surface. The local friction factor, C_f, is correlated in terms of Re_x and therefore it is useful to transform the coordinate of integration in Eq. (8.16) from x to Re_x:

$$x = \frac{\mu \, Re_x}{\rho \, u_\infty} \tag{8.17}$$

so that

$$dx = \frac{\mu}{\rho \, u_\infty} d Re_x. \tag{8.18}$$

Substituting Eqs. (8.17) and (8.18) into Eq. (8.16) leads to:

$$\bar{C}_f = \frac{1}{L} \int_0^{Re_L} C_f \frac{\mu}{\rho \, u_\infty} dRe_x = \frac{\mu}{\rho \, u_\infty \, L} \int_0^{Re_L} C_f dRe_x, \tag{8.19}$$

or

$$\bar{C}_f = \frac{1}{Re_L} \int_0^{Re_L} C_f dRe_x, \tag{8.20}$$

where Re_L is the Reynolds number defined based on the length of the plate. Equation (8.20) is valid regardless of the specific correlation(s) that are used to express the local friction coefficient C_f as a function of Re_x.

In the laminar region, Eq. (8.6) can be substituted into Eq. (8.20) to provide:

$$\bar{C}_f = \frac{1}{Re_L} \int_0^{Re_L} \frac{0.664}{\sqrt{Re_x}} dRe_x = \frac{0.664}{Re_L} \left[2 \, Re_x^{0.5} \right]_0^{Re_L} \tag{8.21}$$

or

$$\bar{C}_f = \frac{1.328}{\sqrt{Re_L}} \text{ for } Re_L < Re_{crit}. \tag{8.22}$$

The average friction coefficient including both the laminar and the turbulent regions is obtained by integrating Eq. (8.20) in two parts: the laminar region for $0 < Re_x < Re_{crit}$ and the turbulent region for $Re_{crit} < Re_x < Re_L$.

$$\bar{C}_f = \frac{1}{Re_L} \left[\int_0^{Re_{crit}} C_{f, lam}(Re_x) dRe_x + \int_{Re_{crit}}^{Re_L} C_{f, turb}(Re_x) dRe_x \right]. \tag{8.23}$$

For a smooth plate, Eq. (8.7) can be used in the turbulent region:

$$\bar{C}_f = \frac{1}{Re_L} \left[\int_0^{Re_{crit}} \frac{0.664}{\sqrt{Re_x}} dRe_x + \int_{Re_{crit}}^{Re_L} \frac{0.027}{Re_x^{1/7}} dRe_x \right] \tag{8.24}$$

or

$$\bar{C}_f = \frac{1}{Re_L} \left[1.328 Re_{crit}^{0.5} + 0.0315 \left(Re_L^{6/7} - Re_{crit}^{6/7} \right) \right] \text{ for } Re_{crit} < Re_L < 1 \times 10^{10}. \tag{8.25}$$

For a rough plate the integration associated with the second part of Eq. (8.23) can be accomplished numerically using the implicit solution to Eq. (8.9). Figure 8.3 illustrates the average friction coefficient as a function of the Reynolds number based on the plate length for various values of the relative roughness, e/L.

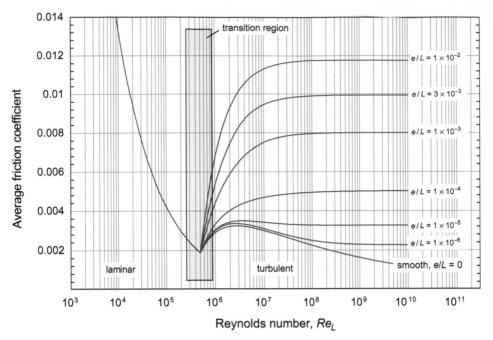

Figure 8.3 Average friction coefficient as a function of the Reynolds number Re_L for various values of the relative roughness, e/L.

Notice that there is no longer a discontinuity at the critical Reynolds number because the local friction factor shown in Figure 8.2 is being integrated with respect to Reynolds number.

Example 8.1

A car has a sign mounted on its roof that is $L = 4$ ft long and $H = 1$ ft high. The sign is thin and mounted so that it is parallel to the direction of the car's motion. Estimate the additional drag force experienced by the car due to the sign when it is travelling at $u_\infty = 55$ mph.

Known Values

Figure 1 illustrates a sketch of the problem with the known values converted to SI units.

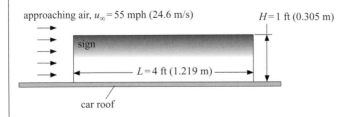

Figure 1 Sketch of the problem with the known values indicated.

Example 8.1 (cont.)

Assumptions

- The correlations for a flat plate can be used to estimate the shear force.
- The sign is sufficiently thin that form drag can be neglected.
- No temperature or pressure values are provided; therefore, reasonable values will be assumed to obtain the properties of air.
- The sign surface is assumed to be smooth.

Analysis and Solution

The steps associated with using a convection coefficient are outlined in Section 8.1. The flow situation is forced convection over a flat plate and the average friction coefficient is required in order to determine the total force. The properties of air are obtained from Appendix C assuming near room temperature and atmospheric pressure conditions: $\rho = 1.177$ kg/m^3 and $\mu = 18.54 \times 10^{-6}$ Pa-s. The dimensionless number required by the correlation is the Reynolds number:

$$Re_L = \frac{\rho \, u_\infty \, L}{\mu} = \frac{1.177 \text{ kg}}{\text{m}^3} \left| \frac{24.6 \text{ m}}{\text{s}} \right| \frac{1.219 \text{ m}}{} \left| \frac{}{18.54 \times 10^{-6} \text{ Pa s}} \right| \left| \frac{\text{Pa m}^2}{\text{N}} \right| \frac{\text{N s}^2}{\text{kg m}} = 1.90 \times 10^6. \tag{1}$$

The Reynolds number is above the critical Reynolds number and therefore Eq. (8.25) must be used to determine the average friction coefficient.

$$\begin{aligned}
\bar{C}_f &= \frac{1}{Re_L} \left[1.328 \, Re_{crit}^{0.5} + 0.0315 \left(Re_L^{6/7} - Re_{crit}^{6/7} \right) \right] \\
&= \frac{1}{1.90 \times 10^6} \left[1.328 \left(5 \times 10^5 \right)^{0.5} + 0.0315 \left(\left(1.90 \times 10^6 \right)^{6/7} - \left(5 \times 10^5 \right)^{6/7} \right) \right] = 0.00322.
\end{aligned} \tag{2}$$

Alternatively, this value could be obtained less precisely using Figure 8.3 at $Re_L = 1.90 \times 10^6$ and $e/L = 0$. The average shear stress experienced by the sign is obtained using the definition of the average friction coefficient:

$$\bar{\tau}_s = \bar{C}_f \frac{\rho \, u_\infty^2}{2} = \frac{0.00322}{2} \left| \frac{1.177 \text{ kg}}{\text{m}^3} \right| \frac{(24.6)^2 \text{ m}^2}{\text{s}^2} \left| \frac{\text{N s}^2}{\text{kg m}} \right| \frac{\text{Pa m}^2}{\text{N}} = 1.14 \text{ Pa}. \tag{3}$$

The total drag force added to the car is therefore:

$$F = 2 \, \bar{\tau}_s \, H \, L = \frac{2}{} \left| \frac{1.14 \text{ Pa}}{} \right| \frac{0.305 \text{ m}}{} \left| \frac{1.219 \text{ m}}{} \right| \frac{\text{N}}{\text{Pa m}^2} = \boxed{0.850 \text{ N} \,(0.19 \text{ lb}_f)}. \tag{4}$$

Discussion

The drag added by the sign is clearly insignificant relative to the force required to move the car.

8.2.2 The Nusselt Number

This section provides both local and average Nusselt number for flow over a smooth plate subjected to three different boundary conditions: at constant temperature, with an unheated starting length, and with a constant heat flux.

Constant Temperature

When the Reynolds number is less than Re_{crit}, then the self-similar solution presented in Schlichting (2000) can be used to determine the local Nusselt number for laminar flow over a constant temperature plate.

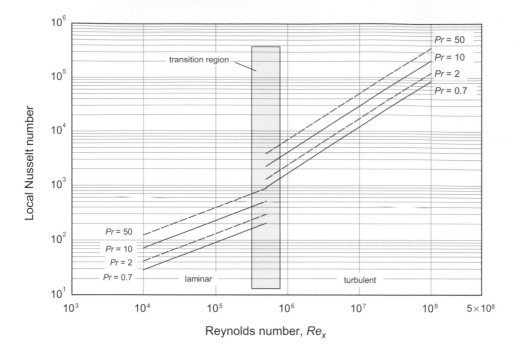

Figure 8.4 Local Nusselt number as a function of Reynolds number for various values of the Prandtl number.

The self-similar solution is not explicit, but the local Nusselt number provided by the solution has been curve fit by Churchill and Ozoe (1973) according to:

$$Nu_x = \frac{0.3387\, Re_x^{1/2}\, Pr^{1/3}}{\left[1 + \left(\dfrac{0.0468}{Pr}\right)^{2/3}\right]^{1/4}} \text{ for } Re_x < Re_{crit}. \tag{8.26}$$

For turbulent flow, correlations are based on experimental data and several alternative forms are available in the literature. For a smooth plate, the local Nusselt number can be computed using the modified Reynolds analogy (i.e., the Colburn analogy), discussed in Sections 7.1.8, together with Eq. (8.7). The Colburn analogy produces reliable results for Reynolds numbers up to approximately 1×10^8 and $0.5 < Pr < 60$:

$$Nu_x = 0.0135\, Re_x^{6/7}\, Pr^{1/3} \text{ for } Re_{crit} < Re_x < 1 \times 10^8 \text{ and } 0.5 < Pr < 60. \tag{8.27}$$

Figure 8.4 illustrates the local Nusselt number in the laminar and turbulent regions for various values of the Prandtl number and shows the large increase in the local Nusselt number that occurs at the transition from laminar to turbulent flow.

Figure 8.4 shows that the Nusselt number increases with Prandtl number in both the laminar and turbulent regions. The conceptual model of the turbulent boundary layer presented in Section 7.2.3 suggested that the dominant resistance to heat transfer in a turbulent boundary layer is associated with conduction through the viscous sublayer. The thermal resistance of the viscous sublayer is substantially affected by the molecular Prandtl number.

Figure 8.4 can be somewhat misleading at first glance. The local Nusselt number increases with Reynolds number (which is proportional to the position along the plate) even though the local heat transfer coefficient itself tends to decrease with x due to the thickening of the boundary layer (in the laminar region) and of the viscous sublayer (in the turbulent region). This apparent discrepancy is related to the fact that the characteristic length used to define the Nusselt number for a flat plate is position x:

$$Nu_x = \frac{h\,x}{k}. \tag{8.28}$$

Solving Eq. (8.28) for h provides:

$$h = \frac{k}{x} Nu_x. \tag{8.29}$$

Recognizing the Reynolds number Re_x is proportional to x and substituting either Eq. (8.26) or Eq. (8.27) into Eq. (8.29) leads to:

$$h \propto x^{-1/2} u_\infty^{1/2} \text{ in the laminar region, } Re < Re_{crit} \tag{8.30}$$

$$h \propto x^{-1/7} u_\infty^{6/7} \text{ in the turbulent region, } Re > Re_{crit}. \tag{8.31}$$

The average Nusselt number $(\overline{Nu})$ was discussed in Section 7.1.9 and defined according to:

$$\overline{Nu_L} = \frac{\bar{h} L}{k}, \tag{8.32}$$

where $(\bar{h})$ is the average heat transfer coefficient, defined according to:

$$\bar{h} = \frac{1}{L} \int_0^L h \, dx. \tag{8.33}$$

Substituting Eqs. (8.33) and (8.29) into Eq. (8.32) leads to:

$$\overline{Nu_L} = \frac{L}{k} \frac{1}{L} \int_0^L \frac{Nu_x \, k}{x} \, dx \tag{8.34}$$

or

$$\overline{Nu_L} = \int_0^L \frac{Nu_x}{x} \, dx. \tag{8.35}$$

Because Nusselt number correlations are expressed in terms of the Reynolds number, it is convenient to change the coordinate of integration in Eq. (8.35) from x to Re_x by substituting Eqs. (8.17) and (8.18) into Eq. (8.35):

$$\overline{Nu_L} = \int_0^{Re_L} \frac{Nu_x}{Re_x} \, dRe_x. \tag{8.36}$$

Equation (8.36) can be integrated using any correlation or set of correlations for the local Nusselt number. In the laminar region, Eq. (8.26) is substituted into Eq. (8.36) in order to obtain:

$$\overline{Nu_L} = \frac{0.3387 \, Pr^{1/3}}{\left[1 + \left(\frac{0.0486}{Pr}\right)^{2/3}\right]^{1/4}} \int_0^{Re_L} Re_x^{-1/2} \, dRe_x = \frac{0.3387 \, Pr^{1/3}}{\left[1 + \left(\frac{0.0468}{Pr}\right)^{2/3}\right]^{1/4}} \left[2 \, Re_x^{1/2}\right]_0^{Re_L} \tag{8.37}$$

or

$$\overline{Nu_L} = \frac{0.6774 \, Pr^{1/3} \, Re_L^{1/2}}{\left[1 + \left(\frac{0.0468}{Pr}\right)^{2/3}\right]^{1/4}} \text{ for } Re_L < Re_{crit}. \tag{8.38}$$

The average Nusselt number for the combined laminar and turbulent regions is obtained by integrating Eq. (8.36) in two parts; Eq. (8.26) is used in the laminar region (from $Re_x < Re_{crit}$) and Eq. (8.27) is used in the turbulent region (from $Re_{crit} < Re_x < Re_L$):

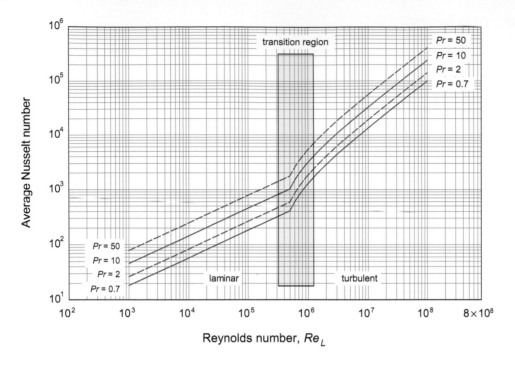

Figure 8.5 Average Nusselt number as a function of Reynolds number for various values of the Prandtl number.

$$\overline{Nu}_L = \frac{0.3387 \, Pr^{1/3}}{\left[1 + \left(\dfrac{0.0468}{Pr}\right)^{2/3}\right]^{1/4}} \int\limits_{0}^{Re_{crit}} Re_x^{-1/2} \, dRe_x + 0.0135 \, Pr^{1/3} \int\limits_{Re_{crit}}^{Re_L} Re_x^{-1/7} \, dRe_x \qquad (8.39)$$

or

$$\overline{Nu}_L = \frac{0.6774 \, Pr^{1/3} \, Re_{crit}^{1/2}}{\left[1 + \left(\dfrac{0.0468}{Pr}\right)^{2/3}\right]^{1/4}} + 0.0158 \, Pr^{1/3} \left(Re_L^{6/7} - Re_{crit}^{6/7}\right)$$

for $Re_{crit} < Re_L < 1 \times 10^8$ and $0.5 < Pr < 60$. $\qquad (8.40)$

The average Nusselt number is shown in Figure 8.5 as a function of the Reynolds number for various values of the Prandtl number.

The local and average correlations for a smooth flat plate at a constant temperature are summarized in Table 8.1. The average correlations for a smooth flat plate at constant temperature that are presented in this section are available as a built-in procedure in EES. To access this procedure, select Function Info from the Options menu and then select Convection from the pull down menu at the lower right corner of the upper window. Select External Flow – Non-dimensional and scroll to the flat plate, as shown in Figure 8.6.

Most convection correlations are provided as procedures rather than as functions. *Procedures* may return multiple outputs as opposed to *functions*, which can return only one. The arguments to the left of the colon in the argument list of a procedure are inputs; these include the Reynolds number and the Prandtl number for the procedure External_Flow_Plate_ND shown in Figure 8.6. The arguments to the right of the colon are outputs; these include the average Nusselt number and average friction coefficient.

The correlations provided in this section are valid for smooth plates. There is no generally acceptable correlation for the Nusselt number associated with flow over a rough plate, but certainly the effect of surface

Table 8.1 Summary of correlations for a smooth, isothermal flat plate.

Flow condition	Parameter	Local value	Average value
laminar, $Re_x < Re_{crit}$ $Re_L < Re_{crit}$	friction coefficient	$C_f = \dfrac{0.664}{\sqrt{Re_x}}$	$\bar{C}_f = \dfrac{1.328}{\sqrt{Re_L}}$
	Nusselt number	$Nu_x = \dfrac{0.3387\, Re_x^{1/2}\, Pr^{1/3}}{\left[1 + \left(\frac{0.0468}{Pr}\right)^{2/3}\right]^{1/4}}$	$\overline{Nu}_L = \dfrac{0.6774\, Pr^{1/3}\, Re_L^{1/2}}{\left[1 + \left(\frac{0.0468}{Pr}\right)^{2/3}\right]^{1/4}}$
turbulent[*], $Re_x > Re_{crit}$ $Re_L > Re_{crit}$	friction coefficient	$C_f = \dfrac{0.027}{Re_x^{1/7}}$	$\bar{C}_f = \dfrac{1}{Re_L}\left[1.328\, Re_{crit}^{0.5} + 0.0315\left(Re_L^{6/7} - Re_{crit}^{6/7}\right)\right]$
	Nusselt number	$Nu_x = 0.0135\, Re_x^{6/7}\, Pr^{1/3}$	$\overline{Nu}_L = \dfrac{0.6774\, Pr^{1/3}\, Re_{crit}^{1/2}}{\left[1 + \left(\frac{0.0468}{Pr}\right)^{2/3}\right]^{1/4}} + 0.0158\, Pr^{1/3}\left(Re_L^{6/7} - Re_{crit}^{6/7}\right)$

[*] Note that the average correlations in the turbulent region include integration of the local correlations for laminar flow through the laminar region.

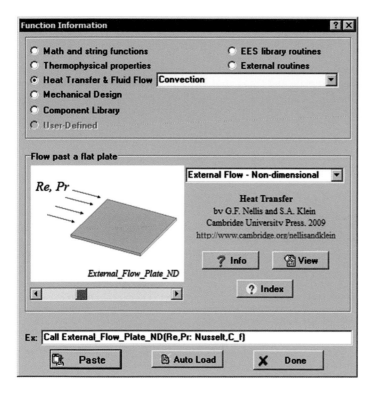

Figure 8.6 Dialog for the EES function that implements the correlation for flow over a flat plate.

roughness can be important in a turbulent flow. The results for the friction coefficient, Figure 8.2, can be used to understand under what conditions the roughness might become important, but not the absolute effect. The Colburn analogy does not hold for a rough plate and it has been shown that generally, the presence of roughness does not improve heat transfer by as much as it increases wall shear stress (Jiji, 2010).

Example 8.2

A flat plate is maintained at a temperature of $T_s = 380$ K while being subjected on one side to a flow of water at $T_\infty = 300$ K and $u_\infty = 3$ m/s. The dimension of the plate in the direction that the water flows is $L = 1$ m and the dimension perpendicular to the water flow is $W = 0.5$ m. Estimate the rate of heat transfer from the plate.

Known Values

Figure 1 illustrates a sketch of the problem with the known values shown.

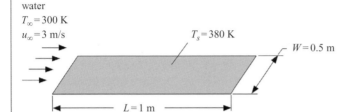

water
$T_\infty = 300$ K
$u_\infty = 3$ m/s

$T_s = 380$ K

$W = 0.5$ m

$L = 1$ m

Figure 1 Sketch of the problem with the known values indicated.

Assumptions

- The correlations for a flat plate can be used to estimate the heat transfer coefficient.
- The plate is smooth.

Analysis and Solution

The flow situation is forced convection over a flat plate and the average Nusselt number is required in order to determine the total heat transfer rate. The film temperature:

$$T_f = \frac{T_\infty + T_s}{2} = \frac{(300 + 380)\ \text{K}}{2} = 340\ \text{K} \tag{1}$$

is used to obtain the properties of water using Appendix D; $\rho = 979.5$ kg/m^3, $\mu = 0.422 \times 10^{-3}$ Pa-s, $k = 0.6608$ W/m-K, and $Pr = 2.675$. The dimensionless numbers required by the correlation are the Prandtl number and the Reynolds number:

$$Re_L = \frac{\rho\, u_\infty L}{\mu} = \frac{979.5\ \text{kg}}{\text{m}^3} \left| \frac{3\ \text{m}}{\text{s}} \right| \frac{1\ \text{m}}{\left| 0.422 \times 10^{-3}\ \text{Pa s} \right|} \left\| \frac{\text{Pa m}^2}{\text{N}} \right| \frac{\text{N s}^2}{\text{kg m}} = 6.96 \times 10^6. \tag{2}$$

The Reynolds number is above the critical Reynolds number and therefore Eq. (8.40) must be used to determine the average Nusselt number:

$$\overline{Nu_L} = \frac{0.6774\, Pr^{1/3}\, Re_{crit}^{1/2}}{\left[1 + \left(\frac{0.0468}{Pr} \right)^{2/3} \right]^{1/4}} + 0.0158\, Pr^{1/3} \left(Re_L^{6/7} - Re_{crit}^{6/7} \right)$$

$$= \frac{0.6774\, (2.675)^{1/3}\, \left(5 \times 10^5 \right)^{1/2}}{\left[1 + \left(\frac{0.0468}{2.675} \right)^{2/3} \right]^{1/4}} + 0.0158(2.675)^{1/3} \left(\left(6.96 \times 10^6 \right)^{6/7} - \left(5 \times 10^5 \right)^{6/7} \right) = 15{,}055. \tag{3}$$

Alternatively, this value could be obtained less precisely using Figure 8.5 at $Re_L = 6.96 \times 10^6$ and $Pr = 2.675$. The average heat transfer coefficient experienced by the plate is obtained using the definition of the average Nusselt number:

Example 8.2 (cont.)

$$\bar{h} = \overline{Nu}\frac{k}{L} = \frac{15,055\left|0.6608\text{ W}\right|}{\left|\text{m K}\right|}\frac{}{1\text{ m}} = 9948\frac{\text{W}}{\text{m}^2\text{ K}}. \tag{4}$$

The total rate of heat transfer from the plate is therefore:

$$\dot{q} = \bar{h}\,W\,L(T_s - T_\infty) = \frac{9948\text{ W}\left|1\text{ m}\right|0.5\text{ m}\left|(380-300)\text{ K}\right|}{\text{m}^2\text{ K}\left|\right|\left|\right|} = \boxed{398,000\text{ W/m}^2\ (398\text{ kW/m}^2)}. \tag{5}$$

Discussion

The Convection Heat Transfer Library in EES could be used to obtain this solution. The required EES program is shown below.

```
$UnitSystem SI Mass J K Pa
T_s=380 [K]                                         "plate surface temperature"
T_infinity = 300 [K]                                "free stream temperature"
L=1 [m]                                             "plate length in the flow direction"
W=0.5 [m]                                           "plate width perpendicular to flow"
u_infinity=3 [m/s]                                  "free stream velocity"
F$='Steam_IAPWS'                                    "fluid"

T_f=(T_s+T_infinity)/2                              "film temperature"
rho=Density(F$,T=T_f,P=Po#)                         "density"
mu=Viscosity(F$,T=T_f,P=Po#)                        "viscosity"
k=Conductivity(F$,T=T_f,P=Po#)                      "conductivity"
Pr=Prandtl(F$,T=T_f,P=Po#)                          "Prandtl number"

Re_L=rho*u_infinity*L/mu                            "Reynolds number"
CALL External_Flow_Plate_ND(Re_L,Pr: Nusselt_bar,C_f_bar)  "access correlations"
h_bar=Nusselt_bar*k/L                               "average heat transfer coefficient"
q_dot=h_bar*W*L*(T_s-T_infinity)                    "heat transfer rate"
```

Solving provides $\dot{q} = 396.5\text{ kW/m}^2$. The small difference in the result compared to the hand calculation arises from evaluating the properties of water at 1 atm rather than using the saturated liquid properties provided in Appendix D.

Example 8.3

A plate has a heater mounted on its underside that provides $\dot{q} = 25$ W. The plate is cooled by a flow of air at $u_\infty = 2.8$ m/s and $T_\infty = 20°$C. The plate is $L = 10$ cm long in the direction that the air is flowing and $W = 10$ cm wide perpendicular to the flow. Determine the surface temperature of the plate.

Known Values

Figure 1 illustrates a sketch of the problem with the known values shown.

Continued

Example 8.3 (cont.)

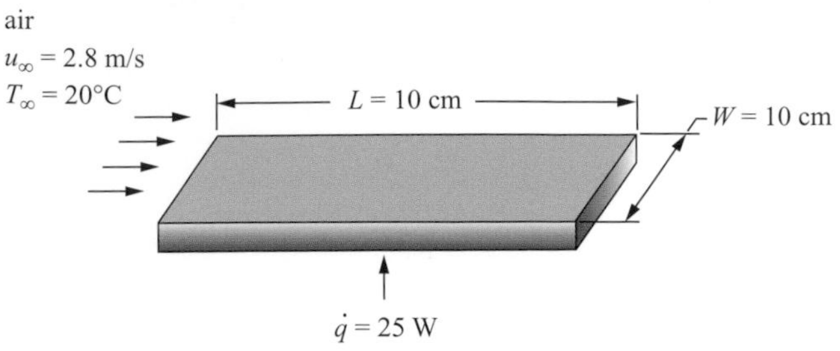

air
$u_\infty = 2.8$ m/s
$T_\infty = 20°C$

Figure 1 Sketch of the problem with the known values indicated.

$L = 10$ cm
$W = 10$ cm
$\dot{q} = 25$ W

Assumptions

- The correlations for a flat plate can be used to estimate the heat transfer coefficient.
- The plate is smooth.
- The plate is sufficiently thick and conductive that it comes to a uniform temperature.
- The pressure of the air is atmospheric.

Analysis

The flow situation is forced convection over a flat plate and the average Nusselt number is required in order to determine the total heat transfer rate. The film temperature is defined as:

$$T_f = \frac{T_\infty + T_s}{2}. \tag{1}$$

The film temperature is used to obtain the properties of air:

$$\rho = \rho(T_f, P_o) \tag{2}$$

$$\mu = \mu(T_f) \tag{3}$$

$$k = k(T_f) \tag{4}$$

$$Pr = Pr(T_f), \tag{5}$$

where P_o is the assumed atmospheric pressure. The dimensionless numbers required by the correlation are the Prandtl number and the Reynolds number:

$$Re_L = \frac{\rho\, u_\infty\, L}{\mu}. \tag{6}$$

The Reynolds number and Prandtl number together are used in the proper correlation in order to determine the average Nusselt number:

$$\overline{Nu}_L = \overline{Nu}_L(Re_L, Pr). \tag{7}$$

The average heat transfer coefficient experienced by the plate is obtained using the definition of the average Nusselt number:

$$\bar{h} = \overline{Nu}\frac{k}{L}. \tag{8}$$

The temperature of the plate is obtained from Newton's Law of Cooling:

$$\dot{q} = \bar{h}\, W\, L(T_s - T_\infty). \tag{9}$$

Equations (1) through (9) are nine equations in the nine unknowns: T_f, T_s, ρ, μ, k, Pr, Re_L, $\overline{Nu}_L$, and $\bar{h}$.

Example 8.3 (cont.)

Solution

Equations (1) through (9) are sufficient to solve the problem but they are not explicit. Rather, the surface temperature appears initially in Eq. (1) and again in Eq. (9), but it is not known; the equations are implicit and must be solved iteratively, ideally using a computer. Here we will program the equations in EES, but we will do so in an *intelligent* way that allows us to control where the iterative process used by EES starts.

The inputs are entered in EES and converted to base SI units. Also, the unit system is specified using the $UnitSystem directive so that the EES property database can be used.

```
$UnitSystem SI Mass J K Pa
T_infinity = ConvertTemp(C,K,20 [C])     "free stream temperature"
L=10 [cm]*Convert(cm,m)                  "plate length in the streamwise direction"
W=10 [cm]*Convert(cm,m)                  "plate width perpendicular to flow"
u_infinity=2.8 [m/s]                     "free stream velocity"
q_dot=25 [W]                             "heater power"
```

In order to proceed with the solution in an explicit manner, a value for the surface temperature is "guessed". Notice that this value need only be *reasonable* as it provides a starting point for the iterative process.

```
T_s=400 [K]                              "guess for T_s"
```

The film temperature is calculated using Eq. (1) and the properties are obtained using the Density, Viscosity, Conductivity, and Prandtl functions in EES.

```
T_f=(T_s+T_infinity)/2                   "film temperature"
rho=Density(Air,T=T_f,P=Po#)             "density"
mu=Viscosity(Air,T=T_f)                  "viscosity"
k=Conductivity(Air,T=T_f)                "conductivity"
Pr=Prandtl(Air,T=T_f)                    "Prandtl number"
```

The Reynolds number is computed using Eq. (6). The procedure External_Flow_Plate_ND is used to access the correlations and determine the average Nusselt number. The heat transfer coefficient is computed using Eq. (8).

```
Re_L=rho*u_infinity*L/mu                                      "Reynolds number"
CALL External_Flow_Plate_ND(Re_L,Pr: Nusselt_bar,C_f_bar)    "access correlations"
h_bar=Nusselt_bar*k/L                                        "average heat transfer coefficient"
```

Implementation of Eq. (9) to compute the surface temperature will over-specify the EES code. At this point, it is best to update the guess values used by EES to converge to a solution; this is done by selecting Update Guesses from the Calculate menu. Then, comment out the guessed value for the variable T_s and implement Eq. (9).

```
{T_s=400 [K]}                            "guess for T_s"
```

```
q_dot=h_bar*W*L*(T_s-T_infinity)         "heat transfer rate"
```

Solving should provide $T_s = 418.7$ K (145.6°C). At this point, the EES code consists of nine coupled, implicit equations in nine unknowns. The advantages of taking this approach are twofold. First, by guessing a value for T_s it became possible to enter equations one at a time and check each one as it was entered; this makes debugging much easier than if many equations had to be entered before a solution could be obtained.

Continued

Example 8.3 (cont.)

Second, at the point where the problem becomes implicit and therefore requires iteration, EES is provided with a good starting point for the iteration.

Discussion

The final version of the EES code is shown below.

```
$UnitSystem SI Mass J K Pa
T_infinity = ConvertTemp(C,K,20 [C])        "free stream temperature"
L=10 [cm]*Convert(cm,m)                      "plate length in the streamwise direction"
W=10 [cm]*Convert(cm,m)                      "plate width perpendicular to flow"
u_infinity=2.8 [m/s]                         "free stream velocity"
q_dot=25 [W]                                 "heater power"

{T_s=400 [K]}                                "guess for T_s"

T_f=(T_s+T_infinity)/2                        "film temperature"
rho=Density(Air,T=T_f,P=Po#)                 "density"
mu=Viscosity(Air,T=T_f)                      "viscosity"
k=Conductivity(Air,T=T_f)                    "conductivity"
Pr=Prandtl(Air,T=T_f)                        "Prandtl number"

Re_L=rho*u_infinity*L/mu                      "Reynolds number"
CALL External_Flow_Plate_ND(Re_L,Pr: Nusselt_bar,C_f_bar)
                                             "access correlations"
h_bar=Nusselt_bar*k/L                        "average heat transfer coefficient"

q_dot=h_bar*W*L*(T_s-T_infinity)             "heat transfer rate"
```

An advantage of using the computer is that it becomes easy to run parametric studies. For example, a Parametric Table can be generated in which the air velocity is varied and the surface temperature is computed and recorded. The resulting plot of surface temperature as a function of air velocity is shown in Figure 2.

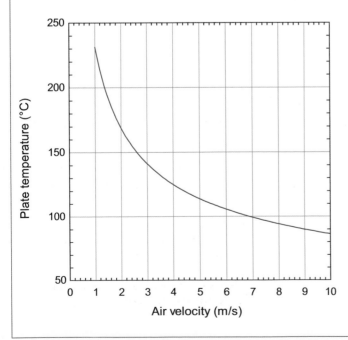

Figure 2 Plate temperature as a function of air velocity.

free stream, u_∞, T_∞

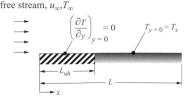

Figure 8.7 Plate with an unheated starting length.

Unheated Starting Length

Consider a flat plate that has an adiabatic section at its leading edge where the flow develops hydrodynamically before it begins to develop thermally, as shown in Figure 8.7. The solution to the local heat transfer problem under laminar conditions can be obtained using integral techniques, as discussed in Nellis and Klein (2009):

$$Nu_x = \frac{Nu_{x, L_{uh}=0}}{\left[1 - \left(\frac{L_{uh}}{x}\right)^{0.75}\right]^{1/3}} \text{ for } Re_x < Re_{crit} \tag{8.41}$$

where $Nu_{x, L_{uh}=0}$ is the local Nusselt number that would be obtained at the same position, x, with no unheated starting length (i.e., with $L_{uh} = 0$); under laminar conditions, $Nu_{x, L_{uh}=0}$ is the result obtained using Eq. (8.26).
  Under turbulent conditions, the solution is

$$Nu_x = \frac{Nu_{x, L_{uh}=0}}{\left[1 - \left(\frac{L_{uh}}{x}\right)^{0.90}\right]^{1/9}} \text{ for } Re_x > Re_{crit}, \tag{8.42}$$

where $Nu_{x, L_{uh}=0}$ is the local Nusselt number obtained using Eq. (8.27). The average Nusselt number coefficient $\left(\overline{Nu_L}\right)$ (averaged over the constant surface temperature portion of the plate) is given by Ameel (1997):

$$\overline{Nu_L} = \overline{Nu_{L, L_{uh}=0}} \frac{L}{(L - L_{uh})} \left[1 - \left(\frac{L_{uh}}{L}\right)^{\frac{(p+1)}{(p+2)}}\right]^{\frac{p}{(p+1)}}, \tag{8.43}$$

where $\overline{Nu_{L, L_{uh}=0}}$ is the average Nusselt number obtained if $L_{uh} = 0$; i.e., the result obtained using Eq. (8.38) or Eq. (8.40) for laminar and turbulent flow, respectively. The parameter p in Eq. (8.43) is 2 for laminar flow and 8 for turbulent flow. Equation (8.43) can be expressed as the ratio of $\overline{Nu_L}$ to $\overline{Nu_{L, L_{uh}=0}}$ as a function of L_{uh}/L:

$$\frac{\overline{Nu_L}}{\overline{Nu_{L, L_{uh}=0}}} = \left(1 - \frac{L_{uh}}{L}\right)^{-1} \left[1 - \left(\frac{L_{uh}}{L}\right)^{\frac{(p+1)}{(p+2)}}\right]^{\frac{p}{(p+1)}}. \tag{8.44}$$

The ratio of Nusselt numbers is shown in Figure 8.8 for laminar and turbulent flow. Notice that the average Nusselt number $\left(\overline{Nu_L}\right)$ with an adiabatic section tends to be larger than the value that would be obtained if there were no unheated starting length $\left(\overline{Nu_{L, L_{uh}=0}}\right)$ because the thermal boundary layer is thinner than it would have otherwise been.

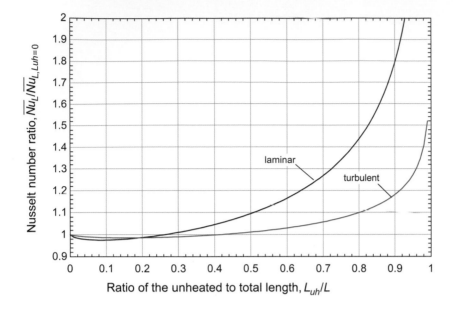

Figure 8.8 The Nusselt number ratio $\left(\overline{Nu}_L/\overline{Nu}_{L,\,L_{uh}=0}\right)$ as a function of the ratio of the unheated length to the plate length (L_{uh}/L).

Example 8.4

The heat transfer fluid Therminol 66 flows over a flat plate with $u_\infty = 3.2$ m/s and $T_\infty = 60°C$. The entire length of the plate is $L = 20$ cm. In the flow direction, the initial $L_{uh} = 10$ cm of the plate is insulated and the remainder of the plate is maintained at $T_s = 15°C$. The width of the plate (perpendicular to the flow direction) is $W = 10$ cm. Determine the rate of heat transfer from the fluid to the plate.

Known Values

Figure 1 illustrates a sketch of the problem with the known values shown.

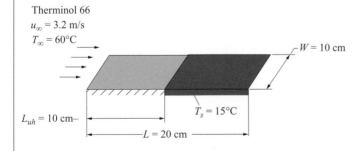

Figure 1 Sketch of the problem with the known values indicated.

Assumptions

- The plate is smooth.
- The properties of Therminol 66 at 20°C from Appendix B can be used at the temperatures in the problem.

Example 8.4 (cont.)

Analysis and Solution

The properties of Therminol 66 are obtained from Appendix B: ρ = 1004 kg/m³, μ = 92.35 × 10⁻³ Pa-s, k = 0.1174 W/m-K, and Pr = 1239. The flow situation is forced convection over a flat plate with an unheated starting length and the average Nusselt number is required in order to determine the total heat transfer rate. The dimensionless numbers required by the correlation are the Prandtl number and the Reynolds number:

$$Re_L = \frac{\rho\,u_\infty\,L}{\mu} = \frac{1004\,\text{kg}}{\text{m}^3}\left|\frac{3.2\,\text{m}}{\text{s}}\right|0.2\,\text{m}\left|\frac{}{92.35\times10^{-3}\,\text{Pa s}}\right|\left\|\frac{\text{Pa m}^2}{\text{N}}\right|\frac{\text{N s}^2}{\text{kg m}} = 6960. \tag{1}$$

The Reynolds number is lower than the critical Reynolds number and therefore the flow is laminar. Equation (8.38) can be used to estimate the average Nusselt number in the absence of the unheated length:

$$\overline{Nu}_{L,\,L_{uh}=0} = \frac{0.6774\,Pr^{1/3}\,Re_L^{1/2}}{\left[1+\left(\dfrac{0.0468}{Pr}\right)^{2/3}\right]^{1/4}} = \frac{0.6774(1239)^{1/3}(6960)^{1/2}}{\left[1+\left(\dfrac{0.0468}{1239}\right)^{2/3}\right]^{1/4}} = 606.7. \tag{2}$$

The ratio of the actual average Nusselt number (i.e., with the unheated length) to the average Nusselt number calculated using Eq. (2) can be obtained using Eq. (8.44) with p = 2 (for laminar flow):

$$\frac{\overline{Nu}_L}{\overline{Nu}_{L,\,L_{uh}=0}} = \left(1-\frac{L_{uh}}{L}\right)^{-1}\left[1-\left(\frac{L_{uh}}{L}\right)^{\frac{(p+1)}{(p+2)}}\right]^{\frac{p}{(p+1)}} = (1-0.5)^{-1}\left[1-(0.5)^{3/4}\right]^{2/3} = 1.096. \tag{3}$$

This value can also be obtained from Figure 8.8. The actual average Nusselt number is therefore:

$$\overline{Nu}_L = \frac{\overline{Nu}_L}{\overline{Nu}_{L,\,L_{uh}=0}}\,\overline{Nu}_{L,\,L_{uh}=0} = 1.096(606.7) = 664.7. \tag{4}$$

The average heat transfer coefficient experienced by the plate is obtained using the definition of the average Nusselt number:

$$\bar{h} = \overline{Nu}\frac{k}{L} = \frac{664.7}{}\left|\frac{0.1174\,\text{W}}{\text{m K}}\right|\frac{}{0.2\,\text{m}} = 390\frac{\text{W}}{\text{m}^2\text{K}}. \tag{5}$$

The total rate of heat transfer to the cooled portion of the plate is therefore:

$$\dot{q} = \bar{h}\,W(L-L_{uh})\,(T_\infty - T_s) = \frac{390\,\text{W}}{\text{m}^2\,\text{K}}\left|\frac{0.1\,\text{m}}{}\right|0.1\,\text{m}\left|\frac{(60-15)\,\text{K}}{}\right| = \boxed{176\,\text{W}}. \tag{6}$$

Constant Heat Flux

The local Nusselt number for a flat plate with a constant heat flux, $\dot{q}''_s$, under laminar conditions is given by Kays *et al.* (2005):

$$Nu_x = 0.453\,Re_x^{1/2}\,Pr^{1/3} \quad \text{for}\ \ Pr > 0.6\ \ \text{and}\ \ Re_x < Re_{crit}. \tag{8.45}$$

The local Nusselt number for turbulent flow over a flat plate with constant heat flux is:

$$Nu_x = 0.0308\,Re_x^{0.8}\,Pr^{1/3} \quad \text{for}\ \ 0.6 < Pr < 60\ \ \text{and}\ \ Re_x < Re_{crit}. \tag{8.46}$$

In the case of a uniform heat flux, the total rate of heat transfer from the surface is known directly from the product of the area and the heat flux. Therefore, the average surface temperature $(\bar{T}_s)$ is used to define an average heat transfer coefficient

$$\bar{h} = \frac{\dot{q}''_s}{(\bar{T}_s - T_\infty)} \tag{8.47}$$

and an average Nusselt number

$$\overline{Nu_L} = \frac{\bar{h}\,L}{k}. \tag{8.48}$$

The average Nusselt number associated with the laminar flow solution, Eq. (8.45), is

$$\overline{Nu_L} = 0.680\,Re_x^{1/2}\,Pr^{1/3}. \tag{8.49}$$

Equation (8.49) is very close to the result associated with a constant surface temperature, Eq. (8.38). The average Nusselt number for a turbulent flow agrees even more closely with the constant temperature result, Eq. (8.40).

Example 8.5

A flat plate experiences a uniform heat flux on its surface of $\dot{q}''_s = 5400\ \mathrm{W/m^2}$. The plate is $L = 1.5$ m long in the flow direction. The free stream temperature is $T_\infty = 20°\mathrm{C}$ and the free stream velocity is $u_\infty = 5$ m/s. The properties of the fluid are $\rho = 850\ \mathrm{kg/m^3}$, $\mu = 0.005$ Pa-s, $k = 0.32$ W/m-K, and $Pr = 5.4$. Determine the average plate temperature as well as the hottest local temperature on the plate.

Known Values

Figure 1 illustrates a sketch of the problem with the known values shown.

free stream
$u_\infty = 5$ m/s, $T_\infty = 20°\mathrm{C}$
$\rho = 850\ \mathrm{kg/m^3}$, $\mu = 0.005$ Pa-s, $k = 0.32$ W/m-K, $Pr = 5.4$

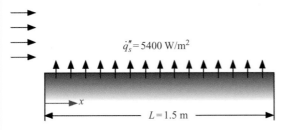

$\dot{q}''_s = 5400\ \mathrm{W/m^2}$

x

$L = 1.5$ m

Figure 1 Sketch of the problem with the known values indicated.

Assumption

• The plate is smooth.

Analysis and Solution

The flow situation is forced convection over a flat plate with a constant heat flux. The average Nusselt number is required in order to determine the average temperature and the local Nusselt number will be required to determine the local temperature.

Example 8.5 (cont.)

The dimensionless numbers required by the average Nusselt number correlation are the Prandtl number and the Reynolds number:

$$Re_L = \frac{\rho \, u_\infty L}{\mu} = \frac{850 \text{ kg}}{\text{m}^3} \left| \frac{5 \text{ m}}{\text{s}} \right| 1.5 \text{ m} \left| \frac{}{0.005 \text{ Pa s}} \right| \left| \frac{\text{Pa m}^2}{\text{N}} \right| \frac{\text{N s}^2}{\text{kg m}} = 1.28 \times 10^6. \tag{1}$$

The Reynolds number is greater than the critical Reynolds number ($Re_{crit} = 5 \times 10^5$) and therefore the flow will transition to turbulence. Equation (8.40) can be used to estimate the average Nusselt number with little error for a constant heat flux case:

$$\overline{Nu}_L = \frac{0.6774 \, Pr^{1/3} \, Re_{crit}^{1/2}}{\left[1 + \left(\frac{0.0468}{Pr} \right)^{2/3} \right]^{1/4}} + 0.0158 \, Pr^{1/3} \left(Re_L^{6/7} - Re_{crit}^{6/7} \right)$$

$$= \frac{0.6774(5.4)^{1/3} \left(5 \times 10^5 \right)^{1/2}}{\left[1 + \left(\frac{0.0468}{5.4} \right)^{2/3} \right]^{1/4}} + 0.0158(5.4)^{1/3} \left[\left(1.28 \times 10^6 \right)^{6/7} - \left(5 \times 10^5 \right)^{6/7} \right] = 3449. \tag{2}$$

The average heat transfer coefficient experienced by the plate is obtained using the definition of the average Nusselt number:

$$\bar{h} = \overline{Nu} \frac{k}{L} = \frac{3449}{} \left| \frac{0.32 \text{ W}}{\text{m K}} \right| \frac{}{1.5 \text{ m}} = 736 \frac{\text{W}}{\text{m}^2 \text{ K}}. \tag{3}$$

The average plate temperature is obtained by rearranging Eq. (8.47):

$$\bar{T}_s = T_\infty + \frac{\dot{q}_s''}{\bar{h}} = 20°\text{C} + \frac{5400 \text{ W}}{\text{m}^2} \left| \frac{\text{m}^2 \text{ K}}{736 \text{ W}} \right| = \boxed{27.3°\text{C}}. \tag{4}$$

The local plate temperature must be computed using the local heat transfer coefficient according to:

$$T_s = T_\infty + \frac{\dot{q}_s''}{h}, \tag{5}$$

Examination of Eq. (5) shows that the surface temperature of the plate will be largest when the local heat transfer coefficient is smallest. According to Eq. (8.30), the local heat transfer coefficient in a laminar flow decreases relatively quickly with position due to the thickening of the thermal boundary layer. Also, we know that the heat transfer coefficient increases substantially as the flow becomes turbulent. Therefore, it is likely that the heat transfer coefficient is lowest at the transition from laminar to turbulent flow. The position of this transition is:

$$x_{crit} = \frac{Re_{crit} \, \mu}{\rho \, u_\infty} = \frac{5 \times 10^5}{} \left| \frac{0.005 \text{ Pa s}}{} \right| \frac{\text{m}^3}{850 \text{ kg}} \left| \frac{\text{s}}{5 \text{ m}} \right| \frac{\text{N}}{\text{Pa m}^2} \left| \frac{\text{kg m}}{\text{N s}^2} \right| = 0.588 \text{ m}. \tag{6}$$

The local Nusselt number just prior to transition is given by Eq. (8.45):

$$Nu_{x_{crit}} = 0.453 \, Re_{crit}^{1/2} \, Pr^{1/3} = 0.453 \left(5 \times 10^5 \right)^{1/2} (5.4)^{1/3} = 562. \tag{7}$$

The corresponding local heat transfer coefficient is:

$$h_{x = x_{crit}} = Nu_{x_{crit}} \frac{k}{x_{crit}} = \frac{562}{} \left| \frac{0.32 \text{ W}}{\text{m K}} \right| \frac{}{0.588 \text{ m}} = 306 \frac{\text{W}}{\text{m}^2 \text{K}}. \tag{8}$$

The local plate temperature at the end of the laminar region is given by:

$$T_{s, x = x_{crit}} = T_\infty + \frac{\dot{q}_s''}{h_{x = x_{crit}}} = 20°\text{C} + \frac{5400 \text{ W}}{\text{m}^2} \left| \frac{\text{m}^2 \text{ K}}{306 \text{ W}} \right| = \boxed{37.7°\text{C}}. \tag{9}$$

Continued

Example 8.5 (cont.)

Discussion

To be complete, we should also check the local temperature at the trailing edge of the plate. Equation (8.31) indicates that the local heat transfer coefficient in a turbulent flow also decreases with position, but not very quickly because the viscous sublayer thickness does not change rapidly. However, it is easy to estimate the local heat transfer coefficient at $x = L$ to verify that it is higher than the value computed at $x = x_{crit}$. The local Nusselt number at the trailing edge of the plate is given by Eq. (8.46):

$$Nu_L = 0.0308\ Re_L^{0.8}\ Pr^{1/3} = 0.0308\left(1.28 \times 10^6\right)^{0.8}(5.4)^{1/3} = 4140. \tag{10}$$

The corresponding local heat transfer coefficient is:

$$h_{x=L} = Nu_L \frac{k}{L} = \frac{4140}{} \left|\frac{0.32\ \text{W}}{\text{m K}}\right|\frac{1}{1.5\ \text{m}} = 883\ \frac{\text{W}}{\text{m}^2\ \text{K}}. \tag{11}$$

The local heat transfer coefficient computed by Eq. (11) is much larger than the value computed in Eq. (8) and therefore the highest plate temperature is indeed at $x = x_{crit}$.

8.3 Flow across a Cylinder

Flows over an object other than a flat plate are complicated by the fact that the boundary layer will be subjected to pressure gradients that are generated as the free stream flow decelerates and accelerates. In regions where the pressure gradient is very large and positive, the boundary layer may separate completely from the object and a wake region will form. For example, Figure 8.9 illustrates conceptually cross-flow over a cylinder. The cylinder is exposed to a uniform upstream velocity, u_f. However, the free stream velocity that is experienced at the outer edge of the boundary layer that forms at the cylinder surface, u_∞, changes as it moves along the surface. Therefore, the free stream pressure, p_∞, varies along the surface as well. The pressure gradient term in the momentum equation is not zero as it was for the flat plate and the free stream pressure variation will affect the boundary layer growth. A large adverse (i.e., positive) free stream pressure gradient will cause the boundary layer to separate from the surface, resulting in a wake. This is understandable as fluid does not like to flow from low pressure to high pressure.

Figure 8.9(a) illustrates, qualitatively, the streamlines associated with flow around the cylinder at very low velocity. A **stagnation point** will occur at $\theta = 0°$, where the fluid is initially brought to rest. The reduction in the fluid velocity is accompanied by an increase in the pressure at the stagnation point. The velocity of the fluid in the free stream that is adjacent to the cylinder surface will then increase from $\theta = 0°$ to $\theta = 90°$ (notice that the streamlines are getting closer together) resulting in a decrease in pressure. The free stream pressure gradient experienced by the boundary layer in this region is negative; pressure decreases in the direction of flow from $\theta = 0°$ to $\theta = 90°$. Because fluid likes to flow from high pressure to low pressure, a negative pressure gradient is referred to as a *favorable* pressure gradient. The boundary layer develops from the stagnation point and thickens in the downstream direction in a manner that is similar to its behavior on a flat plate. The favorable pressure gradient results in a very stable boundary layer that is somewhat thinner than it would be at the same distance from the leading edge of a flat plate.

The flow will then decelerate from $\theta = 90°$ to $\theta = 180°$ and therefore the pressure will increase, the pressure gradient experienced by the boundary layer in this region will be positive. Fluid does not easily flow from low pressure to high pressure and therefore a positive pressure gradient is referred to as an *adverse* pressure gradient. An adverse pressure gradient results in an unstable boundary layer and as a result the flow will eventually separate from the cylinder's surface, as shown in Figure 8.9(b). The effect of the adverse pressure gradient and separation causes the behavior of external flow around an object such as a cylinder to deviate substantially from the behavior of a boundary layer on a simple flat plate. However, the thermal fluid behavior can still be correlated using an appropriately defined set of nondimensional parameters.

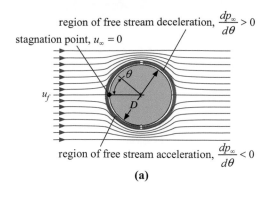

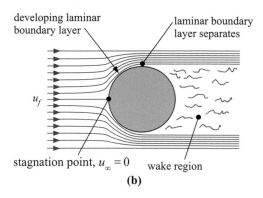

(a)

(b)

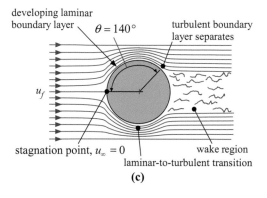

(c)

Figure 8.9 Flow over a cylinder, (a) at very low velocity, (b) under conditions where the boundary layer remains laminar and separates, and (c) under conditions where the boundary layer transitions to turbulence and then separates.

The Reynolds number for flow over a cylinder is defined according to:

$$Re_D = \frac{\rho \, u_f \, D}{\mu}, \tag{8.50}$$

where D is the diameter of the cylinder, u_f is the upstream velocity, and ρ and μ are the density and viscosity of the fluid evaluated at the film temperature, T_{film}, which is the average of the free stream and surface temperatures.

The drag force experienced by the cylinder (F_D) is correlated by the drag coefficient (C_D):

$$C_D = \frac{2F_D}{\rho \, u_f^2 \, LD}, \tag{8.51}$$

where L is the length of the cylinder and therefore the product of L and D is the projected area of the cylinder, as seen from the direction of the approaching flow. The local heat transfer coefficient (h) is correlated by the Nusselt number

$$Nu_D = \frac{h \, D}{k}, \tag{8.52}$$

where k is the conductivity of the fluid, also evaluated at the film temperature. The local Nusselt number is a function of θ, the position on the surface of the cylinder. More typically, correlations for an immersed object are given in terms of the average Nusselt number, defined according to:

$$\overline{Nu_D} = \frac{\bar{h} \, D}{k}. \tag{8.53}$$

560

8.3.1 The Drag Coefficient

Figure 8.10 illustrates the drag coefficient for a cylinder in cross-flow as a function of the Reynolds number (Schlichting, 2000). The behavior exhibited in Figure 8.10 can be understood by returning to Figure 8.9. At small Reynolds numbers **creeping flow** occurs in which the flow does not separate, as shown in Figure 8.9 (a), and therefore the drag force is primarily related to viscous shear experienced at the surface. This is the same phenomenon that leads to a shear force on the flat plate considered in Section 8.2. The force associated with viscous shear will scale according to the area exposed to shear and the product of the viscosity and the velocity gradient at the wall. For laminar flow, the velocity gradient scales as the ratio of the free stream velocity to the momentum boundary layer thickness. To first order, therefore, the shear force is given by:

$$F_D \approx \underbrace{\mu \frac{u_f}{\delta_m}}_{\approx \tau_s} \underbrace{D L}_{\approx A_s} . \tag{8.54}$$

Substituting Eq. (8.54) into Eq. (8.51) shows that the drag coefficient will scale according to:

$$C_D \approx \frac{\mu \frac{u_f}{\delta_m} D L}{\rho\, u_f^2\, D L} = \frac{\mu}{\rho\, u_f\, \delta_m} = \frac{\mu}{\rho\, u_f\, D} \frac{D}{\delta_m} = \frac{1}{Re_D} \frac{D}{\delta_m} . \tag{8.55}$$

The ratio of the momentum boundary layer thickness to the diameter should scale approximately as $Re_D^{-0.5}$ according to flat plate theory and therefore the drag coefficient should scale approximately as:

$$C_D \approx \frac{1}{\sqrt{Re_D}} . \tag{8.56}$$

In fact, Figure 8.10 shows that the drag coefficient scales somewhere between $Re_D^{-0.5}$ and Re_D^{-1} at very low Reynolds number.

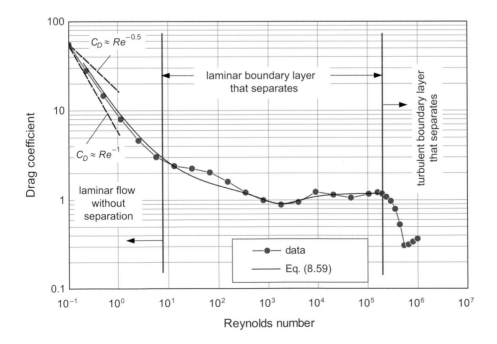

Figure 8.10 Drag coefficient as a function of Reynolds number for a cylinder, based on Schlichting (2000).

When the Reynolds number reaches approximately 10, the laminar boundary layer detaches and a **wake region** forms, as shown in Figure 8.9(b). The laminar boundary layer is thick and has relatively low momentum throughout. Therefore, it is particularly susceptible to adverse pressure gradients and will detach almost immediately at the apex of the cylinder so that the wake region covers essentially all of the entire downstream side of the cylinder. The drag force experienced on the cylinder after the laminar boundary layer separates is primarily due to form drag. The pressure associated with the fluid deceleration exerted on the upstream side of the cylinder is, approximately, $\rho\, u_f^2/2$ greater than ambient pressure while the pressure in the wake region is nearly equal to the ambient pressure. Therefore, the drag force experienced by the cylinder under the condition shown in Figure 8.9(b) is approximately:

$$F_D \approx \frac{\rho\, u_f^2}{2} D\, L.$$

(8.57)

Substituting Eq. (8.57) into Eq. (8.51) leads to:

$$C_D \approx \frac{\dfrac{\rho\, u_f^2}{2} D\, L}{\dfrac{\rho\, u_f^2}{2} D\, L} = 1.$$

(8.58)

Figure 8.10 shows that the drag coefficient becomes relatively insensitive to Reynolds number and remains close to unity for a wide range of Reynolds number. This behavior persists until the Reynolds number approaches the critical Reynolds number for transition to turbulence ($Re_{crit} \approx 5 \times 10^5$ according to flat plate theory); notice the sharp drop in the drag coefficient that occurs at this point. This phenomenon occurs because the turbulent boundary layer has a higher momentum than the laminar boundary layer since the region of low velocity, the viscous sublayer, is very thin. As a result, a turbulent boundary layer has sufficient momentum to overcome the adverse pressure gradient and separation is delayed until approximately $\theta = 140°$, as shown in Figure 8.9(c). A smaller area of the cylinder is exposed to the low pressure wake region and the form drag is substantially smaller. A similar phenomenon is observed for other smooth bodies, such as spheres. The reason that golf balls are dimpled is to promote transition to turbulence at lower values of Reynolds number and therefore reduce the drag force that is experienced by the golf ball during flight.

White (1991) suggests the equation below as a curve fit to the drag coefficient data over a wide range of Reynolds number:

$$C_D = 1.18 + \frac{6.8}{Re_D^{0.89}} + \frac{1.96}{Re_D^{0.5}} - \frac{0.0004\, Re_D}{1 + 3.64 \times 10^{-7}\, Re_D^2}.$$

(8.59)

Equation (8.59) has been entered in EES and is accessible using the built-in procedure External_Flow_Cylinder_ND; this procedure also returns the average Nusselt number, discussed in the subsequent section.

8.3.2 The Nusselt Number

The behavior of the local Nusselt for flow past a cylinder is shown qualitatively in Figure 8.11 for various values of the Reynolds number. Detailed data for this situation can be found in Giedt (1949) and elsewhere. At very low Reynolds number, the local Nusselt number decreases with angular position as the laminar thermal boundary layer thickens in the downstream direction and never detaches. This situation corresponds to Figure 8.9(a). At higher Reynolds number, the local Nusselt number decreases from $\theta = 0°$ to $90°$ as the laminar thermal boundary layer thickens. At approximately $90°$, the boundary layer separates and the local heat transfer coefficient increases in the wake region. This situation corresponds to Figure 8.9(b). For still higher Reynolds number, the Nusselt number decreases as the laminar boundary layer thickens and then, at some critical value of θ (similar to x_{crit} on a flat plate), the boundary layer transitions to turbulence with a corresponding jump in the local Nusselt number. The local Nusselt number associated with the turbulent boundary layer decreases in the downstream direction until finally the turbulent boundary layer detaches and a second jump in the local Nusselt number occurs in the wake region. This situation corresponds to Figure 8.9(c).

The behavior of the local Nusselt number in Figure 8.11 is a complex function of both position and Reynolds number. However, the average value of the Nusselt number increases monotonically with Reynolds number

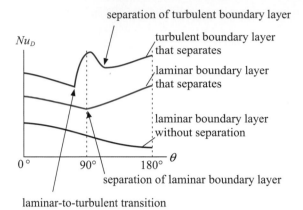

Figure 8.11 The qualitative behavior of the local Nusselt number as a function of position for very low Reynolds number (black), moderate Reynolds number (blue), and very high Reynolds number (red).

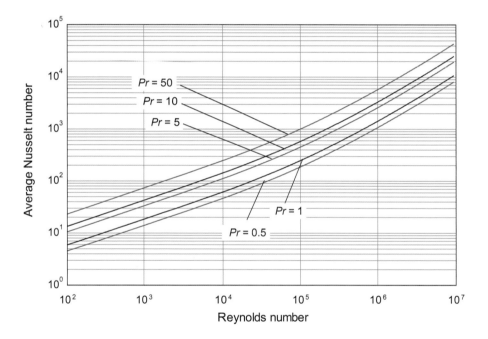

Figure 8.12 Average Nusselt number as a function of Reynolds number for various values of Prandtl number for a cylinder in cross-flow.

even as the characteristics of the flow change substantially. The average Nusselt number is correlated over a wide range of Reynolds number and Prandtl number by Churchill and Bernstein (1977):

$$\overline{Nu}_D = 0.3 + \frac{0.62\, Re_D^{0.5}\, Pr^{1/3}}{\left[1 + \left(\frac{0.4}{Pr}\right)^{2/3}\right]^{0.25}} \left[1 + \left(\frac{Re_D}{2.82 \times 10^5}\right)^{0.625}\right]^{0.80} \quad \text{for } Re_D\, Pr > 0.2. \qquad (8.60)$$

The average Nusselt number predicted by Eq. (8.60) is shown in Figure 8.12 and can be accessed using the procedure External_Flow_Cylinder_ND in EES.

Example 8.6

An uninsulated pipe is carrying water at an average temperature of $T_w = 160°F$ outdoors on a day when the wind is blowing with velocity $u_f = 5$ mph and temperature $T_\infty = -5°F$. The pipe has an outer diameter of $D = 0.5$ inch and a length of $L = 10$ ft. Estimate the rate of heat loss from the pipe.

Known Values

Figure 1 illustrates a sketch of the problem with the known values shown and converted to SI units.

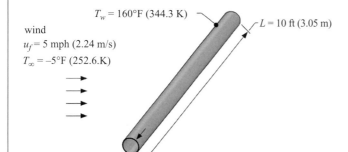

$T_w = 160°F$ (344.3 K)

wind
$u_f = 5$ mph (2.24 m/s)
$T_\infty = -5°F$ (252.6.K)

$L = 10$ ft (3.05 m)

$D = 0.5$ in (0.0127 m)

Figure 1 Sketch of the problem with the known values indicated.

Assumptions

- The pipe surface is smooth.
- The wind is blowing in cross-flow over the pipe.
- The surface of the pipe is at the same temperature as the water.
- The average temperature of the water can be used to compute the heat transfer rate (i.e., the water temperature does not change much as it flows through the pipe).

Analysis and Solution

The flow situation is external forced convection over a cylinder. The average Nusselt number is required in order to determine the total rate of heat transfer. The properties should be evaluated at the film temperature:

$$T_f = \frac{T_w + T_\infty}{2} = \frac{344.3 \text{ K} + 252.6 \text{ K}}{2} = 298.4 \text{ K}. \tag{1}$$

The properties of air are obtained from Appendix C: $\rho = 1.177$ kg/m³, $\mu = 18.54 \times 10^{-6}$ Pa-s, $k = 0.02638$, and $Pr = 0.7072$. The dimensionless numbers required by the average Nusselt number correlation are the Prandtl number and the Reynolds number:

$$Re_D = \frac{\rho u_f D}{\mu} = \frac{1.177 \text{ kg}}{\text{m}^3} \left| \frac{2.24 \text{ m}}{\text{s}} \right| 0.0127 \text{ m} \left| \frac{1}{18.54 \times 10^{-6} \text{ Pa s}} \right| \left| \frac{\text{Pa m}^2}{\text{N}} \right| \frac{\text{N s}^2}{\text{kg m}} = 1802. \tag{2}$$

Equation (8.60) can be used to estimate the average Nusselt number:

$$\overline{Nu}_D = 0.3 + \frac{0.62 \, Re_D^{0.5} \, Pr^{1/3}}{\left[1 + \left(\frac{0.4}{Pr} \right)^{2/3} \right]^{0.25}} \left[1 + \left(\frac{Re_D}{2.82 \times 10^5} \right)^{0.625} \right]^{0.80}$$

$$= 0.3 + \frac{0.62(1802)^{0.5}(0.7072)^{1/3}}{\left[1 + \left(\frac{0.4}{0.7072} \right)^{2/3} \right]^{0.25}} \left[1 + \left(\frac{1802}{2.82 \times 10^5} \right)^{0.625} \right]^{0.80} = 21.6. \tag{3}$$

Continued

Example 8.6 (cont.)

The average heat transfer coefficient experienced by the pipe is obtained using the definition of the average Nusselt number:

$$\bar{h} = \overline{Nu_D}\,\frac{k}{D} = \frac{21.6}{}\left|\frac{0.0264\text{ W}}{\text{m K}}\right|\frac{}{0.0127\text{ m}} = 44.8\,\frac{\text{W}}{\text{m}^2\text{ K}}. \tag{4}$$

The total rate of heat transfer from the pipe to the air is given by:

$$\dot{q} = \bar{h}\,\pi\,D\,L\,(T_w - T_\infty) = \frac{44.8\text{ W}}{\text{m}^2\text{ K}}\left|\pi\right|\frac{0.0127\text{ m}}{}\left|3.05\text{ m}\right|\frac{(344.3 - 252.6)\text{ K}}{} = \boxed{500\text{ W}}. \tag{5}$$

Discussion

The problem could be done using EES, as shown below.

```
$UnitSystem SI Mass J K Pa
D=0.5 [inch]*Convert(inch,m)                                    "diameter"
L=10 [ft]*Convert(ft,m)                                         "length"
T_w=ConvertTemp(F,K,160 [F])                                    "water temperature"
T_infinity=ConvertTemp(F,K,-5 [F])                             "air temperature"
u_f=5 [mph]*Convert(mph,m/s)                                    "wind velocity"

T_f=(T_w+T_infinity)/2                                          "film temperature"
rho=Density(Air,T=T_f,P=Po#)                                    "density"
mu=Viscosity(Air,T=T_f)                                         "viscosity"
k=Conductivity(Air,T=T_f)                                       "conductivity"
Pr=Prandtl(Air,T=T_f)                                           "Prandtl number"

Re_D=D*rho*u_f/mu                                               "Reynolds number"
Call External_Flow_Cylinder_ND(Re_D,Pr: Nusselt_bar_D,C_D)    "access correlations"
h_bar=Nusselt_bar_D*k/D                                         "average heat transfer coefficient"
q_dot=h_bar*pi*D*L*(T_w-T_infinity)                            "rate of heat transfer"
```

Solving provides $\dot{q} = 491.0$ W which is within 2 percent of the value computed manually (the small difference occurs because Appendix C uses nonideal gas properties for air whereas the fluid 'Air' in this EES program models air as an ideal gas). Solving the problem in EES allows a parametric study to be carried out in which the effect of the wind velocity is examined. The rate of heat transfer as a function of wind velocity predicted by this model is shown in Figure 2.

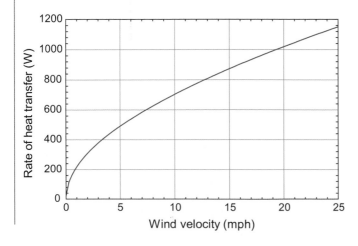

Figure 2 Rate of heat transfer from pipe as a function of the wind velocity.

Example 8.6 (cont.)

Notice that the rate of heat transfer predicted by this model approaches zero as the wind velocity approaches zero. In fact, at very low wind velocity this problem would become a natural convection problem where the air motion would be driven by buoyancy effects, as discussed in Chapter 10. The forced convection correlations that are used in this example do not capture this effect.

Example 8.7

A hot-wire anemometer is used to measure the velocity of air flowing in the duct of a heating system. The anemometer consists of an electrically heated tungsten wire with diameter $D = 10$ μm and length $L = 0.75$ cm. The air temperature is $T_\infty = 20°C$. The anemometer is operated in constant temperature mode, which means that the current passing through the wire is controlled so that the wire temperature remains always $\Delta T = 40$ K above the ambient temperature. As the air velocity increases, the heat transfer coefficient increases and therefore more current is required to maintain this temperature difference. The voltage measured across the wire is therefore related to the air velocity.

If the air velocity is $u_f = 5$ m/s determine the voltage across the wire. Prepare a plot showing the voltage as a function of air velocity.

Known Values

Figure 1 illustrates a sketch of the problem with the known values indicated.

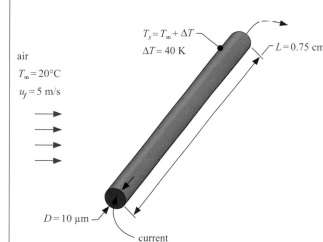

air

$T_\infty = 20°C$

$u_f = 5$ m/s

$T_s = T_\infty + \Delta T$

$\Delta T = 40$ K

$L = 0.75$ cm

$D = 10$ μm

current

Figure 1 Sketch of problem with known values indicated.

Assumptions

- The wire surface is smooth.
- The wire is at a uniform temperature. This implies that there are minimal temperature gradients in the r-direction which is consistent with the Biot number being small so that the extended surface approximation can be made. It also implies that the effect of conduction axially to the mounting structure is neglected.

Continued

Example 8.7 (cont.)

- Radiation is neglected.
- The air is at ambient pressure.

Analysis

The surface temperature of the wire can be computed using the known ΔT:

$$T_s = T_\infty + \Delta T. \tag{1}$$

The film temperature is computed according to:

$$T_f = \frac{T_\infty + T_s}{2}. \tag{2}$$

The properties of air are obtained using the film temperature and the assumed atmospheric pressure:

$$\rho = \rho\left(T_f, P_o\right) \tag{3}$$

$$\mu = \mu\left(T_f\right) \tag{4}$$

$$k = k\left(T_f\right) \tag{5}$$

$$Pr = Pr\left(T_f\right). \tag{6}$$

The dimensionless numbers required by the correlation are the Prandtl number and the Reynolds number:

$$Re_D = \frac{\rho\, u_f\, D}{\mu}. \tag{7}$$

The Reynolds number and Prandtl number together can be used to determine the average Nusselt number:

$$\overline{Nu}_D = \overline{Nu}_D\left(Re_D, Pr\right). \tag{8}$$

The average heat transfer coefficient experienced by the hot wire is obtained using the definition of the average Nusselt number:

$$\bar{h} = \overline{Nu}_D \frac{k}{D}. \tag{9}$$

The rate of heat transfer from the wire can be obtained from Newton's Law of Cooling:

$$\dot{q} = \bar{h}\,\pi\, D\, L\left(T_s - T_\infty\right). \tag{10}$$

The electrical resistivity of tungsten is a function of its temperature:

$$\rho_e = \rho_e(T_s). \tag{11}$$

The electrical resistance of the tungsten wire can be computed according to:

$$R_e = \frac{\rho_e\, L}{A_c}, \tag{12}$$

where A_c is the cross-sectional area for current flow

$$A_c = \frac{\pi}{4}\, D^2. \tag{13}$$

An energy balance on the wire requires that the ohmic dissipation must equal the rate of convection heat transfer at steady state

$$\dot{q} = R_e\, I_c^2, \tag{14}$$

Example 8.7 (cont.)

where I_c is the current. The voltage measured across the wire is given by Ohm's Law:

$$V = R_e I_c. \tag{15}$$

Equations (1) through (15) are 15 equations in the 15 unknowns T_s, T_f, ρ, μ, k, Pr, Re_D, $\overline{Nu}_D$, $\bar{h}$, $\dot{q}$, ρ_e, R_e, A_c, I_c, and V.

Solution

Equations (1) through (15) are sufficient to solve the problem. The solution will be implemented using EES in order to access the property database for Eqs. (2) through (6) and Eq. (11) and the Heat Transfer Library for Eq. (8). The use of EES also makes it possible to make the requested plot.

The inputs are entered in EES and converted to base SI units. Also, the unit system is specified using the $UnitSystem directive so that the EES property database can be used.

```
$UnitSystem SI Mass J K Pa
D=10 [micron]*Convert(micron,m)          "wire diameter"
L=0.75 [cm]*Convert(cm,m)                "wire length"
T_infinity=ConvertTemp(C,K,20 [C])       "free stream temperature"
DT=40 [K]                                "temp. diff. between wire and air"
u_f=5 [m/s]                              "velocity"
```

Equations (1) through (15) are then entered.

```
T_s=T_infinity+DT                                       "surface temperature"
T_f=(T_s+T_infinity)/2                                  "film temperature"
rho=Density(Air,T=T_f,P=Po#)                            "density of air"
mu=Viscosity(Air,T=T_f)                                 "viscosity of air"
k=Conductivity(Air,T=T_f)                               "conductivity of air"
Pr=Prandtl(Air,T=T_f)                                   "Prandtl number of air"
Re_D=u_f*D*rho/mu                                       "Reynolds number"
CALL External_Flow_Cylinder_ND(Re_D,Pr: Nusselt_bar_D,) "access correlation"
h_bar=Nusselt_bar_D*k/D                                 "average heat transfer coefficient"
q_dot=h_bar*pi*D*L*(T_s-T_infinity)                     "rate of heat transfer"
rho_e=ElectricalResistivity(Tungsten, T=T_s)           "electrical resistivity"
A_c=pi*D^2/4                                            "cross-sectional area"
R_e=rho_e*L/A_c                                         "electrical resistance"
q_dot=R_e*I_c^2                                         "ohmic dissipation in wire"
V=R_e*I_c                                               "voltage"
```

Solving provides $I_c = 0.068$ amp and $V = 0.420$ V. A Parametric Table is set up that includes the variables u_f and V and u_f is varied from 0.5 m/s to 10 m/s. The value of u_f in the Equations Window is commented out and the Parametric Table is run. The resulting information is used to generate Figure 2 which shows the voltage as a function of air velocity (labeled constant temperature mode). The measured voltage increases with the air velocity in a predictable way allowing the hot wire to be used to measure velocity.

Continued

Example 8.7 (cont.)

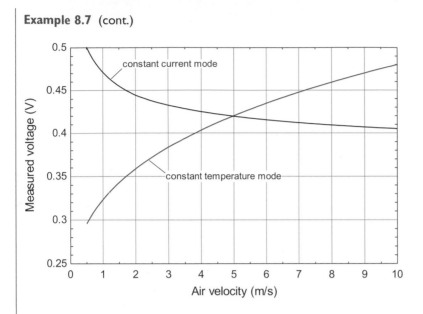

Figure 2 Measured voltage as a function of the air velocity for the hot wire run in constant temperature and constant current modes.

Discussion

Although most hot-wire anemometers are operated in constant temperature mode it is also possible to operate them in constant current mode. In this case, a fixed current is applied to the hot wire and the wire then changes temperature in response to the air velocity. Because the resistivity of tungsten is a function of temperature, the change in temperature will lead to a change in the measured voltage. The model that has been developed in EES can be modified to simulate this type of operation. First, update the guess values as the modifications will lead to a set of implicit equations. Then, remove the equation that specifies the temperature difference.

| //DT=40 [K] | "temp. diff. between wire and air" |

This leaves us one equation short. Therefore, add an equation that specifies the current to be used, for example $I_c = 0.068$ amp.

| I_c=0.068 [amp] | "current" |

Solving with $u_f = 5$ m/s should provide a voltage of 0.420 V. Running the Parametric Table allows the measured voltage as a function of air velocity to be determined and the result is also shown in Figure 2. Notice that the voltage does not vary as much in constant current mode as it does in constant temperature mode for the same change in air velocity. As a result, the instrument is less sensitive when run in this mode.

8.4 Flow across other Extrusions

Cylinders are circular extrusions; noncircular extrusions (e.g., square channels) appear often in engineering applications and their behavior can be correlated using a similar set of dimensionless parameters. The average Nusselt number and Reynolds number are defined according to:

$$\overline{Nu}_W = \frac{\bar{h}\,W}{k}$$

(8.61)

and

$$Re_W = \frac{\rho\,u_f\,W}{\mu},$$

(8.62)

where W is the width of the extrusion, as shown in Table 8.2.

Jakob (1949) has correlated the average Nusselt number for flow across several common extrusions as a function of Reynolds number and Prandtl number according to:

$$\overline{Nu}_W = C\,Re_W^n\,Pr^{1/3}$$

(8.63)

where C and n are dimensionless constants that are specific to the shape of the extrusion. The constants C and n as well as the correct definition of W are provided in Table 8.2. Also shown in Table 8.2 are the drag coefficients for each shape in the region $Re_W > 1 \times 10^4$. Care should be taken not to apply the correlations beyond their

Table 8.2 Average Nusselt number correlations and drag coefficient for external flow over noncircular extrusions.

Geometry	Average Nusselt number			Drag coefficient for $Re_W > 1 \times 10^4$
	Range of applicability	Constants for Eq. (8.63)		
		C	n	
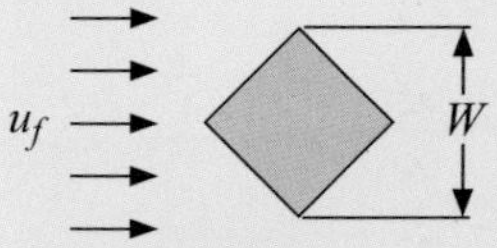 u_f — diamond (square), W	$5 \times 10^3 < Re_W < 1 \times 10^5$ $Pr > 0.7$	0.246	0.588	1.6
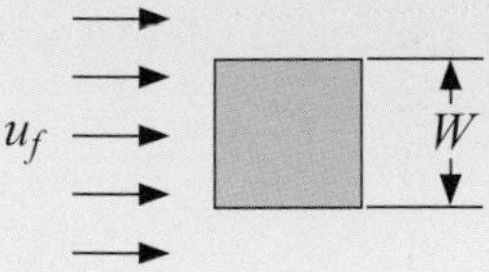 u_f — square, W	$5 \times 10^3 < Re_W < 1 \times 10^5$ $Pr > 0.7$	0.102	0.675	2.1
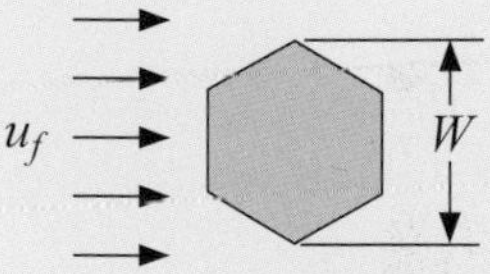 u_f — hexagon (flat top), W	$5 \times 10^3 < Re_W < 1.95 \times 10^4$ $Pr > 0.7$	0.160	0.668	0.7
	$1.95 \times 10^4 < Re_W < 1 \times 10^5$ $Pr > 0.7$	0.0385	0.782	
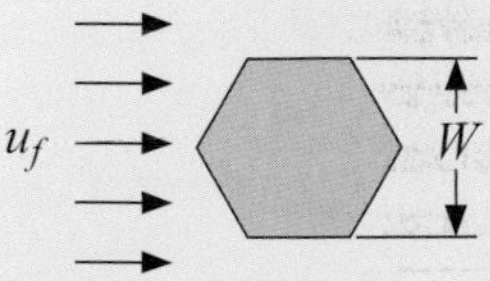 u_f — hexagon (point), W	$5 \times 10^3 < Re_W < 1 \times 10^5$ $Pr > 0.7$	0.153	0.638	1.0
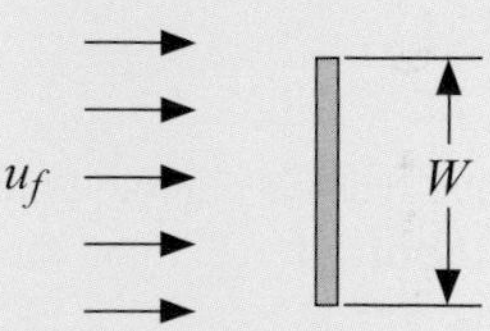 u_f — flat plate, W	$4 \times 10^3 < Re_W < 1.5 \times 10^4$ $Pr > 0.7$	0.228	0.731	2.0

range of validity, which is also indicated the table. The correlations are provided by procedures in the EES Heat Transfer Library (select Convection and then External Flow and scroll to the appropriate shape).

Example 8.8

A fin with a square cross-section is being used to heat R134a flowing through a large duct. The side length of the fin is $H = 4$ mm and the length of the fin is $L = 2.5$ cm. The fin is made from material with $k_m = 320$ W/m-K. The base temperature of the fin is $T_b = 280$ K and the free stream temperature and velocity are $T_\infty = 240$ K and $u_f = 3.5$ m/s. The pressure of the R134a is $p = 500$ kPa and therefore the refrigerant is sub-cooled.

Determine the rate of heat transfer from the fin to the refrigerant and the drag force experienced by the fin. Carry out your calculations for the two cases where the flow is directed towards the side of the fin and towards the corner of the fin (i.e., the first two entries in Table 8.2).

Known Values

Figure 1 illustrates a sketch of the problem with the known values shown. The properties of R134a are estimated using Appendix D at the film temperature of 260 K: $\rho = 1337$ kg/m³, $k = 0.1001$ W/m-K, $\mu = 0.3153 \times 10^{-3}$ Pa-s, and $Pr = 4.121$. (Note that the properties of sub-cooled R134a are approximated with the saturated liquid properties provided in Appendix D.)

R134a
$u_f = 3.5$ m/s, $T_\infty = 240$ K
$\rho = 1337$ kg/m³, $k = 0.1001$ W/m-K, $\mu = 0.3153 \times 10^{-3}$ Pa-s, $Pr = 4.121$

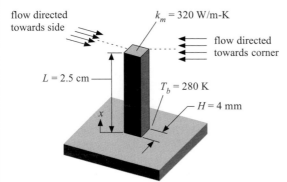

Figure I Sketch of the problem with the known values indicated.

Assumptions

- The properties of R134a can be evaluated using saturated liquid values in Appendix D.
- The extended surface approximation can be made to analyze the fin. This implies that the temperature varies in the x-direction only so that the fin efficiency solutions discussed in Section 3.3 can be applied. This assumption should be checked by calculation of an appropriately defined Biot number.
- Radiation is neglected.

Analysis and Solution

The flow situation is external forced convection over a square extrusion. The average Nusselt number and drag coefficient are required in order to determine the total rate of heat transfer and the drag force. First, the

Example 8.8 (cont.)

situation where the flow direction is perpendicular to the side of the fin is analyzed. In this case, the characteristic dimension W is equal to the side length H:

$$W = H = 0.004 \text{ m} \quad (4 \text{ mm}).\tag{1}$$

The dimensionless numbers required by the average Nusselt number correlation are the Prandtl number and the Reynolds number:

$$Re_W = \frac{\rho \, u_f \, W}{\mu} = \frac{1337 \text{ kg}}{\text{m}^3}\left|\frac{3.5 \text{ m}}{\text{s}}\right|\frac{0.004 \text{ m}}{\left|0.3153 \times 10^{-3} \text{ Pa s}\right|}\left|\frac{\text{Pa m}^2}{\text{N}}\right|\frac{\text{N s}^2}{\text{kg m}} = 5.94 \times 10^4.\tag{2}$$

Equation (8.63) can be used to estimate the average Nusselt number with the constants $C = 0.102$ and $n = 0.675$ from Table 8.2:

$$\overline{Nu}_W = C \, Re_W^n \, Pr^{1/3} = 0.102 \left(5.94 \times 10^4\right)^{0.675} (4.121)^{(1/3)} = 272.7.\tag{3}$$

The average heat transfer coefficient experienced by the fin is obtained using the definition of the average Nusselt number:

$$\bar{h} = \overline{Nu}_W \frac{k}{W} = \frac{272.7}{}\left|\frac{0.1001 \text{ W}}{\text{m K}}\right|\frac{1}{0.004 \text{ m}} = 6825 \frac{\text{W}}{\text{m}^2 \text{ K}}.\tag{4}$$

The fin efficiency can be computed using the solution for a constant cross-section, adiabatic tip derived in Chapter 3. The fin constant is:

$$mL = \sqrt{\frac{\bar{h} \, per}{k_m \, A_c}} L = \sqrt{\frac{\bar{h} \, 4 \, H}{k_m \, H^2}} L = \sqrt{\frac{\bar{h} \, 4}{k_m \, H}} L$$

$$= \sqrt{\frac{6825 \text{ W}}{\text{m}^2 \text{ K}}\left|\frac{4}{}\right|\frac{\text{m K}}{320 \text{ W}}\left|\frac{1}{0.004 \text{ m}}\right.}(0.025 \text{ m}) = 3.651.\tag{5}$$

The fin efficiency is given by:

$$\eta_{fin} = \frac{\tanh \, (mL)}{mL} = \frac{\tanh \, (3.651)}{3.651} = 0.2735.\tag{6}$$

The total rate of heat transfer from the fin to the refrigerant is given by:

$$\dot{q} = \bar{h} \, 4 \, H \, L \, \eta_{fin} \, (T_b - T_\infty) = \frac{6825 \text{ W}}{\text{m}^2 \text{ K}}\left|\frac{4}{}\right|\frac{0.004 \text{ m}}{}\left|\frac{0.025 \text{ m}}{}\right|\frac{0.2735}{}\left|\frac{(280 - 240) \text{ K}}{}\right. = \boxed{29.87 \text{ W}}.\tag{7}$$

The drag coefficient is $C_D = 2.1$ according to Table 8.2. The projected area, seen from the direction of the approaching flow, is:

$$A_p = H \, L = (0.004 \text{ m}) \, (0.025 \text{ m}) = 0.001 \text{ m}^2.\tag{8}$$

The drag force is:

$$F_D = \frac{C_D \, \rho \, u_f^2 \, A_p}{2} = \frac{2.1}{2}\left|\frac{1337 \text{ kg}}{\text{m}^3}\right|\frac{(3.5)^2 \text{ m}^2}{\text{s}^2}\left|\frac{0.0001 \text{ m}^2}{}\right|\left|\frac{\text{N s}^2}{\text{kg m}}\right. = \boxed{1.72 \text{ N}}.\tag{9}$$

The calculations can be repeated for the case in which the flow is directed into the corner of the fin, taking care to use the appropriate definition of W:

$$W = 2 \, H \, \cos\left(\frac{\pi}{4}\right) = 2(0.004 \text{ m}) \, \cos\left(\frac{\pi}{4}\right) = 0.00566 \text{ m} \quad (5.66 \text{ mm}).\tag{10}$$

Continued

Example 8.8 (cont.)

The Reynolds number is computed using this new value of W:

$$Re_W = \frac{\rho\, u_f\, W}{\mu} = \frac{1337\ \text{kg}}{\text{m}^3}\left|\frac{3.5\ \text{m}}{\text{s}}\right|\frac{0.00566\ \text{m}}{}\left|\frac{}{0.3153 \times 10^{-3}\ \text{Pa s}}\right|\left\|\frac{\text{Pa m}^2}{\text{N}}\right|\frac{\text{N s}^2}{\text{kg m}} = 8.40 \times 10^4. \tag{11}$$

Equation (8.63) can be used to estimate the average Nusselt number with the constants $C = 0.246$ and $n = 0.588$ from Table 8.2:

$$\overline{Nu}_W = C\, Re_W^n\, Pr^{1/3} = 0.246\left(8.40 \times 10^4\right)^{0.588}(4.121)^{(1/3)} = 309.9. \tag{12}$$

The average heat transfer coefficient experienced by the fin is obtained using the definition of the average Nusselt number, again taking care to use the correct value of W:

$$\bar{h} = \overline{Nu}_W\,\frac{k}{W} = \frac{309.9}{}\left|\frac{0.1001\ \text{W}}{\text{m K}}\right|\frac{}{0.00566\ \text{m}} = 5484\,\frac{\text{W}}{\text{m}^2\ \text{K}}. \tag{13}$$

The fin efficiency is computed in the same way.

$$mL = \sqrt{\frac{\bar{h}\, per}{k_m\, A_c}}L = \sqrt{\frac{\bar{h}\, 4\, H}{k_m\, H^2}}L = \sqrt{\frac{\bar{h}\, 4}{k_m\, H}}L \tag{14}$$

$$= \sqrt{\frac{5484\ \text{W}}{\text{m}^2\ \text{K}}\left|\frac{4}{}\right|\frac{\text{m K}}{320\ \text{W}}\left|\frac{}{0.004\ \text{m}}\right|}(0.025\ \text{m}) = 3.273$$

$$\eta_{fin} = \frac{\tanh\,(mL)}{mL} = \frac{\tanh\,(3.273)}{3.273} = 0.3047. \tag{15}$$

The total rate of heat transfer from the fin to the refrigerant is given by:

$$\dot{q} = \bar{h}\, 4\, H\, L\, \eta_{fin}\,(T_b - T_\infty)$$

$$= \frac{5484\ \text{W}}{\text{m}^2\ \text{K}}\left|\frac{4}{}\right|\frac{0.004\ \text{m}}{}\left|\frac{0.025\ \text{m}}{}\right|\frac{0.3047}{}\left|\frac{(280 - 240)\ \text{K}}{}\right| = \boxed{26.73\ \text{W}}. \tag{16}$$

The drag coefficient is $C_D = 1.6$ according to Table 8.2. The projected area, seen from the direction of the approaching flow, is:

$$A_p = W\, L = (0.00566\ \text{m})\,(0.025\ \text{m}) = 0.000141\ \text{m}^2. \tag{17}$$

The drag force is:

$$F_D = \frac{C_D\,\rho\, u_f^2\, A_p}{2} = \frac{1.6}{2}\left|\frac{1337\ \text{kg}}{\text{m}^3}\right|\frac{(3.5)^2\ \text{m}^2}{\text{s}^2}\left|\frac{0.000141\ \text{m}^2}{}\right|\left\|\frac{\text{N s}^2}{\text{kg m}} = \boxed{1.85\ \text{N}}. \tag{18}$$

Discussion

Because the problem involved two conductivity values (k for the refrigerant and k_m for the fin material) it is important to ensure that the correct conductivity is used in the calculations. The refrigerant conductivity is used to determine the heat transfer coefficient in Eqs. (4) and (13) while the fin material conductivity should be used to determine the fin constant in Eqs. (5) and (14).

The extended surface approximation should be checked by calculating a Biot number using the definition provided below. The largest value of $\bar{h}$ is used in order to establish the upper limit of the Biot number:

$$Bi = \frac{H\,\bar{h}}{2\, k_m} = \frac{0.004\ \text{m}}{2}\left|\frac{6825\ \text{W}}{\text{m}^2\ \text{K}}\right|\frac{\text{m K}}{320\ \text{W}} = 0.043. \tag{19}$$

This value is sufficiently less than unity that the extended surface approximation is valid.

8.5 Flow past a Sphere

Figure 8.13 illustrates external flow past a sphere. The behavior of flow past a sphere is correlated in terms of the Prandtl number and the Reynolds number based on the sphere diameter:

$$Re_D = \frac{\rho\, u_f\, D}{\mu}. \tag{8.64}$$

The drag force, F_D, is correlated with a drag coefficient

$$C_D = \frac{8\, F_D}{\rho\, u_f^2\, \pi\, D^2}, \tag{8.65}$$

and the average heat transfer coefficient is correlated with an average Nusselt number

$$\overline{Nu}_D = \frac{\bar{h}\, D}{k}. \tag{8.66}$$

Figure 8.14 illustrates the drag coefficient as a function of Reynolds number for a sphere (from Schlichting, 2000); the drag coefficient exhibits essentially the same behavior that is discussed in Section 8.3.1 in the context of flow across a cylinder.

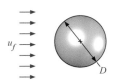

Figure 8.13 Flow past a sphere.

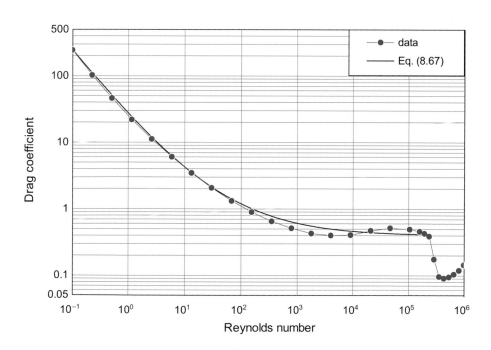

Figure 8.14 Drag coefficient as a function of Reynolds number for flow past a sphere.

White (1991) provides the following expression for the drag coefficient that is valid for $Re_D < 2 \times 10^5$:

$$C_D = \frac{24}{Re_D} + \frac{6}{\left(1 + \sqrt{Re_D}\right)} + 0.40. \tag{8.67}$$

Whitaker (1972), as provided by Mills (1995), has proposed the correlation for average Nusselt number that is valid for $3.5 < Re_D < 8 \times 10^4$ and $0.7 < Pr < 380$:

$$\overline{Nu}_D = 2 + \left(0.4\, Re_D^{0.5} + 0.06\, Re_D^{2/3}\right) Pr^{0.4}. \tag{8.68}$$

The drag coefficient and the average Nusselt number predicted by Eqs. (8.67) and (8.68) can be accessed using the built-in EES function External_Flow_Sphere_ND.

Example 8.9

A copper sphere is falling through a bath of glycerin. The diameter of the sphere is $D = 20$ mm and the glycerin temperature is $T_\infty = 20°C$. Initially, the sphere is falling at a rate of $u = 0.5$ m/s and has a temperature of $T = 200°C$. Determine the rate of change of the temperature and the rate of change of the velocity of the sphere.

Known Values

Figure 1 illustrates a sketch of the problem with the known values indicated.

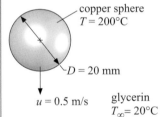

Figure I Sketch of problem with known values indicated.

Assumptions

- The sphere surface is smooth.
- The sphere is at a uniform temperature. This implies that there are minimal temperature gradients within the copper, and is consistent with the Biot number being small so that the lumped capacitance approximation can be made.

Analysis

The properties of the sphere material are obtained using its temperature:

$$\rho_s = \rho_s(T) \tag{1}$$

$$c_s = c_s(T). \tag{2}$$

The film temperature is computed according to:

$$T_f = \frac{T_\infty + T}{2}. \tag{3}$$

Example 8.9 (cont.)

The properties of the fluid are obtained using the film temperature:

$$\rho = \rho\left(T_f\right) \tag{4}$$

$$\mu = \mu\left(T_f\right) \tag{5}$$

$$k = k\left(T_f\right) \tag{6}$$

$$Pr = Pr\left(T_f\right). \tag{7}$$

The problem requires both the average Nusselt number and the drag coefficient. The dimensionless numbers required by the correlation are the Prandtl number and the Reynolds number:

$$Re_D = \frac{\rho u D}{\mu}. \tag{8}$$

The Reynolds number can be used to obtain the drag coefficient:

$$C_D = C_D\left(Re_D\right). \tag{9}$$

The Reynolds number and Prandtl number together can be used to determine the average Nusselt number:

$$\overline{Nu}_D = \overline{Nu}_D\left(Re_D, Pr\right) \tag{10}$$

The average heat transfer coefficient experienced by the sphere is obtained using the definition of the average Nusselt number:

$$\bar{h} = \overline{Nu}_D\,\frac{k}{D}. \tag{11}$$

The volume and surface area of the sphere are computed according to:

$$V = \frac{4\,\pi}{3}\left(\frac{D}{2}\right)^3 \tag{12}$$

$$A_s = 4\,\pi\left(\frac{D}{2}\right)^2. \tag{13}$$

An energy balance on the sphere provides the rate of temperature change:

$$0 = \rho_s\,c_s\,V\frac{dT}{dt} + \bar{h}\,A_s(T - T_\infty). \tag{14}$$

The projected area of the sphere is computed according to:

$$A_p = \pi\left(\frac{D}{2}\right)^2. \tag{15}$$

The drag force is given by:

$$F_D = C_D\,A_p\,\frac{\rho\,u^2}{2}. \tag{16}$$

A momentum balance on the sphere provides the rate of change of the velocity:

$$\rho_s\,V\frac{du}{dt} = \rho_s\,V\,g - F_D. \tag{17}$$

Equations (1) through (17) are 17 equations in the 17 unknowns: ρ_s, c_s, T_f, ρ, μ, k, Pr, Re_D, C_D, $\overline{Nu}_D$, $\bar{h}$, V, A_s, $\frac{dT}{dt}$, A_p, F_D, and $\frac{du}{dt}$.

Continued

Example 8.9 (cont.)

Solution

Equations (1) through (17) are sufficient to solve the problem. The solution will be implemented using EES in order to access the property database for Eqs. (1), (2), and (4) through (7) and the Heat Transfer Library for Eqs. (9) and (10). The use of EES also makes it possible to simulate the dynamic behavior of the sphere.

The inputs are entered in EES and converted to base SI units. The unit system is specified using the $UnitSystem directive so that the EES property database can be used and the type of material used for the sphere and the fluid are stored in strings so that these can be easily changed if needed.

```
$UnitSystem SI Mass J K Pa
D=20 [mm]*Convert(mm,m)               "diameter"
S$='Copper'                          "sphere material"
F$='Glycerin'                        "fluid"
T_infinity=ConvertTemp(C,K,20 [C])   "fluid temperature"
T=ConvertTemp(C,K,200 [C])           "sphere temperature"
u=0.5 [m/s]                          "sphere velocity"
```

Equations (1) through (17) are then entered.

```
"Sphere properties"
rho_s=Density(S$,T=T)                              "density"
c_s=cP(S$,T=T)                                     "specific heat capacity"

"Fluid properties"
T_f=(T+T_infinity)/2                               "film temperature"
rho=Density(F$,T=T_f)                              "density"
mu=Viscosity(F$,T=T_f)                             "viscosity"
k=Conductivity(F$,T=T_f)                           "conductivity"
Pr=Prandtl(F$,T=T_f)                               "Prandtl number"

Re_D=rho*D*u/mu                                    "Reynolds number"
CALL External_Flow_Sphere_ND(Re_D,Pr: Nusselt_bar,C_D)  "access correlations"
h_bar=Nusselt_bar*k/D                              "average heat transfer coefficient"
V=4*pi*(D/2)^3/3                                   "volume of sphere"
A_s=4*pi*(D/2)^2                                   "surface area of sphere"
0=rho_s*c_s*V*dTdt+h_bar*A_s*(T-T_infinity)        "energy balance"
A_p=pi*D^2/4                                       "projected area"
F_D=C_D*A_p*rho*u^2/2                              "drag force"
rho_s*V*dudt=rho_s*V*g#-F_D                        "momentum balance"
```

Solving provides $\dfrac{dT}{dt} = -30.6 \, \text{K/s}$ and $\dfrac{du}{dt} = 9.033 \, \text{m/s}$.

Discussion

The lumped capacitance approximation should be checked by calculating a Biot number using the definition provided below:

$$Bi = \frac{L_{cond}\,\bar{h}}{k_s}, \tag{18}$$

Example 8.9 (cont.)

where L_{cond} is the conduction length, estimated according to:

$$L_{cond} = \frac{V}{A_s}.$$ (19)

```
k_s=Conductivity(S$,T=T)          "conductivity"
L_cond=V/A_s                      "conduction length"
Bi=h_bar*L_cond/k_s               "Biot number"
```

Solving provides $Bi = 0.016$ which is sufficiently less than one to justify the lumped capacitance assumption.

It is relatively easy to use the Integral command that was introduced in Section 5.3.5 and revisited in Section 6.3.6 to numerically integrate these derivatives through time. For this problem, the variables u and T are the state variables and our model implements the state equations, returning their derivatives. A third state variable, the position of the sphere (x), can be easily considered as well since its derivative is simply velocity (u).

Update the guess values for the problem and then comment out the specified values of u and T. Add the required three simultaneous calls to the Integral function in order to integrate the quantities $\frac{dx}{dt}$, $\frac{du}{dt}$, and $\frac{dT}{dt}$ through time from an initial value. The $IntegralTable directive is used to store the trajectory of these variables. The modified EES code is shown below.

```
$UnitSystem SI Mass J K Pa
D=20 [mm]*Convert(mm,m)                     "diameter"
S$='Copper'                                 "substance of sphere"
F$='Glycerin'                               "fluid"
T_infinity=ConvertTemp(C,K,20 [C])          "fluid temperature"
{T=ConvertTemp(C,K,200 [C])                 "sphere temperature"
u=0.5 [m/s]                                 "sphere velocity"}

"Sphere properties"
rho_s=Density(S$,T=T)                       "density"
c_s=cP(S$,T=T)                              "specific heat capacity"

"Fluid properties"
T_f=(T+T_infinity)/2                        "film temperature"
rho=Density(F$,T=T_f)                       "density"
mu=Viscosity(F$,T=T_f)                      "viscosity"
k=Conductivity(F$,T=T_f)                    "conductivity"
Pr=Prandtl(F$,T=T_f)                        "Prandtl number"

Re_D=rho*D*u/mu                             "Reynolds number"
CALL External_Flow_Sphere_ND(Re_D,Pr: Nusselt_bar,C_D)   "access correlations"
h_bar=Nusselt_bar*k/D                       "average heat transfer coefficient"
V=4*pi*(D/2)^3/3                            "volume of sphere"
A_s=4*pi*(D/2)^2                            "surface area of sphere"
0=rho_s*c_s*V*dTdt+h_bar*A_s*(T-T_infinity) "energy balance"
A_p=pi*D^2/4                                "projected area"
F_D=C_D*A_p*rho*u^2/2                       "drag force"
rho_s*V*dudt=rho_s*V*g#-F_D                 "momentum balance"
```

Continued

Example 8.9 (cont.)

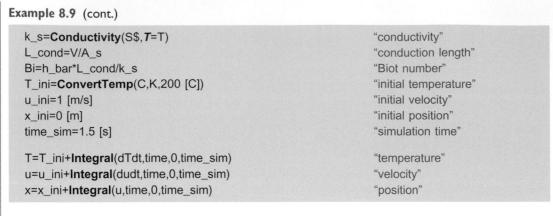

k_s=**Conductivity**(S$,*T*=T)	"conductivity"
L_cond=V/A_s	"conduction length"
Bi=h_bar*L_cond/k_s	"Biot number"
T_ini=**ConvertTemp**(C,K,200 [C])	"initial temperature"
u_ini=1 [m/s]	"initial velocity"
x_ini=0 [m]	"initial position"
time_sim=1.5 [s]	"simulation time"
T=T_ini+**Integral**(dTdt,time,0,time_sim)	"temperature"
u=u_ini+**Integral**(dudt,time,0,time_sim)	"velocity"
x=x_ini+**Integral**(u,time,0,time_sim)	"position"

Figure 2 illustrates the velocity and temperature as a function of position.

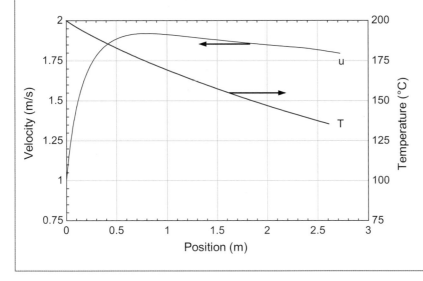

Figure 2 Velocity and temperature as a function of position.

8.6 Conclusions and Learning Objectives

Chapter 7 discussed boundary layers, which are the important flow phenomenon governing convection problems. The behavior of laminar and turbulent boundary layers was interpreted in terms of their effect on engineering quantities like heat transfer coefficient, shear, and drag. The mathematical equations that must be solved in order to develop a compete description of a boundary layer are complex, even when simplified by using the boundary layer assumptions. Therefore, these solutions (or experiments) are presented in terms of *correlations* that describe relationships between dimensionless parameters.

This chapter provides some specific correlations for external forced convection situations. These correlations are presented in several forms including graphs, equations, and procedures in the EES software. However, all accomplish the same thing as they provide the Nusselt number in terms of the Reynolds number and Prandtl number for flow over a specific geometry. The appropriate steps required to apply such correlations to an engineering problem are laid out in Section 8.1 and these steps are then applied in numerous examples throughout the remainder of the chapter.

Some specific concepts that you should understand are listed below.

- The methodology for using a convection correlation discussed in Section 8.1 should be understood well as these correlations are used often in engineering calculations.

- The difference between a local and an average Nusselt number should be understood.
- The definition of the important dimensionless numbers should be understood: Nusselt number (local and average), friction coefficient (local and average), drag coefficient, Reynolds number, and Prandtl number.
- The linkage between the behavior of the Nusselt number and friction coefficient and the underlying physics of the boundary layer should be understood. It is not enough to simply say that "this answer is what the correlation predicts ..." but rather a good engineer should also be able to say "... and I believe it because it agrees with my physical understanding of boundary layer behavior."
- The different regimes that characterize the behavior of the drag coefficient for flow over a cylinder or sphere should be understood.

References

Ameel, T. A., "Average effects of forced convection over a flat plate with an unheated starting length," *Int. Comm. Heat Mass Transfer,* Vol. 24, No. 8, pp. 1113–1120 (1997).

Churchill, S. W. and M. Bernstein, "A correlating equation for forced convection from gases and liquids to a circular cylinder in crossflow," *J. Heat Transfer*, Vol. 99, pp. 300–306 (1977).

Churchill, S. W. and H. Ozoe, "Correlations for laminar forced convection with uniform heating in flow over a plate and in developing and fully developed flow in a tube," *J. Heat Transfer*, Vol. 95, p. 78 (1973).

Giedt, W. H., "Investigation of variation of point unit-heat transfer coefficient around a cylinder normal to an air stream," *Trans. ASME*, Vol. 71, pp. 375–381 (1949).

Jakob, M., *Heat Transfer*, John Wiley and Sons, New York (1949).

Jiji, L. M., *Heat Convection,* Second Edition, Springer Publishing, New York (2010).

Kays, W. M., M. E. Crawford, and B. Weigand, *Convective Heat and Mass Transfer*, Fourth Edition, McGraw-Hill, Boston (2005).

Mills, A. F., *Basic Heat and Mass Transfer*, Irwin, Inc., Chicago (1995).

Mitchell, J. W., "Heat transfer from spheres and other animal forms," *Biophys J.*, Vol. 16, pp. 561–569 (1976).

Nellis, G. F. and S. A. Klein, *Heat Transfer*, Cambridge University Press, New York (2009).

Schlichting, H., *Boundary Layer Theory*, Springer Publishing, New York (2000).

Whitaker, S., "Forced convection heat-transfer correlations for flow in pipes, past flat plates, single cylinders, single spheres, and flow in packed beds and tube bundles," *AIChE J.*, Vol. 18, p. 361 (1972).

White, F. W., *Viscous Fluid Flow,* Second Edition, McGraw Hill, New York (1991).

Problems

Flow over a Flat Plate

8.1 A hot-film anemometer is used to measure air velocity. The anemometer is a thin plate of metal that is oriented parallel to the oncoming air flow. The free stream temperature (T_∞) and the plate surface temperature (T_s – assume that the plate is at a uniform temperature) are both measured. The plate temperature is controlled by electrically heating the material. The plate temperature is always kept $\Delta T = 5$ K higher than the free stream temperature by controlling the electrical power provided, $\dot{q}_e$. The heat transfer coefficient between the surface of the plate and the air flow depends on air velocity and therefore it is possible to relate $\dot{q}_e$ to the free stream velocity, u_∞. The length of the plate in the direction of flow is $L = 10$ mm and the width of the plate is $W = 25$ mm. The plate is placed in an air flow at pressure $p_\infty = 1$ atm and $T_\infty = 20°C$ and the power required by the hot film is $\dot{q}_e = 0.20$ W. You can neglect radiation for this problem. Note that both sides of the plate (the top and the bottom in the figure) convect to the free stream.

(a) What is the air velocity, u_∞?
(b) Plot the air velocity as a function of $\dot{q}_e$. This is the calibration curve for the anemometer; if you were using the anemometer you would measure $\dot{q}_e$ and use this curve to determine u_∞.
(c) Assess the limits of the operating temperature range for the anemometer. If you change the operating temperature (T_∞) by ±50 K will the calibration curve from (b) still be applicable? Justify your answer with a plot.

8.2 The figure illustrates the wall of a house. When the wind is blowing hard in the middle of winter, it *feels* like it is colder in your house. This problem examines whether the amount of heat loss from your house is really strongly affected by the out-door wind velocity.

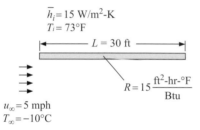

$$\bar{h}_i = 15 \text{ W/m}^2\text{-K}$$
$$T_i = 73°F$$
$$L = 30 \text{ ft}$$
$$R = 15 \frac{\text{ft}^2\text{-hr-}°F}{\text{Btu}}$$
$$u_\infty = 5 \text{ mph}$$
$$T_\infty = -10°C$$

The height of the wall is $H = 10$ ft and the length is $L = 30$ ft. The velocity of the wind is $u_\infty = 5$ mph and the temperature of the air outside is $T_\infty = -10°C$. The indoor air temperature is $T_i = 73°F$ and the natural convection heat transfer coefficient on the inside of the wall is $\bar{h}_i = 15$ W/m^2 – K. The R-value of the wall is $R = 15$ ft^2-hr-F/Btu; this is the area-specific thermal resistance to conduction through the wall.
(a) What is the rate of energy transfer from the wall of the house?
(b) Plot the rate of heat loss as a function of the wind velocity. Comment on any interesting features of your plot. Also comment on whether you feel that the heat loss predicted as u_∞ approaches zero is accurate.

8.3 A solar photovoltaic panel is mounted on a mobile traffic sign in order to provide power without being connected to the grid. The panel is $W = 0.75$ m wide by $L = 0.5$ m long. The wind blows across the panel with velocity $u_\infty = 5$ mph and temperature $T_\infty = 90°F$. The back side of the panel is insulated. The panel surface has an emissivity of $\varepsilon = 1.0$ and radiates to surroundings at T_∞. The PV panel receives a solar flux of

$\dot{q}_s'' = 490$ W/m^2. The panel produces electricity with an efficiency η (that is, the amount of electrical energy produced by the panel is the product of the efficiency, the solar flux, and the panel area). The efficiency of the panel is a function of surface temperature; at 20°C the efficiency is 15 percent and the efficiency drops by 0.25 percent/K as the surface temperature increases (i.e., if the panel surface is at 40°C then the efficiency has been reduced to 10 percent). All of the solar radiation absorbed by the panel and not transformed into electrical energy must be either radiated or convected to its surroundings.

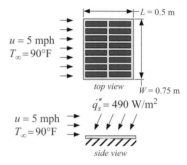

$$L = 0.5 \text{ m}$$
$$u = 5 \text{ mph}$$
$$T_\infty = 90°F$$
top view
$$W = 0.75 \text{ m}$$
$$\dot{q}_s'' = 490 \text{ W/m}^2$$
$$u = 5 \text{ mph}$$
$$T_\infty = 90°F$$
side view

(a) Determine the panel surface temperature, T_s, and the amount of electrical energy generated by the panel.
(b) Prepare a plot of the electrical energy generated by the panel as a function of the solar flux for $\dot{q}_s''$ ranging from 100 W/m^2 to 700 W/m^2. Your plot should show that there is an optimal value for the solar flux – explain this result.
(c) Prepare a plot of the electrical energy generated by the panel as a function of the wind velocity (with $\dot{q}_s'' = 490$ W/m^2) for u_∞ ranging from 5 mph to 50 mph – explain any interesting aspects of your plot.
(d) Prepare a plot of the shear force experienced by the panel due to the wind as a function of wind velocity for u_∞ ranging from 5 mph to 50 mph – explain any interesting aspects of your plot.

8.4 A PV panel is composed of two separate modules.

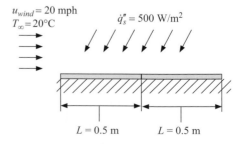

$$u_{wind} = 20 \text{ mph}$$
$$T_\infty = 20°C$$
$$\dot{q}_s'' = 500 \text{ W/m}^2$$
$$L = 0.5 \text{ m} \qquad L = 0.5 \text{ m}$$

The panel is typically cooled by forced convection due to the wind. The ambient temperature is $T_\infty = 20°C$ and the wind velocity is $u_{wind} = 20$ mph. The length of each module in the flow direction is $L = 0.5$ m. The width is $W = 1$ m. Each module is mounted on a thick and conductive substrate and therefore comes to a uniform temperature (T_1 and T_2). The solar flux on the panel is $\dot{q}''_s = 500$ W/m². The efficiency of each module depends on its temperature according to:

$$\eta = \eta_{max} - a(T - T_\infty),$$

where $\eta_{max} = 0.14$ and $a = 0.0005$ K^{-1}.

(a) Determine the temperature of each module and the total power produced by the PV panel.

(b) Plot the temperature of the two modules as a function of the wind velocity. Explain the characteristics of the plot.

(c) Plot the power produced by the panel as a function of the wind velocity.

You are exploring the idea of adding a fan to the system in order to augment the cooling. The fan has the effect of adding $u_{fan} = 10$ m/s to the free stream velocity associated with the wind but requires $\dot{w}_{fan} = 15$ W of power.

(d) Overlay on your plot from (c) the total power produced as a function of wind velocity with the fan turned on. At what wind velocity would you turn on the fan?

8.5 Air at $u_\infty = 1.8$ m/s and $T_\infty = 450$ K flows over a flat plate in a wind tunnel. The plate is $L = 1.6$ m in length in the flow direction and 1 m wide. Measurements are made of the air temperature at various positions above the plate (various values of y) using a small thermocouple that is designed to minimize the disturbance to the flow. The measurements, made at position $x = 0.7$ m from the leading edge of the plate (as shown in the figure), are summarized in the table.

y (m)	T (K)
0	350
0.0012	362
0.0045	400
0.008	425
0.012	440
0.015	447
0.020	450

(a) Estimate the thermal boundary layer thickness from the experimental data.

(b) Estimate the local heat transfer coefficient at $x = 0.7$ m using the experimental data.

(c) What it the local heat transfer coefficient according to accepted correlations?

8.6 In a solar collector, water at 25°C is heated as it flows with a velocity of 0.10 m/s over a black absorber plate. The plate is 1 m in width and 2 m in the flow direction and well insulated on its backside. It is made of copper and it is at a uniform temperature of 28°C. The incident radiation is 750 W/m².

(a) Plot the local heat transfer coefficient as a function of position in the flow direction.

(b) Determine the average heat transfer coefficient assuming a critical Reynold's number of 5×10^5.

(c) Determine the efficiency of the solar collector, defined as the ratio of the heat transfer rate to the water to the incident solar radiation.

8.7 Air at 20°C, 1 atm is blown over a flat plate at 3 m/s. The plate is 60 cm wide and 640 cm long and it is maintained at 90°C.

(a) Determine the location in the flow direction that the flow transitions from laminar to turbulent.

(b) Determine the rate of heat transfer from the plate.

(c) There is some uncertainty in the critical Reynold's number. Determine what the rate of heat transfer would be if the critical Reynold's number is 4.5×10^5 and 5.5×10^5.

8.8 Hydrogen gas at 300 K and 5 atm flows over one side of a flat plate at 18 m/s. The plate is 1 m wide and 2 m long in the flow direction (i.e., the x-direction).

(a) Plot the local friction factor as a function of position x and identify the point at which a transition from laminar to turbulent flow occurs.

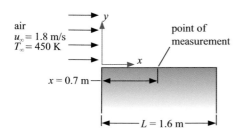

(b) Determine the force exerted by the hydrogen on the plate.

8.9 A very-high-conductivity plate is being dragged through the water at with a velocity of 5 m/s; half of the plate is exposed to air at 30°C and the other half is exposed to water at 5°C. The water and air are both stagnant (i.e., there is no wind or current to speak of). The length of the plate is 1 m and the half-width is 0.5 m, as shown.

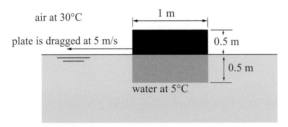

air at 30°C

plate is dragged at 5 m/s

1 m

0.5 m

0.5 m

water at 5°C

(a) Determine the total force exerted on the plate (N); how much of this force is related to the drag from the water and how much from the air?
(b) Determine the average heat transfer coefficient between the plate and the water and the plate and the air (W/m²-K).
(c) If the plate has a very high thermal conductivity then what is the temperature of the plate (°C)?

8.10 An industrial building has a south-facing vertical metal surface that is 3.5 m high and 5.5 m wide. A net energy flux of 280 W/m² is absorbed on the surface as a result of solar radiation. The average wind speed parallel to the surface is 5 m/s with air at 16°C.
(a) Determine the steady-state temperature of the surface.
(b) Owing to the natural turbulence in the air, it is likely that flow is turbulent over the entire surface. Compare the results obtained assuming the flow is entirely turbulent with that obtained assuming a typical laminar to turbulent transition.

8.11 Air at 23°C, 1 atm flows at a velocity of 45 m/s over a flat plate that is 0.5 m wide and 1.5 m long. The plate is maintained at 88°C.
(a) Assume the boundary layer is laminar over the entire length of the plate, compute the total drag force and the total rate of heat transfer.
(b) Assume the boundary layer is turbulent over the entire length of the plate, compute the

total drag force and the total rate of heat transfer.
(c) Assume that the flow transitions from laminar to turbulent when $Re = Re_{crit} = 5 \times 10^5$, compute the total drag force and the total rate of heat transfer.

8.12 The figure illustrates a series of plates inserted into a stream of water in order to provide cooling to a power electronics system.

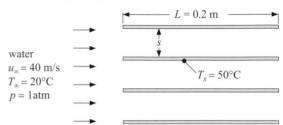

$L = 0.2$ m

water
$u_\infty = 40$ m/s
$T_\infty = 20$°C
$p = 1$atm

s

$T_s = 50$°C

The plates are $L = 0.2$ m long and $W = 0.05$ m wide (into the page). The plates are spaced far enough apart that the boundary layers from adjacent plates do not meet; therefore, the flow over the plates can be treated as an external flow rather than an internal flow. The free stream velocity is $u_\infty = 40$ m/s, the free stream temperature and pressure are $T_\infty = 20$°C and $p = 1$ atm, respectively. The surface of the plates are maintained at $T_s = 50$°C. Assume that the plates are smooth.
(a) What is the total rate of heat transfer from each plate?
(b) Estimate the minimum spacing (s) that can be used before the boundary layers from adjacent plates meet and the problem becomes an internal flow rather than an external flow problem.
(c) Surface roughness begins to affect a turbulent flow when the size of the roughness elements (e) is comparable to the size of the viscous sublayer. Estimate the maximum size of the roughness elements that can be present on the plate before the plate roughness will begin to affect your answer from (a). The viscous sublayer thickness may be estimated by using the equation

$$\delta_{vs} \approx \frac{34.9\,L}{Re^{0.9}}.$$

8.13 You and your friend are looking for an apartment in a high-rise building. You have your choice of four different south-facing units (units

#2 through #5 in the figure) in a city where the wind is predominately from west to east, as shown. You are responsible for paying the heating bill for your apartment and you have noticed that the exterior wall is pretty cheaply built. Your friend has taken heat transfer and therefore is convinced that you should take unit #5 in order to minimize the cost of heating because the boundary layer will be thickest and heat transfer coefficient smallest for the exterior wall of that unit. Prepare an analysis that can predict the cost of heating each of the four units so that you can (a) decide whether the difference is worth considering, and (b) if it is, choose the optimal unit. Assume that the heating season is *time* = 4 months long (120 days), the average outdoor air temperature during that time is $T_\infty = 0°C$ and the average wind velocity is $u_\infty = 5$ mph. The dimensions of the external walls are provided in the figure; assume that no heat loss occurs except through the external walls. Further, assume that the walls have a total thermal resistance on a unit area basis (not including convection) of $R_w'' = 1$ K-m²/W. The internal heat transfer coefficient is $\bar{h}_{in} = 10$ W/m²-K. You like to keep your apartment at $T_{in} = 22°C$ and use electric heating at a cost of $ec = 0.15\$/kW-hr$. You may use the properties of air at T_∞ for your analysis and neglect the effect of any windows.

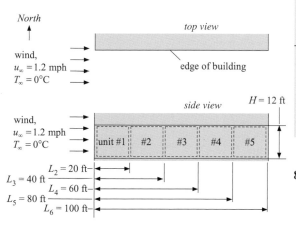

(a) Determine the average yearly heating cost for each of the four units and discuss which apartment is best and why.

(b) Prepare a plot showing the heating cost for unit #2 and unit #5 as a function of the wind velocity for the range 0.5 mph to 5.0 mph.

Explain any interesting characteristics that you observe.

Flow over Cylinders and other Extrusions

8.14 A pipe with outer diameter $D = 12$ inch and length $L = 20$ ft runs outside where it is exposed to a cross-flow of wind with velocity $u_\infty = 5$ m/s. The surface temperature of the pipe is $T_s = 100°C$ and the temperature of the air is $T_\infty = 10°C$.

(a) Determine the heat transfer coefficient and heat transfer rate from the pipe to the air.

(b) Determine the drag force exerted by the wind onto the pipe.

(c) Plot the heat transfer rate and drag force as a function of wind velocity.

8.15 A hollow stainless steel cylinder, 0.5 m in length and 12 cm in diameter is electrically heated with a variable voltage source. The cylinder is horizontal and placed in a wind tunnel with air flowing at a controlled velocity and temperature across it. The outside temperature of the cylinder is measured together with the current and voltage provided to the heater. For the different conditions shown in the table, calculate the average heat transfer coefficient and compare the results with an established correlation.

Air velocity (m/s)	Air temperature (°C)	Cylinder surface temperature (°C)	Heater voltage (V)	Heater current (amp)
1.3	21.1	28.2	42.22	0.422
2.6	24.2	34.6	61.2	0.612
3.8	26.6	30.2	39.7	0.397
4.2	17.3	44.8	112.7	1.127
6.3	19.4	40.2	108.8	1.088

8.16 A long cylindrical container with outer diameter 0.25 m and length 4.0 m is used in a factory to process chemicals. Assume that the ends of the vat are perfectly insulated. The chemicals within the container are maintained at 400°C via an electrical heater that requires 5 kW of power. Although the walls of the cylinder are coated with a thick layer of ceramic in an effort to reduce heat loss, the outer surface is still at 100°C, which is too hot for the safety of workers and poses a fire hazard. The ambient air in the factory is 20°C.

(a) Estimate the heat transfer coefficient between the external surface of the cylinder and the ambient air. Also estimate the thermal resistance (K/W) between the external surface of the cylinder and the chemicals. Note that you must estimate these values based on the information (temperatures and heat transfer rate) given in the problem since you are not given either an air velocity or the internal geometry of the cylinder.

In order to reduce the external surface temperature, you install a large fan that blows air across the cylinder with a velocity of 15 m/s. After the fan is turned on, it was found that the heater power had to be increased in order to maintain the chemicals at the required 400°C. Assume the air properties are: $\rho = 0.97$ kg/m³, $k = 0.031$ W/m-K, $\mu = 212 \times 10^{-7}$ Pa-s, and $Pr = 0.70$.

(b) Estimate the heat transfer coefficient between the air and the cylinder surface.

(c) Estimate the surface temperature of the vat after the fan is turned on (°C) and determine the increase in the heater power (W).

8.17 Assume that the heat transfer characteristics of a person exposed to wind can be adequately approximated as a cylinder. In other words, the heat transfer coefficient between the surface of your winter coat and the winter air can be estimated by assuming you "look like" (at least thermally) a cylinder with a diameter of 0.25 m and length 1.0 m. When you are exposed to a gentle breeze of 5 mph (2.24 m/s) you find that your skin temperature is 30°C. The temperature of the air is –10°C and your rate of metabolic heat generation is 100 W.

(a) Estimate the thermal resistance of your coat (K/W). Neglect any heat loss from your top and bottom surfaces (remember you are a cylinder).

(b) If the wind picks up to 25 mph (11.2 m/s) how much colder does it feel (how much does your skin temperature drop)? Assume that your metabolic heat generation remains at 100 W and the air temperature does not change.

8.18 A piece of metal has a square cross-section with side width $W = 0.005$ m and length

$L = 0.07$ m. The top and bottom surfaces are insulated. The metal has properties $\rho_m = 2300$ kg/m³, $c_m = 750$ J/kg-K, and $k_m = 52$ W/m-K. The sides of the metal are coated with a very thin layer of plastic. The thickness of the plastic coating is $th_c = 0.0005$ m and the conductivity of the plastic is $k_c = 1.1$ W/m-K. The outer surface of the plastic is exposed to a cross-flow of air with velocity $u_f = 50$ m/s and temperature $T_\infty = 300$ K. The properties of air are $k_a = 0.025$ W/m-K, $\mu_a = 0.000019$ Pa-s, $\rho_a = 0.9$ kg/m³, and $Pr_a = 0.7$. Alternating current flows through the metal causing a total generation of thermal energy within the material that varies with time according to:

$$\dot{g}(t) = \dot{g}_{max} \sin^2(2\pi f\, t),$$

where $\dot{g}_{max} = 100$ W and f is the frequency.

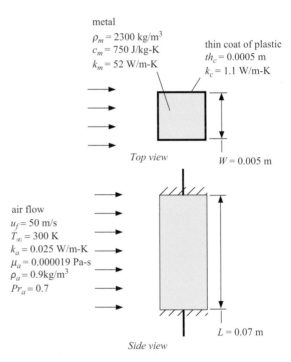

air flow
$u_f = 50$ m/s
$T_\infty = 300$ K
$k_a = 0.025$ W/m-K
$\mu_a = 0.000019$ Pa-s
$\rho_a = 0.9$kg/m³
$Pr_a = 0.7$

(a) Determine the average heat transfer coefficient and drag force.

(b) Is a lumped capacitance model of the metal appropriate? Justify your answer with a calculation.

For the remainder of this problem, assume that a lumped capacitance model of the metal is appropriate.

(c) Determine the lumped capacitance time constant.

(d) Derive the ordinary differential equation that governs the behavior of metal.

8.19 An experiment is run to measure the convective heat transfer coefficient for forced air flow past a cylinder. The free stream air is at 72°F and 1 atm. The cylinder is 4 inches long with a diameter of 0.5 inch. It is polished to minimize radiation. An electrical heater is embedded within the cylinder and the power dissipation has been measured as a function of the air velocity. Also measured is the temperature difference between the cylinder surface and the free stream air. The results are provided in the table.

Velocity (ft/s)	Power (W)	Temperature difference (°F)
41.7	8.18	29.0
54.9	8.15	26.1
70.9	8.20	22.8
95.6	8.22	19.3
126.5	8.15	15.5

(a) Using the experimental data, prepare a plot of Nusselt number vs. Reynolds number.

(b) Compare the experimental plot with an accepted correlation.

(c) List three likely causes for the observed discrepancies.

8.20 Figure 8.10 in the chapter illustrates the drag coefficient for a cylinder as a function of Reynolds number.

(a) Using this figure, discuss briefly (in one or two sentences) why it might make sense to add dimples to a baseball bat.

(b) Using this figure of your text, estimate how fast you would have to be able to swing a bat in order for it to make sense to think about adding dimples (the estimate can be rough but should be explained well). Assume that a bat has diameter $D = 0.04$ m and air has properties $\rho = 1$ kg/m^3 and $\mu = 0.00002$ Pa-s.

8.21 A soldering iron tip can be approximated as a cylinder of metal with radius $r_{out} = 5.0$ mm

and length $L = 20$ mm. The metal is carbon steel; assume that the steel has constant density $\rho = 7854$ kg/m^3 and constant conductivity $k = 50.5$ W/m-K, but a specific heat capacity that varies with temperature (in K) according to:

$$c = 374.9 \left[\frac{J}{kg\text{-}K}\right] + 0.0992 \left[\frac{J}{kg\text{-}K^2}\right] T + 3.596$$

$$\times 10^{-4} \left[\frac{J}{kg\text{-}K^3}\right] T^2.$$

The surface of the iron radiates and convects to surroundings that have temperature $T_{amb} = 20°C$. Radiation and convection occur from the sides of the cylinder (the top and bottom are insulated). The soldering iron is exposed to an air flow (across the cylinder) with a velocity $V = 3.5$ m/s at T_{amb} and $P_{amb} = 1$ atm. The surface of the iron has an emissivity $\varepsilon = 1.0$. The iron is heated electrically by ohmic dissipation; the rate at which electrical energy is added to the iron is $\dot{g} = 35$ W.

(a) Assume that the soldering iron tip can be treated as a lumped capacitance. Develop a numerical model using the Euler technique that can predict the temperature of the soldering iron as a function of time after it is activated. Assume that it is activated when the tip is at ambient temperature. Be sure to account for the fact that the heat transfer coefficient, the radiation resistance, and the heat capacity of the soldering iron tip are all a function of the temperature of the tip.

(b) Plot the temperature of the soldering iron as a function of time. Make sure that your plot covers sufficient time that your soldering iron has reached steady state.

(c) Verify that the soldering iron tip can be treated as a lumped capacitance.

Flow over a Sphere

8.22 In a lead shot tower such as the one at Tower Hill State Park in Wisconsin, spherical shots having a diameter of 0.95 cm are formed from drops of molten lead that solidify as they descend in 20°C air. The average shot temperature is 170°C. Estimate the convective heat transfer coefficient (excluding radiative effects) for the spherical shot at its terminal velocity.

8.23 In his paper "Heat transfer from spheres and other animal forms," *Biophys J.*, Vol. 16, pp. 561–569 (1976), John W. Mitchell proposed that the heat transfer coefficient for all animals can be accurately estimated using the correlations for a sphere with the significant length being the cube root of the volume of the animal. Using this concept, estimate that rate of convective heat loss from a 175 lb_m jogger moving at a pace of 7 min/mile. Assume there is no wind and the density of the jogger is the same as that of liquid water at 25°C. The air temperature is 80°F and the temperature of the jogger's skin is 94°F.

8.24 Two spherical thermocouples are installed in the cylinder of an air compressor: a large one (1.0 mm diameter) and a small one (0.1 mm in diameter). The thermocouple material has properties ρ_{tc} = 8920 kg/m³, c_{tc} = 385 J/kg-K, and k_{tc} = 100 W/m-K. The compressor piston has a diameter of 0.05 m and reciprocates at 600 cycles/min. The piston stroke is 0.05 m. The air in the cylinder has properties ρ = 5.8 kg/m³, μ = 1.86 × 10⁻⁵ Pa-s, k = 0.026 W/m-K, and Pr = 0.73.

(a) Assume that the thermocouples can be treated using a lumped capacitance model. Estimate the time constant of each of the two thermocouples. Assume that the heat transfer coefficient can be estimated using correlations for a sphere in cross-flow with a free stream velocity that is equal to the maximum velocity of the piston.

(b) If the actual temperature of the air in the piston varies sinusoidally with the same frequency as the compressor, then sketch the actual temperature as well as the temperature measured by each of the two thermocouples.

(c) Justify modeling the thermocouples using a lumped capacitance model.

8.25 The wind chill temperature is loosely defined as the temperature that it "feels like" outside when the wind is blowing. More precisely, the wind chill temperature is the temperature of still air that would produce the same bare skin temperature that you experience on a windy day. If you are alive then you are always transferring thermal energy (at rate $\dot{q}$) from your skin (at

temperature T_{skin}). On a windy day, this heat loss is resisted by a convection resistance where the heat transfer coefficient is related to forced convection ($R_{conv,fc}$), as shown in the left-hand figure. The skin temperature is therefore greater than the air temperature (T_{air}). On a still day, this heat loss is resisted by a higher convection resistance because the heat transfer coefficient is related to natural convection ($R_{conv,nc}$), as shown in the right-hand figure. For a given heat loss, air temperature, and wind velocity, the wind chill temperature (T_{WC}) is the temperature of still air that produces the same skin temperature.

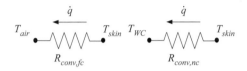

It is surprisingly complicated to compute the wind chill temperature because it requires that you know the rate at which the body is losing heat and the heat transfer coefficient between a body and air on both a windy and still day. At the same time, the wind chill temperature is important and controversial because it affects winter tourism in many places. The military and other government agencies that deploy personnel in extreme climates are also very interested in the wind chill temperature in order to establish allowable exposure limits. This problem looks at the wind chill temperature using your heat transfer background. It has been shown that the heat transfer coefficient for most animals can be obtained by treating them as if they were spherical with an equivalent volume.

(a) What is the diameter of a sphere that has the same volume as a person weighing M = 170 lb_m (assume that the density of human flesh is ρ_f = 64 lb_m/ft^3)?

(b) Assuming that the person can be treated as a sphere, compute their skin temperature on a day when the wind blows at V = 10 mph and the air temperature is T_{air} = 0°F. Assume that the metabolic heat generation is $\dot{q}$ = 150 W.

(c) Assume that the natural convection heat transfer coefficient that would occur on a day with no wind is h_{nc} = 8.0 W/m²-K. What is the wind chill temperature?

According to the National Weather Service, the wind chill temperature can be computed according to:

$$T_{WC} = 35.74 + 0.6215$$
$$T_{air} - 35.75 V^{0.16} + 0.4275 T_{air} V^{0.16},$$

where T_{air} is the air temperature in °F and V is the wind velocity in mph.

(d) Use the National Weather Service equation to compute T_{WC} on a day when $T_{air} = 0$°F and $V = 10$ mph.

(e) Plot the wind chill temperature on a day with $T_{air} = 0$°F as a function of the wind velocity; show the value predicted by your model and by the National Weather Service equation for wind velocities ranging from 5 mph to 30 mph.

8.26 Experimental data for forced convection past spheres of different sizes are to be correlated with Nusselt number as a function of the Reynolds and Prandtl number. The correlation has the form $Nu = a + b\, Re^n Pr^m$. The coefficients a, b, n, and m are parameters that are to be determined in the correlation process. Indicate what value you expect for a and provide a detailed explanation.

8.27 Molten metal droplets must be injected into a plasma for an extreme ultraviolet radiation source.

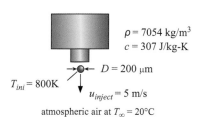

$\rho = 7054$ kg/m³
$c = 307$ J/kg-K

$D = 200$ μm

$T_{ini} = 800$K

$u_{inject} = 5$ m/s

atmospheric air at $T_\infty = 20$°C

The fuel droplets have a diameter of $D = 200$ μm and are injected at a velocity $u_{inject} = 5$ m/s with temperature $T_{ini} = 800$ K. The density of the droplet is $\rho = 7054$ kg/m³ and the specific heat capacity is $c = 307$ J/kg-K. You may assume that the droplet can be treated as a lumped capacitance. The droplet is exposed to still air at $T_\infty = 20$°C.

(a) Develop a numerical model in EES using the Integral command that can predict the

velocity, temperature, and position of the droplet as a function of time.

(b) Plot the velocity as a function of time and the temperature as a function of time.

(c) Plot the temperature as a function of position. If the temperature of the droplet must be greater than 500 K when it reaches the plasma then what is the maximum distance that can separate the plasma from the injector?

Projects

8.28 Fruit should be frozen as soon as possible after it has been harvested and the temperature should be reduced very rapidly in order to maintain its quality for shipping. In order to accomplish this, large "flash freezing" chambers are often utilized at or near the orchard. In one flash freezing operation, spherical fruit with diameter $D = 1.2$ cm is exposed to a high-velocity flow of very cold air. The velocity of the air is $v_f = 12$ ft/s and the air temperature is $T_\infty = -40$°F. The fruit is initially at $T_{ini} = 85$°F and can be treated as a lumped capacitance for this problem. The density of the fruit is $\rho_f = 870$ kg/m³. The specific heat capacity of the fruit varies strongly with temperature, as shown in the figure.

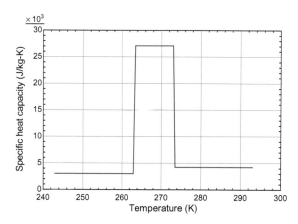

The specific heat capacity of fruit before it freezes is $c_{f,l} = 4200$ J/kg-K and after it is completely frozen is $c_{f,s} = 3000$ J/kg-K. Because the liquid in fruit is not pure water, the freezing process occurs over a temperature range rather than at a single temperature. The fruit begins to freeze at $T_{f,max} = 273.2$ K and is completely frozen at $T_{f,min} = 263.2$ K. During the freezing process, the latent heat of fusion, $\Delta h_{fus} = 235 \times 10^3$ J/kg, is released. The specific heat capacity is given approximately by:

$$c_f = \begin{cases} c_{f,l} & T > T_{f,max} \\ \dfrac{(c_{f,s}+c_{f,l})}{2} + \dfrac{\Delta h_{fus}}{(T_{f,max}-T_{f,min})} & T_{f,min} < T < T_{f,max} \,. \\ c_{f,s} & T < T_{f,min} \end{cases}$$

The logic required to implement the specific heat capacity function provided by the equation above should be accomplished using if-then-else statements in a function.

(a) Determine the rate at which the temperature of the fruit is changing when the temperature of the fruit is $T = 290$ K.

(b) Comment out your assumed value of temperature from (a) (i.e., remove the specification that $T = 290$ K) and use the Integral command in EES in order to predict the temperature of the fruit as a function of time for $0 < t < 10$ min. You will likely need to specify relatively small time steps in the Integral command in order to make the problem converge due to the sharp discontinuities in specific heat capacity that are shown in the plot of specific heat capacity vs. temperature.

(c) Plot the temperature of the fruit as a function of time. Point out and explain any interesting characteristics. How long does it take for the fruit to become completely frozen?

(d) Overlay on your plot from (c) the temperature of the fruit as a function of time for $v_f = 15$ ft/s and $v_f = 9$ ft/s.

8.29 Many of us have made the mistake of trying to use the metal slides on a playground on a hot summer day when the Sun is shining brightly. This problem will predict the temperature of a slide; the figure illustrates a simplified slide geometry.

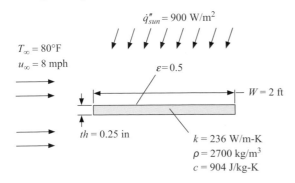

The slide is $L = 10$ ft long and $W = 2$ ft wide. Wind blows across the slide as shown. The air temperature is $T_\infty = 80°$F and the wind velocity is $u_\infty = 8$ mph. Both the top and bottom surfaces of the slide are exposed to convection to the wind as well as radiation. The emissivity of the surface is $\varepsilon = 0.5$. The slide is $th = 0.25$ inch thick and made of material with conductivity $k = 236$ W/m-K, density $\rho = 2700$ kg/m^3, and specific heat capacity $c = 904$ J/kg-K. The top surface of the slide is exposed to solar radiation at a rate of $\dot{q}''_{sun} = 900$ W/m^2. Note that the amount of solar radiation that is absorbed is $c\,\dot{q}''_{sun}$ (the remainder of the radiation is reflected).

(a) Determine the temperature of the slide.

(b) What is the force exerted on the slide by the wind?

(c) Plot the slide temperature as a function of the slide emissivity. Explain the shape of the graph.

(d) Plot the temperature of the slide as a function of the wind velocity from 1 mph to 50 mph. Explain any interesting characteristics in your graph.

(e) You are interested in evaluating the transient response of the slide as the Sun goes behind a cloud causing the solar flux to approach zero suddenly. Is a lumped capacitance model of the slide appropriate? Justify your answer.

(f) Estimate the time constant associated with the slide cooling down after the Sun goes behind a cloud.

(g) Use the Integral command in EES to generate a numerical model of the transient process. Assume the slide is initially at $T_{ini} = 120°$F when the Sun goes behind a cloud, causing the solar flux to drop to zero. Plot the temperature as a function of time.

8.30 A hot-film anemometer is used to measure air velocity. The anemometer is a thin plate of metal that is oriented parallel to the oncoming air flow at $u_\infty = 25$ m/s. The width of the plate is $W = 25$ mm, the length (in the flow direction) is $L = 10$ mm, and the thickness of the plate is $th = 0.4$ mm. The plate is made of material with electrical resistivity $\rho_e = 1.8 \times 10^{-4}$ ohm-m, density $\rho_m = 7500$ kg/m^3, specific heat capacity $c_m = 890$ J/kg-K, and conductivity $k_m = 55$ W/m-K. A voltage difference of $V = 1$ V is maintained across the two edges of the plate (at $y = 0$ and $y = W$) in order to provide a thermal energy generation in the plate material which causes its temperature to rise relative to the temperature of the free stream, $T_\infty = 20°C$. The temperature of the trailing edge of the plate (at $x = L$) is measured. The difference between the trailing edge temperature and the free stream temperature can be correlated to the free stream velocity. You can neglect radiation for this problem. Note that both sides of the plate (the top and the bottom in the figure) experience convection with the free stream. Assume that the temperature distribution in the plate is 1-D (in x).

air at u_∞
$T_\infty = 20°C$
$p_\infty = 1$ atm

$W = 25$ mm
ground
+1 V
$L = 10$ mm

$u_\infty = 25$ m/s
$th = 0.4$ mm
$L = 10$ mm

(a) Develop a 1-D numerical model in EES that can predict the temperature distribution in the plate. Note that you will need to compute the average heat transfer coefficient that exists within each of the control volumes. You may use the properties of air evaluated at the free stream temperature. Plot the temperature as a function of position.

(b) Prepare a plot of the temperature difference between the trailing edge of the plate and the free stream temperature as a function of the free stream velocity for 5 m/s $< u_\infty <$ 50 m/s.

(c) Based on your plot from (b), if the uncertainty in your measurement of the temperature difference is $\delta\Delta T = 0.25$ K then estimate the associated uncertainty in the velocity measurement at the middle of the range (25 m/s).

The next part of problem examines the transient response of the hot film as it is exposed to time-varying velocity.

(d) Estimate the two time constants that will help you understand the transient behavior of the device.

(e) Prepare a 1-D transient numerical model using the Integral command in EES of the device start-up process. Initially, the device is

at a uniform temperature of T_∞ when the voltage is applied. Plot the temperature difference between the trailing edge of the plate and the free stream temperature as a function of time.

8.31 In climates where the night-time temperatures are much less than the day-time temperatures it is possible to reduce the day-time cooling load by using night-time air to cool some thermal storage medium. For example, the figure illustrates a schematic of a system where night-time air is passed over an array of cylinders (one of these is shown in the figure) that are filled with phase change material.

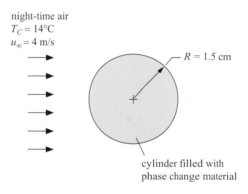

night-time air
$T_C = 14°C$
$u_\infty = 4$ m/s

$R = 1.5$ cm

cylinder filled with
phase change material

The cold night-time air at $T_C = 14°C$ and $u_\infty = 4$ m/s is passed over the cylinder for $t_c = 5$ hr, causes the liquid in the cylinder to solidify; this process releases thermal energy to the air. The phase change material is initially at a uniform temperature T_H. In a subsequent process, hot day-time air is passed over the cylinders at $T_H = 28°C$ causing the reverse process. The phase change material melts, absorbing thermal energy from the air that causes it to be cooled, thus reducing the cooling load. This problem will model the process in which the phase change material is solidified. The outer radius of the phase change material is $R = 1.5$ cm; you may assume that the casing presents negligible thermal resistance and has negligible heat capacity. The density of the phase change material is $\rho = 976$ kg/m^3 and independent of temperature. You may model the freezing process as if it occurs over a small range of temperatures centered at $T_f = 18°C$. The freezing range interval is $\Delta T = 2$ K. The specific heat capacity and conductivity of liquid (i.e., at temperatures above the freezing range) are $c_l = 4500$ J/kg-K and $k_l = 0.25$ W/m-K. The specific heat capacity and conductivity of solid (i.e., at temperatures below the freezing range) are $c_s = 2800$ J/kg-K and $k_s = 0.6$ W/m-K. Within the freezing range, the specific heat capacity is augmented by the latent heat of fusion, $\Delta h_{fus} = 190 \times 10^3$ J/kg. Expressions for these quantities are:

$$c = \begin{cases} c_l \text{ for } T > T_f + \dfrac{\Delta T}{2} \\[2mm] \dfrac{(c_s + c_l)}{2} + \dfrac{\Delta h_{fus}}{\Delta T} \text{ for } T_f - \dfrac{\Delta T}{2} < T < T_f + \dfrac{\Delta T}{2} \\[2mm] c_s \text{ for } T < T_f - \dfrac{\Delta T}{2} \end{cases}$$

$$k = \begin{cases} k_l \text{ for } T > T_f + \dfrac{\Delta T}{2} \\[2mm] \dfrac{(k_s + k_l)}{2} k \text{ for } T_f - \dfrac{\Delta T}{2} < T < T_f + \dfrac{\Delta T}{2} \\[2mm] k_s \text{ for } T < T_f - \dfrac{\Delta T}{2} \end{cases}.$$

(a) Estimate the heat transfer coefficient between the air and the surface of the phase change material. Plot the heat transfer coefficient as a function of the temperature of the surface of the phase change material for values between T_C and T_H.

(b) Develop a 1-D, transient numerical model of the process where the phase change material is frozen during the night. Use EES and use the fully implicit method with one modification – evaluate the properties of the phase change material using the temperatures at the beginning rather than the end of the time step. Your plot from part (a) should have shown that the heat transfer coefficient can be assumed to be constant. Plot the temperature at the surface, the mid-radius, and the center of the phase change material as a function of time.

(c) Plot the center temperature as a function of time for various values of the air velocity.

9 | Internal Forced Convection

Chapter 8 provides correlations that can be used to solve *external flow* forced convection problems where an external flow is defined as one where the boundary layer can grow without bound. For flow over a flat plate located sufficiently far from any other surface, the boundary layer is never confined by the presence of another object and therefore continues to grow from the leading edge to the trailing edge. An **internal flow** is defined as a flow situation where the growth of the boundary layer is confined; that is, the boundary layers can only grow to a certain thickness before being constrained. Internal flows are often encountered in engineering applications (e.g., the flow through tubes or ducts).

Section 9.1 discusses the qualitative behavior of internal flows. Many of the concepts that are discussed in Chapters 7 and 8 for external flows can be extended to understand the behavior of an internal flow. Section 9.2 provides specific correlations that can be used to estimate the pressure gradient and heat transfer coefficient for an internal flow situation. In an external flow, the outer edge of the boundary layer corresponds to the *free stream*, which is fluid that remains unaffected by the surface; the temperature of the free stream is always at the same value (T_∞) regardless of the position on the surface. In an internal flow the boundary layer does not end in a free stream and therefore the idea of a free stream temperature is no longer useful. As a result, it becomes necessary to solve an energy balance when examining an internal flow as discussed in Section 9.3.

9.1 Internal Flow Concepts

Together, the correlations and energy balance discussed in Sections 9.2 and 9.3, respectively, provide the tools that are required to obtain accurate solutions to a variety of internal flow problems. This section discusses the behavior of internal flows at a conceptual level so that these solutions can be understood more completely.

9.1.1 Velocity and Momentum Considerations

Internal vs. External Flow

Figure 9.1(a) illustrates, qualitatively, the momentum boundary layer and velocity distribution that result from a laminar external flow over a plate. The picture should be familiar from the discussion in Chapter 7. The velocity changes from zero at the surface of the plate to the free stream velocity at the edge of the boundary layer. The momentum boundary layer grows in the flow direction due to molecular diffusion.

Figure 9.1(b) illustrates laminar flow through a passage that is formed between two parallel plates. Near the inlet of the passage, the behavior shown in Figure 9.1(b) for an internal flow looks quite similar to the behavior shown in Figure 9.1(a) for an external flow; the momentum boundary layer grows in the flow direction. However, notice that eventually (i.e., at sufficiently large values of x), the momentum boundary layer stops growing because it meets a momentum boundary layer that is growing from the opposite side of the channel. Therefore, at larger values of x the momentum boundary layer becomes bounded. This behavior is the fundamental difference between external and internal flows.

The Developing Region vs. the Fully Developed Region

The momentum boundary layers growing from the upper and lower plates in Figure 9.1(b) meet at some distance from the inlet; this distance defines the **hydrodynamic entry length**, $x_{fd,h}$. Beyond this hydrodynamic entry length, the momentum boundary layer thickness must remain constant as the fluid moves further down the

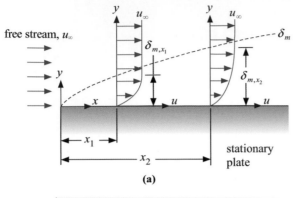

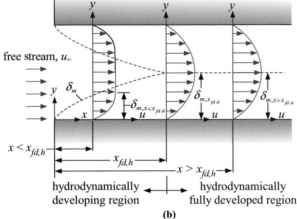

Figure 9.1 (a) Laminar external flow over a flat plate, (b) laminar internal flow between two parallel plates.

flow passage (i.e., for $x > x_{fd,h}$). The region close to the inlet where the boundary layers are growing (i.e., for $x < x_{fd,h}$) is referred to as the **hydrodynamically developing region**. The behavior of an internal flow in the developing region is similar to that of an external flow. The region of the flow after the momentum boundary layer becomes bounded (i.e., for $x > x_{fd,h}$) is referred to as the **hydrodynamically fully developed region**.

Recall from Chapter 7 that the shear stress at the surface, τ_s, is related to the velocity gradient according to:

$$\tau_s = \mu \left. \frac{\partial u}{\partial y} \right|_{y=0}. \tag{9.1}$$

For a laminar flow, the shear stress can be expressed, approximately, as

$$\tau_s \approx \mu \frac{u_\infty}{\delta_m}, \tag{9.2}$$

where δ_m is the momentum boundary layer thickness. Figure 9.2 illustrates, qualitatively, the variation of the shear stress as a function of position for the external and internal flow cases shown in Figure 9.1(a) and (b), respectively.

The shear stress associated with the external flow continues to decrease with position as the boundary layer continues to grow. Eventually, the viscous shear will become sufficiently small that the flow will transition to turbulence, as discussed in Section 7.2. In the hydrodynamically developing region of an internal flow, the shear stress will also decrease as the boundary layer grows and the velocity gradient at the wall is reduced. Notice that the shear stress in the developing region of the internal flow is somewhat larger than the value associated with an external flow at the corresponding position. This occurs because the velocity at the outer edge of the boundary layer (i.e., near the center of the duct) must increase to satisfy continuity for an internal flow; the total flow through the channel at any position is constant and so the velocity at the center of the duct must increase as the flow near the edges is retarded.

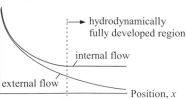

Figure 9.2 Shear stress as a function of position for a laminar external flow and a laminar internal flow.

Because the momentum boundary layer thickness does not change with position (x) in the hydrodynamically fully developed region, the velocity gradient at the wall (and therefore the shear stress) will also be independent of position in this region, as shown in Figure 9.2. If the duct has a constant cross-sectional area and the flow is incompressible, then the velocity distribution associated with an internal flow will not change with x in the hydrodynamically fully developed region.

The Mean Velocity, Hydraulic Diameter, and Reynolds Number

As discussed in Section 7.1.6, the Reynolds number for any flow situation is defined as

$$Re = \frac{u_{char} \, L_{char} \, \rho}{\mu}, \tag{9.3}$$

where u_{char} and L_{char} are the characteristic velocity and length, respectively, associated with the problem. The **mean velocity** (u_m, sometimes referred to as the **bulk velocity**) is the characteristic velocity associated with an internal flow. The mean velocity can be thought of as the single velocity that represents the mass flow rate ($\dot{m}$) carried by the actual velocity distribution that exists within the duct:

$$u_m = \frac{\dot{m}}{\rho \, A_c}, \tag{9.4}$$

where ρ is the density of the flow and A_c is the cross-sectional area of the flow. Equation (9.4) can also be written in terms of the volumetric flow rate ($\dot{V}$):

$$u_m = \frac{\dot{V}}{A_c}. \tag{9.5}$$

Figure 9.1(b) shows that the velocity varies within the cross-section of the duct and therefore the volumetric flow rate must be obtained by integration of the velocity distribution across the duct cross-sectional area, A_c:

$$\dot{V} = \int_{A_c} u \, dA_c, \tag{9.6}$$

where u is the velocity in the x-direction. Substituting Eq. (9.6) into Eq. (9.5) leads to:

$$u_m = \frac{1}{A_c} \int_{A_c} u \, dA_c. \tag{9.7}$$

The mass flow rate or volumetric flow rate in a duct is typically known and therefore Eqs. (9.4) or (9.5) can be used to compute the bulk velocity. Notice that continuity requires that u_m be constant with position (x) for a steady internal flow with constant density, provided that the cross-sectional area of the duct remains constant. Therefore, u_m must be identical to u_∞, the free stream velocity at the entrance to the duct.

The characteristic length for an internal flow is the **hydraulic diameter** (D_h). The hydraulic diameter is defined according to:

$$D_h = \frac{4 \, A_c}{per}, \tag{9.8}$$

where A_c is the cross-sectional area and *per* is the wetted perimeter of the duct, as viewed from the flow direction. For a circular tube with internal diameter D, the hydraulic diameter is equal to the diameter, this is consistent with Eq. (9.8):

$$D_h = \frac{4\,A_c}{per} = \frac{4\pi\,D^2}{4\pi\,D} = D. \tag{9.9}$$

For other duct shapes, Eq. (9.8) should be used to determine the hydraulic diameter. For example, a rectangular duct of dimension $a \times b$ would have a hydraulic diameter

$$D_h = \frac{4\,A_c}{per} = \frac{4ab}{2(a+b)}. \tag{9.10}$$

The Reynolds number that characterizes an internal flow is based on u_m and the hydraulic diameter, D_h:

$$Re_{D_h} = \frac{u_m\,D_h\,\rho}{\mu}. \tag{9.11}$$

The Laminar Hydrodynamic Entry Length

The hydrodynamic entry length is the distance required for the momentum boundary layers to join, as shown in Figure 9.1(b). A conceptual model for momentum boundary layer growth in a laminar external flow was discussed in Section 7.1.3 and provides:

$$\delta_{m,\,lam} \approx 2\sqrt{\upsilon t} = 2\sqrt{\upsilon\frac{x}{u_m}}. \tag{9.12}$$

The laminar hydrodynamic entry length ($x_{fd,h,lam}$) is approximately equal to the axial position at which the momentum boundary layer spans the half-width of the duct. Substituting $\delta_{m,lam} = D_h/2$ and $x = x_{fd,h,lam}$ into Eq. (9.12) leads to:

$$\frac{D_h}{2} \approx 2\sqrt{\upsilon\frac{x_{fd,\,h,\,lam}}{u_m}}. \tag{9.13}$$

Squaring both sides of Eq. (9.13) and substituting the definition of kinematic viscosity leads to:

$$\frac{D_h^2}{4} \approx 4\frac{\mu\,x_{fd,\,h,\,lam}}{\rho\,u_m}. \tag{9.14}$$

Equation (9.14) is solved for the ratio of the hydrodynamic entry length to the hydraulic diameter:

$$\frac{x_{fd,\,h,\,lam}}{D_h} \approx \frac{1}{16}\frac{\rho\,u_m\,D_h}{\mu}. \tag{9.15}$$

Substituting the definition of the Reynolds number for an internal flow, Eq. (9.11), into Eq. (9.15) provides:

$$\frac{x_{fd,\,h,\,lam}}{D_h} \approx 0.063\,Re_{D_h}. \tag{9.16}$$

Equation (9.16) is the accepted correlation for the hydrodynamic entry length of a laminar internal flow (see, for example, White (1991)). Clearly, our conceptual understanding of the characteristics of external flows can be used to understand the behavior of internal flows as well.

Example 9.1

A circular pipe with inner diameter $D = 2$ inch carries $\dot{V} = 1.5$ gal/min of water at atmospheric pressure and room temperature. Estimate the hydrodynamic entry length associated with the flow.

Example 9.1 (cont.)

Known Values

Figure 1 illustrates a sketch of the problem with the known values converted to SI units.

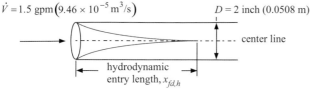

water
$\dot{V} = 1.5$ gpm $\left(9.46 \times 10^{-5}\,\mathrm{m^3/s} \right)$ $D = 2$ inch (0.0508 m)

center line

hydrodynamic
entry length, $x_{fd,h}$

Figure I Sketch of the problem with the known values indicated.

Assumptions

- The flow remains laminar in the pipe.
- The properties of saturated liquid water are used.

Analysis and Solution

The properties of water are obtained from Appendix D, $\rho = 998.2$ kg/m^3, $\mu = 1.002 \times 10^{-3}$ Pa-s. The cross-sectional area and perimeter of the pipe are computed:

$$A_c = \pi \frac{D^2}{4} = \pi \frac{(0.0508)^2\,\mathrm{m^2}}{4} = 0.00203\,\mathrm{m^2} \tag{1}$$

$$per = \pi D = \pi(0.0508\,\mathrm{m}) = 0.160\,\mathrm{m}. \tag{2}$$

The hydraulic diameter is computed according to Eq. (9.8)

$$D_h = \frac{4\,A_c}{per} = \frac{4}{-}\left|\frac{0.00203\,\mathrm{m^2}}{0.160\,\mathrm{m}}\right| = 0.0508\,\mathrm{m}. \tag{3}$$

The mean velocity is computed according to Eq. (9.5).

$$u_m = \frac{\dot{V}}{A_c} = \frac{9.46 \times 10^{-5}\,\mathrm{m^3}}{\mathrm{s}}\left|\frac{}{0.00203\,\mathrm{m^2}}\right| = 0.0467\,\mathrm{m/s}. \tag{4}$$

The Reynolds number is computed according to Eq. (9.11):

$$Re_{D_h} = \frac{\rho\,u_m\,D_h}{\mu} = \frac{998.2\,\mathrm{kg}}{\mathrm{m^3}}\left|\frac{0.0467\,\mathrm{m}}{\mathrm{s}}\right|\left|\frac{0.0508\,\mathrm{m}}{1.002 \times 10^{-3}\,\mathrm{Pa\,s}}\right|\left|\frac{\mathrm{Pa\,m^2}}{\mathrm{N}}\right|\left|\frac{\mathrm{N\,s^2}}{\mathrm{kg\,m}}\right| = 2360. \tag{5}$$

The hydrodynamic entry length is estimated using Eq. (9.16), which assumes laminar flow:

$$x_{fd,\,h,\,lam} \approx 0.063\,Re_{D_h}\,D_h = 0.063(2360)(0.0508\,\mathrm{m}) = \boxed{7.56\,\mathrm{m}\ (24.8\,\mathrm{ft})}. \tag{6}$$

Discussion

The Reynolds number should be checked to ensure that the flow remains laminar. The subsequent section will show that $Re_{D_h} = 2360$ is near the transition point between laminar and turbulent internal flow.

Turbulent Internal Flow

An internal flow will be turbulent provided that it transitions to turbulence before it becomes hydrodynamically fully developed. Once the boundary layers join, the viscous shear stress is constant with position and therefore the flow is not likely to become turbulent unless it experiences a disturbance or there is a change in the cross-sectional area or fluid properties that causes an increase in the Reynolds number. We can use our understanding of external flow behavior to approximately identify the flow conditions that would lead to a turbulent internal flow.

The critical Reynolds number based on position in the flow direction ($Re_{x,crit}$) at which an external flow becomes turbulent depends significantly on the free stream conditions and other characteristics of the flow but typically ranges from 3×10^5 to 6×10^6, as discussed in Section 7.2.2. A value of $Re_{x,crit} = 5 \times 10^5$ is often used to characterize this transition:

$$Re_{x,\,crit} = \frac{\rho\,u_\infty\,x_{crit}}{\mu}.$$ (9.17)

Solving Eq. (9.17) for x_{crit} leads to:

$$x_{crit} = Re_{x,\,crit}\,\frac{\mu}{\rho\,u_\infty}.$$ (9.18)

Equation (9.12) is used to estimate the laminar boundary layer thickness at the transition from laminar to turbulent flow, $\delta_{m,crit}$:

$$\delta_{m,\,crit} \approx 2\sqrt{\frac{x_{crit}\,\mu}{\rho\,u_m}}.$$ (9.19)

Substituting Eq. (9.18) into Eq. (9.19) leads to an expression for the critical boundary layer thickness:

$$\delta_{m,\,crit} \approx \frac{2\,\mu}{\rho\,u_\infty}\sqrt{Re_{x,\,crit}}.$$ (9.20)

If the critical boundary layer thickness is less than the half-width of the duct, $D_h/2$, then the flow will likely transition to turbulence before becoming fully developed. Therefore, the condition at which an internal flow will become turbulent is, approximately:

$$\frac{D_h}{2} \approx \frac{2\,\mu}{\rho\,u_\infty}\sqrt{Re_{x,\,crit}}.$$ (9.21)

Rearranging Eq. (9.21) and recognizing that $u_\infty = u_m$ leads to:

$$\underbrace{\frac{\rho\,u_m\,D_h}{\mu}}_{Re_{D_h,\,crit}} \approx 4\sqrt{Re_{x,\,crit}}.$$ (9.22)

Assuming that $Re_{x,crit} = 5 \times 10^5$ and recognizing that the left side of Eq. (9.22) is the Reynolds number based on the hydraulic diameter leads to:

$$Re_{D_h,\,crit} \approx 2800.$$ (9.23)

Equation (9.23) suggests that an internal flow will be turbulent provided that the Reynolds number based on the hydraulic diameter (Re_{D_h}) is greater than a critical value ($Re_{D_h,\,crit}$). Based on more detailed studies, the critical Reynold's number is typically assumed to be $Re_{D_h,\,crit} = 2300$ but can be as low as 2100 or as high as 10,000 for a carefully controlled experiment (Lienhard and Lienhard (2005)).

Figure 9.3 illustrates flow through the passage formed by two parallel plates. A laminar momentum boundary layer develops from the leading edge of the plate and the boundary layer grows until it reaches a critical value, $\delta_{m,crit}$ given by Eq. (9.20), at which point the flow transitions to turbulence. The turbulent, internal flow has many of the same characteristics as turbulent external flow discussed in Section 7.2.3. Throughout the bulk of the flow, turbulent eddies provide an efficient mechanism for momentum transport and therefore the fluid has a very large, effective viscosity. As a result, the turbulent boundary layer grows quickly and the flow becomes fully developed soon after it transitions to turbulence. The turbulent hydrodynamic entry length is relatively short for any reasonable Reynolds number. A typical rule of thumb is that a turbulent flow will become fully developed in 10–50 diameters, which is a small fraction of the length of a typical pipe.

The turbulent eddies are suppressed in the thin viscous sublayer that exists next to the wall and therefore the velocity gradient occurs primarily across the viscous sublayer. The shear stress jumps dramatically upon transitioning to turbulence. The viscous sublayer thickens very slowly in the downstream direction and therefore

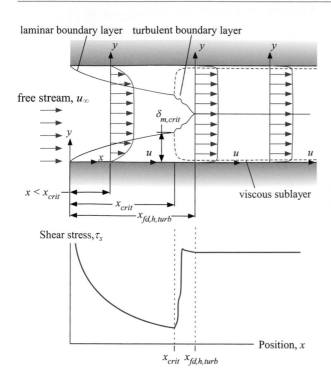

laminar boundary layer   turbulent boundary layer

free stream, u_∞

$\delta_{m,crit}$

$x < x_{crit}$

x_{crit}

$x_{fd,h,turb}$

viscous sublayer

Shear stress, τ_s

Position, x

x_{crit} $x_{fd,h,turb}$

Figure 9.3 Qualitative characteristics of a turbulent internal flow.

the shear stress drops slightly until the flow becomes fully developed. The velocity distribution and therefore the shear stress remain unchanged after the flow is fully developed.

The Pressure Gradient

The pressure gradient is an important engineering quantity for an internal flow because the pressure drop experienced by the fluid must be overcome by a fan or pump that consumes energy and therefore costs money to operate. The pressure gradient is related to the shear stress at the wall as well as to the acceleration of the fluid through a momentum balance. Figure 9.4 illustrates a momentum balance carried out on a control volume that is defined as being differentially small in the x-direction but extending across the entire cross-section of the duct.

Recall from fluid mechanics that momentum is not a conserved quantity. Rather, a momentum balance in the x-direction can be expressed as:

$$\sum F_x + \text{rate of } x\text{-momentum in} = \text{rate of } x\text{-momentum out} + \frac{d}{dt}(x\text{-momentum}). \tag{9.24}$$

The first term in Eq. (9.24) is the sum of the forces acting in the positive x-direction. The remaining terms correspond to the rate of x-momentum entering or leaving the system and being stored within the system. At steady state, the final term in Eq. (9.24) must be zero and so we find that any positive forces in the x-direction must act to increase the rate of x-momentum leaving the system relative to the amount entering:

$$\sum F_x = \text{rate of } x\text{-momentum out} - \text{rate of } x\text{-momentum in}. \tag{9.25}$$

For the control volume shown in Figure 9.4, the x-forces include a pressure force acting on either side of the control volume (i.e., at x and at $x + dx$) as well as the viscous shear acting on the wall. Notice that the viscous shear force tends to retard the fluid flow and therefore acts in the negative x-direction. The rate of momentum flowing across a boundary is obtained by integration of the differential mass flow rate ($\rho\, u\, dA_c$) multiplied by its x-velocity (u). The momentum balance suggested Figure 9.4 is therefore:

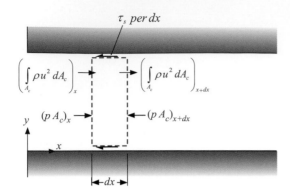

Figure 9.4 A differential momentum balance on a steady internal flow.

$$\underbrace{(p\,A_c)_x - \tau_s\,per\,dx - (p\,A_c)_{x+dx}}_{\sum F_x} + \underbrace{\left(\int_{A_c}\rho\,u^2\,dA_c\right)_x}_{\substack{\text{rate of}\\ x\text{-momentum in}}} = \underbrace{\left(\int_{A_c}\rho\,u^2\,dA_c\right)_{x+dx}}_{\substack{\text{rate of}\\ x\text{-momentum out}}}. \tag{9.26}$$

Expressing the $x + dx$ terms in Eq. (9.26) in terms of their derivatives leads to:

$$(p\,A_c)_x - \tau_s\,per\,dx - (p\,A_c)_x - \frac{d(p\,A_c)}{dx}\,dx + \left(\int_{A_c}\rho\,u^2\,dA_c\right)_x = \left(\int_{A_c}\rho\,u^2\,dA_c\right)_x + \frac{d}{dx}\left(\int_{A_c}\rho\,u^2\,dA_c\right)_x dx, \tag{9.27}$$

which can be simplified to:

$$-\frac{d(p\,A_c)}{dx} = \tau_s\,per + \frac{d}{dx}\left(\int_{A_c}\rho\,u^2\,dA_c\right)_x. \tag{9.28}$$

Equation (9.28) shows that the pressure force must balance the shear force at the wall as well as any change in the momentum of the fluid. For a constant cross-sectional area duct, Eq. (9.28) reduces to:

$$-\frac{dp}{dx} = \tau_s\,\frac{per}{A_c} + \frac{d}{dx}\left[\frac{1}{A_c}\int_{A_c}\rho\,u^2\,dA\right]. \tag{9.29}$$

Equation (9.29) suggests that the pressure gradient will be largest (i.e., most negative) in the developing region, both because the shear stress is higher (see Figure 9.2) as well as because the core of the flow is accelerating. The second term on the right side of Eq. (9.29) is related to the change in the momentum of the flow. The flow enters the duct with a uniform velocity that becomes nonuniform due to the effect of the shear applied by the wall. The velocity at the center of the duct increases while the velocity at the edge of the duct decreases; the integral of the velocity squared must therefore increase as the flow develops hydrodynamically. This effect also contributes to an increased negative pressure gradient in the developing region.

In the hydrodynamically fully developed region (i.e., $x > x_{fd,h}$), the velocity distribution does not change with x and therefore the last term in Eq. (9.29) will be zero (i.e., the momentum of the flow does not change in the x direction). Furthermore, in the hydrodynamically fully developed region the shear stress at the wall is independent of x (as indicated in Figure 9.2) and therefore, the pressure gradient will be constant:

$$\left(-\frac{dp}{dx}\right)_{x>x_{fd,h}} = \tau_s\,\frac{per}{A_c}. \tag{9.30}$$

Figure 9.5 illustrates, qualitatively, the variation of the shear stress and pressure in the entrance region of an internal flow. The pressure initially falls sharply due to the larger shear stress and, to a lesser extent, the fact that

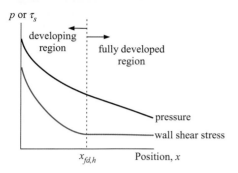

Figure 9.5 Wall shear stress and pressure as a function of position for an internal flow.

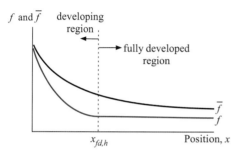

Figure 9.6 Local and average friction factor as a function of position.

the core of the flow is being accelerated. However, for $x > x_{fd,h}$ the pressure gradient is constant and therefore the pressure decreases linearly with position.

The Friction Factor

For an external flow over a plate, the local shear stress was correlated with the local friction coefficient, C_f. For an internal flow, the local pressure gradient is correlated using the **local Moody friction factor** (sometimes referred to as the Darcy friction factor), f, defined according to:

$$f = \left(-\frac{dp}{dx}\right)\frac{2 D_h}{\rho u_m^2}.$$ (9.31)

The local friction factor will be large in the hydrodynamically developing region and then become constant in the fully developed region, as shown in Figure 9.6.

It is important that the Moody friction factor (f) not be confused with the Fanning friction factor. The Fanning friction factor is defined based on the wall shear stress, according to:

$$f_{Fanning} = \frac{2 \tau_s}{\rho u_m^2}.$$ (9.32)

Combining Eqs. (9.30), (9.31), and (9.32) with the definition of the hydraulic diameter shows that the Moody friction factor (f) is four times larger than the Fanning friction factor ($f_{Fanning}$). Therefore, care should be taken so that these friction factor definitions are not accidentally interchanged. This book deals exclusively with the Moody friction factor, defined in Eq. (9.31).

The friction factor defined by Eq. (9.31) is a local friction factor that is based on the local pressure gradient. The **average** or **apparent friction factor** ($\bar{f}$) is defined based on the total change in the pressure:

$$\bar{f} = (p_{x=0} - p_{x=L})\frac{2 D_h}{L \rho u_m^2} = \Delta p \frac{2 D_h}{L \rho u_m^2},$$ (9.33)

where L is the length of the pipe. It is almost always necessary to calculate the average rather than the local friction factor in order to solve an engineering problem because the total pressure change across a duct that must be overcome by a pump or fan is usually the quantity of interest. Equation (9.33) can be rewritten in terms of the local friction factor:

$$\bar{f} = -\frac{2D_h}{L\rho u_m^2}\int_0^L \frac{dp}{dx}dx = \frac{2D_h}{L\rho u_m^2}\int_0^L \frac{\rho u_m^2}{2 D_h}f\,dx = \frac{1}{L}\int_0^L f\,dx. \tag{9.34}$$

The average friction factor will approach the local friction factor in the fully developed region. However, the average friction factor has some memory of the developing region and therefore it will always be somewhat larger than the local value, as shown in Figure 9.6.

Specific correlations for the behavior of the friction factor in an internal flow are provided in Section 9.2.3. However, some of the characteristics of these correlations can be anticipated based upon our knowledge of laminar and turbulent flows. For example, the velocity in a laminar flow changes smoothly across the entire momentum boundary layer which, in a fully developed internal flow, corresponds to the entire cross-section of the duct. Therefore, we should expect that the friction factor for a laminar internal flow will be quite sensitive to the shape of the duct itself (e.g., round vs. square) but insensitive to small-scale roughness at the duct surface. In contrast, the velocity gradient in a turbulent flow is primarily confined to a very thin region referred to as the viscous sublayer and the core of the flow is very nearly at a uniform velocity. As a result, the shear stress at the wall and the associated friction factor will be insensitive to the shape of the duct but strongly affected by the presence of roughness at the duct surface when the flow is turbulent. In Section 9.2, we will see that these intuitive characteristics are reflected in the specific correlations that are used to carry out engineering calculations.

Example 9.2

A duct with a rectangular cross-section is carrying a flow of helium gas at room temperature and atmospheric pressure. The cross-section of the duct is $W = 1$ cm $\times$ $H = 0.8$ cm and the length of the duct is $L = 10$ m. The mass flow rate of helium is equal to $\dot{m} = 1$ g/s and the pressure drop across the duct (i.e., the pressure at the inlet less the pressure at the outlet) is $\Delta p = 17$ kPa.

Determine the Reynolds number and the average friction factor associated with this situation.

Known Values

Figure 1 illustrates a sketch of the problem with the known values indicated.

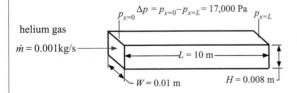

Figure 1 Sketch of the problem with the known values indicated.

Assumption

. The properties of helium at atmospheric temperature and pressure can be used.

Analysis and Solution

The properties of helium are obtained from Appendix C, $\rho = 0.1625$ kg/m^3, $\mu = 19.93 \times 10^{-6}$ Pa-s. The cross-sectional area and perimeter of the duct are computed:

Example 9.2 (cont.)

$$A_c = W\,H = (0.01\text{ m})(0.008\text{ m}) = 8 \times 10^{-5}\text{ m}^2 \tag{1}$$

$$per = 2(W + H) = 2(0.01\text{m} + 0.008\text{ m}) = 0.036\text{ m.} \tag{2}$$

The hydraulic diameter is computed according to Eq. (9.8)

$$D_h = \frac{4\,A_c}{per} = \frac{4\left|8 \times 10^{-5}\text{ m}^2\right|}{\left|0.036\text{ m}\right|} = 0.0089\text{ m.} \tag{3}$$

The mean velocity is computed according to Eq. (9.4):

$$u_m = \frac{\dot{m}}{\rho\,A_c} = \frac{0.001\text{ kg}}{\text{s}}\left|\frac{\text{m}^3}{0.1625\text{ kg}}\right|\frac{}{8 \times 10^{-5}\text{ m}^2} = 76.9\text{ m/s.} \tag{4}$$

The Reynolds number is computed according to Eq. (9.11):

$$Re_{D_h} = \frac{\rho\,u_m\,D_h}{\mu} = \frac{0.1625\text{ kg}}{\text{m}^3}\left|\frac{76.9\text{ m}}{\text{s}}\right|\frac{0.0089\text{ m}}{}\left|\frac{}{19.93 \times 10^{-6}\text{ Pa s}}\right|\left|\frac{\text{Pa m}^2}{\text{N}}\right|\frac{\text{N s}^2}{\text{kg m}} = \boxed{5575}. \tag{5}$$

The average friction factor is computed using Eq. (9.33):

$$\bar{f} = \Delta p\,\frac{2D_h}{L\rho\,u_m^2}$$

$$= \frac{17{,}000\text{ Pa}}{}\left|\frac{2\left|0.0089\text{ m}\right|}{10\text{ m}}\right|\frac{\text{m}^3}{0.1625\text{ kg}}\left|\frac{\text{s}^2}{(76.9)^2\text{ m}^2}\right|\left|\frac{\text{N}}{\text{Pa m}^2}\right|\frac{\text{kg m}}{\text{N s}^2} = \boxed{0.0314}. \tag{6}$$

Discussion

The correlations provided in Section 9.2 can be used to predict the average friction factor given the Reynolds number and duct shape. These correlations were developed based on actual measurements such as the one described in this example.

The flow is likely to be turbulent since $Re_{D_h} > 2300$ and therefore the hydrodynamic entry length will be on the order of 10–50 hydraulic diameters. Because the tube length is more than 1000× the hydraulic diameter (L/D_h = 10 m/0.0089 m = 1125), the effect of the entrance region on the average friction factor is not likely to be large and the friction factor can be interpreted as the fully developed friction factor.

9.1.2 Thermal Considerations

The Developing Region vs. the Fully Developed Region
Figure 9.7(a) illustrates laminar external flow over a plate and shows, qualitatively, the thermal boundary layer and temperature distribution that results. Figure 9.7(b) illustrates laminar internal flow through the passage that is formed by two parallel plates. Notice that at some location the thermal boundary layer becomes bounded in the same way that the momentum boundary layer does in Figure 9.1(b).

The position where the thermal boundary layers that are growing from the upper and lower plates in Figure 9.7(b) meet is referred to as the **thermal entry length**, $x_{fd,t}$. The thermal boundary layer cannot grow further as the fluid moves downstream (i.e., for $x > x_{fd,t}$); this region is referred to as the **thermally fully developed region**. The region where the thermal boundary layers are growing (i.e., $x < x_{fd,t}$) is referred to as the **thermally developing region**.

The Mean Temperature and the Heat Transfer Coefficient
The velocity distribution for an internal flow does not change with x in the hydrodynamically fully developed region as shown in Figure 9.1(b), provided that the flow has constant properties and the cross-section of the duct does not change. However, Figure 9.7(b) illustrates that the temperature distribution in the thermally fully

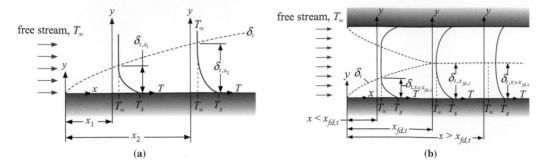

Figure 9.7 (a) Laminar external flow over a flat plate, and (b) laminar internal flow between two parallel plates.

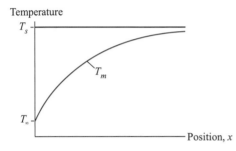

Figure 9.8 Mean temperature as a function of position in a duct with a constant surface temperature.

developed region *does* continue to change even though the thermal boundary layer thickness does not. In general, energy continues to be added or removed from the fluid in the thermally fully developed region and therefore the fluid temperature must change in order to conserve energy. As a result, the free stream temperature at the duct inlet, T_∞ in Figure 9.7(b), is not a useful reference temperature for the flow, particularly after the flow becomes thermally fully developed. Instead, the **mean temperature** (T_m, sometimes referred to as the **bulk temperature**) is the reference temperature for internal flows. The mean temperature is the single temperature that represents the thermal energy carried by the flow at a particular axial position:

$$\dot{m}\, c\, T_m(x) = \rho\, c \int_{A_c} T\, u\, dA_c. \tag{9.35}$$

where T is the local temperature of the flow. Equation (9.35) can be written in terms of the volumetric flow rate ($\dot{V}$):

$$T_m(x) = \frac{1}{\dot{V}} \int_{A_c} T\, u\, dA_c. \tag{9.36}$$

Note that both Eqs. (9.35) and (9.36) are only valid for incompressible fluids with constant specific heat capacity.

The mean temperature is the single temperature at position x that can be used in an energy balance on the flow, as discussed in Section 9.3. You may have solved thermodynamics problems where the temperature of a fluid flowing into or out of a system was specified. You now know that the temperature of the flow actually varied across the cross-section of the duct that was carrying it; therefore, the given temperature was actually the mean temperature that could be used to determine an appropriate energy transfer associated with the flowing fluid. Figure 9.8 illustrates the variation of the mean temperature of the fluid with position in the constant surface temperature duct shown in Figure 9.7(b).

The mean temperature is the most meaningful reference temperature for an internal flow and is therefore used in place of the free stream temperature to define the local heat transfer coefficient:

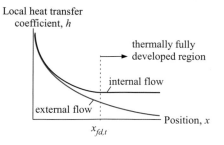

Figure 9.9 Local heat transfer coefficient as a function of position for a laminar external and internal flow.

$$h = \frac{\dot{q}''_s}{(T_s - T_m)},$$ (9.37)

where $\dot{q}''_s$ is the local surface heat flux and T_s and T_m are the local surface and mean temperatures, respectively.

For a laminar flow, the thermal boundary layer can be thought of as a conduction resistance to heat transfer, as discussed in Section 7.1.5. Therefore, the local heat transfer coefficient is inversely proportional to the laminar boundary layer thickness:

$$h \approx \frac{k}{\delta_t}.$$ (9.38)

Figure 9.9 illustrates the local heat transfer coefficient as a function of position for an external flow; notice that h continues to decrease as δ_t grows without bound. Eventually, the flow would transition to turbulence causing h to increase. Figure 9.9 also shows the local heat transfer associated with an internal flow. The heat transfer coefficient is large in the thermally developing region where the thermal boundary layer is small and decreases until the thermal boundary layer becomes bounded, at which point it reaches a constant value in the thermally fully developed region (assuming property changes are insignificant).

Example 9.3

Fluid flows through a duct with length L that has a uniform heat flux $\dot{q}''_s$ applied to its surface. Assume that the flow remains laminar and becomes thermally fully developed at $L/2$. Flow enters the tube at uniform temperature T_{in}. Sketch the mean temperature of the fluid and the surface temperature of the duct as a function of position.

Known Values

Figure 1 illustrates a sketch of the problem.

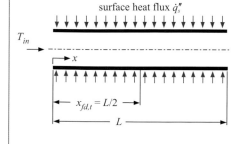

Figure 1 Sketch of the problem with the known values indicated.

Continued

Example 9.3 (cont.)

Assumptions

- The flow remains laminar.
- The specific heat capacity of the flow is constant.

Analysis and Solution

The energy carried by the flow increases at a constant rate as it flows through the tube because the heat flux at the surface is constant. Therefore, the enthalpy of the fluid must increase linearly; if the specific heat capacity is constant then the mean temperature will also increase linearly, as shown in Figure 2.

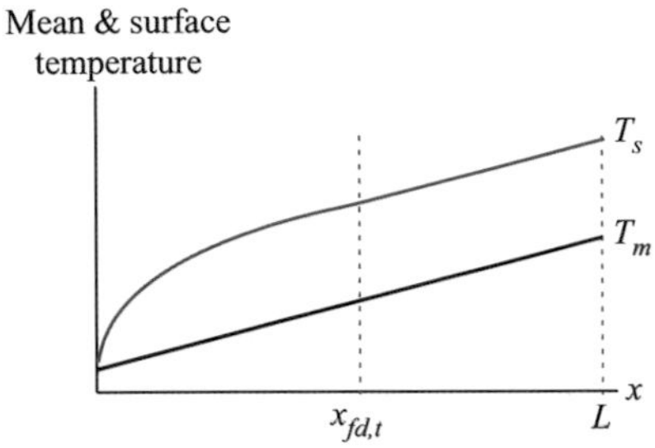

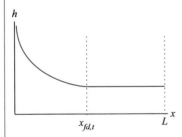

Figure 2 Sketch of the mean temperature and the surface temperature as a function of position.

The local heat transfer coefficient dictates the difference between the mean temperature and the surface temperature according to Eq. (9.37), rearranged below:

$$T_s = T_m + \frac{\dot{q}_s''}{h}. \qquad (1)$$

The local heat transfer coefficient varies with position according to Figure 3. Where the flow is developing the heat transfer coefficient is large and when it becomes fully developed at $L/2$ the heat transfer coefficient becomes constant.

Figure 3 Sketch of the local heat transfer coefficient as a function of position.

Equation (1) together with Figure 3 allows us to sketch the surface temperature. Where the local heat transfer coefficient is large, the surface temperature is close to the mean temperature. When the heat transfer coefficient becomes constant, the difference between the surface and mean temperatures is constant. The surface temperature is overlaid onto the sketch of the mean temperature in Figure 2.

Discussion

Section 9.3 provides a more rigorous treatment of a duct with a specified heat flux based on a formal energy balance.

The Laminar Thermal Entry Length

The thermal entry length is the distance required for the thermal boundary layers to join. The thermal boundary layer growth for a laminar, external flow is discussed conceptually in Section 7.1.2. A simple model based on the diffusion of energy into the fluid provides:

$$\delta_{t,\,lam} \approx 2\sqrt{\alpha\,t} \approx 2\sqrt{\alpha\frac{x}{u_m}},\tag{9.39}$$

where t is the time that the energy has to diffuse, which is related to the position and velocity in the flow direction. The thermal entry length is the position at which the thermal boundary layer extends across the half-width of the duct. Substituting $\delta_{t,lam} = D_h/2$ and $x = x_{fd,t,lam}$ into Eq. (9.39) leads to:

$$\frac{D_h}{2} \approx 2\sqrt{\alpha\frac{x_{fd,\,t,\,lam}}{u_m}},\tag{9.40}$$

where $x_{fd,t,lam}$ is the thermal entry length for a laminar internal flow. Squaring both sides of Eq. (9.40) leads to:

$$\frac{D_h^2}{4} \approx 4\,\alpha\,\frac{x_{fd,\,t,\,lam}}{u_m}\tag{9.41}$$

Equation (9.41) is solved for the ratio of the thermal entry length to the hydraulic diameter:

$$\frac{x_{fd,\,t,\,lam}}{D_h} \approx \frac{D_h\,u_m}{16\,\alpha}.\tag{9.42}$$

Multiplying and dividing the right side of Eq. (9.42) by the kinematic viscosity and substituting the definition of Prandtl number and Reynolds number for an internal flow provides:

$$\frac{x_{fd,\,t,\,lam}}{D_h} \approx \frac{D_h\,u_m}{16\,\alpha}\frac{\upsilon}{\upsilon} = \frac{1}{16}\underbrace{\frac{D_h\,u_m}{\upsilon}}_{Re_{D_h}}\underbrace{\frac{\upsilon}{\alpha}}_{Pr}\tag{9.43}$$

or

$$\frac{x_{fd,\,t,\,lam}}{D_h} \approx 0.063\,Re_{D_h}\,Pr.\tag{9.44}$$

It is interesting to note that the ratio of the thermal to the hydrodynamic laminar entry lengths ($x_{fd,t,lam}/x_{fd,h,lam}$) is only related to the Prandtl number; dividing Eq. (9.44) by Eq. (9.16) provides:

$$\frac{x_{fd,\,t,\,lam}}{x_{fd,\,h,\,lam}} \approx Pr.\tag{9.45}$$

Equation (9.45) makes sense given that the relative size of the momentum to the thermal boundary layer thickness in a laminar external flow was shown in Section 7.1.4 to be related to the Prandtl number. For example, engine oil is viscous but not conductive which results in its very large Prandtl number (much greater than unity). The relative growth of the boundary layers for a high Prandtl number fluid such as engine oil is shown in Figure 9.10(a). The momentum boundary layer will grow more quickly than the thermal boundary layer and the flow will become hydrodynamically fully developed much sooner than it will become thermally developed. The ratio of $x_{fd,t,lam}$ to $x_{fd,h,lam}$ should be much greater than unity, as predicted by Eq. (9.45).

The opposite behavior occurs for a fluid such as a liquid metal that has a Prandtl number that is much less than unity. The thermal boundary layer develops quickly due to the conductive nature of the liquid metal and therefore $x_{fd,t,lam}$ is small. The momentum boundary layer takes longer to develop and so the ratio of $x_{fd,t,lam}$ to $x_{fd,h,lam}$ should be much less than unity, as predicted by Eq. (9.45) and shown in Figure 9.10(b).

Turbulent Internal Flow

As discussed in Section 9.1.1, an internal flow will be turbulent provided that it transitions before it becomes fully developed. This condition is consistent with a critical Reynolds number based on the hydraulic diameter, Re_{D_h}, that is nominally equal to $Re_{D_h,\,crit} = 2300$. A turbulent internal flow will exhibit the same characteristics

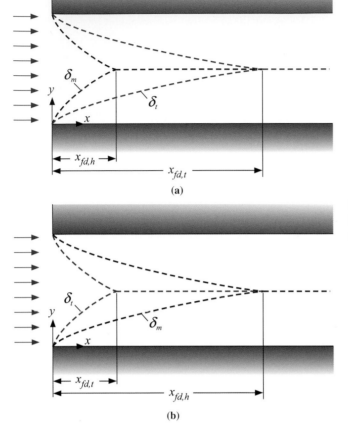

Figure 9.10 Thermal and momentum boundary layer growth for (a) a high Prandtl number, and (b) a low Prandtl number fluid.

that were previously discussed for a turbulent external flow. The dominant thermal resistance is a very thin, viscous sublayer while the remainder of the flow has an extremely high, effective thermal conductivity due to the presence of random fluid motion associated with turbulent eddies. Figure 9.11 illustrates, qualitatively, the temperature distribution expected for a fully developed laminar and turbulent internal flow that each have the same mean temperature and surface temperature.

The energy transport in the laminar internal flow shown in Figure 9.11(a) is primarily diffusive (i.e., related to conduction rather than to fluid motion) across the entire cross-section because no turbulent eddies are present. It is completely diffusive in the fully developed region because there is no velocity component at all in the y-direction. As a result, the temperature gradient extends over the entire cross-section of the duct. The energy transport in the turbulent internal flow shown in Figure 9.11(b) is primarily due to the macroscopic fluid motion induced by turbulent eddies. The effective conductivity associated with this turbulent condition, k_{turb}, is much higher than the conductivity of the fluid itself. Only in the viscous sublayer very near the wall will the energy transport be diffusive and, due to the fact that $k_{turb} \gg k$, the temperature gradient is primarily confined to the viscous sublayer.

The heat transfer coefficients for the laminar and turbulent flows are, approximately,

$$h_{lam} \approx \frac{k}{\delta_{t,\,lam}} \tag{9.46}$$

$$h_{turb} \approx \frac{k}{\delta_{vs}}, \tag{9.47}$$

where δ_{vs} is the thickness of the viscous sublayer. Comparing Eqs. (9.46) and (9.47) indicates that the heat transfer coefficient will be substantially higher in a turbulent flow because δ_{vs} is much less than $\delta_{t,\,lam}$.

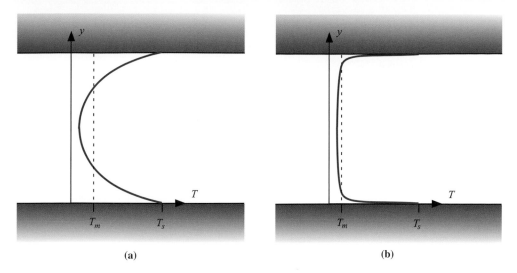

Figure 9.11 Qualitative fully developed temperature distribution expected in (a) a laminar and (b) a turbulent internal flow.

The Nusselt Number

The local heat transfer coefficient for an internal flow is correlated using a local Nusselt number that is based on the hydraulic diameter:

$$Nu_{D_h} = \frac{h\,D_h}{k},$$ (9.48)

where k is the thermal conductivity of the fluid.

The average heat transfer coefficient $(\bar{h})$ for an internal flow is defined according to:

$$\bar{h} = \frac{1}{L}\int_0^L h\,dx.$$ (9.49)

The average Nusselt number for an internal flow is defined based on the average heat transfer coefficient:

$$\overline{Nu}_{D_h} = \frac{\bar{h}\,D_h}{k}.$$ (9.50)

Substituting Eqs. (9.49) and (9.48) into Eq. (9.50) leads to an expression for the average Nusselt number in terms of the local Nusselt number:

$$\overline{Nu}_{D_h} = \frac{D_h}{k}\frac{1}{L}\int_0^L \frac{k\,Nu_{D_h}}{D_h}\,dx = \frac{1}{L}\int_0^L Nu_{D_h}dx.$$ (9.51)

Figure 9.12 illustrates the local and average Nusselt number for an internal flow. The average Nusselt number will approach the local Nusselt number in the thermally fully developed region. However, because the average Nusselt number has some memory of the developing region, it will always be somewhat larger than the local value.

Specific correlations for the Nusselt number in an internal flow are provided in Section 9.2. Some of the characteristics of these correlations can be anticipated based on the discussion in this section. For example, the temperature gradient in a laminar flow encompasses the entire boundary layer which, in a fully developed internal flow, corresponds to the entire cross-section of the duct. Therefore the Nusselt number for a laminar internal flow will be quite sensitive to the shape of the duct (e.g., round vs. square) as well as the boundary conditions on the flow (e.g., constant temperature or constant heat flux) but insensitive to small-scale roughness. The temperature gradient in a turbulent flow is confined to the viscous sublayer and therefore the turbulent

Nu and $\overline{Nu}$

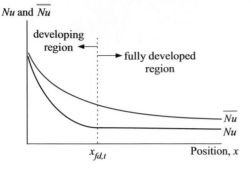

developing
region → ← fully developed
region

$\overline{Nu}$
Nu

$x_{fd,t}$ Position, *x*

Figure 9.12 Local and average friction factor as a function of position.

Nusselt number will be insensitive to the shape of the duct or the thermal boundary conditions but will be strongly affected by the presence of surface roughness.

Example 9.4

The heat transfer fluid Dowtherm A flows through a tube with inner diameter $D = 1.5$ cm and length $L = 12$ m. The mass flow rate of the fluid is $\dot{m} = 0.04$ kg/s. Determine the hydrodynamic and thermal entry lengths for this situation.

Known Values

Figure 1 illustrates a sketch of the problem. The properties of Dowtherm A are obtained from Appendix B.

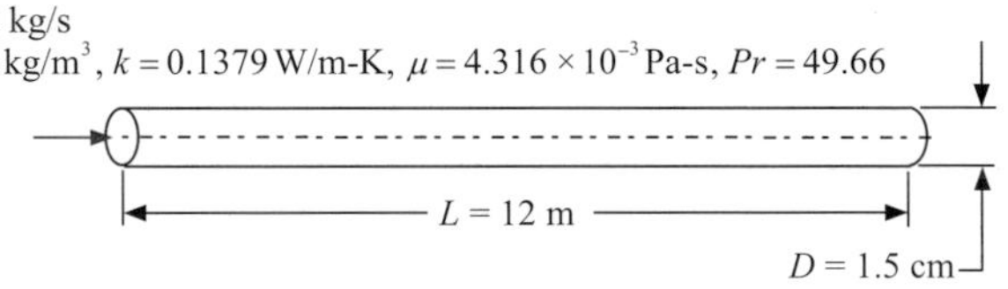

Dowtherm A
$\dot{m} = 0.04$ kg/s
$\rho = 1056 \, \text{kg/m}^3$, $k = 0.1379$ W/m-K, $\mu = 4.316 \times 10^{-3}$ Pa-s, $Pr = 49.66$

Figure 1 Sketch of the problem with the known values indicated.

$L = 12$ m

$D = 1.5$ cm

Assumption

• The properties from Appendix B at room temperature can be used.

Analysis and Solution

The cross-sectional area of the tube is computed:

$$A_c = \frac{\pi}{4}D^2 = \frac{\pi}{4}(0.015 \text{ m})^2 = 1.77 \times 10^{-4} \text{ m}^2. \tag{1}$$

The mean velocity is computed according to Eq. (9.4):

$$u_m = \frac{\dot{m}}{\rho \, A_c} = \frac{0.04 \text{ kg}}{\text{s}} \left| \frac{\text{m}^3}{1056 \text{ kg}} \right| \frac{1}{1.77 \times 10^{-4} \text{ m}^2} = 0.214 \text{ m/s}. \tag{2}$$

The Reynolds number is computed according to Eq. (9.11):

$$Re_{D_h} = \frac{\rho \, u_m D_h}{\mu} = \frac{1056 \text{ kg}}{\text{m}^3} \left| \frac{0.214 \text{ m}}{\text{s}} \right| \frac{0.015 \text{ m}}{\text{s}} \left| \frac{1}{4.316 \times 10^{-3} \text{ Pa s}} \right| \left| \frac{\text{Pa m}^2}{\text{N}} \right| \frac{\text{N s}^2}{\text{kg m}} = 787. \tag{3}$$

Example 9.4 (cont.)

Because the Reynolds number is below the critical Reynolds number for internal flow (2300), the flow is likely to be laminar. The hydrodynamic entry length is estimated using Eq. (9.16):

$$x_{fd, h, lam} \approx 0.063 \, Re_{D_h} D = 0.063(787)(0.015 \, \text{m}) = \boxed{0.74 \, \text{m}}. \tag{4}$$

The thermal entry length is estimated using Eq. (9.44):

$$x_{fd, t, lam} \approx 0.063 \, Re_{D_h} Pr \, D = 0.063(787)(49.66)(0.015 \, \text{m}) = \boxed{36.9 \, \text{m}}. \tag{5}$$

Discussion

Because the fluid has a large Prandtl number and the flow is laminar, the hydrodynamic entry length is quite small relative to the thermal entry length. In fact, the hydrodynamically developing region occupies less than 10 percent of the tube while the fluid *never* becomes thermally fully developed within the given tube length. You can expect that the local friction factor at the exit of the tube will not be very different than the fully developed value. However, the local Nusselt number at the exit of the tube will be much higher than the fully developed value.

9.2 Internal Flow Correlations

9.2.1 Introduction

Section 9.1 discusses the behavior of laminar and turbulent internal flow at a conceptual level, without providing specific information that could be used to obtain a precise solution to an internal flow problem. This section presents a set of useful correlations that are based on analytical solutions and experimental data obtained by many researchers over a long period of time. These correlations can be used to solve engineering problems and are widely used for the thermal-fluid design of various components. A relatively complete set of these correlations has been provided in a library of functions that are available in EES. The correlations are examined as they are presented in order to demonstrate that they agree with our physical understanding of internal flow problems.

9.2.2 Flow Classification

In order to select the appropriate correlation for an internal flow problem, it is necessary to classify the salient features of the flow.

1. *Laminar vs. Turbulent Flow*

   The most critical classification is the flow condition, which is determined by the Reynolds number based on the hydraulic diameter (D_h):

$$Re_{D_h} = \frac{\rho \, u_m \, D_h}{\mu}, \tag{9.52}$$

   where the hydraulic diameter is defined according to:

$$D_h = \frac{4 \, A_c}{per}. \tag{9.53}$$

   As discussed in Section 9.1, the flow will be laminar if the Reynolds number based on the hydraulic diameter is less than approximately 2300 and otherwise it will be turbulent unless other information relating to the transition is available.

2. *Developing vs. Fully Developed*

   Both the Nusselt number and friction factor depend on axial position in the developing region, as discussed in Section 9.1. The behavior of an internal flow is different in the developing as opposed to the fully developed regions. The friction factor depends only on whether the flow is hydrodynamically fully developed or not. However, the Nusselt number depends on whether the flow is developing both thermally and hydrodynamically (referred to as *simultaneously developing*) or is thermally developing after it has previously become hydrodynamically fully developed (referred to as *thermally developing/hydrodynamically developed*). The flow will be simultaneously developing at the inlet to a pipe whereas it will be thermally developing/hydrodynamically developed when heating or cooling is applied at some location on a pipe surface that is far downstream from the inlet.

3. *Duct Shape and Boundary Conditions*

   The heat transfer coefficient and friction factor for a laminar flow will be affected by the large-scale features of the flow, for example the shape of the duct. Correlations are presented in this section for circular, rectangular, and annular passages. The Nusselt number for a laminar flow also depends on the thermal boundary conditions. The two most common boundary conditions are constant wall temperature and constant heat flux; these two boundary conditions provide lower and upper bounds, respectively, for more realistic boundary conditions that are not as well defined.

4. *Surface Roughness*

   The roughness at the duct surface will affect the Nusselt number and friction factor for a turbulent flow but not a laminar flow.

The engineer must carefully classify an internal flow problem according to the criteria discussed above in order to identify the most appropriate correlation. The Internal Flow Convection Library in EES automatically accomplishes this classification and implements the correct correlation. However, it is incumbent on the engineer to understand this process so that the result can be critically assessed and also so that correlations for other geometries and conditions can be correctly applied.

9.2.3 The Friction Factor

The local friction factor in the fully developed region is shown in Figure 9.13 as a function of Reynolds number for various duct shapes (in the laminar region) and various values of surface roughness (in the turbulent region).

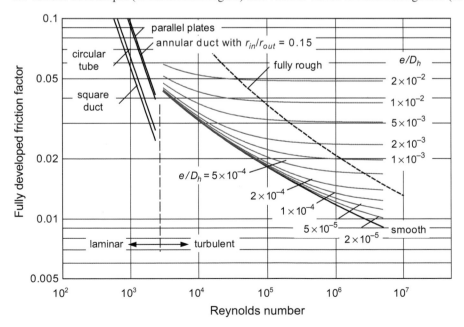

Figure 9.13 Fully developed friction factor as a function of Reynolds number for various duct shapes (in the laminar region) and relative roughness, e/D_h (in the turbulent region).

The surface roughness is quantified as the ratio of the average height of the surface disparities (e) to the hydraulic diameter of the duct. The specific correlations that correspond to the information in Figure 9.13, as well as the increase in friction factor related to the developing region for finite length ducts, are discussed in this subsection.

Laminar Flow

Figure 9.13 shows that the friction for a laminar flow is affected by the duct shape but not by the surface roughness. Therefore, this section presents correlations for various duct shapes. In each case, the local friction factor for the hydrodynamically fully developed region is presented. Correlations for the average friction factor (sometimes referred to as the *apparent* friction factor) that account for the effect of the developing region are also presented for each duct shape. As discussed in Section 9.1.1, the laminar hydrodynamic entry length is approximately equal to:

$$x_{fd, h, lam} \approx 0.063 \, Re_{D_h} D_h. \tag{9.54}$$

Therefore correlations that include the effect of the developing region are presented in terms of a dimensionless tube length, defined according to:

$$\tilde{L} = \frac{L}{D_h Re_{D_h}}. \tag{9.55}$$

The local friction factor will approach the fully developed value when length becomes greater than $x_{fd, h, lam}$, which corresponds to $\tilde{L} > 0.063$. The average friction factor will take longer to approach the fully developed value as it has some memory of the developing region.

Circular Tubes

The local friction factor for a laminar, hydrodynamically fully developed flow in a circular tube is given by:

$$f_{fd} = \frac{64}{Re_{D_h}} \quad \text{for } Re_{D_h} < 2300 \text{ and } x > x_{fd, h, lam}. \tag{9.56}$$

Equation (9.56) is obtained by solving the momentum equation in radial coordinates in the fully developed region. The ratio of the average to the fully developed friction factor for this situation is given by Shah and London (1978):

$$\frac{\bar{f}}{f_{fd}} = 0.215 \, \tilde{L}^{-0.5} + \frac{0.0196\tilde{L}^{-1} + 1 - 0.215\tilde{L}^{-0.5}}{1 + 0.00021\tilde{L}^{-2}}, \tag{9.57}$$

where f_{fd} is the fully developed friction factor computed using Eq. (9.56) and $\tilde{L}$ is calculated according to Eq. (9.55). Notice that in the limit that $\tilde{L} \to \infty$, Eq. (9.57) approaches 1, as it should.

Figure 9.14 illustrates the ratio of the average friction factor to the fully developed friction factor as a function of the dimensionless tube length. Notice that the average friction factor approaches the fully developed friction factor when $\tilde{L}$ reaches a value somewhat larger than 0.063.

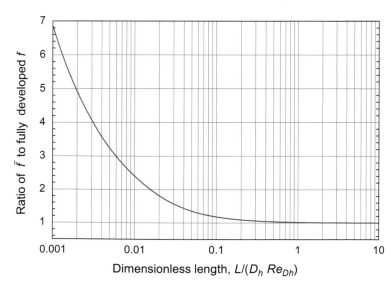

Figure 9.14 The ratio of the apparent friction factor to the fully developed friction factor as a function of $\tilde{L} = L/(D_h Re_{D_h})$.

Example 9.5

A drug delivery system must deliver steady flow at a rate of $\dot{V}$ = 24 cm³/min through a tube with inner diameter D = 500 μm and length L = 2.5 cm. The drug has properties ρ = 1030 kg/m³ and μ = 0.004 Pa-s. Determine the pressure elevation at the inlet of the tube (Δp) that is required to provide this flow.

Known Values

Figure 1 illustrates a sketch of the problem.

drug

$\dot{V} = 4 \times 10^{-7}$ m³/s (24 cm³/min)

ρ = 1030 kg/m³, μ = 0.004 Pa-s

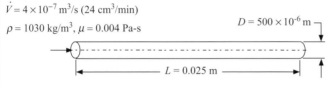

$D = 500 \times 10^{-6}$ m

$L = 0.025$ m

Figure I Sketch of the problem with the known values indicated.

Assumption

- The flow is steady state.

Analysis and Solution

The cross-sectional area of the tube is computed:

$$A_c = \frac{\pi}{4} D^2 = \frac{\pi}{4} \left(500 \times 10^{-6} \text{ m}\right)^2 = 1.96 \times 10^{-7} \text{ m}^2. \tag{1}$$

The mean velocity is computed:

$$u_m = \frac{\dot{V}}{A_c} = \frac{4 \times 10^{-7} \text{ m}^3}{\text{s}} \left| \frac{}{1.96 \times 10^{-7} \text{ m}^2} \right. = 2.04 \text{ m/s}. \tag{2}$$

The Reynolds number is computed:

$$Re_{D_h} = \frac{\rho \, u_m D_h}{\mu} = \frac{1030 \text{ kg}}{\text{m}^3} \left| \frac{2.04 \text{ m}}{\text{s}} \right| \frac{500 \times 10^{-6} \text{ m}}{} \left| \frac{}{0.004 \text{ Pa s}} \right| \left| \frac{\text{Pa m}^2}{\text{N}} \right| \frac{\text{N s}^2}{\text{kg m}} = 262. \tag{3}$$

Because the Reynolds number is significantly below the critical Reynolds number for internal flow (2300), the flow will be laminar. We therefore will need the average friction factor correlation associated with laminar flow through a round tube. The hydrodynamic entry length is estimated using Eq. (9.16):

$$x_{fd, h, lam} \approx 0.063 \, Re_{D_h} D = 0.063(262) \left(500 \times 10^{-6} \text{ m}\right) = 0.0083 \text{ m} \ (0.83 \text{ cm}). \tag{4}$$

Because the hydrodynamic entry length is significant relative to the tube length it will be necessary to include the impact of the developing region. Equation (9.56) is used to obtain the fully developed friction factor:

$$f_{fd} = \frac{64}{Re_{D_h}} = \frac{64}{262} = 0.244. \tag{5}$$

Equation (9.55) is used to obtain the dimensionless length of the tube:

$$\tilde{L} = \frac{L}{D_h \, Re_{D_h}} = \frac{0.025 \text{ m}}{500 \times 10^{-6} \text{ m}} \left| \frac{}{262} \right. = 0.191. \tag{6}$$

Example 9.5 (cont.)

Equation (9.57) is used to determine the ratio of the average friction factor to the fully developed friction factor:

$$\frac{\bar{f}}{f_{fd}} = 0.215\,\tilde{L}^{-0.5} + \frac{0.0196\,\tilde{L}^{-1} + 1 - 0.215\,\tilde{L}^{-0.5}}{1 + 0.00021\,\tilde{L}^{-2}}$$

$$= 0.215(0.191)^{-0.5} + \frac{0.0196(0.191)^{-1} + 1 - 0.215(0.191)^{-0.5}}{1 + 0.00021(0.191)^{-2}} = 1.099.$$

(7)

The average friction factor is computed according to:

$$\bar{f} = \frac{\bar{f}}{f_{fd}} f_{fd} = 1.099(0.244) = 0.268.$$

(8)

The pressure drop that occurs in the tube due to the flow is given by:

$$\Delta p = \bar{f} \frac{L}{D} \left(\frac{\rho\, u_m^2}{2} \right)$$

$$= \frac{0.268}{2} \left| \frac{0.025\ \mathrm{m}}{500 \times 10^{-6}\ \mathrm{m}} \right| \left| \frac{1030\ \mathrm{kg}}{\mathrm{m}^3} \right| \frac{(2.04)^2\ \mathrm{m}^2}{\mathrm{s}^2} \left| \frac{\mathrm{N\,s}^2}{\mathrm{kg\,m}} \right| \frac{\mathrm{Pa\,m}^2}{\mathrm{N}}$$

$$= \boxed{28{,}700\ \mathrm{Pa}\ (28.7\ \mathrm{kPa})}.$$

(9)

Discussion

Because the flow is laminar, the surface roughness of the tube is not an important factor in the calculation.

Rectangular Ducts

In the laminar, fully developed region, the product of the Reynolds number and the friction factor will be a constant that depends on the shape of the duct. For example, in a circular duct this constant is 64 according to Eq. (9.56). For a rectangular duct, the product $f_{fd}\,Re_{D_h}$ is a function of the aspect ratio, AR, which is defined as the ratio of the minimum to the maximum dimensions of the duct cross-section (as viewed in the direction of the flow). The value of $f_{fd}\,Re_{D_h}$ can be obtained from Eq. (9.58) or by using Figure 9.15:

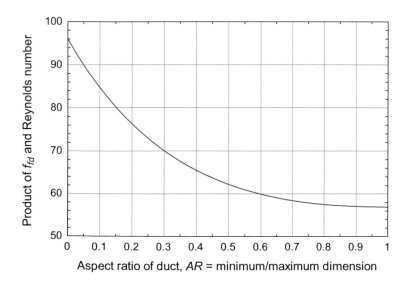

Figure 9.15 The product of the fully developed friction factor and the Reynolds number as a function of the aspect ratio for a rectangular duct.

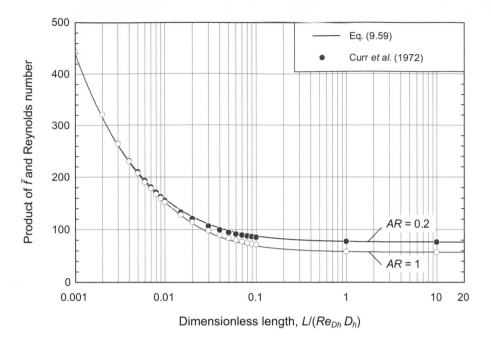

Figure 9.16 The product of the average friction factor in a rectangular duct and the Reynolds number as a function of $\tilde{L}$ for two values of the aspect ratio. The results predicted by Eqs. (9.58) and (9.59) are shown, as well as the numerical solution provided by Curr *et al.* (1972).

$$f_{fd}Re_{D_h} = 96\left(1 - 1.3553\,AR + 1.9467\,AR^2 - 1.7012\,AR^3 + 0.9564\,AR^4 - 0.2537\,AR^5\right). \tag{9.58}$$

As the aspect ratio approaches 0, the solution provided by Eq. (9.58) limits to $f_{fd} = 96/Re_{D_h}$ which is the solution for flow between two infinite parallel plates. The average friction factor (considering the developing region) for laminar flow in rectangular ducts can be approximately computed using the dimensionless tube length, $\tilde{L}$, defined by Eq. (9.55) and the fully developed friction factor calculated using Eq. (9.58):

$$\bar{f}\,Re_{D_h} \approx 13.76\,\tilde{L}^{-0.5} + \frac{1.25\,\tilde{L}^{-1} + \left(f_{fd}\,Re_{D_h}\right) - 13.76\,\tilde{L}^{-0.5}}{1 + 0.00021\,\tilde{L}^{-2}}. \tag{9.59}$$

Figure 9.16 illustrates the product of the apparent friction factor and the Reynolds number as a function of $\tilde{L}$ for two values of the aspect ratio. Also shown are the numerical results obtained by Curr *et al.* (1972).

Annular Duct

Flow in the space between concentric inner and outer tubes is often encountered in engineering applications. For an annular duct, the product of the fully developed friction factor and the Reynolds number is a function of the radius ratio, RR, defined as the ratio of the inner radius to the outer radius of the flow passage (Shah and London, 1978):

$$f_{fd}\,Re_{D_h} = \frac{64(1 - RR)^2}{1 + RR^2 - \left[\dfrac{1 - RR^2}{\ln\left(RR^{-1}\right)}\right]}. \tag{9.60}$$

Notice that as the radius ratio approaches 0, the solution provided by Eq. (9.60) limits to $64/Re_{D_h}$, which is the correct solution for flow through a circular duct. The solution provided by Eq. (9.60) limits to $96/Re_{D_h}$ as the radius ratio approaches unity, which is the correct solution for flow through parallel plates. Figure 9.17 illustrates $f_{fd}\,Re_{D_h}$ for an annular duct as a function of radius ratio predicted by Eq. (9.60).

The apparent friction factor can be obtained using the same approximation previously presented for a rectangular duct, Eq. (9.59), with f_{fd} obtained from Eq. (9.60). Figure 9.18 illustrates the product of the

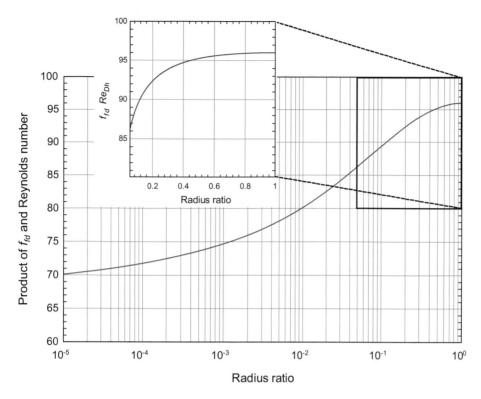

Figure 9.17 The product of the fully developed friction factor and the Reynolds number as a function of the radius ratio for an annular duct.

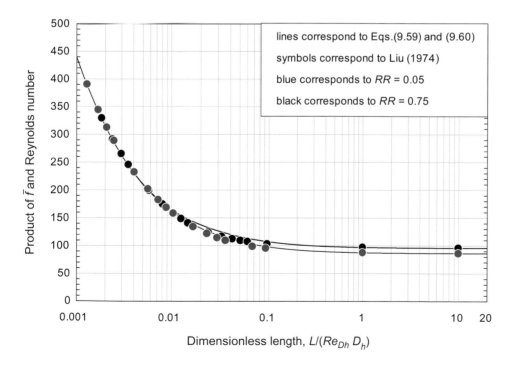

Figure 9.18 The product of the apparent friction factor for an annular duct and the Reynolds number as a function of $\tilde{L}$ for two values of the radius ratio. The results predicted by Eqs. (9.59) and (9.60) are shown, as well as the numerical solution provided by Liu (1974).

apparent friction factor and the Reynolds number as a function of $\tilde{L}$ for two values of the radius ratio. Also shown are the numerical results obtained by Liu (1974).

Example 9.6

A duct with an annular cross-section is formed by sliding an inner tube inside of an outer tube. Engine oil (10W) is forced to flow through this annular gap in a heat exchanger. The outer radius of the annular space is $r_{out} = 1.25$ inch and the inner radius is $r_{in} = 1.0$ inch. The total length of the passage is $L = 40$ ft. The oil pump that provides the flow is able to produce a pressure rise of $\Delta p = 5$ psi. Determine the volumetric flow rate of engine oil that will result.

Known Values

Figure 1 illustrates a sketch of the problem with the known values indicated. The properties of 10 W engine oil are obtained from Appendix B.

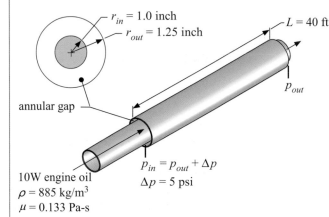

Figure I Sketch of the problem with the known values indicated.

$r_{in} = 1.0$ inch
$r_{out} = 1.25$ inch
$L = 40$ ft
P_{out}
annular gap
$p_{in} = p_{out} + \Delta p$
$\Delta p = 5$ psi
10W engine oil
$\rho = 885$ kg/m^3
$\mu = 0.133$ Pa-s

Assumptions

- The flow is steady state.
- The flow remains laminar.
- The properties from Appendix B at room temperature can be used.

Analysis

The cross-sectional area of the annular space is computed:

$$A_c = \pi\left(r_{out}^2 - r_{in}^2\right). \tag{1}$$

The perimeter of the annular space is determined:

$$per = 2\pi(r_{out} + r_{in}). \tag{2}$$

The hydraulic diameter of the passage is:

$$D_h = \frac{4\,A_c}{per}. \tag{3}$$

Example 9.6 (cont.)

The mean velocity is computed from

$$u_m = \frac{\dot{V}}{A_c},\qquad(4)$$

where $\dot{V}$ is the (unknown) volumetric flow rate. The Reynolds number is computed from the mean velocity and the hydraulic diameter:

$$Re_{D_h} = \frac{\rho\, u_m\, D_h}{\mu}.\qquad(5)$$

The radius ratio of the passage is given by

$$RR = \frac{r_{in}}{r_{out}}.\qquad(6)$$

Assuming the flow is laminar, the fully developed friction factor may be computed using Eq. (9.60):

$$f_{fd}\, Re_{D_h} = \frac{64(1-RR)^2}{1+RR^2 - \left[\dfrac{1-RR^2}{\ln\left(RR^{-1}\right)}\right]}.\qquad(7)$$

The average friction factor depends on the dimensionless length

$$\tilde{L} = \frac{L}{Re_{D_h} D_h}\qquad(8)$$

and the fully developed friction factor according to Eq. (9.59):

$$\bar{f}\, Re_{D_h} \approx 13.76\,\tilde{L}^{-0.5} + \frac{1.25\,\tilde{L}^{-1} + \left(f_{fd}\, Re_{D_h}\right) - 13.76\,\tilde{L}^{-0.5}}{1+0.00021\,\tilde{L}^{-2}}.\qquad(9)$$

The pressure drop across the pipe is related to the average friction factor according to:

$$\Delta p = \bar{f}\,\frac{L}{D_h}\frac{\rho\, u_m^2}{2}.\qquad(10)$$

Solution

Equations (1) through (10) are ten equations in the ten unknowns: A_c, per, D_h, u_m, $\dot{V}$, Re_{D_h}, RR, f_{fd}, $\tilde{L}$, and $\bar{f}$. These equations are not explicit and therefore some iterative technique will be required to provide a solution. This is typical of an internal flow problem; the pressure drop is known and the volumetric flow rate is not. We saw in Example 9.5 that when the flow rate is known, the problem becomes explicit for the pressure drop. Therefore, the best way of solving this problem (either by hand or using a computer program) is to "guess" a value of the flow rate and solve the problem, leading to the calculation of the pressure drop. Finally, the guessed value of the flow rate can be adjusted in order to obtain the required pressure drop. In this example, the EES software is used in order to facilitate the iterative process of finding a solution.

The inputs are entered in EES and converted to base SI units.

```
$UnitSystem SI Mass J K Pa
r_out=1.25 [inch]*Convert(inch,m)        "outer radius of annular space"
r_in=1.0 [inch]*Convert(inch,m)          "inner radius of annular space"
L=40 [ft]*Convert(ft,m)                  "length of duct"
DELTAp= 5 [psi]*Convert(psi,Pa)          "pressure drop across the duct"
rho=885 [kg/m^3]                         "Density"
mu= 0.133 [Pa-s]                         "Viscosity"
```

Continued

Example 9.6 (cont.)

A reasonable value for the volumetric flow rate is assumed; note that it will eventually be necessary to remove this line of code to iterate to the correct solution.

V_dot=1e-3 [m^3/s]	"assumed volumetric flow rate"

Equations (1) through (10) are entered, leading to the calculation of the pressure drop that corresponds to the assumed value of flow rate (Δp_c).

A_c=pi*(r_out^2-r_in^2)	"cross-sectional area"
per=2*pi*(r_out+r_in)	"perimeter"
D_h=4*A_c/per	"hydraulic diameter"
u_m=V_dot/A_c	"mean velocity"
Re_Dh=rho*D_h*u_m/mu	"Reynolds number"
RR=r_in/r_out	"radius ratio"
f_fd*Re_Dh=64*(1-RR)^2/(1+RR^2-((1-RR^2)/Ln(1/RR)))	"fully developed friction factor"
L_tilde=L/(Re_Dh*D_h)	"dimensionless length"
f_bar*Re_Dh=13.76/**Sqrt**(L_tilde)+(1.25/L_tilde+f_fd*Re_Dh-&	
13.76/**Sqrt**(L_tilde))/(1+0.00021/L_tilde^2)	"average friction factor"
DELTAp_c=f_bar*L/D_h*rho*u_m^2/2	"calculated pressure drop"

Solving provides $\Delta p_c = 423,348$ Pa (61.4 psi) which is higher than the pressure drop specified in the problem statement, $\Delta p = 34,500$ Pa (5 psi). Therefore, the guessed value for the volumetric flow rate is somewhat higher than the actual value. Update the guess values used for the calculation process by selecting Update Guesses from the Calculate menu. Comment out the assumed value of $\dot{V}$ and replace this equation with the requirement that $\Delta p = \Delta p_c$. The resulting code is shown below.

$UnitSystem SI Mass J K Pa	
r_out=1.25 [inch]***Convert**(inch,m)	"outer radius of annular space"
r_in=1.0 [inch]***Convert**(inch,m)	"inner radius of annular space"
L=40 [ft]***Convert**(ft,m)	"length of duct"
DELTAp= 5 [psi]***Convert**(psi,Pa)	"pressure drop across the duct"
rho=885 [kg/m^3]	"Density"
mu= 0.133 [Pa-s]	"Viscosity"
//V_dot=1e-3 [m^3/s]	"assumed volumetric flow rate"
A_c=pi*(r_out^2-r_in^2)	"cross-sectional area"
per=2*pi*(r_out+r_in)	"perimeter"
D_h=4*A_c/per	"hydraulic diameter"
u_m=V_dot/A_c	"mean velocity"
Re_Dh=rho*D_h*u_m/mu	"Reynolds number"
RR=r_in/r_out	"radius ratio"
f_fd*Re_Dh=64*(1-RR)^2/(1+RR^2-((1-RR^2)/**Ln**(1/RR)))	"fully developed friction factor"
L_tilde=L/(Re_Dh*D_h)	"dimensionless length"
f_bar*Re_Dh=13.76/**Sqrt**(L_tilde)+(1.25/L_tilde+f_fd*Re_Dh-&	
13.76/**Sqrt**(L_tilde))/(1+0.00021/L_tilde^2)	"average friction factor"
DELTAp_c=f_bar*L/D_h*rho*u_m^2/2	"calculated pressure drop"
DELTAp_c=DELTAp	"set calculated Dp to specified Dp"

Example 9.6 (cont.)

Solving provides $\boxed{\dot{V} = 8.15 \times 10^{-5}\,\text{m}^3/\text{s}\,(1.29\,\text{gal/min})}$.

Discussion

The Reynolds number associated with the solution is $Re_{D_h} = 6.0$ which indicates that the assumption of laminar flow is justified. One advantage of using a computer program is that it is possible to run parametric studies once the problem has been solved. For example, Figure 2 illustrates the pressure drop as a function of flow rate for the annular duct analyzed here.

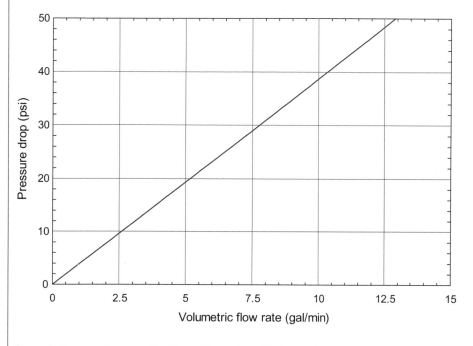

Figure 2 Pressure drop as a function of the volumetric flow rate.

In many engineering problems, neither the pressure drop nor the flow rate is provided. Rather, information is known about the pump that is energizing the system, typically in the form of a pump curve. The **pump curve** relates pressure drop and flow rate for the pump and the engineer must find the combination of these quantities that satisfies both the behavior of the system that the pump is connected to (i.e., the **system resistance curve** shown in Figure 2) as well as the pump itself. This situation is illustrated in Example 9.7.

Turbulent Flow

Figure 9.13 shows that the fully developed friction factor for a turbulent flow is not strongly affected by the duct shape but is sensitive to the scale of the surface roughness (e). The friction factor for fully developed turbulent flow in an aerodynamically smooth duct is provided by Petukhov *et al.* (1970):

$$f_{fd} = \frac{1}{[0.790\,\ln(Re_{D_h}) - 1.64]^2} \text{ for } 3000 < Re_{D_h} < 5.0 \times 10^6 \text{ with } \frac{e}{D_h} = 0. \quad (9.61)$$

More recently, Li *et al.* (2011) provided a more accurate, explicit correlation for turbulent flow in smooth tubes:

$$f_{fd} = 4 \left[\frac{C_1}{\ln (Re_{D_h})} + \frac{C_2}{(\ln (Re_{D_h}))^2} + \frac{C_3}{(\ln (Re_{D_h}))^3} \right]$$

$$C_1 = -0.0015702, \ C_2 = 0.3942031, \ C_3 = 2.5341533$$

$$\text{for } 4000 < Re_{D_h} < 1 \times 10^7 \text{ with } \frac{e}{D_h} = 0. \tag{9.62}$$

Colebrook (1939) presents an implicit expression for the fully developed, turbulent friction factor in a duct that has significant surface roughness, e (referred to as a *rough* duct):

$$\frac{1}{\sqrt{f_{fd}}} = -2 \log_{10} \left(\frac{e}{3.71 \, D_h} + \frac{2.51}{Re_{D_h} \sqrt{f_{fd}}} \right). \tag{9.63}$$

Zigrang and Sylvester (1982) present an explicit and therefore more convenient correlation for the fully developed, turbulent friction factor in a rough duct:

$$f_{fd} = \left\{ -2.0 \log_{10} \left[\frac{2e}{7.54 \, D_h} - \frac{5.02}{Re_{D_h}} \log_{10} \left(\frac{2e}{7.54 \, D_h} + \frac{13}{Re_{D_h}} \right) \right] \right\}^{-2}$$

$$4 \times 10^3 < Re_{D_h} < 1 \times 10^8, \ 4 \times 10^{-5} < \frac{e}{D_h} < 5 \times 10^{-2}. \tag{9.64}$$

Offor and Alabi (2016) present a more accurate but still explicit correlation:

$$f_{fd} = \left\{ -2.0 \log_{10} \left[\frac{e}{C_1 D_h} - \frac{C_2}{Re_{D_h}} \left(\ln \left(\left(\frac{e/D_h}{C_3} \right)^{C_4} + \left(\frac{C_5}{Re_{D_h} + C_6} \right) \right) \right) \right] \right\}^{C_7}$$

$$C_1 = 3.71, \ C_2 = -1.975, \ C_3 = 3.93, \ C_4 = 1.092, \ C_5 = 7.627, \ C_6 = 395.9, \ C_7 = -2$$

$$4 \times 10^3 < Re_{D_h} < 1 \times 10^8, \ 1 \times 10^{-6} < \frac{e}{D_h} < 5 \times 10^{-2}. \tag{9.65}$$

Because the turbulent entry length is usually very short, as discussed in Section 9.1.1, most correlations for turbulent flow ignore the additional pressure drop incurred in the hydrodynamically developing region and assume that the average friction factor is approximately equal to the fully developed value, $\bar{f} \approx f_{fd}$. However, the effect of the developing region on the apparent friction factor can be approximately accounted for by using:

$$\bar{f} \approx f_{fd} \left(1 + \left(\frac{D_h}{L} \right)^{0.7} \right). \tag{9.66}$$

Notice that the average friction factor is only slightly higher than the fully developed friction factor and quickly approaches f_{fd} as L/D_h increases.

The effect of surface roughness on a turbulent flow is consistent with its impact on the external flow over a plate, discussed in Section 8.2.1. The velocity gradient is concentrated in a region very near the wall referred to as the viscous sublayer. The viscous sublayer, δ_{vs}, is quite thin and therefore it is often the case that engineering materials can have roughness elements that are on this order or larger. Notice there is a line in Figure 9.13 that delineates the **fully rough region**. Within the fully rough region the friction factor is independent of Reynolds number and only depends on the size of the roughness; the friction factor in this region can be obtained from the correlation below (Nikuradse, 1933):

$$f_{fd} = \frac{1}{\left[1.74 + 2.0 \log_{10} \left(\frac{D_h}{2 \, e} \right) \right]^2} \quad \text{fully rough region.} \tag{9.67}$$

In the transitionally rough region both roughness and Reynolds number affect the pressure drop. Typical roughness values for various surfaces are listed in Table 9.1.

Table 9.1 Typical roughness of commercial pipes (various sources).

Material	Roughness, e
Riveted steel	0.9–9.0 mm
Galvanized steel	0.15 mm
Forged steel	0.045 mm
New cast iron	0.26–0.80 mm
Rusty cast iron	1.5–2.5 mm
Drawn tubing	0.0015 mm
Concrete	0.3–3.0 mm
Glass	smooth
PVC and plastic pipes	0.0015–0.007 mm
Wood	0.5 mm
Rubber	0.01 mm

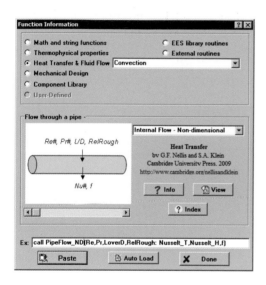

Figure 9.19 Function dialog for Internal Flow Convection Library.

EES' Internal Flow Convection Library

The friction factor correlations discussed in this section (as well as others) have been implemented in EES in order to facilitate solving internal flow problems. The Internal Flow Convection Library can be accessed by selecting Function Information from the Options menu and then selecting Convection from the list of options in the Heat Transfer pulldown menu. Select Internal Flow – Non-dimensional and scroll through the available duct shapes, as shown in Figure 9.19.

The Internal Flow – Non-dimensional Library implements correlations for the average friction factor in ducts of various shapes (as well as the Nusselt number for constant temperature and constant heat flux boundary conditions, discussed in Section 9.2.4). The companion library, Internal Flow – Non-dim (local), includes procedures that return the local friction factor (and Nusselt numbers) for the same duct geometries. Procedures are accessed using the Call command. For example, to access the correlations for the average friction factor for circular ducts, it is necessary to call the procedure PipeFlow_ND as shown below.

```
$UnitSystem SI MASS RAD PA K J
Re = 1000 [-]                              "Reynolds number"
Pr = 1 [-]                                 "Prandtl number"
```

LoverD = 100 [-]   "length to diameter ratio"
RelRough = 1e-4 [-]   "relative roughness"
CALL **PipeFlow_ND**(Re,Pr,LoverD,RelRough: Nusselt_T_bar,Nusselt_H_bar,f_bar)
                                                                   "access correlation"

Solving leads to $\bar{f} = 0.0758$; this result is approximately consistent with the value indicated in Figure 9.13 (it is slightly higher, due to the effect of the hydrodynamically developing region). This value could also be obtained manually by using Eqs. (9.56) and (9.57). The EES library procedures identify whether the flow is laminar or turbulent and which of the correlations should be used. The arguments to the left of the colon in the procedure are inputs and those to the right are outputs; however, it is possible to specify an output, such as the friction factor, and have the procedure determine one of the inputs (provided that the guess values and limits for the variables are appropriately set). The inputs to the procedure include Reynolds number, Prandtl number, length to diameter ratio (L/D_h), and the relative roughness (e/D_h). The outputs are the average Nusselt numbers assuming constant wall temperature and constant heat flux (which provide lower and upper bounds, as discussed in the subsequent section), and the average friction factor. The procedure models the transition from laminar to turbulence (i.e., the region between a Reynolds number of 2300 and 3000) by determining the fully laminar and fully turbulent results and interpolating between these values; this gradual rather than abrupt transition will help prevent instability and convergence problems in many solutions. Further details regarding the procedure can be obtained by examining the Help information in EES.

Example 9.7

A pump is used to fill an open tank with water. The inlet to the pump pulls water directly from a reservoir at $T_o = 70°F$ and $p_o = 1$ atm. The pump discharge is connected to a length of new cast iron pipe with inner diameter $D = 1.5$ inch and length $L = 500$ ft. The exit of the pipe discharges into the tank at approximately the same elevation as the reservoir. The volume of the tank is $V = 600$ gal. The pump behavior is characterized by the pump curve in Figure 1, which shows the pressure rise produced by the pump (Δp_{pump}) as a function of the flow rate that it provides ($\dot{V}$).

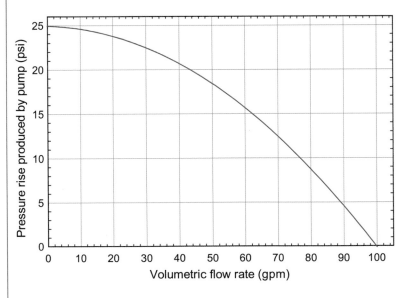

Figure 1 Pump curve.

Example 9.7 (cont.)

The pump curve shown in Figure 1 is described by the equation:

$$\frac{\Delta p_{pump}}{\Delta p_{dh}} = 1 - 0.0353 \left(\frac{\dot{V}}{\dot{V}_{oc}}\right) - 0.9647 \left(\frac{\dot{V}}{\dot{V}_{oc}}\right)^2, \tag{1}$$

where $\Delta p_{dh} = 25$ psi is the pressure rise produced by the pump when it is "dead-headed" (that is, when it is producing no net flow) and $\dot{V}_{oc} = 100$ gal/min is the flow rate produced by the pump in an open circuit condition (i.e., when it is producing no pressure rise).

Determine the time required to fill the tank with this pump.

Known Values

Figure 2 illustrates a sketch of the problem with the known values indicated.

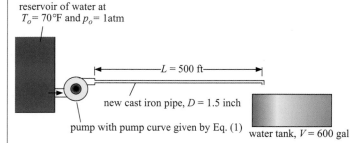

reservoir of water at
$T_o = 70°F$ and $p_o = 1$ atm

$-L = 500$ ft$-$

new cast iron pipe, $D = 1.5$ inch

pump with pump curve given by Eq. (1)

water tank, $V = 600$ gal

Figure 2 Sketch of the problem with the known values indicated.

Assumptions

- The flow is steady.
- The properties of water at T_o and p_o can be used; this implies that the properties do not change significantly as the water passes through the pump and the pipe.
- The pressure drop caused by any minor losses (i.e., bends, contractions and expansions) are neglected.
- The roughness of new cast iron pipe can be estimated from Table 9.1.
- Water is treated as being incompressible.

Analysis

The density and viscosity of water are determined at the reservoir temperature and pressure:

$$\rho = \rho(T_o, p_o) \tag{2}$$
$$\mu = \mu(T_o, p_o). \tag{3}$$

The roughness of new cast iron pipe is estimated from Table 9.1 to be $e = 0.5$ mm. The cross-sectional area of the pipe is computed:

$$A_c = \pi \frac{D^2}{4}. \tag{4}$$

The mean velocity is related to the flow rate through the pipe according to:

$$u_m = \frac{\dot{V}}{A_c}. \tag{5}$$

The Reynolds number is computed from the mean velocity and the diameter:

$$Re_{D_h} = \frac{\rho u_m D}{\mu}. \tag{6}$$

Continued

Example 9.7 (cont.)

The average friction factor for this circular duct is a function of the Reynolds number, length to diameter ratio, and relative roughness:

$$\bar{f} = \bar{f}\left(Re_{D_h}, \frac{L}{D_h}, \frac{e}{D_h} \right). \tag{7}$$

The pressure drop across the pipe is computed from the average friction factor:

$$\Delta p = \bar{f} \frac{L}{D_h} \frac{\rho\, u_m^2}{2}. \tag{8}$$

The reservoir and the open tank are both at atmospheric pressure. Therefore, the pressure rise produced by the pump must equal the pressure drop through the pipe:

$$\Delta p_{pump} = \Delta p. \tag{9}$$

The time required to fill the tank is given by:

$$t = \frac{V}{\dot{V}}. \tag{10}$$

Solution

Equations (1) through (10) are ten equations in the ten unknowns: Δp_{pump}, $\dot{V}$, ρ, μ, A_c, u_m, Re_{D_h}, $\bar{f}$, Δp, and t. These equations are not explicit and therefore some iteration will be required to provide a solution. As in Example 9.6, the best way of solving this problem is to "guess" a value of the flow rate and solve the problem, leading to the calculation of the pressure drop. Finally, the guessed value of the flow rate can be adjusted in order to obtain the required pressure drop.

The inputs are entered in EES and converted to base SI units.

```
$UnitSystem SI MASS RAD PA K J
DELTAp_dh=25 [psi]*Convert(psi,Pa)          "dead head pressure rise for pump"
V_dot_oc=100 [gal/min]*Convert(gal/min,m^3/s)  "open circuit flow rate for pump"
L=500 [ft]*Convert(ft,m)                     "length of pipe"
D=1.5 [inch]*Convert(inch,m)                 "diameter of pipe"
V=600 [gal]*Convert(gal,m^3)                 "volume of tank"
To=ConvertTemp(F,K,70 [F])                   "temperature of water"
```

A reasonable value for the volumetric flow rate is assumed; note that it will eventually be necessary to remove this line of code in order to iterate to the correct solution.

```
V_dot=20 [gal/min]*Convert(gal/min,m^3/s)    "guess for the flow rate"
```

Equations (1) through (8) are entered, leading to the calculation of the pressure rise produced by the pump and also the pressure drop across the pipe; both of these values correspond to the assumed value of flow rate.

```
DELTAp_pump=DELTAp_dh*(1-0.0353*(V_dot/V_dot_oc)-0.9647*(V_dot/V_dot_oc)^2)
                                             "pump curve"
rho=Density(Water,T=To,P=Po#)                "density"
mu=Viscosity(Water,T=To,P=Po#)               "viscosity"
e=0.5 [mm]*Convert(mm,m)                      "roughness"
A_c=pi*D^2/4                                  "area for flow"
u_m=V_dot/A_c                                 "mean velocity"
```

Example 9.7 (cont.)

```
Re_Dh=rho*u_m*D/mu                              "Reynolds number"
CALL PipeFlow_ND(Re_Dh,1 [-],L/D,e/D: , ,f_bar)  "access correlation"
DELTAp=f_bar*L/D*rho*u_m^2/2                     "pressure drop through pipe"
```

Solving provides Δp_{pump} = 164,500 Pa (23.9 psi) and Δp = 104,800 Pa (15.2 psi). The pump pressure rise is higher than the pipe pressure drop, suggesting that the guessed value of the flow rate is too low. The flow rate can now be adjusted in order to bring these two quantities into agreement. Physically, this process corresponds to finding the intersection between the pump curve shown in Figure 1 and the system resistance curve which relates the flow rate through the pipe to its pressure drop. Figure 3 overlays the system resistance curve onto the pump curve and shows that the operating point corresponds approximately to a flow rate of 25 gal/min and a pump pressure rise of 23.5 psi.

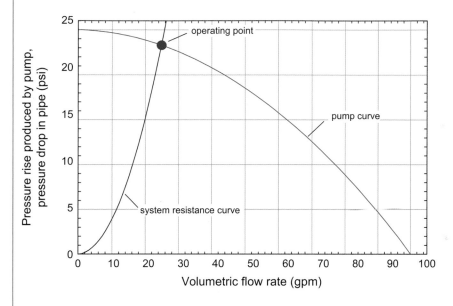

Figure 3 System resistance curve overlaid onto the pump curve from Figure 1 in order to show the operating point.

Update the guess values used for the calculation process by selecting Update Guesses from the Calculate menu. Comment out the assumed value of $\dot{V}$ and replace this equation with Eq. (9). The resulting code is shown below.

```
$UnitSystem SI MASS RAD PA K J
DELTAp_dh=25 [psi]*Convert(psi,Pa)              "dead head pressure rise for pump"
V_dot_oc=100 [gal/min]*Convert(gal/min,m^3/s)   "open circuit flow rate for pump"
L=500 [ft]*Convert(ft,m)                        "length of pipe"
D=1.5 [inch]*Convert(inch,m)                    "diameter of pipe"
V=600 [gal]*Convert(gal,m^3)                    "volume of tank"
To=ConvertTemp(F,K,70 [F])                      "temperature of water"

{V_dot=20 [gal/min]*Convert(gal/min,m^3/s)      "guess for the flow rate"}
```

Continued

Example 9.7 (cont.)

```
DELTAp_pump=DELTAp_dh*(1-0.0353*(V_dot/V_dot_oc)-0.9647*(V_dot/V_dot_oc)^2)
                                                    "pump curve"
rho=Density(Water,T=To,P=Po#)                       "density"
mu=Viscosity(Water,T=To,P=Po#)                      "viscosity"
e=0.5 [mm]*Convert(mm,m)                             "roughness"
A_c=pi*D^2/4                                         "area for flow"
u_m=V_dot/A_c                                        "mean velocity"
Re_Dh=rho*u_m*D/mu                                   "Reynolds number"
CALL PipeFlow_ND(Re_Dh,1 [-],L/D,e/D: , ,f_bar)     "access correlation"
DELTAp=f_bar*L/D*rho*u_m^2/2                         "pressure drop through pipe"
DELTAp=DELTAp_pump                                  "find the intersection point"
time=V/V_dot                                         "time to fill tank"
```

Solving the code shows that the time required to fill the tanks is $\boxed{t = 1451 \text{ s} \, (24.2 \text{ min})}$.

Discussion

The Reynolds number associated with the solution is $Re_{D_h} = 43{,}100$ which indicates that the flow is turbulent and therefore the roughness of the cast iron pipe wall is probably an important parameter.

Again, an advantage of using a computer to solve the problem is that it is possible to use this model for design purposes. For example, Figure 4 illustrates the volumetric flow rate produced as a function of the diameter of the pipe. At small diameters, the system is very restrictive which corresponds to moving the system resistance curve in Figure 3 more towards the left. As a result the flow approaches zero. At large diameters, the system is not restrictive and the system resistance curve in Figure 3 moves towards the right. Eventually, the flow approaches the open circuit value associated with the pump of 100 gal/min.

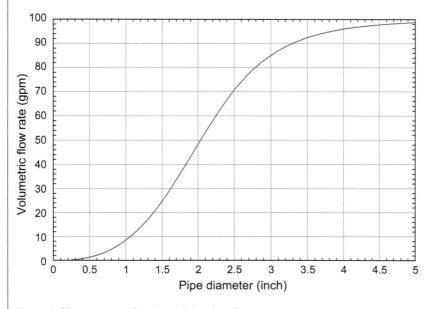

Figure 4 Flow rate as a function of the pipe diameter.

9.2.4 The Nusselt Number

The local Nusselt number in the fully developed region is shown in Figure 9.20 as a function of Reynolds number for various duct shapes and thermal boundary conditions (in the laminar region) and as a function of wall roughness and Prandtl number (in the turbulent region).

Laminar Flow

Figure 9.20 shows that the Nusselt number (and thus the heat transfer coefficient) for a fully developed laminar flow is affected by the duct shape and thermal boundary conditions but not by the surface roughness, Prandtl number, or even the Reynolds number. This behavior is consistent with the discussion of the heat transfer coefficient in Section 9.1.2. The heat transfer coefficient in a laminar flow is approximately equal to the ratio of the thermal conductivity of the fluid to the thermal boundary layer thickness, $h \approx k/\delta_t$. In the fully developed, laminar region of a duct, the thermal boundary layer is, approximately, half the duct width, $\delta_t \approx D_h/2$. Substituting these approximate relationships into the definition of the Nusselt number leads to:

$$Nu_{D_h} = \frac{h\,D_h}{k} \approx \frac{2\,k\,D_h}{D_h\,k} \approx 2. \tag{9.68}$$

In fact, Figure 9.20 shows that the fully developed Nusselt number can range anywhere from 3.66 (for a circular tube with a constant temperature) to 8.24 (for flow between infinite parallel plates with constant heat flux). However, Eq. (9.68) does explain why the Nusselt number is independent of Reynolds number and Prandtl number and also why its order of magnitude is not very different than 1.

Circular Tubes

The local Nusselt number for a laminar, hydrodynamically and thermally fully developed flow in a circular tube depends on the thermal boundary condition. For a uniform heat flux (indicated by the subscript H), the Nusselt number is:

$$Nu_{D_h, H, fd} = 4.36 \ \ \text{for } Re_{D_h} < 2300 \text{ and } x > x_{fd, t, lam}. \tag{9.69}$$

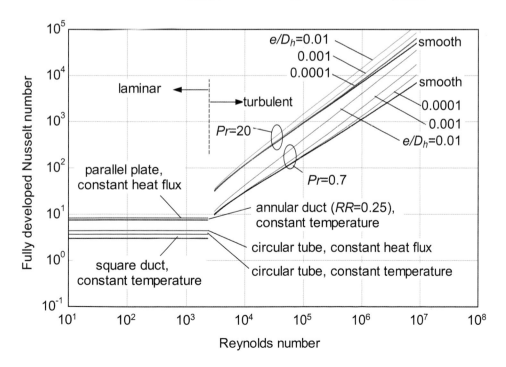

Figure 9.20 Local, fully developed Nusselt number as a function of Reynolds number for various duct shapes and thermal boundary conditions (in the laminar region) and relative roughness and Prandtl number (in the turbulent region).

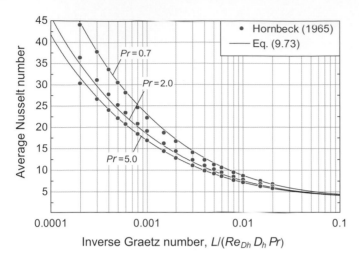

Figure 9.21 Average Nusselt number in a simultaneously developing laminar flow within a circular pipe with a constant surface temperature as a function of the inverse Graetz number ($\tilde{L}/Pr$) for various values of the Prandtl number. The graphical results presented by Hornbeck (1965) are shown, as well as the correlation provided by Eq. (9.72).

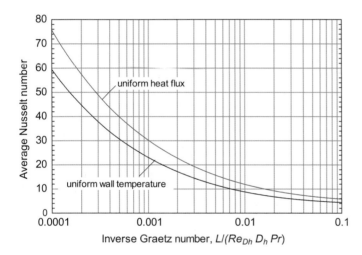

Figure 9.22 Average Nusselt number in a simultaneously developing laminar flow within a circular pipe with a constant surface temperature and a constant heat flux boundary condition as a function of the inverse Graetz number ($\tilde{L}/Pr$) for $Pr = 0.70$.

For a uniform wall temperature (indicated by the subscript T), the fully developed Nusselt number is (Kays and Crawford, 1993):

$$Nu_{D_h, T, fd} = 3.66 \text{ for } Re_{D_h} < 2300 \text{ and } x > x_{fd, t, lam}. \tag{9.70}$$

The average Nusselt number depends on the entrance conditions. Figure 9.21 illustrates the average Nusselt number for simultaneously developing flow (i.e., the thermal and momentum boundary layers are both developing together) with a constant wall temperature for various values of the Prandtl number as a function of the inverse of the Graetz number, Gz. The Graetz number is defined as:

$$Gz = \frac{Pr}{\tilde{L}} = \frac{D_h \, Re_{D_h} Pr}{L}. \tag{9.71}$$

The symbols in Figure 9.21 correspond to graphical results presented by Hornbeck (1965). The correlation provided by Eq. (9.72) is also shown in Figure 9.21 and provides an adequate fit to these results. Notice that the correlation limits to the fully developed Nusselt number as Gz approaches 0 (i.e., as $\tilde{L}/Pr$ becomes large):

$$\overline{Nu}_{D_h, T} = 3.66 + \frac{\left(0.049 + \dfrac{0.020}{Pr}\right) Gz^{1.12}}{\left(1 + 0.065 \, Gz^{0.7}\right)} \text{ for } Re_{D_h} < 2300. \tag{9.72}$$

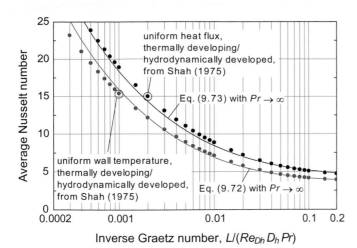

Figure 9.23 Average Nusselt number as a function of the inverse Graetz number ($\tilde{L}/Pr$) for a thermally developing/hydrodynamically developed flow from Shah (1975). Also shown are the results predicted by the correlations for the simultaneously developing flow, Eqs. (9.72) and (9.73), in the limit that $Pr \rightarrow \infty$.

The correlation

$$\overline{Nu}_{D_h, H} = 4.36 + \frac{\left(0.1156 + \dfrac{0.08569}{Pr^{0.4}}\right) Gz}{\left(1 + 0.1158\, Gz^{0.6}\right)} \quad \text{for } Re_{D_h} < 2300 \tag{9.73}$$

provides the average Nusselt number for simultaneously developing flow in a duct that is exposed to a uniform heat flux (see also Figure 9.22). Equation (9.73) is based on integrating and fitting data provided by Hornbeck (1965) and presented in Shah and London (1978). The average heat transfer coefficient for the constant temperature and constant heat flux conditions provide natural bounding cases for arbitrary boundary conditions and therefore it is generally useful to compute both values. Figure 9.23 illustrates the average Nusselt number as a function of the inverse Graetz number calculated using Eqs. (9.72) and (9.73) for a Prandtl number of 0.70. The actual heat transfer coefficient is likely to fall between these bounds.

In some situations, the flow will be hydrodynamically fully developed at the point where it begins to develop thermally; for example, if a heated section is placed far downstream of the inlet to a pipe. Figure 9.23 shows the average Nusselt number for a thermally developing/hydrodynamically developed flow as a function of the inverse Graetz number. Both the constant temperature and constant heat flux boundary condition solutions are shown based on the results presented in Shah (1975). Notice that the Prandtl number does not affect the solution because the momentum boundary layer does not change with position in a hydrodynamically fully developed flow. In the limit that the Prandtl number is very large, the simultaneously developing solution will approach the thermally developing/hydrodynamically developed result because the momentum boundary layer grows much faster than the thermal boundary layer for a high Prandtl number fluid. The average Nusselt number predicted using Eqs. (9.72) and (9.73) for a simultaneously developing flow in the limit that $Pr \rightarrow \infty$ is also shown in Figure 9.23 to illustrate this behavior.

The EES library of internal forced convection procedures provides the Nusselt number for simultaneously developing flow under conditions of both constant heat flux and temperature; these provide upper and lower bounds on the result, respectively. The average Nusselt number for a circular duct may be accessed using the PipeFlow_ND procedure and the local Nusselt number may be obtained using the PipeFlow_N_local procedure (discussed previously in Section 9.3.3). The local Nusselt number provided by the EES procedure is obtained by numerically differentiating Eqs. (9.72) and (9.73) according to Eq. (9.51). It is possible to obtain results for laminar flow that are consistent with a thermally developing/hydrodynamically developed flow by calling EES' internal convection procedures with a very large Prandtl number.

Example 9.8

A circular pipe carries an ethylene glycol/water solution (40 percent by volume). The pipe is wrapped in electrical heat tape that provides approximately a constant heat flux at the surface. The inner diameter of

Continued

Example 9.8 (cont.)

the pipe is $D = 1$ cm and the length is $L = 5$ m. The mass flow rate of the ethylene glycol solution is $\dot{m} = 0.035$ kg/s. Determine the average heat transfer coefficient between the fluid and the inner surface of the pipe.

Known Values

Figure 1 illustrates a sketch of the problem. The properties of a 40 percent ethylene glycol solution are obtained from Appendix B: $\rho = 1049$ kg/m³, $k = 0.4291$ W/m-K, $\mu = 0.002424$ Pa-s, and $Pr = 19.99$.

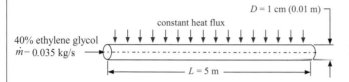

Figure 1 Sketch of the problem with the known values indicated.

Assumptions

- The flow is steady state.
- The properties in Appendix B can be used.

Analysis and Solution

The cross-sectional area of the tube is computed:

$$A_c = \frac{\pi}{4} D^2 = \frac{\pi}{4} (0.01 \text{ m})^2 = 7.85 \times 10^{-5} \text{ m}^2. \tag{1}$$

The mean velocity is computed:

$$u_m = \frac{\dot{m}}{\rho A_c} = \frac{0.035 \text{ kg}}{\text{s}} \left| \frac{\text{m}^3}{1049 \text{ kg}} \right| \frac{1}{7.85 \times 10^{-5} \text{ m}^2} = 0.425 \text{ m/s}. \tag{2}$$

The Reynolds number is computed:

$$Re_{D_h} = \frac{\rho u_m D_h}{\mu} = \frac{1049 \text{ kg}}{\text{m}^3} \left| \frac{0.425 \text{ m}}{\text{s}} \right| 0.01 \text{ m} \left| \frac{1}{0.002424 \text{ Pa s}} \right| \left| \frac{\text{Pa m}^2}{\text{N}} \right| \frac{\text{N s}^2}{\text{kg m}} = 1840. \tag{3}$$

Because the Reynolds number is below the critical Reynolds number for internal flow (2300), the flow will likely be laminar. We therefore will need the average Nusselt number correlation associated with simultaneously developing flow through a round tube exposed to a constant heat flux. The Graetz number is computed according to Eq. (9.71) and the average heat transfer coefficient is computed using Eq. (9.73):

$$Gz = \frac{D_h Re_{D_h} Pr}{L} = \frac{0.01 \text{ m}}{} \left| \frac{1840}{} \right| \frac{19.99}{5 \text{ m}} = 73.5 \tag{4}$$

$$\overline{Nu}_{D_h, H} = 4.36 + \frac{\left(0.1156 + \dfrac{0.08569}{Pr^{0.4}} \right) Gz}{\left(1 + 0.1158 \, Gz^{0.6} \right)} = 4.36 + \frac{\left[0.1156 + \dfrac{0.08569}{(19.99)^{0.4}} \right] 73.5}{\left[1 + 0.1158 (73.5)^{0.6} \right]} = 8.48. \tag{5}$$

The average heat transfer coefficient is calculated according to:

$$\bar{h} = \frac{\overline{Nu}_{D_h, H} k}{D} = \frac{8.48}{} \left| \frac{0.4291 \text{ W}}{\text{m K}} \right| \frac{1}{0.01 \text{ m}} = \boxed{364 \frac{\text{W}}{\text{m}^2 \text{ K}}}. \tag{6}$$

Discussion

Because the flow is laminar, the surface roughness of the tube is not an important factor in the calculation.

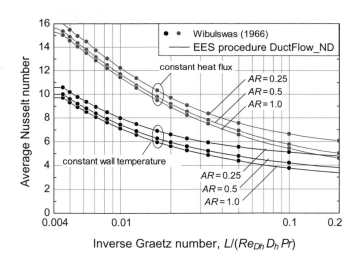

Figure 9.24 Average Nusselt number in a rectangular duct with simultaneously developing laminar flow as a function of the inverse Graetz number ($\tilde{L}/Pr$) for various values of aspect ratio and $Pr = 0.72$. Results are shown for uniform heat flux and uniform wall temperature from Wibulswas (1966) and also from the EES procedure DuctFlow_ND.

Rectangular Ducts

Shah and London (1978) provide the local Nusselt number for a laminar, hydrodynamically and thermally fully developed flow in a rectangular duct. The constant temperature result is correlated by

$$Nu_{D_h,\,T,fd} = 7.541\left(1 - 2.610\,AR + 4.970\,AR^2 - 5.119\,AR^3 + 2.702\,AR^4 - 0.548\,AR^5\right) \qquad (9.74)$$

and the constant heat flux result is correlated by

$$Nu_{D_h,\,H,fd} = 8.235\left(1 - 2.042\,AR + 3.085\,AR^2 - 2.477\,AR^3 + 1.058\,AR^4 - 0.186\,AR^5\right), \qquad (9.75)$$

where AR is the aspect ratio of the duct (the ratio of the minimum to the maximum dimensions of the duct cross-section).

Figure 9.24 illustrates the average Nusselt number for a simultaneously developing flow in a rectangular duct exposed to a constant wall temperature and constant heat flux with $Pr = 0.72$ for various values of the aspect ratio as a function of the inverse Graetz number (Wibulswas, 1966). Also shown in Figure 9.24 are the predictions obtained from the EES procedure DuctFlow_ND that is part of the Heat Transfer Library. The procedure DuctFlow_ND interpolates a table of data provided by Kakaç et al. (1987) and corrects for Prandtl number effects by applying the correction associated with a square duct, also from Kakaç et al. (1987).

Figure 9.25 illustrates the average Nusselt number in a rectangular duct exposed to a constant heat flux that is thermally developing but hydrodynamically fully developed for various values of the aspect ratio (Wibulswas, 1966). Also shown in Figure 9.25 are the results from the DuctFlow_ND procedure in EES called with a large Prandtl number; note that the results do not match exactly, but are sufficiently accurate for most engineering calculations.

More complex boundary conditions can be considered in which one or more of the duct walls are adiabatic. The solutions for alternative boundary conditions are presented in various references, including Shah and London (1978) and Rohsenow et al. (1998).

Annular Ducts

The fully developed Nusselt number for an annular duct with an adiabatic external surface and an internal surface that is subjected to a constant temperature and a constant heat flux boundary condition is provided by Rohsenow et al. (1998) and shown in Figure 9.26 as a function of the radius ratio (RR, the ratio of the inner to the outer radii of the duct).

The AnnularFlow_ND procedure provides the average Nusselt number for simultaneously developing flow in an annular duct with either a uniform heat flux or a uniform wall temperature boundary conditions.

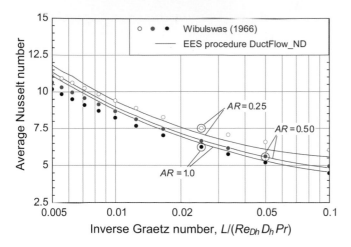

Figure 9.25 Average Nusselt number in a rectangular duct with laminar, thermally developing but hydrodynamically fully developed flow as a function of the inverse Graetz number ($\tilde{L}/Pr$) for various values of aspect ratio; results are shown for uniform heat flux from Wibulswas (1966) and the EES procedure DuctFlow_ND called with $Pr \rightarrow \infty$.

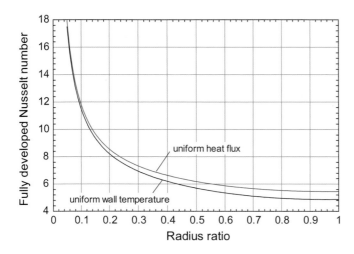

Figure 9.26 Fully developed Nusselt number in an annular duct with an adiabatic external surface and a constant temperature and constant heat flux boundary condition on the internal surface as a function of the radius ratio.

Turbulent Flow

Figure 9.20 shows that the Nusselt number for a turbulent flow is not strongly affected by the duct shape (e.g., round or rectangular) or the thermal boundary conditions (e.g., constant temperature or constant heat flux), but it is sensitive to the scale of the surface roughness (e), the Prandtl number, and the Reynolds number. This behavior occurs because the characteristics of the viscous sublayer, which is the primary thermal resistance, are affected by both Reynolds number and Prandtl number. Further, the viscous sublayer is on the same scale as the surface roughness encountered in many ducts. The Nusselt number for fully developed turbulent flow can be estimated using the correlation suggested by Gnielinski (1976):

$$Nu_{Dh,fd} = \frac{\left(\frac{f_{fd}}{8}\right)(Re_{D_h} - 1000)\,Pr}{1 + 12.7\left(Pr^{2/3} - 1\right)\sqrt{\frac{f_{fd}}{8}}}$$

(9.76)

$$\text{for } 0.5 < Pr < 2000 \text{ and } 2300 < Re_{D_h} < 5 \times 10^6,$$

where f_{fd} is the fully developed friction factor, obtained using one of the appropriate correlations presented in Section 9.3.3. The effect of roughness on the Nusselt number in Eq. (9.76) is included through its effect on the friction factor. The Nusselt number for smooth tubes at very low Prandtl number can be calculated according to (Notter and Sleicher, 1972):

$$Nu_{D_h, T, fd} = 4.8 + 0.0156 Re_{D_h}^{0.85} Pr^{0.93} \quad \text{for } 0.004 < Pr < 0.1 \text{ and } 1 \times 10^4 < Re_{D_h} < 1 \times 10^6 \tag{9.77}$$

$$Nu_{D_h, H, fd} = 6.3 + 0.0167 Re_{D_h}^{0.85} Pr^{0.93} \quad \text{for } 0.004 < Pr < 0.1 \text{ and } 1 \times 10^4 < Re_{D_h} < 1 \times 10^6. \tag{9.78}$$

Note that the special case of turbulent flow under conditions of very low Prandtl number does depend somewhat on the thermal boundary condition.

The average Nusselt number can be approximately computed according to (Kakaç et al., 1987):

$$\overline{Nu}_{D_h} \approx Nu_{D_h, fd} \left[1 + C \left(\frac{x}{D_h} \right)^{-m} \right] \quad \text{for } Re > 2300, \tag{9.79}$$

where reasonable values of the constants C and m are 1.0 and 0.7.

Example 9.9

A device is being designed to heat a flow of air from an inlet temperature of $T_{in} = 20°C$ to a mean outlet temperature of $T_{out} = 80°C$. The inlet pressure is $p_{in} = 100$ psia and the mass flow rate of air is $\dot{m} = 0.01$ kg/s. The preliminary design concept is a drawn copper tube wrapped with a heater that provides a uniform heat flux to the air of $\dot{q}_s''$. The thickness of the tube wall is $th = 0.035$ inch, the length is $L = 4$ ft long, and the outer diameter is $D_o = 0.25$ inch. Determine the pressure drop, the required heat flux on the outer surface of the tube, and the maximum temperature of the tube material.

Known Values

Figure 1 illustrates a sketch of the problem with the known values indicated.

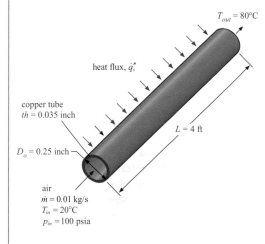

$T_{out} = 80°C$

heat flux, $\dot{q}_s''$

copper tube
$th = 0.035$ inch

$L = 4$ ft

$D_o = 0.25$ inch

air
$\dot{m} = 0.01$ kg/s
$T_{in} = 20°C$
$p_{in} = 100$ psia

Figure 1 Sketch of the problem with the known values indicated.

Assumptions

- The flow is steady.
- The properties of air at the average temperature and the inlet pressure can be used.
- The roughness of drawn copper pipe can be estimated from Table 9.1.

Continued

Example 9.9 (cont.)

- The thermal resistance of the copper tube wall is negligible.

Analysis

The properties of air are computed using the average bulk temperature and, when necessary, the inlet pressure:

$$\bar{T} = \frac{T_{in} + T_{out}}{2} \tag{1}$$

$$\rho = \rho\,(\bar{T}, p_{in}) \tag{2}$$

$$\mu = \mu(\bar{T}) \tag{3}$$

$$c = c\,(\bar{T}) \tag{4}$$

$$k = k\,(\bar{T}) \tag{5}$$

$$Pr = Pr\,(\bar{T}). \tag{6}$$

The roughness of drawn pipe is estimated from Table 9.1 to be $e = 0.0015$ mm. An energy balance on the pipe provides the total rate of heat transfer that must be delivered to heat the air from T_{in} to T_{out} :

$$\dot{q} = \dot{m}\,c(T_{out} - T_{in}). \tag{7}$$

The heat flux that must be applied is:

$$\dot{q}'' = \frac{\dot{q}}{\pi D_o L}. \tag{8}$$

The inner diameter of the tube is computed according to:

$$D_i = D_o - 2\,th. \tag{9}$$

The cross-sectional area for flow is:

$$A_c = \pi \frac{D_i^2}{4}. \tag{10}$$

The mean velocity is related to the flow rate through the pipe according to:

$$u_m = \frac{\dot{m}}{\rho\,A_c}. \tag{11}$$

The Reynolds number is computed from the mean velocity and the diameter:

$$Re_{D_h} = \frac{\rho\,u_m\,D_i}{\mu}. \tag{12}$$

The average friction factor for this circular duct is a function of the Reynolds number, length-to-diameter ratio, and relative roughness:

$$\bar{f} = \bar{f}\left(Re_{D_h}, \frac{L}{D_i}, \frac{e}{D_i}\right). \tag{13}$$

The pressure drop across the pipe is computed from the average friction factor:

$$\Delta p = \bar{f}\,\frac{L}{D_i}\,\frac{\rho\,u_m^2}{2}. \tag{14}$$

The hottest location in the tube will be at the exit to the pipe where the mean temperature is the highest and the heat transfer coefficient is the lowest. The local Nusselt number at $x = L$ under a constant heat flux condition is a function of the Reynolds number, length-to-diameter ratio, relative roughness, and the Prandtl number:

$$Nu_H = Nu_H\left(Re_{D_h}, \frac{L}{D_i}, \frac{e}{D_i}, Pr\right). \tag{15}$$

Example 9.9 (cont.)

The local heat transfer coefficient at the outlet to the tube is estimated according to:

$$h_{x=L} = \frac{Nu_H k}{D_i}.$$ (16)

The maximum tube temperature is therefore:

$$T_{s,\,x=L} = T_{out} + \frac{\dot{q}''_s}{h_{x=L}}.$$ (17)

Solution

Equations (1) through (17) are 17 equations in the 17 unknowns: $\bar{T}$, ρ, μ, c, k, Pr, $\dot{q}$, $\dot{q}''$, D_i, A_c, u_m, Re_{D_h}, $\bar{f}$, Δp, Nu_H, $h_{x=L}$, and $T_{s,x=L}$. These equations are programmed in EES below. Note that the property database in EES is used to obtain the required properties in Eqs. (2) through (6). The PipeFlow_ND procedure from the internal flow section of the Heat Transfer Convection Library is used to determine the average friction factor in Eq. (13). The PipeFlow_Local_ND procedure is used to determine the local Nusselt number in Eq. (15).

```
$UnitSystem SI MASS RAD PA K J

"Inputs"
T_in=ConvertTemp(C,K,20[C])            "inlet air temperature"
T_out=ConvertTemp(C,K,80[C])           "outlet mean air temperature"
m_dot=0.01 [kg/s]                      "mass flow rate"
p_in=100 [psi]*Convert(psi,Pa)         "inlet pressure"
th=0.035 [inch]*Convert(inch,m)        "thickness"
L=4 [ft]*Convert(ft,m)                 "tube length"
D_o=0.25 [inch]*Convert(inch,m)        "tube diameter"


"Calculations"
T_bar=(T_in+T_out)/2                   "average air temperature"
rho=Density(Air,T=T_bar,p=p_in)        "density of air"
mu=Viscosity(Air,T=T_bar)              "viscosity of air"
c=cP(Air,T=T_bar)                      "specific heat capacity of air"
k=Conductivity(Air,T=T_bar)            "conductivity of air"
Pr=Prandtl(Air,T=T_bar)                "Prandtl number of air"
e=0.0015 [mm]*Convert(mm,m)            "roughness of tube"
q_dot=m_dot*c*(T_out-T_in)             "heater power required"
q``_dot_s=q_dot/(pi*D_o*L)             "heat flux"
D_i=D_o-2*th                           "inner diameter of the tube"
A_c=pi*D_i^2/4                         "cross-sectional area for flow"
u_m=m_dot/(rho*A_c)                    "mean velocity"
Re=rho*u_m*D_i/mu                      "Reynolds number"
CALL PipeFlow_ND(Re,Pr,L/D_i,e/D_i: Nusselt_bar_T,Nusselt_bar_H,f_bar)
                                       "correlations for f_bar"
DELTAp=(rho*u_m^2/2)*(f_bar*L/D_i)     "pressure drop"
```

Continued

Example 9.9 (cont.)

```
CALL PipeFlow_ND_local(Re,Pr,L/D_i,e/D_i: Nusselt_T,Nusselt_H,f)
                                            "correlations for local Nusselt numbers"
h=Nusselt_H*k/D_i                           "local heat transfer coefficient at the outlet"
T_s_max=T_out+q``_dot_s/h                   "maximum tube temperature"
```

Executing the EES program provides the solution $\boxed{\Delta p = 126{,}600 \text{ Pa}(18.4 \text{ psi})}$, $\boxed{\dot{q}_s'' = 24{,}800 \text{ W/m}^2 \; (24.8 \text{ kW/m}^2)}$, and $\boxed{T_{s,\,max} = 368.5 \text{ K } (95.31°\text{C})}$.

Discussion

One advantage of solving the problem using a computer program is that the resulting model can be used for design purposes. Typically, a model of a system is developed using a natural set of inputs, in this case the diameter and length of the tube, in order to calculate the performance, in this case pressure drop, heat flux, and maximum surface temperature. Once the model works, the design process manipulates the inputs in order to optimize some figure of merit or satisfy some design constraint. In this case, we can adjust the diameter and length in order to meet a set of design requirements, for example Δp = 10 psi and $T_{s,max}$ = 100°C. In EES, this process requires that we remove the specified values of D_o and L and instead specify the two outputs Δp and $T_{s,max}$. This will lead to a set of equations that are implicit rather than explicit, and therefore it is important to provide good guess values before this change is made. Select Update Guess Values from the Calculate menu and then change the code according to the following.

```
$UnitSystem SI MASS RAD PA K J

"Inputs"
T_in=ConvertTemp(C,K,20[C])                 "inlet air temperature"
T_out=ConvertTemp(C,K,80[C])                "outlet mean air temperature"
m_dot=0.01 [kg/s]                           "mass flow rate"
p_in=100 [psi]*Convert(psi,Pa)              "inlet pressure"
th=0.035 [inch]*Convert(inch,m)             "thickness"
//L=4 [ft]*Convert(ft,m)                    "tube length"
//D_o=0.25 [inch]*Convert(inch,m)           "tube diameter"


"Calculations"
T_bar=(T_in+T_out)/2                        "average air temperature"
rho=Density(Air,T=T_bar,p=p_in)            "density of air"
mu=Viscosity(Air,T=T_bar)                   "viscosity of air"
c=cP(Air,T=T_bar)                           "specific heat capacity of air"
k=Conductivity(Air,T=T_bar)                 "conductivity of air"
Pr=Prandtl(Air,T=T_bar)                     "Prandtl number of air"
e=0.0015 [mm]*Convert(mm,m)                 "roughness of tube"
q_dot=m_dot*c*(T_out-T_in)                  "heater power required"
q``_dot_s=q_dot/(pi*D_o*L)                  "heat flux"
D_i=D_o-2*th                                "inner diameter of the tube"
A_c=pi*D_i^2/4                              "cross-sectional area for flow"
u_m=m_dot/(rho*A_c)                         "mean velocity"
Re=rho*u_m*D_i/mu                           "Reynolds number"
```

Example 9.9 (cont.)

> *CALL* **PipeFlow_ND**(Re,Pr,L/D_i,e/D_i: Nusselt_bar_T,Nusselt_bar_H,f_bar)
>   "correlations f_bar"
> DELTAp=(rho*u_m^2/2)*(f_bar*L/D_i)   "pressure drop"
> *CALL* **PipeFlow_ND_local**(Re,Pr,L/D_i,e/D_i: Nusselt_T,Nusselt_H,f)
>   "correlations for local Nusselt numbers"
> h=Nusselt_H*k/D_i   "local heat transfer coefficient at the outlet"
> T_s_max=T_out+q``_dot_s/h   "maximum tube temperature"
> DELTAp=10 [psi]***Convert**(psi,Pa)   "design value of pressure drop"
> T_s_max=**ConvertTemp**(C,K,100 [C])   "design value of surface temperature"

Solving provides the design that satisfies the specified pressure drop and surface temperature requirements: $D_o = 0.00679$ m (0.267 inch), and $L = 1.04$ m (3.42 ft).

9.3 The Energy Balance

9.3.1 Introduction

In the external flow problems that were investigated in Chapter 8, the temperature of the fluid far from the surface, the free stream temperature T_∞, was the same everywhere and unaffected by the energy transfer to or from the surface. As a result, the heat transfer coefficient was defined using the free stream temperature as the appropriate reference temperature:

$$h = \frac{\dot{q}_s''}{(T_s - T_\infty)} \text{ for an external flow.} \tag{9.80}$$

The mean temperature associated with an internal flow will vary in the direction of the flow as energy is added or removed; for example see Figure 9.8, which shows the mean temperature variation when the surface temperature is constant. The free stream temperature (or inlet temperature) is no longer an appropriate reference at every location along the duct surface. In fact, far from the inlet to the duct, the free stream temperature is essentially irrelevant to the local heat transfer process. Therefore, the local heat transfer coefficient for an internal flow is defined using the *mean temperature*, defined in Section 9.1.1. Recall that the mean temperature is the mass-average temperature over the cross-sectional area of the duct at a specified position in the flow direction. The local heat transfer coefficient for an internal flow is defined as:

$$h = \frac{\dot{q}_s''}{(T_s - T_m)} \text{ for an internal flow.} \tag{9.81}$$

In order to solve an internal flow problem, it is not only necessary to determine the heat transfer coefficient but you must also establish the variation in the mean temperature with position. This is accomplished using an energy balance. For a single internal flow interacting with a prescribed boundary condition (e.g., a constant temperature duct surface), the energy balance will result in an ordinary differential equation that can be solved analytically or numerically provided information is known about the heat transfer coefficient. Heat exchangers are devices in which two (or more) internal flows interact, causing both of their mean temperatures to vary; the understanding and modeling of this interaction lies at the heart of much of Chapter 12.

9.3.2 The Energy Balance

Figure 9.27 illustrates an energy balance on a control volume that is differential in the flow direction, x, but extends across the entire cross-section of the duct.

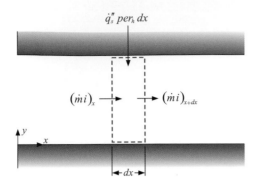

$\dot{q}_s'' \, per_h \, dx$

$(\dot{m}i)_x$ → → $(\dot{m}i)_{x+dx}$

dx

Figure 9.27 An energy balance on a differential control volume.

The heat flux from the duct wall to the fluid ($\dot{q}_s''$) must be balanced by the change in the enthalpy carried by the fluid. The energy balance suggested by Figure 9.27 is

$$\dot{q}_s'' \, per_h \, dx + (\dot{m}i)_x = (\dot{m}i)_{x+dx}, \tag{9.82}$$

where i is the mean specific enthalpy of the fluid (the specific enthalpy evaluated at the mean temperature at position x) and per_h is the perimeter associated with the heat flux; note that per_h may be different from the entire wetted perimeter used to calculate the hydraulic diameter (per) depending on whether the heat flux is uniformly applied. Expanding the term on the right side of Eq. (9.82) and substituting the definition of a derivative provides:

$$\dot{q}_s'' \, per_h \, dx + (\dot{m}i)_x = (\dot{m}i)_x + \frac{d(\dot{m}i)}{dx} dx. \tag{9.83}$$

Recognizing that the mass flow rate is constant allows Eq. (9.83) to be written as:

$$\dot{q}_s'' \, per_h = \dot{m} \frac{di}{dx}. \tag{9.84}$$

Typically, the impact of the pressure gradient on the enthalpy is small and therefore the rate of change of specific enthalpy can be written in terms of the specific heat capacity and the rate of change of the mean temperature:

$$\dot{q}_s'' \, per_h = \dot{m} \, c \frac{dT_m}{dx}. \tag{9.85}$$

Equation (9.85) neglects axial conduction and volumetric thermal energy generation in the fluid, both of which are typically negligible for an internal convection problem. If the heat flux at the wall is prescribed, then Eq. (9.85) can be integrated directly to obtain T_m as a function of x. If the wall temperature, T_s, is prescribed, then the definition of the local heat transfer coefficient, Eq. (9.81), must be substituted into Eq. (9.85) to provide:

$$per_h \, h \, (T_s - T_m) = \dot{m} \, c \frac{dT_m}{dx}. \tag{9.86}$$

The energy balances provided by either Eq. (9.85) or Eq. (9.86) are 1-D ordinary differential equations for T_m, whereas the actual flow situation in the duct is clearly 2-D or 3-D. The use of the 1-D energy balance is facilitated by the definition of the mean temperature and the associated definition of the local heat transfer coefficient in terms of the mean temperature. This is an engineering approach that simplifies the solution to internal flow problems. The mean temperature represents the appropriate local energy flow rate associated with the fluid while the local heat transfer coefficient encompasses the details of the local temperature distribution and flow conditions. The heat transfer coefficient for many practical problems is correlated in the form of Nusselt number as a function of Reynolds number and Prandtl number (as discussed in Section 9.2) and therefore many engineering problems can be solved without ever considering the 2-D or 3-D details of the flow. However, it is important to realize that by using these correlations you are relying on the prior work (either theoretical or experimental) of other researchers to characterize these 2-D or 3-D flows and summarize the result in a convenient format: the heat transfer coefficient.

Two common wall conditions that are encountered in practice are the constant wall temperature and constant surface temperature conditions. Analytical solutions to the energy balance can be obtained for these limiting cases. However the energy balance equations derived in this section, Eqs. (9.85) and (9.86), are not limited to these simple wall conditions.

9.3.3 Specified Heat Flux

In a duct with a specified heat flux variation with position, $\dot{q}_s''(x)$, the energy balance provided by Eq. (9.85) can be separated and integrated to determine the variation in the mean temperature with position. This situation might occur due to the electric dissipation in a heater wire that is wrapped on the surface of the duct. Assuming that the specific heat capacity of the fluid is constant, this results in

$$\int_{T_{in}}^{T_m} dT_m = \frac{per_h}{\dot{m}\,c}\int_0^x \dot{q}_s''(x)\,dx \tag{9.87}$$

or

$$T_m = T_{in} + \frac{per_h}{\dot{m}\,c}\int_{x=0}^x \dot{q}_s''(x)\,dx, \tag{9.88}$$

where T_{in} is the mean temperature of the flow at the inlet. Equation (9.88) shows that the mean temperature of the flow increases at a rate that is proportional to the applied heat flux and is inversely proportional to the total capacitance rate of the fluid. The **total capacitance rate**, $\dot{C}$, is the product of the mass flow rate and specific heat capacity ($\dot{m}\,c$) and represents the rate at which energy must be added to the fluid in order to increase its temperature. Equation (9.88) can be solved either analytically (for simple variations in the heat flux) or numerically using any of the numerical integration techniques discussed in Section 5.3.

Equation (9.88) indicates that the variation in the mean temperature of the fluid is insensitive to the local heat transfer coefficient for situations where the heat flux is specified. The surface temperature (i.e., the temperature of the wall), however, does depend on the local heat transfer coefficient according to Eq. (9.81):

$$T_s = T_m + \frac{\dot{q}_s''}{h}. \tag{9.89}$$

Constant Heat Flux

In the limit that a constant heat flux is applied at the surface of the duct, Eq. (9.88) provides:

$$T_m = T_{in} + \frac{per_h\,\dot{q}_s''}{\dot{m}\,c}\,x. \tag{9.90}$$

Figure 9.28 illustrates the mean fluid temperature and the surface temperature as a function of position for a duct that is exposed to a constant heat flux.

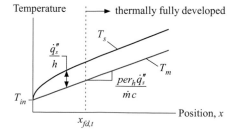

Figure 9.28 The mean fluid temperature and the surface temperature as a function of position for a duct exposed to a constant heat flux.

Notice that the wall-to-mean temperature difference becomes constant in the thermally fully developed region (i.e., for $x > x_{fd,t}$) because the heat transfer coefficient becomes constant in this region. Further, the surface temperature approaches the mean temperature at the inlet because the heat transfer coefficient is large in the developing region.

Example 9.10

A single fuel rod is being cooled by a molten salt. The salt being used corresponds to the fluid 'Salt (58NaCl_42MgCl2)' in the EES property database. The fuel rod is a solid cylinder of radius $r_{in} = 0.25$ inch and the molten salt flows through an annular gap formed by a smooth outer tube that is concentric to the rod. The outer radius of the gap formed between the fuel rod and the outer tube is $r_{out} = 0.375$ inch. The mass flow rate of molten salt is $\dot{m} = 0.27$ kg/s and the inlet temperature is $T_{in} = 500°C$. The length of the fuel rod is $L = 6$ ft. The heat flux at the surface of the fuel rod varies with position according to the equation

$$\dot{q}_s''(x) = \dot{q}_{s,max}'' \sin\left(\pi \frac{x}{L}\right), \tag{1}$$

where $\dot{q}_{s,max}'' = 395$ kW/m^2.

Determine the pressure drop associated with the molten salt as it flows through the tube. Plot the mean temperature and the fuel rod surface temperature as a function of position.

Known Values

Figure 1 illustrates a sketch of the problem with the known values indicated.

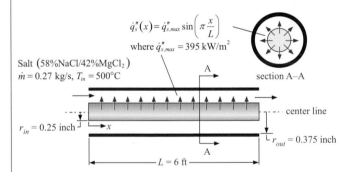

Figure 1 Sketch of the problem with the known values indicated.

Assumptions

- The flow is steady.
- The properties of molten salt at the average of the inlet and outlet temperatures can be used and assumed to be constant.
- The surfaces are smooth.
- The outer surface of the outer tube is insulated.

Analysis

The perimeter of the heated portion of the duct is:

$$per_h = 2\pi r_{in}. \tag{2}$$

Example 9.10 (cont.)

The total wetted perimeter is:

$$per = 2\pi(r_{in} + r_{out}). \tag{3}$$

The cross-sectional area for flow is:

$$A_c = \pi\left(r_{out}^2 - r_{in}^2\right). \tag{4}$$

The hydraulic diameter of the duct is:

$$D_h = \frac{4\,A_c}{per}. \tag{5}$$

The properties of the molten salt are computed using the average bulk temperature:

$$\bar{T} = \frac{T_{in} + T_{out}}{2} \tag{6}$$

$$\rho = \rho\left(\bar{T}\right) \tag{7}$$

$$\mu = \mu\left(\bar{T}\right) \tag{8}$$

$$c = c\left(\bar{T}\right) \tag{9}$$

$$k = k\left(\bar{T}\right) \tag{10}$$

$$Pr = Pr\left(\bar{T}\right). \tag{11}$$

The mean temperature as a function of position can be computed by substituting Eq. (1) into Eq. (9.88):

$$T_m = T_{in} + \frac{per_h}{\dot{m}\,c}\int_{x=0}^{x}\dot{q}_{s,\,max}''\sin\left(\pi\frac{x}{L}\right)dx = T_{in} + \frac{per_h\,L\,q_{s,\,max}''}{\dot{m}\,c\,\pi}\left[1 - \cos\left(\pi\frac{x}{L}\right)\right]. \tag{12}$$

Substituting $x = L$ into Eq. (12) provides the mean temperature at the outlet:

$$T_{out} = T_{in} + \frac{2per_h\,L\,\dot{q}_{s,\,max}''}{\dot{m}\,c\,\pi}. \tag{13}$$

The mean velocity is computed according to:

$$u_m = \frac{\dot{m}}{\rho\,A_c}. \tag{14}$$

The Reynolds number is computed from the mean velocity and the hydraulic diameter:

$$Re_{D_h} = \frac{\rho\,u_m\,D_h}{\mu}. \tag{15}$$

The average friction factor for this annular duct is a function of the Reynolds number, length to hydraulic diameter ratio, radius ratio, and relative roughness (where e is assumed to be zero):

$$\bar{f} = \bar{f}\left(Re_{D_h}, \frac{r_{in}}{r_{out}}, \frac{L}{D_h}, \frac{e}{D_h}\right). \tag{16}$$

The pressure drop is computed from the average friction factor:

$$\Delta p = \bar{f}\,\frac{L}{D_h}\frac{\rho\,u_m^2}{2}. \tag{17}$$

The local Nusselt number is a function of the Reynolds number, position to hydraulic diameter ratio, radius ratio, relative roughness, and the Prandtl number:

$$Nu_H = Nu_H\left(Re_{D_h}, \frac{r_{in}}{r_{out}}, \frac{x}{D_h}, \frac{e}{D_h}, Pr\right). \tag{18}$$

Continued

Example 9.10 (cont.)

The local heat transfer coefficient is estimated according to:

$$h = \frac{Nu_H k}{D_h}. \tag{19}$$

The surface temperature of the fuel rod is therefore:

$$T_s = T_m + \frac{\dot{q}_s''}{h}. \tag{20}$$

Solution

Given a position x, Eqs. (2) through (20) are 19 equations in the 19 unknowns: per_h, per, A_c, D_h, $\bar{T}$, ρ, μ, c, k, Pr, T_m, T_{out}, u_m, Re_{D_h}, $\bar{f}$, Δp, Nu_H, h, and T_s. These equations are programmed in EES below. Note that the property database in EES is used to obtain the required properties in Eqs. (7) through (11). The AnnularFlow_ND procedure from the internal flow section of the Heat Transfer Convection Library is used to determine the average friction factor in Eq. (16). The AnnularFlow_Local_ND procedure is used to determine the local Nusselt number in Eq. (18).

```
$UnitSystem SI Mass J K Pa rad

L=6 [ft]*Convert(ft,m)                          "length"
r_in=0.25 [inch]*Convert(inch,m)                "inner radius of annular passage"
r_out=0.375 [inch]*Convert(inch,m)              "outer radius of annular passage"
m_dot=0.27 [kg/s]                               "mass flow rate"
T_in=ConvertTemp(C,K,500 [C])                   "inlet temperature"
q``_dot_max=395000 [W/m^2]                      "amplitude of sine function"
F$='Salt(58NaCl_42MgCl2)'

per_h=2*pi*r_in                                 "perimeter exposed to heat flux"
per=2*pi*(r_in+r_out)                           "total wetted perimeter"
A_c=pi*(r_out^2-r_in^2)                         "cross-sectional area for flow"
D_h=4*A_c/per                                   "hydraulic diameter"

T_bar=(T_in+T_out)/2                            "average temperature for properties"
rho=Density(F$, T=T_bar)                        "density"
mu=Viscosity(F$, T=T_bar)                       "viscosity"
k=Conductivity(F$, T=T_bar)                     "conductivity"
c=cP(F$, T=T_bar)                               "specific heat capacity"
Pr=Prandtl(F$, T=T_bar)                         "Prandtl number"

T_out=T_in+2*per_h*L*q``_dot_max/(m_dot*c*pi)   "outlet temperature"

u_m=m_dot/(rho*A_c)                             "mean velocity"
Re_Dh=u_m*D_h*rho/mu                            "Reynolds number"
```

Example 9.10 (cont.)

> **CALL AnnularFlow_ND**(Re_Dh, Pr, L/D_h, r_in/r_out, 0 [-]: , , f_bar)
> "average friction factor"
>
> DELTAp=f_bar*L/D_h*rho*u_m^2/2   "pressure drop"
>
> x=L/2   "location"
> q``_dot_s=q``_dot_max***Sin**(pi*x/L)   "heat flux"
> T_m=T_in+per_h*L*q``_dot_max*(1-**Cos**(pi*x/L))/(m_dot*c*pi) "mean temperature"
> **CALL AnnularFlow_ND_Local**(Re_Dh, Pr, x/D_h, r_in/r_out, 0 [-]: , Nusselt_H,)
> "local Nusselt number"
> h=Nusselt_H*k/D_h   "local heat transfer coefficient"
> T_s=T_m+q``_dot_s/h   "surface temperature"

Solving provides $\Delta p = 4180 \text{ Pa } (0.606 \text{ psi})$. A Parametric Table is created that includes x, T_m, and T_s. The value of x is varied from 0 to L and the result is used to generate the plot of mean and surface temperature as a function of position shown in Figure 2.

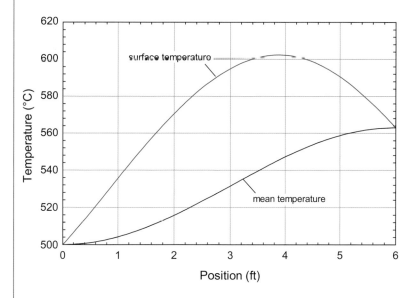

Figure 2 Surface temperature and mean temperature as a function of position

Discussion

Notice in Figure 2 that the rate at which the mean temperature increases is entirely related to the rate at which energy is being added due to the heat flux at the wall, given by Eq. (1). The variation of the mean temperature has nothing to do with the heat transfer coefficient. The surface temperature is elevated relative to the mean temperature by an amount that depends on both the heat flux and the local heat transfer coefficient according to Eq. (20).

9.3.4 Specified Wall Temperature

In a duct with a specified wall temperature variation with position, $T_s(x)$, the energy balance provided by Eq. (9.86) must be used:

$$\frac{dT_m}{dx} + \frac{per_h\,h}{\dot{m}\,c}T_m = \frac{per_h\,h}{\dot{m}\,c}T_s(x). \tag{9.91}$$

In general, the local heat transfer coefficient may be a function of position, particularly in the developing region, and therefore Eq. (9.91) must be solved numerically using one of the integration techniques discussed in Section 5.3. Notice that Eq. (9.91) provides the state equation for the mean temperature and therefore numerical integration can be applied relatively easily if a correlation is available for the heat transfer coefficient. If the local heat transfer coefficient, perimeter, mass flow rate and specific heat capacity are assumed to be constant then Eq. (9.91) is a linear, first-order ordinary differential equation that may be solved analytically for simple variations in the duct surface temperature. The solution proceeds by dividing the mean temperature into its homogeneous and particular components in the same way that lumped capacitance problems were solved in Section 5.2.

Constant Wall Temperature

In the limit that the duct has a constant wall temperature ($T_s(x) = T_s$) the energy balance, Eq. (9.86), can be separated and integrated:

$$\int_{T_{in}}^{T_m} \frac{dT_m}{(T_s - T_m)} = \frac{per_h}{\dot{m}\,c}\int_0^x h\,dx. \tag{9.92}$$

Substituting the definition of the average heat transfer coefficient, Eq. (9.49), into Eq. (9.92) leads to:

$$\int_{T_{in}}^{T_m} \frac{dT_m}{(T_s - T_m)} = \frac{per_h\,x\,\bar{h}}{\dot{m}\,c}. \tag{9.93}$$

Equation (9.93) is integrated:

$$\ln\left(\frac{T_s - T_m}{T_s - T_{in}}\right) = -\frac{per_h\,x\,\bar{h}}{\dot{m}\,c}. \tag{9.94}$$

Solving for the mean temperature provides:

$$T_m = T_s - (T_s - T_{in})\exp\left(-\frac{per_h\,x\,\bar{h}}{\dot{m}\,c}\right). \tag{9.95}$$

Equation (9.95) shows that the mean fluid temperature will approach the wall temperature approximately exponentially (exactly exponentially if $\bar{h}$ is constant). This behavior is similar to the transient problem associated with a lumped capacitance subjected to a step change in its ambient temperature; the analytical solution for this case is given by

$$T = T_\infty + (T_{ini} - T_\infty)\exp\left(-\frac{\bar{h}\,A_s}{m\,c}t\right). \tag{9.96}$$

Equation (9.96) can be thought of as the mean temperature experienced by a differential mass of fluid as a function of time as it moves through the tube (i.e., the temperature expressed in a Lagrangian frame of reference, where the observer moves with the fluid), whereas Eq. (9.95) is the mean temperature expressed as a function of axial position (i.e., the temperature expressed in an Eulerian frame of reference, where the

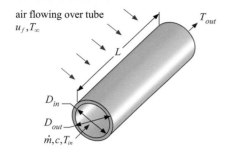

Figure 9.29 Fluid in a tube exposed to a flow of air

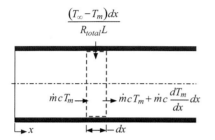

Figure 9.30 Differential control volume for a fluid subjected to a constant external temperature rather than a constant duct wall temperature.

observer is stationary). The mean temperature of the fluid leaving the duct, T_{out}, is obtained by evaluating Eq. (9.95) at $x = L$:

$$T_{out} = T_s - (T_s - T_{in}) \exp\left(-\frac{per_h L \bar{h}}{\dot{m} c}\right). \tag{9.97}$$

9.3.5 Specified External Temperature

It is more often the case that a temperature *external* to the duct will be prescribed, rather than the temperature of the duct wall itself. For example, Figure 9.29 illustrates fluid running through a tube with length L and inner and outer diameters D_{in} and D_{out}, respectively. The tube is placed in an air flow at T_∞. The fluid enters the tube at T_{in}. In this case, the temperature of the inner wall of the tube is not constant or even directly specified and therefore the solution derived in Section 9.3.4 does not apply.

In this situation, it is necessary to write a differential energy balance on the fluid in terms of the total thermal resistance that separates the prescribed external temperature (T_∞) from the mean temperature of the fluid (T_m, which does vary with position). For the situation shown in Figure 9.29, the total thermal resistance will include a resistance associated with convection from the tube wall to the fluid as well as the resistance to conduction through the tube wall and convection between the outside surface of the duct and the air:

$$R_{total} = \underbrace{\frac{1}{\bar{h}_{in} \pi D_{in} L}}_{\text{internal convection}} + \underbrace{\frac{\ln\left(\dfrac{D_{out}}{D_{in}}\right)}{2 \pi k_{tube} L}}_{\text{conduction through tube}} + \underbrace{\frac{1}{\bar{h}_{out} \pi D_{out} L}}_{\text{external convection}}. \tag{9.98}$$

Additional resistances could be added to represent fins, contact resistance, radiation, or fouling (the build-up of scale or deposit on the inside or outside surfaces of a tube subjected to a flowing fluid, discussed in Chapter 12).

A differential energy balance is shown in Figure 9.30 for a duct that is exposed to a constant external temperature.

The energy balance suggested by Figure 9.30 is

$$\dot{m} c T_m + \frac{(T_\infty - T_m)dx}{R_{total}L} = \dot{m} c T_m + \dot{m} c \frac{dT_m}{dx} dx, \tag{9.99}$$

which can be simplified to

$$\frac{(T_\infty - T_m)}{R_{total}L} = \dot{m} c \frac{dT_m}{dx}. \tag{9.100}$$

The difference between Eq. (9.99) and Eq. (9.86) is that the heat transfer rate is written in terms of total resistance rather than the internal heat transfer coefficient and in terms of an external temperature (T_∞) rather than the duct surface temperature (T_s). Assuming the total resistance can be assumed to be a constant, Eq. (9.100) can be separated and integrated to provide:

$$T_{out} = T_\infty - (T_\infty - T_{in}) \exp\left(-\frac{1}{R_{total} \dot{m} c}\right). \tag{9.101}$$

The inverse of the total resistance is referred to as the **total conductance** (UA) in heat exchanger analysis:

$$UA = \frac{1}{R_{total}}. \tag{9.102}$$

Substituting the definition of conductance, Eq. (9.102), into Eq. (9.101) leads to:

$$T_{out} = T_\infty - (T_\infty - T_{in}) \exp\left(-\frac{UA}{\dot{m} c}\right), \tag{9.103}$$

which is the same as the solution for the constant duct temperature, Eq. (9.97), with the terms $per\, L\bar{h}$ replaced by UA and T_s replaced by T_∞.

The situation shown in Figure 9.29 is an example of a **heat exchanger** in that the fluid flowing within the tube is exchanging heat with the air flowing over the tube. However, it is a simple heat exchanger because the air temperature seen by the tube is the same everywhere. In Chapter 12, more complicated heat exchangers are introduced where two fluid streams interact thermally and both change temperature substantially. In a heat exchanger analysis, two energy balances of the type shown in Figure 9.30 must be written and the solution becomes more complex. However, the idea is the same and the concept of a total conductance is very useful for a heat exchanger problem.

Example 9.11

The heat rejection in a water source heat pump occurs by pumping a 20 percent ethylene glycol solution through a drawn copper tube that is $L = 10$ m long with outer diameter $D_{out} = 1.7$ cm and wall thickness $th = 2$ mm. The ethylene glycol enters the tube at $T_{in} = 17°C$ with mass flow rate $\dot{m}_f = 0.25$ kg/s. The tube is submerged in a small creek that flows with average velocity $u_w = 3.2$ m/s and temperature $T_w = 5°C$. Estimate the outlet temperature of the ethylene glycol and the heat transfer rate assuming that (a) the internal surface temperature of the tube is equal to the water temperature, and (b) including the additional resistance associated with the conduction through the tube and convection from the outer surface of the tube to the water.

Known Values

Figure 1 illustrates a sketch of the problem. The properties of a 20 percent ethylene glycol solution are obtained from Appendix B: $\rho_f = 1022$ kg/m^3, $c_f = 3905$ J/kg-K, $k_f = 0.5132$ W/m-K, $\mu_f = 0.00145$ Pa-s, and $Pr_f = 11.03$. The conductivity of copper is obtained from Appendix A: $k_t = 396$ W/m-K. The properties of water are obtained from Appendix D: $\rho_w = 996$ kg/m^3, $k_w = 0.6111$ W/m-K, $\mu_w = 0.854 \times 10^{-3}$ Pa-s, and $Pr_w = 5.841$.

Example 9.11 (cont.)

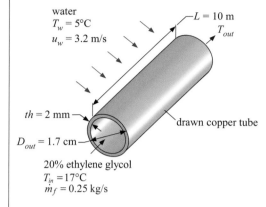

water
$T_w = 5°C$
$u_w = 3.2$ m/s

$L = 10$ m
T_{out}

$th = 2$ mm
drawn copper tube

$D_{out} = 1.7$ cm

20% ethylene glycol
$T_{in} = 17°C$
$\dot{m}_f = 0.25$ kg/s

Figure 1 Sketch of the problem with the known values indicated.

Assumptions

- The flow is steady state.
- The properties in Appendices A, B, and D can be used.
- The surface of the pipe is smooth (i.e., the roughness is negligibly small).
- The effect of the developing region is small for the internal flow.
- The external flow can be modeled as a cylinder in cross-flow.

Analysis and Solution

The inner diameter, cross-sectional area, and inside perimeter of the tube are computed:

$$D_{in} = D_{out} - 2\,th = 0.017\text{ m} - 2(0.002\text{ m}) = 0.013\text{ m } (1.3\text{ cm}) \tag{1}$$

$$A_c = \frac{\pi}{4} D_{in}^2 = \frac{\pi}{4}(0.013\text{ m})^2 = 1.33 \times 10^{-4}\text{ m}^2 \tag{2}$$

$$per = \pi\, D_{in} = \pi(0.013\text{ m}) = 0.0408\text{ m}. \tag{3}$$

The mean velocity of the ethylene glycol solution is computed:

$$u_{m,f} = \frac{\dot{m}_f}{\rho_f A_c} = \frac{0.25\text{ kg}}{\text{s}}\left|\frac{\text{m}^3}{1022\text{ kg}}\right|\frac{1}{1.33 \times 10^{-4}\text{ m}^2} = 1.843\text{ m/s}. \tag{4}$$

The Reynolds number associated with the internal flow of ethylene glycol is computed:

$$Re_{D_h,f} = \frac{\rho_f u_{m,f} D_{in}}{\mu_f} = \frac{1022\text{ kg}}{\text{m}^3}\left|\frac{1.843\text{ m}}{\text{s}}\right|\frac{0.013\text{ m}}{0.00145\text{ Pa s}}\left|\left|\frac{\text{Pa m}^2}{\text{N}}\right|\frac{\text{N s}^2}{\text{kg m}}\right| = 1.69 \times 10^4. \tag{5}$$

Because the Reynolds number is above the critical Reynolds number for internal flow (2300), the flow will likely be turbulent. The tube is very long relative to its diameter and the flow is turbulent so we will assume that the fully developed friction factor and Nusselt number can be used. The roughness of drawn tube is very small and therefore Eq. (9.61) can be used to estimate the fully developed, turbulent friction factor and Eq. (9.76) can be used to estimate the fully developed Nusselt number:

$$f_{fd} = \frac{1}{[0.790\ln(Re_{D_h,f}) - 1.64]^2} = \frac{1}{[0.790\ln(1.69 \times 10^4) - 1.64]^2} = 0.0273 \tag{6}$$

Continued

Example 9.11 (cont.)

$$\overline{Nu}_{D_h,f} = \frac{\left(\frac{f_{fd}}{8}\right)(Re_{D_h,f} - 1000)\,Pr_f}{1 + 12.7\left(Pr_f^{2/3} - 1\right)\sqrt{\frac{f_{fd}}{8}}} = \frac{\left(\frac{0.0273}{8}\right)(1.69 \times 10^4 - 1000)(11.03)}{1 + 12.7\left[(11.03)^{2/3} - 1\right]\sqrt{\frac{0.0273}{8}}} = 152.1. \tag{7}$$

The convection heat transfer coefficient between the ethylene glycol and the internal surface of the tube is determined:

$$\overline{h}_f = \frac{\overline{Nu}_{D_h,f}\,k_f}{D_{in}} = \frac{152.1}{} \left|\frac{0.5132\,\text{W}}{\text{m K}}\right| \frac{1}{0.013\,\text{m}} = 6003\,\frac{\text{W}}{\text{m}^2\,\text{K}}. \tag{8}$$

If the internal surface temperature of the tube is assumed to be at the water temperature then Eq. (9.97) can be used to determine the ethylene glycol outlet temperature:

$$T_{out} = T_w - (T_w - T_{in})\exp\left(-\frac{per\,L\,\overline{h}_f}{\dot{m}_f\,c_f}\right)$$

$$= 5°\text{C} - (5°\text{C} - 17°\text{C})\exp\left(-\frac{0.0408\,\text{m}}{}\left|\frac{10\,\text{m}}{}\right|\frac{6003\,\text{W}}{\text{m}^2\,\text{K}}\right|\frac{\text{s}}{0.25\,\text{kg}}\left|\frac{\text{kg K}}{3905\,\text{J}}\right|\left|\frac{\text{J}}{\text{W s}}\right|\right). \tag{9}$$

$$= \boxed{5.97°\text{C}}$$

The rate of heat transfer from the ethylene glycol to the water is obtained from an energy balance on the ethylene glycol:

$$\dot{q} = \dot{m}_f\,c_f(T_{in} - T_{out}) = \frac{0.25\,\text{kg}}{\text{s}}\left|\frac{3905\,\text{J}}{\text{kg K}}\right|\frac{(17 - 5.97)\,\text{K}}{} = \boxed{10,760\,\text{W (10.76 kW)}}. \tag{10}$$

If the additional thermal resistance associated with conduction through the tube and convection from the outer surface to the water is taken into account then it becomes necessary to use Eq. (9.98) to determine the total resistance. The resistance to convection within the tube is obtained from:

$$R_{conv,f} = \frac{1}{\overline{h}_f\,per\,L} = \frac{\text{m}^2\,\text{K}}{6003\,\text{W}}\left|\frac{1}{0.0408\,\text{m}}\right|\frac{1}{10\,\text{m}} = 4.08 \times 10^{-4}\,\frac{\text{K}}{\text{W}}. \tag{11}$$

The resistance to conduction through the tube is determined from:

$$R_{cond} = \frac{\ln\left(\frac{D_{out}}{D_{in}}\right)}{2\pi k_t L} = \frac{\ln\left(\frac{0.017}{0.013}\right)}{2\pi}\left|\frac{\text{m K}}{396\,\text{W}}\right|\frac{1}{10\,\text{m}} = 1.08 \times 10^{-5}\,\frac{\text{K}}{\text{W}}. \tag{12}$$

In order to determine the resistance due to convection from the outside surface of the tube it is necessary to solve the external flow problem associated with the flow of water over the cylinder. The Reynolds number associated with the external flow is calculated:

$$Re_{D_{out},w} = \frac{\rho_w\,u_w\,D_{out}}{\mu_w}$$

$$= \frac{996\,\text{kg}}{\text{m}^3}\left|\frac{3.2\,\text{m}}{\text{s}}\right|0.017\,\text{m}\left|\frac{1}{0.854 \times 10^{-3}\,\text{Pa s}}\right|\left|\frac{\text{Pa m}^2}{\text{N}}\right|\frac{\text{N s}^2}{\text{kg m}} = 6.34 \times 10^4. \tag{13}$$

The correlation for external flow over a cylinder from Section 8.3.2 is used to determine the average Nusselt number:

Example 9.11 (cont.)

$$\overline{Nu}_{D_{out,\,w}} = 0.3 + \frac{0.62\,Re_{D_{out,\,w}}^{0.5}Pr_w^{1/3}}{\left[1 + \left(\frac{0.4}{Pr_w}\right)^{2/3}\right]^{0.25}}\left[1 + \left(\frac{Re_{D_{out,\,w}}}{2.82 \times 10^5}\right)^{0.625}\right]^{0.80}$$

$$= 0.3 + \frac{0.62\left(6.34 \times 10^4\right)^{0.5}(5.841)^{1/3}}{\left[1 + \left(\frac{0.4}{5.841}\right)^{2/3}\right]^{0.25}}\left[1 + \left(\frac{6.34 \times 10^4}{2.82 \times 10^5}\right)^{0.625}\right]^{0.80} = 353.2. \qquad (14)$$

The heat transfer coefficient is:

$$\bar{h}_w = \frac{\overline{Nu}_{D_{out,\,w}}\,k_w}{D_{out}} = \frac{353.2\left|0.6111\text{ W}\right|}{\left|\text{m K}\right|0.017\text{ m}} = 12{,}700\,\frac{\text{W}}{\text{m}^2\text{ K}}. \qquad (15)$$

The resistance to convection from the outer surface is:

$$R_{conv,\,w} = \frac{1}{\bar{h}_w\,\pi\,D_{out}\,L} = \frac{\text{m}^2\text{ K}}{12{,}700\text{ W}}\left|\frac{1}{\pi}\right|\frac{1}{0.017\text{ m}}\left|\frac{1}{10\text{ m}}\right| = 1.48 \times 10^{-4}\,\frac{\text{K}}{\text{W}}. \qquad (16)$$

The total resistance between the ethylene glycol and the water is computed:

$$R_{total} = R_{conv,\,f} + R_{cond} + R_{conv,\,w}$$

$$= 4.08 \times 10^{-4}\,\frac{\text{K}}{\text{W}} + 1.08 \times 10^{-5}\,\frac{\text{K}}{\text{W}} + 1.48 \times 10^{-4}\,\frac{\text{K}}{\text{W}} = 5.66 \times 10^{-4}\,\frac{\text{K}}{\text{W}}. \qquad (17)$$

The conductance of the heat exchanger is given by Eq. (9.102):

$$UA = \frac{1}{R_{total}} = \frac{\text{W}}{5.66 \times 10^{-4}\text{K}} = 1770\,\frac{\text{W}}{\text{K}}. \qquad (18)$$

The outlet temperature is computed using Eq. (9.103):

$$T_{out} = T_w - (T_w - T_{in})\exp\left(-\frac{UA}{\dot{m}_f\,c_f}\right)$$

$$= 5°\text{C} - (5°\text{C} - 17°\text{C})\exp\left(-\frac{1770\text{ W}}{\text{K}}\left|\frac{\text{s}}{0.25\text{ kg}}\right|\frac{\text{kg K}}{3905\text{ J}}\left|\frac{\text{J}}{\text{W s}}\right|\right). \qquad (19)$$

$$= \boxed{6.97°\text{C}}$$

The rate of heat transfer is computed using an energy balance on the ethylene glycol:

$$\dot{q} = \dot{m}_f\,c_f(T_{in} - T_{out}) = \frac{0.25\text{ kg}}{\text{s}}\left|\frac{3905\text{ J}}{\text{kg K}}\right|\frac{(17 - 6.97)\text{K}}{} = \boxed{9800\text{ W }(9.80\text{ kW})}. \qquad (20)$$

Discussion

As expected, the inclusion of the additional resistances causes the outlet temperature to rise (from 5.97°C to 6.97°C) and the rate of heat transfer to drop (from 10.76 kW to 9.8 kW). However, the change is not large because the additional resistances are small relative to the internal convection resistance. The resistance to conduction is negligibly small relative to the convection resistances. The resistance to convection on the water side is about 2.5× smaller than the resistance to convection on the ethylene glycol side.

9.4 Conclusions and Learning Objectives

This chapter provides a discussion of internal forced convection flow and describes how this situation is fundamentally different from the external forced convection flow examined in Chapters 7 and 8. The fact that the boundary layers are confined in an internal flow leads to some interesting behaviors in the fully developed region and the reference temperature for the flow must be taken to be the mean temperature as compared to the free stream temperature. A number of useful correlations were presented for different flow situations; it is incumbent on the engineer to be able to classify the flow correctly in order to then identify the most appropriate correlation. Finally, the concept of an energy balance was explored in order to understand and predict the variation of the fluid mean temperature as a function of position in the flow direction.

Some specific concepts that you should understand are listed below.
- The fundamental differences between internal and external flows in terms of the boundary layer development.
- The importance of the fully developed region and the behavior of the pressure gradient in the hydrodynamically fully developed region and the heat transfer coefficient in the thermally fully developed region.
- Methods to estimate the hydrodynamic and thermal entry length for laminar flow.
- The relationship between pressure gradient and shear stress.
- The definition of the friction factor and Nusselt number for an internal flow.
- The difference between average and local friction factor and Nusselt number for an internal flow.
- The methodology for classifying an internal flow problem in terms of laminar vs. turbulent, developing (thermally, hydrodynamically, or both) vs. fully developed, duct shape, thermal boundary condition, etc.
- The methodology for carefully and correctly applying a convection correlation to a particular flow problem.
- The behavior of the friction factor and Nusselt number should be understood in terms of the fundamental characteristics of laminar and turbulent boundary layers.
- The difference between a mean temperature and a free stream temperature.
- The general form of the energy balance for an internal flow and its application to various boundary conditions including specified heat flux, specified surface temperature, and specified external temperature.
- The definition of the conductance between a fluid and its surroundings.

References

Colebrook, C. F., "Turbulent flow in pipes with particular reference to the transitional region between the smooth and the rough pipe laws," *J. Inst. Civil Eng.*, Vol. 11, pp. 133–156 (1939).

Curr, R. M., D. Sharma, and D. G. Tatchell, "Numerical predictions of some three-dimensional boundary layers in ducts," *Comput. Methods Appl. Mech. Eng.*, Vol. 1, pp. 143–158 (1972).

Gnielinski, V., "New equations for heat and mass transfer in turbulent pipe and channel flow," *Int. Chem. Eng.*, Vol. 16, pp. 359–368 (1976).

Hornbeck, R. W., "An all-numerical method for heat transfer in the inlet of a tube," *Am. Soc. Mech. Eng.*, Paper 65-WA/HT-36 (1965).

Kakaç, S., R. K. Shah, and W. Aung, eds., *Handbook of Single-Phase Convective Heat Transfer*, Wiley-Interscience (1987).

Kays, W. M. and M. E. Crawford, *Convective Heat and Mass Transfer*, Third Edition, McGraw-Hill, New York (1993).

Li, P., J. Seem, and Y. Li, "A new explicit equation for accurate friction factor calculation of smooth pipes," *Int. J. Refrig.*, Vol. 34(6), pp. 1535–1541 (2011).

Lienhard, J. H., IV and J. H. Lienhard V, *A Heat Transfer Textbook*, Third Edition, Phlogiston Press, Cambridge, MA (2005).

Liu, J., *Flow of a Bingham Fluid in the Entrance Region of an Annular Tube*, M.S. Thesis, University of Wisconsin at Milwaukee (1974).

Nikuradse, J., "Strömungsgesetze in rauhen Rohren," *VDI-Forschungsh.*, Vol. 361, pp. 1–22 (1933).

Notter, R. H. and C. A. Sleicher, "A solution to the turbulent Graetz problem III. Fully developed and entry region heat transfer rates," *Chem. Eng. Sci.*, Vol. 27 (1972).

Offor, U. H. and S. B. Alabi, "An accurate and computationally efficient explicit friction factor model," *Adv. In Chem. Eng. and Science*, Vol. 6, pp. 237–245 (2016).

Petukhov, B. S., in *Advances in Heat Transfer*, Vol. 6, T. F. Irvine and J. P. Hartnett eds., Academic Press, New York (1970).

Rohsenow, W. M., J. P. Hartnett, and Y.I. Cho, eds., *Handbook of Heat Transfer,* Third Edition, McGraw-Hill, New York (1998).

Shah, R. K., "Thermal entry length solutions for the circular tube and parallel plates," *Proc. 3rd Natl. Heat Mass Transfer Conf.,* Indian Inst. Technology., Bombay, Vol. I, Paper No. HMT-11-75 (1975).

Shah, R. K. and A. L. London, *Laminar Flow Forced Convection in Ducts*, Academic Press, New York (1978).

White, F. M., *Viscous Fluid Flow,* McGraw-Hill, New York (1991).

Wibulswas, P., *Laminar Flow Heat Transfer in Non-Circular Ducts*, Ph.D. thesis, London University, London (1966).

Zigrang, D. J. and N. D. Sylvester, "Explicit approximations to the solution of Colebrook's friction factor equation," *AIChE Journal*, Vol. 28, pp. 514–515 (1982).

Problems

Internal Flow Concepts

9.1 Liquid metal enters a tube in a laminar flow condition. Liquid metal has a very large thermal diffusivity (α) relative to its kinematic viscosity (v).

(a) Sketch the thermal and momentum boundary layers that develop in the tube; make sure that the relative magnitude of these boundary layers is approximately correct.

(b) Do you expect the hydrodynamic or thermal entry length to be larger?

9.2 Oil at temperature T_{in} and pressure P_{in} flows into a round tube with radius r_o. The first half of the tube length (L) is insulated and the surface of the remainder of the tube is exposed to a uniform heat flux.

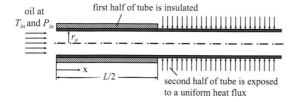

(a) Sketch the thermal boundary layer thickness (δ_t) as a function of position (x). Assume that the flow becomes thermally fully developed within the tube.

(b) Sketch the mean temperature of the fluid (T_m) and the surface temperature of the tube (T_s) as a function of position (x). Your sketch should be consistent with the thermal boundary layer thickness from (a).

9.3 Sketch the fully developed, time-averaged temperature distribution (i.e., temperature as a function of radius) in a pipe of radius 1 cm at a location where the surface temperature of the pipe is 100°C and the mean temperature of the fluid is 60°C under (a) laminar and (b) turbulent conditions. You may want to add a few words justifying your answer.

9.4 Fluid enters a pipe that has a uniform heat flux at its surface. The velocity and temperature distribution at the inlet to the pipe are uniform at u_{in} and T_{in}, respectively. The length of the pipe is L and its inner radius is R. The flow becomes hydrodynamically fully developed at location (1) and thermally fully developed at location (2), as shown in the figure. The flow is laminar.

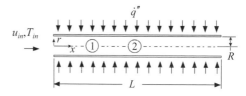

(a) Is the Prandtl number of the fluid less than, equal to, or greater than 1.0? Justify your answer.

(b) Sketch the pressure as a function of x. You need only get the qualitative features of your sketch correct. You may add justification for your sketch.

(c) Sketch the heat transfer coefficient as a function of x. You need only get the qualitative features of your sketch correct. You may add justification for your sketch.

(d) Sketch the temperature as a function of radius at locations 1 and 2. You need only get the qualitative features of your sketch correct. You may add justification for your sketch.

9.5 The figure illustrates flow through a duct. The flow enters at a uniform temperature, T_{in}. The first half of the duct (from $0 < x < L/2$) experiences a constant heat flux and the second half (from $L/2 < x < L$) is adiabatic. The flow transitions from laminar to turbulent at $x = L/4$ and becomes fully developed very quickly after it transitions. The properties of the fluid are constant.

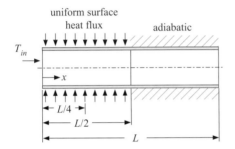

(a) Sketch the mean temperature of the fluid as a function of position from $0 < x < L$. Your sketch should capture the qualitative features of the flow correctly.

(b) Overlay on your sketch from (a) the surface temperature of the duct as a function of position from $0 < x < L$. Your sketch should capture the qualitative features of the flow correctly.

(c) If the roughness of the duct wall increased substantially then how would your sketch from (a) and (b) would change? Clearly justify your answer.

9.6 Fluid with a small Prandtl number, $Pr < 1$, enters a pipe. The fluid becomes thermally fully developed at about $x = b$ on the figure. The flow of the fluid is laminar.

flow of fluid
with $Pr < 1$
entering a pipe

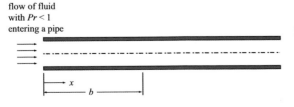

(a) Sketch on the thermal and momentum boundary layer thickness as a function of position (δ_t and δ_m – be sure to clearly show which is which). Label the hydrodynamic and thermal entry length, $x_{fd,t}$ and $x_{fd,h}$.

(b) Sketch the local and average heat transfer coefficient, h and $\bar{h}$, as a function of x.

(c) Sketch the local and average friction factor, f and $\bar{f}$, as a function of x.

A nonuniform heat flux is applied to the surface of the pipe. The fluid enters the pipe with mean temperature, $T_{m,in}$.

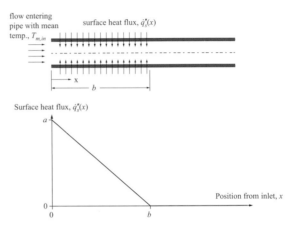

The heat flux as a function of position, x, is given by:

$$\dot{q}_s''(x) = \begin{cases} a\left(1 - \dfrac{x}{b}\right) \text{ for } x < b \\ 0 \text{ for } x > b \end{cases},$$

where a and b are constants. The heat flux as a function of position is shown in the figure; notice that it decreases linearly to 0 at $x = b$ and remains at 0 for all subsequent x.

(d) Sketch the mean temperature of the fluid as a function of position.

(e) On the same plot as (d), sketch the surface temperature of the pipe as a function of position.

9.7 Consider flow between two parallel plates with separation distance H, width W and length L. The dimension W (into the paper) is very large. The top side is subject to a constant heat flux and the bottom side is insulated.

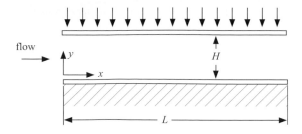

flow

(a) Carefully sketch and label the bulk temperature and the upper plate temperature as a function of distance from the entrance for both laminar and turbulent flow. The developing nature of the flow should be evident in the curves.

(b) Sketch and label the fluid temperature profile at $x = L$ for the two conditions.

9.8 Fluid flows through a tube with inner diameter D. You have been asked to help interpret some measured data for flow through the tube. Specifically, the heat transfer coefficient, h, has been measured as a function of the tube diameter. The figure illustrates the heat transfer coefficient measured in the thermally fully developed region of the tube as a function of the tube diameter; note that the mass flow rate of fluid ($\dot{m}$), the type of fluid, and all other aspects of the experiment are not changed for these measurements.

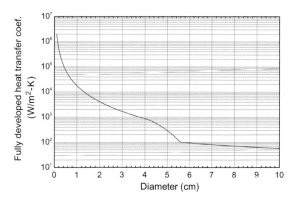

(a) Explain in a few sentences the abrupt change in the heat transfer coefficient that occurs at approximately $D = 5.5$ cm.

(b) Explain in a few sentences why the heat transfer is inversely proportional to diameter for diameters above about $D = 5.5$ cm; that is, why is it true that $h \propto D^{-1}$ for $D > 5.5$ cm?

(c) Sketch your expectation for how the figure would change if the roughness of the tube wall is increased dramatically.

Internal Flow Correlations

9.9 A 20 ft long plastic pipe has outer diameter 0.5 inch and wall thickness 0.065 inch. The conductivity of the plastic is 0.2 W/m-K and the surface roughness is very close to zero. Water flows through the pipe with volumetric flow rate 4.2 gal/min. The outer surface of the pipe is wrapped with a heater that applies a uniform heat flux of 2.5 kW/m^2. At the pipe exit the water has pressure 5 psig (gage) and mean temperature 12°C.

(a) Estimate the gage pressure of the water at the inlet (psig).

(b) Determine the temperature of the inner surface of the plastic pipe at the exit.

(c) Determine the temperature of the outer surface of the plastic pipe at the exit.

9.10 Repeat Problem 9.9 but change the following parameters in order to allow a higher heat flux, 25 kW/m^2. The pipe is made of steel with conductivity 30 W/m-K and roughness 0.15 mm.

(a) Estimate the gage pressure of the water at the inlet (psig).

(b) Determine the temperature of the inner surface of the pipe at the exit.

(c) Determine the temperature of the outer surface of the pipe at the exit.

(d) Plot the inlet gage pressure and the inner surface temperature of the pipe at the exit as a function of the surface roughness (for 0 mm < e < 0.4 mm). Explain the shape of these plots.

9.11 A geothermal bore field consists of many individual wells each 100 m deep. Within each well there is a plastic U-tube that allows fluid to travel to the bottom of the well, turn around and return to the top. The contact between the fluid and the surrounding soil either cools the fluid (in the summer) or heats it (in the winter). The fluid is a 20 percent ethylene-glycol in water mixture (the properties of such a mixture can be obtained using the brines property database in EES). The average temperature of the fluid is 20°C. The tube has an outer diameter of 0.625 inch and a wall thickness of 0.035 inch. The flow rate of the fluid is 3 gal/min.

(a) Determine the average heat transfer coefficient between the fluid and the tube.

(b) Determine the pressure drop associated with the flow through the U-tube.

(c) Plot the average heat transfer coefficient as a function of flow rate. Comment on any characteristics you see in your plot; is there a certain range of flows that should be avoided? Why?

9.12 A schedule 40 0.5 inch PVC pipe (0.84 inch outer diameter and 0.109 inch wall) has conductivity 0.19 W/m-K. Water is flowing through the pipe with flow rate 4.2 gal/min, pressure 15 psig and mean temperature 40°F. The outer surface of the pipe is exposed to very cold air at $\bar{h}_{out}$ = 50 W/m²-K and T_∞ = –30°F.

(a) Estimate the heat transfer per unit length from the water.

(b) Estimate the inner and outer surface temperatures of the pipe. Do you expect any freezing to occur for the water?

(c) Plot the inner surface temperature of the pipe as a function of flow rate. Is there a flow rate below which freezing might occur? Why?

9.13 You are designing a solar-powered livestock watering system for an off-grid application. The system must pump water from a L = 45 ft deep well through a D_{tube} = 0.5 inch tube. The pump has a dead-head pressure rise (i.e., the pressure rise produced in the absence of flow) of Δp_{dh} = 40 psi and an open circuit flow rate (i.e., the flow produced in the absence of any pressure rise) of $\dot{V}_o$ = 10 gal/min. The pump curve is linear between these two points. The water temperature is nominally T_w = 50°F. The pump efficiency is η_p = 0.65.

(a) What is the volumetric flow rate produced by the system?

(b) What is the power required by the pump?

The solar flux is on average $\dot{q}''_s$ = 650 W/m² over t_s = 8 hr per day and the photovoltaic panel has an efficiency of η_s = 0.08. The solar panel costs $3800/m².

(c) Determine the area and cost of the collector required.

(d) If a cow requires approximately 25 gallons of water per day then how many cows can your system service?

(e) Plot the number of cows that can be serviced and the cost of the collector as a function of the depth of the well, L. Explain the shape of these curves.

(f) Plot the cost of the collector per cow (i.e., the cost of the collector divided by the number of cows that can be serviced) as a function of the depth of the well.

(g) For a L = 45 ft deep well, plot the cost of the collector per cow as a function of the diameter of the tube. Explain the shape of this curve.

9.14 Air is being blown across a plastic tube through which water is flowing in order to cool the water. The tube has outer diameter 0.375 inch and wall thickness 0.035 inch. The conductivity of the tube material is 0.15 W/m-K and it has a surface roughness very close to zero. The water flow is hydrodynamically and thermally fully developed at this location. The water has bulk temperature 42°C and pressure 300 kPa and flows at a rate of 3.0 gal/min. The air has temperature 20°C and free stream velocity 15 m/s.

(a) Determine the rate of heat transfer from the water to the air per unit length of tube.

(b) If you were asked to improve the performance (i.e., increase the rate of heat transfer) what design modifications would you suggest? Justify your suggestions.

9.15 The figure illustrates a system designed to fill medicine bladders. The medicine bladders are balloons. Once filled, the balloon pressure can be used to passively deliver a small amount of medicine for a long period of time.

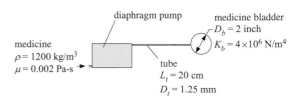

The system is energized by a small diaphragm pump. The pump curve data are provided in the table.

Flow rate (liter/min)	Pressure rise (psi)
0	24
0.2	18
0.4	14
0.6	10
0.8	6.5
1.0	4.6
1.2	3.5
1.6	1.2
1.9	0

The pressure within the balloon is related to its diameter according to:

$$\Delta p_b = K_b D_b^2,$$

where $K_b = 4 \times 10^6$ N/m^2. The diameter of the balloon is $D_b = 2$ inch. The density of the medicine is $\rho = 1200$ kg/m^3 and the viscosity of the medicine is $\mu = 0.002$ Pa-s. The diaphragm pump is connected to the balloon by a tube with length $L_t = 20$ cm and inner diameter $D_t = 1.25$ mm.

(a) Determine the volumetric flow rate provided to the balloon.

(b) What is the rate that the balloon diameter is growing?

(c) If the initial balloon diameter is $D_{b,ini} = 0.25$ inch, use the Integral command in EES to determine and plot the balloon diameter as a function of time. Also show the pressure rise across the pump, pressure rise across the balloon and volumetric flow rate as a function of time.

9.16 A heat exchanger cools water that flows through $N = 25$ tubes in a tube bundle. The water enters with mass flow rate $\dot{m}$ and a pressure of 80 psig. The nominal temperature of the water is 65°F. The water splits equally and flows through each of the tubes in parallel. The external surface of the tube is exposed to an evaporating refrigerant causing the water to cool. Let's focus on analyzing a pump for this application. Each tube has inner diameter $D = 10$ mm and length $L = 2.5$ m.

You have selected a pump for this application – the manufacturer's pump curve is shown in the figure (you have selected the 7.75 inch impeller

model). Note that the y-axis is pressure rise (in feet of water) and the x-axis is flow rate (in gallons per minute).

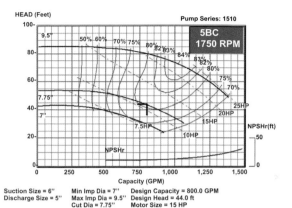

Suction Size = 6"   Min Imp Dia = 7"   Design Capacity = 800.0 GPM
Discharge Size = 5"   Max Imp Dia = 9.5"   Design Head = 44.0 ft
                      Cut Dia = 7.75"   Motor Size = 15 HP

(a) Estimate the operating point that will result if you connect this pump to your heat exchanger. What flow rate, pressure rise and efficiency will result? What power will the pump consume?

(b) What is the average heat transfer coefficient between the water and the inner surface of the tube?

9.17 Dismounted soldiers and emergency response personnel are routinely exposed to high-temperature/high-humidity environments as well as external energy sources such as flames, motor heat, or solar radiation. The protective apparel required by chemical, laser, biological, and other threats tend to have limited heat removal capability. These and other factors can lead to severe heat stress. One solution is a portable, cooling system integrated with an encapsulating garment to provide metabolic heat removal. A portable metabolic heat removal system that is acceptable for use by a dismounted soldier or emergency response personnel must satisfy a unique set of criteria. The key requirement for such a system is that it should be extremely low mass and very compact in order to ensure that any gain in performance due to active cooling is not offset by fatigue related to an increase in pack load. In order to allow operation for an extended period of time, a system must either be passive (require no consumable energy source), very efficient (require very little consumable energy), or draw energy from a high-energy density power source. One alternative for providing portable metabolic heat removal is with an ice pack.

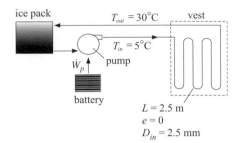

ice pack | $T_{out} = 30°C$ | vest

$T_{in} = 5°C$
pump
$\dot{W}_p$
battery

$L = 2.5$ m
$e = 0$
$D_{in} = 2.5$ mm

The pump forces a liquid antifreeze solution to flow through plastic tubes in the vest in order to transfer the cooling from the ice to the person. Assume that the surface of the plastic is completely smooth, $e = 0$, the total length of the tube is $L = 2.5$ m and the inner diameter of the tube is $D_{in} = 2.5$ mm. There are $N_b = 20$ bends in the vest; the loss coefficient associated with each bend is $C_b = 1.0$. The fluid that is being circulated through the vest has properties $\rho_f = 1110$ kg/m³, $c_f = 2415$ J/kg-K, $\mu_f = 0.0157$ Pa-s, and $Pr_f = 151$. The fluid enters the vest at $T_{in} = 5.0°C$ and leaves the vest at $T_{out} = 30°C$. You may assume that the pressure drop associated with the vest is much greater than the pressure drop associated with any other part of the system.

(a) Assume that the bulk velocity in the tube is $u_m = 1.0$ m/s. Determine the pressure drop required to circulate the fluid through the vest.

A miniature diaphragm pump is used to circulate the fluid. Assume that the pressure rise produced by the pump (Δp_p) varies linearly from the dead head pressure rise $\Delta p_{p,dh} = 30$ psi at no flow ($\dot{V}_p = 0$) to zero at the maximum unrestricted flow rate $\dot{V}_{p, open} = 650$ mL/ min :

$$\Delta p_p = \Delta p_{p, dh} \left(1 - \frac{\dot{V}_p}{\dot{V}_{p, open}} \right).$$

(b) Determine the fluid flow rate through the vest that is consistent with the pump curve given by the equation above; that is, vary the value of the mean velocity, u_m, until the pressure drop across the vest and flow rate through the vest falls on the pump curve.

(c) How much cooling is provided by the vest?
(d) If the pump efficiency is $\eta_p = 0.20$ then how much power is consumed by the pump?
(e) If the system is run for *time* = 1 hour then what is the mass of ice that is consumed? (Assume that the latent heat of fusion associated with melting ice is $i_{fs} = 3.33 \times 10^5$ J/kg and that the only energy transfer to the ice is from the fluid.) What is the mass of batteries that are required, assuming that the energy density of a lead acid battery is $ed_b = 0.05$ kW-hr/kg?

9.18 A heat exchanger cools water that flows through $N = 25$ tubes in a tube bundle. The water enters with mass flow rate $\dot{m} = 24$ kg/s and a pressure of 80 psig. The nominal temperature of the water is 65°F. The water splits equally and flows through each of the tubes in parallel. The external surface of the tube is exposed to an evaporating refrigerant causing the water to cool. Let's focus on selecting a pump for this application. Each tube has inner diameter $D = 10$ mm and length $L = 2.5$ m.

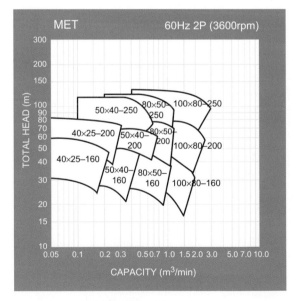

(a) Estimate the pressure drop experienced by the water as it passes through the heat exchanger.
(b) If the pump efficiency is nominally $\eta_p = 0.65$ then what pumping power is required?

(c) Select the most appropriate centrifugal pump model from the manufacturer's selection table shown in the figure.

The Energy Balance

9.19 The figure illustrates flow through a pipe where the surface heat flux decreases linearly with axial position until it reaches zero at the exit, $x = L$.

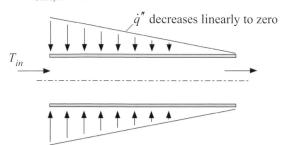

(a) Sketch the mean temperature of the fluid as a function of position, x.

(b) Sketch the surface temperature of the duct as a function of position, x.

9.20 In a biofuel refinery, ethanol enters a tube at temperature $T_{in} = 5°C$ and pressure $p_{in} = 45$ psia. The mass flow rate of the ethanol is $\dot{m} = 0.04$ kg/s. The outer diameter of the tube is $D_o = 0.25$ inch and the length of the tube is $L = 5$ ft. The tube wall thickness is $th = 0.03$ inch. The roughness on the internal surface of the tube is estimated to be $e = 15$ μm. The outer surface of the tube is exposed to saturated steam at $p_s = 5$ psia. You may neglect the convection resistance on the outer surface of the tube as well as the resistance to radial conduction through the tube (therefore, the inner surface of the tube may be assumed to be at a constant temperature, equal to the saturation temperature of the steam).

(a) Determine the outlet pressure and temperature of the ethanol. Calculate the properties of ethanol using the average temperature and pressure of the ethanol as it flows through the tube.

(b) Plot the outlet temperature and pressure drop as a function of the mass flow rate of ethanol for 0.001 kg/s $< \dot{m} < 0.05$ kg/s. Explain your plots.

(c) Determine the tube diameter and length that will lead to the target performance, $T_{out} = 40°C$ and $\Delta p = 20$ kPa. Assume that the tube thickness does not change with tube diameter.

9.21 Cold fluid enters a tube that has a constant surface temperature. The surface temperature of the tube is higher than the fluid inlet temperature (i.e., the fluid is heated as it passes through the tube). If the mass flow rate of the fluid is doubled, answer the following questions.

(a) Will the mean temperature of the fluid leaving the pipe increase or decrease? Assume that the average heat transfer coefficient does not change and justify your answer.

(b) Will the amount of heat transferred to the fluid increase or decrease? Again, assume that the average heat transfer coefficient does not change and justify your answer.

9.22 Your cabin is located adjacent to a stream and you like cold beverage out of a keg. There is no electricity to your cabin, so you cannot put the keg in the refrigerator. Therefore, you have decided to run the warm fluid from the keg through a tube that is submerged in the stream to get cold beverage. You can pressurize the keg to $\Delta P_k = 20$ psi above ambient to force the fluid through the tube. Your initial design for the cooler is a tube, $L = 20$ ft long, that has an outer diameter of $D_o = 0.25$ inch with thickness $th = 0.02$ inch. The conductivity of the tube material is $k_t = 15$ W/m-K. The fluid in the keg is at room temperature, $T_{in} = 20°C$. The properties of the fluid are $\rho_b = 900$ kg/m^3, $k_b = 0.4$ W/m-K, $\mu_b = 0.001$ Pa-s, and $c_b = 3850$ J/kg-K. The temperature of the water in the stream is $T_w = 5°C$ and the velocity of the stream is $u_w = 5$ ft/s. Assume that the tube is exposed to the stream velocity in cross-flow.

(a) Determine the volumetric flow rate and mass flow rate of fluid through the tube.

(b) If your cup is $V_{cup} = 0.5$ liter then how long does it take to fill?

(c) Determine the total conductance between the fluid in the tube and the stream.

(d) What is the outlet temperature of the fluid?

(e) Plot the outlet temperature of the fluid and the time to fill the cup as a function of the length of the tube.

9.23 Air enters a 1 m long pipe having a 25 mm inner diameter at 300 K with a mass flow rate of 0.0041 kg/s. The pipe wall is maintained at 500 K. The outlet air temperature is 400 K.

(a) Determine the rate of heat transfer to the air.

(b) Determine the experimental value of the convection heat transfer coefficient.

(c) Determine the experimental value of the Nusselt number.

(d) Compare the value determined in part (c) to the value predicted by an appropriate correlation.

9.24 Fluid enters a long circular tube with a constant surface temperature, $T_s = 100°C$, with an inlet temperature, $T_{m,in} = 0°C$.

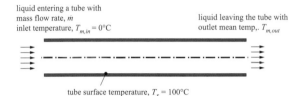

liquid entering a tube with mass flow rate, $\dot{m}$ inlet temperature, $T_{m,in} = 0°C$

liquid leaving the tube with outlet mean temp,. $T_{m,out}$

tube surface temperature, $T_s = 100°C$

You have been asked to help interpret some data that are puzzling. The figure illustrates the measured mean outlet fluid temperature, $T_{m,out}$, as a function of the mass flow rate of the fluid (all other quantities are fixed, including the fluid properties and tube geometry).

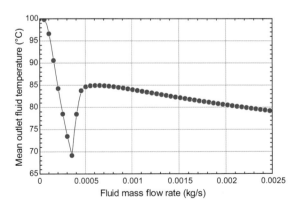

(a) Explain (in a few sentences) the sharp transition that causes the outlet temperature to begin to initially drop and then increase as the flow rate increases.

(b) Explain (in a few sentences) why the outlet temperature at high mass flow rate tends to decrease with increasing mass flow rates.

9.25 A solar collector for water heating is mounted on the roof of a house. The heated water leaving the collector passes through an un-insulated, long tube before re-entering the house through the roof. You need to estimate the heat loss from the pipe in order to decide whether it is worth insulating it.

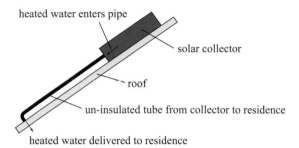

heated water enters pipe

solar collector

roof

un-insulated tube from collector to residence

heated water delivered to residence

The figure below shows the pipe in more detail.

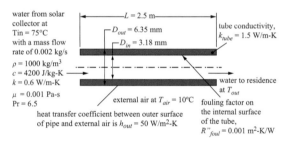

water from solar collector at Tin = 75°C with a mass flow rate of 0.002 kg/s
$\rho = 1000$ kg/m³
$c = 4200$ J/kg-K
$k = 0.6$ W/m-K
$\mu = 0.001$ Pa-s
Pr = 6.5

$L = 2.5$ m
$D_{out} = 6.35$ mm
$D_{in} = 3.18$ mm

tube conductivity, $k_{tube} = 1.5$ W/m-K

water to residence at T_{out}

external air at $T_{air} = 10°C$

fouling factor on the internal surface of the tube, $R''_{foul} = 0.001$ m²-K/W

heat transfer coefficient between outer surface of pipe and external air is $h_{out} = 50$ W/m²-K

The inner diameter of the tube (D_{in}) is 3.18 mm and the outer diameter (D_{out}) is 6.35 mm. The length of the tube (L) is 2.5 m. The outer surface of the pipe is exposed to blowing air at $T_{air} = 10°C$; the average convection coefficient between the outer surface of the pipe and the air (h_{out}) is 50 W/m²-K. The conductivity of the tube material (k_{tube}) is 1.5 W/m-K. The internal surface of the tube is fouled with calcium chloride deposits due to hard water; the fouling factor is $R''_{foul} = 0.001$ m²-K/W. Water enters the tube at a temperature (T_{in}) of 75°C and a mass flow rate ($\dot{m}$) of 0.002 kg/s. The properties of water are $\rho = 1000$ kg/m³, $\mu = 0.001$ Pa-s, $k = 0.6$ W/m-K, $c = 4200$ J/kg-K, Pr = 6.5.

Determine the mean outlet temperature of the water (T_{out}). What is the rate of heat loss from the fluid as it flows through the pipe (W)?

9.26 Fluid flows through a tube that is exposed to a cross-flow of gas on its outer surface.

gas flowing over a tube
$T_g = 10°C$
$u_g = 10$ m/s
$\rho_g = 1$ kg/m³
$c_g = 1000$ J/kg-K
$\mu_g = 2 \times 10^{-5}$ Pa-s
$k_g = 0.026$ W/m-K

$L = 20$ m

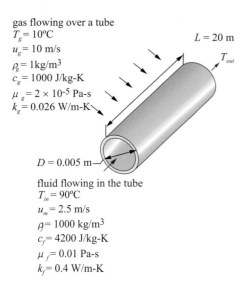

T_{out}

$D = 0.005$ m

fluid flowing in the tube
$T_{in} = 90°C$
$u_m = 2.5$ m/s
$\rho_f = 1000$ kg/m³
$c_f = 4200$ J/kg-K
$\mu_f = 0.01$ Pa-s
$k_f = 0.4$ W/m-K

The fluid enters the tube with mean temperature $T_{in} = 90°C$ and mean velocity $u_m = 2.5$ m/s. The fluid has density $\rho_f = 1000$ kg/m³, specific heat capacity $c_f = 4200$ J/kg-K, viscosity $\mu_f = 0.01$ Pa-s, and conductivity $k_f = 0.4$ W/m-K. Gas flows over the tube with velocity $u_g = 10$ m/s and temperature $T_g = 10°C$. The gas has density $\rho_g = 1$ kg/m³, specific heat capacity $c_g = 1000$ J/kg-K, viscosity $\mu_g = 2 \times 10^{-5}$ Pa-s, and conductivity $k_g = 0.026$ W/m-K. The diameter of the tube is $D = 0.005$ m and its length is $L = 20$ m. The inner and outer surfaces of the tube are smooth. The tube has a very thin wall so that its inner and outer diameters are nearly the same and there is very little resistance to conduction. You may assume that the liquid is fully developed for most of the tube. Determine the outlet temperature of the fluid.

9.27 Water enters a thin wall tube with length $L = 5$ m and diameter $D = 0.25$ inch. The internal roughness of the tube surface is $e = 15$ μm. A uniform heat flux of $\dot{q}'' = 45.1$ kW/m² is applied to the surface of the tube. The flow rate of the water is $\dot{V} = 0.5$ gal/min and the inlet temperature is $T_{in} = 10°C$. The tube exits to atmospheric pressure.
(a) Determine the mean temperature of the flow leaving the tube.
(b) Determine the pressure at the inlet to the tube.
(c) Determine the surface temperature of the tube at the exit.
(d) How long can the tube be before you will see local boiling at the surface of the tube at the exit?

9.28 Water flows through a 1 m long, flat, aluminum strip that has been extruded with 10 "mini-channels" in it. Each of these channels is 3 mm in diameter and is separated from its neighbors by 1 mm. Air flows across the 43 mm wide aluminum strip at a free stream velocity of 5 m/s and a free stream temperature of 300 K. If the total water flow rate is 0.007 kg/s, and the inlet water temperature is 350 K, then estimate the outlet water temperature.

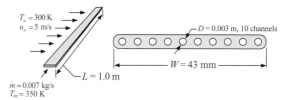

$T_\infty = 300$ K
$u_\infty = 5$ m/s

$D = 0.003$ m, 10 channels

$W = 43$ mm

$L = 1.0$ m

$\dot{m} = 0.007$ kg/s
$T_{in} = 350$ K

9.29 The surface of a tube is exposed to a constant heat flux, $\dot{q}_s'' = 19{,}000$ Btu/hr-ft². The diameter of the tube is $D = 0.25$ inch and the length is $L = 4$ ft. The tube wall roughness is $e = 0.005$ inch. Water flows through the tube with $\dot{m} = 400$ lbm/hr. The water enters the tube at $T_{in} = 50°F$ and $p_{in} = 50$ psi.
(a) Determine the outlet temperature of the water.
(b) Determine the temperature at the pipe surface at the outlet.
(c) Determine the pressure drop through the pipe.

9.30 A fluid with a Prandtl number less than one, $Pr < 1$, flows through a pipe. The heat flux on the surface of the pipe depends on position as shown in the figure.

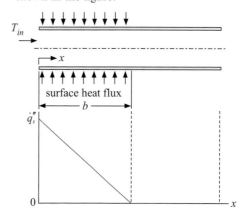

T_{in}

x

surface heat flux

b

$\dot{q}_s''$

0

x

The fluid becomes thermally fully developed at location $x = b$. The flow of the fluid is laminar.
(a) Sketch the thermal and momentum boundary layer thickness as a function of position

(δ_t and δ_m – be sure to clearly show which is which). Label the hydrodynamic and thermal entry lengths, $x_{fd,t}$ and $x_{fd,h}$, in your sketch. Show the location $x = b$ in your sketch.

(b) Sketch the local and average heat transfer coefficient, h and $\bar{h}$, as a function of x; indicate on your sketch the location $x = b$.

(c) Sketch the local and average friction factor, f and $\bar{f}$, as a function of x; indicate on your sketch the location $x = b$.

The figure shows the nonuniform heat flux is applied to the surface of the pipe. The heat flux decreases linearly from $x = 0$ to $x = b$ and remains at 0 for all subsequent x. The fluid enters the pipe with mean temperature, T_{in}.

(d) Sketch the mean temperature of the fluid as a function of position.

(e) Sketch the surface temperature of the pipe as a function of position.

9.31 Engine oil (20W) enters a 50 m long thin-walled smooth copper tube of 30 mm inner diameter at 150°C at a rate of 0.35 kg/s. The tube is suspended in a location in which the air temperature is 20°C and the convection coefficient between the outer wall of the tube and the air is 21 W/m²-K. Estimate the pressure drop, exit oil temperature and heat transfer rate for the tube.

9.32 Water flowing through a 1 inch inner diameter pipe at a rate of 5000 lb$_m$/hr is to be heated from 100°F to 180°F at atmospheric pressure. The outside surface temperature of the pipe is held constant at 200°F and the thermal resistance of the pipe wall is negligible. A test with a 4 ft pipe under the above conditions resulted in a water outlet temperature of 125°F.

(a) Estimate the length of pipe needed to provide the 180°F outlet temperature.

(b) Determine the pressure drop for these conditions.

9.33 Water is to be heated to 150°F by pumping it through a smooth copper tube that has an inner diameter of 0.87 inch. The tube is wrapped on the outside with an electrical resistance heater. The tube and the heater are then covered with a thick layer of insulating material. The water flow rate is 156 lb$_m$/min and the water enters the tube at 50°F.

(a) What heater power is required?

(b) How long must the tube be to heat the water to 150°F if the maximum tube wall temperature cannot exceed 210°F?

(c) What is the minimum pump power requirement?

9.34 An air-cooled condenser for a refrigeration cycle consists of 2000 thin-walled aluminum tubes that are 0.5 m long and 0.015 m in diameter. The total mass flow rate of air through these tubes is 10 kg/s. The air enters at 20°C. Refrigerant at 60°C condenses at constant pressure on the outside surface of the tubes with a heat transfer coefficient of 620 W/m²-K. The enthalpy of vaporization of the refrigerant is 150 kJ/kg.

(a) Determine the air exit temperature.

(b) Determine the mass flow rate of condensate.

9.35 Water at 1 atmosphere flows in a thin-walled pipe with a smooth surface. The inlet temperature of the water is $T_{in} = 20$°C. The pipe has inner diameter $D = 1.0$ cm and length $L = 11.5$ m. The external surface of the pipe is wrapped with a heater that provides a uniform heat flux of $\dot{q}'' = 75$ kW/m². The properties of water are $\rho = 1000$ kg/m³, $c = 4200$ J/kg-K, $k = 0.6$ W/m-K, and $\mu = 0.00085$ Pa-s.

(a) Estimate the heat transfer coefficient between the water and the pipe surface.

(b) Determine whether the water boils anywhere before it exits the pipe.

(c) Will the mean temperature of the water at the outlet of the pipe increase, decrease, or stay the same if the pipe surface roughness increases?

(d) Will the surface temperature at the exit increase, decrease, or stay the same if the pipe surface roughness increases?

9.36 One concept for rapidly launching small satellites involves a rocket-boosted, expendable launch vehicle that is dropped from the cargo bay of a military cargo aircraft. The launch vehicle is propelled by self-pressurized tanks of liquid oxygen and liquid propane. The liquid oxygen fuel tank (referred to as the propellant tank) is at elevated pressure and must be kept full while the aircraft sits on the runway, flies to the launch coordinates, and potentially holds position in order to wait for a strategically appropriate launch time; the design requires that the propellant tank remain full for $time_{wait}$ = 12 hours. The propellant tank contains saturated liquid oxygen at p_{tank} = 215 psia. Saturated liquid oxygen at this pressure has a temperature of T_{tank} = 126.8 K. Because the tank is so cold, it is subjected to a large heat leak, $\dot{q}_{tank}$. Without external cooling, it would be necessary to vent the liquid oxygen that boils off in order to maintain the proper pressure and therefore the tank would slowly be emptied. It is not possible to place a cryogenic refrigerator with 20 kW of cooling capacity in the propellant tanks in order to re-liquefy the oxygen. Rather, an adjacent Dewar of liquid oxygen (referred to as the conditioning tank) is used to remove the parasitic heat transfer and prevent any oxygen in the propellant tank from boiling away. The figure illustrates the proposed system. A pump is used to circulate liquid oxygen from the propellant tank through a cooling coil that is immersed in the conditioning tank. The pump and conditioning tank can be quickly removed from the launch vehicle when it is time for launch. The conditioning tank is maintained at p_{ct} = 14.7 psia and contains saturated liquid oxygen; any oxygen that evaporates due to the heat added by the cooling coil is allowed to escape. The cooling coil is a coiled up tube with total length L = 10 m, inner diameter D_i = 0.8 cm and outer diameter D_o = 1.0 cm. The internal surface of the tube has roughness e = 50 μm and the conductivity of the tube material is k_{tube} = 2.5 W/m-K. The mass flow rate provided by the pump is $\dot{m} = 0.25$ kg/s and the pump efficiency is η_{pump} = 0.45. The heat transfer coefficient associated with the evaporation of the liquid oxygen in the conditioning tank from the external surface of the tube is $h_o = 2 \times 10^4$ W/m²-K. You may assume that the liquid oxygen that is pumped through the cooling coil has constant properties that are consistent with saturated liquid oxygen at the tank pressure.

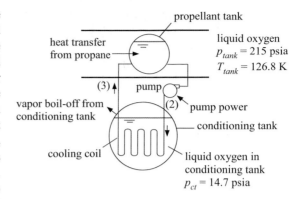

propellant tank

heat transfer from propane

liquid oxygen
p_{tank} = 215 psia
T_{tank} = 126.8 K

(3) pump

(2) pump power

vapor boil-off from conditioning tank

conditioning tank

cooling coil

liquid oxygen in conditioning tank
p_{ct} = 14.7 psia

(a) What is the pressure drop associated with forcing the liquid oxygen through the cooling coil?
(b) What is the power required by the pump?
(c) If all of the pump power ultimately is transferred to the liquid oxygen that is being pumped then what is the temperature of the liquid oxygen leaving the pump (T_2 in the figure)?
(d) What is the heat transfer coefficient between the liquid oxygen flowing through the cooling coil and the internal surface of the tube?
(e) What is the total conductance associated with the cooling coil?
(f) What is the temperature of the liquid oxygen leaving the cooling coil (T_3 in the figure)?
(g) How much cooling is provided to the propellant tank?
(h) Plot the cooling provided to the conditioning tank and the pump power as a function of the mass flow rate. If the parasitic heat leak to the propellant tank is $\dot{q}_{tank}$ = 10 kW then suggest the best mass flow rate to use for the system.

9.37 Air at atmospheric pressure and inlet temperature 10°C is blown through a rectangular channel that forms part of an air-heating solar collector at an inlet volumetric flow rate of 0.0023 m³/s. The channel is 0.9 m wide, 1.8 m in length and 2.5 cm high. The channel walls may be assumed to be at a constant temperature of 82°C.
(a) Determine the required pressure drop.
(b) Determine the outlet air temperature and heat transfer rate to the air.

9.38 Air is blown through a rectangular duct measuring 12 inches × 7 inches in cross-section. Air enters the duct from a residential furnace at 38°C at a volumetric flow rate of 230 cfm. The duct is 24 ft in length. Thermal losses occur to the surroundings that are at 25°C. Heat transfer on the outside surface of the duct occurs with an average heat transfer coefficient of 1.6 W/m²-K. Determine the following quantities:

(a) the average heat transfer coefficient on the inside surface of the duct,

(b) the overall conductance considering the heat transfer resistances on both the inside and outside of the duct,

(c) the temperature of the air exiting the duct,

(d) rate of heat loss from the duct, and

(e) the minimum power required to move the air through this duct.

9.39 Water from a building chilled water system passes through the evaporator of a refrigeration system. The evaporator consists of 100 copper tubes; each tube is 5 m long with a 1.0 cm outer diameter and a 0.8 cm inner diameter. The *total* mass flow rate of water passing through the evaporator is 0.50 kg/s. The water enters the evaporator at 20°C and passes through the tubes in parallel (assume an equal mass flow of water in each tube). The properties of water are: $\rho = 1000$ kg/m³, $\mu = 0.001$ Pa-s, $k = 0.59$ W/m-K, $c = 4183$ J/kg-K, and $Pr = 7.2$. On the outside of the tube, refrigerant is evaporating at a constant temperature of 10°C. The heat transfer coefficient between the tube's external surface and the refrigerant is 500 W/m²-K. Determine the temperature of the water leaving the evaporator (°C) and returning to the building.

9.40 The figure illustrates a system to reject energy from a heat pump to a lake.

Hot ethylene glycol at $T_{eg,in} = 40$°C enters a long tube that is immersed in the lake. The properties of ethylene glycol are $\rho_{eg} = 1114$ kg/m³, $c_{eg} = 2415$ J/kg-K, $\mu_{eg} = 0.0157$ Pa-s, $k_{eg} = 0.252$ W/m-K, and $Pr_{eg} = 151$. The volumetric flow rate of ethylene glycol is $\dot{V}_{eg} = 1 \times 10^{-4}$ m³/s. The tube has length $L = 200$ m and diameter $D = 0.015$ m; you may assume that the tube is very thin walled. The temperature of the lake water is $T_w = 8$°C and the heat transfer coefficient between the lake water and the outer surface of the tube is $\bar{h}_{out} = 75$ W/m²-K.

(a) Determine the heat transfer coefficient between the ethylene glycol and the tube wall.

(b) Determine the total pressure drop associated with the flow of ethylene glycol through the tube.

(c) Determine the temperature of the ethylene glycol leaving the lake, $T_{eg,out}$.

(d) The cooled ethylene glycol leaving the lake enters a counterflow heat exchanger and cools a flow of hot water. The water enters at $T_{w,in} = 60$°C with capacitance rate $\dot{C}_w = 200$ W/K and the ethylene glycol exits at 40°C, as shown in the figure. Determine the exiting water temperature, $T_{w,out}$.

9.41 When you turn on your shower, 0.75 gallons/min of water is taken from the city water supply (at 300 kPa) and directed to your hot water heater. The water leaves your hot water heater at 50°C and 250 kPa and travels through 25 m of half-inch copper tube (1.1 cm inner diameter and 1.3 cm outer diameter) to finally reach your shower head. The copper tube is exposed to ambient air (at 20°C). The heat transfer coefficient between the external surface of the tube and the ambient air is 35 W/m²-K.

(a) Estimate the steady-state temperature of the water when it reaches your shower (°C).

(b) Estimate the water pressure when it reaches your shower (kPa).

(c) If you adjust the shower to increase the mass flow rate of hot water (e.g. 1 gal/min), will the hot water reach the shower warmer or cooler than the answer from part (a)?

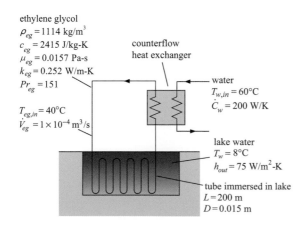

ethylene glycol
$\rho_{eg} = 1114$ kg/m³
$c_{eg} = 2415$ J/kg-K
$\mu_{eg} = 0.0157$ Pa-s
$k_{eg} = 0.252$ W/m-K
$Pr_{eg} = 151$

$T_{eg,in} = 40$°C
$\dot{V}_{eg} = 1 \times 10^{-4}$ m³/s

counterflow
heat exchanger

water
$T_{w,in} = 60$°C
$\dot{C}_w = 200$ W/K

lake water
$T_w = 8$°C
$\bar{h}_{out} = 75$ W/m²-K

tube immersed in lake
$L = 200$ m
$D = 0.015$ m

9.42 The figure illustrates flow through a rectangular duct. The height of the duct is $H = 0.01$ m and the width of the duct is $W = 0.05$ m. The total length of the duct in the flow direction is

$L_{total} = 1$ m. The duct has two sections of equal length, $L = 0.5$ m. The surface temperature of the first section of the duct is maintained at a constant temperature, $T_s = 200$ K. The second section of the duct is insulated and therefore you may assume that the surface of the second section is adiabatic. Fluid enters the duct with a uniform temperature, $T_{in} = 100$ K. The mass flow rate is $\dot{m} = 0.25$ kg/s and the properties of the fluid are $\rho = 1000$ kg/m^3, $c = 100$ J/kg-K, $\mu = 0.01$ Pa-s, and $k = 1.0$ W/m-K.

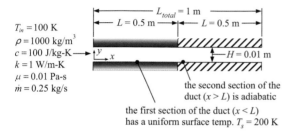

$T_{in} = 100$ K
$\rho = 1000$ kg/m^3
$c = 100$ J/kg-K
$k = 1$ W/m-K
$\mu = 0.01$ Pa-s
$\dot{m} = 0.25$ kg/s

$L_{total} = 1$ m
$L = 0.5$ m $L = 0.5$ m
$H = 0.01$ m

the second section of the duct ($x > L$) is adiabatic

the first section of the duct ($x < L$) has a uniform surface temp. $T_s = 200$ K

$W = 0.05$ m $H = 0.01$ m

view of duct in the direction of flow

(a) Is the flow in the duct laminar or turbulent? Justify your answer.

(b) Estimate the hydrodynamic and thermal entry lengths ($x_{fd,h}$ and $x_{fd,t}$) for the flow.

For the remaining questions, assume that the flow becomes hydrodynamically and thermally fully developed at exactly half-way through the heated portion of the duct (i.e., $x_{fd,h} \approx x_{fd,t} \approx L/2$); note that this may or may not be the correct answer to part (b).

(c) Sketch the pressure as a function of position x in the duct.

(d) The average friction factor in the duct is $\bar{f} = 0.13$; use this value to estimate the pressure drop across the duct (i.e., the pressure drop from $x = 0$ to $x = L_{total}$).

(e) If the roughness of the surface of the duct is increased substantially do you expect that the pressure drop across the duct would change significantly? Justify your answer.

(f) The average Nusselt number in the first section of the duct (i.e., from $x = 0$ to $x = L$) is $\overline{Nu} = 5.2$; use this value to estimate the average heat transfer coefficient in the first section of the duct.

(g) What is the mean temperature of the fluid leaving the first section of the duct?

For the remainder of this problem, assume that the mean temperature of the fluid leaving the first section of the duct is $T_{out} = 175$ K; note that this may or may not be the correct answer to part (g).

(h) Sketch the mean temperature as a function of position x in the duct. Make sure that your sketch extends all the way to the outlet of the duct (i.e., from $x = 0$ to $x = L_{total}$).

(i) Sketch the temperature as a function of y at the five x-locations in the duct that are indicated in the figure. The x-locations correspond to the following axial positions:

1: inlet to the duct,
2: half-way through the thermally developing region (i.e., $x_2 = x_{fd,t}/2$),
3: at the thermal entry length (i.e., $x_3 = x_{fd,t}$),
4: at the end of the 1st section (i.e., $x_4 = L$),
5: at the outlet of the duct (i.e., $x_5 = L_{total}$).

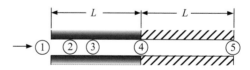

L L

① ② ③   ④   ⑤

9.43 Fluid flows through the annular space formed between a cylindrical heating element and an insulated pipe.

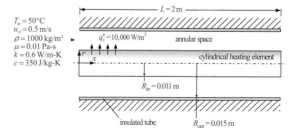

$T_{in} = 50°$C
$u_\infty = 0.5$ m/s
$\rho = 1000$ kg/m^3
$\mu = 0.01$ Pa-s
$k = 0.6$ W/m-K
$c = 350$ J/kg-K

$L = 2$ m
$\dot{q}_s'' = 10{,}000$ W/m^2 annular space
cylindrical heating element

$R_{in} = 0.011$ m

insulated tube $R_{out} = 0.015$ m

The inner and outer radii of the gap are R_{in} = 0.011 m and R_{out} = 0.015 m, respectively. The length of the flow passage is L = 2 m. The outer tube is insulated. The cylindrical heating element provides a uniform heat flux, $\dot{q}_s''$ = 10,000 W/m², at the inner radius. The fluid enters the gap at a uniform temperature, T_{in} = 50°C, and uniform velocity u_∞ = 0.5 m/s. The fluid properties are ρ = 1000 kg/m³, μ = 0.01 Pa-s, k = 0.6 W/m-K, and c = 350 J/kg-K.

(a) Is the flow laminar or turbulent? Justify your answer with a calculation.

For the remainder of this problem assume that your calculation from (a) showed that the flow is laminar.

(b) Estimate the thermal entry length using your conceptual understanding of boundary layer behavior.

For the remainder of this problem assume that your calculation from (b) showed that $x_{fd,t}$ = 1 m.

(c) Estimate the local heat transfer coefficient at the exit to the flow passage (x = 2 m). Do not use a correlation – instead obtain an estimate based on your conceptual understanding of laminar flow.

For the remainder of this problem assume that your calculation from (c) showed that the local heat transfer coefficient at the exit is 175 W/m²-K.

(d) Sketch the local heat transfer coefficient as a function of x.

(e) If the fluid boils at T_{boil} = 100°C then will any boiling occur within the duct at any location? Justify your answer.

(f) Sketch the mean temperature, the temperature at the surface of the element (r = R_{in}), and the temperature at the surface of the insulated tube (r = R_{out}) as a function of position.

(g) Sketch the temperature of the fluid as a function of radius at x = 0, x = 0.5 m, x = 1 m, and x = 2 m.

9.44 A beverage dispenser provides carbonated or noncarbonated drinks that are cooled at the time of dispense by running the fluid through stainless steel tubes that are embedded in a cast aluminum cold plate that is cooled by ice cubes. The ice cubes are either manually loaded into the unit or automatically produced by a top-mounted ice cuber. The cold plate represents the heart of the beverage dispenser; it is the most expensive component within the system and its performance determines whether the beverage dispenser will be capable of meeting the standards for temperature/flow performance that are set by the large distributors. The cold plate must provide a high rate of beverage cooling by channeling the fluid through stainless steel tubes that are embedded within the heavy aluminum plate. The cold plate system is shown schematically in the figure. The melting ice cools the aluminum to a spatially uniform temperature of T_{wall} = 0°C. This process requires a relatively long time (minutes) because the thermal communication between the external surface of the cylinder and the ice is not very high; fortunately, in most convenience stores there is a long time between customers dispensing the same beverage. When the customer activates the dispenser, liquid at T_{in} = 20°C enters the tubes.

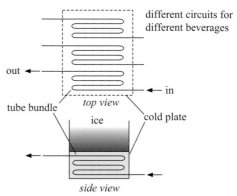

The beverage dispenser actually works as a thermal storage unit; the plate is cooled to the ice temperature between dispense processes, storing "cold" or cooling potential that is transferred very quickly to the beverage during each dispense. The relatively large aluminum cold plate is necessary so that aluminum doesn't change temperature much. For this problem, you may assume that the temperature of the aluminum block does not change at all during the dispense process and therefore the flowing fluid is exposed to a uniform wall temperature.

A V = 0.35 liter beverage must be dispensed in less than t_d = 10.0 s (the customer doesn't want to wait long to fill his cup). Assume that the beverage has the properties of water. The total length of the stainless steel tube bundle embedded in the aluminum block is $L = 1$ m and the inner diameter of the tube is $D_i = 3.1$ mm. The roughness associated with the tube surface is $e = 15$ μm. There are $N_b = 10$ bends of 180° in the tube bundle, each characterized by a loss coefficient of $C_b = 1.7$.

(a) What is the heat transfer coefficient between the fluid and the tube wall?

(b) What is the temperature of the fluid as it exits the cold plate?

(c) Plot the exit temperature of the fluid as a function of the dispense time (which dictates the flow rate) for $t_d = 2.5$ s to 40 s. Explain the characteristics that you observe in your plot.

(d) What is the pressure drop of the fluid as it flows through the tube (for $t_d = 10$ s)?

9.45 You must cool an electrical conductor that carries $I = 25,000$ amp of current from a power supply to a magnetic energy storage device. The figure illustrates the initial design of the conductor.

water at $p_{in} = 50$ psig and $T_{in} = 10°C$

atmospheric pressure

$L = 2$ m

$D_p = 2.0$ cm $th = 0.25$ cm

conductor
$I = 25,000$ amp
$\rho_e = 5 \times 10^{-9}$ ohm-m
$k = 50$ W/m-K

The outer surface of the conductor is cooled by a flow of water. The conductor is cylindrical and the water is contained in an annular channel. The inner diameter of the water jacket is $D_p = 2.0$ cm and the thickness of the gap through which the water flows is $th = 0.25$ cm. The roughness of the jacket and the conductor surface is $e = 20$ μm. The length of the conductor and the jacket is $L = 2$ m. You may assume that the jacket is insulated. Water is provided from the city water source at $T_{in} = 10°C$ and $p_{in} = 50$ psig. The water exits the jacket at atmospheric pressure. The electrical resistivity of the conductor is $\rho_e = 5 \times 10^{-9}$ ohm-m and the conductivity of the conductor is $k = 50$ W/m-K. You may compute the properties of water at the inlet temperature and average pressure within the

jacket and assume that phase change does not occur.

(a) Determine the total rate of ohmic dissipation in the conductor that must be removed by the coolant.

(b) Determine the volumetric flow rate of water through the jacket.

(c) What is the mean temperature of the water at the outlet of the cooling channel?

(d) Determine the surface temperature of the conductor at the exit of the cooling channel.

(e) Determine the center temperature of the conductor at the exit of the conduit.

(f) Plot the maximum value (i.e., the value at the exit) of the fluid, conductor surface, and conductor center temperature as a function of the thickness of the gap for 0.1 cm < th < 0.5 cm. You should see that there is an optimal gap thickness that minimizes the temperature of the conductor – explain why this optimal gap exists.

9.46 Supermarkets will sometimes utilize a secondary loop refrigeration system, as shown in the figure on the right, as opposed to a primary system, shown in the figure on the left.

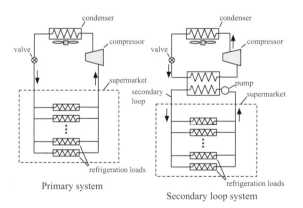

Primary system

Secondary loop system

In either system, the compressors and condensers are located remotely (e.g., in a back room or on the roof) while the refrigeration must be provided at various locations throughout the store (e.g., at several freezer cases). In a primary system, the synthetic refrigerant (e.g., R134a) is circulated throughout the store to each of the refrigeration loads. Such a system may require as much as 5000 lb$_m$ of refrigerant and will lose 30 percent to 50 percen of this charge annually; most states have regulations prohibiting this level of

666

refrigerant release due to global warming concerns. The secondary loop system addresses this problem. In the secondary loop, the synthetic refrigerant is confined to the refrigeration system itself and a secondary, environmentally benign fluid (e.g., carbon dioxide) is pumped from the refrigeration system to the refrigeration loads in the store. This design can reduce the amount of synthetic refrigerant required by an order of magnitude.

The figure illustrates the secondary refrigerant system that will be analyzed in this problem.

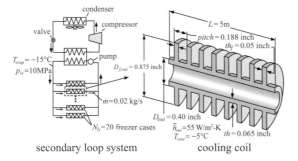

secondary loop system

cooling coil

The secondary refrigerant is carbon dioxide which leaves the evaporator of the refrigeration system at $T_{evap} = -15°C$ and $p_{in} = 10$ MPa (note that you cannot model carbon dioxide as an ideal gas at these conditions). The flow is divided into $N_{case} = 20$ branches, one branch going to each of the freezer cases within the supermarket. The mass flow of carbon dioxide in each of the branches is 0.020 kg/s. The carbon dioxide is provided to a cooling coil within each freezer case. The cooling coil consists of a $L = 5$ m long finned tube. The outer diameter of the tube is $D_{out} = 0.40$ inch and the thickness of the tube wall is $th = 0.065$ inch. The roughness of the inner surface of the tube is $e = 50$ μm. The fins are spaced $pitch = 0.188$ inch apart (i.e., there is one fin every 0.188 inch) and the thickness of each fin is $th_f = 0.05$ inch. The outer diameter of the fins is $D_{f,out} = 0.875$ inch. The conductivity of the fins and the tube is $k_m = 38$ W/m-K. Heat transfer occurs between the air in the freezer case, at $T_{case} = -5°C$, and the carbon dioxide in the finned tube. The heat transfer coefficient between the air in the case and surface of the finned tube is $\bar{h}_{out} = 55$ W/m²-K.

(a) Determine the average heat transfer coefficient between the carbon dioxide and the inner surface of the cooling coil.

(b) Determine the total conductance (UA) of the cooling coil.
(c) Determine the outlet temperature of the carbon dioxide leaving the cooling coil and the refrigeration provided to each freezer case.
(d) Determine the total refrigeration load on the system.
(e) Determine the pumping power required by the system; assume that the pump efficiency is $\eta_p = 0.65$.
(f) Plot the refrigeration provided per case and the total pumping power as a function of the mass flow rate; suggest a reasonable operating point for the system.

9.47 Your father-in-law has asked for your help with the design and installation of a water source heat pump in his cabin in northern Wisconsin, as shown in the figure. In particular, he would like you to analyze the system and possibly optimize the design. During the cooling season, the water source heat pump is, essentially, an air conditioner that rejects heat to a water source rather than to air. The cabin is located next to a lake and therefore you intend to reject heat by running a cheap plastic tube through the lake. Currently, your father-in-law has selected a tube with an outer diameter, $D_{out} = 0.50$ inch and a wall thickness $th = 0.065$ inch. You measure the temperature of the water in the lake to be $T_{lake} = 50°F$ and estimate that the heat transfer coefficient between the external surface of the pipe and the water is $\bar{h}_o = 450$ W/m²-K. The conductivity of the tube material is $k_{tube} = 1.5$ W/m-K.

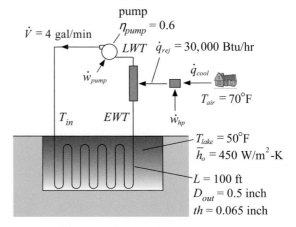

The manufacturer's sheet for the particular heat pump that has been purchased lists many characteristics of the heat pump as a function of the entering water temperature, *EWT*. The manufacturer recommends a fixed flow rate of water through the pipe of $\dot{V} = 4.0$ gal/min, and so you have found an appropriate fixed displacement pump to provide this constant volumetric flow rate of water; the pump has an efficiency, $\eta_{pump} = 0.60$. The data from the manufacturer's sheet have been used to correlate the heat pump power consumption as a function of the entering water temperature according to:

$$\dot{w}\,[\text{kW}] = 0.8513\,[\text{kW}] + 1.347 \times 10^{-3}$$
$$\left[\frac{\text{kW}}{{}^\circ\text{F}}\right] EWT\,[{}^\circ\text{F}] + 9.901 \times 10^{-5}$$
$$\left[\frac{\text{kW}}{{}^\circ\text{F}^2}\right] (EWT\,[{}^\circ\text{F}])^2.$$

You have been asked to determine the length of tube, L, that should be run through the lake in order to maximize the efficiency of the system (defined as the coefficient of performance, *COP*, which is the ratio of the cooling provided to the power consumed by both the pump and the heat pump). This is not a straightforward problem because it is difficult to see where to start. We'll tackle it in small steps as discussed here. We'll start by making a couple of assumptions that will eventually be relaxed; the assumptions are just to get the solution going – it is easier to accomplish a meaningful analysis when you have a working model. Assume that the leaving water temperature is $LWT = 40°C$ and that the length of the tube is $L = 100$ ft.

(a) Calculate the pressure drop required to push the water through the tube in the lake.

(b) Predict the temperature of the water leaving the pump (T_{in} in the figure); assume all of the pump energy goes into the water.

(c) Predict the temperature of the water leaving the lake and entering the heat pump (*EWT* in the figure) by considering the heat transfer coefficient associated with the flow of water in the tube and the energy balance for this flow.

(d) Using your model, adjust the leaving water temperature (that you initially assumed to be 40°C) until the heat rejected to the water is equal to the heat rejection required by the heat pump (i.e., $\dot{q}_{rej} = 30 \times 10^3$ Btu/hr).

(e) Using the manufacturer's data provided by the curve fit, calculate the power required by the

heat pump and, from that, the cooling provided to the cabin and the total *COP* (including both the heat pump and the pump).

(f) Use your model to prepare a single plot that shows how the COP and cooling capacity vary with length of tube. You should see an optimal length of tube that maximizes the *COP*; explain why this optimal value exists.

Projects

9.48 A solar electric generating system (SEGS) uses parabolic solar collectors to heat oil; this thermal energy is transferred to a power generation plant that uses a steam Rankine cycle. The collector field is located in the Mojave Desert in southern California. The parabolic solar collectors are long, parallel rows of curved glass mirrors that focus the Sun's energy onto absorber pipes that are located at the focal point of the mirror. A heat transfer fluid (an oil with $\rho = 825$ kg/m³, $\mu = 0.0087$ Pa-s, $k = 0.134$ W/m-K, $Pr = 152$) is pumped through the absorber pipe. The volumetric flow rate of oil in each pipe is $\dot{V} = 0.012$ m³/s. The inner diameter of the absorber pipe is $D_i = 0.066$ m.

There are a total of $N = 50$ parallel flow loops in the collector field. In each loop, the oil is pumped from the power plant through an absorber tube from the center to the edge of the field and then back from the edge of the field to the center where the oil finally returns to the power generation plant. Therefore, each loop consists of $L = 750$ m of absorber tube length that is exposed to a constant surface heat flux due to concentrated solar radiation. The oil enters the collector field (i.e., leaves the power generation plant) with a mean temperature of $T_{in} = 500$ K and leaves the collector field with an elevated mean temperature.

The intensity of solar radiation onto the outer edge of the Earth's atmosphere has a nearly constant value of 1370 W/m² (it varies by about 3 percent during a year). The radiation that is incident on the collector at the Earth's surface (on a sunny day) is nominally 1000 W/m² (it is reduced by absorption in the atmosphere and scattering). The collector concentrates the radiation so that the heat flux that is incident on the surface of the absorber tube is $q_s'' = 15,000$ W/m². You may assume that the absorber tube is very thin and the heat flux is uniformly distributed.

To restate the problem more concisely; you are going to analyze $N = 50$ tubes, each with an inner diameter of $D_i = 0.066$ m and a length of $L = 750$ m. The outer diameter of the tube is essentially equal to the inner diameter and the outer surface of these tubes is exposed to a uniform heat flux of $q_s'' = 15{,}000$ W/m^2. Each tube has oil flowing through it with a volumetric flow rate (per tube) of $\dot{V} = 0.012$ m^3/s. The oil enters the tube with a mean temperature of $T_{in} = 500$ K.

(a) Assuming that there are no losses from the tube to the ambient (due to convection or radiation), what is the mean temperature of the oil leaving the tube?

(b) What is the total rate at which thermal energy delivered to the power plant if there are no thermal losses from the absorber (i.e., for all of the absorber tubes)?

(c) What is the pressure drop associated with pumping the oil through the collector field? You may assume that the absorber tubes are perfectly smooth and neglect losses in the bends.

(d) What is the total power (i.e., for all of the absorber tubes) that is required to pump the oil through the field? Assume that your pumps have efficiency $\eta_{pump} = 0.50$ and that the oil is incompressible.

(e) If the power plant efficiency is $\eta_{power} = 0.38$ (relative to converting thermal energy to electrical energy) then how much electrical power is produced by the plant? Don't forget to subtract the pump power that is consumed.

(f) What is the maximum temperature of the surface of the absorber tube? Where does it occur?

(g) Estimate the average temperature of the surface of the absorber tube within the collector field?

The absorber tubes are actually vacuum insulated (like a thermos with a transparent outer shell) in order to reduce the loss of heat to the ambient.

(h) Estimate the total amount of heat that would be lost to ambient if the absorber tubes were NOT vacuum insulated but rather just exposed to a gentle breeze with $u_\infty = 5$ mph on a day when the ambient

temperature is $T_{amb} = 20°$C. Assume that the absorber tube thickness is small and neglect radiation in your calculation. Use the average tube surface temperature that you computed in part (g) in this calculation.

(i) On the same day (i.e., one with a 5 mph breeze), estimate the force per unit meter that is exerted on the absorber tube due to wind drag?

(j) What would the electrical output of the plant be considering the thermal losses that you calculated in (h)?

9.49 The figure illustrates a new design for a point-of-use water heater.

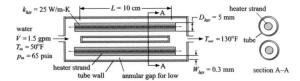

Heater strands with diameter $D_{htr} = 5$ mm are surrounded by a tube in order to form an annular flow passage through which the water passes. Ohmic dissipation in the strand material is used to heat the water. The width of the flow passage is $W_{htr} = 0.3$ mm. The tube can be considered to be perfectly insulated. You can neglect axial conduction within the heater strands and the tube and assume that the heater strand provides a uniform heat flux at its surface. The initial design uses $N = 2$ of these strand/tube assemblies in parallel. Each strand is $L = 10$ cm long. A total flow rate of $\dot{V} = 1.5$ gal/min is split between the parallel flow passages. The water enters at $p_{in} = 65$ psia and $T_{in} = 50°$F and must be heated to $T_{out} = 130°$F. The conductivity of the heater strand is $k_{htr} = 25$ W/m-K. The surface of the strand and the inner surface of the tube may be assumed to be smooth, $e = 0$.

(a) Develop a model of the point-of-use water heater that can predict the pressure drop in the water as well as the maximum temperature anywhere in the heater material.

(b) Plot the pressure drop and the maximum temperature as a function of the water flow rate for flow rates in the range of 0.1 gal/min $< \dot{V} < 5$ gal/min. Explain the shape of your plots and any distinctive features that you notice.

(c) Overlay on your plots from (b) the results obtained if the roughness of the strand and the tube surface is $e = 20$ μm. Explain the results.

(d) The maximum allowable pressure drop across the point-of-use water heater is 30 psi and the maximum allowable temperature of the strand material is 400°F. With the volumetric flow rate set to $\dot{V} = 1.5$ gpm and the roughness set to $e = 20$ μm, adjust the values of L and W_{htr} so that these two constraints are satisfied.

(e) Plot the total length of strand material required as a function of the number of strand/tube assemblies for $2 < N < 11$. Be sure that each point on your plot continues to satisfy the design constraints listed in (d). Explain the shape of your plot.

9.50 You have been hired as a consultant for an eco-resort. The central theme of the resort is the conservation of energy in every possible way. Therefore, the resort is considering the installation of a domestic hot water heating system integrated with its campfire pits, as shown in the figure.

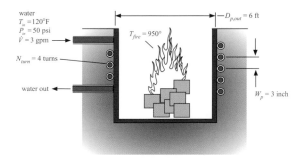

The fire pit container is a cylinder of steel with outer diameter $D_{p,out} = 6$ ft and thickness $th_p = 0.375$ inch. The conductivity of the steel is $k_p = 12.5$ W/m-K. A tube is wrapped around the outer diameter of the pit carrying water to be heated. Water enters the tube at $T_{in} = 120°$F and $P_{in} = 50$ psi. The volumetric flow rate is $\dot{V} = 3$ gpm. The outer diameter of the tube is $D_{t,out} = 0.625$ inch and the tube thickness is $th_t = 0.065$ inch. The conductivity of the tube material is $k_t = 242$ W/m-K. Water is pumped through the tube with a pump that has efficiency $\eta_p = 0.58$. Here we will assume that the inner surface of the pit container experiences only radiation heat

transfer with the fire (i.e., we will neglect radiation and convection to the surroundings). The emissivity of the pit container is $\varepsilon = 0.9$ and the temperature of the fire is $T_{fire} = 950°$C. The figure shows a close-up view of the attachment of the tube to the pit container.

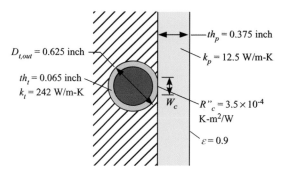

The tube wraps are placed $W_p = 3$ inch apart. There are $N_{turn} = 4$ wraps around the pit. The tube is clamped to the pit container causing a slight compression of the material. You have estimated that the contact distance (W_c) is 20 percent of the tube diameter and that the area-specific contact resistance is $R''_c = 3.5 \times 10^{-4}$ K-m^2/W. Neglect heat loss from the tube to the surrounding soil.

(a) Determine the average heat transfer coefficient between the water and the inner surface of the tube and the pressure drop across the tube.

(b) Estimate the pump power required to run the system.

(c) In order to develop a thermal model of the system we need to understand the thermal resistances that govern the problem. Estimate the following thermal resistances for a single loop of pipe:
  - convection between the water and the inside surface of the tube, R_{conv}
  - conduction along the circumference of the tube, $R_{cond,t}$
  - contact resistance, R_c
  - conduction along the pit container wall in the x-direction (i.e., axially), $R_{cond,p,x}$
  - conduction through the pit container wall in the r-direction (i.e., radially), $R_{cond,p,r}$
  - radiation between the pit container wall surface and the flame surface, R_{rad}.

Your analysis from (c) should indicate a few things. First, it should show that R_{conv} and $R_{cond,t}$ are negligibly small relative to the others and can therefore be ignored. Second, it should show that $R_{cond,r,p}$ is much less than R_{rad} but $R_{cond,x,p}$ is on the same order as R_{rad}. Therefore the fire pit wall can be treated as an extended surface with temperature varying only in the x- and not the r-direction.

(d) Estimate the total thermal resistance between the flame and the water. Treat the fire pit wall as a constant cross-sectional area fin using the concept of a radiation coefficient, calculated according to: $h_{rad} = \varepsilon \sigma \left(T_{in}^2 + T_{fire}^2 \right) \left(T_{in} + T_{fire} \right)$.

(e) Estimate the outlet temperature of the water.

(f) The pump speed is controlled so that the outlet temperature of the water is always $T_{out} = 140°F$). Determine the required volumetric flow rate.

(g) The electricity required to run the pump costs $ec = 0.12$ \$/kW-hr and the natural gas saved by the heating system costs $gc = 1.05$ \$/therm. Determine the net savings associated with running the system per hour.

(h) Plot the net savings per hour as a function of the number of turns for various values of the outer diameter of the tube.

9.51 A house requires 1500 W of heating for an average of 6000 hr/year. Currently, the house uses electrical heating. However, the cost of electricity has risen to $0.08/kW-hr and therefore the home owner would like to install a ground source heat pump to alleviate his energy cost.

The ground source heat pump pumps ethylene glycol ($\rho = 1130$ kg/m^3, $c = 2290$ J/kg-K, $k = 0.24$ W/m-K, and $\mu = 0.065$ Pa-s) through a tube buried in the ground. The tube is made of plastic ($k_{tube} = 0.5$ W/m-K) with a thickness (t_{tube}) of 5.0 mm and it has an inner diameter (D_i) of 25.0 mm. The ground remains at a relatively constant temperature (T_g) of 10°C throughout the year; for this problem, you may assume that the temperature of the external surface of the tube is equal to the ground temperature. The ethylene glycol enters the ground source heat exchanger at a relatively low temperature (T_1) and is heated as it passes through the tube and receives heat from the ground. Thus it leaves the ground at a higher temperature (T_2) and enters the heat pump at this temperature (neglect the effect of the pump on the fluid temperature). The efficiency of the pump (η_p) is 50 percent; you will need to calculate the pumping power ($\dot{w}_p$).

The ground source heat pump can be modeled as having a coefficient of performance (COP, the ratio of the heat supplied to the house, $\dot{q}_h$, to the electrical power required by the heat pump, $\dot{w}_{hp}$) that is 25 percent of the COP for a reversible heat pump operating between T_1 and the temperature of the conditioned space ($T_{sp} = 25°C$). The purpose of this problem is to select the size of the tube (the ground source heat exchanger) and the mass flow rate of glycol that achieves the highest

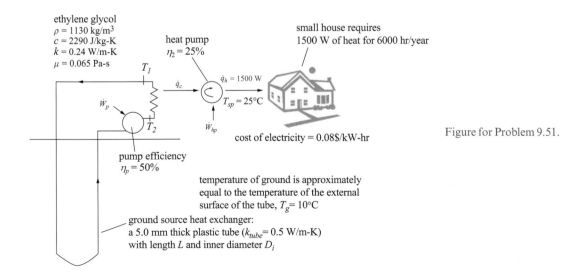

Figure for Problem 9.51.

savings over a five year period. To start the problem, you can assume that the mass flow rate of glycol ($\dot{m}$) is 0.10 kg/s and the tube length (L) is 50 m. The specifications will be relaxed as specified in the problem.

(a) Assume that the leaving temperature of the heat pump (T_1) is 0°C. Determine the entering temperature of the heat pump (T_2) under these conditions.

(b) Determine the pump power required to pump fluid through the ground source heat exchanger under these conditions. Neglect any pressure drop within the heat pump itself.

(c) Determine the heat transfer rate supplied to the house and the electrical power that is required by the heat pump under these conditions.

(d) Relax the assumption that T_1 is 0°C in order to specify that the heat transfer rate supplied to the house is the required 1500 W.

(e) Calculate the energy cost associated with running the heat pump for five years and subtract this value from the cost associated with using electric heat for five years (neglect the time value of money). This is the value of your energy savings for a five year period.

Assume that purchasing and installing the heat pump unit costs $800 and the cost to purchase and install the ground source heat exchanger is $10/m (i.e., the cost of the ground source heat exchanger is 10L). Calculate the total initial outlay of capital associated with installing the heat pump system. Calculate the net savings associated with installing the heat pump and operating it for the next five years.

(f) Prepare a plot showing the net savings as a function of the mass flow rate of glycol for mass flow rates ranging from 0.05 kg/s to 0.5 kg/s. Explain your graph – there should be an optimal mass flow rate. Use EES' built-in optimization capability to automatically select the optimal mass flow rate. Note that this step should be done in such a way that the heat pump is always providing 1500 W of heat to the house.

(g) Prepare a plot showing the net savings as a function of the length of the ground source heat exchanger. Overlay the total energy savings and the total capital cost on this plot. This step should be done so that the heat pump is always providing 1500 W of heat to the house and the mass flow rate is always optimal (you may want to use the Min/Max Table feature in EES). Explain the shape of your plot.

10 | Free Convection

Chapters 7 through 9 discuss *forced* convection problems. In a forced convection problem the fluid is driven externally over a surface (for example by a fan or a pump). **Free (or natural) convection** refers to a problem where, in the absence of a temperature difference between the surface and the fluid, the fluid would be completely quiescent. However, because the density of most fluids depends at least weakly on temperature, the heating or cooling of the fluid leads to density gradients and an imbalance in the **buoyancy forces** (i.e., forces related to the action of gravity) that may cause fluid motion. The fluid motion in a free convection situation is fundamentally driven by density gradients that are induced in the fluid as it is heated or cooled due to the presence of a surface. The velocities induced by these density gradients are typically small and therefore the absolute magnitude of natural convection heat transfer coefficients is also small compared to forced convection values.

10.1 Free Convection Flow

The flow patterns induced by heating or cooling can usually be understood intuitively. Hot fluid tends to have lower density and therefore rise (i.e., flow *against* gravity) while cold fluid with higher density tends to fall (i.e., flow *in the direction of* gravity). The existence of a temperature gradient does not guarantee fluid motion. For example, Figure 10.1 illustrates fluid trapped between two plates that are oriented **horizontally** (i.e., perpendicular to the gravity vector g). In Figure 10.1(a) the lower plate is heated (to temperature T_H) and the upper plate is cooled (to temperature T_C). The heated fluid adjacent to the lower plate will tend to rise and the cooled fluid adjacent to the upper plate will tend to fall; the result is the formation of natural convection "cells" as shown in Figure 10.1(a). In Figure 10.1(b) the lower plate is cooled and the upper plate heated. This situation is stable; the cold fluid cannot fall further and the hot fluid cannot rise further. As a result, the fluid in Figure 10.1(b) will remain stagnant. The heat transfer rate between the two plates shown in Figure 10.1(a) will be substantially higher than for the two plates in Figure 10.1(b) due to the presence of the flowing fluid.

Section 10.2 discusses the natural set of dimensionless parameters that are used to correlate the solutions and experimental data for free convection problems. In Section 10.3, several commonly encountered configurations are examined and correlations are presented that can be used to solve engineering problems. A relatively comprehensive set of correlations is included in EES; the use of these correlations is illustrated with examples.

10.2 Dimensionless Parameters

In Section 7.4 we found that the average Nusselt number for *forced* convection problems is typically correlated using the Reynolds number and the Prandtl number. This section explores the analogous set of dimensionless parameters that are appropriate for a *free* convection problem.

10.2.1 The Characteristic Buoyancy Velocity

There is no obvious reference velocity associated with a free convection problem in the same way that there is for a forced convection situation. This section derives an expression for the approximate magnitude of the velocity that will be induced by buoyancy forces. This is helpful in two ways. First, the characteristic buoyancy velocity forms the basis for the appropriate dimensionless parameter that is analogous to a Reynolds number. Second, the characteristic buoyancy velocity provides a basis for estimating the scale of the velocity that might

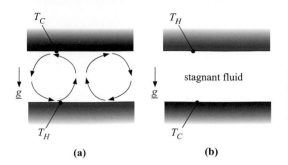

Figure 10.1 Flow patterns induced between horizontal plates in which (a) the upper plate is cooled and the lower plate heated, and (b) the upper plate is heated and the lower plate cooled.

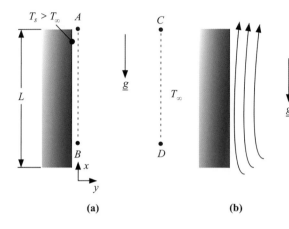

Figure 10.2 (a) A plate heated to T_H in an environment of fluid at T_∞ and (b) the associated flow pattern.

be induced in a free convection problem; this might be useful for checking the results of a more detailed CFD analysis or experimental measurement.

Figure 10.2(a) illustrates a **vertically** oriented (i.e., oriented parallel to the gravity vector g) plate with a surface that is heated to temperature T_s in an environment of fluid that would otherwise be stagnant at temperature T_∞. The governing equations that apply to the flow of a viscous fluid were derived in Section 7.3.2. The momentum conservation equation in the vertical direction (the x-direction with vertical velocity u) is:

$$\rho\left(\frac{\partial u}{\partial t} + u\frac{\partial u}{\partial x} + v\frac{\partial u}{\partial y}\right) = -\frac{\partial p}{\partial x} + \mu\left(\frac{\partial^2 u}{\partial x^2} + \frac{\partial^2 u}{\partial y^2}\right) - \rho g. \tag{10.1}$$

The gravitational term in Eq. (10.1) is negative because the gravity vector is in the negative x-direction. In the absence of any fluid motion (i.e., $u = v = 0$), Eq. (10.1) reduces to:

$$\frac{\partial p}{\partial x} + \rho g = 0. \tag{10.2}$$

If the density of the fluid is everywhere constant, then pressure will only vary with x (i.e., due to hydrostatic effects) and the pressure at any y-position will be the same. However, if density is a function of temperature, then the pressure gradient in the hot fluid that is immediately adjacent to the heated plate, along line A–B in Figure 10.2(a), will be different than the pressure gradient in the cold fluid far from the plate, along line C–D in Figure 10.2(a).

In the absence of any fluid motion, the pressure gradient away from the plate will be

$$\left(\frac{\partial p}{\partial x}\right)_{C-D} = -\rho_{T=T_\infty}\, g, \tag{10.3}$$

whereas the pressure gradient adjacent to the plate will be

$$\left(\frac{\partial p}{\partial x}\right)_{A-B} = -\rho_{T=T_s}\, g. \tag{10.4}$$

If the pressures at points A and C are the same, then Eqs. (10.3) and (10.4) imply that a pressure *difference* will be induced between points B and D (again, in the absence of any fluid motion):

$$p_D - p_B = \Delta p = g\, L\left(\rho_{T=T_\infty} - \rho_{T=T_s}\right). \tag{10.5}$$

The y-directed pressure difference predicted by Eq. (10.5) does not persist – instead, fluid will be pushed towards the plate by the pressure force where it will be heated and rise against gravity, as shown in Figure 10.2(b). However, the pressure difference given by Eq. (10.5) provides the fundamental driving force for fluid motion and allows the definition of a characteristic velocity for the free convection problem. The pressure difference will induce a consistent fluid momentum change, according to

$$g\, L\left(\rho_{T=T_\infty} - \rho_{T=T_s}\right) = \rho\, u_{char}^2, \tag{10.6}$$

where ρ is the average density of the fluid. Therefore, the reference velocity for free convection is:

$$u_{char} = \sqrt{\frac{g\, L}{\rho}\left(\rho_{T=T_\infty} - \rho_{T=T_s}\right)}. \tag{10.7}$$

Equation (10.7) provides an order of magnitude estimate for the velocity that might be induced in a free convection problem. Notice that the size of the characteristic buoyancy velocity increases with the density difference that is induced by the difference in temperature between the fluid and the surface.

10.2.2 The Volumetric Thermal Expansion Coefficient

The density difference in Eq. (10.7) is driven by a temperature difference. Density varies approximately linearly with temperature for the relatively small density changes that are involved in most free convection problems. Therefore, the density difference in Eq. (10.7) can be expressed as:

$$\rho_{T=T_\infty} - \rho_{T=T_s} = \left(\frac{\partial \rho}{\partial T}\right)_p (T_\infty - T_s). \tag{10.8}$$

The partial derivative of density with respect to temperature at constant pressure is related to the **volumetric thermal expansion coefficient**:

$$\beta = -\frac{1}{\rho}\left(\frac{\partial \rho}{\partial T}\right)_p. \tag{10.9}$$

Density is a decreasing function of temperature for most fluids and therefore the volumetric thermal expansion coefficient will be positive. The units of β are K^{-1} and the property can be interpreted as the fractional change in density that occurs per change in temperature. The volumetric thermal expansion coefficient is a thermodynamic property of a fluid, just as density is. The volumetric thermal expansion coefficient of some saturated liquids can be found in Appendix D and the EES function VolExpCoef returns this property for any fluid in the EES database. Figure 10.3 illustrates the volumetric thermal expansion coefficient of several fluids as a function of temperature.

Substituting Eq. (10.9) into Eq. (10.8) leads to

$$\rho_{T=T_\infty} - \rho_{T=T_s} = -\beta\, \rho(T_\infty - T_s), \tag{10.10}$$

where ρ is the nominal density of the fluid (i.e., the *average* density). Substituting Eq. (10.10) into Eq. (10.7) leads to the characteristic buoyancy velocity for a natural convection problem:

$$u_{char} = \sqrt{g\, L\beta(T_s - T_\infty)}. \tag{10.11}$$

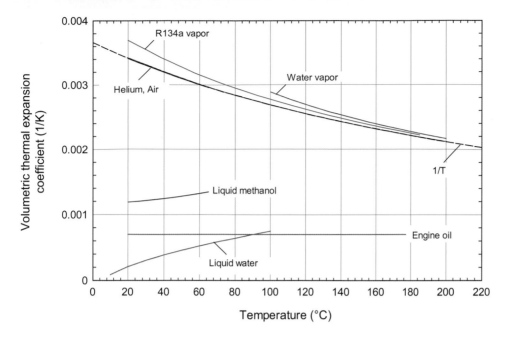

Figure 10.3 Volumetric thermal expansion coefficient of several fluids as a function of temperature at atmospheric pressure.

Equation (10.11) shows that the magnitude of the induced velocity will increase with the temperature difference and also with the volumetric thermal expansion coefficient.

It is worth noting that the magnitude of the velocity that will be induced by a temperature difference under most conditions is quite small. For example, a 20 cm long plate heated to 100°C in room temperature air leads to $u_{char} \approx 0.7$ m/s; a fan or blower can easily provide air flow velocities that are an order of magnitude higher than this value. As a consequence of the very small velocity associated with a free convection problem, the heat transfer coefficient is typically also small compared to a forced convection heat transfer coefficient. Often, radiation is an important phenomenon in a free convection situation whereas it rarely is in a forced convection problem.

The Volumetric Thermal Expansion Coefficient of an Ideal Gas
Notice that the gases shown in Figure 10.3 tend to collapse onto a single line, particularly at higher temperature. This observation can be explained by substituting the Ideal Gas Law into the definition of the volumetric thermal expansion coefficient. The Ideal Gas Law is

$$\rho = \frac{p}{RT},$$
(10.12)

where R is the gas constant. Substituting Eq. (10.12) into Eq. (10.9) leads to:

$$\beta = -\frac{1}{\rho}\left(\frac{\partial \rho}{\partial T}\right)_p = -\frac{RT}{p}\frac{\partial}{\partial T}\left(\frac{p}{RT}\right) = -\frac{RT}{p}\left(-\frac{p}{RT^2}\right) = \frac{1}{T}.$$
(10.13)

The volumetric thermal expansion coefficient of an ideal gas is the inverse of its absolute temperature. The small differences that can be seen in Figure 10.3 result because these gases do not exactly behave according to the Ideal Gas Law at the conditions used to construct the plot.

10.2.3 The Grashof Number and the Rayleigh Number

The forced convection results from Chapters 7 through 9 are correlated using the Prandtl number and the Reynolds number based on a characteristic length and velocity:

$$Re = \frac{\rho \, L_{char} \, u_{char}}{\mu}. \tag{10.14}$$

In forced convection problems, the characteristic velocity in Eq. (10.14) is typically the free stream velocity (u_∞) for external flow problems or the mean velocity (u_m) for internal flow problems. Free convection correlations are also correlated using something like the Reynolds number, but the appropriate characteristic velocity is the velocity induced by the buoyancy forces, derived in Eq. (10.11):

$$Re = \frac{\rho \, L_{char} \, u_{char}}{\mu} = \frac{\rho \, L_{char}}{\mu} \sqrt{g \, L_{char} \, \beta (T_s - T_\infty)}. \tag{10.15}$$

Natural convection correlations are often presented in terms of the **Grashof number** (Gr), which is the Reynolds number defined in Eq. (10.15) squared:

$$Gr = \left(\frac{\rho \, L_{char} \, u_{char}}{\mu} \right)^2 = \frac{g \, L_{char}^3 \, \beta (T_s - T_\infty)}{\upsilon^2}. \tag{10.16}$$

Alternatively, the **Rayleigh number** (Ra) is sometimes used to correlate the results; the Rayleigh number is defined as the product of the Grashof number and the Prandtl number:

$$Ra = Gr \, Pr = \frac{g \, L_{char}^3 \, \beta \, (T_s - T_\infty)}{\upsilon \, \alpha}. \tag{10.17}$$

The Nusselt number provided by natural convection correlations is defined the same way as it is for forced convection correlations:

$$\overline{Nu} = \frac{\bar{h} \, L_{char}}{k}. \tag{10.18}$$

The drag coefficient is not often included in natural convection correlations, primarily because the forces induced by the very small buoyancy-induced velocity are not usually large enough to be of engineering interest.

10.3 External Free Convection Correlations

10.3.1 Introduction

Section 10.2 provides an introduction to natural convection and identifies the important correlating parameters. This section presents a set of correlations that can be used to examine several commonly encountered natural convection situations in which the boundary layers can grow without bound. These situations are analogous to the forced external flow geometries examined in Chapter 8.

10.3.2 Plate

Free convection in the presence of a heated or cooled plate has been thoroughly studied. The behavior of the natural convection flow that is induced depends on the orientation of the plate with respect to gravity. The correlations provided in this section are from Raithby and Hollands (1998).

Heated or Cooled Vertical Plate
Figure 10.4 illustrates, qualitatively, the velocity and temperature distribution associated with natural convection from a heated vertical plate; vertical refers to a plate that is parallel to the gravity vector.

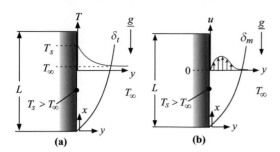

Figure 10.4 (a) Temperature and (b) velocity distributions associated with the natural convection from a vertical plate.

Notice that both the momentum and thermal boundary layers develop from the lower edge of a heated plate in a manner that is similar to external forced convection over a plate. However, the velocity distribution is somewhat different in that the velocity is zero both at $y = 0$ and as $y \to \infty$. Also, the temperature rise associated with the thermal boundary layer drives the velocity.

The solution for natural convection from a vertical plate is correlated using a Rayleigh number and average Nusselt number that are defined based on the length of the plate in the vertical direction (see Figure 10.4):

$$Ra_L = \frac{g\,L^3\,\beta\,(T_s - T_\infty)}{\upsilon\,\alpha} \tag{10.19}$$

$$\overline{Nu_L} = \frac{\bar{h}\,L}{k}. \tag{10.20}$$

The boundary layer will become turbulent at a critical Rayleigh number of approximately $Ra_{crit} \approx 1 \times 10^9$ causing the average Nusselt number to increase. The average Nusselt number associated with an isothermal, vertical heated plate is provided by Churchill and Chu (1975b):

$$\overline{Nu_L} = \left\{ 0.825 + \frac{0.387\,Ra_L^{1/6}}{\left[1 + \left(\dfrac{0.492}{Pr}\right)^{9/16}\right]^{8/27}} \right\}^2. \tag{10.21}$$

The correlation associated with Eq. (10.21) is valid for the entire range of Raleigh number. The correlations for free convection can be accessed in EES by selecting Function Info from the Options menu and then selecting Heat Transfer & Fluid Flow. Select the Convection Library and then the Free Convection – Non-dimensional option in order to scroll through the available correlations. The correlation for a vertical plate is implemented by the procedure FC_plate_vertical_ND in EES. Figure 10.5 illustrates the average Nusselt number as a function of the Rayleigh number for various values of the Prandtl number.

If the vertical plate is cooled rather than heated (i.e., if $T_s < T_\infty$), then the boundary layers shown in Figure 10.4 initiate at the *upper* edge of the plate and grow in the *downward* direction. Therefore, the average Nusselt number for a cooled vertical plate can also be computed using the correlation presented in this section. Note that the Rayleigh and Grashof must be defined based on the absolute value of the temperature difference in this instance.

Specific correlations exist for a plate with a uniform heat flux and other thermal boundary conditions. However, reasonable accuracy can be obtained using the correlation for an isothermal plate with a Rayleigh number based on the average temperature difference between the plate and the ambient air.

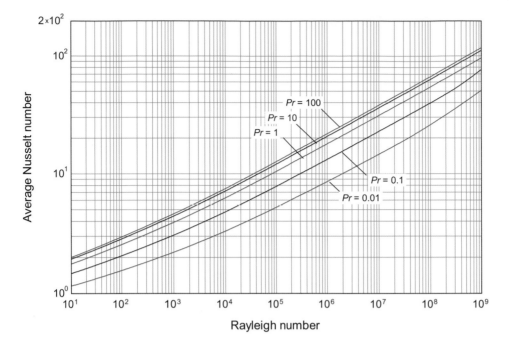

Figure 10.5 The average Nusselt number for a heated vertical plate as a function of Rayleigh number for various values of the Prandtl number.

Example 10.1

A flat plate is mounted vertically in a room filled with still air at atmospheric pressure and $T_\infty = 300$ K. The plate has width $W = 5$ cm (perpendicular to gravity) and length $L = 10$ cm (parallel to gravity). One side of the plate experiences convection. The plate is electrically heated until its surface temperature reaches $T_s = 500$ K. Estimate the steady-state rate of heat transfer required.

Known Values

Figure 1 illustrates a sketch of the problem with the known values indicated.

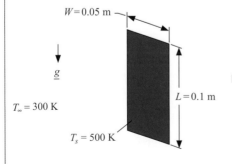

Figure 1 Sketch of the problem with the known values indicated.

Example 10.1 (cont.)

Assumptions

- The problem is steady state.
- Radiation is neglected.
- Air behaves as an ideal gas.

Analysis and Solution

The properties of air are obtained from Appendix C at the film temperature (400 K): $\rho = 0.8821 \ \text{kg/m}^3$, $k = 0.03345 \ \text{W/m-K}$, $\alpha = 37.39 \times 10^{-6} \ \text{m}^2/\text{s}$, and $\upsilon = 26.14 \times 10^{-6} \ \text{m}^2/\text{s}$. The volumetric thermal expansion coefficient is estimated using Eq. (10.13):

$$\beta = \frac{1}{T_{film}} = \frac{1}{400 \ \text{K}} = 0.0025 \ \text{K}^{-1}. \tag{1}$$

The Prandtl number is calculated according to:

$$Pr = \frac{\upsilon}{\alpha} = \frac{26.14 \times 10^{-6}}{37.39 \times 10^{-6}} = 0.699. \tag{2}$$

The Rayleigh number based on the length of the plate parallel to gravity is computed:

$$Ra_L = \frac{g \, L^3 \, \beta \, (T_s - T_\infty)}{\upsilon \, \alpha}$$

$$= \frac{9.81 \ \text{m}}{\text{s}^2} \left| \frac{(0.1)^3 \ \text{m}^3}{} \right| \frac{0.0025}{\text{K}} \left| (500 - 300) \ \text{K} \right| \frac{\text{s}}{26.14 \times 10^{-6} \ \text{m}^2} \left| \frac{\text{s}}{37.39 \times 10^{-6} \ \text{m}^2} \right|$$

$$= 5.02 \times 10^6. \tag{3}$$

The average Nusselt number at $Ra_L = 5.02 \times 10^6$ and $Pr = 0.699$ is approximately $\overline{Nu}_L = 25$ according to Figure 10.5. The average heat transfer coefficient is:

$$\bar{h} = \overline{Nu}_L \frac{k}{L} = \frac{25}{} \left| \frac{0.03345 \ \text{W}}{\text{m K}} \right| \frac{1}{0.1 \ \text{m}} = 8.36 \frac{\text{W}}{\text{m}^2 \ \text{K}}. \tag{4}$$

The rate of convection heat transfer is computed according to:

$$\dot{q} = h \, L \, W \, (T_s - T_\infty) = \frac{8.36 \ \text{W}}{\text{m}^2 \ \text{K}} \left| 0.1 \ \text{m} \right| 0.05 \ \text{m} \left| (500 - 300) \ \text{K} \right| = \boxed{8.36 \ \text{W}}. \tag{5}$$

Discussion

The heat transfer coefficient for this natural convection problem is small relative to forced convection and therefore it is likely that radiation should also be considered. The rate of heat transfer due to radiation cannot be calculated precisely without knowing the emissivity of the surface. However, we can bound its potential impact assuming an emissivity of $\varepsilon = 1$:

$$\dot{q}_{rad} = \sigma \, \varepsilon \, L \, W \, (T_s^4 - T_\infty^4) = \frac{5.67 \times 10^{-8} \ \text{W}}{\text{m}^2 \ \text{K}^4} \left| 1 \right| 0.1 \ \text{m} \left| 0.05 \ \text{m} \right| \frac{(500^4 - 300^4) \ \text{K}^4}{} = 15.4 \ \text{W}. \tag{6}$$

The heat transfer due to radiation could be almost twice as large as the heat transfer due to free convection.

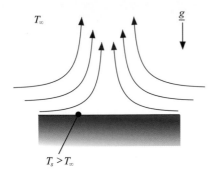

Figure 10.6 Flow induced by a horizontal heated plate facing upwards.

Horizontal Plate – Heated Upward Facing or Cooled Downward Facing

Figure 10.6 illustrates, qualitatively, the flow pattern that is associated with a heated plate oriented horizontally (i.e., perpendicular to gravity) with the heated surface facing upwards and no restriction to flow at the edges of the plate. The solution to the problem shown in Figure 10.6 has been correlated using a Rayleigh number and average Nusselt number that are based on a characteristic length for the plate, L_{char}, defined as the ratio of the surface area to the perimeter:

$$L_{char} = \frac{A_s}{per} \tag{10.22}$$

$$Ra_{L_{char}} = \frac{g\, L_{char}^3\, \beta\, (T_s - T_\infty)}{\upsilon\, \alpha} \tag{10.23}$$

$$\overline{Nu}_{L_{char}} = \frac{\bar{h}\, L_{char}}{k}. \tag{10.24}$$

For a square plate, the characteristic length is one quarter the side length and for a circular plate the characteristic length is one quarter the diameter. The Nusselt number is found by weighting the Nusselt numbers calculated separately assuming laminar and turbulent flow ($\overline{Nu}_{L,\,lam}$ and $\overline{Nu}_{L,\,turb}$, respectively) using the empirical formula proposed by Churchill and Usagi (1972):

$$\overline{Nu}_{L_{char}} = \left(\overline{Nu}_{L_{char},\,lam}^{10} + \overline{Nu}_{L_{char},\,turb}^{10}\right)^{1/10}. \tag{10.25}$$

Note that under laminar condition $\overline{Nu}_{L,\,lam}$ is larger while under turbulent conditions $\overline{Nu}_{L,\,turb}$ is larger; therefore, Eq. (10.25) will limit to the correct value over the range of Ra_L from $1 \times 10^{-1} < Ra_L < 1 \times 10^{12}$.

The laminar Nusselt number is

$$\overline{Nu}_{L_{char},\,lam} = \frac{1.4}{\ln\left(1 + \dfrac{1.4}{0.835\, C_{lam}\, Ra_{L_{char}}^{0.25}}\right)}, \tag{10.26}$$

where C_{lam} is

$$C_{lam} = \frac{0.671}{\left[1 + \left(\dfrac{0.492}{Pr}\right)^{9/16}\right]^{4/9}}. \tag{10.27}$$

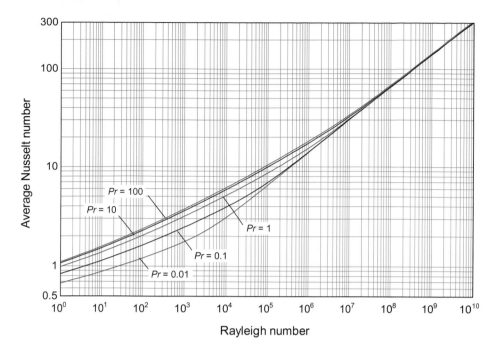

Figure 10.7 The average Nusselt number for a heated horizontal plate facing upward as a function of Rayleigh number for various values of the Prandtl number. Both the Nusselt number and Rayleigh number are based on the characteristic length defined by Eq. (10.22).

The turbulent Nusselt number is

$$\overline{Nu}_{L_{char}, turb} = C_{turb}\, Ra_{L_{char}}^{1/3}, \tag{10.28}$$

where C_{turb} is calculated according to

$$C_{turb} = 0.14\left(\frac{1 + 0.0107\, Pr}{1 + 0.01\, Pr}\right). \tag{10.29}$$

The correlation associated with Eqs. (10.25) through (10.29) is valid for $1.0 < Ra_{L_{char}} < 1 \times 10^{10}$ and is implemented by the procedure **FC_plate_horizontal1_ND** in EES. Figure 10.7 illustrates the average Nusselt number as a function of the Rayleigh number for various values of the Prandtl number.

If the plate were cooled (i.e., $T_s < T_\infty$) and facing downward, then the flow pattern would appear as shown in Figure 10.6, but flipped upside down so that the fluid tends to fall after it is cooled by the plate. The average Nusselt number for a cooled downward facing plate can therefore be computed using the same correlation that is presented in this section; note that the Rayleigh and Grashof numbers must be defined based on the absolute value of the temperature difference in this case.

Horizontal Plate – Heated Downward Facing or Cooled Upward Facing
Figure 10.8 illustrates, qualitatively, the flow induced by a heated plate oriented horizontally with the heated surface facing down and no restriction to flow from the edges. The heated fluid tends to escape from the side of the plate, inducing further fluid flow from the ambient.

The solution to the problem shown in Figure 10.8 is correlated using a Rayleigh number and average Nusselt number that are based on the characteristic length scale L_{char} defined previously in Eq. (10.22). The flow velocity that is induced in this configuration is substantially less than for the vertical or horizontal

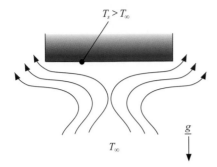

Figure 10.8 Flow induced by a horizontal heated plate facing downward.

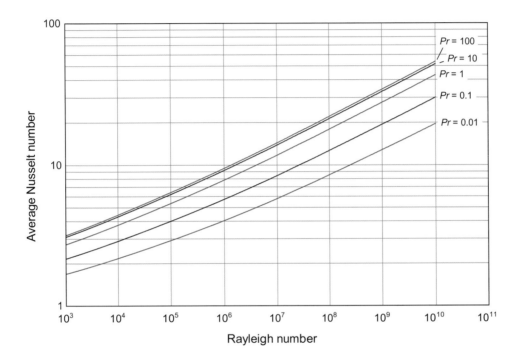

Figure 10.9 The average Nusselt number for a heated horizontal plate facing downward as a function of Rayleigh number for various values of the Prandtl number. Both the Nusselt number and Rayleigh number are based on the characteristic length defined by Eq. (10.22).

upward facing plate because the configuration is nearly stable. Therefore, the flow is laminar even at high values of the Rayleigh number and the average Nusselt number is equal to the laminar Nusselt number, computed according to:

$$\overline{Nu}_{L_{char}} = \frac{2.5}{\ln\left\{1 + \dfrac{2.5}{0.527\, Ra_{L_{char}}^{0.20} \left[1 + \left(\dfrac{1.9}{Pr}\right)^{0.9}\right]^{2/9}}\right\}}.$$ (10.30)

The correlation provided by Eq. (10.30) is valid for $1 \times 10^3 < Ra_{L_{char}} < 1 \times 10^{10}$ and is implemented by the procedure FC_plate_horizontal2_ND in EES. Figure 10.9 illustrates the average Nusselt number as a function of the Rayleigh number for various values of the Prandtl number.

If the plate were cooled (i.e., $T_s < T_\infty$) and facing upward, then the flow pattern would appear as shown in Figure 10.9, but flipped upside down so that the fluid tends to fall after it is cooled by the plate. The average Nusselt number for an upward facing cooled plate can therefore be computed using the same correlation presented in this section, again using the absolute value of the temperature difference to compute the Rayleigh and Grashof numbers.

Example 10.2

A flat plate heater is mounted horizontally in an open tank of otherwise still water at $T_\infty = 20°C$. The plate has dimensions $W = 5$ cm $\times L = 15$ cm. The plate is electrically heated at a rate $\dot{q} = 200$ W. Determine the surface temperature if the plate is mounted so that it is upward facing and also if it is mounted so that it is downward facing.

Known Values

Figure 1 illustrates a sketch of the problem in the upward facing configuration with the known values indicated.

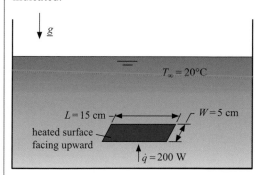

Figure 1 Sketch of the problem with the heated surface facing upward.

Assumptions

- The problem is steady state.
- The pressure in the tank is atmospheric and the temperature of the water in the tank is uniform.
- The tank is large enough that the edges do not affect the flow situation.
- No evaporation occurs.

Analysis

The film temperature is given by:

$$T_f = \frac{(T_s + T_\infty)}{2}. \tag{1}$$

The properties of water are obtained at the film temperature and atmospheric pressure (p_{atm}):

$$\rho = \rho\left(T_f, p_{atm}\right) \tag{2}$$

$$k = k\left(T_f, p_{atm}\right) \tag{3}$$

$$\upsilon = \upsilon\left(T_f, p_{atm}\right) \tag{4}$$

Continued

Example 10.2 (cont.)

$$\alpha = \alpha(T_f, p_{atm}) \qquad (5)$$

$$\beta = \beta(T_f, p_{atm}) \qquad (6)$$

$$Pr = \frac{\upsilon}{\alpha}. \qquad (7)$$

The surface area and perimeter of the plate are used to compute the characteristic length according to Eq. (10.22):

$$A_s = W\,L \qquad (8)$$

$$per = 2\,(W + L) \qquad (9)$$

$$L_{char} = \frac{A_s}{per}. \qquad (10)$$

The Rayleigh number based on the characteristic length is determined:

$$Ra_{L_{char}} = \frac{g\,L_{char}^3\,\beta\,(T_s - T_\infty)}{\upsilon\,\alpha}. \qquad (11)$$

The correlations for an upward facing or downward facing heated plate are used; in either case, the average Nusselt number is a function of the Rayleigh number and Prandtl number:

$$\overline{Nu}_{L_{char}} = \overline{Nu}_{L_{char}}(Ra_{L_{char}}, Pr). \qquad (12)$$

The average heat transfer coefficient is computed using the definition of the Nusselt number:

$$\bar{h} = k\,\frac{\overline{Nu}_{L_{char}}}{L_{char}}. \qquad (13)$$

The surface temperature is calculated from:

$$T_s = T_\infty + \frac{\dot{q}}{\bar{h}\,A_s}. \qquad (14)$$

Equations (1) through (14) are 14 equations in the 14 unknowns: T_s, T_f, ρ, k, υ, α, β, Pr, A_s, per, L_{char}, $Ra_{L_{char}}$, $\overline{Nu}_{L_{char}}$, and $\bar{h}$.

Solution

Equations (1) through (14) are a coupled set of implicit equations because the properties and Rayleigh number all depend on the surface temperature, which is not known at the beginning of the calculations. The inputs are entered in EES.

```
$UnitSystem SI Mass J K Pa Radian
p_atm=1 [atm]*Convert(atm,Pa)          "pressure of water"
T_infinity=ConvertTemp(C,K,20 [C])     "temperature of water"
q_dot=200 [W]                          "heat transfer rate"
W=5 [cm]*Convert(cm,m)                 "width of plate"
L=15 [cm]*Convert(cm,m)                "length of plate"
```

An initial guess for the surface temperature is used in order to allow the equations to be entered and solved sequentially; this enables the units to be set and the equations checked as they are entered. Equations (1) through (14) are entered; note that EES' internal database for properties is used for Eqs. (2) through (6) and the Free Convection Library is used for Eq. (12).

Example 10.2 (cont.)

```
T_s=350 [K]                                      "guess for T_s"
T_f=(T_infinity+T_s)/2                           "film temperature"
rho=Density(Water,T=T_f,p=p_atm)                 "density"
k=Conductivity(Water,T=T_f,p=p_atm)              "thermal conductivity"
nu=KinematicViscosity(Water,T=T_f,p=p_atm)       "kinematic viscosity"
alpha=ThermalDiffusivity(Water,T=T_f,p=p_atm)    "thermal diffusivity"
beta=VolExpCoef(Water,T=T_f,p=p_atm)             "volumetric expansion coefficient"
Pr=nu/alpha                                      "Prandtl number"
A_s=W*L                                          "surface area"
per=2*(W+L)                                      "perimeter"
L_char=A_s/per                                   "characteristic length"
Ra_L_char=g#*L_char^3*beta*(T_s-T_infinity)/(nu*alpha)
                                                 "Rayleigh number"
Call FC_plate_horizontal1_ND(Ra_L_char, Pr: Nusselt_L_char)
                                                 "access correlations for upward facing plate"
h_bar=Nusselt_L_char*k/L_char                    "average heat transfer coefficient"
T_s_c=T_infinity+q_dot/(h_bar*A_s)               "calculated surface temperature"
```

Solving the problem provides $T_{s,c}$ = 313.8 K, which is different from the assumed value of T_s = 350 K. To finish the problem, update the guess values by selecting Update Guesses from the Calculate menu and comment out the assumed value for T_s. Add an equation requiring that T_s must equal $T_{s,c}$. The resulting code is as follows.

```
$UnitSystem SI Mass J K Pa Radian
p_atm=1 [atm]*Convert(atm,Pa)                    "pressure of water"
T_infinity=ConvertTemp(C,K,20 [C])               "temperature of water"
q_dot=200 [W]                                    "heat transfer rate"
W=5 [cm]*Convert(cm,m)                           "width of plate"
L=15 [cm]*Convert(cm,m)                          "length of plate"
//T_s=350 [K]                                    "guess for T_s"
T_f=(T_infinity+T_s)/2                           "film temperature"
rho=Density(Water,T=T_f,p=p_atm)                 "density"
k=Conductivity(Water,T=T_f,p=p_atm)              "thermal conductivity"
nu=KinematicVIscosity(Water,T=T_f,p=p_atm)       "kinematic viscosity"
alpha=ThermalDiffusivity(Water,T=T_f,p=p_atm)    "thermal diffusivity"
beta=VolExpCoef(Water,T=T_f,p=p_atm)             "volumetric expansion coefficient"
Pr=nu/alpha                                      "Prandtl number"
A_s=W*L                                          "surface area"
per=2*(W+L)                                      "perimeter"
L_char=A_s/per                                   "characteristic length"
Ra_L_char=g#*L_char^3*beta*(T_s-T_infinity)/(nu*alpha)
                                                 "Rayleigh number"
Call FC_plate_horizontal1_ND(Ra_L_char, Pr: Nusselt_L_char)
                                                 "access correlations for upward facing plate"
```

Continued

Example 10.2 (cont.)

h_bar=Nusselt_L_char*k/L_char	"average heat transfer coefficient"
T_s_c=T_infinity+q_dot/(h_bar*A_s)	"calculated surface temperature"
T_s=T_s_c	"iterate to find surface temperature"

Solving leads to $\boxed{T_s = 322.8 \text{ K } (49.6°\text{C})}$ for an upward facing heated plate.

In order to solve the problem for the case where the heated plate is facing downward it is only necessary to call a different correlation in Eq. (12). The call to the procedure **FC_plate_horizontal1_ND** is replaced with a call to the procedure **FC_plate_horizontal2_ND**; otherwise the code is unchanged. Note that it is possible to browse through the Free Convection Library by selecting Function Info from the Options menu and then selecting Heat Transfer & Fluid Flow radio button. Navigate to the Convection Library from the drop down menu and then Free Convection – Non-dimensional in the lower pane. Scroll through the various geometries that are available or select Index to access a clickable index of the correlations, as shown in Figure 2.

Free Convection

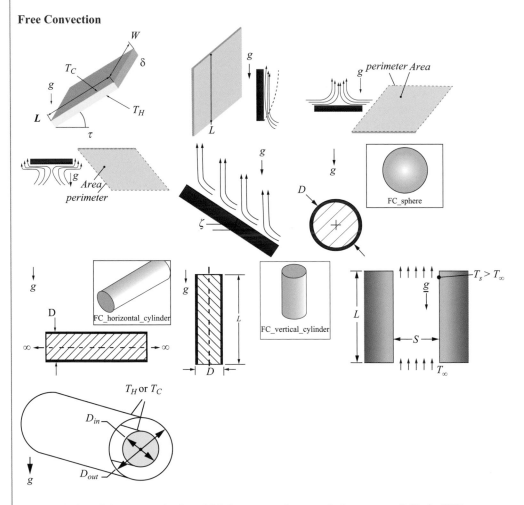

Figure 2 Index of the geometries for which free convection correlations are available in EES.

Example 10.2 (cont.)

The code altered in order to determine the surface temperature for a downward facing heated plate is shown below.

```
$UnitSystem SI Mass J K Pa Radian
p_atm=1 [atm]*Convert(atm,Pa)                       "pressure of water"
T_infinity=ConvertTemp(C,K,20 [C])                  "temperature of water"
q_dot=200 [W]                                       "heat transfer rate"
W=5 [cm]*Convert(cm,m)                              "width of plate"
L=15 [cm]*Convert(cm,m)                             "length of plate"
//T_s=350 [K]                                       "guess for T_s"
T_f=(T_infinity+T_s)/2                              "film temperature"
rho=Density(Water,T=T_f,p=p_atm)                    "density"
k=Conductivity(Water,T=T_f,p=p_atm)                 "thermal conductivity"
nu=KinematicViscosity(Water,T=T_f,p=p_atm)          "kinematic viscosity"
alpha=ThermalDiffusivity(Water,T=T_f,p=p_atm)       "thermal diffusivity"
beta=VolExpCoef(Water,T=T_f,p=p_atm)                "volumetric expansion coefficient"
Pr=nu/alpha                                         "Prandtl number"
A_s=W*L                                             "surface area"
per=2*(W+L)                                         "perimeter"
L_char=A_s/per                                      "characteristic length"
Ra_L_char=g#*L_char^3*beta*(T_s-T_infinity)/(nu*alpha)
                                                    "Rayleigh number"
//Call FC_plate_horizontal1_ND(Ra_L_char, Pr: Nusselt_L_char)
                                                    "access correlations for upward facing plate"
Call FC_plate_horizontal2_ND(Ra_L_char, Pr: Nusselt_L_char)
                                                    "access correlations for downward facing plate"
h_bar=Nusselt_L_char*k/L_char                       "average heat transfer coefficient"
T_s_c=T_infinity+q_dot/(h_bar*A_s)                  "calculated surface temperature"
T_s=T_s_c                                           "iterate to find surface temperature"
```

Solving leads to $\boxed{T_s = 347.5 \text{ K } (74.3°\text{C})}$ for a downward facing heated plate.

Discussion

The surface temperature is much higher for a downward facing plate than it is for an upward facing plate ($74.3°\text{C}$ vs. $49.6°\text{C}$) because the hot fluid adjacent to the plate is above the colder fluid and therefore the situation is very nearly stable with regard to buoyancy effects. The velocity of the buoyancy-induced flow is smaller in this case, resulting in a larger boundary layer and a lower heat transfer coefficient.

Figure 3 illustrates the surface temperature as a function of the rate of heat transfer for both the upward and downward facing cases. Notice that the relationship is more nonlinear than a forced convection situation due to the dependence of the Rayleigh number on the surface to fluid temperature difference. The solution is not valid for surface temperatures that are above $100°\text{C}$ because at that point boiling will begin. Boiling heat transfer is discussed in Chapter 11.

Continued

Example 10.2 (cont.)

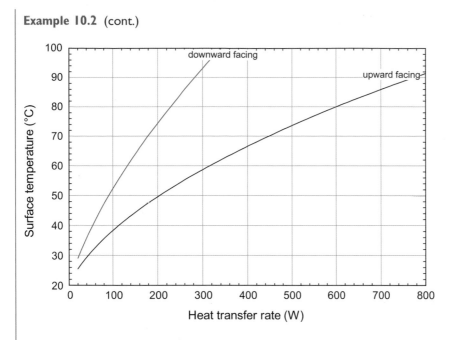

Figure 3 Surface temperature as a function of the rate of heat transfer to the plate if it is upward and downward facing.

Plate at an Arbitrary Angle

Figure 10.10 illustrates a heated plate oriented at an angle ζ relative to horizontal. When $\zeta = 0$ the plate is heated upward facing, when $\zeta = \pi/2$ the plate is vertical, and when $\zeta = \pi$ the plate is heated downward facing; correlations for these situations have been presented previously in this section and are summarized in Figure 10.11.

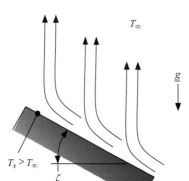

Figure 10.10 Heated plate oriented at an arbitrary angle ζ relative to horizontal.

Raithby and Hollands (1998) present a methodology that uses the correlations for the three situations shown in Figure 10.11 in order to estimate the heat transfer coefficient from a heated plate inclined at an arbitrary angle, $0 < \zeta < \pi$ rad. The procedure is illustrated in Figure 10.12 and requires that two of the three functions shown in Figure 10.11 are called using an appropriate projection of the gravity vector. The maximum of the heat transfer coefficients that is calculated is taken as the best estimate of the actual heat transfer coefficient. (Note that it is *not* correct to take the maximum Nusselt numbers, since different length scales are used for the correlations.) For a cooled plate, the same procedure applies but the tilt angle should be adjusted to be $(\pi - \zeta)$ rad.

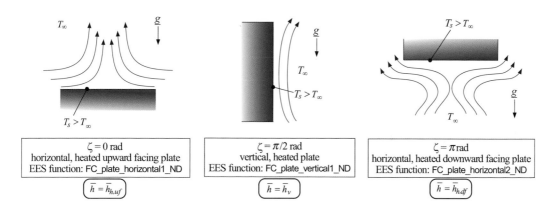

Figure 10.11 Correlations available for heated upward facing plate (left), heated vertical plate (middle), and heated downward facing plate (right).

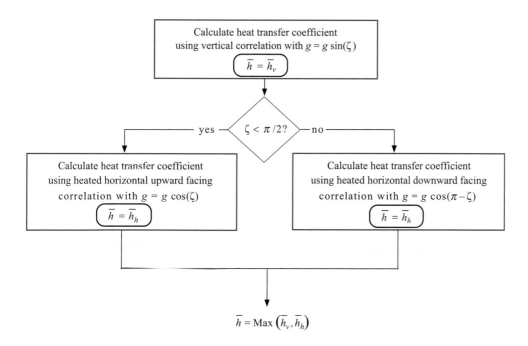

Figure 10.12 Methodology used to calculate the heat transfer coefficient from a heated plate at arbitrary angle ζ relative to horizontal, as shown in Figure 10.10.

Example 10.3

Consider a flat plate heater that is tilted to an angle $\zeta = \pi/4$ relative to horizontal. The plate is in a tank filled with helium at $p = 10$ atm and $T_\infty = 300$ K. The plate is square with side length $L = 10$ cm. Only the top surface of the plate experiences convection and radiation. The plate is electrically heated until its surface temperature reaches $T_s = 400$ K. The emissivity of the plate is $\varepsilon = 0.64$. Estimate the rate of heat transfer required.

Known Values

Figure 1 illustrates a sketch of the problem with the known values indicated.

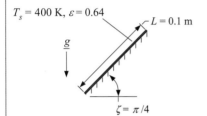

Figure 1 Sketch of the problem with the known values indicated.

Assumptions

- The problem is steady state.
- The tank is sufficiently large that the effect of the tank surface on the flow can be neglected.
- Radiation occurs to surroundings that are at a uniform temperature of T_∞.

Analysis

The film temperature is:

$$T_f = \frac{(T_s + T_\infty)}{2}. \tag{1}$$

The properties of helium are obtained at the film temperature and atmospheric pressure (p_{atm}):

$$\rho = \rho(T_f, p) \tag{2}$$

$$k = k(T_f, p) \tag{3}$$

$$\upsilon = \upsilon(T_f, p) \tag{4}$$

$$\alpha = \alpha(T_f, p) \tag{5}$$

$$\beta = \beta(T_f, p) \tag{6}$$

$$Pr = \frac{\upsilon}{\alpha}. \tag{7}$$

Example 10.3 (cont.)

Following the methodology in Figure 10.12, the vertical plate correlation is used with the component of gravity that is parallel to the plate surface. The associated Rayleigh number is:

$$Ra_{L,v} = \frac{g \sin (\zeta) L^3 \beta (T_s - T_\infty)}{\upsilon \alpha}.$$ (8)

Notice that the characteristic length for the correlation for a vertical plate is the length of the plate. The average Nusselt number is a function of the Rayleigh number and Prandtl number:

$$\overline{Nu}_{L,v} = \overline{Nu}_{L,v}(Ra_{L,v}, Pr).$$ (9)

The average heat transfer coefficient is computed using the definition of the Nusselt number:

$$\bar{h}_v = k \frac{\overline{Nu}_{L,v}}{L}.$$ (10)

The surface area and perimeter of the plate are used to compute the characteristic length for the horizontal plate correlations according to Eq. (10.22):

$$A_s = W L$$ (11)

$$per = 2 (W + L)$$ (12)

$$L_{char} = \frac{A_s}{per}.$$ (13)

Because $\zeta < \pi/2$, the correlation for a horizontal, upward facing heated plate is used with the component of gravity perpendicular to the plate. The Rayleigh number based on the characteristic length is determined:

$$Ra_{L_{char},h} = \frac{g \cos (\zeta) L_{char}^3 \beta (T_s - T_\infty)}{\upsilon \alpha}.$$ (14)

The average Nusselt number is a function of the Rayleigh number and Prandtl number:

$$\overline{Nu}_{L_{char},h} = \overline{Nu}_{L_{char},h}(Ra_{L_{char},h}, Pr).$$ (15)

The average heat transfer coefficient is computed using the definition of the Nusselt number:

$$\bar{h}_h = k \frac{\overline{Nu}_{L_{char},h}}{L_{char}}.$$ (16)

The heat transfer coefficient is estimated as the maximum of the two values calculated:

$$\bar{h} = \text{Max} (\bar{h}_v, \bar{h}_h).$$ (17)

The total heat transfer rate is the sum of convection and radiation:

$$\dot{q}_{conv} = \bar{h} A_s (T_s - T_\infty)$$ (18)

$$\dot{q}_{rad} = \sigma \varepsilon A_s (T_s^4 - T_\infty^4)$$ (19)

$$\dot{q} = \dot{q}_{conv} + \dot{q}_{rad}.$$ (20)

Equations (1) through (20) are 20 equations in the 20 unknowns: $T_f, \rho, k, \upsilon, \alpha, \beta, Pr, Ra_{L,v}, \overline{Nu}_{L, v}, \bar{h}_v, A_s, per, L_{char}, Ra_{L_{char}}, \overline{Nu}_{L_{char}}, \bar{h}_h, \bar{h}, \dot{q}_{conv}, \dot{q}_{rad}$, and $\dot{q}$.

Solution

Equations (1) through (20) are entered in EES. The properties required by Eqs. (2) through (6) are obtained from EES' internal database and the correlations in Eqs. (9) and (15) are obtained from the Heat Transfer Library.

Continued

Example 10.3 (cont.)

```
$UnitSystem SI Mass J K Pa Radian
T_infinity=300 [K]                              "ambient temperature"
T_s=400 [K]                                     "surface temperature"
zeta=pi/4 [radian]                              "angle relative to horizontal"
L = 0.1 [m]                                      "side width of square plate"
e=0.64 [-]                                       "emissivity"
p = 10 [atm]*Convert(atm,Pa)                    "pressure"


T_f=(T_s+T_infinity)/2                          "film temperature"
rho=Density(Helium,T=T_f,P=p)                   "density"
k=Conductivity(Helium,T=T_f,P=p)               "conductivity"
nu=KinematicViscosity(Helium,T=T_f,P=p)        "kinematic viscosity"
alpha=ThermalDiffusivity(Helium,T=T_f,P=p)     "thermal diffusivity"
beta=VolExpCoef(Helium,T=T_f,P=p)              "volumetric expansion coefficient"
Pr=nu/alpha                                     "Prandtl number"

"Vertical correlation"
Ra_L_v=g#*Sin(zeta)*L^3*beta*(T_s-T_infinity)/(nu*alpha)
                                                "Rayleigh number"
CALL FC_Plate_Vertical_ND(Ra_L_v, Pr: Nusselt_L_v)
                                                "Nusselt number"
h_bar_v=Nusselt_L_v*k/L                         "heat transfer coefficient"

"Horizontal correlation"
A_s=L^2                                          "surface area"
per=4*L                                          "perimeter"
L_char=A_s/per                                   "char. length for horizontal correlations"
Ra_Lchar_h=g#*Cos(zeta)*L_char^3*beta*(T_s-T_infinity)/(nu*alpha)
                                                "Rayleigh number"
CALL FC_Plate_Horizontal1_ND(Ra_Lchar_h, Pr: Nusselt_Lchar_h)
                                                "Nusselt number"
h_bar_h=Nusselt_Lchar_h*k/L_char               "heat transfer coefficient"

h_bar=Max(h_bar_v,h_bar_h)                      "heat transfer coefficient"
q_dot_conv=h_bar*A_s*(T_s-T_infinity)          "convection"
q_dot_rad=sigma#*e*A_s*(T_s^4-T_infinity^4)    "radiation"
q_dot=q_dot_conv+q_dot_rad                      "total heat transfer rate"
```

Solving provides $\boxed{\dot{q} = 61.4 \text{ W}}$.

Discussion

Figure 2 illustrates the rate of heat transfer as a function of the angle, ζ. Notice that the heat transfer rate is highest for a horizontal plate and is reduced as the plate becomes more vertical because the situation is more stable relative to buoyancy. The discontinuity at $\zeta \approx 70°$ is related to the vertical plate heat transfer coefficient becoming larger than the horizontal plate heat transfer coefficient in Eq. (17).

Example 10.3 (cont.)

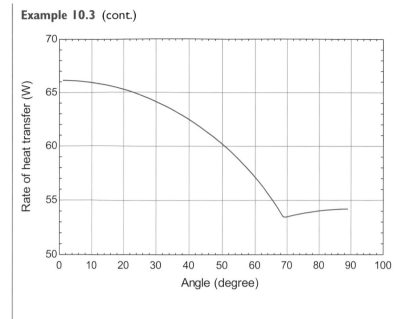

Figure 2 Rate of heat transfer as a function of the angle ζ in Figure 1.

10.3.3 Sphere

Natural convection from a heated or cooled sphere is correlated using a Rayleigh number and average Nusselt number that are based on the diameter of the sphere:

$$Ra_D = \frac{g \, D^3 \, \beta \, (T_s - T_\infty)}{\nu \, \alpha} \tag{10.31}$$

$$\overline{Nu}_D = \frac{\bar{h} \, D}{k}. \tag{10.32}$$

Churchill (1983) recommends the following correlation for this situation:

$$\overline{Nu}_D = 2 + \frac{0.589 \, Ra_D^{0.25}}{\left[1 + \left(\frac{0.469}{Pr} \right)^{9/16} \right]^{4/9}} \quad \text{for } Pr > 0.5 \text{ and } Ra_D < 1 \times 10^{11}. \tag{10.33}$$

Note that the Nusselt number predicted by Eq. (10.33) limits to a value of 2 at very low Rayleigh number. In the limit that the Rayleigh number approaches zero, there will be no buoyancy-induced fluid motion and therefore the problem is reduced to conduction through a stationary medium. One-dimensional, steady-state conduction through a spherical shell is characterized by the spherical conduction resistance derived in Section 2.2.4:

$$R_{sph} = \frac{1}{4 \, \pi \, k} \left[\frac{1}{r_{in}} - \frac{1}{r_{out}} \right]. \tag{10.34}$$

The surface of the sphere at $r_{in} = D/2$ is at T_s and the surrounding medium that is far from the sphere (at $r_{out} \rightarrow \infty$) is at T_∞. The rate of heat transfer from the sphere for this limiting case is

$$\dot{q} = \frac{(T_s - T_\infty)}{R_{sph}} = \frac{4 \, \pi \, k \, (T_s - T_\infty)}{\left[\frac{2}{D} - \frac{1}{\infty} \right]} \tag{10.35}$$

or

$$\dot{q} = 2 \, \pi \, k \, D \, (T_s - T_\infty). \tag{10.36}$$

The heat transfer rate may also be written in terms of the average heat transfer coefficient and the surface area of the sphere:

$$\dot{q} = \bar{h} \, A_s \, (T_s - T_\infty). \tag{10.37}$$

Expressing the average heat transfer coefficient in Eq. (10.37) in terms of an average Nusselt number and substituting the surface area for a sphere leads to:

$$\dot{q} = \underbrace{\overline{Nu_D}\frac{k}{D}}_{\bar{h}} \, \underbrace{\pi D^2}_{A_s} \, (T_s - T_\infty) = \underbrace{\overline{Nu_D}}_{=2} \, \pi k D \, (T_s - T_\infty). \tag{10.38}$$

Comparing Eqs. (10.36) and (10.38) shows that the average Nusselt number for a sphere in a stagnant medium must be 2. The correlation provided by Eq. (10.33) is implemented in EES in the procedure FC_sphere_ND. The average Nusselt number for a cooled sphere may be computed using the correlation provided in this section provided that the absolute value of the temperature difference is used.

10.3.4 Cylinder

Horizontal Cylinder

Natural convection from a long cylinder that is either heated or cooled and is horizontal with respect to gravity (i.e., gravity is perpendicular to the axis of the cylinder) is correlated using a Rayleigh number and average Nusselt number that are based on the diameter of the cylinder:

$$Ra_D = \frac{g \, D^3 \, \beta \, (T_s - T_\infty)}{\upsilon \, \alpha} \tag{10.39}$$

$$\overline{Nu_D} = \frac{\bar{h} \, D}{k}. \tag{10.40}$$

The correlation provided by Churchill and Chu (1975a) is valid for $1 \times 10^{-5} < Ra_D < 1 \times 10^{12}$:

$$\overline{Nu_D} = \left\{ 0.60 + \frac{0.387 \, Ra_D^{1/6}}{\left[1 + \left(\dfrac{0.559}{Pr} \right)^{9/16} \right]^{8/27}} \right\}^2. \tag{10.41}$$

Figure 10.13 illustrates the average Nusselt number for a horizontal cylinder as a function of the Rayleigh number and various values of the Prandtl number. The correlation discussed in this section is implemented in EES as the procedure FC_horizontal_cylinder_ND and can be used for either a heated or cooled cylinder (provided that the absolute value of the temperature difference is used).

Vertical Cylinder

Natural convection from the sides of a heated or cooled cylinder that is vertical with respect to gravity (i.e., gravity is parallel to the axis of the cylinder) leads to the development of a boundary layer that starts at one end (the lower end of a heated cylinder and upper edge of a cooled cylinder) and grows in the direction of the induced fluid motion. The flow pattern is therefore similar to the flow associated with natural convection from a vertical plate, Figure 10.4. The Rayleigh number and average Nusselt number are based on the length of the cylinder rather than its diameter:

$$Ra_L = \frac{g \, L^3 \, \beta \, (T_s - T_\infty)}{\upsilon \, \alpha} \tag{10.42}$$

$$\overline{Nu_L} = \frac{\bar{h} \, L}{k}, \tag{10.43}$$

where L is the length of the cylinder.

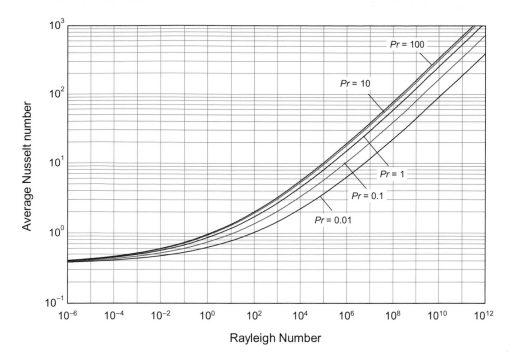

Figure 10.13 Nusselt number for a horizontal cylinder as a function of the Rayleigh number and for various values of the Prandtl number.

A vertical cylinder can be treated as a flat plate provided that the boundary layer thickness (δ_m) is much smaller than the cylinder diameter (D). Section 7.1 showed that the momentum boundary layer thickness for laminar flow over a flat plate in forced convection is approximately:

$$\delta_m \approx \frac{2\,x}{\sqrt{Re_x}}. \tag{10.44}$$

Recall from the discussion in Section 10.2.3 that in a free convection problem, the square root of the Grashof number is analogous to the Reynolds number. Therefore, the boundary layer thickness at the upper edge of a heated cylinder will be approximately:

$$\delta_m = \frac{2\,L}{Gr_L^{0.25}}. \tag{10.45}$$

The ratio of diameter to momentum boundary layer thickness is therefore:

$$\frac{D}{\delta_{m,\,lam}} \approx \frac{D}{2\,L}\,Gr_L^{0.25}. \tag{10.46}$$

Equation (10.46) suggests that it is appropriate to treat free convection from the sides of a vertical cylinder using the vertical flat plate correlation in Section 10.3.2 provided that:

$$\frac{D}{L} >> \frac{2}{Gr_L^{0.25}}. \tag{10.47}$$

Sparrow and Gregg (1956) suggest that natural convection from a vertical cylinder can be treated as a vertical flat plate when the diameter-to-length ratio meets the criteria:

$$\frac{D}{L} > \frac{35}{Gr_L^{0.25}}. \tag{10.48}$$

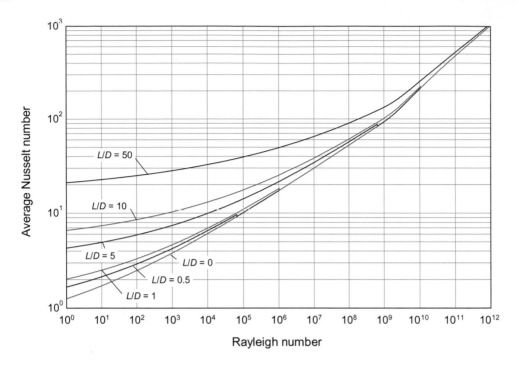

Figure 10.14 Nusselt number for a vertical cylinder as a function of the Rayleigh number for various values of L/D and $Pr = 0.7$.

For smaller diameter-to-length ratios, the Nusselt number is augmented by the effect of the curvature of the surface and can be obtained approximately from

$$\overline{Nu}_L = \overline{Nu}_{L,\,vp} \frac{\zeta}{\ln(1 + \zeta)}, \tag{10.49}$$

where $\overline{Nu}_{L,\,vp}$ is the average Nusselt number neglecting curvature (i.e., obtained using the correlation for a vertical flat plate) and ζ is given by

$$\zeta = \frac{1.8}{\overline{Nu}_{L,\,vp}} \frac{L}{D}. \tag{10.50}$$

Figure 10.14 illustrates the average Nusselt number for a vertical cylinder as a function of the Rayleigh number for various values of the length-to-diameter ratio with $Pr = 0.7$. The augmentation associated with slender cylinders (i.e., cylinders with small diameters and long lengths) is most apparent at low Rayleigh number where the boundary layer is the thickest.

The correlation for a vertical cylinder is implemented in EES by the procedure FC_vertical_cylinder_ND. The correlations in this section can be used for either a heated or a cooled cylinder (provided that the absolute value of the temperature difference is used).

Example 10.4

An electrical heater is installed in a large tank of water. The heater is a cylinder with diameter $D = 1$ cm and length $L = 40$ cm that is oriented vertically in the otherwise quiescent water within the tank. The water is at atmospheric pressure and $T_\infty = 20°C$. The current passing through the heater causes a uniform volumetric

Example 10.4 (cont.)

rate of thermal energy generation equal to $\dot{g}''' = 1 \times 10^7$ W/m^3. The heater material has properties $\rho_h = 9000$ kg/m^3, $c_h = 350$ J/kg-K, and $k_h = 120$ W/m-K. Determine the rate of temperature change of the heater at the instant of time that it is at a uniform temperature of $T_h = 40°C$.

Known Values

Figure 1 illustrates a sketch of the problem with the known values indicated.

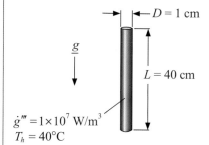

Figure 1 Sketch of the problem with the known values indicated.

water at atmospheric pressure and $T_\infty = 20°C$

Assumptions

- The heater can be modeled as being a lumped capacitance.
- The tank is sufficiently large that the water temperature is constant and the flow is not affected by the presence of the tank walls.
- Convection from the top and bottom surfaces is neglected.
- The convection correlation is applicable for unsteady-state conditions.

Analysis

The film temperature is:

$$T_f = \frac{(T_s + T_\infty)}{2}. \tag{1}$$

The properties of water are obtained at the film temperature and atmospheric pressure (p_{atm}):

$$\rho = \rho\left(T_f, p_{atm}\right) \tag{2}$$

$$k = k\left(T_f, p_{atm}\right) \tag{3}$$

$$v = v\left(T_f, p_{atm}\right) \tag{4}$$

$$\alpha = \alpha\left(T_f, p_{atm}\right) \tag{5}$$

$$\beta = \beta\left(T_f, p_{atm}\right) \tag{6}$$

$$Pr = \frac{v}{\alpha}. \tag{7}$$

The Rayleigh number is:

$$Ra_L = \frac{g\,L^3\,\beta\,(T_s - T_\infty)}{v\,\alpha}. \tag{8}$$

Continued

Example 10.4 (cont.)

Notice that the characteristic length for the correlations for a vertical cylinder is the length of the cylinder. The average Nusselt number is a function of the Rayleigh number, the Prandtl number, and the length-to-diameter ratio:

$$\overline{Nu}_{L,v} = \overline{Nu}_{L,v}\,(Ra_L, Pr, L/D). \tag{9}$$

The average heat transfer coefficient is computed using the definition of the Nusselt number:

$$\bar{h} = k\,\frac{\overline{Nu}_L}{L}. \tag{10}$$

The volume and surface area of the cylinder are computed:

$$V = \pi\,\frac{D^2}{4}\,L \tag{11}$$

$$A_s = \pi\,D\,L. \tag{12}$$

The total rate of thermal energy generation in the heater is given by:

$$\dot{g} = \dot{g}'''\,V. \tag{13}$$

An energy balance on the cylinder provides:

$$\dot{g} = A_s\,\bar{h}\,(T_h - T_\infty) + V\,\rho_h\,c_h\,\frac{dT_h}{dt}. \tag{14}$$

Equations (1) through (14) are 14 equations in the 14 unknowns: T_f, ρ, k, v, α, β, Pr, Ra_L, $\overline{Nu}_L$, $\bar{h}$, V, A_s, $\dot{g}$, and $\frac{dT_h}{dt}$.

Solution

Equations (1) through (14) are entered in EES. The properties in Eqs. (2) through (6) are obtained from EES' internal database and the correlation required by Eq. (9) is accessed from the Heat Transfer Library.

```
$UnitSystem SI Mass J K Pa Radian
D=1 [cm]*Convert(cm,m)                          "diameter"
L=40 [cm]*Convert(cm,m)                         "length"
T_infinity=ConvertTemp(C,K,20[C])               "ambient temperature"
g```_dot=1e7 [W/m^3]                            "volumetric generation of thermal energy"
rho_h=9000 [kg/m^3]                             "density of heater element"
c_h=350 [J/kg-K]                                "specific heat capacity of heater element"
k_h=120 [W/m-K]                                 "thermal conductivity of heater element"
T_h=ConvertTemp(C,K,40 [C])                     "heater element temperature"
T_f=(T_h+T_infinity)/2                          "film temperature"
rho=Density(Water,T=T_f,P=po#)                  "density of water"
k=Conductivity(Water,T=T_f,P=po#)               "conductivity of water"
nu=KinematicViscosity(Water,T=T_f,P=po#)        "kinematic viscosity of water"
alpha=ThermalDiffusivity(Water,T=T_f,P=po#)     "thermal diffusivity of water"
beta=VolExpCoef(Water,T=T_f,P=po#)              "volumetric expansion coefficient of water"
```

Example 10.4 (cont.)

Pr=nu/alpha	"Prandtl number of water"
Ra_L=g#*L^3*beta*(T_h-T_infinity)/(nu*alpha)	"Rayleigh number"
Call **FC_Vertical_Cylinder_ND**(Ra_L, Pr, L/D: Nusselt_bar_L)	
	"access correlations"
h_bar=Nusselt_bar_L*k/L	"heat transfer coefficient"
V=pi*(D^2/4)*L	"volume of heater"
A_s=pi*D*L	"surface area of heater"
g_dot=g```_dot*V	"rate of thermal energy generation"
g_dot=A_s*h_bar*(T_h-T_infinity)+rho_h*c_h*V*dThdt	"energy balance"

Solving provides $\boxed{\dfrac{dT_h}{dt} = 1.282 \; \dfrac{K}{s}}$.

Discussion

The lumped capacitance assumption should be checked by computing a Biot number:

$$Bi = \frac{\bar{h}\,D}{2\,k}. \tag{15}$$

Bi=h_bar*(D/2)/k_h	"Biot number"

Solving provides $Bi = 0.022$ which is sufficiently small to justify this assumption.

It is relatively easy to investigate the transient behavior of the heater by numerically integrating the time derivative forward using EES' Integral command, as discussed previously in Section 5.3.5. For example, we can examine the activation of the heater from some initial temperature, T_{ini}, by carrying out the following steps. Update the guess values by selecting Update Guesses from the Calculate menu. Comment out the specified value of T_h and instead use the Integral command, as shown in the code below.

$UnitSystem SI Mass J K Pa Radian	
D=1 [cm]***Convert**(cm,m)	"diameter"
L=40 [cm]***Convert**(cm,m)	"length"
T_infinity=**ConvertTemp**(C,K,20[C])	"ambient temperature"
g```_dot=1e7 [W/m^3]	"volumetric generation of thermal energy"
rho_h=9000 [kg/m^3]	"density of heater element"
c_h=350 [J/kg-K]	"specific heat capacity of heater element"
k_h=120 [W/m-K]	"thermal conductivity of heater element"
//T_h=**ConvertTemp**(C,K,40 [C])	"heater element temperature"
T_f=(T_h+T_infinity)/2	"film temperature"
rho=**Density**(*Water*,*T*=T_f,*P*=po#)	"density of water"
k=**Conductivity**(*Water*,*T*=T_f,*P*=po#)	"conductivity of water"
nu=**KinematicViscosity**(*Water*,*T*=T_f,*P*=po#)	"kinematic viscosity of water"
alpha=**ThermalDiffusivity**(*Water*,*T*=T_f,*P*=po#)	"thermal diffusivity of water"
beta=**VolExpCoef**(*Water*,*T*=T_f,*P*=po#)	"volumetric expansion coefficient of water"
Pr=nu/alpha	"Prandtl number of water"

Continued

Example 10.4 (cont.)

```
Ra_L=g#*L^3*beta*(T_h-T_infinity)/(nu*alpha)         "Rayleigh number"
Call FC_Vertical_Cylinder_ND(Ra_L, Pr, L/D: Nusselt_bar_L) "access correlations"
h_bar=Nusselt_bar_L*k/L                              "heat transfer coefficient"
V=pi*(D^2/4)*L                                       "volume of heater"
A_s=pi*D*L                                           "surface area of heater"
g_dot=g```_dot*V                                     "rate of thermal energy generation"
g_dot=A_s*h_bar*(T_h-T_infinity)+rho_h*c_h*V*dThdt   "energy balance"
Bi=h_bar*(D/2)/k_h                                   "Biot number"

T_ini=300 [K]                                        "initial temperature"
T_h=T_ini+Integral(dThdt,time,0,60)                  "integrate the state equation"
$IntegralTable time,T_h                              "Integral table stores results"
```

Note that the $IntegralTable directive is used to store the temperature of the heater as a function of time. The temperature of the heater as a function of time is shown in Figure 2.

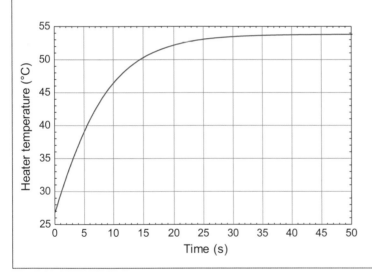

Figure 2 Temperature of the heater as a function of time.

10.4 Internal Free Convection Correlations

10.4.1 Introduction

Section 10.3 explores natural convection problems associated with flow induced around geometries that are immersed in an infinitely large medium of otherwise stagnant fluid. These problems are analogous to external flow forced convection problems in that the boundary layers that form are unbounded. This section reviews correlations for free convection in channels and cavities, flow situations that are more similar to internal forced convection because the boundary layers that form are inherently bounded by the adjacent walls.

10.4.2 Vertical Parallel Plate Channels

Figure 10.15 illustrates a channel formed by two plates that are each oriented parallel to gravity. The flow within this channel will resemble a forced convection, internal flow problem. There is a developing region where

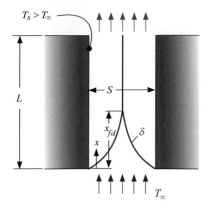

Figure 10.15 Open channel flow induced in a vertical channel formed by two vertical plates.

the boundary layers are growing ($x < x_{fd}$), followed by a fully developed region where the boundary layers are bounded ($x > x_{fd}$). If the plates are sufficiently short such that the flow does not become fully developed (i.e., $L < x_{fd}$) then the flow can be adequately modeled using the correlations for a vertical plate presented in Section 10.3.2. Otherwise the flow will become fully developed and the correlations provided in this section must be used.

The Rayleigh number and average Nusselt number for flow through the channels defined by vertical parallel plates are defined based on the plate spacing, S:

$$Ra_S = \frac{g \, S^3 \, \beta \, (T_s - T_\infty)}{\nu \, \alpha} \tag{10.51}$$

$$\overline{Nu}_S = \frac{\bar{h} \, S}{k}. \tag{10.52}$$

The average heat transfer coefficient in Eq. (10.52) is defined as

$$\bar{h} = \frac{\dot{q}}{A_s \, (T_s - T_\infty)}, \tag{10.53}$$

where $\dot{q}$ is the total rate of heat transfer from one of the plates and A_s is the surface area of the plate. Note that the average heat transfer coefficient in Eq. (10.53) is based upon the difference between the plate surface temperature (T_s) and the temperature of the fluid that is being pulled into the channel (T_∞) rather than on the temperature difference between the wall and the local mean temperature (as was the case for forced, internal flow).

The Nusselt number is correlated against the Rayleigh number and the ratio of the channel length to spacing (L/S) according to Elenbaas (1942):

$$\overline{Nu}_S = \frac{Ra_S}{24} \frac{S}{L} \left[1 - \exp\left(-\frac{35}{Ra_S} \frac{L}{S} \right) \right]^{0.75}. \tag{10.54}$$

The correlation provided by Eq. (10.54) is implemented in the EES procedure **FC_Vertical_Channel_ND**. Correlations for parallel plate channels under other boundary conditions (e.g., uniform heat flux or with one channel insulated) are presented by various researchers including Bar-Cohen and Rohsenow (1984). Correlations for channels with other cross-sections (e.g., circular channels) and arrays of extended surfaces (such as are commonly encountered in heat sink applications) are also available and have been compiled by Raithby and Hollands (1998).

Example 10.5

A passive heat sink is being designed for an electronic component. Several plates of copper with thickness $th = 1.5$ mm extend between two parallel surfaces that are separated by a distance $H = 10$ cm. The space between adjacent plates forms an open channel. The copper plates, and therefore the channels, are $L = 25$ cm long in the direction of gravity. Each plate has area $H \times L$ on one side. The total distance that is available to install plates is $W = 10$ cm wide. Therefore, a large number of plates can be installed but they must be very close together leaving a small space for air flow (S in Figure 1). Alternately, a small number of plates can be installed with a large spacing, S. The heat sink is placed in stagnant air at a nominal temperature of $T_\infty = 20°C$ and the plates are maintained at a temperature of $T_s = 80°C$.

Determine the number of plates that maximizes the rate of heat transfer from the heat sink.

Known Values

Figure 1 provides a sketch of the problem with the known values indicated.

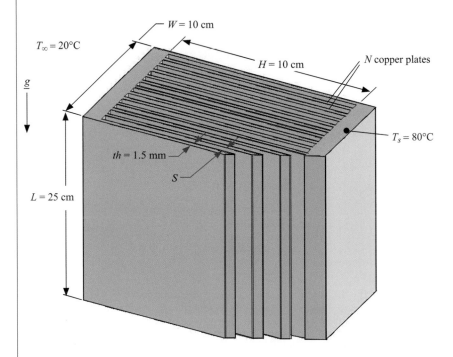

Figure 1 Sketch of the problem with the known values indicated.

Assumptions

- The copper is sufficiently conductive that the plates can be assumed to be isothermal. (This assumption should be confirmed by computing the fin efficiency of the plates.)
- The air pressure is atmospheric.
- Convection from the top and bottom surfaces is neglected.

Example 10.5 (cont.)

- Convection from the edges of the outer plates is neglected.
- Radiation is neglected.

Analysis

In order to carry out the optimization required by the problem we will develop a model capable of computing the total rate of heat transfer from the heat sink given a reasonable value of N, the number of plates. Once completed, the model can be used to parametrically vary N and examine its effect on the total heat transfer rate. The width of the heat sink is taken up by N plates and $(N-1)$ spaces; therefore, the spacing between adjacent plates for a particular value of N is given by:

$$S = \frac{(W - N\,th)}{(N-1)}. \tag{1}$$

The film temperature is:

$$T_f = \frac{(T_s + T_\infty)}{2}. \tag{2}$$

The properties of air are obtained at the film temperature and atmospheric pressure (p_{atm}):

$$\rho = \rho\left(T_f, p_{atm}\right) \tag{3}$$

$$k = k\left(T_f\right) \tag{4}$$

$$\upsilon = \upsilon\left(T_f, p_{atm}\right) \tag{5}$$

$$\alpha = \alpha\left(T_f, p_{atm}\right) \tag{6}$$

$$\beta = \beta\left(T_f\right) \tag{7}$$

$$Pr = \frac{\upsilon}{\alpha}. \tag{8}$$

The Rayleigh number is:

$$Ra_S = \frac{g\,S^3\,\beta\,(T_s - T_\infty)}{\upsilon\,\alpha}. \tag{9}$$

Notice that the characteristic length used for the correlations for a channel formed between two vertical plates is the plate-to-plate spacing. The average Nusselt number is a function of the Rayleigh number and the length-to-spacing ratio:

$$\overline{Nu}_S = \overline{Nu}_S\,(Ra_S, L/S). \tag{10}$$

The average heat transfer coefficient is computed using the definition of the Nusselt number:

$$\bar{h} = k\frac{\overline{Nu}_S}{S}. \tag{11}$$

Continued

Example 10.5 (cont.)

The rate of heat transfer per channel is computed:

$$\dot{q}_{channel} = 2\,W\,L\,\bar{h}\,(T_s - T_\infty). \tag{12}$$

The total rate of heat transfer from the heat sink is:

$$\dot{q}_{total} = (N - 1)\,\dot{q}_{channel}. \tag{13}$$

Equations (1) through (13) are 13 equations in the 13 unknowns: S, T_f, ρ, k, v, α, β, Pr, Ra_S, $\overline{Nu}_S$, $\bar{h}$, $\dot{q}_{channel}$, and $\dot{q}_{total}$.

Solution

Equations (1) through (13) are entered in EES using an arbitrary value of the number of plates, N. The properties required by Eqs. (3) through (7) are obtained from EES' internal database and the correlation required by Eq. (10) is accessed from the Heat Transfer Library.

```
$UnitSystem SI Mass J K Pa
T_infinity=ConvertTemp(C,K,20 [C])          "ambient temperature"
T_s=ConvertTemp(C,K,80[C])                  "surface temperature"
L = 25 [cm]*Convert(cm,m)                   "length of plates in direction of gravity"
H = 10 [cm]*Convert(cm,m)                   "plate height, perpendicular to gravity"
W = 10 [cm]*Convert(cm,m)                   "distance available to install plates"
th = 1.5 [mm]*Convert(mm,m)                 "plate thickness"


N=10 [-]                                    "number of plates"
S=(W-N*th)/(N-1)                            "space for air flow between plates"
T_f=(T_s+T_infinity)/2                      "film temperature"
rho=Density(Air,T=T_f,P=po#)                "density"
k=Conductivity(Air,T=T_f)                   "thermal conductivity"
nu=KinematicViscosity(Air,T=T_f,P=po#)      "kinematic viscosity"
alpha=ThermalDiffusivity(Air,T=T_f,P=po#)   "thermal diffusivity"
beta=VolExpCoef(Air,T=T_f)                  "volumetric expansion coefficient"
Pr=nu/alpha                                 "Prandtl number"
Ra_S=g#*S^3*beta*(T_s-T_infinity)/(nu*alpha) "Rayleigh number based on S"
CALL FC_Vertical_Channel_ND(Ra_S,L/S:Nusselt_bar_S)
                                            "correlation for open channel"
h_bar=Nusselt_bar_S*k/S                     "average heat transfer coefficient"
q_dot_channel=2*h_bar*H*L*(T_s-T_infinity)  "heat transfer per channel"
q_dot_total=q_dot_channel*(N-1)             "total rate of heat transfer from heat sink"
```

The value of N in the Equations Window is commented out and a Parametric Table is created that includes the variables N, $\dot{q}_{channel}$, and $\dot{q}_{total}$. The values of N are varied in integer values in order to generate Figure 2, which shows the rate of heat transfer per channel and the total rate of heat transfer as a function of the number of plates.

Example 10.5 (cont.)

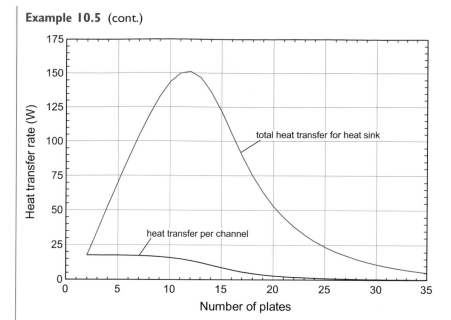

Figure 2 Rate of heat transfer per channel and total rate of heat transfer from the heat sink as a function of the number of plates.

Figure 2 shows that the optimal number of plates is about $\boxed{N = 12}$, resulting in a total heat transfer of approximately $\boxed{\dot{q}_{total} = 150\text{ W}}$.

Discussion

Notice in Figure 2 that the rate of heat transfer per channel is initially unaffected by the number of plates, corresponding to the situation where the boundary layers on the two sides of the channel do not interact. Eventually, the channel spacing becomes sufficiently small that the heat transfer per channel is degraded. As a result, there is a maximum value of the total rate of heat transfer.

The fin efficiency of the plates should be checked to support the assumption that the plates are isothermal at T_s. The conductivity of copper, k_p, is obtained from the Conductivity function in EES. Following the methodology discussed in Section 3.3, the fin efficiency is estimated according to:

$$\eta_{fin} = \frac{\tanh\,(mL)}{mL}, \tag{14}$$

where

$$mL = \sqrt{\frac{\bar{h}\,2\,L}{L\,th}\left(\frac{H}{2}\right)}. \tag{15}$$

```
k_p=Conductivity(Copper,T=T_s)          "conductivity of copper"
mL=Sqrt(h_bar*2*L/(k_p*L*th))*H/2        "fin constant"
eta_fin=tanh(mL)/mL                       "fin efficiency"
```

Carrying out the calculations above for $N = 12$ provides $\eta_{fin} = 0.99$ which suggests that the plates are very nearly isothermal.

10.4.3 Enclosures

The fluid in an enclosed volume may not remain stagnant if it is heated and cooled at different surfaces in the presence of gravity. The rectangular enclosure is encountered often in engineering applications and it has been extensively studied. Rectangular enclosures with large aspect ratios in both directions, as shown in Figure 10.16, are common. Notice that the separation distance between two walls (S) is much less than either of the other two dimensions of the enclosures (L or W). Examples of such high aspect ratio enclosures include the glass covering over solar collectors and multi-pane windows.

The enclosure in Figure 10.16 is tilted with angle ζ relative to the horizontal. If ζ is equal to 0, then the cooled surface (at temperature T_C) lies above the heated surface (at temperature T_H). When the tilt angle is $\zeta = \pi/2$ radian (90°), then the cooled and heated surfaces are both parallel to the gravity vector. At $\zeta = \pi$ radian (180°), the enclosure is again horizontal but with the heated surface over the cooled surface. The free convection flows that are induced within the enclosure depend strongly on the angle of tilt; however, regardless of tilt angle, the results are correlated using a Rayleigh number and average Nusselt number that are defined based on the separation distance, S:

$$Ra_S = \frac{g\,S^3\,\beta\,(T_H - T_C)}{\nu\,\alpha} \tag{10.55}$$

$$\overline{Nu}_S = \frac{\overline{h}\,S}{k}. \tag{10.56}$$

The heat transfer coefficient is defined based on the temperature difference between the two walls of the enclosure:

$$\overline{h} = \frac{\dot{q}}{W\,L\,(T_H - T_C)}, \tag{10.57}$$

where $\dot{q}$ is the total rate of heat transfer from the heated surface to the cooled surface.

At a tilt angle of $\zeta = \pi/2$ radian (90°, where walls of the enclosure are vertical), the fluid adjacent to the heated wall tends to rise until it reaches the top of the cavity and is then forced into contact with the cooled wall and falls. This circulation pattern carries energy from the hot surface to the cold surface more efficiently than simple conduction through the fluid that fills the gap.

If the Rayleigh number is less than a critical value of $Ra_{L,crit} \approx 1000$, then the buoyancy force is insufficient to overcome the viscous force and the fluid remains stagnant. In this limit, the free convection problem reduces to a simple conduction problem and the heat transfer can be calculated according to:

$$\dot{q} = \frac{k\,W\,L}{S}\,(T_H - T_C). \tag{10.58}$$

Substituting Eq. (10.58) into Eq. (10.57) leads to:

$$\overline{h} = \frac{k}{S} \tag{10.59}$$

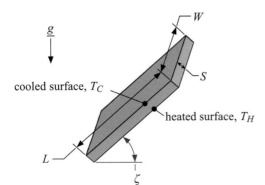

Figure 10.16 A high aspect ratio cavity tilted at angle ζ relative to horizontal.

and therefore, according to Eq. (10.56), the Nusselt number will be 1 when $Ra_L < Ra_{L,crit}$. MacGregor and Emery (1969) recommend the following correlation for an enclosure at $\zeta = \pi/2$ radian at low Rayleigh number with $W >> L$:

$$\overline{Nu}_{S, \zeta = \pi/2} = 0.42 \, Ra_S^{0.25} Pr^{0.012} \left(\frac{L}{S}\right)^{-0.3} \quad \text{for } 10 < \frac{L}{S} < 40, \; 1 < Pr < 2 \times 10^4, \; 1 \times 10^4 < Ra_S < 1 \times 10^7. \quad (10.60)$$

At higher Rayleigh numbers the Nusselt number should be calculated according to:

$$\overline{Nu}_{S, \zeta = \pi/2} = 0.046 \, Ra_S^{1/3} \quad \text{for } 1 < \frac{L}{S} < 40, \; 1 < Pr < 20, \; 1 \times 10^6 < Ra_S < 1 \times 10^9. \quad (10.61)$$

When the tilt angle is reduced below a critical tilt angle, ζ_{crit}, then regularly spaced, convective cells such as those shown in Figure 10.1(a) tend to form. The critical tilt angle is a function of the aspect ratio, L/S; however, for high aspect ratio enclosures ($L/S > 12$), the critical angle is approximately $\zeta_{crit} = 1.22$ rad (or $70°$).

For tilt angles less than $\zeta_{crit} = 1.22$ radian, flow is suppressed unless the Rayleigh number exceeds a critical value, $Ra_{S,crit}$. The critical Rayleigh can be estimated according to:

$$Ra_{S, crit} = \frac{1708}{\cos{(\zeta)}}. \quad (10.62)$$

Below the critical Rayleigh number, the buoyancy force is not sufficient to overcome the viscous forces that suppress fluid motion and therefore the fluid remains stagnant. Above the critical Rayleigh number, the Nusselt number has been correlated by Hollands *et al.* (1976) according to:

$$\overline{Nu}_S = 1 + 1.44 \, \text{Max} \left(0, 1 - \frac{1708}{Ra_S \cos{(\zeta)}}\right) \text{Max} \left[0, 1 - \frac{1708 \, [\sin{(1.8\zeta)}]^{1.6}}{Ra_S \cos{(\zeta)}}\right] + \text{Max} \left[0, \left(\frac{Ra_S \cos{(\zeta)}}{5830}\right)^{1/3} - 1\right],$$

$$\text{for } \frac{L}{S}, \frac{W}{S} > 12 \text{ and } 0 < \zeta < 1.22 \text{ rad.} \quad (10.63)$$

Note that Eq. (10.63) will reduce to 1 when $Ra_S < Ra_{S,crit}$. For tilt angles between the critical angle and vertical (i.e., 1.22 radian $< \zeta < \pi/2$ radian), Ayyaswamy and Catton (1973) recommend:

$$\overline{Nu}_S = \overline{Nu}_{S, \zeta = \pi/2} \, [\sin{(\zeta)}]^{0.25}. \quad (10.64)$$

For tilt angles greater than vertical (i.e., $\pi/2$ radian $< \zeta < \pi$ radian), Arnold *et al.* (1975) recommend:

$$\overline{Nu}_S = 1 + \left[\overline{Nu}_{S, \zeta = \pi/2} - 1\right] \sin{(\zeta)} \quad (10.65)$$

where $\overline{Nu}_{S, \zeta = \pi/2}$ is given by either Eq. (10.60) or Eq. (10.61) depending on the conditions. Notice that if $\zeta = \pi$ radian, then the enclosure is horizontal with the heated side up. This is an unconditionally stable situation and therefore the Nusselt number approaches unity regardless of Rayleigh number.

The correlations discussed in this section are implemented in EES by the procedure Tilted_Rect_Enclosure_ND. Figure 10.17 illustrates the Nusselt number as a function of Rayleigh for various values of the tilt angle with $L/S = 20$. Notice that the Nusselt number increases with Rayleigh number and decreases with tilt angle. As the cooled edge of the enclosure moves from being at the top of the enclosure ($\zeta = 0$) to the bottom of the enclosure ($\zeta = \pi$ radian), the free convection situation becomes progressively more stable and therefore the Nusselt number is reduced.

10.5 Combined Free and Forced Convection

In Chapters 7 through 9, forced convection was analyzed as if the *only* fluid motion that occurred was related to the forced flow. However, we have seen in this chapter that fluid motion can occur even without a mechanical

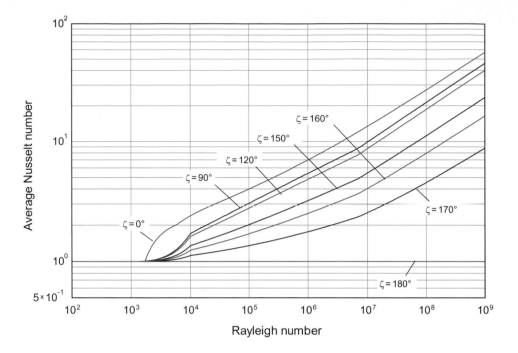

Figure 10.17 Nusselt number as a function Rayleigh number for various values of tilt angle (with $L/S = 20$).

input. Whenever a temperature gradient is present in a fluid in the presence of gravity, there can be some buoyancy-induced fluid motion. The relative importance of the free to the forced fluid motion can be quantified by the ratio of the Grashof number defined in Section 10.2.3 to the square of the Reynolds number defined based on the forced flow velocity. This ratio provides an index that is consistent with the discussion in Section 10.2.3, where the Grashof number was identified as the square of a Reynolds number that is defined based on the buoyancy-induced velocity. In Chapters 7 through 9, we implicitly assumed that the buoyancy-induced flow was so small that it could be neglected relative to the forced flow. This is consistent with the situation where the Grashof number is much less than the Reynolds number squared:

$$\frac{Gr}{Re^2} << 1 \quad \rightarrow \quad \text{consider only forced convection effects.} \tag{10.66}$$

In this chapter, forced flow has been neglected and *only* buoyancy-induced flow is considered. This assumption is valid provided that the Grashof number is much larger than the Reynolds number squared:

$$\frac{Gr}{Re^2} >> 1 \quad \rightarrow \quad \text{consider only free convection effects.} \tag{10.67}$$

However, there will be situations where both free and forced convection are important; in this situation the Grashof number has the same order of magnitude as the Reynolds number squared:

$$\frac{Gr}{Re^2} \approx 1 \quad \rightarrow \quad \text{both free and forced convection are important.} \tag{10.68}$$

There is a substantial body of literature that addresses these **mixed** convection problems. However, an approximate method for dealing with a mixed convection problem is to calculate the heat transfer coefficients using the appropriate forced convection correlation ($\bar{h}_{fc}$) and natural convection correlation ($\bar{h}_{nc}$). A rough approximation of the actual heat transfer coefficient ($\bar{h}$) may be obtained by taking the maximum of these two values:

$$\bar{h} \approx \text{Max}\left[\bar{h}_{fc}, \bar{h}_{nc}\right]. \tag{10.69}$$

A slightly more sophisticated methodology combines the two estimates of the heat transfer coefficient according to

$$\bar{h} = \left[\left(\text{Max} \left[\bar{h}_{fc}, \bar{h}_{nc} \right] \right)^m \pm \left(\text{Min} \left[\bar{h}_{fc}, \bar{h}_{nc} \right] \right)^m \right]^{1/m}, \tag{10.70}$$

where m is typically taken to be 3. The positive sign in Eq. (10.70) is used if the natural and forced convection flows augment one another (i.e., if the velocities are in the same direction or are perpendicular to one another).

Example 10.6

A sphere is heated until its surface temperature reaches $T_s = 500$ K in an environment of air at $T_\infty = 300$ K. Air at T_∞ flows across the sphere (perpendicular to gravity) at a velocity of $u_f = 1.1$ m/s. The diameter of the sphere is $D = 10$ cm and the emissivity of the surface of the sphere is $\varepsilon = 0.42$. Determine the rate at which the sphere must be heated.

Known Values

Figure 1 illustrates a sketch of the problem with the known values indicated.

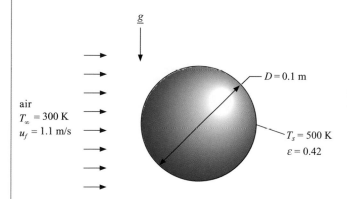

Figure 1 Sketch of the problem with the known values indicated.

Assumptions

- The problem is steady state.
- Air behaves as an ideal gas.

Analysis and Solution

The properties of air are obtained from Appendix C at the film temperature (400 K): $\rho = 0.8821$ kg/m³, $k = 0.03345$ W/m-K, $\alpha = 37.39 \times 10^{-6}$ m²/s, and $\upsilon = 26.14 \times 10^{-6}$ m²/s. The volumetric thermal expansion coefficient is estimated using Eq. (10.13):

$$\beta = \frac{1}{T_{film}} = \frac{1}{400 \text{ K}} = 0.0025 \text{ K}^{-1}. \tag{1}$$

The Prandtl number is calculated according to:

$$Pr = \frac{\upsilon}{\alpha} = \frac{26.14 \times 10^{-6}}{37.39 \times 10^{-6}} = 0.699. \tag{2}$$

Continued

Example 10.6 (cont.)

In order to determine whether forced convection or natural convection fluid motion dominates the problem, the Grashof and Reynolds numbers are calculated. The Grashof number is:

$$Gr_D = \frac{g\,D^3\,\beta\,(T_s - T_\infty)}{v^2}$$

$$= \frac{9.81\text{ m}}{\text{s}^2}\left|\frac{(0.1)^3\text{ m}^3}{}\right|\frac{0.0025}{\text{K}}\left|\frac{(500-300)\text{ K}}{}\right|\frac{\text{s}^2}{(26.14\times10^{-6})^2\text{ m}^4} = 7.18\times10^6. \qquad (3)$$

The Reynolds number is computed according to:

$$Re_D = \frac{\rho\,u_f\,D}{\mu} = \frac{u_f\,D}{v} = \frac{1.1\text{ m}}{\text{s}}\left|\frac{0.1\text{ m}}{}\right|\frac{\text{s}}{26.14\times10^{-6}\text{ m}^2} = 4208. \qquad (4)$$

The ratio of Grashof number to Reynolds number squared is therefore

$$\frac{Gr_D}{Re_D^2} = \frac{7.18\times10^6}{(4208)^2} = 0.405, \qquad (5)$$

which suggests that this is a mixed convection problem where both natural and forced convection are important. The Rayleigh number is computed according to:

$$Ra_D = Gr_D\,Pr = 7.18\times10^6(0.699) = 5.02\times10^6. \qquad (6)$$

The average Nusselt number assuming only free convection is computed using Eq. (10.33):

$$\overline{Nu}_{D,\,nc} = 2 + \frac{0.589\,Ra_D^{0.25}}{\left[1+\left(\dfrac{0.469}{Pr}\right)^{9/16}\right]^{4/9}} = 2 + \frac{0.589\,(5.02\times10^6)^{0.25}}{\left[1+\left(\dfrac{0.469}{0.699}\right)^{9/16}\right]^{4/9}} = 23.47. \qquad (7)$$

The average heat transfer coefficient assuming free convection is:

$$\overline{h}_{nc} = \overline{Nu}_{D,\,nc}\frac{k}{D} = \frac{23.47}{}\left|\frac{0.03345\text{ W}}{\text{m K}}\right|\frac{1}{0.1\text{ m}} = 7.85\frac{\text{W}}{\text{m}^2\text{ K}}. \qquad (8)$$

The average Nusselt number assuming only forced convection is computed using Eq. (8.68):

$$\overline{Nu}_{D,\,fc} = 2 + \left(0.4\,Re_D^{0.5} + 0.06\,Re_D^{2/3}\right)Pr^{0.4}$$

$$= 2 + \left[0.4\,(4208)^{0.5} + 0.06\,(4208)^{2/3}\right](0.699)^{0.4} = 38.04. \qquad (9)$$

The average heat transfer coefficient assuming forced convection is:

$$\overline{h}_{fc} = \overline{Nu}_{D,\,fc}\frac{k}{D} = \frac{38.04}{}\left|\frac{0.03345\text{ W}}{\text{m K}}\right|\frac{1}{0.1\text{ m}} = 12.72\frac{\text{W}}{\text{m}^2\text{ K}}. \qquad (10)$$

The average heat transfer coefficient for the mixed convection problem is computed using Eq. (10.70):

$$\overline{h} = \left[\left(\text{Max}\,[\overline{h}_{fc},\overline{h}_{nc}]\right)^m + \left(\text{Min}\,[\overline{h}_{fc},\overline{h}_{nc}]\right)^m\right]^{1/m} = \left(12.72^3 + 7.852^3\right)^{1/3} = 13.65\frac{\text{W}}{\text{m}^2\text{ K}}. \qquad (11)$$

The rate of convection heat transfer is computed according to:

$$\dot{q}_{conv} = \overline{h}\,\pi\,D^2(T_s - T_\infty) = \frac{13.65\text{ W}}{\text{m}^2\text{ K}}\left|\frac{\pi}{}\right|\frac{(0.1)^2\text{ m}^2}{}\left|\frac{(500-300)\text{ K}}{}\right| = 85.78\text{ W}. \qquad (12)$$

Example 10.6 (cont.)

The rate of radiation heat transfer is computed according to:

$$\dot{q}_{rad} = \sigma \, \varepsilon \, \pi \, D^2 \left(T_s^4 - T_\infty^4 \right)$$

$$= \frac{5.67 \times 10^{-8} \text{ W}}{\text{m}^2 \text{ K}^4} \left| 0.42 \right| \pi \left| (0.1)^2 \text{ m}^2 \right| \frac{\left(500^4 - 300^4 \right) \text{ K}^4}{} = 40.7 \text{ W}. \tag{13}$$

The total rate of heat transfer is:

$$\dot{q} = \dot{q}_{conv} + \dot{q}_{rad} = 85.78 \text{ W} + 40.7 \text{ W} = \boxed{126.5 \text{ W}}. \tag{14}$$

Discussion

The plus sign is used in Eq. (11) because the natural and forced convection flows do not oppose each other. If the forced convection flow were downwards (i.e., in the direction of gravity) then the minus sign would be used as the forced convection flow would tend to be suppressed by the buoyancy-induced flow.

10.6 Conclusions and Learning Objectives

This chapter provides a discussion of natural convection (also known as free convection). Natural convection is fundamentally different from the forced convection situations discussed in Chapters 7 through 9 because the fluid motion occurs due to the heating or cooling processes that are being considered; in the absence of the heated or cooled surfaces, the fluid would be static. Analysis of free convection requires that a characteristic or reference velocity be defined that is consistent with the fluid velocity that is developed by the relevant buoyancy forces. The reference velocity for free convection is subsequently used to define relevant dimensionless parameters: the Grashof and Raleigh numbers. Correlations for various free convection situations are provided that allow the solution of engineering problems. Finally, methods are presented that allow the consideration of mixed convection problems in which both a forced velocity and a buoyancy-induced velocity are important.

Some specific concepts that you should understand are listed below.
- The flow patterns that will be induced under various situations due to buoyancy forces.
- The derivation of the characteristic buoyancy velocity for a free convection problem.
- The relatively small magnitude of the characteristic buoyancy velocity under most situations and the resulting impact of this on the heat transfer coefficient for natural convection problems.
- The definition of the relevant dimensionless parameters for a natural convection problem.
- The correct application of external flow natural convection correlations for a variety of shapes.
- The correct application of internal free convection correlations associated with channels and enclosures.
- The definition of a mixed convection problem.
- The ability to discern the relative importance of free and forced convection effects in a mixed convection problem.
- The ability to combine free and forced convection effects in a meaningful way.

References

Arnold, J. N., I. Catton, and D. K. Edwards, "Experimental Investigation of Natural Convection in Inclined Rectangular Regions of Differing Aspect Ratios," ASME Paper 75-HT-62 (1975).

Ayyaswamy, P. S. and I. Catton, *J. Heat Transfer*, Vol. 95, p. 543 (1973).

Bar-Cohen, A. and W. M. Rohsenow, "Thermally optimum spacing of vertical natural convection cooled, parallel plates," *J. Heat Transfer*, Vol. 106, p. 116 (1984).

Churchill, S. W., "Free convection around immersed bodies," in E. U. Schlünder ed., *Heat Exchanger Design Handbook*, Hemisphere Publishing, New York (1983).

Churchill, S. W. and H. H. S. Chu, "Correlating equations for laminar and turbulent free convection from a horizontal cylinder," *Int. J. Heat Mass Transfer*, Vol. 18, p. 1049 (1975a).

Churchill, S. W. and H. H. S. Chu, "Correlating equations for laminar and turbulent free convection from a vertical plate," *Int. J. Heat Mass Transfer*, Vol. 18, pp. 1323–1329 (1975b).

Churchill, S. W. and R. Usagi, "A general expression for the correlation of rates of transfer and other phenomena," *AIChE J.*, Vol 18, pp. 1121–1128 (1972).

Elenbaas, W., "Heat dissipation of parallel plates by free convection," *Physica*, Vol. 9, p. 1 (1942).

Hollands, K. G. T., S. E. Unny, G. D. Raithby, and L. Konicek, *J. Heat Transfer*, Vol. 98, p. 189 (1976).

MacGregor, R. K. and A. P. Emery, *J. Heat Transfer*, Vol. 91, p. 391 (1969).

Raithby, G. D. and K. G. T. Hollands, *Natural Convection* in *The Handbook of Heat Transfer,* Third Edition, W. M. Rohsenow, J. P. Hartnett, and Y. I. Cho, eds., McGraw-Hill, New York (1998).

Sparrow, E. M. and J. L. Gregg, "Laminar free convection heat transfer from the outer surface of a vertical circular cylinder," *Trans. ASME*, Vol. 78, pp. 1823 (1956).

Problems

Free Convection Concepts

10.1 A vertical plate is electrically heated in air (at temperature, T_{air}) that would otherwise be stagnant (i.e., there is no wind or other forced air motion). The electrical heating results in a uniform heat flux from the surface of the plate to the air. Sketch the surface temperature of the plate as a function of the distance from the bottom of the plate, x (note that the coordinate x runs against gravity from the bottom of the plate). Justify your answer.

10.2 The density of liquid water in kg/m^3 between 290 K and 310 K and atmospheric pressure can be represented by $\rho = 661.47 + 2.506T - 0.00463T^2$, where T is temperature in K.
 (a) Calculate the volumetric coefficient of thermal expansion for liquid water at 300 K and compare it with the value for saturated water vapor at 300 K, which can be represented as an ideal gas.
 (b) Calculate the ratio of the characteristic velocities for liquid water and water vapor at 300 K.

10.3 A horizontal copper rod having a diameter of 12.5 mm is equipped with an electric heater that is controlled to maintain the rod at a uniform temperature of 76°C. Calculate the Raleigh number for the situation in which the rod is immersed in the following fluids at 25°C, 100 kPa:
 (a) water,
 (b) air,
 (c) glycerin, and
 (d) hydrogen.

10.4 The characteristic length used to define the Raleigh number for free convection in a vertical channel is the channel width. Plot the Rayleigh number and the characteristic velocity in the channel as a function of channel width (ranging from 1 cm to 12 cm) for 25°C air entering a channel with its surfaces maintained at 36°C.

Free Convection from Plates, Spheres, and Cylinders

10.5 Clean engine oil enters an $L = 50$ m long thin-walled pipe having an outer diameter $D_{out} = 30$ mm. The mass flow rate is $\dot{m} = 0.25$ kg/s and the inlet temperature is $T_{in} = 150$°C. The pipe is suspended horizontally in a large room in which the air temperature is $T_\infty = 20$°C. Estimate the heat transfer rate from the oil to the room and the outlet oil temperature.

10.6 A horizontal pipe carrying chilled water passes through a room filled with stagnant air at $T_\infty = 75$°F. The surface of the pipe is at $T_s = 40$°F. The pipe has an outer diameter of $D = 0.5$ inch and length $L = 12$ ft. Determine the rate of heat transfer to the pipe.

10.7 A single-paned glass window is $L = 6$ ft high and $W = 4$ ft wide; the glass is $th_g = 0.25$ inch thick and has conductivity $k_g = 1.4$ W/m-K.

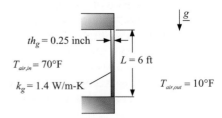

On a typical winter day, the outdoor temperature is $T_{air,out} = 10°F$ and you keep the indoor temperature at $T_{air,in} = 70°F$.

(a) On a still winter day, estimate the rate of heat loss from the window.

(b) Winter lasts $t_{winter} = 90$ days and you are heating with electrical resistance heaters. Electricity costs $e_{cost} = \$0.12/kW$-hr. How much does the heat loss through the window cost you over the course of 1 winter?

(c) Assume that 50 percent of your heat loss in your house is through your windows and that you have $N_{window} = 10$ single-paned windows in your house. Prepare a plot showing the cost of heating your house as a function of the thermostat set point (i.e., the indoor air temperature).

10.8 A French Press coffee pot is a glass cylinder with an inner diameter of 4 inches and height of 9 inches. The glass is Pyrex and it is 5 mm thick. The emissivity of glass is 0.93. The coffee pot is filled with water at 205°F at atmospheric pressure. An insulated cover is placed on top of the coffee pot, but the sides of the cylinder are exposed to air in the room that is at 73°F. Assume that coffee has the same properties as water and the top and bottom of the pot are adiabatic. Note any other assumptions that you employ.

(a) Determine the rate of heat transfer from the sides of the coffee pot when the coffee is at 205°F.

(b) Estimate the time required for the average temperature of the coffee to drop to 195°F. Assume that the coffee is at a uniform temperature at any time for this calculation.

10.9 A homeowner is considering the option of replacing the single-glazed windows currently in use in his home with double-glazed units. These windows are 1.35 m high and 1.0 m in width. The convection coefficient on the outside of the glass, due to free convection and wind is 15 W/m2-K. Air inside the building is maintained at 20°C. Calculate and plot the heat transfer rate through the single-glazed window that is currently installed for outdoor temperatures between –15°C and 15°C. Assume the conductive resistance of the glass material to be negligible relative to the convective resistances and also neglect radiation.

10.10 You have an old apartment with ten single-paned glass windows; each window is 4 ft wide by 6 ft high. During the winter (three months, December through February), the average outdoor air temperature is 12°F. You keep the temperature of the apartment at 70°F. The average wind velocity outside your windows is such that the heat transfer coefficient between the external surface of the window and the outdoor air is $h_{out} = 50$ W/m²-K. The glass window ($k_{glass} = 1.4$ W/m-K) is 0.375 inch thick.

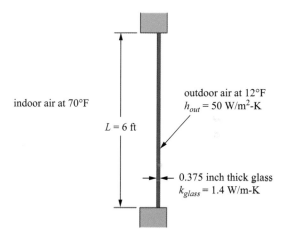

(a) Estimate the rate heat loss through all of your windows on an average winter day (W).

(b) You heat with electricity at an average rate of 0.1\$/kW-hr. Estimate the cost of the heat that is lost through your winters during an average winter (\$/winter).

In order to reduce the heating cost, you put a 0.75 inch thick sheet of polystyrene insulation ($k_{ins} = 0.03$ W/m-K) in the window (it's not an attractive solution).

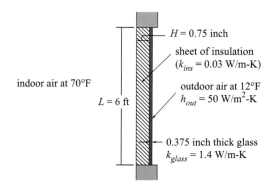

(c) Estimate the heat transfer rate through all of your windows on an average winter day after you've installed the insulation (W).

(d) Estimate the savings per winter associated with adding the insulation ($/winter).

10.11 A flat-screen television measures 44 inches wide by 25 inches high and dissipates 125 W. Estimate the surface temperature of the screen:

(a) neglecting radiation, and

(b) including radiation with a surface emissivity of 0.93.

10.12 A pipe that is 3 m long has an outer diameter $D_{out} = 0.1$ m and is bent in the center to form an "L" shape. One leg is vertical and the other leg is horizontal. The pipe is made of thin-walled copper, and saturated steam at atmospheric pressure is circulating through the pipe. The pipe is in a large room and the air temperature far from the pipe is $T_\infty = 30°C$ at atmospheric pressure. The conduction resistance associated with the pipe wall and the convection resistance associated with steam can be neglected.

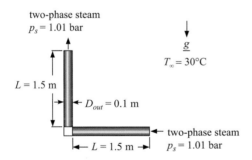

two-phase steam
$p_s = 1.01$ bar

g

$T_\infty = 30°C$

$L = 1.5$ m

$D_{out} = 0.1$ m

two-phase steam
$L = 1.5$ m $p_s = 1.01$ bar

(a) Determine the Grashof, Rayleigh, and Nusselt numbers and the corresponding average heat transfer coefficient for the horizontal section of the pipe.

(b) Determine the Grashof, Rayleigh, and Nusselt numbers and the corresponding average heat transfer coefficient for the vertical section of the pipe.

(c) Calculate the total rate of heat transfer to the air.

10.13 A spherical metal part with diameter $D = 4$ cm is placed in an industrial oven. At a certain time during the heating process the temperature of the surface of the part is $T_p = 200°C$. The part experiences natural convection with the air in the oven at $T_\infty = 70°C$. Assume

that air behaves as an ideal gas with $k_a = 0.033$ W/m-K, $\alpha_a = 3.86 \times 10^{-5}$ m^2/s and $\nu_a = 2.72 \times 10^{-5}$ m^2/s. The part is made of a material with density $\rho = 8000$ kg/m^3, $k = 20$ W/m-K, and $c = 200$ J/kg-K.

(a) Determine the rate of natural convection heat transfer from the part to the air.

(b) Is a lumped capacitance approximation valid for this object? Justify your answer.

(c) Assume your answer from part (b) is yes (this may or may not be the correct answer). Determine the rate at which the temperature of the object is changing.

10.14 An aluminum cylinder 0.75 m long and 0.02 m in diameter is suspended horizontally in still air that is at 20°C. The ends of the cylinder are insulated and the cylinder surface is painted black. The cylinder is heated electrically to a uniform temperature of 60°C. Estimate the steady draw of electrical power required by the heater. Indicate whether or not radiation is an important factor in this calculation.

10.15 A thin aluminum plate measuring 0.6 m in height and 1.8 m in width is suspended vertically in air at 300 K. The plate has been polished so that it has an emissivity of 0.10. The plate is electrically heated so that it is maintained at a temperature of 380 K when the electrical power is 920 W. Convection occurs on both sides of the plate.

(a) Determine the experimental value of the convective heat transfer coefficient.

(b) Determine the experimental value of the Nusselt number.

(c) Determine the Rayleigh number.

(d) Compare the experimental Nusselt number with an appropriate correlation.

10.16 You are given a cold beverage in a can but you cannot drink it now. It would be great if it stayed cold until you are ready to consume it. The can is 4.5 inches high with a 2.5 in diameter. Will it be better to position the can horizontally or vertically while you wait? State and justify the assumptions you use in this analysis.

10.17 The temperature on the outside surface of a 40 W incandescent light bulb is measured to be 197°C in still air that is at 25°C. The bulb is nearly spherical with a diameter of 50 mm.

(a) Determine the convection heat transfer coefficient.

(b) Determine the rate of heat transfer from the bulb to air by free convection.

(c) Assuming that the emissivity of the bulb is 0.9, estimate the rate of heat transfer by radiation.

(d) Based on your results, what is the efficiency of the light bulb, defined as the ratio of the visible light divided by the electrical input?

Free Convection in Channels

10.18 An open vertical channel is formed using two electrically heated metal plates that are each 82 cm high and 47 cm wide. The plates are polished (to reduce radiative heat transfer) and maintained at 75°C. Air enters the channel at 25°C. Prepare a plot of the required electrical power as a function of the plate spacing, S, for values ranging between 1 cm and 8 cm.

10.19 Free convection occurs between two large horizontal parallel flat plates spaced a distance L apart. The temperatures of the plates are constant with the lower plate maintained at a higher temperature than the upper plate. Free convection heat transfer is normally correlated by an equation that provides the Nusselt number as a function of the Grashof and Prandtl numbers. As the temperature difference between the plates is reduced, heat transfer by free convection is also reduced. What is the lowest value of Nusselt number that can occur for this situation?

10.20 A solar "chimney" concept is proposed to heat atmospheric air at 10°C. The chimney consists of black metal walls that form a vertical channel in which the walls are 15 cm apart. The channel is 2.5 m high and 0.5 m wide. The walls of the channel are at 34°C as a result of absorption of solar energy.

(a) Calculate the heat transfer coefficient assuming the chimney can be considered to be a vertical channel.

(b) Calculate the heat transfer coefficient assuming that the chimney surface can be considered to be a vertical wall.

(c) Which of these answers is likely to be more accurate?

Free Convection in Enclosures

10.21 The figure illustrates the bottom of a freezer chamber that is placed in a laboratory environment.

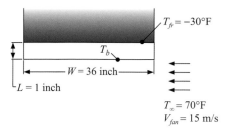

The freezer surface is maintained at $T_{fr} = -30°F$. It is necessary to prevent condensation and freezing due to the water in the room air at $T_\infty = 70°F$ and also to insulate the freezer. Therefore, an air gap is installed on the bottom surface. The bottom of the freezer is square with side width $W = 36$ inch and the width of the air gap is $L = 1$ inch. The outer surface of the air gap is exposed to flow driven by a fan with velocity $V_{fan} = 15$ m/s. You may neglect the conduction resistance of the wall as well as radiation.

(a) Determine the bottom temperature of the freezer (T_b in the figure) and the rate of heat transfer to the freezer.

(b) Plot the bottom surface temperature as a function of the fan velocity. Explain the shape of your plot.

(c) Plot the rate of heat transfer to the freezer as a function of the air gap spacing, L. Explain the shape of your plot. Assume that $V_{fan} = 15$ m/s for this section.

10.22 A homeowner is considering the option of replacing the single-glazed windows currently in use in his home with double-glazed units. These windows are 1.35 m high and 1.0 m in width. The convection coefficient on the outside of the glass, due to free convection and wind is 15 W/m²-K. The double-glazed windows that will be installed have a 1 cm trapped gas space. Air inside the building is maintained at 20°C. Assume the conductive resistance of the glass material to be negligible relative to the convective resistances and also neglect radiation. Calculate and plot the heat transfer rate through double-glazed window for outdoor temperatures between –15°C and 15°C when the trapped gas between the glazings is:

(a) air, and

(b) argon.

10.23 The single-glazed window in Problem 10.7 is replaced with a double-glazed window. Both glass panes are 0.25 inch thick and the gap between the panes is 0.5 inch. The gap contains dry air at atmospheric pressure. All other information is the same as in Problem 10.7. Neglect heat transfer by radiation.

    (a) Repeat the calculations requested in parts (a) and (b) of Problem 10.7.

    (b) Summarize and explain the benefits of the double-glazed window.

10.24 You have seen an advertisement for argon-filled windows. These windows are similar in construction to the window described in Problem 10.23, except that argon, rather than air, is contained in the gap. Neglect heat transfer by radiation.

    (a) Repeat Problem 10.23 assuming that the gap contains argon.

    (b) Are the claims that argon reduces heat loss valid? If so, why does this behavior occur?

    (c) Would nitrogen (which is cheaper) work as well? Why or why not? Can you suggest another gas that would work better than argon?

10.25 The figure illustrates a flat plate solar collector that is mounted at an angle of $\tau = 45°$ on the roof of a house. The collector is used to heat water; a series of tubes are soldered to the backside of a black plate. The collector plate is contained in a case with a glass cover.

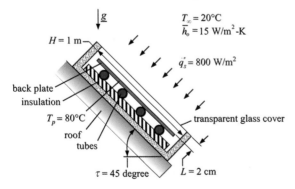

Assume that the solar collector is $H = 1$ m wide by $W = 1$ m long (into the page) and the distance between the heated plate and the glass covering is $L = 2$ cm. The collector receives a solar flux $\dot{q}''_s = 800$ W/m^2 and the collector plate can be assumed to absorb all of the solar energy. The collected energy is either transferred to the water in the pipe (in which case the energy is used to provide useful water heating) or lost due to heat transfer with the environment (either by radiation, which will be neglected in this problem, or convection). The collector plate temperature is $T_p = 80°C$ and the ambient temperature is $T_\infty = 20°C$. The heat transfer coefficient on the external surface of the glass is due to forced convection (there is a slight breeze) and equal to $\bar{h}_o = 15$ W/m^2-K. The glass is thin and can be neglected from the standpoint of providing any thermal resistance between the plate and ambient.

    (a) Determine the rate of heat loss from plate due to convection; you may assume that the insulation on the back of the tubes is perfect so no heat is conducted to the roof and that radiation from the plate is negligible.

    (b) What is the efficiency of the solar collector, $\eta_{collector}$, defined as the ratio of the energy delivered to the water to the energy received from the Sun?

    (c) Prepare a plot showing the collector efficiency as a function of the plate to glass spacing, L. Explain the shape of the plot.

10.26 You are involved in a project to design a solar collector for heating air. Two competing designs are shown in the figure.

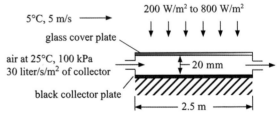

Air flows above the collector plate.

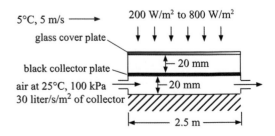

Air flows below the collector plate.

Both designs employ a transparent glass cover plate and a thin metal opaque black collector plate upon which solar radiation is completely absorbed. The glazing is standard safety glass with a thickness of 6 mm. In the first design, shown in the top figure, air is blown through the gap between the cover and plate. In the second design, shown in the bottom figure, the air flows in a second gap that is below the collector plate and free convection occurs between the collector plate and the glass cover plate. The collector is 1 m wide (into the page) and 2.5 m long (in the air flow direction) and oriented horizontally. In both designs, the gaps are 20 mm wide. Air at 25°C, 100 kPa, enters the flow passage at a flow rate of 30 l/s per square meter of collector area (area exposed to solar radiation) in both cases. The outdoor temperature (above the glass cover plate) is 5°C and there is a wind that may be represented as a forced convective flow with a free stream velocity of 5 m/s. Calculate and plot the efficiency of the two collector designs as a function of the solar radiation absorbed on the plate for values between 200 and 800 W/m². Assume that the insulation is adiabatic and neglect radiation in these calculations.

Combined Free and Forced Convection

10.27 On a sunny day in summer, a flat black roof of dimensions 4 m × 3 m absorbs 625 W/m² of solar radiation. The outdoor temperature is 28°C. The backside of the roof is insulated.
   (a) Calculate the roof surface temperature if free convection is the only mode of heat transfer.
   (b) Calculate the roof surface temperature if both free and wind-driven convection occur, with a 4 m/s wind speed parallel to the roof.
   (c) Calculate the roof surface temperature if, in addition to free and wind-driven convection radiation is important and the room has an emissivity of 0.9.

10.28 A vertical wall that is 2.5 m high and 13 m wide absorbs 375 W/m² of solar radiation. The outdoor temperature is 28°C. The backside of the wall is insulated.
   (a) Calculate the wall surface temperature assuming that free convection is the only mode of heat transfer.

   (b) Calculate the wall surface assuming that both free and wind-driven convection occur, with a 4 m/s wind speed parallel to the roof.
   (c) Calculate the wall surface assuming that, in addition to free and wind-driven convection, the wall surface also experiences radiation and has an emissivity of 0.75.

10.29 A spherical object is initially at a uniform temperature of $T_{ini} = 20°C$ when it is placed in a flow of atmospheric air at $T_\infty = -20°C$ with velocity $u_\infty = 0.5$ m/s. The air velocity is in the same direction as gravity, as shown.

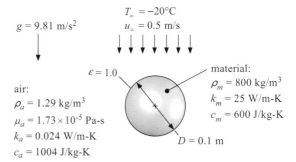

The properties of air are $\rho_a = 1.29$ kg/m³, $\mu_a = 1.73 \times 10^{-5}$ Pa-s, $k_a = 0.024$ W/m-K, $c_a = 1004$ J/kg-K. The object is made of material with $\rho_m = 800$ kg/m³, $c_m = 600$ J/kg-K, and $k_m = 25$ W/m-K. The emissivity of the surface of the object is $\varepsilon = 1.0$ and the object radiates to surroundings that are at T_∞.
   (a) Determine the forced convection heat transfer coefficient associated with the flow over the sphere.
   (b) Do you expect that buoyancy-induced flow will be important for this problem? Justify your answer with a simple calculation.

For the remainder of this problem assume that your answer to (b) was yes.
   (c) Determine the natural convection heat transfer coefficient associated with the sphere.
   (d) Determine a mixed convection heat transfer coefficient. That is, combine your answers from (a) and (c) appropriately in order to determine a heat transfer coefficient that approximately includes the effect of both natural and forced convection.
   (e) Do you expect that radiation plays an important role in this problem? Justify your answer.

For the remainder of the problem, assume that your answer from (e) was yes.

(f) Is a lumped capacitance model of the sphere appropriate? Justify your answer.

For the remainder of the problem, assume that your answer from (f) was yes.

(g) Estimate the time required for the object to cool to T_∞; your answer does not have to be exact but should be justified.

10.30 The receiver of a solar power tower is a vertical metal cylinder with a height of 8 m and a diameter of 6 m. The top and bottom of the receiver are insulated. The emissivity of the exposed surface is 0.74. The receiver is mounted on a tower where it is irradiated from solar radiation that is reflected from heliostats. The surface temperature of the receiver is 800°C. The wind speed at the height of the receiver is 8 m/s and the air temperature is 22°C.

(a) Estimate the average heat transfer coefficient for free convection.

(b) Estimate the average heat transfer coefficient due to the wind.

(c) Determine a combined heat transfer coefficient to account for both wind-driven and free convection.

(d) Estimate the average heat transfer coefficient that characterizes radiation.

(e) Estimate the rate of heat transfer from the receiver considering all modes of heat transfer. What is the dominant heat transfer mechanism?

10.31 A resistance temperature detector (RTD) is inserted into a methane pipeline to measure the gas temperature. The sensor is spherical with a diameter, $D = 5.0$ mm, and is exposed to methane at $p_f = 10$ atm with a fluid temperature of $T_f = 20°C$. The resistance of the sensor is related to its temperature; the resistance is measured by passing a known current through the resistor and measuring the associated voltage drop. The current causes an ohmic dissipation of $\dot{q} = 5.0$ mW. You have been asked to estimate the associated self-heating error as a function of the velocity of the methane in the pipe, V_f. Focus on the very-low-velocity operation (e.g., 0 m/s to 0.1 m/s) where self-heating might be large. The self-heating error is the amount that the temperature sensor surface must

rise relative to the surrounding fluid in order to transfer the heat associated with ohmic dissipation. You may neglect radiation for this problem.

(a) Assume that only forced convection is important and prepare a plot showing the self-heating error as a function of the methane velocity for velocities ranging from 0 m/s to 0.1 m/s.

(b) Assume that only natural convection is important and determine the self-heating error in this limit. Overlay this value on your plot from (a).

(c) Prepare a plot that shows your prediction for the self heating error as a function of velocity considering both natural convection and forced convection effects. Assume that the pipe is mounted horizontally.

Projects

10.32 The figure illustrates a device that is used to measure the thermal conductivity of a sample of material.

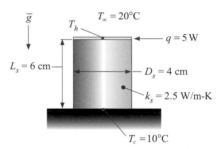

The sample has diameter $D_s = 4$ cm and length $L_s = 6$ cm. The actual conductivity of the sample is $k_s = 2.5$ W/m-K. A heat transfer rate of $\dot{q} = 5$ W is applied to a thin plate that is attached to the top of the sample and the temperature at the top of the sample, T_h, is measured. The temperature of the bottom of the sample is maintained at $T_c = 10°C$. The conductivity of the sample is computed according to:

$$k_{s,\,meas} = \frac{\dot{q}\, L_s\, 4}{\pi D_s^2\, (T_h - T_c)}.$$

This equation is derived by assuming that all of the applied heat passes through the sample. You have been asked to assess the error in the sample conductivity measurement due to heat loss from the top of the sample by natural convection.

(a) Determine the hot end temperature, T_h. You may neglect the resistance to conduction through the heater, radiation from the top surface, convection from the sides of the sample, and contact resistance at the interfaces. Assume that the problem is steady state.

(b) Use your answer from (a) to compute the measured conductivity and the fractional error in the conductivity measurement (i.e., the error normalized by the actual conductivity) that can be attributed to natural convection from the top of the device.

(c) If you can measure the temperature difference with an uncertainty of $\delta T = 1$ K then the fractional uncertainty in the conductivity measurement associated with the temperature measurement is given by:

$$err_{k,\,T} = \frac{\delta T}{(T_h - T_c)}.$$

Determine the fractional uncertainty in the conductivity measurement due to the temperature measurement uncertainty.

(d) Plot the fractional error in the conductivity measurement due to natural convection (your answer from (b)), the fractional error in the conductivity measurement due to uncertainty in the temperature measurement (your answer from (c)), and the total fractional error in the conductivity measurement as a function of $\dot{q}$. Explain the shape of each of your plots.

10.33 A pipe with outer diameter $D = 6$ cm carries steam at $T_p = 100°$C. The heat transfer coefficient between the steam and the inner surface of the pipe is large and the pipe wall is thin and conductive; therefore, the outer surface of the pipe may be assumed to be at T_p. The pipe is surrounded by air at $T_\infty = 20°$C. You are interested in designing an insulation blanket that is economically optimal. The insulation blanket is wrapped on the external surface of the pipe. The insulation has conductivity $k_{ins} = 0.15$ W/m-K and thickness $th_{ins} = 1$ cm. The outer surface of the insulation has emissivity $\varepsilon = 0.5$ and radiates to surroundings that are at $T_{sur} = T_\infty$. Assume that this is a natural convection problem (i.e., the only fluid motion is driven by buoyancy forces).

(a) Determine the rate of heat loss from the pipe per unit length.

(b) Estimate the air velocity that would cause the problem to transition from natural convection to forced convection.

(c) If the value of the lost heat transfer from the pipe is estimated to be 0.2$/therm then estimate the value of the energy lost per unit length of pipe over a five year period.

(d) If the cost of the insulation is 0.02 $/inch^3 then estimate the cost of the insulation per unit length of pipe.

(e) Use your answers from (c) and (d) to determine the total 5 year cost per unit length of pipe.

(f) Plot the rate of heat transfer per unit length as a function of insulation thickness for various values of the emissivity.

(g) For an emissivity of $\varepsilon = 0.5$, plot the energy cost, insulation cost, and total cost as a function of the insulation thickness. You should see that there is an optimal insulation thickness that causes the lowest possible total five year cost.

11 | Boiling and Condensation

Chapters 7 through 10 discuss convection situations involving single-phase fluids. The thermodynamic state of the fluids in these problems is sufficiently far from their vapor dome that they do not undergo a phase change. In this chapter, two-phase convection processes are examined. Two-phase processes occur when the fluid is experiencing heat transfer near the vapor dome so that vapor and liquid are simultaneously present. If the fluid is being transformed from liquid to vapor through heat addition, then the process is referred to as **boiling** or **evaporation**. If vapor is being transformed to liquid by heat removal, then the process is referred to as **condensation**.

Chapter 10 discussed free convection problems involving single-phase fluid and showed that temperature-induced density variations may have a substantial impact on a heat transfer problem because of the resulting buoyancy-induced fluid motion. However, the temperature-induced density gradient that is present in a typical single-phase free convection problem is small and as a result, the buoyancy-induced fluid velocity is also small. The consequence of this is that the heat transfer coefficient that characterizes a free convection process is typically much lower than what is encountered in forced convection processes.

This chapter examines two-phase problems where vapor and liquid are simultaneously present. The density difference between a vapor and a liquid is typically quite large. For example, saturated liquid water at 1 atm has a density of 960 kg/m^3 while saturated water vapor at 1 atm has a density of 0.60 kg/m^3. Such large density differences lead to correspondingly large buoyancy-induced fluid velocities and therefore large heat transfer coefficients. The heat transfer coefficients that characterize boiling and condensation processes are often substantially higher than those encountered in either free convection or forced convection heat transfer with single phase fluids.

11.1 Relevance

Most power and refrigeration cycles operate using thermodynamic cycles that rely on both boiling and condensation. The Rankine cycle used in many power systems includes a boiler where steam (water vapor) is generated from liquid through heat addition and a condenser where that steam is returned to liquid form through heat rejection. The vapor compression refrigeration cycle includes an evaporator where refrigerant is transformed from liquid to vapor at low pressure (accomplishing the heat extraction from the cooled space) and a condenser where the vapor is returned to liquid form at a higher pressure (accomplishing the heat rejection process). Despite the fact that boiling and condensation are present in so many of our thermal energy conversion systems, the fundamental physics associated with these two-phase convection processes are not as well understood as they are for single-phase convection. This lack of understanding is certainly due to the complexity of the phase change process itself; surface tension and other forces that are not important for single-phase convection may play an important role in boiling and condensation processes. However, another reason that phase change heat transfer processes are less-well understood is that boiling and condensation heat transfer coefficients are high. As a result, the thermal resistance that limits the performance of a heat exchanger in a power or a refrigeration device is often unrelated to the boiling and condensation processes. For example, the condenser in a home air conditioner must transfer heat from condensing refrigerant (e.g., R410a) to outdoor air. The total thermal resistance that characterizes the condenser (the inverse of the total *conductance* of the heat exchanger, which is mentioned in Section 9.3.5 and will be studied in more detail in Chapter 12), is likely to be dominated by the single-phase convection with the outdoor air rather than by two-phase convection with the condensing refrigerant. Thus, an engineer focused on improving the performance of this device will be motivated to study and understand the single-phase heat transfer coefficient associated with air flow over a finned surface more carefully than forced condensation of the refrigerant.

Because the heat transfer coefficients that characterize two-phase heat transfer are high, engineers also use boiling and condensation processes to accomplish heat removal or addition in applications where equipment must be compact. The study of two-phase heat transfer processes is an extremely rich and interesting field of research. The complexity of the processes involved makes them difficult to model and therefore careful experimental studies and visualization efforts are important. There are several excellent reviews that discuss these topics in detail, including Collier and Thome (1996), Thome (2006), Carey (1992), and Whalley (1987). In this chapter, a qualitative description of the boiling and condensation processes is presented and a few useful correlations are discussed; these correlations are also available as built-in functions and procedures in the Heat Transfer Library in EES. It should be noted that there is substantially more uncertainty associated with the use of correlations for two-phase heat transfer than there is for the single-phase heat transfer correlations discussed in Chapters 8 through 10 because of the complexity of these two-phase processes. Correlations are typically derived from particular, limited data sets; while these correlations may be appropriate beyond the test conditions associated with the data, they should be applied with some caution.

11.2 Pool Boiling

11.2.1 Introduction

A surface that is heated to a temperature that is greater than the saturation temperature of the surrounding liquid may result in **evaporation** (boiling) of the liquid. **Pool boiling** is analogous to free convection in that there is no external mechanism that would cause fluid motion. However, vigorous fluid motion does occur during pool boiling due to the dramatic difference in the density of the vapor that is generated by the evaporation process in comparison to the density of the surrounding liquid. **Flow boiling** is analogous to forced convection and is discussed in Section 11.3.

If the temperature of the surrounding liquid is lower than the saturation temperature, then the process is referred to as **sub-cooled pool boiling**. If the liquid is at its saturation temperature, then the process is referred to as **saturated pool boiling**. The general behavior observed for pool boiling is discussed in this section and some correlations are presented that predict the behavior of the most commonly encountered mode of pool boiling, **nucleate boiling**, and the limit of this mode which is the **critical heat flux**.

11.2.2 The Boiling Curve

The results of a famous pool boiling experiment are presented in Nukiyama (1934). A platinum wire submerged in a pool of saturated liquid water at ambient pressure is subjected to a controlled level of power (i.e., a controlled heat flux, $\dot{q}_s''$, from the wire surface). The heat flux is measured as a function of the **excess temperature difference**, ΔT_e, defined as the temperature difference between the surface of the wire (T_s) and the saturation temperature of the fluid (T_{sat}):

$$\Delta T_e = T_s - T_{sat}. \qquad (11.1)$$

The results of the experiment are shown in Figure 11.1, where the applied heat flux is shown as a function of the excess temperature difference. Notice that the behavior that is observed as the wire is heated (i.e., as $\dot{q}_s''$ is increased) is substantially different than the behavior observed when the wire is cooled (i.e., as $\dot{q}_s''$ is reduced).

At low power (i.e., low excess temperature), there is no evaporation and therefore heat transfer from the wire is due to single-phase natural convection. Heated liquid near the wire surface tends to rise due to its lower density and cooler liquid from the pool flows in to take its place. The single-phase natural convection correlation presented in Section 10.3.4 can be used to predict this portion of the curve, as shown in Figure 11.1.

As the surface heat flux increases, **nucleate boiling** begins approximately at the point labeled "onset of nucleate boiling" in Figure 11.1. Nucleate boiling is initially characterized by vapor bubbles that form at

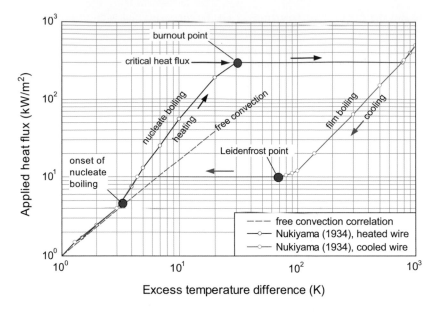

Figure 11.1 The boiling curve measured by Nukiyama (1934) during heating and cooling, observed by controlling the applied heat flux.

nucleation sites on the surface and grow until the buoyancy force is sufficient to cause them to detach from the surface and rise against gravity, as shown in Figure 11.2(a). As the heat flux increases further, more nucleation sites are activated and vapor is generated at a higher rate. Eventually, the bubbles may coalesce and form jets and columns of vapor, as shown in Figure 11.2(b).

Notice that the heat flux in the nucleate boiling region is considerably higher than would be present for natural convection in a single-phase fluid. This enhancement is due to the vigorous fluid motion that is induced by the vapor bubbles leaving the surface; as a vapor bubble leaves, relatively cold liquid rushes in to take its place. Therefore, the heated surface is continuously being exposed to cold fluid. The boiling curve data shown in Figure 11.1 are presented again in Figure 11.3 as the heat transfer coefficient, defined as

$$\bar{h} = \frac{\dot{q}_s''}{\Delta T_e}.$$ (11.2)

Figure 11.3 shows that a sharp increase in the heat transfer coefficient occurs at the onset of nucleate boiling. The heat transfer coefficient rises above the value that would be expected for natural convection from a horizontal cylinder (also shown in Figure 11.3) and continues to increase as the rate of vapor generation increases due to the aforementioned fluid motion related to the vapor escaping from the wire surface.

Eventually, vapor is produced at a rate that is so high that it begins to interfere with the ability of the liquid to re-wet the surface, as shown in Figure 11.2(c). The vapor phase has a substantially lower conductivity than the liquid phase (e.g., for water at 1 atm, the conductivity of saturated vapor is 0.025 W/m-K, whereas the conductivity of saturated liquid is 0.67 W/m-K). The low-conductivity vapor interferes with the heat transfer path causing a reduction in the heat transfer coefficient and, for a fixed heat flux, an increase in the surface temperature. This situation is *unstable* because larger excess temperature differences result in more vapor generation, which further interferes with the flow of liquid and further increases the excess temperature difference. The result is a drastic and sudden increase in the excess temperature difference that occurs at the **burnout point**, which is indicated in both Figure 11.1 and Figure 11.3. The act of reaching the burnout point is

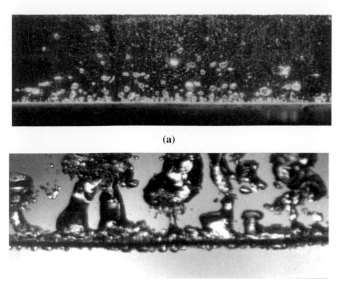

(a)

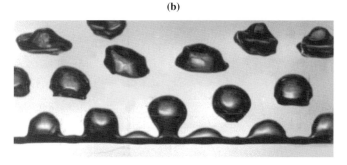

(b)

Figure 11.2 Photographs of pool boiling at (a) low temperature difference, (b) moderate temperature difference, and (c) high temperature difference. Photographs are from Lienhard and Lienhard (2005), available at http://web.mit.edu/lienhard/www/ahtt.html.

(c)

sometimes referred to as the **boiling crisis** because of the large increase in the excess temperature difference that results from a small increase in the applied heat flux. This increase in excess temperature difference is so severe that it will tend to damage or melt most materials. For water at atmospheric pressure, Figure 11.1 indicates that the excess temperature difference will rise from approximately 30 K to almost 800 K when the burnout point is reached. The heat flux at the burnout point is referred to as the peak heat flux or **critical heat flux** ($\dot{q}''_{s,\,crit}$). In most devices, it is important that the heat flux be kept *below* the critical heat flux so that the boiling crisis is avoided and the device operates safely in the nucleate boiling region. For this reason, the correlations available in the literature and discussed in Section 11.2.3 focus on predicting the heat transfer behavior in the nucleate boiling regime and also on predicting the critical heat flux.

After the wire has experienced the boiling crisis (assuming it has not melted), the excess temperature difference will be extremely high because the surface is completely coated with vapor. Heat transfer will primarily occur by a combination of conduction and radiation through the low-conductivity vapor layer. This boiling regime is referred to as **film boiling**. If the wire is subsequently cooled, the excess temperature remains very high even as the heat flux decreases below the critical heat flux. The excess temperature difference during the cooling process is much higher than the excess temperature difference was for the same heat flux during the heating process, as shown for the cooled wire data in Figure 11.1. Film boiling persists until the **Leidenfrost point** is reached. The Leidenfrost point represents the end of vapor-inhibited heat transfer. If the heat flux is decreased still further, then the excess temperature difference drops dramatically (from approximately 70 K to less than 5 K in Figure 11.1) and nucleate boiling is observed again.

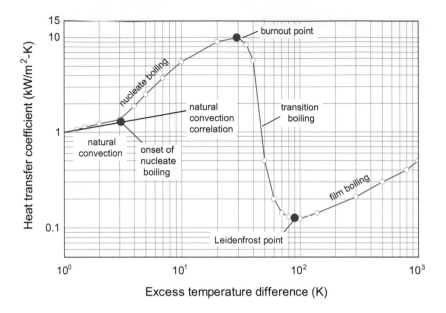

Figure 11.3 Heat transfer coefficient associated with the boiling curve shown in Figure 11.1. Also shown is the heat transfer coefficient predicted for natural convection from a horizontal cylinder.

The hysteresis exhibited by the boiling curve is a consequence of the fact that the surface heat flux is controlled and the surface temperature measured. A single value of the surface heat flux can be associated with two different values of the surface temperature. For example, Figure 11.1 shows that it is possible to provide 100 kW/m^2 under nucleate boiling conditions with a modest excess temperature difference of 14 K. However, the same heat flux of 100 kW/m^2 can also be provided under film boiling conditions, but only with the much larger excess temperature difference of 400 K. Therefore, the excess temperature difference that is measured when the experiment is run in a heat flux controlled mode depends on the history of the test.

Later experiments carried out by Drew and Mueller (1937) controlled the surface temperature and measured the heat flux; they were therefore able to measure the complete boiling curve shown in Figure 11.4. The heat transfer coefficient curve shown in Figure 11.3 is based on data of this type. In these experiments there is a unique value of the surface heat flux corresponding to each value of the excess temperature difference and so the boiling curve measured by Drew and Mueller does not exhibit the same hysteresis that is evident in the data collected by Nukiyama (1934).

In addition to the natural convection, nucleate boiling and film boiling modes already discussed, there is a **transition boiling** regime that joins the burnout point to the Leidenfrost point, as shown in Figure 11.4. Notice that the surface heat flux tends to go down as the excess temperature difference increases in the transition boiling regime. This behavior occurs because the amount of the surface that is completely coated by vapor increases with excess temperature difference so that the applied heat flux (or equivalently, the heat transfer coefficient) decreases. Figure 11.3 shows that the heat transfer coefficient associated with transition and film boiling is actually lower than would be expected for single phase natural convection; these are typically undesirable operating regimes for engineering equipment.

11.2.3 Pool Boiling Correlations

Pool boiling correlations focus on the two most important characteristics of the boiling curve: nucleate boiling and critical heat flux. The heat flux ($\dot{q}''_{s,\,nb}$) in the nucleate boiling region has been correlated by Rohsenow

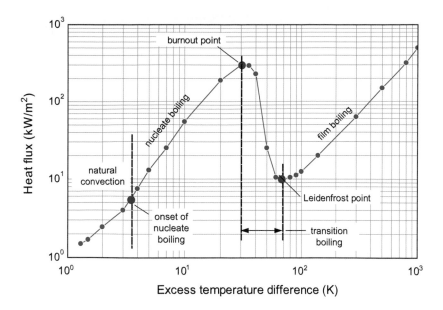

Figure 11.4 Boiling curve measured by Drew and Mueller (1937) by controlling the excess temperature difference.

(1952) to excess temperature difference (ΔT_e), fluid properties, and an empirical constant that is related to the surface–fluid combination (C_{nb}):

$$\dot{q}''_{s,\,nb} = \mu_{l,\,sat}\,\Delta i_{vap}\,\sqrt{\frac{g(\rho_{l,\,sat} - \rho_{v,\,sat})}{\sigma}}\left(\frac{c_{l,\,sat}\,\Delta T_e}{C_{nb}\,\Delta i_{vap}\,Pr^n_{l,\,sat}}\right)^3, \tag{11.3}$$

where $\mu_{l,sat}$, $\rho_{l,sat}$, $c_{l,sat}$, and $Pr_{l,sat}$ are the viscosity, density, specific heat capacity, and Prandtl number of saturated liquid, respectively, $\rho_{v,sat}$ is the density of saturated vapor, Δi_{vap} is the latent heat of vaporization (the difference between the specific enthalpies of the saturated vapor and the saturated liquid), g is the acceleration of gravity, and σ is the liquid–vapor surface tension. The dimensionless exponent n on Prandtl number is equal to 1.0 for water and 1.7 for other fluids.

The dimensionless constant C_{nb} in Eq. (11.3) is an experimentally determined coefficient that depends on the surface–fluid combination. The characteristics of the surface play a role in the nucleate boiling behavior because the number of nucleation sites that are active depends on the surface preparation. Some values of C_{nb} are listed in Table 11.1; note that it is typical to assume that $C_{nb} = 0.013$ if there are no data available for the surface-fluid combination of interest. According to Eq. (11.3), $\dot{q}''_s \propto \Delta T_e^3$; therefore, large errors (as much as 100 percent) can occur when the correlation is used to estimate the heat flux given the temperature difference. On the other hand, $\Delta T_e \propto \dot{q}''^{1/3}_s$ so the error is much smaller when Eq. (11.3) is used to estimate the excess temperature difference associated with a particular heat flux. The nucleate boiling correlation provided by Eq. (11.3) is programmed in EES as the function Nucleate_Boiling, which can be found in the Heat Transfer & Fluid Flow Library under Boiling and Condensation.

The critical heat flux (i.e., the heat flux at the burnout point) is an important engineering quantity for many applications. Lienhard and Dhir (1973) suggest the following correlation for the critical heat flux ($\dot{q}''_{s,\,crit}$):

$$\dot{q}''_{s,\,crit} = C_{crit}\,\Delta i_{vap}\,\rho_{v,\,sat}\left[\frac{\sigma g(\rho_{l,\,sat} - \rho_{v,\,sat})}{\rho^2_{v,\,sat}}\right]^{1/4}, \tag{11.4}$$

Table 11.1 Values of the coefficient C_{nb} in Eq. (11.3) for various surface/fluid combinations, from Rohsenow (1952), Collier and Thome (1996), Vachon et al. (1968).

Fluid	Surface	C_{nb}
water	polished copper	0.0127
	lapped copper	0.0147
	scored copper	0.0068
	ground and polished stainless steel	0.0080
	Teflon-pitted stainless steel	0.0058
	chemically etched stainless steel	0.0133
	mechanically polished stainless steel	0.0132
	brass	0.0060
	nickel	0.006
	platinum	0.0130
n-pentane	polished copper	0.0154
	polished nickel	0.0127
	lapped copper	0.0049
	emery-rubbed copper	0.0074
carbon tetrachloride	polished copper	0.0070
benzene	chromium	0.0101
ethyl alcohol	chromium	0.0027
isopropyl alcohol	copper	0.0023
n-butyl alcohol	copper	0.0030

where C_{crit} is a dimensionless constant that does not depend on the surface characteristics but does depend weakly on the surface size. The size of the surface is characterized by a dimensionless length ($\tilde{L}$), which is defined as the ratio of the characteristic length of the surface (L_{char}, for example the radius of a cylindrical heater) to the length scale associated with the nucleate boiling process (L_{nb}):

$$\tilde{L} = \frac{L_{char}}{L_{nb}}, \tag{11.5}$$

where

$$L_{nb} = \sqrt{\frac{\sigma}{g\left(\rho_{l,\,sat} - \rho_{v,\,sat}\right)}}. \tag{11.6}$$

Table 11.2 summarizes the values of C_{crit} for various heater geometries as well as the definition of the characteristic length. This correlation for critical heat flux is implemented in the Critical_Heat_Flux EES library function.

Table 11.2 Values of C_{crit} for Eq. (11.4) for various heater geometries, from Mills (1992).

Geometry	C_{crit}	Characteristic length	Range of $\tilde{L}$
large flat plate	0.15	width or diameter	$\tilde{L} > 27$
small flat plate	$\dfrac{9\pi L_{nb}^2}{5A_s}$	width or diameter	$9 < \tilde{L} < 20$
large horizontal cylinder	0.12	cylinder radius	$\tilde{L} > 1.2$
small horizontal cylinder	$\dfrac{0.12}{\tilde{L}^{0.25}}$	cylinder radius	$0.15 < \tilde{L} < 1.2$
large sphere	0.11	sphere radius	$4.26 < \tilde{L}$
small sphere	$\dfrac{0.227}{\tilde{L}^{0.5}}$	sphere radius	$0.15 < \tilde{L} < 4.26$
large, finite body	≈ 0.12		

Example 11.1

A power electronics module is immersed in a pool of saturated liquid R134a refrigerant at $p = 550$ kPa. The module is mounted in a horizontal manner (i.e., perpendicular to gravity) with the heated side up. The width of the module is $W = 3.5$ cm wide and it is much longer in length. The heat flux from the surface of the plate is $\dot{q}_s'' = 20$ W/cm². Estimate the temperature of the module.

Known Values

Figure 1 illustrates a sketch of the problem with the known values indicated.

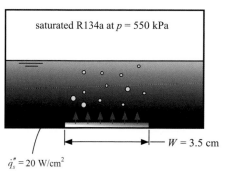

Figure 1 Electronics module immersed in a pool of saturated liquid R134a.

Assumptions

- The problem is at steady state.
- The heat flux is below the critical heat flux.
- The pressure is uniform.
- The pool is large enough that the edges of the container do not affect the flow situation.

Continued

Example 11.1 (cont.)

Analysis

The saturation temperature is a function of the pressure:

$$T_{sat} = T_{sat}(p). \tag{1}$$

The properties of saturated R134a are functions of the saturation pressure:

$$\mu_{l, sat} = \mu_{l, sat}(p_{sat}) \tag{2}$$

$$\rho_{l, sat} = \rho_{l, sat}(p_{sat}) \tag{3}$$

$$c_{l, sat} = c_{l, sat}(p_{sat}) \tag{4}$$

$$Pr_{l, sat} = Pr_{l, sat}(p_{sat}) \tag{5}$$

$$\Delta i_{vap} = \Delta i_{vap}(p_{sat}) \tag{6}$$

$$\rho_{v, sat} = \rho_{v, sat}(p_{sat}) \tag{7}$$

$$\sigma = \sigma(T_{sat}). \tag{8}$$

Because we have no knowledge of the surface preparation of the electronics module, the nucleate boiling constant is set to $C_{nb} = 0.013$. The fluid is not water and therefore the exponent on the liquid Prandtl number in Eq. (11.3) is set to $n = 1.7$:

$$\dot{q}''_{s, nb} = \mu_{l, sat}\,\Delta i_{vap}\,\sqrt{\frac{g(\rho_{l, sat} - \rho_{v, sat})}{\sigma}}\left(\frac{c_{l, sat}\,\Delta T_e}{0.013\,\Delta i_{vap}\,Pr_{l, sat}^{1.7}}\right)^3. \tag{9}$$

Equation (9) can be solved for the excess temperature difference, which can be used to predict the module surface temperature according to:

$$T_s = T_{sat} + \Delta T_e. \tag{10}$$

Solution

Equations (1) through (10) are programmed in EES. The properties required by Eqs. (1) through (8) are obtained using EES' internal property database. The result is shown below.

```
$UnitSystem SI Mass J K Pa Radian
p=550 [kPa]*Convert(kPa,Pa)                          "pressure"
W=3.5 [cm]*Convert(cm,m)                             "width"
q``_s_nb=20 [W/cm^2]*Convert(W/cm^2,W/m^2)           "heat flux"
F$='R134a'                                           "fluid"
T_sat=T_sat(F$,P=p)                                  "saturation temperature"
mu_l_sat=Viscosity(F$,P=p,x=0)                       "saturated liquid viscosity"
rho_l_sat=Density(F$,P=p,x=0)                        "saturated liquid density"
c_l_sat=Cp(F$,P=p,x=0)                               "saturated liquid specific heat capacity"
Pr_l_sat=Prandtl(F$,P=p,x=0)                         "saturated liquid Prandtl number"
DELTAi_vap=Enthalpy_Vaporization(F$,P=p)             "latent heat of vaporization"
rho_v_sat=Density(F$,P=p,x=1)                        "saturated vapor density"
sigma=SurfaceTension(F$,T=T_sat)                     "surface tension"
C_nb=0.013 [-]                                       "nucleate boiling constant"
n=1.7 [-]                                            "exponent on Prandtl number"
```

Example 11.1 (cont.)

q``_s_nb=mu_l_sat*DELTAi_vap***Sqrt**(g#*(rho_l_sat-rho_v_sat)/sigma)*&
   (c_l_sat*DELTAT_e/(C_nb*DELTAi_vap*Pr_l_sat^n))^3 "nucleate boiling correlation"
T_s=T_sat+DELTAT_e   "surface temperature of module"

Solving provides ΔT_e = 22.6 K or $\boxed{T_s = 314.5 \text{ K } (41.36°C)}$.

Discussion

The calculations provided here could have been accomplished using the function Nucleate_Boiling that is contained in the Heat Transfer Library within EES. To access this function select Function Info from the Options menu and then select the Heat Transfer & Fluid Flow radio button. Select Boiling and Condensation from the drop down menu. In the lower pane select Boiling and scroll to the Nucleate_Boiling option.

q``_s_nb=**Nucleate_Boiling**(F$, T_sat, T_s_EES, C_nb) "EES function for nucleate boiling"

Solving provides $T_{s,EES}$ = 314.5 K, which is the same answer obtained by manually entering the correlation.
   It is important to check that the electronics module has not exceeded the critical heat flux, an upper limit on the nucleate boiling regime. The length scale associated with nucleate boiling is computed using Eq. (11.6):

$$L_{nb} = \sqrt{\frac{\sigma}{g(\rho_{l,\,sat} - \rho_{v,\,sat})}}. \tag{11}$$

The dimensionless length is computed according to Eq. (11.5) with $L_{char} = W$ as indicated by the large plate entry in Table 11.2:

$$\tilde{L} = \frac{L_{char}}{L_{nb}}. \tag{12}$$

L_nb=**Sqrt**(sigma/(g#*(rho_l_sat-rho_v_sat)))   "nucleate boiling length scale"
L_tilde=W/L_nb   "dimensionless length"

Solving provides $\tilde{L}$ = 40.4 which is greater than 27 and therefore the critical heat flux constant is taken to be C_{crit} = 0.15. The critical heat flux is estimated using Eq. (11.4):

$$\dot{q}''_{s,\,crit} = C_{crit}\, \Delta i_{vap}\, \rho_{v,\,sat} \left[\frac{\sigma g (\rho_{l,\,sat} - \rho_{v,\,sat})}{\rho^2_{v,\,sat}} \right]^{1/4}. \tag{13}$$

C_crit=0.15   "critical heat flux constant"
q``_s_crit=C_crit*DELTAi_vap*rho_v_sat*(sigma*g#*(rho_l_sat-rho_v_sat)/rho_v_sat^2)^(1/4)
                                                         "critical heat flux"

Solving provides $\dot{q}''_{s,\,crit}$ = 455,000 W/m^2 (45.5 W/cm^2), which is sufficiently higher than the actual surface heat flux that nucleate boiling is likely to exist. The critical heat flux could also be estimated using the Critical_Heat_Flux function in EES.

q``_s_crit_EES=**Critical_Heat_Flux**(F$,'Plate', W, T_sat)   "EES function for critical heat flux"

Figure 2 shows the excess temperature difference as a function of the heat flux with the upper limit of the nucleate boiling region indicated by the vertical line.

Continued

Example 11.1 (cont.)

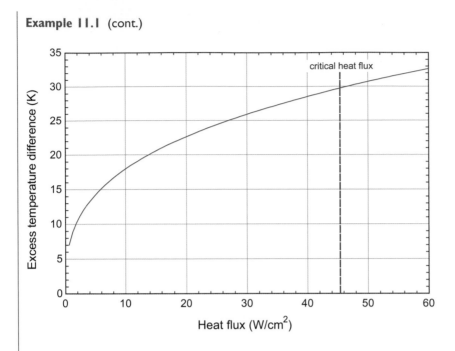

Figure 2 Excess temperature difference as a function of the heat flux with the critical heat flux indicated.

11.3 Flow Boiling

11.3.1 Introduction

Most power generation and refrigeration systems rely on evaporation of a working fluid while the fluid is flowing within a duct. This process is referred to as **flow boiling**. For example, in a direct expansion evaporator for a refrigeration cycle, the refrigerant enters the heat exchanger as a low-quality, two-phase mixture and exits as a saturated or possibly super-heated vapor as a result of heat transfer to the refrigerant. The purpose of this section is to provide methods for estimating the heat transfer coefficient for flow boiling processes.

The physical processes involved in the evaporation of a flowing fluid are more complicated than those associated with the single-phase, internal forced convection heat transfer processes that were examined in Chapter 9. Figure 11.5 shows the qualitative behavior of a sub-cooled liquid entering a horizontal tube in which the wall is heated to a temperature above the saturation temperature of the fluid. Near the inlet, fluid adjacent to the wall is heated and undergoes nucleate boiling. The vapor produced by this boiling process coalesces into vapor bubbles that are distributed in the fluid. The specific volume of the vapor is usually much higher than the specific volume of liquid so that even when the vapor represents a relatively small mass fraction of the fluid (i.e., the two-phase mixture has a low quality) it will still represent a large *volume fraction*. The bubbles tend to concentrate near the center of the tube, forcing the liquid towards the wall. Eventually, the flow enters what is called an **annular flow** regime where the walls are coated with a liquid film and there is a vapor core. The liquid film in contact with the walls continues to produce vapor by nucleate boiling. However, vapor is also produced by evaporation of the liquid at the liquid–vapor interface in a process called **convective boiling**. The presence of the nucleate and convective boiling processes both contribute to the high heat transfer coefficients associated with flow boiling.

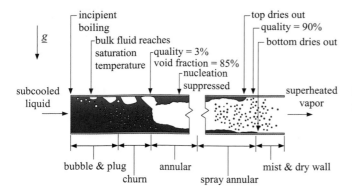

Figure 11.5 Flow regimes that occur during flow boiling in a smooth, horizontal tube.

As boiling continues, the liquid film at the wall is thinned, suppressing nucleate boiling. At some point, the amount of liquid is no longer sufficient to wet the entire perimeter of the tube. Owing to gravity forces, the top part of a horizontal tube will tend to dry out first. The liquid–wall interface has a much higher heat transfer coefficient than the vapor–wall interface that takes its place as dry out progresses. Therefore, the heat transfer coefficient tends to drop precipitously at **dry out**. The fraction of the perimeter that is dry increases as the remaining liquid is vaporized, resulting in a continuous decrease in the heat transfer coefficient until only single-phase vapor exists.

This discussion of flow boiling is simplistic and there continues to be a large number of researchers working towards developing a more complete understanding of this complex process. The interested reader is directed towards books such as Collier and Thome (1996).

11.3.2 Flow Boiling Correlations

Forced convection internal flow correlations for a single-phase fluid (either all liquid or all vapor) are provided in Section 9.2. The physical situation occurring in two-phase heating processes is more complicated and, consequently, the correlations for estimating flow-boiling heat transfer coefficients are also more complicated. These correlations are almost always based on curve fitting some set of measurements rather than a complete model of the physical situation. Therefore, the heat transfer coefficients predicted using flow boiling correlations have larger uncertainty bands than their single-phase counterparts, particularly when the correlations are applied outside of the range of conditions associated with the data. It is fortunate that the heat transfer coefficients resulting from evaporation are ordinarily large and therefore the thermal resistance associated with flow boiling is not typically the limiting thermal resistance in a heat exchanger. A large uncertainty in the heat transfer coefficient that contributes to the smallest thermal resistance may not have a significant effect on predicting the performance of a heat exchanger.

There have been literally hundreds of correlations proposed for flow boiling heat transfer coefficients. A review of the most widely used and well accepted correlations has been prepared by Shah (2006). The review concludes that the correlation proposed by Shah (1976, 1982) provides the most consistent agreement with the available experimental data, with a mean deviation of less than 20 percent. The Shah correlation was developed for saturated flow boiling at sub-critical heat fluxes and can be used for a wide range of vapor qualities, from saturated liquid ($x = 0$) to the liquid-deficient and dry-out regimes that occur at qualities of 0.8 or higher.

The Shah correlation was selected from the available correlations for presentation in this section because it is applicable to any fluid in both horizontal and vertical tubes and it has been compared to a large database. The Shah correlation relates a dimensionless heat transfer coefficient, $\tilde{h}$, to three dimensionless parameters, the convection number (Co), boiling number (Bo), and Froude number (Fr):

$$\tilde{h} = \tilde{h}\,(Co, Bo, Fr). \tag{11.7}$$

The dimensionless heat transfer coefficient for flow boiling is defined as the ratio of the local heat transfer coefficient (h) to the local heat transfer coefficient that would occur if *only* the liquid phase of the two-phase flow were present (h_l, sometimes referred to as the **superficial heat transfer coefficient of the liquid phase**):

$$\tilde{h} = \frac{h}{h_l}. \tag{11.8}$$

The superficial heat transfer coefficient of the liquid phase can be determined using the correlation developed by Gnielinski (1976) and presented in Section 9.2 for the Nusselt number. The heat transfer coefficient for fully developed, single-phase turbulent flow under turbulent conditions is:

$$h_l = \left[\frac{\left(\frac{f_l}{8} \right) (Re_{D_h,\,l} - 1000) Pr_{l,\,sat}}{1 + 12.7 \left(Pr_{l,\,sat}^{2/3} - 1 \right) \sqrt{\frac{f_l}{8}}} \right] \frac{k_{l,\,sat}}{D_h}. \tag{11.9}$$

The Reynolds number appearing in Eq. (11.9), $Re_{D_h,\,l}$, is the liquid superficial Reynolds number, which is calculated using the hydraulic diameter D_h of the tube and evaluated using the mass flow rate *of the liquid only*:

$$Re_{D_h,\,l} = \frac{G(1-x)D_h}{\mu_{l,\,sat}}, \tag{11.10}$$

where G is the **mass velocity** (also called the **mass flux**) and x is the quality (defined as the mass fraction of the fluid that is vapor). The mass velocity of the flow is equal to the total mass flow rate of the two-phase flow ($\dot{m}$, the sum of the liquid and vapor flow rates) divided by the cross-sectional area of the tube (A_c):

$$G = \frac{\dot{m}}{A_c}. \tag{11.11}$$

The properties $\mu_{l,sat}$, $k_{l,sat}$, and $Pr_{l,sat}$ in Eqs. (11.9) and (11.10) are the dynamic viscosity, thermal conductivity, and Prandtl number, respectively, of the saturated liquid. The parameter f_l in Eq. (11.9) is the friction factor associated with the flow of the liquid alone. The correlation developed by Petukhov (1970), is presented in Section 9.2 for fully developed single-phase flow under turbulent conditions in a smooth passage; this correlation is typically used to evaluate f_l in Eq. (11.9):

$$f_1 = \frac{1}{[0.790 \, \ln (Re_{D_h,\,l}) - 1.64]^2}. \tag{11.12}$$

The **convection number** is defined according to

$$Co = \left(\frac{1}{x} - 1 \right)^{0.8} \sqrt{\frac{\rho_{v,\,sat}}{\rho_{l,\,sat}}}, \tag{11.13}$$

where $\rho_{l,\,sat}$ and $\rho_{v,\,sat}$ are the densities of saturated liquid and vapor, respectively. The **boiling number** is defined as

$$Bo = \frac{\dot{q}_s''}{G\,\Delta i_{vap}}, \tag{11.14}$$

where $\dot{q}''$ is the heat flux at the wall and Δi_{vap} is the enthalpy of vaporization (i.e., the difference between the specific enthalpies of saturated vapor and liquid, $i_{v,sat} - i_{l,sat}$). The **Froude number** is related to the ratio of the inertial force of the fluid to the gravitational force

$$Fr = \frac{G^2}{\rho_{l,\,sat}^2 \, g \, D_h}, \tag{11.15}$$

where g is the acceleration of gravity. Note that, according to Shah (1982), the Reynolds number in Eq. (11.10) should be evaluated using the liquid mass velocity, i.e., $G(1-x)$, while the Froude number in Eq. (11.15) should be evaluated using the total mass velocity, G.

The correlation for $\bar{h}$ in terms of Co, Bo, and Fr is facilitated by defining one additional dimensionless parameter, N, in terms of the others:

$$N = \begin{cases} Co & \text{for vertical tubes or horizontal tubes with } Fr > 0.04 \\ 0.38 \, Co \, Fr^{-0.3} & \text{for horizontal tubes with } Fr \leq 0.04. \end{cases} \tag{11.16}$$

The Shah correlation is expressed by Eqs. (11.17) through (11.21):

$$\tilde{h}_{cb} = 1.8 N^{-0.8} \tag{11.17}$$

$$\tilde{h}_{nb} = \begin{cases} 230\sqrt{Bo} & \text{if } Bo \geq 0.3 \times 10^{-4} \\ 1 + 46\sqrt{Bo} & \text{if } Bo < 0.3 \times 10^{-4} \end{cases} \tag{11.18}$$

$$\tilde{h}_{bs,\,1} = \begin{cases} 14.70\sqrt{Bo}\exp\left(2.74N^{-0.1}\right) & \text{if } Bo \geq 11 \times 10^{-4} \\ 15.43\sqrt{Bo}\exp\left(2.74N^{-0.1}\right) & \text{if } Bo < 11 \times 10^{-4} \end{cases} \tag{11.19}$$

$$\tilde{h}_{bs,\,2} = \begin{cases} 14.70\sqrt{Bo}\exp\left(2.74N^{-0.15}\right) & \text{if } Bo \geq 11 \times 10^{-4} \\ 15.43\sqrt{Bo}\exp\left(2.74N^{-0.15}\right) & \text{if } Bo < 11 \times 10^{-4} \end{cases} \tag{11.20}$$

$$\tilde{h} = \begin{cases} \text{Max}(\tilde{h}_{cb}, \tilde{h}_{bs,\,2}) & \text{if } N \leq 0.1 \\ \text{Max}(\tilde{h}_{cb}, \tilde{h}_{bs,\,1}) & \text{if } 0.1 < N \leq 1.0 \, . \\ \text{Max}(\tilde{h}_{cb}, \tilde{h}_{nb}) & \text{if } N > 1.0 \end{cases} \tag{11.21}$$

The Shah correlation is implemented in the Flow_Boiling procedure in EES. There are a few practical implementation details associated with the procedure Flow_Boiling that are not directly addressed by Shah (1982). For example, as the quality increases to 1, the Reynolds number calculated using Eq. (11.10) tends toward 0; however, both the Petukhov and Gnielinski correlations, Eqs. (11.12) and (11.9), respectively, require that the Reynolds number be sufficiently high for turbulent flow (greater than about 2300). In addition, it is necessary to ensure that the heat transfer coefficient provided by the procedure for saturated vapor (i.e., at $x = 1$) is consistent with the heat transfer coefficient for single-phase vapor (i.e., the heat transfer coefficient predicted by Eq. (11.9) evaluated using the properties of the saturated vapor). Therefore, if the Reynolds number determined by Eq. (11.10) is less than 2300, then the Flow_Boiling procedure determines the quality at which Reynolds number is 2300 and computes the heat transfer coefficient at this quality. The heat transfer coefficient at a quality of 1 (i.e., for pure vapor) is also computed. The heat transfer coefficient is found by linear interpolation between these two values.

Example 11.2

Refrigerant R134a flows in a horizontal tube of inner diameter $D = 1$ cm. The refrigerant has mass flow rate $\dot{m} = 0.05$ kg/s, quality $x = 0.2$ and temperature $T_{sat} = 260$ K. The heat flux at the surface of the tube is $\dot{q}_s'' = 13.5$ kW/m^2. Estimate the local heat transfer coefficient and tube wall temperature.

Continued

Example 11.2 (cont.)

Known Values

Figure 1 illustrates a sketch of the problem with the known values indicated.

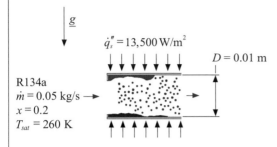

g

$\dot{q}''_s = 13,500\,\text{W/m}^2$

$D = 0.01$ m

R134a
$\dot{m} = 0.05$ kg/s
$x = 0.2$
$T_{sat} = 260$ K

Figure 1 Sketch of problem.

Assumption

- The problem is at steady state.

Analysis and Solution

The properties of the refrigerant are obtained from Appendix D: $\rho_{l,sat} = 1337$ kg/m^3, $\rho_{v,sat} = 8.912$ kg/m^3, $k_{l,sat} = 0.1001$ W/m-K, $\mu_{l,sat} = 0.3153 \times 10^{-3}$ Pa-s, $Pr_{l,sat} = 4.121$, $\Delta i_{vap} = 208.2 \times 10^3$ J/kg, and $\sigma = 0.0133$ N/m. The cross-sectional area of the tube for flow is given by:

$$A_c = \pi \frac{D^2}{4} = \frac{\pi}{4}(0.01)^2\,\text{m}^2 = 7.854 \times 10^{-5}\,\text{m}^2. \tag{1}$$

The mass velocity is:

$$G = \frac{\dot{m}}{A_c} = \frac{0.05\,\text{kg}}{\text{s}} \left| \frac{}{7.854 \times 10^{-5}\,\text{m}^2} \right. = 636.6\,\frac{\text{kg}}{\text{m}^2\,\text{s}}. \tag{2}$$

The liquid superficial Reynolds number is given by Eq. (11.10):

$$Re_{D,l} = \frac{G(1-x)D}{\mu_{l,\,sat}}$$

$$= \frac{636.6\,\text{kg}}{\text{m}^2\,\text{s}} \left| \frac{(1-0.2)}{} \right| \left| \frac{0.01\,\text{m}}{} \right| \left| \frac{}{0.3153 \times 10^{-3}\,\text{Pa s}} \right| \left| \frac{\text{Pa m}^2}{\text{N}} \right| \left| \frac{\text{N s}^2}{\text{kg m}} \right. = 1.62 \times 10^4. \tag{3}$$

The liquid-only friction factor is estimated using Eq. (11.12):

$$f_1 = \frac{1}{[0.790\,\ln\,(Re_{D,l}) - 1.64]^2} = \frac{1}{[0.790\,\ln\,(1.62 \times 10^4) - 1.64]^2} = 0.0276. \tag{4}$$

The superficial heat transfer coefficient of the liquid phase is determined using Eq. (11.9):

$$h_l = \left[\frac{\left(\dfrac{f_l}{8}\right)(Re_{D,l} - 1000)Pr_{l,\,sat}}{1 + 12.7(Pr_{l,\,sat}^{2/3} - 1)\sqrt{\dfrac{f_l}{8}}} \right] \frac{k_{l,\,sat}}{D} \tag{5}$$

$$= \left[\frac{\left(\dfrac{0.0276}{8}\right)(1.61 \times 10^4 - 1000)(4.121)}{1 + 12.7\left((4.121)^{2/3} - 1\right)\sqrt{\dfrac{0.0276}{8}}} \right] \frac{0.1001\,\text{W}}{\text{m K}} \left| \frac{}{0.01\,\text{m}} \right. = 994.1\,\frac{\text{W}}{\text{m}^2\,\text{K}}.$$

Example 11.2 (cont.)

The convection, boiling, and Froude numbers are computed using Eqs. (11.13) through (11.15):

$$Co = \left(\frac{1}{x} - 1\right)^{0.8} \sqrt{\frac{\rho_{v,\,sat}}{\rho_{l,\,sat}}} = \left(\frac{1}{0.2} - 1\right)^{0.8} \sqrt{\frac{8.912}{1337}} = 0.2475 \tag{6}$$

$$Bo = \frac{\dot{q}_s''}{G\,\Delta i_{vap}} = \frac{13{,}500\ \text{W}}{\text{m}^2} \left|\frac{\text{m}^2\ \text{s}}{636.6\ \text{kg}}\right|\frac{\text{kg}}{208{,}200\ \text{J}}\left|\frac{\text{J}}{\text{W s}}\right| = 0.102 \times 10^{-3} \tag{7}$$

$$Fr = \frac{G^2}{\rho_{l,\,sat}^2\, g\, D} = \frac{(636.6)^2\ \text{kg}^2}{\text{m}^4\ \text{s}^2}\left|\frac{\text{m}^6}{(1337)^2\ \text{kg}^2}\right|\frac{\text{s}^2}{9.81\ \text{m}}\left|\frac{}{0.01\ \text{m}}\right| = 2.312. \tag{8}$$

Because the Froude number is greater than 0.04, the parameter N is equal to the convection number according to Eq. (11.16). Because $N > 0.1$, the parameter $\tilde{h}$ is equal to the maximum of $\tilde{h}_{cb}$ and $\tilde{h}_{bs,\,1}$ according to Eq. (11.21); the parameters $\tilde{h}_{cb}$ and $\tilde{h}_{bs,\,1}$ are computed using Eqs. (11.17) and (11.19):

$$\tilde{h}_{cb} = 1.8 N^{-0.8} = 1.8(0.2475)^{-0.8} = 5.501 \tag{9}$$

$$\tilde{h}_{bs,\,1} = 15.43 \sqrt{0.102 \times 10^{-3}} \exp\left[2.74(0.2475)^{-0.1}\right] = 3.636 \tag{10}$$

$$\tilde{h} = \text{Max}\left(\tilde{h}_{cb}, \tilde{h}_{bs,\,1}\right) = \text{Max}\,(5.501, 3.636) = 5.501. \tag{11}$$

The local heat transfer coefficient is computed using Eq. (11.8):

$$h = \tilde{h}\, h_l = (5.501)\left(994.1\ \frac{\text{W}}{\text{m}^2\ \text{K}}\right) = \boxed{5468\ \frac{\text{W}}{\text{m}^2\ \text{K}}}. \tag{12}$$

The wall temperature is computed according to:

$$T_s = T_{sat} + \frac{\dot{q}_s''}{h} = 260\ \text{K} + \frac{13{,}500\ \text{W}}{\text{m}^2}\left|\frac{\text{m}^2\ \text{R}}{5468\ \text{W}}\right| = \boxed{262.5\ \text{K}\ (-10.68\,^\circ\text{C})}. \tag{13}$$

Discussion

The calculations provided here could have been accomplished using the procedure **Flow_Boiling** that is contained in the Heat Transfer Library within EES. To access this function select Function Info from the Options menu and then select the Heat Transfer & Fluid Flow radio button. Select Boiling and Condensation from the drop down menu. In the lower pane select Boiling and scroll to the **Flow_Boiling** option. The code below solves this problem.

```
$UnitSystem SI Mass J K Pa
T_sat=260 [K]                                    "saturation temperature"
D=0.01 [m]                                       "diameter"
m_dot=0.05 [kg/s]                                "mass flow rate"
x=0.2 [-]                                         "quality"
q``_dot_s=13500 [W/m^2]                          "surface heat flux"

A_c=pi*D^2/4                                      "cross-sectional area"
G=m_dot/A_c                                       "mass velocity"
CALL Flow_Boiling('R134a', T_sat, G, D, x, q``_dot_s, 'Horizontal': h, T_s)
                                                 "flow boiling correlation"
```

Continued

Example 11.2 (cont.)

Solving provides $h = 5468$ W/m²-K and $T_s = 262.5$ K, which are the same answers obtained by manually entering the correlation. Figure 2 illustrates the heat transfer coefficient predicted as a function of the thermodynamic quality. Notice that the two-phase flow boiling heat transfer coefficient is higher than either single-phase liquid ($x = 0$) or single-phase vapor ($x = 1$) conditions.

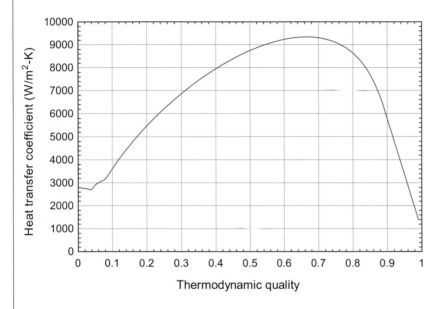

Figure 2 Heat transfer coefficient as a function of the thermodynamic quality.

11.4 Film Condensation

11.4.1 Introduction

Condensation is the reverse of the boiling processes discussed in Sections 11.2 and 11.3; vapor changes phase to liquid due to the presence of a surface that is cooled below the saturation temperature. The condensation processes referred to as **film** and **drop condensation** processes are analogous to pool boiling for evaporation in that fluid motion is induced by density differences between the liquid and vapor and not driven externally. Gravity causes the liquid that condenses on the cooled surface to flow, if that is possible. Therefore, film and drop condensation, like pool boiling, are analogous to natural convection albeit with much larger density differences and therefore larger heat transfer coefficients.

Film condensation occurs when the liquid forms a contiguous film that is pulled downwards by gravity, as shown in Figure 11.6 (on the right). If the condensate does not wet the wall then it will form droplets that will eventually detach under the force of gravity and then roll down the wall, as shown in Figure 11.6 (on the left).

In film condensation, the thermal resistance between the surface of the wall and the surrounding vapor is primarily related to conduction through the thin film of liquid condensate that forms on the wall. If the thickness of the film is δ, then the rate of heat transfer to the wall from the vapor is, approximately:

$$\dot{q}_s'' \approx \frac{k_{l,\,sat}}{\delta}(T_{sat} - T_s) \tag{11.22}$$

Figure 11.6 Photograph of drop condensation (on the left) and film condensation (on the right). Photograph from J. F. Welch and J. W. Westwater, Department of Chemical Engineering, University of Illinois, Urbana.

where $k_{l,sat}$ is the conductivity of saturated liquid, T_{sat} is the saturation temperature of the vapor, and T_s is the temperature of the surface. Comparing Eq. (11.22) with Newton's Law of Cooling

$$\dot{q}_s'' = h(T_{sat} - T_s) \tag{11.23}$$

suggests that the heat transfer coefficient for film condensation is approximately

$$h \approx \frac{k_{l,sat}}{\delta}. \tag{11.24}$$

Equation (11.24) indicates that the heat transfer coefficient is inversely proportional to the film thickness. The film thickness (δ) grows in the direction of flow because more liquid accumulates. Therefore, surfaces with short vertical lengths like horizontal tubes are often used for condensers.

Because of the sweeping action of the droplets that occurs in drop condensation, the average thickness of the liquid film on the wall tends to be smaller than it is in film condensation. Also, the droplets themselves are small and well-mixed due to their motion. Therefore, the heat transfer coefficient for drop condensation processes tends to be higher than it is for film condensation processes. This observation has led heat exchanger designers to strive for drop condensation by applying coatings and treatments that make the heat transfer surfaces hydrophobic, promoting the formation of droplets. However, surface treatments tend to lose their effectiveness over prolonged periods of operation and therefore drop condensation may revert to film condensation over time. Most design calculations assume film condensation in order to capture the conservative, long-term performance of the heat exchanger. The correlations presented in this section are for film condensation in a few geometries.

11.4.2 Correlations for Film Condensation

Vertical Wall

Condensation on a vertical wall is shown in Figure 11.7. Correlations for condensation on a vertical wall are typically expressed in terms of a condensate film Reynolds number, Re_c, defined according to

$$Re_c = \frac{4\dot{m}_l}{W\,\mu_{l,sat}}, \tag{11.25}$$

T_{sat}

g

Figure 11.7 Film condensation on a vertical wall.

where $\mu_{l,sat}$ is the viscosity of saturated liquid and $\dot{m}_l/W$ is the mass flow rate of condensate per unit width of plate (into the page). For small values of Re_c, the solution is obtained from an analysis of a laminar, smooth condensate film:

$$\frac{\bar{h}}{k_{l,sat}}\left[\frac{\mu_{l,sat}^2}{\rho_{l,sat}\left(\rho_{l,sat}-\rho_{v,sat}\right)g}\right]^{1/3}=1.47\,Re_c^{-1/3}\qquad \text{for } Re_c < 30, \tag{11.26}$$

where $\rho_{v,sat}$ and $\rho_{l,sat}$ are the densities of saturated vapor and saturated liquid, respectively, and g is the acceleration of gravity. As the film Reynolds number increases, the film becomes unstable and waves appear at the liquid–vapor interface. At very high values of the film Reynolds number, the film may become turbulent. For Reynolds numbers greater than 30, Butterworth (1981) suggests the correlations:

$$\frac{\bar{h}}{k_{l,sat}}\left[\frac{\mu_{l,sat}^2}{\rho_{l,sat}\left(\rho_{l,sat}-\rho_{v,sat}\right)g}\right]^{1/3}=\frac{Re_c}{1.08\,Re_c^{1.22}-5.2}\qquad \text{for } 30 < Re_c < 1600 \tag{11.27}$$

$$\frac{\bar{h}}{k_{l,sat}}\left[\frac{\mu_{l,sat}^2}{\rho_{l,sat}\left(\rho_{l,sat}-\rho_{v,sat}\right)g}\right]^{1/3}=\frac{Re_c}{8750+\dfrac{58}{\sqrt{Pr_{l,sat}}}\left(Re_c^{0.75}-253\right)}\qquad \text{for } Re_c > 1600. \tag{11.28}$$

The calculation of the Reynolds number depends on the mass flow rate of condensate which depends on the heat transfer coefficient; therefore, the calculation is inherently iterative. The mass flow rate of condensate can be computed according to:

$$\dot{m}_c\,\Delta i'_{vap}=\bar{h}\,W\,L(T_{sat}-T_s). \tag{11.29}$$

The parameter $\Delta i'_{vap}$ is the latent heat of vaporization corrected to account for the sensible cooling of the condensate:

$$\Delta i'_{vap}=\Delta i_{vap}+0.68c_{l,sat}(T_{sat}-T_s), \tag{11.30}$$

where $c_{l,sat}$ is the specific heat capacity of saturated liquid vapor. The properties of saturated liquid ($\rho_{l,sat}$, $\mu_{l,sat}$, $k_{l,sat}$, and $c_{l,sat}$) should be computed using the average of the wall and saturation temperatures. This correlation for film condensation from a vertical wall is implemented in the EES procedure Cond_Vertical_Plate.

Example 11.3

A simple water purification system consists of a pressure vessel that is partially filled with brackish water. A heater element is submerged in the water causing it to boil. A cooled plate is mounted vertically to the side of the vessel in the vapor space. The plate is cooled to a uniform temperature of $T_s = 40°C$, causing

Example 11.3 (cont.)

condensation on its surface. The water that condenses on the cold plate is relatively pure and can be collected for drinking. The length of the cold plate is $L = 4.0$ cm in the vertical direction and the plate is $W = 10$ cm wide. The temperature of the saturated water vapor in the vessel is $T_{sat} = 110°C$. Determine the rate at which clean water is produced by the device.

Known Values

Figure 1 illustrates a sketch of the problem with the known values indicated.

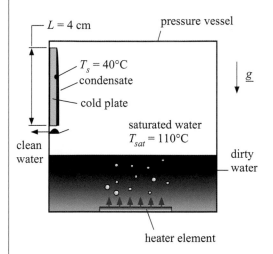

Figure 1 Water purification system.

Assumptions

- Noncondensable air in the vessel is neglected.
- The tank is large enough that the edges do not affect the flow situation.

Analysis

The film temperature is used to compute the saturated liquid properties:

$$T_f = \frac{(T_{sat} + T_s)}{2}. \tag{1}$$

The properties of water are obtained:

$$\mu_{l, sat} = \mu_{l, sat}(T_f) \tag{2}$$

$$k_{l, sat} = k_{l, sat}(T_f) \tag{3}$$

$$\rho_{l, sat} = \rho_{l, sat}(T_f) \tag{4}$$

$$c_{l, sat} = c_{l, sat}(T_f) \tag{5}$$

$$\rho_{v, sat} = \rho_{v, sat}(T_{sat}) \tag{6}$$

$$\Delta i_{vap} = \Delta i_{vap}(T_{sat}). \tag{7}$$

The condensate film Reynolds number is computed according to Eq. (11.25):

$$Re_c = \frac{4\,\dot{m}_l}{W\,\mu_{l, sat}}, \tag{8}$$

Continued

Example 11.3 (cont.)

where $\dot{m}_l$ is the mass flow rate of condensate. The heat transfer coefficient is computed using Eq. (11.27), which assumes that $30 < Re_c < 1600$:

$$\bar{h}_c = \frac{Re_c \, k_{l,\,sat}}{(1.08 \, Re_c^{1.22} - 5.2)} \left[\frac{\mu_{l,\,sat}^2}{\rho_{l,\,sat}(\rho_{l,\,sat} - \rho_{v,\,sat})g} \right]^{-1/3}. \tag{9}$$

The rate of heat transfer to the condensate plate is given by:

$$\dot{q}_c = \bar{h}_c \, L \, W \, (T_{sat} - T_s). \tag{10}$$

The latent heat of vaporization is corrected according to Eq. (11.30) in order to approximately account for sensible cooling of the condensate:

$$\Delta i'_{vap} = \Delta i_{vap} + 0.68 c_{l,\,sat}(T_{sat} - T_s). \tag{11}$$

The mass flow rate of condensate is then computed according to:

$$\dot{m}_c = \frac{\dot{q}_c}{\Delta i'_{vap}}. \tag{12}$$

The volumetric flow rate of condensate is:

$$\dot{V}_c = \frac{\dot{m}_c}{\rho_{l,\,sat}}. \tag{13}$$

Solution

Equations (1) through (13) are 13 equations in the 13 unknowns T_f, $\mu_{l,sat}$, $k_{l,sat}$, $\rho_{l,sat}$, $c_{l,sat}$, $\rho_{v,sat}$, Δi_{vap}, Re_c, $\bar{h}_c$, $\dot{q}_c$, $\Delta i'_{vap}$, $\dot{m}_c$, and $\dot{V}_c$. Because these are an implicit set of equations it is advisable to enter a guess for $\dot{m}_c$ and then enter Eqs. (1) through (11), as shown below.

```
$UnitSystem SI Mass J K Pa
T_s=ConvertTemp(C,K,40 [C])                      "surface temperature of cold plate"
L = 4 [cm]*Convert(cm,m)                          "length of cooling plate in the direction of gravity"
W = 10 [cm]*Convert(cm,m)                         "width of heater and cooling plate"
T_sat = ConvertTemp(C,K,110 [C])                  "saturation temperature"
m_dot_c=0.001 [kg/s]                              "guess for condensation rate"
T_f=(T_sat+T_s)/2
mu_l_sat=Viscosity(Water,T=T_f,x=0)               "viscosity of saturated liquid"
k_l_sat=Conductivity(Water,T=T_f,x=0)             "conductivity of saturated liquid"
rho_l_sat=Density(Water,T=T_f,x=0)                "density of saturated liquid"
c_l_sat=cP(Water,T=T_f,x=0)                       "specific heat capacity of saturated liquid"
rho_v_sat=Density(Water,T=T_sat,x=1)              "density of saturated vapor"
DELTAi_vap=Enthalpy_Vaporization(Water,T=T_sat)   "latent heat of vaporization"
Re_c=4*m_dot_c/(W*mu_l_sat)                        "condensate film Reynolds number"
h_bar_c=Re_c*k_l_sat*(mu_l_sat^2/(rho_l_sat*(rho_l_sat-rho_v_sat)*g#))^(-1/3)/(1.08*Re_c^1.22-5.2)
                                                  "film condensation correlation"
q_dot_c=h_bar_c*L*W*(T_sat-T_s)                   "cooling provided by condensate plate"
DELTAi_vap`=DELTAi_vap+0.68*c_l_sat*(T_sat-T_s)   "corrected latent heat of vaporization"
```

The program is solved and the guess values are updated. The guessed value of $\dot{m}_c$ is removed and Eqs. (12) and (13) are added, as shown below.

Example 11.3 (cont.)

```
$UnitSystem SI Mass J K Pa
T_s=ConvertTemp(C,K,40 [C])                    "surface temperature of cold plate"
L = 4 [cm]*Convert(cm,m)                        "length of cooling plate in the direction of gravity"
W = 10 [cm]*Convert(cm,m)                       "width of heater and cooling plate"
T_sat = ConvertTemp(C,K,110 [C])               "saturation temperature"

//m_dot_c=0.001 [kg/s]                          "guess for condensation rate"
T_f=(T_sat+T_s)/2
mu_l_sat=Viscosity(Water,T=T_f,x=0)            "viscosity of saturated liquid"
k_l_sat=Conductivity(Water,T=T_f,x=0)          "conductivity of saturated liquid"
rho_l_sat=Density(Water,T=T_f,x=0)             "density of saturated liquid"
c_l_sat=cP(Water,T=T_f,x=0)                     "specific heat capacity of saturated liquid"
rho_v_sat=Density(Water,T=T_sat,x=1)           "density of saturated vapor"
DELTAi_vap=Enthalpy_Vaporization(Water,T=T_sat) "latent heat of vaporization"
Re_c=4*m_dot_c/(W*mu_l_sat)                     "condensate film Reynolds number"
h_bar_c=Re_c*k_l_sat*(mu_l_sat^2/(rho_l_sat*(rho_l_sat-rho_v_sat)*g#))^(-1/3)/(1.08*Re_c^1.22-5.2)
                                                "film condensation correlation"
q_dot_c=h_bar_c*L*W*(T_sat-T_s)                 "cooling provided by condensate plate"
DELTAi_vap`=DELTAi_vap+0.68*c_l_sat*(T_sat-T_s) "corrected latent heat of vaporization"
q_dot_c=m_dot_c*DELTAi_vap`                     "mass flow rate of condensate"
V_dot_c=m_dot_c/rho_l_sat                       "volumetric flow rate of water"
```

Solving provides $\boxed{\dot{m}_c = 0.00102 \text{ kg/s and } \dot{V}_c = 1.04 \times 10^{-6} \text{ m}^3/\text{s } (3.76 \text{ liter/hr})}$. The calculated value of $Re_c = 107.8$, which is within the range of applicability for Eq. (11.27).

Discussion

The calculation is necessarily iterative since the value of Re_c depends on $\dot{m}_c$ which itself depends on the heat transfer coefficient. The function Cond_Vertical_Plate could have been used to accomplish the same iteration.

```
$UnitSystem SI Mass J K Pa
T_s=ConvertTemp(C,K,40 [C])                    "surface temperature of cold plate"
L = 4 [cm]*Convert(cm,m)                        "length of cooling plate in the direction of gravity"
W = 10 [cm]*Convert(cm,m)                       "width of heater and cooling plate"
T_sat = ConvertTemp(C,K,110 [C])               "saturation temperature"
CALL Cond_Vertical_Plate('Water', L, W, T_s, T_sat :h_bar_c, Re_c, q_dot_c, m_dot_c)
                                                "EES function for condensation correlation"
```

Solving provides $\dot{m}_c - 0.00102$ kg/s.

Horizontal, Downward Facing Plate

Gerstmann and Griffith (1967) present correlations for condensation on a horizontal, downward facing plate where the condensate is removed in the form of droplets that form, grow, and detach:

$$\frac{\bar{h}}{k_{l,sat}}\sqrt{\frac{\sigma}{(\rho_{l,sat} - \rho_{v,sat})g}} = 0.69\, Ra^{0.20} \qquad \text{for } 1 \times 10^6 < Ra < 1 \times 10^8 \qquad (11.31)$$

$$\frac{\bar{h}}{k_{l,\,sat}}\sqrt{\frac{\sigma}{\left(\rho_{l,\,sat}-\rho_{v,\,sat}\right)g}}=0.81\,Ra^{0.193} \qquad \text{for } 1\times10^{8} < Ra < 1\times10^{10}, \tag{11.32}$$

where

$$Ra = \frac{g\,\rho_{l,\,sat}\left(\rho_{l,\,sat}-\rho_{v,\,sat}\right)\Delta i'_{vap}}{\mu_{l,\,sat}(T_{sat}-T_{s})k_{l,\,sat}}\left[\frac{\sigma}{\left(\rho_{l,\,sat}-\rho_{v,\,sat}\right)g}\right]^{3/2}, \tag{11.33}$$

where $\Delta i'_{vap}$ is computed using Eq. (11.30). For slightly inclined surfaces (less than 20°), it is possible to use Eqs. (11.31) through (11.33) provided that g is replaced by $g\cos(\theta)$ where θ is the angle of inclination of the plate from horizontal. The correlation for condensation from a horizontal, downward facing plate is implemented in EES procedure Cond_Horizontal_Down.

Horizontal, Upward Facing Plate
Nimmo and Leppert (1970) present a correlation for condensation on a horizontal, upward facing plate that is infinite in one direction and has length L in the other. The condensate drains from the side under the influence of the hydrostatic pressure gradient related to the film thickness variation between the side and center of the plate:

$$\frac{\bar{h}\,L}{k_{l,\,sat}}=0.82\left[\frac{\rho_{l,\,sat}^{2}\,g\,\Delta i'_{vap}\,L^{3}}{\mu_{l,\,sat}(T_{sat}-T_{s})k_{l,\,sat}}\right]^{1/5}. \tag{11.34}$$

The correlation for condensation on a horizontal, upward facing plate is implemented in the procedure Cond_Horizontal_Up.

Single Horizontal Cylinder
Marto (1998) presents a solution for laminar film condensation on the outer surface of a single cylinder of diameter D:

$$\frac{\bar{h}\,D}{k_{l,\,sat}}=0.728\left[\frac{\rho_{l,\,sat}\left(\rho_{l,\,sat}-\rho_{v,\,sat}\right)g\,\Delta i'_{vap}\,D^{3}}{\mu_{l,\,sat}(T_{sat}-T_{s})k_{l,\,sat}}\right]^{1/4}. \tag{11.35}$$

The correlation for condensation on a single horizontal cylinder is implemented in EES procedure Cond_Horizontal_Cylinder.

Bank of Horizontal Cylinders
Film condensation in bundles of tubes is common in industrial condenser applications. The heat transfer coefficient associated with the tubes inside the bundle may be substantially less than the heat transfer coefficient calculated for a single tube using Eq. (11.35). Kern (1958), as presented in Kakaç and Liu (1998), suggests that the average heat transfer coefficient for a single tube $(\bar{h}_{1-tube})$ should be modified based on the number of rows of tubes in the vertical direction $(N_{tube,\,vert})$ according to:

$$\bar{h} = \bar{h}_{1-tube}\,N_{tube,\,vert}^{-1/6}. \tag{11.36}$$

The correlation for condensation on a bank of horizontal cylinders is implemented in the procedure Cond_Horizontal_N_Cylinders.

Single Horizontal Finned Tube
Film condensation on a horizontal finned tube, shown in Figure 11.8, was studied by Beatty and Katz (1948). The following correlation is suggested:

$$\bar{h}=0.689\left[\frac{\rho_{l,\,sat}^{2}\,k_{l,\,sat}^{3}\,g\,\Delta i'_{vap}}{\mu_{l,\,sat}(T_{sat}-T_{s})D_{eff}}\right]^{1/4}, \tag{11.37}$$

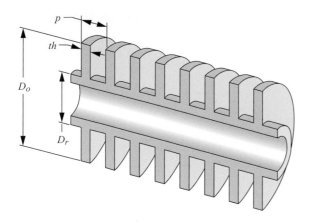

Figure 11.8 A finned tube.

where D_{eff} is an effective diameter, defined according to

$$D_{eff} = \left(1.30\eta_{fin} \frac{A_f}{A_{eff}\, \tilde{L}^{1/4}} + \frac{A_{uf}}{A_{eff}\, D_r^{1/4}} \right)^{-4},$$ (11.38)

where D_r is the root diameter of the finned tube, shown in Figure 11.8. The parameter η_{fin} is the fin efficiency, calculated using the solution presented in Section 3.3.1 and implemented using the EES function Eta_Fin_Annular_Rect. The surface area of the sides of a single fin, A_f in Eq. (11.38), is calculated according to:

$$A_f = 2\frac{\pi}{4}\left(D_o^2 - D_r^2\right),$$ (11.39)

where D_o is the outer diameter of the fins. The area of the exposed tube between adjacent fins, A_{uf} in Eq. (11.38), is calculated according to:

$$A_{uf} = \pi D_o (p - th),$$ (11.40)

where p is the fin pitch and th is the fin thickness (again, see Figure 11.8). The effective area of a fin, A_{eff} in Eq. (11.38), is calculated according to:

$$A_{eff} = \eta_f A_f + A_{uf}.$$ (11.41)

The parameter $\tilde{L}$ in Eq. (11.38) is calculated according to:

$$\tilde{L} = \frac{\pi\left(D_o^2 - D_r^2\right)}{4\, D_o}.$$ (11.42)

Note that the fin efficiency used in Eqs. (11.38) and (11.41) depends on the heat transfer coefficient and therefore implementation of the correlation is necessarily implicit and iterative. The correlation for condensation from a finned tube is implemented by the EES function Cond_Finned_Tube.

11.5 Flow Condensation

11.5.1 Introduction

Flow condensation processes are analogous to the flow boiling processes that were discussed in Section 11.3. When the condensation phase change process occurs while the fluid is flowing within a duct, then it is referred to as flow condensation. Flow condensation occurs within the condenser of vapor compression refrigeration cycles and Rankine power cycles. A flow condensation process is likely to experience several different flow regimes that are similar to those shown in Figure 11.5 for flow boiling, but occur in reverse. Flow condensation, like flow boiling, is complicated and there have been many researchers involved in developing a more complete understanding of this process; the interested reader is directed towards books such as Collier and Thome (1996). The next section provides one correlation that can be used to estimate the heat transfer coefficient for flow condensation processes.

11.5.2 Flow Condensation Correlations

The correlation suggested by Dobson and Chato (1998) is presented in this section and is implemented in EES as the procedure Cond_HorizontalTube. This correlation has been experimentally validated by Smit *et al.* (2002) and others. However, it should be noted that the correlation over-predicts the heat transfer coefficient for refrigerants that condense at high pressure (e.g., R125, R32, and R410a). Alternative condensation correlations from Shah (2016) are available in EES as the procedure Cond_Tube.

The correlation by Dobson and Chato is divided by flow regime into wavy or annular depending on the mass flux and the modified Froude number. The total mass flux is

$$G = \frac{4\,\dot{m}}{\pi\,D^2},\tag{11.43}$$

where D is the inner diameter of the tube and $\dot{m}$ is the total mass flow rate of the fluid (vapor and liquid). If the mass flux is greater than 500 kg/m²-s, then the flow is assumed to be annular regardless of the Froude number and the local heat transfer coefficient is computed according to:

$$h = \underbrace{\frac{k_{l,\,sat}}{D}\,0.023\,Re_{D,\,l}^{0.8}\,Pr_{l,\,sat}^{0.4}}_{\text{Dittus–Boelter equation}}\,\underbrace{\left(1 + \frac{2.22}{X_{tt}^{0.89}}\right)}_{\text{two-phase multiplier}} \qquad \text{for } G > 500 \text{ kg/m}^2,\tag{11.44}$$

where X_{tt} is the **Lockhart Martinelli parameter** and $Re_{D,\,l}$ is the superficial liquid Reynolds number. The Lockhart Martinelli parameter is computed according to:

$$X_{tt} = \sqrt{\frac{\rho_{v,\,sat}}{\rho_{l,\,sat}}}\left(\frac{\mu_{l,\,sat}}{\mu_{v,\,sat}}\right)^{0.1}\left[\frac{(1-x)}{x}\right]^{0.9},\tag{11.45}$$

where x is the local thermodynamic quality. The superficial liquid Reynolds number was defined previously in Eq. (11.10) and is the Reynolds number that is consistent with the liquid flowing alone in the tube:

$$Re_{D,\,l} = \frac{G\,D(1-x)}{\mu_{l,\,sat}}.\tag{11.46}$$

Comparing Eq. (11.44) to the Dittus–Boelter equation for fully developed turbulent flow shows that the parameter in parentheses is a two-phase multiplier on a single-phase heat transfer coefficient prediction.

If the mass flux is less than 500 kg/m²-s, then the flow is either annular or wavy depending on the value of the modified Froude number. The modified Froude number is computed according to:

$$Fr_{mod} = \begin{cases} \dfrac{0.025\,Re_{D,\,l}^{1.59}}{Ga^{0.5}}\left(\dfrac{1 + 1.09\,X_{tt}^{0.039}}{X_{tt}}\right)^{1.5} & \text{for } Re_{D,\,l} \le 1250 \\[4mm] \dfrac{1.26\,Re_{D,\,l}^{1.04}}{Ga^{0.5}}\left(\dfrac{1 + 1.09\,X_{tt}^{0.039}}{X_{tt}}\right)^{1.5} & \text{for } Re_{D,\,l} > 1250, \end{cases}\tag{11.47}$$

where Ga is the **Galileo number**, defined as

$$Ga = \frac{g\,\rho_{l,\,sat}\left(\rho_{l,\,sat} - \rho_{v,\,sat}\right)D^3}{\mu_{l,\,sat}^2}.\tag{11.48}$$

If the modified Froude number is greater than 20, then the flow is assumed to be annular and the local heat transfer coefficient is computed according to Eq. (11.44). However, if the modified Froude number is less than 20, then the flow is assumed to be wavy and the local heat transfer coefficient is computed according to:

$$h = \left(\frac{k_{l,\,sat}}{D}\right)\left[\left(\frac{0.23}{1 + 1.11 X_{tt}^{0.58}}\right)\left(\frac{G\,D}{\mu_{v,\,sat}}\right)^{0.12}\left(\frac{\Delta i'_{vap}}{c_{l,\,sat}(T_{sat} - T_s)}\right)^{0.25} Ga^{0.25}\, Pr_{l,\,sat}^{0.25} + A\, Nu_{fc}\right].$$

(11.49)

$$\text{if } G < 500 \text{ kg/m}^2 \text{ and } Fr_{mod} < 20$$

The parameter A in Eq. (11.49) is related to the angle from the top of the tube to the liquid level:

$$A = \frac{\cos^{-1}(2\,vf - 1)}{\pi},$$

(11.50)

where vf is the **void fraction** (the fraction of the volume occupied by vapor), estimated using the correlation provided by Zivi (1964):

$$vf = \left[1 + \frac{(1 - x)}{x}\left(\frac{\rho_{v,\,sat}}{\rho_{l,\,sat}}\right)^{2/3}\right]^{-1}.$$

(11.51)

The parameter Nu_{fc} in Eq. (11.49) is a Nusselt number related to forced convection in the bottom pool, evaluated according to:

$$Nu_{fc} = 0.0195\, Re_{D,\,l}^{0.8}\, Pr_{l,\,sat}^{0.4}\sqrt{1.376 + \frac{C_1}{X_{tt}^{C_2}}}.$$

(11.52)

The parameters C_1 and C_2 in Eq. (11.52) are evaluated based on the Froude number, Fr (note that this is not the modified Froude number):

$$Fr = \frac{G^2}{\rho_{l,\,sat}^2\, g\, D}.$$

(11.53)

If the Froude number is greater than 0.7 then:

$$C_1 = 7.242 \text{ and } C_2 = 1.655 \qquad \text{for } Fr > 0.7$$

(11.54)

otherwise:

$$C_1 = 4.172 + 5.48\, Fr - 1.564\, Fr^2 \text{ and } C_2 = 1.773 - 0.169\, Fr \qquad \text{for } Fr \leq 0.7.$$

(11.55)

The Dobson and Chato correlation for condensation in a horizontal tube discussed in this section is implemented in the procedure Cond_HorizontalTube. Note that the EES procedure slightly modifies the correlation described in this section in order to smooth out the sharp transition and the associated discontinuity that otherwise occurs at $Fr_{mod} = 20$ according to Eq. (11.47). Also, the Cond_HorizontalTube procedure provides the film condensation correlation recommended by Chato (1962) as reported in Incropera and DeWitt (2002) when the mass flow rate of fluid in the tube is set to zero.

11.6 Conclusions and Learning Objectives

This chapter provides a discussion of two-phase heat transfer. The behavior of pool boiling is presented qualitatively and the regimes of pool boiling are discussed. Correlations are presented for the pool boiling heat transfer coefficient in the nucleate boiling regime, which is of the greatest engineering interest. A method to estimate the critical heat flux is also presented as it represents an important upper limit in most applications. Flow boiling is discussed and a correlation for flow boiling is presented that allows an estimate of the heat transfer coefficient in this situation. Film and dropwise condensation are discussed but only correlations for film condensation are presented as this is the typical mode of condensation heat transfer for most surfaces. Flow condensation is also discussed and a correlation is presented for this situation.

Some specific concepts that you should understand are listed below.

- The qualitative behavior of the pool boiling curve including nucleate boiling, the critical heat flux, film boiling, and the Leidenfrost point.
- The hysteretic behavior of pool boiling and why this behavior occurs.
- The relative magnitude of two-phase heat transfer processes compared to single-phase heat transfer processes.
- Methods for estimating nucleate pool boiling behavior and the critical heat flux.
- Flow boiling regimes including at least bubbly and annular flow and dry out. The qualitative variation of the flow boiling heat transfer coefficient with quality.
- The difference between film and dropwise condensation.

References

Beatty, K. O. and D. L. Katz, "Condensation of vapors on outside of finned tubes," *Chem. Eng. Prog.*, Vol. 44, pp. 55–70 (1948).

Butterworth, D., "Condensers: Basic heat transfer and fluid flow," in *Heat Exchanger*, S. Kakac, A. E. Bergles, and F. Mayinger, eds., Hemisphere Publishing Corp., New York, pp. 289–314 (1981).

Carey, V. P., *Liquid Vapor Phase Change Phenomena*, Hemisphere Publishing Corporation, Washington, D.C. (1992).

Chato, J. C., "Laminar condensation inside horizontal and inclined tubes," *ASHRAE J.*, 4, pp. 52–60 (1962).

Collier, J. G. and J. R. Thome, *Convective Boiling and Condensation,* Third Edition, Oxford University Press, United Kingdom (1996).

Dobson, M. K. and J. C. Chato, "Condensation in smooth horizontal tubes," *J. Heat Transfer*, Vol. 120, pp. 193–213 (1998).

Drew, T. B. and C. Mueller, "Boiling," *Trans. AIChE*, Vol. 33, pp. 449 (1937).

Gerstmann, J. and P. Griffith, "Laminar film condensation on the underside of horizontal and inclined surfaces," *Int. J. Heat Mass Transfer*, Vol. 10, pp. 567–580 (1967).

Gnielinski, V., "New equations for heat and mass transfer in turbulent pipe and channel flow," *Int. Chem. Eng.*, Vol. 16, pp. 359–368 (1976).

Incropera, F. P and DeWitt, D. P. *Introduction to Heat Transfer*, Fourth Edition, John Wiley & Sons, New York (2002).

Kakaç, S. and H. Liu, *Heat Exchangers: Selection, Rating, and Thermal Design*, CRC Press, New York (1998).

Kern, D. Q., "Mathematical development of loading in horizontal condensers," *AIChE J.*, Vol. 4, pp. 157–160 (1958).

Lienhard, J. H. and V. K. Dhir, "Hydrodynamic prediction of peak pool-boiling heat fluxes from finite bodies," *J. Heat Transfer*, Vol. 95, pp. 152–158 (1973).

Marto, P. J, "Condensation", in *Handbook of Heat Transfer,* Third Edition, W. M. Rohsenow, J. P. Hartnett, and Y. I. Cho, eds., McGraw-Hill, New York (1998).

Mills, A. F., *Heat Transfer*, Irwin Publishing, Homewood, IL (1992).

Nimmo, B.G. and G. Leppert, "Laminar film condensation on a finite horizontal surface," In *Proc. 4th Int. Heat Transfer Conf.,* Paris, Vol. 6, Cs2.2 (1970).

Nukiyama, S., "The maximum and minimum values of heat transmitted from metal to boiling water under atmospheric pressure," *J. Japan Soc. Mech. Eng.*, Vol. 37, pp. 367–374 (1934) (in Japanese); translated in *Int. J. Heat Mass Transfer*, Vol. 9, pp. 1419–1433 (1966).

Petukhov, B. S., in *Advances in Heat Transfer*, Vol. 6, T. F. Irvine and J. P. Hartnett eds., Academic Press, New York (1970).

Rohsenow, W. M., "A method of correlating heat transfer data for surface boiling of liquids," *Transactions of the ASME*, Vol. 74, pp. 969 (1952).

Shah, M. M., "A new correlation for heat transfer during boiling flow through pipes,", *ASHRAE Trans.*, Vol 82(2), pp. 66–86 (1976).

Shah, M. M., "Chart correlation for saturated boiling heat transfer: Equations and further study," *ASHRAE Trans.*, Vol 88(1), pp. 185–186 (1982).

Shah, M. M., "Evaluation of general correlations for heat transfer during boiling of saturated liquids in tubes and annuli," *HVAC&R Journal*, Vol. 12 (4), pp. 1047–1065 (2006).

Shah, M. M., "Comprehensive correlations for heat transfer during condensation in conventional and mini/micro channels in all orientations," *Int. J. Refrigeration*, Vol. 67, pp. 22–41 (2016).

Smit, F. J., J. R. Thome, and J. P. Meyer, "Heat transfer coefficients during condensation of the zeotropic refrigerant mixture HCFC-22/HCFC-142b," *J. Heat Transfer*, Vol. 124, pp. 1137–1146 (2002).

Thome, J. R., *Engineering Data Book III*, Wolverine Tube, Huntsville, AL (2006); accessible online at http://www.thermalfluidscentral.org/e-books/pdf/edb3/1.pdf.

Vachon, R. I., G. H. Nix, and G. E. Tanger, "Evaluation of constants for the Rohsenow pool-boiling correlation," *Journal of Heat Transfer*, Vol. 90, pp. 239–246 (1968).

Whalley, P. B., *Boiling, Condensation and Gas-Liquid Flow,* Clarendon Press, Oxford, United Kingdom (1987).

Zivi, S. M., "Estimation of steady-state steam void-fraction by means of the principle of minimum entropy production," *J. Heat Transfer*, Vol. 86, pp. 247–252 (1964).

Problems

Pool Boiling

11.1 It is necessary to determine the power requirements of a heater that will be immersed in boiling water at atmospheric pressure. The heater is a thin square copper plate, 10 cm on a side that is maintained at a set temperature, T_s, with embedded electrical heaters. Prepare a plot of the electrical power required as a function of the plate temperature. At what plate temperature will the critical heat flux occur?

11.2 A computer manufacturer is reviewing alternative ways to remove heat from electronic components. The electronic circuit board can be assumed to be a thin horizontal plate with a width of 8 cm and a length of 16 cm. Currently, air is blown over the top of the circuit board at a velocity of 10 m/s. Additional cooling could be obtained by a higher air velocity, but the increased noise associated with the larger fan required is judged to be unacceptable. Another alternative is to immerse the board in a fluid at atmospheric pressure that is undergoing nucleate boiling. The fluid R245fa has been chosen as a possibility. The surface tension of R245fa at atmospheric pressure is 0.0153 N/m. Other thermodynamic and transport properties are available from EES. Prepare a plot that shows the surface temperature of the plate as a function of the heat flux using air at 10 m/s and nucleate boiling at atmospheric pressure with R245fa for heat fluxes ranging from 100 to 10,000 W/m².

11.3 A tungsten wire having a diameter of 1 mm and a length of 0.45 m is suspended in saturated carbon dioxide liquid maintained at 3.25 MPa. The fluid–surface coefficient needed in the nucleate boiling relation, C_{nb}, is estimated to be 0.01 and the emissivity of the tungsten wire is 0.4. Prepare a plot of the electrical power dissipated in the wire versus the excess temperature for power levels ranging from 10 W to the power corresponding to the critical heat flux for the nucleate boiling regime. What is your estimate of the excess temperature at the burnout point?

11.4 A kettle having a flat copper bottom 25 cm in diameter is electrically heated and used to produce steam from saturated liquid water at atmospheric pressure. The electrical power can be adjusted from 1.5 kW to 3.0 kW.
(a) Plot the rate of steam produced as a function of the power.
(b) Plot the temperature of the copper surface as a function of the power.
(c) At what power will the critical heat flux occur?

11.5 Determine the heat flux and excess temperature difference at the burnout point for boiling benzene at atmospheric pressure on a horizontal chromium cylinder having a 1.25 cm outer diameter.

11.6 You are preparing a spaghetti dinner for guests when you realize that your heat transfer training can be used to answer some fundamental questions about the process. The pot you are using holds 4 liters of water. The atmospheric pressure is 101 kPa. When on its high setting, the electric stove heating unit consumes 1.8 kW of electrical power of which 20 percent is transferred to the surroundings, rather than to the water. The pot is made of 4 mm thick polished AISI 304 stainless steel and it has a diameter of 0.25 m. The burner diameter is also 0.25 m.

(a) How much time is required to heat the water from 15°C to its boiling temperature?

(b) What are the temperatures of the outside and inside surfaces of the bottom of pot while the water is boiling?

(c) What would the burner electrical power requirement have to be to achieve the critical heat flux? Compare the actual heat flux during the boiling process to the critical heat flux.

(d) How much water is vaporized during the 10 min required to cook the spaghetti?

Flow Boiling

11.7 The refrigerant R134a is evaporating as it flows through a pipe with inner diameter $D = 0.5$ inch. The mass flow rate is $\dot{m} = 40$ lbm/min and the saturation temperature is $T_{sat} = 54°F$. The heat flux at the surface of the tube is $\dot{q}''_s = 5000$ Btu/hr-ft^2.

(a) Plot the local heat transfer coefficient as a function of the quality of the R134a.

(b) Determine the average heat transfer coefficient in the pipe if the R134a enters the pipe as saturated liquid and leaves as saturated vapor. Use the function Flow_Boiling_avg.

11.8 An industrial boiler produces steam from saturated liquid water at 2.4 atm flowing at 0.02 kg/s within a serpentine 1.0 cm inner diameter tube that is electrically heated. The heat flux is maintained at 58,000 W/m^2.

(a) Prepare a plot of the wall temperature of the tube versus quality, ranging from 0 to 1. Provide an explanation for the shape of the plot.

(b) Determine the total length of tube required to change the quality from 0 to 1.

11.9 When one fluid is changing phase in a heat exchanger, it is commonly assumed to be at a uniform temperature. However, there is a pressure drop in the evaporating fluid, which affects its saturation temperature. In a particular case, a

2 m long horizontal concentric tube heat exchanger made of copper is used to evaporate 0.028 kg/s of refrigerant R134a with an entering state of 300 kPa and a quality of 0.35. Heat transfer is provided by a flow of water that enters the heat exchanger at 12°C and 1.10 bar, with a mass flow rate of 0.20 kg/s. The refrigerant passes through the central tube of the heat exchanger, which has an inner diameter of 1.25 cm and a wall thickness of 2 mm. The water flows through the annulus; the inner diameter of the outer tube is 2.5 cm. Estimate the outlet temperature of the water and the outlet temperature and quality of the refrigerant.

11.10 A circular finned tube evaporator designed for cooling air is made from aluminum. The outer diameter of the tubes is 10.21 mm, with a tube wall thickness of 1 mm. The evaporator is plumbed such that there are 12 parallel circuits of tubes with each circuit having a length of 0.6 m. Refrigerant R134a enters the evaporator at a mass flow rate of 0.15 kg/s. The refrigerant enters the throttle valve upstream of the evaporator as 35°C saturated liquid. The pressure in the evaporator is 240 kPa. The refrigerant exits as a saturated vapor.

(a) What is the rate of heat transfer to the air for this evaporator?

(b) Estimate the average heat transfer coefficient between the R134a and the tube wall.

(c) Estimate the pressure drop of the R134a as it passes through the evaporator. Does this pressure drop significantly affect the saturation temperature?

11.11 A test facility for an evaporator consists of a 0.01 m inner diameter tube, 0.5 m in length. The tube is insulated on its outer surface. Saturated liquid ammonia at 850 kPa enters the tube with a mass flow rate of 0.0065 kg/s. The tube is electrically heated at a rate that provides a heat flux of 12,500 W/m^2 to the inside surface of the tube.

(a) Determine the electrical power required.

(b) Determine the quality of the ammonia at the tube exit.

(c) Determine the average heat transfer coefficient on the inside surface of the tube.

(d) Estimate the average tube wall temperature.

11.12 Problem 9.48 describes a solar electric generating system (SEGS) plant that uses solar collectors to heat oil. Use of the oil is problematic since a heat exchanger is required to transfer energy between the oil and the steam that must be generated to operate the power cycle. In this problem, we wish to investigate the option of boiling water directly within the tubes of the parabolic collectors. As in Problem 9.48, we will analyze $N = 50$ tubes, each with an inner diameter of $D_i = 0.066$ m and a length of $L = 750$ m. The outer diameter of the tube is essentially equal to the inner diameter and the outer surface of these tubes is exposed to a uniform heat flux of $\dot{q}_s'' = 15{,}000$ W/m². In Problem 9.48, each tube has oil (with specific heat of 2341 J/kg-K and a density of 825 kg/m³) flowing through it with a volumetric flow rate (per tube) of $\dot{V} = 0.012$ m³/s. The oil enters the tube with a mean temperature of $T_{in} = 500$ K and exits at 600 K. The pump power for all 50 tubes with a pump having an efficiency of 0.5 is about 0.9 MW. Assume that, instead of the oil, saturated liquid water at 600 K enters the tubes and exits as saturated vapor.

(a) Compute the necessary mass flow rate of water so that the rate of energy transfer to the water is the same as that for the oil.

(b) Calculate the average heat transfer coefficient on the inside surface of the tube when water is used as the heat transfer fluid and compare it to the heat transfer coefficient using oil.

(c) Determine the pressure drop across each of the 50 tubes.

(d) Calculate the outlet temperature of the steam.

(e) Estimate the minimum pump power required for all 50 tubes and compare the result with the minimum pump power needed for the oil. Note that the pump that provides the pressure operates with liquid water at 35°C.

Film Condensation

11.13 In the condenser in a lithium bromide–water absorption refrigeration system, water vapor at 16 kPa is condensed on the outside surface of horizontal tubes, thereby transferring energy to liquid water flowing inside of the tubes at atmospheric pressure. The tubes are made of copper with an inner diameter of 2.5 cm and a wall thickness of 3 mm. The tubes are arranged in banks with N tubes, stacked vertically, in each bank. The N tubes are plumbed in parallel, with water at 30°C and a total mass flow rate of 4 kg/s split between the tubes. The cooling water exits at a mean temperature of 41.4°C.

(a) Determine the rate of heat transfer resulting from this condensation process.

(b) Estimate the flow rate of condensate.

(c) Calculate and plot the total length of all tubes required for the condensation process as a function of the number of vertical tubes plumbed in parallel for $N = 2$ to 12. Also, calculate and plot the pressure drop for the cooling water in the tube bank as a function of N.

11.14 Evacuated tubular solar collectors may employ a heat pipe to transfer collected solar energy for water heating. Heat transfer between the water that is being heated and the solar collector occurs at the condenser of the heat pipe, which is a thin-walled cylinder made of copper with a length of 6 cm and a diameter of 1 cm as shown in the figure. Water at 40°C and 1 atm flows past the condenser at a velocity that can be specified by the flow rate and duct diameter. The fluid inside the heat pipe is also water and it condenses at a pressure of 100 kPa. The heat transfer situation of the condensing water within the heat pipe is not known, but here it will be assumed that it can be represented with the same relations as used for film condensation on the inside surface of a cylinder. This heat transfer coefficient for film condensation is provided by the Cond_Horizontal_Cylinder procedure when the mass flow rate is set to 0. Plot the rate of heat transfer from the solar collector to the water that is being heated as a function of the flow velocity of the water for velocities between 1 m/s and 10 m/s.

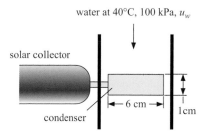

water at 40°C, 100 kPa, u_w

solar collector

← 6 cm →

condenser

1 cm

11.15 The condenser in steam power cycle utilizes a shell and tube heat exchanger that consists of 1200 nominal 1.5 inch schedule 40 tubes made of brass. Each tube is 8 ft long and internally smooth. Cooling water enters each of the tubes at 68°F and exits at 74°F. Saturated steam at 1 psia having a quality of 91 percent enters the condenser at a low velocity and is condensed on the tubes. Estimate the rate of condensate formation and the associated water flow rate at steady-state conditions.

11.16 Problem 11.2 describes an electronics cooling system that removes the heat dissipated in an electronic circuit board by submersing the board in R245fa. The circuit board is maintained at a relatively low and uniform temperature over a range of heat fluxes by boiling R245fa. However, a problem now arises in dealing with the vapor produced by the evaporation. One possibility is to condense the vapor on the bottom side of a vertical plate that is cooled by chilled water on its top side, as shown in the figure in a sealed container. The top of the enclosure is made of metal and it can be considered to be isothermal. The chilled water is at 1 atm and has a free stream velocity of 10 m/s and a free stream temperature of 10°C. The circuit board is 8 cm wide and 16 cm long. The saturation pressure (and thus temperature) of the R245fa in the enclosure should vary with the heat flux.

water at
10°C, 1 atm, 10 m/s

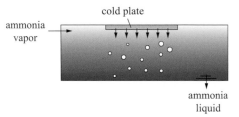

— 16 cm —
saturated R245fa

circuit board
8 cm × 16 cm

(a) Prepare a plot of the saturation pressure and circuit board surface temperature as a function of the heat flux for heat fluxes ranging from 100 W/m² to 10000 W/m².
(b) How sensitive are the results to the velocity of the chilled water?

11.17 Calculate the heat flux for a square plate 1 m on each side that is used to condense steam at 6 kPa. Consider three plate orientations: (1) horizontal

facing downward; (2) horizontal facing upward; and (3) vertical (one side only is active).
(a) Plot the heat flux for each orientation as a function of plate surface temperature.
(b) Which geometry provides the highest rate of condensation per unit surface area?
(c) How do the answers to (a) and (b) change if the plate dimensions are reduced to 0.5 m per side?

11.18 A condenser designed for an ammonia refrigeration cycle includes a flat square cold plate 12 cm on a side mounted to the top of an insulated enclosure. The plate is maintained at 30°C. Saturated ammonia vapor is maintained in the enclosure by a pressure-sensitive valve. Ammonia condenses on the plate and drops to the bottom of the enclosure where it passes through a liquid trap. As liquid ammonia exits, vapor enters to maintain steady conditions in the enclosure. Plot the rate of liquid ammonia produced as a function of the ammonia pressure from 1500 kPa and 4000 kPa.

cold plate

ammonia
vapor

ammonia
liquid

11.19 Saturated steam at 4 kPa condenses on the outside surface of a horizontal copper tube having a 26.6 mm inner diameter and 33.4 mm outer diameter. Water at an average temperature of 15°C flows through the smooth tube at a flow rate of 2.45 kg/s.
(a) Estimate the rate of condensate produced per meter of tube assuming the wall temperature is the average of the water and steam temperatures.
(b) Refine the calculation by estimating the wall temperature based on knowledge of the heat transfer coefficients on the inside and outside of the pipe wall.

11.20 A heat pipe has been instrumented in order to measure its ability to transfer thermal energy. The heat pipe consists of a sealed vertical thin-walled copper tube that is 1.5 m in length and 2.5 cm in diameter. The heat pipe contains liquid and vapor toluene. The bottom 5 cm of the tube are wrapped with heater tape that provides 100 W of heat input to the toluene. The toluene evaporates at the lower end of the tube and the vapor rises to the top where it is condensed by contact with the cold top surface of the tube. The top 6 cm of the heat pipe are maintained at 29°C by a flow of liquid water at 25°C and 1 atm. Toluene condensate flows back to the bottom of the tube; the flow is assisted by surface tension due to the presence of a wicking material on the inner surface of the copper tube. The heat pipe tube is well insulated except for the bottom part that is in contact with the heater and the top part that is in contact with the water.

(a) Estimate the saturation temperature and pressure of the toluene in the heat pipe.

(b) Estimate the surface temperature of the tube that is in contact with the heater.

(c) Estimate the velocity of the cooling water provided at 25°C needed to maintain the top surface of the heat pipe at 29°C.

(d) Compare the heat transfer rate provided by the heat pipe to the heat transfer rate that would occur if the tube were replaced with 2.5 cm diameter solid copper rod with the same temperatures imposed at the hot and cold ends.

(e) What is the effective thermal conductivity of the heat pipe? What do you see as advantages of the heat pipe?

11.21 A vertical cylindrical container is made of aluminum having a wall thickness of 2.5 mm. The cylinder is 0.24 m in height and it has an outer diameter of 7.5 cm. Liquid water is placed in the cylinder and the bottom is heated, evaporating the liquid. The vapor that is produced escapes through a vent at the top of the cylinder. The flow of vapor drives out air that was originally in the cylinder. When all of the liquid has been boiled, the heating is stopped and the vent at the top of the cylinder is closed. The aluminum surfaces are nearly at a uniform temperature of 100°C. The cylinder is allowed to stand in a large room and it transfers energy by free convection to the 25°C air.

Calculate and plot the pressure inside the cylinder as a function of time for a 5 minute period after the vent is closed. State and justify

any assumptions that you employ. (Note that the heat transfer coefficient for film condensation on the inside surfaces of the cylinder can be estimated using the Cond_HorizontalTube procedure with a mass flow rate of zero.)

11.22 The tube in Problem 11.19 is replaced with a tube having annular fins. The fins have an outside diameter of 42 mm. They are made of aluminum with a thickness of 1 mm and a spacing of 4 mm. Calculate the rate of condensate production and compare it with the result from Problem 11.19.

11.23 One method of removing water and other contamination from a gas is to pass it through a cooled tube so that contaminants with high freezing and liquefaction points (e.g., water) tend to be collected at the wall. A quick and easy liquid nitrogen trap for methane is constructed by placing a tube in a Styrofoam cooler that is filled with liquid nitrogen, as shown in the figure.

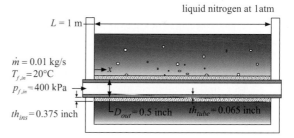

The length of the tube is $L = 1$ m. The outer diameter of the tube is $D_{out} = 0.5$ inch and the tube thickness is $th_{tube} = 0.065$ inch. The tube conductivity is $k_{tube} = 150$ W/m-K. The tube is wrapped in insulation (to avoid liquefying the methane). The thickness of the insulation is $th_{ins} = 0.375$ inch and the insulation conductivity is $k_{ins} = 1.5$ W/m-K. Methane enters the tube at $\dot{m} = 0.01$ kg/s with temperature $T_{f,in} = 20$°C and pressure $p_{f,in} = 400$ kPa. The liquid nitrogen that fills the container is at 1 atm and is undergoing nucleate boiling on the external surface of the insulation. You may neglect axial conduction through the tube.

(a) Set up an EES program that can evaluate the state equations for this problem. That is, given a value of position, x, methane temperature, T_f, and methane pressure, p_f, your program should be able to compute $\frac{dT_f}{dx}$ and $\frac{dp_f}{dx}$.

(b) Use the Integral command in EES to integrate the state equations from $x = 0$ to $x = L$. Plot the fluid temperature and pressure as a function of position.

(c) Plot the heat flux at the insulation surface and the critical heat flux as a function of position.

(d) Plot the temperature of the methane at the surface of the tube as a function of position.

(e) Plot the lowest temperature experienced by the methane in the trap as a function of the insulation thickness. If the methane temperature must be maintained at above its liquefaction point (131.4 K at 400 kPa) then what should the insulation thickness be?

11.24 A condenser consists of a tube bank within a large cylindrical shell. Saturated steam vapor at 4 kPa exists within the shell. The tube bank consists of 12 unfinned copper tubes in the horizontal direction and 10 tubes in the vertical direction. Each tube has an outer diameter of 33.4 mm and an inner diameter of 26.6 mm and a length of 2 m. The tubes are internally smooth. The tubes are spaced so there is a gap of 33.4 mm between each tube in both the horizontal and vertical directions. Water enters the tube bank at 15°C and a total flow rate of 210 kg/s.

(a) Determine the average heat transfer coefficient for the tube bank.

(b) Determine the average heat transfer coefficient on the water side.

(c) Estimate the rate of condensate produced.

(d) Determine the water exit temperature.

11.25 The left and right side walls of an old refrigerator are poorly insulated relative to current standards and the electrical heaters have been disabled. Consequently, the wall temperature is 18.8°C on a day in which the temperature in the kitchen where the refrigerator is located is 27°C and the relative humidity is 62 percent. The side walls are each 60 inches high and 28 inches deep. Determine the rate at which condensate will be generated at these conditions, assuming that the diffusional resistance resulting from water vapor transferring through air is negligible and conditions are steady.

Flow Condensation

11.26 A condenser for a refrigeration cycle consists of a bank of 16 copper tubes with 1 cm inner

diameter, each having a wall thickness of 1.5 mm and a length of 30 cm. The tubes are plumbed in parallel. Saturated refrigerant R134a flows inside of the tubes at a pressure of 1200 kPa with a total mass flow rate of 0.85 kg/s. Atmospheric air at 35°C is blown past the tubes at a velocity of 3.5 m/s.

(a) Estimate the heat transfer coefficient between the air and the outside of the tubes.

(b) Plot the heat transfer coefficient between the R134a and inside of the tubes for quality ranging between 0.95 and 0.05.

(c) Comment on the importance of accurately knowing the condensation heat transfer coefficient for estimating the rate of condensation in this condenser.

11.27 You have fabricated an inexpensive condenser for your air conditioner by running the refrigerant through a plastic tube that you have submerged in a lake. The outer diameter of the tube is $D_o = 7.0$ mm and the inner diameter is $D_i = 5.0$ mm. The tube conductivity is $k_{tube} = 1.4$ W/m-K. The refrigerant is R134a and it enters the tube with quality $x = 0.97$, temperature $T_{sat} = 35°C$ and mass flow rate $\dot{m}_r = 0.01$ kg/s. The water in the lake has temperature $T_\infty = 10°C$.

(a) Determine the heat transfer rate per unit length from the refrigerant to the lake at the tube inlet.

(b) If the length of the tube is $L = 6$ m, then what is the quality of the refrigerant leaving the tube? Plot the quality and refrigerant heat transfer coefficient as a function of position in the tube.

11.28 Condensation and boiling are analogous processes in that both a involve phase change. Heat exchangers that provide condensation and boiling are often designed in a similar manner. In a particular case, a phase change of R134a takes place within horizontal tubes having an inner diameter of 1 cm. The mass velocity is 300 kg/s-m².

(a) Prepare a plot of the heat transfer coefficient for condensation and boiling as a function of quality at a saturation temperature of 10°C for heat fluxes of 5000 W/m² and 10,000 W/m².

(b) Plot the excess temperature for condensation and boiling as a function of quality.

(c) What conclusion can you draw from the results?

11.29 Absorption refrigeration cycles are often used to operate small refrigerators in hotel rooms because they do not require compressors and thus operate quietly. In a particular case, an absorption refrigeration system uses an ammonia–water mixture. The ammonia is separated from the water and passes through a condenser where it is isobarically changed from saturated vapor at 76°C to subcooled liquid at 40°C. The condenser consists of a single unfinned thin-walled copper tube with a 1 cm inner diameter. The thermal energy released from the ammonia in this process is transferred to the 25°C room air by free convection. The subcooled ammonia is throttled to a saturation temperature of –5°C and evaporated to saturated vapor to produce a refrigeration capacity of 110 W.

(a) Determine the mass flow rate of ammonia through the evaporator and condenser.

(b) Estimate the length of piping required to condense the ammonia from saturated vapor to saturated liquid at 76°C.

(c) Estimate the additional length of piping required to subcool the ammonia to 40°C.

Projects

11.30 An industrial boiler generates steam by heat exchanging combustion gases with saturated water at 125 kPa through mechanically polished AISI 302 stainless steel tubing having an inside diameter of 5.48 cm with a wall thickness of 2.7 mm and a total submerged length of 10 m. The combustion gases enter the tubing at 750°C with a mass flow rate of 0.0115 kg/s. The gases exhaust at ambient pressure. Assume that the combustion gases have the same thermodynamic properties as air.

(a) Identify the state equation for this problem; the differential equation that can be used to determine the rate of change of the temperature of the combustion gas with respect to position.

(b) Integrate the state equations developed in part (a) in order to determine the outlet temperature of the combustion gases

(c) Calculate the rate at which steam is generated in this boiler.

11.31 A cross-section of one type of evacuated solar collector is shown in the figure. The collector consists of a cylindrical glass tube with an outer diameter of 7.5 cm and wall thickness of 5 mm. In the center of the tube is a heat pipe, which is a copper tube with an outer diameter of 2 cm and wall thickness of 1.5 mm. The heat pipe contains a small amount of water at a pressure of 200 kPa that experiences nucleate boiling as solar radiation is absorbed on the outside surface of the copper tube at a rate of 745 W/m². You may assume that the glass is transparent to solar radiation and that the absorptivity of the copper tube with respect to solar radiation is 1.0. The surface of the copper tube has an emissivity of 0.13 with respect to its radiative interaction with the inner surface of the glass tube. The glass may be assumed to be opaque to thermal radiation from the copper tube with an emissivity of 1.0 on both its inner and outer surfaces. The outside surface of the glass interacts with the 25°C, 101.3 kPa, surroundings through radiation and free convection.

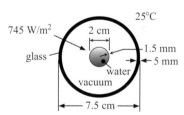

(a) Calculate the net rate of energy transfer to the water per unit length.

(b) Calculate the efficiency of the solar collector, defined as the ratio of the rate of energy transfer to the water in the heat pipe to the incident solar radiation.

12 | Heat Exchangers

A **heat exchanger** is a device that is designed to transfer thermal energy from one fluid to another. Heat exchangers are everywhere in our modern society. Nearly all thermal systems employ at least one and usually several heat exchangers. The background material related to conduction and convection, presented in Chapters 2 through 11, is required to analyze and design heat exchangers. Section 12.1 reviews the applications and types of heat exchangers that are commonly encountered. The subsequent sections provide the theory and tools required to predict and understand the performance of these devices.

12.1 Introduction to Heat Exchangers

12.1.1 Applications of Heat Exchangers

You may be unaware of just how common heat exchangers are in both residential and industrial applications. For example, you probably live in a building that is heated to a comfortable temperature in winter and cooled in the summer. Heating is usually accomplished by combusting a fuel (e.g., natural gas, propane, wood, or oil) that provides the desired thermal energy, but also produces combustion gases that can be harmful. Therefore, your heating system includes a heat exchanger that transfers thermal energy from the combustion gases to an air or water stream that can be safely circulated through the building.

The water used for showering and washing clothes and dishes is much warmer than the temperature of the water supplied by the city or drawn from a well. Some water heaters use electrical heaters, but many rely on the combustion of fuel. The hot combustion gases are heat exchanged with the city or well water in an insulated tank in order to produce domestic hot water that may be stored for later use.

Your food is preserved in a refrigerator that includes at least two heat exchangers. One of the heat exchangers, the evaporator, transfers thermal energy from the air inside the refrigerator to the evaporating refrigerant (typically a synthetic fluid such as R134a contained in a hermetically sealed circuit). The second heat exchanger, the condenser, transfers the thermal energy removed from the refrigerator (plus the thermal equivalent of the required mechanical work added by the compressor) from the condensing refrigerant to the surroundings. Additional heat exchangers are sometimes used in refrigerators to improve their efficiency.

The electricity that is provided to your home and used to run the refrigerator as well as other appliances is most likely produced using a power cycle. Some modification of a Rankine cycle is a common choice for large power plants. The Rankine cycle may use hot combustion gases to boil water in a heat exchanger called the boiler. After passing through a turbine that produces work, the spent steam is condensed in another heat exchanger called the condenser. A variety of additional heat exchangers are employed in the power plant to pre-heat the air used for combustion and to pre-heat the water returning to the boiler.

You may drive to work in an automobile that requires a variety of heat exchangers for thermal management. Heat exchangers are provided for window defrosting and cabin comfort. Also, the thermal energy that is transferred to the engine by the combustion process must be discharged to the surroundings in order to avoid over-heating. This heat exchange process is accomplished by circulating an antifreeze solution of water and ethylene glycol through passages in the engine block and then to a heat exchanger that is located in the front of the vehicle and cooled by ambient air. This heat exchanger is often called the "radiator," but this is a misnomer as radiation heat transfer actually plays a small role in its operation. The vehicle may also provide additional heat exchangers to cool the transmission fluid and the lubricating oil.

The efficiency of most energy systems is a strong function of the performance of the heat exchangers that they employ. The heat exchangers are often physically large and expensive. Therefore, proper design of the heat exchangers is critical to successful operation of energy systems. The heat exchanger analysis tools that are provided in this chapter will be helpful in identifying an optimum design for a heat exchanger application.

12.1.2 Heat Exchanger Classifications and Flow Configurations

Heat exchangers can be classified as either **direct transfer** or **indirect transfer** devices. A direct transfer heat exchanger transfers heat *directly* from one stream of fluid to another, typically through a separating wall that forms a pressure boundary. For example, Figure 12.1 illustrates a simple direct transfer heat exchanger formed by placing a smaller tube within a larger one. In this case, cold fluid passes through the smaller, inner tube while hot fluid passes through the annular space formed between the tubes. A direct transfer heat exchanger will usually operate at steady state or at least under quasi-steady conditions.

Indirect transfer heat exchangers (often referred to as **regenerators**) do not transfer thermal energy directly from one stream to the other but rather utilize a secondary, intermediate substance to accomplish this transfer process. In a regenerator, the fluid flows through a matrix of material with high heat capacity (e.g., a packed bed of lead spheres). When the hot stream flows in one direction, energy is transferred from the fluid to the matrix and when the cold stream flows in the opposite direction, the energy is transferred from the matrix to the fluid. Thus the matrix is a secondary medium in an indirect transfer device.

The rotary regenerator shown in Figure 12.2 can be used for energy recovery in buildings. The regenerator matrix in the rotary regenerator is a large wheel that has many small passages for fluid flow. The regenerator wheel rotates so that the material is exposed to hot air in one duct and cold air in the other. The regenerator material therefore warms during half of its cycle and cools during the other half. The heat transfer ultimately passes from the hot fluid to the cold fluid, but only after being stored for a time in the matrix. This device is useful for preheating outdoor cold air in the winter, as shown in Figure 12.2. It can also be used to precool outdoor hot air in the summer. See Nellis and Klein (2009) or Ackermann (1997) for more information on the analysis of regenerators.

This chapter is confined to the analysis of direct transfer heat exchangers. There are many different types of direct transfer heat exchanger configurations that exist in order to accommodate different fluid properties and operating requirements. Each of these configurations is classified based on the flow path taken by the two fluids (hot and cold) relative to one another. The temperature of the hot fluid decreases and the temperature of the cold fluid increases as they flow through the device. As a result, the local temperature difference between the hot and cold fluid within the heat exchanger and therefore the performance depends on flow configuration.

One of the simplest configurations is the **tube-in-tube** heat exchanger shown in Figure 12.1. In the specific case that is shown in Figure 12.1, the heat exchanger is operating in a **parallel flow** configuration. In parallel flow, the

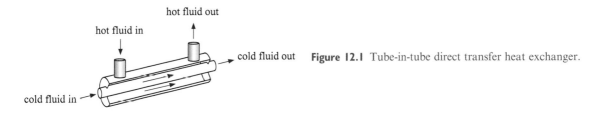

Figure 12.1 Tube-in-tube direct transfer heat exchanger.

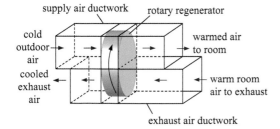

Figure 12.2 Schematic of a rotary regenerator system.

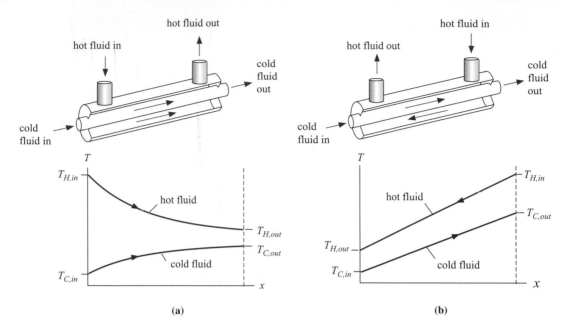

Figure 12.3 Tube-in-tube heat exchanger and the qualitative variation of the mean temperature of the hot and cold fluids when the device is plumbed in parallel flow (a) and counter flow (b).

two fluids flow in the *same direction* (i.e., parallel to each other, in this case from left to right). Figure 12.3(a) shows, qualitatively, the variation of the mean temperature of the hot and cold fluids that might result from a parallel flow configuration.

An alternative (and usually more effective) flow arrangement is **counter flow**. In a counter flow arrangement, the two fluids flow in *exactly opposite* directions. The tube-in-tube heat exchanger shown in Figure 12.3(a) can easily be reconfigured to be in counter flow; the hot fluid inlet and outlet ports are switched. The result is shown in Figure 12.3(b). Notice that the temperature distribution within the heat exchanger is strongly affected by changing from a parallel flow (a) to counter flow (b) configuration. The same heat exchanger, when configured in counter flow can perform better, as evident by the fact that the hot outlet temperature is lower and the cold outlet temperature higher. Flow configuration matters for heat exchangers because of the change in the fluid temperatures that occurs as heat is transferred to or from them.

The **plate heat exchanger** is another common type of direct transfer heat exchanger that is used for thermal energy exchange between two liquid streams. A plate heat exchanger consists of many individual plates that are stacked together, as shown in Figure 12.4. The plates are corrugated in order to form flow channels between adjacent plates. The stack of plates is manifolded (via the large holes at the ends) so that one fluid flows on one side of each plate while a second fluid flows (usually in the opposite direction, resulting in counter flow) on the other side. The plates are bolted together and sealed with gaskets. Plate heat exchangers offer several advantages. They are compact, easy to disassemble for cleaning and it is relatively easy to increase or decrease their size as needed by adding or removing plates.

Figure 12.5(a) is a cutaway view of a **shell-and-tube** heat exchanger and Figure 12.5(b) is a schematic of the shell-and-tube heat exchanger configuration. One fluid flows through a bank of tubes that is situated within a large shell while the other fluid flows within the shell and around the outer surface of the tubes. Baffles are typically placed in the shell in order to force the shell-side flow to pass *across* the tube bundle as it makes its way through the shell in a serpentine pattern, as shown in Figure 12.5(b).

Figure 12.5(b) shows a shell-and-tube heat exchanger with a single tube pass (i.e., the fluid in the tubes flows along the length of the heat exchanger just one time) and a single shell pass (i.e., the fluid in the shell passes along the length of the heat exchanger just one time). Shell-and-tube heat exchangers can be configured in a

Figure 12.4 Plate heat exchanger, from Turns (2006).

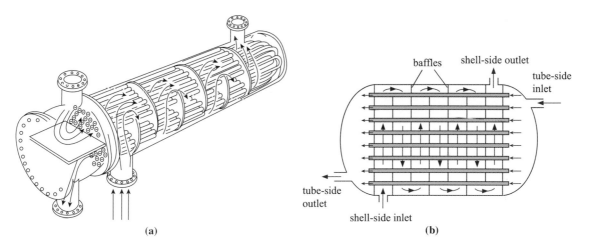

(a) (b)

Figure 12.5 (a) Cutaway view of a shell-and-tube heat exchanger from Turns (2006), and (b) a schematic of a shell-and-tube heat exchanger with a single tube pass and a single shell pass.

number of ways, depending on how the fluid is directed through the tubes and the shell. For example, Figure 12.6(a) shows a shell-and-tube heat exchanger with two tube passes and a single shell pass; a manifold forces the fluid to flow through half of the tubes in one direction (from right to left) before turning around and flowing in the opposite direction (from left to right) through the other half of the tubes. Multiple shell passes require the fluid in the shell to pass back and forth along the length of the heat exchanger. For example, Figure 12.6(b) shows a shell-and-tube heat exchanger with a single tube pass and two shell passes. Note that the shell-and-tube heat exchanger is neither a counter flow nor a parallel flow configuration, but rather has some characteristics of both. Heat exchange processes in which one of the fluids changes phase, such as during boiling or condensation, often use shell-and-tube heat exchanger designs.

The heat transfer coefficient associated with the flow of gas is usually much lower than what is associated with liquid flow. Therefore, fins are often placed on the gas side of a gas-to-liquid heat exchanger in order to increase the heat

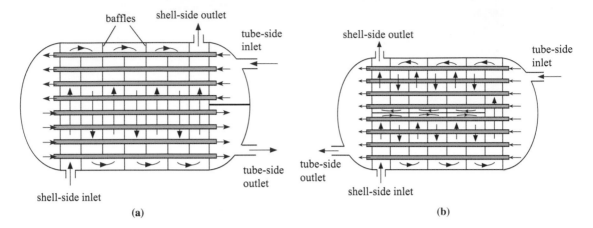

Figure 12.6 Schematic of a shell-and-tube heat exchanger with (a) two tube passes and one shell pass, and (b) one tube pass and two shell passes

Figure 12.7 Examples of cross-flow heat exchangers.

exchanger surface area and therefore compensate for the low convection heat transfer coefficient. Gas-to-liquid heat exchangers are usually configured in a **cross-flow** arrangement in which the direction of the gas flow is *perpendicular* to that of the liquid flow. An automobile "radiator" that is designed to transfer thermal energy to the surrounding air is a familiar example of a cross-flow heat exchanger. Another example of a cross-flow heat exchanger is the evaporator in a refrigeration system. Figure 12.7 shows two photographs of cross-flow heat exchangers.

The behavior of a cross-flow heat exchanger depends on how well the fluid at any position along the flow path has mixed together; that is, how well the fluid can mix in the direction *perpendicular* to the main flow. There are two limiting behaviors in this regard, referred to as **mixed** and **unmixed**; the actual behavior of a cross-flow heat exchanger will fall somewhere between these limits. However, the behavior can approach one of these two limits depending on the geometric configuration. For example, the cross-flow heat exchanger shown in Figure 12.8(a) is likely to behave as if both fluids are unmixed since the tubes (for fluid 2) and the fins (for fluid 1) limit the amount of mixing that can occur in the transverse direction. Figure 12.8(b) shows a cross-flow heat exchanger consisting of an un-finned bank of tubes. Fluid 1 in Figure 12.8(b) will approach mixed behavior as there are no geometric barriers external to the tubes that would prevent mixing, while fluid 2 will be unmixed as the tubes themselves prevent lateral mixing.

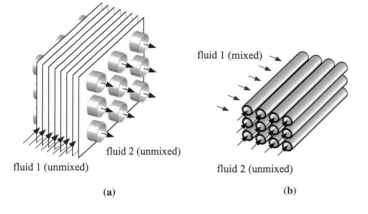

Figure 12.8 Cross-flow heat exchangers for (a) both fluids unmixed and (b) one fluid mixed and one unmixed.

fluid 1 (mixed)

fluid 2 (unmixed)

fluid 1 (unmixed)

fluid 2 (unmixed)

(a)

(b)

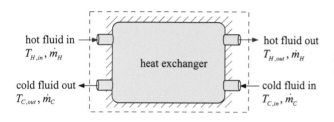

hot fluid in
$T_{H,in}, \dot{m}_H$

heat exchanger

hot fluid out
$T_{H,out}, \dot{m}_H$

cold fluid out
$T_{C,out}, \dot{m}_C$

cold fluid in
$T_{C,in}, \dot{m}_C$

Figure 12.9 An overall energy balance on a heat exchanger.

There are other heat exchanger geometries and flow configurations. This section has only reviewed the most common types. Additional information can be found in reference texts on this subject, such as Kakaç and Liu (1998) and Rohsenow *et al.* (1998).

12.1.3 Overall Energy Balance

Regardless of the geometry or flow configuration, all heat exchangers can be analyzed using an **overall energy balance**. An overall energy balance does not consider the details of the internal heat transfer process and requires only that the rate of energy transfer from the hot fluid must equal the rate of energy transfer to the cold fluid (at steady state and in the absence of losses to the surrounding). Figure 12.9 illustrates a heat exchanger and shows the system that is being considered in an overall energy balance. The hot fluid (indicated by subscript H) enters with mean temperature $T_{H,in}$ and leaves with mean temperature of $T_{H,out}$. The hot fluid flows continuously with a mass flow rate of $\dot{m}_H$. The cold fluid (indicated by subscript C) enters with mean temperature $T_{C,in}$ and leaves with mean temperature of $T_{C,out}$. The mass flow rate of the cold fluid is $\dot{m}_c$.

An overall energy balance is written for a control volume that encloses the heat exchanger, as shown in Figure 12.9. A few assumptions are typically made in a heat exchanger analysis. The kinetic and potential energy of the fluid streams are generally negligible relative to the change in the enthalpy of the fluid. Also, the heat exchanger is assumed to be operating at steady state. Finally, it is typical to assume that the heat exchanger is well insulated so that it is externally adiabatic; that is, the heat transfer with the surroundings is negligible. With these assumptions, the energy balance simplifies to

$$\dot{m}_H \, i_{H,in} + \dot{m}_C \, i_{C,in} - \dot{m}_H \, i_{H,out} - \dot{m}_C \, i_{C,out} = 0, \tag{12.1}$$

where i is the specific enthalpy. Specific enthalpy is nearly independent of pressure for liquids and totally independent of pressure for gases that conform to the Ideal Gas Law. In either case, if the specific heat capacity (c) can be assumed to be constant, then the enthalpy of the fluid can be written as the product of the constant pressure specific heat capacity and the temperature (T) relative to an arbitrary reference temperature (T_{ref}):

$$i = c\left(T - T_{ref}\right). \tag{12.2}$$

The specific heat capacity for most fluids is a function of temperature. However, in many cases the temperature dependence of specific heat capacity is small and can be neglected over the temperature range spanned by the heat exchanger. The specific heat capacity for each fluid used in the energy balance should be the average value over the range of temperatures encountered in the heat exchanger. Substituting Eq. (12.2) into Eq. (12.1) provides:

$$\dot{m}_H\, c_H(T_{H,in} - T_{H,out}) = \dot{m}_C c_C(T_{C,out} - T_{C,in}). \tag{12.3}$$

The heat exchanger is externally adiabatic, but internally heat is transferred from the hot fluid to the cold fluid at a rate $\dot{q}$:

$$\dot{q} = \dot{m}_H c_H(T_{H,in} - T_{H,out}) = \dot{m}_C c_C(T_{C,out} - T_{C,in}). \tag{12.4}$$

The product of the mass flow rate and specific heat capacity appears in the energy balance and is referred to as the **capacitance rate** of the fluid. The capacitance rates of the hot and cold fluids, $\dot{C}_H$ and $\dot{C}_C$, respectively, are given by:

$$\dot{C}_H = \dot{m}_H\, c_H \tag{12.5}$$

$$\dot{C}_C = \dot{m}_C\, c_C. \tag{12.6}$$

The capacitance rate has units of energy per time per temperature change (e.g., W/K) and can be thought of as the rate of heat transfer required to change the temperature of the stream by 1 K. A large heat transfer rate is required to change the temperature of a stream with a large capacitance rate. Substituting Eqs. (12.5) and (12.6) into Eq. (12.4) leads to:

$$\dot{q} = \dot{C}_H(T_{H,in} - T_{H,out}) \tag{12.7}$$

$$\dot{q} = \dot{C}_C(T_{C,out} - T_{C,in}). \tag{12.8}$$

Notice that Eqs. (12.7) and (12.8) alone are not sufficient to solve a heat transfer problem. In a typical heat transfer simulation problem, the inlet temperatures and flow rates are given and the outcome of the calculation is the heat transfer rate and outlet temperatures. Therefore, $T_{H,in}$, $T_{C,in}$, $\dot{C}_H$, and $\dot{C}_C$ should all be known and Eqs. (12.7) and (12.8) are two equations in three unknowns, $\dot{q}$, $T_{H,out}$, and $T_{C,out}$. In order to solve a heat exchanger problem, it is necessary to look *inside* the device and understand both the thermal resistance between the streams as well as the flow configuration and associated temperature changes. The first step in this process is the calculation of the total thermal resistance that couples the fluids as they pass through the heat exchanger. The total thermal resistance is the inverse of the heat exchanger conductance, discussed in the next section.

Example 12.1

A heat exchanger is used to cool engine oil using a flow of 40 percent ethylene glycol. The engine oil enters at $T_{H,in} = 85°C$ with mass flow rate $\dot{m}_H = 0.023$ kg/s. The ethylene glycol enters at temperature $T_{C,in} = 22°C$ with mass flow rate $\dot{m}_c = 0.012$ kg/s. The oil is cooled to $T_{H,out} = 35°C$. Determine the rate of heat transfer from the oil to the ethylene glycol and the outlet temperature of the ethylene glycol.

Known Values

Figure 1 illustrates a sketch of the problem with the known values indicated.

Example 12.1 (cont.)

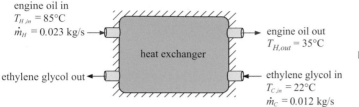

engine oil in
$T_{H,in} = 85°C$
$\dot{m}_H = 0.023$ kg/s

engine oil out
$T_{H,out} = 35°C$

heat exchanger

ethylene glycol out

ethylene glycol in
$T_{C,in} = 22°C$
$\dot{m}_C = 0.012$ kg/s

Figure 1 Heat exchanger.

Assumptions

- The heat exchanger is at steady state.
- The specific heat capacities of both fluids are assumed to be independent of temperature.
- The heat exchanger is well insulated so that the heat transfer through the jacket is minimal.

Analysis and Solution

The specific heat capacity of engine oil and 40 percent ethylene glycol are obtained from Appendix B, $c_H = 1901$ J/kg-K and $c_C = 3539$ J/kg-K, respectively. The capacitance rates of the two fluids are calculated.

$$\dot{C}_H = \dot{m}_H\, c_H = \frac{0.023\ \text{kg}}{\text{s}}\left|\frac{1901\ \text{J}}{\text{kg K}}\right|\left|\frac{\text{W s}}{\text{J}}\right| = 43.72\frac{\text{W}}{\text{K}} \tag{1}$$

$$\dot{C}_C = \dot{m}_C\, c_C = \frac{0.012\ \text{kg}}{\text{s}}\left|\frac{3539\ \text{J}}{\text{kg K}}\right|\left|\frac{\text{W s}}{\text{J}}\right| = 42.47\frac{\text{W}}{\text{K}}. \tag{2}$$

Equation (12.7) is used to compute the rate of heat transfer within the heat exchanger.

$$\dot{q} = \dot{C}_H(T_{H,in} - T_{H,out}) = \frac{43.72\ \text{W}}{\text{K}}\left|\frac{(85-35)\text{K}}{}\right| = \boxed{2186\ \text{W (2.186 kW)}}. \tag{3}$$

Equation (12.8) is rearranged and used to determine the cold fluid outlet temperature:

$$T_{C,out} = T_{C,in} + \frac{\dot{q}}{\dot{C}_C} = 22°C + \frac{2186\ \text{W}}{}\left|\frac{\text{K}}{42.47\ \text{W}}\right| = \boxed{73.48°C}. \tag{4}$$

Discussion

Notice that the geometry and flow arrangement of the heat exchanger did not need to be specified for this problem. Because three of the four fluid temperatures were specified, it was possible to use an overall energy balance to determine the rate of heat transfer and the final fluid temperature. In a more common heat exchanger calculation, it would also be necessary to determine the geometry of the heat exchanger that is required to actually provide this level of performance.

12.1.4 Heat Exchanger Conductance

In order to analyze a heat exchanger, it is necessary to characterize the total thermal resistance that separates the two fluids. The inverse of the total thermal resistance is commonly referred to as the heat exchanger **conductance** and given the symbol UA. The process of estimating the conductance relies on the thermal resistance concepts introduced in Section 2.2. The fin efficiency functions discussed in Section 3.3 will be needed for finned heat exchangers. The correlations provided for internal flow in Chapter 9 (for single-phase flow) and Chapter 11 (for two-phase flow) will likely be required to evaluate the heat transfer coefficients. It is often the case that the geometry of the heat exchanger is complicated relative to the simple ducts that are considered in the internal flow correlations; in this case the reader is referred to the tube bank and compact heat exchanger correlations discussed in Sections 12.1.5 and 12.1.6, respectively.

The conductance is computed by considering all of the thermal resistances that separate the two fluids. In the absence of any fluid temperature change (e.g., for condensing or evaporating flows or flows with very high capacitance rates), the conductance could be used in conjunction with the inlet temperatures directly to determine the heat transfer rate. However, in most heat exchangers the temperature change of one or both fluids in the flow direction is significant. (Changing fluid temperatures is, after all, the purpose of most heat exchangers.) Therefore, the conductance must be determined and used in conjunction with one of the heat exchanger solutions discussed in Sections 12.2 and 12.3.

Fouling Resistance

Heat exchangers tend to operate for prolonged periods of time and eventually heat transfer surfaces that are clean when a heat exchanger is installed will become "fouled". Fouling refers to any type of build up or contamination that has the effect of increasing the thermal resistance between the underlying surface and the adjacent fluid. One example of fouling is the scale that builds up when untreated water flows in contact with metal surfaces for a long period of time. Fouling of heat transfer surfaces can cause a dramatic reduction in the performance of a heat exchanger relative to its installed performance.

The effect of fouling can be represented by the addition of a **fouling resistance** to the thermal resistance network that is used to estimate the conductance of the heat exchanger. The fouling resistance (R_f) is calculated using a **fouling factor** (R_f'') according to:

$$R_f = \frac{R_f''}{A_s}, \tag{12.9}$$

where A_s is the area of the surface that is being fouled. Notice the similarity between the fouling resistance and the contact resistance discussed in Section 2.2.5. The fouling factor, like the area-specific contact resistance, depends on a variety of factors and therefore is typically tabulated for different situations using the results of experiments. Extensive tables of fouling factor values are provided in Kakaç and Liu (1998) and Rohsenow et al. (1998). Much of the information in these sources has been compiled in the EES FoulingFactor function.

Example 12.2

A tube-in-tube heat exchanger like the one shown in Figure 12.1 is used to cool engine oil by transferring heat to a flow of 40 percent ethylene glycol-water solution. The engine oil enters the heat exchanger at $T_{H,in} = 85°C$ and flows through the annular space that is formed between the inner and outer tubes while the ethylene glycol flows through the inner tube. The total length of the heat exchanger is $L = 12.1$ m and the inner tube is made of 304 stainless steel with outer diameter $D_o = 1.0$ cm and wall thickness $th = 1.1$ mm. The heat transfer coefficient between the engine oil and the inner tube surface is $\bar{h}_H = 350$ W/m²-K. The ethylene glycol enters the heat exchanger at $T_{C,in} = 22°C$ and flows through the center tube in a counterflow configuration with mass flow rate of $\dot{m}_c = 0.012$ kg/s. The fouling factor associated with the flow of the engine oil is $R_{f,H}'' = 0.000179$ K-m²/W and the fouling factor associated with the ethylene glycol is $R_{f,C}'' = 0.00043$ K-m²/W. Determine the conductance of the heat exchanger.

Known Values

Figure 1 illustrates a sketch of the problem with the known values indicated.

Example 12.2 (cont.)

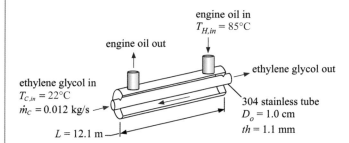

Figure 1 Heat exchanger.

engine oil heat transfer coefficient, $\bar{h}_H = 350$ W/m²-K
engine oil fouling factor, $R''_{f,H} = 0.000179$ K-m²/W
ethylene glycol fouling factor, $R''_{f,C} = 0.00043$ K-m²/W

Assumption

- The properties of the tube and the ethylene glycol at room temperature from Appendices A and B can be used.

Analysis and Solution

The conductance is the inverse of the total thermal resistance separating the two fluids. The resistance is composed of a convection resistance between the oil and the outer surface of the inner tube ($R_{conv,H}$), a fouling resistance on the oil side ($R_{f,H}$), a conduction resistance through the tube (R_{cond}), a fouling resistance on the ethylene glycol side ($R_{f,C}$), and a convection resistance between the ethylene glycol and the inner surface of the inner tube ($R_{conv,C}$). These resistances are all in series.

The resistance to convection on the hot side can be computed according to:

$$R_{conv,H} = \frac{1}{\pi \bar{h}_H D_o L} = \frac{1}{\pi} \left| \frac{m^2\ K}{350\ W} \right| \frac{1}{0.01\ m} \left| \frac{1}{12.1\ m} \right| = 0.00752 \frac{K}{W}. \tag{1}$$

The resistance to fouling on the hot side is computed according to:

$$R_{f,H} = \frac{R''_{f,H}}{\pi D_o L} = \frac{1}{\pi} \left| \frac{0.000179\ m^2\ K}{W} \right| \frac{1}{0.01\ m} \left| \frac{1}{12.1\ m} \right| = 0.000471 \frac{K}{W}. \tag{2}$$

The conductivity of 304 stainless steel is obtained from Appendix A, $k_t = 14.9$ W/m-K. The inner diameter of the tube and the resistance to conduction radially through the tube are computed:

$$D_i = D_o - 2\,th = 0.01\ m - 2\,(0.0011\ m) = 0.0078\ m \tag{3}$$

$$R_{cond} = \frac{\ln\left(\dfrac{D_o}{D_i}\right)}{2\pi L k_t} = \frac{\ln\left(\dfrac{0.010}{0.0078}\right)}{2\pi} \left| \frac{1}{12.1\ m} \right| \frac{m\ K}{14.9\ W} = 0.000219 \frac{K}{W}. \tag{4}$$

The resistance to fouling on the cold side is computed according to:

$$R_{f,C} = \frac{R''_{f,C}}{\pi D_i L} = \frac{1}{\pi} \left| \frac{0.00043\ m^2\ K}{W} \right| \frac{1}{0.0078\ m} \left| \frac{1}{12.1\ m} \right| = 0.00145 \frac{K}{W}. \tag{5}$$

The heat transfer coefficient on the cold side must be evaluated based on the flow conditions and geometry using correlations presented in Chapter 9. The properties of a 40 percent ethylene glycol–water solution (for

Continued

Example 12.2 (cont.)

solution at 25°C) are obtained from Appendix B: $\rho_C = 1049$ kg/m^3, $\mu_C = 0.002424$ Pa-s, $k_C = 0.4291$, $Pr_C = 19.99$. The cross-sectional area for flow on the cold side is:

$$A_{c,C} = \pi \frac{D_i^2}{4} = \frac{\pi}{4}(0.0078 \text{ m})^2 = 4.78 \times 10^{-5} \text{ m}^2. \tag{6}$$

The mean velocity on the cold side is:

$$u_{m,C} = \frac{\dot{m}_C}{A_{c,C}\rho_C} = \frac{0.012 \text{ kg}}{\text{s}} \left| \frac{}{4.78 \times 10^{-5} \text{ m}^2} \right| \frac{\text{m}^3}{1049 \text{ kg}} = 0.239 \frac{\text{m}}{\text{s}}. \tag{7}$$

The Reynolds number associated with the cold side flow is:

$$Re_C = \frac{u_{m,C} \, D_i \rho_C}{\mu_C} = \frac{0.239 \text{ m}}{\text{s}} \left| \frac{0.0078 \text{ m}}{} \right| \frac{1049 \text{ kg}}{\text{m}^3} \left| \frac{}{0.002424 \text{ Pa s}} \right| \left| \frac{\text{Pa m}^2}{\text{N}} \right| \frac{\text{N s}^2}{\text{kg m}} = 808. \tag{8}$$

Because the Reynolds number is less than 2300, the flow will be laminar. The Graetz number is computed according to:

$$Gz_C = \frac{D_i \, Re_C \, Pr_C}{L} = \frac{0.0078 \text{ m}}{} \left| \frac{808}{} \right| \frac{19.99}{} \left| \frac{}{12.1 \text{ m}} \right| = 10.41. \tag{9}$$

The average Nusselt number is computed using the correlation for laminar, developing flow in a round tube with a constant surface temperature, Eq. (9.72):

$$\overline{Nu}_C = 3.66 + \frac{\left(0.049 + \dfrac{0.020}{Pr_C}\right) Gz_C^{1.12}}{(1 + 0.065 \, Gz_C^{0.7})} = 3.66 + \frac{\left(0.049 + \dfrac{0.020}{19.99}\right)(10.41)^{1.12}}{\left[1 + 0.065(10.41)^{0.7}\right]} = 4.18. \tag{10}$$

The average heat transfer coefficient on the cold side is:

$$\bar{h}_C = \frac{\overline{Nu}_c \, k_C}{D_i} = \frac{4.18}{} \left| \frac{0.4291 \text{ W}}{\text{m K}} \right| \frac{}{0.0078 \text{ m}} = 229.8 \frac{\text{W}}{\text{m}^2 \text{ K}}. \tag{11}$$

The resistance to convection on the cold side is:

$$R_{conv,C} = \frac{1}{\pi \bar{h}_C \, D_i \, L} = \frac{1}{\pi} \left| \frac{\text{m}^2 \text{ K}}{229.7 \text{ W}} \right| \frac{}{0.0078 \text{ m}} \left| \frac{}{12.1 \text{ m}} \right| = 0.0147 \frac{\text{K}}{\text{W}}. \tag{12}$$

The total resistance is:

$$R_{total} = R_{conv,H} + R_{f,H} + R_{cond} + R_{f,C} + R_{conv,C}$$

$$= (0.00752 + 0.000471 + 0.000219 + 0.00145 + 0.0147)\frac{\text{K}}{\text{W}} = 0.0243 \frac{\text{K}}{\text{W}}. \tag{13}$$

The conductance is:

$$UA = \frac{1}{R_{total}} = \frac{W}{0.0243 \text{ K}} = \boxed{41.09 \frac{\text{W}}{\text{K}}}. \tag{14}$$

Discussion

The size of the individual resistances that together make up the total resistance can be used to understand their relative importance. In this problem, the two convection resistances dominate and the others are much less important.

Example 12.2 (cont.)

The mean fluid temperature should be used to determine fluid properties. In this example, the outlet temperature of the glycol solution was not specified and property data at 25°C were used. A more accurate solution would use the estimate of the conductance obtained with these data to calculate the outlet temperature of the glycol. Then, the calculations would need to be repeated using the new outlet temperature to determine the conductance. Calculations of this type are necessarily iterative.

The conductance is not, by itself, sufficient to solve this problem because the driving temperature difference within the heat exchanger is not constant. The counter-flow heat exchanger will have a temperature distribution similar to what is shown in Figure 12.3(b). For example, it would **NOT** be correct to evaluate the heat transfer according to:

$$\dot{q} = \frac{(T_{H,in} - T_{C,in})}{R_{total}} = \frac{(85 - 22)\ \text{K}}{0.0243\ \text{K}} \left| \frac{\text{W}}{} \right. = 2589\ \text{W}. \tag{15}$$

Careful examination of Figure 12.3(b) shows that the stream-to-stream temperature difference corresponding to the inlet temperatures, $85 - 22 = 63$ K, does not exist *anywhere* within the heat exchanger. The temperature difference everywhere in the heat exchanger is less than this value and therefore 2589 W is an unrealistic overestimate of the rate of heat transfer. In order to obtain a solution to the problem it is necessary to simultaneously solve the differential equations corresponding to energy balances on both the hot and cold flows; these energy balances are coupled via a rate equation that involves the total resistance (or conductance). Methods to do these calculations are provided in following sections.

12.1.5 Flow across Tube Banks

The geometry associated with Example 12.2 is straightforward; one fluid flows in a tube and the other in the annular space between two tubes. The internal flow correlations discussed in Chapter 9 could be applied to determine the heat transfer coefficients and therefore the conductance. However, most heat exchanger cores are not as straightforward to analyze and involve flow through complex passages that are not well characterized by the correlations discussed in either Chapters 8 or 9. One common example is flow across a bank of tubes through which a second tube is flowing.

Flow across a bank of tubes is shown in Figure 12.10; the correlations in this section specifically refer to the flow of the fluid passing over the outer surface of the tubes. The flow through the tubes themselves can be

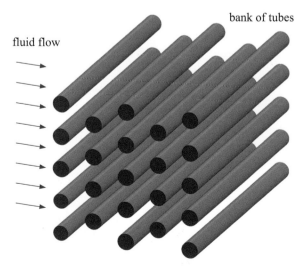

bank of tubes

fluid flow

Figure 12.10 Flow across a bank of tubes.

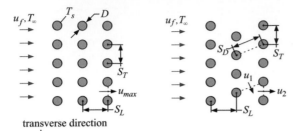

Figure 12.11 Bank of tubes in an inline (left) and staggered (right) arrangement.

analyzed using the internal flow correlations found in Chapter 9. Flow across a bank of tubes is encountered often in the analysis of shell-and-tube heat exchangers in which one fluid flows through the tube and a second one around them.

A bank of tubes is characterized by the tube diameter (D) as well as the longitudinal pitch (S_L, the tube-to-tube spacing in the flow direction) and the transverse pitch (S_T, the tube-to-tube spacing perpendicular to the flow direction). The tubes may be placed in an inline arrangement, as shown in Figure 12.11 (left), where each tube is directly behind the tube in the adjacent row. Alternatively, the tubes may be placed in a staggered arrangement, as shown in Figure 12.11 (right), where each row is offset by half of the transverse pitch (S_T) relative to its neighbors. The flow approaches the tube bundle with a uniform velocity (u_f) and temperature (T_∞).

The characteristic length scale used for the tube bank correlations is the diameter of the tube. The characteristic velocity is the maximum velocity (u_{max}) that is attained by the flow (*not* the approach velocity). For the inline array of tubes, the maximum velocity must occur between adjacent tubes, as shown in Figure 12.11 (left). Assuming that the flow is incompressible, the maximum velocity is given by:

$$u_{max} = u_f \frac{S_T}{(S_T - D)} \text{ for inline array.} \tag{12.10}$$

For the staggered array of tubes, the maximum velocity may either occur in the plane that separates tubes in adjacent rows, u_1 in Figure 12.11 (right), or in the same row, u_2 in Figure 12.11 (right) depending on the spacing. If the spacing is such that

$$(S_D - D) < \frac{(S_T - D)}{2} \tag{12.11}$$

then the maximum velocity will be u_1:

$$u_{max} = u_1 = u_f \frac{S_T}{2(S_D - D)} \text{ for staggered array with } (S_D - D) < \frac{(S_T - D)}{2}. \tag{12.12}$$

Otherwise, the maximum velocity will be u_2:

$$u_{max} = u_2 = u_f \frac{S_T}{(S_T - D)} \text{ for staggered array with } (S_D - D) > \frac{(S_T - D)}{2}, \tag{12.13}$$

where S_D is the center-to-center distance of tubes in adjacent rows:

$$S_D = \sqrt{S_L^2 + \left(\frac{S_T}{2}\right)^2}. \tag{12.14}$$

The Reynolds number is computed according to:

$$Re_D = \frac{\rho D u_{max}}{\mu}. \tag{12.15}$$

Note that the fluid properties for the correlations presented here should be computed at the average free stream temperature of the fluid passing through the array rather than the film temperature.

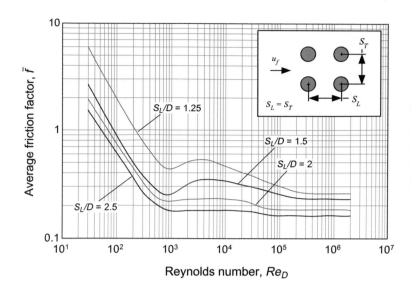

Figure 12.12 Average friction factor for square array ($S_L = S_T$) of inline tubes as a function of Reynolds number for various values of normalized longitudinal pitch, S_L/D (from Zukauska, 1987).

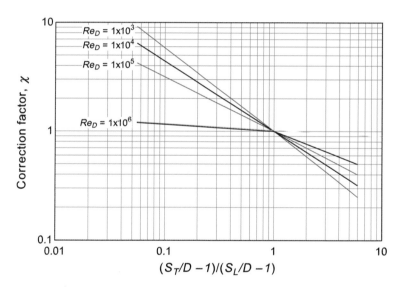

Figure 12.13 Correction factor for nonsquare inline arrays as a function of $(S_T/D - 1)/(S_L/D - 1)$ for various values of the Reynolds number (from Zukauska, 1987).

The Friction Factor

The pressure drop (Δp) associated with the flow as it moves progressively through rows of tubes is typically a critical parameter of interest as this pressure drop must be overcome using a fan, blower, or pump of some type and therefore contributes to the operating cost of the device. The pressure drop is correlated using the average friction factor ($\bar{f}$) according to:

$$\bar{f} = \frac{2 \, \Delta p}{\rho \, u_{max}^2 \, N_L},$$

(12.16)

where N_L is the number of tube rows in the longitudinal direction. The average friction factor for a square array (i.e., one where $S_L = S_T$) of inline tubes is shown in Figure 12.12 as a function of Reynolds number for various values of the normalized longitudinal pitch (S_L/D). For inline arrays that are not square, the pressure drop should be multiplied by the correction factor obtained from Figure 12.13.

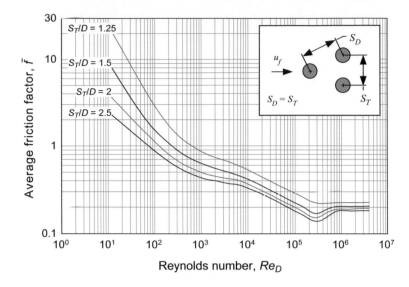

Figure 12.14 Average friction factor for staggered array of tubes with $S_T = S_D$ (i.e., a hexagonal close packed array) is shown in Figure 12.14 as a function of Reynolds number for various values of normalized transverse pitch, S_T/D (from Zukauska, 1987).

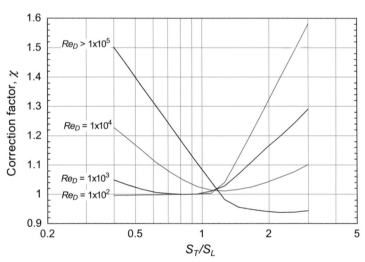

Figure 12.15 Correction factor for staggered arrays as a function of S_T/S_L for various values of the Reynolds number (from Zukauska, 1987).

The average friction factor for a staggered array with $S_T = S_D$ (i.e., a hexagonal close packed array) is shown in Figure 12.14 as a function of Reynolds number for various values of the normalized longitudinal pitch (S_L/D). The correction factor for cases where $S_T \neq S_D$ is given in Figure 12.15.

The Nusselt Number

The average heat transfer coefficient is correlated with an average Nusselt number, defined as:

$$\overline{Nu}_D = \frac{\bar{h}D}{k}.$$ (12.17)

The average Nusselt number for an array of tubes with $N_L >= 20$ (i.e., with at least 20 rows of tubes in the longitudinal direction) is given by (Zukauskas, 1972, as presented in Bergman *et al.*, 2011):

$$\overline{Nu}_D = C\,Re^m\,Pr^{0.36}\left(\frac{Pr}{Pr_s}\right)^{1/4} \text{ for } 10 < Re_D < 2 \times 10^6 \text{ and } 0.7 < Pr < 500,$$ (12.18)

Table 12.1 Constants for tube bank correlations with $N_L \geq 20$ to be used in Eq. (12.18).

Staggered array of tubes				Inline array of tubes		
Reynolds number range	C		m	Reynolds number range	C	m
$10 < Re_D < 1 \times 10^2$	0.90		0.40	$10 < Re_D < 1 \times 10^2$	0.80	0.40
$1 \times 10^2 < Re_D < 1 \times 10^3$	treat as single cylinder			$1 \times 10^2 < Re_D < 1 \times 10^3$	treat as single cylinder	
$1 \times 10^3 < Re_D < 2 \times 10^5$	$0.35\left(\dfrac{S_T}{S_L}\right)^{0.2}$ for $\left(\dfrac{S_T}{S_L}\right) < 2$		0.60	$1 \times 10^3 < Re_D < 2 \times 10^5$	0.27 for $\left(\dfrac{S_T}{S_L}\right) > 0.7$	0.63
	0.40 for $\left(\dfrac{S_T}{S_L}\right) > 0.2$					
$2 \times 10^5 < Re_D < 2 \times 10^6$	$0.027\left(\dfrac{S_T}{S_L}\right)^{0.2}$ for $Pr \approx 0.7$		0.80	$2 \times 10^5 < Re_D < 2 \times 10^6$	0.021	0.84
	$0.031\left(\dfrac{S_T}{S_L}\right)^{0.2}$ for $Pr > 0.7$					

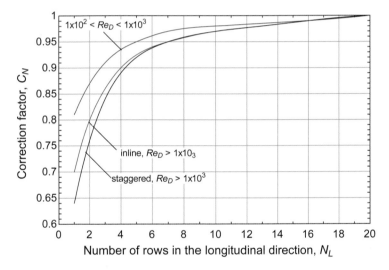

Figure 12.16 Correction factor for the average Nusselt number in arrays with $N_L < 20$.

where the constants C and m depend on the Reynolds number and whether the tubes are arranged in an inline or staggered array, as given by Table 12.1. The correction factor $(Pr/Pr_s)^{0.25}$ is the ratio of the Prandtl number evaluated at the bulk temperature of the fluid to the value evaluated at the surface temperature of the tube.

For tube arrays that have fewer than 20 rows of tubes in the longitudinal direction, the average Nusselt number will be less than the value computed according to Eq. (12.18) and therefore should be corrected using the correction factor found in Figure 12.16.

Procedures that return the average heat transfer and pressure drop associated with flow across inline and staggered tube banks are available in EES based on the correlations presented in this section as well as other information. These are provided in the External Flow – Dimensional portion of the Convection Library as the procedures External_Flow_Inline_Bank and External_Flow_Staggered_Bank.

The fluid passing over the tubes will change temperature as it either gains or loses thermal energy through heat transfer with the tube walls. Therefore, it is not generally appropriate to use the inlet temperature of the fluid to compute the heat transfer rate. Instead the heat exchanger solutions presented in subsequent sections should be used.

Example 12.3

A bank of cylindrical electrical heaters is used to heat a flow of air. The air flow is perpendicular to the axes of the heaters. The power to each heater is adjusted so that the surface temperature is $T_s = 180°C$. The air enters the bank with a uniform velocity $u_f = 22$ m/s at atmospheric pressure and temperature $T_\infty = 20°C$. The heaters are arranged in a staggered array. There are $N_L = 8$ heaters in the longitudinal direction and $N_T = 10$ heaters in the transverse direction for the first row and every subsequent odd row. Even rows have 9 heaters. The heaters have length $L = 10$ cm and diameter $D = 1$ cm. The transverse pitch is $S_T = 2.5$ cm and the longitudinal pitch is $S_L = 1.5$ cm.

   Determine the pressure drop experienced by the air as it passes through the heater bank and estimate the total rate of heat transfer from the heaters to the air.

Known Values

Figure 1 illustrates a sketch of the problem with the known values shown.

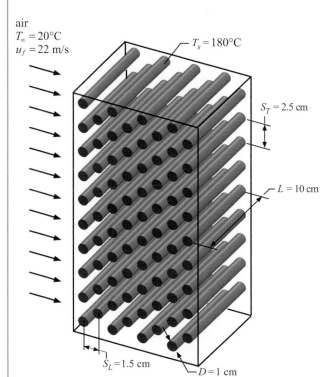

air
$T_\infty = 20°C$
$u_f = 22$ m/s

$T_s = 180°C$

$S_T = 2.5$ cm

$L = 10$ cm

$S_L = 1.5$ cm

$D = 1$ cm

Figure 1 Sketch of the problem with the known values indicated.

Assumptions

- The temperature of the air does not change substantially as it flows through the heater.
- The effect of the walls can be neglected.
- The tubes are smooth.

Example 12.3 (cont.)

Analysis and Solution

The properties of air are estimated using Appendix C at the inlet temperature: $\rho = 1.20$ kg/m^3, $k = 0.025$ W/m-K, $\mu = 1.83 \times 10^{-5}$ Pa-s, and $Pr = 0.73$. The center-to-center distance between tubes in adjacent rows is computed:

$$S_D = \sqrt{S_L^2 + \left(\frac{S_T}{2}\right)^2} = \sqrt{(0.0015 \text{ m})^2 + \left(\frac{0.0025 \text{ m}}{2}\right)^2} = 0.0195 \text{ m } (1.95 \text{ cm}).\tag{1}$$

The two sides of the criteria provided in Eq. (12.11) are evaluated in order to determine where the maximum velocity occurs within the array:

$$S_D - D = 0.0195 \text{ m} - 0.01 \text{ m} = 0.0095 \text{ m } (0.95 \text{ cm})\tag{2}$$

and

$$\frac{(S_T - D)}{2} = \frac{(0.025 \text{ m} - 0.01 \text{ m})}{2} = 0.0075 \text{ m } (0.75 \text{ cm}).\tag{3}$$

Because Eq. (2) provides a larger value than Eq. (3), the maximum velocity will occur between adjacent tubes in the same row, location 2 in Figure 12.11 (right):

$$u_{max} = u_2 = u_f \frac{S_T}{(S_T - D)} = (22 \text{ m/s}) \frac{0.025 \text{ m}}{(0.025 \text{ m} - 0.01 \text{ m})} = 36.7 \text{ m/s}.\tag{4}$$

The Reynolds number associated with the flow situation is:

$$Re_D = \frac{\rho D u_{max}}{\mu} = \frac{1.20 \text{ kg}}{\text{m}^3} \left|\frac{0.01 \text{ m}}{}\right|\frac{36.7 \text{ m}}{\text{s}}\left|\frac{}{18.3 \times 10^{-5} \text{ Pa s}}\right|\left|\frac{\text{Pa m}^2}{\text{N}}\right|\frac{\text{N s}^2}{\text{kg m}} = 2.4 \times 10^4.\tag{5}$$

According to Figure 12.14, the average friction factor is approximately $\bar{f} = 0.25$ and according to Figure 12.15, the correction factor is approximately $\chi = 1.06$. The pressure drop is computed using Eq. (12.16) corrected by χ:

$$\Delta p = \chi \bar{f} N_L \frac{\rho u_{max}^2}{2} = \frac{(1.06)(0.25)(8)}{2}\left|\frac{1.20 \text{ kg}}{\text{m}^3}\right|\frac{(36.7)^2 \text{ m}^2}{\text{s}^2}\left|\frac{\text{N s}^2}{\text{kg m}}\right|\frac{\text{Pa m}^2}{\text{N}} = \boxed{1720 \text{ Pa}}.\tag{6}$$

According to Table 12.1, the constant for a staggered array can be calculated according to:

$$C = 0.35\left(\frac{S_T}{S_L}\right)^{0.2} = 0.35\left(\frac{0.025 \text{ m}}{0.015 \text{ m}}\right)^{0.2} = 0.388\tag{7}$$

The exponent should be $m = 0.60$. Using Figure 12.16, the correction factor associated with the number of rows is $C_N = 0.96$. The Prandtl number of air at $T_s = 180°C$ is obtained from Appendix C, $Pr_s = 0.701$. The average Nusselt number is obtained using Eq. (12.18) corrected by C_N:

$$\overline{Nu}_D = C_N \, C \, Re^m \, Pr^{0.36}\left(\frac{Pr}{Pr_s}\right)^{1/4}$$

$$= (0.96)(0.388)\left(2.42 \times 10^4\right)^{0.6}(0.729)^{0.36}\left(\frac{0.729}{0.701}\right)^{0.25} = 143.2.\tag{8}$$

Continued

Example 12.3 (cont.)

The average heat transfer coefficient is computed according to:

$$\bar{h} = \frac{\overline{Nu}_D \, k}{D} = \frac{143.2 \left|0.0251 \text{ W}\right|}{\left| \text{m K} \right| 0.01 \text{ m}} = 359.9 \frac{\text{W}}{\text{m}^2 \text{ K}}. \tag{9}$$

The total number of tubes is calculated according to:

$$N_{tubes} = N_L N_T - \frac{N_L}{2} = (8)(10) - \frac{8}{2} = 76 \text{ tubes}. \tag{10}$$

where the second term recognizes that every odd row has $N_T - 1$ tubes in it. The rate of heat transfer is given by:

$$\dot{q} - \bar{h} \, N_{tubes} \, \pi \, D \, L \, (T_s - T_\infty)$$

$$= \frac{359.9 \text{ W}}{\text{m}^2 \text{ K}} \left|76\right| \frac{\pi}{-} \left|0.01 \text{ m}\right| 0.10 \text{ m} \left|(180 - 20) \text{ K}\right| = \boxed{13{,}750 \text{ W} (13.75 \text{ kW})}. \tag{11}$$

Discussion

The problem could be accomplished using the Heat Transfer Library in EES. The inputs are entered and then the procedure External_Flow_Staggered_Bank is accessed by selecting Function Info from the Options menu and then selecting Convection from the Heat Transfer and Fluid Flow Library. In the bottom panel select External Flow – Dimensional from the drop down menu and scroll to the correlation for a staggered bank of tubes, as shown in Figure 2.

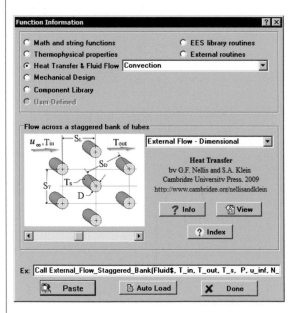

Figure 2 External_Flow_Staggered_Bank function dialog.

The procedure provides the pressure drop and heat transfer coefficient. Equations (10) and (11) are used to compute the rate of heat transfer.

```
$UnitSystem SI Mass J K Pa
L = 10 [cm]*Convert(cm,m)          "length of heater"
D = 1 [cm]*Convert(cm,m)           "diameter of heater"
```

Example 12.3 (cont.)

> S_T=2.5 [cm]***Convert**(cm,m)   "transverse pitch"
> S_L = 1.5 [cm]***Convert**(cm,m)   "longitudinal pitch"
> N_L = 8 [-]   "number of rows in transverse direction"
> N_T= 10 [-]   "number of rows in long. direction in odd rows"
> u_f=22 [m/s]   "approach velocity of air"
> T_infinity=**ConvertTemp**(C,K,20 [C])   "inlet temperature of air"
> T_s=**ConvertTemp**(C,K,180 [C])
>
>
> *CALL* **External_Flow_Staggered_Bank**('Air', T_infinity, T_infinity, T_s,&
>   Po#, u_f, N_L, D ,S_T,S_L: h_bar, Dp, Nusselt_bar, Re_D)
>
>
> N_tubes = N_L*N_T-N_L/2   "number of tubes"
> q_dot=h_bar*pi*D*L*N_tubes*(T_s-T_infinity)   "rate of heat transfer"

Solving provides $\Delta p = 1780$ Pa and $\dot{q} = 13{,}740$ W; the slight differences from the values calculated manually are related to reading values off of charts.

It is a good idea to check the assumption that the air temperature does not change substantially, allowing Eq. (11) to be used to compute the rate of heat transfer. The mass flow rate of the air is estimated according to:

$$\dot{m} = u_f S_T N_T L \rho = \frac{22 \text{ m}}{\text{s}} \left| \frac{0.025 \text{ m}}{} \right| 10 \left| \frac{0.10 \text{ m}}{} \right| \frac{1.20 \text{ kg}}{\text{m}^3} = 0.662 \frac{\text{kg}}{\text{s}}. \tag{12}$$

The constant pressure specific heat capacity of the air is estimated from Appendix C, $c = 1004$ J/kg-K. An energy balance on the air provides the outlet temperature:

$$T_{out} = T_\infty + \frac{\dot{q}}{\dot{m} \, c} = 20°C + \frac{13{,}750 \text{ W}}{\text{s}} \left| \frac{\text{s}}{0.662 \text{ kg}} \right| \frac{\text{kg K}}{1004 \text{ J}} \left\| \frac{\text{J}}{\text{W s}} \right. = 40.67°C. \tag{13}$$

The temperature of the air increases by more than 20 K which reduces the temperature difference driving the convection process from 160 K at the inlet to only 140 K at the outlet. This effect will tend to reduce the heat transfer rate relative to what was calculated using Eq. (11). The proper way to account for this type of fluid temperature change that inevitably occurs in a heat exchanger is discussed in the subsequent sections of this chapter.

12.1.6 Compact Heat Exchanger Correlations

Most heat exchanger cores are quite complex geometrically. For example, consider the finned tube heat exchanger shown in Figure 12.8(a). The flow path associated with fluid 1 is complicated because it resembles internal flow between adjacent fins but it also has aspects of external flow over a tube bank. No correlation provided in either Chapter 9 or Section 12.1.5 directly corresponds to this situation. This type of heat exchanger is referred to as a compact heat exchanger because of the large surface area to volume ratio on the finned side of the heat exchanger.

Compact heat exchangers are often used in applications in which liquid and gas are transferring energy. The compact heat exchanger geometry can be found on the gas side of such a device because the gas heat transfer coefficient is typically small and therefore the heat exchanger performance is dramatically improved by the additional surface area associated with the fins.

Experimental data are required to provide accurate models of compact heat exchanger geometries. Cross-flow heat exchangers are common and therefore there exists a great deal of data for a set of specific compact heat exchanger cores. Experimental measurements of gas-side heat transfer coefficient and pressure drop data have been correlated by Kays and London (1984) for many of these compact heat exchanger core geometries. These heat transfer data are typically presented in terms of the Colburn j_H factor, defined as

$$j_H = St\,Pr^{2/3}, \tag{12.19}$$

where St is the Stanton number and Pr is the Prandtl number. The Stanton number is defined as

$$St = \frac{\bar{h}_{out}}{G\,c}, \tag{12.20}$$

where c is the specific heat capacity and G is the mass velocity. The mass velocity is defined as the mass flow rate per free flow area (A_{ff}):

$$G = \frac{\dot{m}}{A_{ff}}. \tag{12.21}$$

Note that A_{ff} in Eq. (12.21) is the free flow area available to the fluid after it enters the core. The Colburn j_H factor is correlated in terms of the Reynolds and Prandtl numbers, defined as

$$Re = \frac{G\,D_h}{\mu} \tag{12.22}$$

$$Pr = \frac{c\,\mu}{k}, \tag{12.23}$$

where μ is the kinematic viscosity and k is the thermal conductivity of the gas. The hydraulic diameter of the flow channels (D_h) as well as other details of the geometry such as the outer diameter of the tubes (D_{out}), number of fins per length (fpl), fin thickness (th_{fin}), ratio of free flow to frontal area ($\sigma = A_{ff}/A_{front}$), ratio of gas-side heat transfer area to total volume ($\alpha = A_{s,out}/V$), and the ratio of finned to total surface area on the gas-side ($A_{s,fin}/A_{s,out}$) are provided with each correlation for a specific core geometry.

A gas-side friction factor, f, is also provided in the Kays and London compact heat exchanger database. Using the friction factor, Kakaç and Liu (1998) show that the gas-side pressure drop, Δp, for finned heat exchangers can be estimated according to:

$$\Delta p = \frac{G^2}{2\rho_{in}}\left[f\frac{4L_{flow}}{D_h}\frac{\rho_{in}}{\bar{\rho}} + (1+\sigma^2)\left(\frac{\rho_{in}}{\rho_{out}} - 1\right)\right], \tag{12.24}$$

where L_{flow} is the length of flow passage in the direction of the gas flow, ρ_{in} and ρ_{out} are the gas densities at the inlet and outlet of the heat exchanger, respectively, and $\bar{\rho}$ is the average gas density, defined as:

$$\frac{1}{\bar{\rho}} = \frac{1}{2}\left(\frac{1}{\rho_{in}} + \frac{1}{\rho_{out}}\right). \tag{12.25}$$

The second term in the brackets in Eq. (12.24) accounts for the effects of density-induced flow acceleration on the pressure drop. For plate-fin exchangers, Kakaç and Liu recommend including additional terms for entrance and exit effects, so that the total pressure drop is calculated according to:

$$\Delta p = \frac{G^2}{2\rho_{in}}\left[(k_c + 1 - \sigma^2) + 2\left(\frac{\rho_{in}}{\rho_{out}} - 1\right) + f\frac{4L_{flow}}{D_h}\frac{\rho_{in}}{\bar{\rho}} - (1 - k_e - \sigma^2)\left(\frac{\rho_{in}}{\rho_{out}}\right)\right], \tag{12.26}$$

where k_c and k_e are the contraction and expansion loss coefficients. The frictional term in Eq. (12.26) is ordinarily responsible for most of the total pressure drop. The entrance and exit losses are important for heat exchangers with short flow lengths.

Kays and London provide the heat exchanger correlations in graphical form; an example is shown in Figure 12.17. A library of EES procedures has been developed in order to provide the information contained in the Compact Heat Exchanger correlations in the most convenient form. There are four categories of compact heat exchanger procedures that can be found in EES: Geometry, Non-dimensional, Coefficient of heat transfer, and Pressure drop. Within each category, different heat exchanger core configurations are available (e.g., Finned circular tubes) and within each core configuration it is possible to use the scroll bar to select a particular geometry (e.g., surface CF-8.8-1.0J). The Geometry procedures return the geometric characteristics of the core (e.g., the hydraulic diameter) that allow the user to compute the Reynolds number and heat transfer area. The nondimensional functions provide the Colburn j_H and friction factors as a function of the Reynolds number

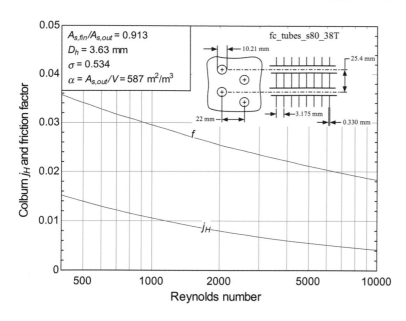

Figure 12.17 Colburn j_H and friction factors for heat exchanger surface 8.0-3/8T; based on data from Kays and London (1984).

(i.e., the information that is directly provided by the Compact Heat Exchanger correlations in charts such as that in Figure 12.17). The Coefficient of heat transfer and Pressure drop procedures carry out the additional calculations that are required to determine the dimensional heat transfer coefficient and pressure drop.

Example 12.4

A heat exchanger with the surface 8.0-3/8T is used to transfer thermal energy from a fluid (in the tubes) to air flowing across the tubes and through the fins. The length of the heat exchanger in the air flow direction is $L = 0.08$ m. The frontal area (seen from the direction of the air flow) is $W = 0.2$ m wide and $H = 0.26$ m high. Ambient air enters the heat exchanger with volumetric flow rate $\dot{V} = 0.25$ m³/s and at temperature $T_{in} = 20°C$. Estimate the air-side pressure drop and heat transfer coefficient and determine the thermal resistance to heat transfer associated with convection on the air side.

Known Values

Figure 1 illustrates a sketch of the problem with the known values indicated.

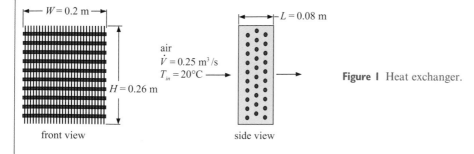

Figure 1 Heat exchanger.

Continued

Example 12.4 (cont.)

Assumptions

- The properties of air can be obtained from Appendix C.
- The contraction and expansion pressure changes are not significant.
- The fin efficiency associated with the plates is 100 percent.

Analysis and Solution

The properties of air are estimated using Appendix C at 300 K: $\rho = 1.177$ kg/m³, $c = 1007$ J/kg-K, $k = 0.02638$ W/m-K, $\mu = 18.54 \times 10^{-6}$ Pa-s, and $Pr = 0.7072$. The geometric properties of the heat transfer surface are obtained from Figure 12.17: $\sigma = 0.534$, $D_h = 3.63$ mm, $\alpha = 587$ m²/m³, and $A_{s,fin}/A_{s,out} = 0.913$.

The frontal area is given by:

$$A_{frontal} = W\,H = (0.2\text{ m})(0.26\text{ m}) = 0.052\text{ m}^2. \tag{1}$$

The free flow area is therefore:

$$A_{ff} = \sigma\,A_{frontal} = (0.534)\,0.052\text{ m}^2 = 0.0278\text{ m}^2. \tag{2}$$

The mass flow rate of air is:

$$\dot{m} = \dot{V}\rho = \frac{0.25\text{ m}^3}{\text{s}}\left|\frac{1.177\text{ kg}}{\text{m}^3}\right. = 0.294\text{ kg/s}. \tag{3}$$

The mass velocity is then:

$$G = \frac{\dot{m}}{A_{ff}} = \frac{0.294\text{ kg}}{\text{s}}\left|\frac{}{0.0278\text{ m}^2}\right. = 10.6\frac{\text{kg}}{\text{m}^2\,\text{s}}. \tag{4}$$

The Reynolds number is determined using Eq. (12.22):

$$Re = \frac{G\,D_h}{\mu} = \frac{10.6\text{ kg}}{\text{m}^2\,\text{s}}\left|\frac{0.00363\text{ m}}{}\right|\frac{}{18.54\times10^{-6}\text{ Pa s}}\left|\frac{\text{Pa m}^2}{\text{N}}\right|\frac{\text{N s}^2}{\text{kg m}} = 2075. \tag{5}$$

Using Figure 12.17, the Colburn j_H factor and friction factor are approximately $j_H = 0.0078$ and $f = 0.025$. The Stanton number is determined using Eq. (12.19):

$$St = \frac{j_H}{Pr^{2/3}} = \frac{0.0078}{(0.7072)^{2/3}} = 0.00982. \tag{6}$$

The heat transfer coefficient on the air side is determined using Eq. (12.20):

$$\bar{h} = G\,c\,St = \frac{10.6\text{ kg}}{\text{m}^2\,\text{s}}\left|\frac{1007\text{ J}}{\text{kg K}}\right|\frac{0.00982}{}\left|\frac{\text{W s}}{\text{J}}\right| = \boxed{104.8\frac{\text{W}}{\text{m}^2\,\text{K}}}. \tag{7}$$

The total surface area on the air side is:

$$A_{s,out} = \alpha V = \alpha H\,W\,L = \frac{587\text{ m}^2}{\text{m}^3}\left|\frac{0.26\text{ m}}{}\right|\frac{0.2\text{ m}}{}\left|\frac{0.08\text{ m}}{}\right. = 2.442\text{ m}^2. \tag{8}$$

The convection resistance on the air side is:

$$R_{conv} = \frac{1}{\bar{h}\,A_{s,out}} = \frac{\text{m}^2\text{ K}}{104.8\text{ W}}\left|\frac{}{2.442\text{ m}^2}\right. = \boxed{0.00391\text{ K/W}}. \tag{9}$$

If the change in air density and the inlet and outlet losses are neglected then the pressure drop on the air side can be estimated using Eq. (12.24), simplified to:

Example 12.4 (cont.)

$$\Delta p = \frac{2LG^2 f}{\rho \, D_h}$$

$$= \frac{2}{}\left|0.08 \text{ m}\right|\frac{(10.6)^2 \text{ kg}^2}{\text{m}^4 \text{ s}^2}\left|0.025\right|\frac{\text{m}^3}{1.177 \text{ kg}}\left|\frac{1}{0.00363 \text{ m}}\right|\left|\frac{\text{N s}^2}{\text{kg m}}\right|\frac{\text{Pa m}^2}{\text{N}} = \boxed{105.1 \text{ Pa}}. \quad (10)$$

Discussion

The procedures in EES can be used to determine the heat transfer and pressure drop for this heat exchanger core.

```
$UnitSystem SI Mass J K Pa

W = 0.2 [m]                          "width of heat exchanger"
H = 0.26 [m]                         "height of heat exchanger"
L = 0.08 [m]                         "length of heat exchanger"
V_dot=0.25 [m^3/s]                   "volumetric flow rate of air"
T_in=ConvertTemp(C,K,20[C])          "air temperature"

rho=1.177 [kg/m^3]                   "density"
m_dot=rho*V_dot                      "mass flow rate"
A_frontal=W*H                        "frontal area"
CALL CHX_h_finned_tube('fc_tubes_s80-38T', m_dot, A_frontal, 'Air',T_in, Po#: h_bar)
CALL CHX_DELTAP_finned_tube('fc_tubes_s80-38T', m_dot, A_frontal,L, 'Air', T_in,&
     T_in, Po#: DELTAp)
```

Solving provides $\bar{h} = 102.7$ W/m²-K and $\Delta p = 103.2$ Pa; the slight differences from the solution obtained manually are from reading the values of j_H and f from Figure 12.17.

The pressure changes associated with the contraction and expansion of the air into and out of the heat exchanger are neglected. In order to understand how important these effects might be for this situation the contraction loss should be estimated using a reasonable contraction coefficient (e.g., $k_c = 0.5$) according to:

$$\Delta p_c = \frac{G^2}{2\rho}\left(k_c + 1 - \sigma^2\right) = \frac{(10.6)^2 \text{ kg}^2}{\text{m}^4 \text{ s}^2}\left|\frac{(0.5 + 1 - 0.534^2)}{2}\right|\frac{\text{m}^3}{1.177 \text{ kg}}\left|\frac{\text{N s}^2}{\text{kg m}}\right|\frac{\text{Pa m}^2}{\text{N}} = 57.95 \text{ Pa}. \quad (11)$$

Because Δp_c is a significant fraction of Δp it is likely that the contraction and expansion losses will make up a significant fraction of the total pressure loss and these effects should not be neglected.

The fins were assumed to be 100 percent efficient when computing the resistance to convection using Eq. (9). It is possible to check this assumption by modeling the fins approximately as annular fins with thickness $th_f = 0.33$ mm and inner and outer radii equal to the tube radius and half the tube-to-tube distance, respectively. The tube-to-tube distance is given by:

$$s_{tt} = \sqrt{(0.0254 \text{ m})^2 + (0.022 \text{ m})^2} = 0.0336 \text{ m}. \quad (12)$$

The conductivity of copper is obtained from Appendix A, $k_f = 396$ W/m-K. The efficiency is computed using the EES function Eta_Fin_Annular_Rect which implements the solution for an annular fin recorded in Table 3.2.

Continued

Example 12.4 (cont.)

th_fin=0.33 [mm]*Convert(mm,m)	"fin thickness"
r_in=10.21 [mm]*Convert(mm,m)/2	"tube radius"
s_tt=0.0336 [m]	"tube-to-tube distance"
r_out=s_tt/2	"outer radius of fin"
k_f=396 [W/m-K]	"conductivity"
eta_fin=Eta_Fin_Annular_Rect(th_fin, r_in, r_out, h_bar, k_f)	"efficiency of fin"

Solving provides $\eta_{fin} = 0.884$, which is different enough from the assumed value of 1.0 that the effect of the fin efficiency should be accounted for in the calculation of the air side convection resistance. Using Eq. (3.101), the overall surface efficiency is computed:

$$\eta_o = 1 - \frac{A_{s,fin}}{A_{s,out}} \left(1 - \eta_{fin}\right) = 1 - 0.913 \left(1 - 0.884\right) = 0.894. \tag{13}$$

The convection resistance is computed using Eq. (3.102):

$$R_{conv} = \frac{1}{\eta_o \bar{h} A_{s,out}} = \frac{1}{0.894} \left| \frac{\text{m}^2 \text{ K}}{104.8 \text{ W}} \right| \frac{1}{2.442 \text{ m}^2} = 0.00437 \text{ K/W}. \tag{14}$$

The effect of the fin efficiency is to increase the resistance by approximately 10 percent relative to the value that was determined by neglecting this effect.

12.2 The Heat Exchanger Problem

12.2.1 Introduction

Section 12.1 showed that an overall energy balance on a heat exchanger alone is not sufficient to predict its performance. In order to solve a heat exchanger problem it is necessary to solve a differential energy balance on each of the flows. The rate of heat transfer from one fluid to the other is driven by the local difference in the mean temperatures of the two fluids and the thermal resistance is related to the heat exchanger conductance, discussed in Section 12.1.4. The governing equations result in a set of coupled differential equations that must be solved. These may be ordinary differential equations or partial differential equations depending on the heat exchanger configuration.

The solutions to these differential equations for most common heat exchanger configurations have been derived and are presented in two different forms: the log-mean temperature difference solutions (*LMTD*) and the effectiveness–*NTU* (*ε-NTU*) solutions. Although these methods appear to be different, they are actually algebraically identical and represent different presentations of the same information. The following section presents the process of deriving the heat exchanger problem in the context of a counter-flow heat exchanger. Sections 12.3 and 12.4 discuss the presentation of this solution in the *LMTD* and *ε-NTU* forms, respectively.

12.2.2 The Counter-Flow Heat Exchanger Solution

Consider the counter-flow heat exchanger shown in Figure 12.18. The hot fluid enters on the left (at $x = 0$) with mass flow rate $\dot{m}_h$ and temperature $T_{H,in}$. The cold fluid enters on the right (at $x = L$) with mass flow rate $\dot{m}_c$ and temperature $T_{C,in}$. In order to solve the heat exchanger problem it is necessary to determine the exiting fluid temperatures ($T_{H,out}$ and $T_{C,out}$) and the heat transfer rate $\dot{q}$ between the two fluid streams.

Two differential control volumes are defined within the heat exchanger, one for each fluid as shown in Figure 12.18. For this configuration, the mean temperature of the fluids is only a function of x and therefore the control volumes are differential in the x-direction. The energy balance on the hot fluid for the differential control volume is

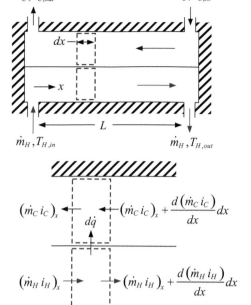

Figure 12.18 A counter-flow heat exchanger; the differential control volumes used to derive the governing differential equation are also shown.

$$(\dot{m}_H i_H)_x = (\dot{m}_H i_H)_x + \frac{d(\dot{m}_H i_H)}{dx} dx + d\dot{q}, \qquad (12.27)$$

where i_H is the specific enthalpy of the hot stream and $d\dot{q}$ is the differential heat transfer rate between the fluids that occurs across the control surface used to define the control volume. Notice that Eq. (12.27) assumes that the jacket of the heat exchanger is adiabatic. Provided that the heat exchanger has no leakage, the mass flow rate will not vary with x and Eq. (12.27) can be simplified to:

$$0 = \dot{m}_H \frac{di_H}{dx} dx + d\dot{q}. \qquad (12.28)$$

In general, the change in specific enthalpy may be driven by both pressure and temperature changes in the fluid. Therefore, Eq. (12.28) may be written as:

$$0 = \dot{m}_H \left[\underbrace{\left(\frac{\partial i_H}{\partial T}\right)_P}_{c_H} \frac{dT_H}{dx} + \underbrace{\left(\frac{\partial i_H}{\partial P}\right)_T \frac{dp_H}{dx}}_{\text{neglected}} \right] dx + d\dot{q}. \qquad (12.29)$$

The partial derivative of enthalpy with respect to temperature at constant pressure in Eq. (12.29) is the constant pressure specific heat capacity of the hot fluid, c_H. The pressure driven change in the enthalpy (the second term in brackets) is typically neglected because both the pressure gradient and the partial derivative of enthalpy with respect to pressure at constant temperature are usually small:

$$0 = \dot{m}_H \, c_H \frac{dT_H}{dx} dx + d\dot{q}. \qquad (12.30)$$

Rearranging Eq. (12.30) leads to

$$d\dot{q} = - \dot{C}_H \frac{dT_H}{dx} dx, \qquad (12.31)$$

where $\dot{C}_H$ is the capacitance rate of the hot fluid. A similar energy balance for the differential control volume on the cold fluid (see Figure 12.18) leads to

$$d\dot{q} = -\dot{C}_C \frac{dT_C}{dx} dx, \tag{12.32}$$

where $\dot{C}_C$ is the capacitance rate of the cold fluid.

The differential energy transfer between the streams within the control volume is related to the conductance according to

$$d\dot{q} = \underbrace{(T_H - T_C)}_{\substack{\text{local temp.}\\\text{difference}}} \underbrace{UA \frac{dx}{L}}_{\substack{\text{amount of}\\\text{conductance}\\\text{in segment } dx}}, \tag{12.33}$$

where L is the total length of the heat exchanger. Note that Eq. (12.33) is the product of the local driving temperature difference and the amount of the total conductance that is contained in the differential segment. Substituting Eq. (12.33) into Eqs. (12.31) and (12.32) leads to

$$UA(T_H - T_C)\frac{dx}{L} = -\dot{C}_H \frac{dT_H}{dx} dx \tag{12.34}$$

$$UA(T_H - T_C)\frac{dx}{L} = -\dot{C}_C \frac{dT_C}{dx} dx. \tag{12.35}$$

Equations (12.34) and (12.35) can be rearranged to provide the two coupled ordinary differential equations that must be solved for a counter-flow heat exchanger:

$$\frac{dT_H}{dx} = -\frac{UA}{L\dot{C}_H}(T_H - T_C) \tag{12.36}$$

$$\frac{dT_C}{dx} = -\frac{UA}{L\dot{C}_C}(T_H - T_C). \tag{12.37}$$

The technique used to obtain the solution to the set of differential equations that describe the heat exchanger will vary. For example, the solution can be obtained numerically. In this problem, Eqs. (12.36) and (12.37) are the state equations; they provide the rate of change of the state variables, T_H and T_C, in terms of the state variables and therefore can be integrated using any of the numerical integration techniques that are discussed in Section 5.3. The counter-flow heat exchanger problem is sufficiently simple that an analytical solution can be obtained provided that the capacitance rates of the two fluids are not temperature dependent. Equation (12.37) is subtracted from Eq. (12.36) in order to obtain

$$\frac{d(T_H - T_C)}{dx} = -\frac{UA}{L}(T_H - T_C)\left(\frac{1}{\dot{C}_H} - \frac{1}{\dot{C}_C}\right), \tag{12.38}$$

which is a single ordinary differential equation in terms of the local temperature difference, θ, defined as:

$$\theta = T_H - T_C. \tag{12.39}$$

Substituting Eq. (12.39) into Eq. (12.38) leads to:

$$\frac{d\theta}{dx} = -\frac{UA}{L}\theta\left(\frac{1}{\dot{C}_H} - \frac{1}{\dot{C}_C}\right). \tag{12.40}$$

Equation (12.40) can be separated:

$$\frac{d\theta}{\theta} = -\frac{UA}{L}\left(\frac{1}{\dot{C}_H} - \frac{1}{\dot{C}_C}\right)dx, \tag{12.41}$$

and integrated from $x = 0$ to $x = L$:

$$\int_{\theta_{x=0}}^{\theta_{x=L}} \frac{d\theta}{\theta} = -\frac{UA}{L}\left(\frac{1}{\dot{C}_H} - \frac{1}{\dot{C}_C}\right)\int_{x=0}^{x=L} dx \tag{12.42}$$

$$\ln \frac{\theta_{x=L}}{\theta_{x=0}} = - UA \left(\frac{1}{\dot{C}_H} - \frac{1}{\dot{C}_C} \right). \tag{12.43}$$

Examination of Figure 12.18 shows that $\theta_{x=L} = T_{H,out} - T_{C,in}$ and $\theta_{x=0} = T_{H,in} - T_{C,out}$:

$$\ln \left(\frac{T_{H,out} - T_{C,in}}{T_{H,in} - T_{C,out}} \right) = - UA \left(\frac{1}{\dot{C}_H} - \frac{1}{\dot{C}_C} \right). \tag{12.44}$$

Equation (12.44) is a fundamental relationship between the exit temperatures, conductance, and capacitance rates for a counter-flow heat exchanger; this equation provides the missing piece of information that can be used in conjunction with the overall energy balances on each stream:

$$\dot{q} = \dot{C}_H (T_{H,in} - T_{H,out}) \tag{12.45}$$

$$\dot{q} = \dot{C}_C (T_{C,out} - T_{C,in}) \tag{12.46}$$

to solve a counter-flow heat exchanger problem. Assuming that the operating conditions ($T_{H,in}$, $T_{C,in}$, $\dot{C}_H$, and $\dot{C}_C$) and heat exchanger conductance (UA) are known, Eqs. (12.44) through (12.46) are three equations in the three unknowns $\dot{q}$, $T_{H,out}$, and $T_{C,out}$. These equations can be solved simultaneously using an equation solver such as EES. However, the *LMTD* and *ε-NTU* methods presented in the following sections rearrange these equations into more accessible and easy-to-use forms.

Example 12.5

Example 12.2 analyzed the tube-in-tube heat exchanger shown in Figure 1. The heat exchanger is used to cool engine oil by transferring heat to a flow of 40 percent ethylene glycol–water solution. The engine oil enters the heat exchanger at $T_{H,in} = 85°C$ with a mass flow rate of $\dot{m}_H = 0.023$ kg/s and flows through the annular space that is formed between the inner and outer tubes. The ethylene glycol enters the heat exchanger at $T_{C,in} = 22°C$ and flows through the center tube in a counter-flow configuration with mass flow rate of $\dot{m}_c = 0.012$ kg/s. The analysis in Example 12.2 indicates that the conductance of the heat exchanger under these conditions is $UA = 41.09$ W/K. Determine the rate of heat transfer and the exit temperatures of the two fluids.

Known Values

Figure 1 illustrates a sketch of the problem with the known values indicated.

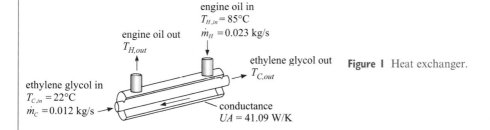

engine oil in
$T_{H,in} = 85°C$
engine oil out
$T_{H,out}$
$\dot{m}_H = 0.023$ kg/s

ethylene glycol out
$T_{C,out}$

Figure 1 Heat exchanger.

ethylene glycol in
$T_{C,in} = 22°C$
$\dot{m}_c = 0.012$ kg/s

conductance
$UA = 41.09$ W/K

Assumptions

- The specific heat capacities of the ethylene glycol solution and engine oil are constant.
- The jacket of the heat exchanger is adiabatic.

Continued

Example 12.5 (cont.)

Analysis

The specific heat capacities of the ethylene glycol solution and engine oil are obtained from Appendix B: $c_C = 3539$ J/kg-K and $c_H = 1901$ J/kg-K, respectively. The capacitance rates of the two flows are computed according to

$$\dot{C}_C = \dot{m}_C \, c_C \tag{1}$$

$$\dot{C}_H = \dot{m}_H \, c_H. \tag{2}$$

Equation (12.44) together with Eqs. (12.45) and (12.46) are three equations in the three unknowns $T_{H,out}$, $T_{C,out}$, and $\dot{q}$:

$$\ln\left(\frac{T_{H,out} - T_{C,in}}{T_{H,in} - T_{C,out}}\right) = -UA\left(\frac{1}{\dot{C}_H} - \frac{1}{\dot{C}_C}\right) \tag{3}$$

$$\dot{q} = \dot{C}_H(T_{H,in} - T_{H,out}) \tag{4}$$

$$\dot{q} = \dot{C}_C(T_{C,out} - T_{C,in}). \tag{5}$$

Solution

Equations (1) through (5) are programmed in EES.

```
$UnitSystem SI Mass J K Pa
T_C_in=22 [C]                           "cold fluid inlet temperature"
m_dot_C=0.012 [kg/s]                    "cold fluid mass flow rate"
T_H_in=85 [C]                           "hot fluid inlet temperature"
m_dot_H=0.023 [kg/s]                    "hot fluid mass flow rate"
UA=41.09 [W/K]                          "conductance"

c_C=3539 [J/kg-K]                       "specific heat capacity of the cold fluid"
c_H=1901 [J/kg-K]                       "specific heat capacity of the hot fluid"

C_dot_C=m_dot_C*c_C                     "capacitance rate of cold fluid"
C_dot_H=m_dot_H*c_H                     "capacitance rate of hot fluid"

Ln((T_H_out-T_C_in)/(T_H_in-T_C_out))=-UA*(1/C_dot_H-1/C_dot_C)
                                        "heat exchanger solution"
q_dot=C_dot_H*(T_H_in-T_H_out)          "hot fluid energy balance"
q_dot=C_dot_C*(T_C_out-T_C_in)          "cold fluid energy balance"
```

It will be necessary to set reasonable guess values for the outlet temperatures in order to obtain a valid solution: $T_{H,out} = 54.7°C$, $T_{C,out} = 53.2°C$, $\dot{q} = 1325$ W.

Discussion

The *LMTD* and *ε-NTU* solutions presented in the next two sections are simply different ways of representing the solution that was derived and solved here for the particular case of a counter-flow heat exchanger.

12.3 The Log-Mean Temperature Difference Method

12.3.1 Introduction

The log-mean temperature difference (*LMTD*) method expresses the heat transfer rate between the two fluid streams in a heat exchanger as the product of an appropriately defined temperature difference (referred to as the **log-mean temperature difference**, ΔT_{lm}) and the conductance of the heat exchanger (*UA*):

$$\dot{q} = UA\,\Delta T_{lm}. \tag{12.47}$$

The major consideration in applying Eq. (12.47) is how to define the temperature difference ΔT_{lm}. Both the hot and cold fluid temperatures will change by different amounts (and usually substantially) as the fluid passes through the heat exchanger as seen, for example, in Figure 12.3. Because the temperature distribution in the heat exchanger depends on flow configuration, the necessary form for ΔT_{lm} also depends upon the flow configuration.

12.3.2 Counter-Flow and Parallel-Flow Heat Exchangers

An analytical expression for ΔT_{lm} can be derived for the counter-flow and parallel-flow configurations. The solution for a counter-flow heat exchanger was derived in Section 12.2.2 and is repeated below:

$$\ln\left(\frac{T_{H,out} - T_{C,in}}{T_{H,in} - T_{C,out}}\right) = -UA\left(\frac{1}{\dot{C}_H} - \frac{1}{\dot{C}_C}\right). \tag{12.48}$$

Equation (12.44) can be expressed in log-mean temperature difference form by solving the overall heat exchanger energy balances:

$$\dot{q} = \dot{C}_H(T_{H,in} - T_{H,out}) \tag{12.49}$$

$$\dot{q} = \dot{C}_C(T_{C,out} - T_{C,in}) \tag{12.50}$$

for the capacitance rates:

$$\dot{C}_H = \frac{\dot{q}}{(T_{H,in} - T_{H,out})} \tag{12.51}$$

$$\dot{C}_C = \frac{\dot{q}}{(T_{C,out} - T_{C,in})}. \tag{12.52}$$

Equations (12.51) and (12.52) are substituted into Eq. (12.48):

$$\ln\left[\frac{(T_{H,out} - T_{C,in})}{T_{H,in} - T_{C,out}}\right] = -UA\left[\frac{(T_{H,in} - T_{H,out}) - (T_{C,out} - T_{C,in})}{\dot{q}}\right]. \tag{12.53}$$

Equation (12.53) is then rearranged so that it resembles Eq. (12.47):

$$\dot{q} = -UA\left[\frac{(T_{H,in} - T_{H,out}) - (T_{C,out} - T_{C,in})}{\ln\left[\frac{(T_{H,out} - T_{C,in})}{(T_{H,in} - T_{C,out})}\right]}\right] \tag{12.54}$$

or

$$\dot{q} = UA\underbrace{\left[\frac{(T_{H,out} - T_{C,in}) - (T_{H,in} - T_{C,out})}{\ln\left[\frac{(T_{H,out} - T_{C,in})}{(T_{H,in} - T_{C,out})}\right]}\right]}_{\Delta T_{lm}}. \tag{12.55}$$

Comparing Eq. (12.55) with Eq. (12.47) shows that the log-mean temperature difference for a counter-flow heat exchanger is

$$\Delta T_{lm} = \frac{(T_{H,out} - T_{C,in}) - (T_{H,in} - T_{C,out})}{\ln\left[\frac{(T_{H,out} - T_{C,in})}{(T_{H,in} - T_{C,out})}\right]} \quad \text{for counter-flow configuration.} \tag{12.56}$$

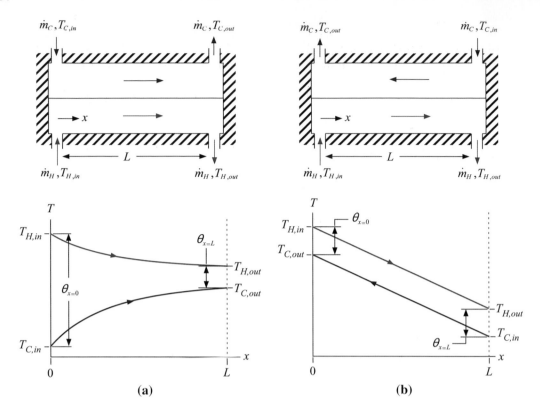

Figure 12.19 (a) Parallel-flow and (b) counter-flow heat exchangers and their associated temperature distributions. Also shown are the temperature differences that are required to compute the log-mean temperature difference.

The log-mean temperature difference for a parallel-flow heat exchanger is derived using a similar procedure:

$$\Delta T_{lm} = \frac{(T_{H,in} - T_{C,in}) - (T_{H,out} - T_{C,out})}{\ln\left[\dfrac{(T_{H,in} - T_{C,in})}{(T_{H,out} - T_{C,out})}\right]} \quad \text{for parallel-flow configuration.} \tag{12.57}$$

Note that the log-mean temperature difference equations for the counter-flow and parallel-flow configurations are identical if they are expressed in terms of the temperature difference at the two ends of the heat exchanger (see Figure 12.19):

$$\Delta T_{lm} = \frac{\theta_{x=0} - \theta_{x=L}}{\ln\left[\dfrac{\theta_{x=0}}{\theta_{x=L}}\right]}. \tag{12.58}$$

It does not matter which end of the heat exchanger is defined as being $x = 0$ and which end is $x = L$; changing the order in the numerator and denominator of Eq. (12.58) both result in a negative sign, which cancels:

$$\Delta T_{lm} = \frac{\theta_{x=0} - \theta_{x=L}}{\ln\left[\dfrac{\theta_{x=0}}{\theta_{x=L}}\right]} = \frac{-(\theta_{x=L} - \theta_{x=0})}{-\ln\left[\dfrac{\theta_{x=L}}{\theta_{x=0}}\right]} = \frac{\theta_{x=L} - \theta_{x=0}}{\ln\left[\dfrac{\theta_{x=L}}{\theta_{x=0}}\right]}. \tag{12.59}$$

Example 12.6

Example 12.5 analyzed a counter-flow heat exchanger where engine oil enters at $T_{H,in} = 85°C$ with a capacitance rate of $\dot{C}_H = 43.72$ W/K and an ethylene glycol solution enters at $T_{C,in} = 22°C$ with

Example 12.6 (cont.)

a capacitance rate of $\dot{C}_C = 42.47$ W/K. Determine the conductance of the heat exchanger that would be required in order to achieve a heat transfer rate of $\dot{q} = 1500$ W.

Known Values

Figure 1 illustrates a sketch of the problem with the known values indicated.

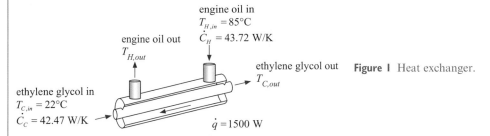

engine oil in
$T_{H,in} = 85°C$
$\dot{C}_H = 43.72$ W/K

engine oil out
$T_{H,out}$

ethylene glycol out
$T_{C,out}$

Figure 1 Heat exchanger.

ethylene glycol in
$T_{C,in} = 22°C$
$\dot{C}_C = 42.47$ W/K

$\dot{q} = 1500$ W

Assumptions

- The specific heat capacities of the ethylene glycol solution and engine oil are constant.
- The jacket of the heat exchanger is adiabatic.

Analysis and Solution

The outlet temperatures are calculated using the overall energy balances:

$$T_{H,out} = T_{H,in} - \frac{\dot{q}}{\dot{C}_H} = 85°C - \frac{1500\text{ W}}{43.72\text{ W}}\bigg|\frac{\text{K}}{} = 50.69°C \tag{1}$$

$$T_{C,out} = T_{C,in} + \frac{\dot{q}}{\dot{C}_C} = 22°C - \frac{1500\text{ W}}{42.47\text{ W}}\bigg|\frac{\text{K}}{} = 57.32°C. \tag{2}$$

The log-mean temperature difference is computed using Eq. (12.56):

$$\Delta T_{lm} = \frac{(T_{H,out} - T_{C,in}) - (T_{H,in} - T_{C,out})}{\ln\left[\frac{(T_{H,out} - T_{C,in})}{(T_{H,in} - T_{C,out})}\right]} = \frac{(50.69 - 22) - (85 - 57.32)}{\ln\left[\frac{(50.69 - 22)}{(85 - 57.32)}\right]} = 28.18\text{ K}. \tag{3}$$

The conductance is computed using Eq. (12.47):

$$UA = \frac{\dot{q}}{\Delta T_{lm}} = \frac{1500\text{ W}}{28.18\text{ K}}\bigg| = \boxed{53.22\frac{\text{W}}{\text{K}}}. \tag{4}$$

Discussion

The rate of heat transfer calculated in Example 12.4 is $\dot{q} = 1325$ W with a conductance of $UA = 41.09$ W/K. It is not surprising that a larger conductance is required ($UA = 53.22$ W/K) to achieve a larger heat transfer rate ($\dot{q} = 1500$ W).

The log-mean temperature difference method is particularly well suited to problems where the required performance is known (i.e., the rate of heat transfer) and the size of the heat exchanger must be determined (i.e., the conductance). The method is much less convenient if the conductance is known and the performance must be determined. The ε-NTU method presented in Section 12.4 is equally useful for both situations.

12.3.3 Shell-and-Tube and Cross-Flow Heat Exchangers

The appropriate form of the log-mean temperature difference, ΔT_{lm}, that should be used with Eq. (12.47) for configurations other than parallel- and counter-flow configurations is more difficult to derive. Although an analytical solution for ΔT_{lm} can be obtained for some cases, the resulting expression is algebraically complicated and therefore inconvenient. An alternative approach recognizes that ΔT_{lm} for these heat exchanger arrangements will always be less than the log-mean temperature difference for the counter-flow arrangement, $\Delta T_{lm,cf}$ given by Eq. (12.56) and repeated here:

$$\Delta T_{lm,cf} = \frac{(T_{H,out} - T_{C,in}) - (T_{H,in} - T_{C,out})}{\ln\left[\frac{(T_{H,out} - T_{C,in})}{(T_{H,in} - T_{C,out})}\right]}. \tag{12.60}$$

Therefore, ΔT_{lm} for any configuration can be expressed as

$$\Delta T_{lm} = F \Delta T_{lm,cf}, \tag{12.61}$$

where F is a correction factor that has a value that is less than unity. For a given heat exchanger configuration, the value of F depends on the capacitance rates of the two fluids and the inlet and outlet temperatures. The effect of these factors can be expressed in terms of two non-dimensional numbers, P and R, defined according to:

$$P = \frac{(T_{C,out} - T_{C,in})}{(T_{H,in} - T_{C,in})} \tag{12.62}$$

$$R = \frac{\dot{C}_C}{\dot{C}_H} = \frac{(T_{H,in} - T_{H,out})}{(T_{C,out} - T_{C,in})}. \tag{12.63}$$

The quantity P is sometimes referred to as the *LMTD* effectiveness and R is as the *LMTD* capacitance ratio. These quantities are related to, but different from, the effectiveness (ε) and capacitance ratio (C_R) presented in Section 12.4 for the ε-*NTU* method.

The correction factor F has been derived for most common heat exchanger configurations. For example, Figure 12.20 provides the correction factor for a cross-flow heat exchanger with both fluids unmixed as a function of P for a range of values of R.

Libraries of functions have been developed and made available in EES in order to provide F as a function of P and R for a variety of heat exchanger geometries. These functions can be accessed from the Function Information dialog by selecting the category Heat Exchangers and the sub-category F for LMTD and then scrolling to find the configuration of interest.

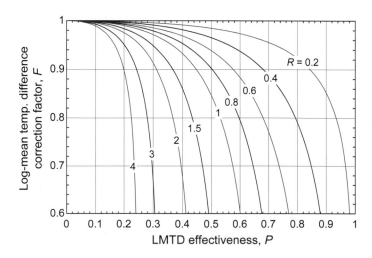

Figure 12.20 The correction factor F for a crossflow heat exchanger with both fluids unmixed as a function of P for various values of R.

Notice in Figure 12.20 that the value of the parameter P is limited to values between 0 and an upper bound that depends on R and is less than 1. For example, if $R = 3$ then P cannot exceed about 0.31. The $LMTD$ effectiveness, P, is the ratio of the temperature change of the cold fluid to the maximum possible temperature change defined as the difference between the hot inlet and cold inlet temperatures. The ratio of the capacitance rates of the two streams, the $LMTD$ capacity ratio R, determines how closely the cold stream can approach the hot stream and therefore dictates the allowable values of P and the associated value of F. It is difficult to accurately determine values of F in terms of P and R when F is less than about 0.70 because of the strong sensitivity of F to P, as shown in Figure 12.20. The effectiveness–NTU method, discussed in Section 12.4, does not have this problem and in general, is superior to the $LMTD$ method.

12.4 The Effectiveness–NTU Method

12.4.1 Introduction

This section presents the **effectiveness–NTU** method for solving heat exchanger problems. The ε-NTU method is more flexible and easier to use than the $LMTD$ method presented in Section 12.3. For example, the effectiveness–NTU method can be used either to directly determine the outlet temperatures of a heat exchanger when the heat exchanger conductance is known (i.e., to carry out a *simulation*-type of problem) or directly determine the conductance if the desired outlet temperatures are known (i.e., to carry out a *design*-type of problem). The $LMTD$ method is difficult to apply to a simulation-type of problem where the outlet temperatures are unknown because it requires iteratively solving a set of nonlinear equations that involve these temperatures. The effectiveness–NTU method is algebraically equivalent to the $LMTD$ method and provides exactly the same results. However, the formulation is much better suited for a wide variety of heat exchanger design problems.

12.4.2 Effectiveness, Number of Transfer Units, and Capacitance Ratio

The ε-NTU method presents the solution for each heat exchanger configuration as an algebraic relationship between three dimensionless numbers: the effectiveness (ε), the number of transfer units (NTU), and the capacitance ratio (C_R).

The **effectiveness** reflects the performance of the heat exchanger and is defined as the ratio of the actual rate of heat transfer between the two fluids ($\dot{q}$) to the maximum possible heat transfer rate ($\dot{q}_{max}$):

$$\varepsilon = \frac{\dot{q}}{\dot{q}_{max}}. \tag{12.64}$$

The value of the maximum possible heat transfer rate is not immediately obvious. An energy balance on the cold fluid provides:

$$\dot{q} = \dot{C}_C(T_{C,out} - T_{C,in}). \tag{12.65}$$

A higher rate of heat transfer will therefore correspond to increasing the outlet temperature of the cold fluid. The absolute highest temperature that the cold fluid can achieve in the heat exchanger is $T_{H,in}$ and therefore, from the standpoint of the cold fluid, the maximum possible rate of heat transfer is:

$$\dot{q}_{max,C} = \dot{C}_C(T_{H,in} - T_{C,in}). \tag{12.66}$$

An energy balance on the hot fluid provides:

$$\dot{q} = \dot{C}_H(T_{H,in} - T_{H,out}). \tag{12.67}$$

A higher rate of heat transfer will correspond to reducing the outlet temperature of the hot fluid; the coldest temperature that the hot fluid can achieve is $T_{C,in}$. Therefore, the maximum possible rate of heat transfer that can be achieved from the standpoint of the hot fluid is:

$$\dot{q}_{max,H} = \dot{C}_H(T_{H,in} - T_{C,in}). \tag{12.68}$$

The actual limiting performance is the smaller of the two values computed using Eqs. (12.66) and (12.68):

$$\dot{q}_{max} = \text{Min} \left[\dot{q}_{max,C}, \dot{q}_{max,H} \right] = \text{Min} \left[\dot{C}_C(T_{H,in} - T_{C,in}), \dot{C}_H(T_{H,in} - T_{C,in}) \right]. \tag{12.69}$$

A more compact manner of expressing $\dot{q}_{max}$ is:

$$\dot{q}_{max} = \dot{C}_{min}(T_{H,in} - T_{C,in}), \tag{12.70}$$

where $\dot{C}_{min}$ is the minimum of the fluid capacitance rates

$$\dot{C}_{min} = \text{Min} \left(\dot{C}_C, \dot{C}_H \right). \tag{12.71}$$

Substituting Eq. (12.70) into Eq. (12.64) provides the definition of effectiveness:

$$\varepsilon = \frac{\dot{q}}{\dot{C}_{min}(T_{H,in} - T_{C,in})}. \tag{12.72}$$

The **number of transfer units** is defined as the ratio of the conductance to the minimum capacitance rate:

$$NTU = \frac{UA}{\dot{C}_{min}}. \tag{12.73}$$

The number of transfer units represents the dimensionless size of the heat exchanger. If the physical size of the heat exchanger is increased then the conductance of the heat exchanger will tend to increase. If the minimum capacitance rate is decreased, then the heat exchanger will need to process less fluid. In either case, its dimensionless size, NTU, will increase.

The **capacitance ratio** is the ratio of the minimum to the maximum capacitance rate

$$C_R = \frac{\dot{C}_{min}}{\dot{C}_{max}}, \tag{12.74}$$

where $\dot{C}_{min}$ is the minimum capacitance rate, defined according to Eq. (12.71) and $\dot{C}_{max}$ is the maximum capacitance rate:

$$\dot{C}_{max} = \text{Max} \left(\dot{C}_C, \dot{C}_H \right). \tag{12.75}$$

Heat exchangers with capacitance ratios near one are said to be **balanced**; the temperature change experienced by the hot and cold fluids will be similar in this case. As the capacitance ratio approaches zero one fluid will experience very little temperature change as it has a much larger capacitance rate than the other. In the case where one fluid is a pure substance that is experiencing a phase change (boiling or condensing) at constant pressure, it will experience a change in quality but no temperature change and the heat exchanger effectively will have a capacitance ratio of zero.

The ε-NTU method algebraically rearranges the solution for each heat exchanger configuration so that is it a relationship between ε, NTU, and C_R. This process will be illustrated in the context of the counter-flow heat exchanger solution derived in Section 12.2.2.

12.4.3 Effectiveness–NTU Solution for a Counter-Flow Heat Exchanger

The ε-NTU method is an algebraic manipulation of the general solution for a particular heat exchanger configuration. In Section 12.2.2, the solution for a counter-flow heat exchanger is derived:

$$\ln \left(\frac{T_{H,out} - T_{C,in}}{T_{H,in} - T_{C,out}} \right) = - UA \left(\frac{1}{\dot{C}_H} - \frac{1}{\dot{C}_C} \right). \tag{12.76}$$

The $LMTD$ method provided one method of rearranging Eq. (12.76) using the overall energy balances:

$$\dot{q} = \dot{C}_C(T_{C,out} - T_{C,in}) \tag{12.77}$$

$$\dot{q} = \dot{C}_H(T_{H,in} - T_{H,out}). \tag{12.78}$$

The ε-NTU relationship for a counter-flow heat exchanger is an alternative arrangement of Eq. (12.76) that leads to a relationship between the three dimensionless parameters introduced in Section 12.4.2: the effectiveness (ε), the number of transfer units (NTU) and the capacity ratio (C_R). The energy balances on the two streams are solved for the fluid outlet temperatures:

$$T_{C,out} = T_{C,in} + \frac{\dot{q}}{\dot{C}_C} \tag{12.79}$$

$$T_{H,out} = T_{H,in} - \frac{\dot{q}}{\dot{C}_H}. \tag{12.80}$$

Substituting the definition of effectiveness, Eq. (12.72), into Eqs. (12.79) and (12.80) leads to:

$$T_{C,out} = T_{C,in} + \frac{\varepsilon\,\dot{C}_{min}(T_{H,in} - T_{C,in})}{\dot{C}_C} \tag{12.81}$$

$$T_{H,out} = T_{H,in} - \frac{\varepsilon\,\dot{C}_{min}(T_{H,in} - T_{C,in})}{\dot{C}_H}. \tag{12.82}$$

Substituting Eqs. (12.81) and (12.82) into Eq. (12.76) leads to:

$$\ln\left(\frac{T_{H,in} - \varepsilon\dfrac{\dot{C}_{min}}{\dot{C}_H}(T_{H,in} - T_{C,in}) - T_{C,in}}{T_{H,in} - T_{C,in} - \varepsilon\dfrac{\dot{C}_{min}}{\dot{C}_C}(T_{H,in} - T_{C,in})}\right) = -UA\left(\frac{1}{\dot{C}_H} - \frac{1}{\dot{C}_C}\right), \tag{12.83}$$

which can be rearranged:

$$\ln\left(\frac{(T_{H,in} - T_{C,in})\left(1 - \varepsilon\dfrac{\dot{C}_{min}}{\dot{C}_H}\right)}{(T_{H,in} - T_{C,in})\left(1 - \varepsilon\dfrac{\dot{C}_{min}}{\dot{C}_C}\right)}\right) = -UA\left(\frac{1}{\dot{C}_H} - \frac{1}{\dot{C}_C}\right) \tag{12.84}$$

and simplified:

$$\ln\left(\frac{1 - \varepsilon\dfrac{\dot{C}_{min}}{\dot{C}_H}}{1 - \varepsilon\dfrac{\dot{C}_{min}}{\dot{C}_C}}\right) = -UA\left(\frac{1}{\dot{C}_H} - \frac{1}{\dot{C}_C}\right). \tag{12.85}$$

To proceed with the derivation, we will assume that $\dot{C}_{min} = \dot{C}_C$ and therefore $\dot{C}_{max} = \dot{C}_H$. Equation (12.85) becomes:

$$\ln\left(\frac{1 - \varepsilon\dfrac{\dot{C}_{min}}{\dot{C}_{max}}}{1 - \varepsilon\dfrac{\dot{C}_{min}}{\dot{C}_{min}}}\right) = -UA\left(\frac{1}{\dot{C}_{max}} - \frac{1}{\dot{C}_{min}}\right). \tag{12.86}$$

Equation (12.86) can be rearranged:

$$\ln\left(\frac{1 - \varepsilon\,\overbrace{\dfrac{\dot{C}_{min}}{\dot{C}_{max}}}^{C_R}}{1 - \varepsilon}\right) = -\underbrace{\frac{UA}{\dot{C}_{min}}}_{NTU}\left(\underbrace{\frac{\dot{C}_{min}}{\dot{C}_{max}}}_{C_R} - 1\right). \tag{12.87}$$

Substituting Eqs. (12.73) and (12.74), the definitions of NTU and C_R, into Eq. (12.87) provides:

$$\ln\left(\frac{1 - \varepsilon\, C_R}{1 - \varepsilon}\right) = -NTU(C_R - 1). \tag{12.88}$$

It was assumed that $\dot{C}_{max} = \dot{C}_H$ and $\dot{C}_{min} = \dot{C}_C$ in order to derive Eq. (12.88). However, if we had instead assumed that $\dot{C}_{max} = \dot{C}_C$ and $\dot{C}_{min} = \dot{C}_H$, then the same result would be obtained.

Equation (12.88) is the effectiveness–NTU relationship for a counter-flow heat exchanger that relates the three dimensionless parameters ε, NTU, and C_R. Equation (12.88) can be rearranged to provide the effectiveness in terms of NTU and C_R (which is useful for simulation-type problems) or the number of transfer units in terms of the ε and C_R (which is useful for design-type problems). Solving Eq. (12.88) for the effectiveness leads to:

$$\varepsilon = \frac{1 - \exp[-NTU(1 - C_R)]}{1 - C_R \exp[-NTU(1 - C_R)]} \text{ for } C_R < 1. \tag{12.89}$$

Equation (12.89) is indeterminate if $C_R = 1$. However, taking the limit of Eq. (12.89) as C_R approaches 1 leads to:

$$\varepsilon = \frac{NTU}{1 + NTU} \text{ for } C_R = 1. \tag{12.90}$$

Solving Eqs. (12.89) and (12.90) for number of transfer units leads to:

$$NTU = \begin{cases} \dfrac{\ln\left[\dfrac{1 - \varepsilon\, C_R}{1 - \varepsilon}\right]}{1 - C_R} & \text{for } C_R < 1 \\[12pt] \dfrac{\varepsilon}{1 - \varepsilon} & \text{for } C_R = 1 \end{cases}. \tag{12.91}$$

Example 12.7

Redo Example 12.6 using the ε-NTU method. Engine oil enters a counter-flow heat exchanger at $T_{H,in} = 85°C$ with a capacitance rate of $\dot{C}_H = 43.72 \text{ W/K}$ and an ethylene glycol solution enters at $T_{C,in} = 22°C$ with a capacitance rate of $\dot{C}_C = 42.47 \text{ W/K}$. Determine the conductance of the heat exchanger that would be required in order to achieve a heat transfer rate of $\dot{q} = 1500 \text{ W}$.

Known Values

Figure 1 in Example 12.6 illustrates a sketch of the problem with the known values indicated.

Assumptions

- The specific heat capacities of the ethylene glycol solution and engine oil are constant.
- The jacket of the heat exchanger is adiabatic.

Analysis and Solution

The problem corresponds to a design-type problem in that the performance is specified and the required conductance must be calculated. Therefore, it is necessary to compute the effectiveness and capacitance ratio and use the correct form of the ε-NTU relation, Eq. (12.91), to determine the number of transfer units and, from that, the conductance. For this problem, the minimum capacitance rate is

$$\dot{C}_{min} = \dot{C}_C = 42.47\frac{\text{W}}{\text{K}} \tag{1}$$

and the maximum capacitance rate is

$$\dot{C}_{max} = \dot{C}_H = 43.72\frac{\text{W}}{\text{K}}. \tag{2}$$

Example 12.7 (cont.)

The capacitance ratio is

$$C_R = \frac{\dot{C}_{min}}{\dot{C}_{max}} = \frac{42.47}{43.72} = 0.9714. \tag{3}$$

The effectiveness is given by Eq. (12.72):

$$\varepsilon = \frac{\dot{q}}{\dot{C}_{min}(T_{H,in} - T_{C,in})} = \frac{1500\ \text{W}}{42.47\ \text{W}} \left| \frac{\text{K}}{(85 - 22)\ \text{K}} \right. = 0.5606. \tag{4}$$

The number of transfer units may be computed using Eq. (12.91):

$$NTU = \frac{\ln\left[\dfrac{1 - \varepsilon C_R}{1 - \varepsilon}\right]}{1 - C_R} = \frac{\ln\left[\dfrac{1 - 0.5606(0.9714)}{1 - 0.5606}\right]}{1 - 0.9714} = 1.253. \tag{5}$$

The required conductance is given by:

$$UA = NTU\ \dot{C}_{min} = 1.253\left(42.47\ \frac{\text{W}}{\text{K}}\right) = \boxed{53.22\ \frac{\text{W}}{\text{K}}}. \tag{6}$$

Discussion

The conductance calculated using the ε-NTU method is identical to the value determined using the $LMTD$ method in Example 12.6 as the two techniques are algebraically identical. However, the ε-NTU technique is equally easy to use for simulation or design type problems.

 Example 12.5 analyzed the same counter-flow exchanger, but specified that the conductance was $UA = 41.09$ W/K and asked for the rate of heat transfer. This is an example of a simulation type problem that would be difficult to do with the $LMTD$ method. In a simulation problem the capacitance ratio and number of transfer units can be used to determine the effectiveness and, from that, the rate of heat transfer and outlet temperatures. The capacitance ratio remains unchanged, $C_R = 0.9714$. The number of transfer units is computed from the given value of UA:

$$NTU = \frac{UA}{\dot{C}_{min}} = \frac{41.09\ \text{W}}{\text{K}} \left| \frac{\text{K}}{42.47\ \text{W}} \right. = 0.9675. \tag{7}$$

Equation (12.89) is used to compute the effectiveness from the NTU and C_R:

$$\varepsilon = \frac{1 - \exp[-NTU(1 - C_R)]}{1 - C_R \exp[-NTU(1 - C_R)]} = \frac{1 - \exp[-0.9675(1 - 0.9714)]}{1 - 0.9714\ \exp[-0.9675(1 - 0.9714)]} = 0.4952. \tag{8}$$

The rate of heat transfer is computed according to Eq. (12.72):

$$\dot{q} = \varepsilon\ \dot{C}_{min}\ (T_{H,in} - T_{C,in}) = \frac{0.4952|42.47\ \text{W}}{\text{K}} \left| (85 - 22)\ \text{K} \right. = \boxed{1325\ \text{W}}, \tag{9}$$

which is the same as the answer obtained in Example 12.5.

12.4.4 Effectiveness–NTU Solutions

Effectiveness–NTU relations (i.e., algebraic equations that relate the quantities ε, NTU, and C_R) have been derived for the most common heat exchanger configurations; analytical or graphical presentations of these solutions are available in Kays and London (1984) and elsewhere. The ε-NTU solutions are presented in Table 12.2 in the form $\varepsilon\ (NTU, C_R)$ and in Table 12.3 in the form of $NTU\ (\varepsilon, C_R)$.

Table 12.2 Effectiveness–*NTU* relations for various heat exchanger configurations in the form effectiveness as a function of number of transfer units and capacity ratio.

Flow arrangement		$\varepsilon\,(NTU,\,C_R)$
One fluid (or any configuration with $C_R = 0$)		$\varepsilon = 1 - \exp(-NTU)$
Counter-flow		$\varepsilon = \begin{cases} \dfrac{1 - \exp[-NTU(1 - C_R)]}{1 - C_R \exp[-NTU(1 - C_R)]} & \text{for } C_R < 1 \\[2ex] \dfrac{NTU}{1 + NTU} & \text{for } C_R = 1 \end{cases}$
Parallel-flow		$\varepsilon = \dfrac{1 - \exp[-NTU(1 + C_R)]}{1 + C_R}$
Cross-flow	both fluids unmixed	$\varepsilon = 1 - \exp\left[\dfrac{NTU^{0.22}}{C_R}\left\{\exp\left(-C_R NTU^{0.78}\right) - 1\right\}\right]$
	both fluids mixed	$\varepsilon = \left[\dfrac{1}{1 - \exp(-NTU)} + \dfrac{C_R}{1 - \exp(-C_R NTU)} - \dfrac{1}{NTU}\right]^{-1}$
	$\dot{C}_{max}$ mixed and $\dot{C}_{min}$ unmixed	$\varepsilon = \dfrac{1 - \exp[C_R\{\exp(-NTU) - 1\}]}{C_R}$
	$\dot{C}_{min}$ mixed and $\dot{C}_{max}$ unmixed	$\varepsilon = 1 - \exp\left[-\dfrac{1 - \exp(-C_R NTU)}{C_R}\right]$
Shell-and-tube	one shell pass and an even number of tube passes	$\varepsilon_1 = 2\left[1 + C_R + \sqrt{1 + C_R^2}\left\{\dfrac{1 + \exp\left(-NTU_1\sqrt{1 + C_R^2}\right)}{1 - \exp\left(-NTU_1\sqrt{1 + C_R^2}\right)}\right\}\right]^{-1}$
	N shell passes and an even number of tube passes	$\varepsilon = \dfrac{\left(\dfrac{1 - \varepsilon_1 C_R}{1 - \varepsilon_1}\right)^N - 1}{\left(\dfrac{1 - \varepsilon_1 C_R}{1 - \varepsilon_1}\right)^N - C_R} \quad \text{where } \varepsilon_1 \text{ is for one shell pass}$ $NTU = N \times NTU_1$

[1] ε and NTU are the overall effectiveness and number of transfer units, while ε_1 and NTU_1 are the effectiveness and number of transfer units associated with each shell pass.

EES functions are available that implement the solutions listed in Table 12.2 and Table 12.3 as well as other heat exchanger solutions that are less easily expressed analytically. These functions are implemented in the two forms that are most useful: $\varepsilon\,(NTU,\,C_R)$ and $NTU\,(\varepsilon,\,C_R)$. The ε-NTU solutions in EES can be accessed from the Function Information Window by selecting Heat Exchangers and then either NTU $\rightarrow$ Effectiveness or Effectiveness $\rightarrow$ NTU, respectively. The first form assumes that UA and the capacitance rates of the two streams are known (and therefore NTU and C_R can be directly computed) and this information is used to determine the effectiveness (which can be used to provide the outlet temperatures). The second form assumes that the outlet temperatures and the capacitance rates of the two streams are known (and therefore ε and C_R can be directly computed) and this information is used to determine the NTU (which can be used to provide the UA that is required).

12.4.5 Further Discussion of Heat Exchanger Effectiveness

Each of the different flow configurations leads to different relationships between ε, NTU, and C_R because the flow configuration influences the temperature distribution and the associated stream-to-stream temperature

Table 12.3 Effectiveness–*NTU* relations for various heat exchanger configurations in the form of number transfer units as a function of effectiveness and capacity ratio.

Flow Arrangement		$NTU\,(\varepsilon,\,C_R)$
One fluid (or any configuration with $C_R = 0$)		$NTU = -\ln(1-\varepsilon)$
Counter-flow		$NTU = \begin{cases} \dfrac{\ln\left[\dfrac{1-\varepsilon\,C_R}{1-\varepsilon}\right]}{1-C_R} & \text{for } C_R < 1 \\[4mm] \dfrac{\varepsilon}{1-\varepsilon} & \text{for } C_R = 1 \end{cases}$
Parallel-flow		$NTU = -\dfrac{\ln\left[1-\varepsilon(1+C_R)\right]}{1+C_R}$
Cross-flow	$\dot{C}_{max}$ mixed and $\dot{C}_{min}$ unmixed	$NTU = -\ln\left[1 + \dfrac{\ln(1-\varepsilon\,C_R)}{C_R}\right]$
	$\dot{C}_{min}$ mixed and $\dot{C}_{max}$ unmixed	$NTU = -\dfrac{\ln\left[C_R \ln(1-\varepsilon) + 1\right]}{C_R}$
Shell-and-tube	one shell pass and an even number of tube passes	$NTU_1 = \dfrac{\ln\left(\frac{E+1}{E-1}\right)}{\sqrt{1+C_R^2}}$ where $E = \dfrac{2 - \varepsilon_1(1+C_R)}{\varepsilon_1\sqrt{1+C_R^2}}$
	N shell passes and an even number of tube passes	use solution for one shell pass with: $\varepsilon_1 = \dfrac{F-1}{F-C_R}$ with $F = \left(\dfrac{\varepsilon\,C_R - 1}{\varepsilon - 1}\right)^{\frac{1}{N}}$ $NTU = N \times NTU_1$

[1] ε and NTU are the overall effectiveness and number of transfer units, while ε_1 and NTU_1 are the effectiveness and number of transfer units associated with each shell pass.

difference; that is why there are multiple entries in Table 12.2 and Table 12.3. For example, Figures 12.21(a) through (c) illustrate ε as a function of NTU for various values of C_R for the counter-flow, parallel-flow, and cross-flow (with both fluids unmixed) configurations, respectively.

Behavior as C_R Approaches Zero
The effect of the heat exchanger configuration on performance is related to the interaction between the temperature changes of the two fluid streams. In the limit that the capacity ratio approaches zero (i.e., one fluid stream has a capacitance rate that is much greater than the other), then there is no such interaction because one fluid stream does not change temperature significantly. For example, Figures 12.22(a) and (b) show the temperature distributions that result within a parallel-flow and counter-flow heat exchanger, respectively, when C_R goes to zero because $\dot{C}_C \gg \dot{C}_H$. In this case, the temperature of the cold fluid does not change significantly. Notice that the local temperature difference between the hot and cold streams for the counterflow and parallel flow heat exchangers are nearly identical, which demonstrates that flow configuration is not important in the limit of C_R approaching zero.

In the limit that $C_R \rightarrow 0$, the effectiveness–NTU relationship for any heat exchanger configuration reduces to:

$$\lim_{C_R \to 0} \varepsilon = 1 - \exp(-NTU). \tag{12.92}$$

Equation (12.83) is identical to the result obtained in Section 9.3.5 for an internal flow problem with a prescribed external temperature. In Figure 12.21, notice that the $C_R = 0$ curves (the upper-most curve in each figure) are identical even though the configuration changes.

The limit of $C_R \rightarrow 0$ is approached in many practical heat exchangers; e.g., the interaction of a flowing fluid with a constant temperature solid (as in a cold-plate) or a well-mixed tank of fluid or the situation that occurs when one of the two streams is undergoing constant pressure evaporation or condensation.

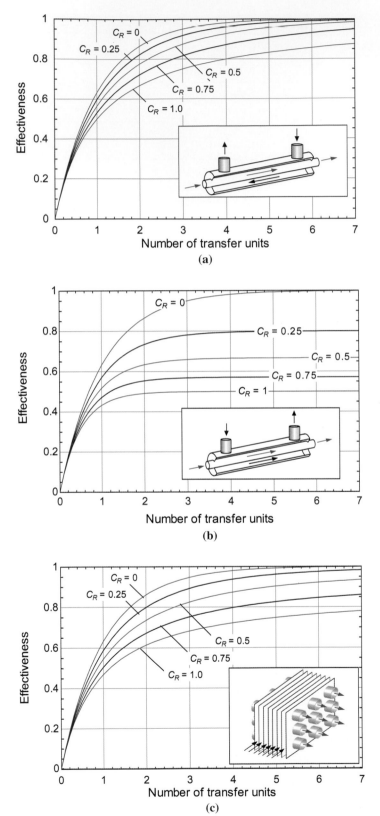

Figure 12.21 Effectiveness as a function of the number of transfer units for various values of capacity ratio for a (a) counter-flow, (b) parallel-flow, and (c) cross-flow (with both fluids unmixed) heat exchanger.

Behavior as NTU Approaches Zero

In the limit that NTU approaches zero, the configuration of the heat exchanger again does not matter. Figures 12.23(a) and (b) illustrate the temperature distribution within a parallel-flow and counter-flow heat exchanger, respectively, for a small NTU.

If NTU is small, then the heat exchanger is under-sized and the rate of heat transfer is not sufficient to change the temperature of either fluid substantially. As a result, the temperature difference in the heat exchanger is approximately constant and equal to $T_{H,in} - T_{C,in}$ throughout the entire heat exchanger. The configuration (e.g., counter-flow vs. parallel-flow) is unimportant and does not impact the local temperature difference. In the limit that $NTU \to 0$, the heat transfer rate can be written as:

$$\dot{q} = UA(T_{H,in} - T_{C,in}) \tag{12.93}$$

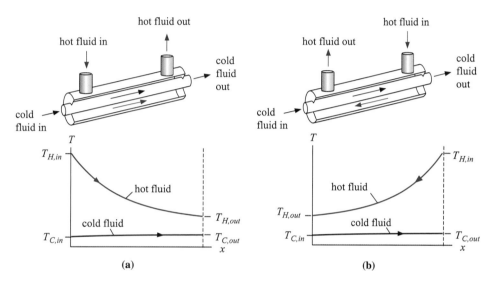

Figure 12.22 Temperature distribution in (a) parallel-flow and (b) counter-flow heat exchanger as the capacity ratio approaches zero because $\dot{C}_C \gg \dot{C}_H$.

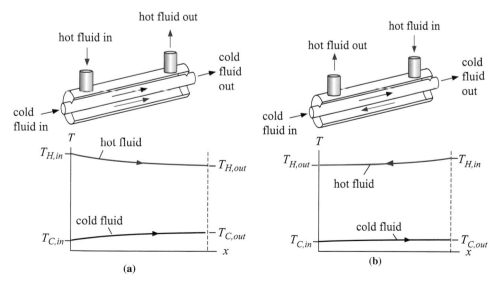

Figure 12.23 Temperature distribution within (a) parallel-flow and (b) counter-flow heat exchanger as NTU approaches zero.

because the temperature difference is constant. The definition of effectiveness is:

$$\varepsilon = \frac{\dot{q}}{\dot{C}_{min}(T_{H,in} - T_{C,in})}. \tag{12.94}$$

Substituting Eq. (12.93) into (12.94) leads to:

$$\varepsilon = \frac{UA(T_{H,in} - T_{C,in})}{\dot{C}_{min}(T_{H,in} - T_{C,in})} \tag{12.95}$$

or

$$\lim_{NTU \to 0} \varepsilon = NTU. \tag{12.96}$$

In Figure 12.21, notice that the curves for all values of C_R collapse near $NTU - 0$ and the slope of this single line is 1.0, regardless of configuration.

Behavior as NTU Becomes Infinite

The configuration is not important when either the capacity ratio or the number of transfer units is small. The heat exchanger configuration becomes increasingly important as the number of transfer units and capacity ratio increase. Consider a "balanced" heat exchanger; that is, a heat exchanger with $C_R = 1$ so that $\dot{C}_C = \dot{C}_H$. Figures 12.24(a) and (b) illustrate the temperature distributions within a parallel-flow and counter-flow heat exchanger, respectively, for the case where $C_R = 1$. Notice that the temperature change experienced by both of

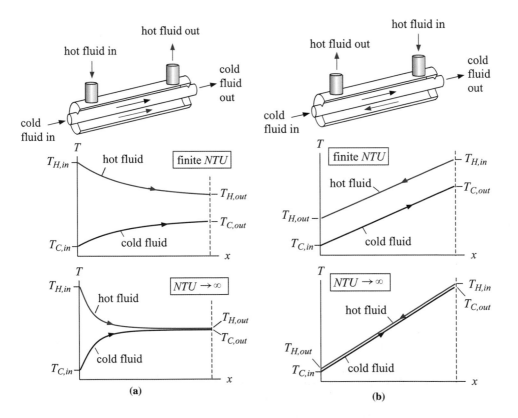

Figure 12.24 Temperature distribution within a balanced (i.e., $C_R = 1$) (a) parallel-flow and (b) counter-flow heat exchanger for a finite NTU (upper) and as $NTU \to \infty$ (lower).

the fluids is the same (consistent with a balanced heat exchanger), but that the temperature distributions are quite different for the two configurations. The temperature distributions shown in the upper sections of Figure 12.24 are consistent with a finite NTU. As NTU increases, the temperature difference somewhere within the heat exchanger will approach zero. This limit is shown in the lower sections of Figure 12.24. The location where the temperature difference approaches zero is referred to as the **pinch point**.

Because the heat exchanger is balanced, the temperature change of the hot and cold streams must be equal regardless of configuration or NTU. As a result, energy balances constrain the effectiveness that can be achieved. For the counter-flow configuration shown in Figure 12.24(b), the effectiveness can approach 1.0 as $NTU \rightarrow \infty$; this is evident because the cold-fluid outlet temperature can approach the hot-fluid inlet temperature (and, for the balanced case, the hot-fluid outlet temperature can approach the cold-fluid inlet temperature). Notice in Figure 12.21(a) that the $C_R = 1$ curve for a counter-flow heat exchanger asymptotically approaches 1.0 as $NTU \rightarrow \infty$. However, the effectiveness approaches 1.0 very slowly because the temperature difference everywhere in the heat exchanger is being driven towards zero.

The same behavior does not occur for any other configuration. For values of $C_R > 0$, the effectiveness of the parallel-flow configuration shown in Figure 12.24(a) does not approach 1.0 even as $NTU \rightarrow \infty$. Regardless of the size of the heat exchanger, the cold-fluid outlet temperature cannot approach the hot-fluid inlet temperature. For the balanced case, the hot-fluid outlet temperature can only get half-way there, corresponding to a maximum possible effectiveness of 0.5. Notice in Figure 12.21(b) that the $C_R = 1$ curve for a parallel-flow heat exchanger asymptotically approaches 0.5 as $NTU \rightarrow \infty$. However, the effectiveness approaches 0.5 quickly because the temperature difference in the heat exchanger is larger than in the counter-flow case. The limit on the effectiveness of a parallel flow heat exchanger as $NTU \rightarrow \infty$ depends only on the capacity ratio. For the balanced flow case shown in Figure 12.24(a), the two streams experience the same temperature change and the maximum effectiveness is 0.5. For smaller capacity ratios, one stream will change temperature by more than the other and therefore the effectiveness that can be achieved will increase. This result is evident by examining the behavior of Figure 12.21(b) as $NTU \rightarrow \infty$. In general, the limit to the performance of a parallel-flow heat exchanger is given by:

$$\lim_{NTU \to \infty} \varepsilon = \frac{1}{C_R + 1} \text{ for a parallel-flow configuration.} \tag{12.97}$$

Notice that if $C_R = 0$ then Eq. (12.97) limits to 1.0, which is consistent with Eq. (12.92).

Other heat exchanger configurations fall between the parallel-flow and counter-flow configurations in terms of the effect of the capacity ratio on their performance at large values of NTU. For example, examine the $C_R = 0.5$ curves for counter-flow, parallel-flow, and cross-flow configurations shown in Figures 12.21(a) through (c), respectively. At any value of NTU, the counter-flow configuration provides the highest effectiveness and the parallel-flow configuration the lowest; other configurations such as cross-flow fall between these limits. This result is illustrated in Figure 12.25, which shows the effectiveness as a function of NTU for various configurations at a constant capacity ratio of (a) $C_R = 1$, (b) $C_R = 0.5$, and (c) $C_R = 0.25$. The difference between heat exchanger configurations is largest for the $C_R = 1$ condition shown in Figure 12.25(a). As C_R is reduced, the effect of configuration diminishes and eventually all configurations collapse onto the same curve, given by Eq. (12.92), as $C_R \rightarrow 0$.

Heat Exchanger Design

The optimal design of a heat exchanger is a complex process that depends on the intended purpose of the heat exchanger and its cost. The design process will inevitably rely on an economic analysis in which the first cost of the heat exchanger is balanced against the operating cost. The first cost of the heat exchanger is directly related to its size and therefore its conductance, UA. For a given operating condition, the conductance dictates the NTU and effectiveness. The operating cost will be inversely related to the effectiveness; a higher effectiveness heat exchanger will reduce the operating cost. Examination of any of the ε-NTU curves that are shown in Figure 12.21 makes it clear that an optimally designed heat exchanger will typically be characterized by an NTU that is in the range of 1 to 2. At low values of NTU, the effectiveness can be increased (and therefore operating cost reduced) substantially by making the heat exchanger larger; a little additional first cost will greatly reduce

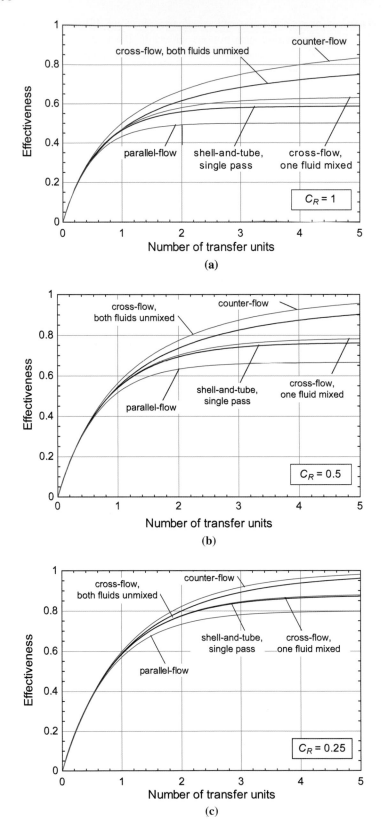

Figure 12.25 Effectiveness as a function of NTU for various configurations with a constant capacity ratio of (a) $C_R = 1$, (b) $C_R = 0.5$, and (c) $C_R = 0.25$.

the operating cost. However, at high values of NTU, the effectiveness becomes insensitive to NTU so that it will take a large increase in the first cost to change the operating cost.

In general, a detailed economic analysis is required to determine an optimal heat exchanger design that minimizes life cycle cost. However, a quick check on a heat exchanger design will involve calculating the NTU. If the NTU is large (say, greater than 10) then it is at least worth asking: what in the economic analysis is pushing you to this extreme? Is the first cost of the heat exchanger somehow very low (perhaps unrealistically) or is the operating cost very large. Addressing these questions will at least lead to greater understanding of the situation and may uncover some flaws in the design process.

Example 12.8

In a food processing plant, an oven exhausts hot air near a fryer that is continuously reheating oil. The hot air leaving the oven is at ambient pressure and high temperature, $T_{H,in} = 450°F$. The volumetric flow rate of hot air is $\dot{V}_{H,in} = 40$ cfm $\left(\text{ft}^3 / \text{min}\right)$. The oil leaves the fryer at $T_{C,in} = 70°F$ with flow rate $\dot{V}_C = 0.75$ gpm $\left(\text{gal}/\text{min}\right)$. The oil leaving the fryer has to be reheated to $T_{fryer} = 350°F$ using an electric heater before it is returned to the fryer. The electricity for the heater costs $ec = 0.15$ \$/kW-hr. You have been asked to look at the potential savings that could be achieved by using a heat exchanger to recover energy from the hot air leaving the oven in order to preheat the oil prior to entering the electric heater. The heat exchanger you have selected is a cross-flow heat exchanger in which both fluids are unmixed and it has an approximate cost per conductance of $UAc = 50$\$-K/W. The food processing plant runs approximately 80 hours a week and your economic time frame is 5 years of operation. The properties of oil are $\rho_c = 930$ kg/m^3 and $c_C = 1670$ J/kg-K.

Known Values

Figure 1 illustrates a sketch of the problem with the known values indicated.

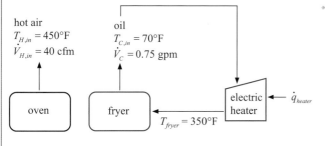

hot air
$T_{H,in} = 450°F$
$\dot{V}_{H,in} = 40$ cfm

oil
$T_{C,in} = 70°F$
$\dot{V}_C = 0.75$ gpm

oven

fryer

$T_{fryer} = 350°F$

electric heater ← $\dot{q}_{heater}$

Figure 1 System without energy recovery heat exchanger.

Assumptions

- The specific heat capacities of the oil and air are constant.
- The jacket of the heat exchanger is adiabatic.
- The oil is modeled as being incompressible.
- The time value of money is neglected.
- The heater is 100 percent efficient relative to converting electrical power into thermal energy provided to the oil.

Analysis

Figure 2 illustrates the system with the energy recovery heat exchanger used to recover energy from the air in order to reduce the electricity consumed by the heater.

Continued

Example 12.8 (cont.)

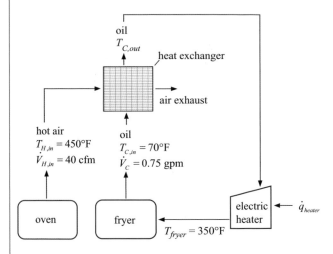

Figure 2 System with energy recovery heat exchanger.

The specific heat capacity and density of air are evaluated at the oven exhaust conditions:

$$c_H = c(T_{H,in}) \tag{1}$$

$$\rho_{H,in} = \rho(T_{H,in}, p_{atm}) \tag{2}$$

where p_{atm} is atmospheric pressure. The capacitance rates of the oil $(\dot{C}_C)$ and the air $(\dot{C}_H)$ are computed according to:

$$\dot{C}_C = \dot{V}_C \rho_C c_C \tag{3}$$

$$\dot{C}_H = \dot{V}_{H,in} \rho_{H,in} c_H. \tag{4}$$

The minimum and maximum capacitance rates are calculated:

$$\dot{C}_{min} = \text{Min}\left(\dot{C}_H, \dot{C}_C\right) \tag{5}$$

$$\dot{C}_{max} = \text{Max}\left(\dot{C}_H, \dot{C}_C\right). \tag{6}$$

The number of transfer units and capacitance ratio are computed:

$$NTU = \frac{UA}{\dot{C}_{min}} \tag{7}$$

$$C_R = \frac{\dot{C}_{min}}{\dot{C}_{max}}. \tag{8}$$

The effectiveness for a cross-flow heat exchanger with both fluids unmixed is a function of NTU and C_R:

$$\varepsilon = \varepsilon(NTU, C_R). \tag{9}$$

The rate of heat transfer in the heat exchanger is computed according to:

$$\dot{q} = \varepsilon \dot{C}_{min}(T_{H,in} - T_{C,in}). \tag{10}$$

The temperature of the oil leaving the heat exchanger, $T_{C,out}$ in Figure 2, is computed:

$$T_{C,out} = T_{C,in} + \frac{\dot{q}}{\dot{C}_C}. \tag{11}$$

The rate of heat transfer that must be obtained from electrical power is:

$$\dot{q}_{heater} = \dot{C}_C(T_{fryer} - T_{C,out}). \tag{12}$$

Example 12.8 (cont.)

The cost of the energy required to run the electric heater is calculated according to:

$$EnergyCost = ec\ time\ \dot{q}_{heater},\qquad(13)$$

where *time* is the time associated with running 80 hours per week for 5 years. The cost of the heat exchanger is:

$$FirstCost = UAc\,UA.\qquad(14)$$

The total cost is:

$$TotalCost = EnergyCost + FirstCost.\qquad(15)$$

Solution

Equations (1) through (15) are 15 equations in the 16 unknowns UA, $\rho_{H,in}$, c_H, $\dot{C}_C$, $\dot{C}_H$, $\dot{C}_{min}$, $\dot{C}_{max}$, NTU, C_R, ε, $\dot{q}$, $T_{C,out}$, $\dot{q}_{heater}$, $EnergyCost$, $FirstCost$, and $TotalCost$. The equations can therefore be programmed in EES and used to optimize the conductance, UA, in order to achieve the minimum value of $TotalCost$. Note that the EES functions **Density** and **Cp** are used for Eqs. (1) and (2) and Function **HX** is used for Eq. (9). Initially a conductance is specified in order to verify that the program is functioning. Then, the conductance is commented out and a Parametric Table is generated that includes UA and $TotalCost$. The code is shown below and the results of the parametric study are shown in Figure 3.

```
$UnitSystem SI Mass J K Pa
T_H_in=ConvertTemp(F,K, 450 [F])              "oven exhaust temperature"
V_dot_H_in=40 [cfm]*Convert(cfm,m^3/s)        "oven exhaust volumetric flow rate"
T_C_in=ConvertTemp(F,K,70 [F])                "fryer oil outlet temperature"
V_dot_C=0.75 [gal/min]*Convert(gal/min,m^3/s) "fryer oil volumetric flow rate"
T_fryer=ConvertTemp(F,K,350 [F])              "fryer oil inlet temperature"
c_C=1670 [J/kg-K]                             "specific heat of oil"
rho_c=930 [kg/m^3]                            "density of oil"
ec=0.15 [$/kW-hr]*Convert($/kW-hr,$/J)        "electricity cost"
time=80*52*5 [hr]*Convert(hr,s)               "time"
UAc=50 [$-K/W]                                "cost per conductance"


//UA=20 [W/K]                                 "arbitrary value of conductance"


rho_H_in=Density(Air,T=T_H_in,P=Po#)          "density of oven exhaust flow"
c_H=Cp(Air,T=T_H_in)                          "specific heat capacity of air"
C_dot_H=V_dot_H_in*rho_H_in*c_H               "capacitance rate of oven exhaust"
C_dot_C=V_dot_C*rho_C*c_C                      "capacitance rate of oil"
C_dot_min=Min(C_dot_H,C_dot_C)                "minimum capacitance rate"
C_dot_max=Max(C_dot_H,C_dot_C)                "maximum capacitance rate"
NTU=UA/C_dot_min                              "number of transfer units"
epsilon=HX('crossflow_both_unmixed', NTU, C_dot_H, C_dot_C, 'epsilon')
                                              "effectiveness"
q_dot=epsilon*C_dot_min*(T_H_in-T_C_in)       "heat transfer rate"
T_C_out=T_C_in+q_dot/C_dot_C                   "temperature of oil leaving heat exchanger"
q_dot_heater=C_dot_C*(T_fryer-T_C_out)        "rate of heat transfer in heater"
EnergyCost=ec*time*q_dot_heater               "energy cost"
FirstCost=UA*UAc                              "capital cost"
TotalCost=EnergyCost+FirstCost                "total cost"
```

Continued

Example 12.8 (cont.)

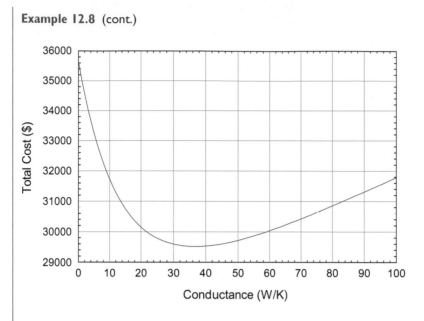

Figure 3 Total cost as a function of heat exchanger conductance.

Figure 3 suggests that a heat exchanger conductance of around 35 W/K would minimize the 5 year cost of operating the food processing plant, resulting in a savings of about $6000 (relative to the $UA = 0$ operating point).

Discussion

Figure 4 illustrates *TotalCost*, *FirstCost*, and *EnergyCost* as a function of *NTU* and shows that the first cost increases linearly while the energy cost drops quickly at first and then more slowly as the heat exchanger becomes oversized. This leads to an optimal *NTU* that is around 2.5.

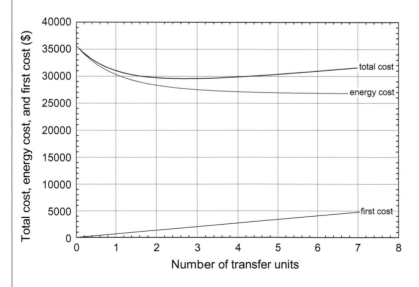

Figure 4 Total cost, 5-year energy cost, and first cost as a function of *NTU*.

12.5 Conclusions and Learning Objectives

This chapter provides a discussion of heat exchangers. Heat exchangers are critical components in most thermal fluid systems and their design and analysis includes concepts from all of the earlier chapters of this book. The basic purpose of a heat exchanger is to transfer heat from one fluid (the hot fluid) to another (the cold fluid). Their operation is described by an overall energy balance, which requires that the rate of heat transfer from the hot fluid be equal to the rate of heat transfer to the cold fluid and this overall energy balance applies to all heat exchangers that are properly insulated. The size of the heat exchanger is measured by its conductance, which is the inverse of the total thermal resistance separating the fluids. A large conductance indicates a small thermal resistance, which is consistent with a good heat exchanger.

In order to obtain a heat exchanger solution it becomes necessary to understand the internal configuration that dictates how the fluids flow with respect to one another. The solution is obtained by solving a set of coupled differential equations that are derived based on energy balances on the two fluids and therefore depend on the flow configuration. The coupling between the energy balances is related to the conductance. The result is a heat exchanger solution that depends on configuration. These solutions can be expressed in either the log-mean temperature difference ($LMTD$) form or the effectiveness–NTU form; both forms are algebraically identical but the ε-NTU form is usually more convenient.

Some specific concepts that you should understand are listed below.
- The difference between direct transfer and indirect transfer heat exchangers.
- The different flow configurations including: counter-flow, parallel-flow, cross-flow, and shell-and-tube configurations.
- The difference between mixed and unmixed geometries for a cross-flow heat exchanger.
- The meaning of multiple tube passes and multiple shell passes for a shell-and-tube configuration.
- The overall energy balance and the assumptions used to derive it.
- The meaning of capacitance rate.
- The meaning of conductance and how to calculate it.
- The concept of fouling and fouling factor and how to use the fouling factor to modify the conductance.
- When and how to use the tube bank and compact heat exchanger correlations.
- The process used to develop the differential equations that must be solved for different flow configurations.
- The ability to access the solutions to different heat exchanger configurations using either the LMTD or the effectiveness–NTU forms.
- The definition of ε, NTU, and C_R.
- The limiting form of the ε-NTU solution as the capacitance ratio approaches zero.
- The limiting form of the ε-NTU solution as the number of transfer units approaches zero.
- The reason for the limitations on ε related to configuration that occurs as capacitance ratio and NTU increase.
- The trade-off between first cost and operating cost that dictate the optimal size of a heat exchanger.

References

Ackermann, R. A., *Cryogenic Regenerative Heat Exchangers*, Plenum Press, New York (1997).

Bergman, T. L, A. S. Lavine, F. P. Incropera, and D. P. DeWitt, *Introduction to Heat Transfer*, Sixth Edition, John Wiley & Sons, Hoboken (2011).

Kakaç, S. and H. Liu, *Heat Exchangers – Selection, Rating and Thermal Design*, CRC Press (1998).

Kays, W. M. and A. L. London, *Compact Heat Exchangers*, Third Edition, McGraw-Hill (1984).

Nellis, G. F. and Klein, S. A., *Heat Transfer*, Second Edition, Cambridge University Press (2009).

Rohsenow, W. M., J. P. Hartnett, and Y. I. Cho, *Handbook of Heat Transfer*, Third Edition, McGraw-Hill (1998).

Turns, S. R., *Thermal-Fluid Sciences, An Integrated Approach*, Cambridge University Press, New York (2006).

Zukauskas, A., "Heat transfer from tubes in cross flow," in *Advances in Heat Transfer*, *Volume 8*, J. P. Hartnett and T. F. Irvine, Jr., eds., Academic Press, New York (1972).

Zukauskas, A., "Convective heat transfer in cross-flow," in *Handbook of Single-Phase Convective Heat Transfer*, Kakaç, S, R. K. Shah, and W. Aung, eds., John Wiley and Sons, New York (1987).

Problems

Heat Exchanger Conductance, Tube Banks, and Compact Heat Exchangers

12.1 A counter-flow concentric pipe heat exchanger used in a solar energy system is made of smooth thin-walled copper and it is well insulated on the outside. The diameter of the outer tube is 2 cm and the diameter of the inner tube is 1 cm. A 50 percent water–propylene glycol solution enters the annulus at one end at 65°C and 0.20 kg/s. Water enters the center pipe at the other end at 25°C and 0.10 kg/s. Estimate the overall conductance for this heat exchanger per meter of length.

12.2 A heat exchanger core consists of finned flat tubes with geometry that is consistent with the compact heat exchanger core: ff_tubes_s91–0737-S. The frontal area of the heat exchanger has width $W = 5$ ft and height $H = 3$ ft. The length of the heat exchanger in the flow direction is $L = 1.5$ ft. The fan that is used to push air through the heat exchanger provides a volumetric flow rate of $\dot{V}_{open} = 30{,}000$ ft^3/ min when there is no pressure rise and a pressure rise of $\Delta p_{dh} = 2$ in H$_2$O when there is no flow. The fan curve is linear between these points. The fan draws in air at $T_{in} = 80°$F and passes it across the coil.
  (a) Determine the total surface area of the coil that is exposed to air.
  (b) Determine the volumetric flow rate of air that will be provided to the coil.
  (c) Determine the heat transfer coefficient on the air side.

12.3 An air-to-air heat exchanger is used in a small airplane as part of a reverse-Brayton refrigeration system that provides conditioned air for cabin comfort. The heat exchanger is composed of a bank of $N_t = 500$ tubes each with inner diameter $D_i = 2.0$ mm and thickness $th = 0.1$ mm. The tubes are $L = 35$ cm long and are made of 302 stainless steel with roughness $e = 100$ μm. The hot air from the bleed line flows over the outer surface of the tubes with a heat transfer coefficient $\bar{h}_h = 75$ W/m^2-K. The mass flow rate of the hot air is $\dot{m}_h = 0.05$ kg/s and the inlet temperature and pressure of the hot air are $T_{h,in} = 275°$C and $P_h = 3.5$ atm, respectively. The cold air from the ambient has a mass flow rate of $\dot{m}_c = 0.20$ kg/s, inlet temperature $T_{c,in} = 35°$C and pressure $P_c - 1$ atm. The hot gas tends to foul the surface of the heat exchanger and can be treated as "Engine Exhaust Gas" from the fouling factor function in EES in terms of predicting the fouling factor. There is no fouling associated with the cold gas.
  (a) Determine the average heat transfer coefficient of the cold gas flowing in the tubes.
  (b) Determine the resistances that separate the hot and cold gas flows. Which resistance dominates this problem?
  (c) Determine the total conductance of the heat exchanger.

12.4 A heat exchanger is built using an aluminum tube ($k = 236$ W/m-K) that is 1 m long with an inner diameter of 2.5 cm and an outer diameter of 2.7 cm. Water flows inside the tube at a velocity of 1.5 m/s and an average temperature of 40°C. Air is blown perpendicular to the tube and tests have shown that the average heat transfer coefficient on the air side is 30 W/m^2-K.
  (a) Determine the conductance for this heat exchanger.
  (b) Circular fins are soldered around the tube at an even spacing of 4 fins per cm. The fins are made of 0.5 mm thick aluminum and they protrude 1.25 cm from the tube surface. Assuming that the air heat transfer coefficient remains the same, determine conductance for the finned heat exchanger and compare it to the unfinned case determined in part (a).

12.5 Dry air at $T_{a,in} = 30°$C, and atmospheric pressure is blown at $\dot{V}_a = 1.0$ m^3/s through a cross-flow heat exchanger in which refrigerant R134a is evaporating at a constant pressure of $p_R = 345$ kPa. The air exits the heat exchanger at $T_{a,out} = 13°$C. The tubes and fins of the heat exchanger are both made of copper. The tubes have an outer diameter of $D_{out,t} = 1.64$ cm and $th_t = 1.5$ mm tube wall thickness. The fins are circular with a spacing that leads to 275 fins per meter, an outer diameter of $D_{out,f} = 3.1$ cm and a thickness of $th_f = 0.25$ mm. The heat transfer coefficient between the R134a and the inner tube wall is estimated to be $\bar{h}_R = 2{,}500$ W/m^2-K. The heat transfer coefficient between the air and tubes and fins is estimated to be $\bar{h}_a = 70$ W/m^2-K. The total length of finned tubes is $L = 110$ m.

(a) Determine the rate of heat transfer from the air.

(b) Determine the value of the heat exchanger conductance for this heat exchanger.

12.6 The cross-flow heat exchanger described in Problem 12.5 has geometry similar to that for compact heat exchanger "fc_tubes_sCF-70-58J". The frontal area of the heat exchanger is $A_f = 0.5$ m^2 and the length of the heat exchanger in the flow direction is $W = 0.25$ m.

(a) Use the Compact Heat Exchanger Library to estimate the air-side conductance and the overall heat exchanger conductance assuming that the heat transfer coefficient between the R134a and the inner tube wall is $\bar{h}_R = 2,500$ W/m^2-K.

(b) Compare the result to the value determined in Problem 12.5.

12.7 A decision has been made to use chilled water, rather than R134a in the heat exchanger described in Problems 12.5 and 12.6. The mass flow rate of chilled water has been chosen so that the temperature rise of the water is $\Delta T_w = 4°$C as it passes through the heat exchanger. The water-side is configured so that the chilled water flows through $N_c = 10$ parallel circuits.

(a) Estimate the overall heat transfer conductance and compare the result to your answers from Problems 12.5 and 12.6.

(b) Estimate how much the overall heat transfer coefficient can be expected to drop over time due to fouling of the closed chilled water loop.

12.8 A concentric tube heat exchanger is to be used to cool oil entering at a flow rate of 8.5 kg/hr from 85°C to 55°C using water flowing at 12.0 kg/hr that enters at 30°C. The specific capacity of the oil is $c_o = 2090$ J/kg-K. The inner tube has an inner diameter of 1.0 cm. The pipe wall thickness is 1 mm and the larger diameter of the annulus is 1.6 cm. The tubes are smooth and initially clean.

(a) Determine the outlet temperature of the water.

(b) Estimate the overall conductance (UA) for this heat exchanger.

(c) After one year, the heat exchanger surfaces have become fouled. The fouling factors are estimated to be 0.000179 K-m^2/W on the oil side and 0.000175 K-m^2/W on the water side. Determine the percent reduction in the overall conductance resulting from the fouling compared to the clean surfaces.

12.9 The conductance of the heat exchanger described Problem 12.8 could perhaps be improved by changing the diameter of the inner tube. Prepare plots of the heat transfer coefficients on the oil and water sides, the pressure drop on the oil side and the overall conductance as a function of the inner tube diameter. Explain the results that you obtain and indicate your recommendation for the optimum inner tube diameter.

12.10 A tube bank is used in the condenser for a refrigeration system. The tube bank has an inline geometry with the outer tube diameter $D = 6.33$ mm, and equal center-to-center tube spacing in the horizontal and vertical directions such that $S_L = S_T = 1.26$ cm. Eight rows of tubes are employed in the flow direction with 24 tubes in each row. The surface temperature of the tubes is 48°C, which is maintained by the condensing refrigerant. Atmospheric air at 30°C is blown over the tubes at an inlet velocity of 5 m/s at the tube-bank inlet. Assume an air outlet temperature of 37°C.

(a) Calculate the pressure drop in the air for this tube bank.

(b) Calculate the average heat transfer coefficient between the air and the tube bank.

(c) Calculate the heat transfer coefficient for a single tube and compare it with your result for part (b).

12.11 Calculate the pressure drop and average heat transfer coefficient for a staggered tube bank that otherwise has the same geometry as the inline tube bank described in Problem 12.10.

12.12 A tube bank designed for the evaporator of a refrigeration system uses AISI 304 stainless steel tubes having an inner diameter of 8 mm with a tube wall thickness of 1.2 mm. The tube bank is arranged in an equilateral triangular array $S_D = S_T = 1.6$ cm. Six rows of tubes are employed in the flow direction with 24 tubes in each row. The refrigerant in the tubes is evaporating at –5°C. The heat transfer coefficient between the inner tube wall and the evaporating fluid is estimated to be 1850 W/m^2-K. Dry atmospheric air at 5°C is blown over the tubes with a velocity of 4.5 m/s at the tube-bank inlet. The outlet air temperature is 0°C.

(a) Calculate the average heat transfer coefficient between the air and the outside tube surface.

(b) Calculate the overall heat transfer coefficient between the air and the refrigerant, including the resistance of the tube wall and the thermal resistance on the inside surface of the tube. How does the outside surface resistance compare with the other thermal resistances between the air and the refrigerant?

Log-Mean Temperature Difference Solution

12.13 The heat exchanger described in Problem 12.1 is to be used to heat water to 50°C.

(a) Determine the log-mean temperature difference.

(b) Determine the required heat exchanger length.

12.14 You are specifying a plate frame heat exchanger with a counter-flow configuration to transfer heat from hot water to cooling water. The hot water enters the heat exchanger at $T_{H,in} = 120°F$ with a flow rate of $\dot{V}_H = 0.85$ gal/min. The cooling water enters at $T_{C,in} = 35°F$ with flow rate $\dot{V}_C = 0.65$ gal/min. If the hot water must be cooled to $T_{H,out} = 75°F$ then what is the required conductance of the heat exchanger?

12.15 Dry air at 84°F and 1 atm is blown at 150 cfm past a cooling coil in which R-134a is evaporating at a constant pressure of 50 psia. The bulk temperature of the air leaving the coil is 56°F. Calculate:

(a) the heat transfer rate,

(b) the log-mean temperature difference, and

(c) the heat exchanger UA.

12.16 Air at 32°C, 1 atm at a volumetric flow rate of 15 m³/s is blown past the condenser of a large commercial refrigeration system. The refrigerant is ammonia and at design conditions, it should condense at a temperature of 12°C above the inlet air temperature. Ammonia enters the condenser at 0.025 kg/s at a temperature of 22°C above the saturation temperature and exits as a saturated liquid. Calculate:

(a) the ammonia saturation pressure,

(b) the heat transfer rate,

(c) the air outlet temperature,

(d) the log-mean temperature difference (neglect the superheat for this calculation), and

(e) the heat exchanger, UA.

12.17 In Problem 12.5, the inlet volumetric flow rate and the inlet and outlet temperatures of the air were known and therefore it was possible to determine the heat transfer rate without a heat exchanger analysis. However, you have just learned that the outlet air temperature was measured with a thermocouple in only one location in the duct and it is not necessarily an accurate measurement of the mixed average outlet air temperature. Use the log-mean temperature difference method to estimate the average air outlet temperature based on the UA value.

12.18 The table provides heat transfer data from a manufacturer's catalog for a counter-flow oil cooler. The table provides the heat transfer rate for three different oil flow rates (expressed in gpm, gallons per minute). The values in the table are the heat transfer rate between the oil and water in units of Btu/min-ETD, where ETD is the entering hot to cold fluid temperature difference (in °F). The density and specific heat of the oil are $\rho_o = 830$ kg/m³ and $c_o = 2.3$ kJ/kg-K, respectively. The water enters at $\dot{V}_w = 35$ gallons per minute at $T_{w,in} = 180°F$ and atmospheric pressure. The oil enters at $T_{o,in} = 240°F$.

	Oil flow rate		
	1 gpm	3 gpm	5 gpm
Model 1	2.5 Btu/ min-°F	4.9 Btu/ min-°F	
Model 2	2.9 Btu/ min-°F	6.1 Btu/ min-°F	8.1 Btu/ min-°F
Model 3	3.1 Btu/ min-°F	6.8 Btu/ min-°F	9.7 Btu/ min-°F

(a) Determine the outlet oil temperature, the log-mean temperature difference, and the overall conductance for Model 2 at oil flow rates 1, 3, and 5 gpm.

(b) Plot the conductance as a function of the oil flow rate and provide an explanation for the observed variation.

(c) Estimate the oil outlet temperature if the oil enters the heat exchanger at $T_{o,in} = 225°F$

with a flow rate of $\dot{V}_o = 4$ gpm and the water enters with temperature $T_{w,in} = 180°F$ at a flow rate of $\dot{V}_w = 35$ gpm.

12.19 Saturated steam at 140 kPa is used in a heat exchanger to heat atmospheric air entering at 18°C and 5 m/s. The heat exchanger consists of a tube bundle with 13.7 mm outside diameter copper tubes. It is configured in an inline arrangement with a transverse spacing of 25 mm and a longitudinal spacing of 32 mm. There are eight rows of tubes in the longitudinal (flow) direction and each row has 16 tubes in the transverse direction. The tubes are 0.84 m in length.

(a) Determine the overall conductance of the tube bundle.

(b) Determine the pressure drop in the air stream

(c) Using the log-mean temperature difference method, determine the rate of heat transfer to the air and the air outlet temperature

(d) Determine the rate of condensation of the steam.

12.20 Water at atmospheric pressure enters a tube bank at 95°C and 7.5 kg/s in order to heat atmospheric air from 18°C to 32°C with an entering velocity of 5 m/s. The heat exchanger consists of a tube bundle with 11.2 mm inner diameter and 13.7 mm outside diameter copper tubes. It is configured in an inline arrangement with a transverse spacing of 25 mm and a longitudinal spacing of 32 mm. There are 12 rows of tubes in the longitudinal (flow) direction and each row has 24 tubes in the transverse direction. All of the water tubes are plumbed in parallel.

(a) Determine the heat transfer coefficient on the air side of the tube bundle. (Assume that the dominant thermal resistance is related to convection to the air on the outside surface of the tube bank.)

(b) Using the log-mean temperature difference method, determine the rate of heat transfer.

(c) Determine the required length of the tubes.

Effectiveness–NTU Solution

12.21 A counter-flow heat exchanger is operating under balanced flow conditions with the capacity rate of the cold fluid equal to 4000 W/K. The cold fluid enters at 60°C and the hot fluid enters at 200°C. The overall conductance of the heat exchanger is 4000 W/K.

(a) What is the effectiveness of this heat exchanger?

(b) What are the outlet temperatures of the hot and cold fluids?

12.22 A parallel-flow heat exchanger is operating under balanced flow conditions with the capacity rate of the cold fluid equal to 4000 W/K. The cold fluid enters at 60°C and the hot fluid enters at 200°C. The overall conductance of the heat exchanger is 4000 W/K.

(a) What is the effectiveness of this heat exchanger?

(b) What are the outlet temperatures of the hot and cold fluids?

12.23 Water at a flow rate of 0.7 lb$_m$/s enters a thin-walled copper tube with 0.75 inch internal diameter at 54°F. Steam is condensing on the outside of the tube at 250°F. The water is to be heated to 140°F. What is the length of the tube required for this application?

12.24 A tube-in-tube heat exchanger is used to cool oil (flowing through the center tube) using a flow of water (flowing through the outer annulus) as shown. The oil has properties: $\rho_o = 850$ kg/m³, $\mu_o = 0.10$ Pa-s, $c_o = 2000$ J/kg-K, $k_o = 0.14$ W/m-K, $Pr_o = 1430$. The mass flow rate of the oil is $\dot{m}_o = 0.01$ kg/s. The inner tube has inner diameter, $d_{i,in} = 0.01$ m and the length of the heat exchanger is $L = 1.0$ m.

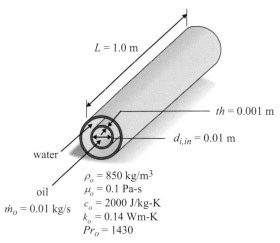

$L = 1.0$ m

$th = 0.001$ m

$d_{i,in} = 0.01$ m

water

oil

$\dot{m}_o = 0.01$ kg/s

$\rho_o = 850$ kg/m³
$\mu_o = 0.1$ Pa-s
$c_o = 2000$ J/kg-K
$k_o = 0.14$ Wm-K
$Pr_o = 1430$

(a) Is the flow of oil laminar or turbulent? Justify your answer.

(b) Will the flow of oil become hydrodynamically fully developed within the heat exchanger? Justify your answer.

(c) Will the flow of oil become thermally fully developed within the heat exchanger? Justify your answer.

(d) The average Nusselt number associated with a fully developed flow of the oil under these conditions is approximately 3.66. Do you expect the Nusselt number for the oil in the heat exchanger to be significantly greater than, less than, or approximately equal to this value? Justify your answer.

For the remainder of the problem assume that $\bar{h}_o = 700 \text{ W/m}^2\text{-K}$ and $\bar{h}_w = 10{,}000 \text{ W/m}^2\text{-K}$. The capacity rate of the water is $C_w = 2100 \text{ W/K}$. The conductivity of the tube is $k_t = 400 \text{ W/m-K}$. The inlet temperatures are $T_{w,in} = 5°C$ and $T_{o,in} = 90°C$.

(e) Calculate the total conductance, UA, of the heat exchanger.

(f) What is the outlet temperature of the oil?

(g) If you reversed the flow of water so that the heat exchanger was counterflow as opposed to parallel flow do you expect the outlet temperature of the oil to increase substantially, decrease substantially, or remain approximately constant? Justify your answer.

(h) If the internal surface of the inner tube goes from being smooth to relatively rough (for example 100 μm) then what effect do you expect that to have on the oil exit temperature? Justify your answer.

12.25 A parallel-flow heat exchanger has a total conductance $UA = 10 \text{ W/K}$. The hot fluid enters at $T_{h,in} = 400 \text{ K}$ and has a capacity rate $\dot{C}_h = 10 \text{ W/K}$. The cold fluid enters at $T_{c,in} = 300 \text{ K}$ and has a capacity rate $\dot{C}_c = 5 \text{ W/K}$.

(a) Determine the number of transfer units (NTU), effectiveness (ε), heat transfer rate ($\dot{q}$), and exit temperatures ($T_{h,out}$ and $T_{c,out}$) for the heat exchanger.

(b) Sketch the temperature distribution within the heat exchanger.

(c) Sketch the temperature distribution within the heat exchanger if the conductance of the heat exchanger is very large; that is, what is the temperature distribution in the limit that $UA \rightarrow \infty$.

(d) Sketch how the hot exit temperature will change as the total conductance (UA) is varied, with all other quantities held constant at the values listed in the problem statement. Be sure to indicate how your plot behaves as UA approaches zero and as UA approaches infinity.

12.26 An air-cooled refrigerant condenser consists of 2000 thin-walled copper tubes that are 0.5 m long and 0.015 m in diameter. The total air mass flow rate through all of the tubes is 10 kg/s and the air enters at 20°C. The heat of vaporization of the refrigerant is 150 kJ/kg and the refrigerant condensing temperature is 60°C. The heat transfer coefficient associated with the condensing refrigerant is 650 W/m²-K. Determine the mass flow rate of the condensed refrigerant.

12.27 Find the length of a concentric tube, counterflow HX that is needed in order to cool oil entering at a flow rate of 0.1 kg/s from 100°C to 60°C using water flowing at 0.1 kg/s that enters at 30°C. Determine the outlet temperature of the water. The total heat transfer coefficient per unit area, U, is 60 W/m²-K, based on the inner tube area. This means that the conductance, UA, is equal to U multiplied by the inner tube surface area. The inner tube has a diameter of 2.5 cm. The specific capacity of the oil is $c_o = 2.13 \text{ kJ/kg-K}$ and the specific heat capacity of the water is $c_w = 4.18 \text{ kJ/kg-K}$.

12.28 The figure illustrates a system to reject energy from a heat pump to a nearby lake.

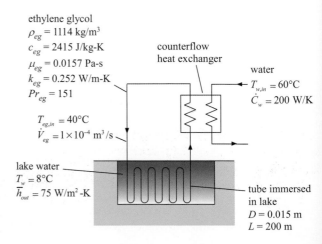

ethylene glycol
$\rho_{eg} = 1114 \text{ kg/m}^3$
$c_{eg} = 2415 \text{ J/kg-K}$
$\mu_{eg} = 0.0157 \text{ Pa-s}$
$k_{eg} = 0.252 \text{ W/m-K}$
$Pr_{eg} = 151$

$T_{eg,in} = 40°C$
$\dot{V}_{eg} = 1 \times 10^{-4} \text{ m}^3/\text{s}$

counterflow
heat exchanger

water
$T_{w,in} = 60°C$
$\dot{C}_w = 200 \text{ W/K}$

lake water
$T_w = 8°C$
$\bar{h}_{out} = 75 \text{ W/m}^2\text{-K}$

tube immersed in lake
$D = 0.015 \text{ m}$
$L = 200 \text{ m}$

Hot ethylene glycol at $T_{eg,in} = 40°C$ enters a long tube that is immersed in the lake. The properties of ethylene glycol are $\rho_{eg} = 1114$ kg/m^3, $c_{eg} = 2415$ J/kg-K, $\mu_{eg} = 0.0157$ Pa-s, $k_{eg} = 0.252$ W/m-K, and $Pr_{eg} = 151$. The volumetric flow rate of ethylene glycol is $\dot{V}_{eg} = 1 \times 10^{-4}$m^3/s. The tube has length $L = 200$ m and diameter $D = 0.015$ m; you may assume that the tube is very thin walled. The temperature of the lake water is $T_w = 8°C$ and the heat transfer coefficient between the lake water and the outer surface of the tube is $\bar{h}_{out} = 75$ W/m^2-K.

(a) Determine the heat transfer coefficient between the ethylene glycol and the tube wall. Assume that the flow is thermally fully developed.

(b) Determine the total pressure drop associated with the flow of ethylene glycol through the tube. Assume that the flow is hydrodynamically fully developed.

(c) Determine the temperature of the ethylene glycol leaving the tube, $T_{eg,out}$.

(d) The cooled ethylene glycol leaving the lake enters a counterflow heat exchanger and cools a flow of hot water. The water enters at $T_{w,in} = 60°C$ with capacitance rate $\dot{C}_w = 200$ W/K and the ethylene glycol exits at 40°C, as shown in the figure. Determine the exiting water temperature, $T_{w,out}$, and the conductance of the heat exchanger, UA_{HX}.

12.29 Water at 35°C at a rate of 230 kg/hr is available for use as a coolant in a concentric-tube heat exchanger whose total surface area is 1.4 m^2. The water is to be used to cool oil ($c_{oil} = 2100$ J/kg-K) from an initial temperature of 120°C. The water exit temperature is not important, but it must be less than 99°C to prevent boiling. The exit temperature of the oil must no higher than 60°C. The overall conductance is 280 W/K.

(a) Assuming a counter flow arrangement, determine the maximum flow rate of oil.

(b) Repeat for a parallel flow arrangement.

12.30 In a power plant, a cross-flow heat exchanger (often called an economizer) is used to heat 8.8×10^5 kg/hr of pressurized liquid water from 120°C to 150°C by the use of hot gases that

enter the heat exchanger at 400°C and exit at 260°C. In this heat exchanger the water steam is unmixed and the gas is mixed. The overall conductance per unit surface area of this heat exchanger is 20 W/m^2-K. The average specific heat of the combustion gas is 1007 J/kg-K.

(a) Determine the capacitance rates of both streams.

(b) Determine the effectiveness of this heat exchanger.

(c) Determine the required heat exchanger surface area.

12.31 Air at a rate of 250 kg/hr is heated from 30°C to 68°C in a single-pass tubular heat exchanger by condensing saturated steam at 110°C on the outside surface of the tube. The tube is made of copper with an inner diameter of 68.8 mm and an outer diameter of 73.0 mm. The convection coefficient due to condensation on the outside of the tube is estimated to be 4500 W/m^2-K.

(a) Determine the effectiveness for this heat exchanger.

(b) Determine the heat exchanger conductance.

(c) Calculate the required length of the heat exchanger tube.

(d) Because there is uncertainty in the value of the heat transfer coefficient on the outside surface, plot the answer to part (c) for values of the heat transfer coefficient ranging from 3000 to 6000 W/m^2-K.

12.32 The water heating system in a commercial building is an inline tube bank consisting of 12 rows of tubes in the flow direction, with each row having 30 tubes for a total of 360 tubes, each 6 ft in length with a 1 inch outer diameter. The longitudinal and transversal spacing of the tubes is 2.4 inch. Steam is condensing in the tubes at a pressure of 24.4 psia. The heat transfer coefficient between the steam and the inside tube surface is high so the internal convection resistance is negligible. Also, neglect the thermal resistance of the tube wall. Water enters the heater at 55°F at 160 lb$_m$/s.

(a) Determine the overall conductance for this water heating system.

(b) Determine the heat exchanger effectiveness.

(c) Determine the heat transfer rate and water outlet temperature.

12.33 A simple heat exchanger is made by soldering two tubes together.

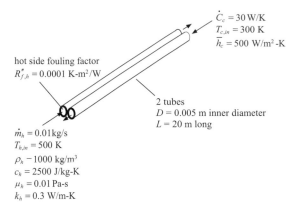

The two tubes each have inner diameter $D = 0.005$ m and length $L = 20$ m. The resistance to conduction through the tube walls and the solder joint can be neglected. The hot fluid has mass flow rate $\dot{m}_h = 0.01$ kg/s and enters with $T_{h,in} = 500$ K. The properties of the hot flow are density $\rho_h = 1000$ kg/m³, specific heat capacity $c_h = 2500$ J/kg-K, viscosity $\mu_h = 0.01$ Pa-s, and conductivity $k_h = 0.3$ W/m-K. There is a fouling factor on the hot fluid side of $R''_{f,h} = 0.0001$ K-m²/W. There is no fouling on the cold side. The cold fluid enters with $T_{c,in} = 300$ K and has capacitance rate $\dot{C}_c = 30$ W/K. The average heat transfer coefficient on the cold fluid side is $\bar{h}_c = 500$ W/m²-K. Assume that the internal surface of both tubes is smooth. You may assume that the tubes are long enough that the fully developed friction factor and Nusselt number represent the average friction factor and Nusselt number.

(a) Estimate the pressure drop associated with the flow of the hot fluid through the heat exchanger.

(b) What is the minimum possible pumping power required to move the hot fluid through the heat exchanger?

(c) Estimate the average heat transfer coefficient on the hot side of the heat exchanger.

(d) Explain briefly why the fully developed Nusselt number at low Reynolds number is independent of both Reynolds number and Prandtl number.

(e) Determine the hot outlet temperature. Assume that your answer from part (c)

was $\bar{h}_h = 150 \frac{W}{m^2\text{-}K}$ (this may or may not be the correct answer).

(f) Your boss has suggested that the fouling on the hot side controls the performance of the heat exchanger and therefore extensive testing should be carried out in order to understand this phenomenon. Do you agree? Why or why not?

(g) If the heat exchanger were re-plumbed so that it was in a parallel flow rather than a counterflow configuration do you expect the hot outlet temperature to change much? Will it go up or down? Justify your answers.

(h) If the tube on the hot side were replaced by one that was not perfectly smooth then what would the effect be on the hot outlet temperature and the pressure drop? Justify your answers.

12.34 A twin-tube heat exchanger arranged in a counter flow configuration is required to preheat a 1 g/s flow of helium entering at $T_{c,in} = 300$ K ($c_c = 5200$ J/kg-K). The average heat transfer coefficient between the helium and inner surface of the tube is $\bar{h}_c = 50$ W/m²-K. A 3.0 g/s flow of steam enters at 500 K ($c_h = 2000$ J/kg-K, $k_h = 0.036$ W/m-K, $\mu_h = 1.7 \times 10^{-5}$ Pa-s, $\rho_h = 0.44$ kg/m³). The heat exchanger is constructed by brazing two smooth copper tubes together as shown. The tubes both have an inner diameter (D_i) of 1.2 cm and a length (L) of 3.0 m.

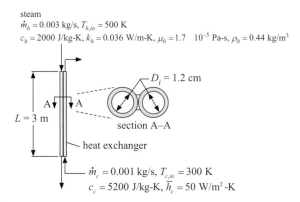

You may ignore the thermal resistance associated with conduction through the tube wall and the braze joint. Assume that the steam does not change phase in the heat exchanger. Estimate the effectiveness of the heat exchanger and the exit temperature of the helium.

12.35 The power delivered to the wheels of a vehicle ($\dot{w}$) as a function of vehicle speed (V) is given by: $\dot{w} = -0.3937[\text{hp}] + 0.6300\left[\frac{\text{hp}}{\text{mph}}\right]V + 0.01417\left[\frac{\text{hp}}{\text{mph}^2}\right]V^2$, where power is in horsepower and velocity is in mph. The rate of heat rejected from the engine block ($\dot{q}_b$) is approximately equal to the power delivered to the wheels (the rest of the energy from the fuel leaves with the exhaust gas). The heat is removed from the engine by pumping water through the engine block with a mass flow rate of $\dot{m} = 0.80$ kg/s. The thermal communication between the engine block and the cooling water is very good, therefore you may assume that the water will exit the engine block at the engine block temperature (T_b). For the purpose of this problem you may model the water as having constant properties that are consistent with liquid water at 70°C. The heat is rejected from the water to the surrounding air using a radiator. When the car is moving, air is forced through the radiator due to the dynamic pressure associated with the relative motion of the car with respect to the air. That is, the air is forced through the radiator by a pressure difference that is equal to equal to $\rho_a V^2/2$, where ρ_a is the density of air. Assume that the temperature of the ambient air is $T_\infty = 35°C$ and model the air in the radiator assuming that it has constant properties consistent with this temperature.

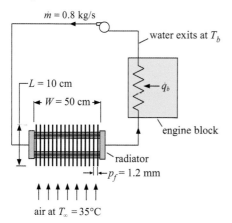

$\dot{m} = 0.8$ kg/s

water exits at T_b

$L = 10$ cm

$W = 50$ cm

$\dot{q}_b$

engine block

radiator

$p_f = 1.2$ mm

air at $T_\infty = 35°C$

The radiator has a plate-fin geometry. There are a series of tubes installed in closely spaced metal plates that serve as fins. The fin pitch is $p_f = 1.2$ mm and therefore there are W/p_f plates available for heat transfer. The heat exchanger core has overall width $W = 50$ cm, height $H = 30$ cm (into the page), and length (in the flow direction) of $L = 10$ cm. For the purpose of modeling the air-side of the core you may assume that the air flow is consistent with an internal flow through rectangular ducts with dimension $H \times p_f$. Assume that the fins are 100 percent efficient and neglect the thermal resistance between the fluid and the internal surface of the tubes. Also neglect convection from the external surfaces of the tubes as well as the reduction in the area of the plates associated with the presence of the tubes.

(a) Using the information above, develop an EES model that will allow you to predict the engine block temperature as a function of vehicle velocity. Prepare a plot showing T_b vs. V and explain the shape of the plot (if necessary, produce additional plots to help with your explanation). If the maximum allowable temperature for the engine block is 100°C (in order to prevent vaporization of the water) then what range of vehicle speeds is allowed? You should see both a minimum and maximum limit.

It is not easy to overcome the maximum speed limit identified (a); however, to overcome the minimum speed limit (so that you can pull up to a stop sign without your car overheating) you decide to add a fan. The fan can provide at most 500 cfm ($\dot{V}_o$ – the open circuit flow) and can produce at most 2.0 inch H_2O (Δp_{dh} – the dead-head pressure). The transition from open circuit to dead-head is linear. The fan curve is given by: $\Delta p_{fan} = \Delta p_{dh}\left(1 - \frac{\dot{V}}{\dot{V}_o}\right)$.

(b) Modify your code to simulate the situation where the air is provided by the fan rather than the vehicle motion. Overlay a plot showing T_b vs. V for this configuration on the one from (a); have you successfully overcome the lower speed limitation?

12.36 You are to design a heat exchanger to condense low-pressure steam from a steam turbine. The heat exchanger is to be of the shell and tube type with two tube passes of 125 tubes per pass. The tubes are to be made of brass with an inner diameter of 0.46 inch and an outer diameter of 5/8 inch with a smooth surface. Cooling water at 75°F enters the tubes at a rate of 40 lb$_m$/s. The condensing steam pressure is 2 psia and the average heat transfer coefficient between the steam and outer tube surface is estimated to be 2000 Btu/hr-ft^2-R.

(a) A well-designed steam condenser will have an effectiveness of 0.70. Using this rule of thumb, calculate the heat transfer rate.

(b) What will the temperature of the cooling water be at the heat exchanger exit?

(c) Determine the overall conductance of this heat exchanger.

(d) Determine the heat transfer coefficient between the water and inner tube walls.

(e) Assuming clean tubes, estimate the length of the tubes required for this condenser.

(f) Assume that the heat exchanger has been in service for an extended period of time such that the fouling factor for the condenser cooling water is 0.000175 m^2-K/W. Determine the percentage reduction in the rate of heat transfer resulting from the fouling.

12.37 A one-shell pass, two-tube pass heat exchanger is used to heat the shell-side fluid from 30°C to 95°C. The tube-side fluid is cooled from 150°C to 105°C.

(a) What is the log-mean temperature difference for an equivalent counter-flow heat exchanger?

(b) What is the heat exchanger effectiveness?

(c) What is the NTU for this heat exchanger?

(d) If the overall heat transfer conductance is reduced by 20 percent because of fouling, what will be the percentage reduction in the heat transfer rate?

12.38 A parallel-flow heat exchanger is shown schematically in the figure. The heat exchanger is being used to heat air from $T_{a,in} = 10°C$ using a flow of water at $T_{w,in} = 80°C$. The capacity rate of the air is $\dot{C}_a = 0.1\,W/K$ and the capacity rate of the water is $\dot{C}_w = 10\,W/K$. The conductance of the heat exchanger is $UA = 0.1\,W/K$.

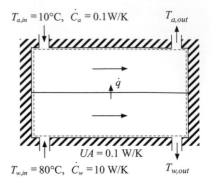

$T_{a,in} = 10°C, \ \dot{C}_a = 0.1\,W/K$ $T_{a,out}$

$UA = 0.1\,W/K$

$T_{w,in} = 80°C, \ \dot{C}_w = 10\,W/K$ $T_{w,out}$

(a) Sketch the temperature distributions in the heat exchanger (i.e., the temperature of the air and the water as a function of position, x); make sure that the qualitative features of your sketch are correct based on the operating conditions.

(b) Determine the heat transfer rate in the heat exchanger ($\dot{q}$) as well as the temperature of the air leaving the heat exchanger ($T_{a,out}$).

(c) Sketch the heat transfer rate ($\dot{q}$) as a function of the heat exchanger conductance (UA, assume that all other conditions remain the same as shown in the figure). Make sure that you indicate the heat transfer rate that you expect at $UA = 0$ and $UA \rightarrow \infty$.

(d) Your boss has heard that it is always better to use a counter-flow as opposed to a parallel-flow heat exchanger and so he insists that you should re-plumb the heat exchanger so that it is in a counter-flow configuration. Do you expect this change to substantially improve the performance (i.e., increase the rate of heat transferred in the heat exchanger)? Clearly justify your result qualitatively.

12.39 An oil cooler for a small engine is a counter-flow, tube-in-tube type heat exchanger that transfers heat from hot oil leaving the engine to cold water. The water enters the annular space with temperature $T_{c,in} = 10°C$ and mass flow rate $\dot{m}_w = 0.02\,kg/s$. The oil enters the center tube with temperature $T_{o,in} = 90°C$ and mass flow rate $\dot{m}_o = 0.004\,kg/s$. The outer diameter of the outer tube is $D_{ot,o} = 0.75$ inch and the outer diameter of the inner tube is $D_{ot,i} = 0.375$ inch. The length and wall thickness of both tubes are $L = 48$ inch and

th = 0.035 inch, respectively. The tubes are 304 stainless steel.

(a) Determine the outlet temperature of the oil as well as the rate of heat transfer in the oil cooler using the solution derived in Section 12.2.2.

(b) Determine the performance of the heat exchanger using the ε-NTU solution and show that the result is consistent with (a).

(c) Plot the rate of heat transfer as a function of the oil inlet temperature. If the engine rejects 200 W to the oil then what will the oil outlet temperature be?

12.40 A counter-flow heat exchanger is operating with a hot fluid inlet temperature of $T_{h,in}$ = 400 K and a cold fluid inlet temperature of $T_{c,in}$ = 300 K. The hot fluid flow has a capacity rate $\dot{C}_h$ = 100 W/K and the cold flow has a capacity rate $\dot{C}_c$ = 50 W/K. The total conductance of the heat exchanger is UA = 200 W/K.

(a) What is the heat transfer rate in the heat exchanger, $\dot{q}$?

(b) What is the exit temperature of the cold fluid, $T_{c,out}$?

(c) Sketch the temperature distribution within the counter-flow heat exchanger.

(d) The heat exchanger is re-plumbed so that the cold and hot fluids flow in the same direction (i.e., in a parallel-flow configuration). What is the exit temperature of the cold fluid? All other aspects of the problem remain the same.

(e) Sketch the temperature distribution within the parallel-flow heat exchanger.

12.41 A parallel-flow heat exchanger is constructed by clamping two metal extrusions together, as shown in the figure.

Each extrusion is L = 0.4 m long and has N = 120 channels for fluid flow (only a few channels are shown in the figure). Each channel is square with dimension W = 2 mm. The thickness of the material separating each channel is th = 0.25 mm. The conductivity of the metal used to make the extrusions is k_m = 15 W/m-K. The heat exchanger is configured to be parallel-flow. The hot and cold fluids have the same mass flow rate, $\dot{m}$ = 0.03 kg/s (this is the total mass flow rate, the mass flow rate per channel is $\dot{m}/N$), with the same properties, ρ = 1000 kg/m³, k = 0.3 W/m-K, μ = 0.01 Pa-s, and c = 4200 J/kg-K. The hot fluid enters at $T_{h,in}$ = 50°C and the cold fluid enters at $T_{c,in}$ = 10°C.

(a) Assume that the flow in the channels is fully developed to estimate the average heat transfer coefficient between the fluid and the channels.

(b) Will the actual average heat transfer coefficient be: substantially higher than, slightly higher than, the same as, slightly less than, or substantially less than, the value that you calculated in (a)? Justify your answer with a calculation.

For the remainder of the problem, assume that the average heat transfer coefficient on both sides of the heat exchanger is h = 300 W/m²-K. This may or may not be the correct answer to (a).

(c) Do you expect that there will be a significant resistance associated with conduction through the material that separates each channel (i.e., the th = 0.25 mm web)? Justify your answer with a calculation.

For the remainder of the problem, assume that your answer (c) is no; that is, assume that the

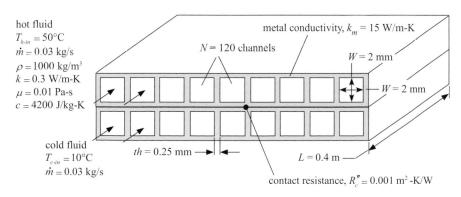

hot fluid
T_{h-in} = 50°C
$\dot{m}$ = 0.03 kg/s
ρ = 1000 kg/m³
k = 0.3 W/m-K
μ = 0.01 Pa-s
c = 4200 J/kg-K

metal conductivity, k_m = 15 W/m-K

N = 120 channels

W = 2 mm

W = 2 mm

Figure for Problem 12.41.

cold fluid
T_{c-in} = 10°C
$\dot{m}$ = 0.03 kg/s

th = 0.25 mm

L = 0.4 m

contact resistance, R''_c = 0.001 m²-K/W

metal in each extrusion will be isothermal and conduction resistance is not important.

(d) Determine the number of transfer units that characterizes this heat exchanger.

For the remainder of this problem, assume that you answer from (d) is $NTU = 0.2$ (this may or may not be correct).

(e) Estimate the effectiveness of the heat exchanger and the hot-fluid exit temperature (you should not need to use any figure or equation from your book to do this).

(f) Do you expect that changing the flow so that the heat exchanger operates in counter rather than parallel flow will have a noticeable effect on the hot exit temperature? Clearly justify your answer.

(g) Do you expect that the roughness of the channel surface will have a noticeable effect on the hot exit temperature? Clearly justify your answer.

12.42 A cross-flow heat exchanger with both fluids unmixed is used to cool a flow of air using a flow of cold water. The air enters the heat exchanger with temperature $T_{a,in} = 35°C$ and mass flow rate $\dot{m}_a = 0.002$ kg/s. The specific heat capacity of the air is $c_a = 1000$ J/kg-K. The water enters the heat exchanger with temperature $T_{w,in} = 5°C$ and mass flow rate $\dot{m}_w = 0.01$ kg/s. The specific heat capacity of the water is $c_w = 4200$ J/kg-K. The total conductance of the heat exchanger is $UA = 4$ W/K.

(a) Determine the temperature of the air leaving the heat exchanger.

(b) The cross-flow heat exchanger is replaced by a parallel flow heat exchanger with the same conductance. Do you expect the air outlet temperature to go up or down? Do you expect the change to be large or small? Clearly justify your answer.

(c) The water flow rate is doubled. The conductance of the heat exchanger remains the same. Do you expect the air outlet temperature to go up or down? Do you expect the change to be large or small? Clearly justify your answer.

12.43 A steam coil is used to provide heating to a zone of a building. The coil has geometry consistent with finned tube fc_tubes_sCF-775-58T in the Compact Heat Exchanger Library. The coil is installed in a square air duct with width

$W = 18$ inch and height $H = 18$ inch. The coil can be installed in modules that are placed in series. The length of each module in the flow direction is $L_{mod} = 6$ inch. The volumetric flow rate of air entering the coil is $\dot{V}_{in} = 900$ cfm. The air enters the coil at $T_{a,in} = -5°C$. The condensing steam in the coil is two-phase throughout and at pressure $p_s = 10$ psi. The temperature of the zone is $T_{zone} = 20°C$. Air is heated in the steam coil and enters the zone at a temperature greater than T_{zone}. The air leaves the zone at T_{zone} and therefore provides heating to the zone. The blower that moves the air through the coil has an efficiency of $\eta_b = 0.55$. You may neglect fouling and assume that the efficiency of the fins on the air side is 100 percent. Further, you may neglect the convection resistance associated with the condensing steam and conduction through the tubes.

(a) If only one module is installed, determine the power required by the blower, the temperature of the air leaving the steam coil, and the heat transfer rate provided to the zone.

(b) Plot the heat transfer rate provided to the zone and the blower power as a function of the number of modules. Explain the shape of your plot. How many modules would you suggest using for the steam coil? Why?

12.44 A schematic of a solar water heating system is shown in the figure. When solar radiation is available, a 30 percent (by mass) propylene glycol solution is pumped at volumetric flow rate $\dot{V}$ through the collectors and through the tubes in a shell-and-tube heat exchanger. The heat exchanger is composed of 40 tubes each having a diameter of 0.5 cm in a shell that has a diameter of 4 cm. The copper pipe walls are thin enough to neglect in this analysis. The tubes are 1.25 m in length. A second pump circulates water at the same volumetric flow rate from the storage tank through the shell of the heat exchanger in a counter-flow direction to the glycol solution flow. The heat transfer coefficient between the water and the outer surface of the tubes has been estimated as $\bar{h}_o = 950$ W/m²-K. An average temperature of 50°C can be assumed to evaluate properties. The purpose of this problem is to determine the effectiveness of the heat exchanger as a function of the volumetric flowrate.

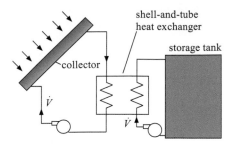

Calculate and plot the effectiveness of the heat exchanger for volumetric flowrates between 0.04 and 0.40 l/s. Explain the behavior of observed in this plot.

12.45 A concentric tube heat exchanger is built and operated as shown in the figure. The hot stream is a heat transfer fluid with specific heat capacity c_H = 2.5 kJ/kg-K. The hot stream enters at the center of the annulus at $T_{H,in}$ = 110°C with mass flow rate $\dot{m}_H$ = 0.64 kg/s and then splits and an equal amount flows in both directions. The cold stream has specific heat capacity c_C = 4.0 kJ/kg-K. The cold enters the center pipe at $T_{C,in}$ = 10°C with mass flow rate $\dot{m}_C$ = 0.2 kg/s. The outlet temperature of the hot-fluid that flows towards the left is $T_{H,out,x=0}$ = 45°C. The two sections of the heat exchanger have the same conductance.

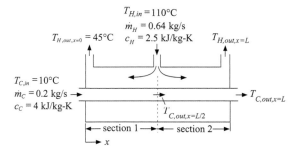

(a) Determine the temperature of the cold stream at the midpoint of the center tube and the temperature of the cold stream leaving the heat exchanger.
(b) Calculate the overall effectiveness of this heat exchanger considering both sections.
(c) How will the overall effectiveness be affected if the inlet temperature is increased to 150°C. (Assume that the properties of the heat transfer fluid are independent of temperature.)

(d) Is the overall effectiveness of this heat exchanger higher, lower, or the same as a counter-flow heat exchanger having the same inlet conditions? Justify your answer.

12.46 Buildings that have high ventilation rates can significantly reduce their heating load by recovering energy from the exhaust air stream. One way that this can be done is by use of a run-around loop. As shown in the figure, a run-around loop consists of two conventional liquid-to-air cross-flow heat exchangers. An ethylene glycol solution with 35 mass percent glycol is pumped at a rate $\dot{m}_g$ = 1 kg/s through both heat exchangers. The specific heat of this glycol solution is c_g = 3.58 kJ/kg-K. (Note that the properties of glycol solutions can be determined using the brine properties in EES.) During winter operation, the glycol solution is heated by the warm air exiting in the exhaust duct. The warm glycol solution is then used to pre-heat cold air entering from outdoors through the ventilation duct.

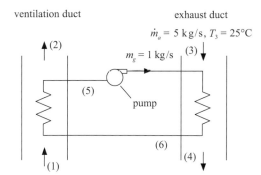

In the present case, outdoor air is blown into the building at a rate of $\dot{m}_a$ = 5 kg/s from outdoors. The outdoor temperature is T_1 = –10°C. The building is tightly constructed so the exhaust air flow rate may be assumed to be equal to the ventilation air flow rate ($\dot{m}_a$ = 5 kg/s). The air leaving the building through the exhaust duct is at T_3 = 25°C. The cross-flow heat exchangers in the exhaust and ventilation streams are identical, each having a finned coil configuration and an estimated conductance UA = 10 kW/K.

(a) Determine the effectiveness of the ventilation and exhaust heat exchangers.

(b) Determine the temperatures of the glycol solution at states (5) and (6).

(c) Determine the overall effectiveness of the run-around loop.

(d) It has been suggested that the performance of the run-around loop can be improved by optimizing the glycol flow rate. Plot the run-around loop overall effectiveness as a function of the glycol solution flow rate for $0.1 \, \text{kg/s} < \dot{m}_g < 4 \, \text{kg/s}$. Assume that the conductance of the heat exchangers vary with glycol solution flow rate to the 0.4 power based on a value of $UA = 10 \, \text{kW/K}$ at $\dot{m}_g = 1 \, \text{kg/s}$. What flow rate do you recommend?

12.47 A heat exchanger has a core geometry that corresponds to finned circular tube core "fc_tubes_s80_38T" in the Compact Heat Exchanger Library. The frontal area of the core has dimensions $W = 7.75$ inch and $H = 7.75$ inch. The length of the core is $L = 1.5$ inch. The core is integrated with a fan that has a head-flow curve given by: $\Delta p = a - b\dot{V}_a$, where Δp is the pressure rise across the fan, $\dot{V}_a$ is the volumetric flow rate of air, and $a = 0.3927$ inch H_2O and $b = 0.0021$ inch H_2O/cfm are the coefficients of the fan curve. The manufacturer has tested the heat exchanger with atmospheric air at $T_{a,in} = 20°C$ and water at $T_{w,in} = 75°C$, $p_w = 65$ psia flowing through the tubes. The test data are shown in the table. The tubes are plumbed in series (i.e., all of the water flows through each tube) and the tube thickness is $th_t = 0.035$ inch.

Water flow rate	Water outlet temperature
0.13 gpm	44.3°C
0.25 gpm	51.1°C
0.5 gpm	60.1°C
1 gpm	66.9°C
2 gpm	70.8°C
4 gpm	72.9°C

(a) Develop a model using the effectiveness–*NTU* technique that can predict the outlet temperature of the water for a water flow rate of the water $\dot{V}_w$.

(b) Plot the outlet temperature of the water as a function of the water flow rate and overlay the manufacturer's data onto your plot.

12.48 The figure illustrates a counter-flow, tube-in-tube heat exchanger that is being used to cool a stream of warm water using a flow of cold ethylene glycol solution.

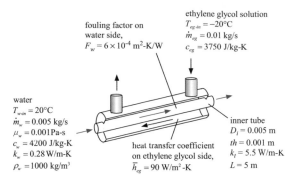

The water enters the heat exchanger and flows through the inner tube. The water enters at $T_{w,in} = 20°C$ and has properties $\rho_w = 1000 \, \text{kg/m}^3$, $\mu_w = 0.001 \, \text{Pa-s}$, $c_w = 4200 \, \text{J/kg-K}$, and $k_w = 0.28 \, \text{W/m-K}$. The mass flow rate of water is $\dot{m}_w = 0.005 \, \text{kg/s}$. The ethylene glycol solution flows through the annular space that is formed between the inner and outer tubes. The ethylene glycol solution enters with temperature $T_{eg,in} = -20°C$ and mass flow rate $\dot{m}_{eg} = 0.01 \, \text{kg/s}$. The specific heat capacity of the ethylene glycol is $c_{eg} = 3750 \, \text{J/kg-K}$. The heat transfer coefficient between the ethylene glycol solution and the outer surface of the inner tube is $\bar{h}_{eg} = 90 \, \text{W/m}^2\text{-K}$. The inner diameter of the inner tube is $D_i = 0.005 \, \text{m}$ and the tube thickness is $th = 0.001 \, \text{m}$. The length of the tube is $L = 5 \, \text{m}$. The inner surface of the tube is smooth (i.e., the surface roughness is zero). The conductivity of the tube material is $k_t = 5.5 \, \text{W/m-K}$. The area-specific fouling factor associated with the water is $F_w = 6.0 \times 10^{-4} \, \text{m}^2\text{-K/W}$. There is no fouling associated with the ethylene glycol.

(a) Is the flow on the water side of the heat exchanger laminar or turbulent? Justify your answer with a calculation.

(b) For the remainder of this problem, assume that your calculation in part (a) showed that the flow is laminar. Further, assume that the tube is sufficiently long that the fully developed Nusselt number can be used to compute the heat transfer coefficient. Determine the bulk temperature of the water leaving the heat exchanger, $T_{w,out}$.

The labels in the figure read:

fouling factor on water side, $F_w = 6 \times 10^{-4} \, \text{m}^2\text{-K/W}$

ethylene glycol solution
$T_{eg,in} = -20°C$
$\dot{m}_{eg} = 0.01 \, \text{kg/s}$
$c_{eg} = 3750 \, \text{J/kg-K}$

water
$T_{w,in} = 20°C$
$\dot{m}_w = 0.005 \, \text{kg/s}$
$\mu_w = 0.001 \, \text{Pa-s}$
$c_w = 4200 \, \text{J/kg-K}$
$k_w = 0.28 \, \text{W/m-K}$
$\rho_w = 1000 \, \text{kg/m}^3$

heat transfer coefficient on ethylene glycol side, $\bar{h}_{eg} = 90 \, \text{W/m}^2\text{-K}$

inner tube
$D_i = 0.005 \, \text{m}$
$th = 0.001 \, \text{m}$
$k_t = 5.5 \, \text{W/m-K}$
$L = 5 \, \text{m}$

For the remainder of this problem, assume that the water outlet temperature that you calculated in (b) is $T_{w,out} = 5°C$; this may or may not be the correct answer from (b).

(c) Do you expect any ice formation within the heat exchanger? That is, does the temperature of the water anywhere in the heat exchanger go below the freezing temperature of water?

(d) What is the relative importance of fouling, conduction through the tube, convection on the water side, and convection on the ethylene glycol side? Justify your answer.

(e) If the tube were not smooth, how do you expect the presence of significant surface roughness to affect the water outlet temperature?

(f) You used the Nusselt number for fully developed flow in a round tube. Do you anticipate that the presence of the thermally developing region will substantially change the water outlet temperature relative to your answer from (b)? Do you expect the outlet temperature to increase or decrease?

(g) The heat exchanger is re-plumbed so that it is in a parallel-flow configuration rather than a counter-flow configuration. Do you expect the water outlet temperature to change significantly? Do you expect it to increase or decrease? Justify your answers.

12.49 Air at atmospheric pressure and $T_{a,in} = 55°C$ flows with a velocity of $u_a = 15$ m/s through a finned circular tube heat exchanger having a geometry corresponding to compact heat exchanger core fc_tubes_s80–38T. The fins are made of copper. The length of the heat exchanger in the flow direction is $L = 0.4$ m. R134a is flowing through the tubes at a saturation temperature of $T_r = -5°C$. The R134a enters the heat exchanger at a quality of $x_{in} = 0.45$ and exits at a quality of $x_{out} = 0.65$. The frontal area of the heat exchanger is $A_f = 0.2$ m^2 and the thickness of the tube wall is $th_w = 0.9$ mm. The tubes are plumbed in $N = 10$ parallel circuits.

(a) Determine the heat transfer coefficient on the air side.

(b) Determine the frictional pressure drop on the air side.

(c) Estimate the fin efficiency.

(d) Determine the average heat transfer coefficient on the R134a side.

(e) Determine the effectiveness of the heat exchanger and the rate of heat transfer.

Projects

12.50 Your company manufactures heat exchangers for building energy recovery systems based on the "run-around loop" concept shown in the figure.

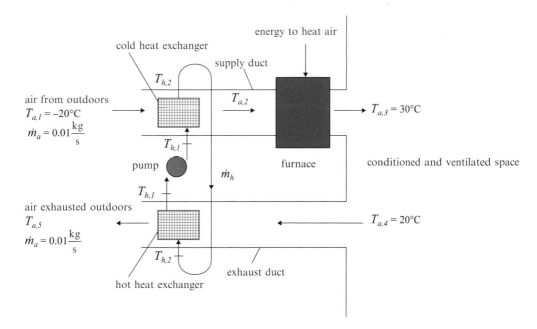

Figure for Problem 12.50.

Ventilation for the building requires that $\dot{m}_a = 0.01$ kg/s of fresh air be continuously supplied to the conditioned space with an equal flow of stale air removed from the space. This fresh air is pulled in from outdoors through the supply duct. During the winter, this air can be quite cold ($T_{a,1} = -20°C$). The cold air passes through a cold heat exchanger that is placed in the duct where it is heated by a stream of heat transfer fluid that has antifreeze in it to prevent freezing. The heat transfer fluid has specific heat capacity of 4228 J/kg-K. The mass flow rate of the heat transfer fluid is $\dot{m}_h$ and the temperatures of the fluid entering and leaving the cold heat exchanger are $T_{h,1}$ and $T_{h,2}$, respectively. The air, having been preheated somewhat by the cold heat exchanger to $T_{a,2}$ enters the furnace where it is heated to $T_{a,3} = 30°C$ and supplied to the conditioned building space.

The air leaving the building is at $T_{a,4} = 20°C$. This air passes through the hot heat exchanger placed in the exhaust duct where it is cooled by the stream of heat transfer fluid coming from the cold heat exchanger. The mass flow rate of the heat transfer fluid is $\dot{m}_h$ and the temperatures of the fluid entering and leaving the hot heat exchanger are $T_{h,2}$ and $T_{h,1}$, respectively.

Assume that all air streams are dry (i.e., neglect the effect of the latent heat load) and that the specific heat capacity of air is 1006 J/kg-K. Neglect the work required to drive the pump, both in terms of the energy cost associated with it as well as the temperature rise of the fluid as it passes through the pump.

The heat exchanger that your company makes is a cross-flow heat exchanger in which both fluids are unmixed. The heat exchanger is designed so that it is easy to change the size of the heat exchanger in order to match the application by laminating different sizes and numbers of plates together. You are asked to perform an economic analysis of this problem in which you select the optimal total surface area of the two heat exchangers (A_{hhx} and A_{chx}) and the optimal mass flow rate of the heat transfer fluid. The overall heat transfer coefficient (U) for both heat exchangers is nominally equal to 10 W/m²-K and the cost per unit of surface area is 50.0 \$/m². The cost per unit of energy supplied to the furnace is 0.25 \$/kW-hr.

This optimization should be based on maximizing the amount of money saved over five winters assuming that each winter can be modeled as 120 days of $-20°C$ weather. Neglect the time-value of money for this problem.

(a) Determine the optimal heat transfer fluid mass flow rate and surface areas for this application. How much money can be saved over five winters?

(b) Prepare a plot showing the net money saved as a function of the cost of the heat exchanger per area (\$/m²). Re-optimize the system for each point on the plot.

12.51 The plant where you work includes a process that results in a stream of hot combustion products at moderate temperature $T_{hg,in} = 150°C$ with mass flow rate $\dot{m}_{hg} = 0.25$ kg/s. The properties of the combustion products can be assumed to be the same as those for air. You would like to recover the energy associated with this flow in order to reduce the cost of heating the plant and therefore you are evaluating the use of the air-to-air heat exchanger shown in the figure.

The air-to-air heat exchanger is a cross-flow configuration. The length of the heat exchanger parallel to the two flow directions is $L = 10$ cm. The width of the heat exchanger in the direction perpendicular to the two flow directions is $W = 20$ cm. Cold air enters the heat exchanger from outdoors at $T_{cg,in} = -5°C$ with mass flow rate $\dot{m}_{cg} = 0.50$ kg/s and is heated by the combustion products to $T_{cg,out}$. The hot and cold air flows through channels that are rectangular (both sides of the heat exchanger have the same geometry). The width of the channels is $h_c = 1.0$ mm. There are fins placed in the channel in order to provide structural support and also increase the surface area for heat transfer. The fins can be assumed to run the complete length of the heat exchanger and are 100 percent efficient. The fins are spaced with pitch, $p_f = 0.5$ mm and the fins are $th_f = 0.10$ mm thick. The thickness of the metal separating the cold channels from the hot channels is $th_w = 0.20$ mm and the conductivity of this metal is $k_m = 15$ W/m-K. Both the hot and cold flows are at nominally atmospheric pressure. The fouling factor associated with the flow of combustion gas through the heat exchanger is $R''_f = 0.0018$ K-m²/W.

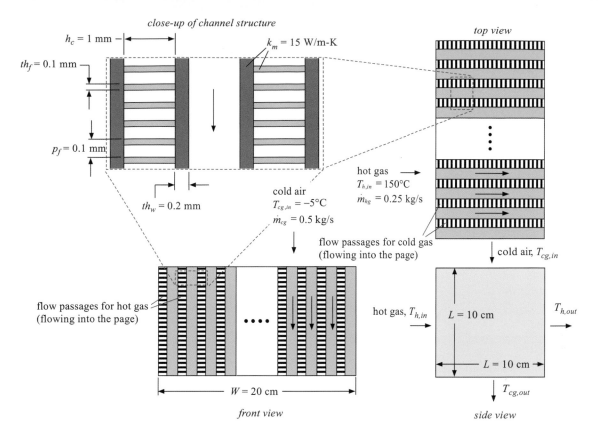

Figure for Problem 12.51.

There is no fouling associated with the flow of outdoor air through the heat exchanger.

(a) Compute the heat transfer coefficient between the hot air and the channel walls and the cold gas and the channel walls. Use the inlet temperatures of the air flows to compute the properties.

(b) Compute the total conductance of the heat exchanger.

(c) Determine the heat transferred in the heat exchanger and the temperature of the cold gas leaving the heat exchanger.

(d) Blowers are required to force the hot and cold flows through the heat exchanger. Assume that you have blowers that are $\eta_{blower} = 0.65$ efficient. Estimate the total blower power required to operate the energy recovery unit.

(e) If you pay $ec = 0.08\$/kW\text{-}hr$ for electricity (to run the blowers) and $1.50\$/therm$ for gas (to heat the plant) then estimate the net savings associated with the energy recovery system

(neglect capital investment cost) for *time* = 1 year; this is the savings associated with the heat transferred in the heat exchanger less the cost to run the blower for a year. Assume that the plant runs continuously.

(f) Plot the net savings per year as a function of the mass flow rate of the cold air that is being heated. Your plot should show a maximum; explain why this maximum exists.

(g) Determine the optimal values of the mass flow rate of combustion gas and cold gas ($\dot{m}_{hg}$ and $\dot{m}_{cg}$) which maximize the net savings per year. You should use the Min/Max capability in EES to accomplish this. What is the maximum savings/year? This is the most you could afford to pay for the blowers and heat exchanger if you wanted a 1 year payback.

12.52 A gas turbine engine is used onboard a small ship to drive the propulsion system. The engine consists of a compressor, turbine, and combustor as shown in the figure. Ambient air is pulled through the gas turbine engine with a mass flow rate of $\dot{m} = 0.1\,\text{kg/s}$ and enters the compressor at $T_1 = 20°C$ and $P_1 = 1$ atm. The exit pressure of the compressor is $P_2 = 3.5$ atm and $T_2 = 167°C$. The air enters a combustor where it is heated to $T_3 = 810°C$ with very little loss of pressure so that $P_3 = 3.5$ atm. The hot air leaving the combustor enters a turbine where it is expanded to $P_4 = 1$ atm. The temperature of the air leaving the turbine is $T_4 = 522°C$. You may assume that the turbine and compressor are well insulated and that the specific heat capacity of air is constant and equal to $c = 1000$ J/kg-K. The difference between the power produced by the turbine and required by the compressor is used to drive the ship.

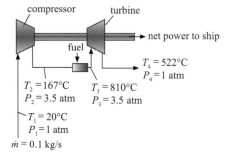

compressor   turbine

fuel

→ net power to ship

$T_4 = 522°C$
$P_4 = 1$ atm

$T_2 = 167°C$
$P_2 = 3.5$ atm

$T_3 = 810°C$
$P_3 = 3.5$ atm

$T_1 = 20°C$
$P_1 = 1$ atm

$\dot{m} = 0.1$ kg/s

(a) Determine the efficiency of the gas turbine engine (the ratio of the net power to the ship to the heat transferred in the combustor).

(b) The combustor runs on fuel with a heating value of $HV = 44 \times 10^6$ J/kg and a mission lasts $t = 2$ days. What is the mass of fuel that the ship must carry?

In order to reduce the amount of fuel required, you have been asked to look at the option of adding a recuperative heat exchanger to the gas turbine cycle, as shown in the figure. You are considering the air-to-air heat exchanger that was evaluated in Problem 12.51. The air-to-air heat exchanger is a cross-flow configuration. The length of the heat exchanger parallel to the two flow directions is $L = 10$ cm. The width of the heat exchanger in the direction perpendicular to the two flow directions is $W = 20$ cm but this can easily be adjusted by adding additional plates. Air enters the heat exchanger from the compressor and is heated by the air leaving the turbine. The hot and cold air flows through

channels that are rectangular (both sides of the heat exchanger have the same geometry). The width of the channels is $h_c = 1.0$ mm. There are fins placed in the channel in order to provide structural support and also increase the surface area for heat transfer. The fins can be assumed to run the complete length of the heat exchanger and are 100 percent efficient. The fins are spaced with pitch, $p_f = 0.5$ mm and the fins are $th_f = 0.10$ mm thick. The thickness of the metal separating the cold channels from the hot channels is $th_w = 0.20$ mm and the conductivity of this metal is $k_m = 15$ W/m-K. The hot and cold flows are at nominally atmospheric pressure and the compressor discharge pressure, respectively. The fouling factor associated with the flow of the combustion products leaving the turbine is $R''_f = 0.0018\ \text{K-m}^2/\text{W}$. There is no fouling associated with the flow of the air leaving the compressor.

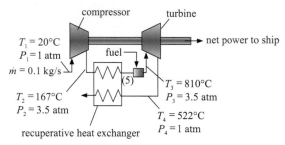

compressor   turbine

$T_1 = 20°C$
$P_1 = 1$ atm
$\dot{m} = 0.1$ kg/s

fuel

→ net power to ship

(5)

$T_3 = 810°C$
$P_3 = 3.5$ atm

$T_2 = 167°C$
$P_2 = 3.5$ atm

$T_4 = 522°C$
$P_4 = 1$ atm

recuperative heat exchanger

(c) Compute the heat transfer coefficient between the hot air from the turbine and the channel walls and the colder air from the compressor and the channel walls. You may use the inlet temperatures of the air flows to compute the properties.

(d) Compute the total conductance of the heat exchanger.

(e) Determine the heat transferred in the heat exchanger and the temperature of the air entering the combustor.

(f) What is the efficiency of the recuperated gas turbine engine?

(g) What is the mass of fuel that must be carried by the ship for the 2-day mission if it uses a recuperated gas turbine engine?

(h) The density of the metal separating the cold channels from the hot channels is $\rho_m = 8000$ kg/m³ and the density of the fins is $\rho_f = 7500$ kg/m³. What is the mass of the heat exchanger?

(i) What is the net savings in mass associated with using the air-to-air recuperated gas turbine engine for the 2-day mission.

(j) Plot the net savings in mass as a function of the width of the heat exchanger (W). Your plot should show a maximum; explain why.

12.53 A cryosurgical probe uses a Joule–Thomson (isenthalpic) expansion of argon gas to produce cold temperatures, as shown in the figure. The argon is compressed to p_2 = 200 atm and cooled at constant pressure to T_3 = 300 K where it enters a recuperative counter-flow heat exchanger. The gas is cooled to state (4) in the recuperator using the return gas that enters the recuperator at T_6 = 140 K. After leaving the recuperator at state (4), the argon is throttled to a low pressure state (5) where it used to cool tissue. The refrigeration load provided by the cryoprobe ($\dot{q}_{load}$) warms the gas to T_6 = 140 K. The argon returning to the compressor is at a pressure of p_1 = 100 kPa; pressure losses occur in the recuperator, but the pressure difference between states (5) and (6) is negligible.

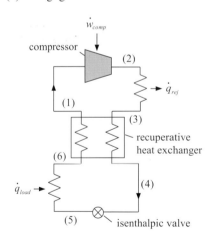

The recuperator is a smooth thin-walled stainless steel concentric tube heat exchanger. The high-pressure argon passes through the center tube and the low-pressure argon returns through the annulus. The length of the tubes is L = 0.35 m. The inner diameter of the outer tube diameter is D_{out} = 1.75 mm. The wall thickness of the tube is th = 180 μm. The inner diameter of the inner tube is D_{in} = 600 μm. The mass flow rate of argon is $\dot{m}$ = 0.3 g/s.

(a) Determine the conductance of the recuperative heat exchanger.

(b) Determine the heat exchanger effectiveness.

(c) Determine the pressures at states (4) and (6).

(d) Determine the temperatures at states (1), (4), (5), and (6).

(e) Determine the refrigeration load provided by the cryoprobe, $\dot{q}_{load} = \dot{m}\,(i_6 - i_5)$, where i is the specific enthalpy.

(f) Plot refrigeration load as a function of mass flow rate. You should see an optimal mass flow rate, why does such an optimal flow exist?

(g) Plot the recuperator effectiveness and the refrigeration load as a function of the inner diameter of the inner tube for 400 μm < D_{in} < 1.1 mm and 0.3 g/s mass flow rate. Is there an optimum inner diameter for the mass flow rate of 0.3 g/s? Why?

12.54 Ammonia is an ideal refrigerant in most respects, although it is an irritant at very low exposure levels and toxic above 1000 ppm. It is also weakly flammable. For these reasons, ammonia has not been used in supermarket refrigeration systems. It is possible, however, to confine the ammonia to the equipment room if a heat exchanger is employed to cool a secondary refrigerant. A 50 percent propylene glycol solution is heat exchanged with the ammonia. The propylene glycol solution is distributed to food cases in the supermarket. The glycol returns from the food cases at T_5 = –9°C and leaves the heat exchanger at T_6 = –15°C. The anticipated load on the food cases is 45 kW. The propylene glycol solution is non-toxic and nonflammable, eliminating the concerns with ammonia used in the supermarket. However, the use of the heat exchanger results in a reduction in the efficiency of the refrigerating equipment and extra pumping power. The ammonia flow rate is controlled with a thermostatic expansion valve to provide saturated vapor leaving the heat exchanger at state (1). The condenser provides saturated liquid at 40°C at state (3). The isentropic efficiency of the compressor is 0.72 and it operates adiabatically. Neglect pressure losses in the refrigeration piping. A shell-and-tube heat exchanger with two shell passes is employed, with the ammonia condensing in the shell and the glycol solution flowing in the tubes.

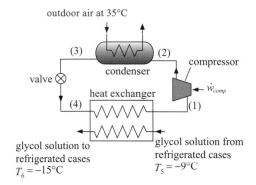

outdoor air at 35°C

(3) (2)
compressor
condenser
valve

heat exchanger

(4) (1)

glycol solution to refrigerated cases
$T_6 = -15°C$

glycol solution from refrigerated cases
$T_5 = -9°C$

(a) Prepare a plot of the power required to operate the compressor and the temperature at state (1) as a function of the conductance of the heat exchanger for values ranging from 1.5 kW/K $< UA <$ 30 kW/K.

(b) The shell-and-tube glycol–ammonia heat exchanger considered for this system has 12 parallel circuits of tubing with an inner diameter of 0.0125 m. The major resistance is due to the heat transfer coefficient on the glycol side of the heat exchanger. Calculate the required length of tubing for each circuit and the minimum pump power required to move the glycol solution through the heat exchanger as a function of the UA. What UA value would you recommend?

12.55 A factory requires 5000 gallons of hot (95°C) water each day. The water is not consumed; rather, it is used to provide heating in one part of a process and therefore goes down the drain at nominally 75°C. The process currently used to supply hot water to the factory is shown in Figure 1. The factory runs a single, 10-hour shift each day and therefore provides the hot water using a 5000 gallon supply tank. The supply tank is empty at the end of the shift and is subsequently filled with city water at 10°C. The cold city water is heated electrically during the 14 hr between the end of the shift and the beginning of the subsequent shift using a 150 kW heater. Therefore, the tank is full of 95°C water at the start of the shift and is emptied over the course of the next shift.

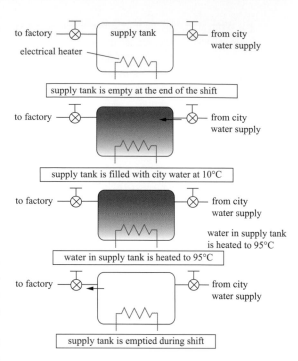

to factory — supply tank — from city water supply
electrical heater

supply tank is empty at the end of the shift

to factory — from city water supply

supply tank is filled with city water at 10°C

to factory — from city water supply

water in supply tank is heated to 95°C

water in supply tank is heated to 95°C

to factory — from city water supply

supply tank is emptied during shift

Figure 1 for Problem 12.55.

You have been asked to replace the current hot water system with an alternative system that utilizes an energy recovery system in order to reduce the cost of the energy required to heat the water. In the proposed system, a second 5000 gallon tank (the drain tank) will be used to store the 75°C water that would otherwise go down the drain, as shown in the figure. At the conclusion of the shift, the drain tank will contain 5000 gallons of 75°C water. This water has been contaminated and therefore cannot be re-used directly. Therefore, a two-part water preparation process is required. During the first process, the water in the drain tank is used to pre-heat the city water using a counterflow heat exchanger. The city water is heated in the counterflow heat exchanger and enters the supply tank at some intermediate temperature that is between 10°C and 75°C, depending on the total conductance (UA), of the counterflow heat exchanger. During the second part of the water preparation process, the electrical heater is activated and used to heat the water to its final temperature, 95°C, so that the next shift can start.

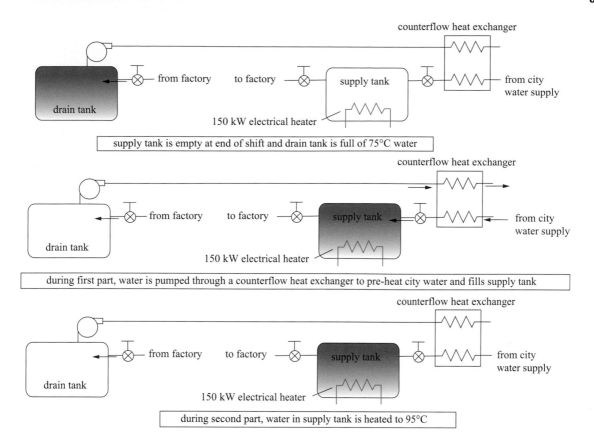

Figure 2 for Problem 12.55.

Assume that the total conductance of the counterflow heat exchanger is $UA = 6000$ W/K and the duration of the first phase of water preparation is $t_{phaseI} = 4$ hr. Neglect any energy loss from either the supply or drain tanks.

(a) What is the volumetric flow rate of city water required to completely fill the supply tank and empty the drain tank (gal/min) during the first phase of water preparation?

(b) What is the temperature of the water in the supply tank at the end of the first phase of water preparation (°C)?

(c) If the 150 kW heater is activated at the conclusion of the first phase of water preparation and remains on until the next shift starts (recall that the time between shifts is 14 hr), then what will the final temperature of the water in the supply tank be (°C)?

(d) Prepare a plot showing the final temperature of the water in the supply tank (°C) as a function of duration of the first phase of water preparation (t_{phaseI}). What is the optimal duration of the first phase of water preparation (i.e., what value of t_{phaseI} provides water at exactly 95°C for the next shift)?

(e) If the cost of electricity is 0.10$/kW-hr, then how much money will the proposed system save per day as compared with the original system ($/day)? Make sure you set the duration of the first part of water preparation to the optimal value you obtained from (d).

(f) The cost to install a new tank and pump and to modify the plumbing is estimated to be $25,000. The cost of the counterflow heat exchanger is proportional to its total conductance; the constant of proportionality is 10$/(W/K). Therefore, the heat exchanger with a total conductance of 6000 W/K will cost $60,000. Determine the simple payback (years) of the proposed system. That is, neglecting the time value of money, how many years of savings are required in order to exactly equal the initial outlay of capital cost for the system?

(g) Prepare a plot showing the payback (years) as a function of the total conductance of the heat exchanger (W/K). Be sure that you are adjusting both the duration of the first part of the water preparation process and the cost of the system as you change UA. What is the optimal value of UA?

12.56 The figure shows a portion of a solar water heating system that employs a concentric tube heat exchanger between the solar collector and the storage tank. The heat exchanger is 6 m in total flow length (with bends so that its actual length is a manageable 1 m). The outer diameter of the heat exchanger is 1.6 cm and in the inner diameter is 1 cm. The heat exchanger is made of thin-walled copper and is well insulated on the outside. A small pump circulates a 50 percent water–propylene glycol solution through the collector and through the center pipe of the heat exchanger at a rate of 0.12 liter/s. A second pump circulates water at 0.08 liter/s from the storage tank through the heat exchanger annulus. Both fluids are pressurized to approximately 1.5 bar. The water at the bottom of the storage tank (T_1) is 32°C. The rate of energy collection from the solar collector (Q_u) is:

$$Q_u = \dot{m}c_g(T_4 - T_3) = A_c(S - U_L(T_3 - T_{amb}))$$

where:

$\dot{m}$ is the mass flow rate of propylene glycol

c_g is the specific heat of the glycol solution

T_i is the temperature at point i

A_c is the collector area, which is 4.5 m² in this system

S is the absorbed solar radiation, which is 800 W/m²

U_L is the collector loss coefficient = 4.2 W/m²-K

T_{amb} is the ambient temperature, which is 12°C.

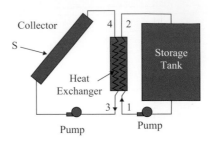

You may assume that there are no pipe losses so the rate of energy transferred to the collector is the rate of energy transfer through the heat exchanger to the water in the storage tank. You may also assume that there is no change in temperature of the streams due to the pumps. State and justify any other assumptions you employ.

(a) Estimate the average heat transfer coefficients in W/m²-K for the water and glycol solutions in the concentric pipe heat exchanger assuming fully developed flow conditions. Propylene glycol properties are available in EES through the brine property functions. To obtain information on brine properties, select the Function Info menu item in the Options menu and click the Thermodynamic properties radio button. Click on Brines and then click the Fluid Info button.

(b) Calculate the heat exchanger effectiveness for this 6 m heat exchanger.

(c) Calculate the rate of useful energy collection (Q_u) and temperatures T_2, T_3, and T_4 for this 6 m heat exchanger.

(d) Calculate the power required to run both pumps, assuming that the pumps each have an efficiency of 0.5.

(e) A contractor has proposed that the performance of the heat exchanger could be improved by using a more dilute solution of propylene glycol. Prepare plots of the heat exchanger effectiveness, rate of useful energy gain, and pump power as a function of the glycol concentration from percentages of 20 percent to 50 percent. Comment on the trends in the plots and indicate advantages and disadvantages of using a more dilute solution.

12.57 Injection molding machines in a factory require cooling. The cooling is provided by a flow of closed-loop treated water. The treated water is cooled using a plate frame heat exchanger. Energy is transferred in the heat exchanger to untreated water that flows through a cooling tower located external to the plant. The system is shown in the figure.

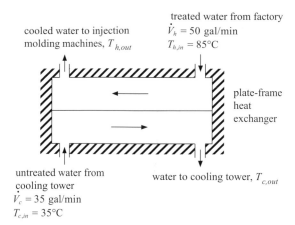

cooled water to injection molding machines, $T_{h,out}$

treated water from factory
$\dot{V}_h = 50$ gal/min
$T_{h,in} = 85°C$

plate-frame heat exchanger

untreated water from cooling tower
$\dot{V}_c = 35$ gal/min
$T_{c,in} = 35°C$

water to cooling tower, $T_{c,out}$

The volumetric flow rate of the treated water is $\dot{V}_h = 50$ gal/ min while the volumetric flow rate of the cooling tower water is $\dot{V}_c = 35$ gal/ min. The flow from the cooling tower is at $T_{c,in} = 35°C$. The cooling tower water is untreated and therefore has a fouling factor of $R''_c = 0.0004$ K-m^2/W. The nominal pressure of both streams in the heat exchanger is $p = 3$ atm. Assume that the water does not change phase in the heat exchanger. The temperature of the treated water returning from the factory is $T_{h,in} = 85°C$.

The plate frame heat exchanger is constructed by bolting together plates in order to alternately form channels for the hot and cold flows, as shown in the figure.

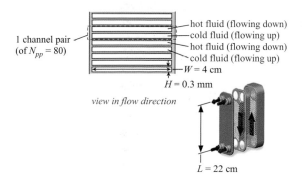

1 channel pair (of $N_{pp} = 80$)

— hot fluid (flowing down)
— cold fluid (flowing up)
— hot fluid (flowing down)
— cold fluid (flowing up)
— $W = 4$ cm
$H = 0.3$ mm

view in flow direction

$L = 22$ cm

There are $N_{pp} = 80$ pairs of plates (i.e., there are 80 hot flow channels and 80 cold flow channels). Each channel is $W = 4$ cm wide and $H = 0.3$ mm high. The plates are $L = 22$ cm long in the flow direction. Assume that the surface of the plates is smooth, $e = 0$. The hot and cold fluids flow in opposite directions; therefore, this is a counterflow heat exchanger.

(a) Determine the total conductance of the heat exchanger. You may neglect the conduction resistance of the plates separating the channels but you should consider fouling and convection on both sides.

(b) Determine the effectiveness of the heat exchanger and the rate of heat transfer from the treated water to the water from the cooling tower.

(c) The temperature of the cooling water returning from the plant, $T_{h,in}$, will change depending on the amount of cooling required by the injection molding machines. If the plant requires $\dot{q}_{plant} = 105$ kW of cooling then what is the value of $T_{h,in}$?

(d) The operating temperature of the injection molding machines is dictated by the temperature of the cooling water. For a fixed cooling load, $\dot{q}_{plant} = 105$ kW, you can reduce the operating temperature by using a larger heat exchanger (i.e., by adding plate pairs to the plate frame heat exchanger). Plot the temperature of the cooling water returning from the plant, $T_{h,in}$, as a function of the number of plate pairs, N_{pp}.

12.58 A relatively large experiment must be maintained at cryogenic temperatures. The experiment is placed some distance from the cryocooler that will provide the refrigeration. Therefore, it is not possible to conductively interface the experiment to the cryocooler. Instead, a system of three heat exchangers, shown in the figure, is used to transfer cooling from the cryocooler to the experiment.

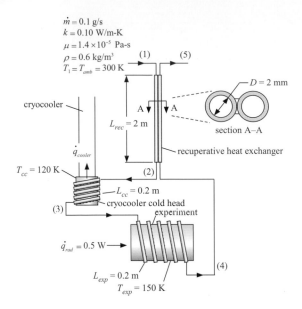

$\dot{m} = 0.1$ g/s
$k = 0.10$ W/m-K
$\mu = 1.4 \times 10^{-5}$ Pa-s
$\rho = 0.6$ kg/m³
$T_1 = T_{amb} = 300$ K

(1)   (5)

$D = 2$ mm

cryocooler

A   A

$L_{rec} = 2$ m

section A–A

recuperative heat exchanger

$\dot{q}_{cooler}$

$T_{cc} = 120$ K

(2)

$L_{cc} = 0.2$ m
cryocooler cold head
experiment

(3)

$\dot{q}_{rad} = 0.5$ W

(4)

$L_{exp} = 0.2$ m
$T_{exp} = 150$ K

Helium at $T_1 = T_{amb} = 300$ K is provided at state (1) to the warm end of a recuperative (counter-flow) heat exchanger. The mass flow rate of the helium is $\dot{m} = 0.1$ g/s and you may model helium as having constant properties, $k = 0.1$ W/m-K, $\mu = 1.4 \times 10^{-5}$ Pa-s, $c = 5200$ J/kg-K, and $\rho = 0.6$ kg/m³. The recuperative heat exchanger is fabricated from two tubes with length $L_{rec} = 2$ m and inner diameter $D = 2$ mm. The tubes are laid side-by-side and soldered together along their length. You may assume that the only resistance to heat transfer is convection from the inner surfaces to the helium. You may also neglect any fouling for this problem as the helium is high purity.

The helium leaving the recuperative heat exchanger at state (2) flows through a second heat exchanger that interfaces the helium with the cold head of the cryocooler. The cold head of the cryocooler may be modeled as an isothermal surface. The cold head heat exchanger consists of $L_{cc} = 0.2$ m of tube with inner diameter $D = 2$ mm that is wrapped around

the cold head and soldered in place; again, assume that convection is the only thermal resistance for this heat exchanger.

The helium leaving the cold head heat exchanger at state (3) flows through the third heat exchanger that interfaces the helium with the experiment. The experiment may be modeled as an isothermal surface. The experiment heat exchanger consists of $L_{exp} = 0.2$ m of tube with inner diameter $D = 2$ mm that is wrapped around the experiment and soldered in place; again, assume that convection is the only thermal resistance for this heat exchanger. The helium leaving the experiment heat exchanger at state (4) enters the cold end of the recuperative heat exchanger, pre-cooling the warm helium.

(a) Determine the effectiveness of the three heat exchangers.

At a particular time, the temperature of the cryocooler cold head is $T_{cc} = 120$ K and the temperature of the experiment is $T_{exp} = 150$ K.

(b) Determine the temperatures at states (2), (3), (4), and (5).

The cooling power of the cryocooler is a function of the cold head temperature, according to:

$$\dot{q}_{cooler} = \dot{q}_{coller,\,max}\left(1 - \frac{T_{nl}}{T_{cc}}\right)$$

where $\dot{q}_{cooler,\,max} = 60$ W and $T_{nl} = 70$ K. The total heat capacity of the cold head and experiment are $C_{cc} = 500$ J/K and $C_{exp} = 1200$ J/K.

(c) Determine the rate of temperature change of the cold head and experiment.

Use the Integral command in EES to determine the temperature of the cryocooler cold head and the experiment as a function of time. Initially, both the cold head and the experiment are at ambient temperature.

(d) Plot the cold head and experiment temperature as a function of time.

13 | Mass Transfer

Mass transfer occurs whenever fluid flows; that is, some mass is transferred from one place to another. However, the focus in this chapter is on the transport of one chemical species (or component) within a mixture that occurs as a direct result of a concentration gradient, independent of a pressure gradient. This type of mass transfer is called **diffusion**. A familiar example of diffusion mass transfer is the humidification process that occurs when an open container of water is allowed to sit out in a room. The gas in the room is a mixture of air (which is itself a mixture of oxygen, nitrogen, and other gases) and water vapor. The air–water mixture in contact with the surface of the liquid water is nearly saturated with water vapor and so it has a relatively high concentration of water vapor. The air–water mixture further from the liquid has a lower concentration of water vapor. Therefore, there is a concentration gradient that drives a mass transfer process causing water to be transported from the liquid surface to the air in the room, thereby humidifying it.

This textbook is focused on heat transfer; therefore, it is appropriate to ask why a chapter on mass transfer is included. There are two answers to this question. First, mass transfer, like momentum transfer, plays a key role in many important heat exchange processes and devices. For example, mass transfer is critical to the operation of the cooling coils, cooling towers, and evaporative coolers and condensers that are commonly used in refrigeration and power systems. The energy transfer that occurs as a result of mass transfer can significantly improve the performance of these heat transfer devices. Second, the processes of heat and mass transfer are analogous. The governing equations for heat and mass transfer are similar and therefore many of the relations and solution techniques that have been developed for heat transfer can be directly applied to mass transfer processes as well.

13.1 Composition Relationships

Diffusion involves a mixture of two or more distinct chemical species. In this chapter, we will assume that these species are *chemically inert* so that no chemical reactions occur. The methods that are used to quantify the relative amounts of the different species in the mixture are summarized in this section. These relationships can be equivalently expressed on either a mass or a molar basis.

The total mass of the mixture (m) is the sum of the mass of each constituent of the mixture (m_i, where i identifies the component):

$$m = \sum_{i=1}^{N_s} m_i,$$ (13.1)

where N_s is the number of species that are present in the mixture. The **mass fraction** of species i (mf_i) is the ratio of the mass of species i to the total mass of the mixture:

$$mf_i = \frac{m_i}{m}.$$ (13.2)

Substituting Eq. (13.2) into Eq. (13.1) leads to the requirement that the mass fractions of all of the species in a mixture must add to unity:

$$\sum_{i=1}^{N_s} mf_i = 1.$$ (13.3)

Since diffusion is a molecular phenomenon, it is often useful to specify composition on a molar rather than a mass basis. A **mole** of species i is defined as the amount of mass that is equal to the molar mass of species i (MW_i). Therefore, the mass units need to be specified in order to precisely define a mole. If the mass unit is chosen to be kg, then the corresponding mole is called a kmol (or kgmol). For example, the molar mass of helium is 4 and therefore a kmol of helium has a mass of 4 kg. A pound mole (lbmol) of helium has a mass of 4 pounds. A gram mole (gmol) of helium has a mass of 4 grams, and so on. Often the term "mole" is expressed without reference to mass units; in this case, mole usually refers to a gmol.

The total number of moles in a mixture (n) is the sum of the moles of each species (n_i):

$$n = \sum_{i=1}^{N_s} n_i. \tag{13.4}$$

The **mole fraction** of species i (y_i) is the number of moles of that species divided by the total moles of mixture:

$$y_i = \frac{n_i}{n}. \tag{13.5}$$

Substituting Eq. (13.5) into Eq. (13.4) provides the requirement that the mole fractions of all of the species in a mixture must sum to unity:

$$\sum_{i=1}^{N_s} y_i = 1. \tag{13.6}$$

The number of moles and mass of species i are related according to the molar mass of that species (MW_i):

$$n_i = \frac{m_i}{MW_i}. \tag{13.7}$$

Note that the total mass (m) and total number of moles (n) of a mixture are related by MW, the molar mass of the mixture:

$$MW = \frac{m}{n}. \tag{13.8}$$

The molar mass of the mixture can be computed according to:

$$MW = \sum_{i=1}^{N_s} y_i MW_i. \tag{13.9}$$

The mass fraction and mole fraction are related according to:

$$MW_i = \frac{m_i}{n_i} = \frac{mf_i\, m}{y_i\, n} = \frac{mf_i}{y_i} MW. \tag{13.10}$$

Diffusion mass transfer of a component is driven by gradients in the concentration of the component in a mixture. **Concentration** is related to the amount of the component that is present per unit volume. Concentration can be expressed on either a mass or a molar basis. On a mass basis, the **mass concentration** of species i (ρ_i, also called the density of species i) is the mass of species i per unit volume, which can be expressed as:

$$\rho_i = \frac{m_i}{V} = \frac{mf_i\, m}{V} = mf_i\, \rho \tag{13.11}$$

where V is the volume and ρ is the mass density of the mixture. The **molar concentration** (also called molar density) of species i (c_i) is the number of moles of species i per unit volume, which can be expressed as

$$c_i = \frac{n_i}{V} = \frac{y_i\, n}{V} = y_i\, c, \tag{13.12}$$

where c is the molar density (n/V) of the mixture.

Example 13.1

A mixture contains carbon dioxide, oxygen, and helium. The mole fractions of the three components are $y_1 = 0.25$ (CO_2), $y_2 = 0.25$ (O_2), and $y_3 = 0.5$ (He). Determine the molar mass of the mixture and the mass fraction of the three components. If the total mixture density is $\rho = 1.788$ kg/m^3 determine the (mass) concentration of each of the components.

Known Values

Figure 1 illustrates a sketch of the problem with the known values indicated.

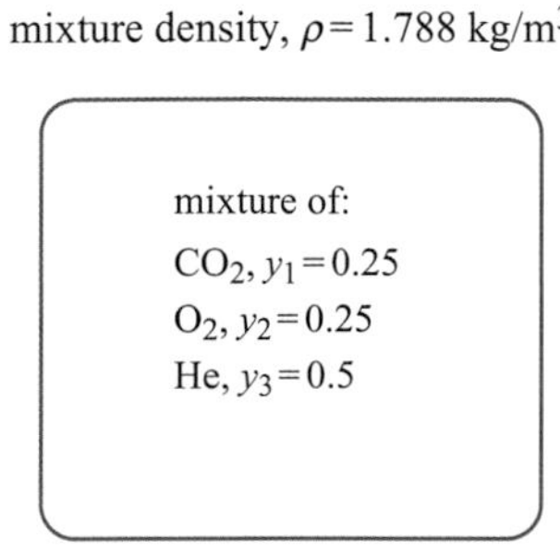

mixture density, $\rho = 1.788$ kg/m^3

mixture of:
$CO_2, y_1 = 0.25$
$O_2, y_2 = 0.25$
He, $y_3 = 0.5$

Figure 1 Mixture.

Analysis and Solution

The molar mass of each component is obtained from Appendix C, $MW_1 = 44.01$ kg/kmol, $MW_2 = 32.00$ kg/kmol, and $MW_3 = 4.003$ kg/kmol. The molar mass of the mixture is calculated using Eq. (13.9).

$$MW = \sum_{i=1}^{N_s} y_i \, MW_i = 0.25\,(44.01) + 0.25\,(32) + 0.55\,(4.003) = \boxed{21.00 \text{ kg/kmol}}. \tag{1}$$

The mass fraction of each of the components is computed using Eq. (13.10):

$$mf_1 = y_1 \frac{MW_1}{MW} = 0.25 \frac{44.01}{21} = \boxed{0.5238} \tag{2}$$

$$mf_2 = y_2 \frac{MW_2}{MW} = 0.25 \frac{32.00}{21} = \boxed{0.3809} \tag{3}$$

$$mf_3 = y_3 \frac{MW_3}{MW} = 0.5 \frac{4.003}{21} = \boxed{0.0953}. \tag{4}$$

The mass concentration of each species is obtained using Eq. (13.11):

$$\rho_1 = mf_1 \, \rho = 0.5238 \left(1.788 \frac{\text{kg}}{\text{m}^3} \right) = \boxed{0.9366 \frac{\text{kg}}{\text{m}^3}} \tag{5}$$

$$\rho_2 = mf_2 \, \rho = 0.3809 \left(1.788 \frac{\text{kg}}{\text{m}^3} \right) = \boxed{0.6810 \frac{\text{kg}}{\text{m}^3}} \tag{6}$$

$$\rho_3 = mf_3 \, \rho = 0.0953 \left(1.788 \frac{\text{kg}}{\text{m}^3} \right) = \boxed{0.1704 \frac{\text{kg}}{\text{m}^3}}. \tag{7}$$

Continued

Example 13.1 (cont.)

Discussion

Notice that the mass fraction of helium is much smaller than its mole fraction due to its relatively low molar mass compared to the other components. Helium makes up 50 percent of the mixture by mole but only 9.53 percent by mass.

13.1.1 Ideal Gas Mixtures

Diffusing species are often low-density gases that obey the Ideal Gas Law. Some of the most important mass transfer applications involve a mixture of dry air and water vapor at near atmospheric pressure and temperature that behaves as an ideal gas mixture. When applicable, the Ideal Gas Law may be used to compute the mass or molar concentrations by defining a **partial pressure** for each component, p_i, according to:

$$p_i = y_i \, p, \tag{13.13}$$

where p is the total pressure of the gas mixture. Substituting Eq. (13.10) into Eq. (13.13) provides:

$$p_i = mf_i \left(\frac{MW}{MW_i} \right) p. \tag{13.14}$$

Application of the Ideal Gas Law allows the mass and molar concentrations of species to be expressed in terms of the partial pressure of the species:

$$\rho_i = \frac{m_i}{V} = \frac{p_i}{R_i \, T} = \frac{y_i \, p}{R_i \, T} \tag{13.15}$$

$$c_i = \frac{n_i}{V} = \frac{p_i}{R_{univ} \, T} = \frac{y_i \, p}{R_{univ} \, T}, \tag{13.16}$$

where R_i is the gas constant for species i (on a mass basis) and R_{univ} is the universal gas constant (8314.3 J/kmol-K); note that R_{univ} and R_i are related according to:

$$R_i = \frac{R_{univ}}{MW_i}. \tag{13.17}$$

13.2 Mass Diffusion

13.2.1 Fick's Law

Mass diffusion refers to the diffusive transport of a species due to a concentration gradient in a mixture. Mass transport by diffusion is analogous to conduction, which is the diffusive transport of energy due to temperature gradients. Therefore, the molecular description of diffusion is similar to the molecular description of conduction that is presented in Section 2.1. Fourier's Law relates conduction heat flux to the temperature gradient using the thermal conductivity of the material:

$$\dot{q}'' = - k \frac{\partial T}{\partial x}. \tag{13.18}$$

Fick's Law relates the diffusive molar flux (or mass flux) of one species in a mixture to the gradient in the concentration of that species. Fick's Law written on a molar basis is

$$\dot{n}_i'' = - D_{i,m} \frac{\partial c_i}{\partial x} = - c \, D_{i,m} \frac{\partial y_i}{\partial x}, \tag{13.19}$$

where $\dot{n}_i''$ is the molar flow rate of species i in a mixture due to diffusion in the x-direction per unit area. On a mass basis, Fick's Law provides:

$$\dot{m}_i'' = - D_{i,m} \frac{\partial \rho_i}{\partial x} = - \rho \, D_{i,m} \frac{\partial mf_i}{\partial x}, \tag{13.20}$$

where $\dot{m}_i''$ is the diffusive mass flow rate in the x-direction per unit area of species i in a mixture. The quantity $D_{i,m}$ is the **diffusion coefficient** for species i in the mixture. Notice in Eqs. (13.19) and (13.20) that diffusive mass transfer is driven by a concentration gradient just as conduction heat transfer is driven by a temperature gradient. The diffusion coefficient, $D_{i,m}$, represents the underlying characteristics of the molecules that are responsible for the diffusive mass transfer. Notice that the diffusion coefficient has units m^2/s (in the SI system), as does thermal diffusivity (α) and kinematic viscosity (v). We will see that the diffusion coefficient plays the same role in diffusion mass transfer processes that α and v play in diffusion energy and momentum transfer processes, respectively.

In order to solve mass transfer processes using Fick's Law it will be necessary to estimate the appropriate diffusion coefficient as well as the concentration boundary conditions that represent the problem. Techniques for accomplishing this are presented in the next two sections.

13.2.2 The Diffusion Coefficient for Binary Mixtures

The diffusion coefficient, $D_{i,m}$, is a transport property that represents the ability of species i to diffuse through a medium m. The medium can be a gas, liquid, or even a solid. Typically, diffusion coefficients are largest for gases, lower for liquids, and lowest for solids. The diffusion coefficient is a mixture property that depends on the properties of all of the interacting species as well as on pressure and temperature. This is also true of the thermal conductivity and viscosity of a gas or liquid mixture.

Diffusion problems are often concerned with the mass transfer of a single species, such as water vapor, within an otherwise homogeneous phase, such as a pure gas or a homogeneous gas mixture (e.g., air). In this case, the mixture can be treated as a binary (i.e., two-component) system. The **binary diffusion coefficient** for species 1 through another species 2 is given the symbol $D_{1,2}$. It is possible to show that $D_{1,2}$ must be equal to $D_{2,1}$.

Gas Mixtures

Table 13.1 provides some values of the diffusion coefficients for various binary gas mixtures at $T_o = 25°C$ and $p_o = 1$ atm. Theoretical methods exist for estimating the binary diffusion coefficient of ideal gas mixtures based on kinetic theory. These are beyond the scope of this text but they suggest that the diffusion coefficient will be proportional to $T^{1.5}/p$ and therefore the diffusion coefficients listed in Table 13.1 at T_o and p_o can be approximately scaled to other temperatures and pressures according to:

$$D \approx D_o \left(\frac{T}{T_o}\right)^{1.5} \left(\frac{p_o}{p}\right), \tag{13.21}$$

where D_o is the value reported at temperature T_o and pressure p_o. The Chapman–Enskog method to estimate the binary diffusion coefficient for gases (Poling *et al.*, 2000) has been programmed in EES in the **D_12_gas** function that is part of the Diffusion Library in EES library routines. The **D_12_gas** function can be used to estimate the diffusion coefficient for any two gases that are available in the EES database.

Table 13.1 Selected binary diffusion coefficients for gas mixtures, from Bolz and Tuve (1976). All values are at $T_o = 25°C$ (298.1 K) and $p_o = 1$ atm (101,325 Pa).

Gas mixture	Diffusion coefficient (cm^2/s)	Gas mixture	Diffusion coefficient (cm^2/s)
Air–H_2	2.76	Air–formic acid	0.615
Air–NH_3	0.886	Air–benzene	0.341
Air–O_2	0.80	Air–xylene	0.275
Air–CO_2	0.635	Air–naphthalene	0.20

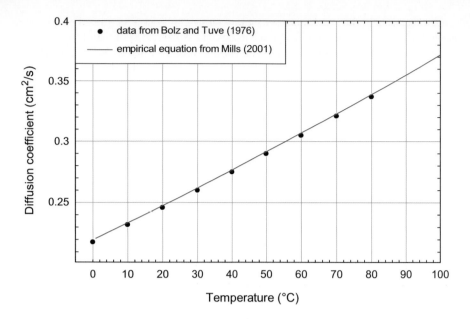

Figure 13.1 Diffusion coefficient for water vapor in air as a function of temperature at atmospheric pressure, from Bolz and Tuve (1976). Also shown is the empirical formula from Mills (2001).

One of the most important diffusion processes is associated with an air–water vapor mixture. The diffusion coefficient for this situation, $D_{a,w}$, is shown as a function of temperature in Figure 13.1 based on data from Bolz and Tuve (1976). The empirical formula from Mills (2001) is often used to estimate the diffusion coefficient of water vapor in air:

$$D_{a,w} = 1.97 \times 10^{-5} \left(\frac{T}{T_o} \right)^{1.685} \left(\frac{p_o}{p} \right), \tag{13.22}$$

where $T_o = 256$ K, $p_o = 1$ atm; note that $D_{a,w}$ is in m²/s, T is the temperature in K, and p is the pressure in atm. Equation (13.22) is also shown in Figure 13.1.

Example 13.2

Plot the diffusion coefficient of water vapor in air as a function of temperature estimated using the empirical formula given by Eq. (13.22) and the kinetic theory model predicted by the EES function **D_12_gas**.

Assumption

• Atmospheric pressure is assumed.

Analysis and Solution

The EES code below provides the two estimates requested in the problem statement. Figure 1 shows the diffusion coefficient estimated both ways as a function of temperature; note that the diffusion coefficient

Example 13.2 (cont.)

calculated by EES each way has units of m²/s but the plot is generated in cm²/s using the secondary units feature in EES.

```
$UnitSystem SI Mass J K Pa
//T=ConvertTemp(C,K,20 [C])              "Temperature"
p=1 [atm]*Convert(atm,Pa)               "Pressure"
T_o=256 [K]                             "Reference temperature"
p_o=1 [atm]*Convert(atm,Pa)             "Reference pressure"
D_a_w=1.97e-5 [m^2/s]*(T/T_o)^1.685*(p_o/p)  "empirical equation"
D_a_w_EES=D_12_gas('H2O','Air',T,P)     "kinetic theory"
```

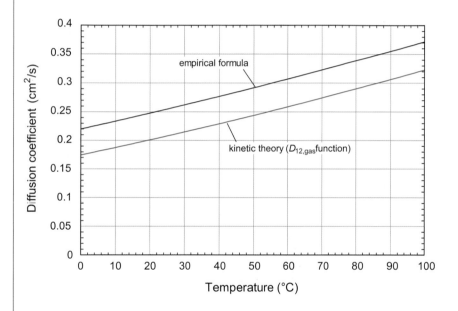

Figure 1 Diffusion coefficient for water in air estimated using Eq. (13.22) and using kinetic theory.

Discussion

Kinetic theory appears to under-predict the diffusion coefficient by approximately 20 percent relative the more accurate empirical formula based on data.

Liquid Mixtures

Table 13.2 provides some values of the diffusion coefficients for species in a dilute solution of water at $T_o = 20°C$ and $p_o = 1$ atm. The diffusion coefficients from Table 13.2 can be estimated at other conditions using the theoretically based scaling relationship

$$D = D_o \left(\frac{T}{T_o}\right) \left(\frac{\mu_o}{\mu}\right),$$

(13.23)

Table 13.2 Selected diffusion coefficients for dilute solutions of water, from Bolz and Tuve (1976). All values are at $T_o = 20°C$ (293.1 K) and $p_o = 1$ atm (101325 Pa), which corresponds to a water viscosity of $\mu_o = 0.001003$ Pa-s.

Mixture	Diffusion coefficient (m²/s)	Ref.	Mixture	Diffusion coefficient (m²/s)	Ref.
water–H_2	5.13×10^{-9}	2	water–NaCl	1.35×10^{-9}	2
water–O_2	1.77×10^{-9}	1	water–acetic acid	0.88×10^{-9}	2
water–CO_2	1.77×10^{-9}	2	water–glycerol	0.72×10^{-9}	2
water–NH_3	1.51×10^{-9}	1	water–sucrose	0.45×10^{-9}	2
water–methane	1.54×10^{-9}	1	water–benzene	0.93×10^{-9}	1

1. From the EES function D_12|o_liquid which uses a modification of the Tyn and Calus method described in Poling *et al.* (2000).
2. From Bolz and Tuve (1976).

Table 13.3 Selected coefficients for Eq. (13.24) from Mills (2001) that can be used to estimate the diffusion coefficient for solids.

Mixture	D_o (m²/s)	E_a (kJ/kmol)	Mixture	D_o (m²/s)	E_a (kJ/kmol)
O_2–Pyrex	6.19×10^{-8}	4.69×10^4	He–Pyrex	4.76×10^{-8}	2.72×10^4
O_2–titanium	5.0×10^{-3}	2.13×10^5	He–borosilicate glass	1.94×10^{-9}	2.34×10^4
H_2–iron	7.60×10^{-8}	5.60×10^3	Ne–borosilicate glass	1.02×10^{-10}	3.77×10^4
H_2–zirconium	1.09×10^{-7}	4.81×10^4	C–FCC iron	2.3×10^{-5}	1.378×10^5

where D_o is the value reported at temperature T_o and water viscosity μ_o. There are theoretical methods for estimating the binary diffusion coefficient of dilute solutions that are beyond the scope of this text. The method described in Poling *et al.* (2000) has been programmed in EES in the D_12|o_liquid function that is part of the Diffusion Library in EES library routines.

Solids

Table 13.3 provides the coefficients D_o and E_a that can be used to estimate the diffusion coefficients for some solids with the equation

$$D = D_o \exp\left(-\frac{E_a}{R_{univ} T}\right).$$ (13.24)

13.2.3 Concentrations at Interfaces

The boundary conditions for diffusion problems are written in terms of concentration in the same way that the boundary conditions for heat transfer problems are expressed in terms of temperatures. This section provides some methods for estimating the concentration boundary condition that should be applied at various types of interfaces.

Gas Mixture in Contact with Pure Liquid or Solid

One situation that often occurs in mass transfer processes is a gas mixture that is in contact with a nearly pure liquid or solid. For example, an air–water vapor mixture may be in contact with pure liquid water as shown in Figure 13.2.

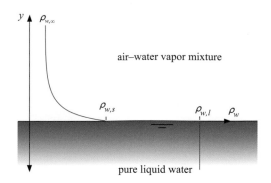

Figure 13.2 Liquid water in contact with an air–water vapor mixture.

In this situation the water vapor at the interface must be in thermodynamic equilibrium with the pure liquid at the interface. This implies that the partial pressure of the water vapor in the gas mixture is equal to the saturation pressure of water evaluated at the interface temperature. The mass concentration of water in the gas mixture at the interface can then be calculated using Eq. (13.15), assuming the Ideal Gas Law holds:

$$\rho_{w,s} = \frac{p_{sat}(T_s)}{R_w T_s}.$$ (13.25)

Liquid Mixture in Contact with Pure Solid
Another situation that can occur is a liquid mixture that is in contact with a nearly pure solid. For example, a solid block of sodium bicarbonate may be dissolving in water as shown in Figure 13.3. In this situation, thermodynamic equilibrium at the interface is specified in terms of **solubility**, a quantity that can be found in various handbooks, for example Perry *et al.* (1963). Table 13.4 lists the solubility of some compounds in water at 20°C.

The units of the solubility constant, S, are g per 100 g of liquid. Therefore, the mass fraction of the solute (i) at the interface is

$$mf_{i,s} = \frac{S}{S + 100.}$$ (13.26)

The mass concentration of the solute at the interface can be computed using Eq. (13.11):

$$\rho_{i,s} = mf_{i,s} \, \rho,$$ (13.27)

where ρ is the density of the mixture.

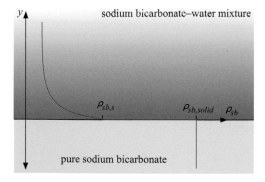

Figure 13.3 Sodium bicarbonate in contact with a sodium bicarbonate–water mixture.

Table 13.4 Solubility of some compounds in water at 20°C, from Perry *et al.* (1963).			
Substance	S (g/100 g of water)	Substance	S (g/100 g of water)
Calcium bicarbonate	16.60	Nickel chloride	64.2
Iodine	0.029	Potassium carbonate	110.5
Lead bromide	0.85	Sodium bicarbonate	9.6
Magnesium chloride	54.5	Sodium chloride	36.0

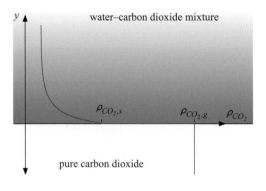

Figure 13.4 Carbon dioxide in contact with a water–carbon dioxide mixture.

Liquid Mixture in Contact with Gas

Another situation that can occur in a mass transfer process is a liquid mixture that is in contact with a gas in a situation where we are interested in the mass transfer occurring in the liquid. Bubbling CO_2 gas through a water–carbon dioxide mixture, as shown in Figure 13.4, is an example of this transport process.

In the situation where the gas is only slightly soluble in the liquid, thermodynamic equilibrium at the interface is specified in terms of **Henry's Law**, which states that the mole fraction of species i in the liquid ($y_{i,s}$) is equal to the mole fraction in the gas ($y_{i,g}$) divided by the **Henry number** (He_i):

$$y_{i,s} = \frac{y_{i,g}}{He_i}. \tag{13.28}$$

The Henry number is given by the ratio of the **Henry constant** ($C_{He,i}$) to the total pressure in the gas:

$$He_i = \frac{C_{He,i}}{p}. \tag{13.29}$$

The Henry constant for a few species in an aqueous solution are listed in Table 13.5.

The mass fraction of species i in the liquid can be computed using Eq. (13.10) and the concentration from Eq. (13.11). The EES function HenryConstant_Water can also be used to provide an estimate of Henry's constant as a function of temperature for many substances in the EES database.

Table 13.5 Henry's constant for a few species in an aqueous solution at 300 K, from Sander (2015).			
Substance	C_{He} (atm)	Substance	C_{He} (atm)
CO_2	1762	CO	58,595
O_2	47,757	Air	75,521
H_2	71,715	N_2	89,359

Example 13.3

Bubbles of carbon dioxide are being passed through water at $p = 5$ atm and $T = 47°C$. Estimate the diffusion coefficient associated with the water–carbon dioxide mixture and the concentration of carbon dioxide at the liquid interface.

Known Values

Figure 1 illustrates a sketch of the problem with the known values indicated.

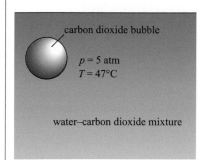

carbon dioxide bubble

$p = 5$ atm
$T = 47°C$

water–carbon dioxide mixture

Figure 1 Carbon dioxide gas passing through water in a carbonation process.

Assumptions

- The concentration of carbon dioxide in the water is sufficiently small that Henry's Law applies.
- The saturated liquid properties of liquid water can be used.

Analysis and Solution

The diffusion coefficient for carbon dioxide in water at $T_o = 20°C$ (293.1 K) and $p_o = 1$ atm (where the viscosity of water is $\mu_o = 0.001003$ Pa-s) is obtained from Table 13.2, $D_o = 1.77 \times 10^{-9}$ m²/s. The viscosity of water at $T = 47°C$ (320.2 K) is obtained from Appendix D, $\mu = 0.577 \times 10^{-3}$ Pa-s. Equation (13.23) is used to estimate the diffusion coefficient at 47°C:

$$D = D_o \left(\frac{T}{T_o}\right)\left(\frac{\mu_o}{\mu}\right) = 1.770 \times 10^{-9} \frac{m^2}{s} \left(\frac{320.2}{293.1}\right)\left(\frac{0.001003}{0.577 \times 10^{-3}}\right) = \boxed{3.361 \times 10^{-9} \frac{m^2}{s}}. \tag{1}$$

Henry's constant for carbon dioxide in water at 300 K is obtained from Table 13.5, $C_{He} = 1762$ atm. (The value returned by the EES HenryConstant_Water function is 1739 atm.) The Henry number is obtained from Eq. (13.29):

$$He = \frac{C_{He}}{p} = \frac{1762 \text{ atm}}{5 \text{ atm}} = 352. \tag{2}$$

The mole fraction of carbon dioxide on the liquid side of the interface is estimated using Henry's Law, Eq. (13.28):

$$y_{CO_2,s} = \frac{y_{i,g}}{He_i} = \frac{1}{352} = 0.00284. \tag{3}$$

The mass fraction is obtained using Eq. (13.10) where the molar mass of the mixture is taken to be the molar mass of water (obtained from Appendix D, $MW = 18.02$ kg/kmol) and the molar mass of carbon dioxide is obtained from Appendix C, $MW_{CO_2} = 44.01$ kg/kmol:

Continued

Example 13.3 (cont.)

$$mf_{CO_2} = y_{CO_2}\frac{MW_{CO_2}}{MW} = 0.00284\left(\frac{44.01}{18.02}\right) = 0.00693. \tag{4}$$

The mass concentration of carbon dioxide on the liquid side of the interface is obtained using Eq. (13.11) where the density of the mixture is taken to be the density of water from Appendix D, $\rho = 989.4$ kg/m³:

$$\rho_{CO_2} = mf_{CO_2}\,\rho = 0.00693\left(989.4\,\frac{\text{kg}}{\text{m}^3}\right) = \boxed{6.857\,\frac{\text{kg}}{\text{m}^3}}. \tag{5}$$

Discussion

The situation analyzed here is one method that can be used to carbonate water.

13.3 Transient Diffusion through a Stationary Solid

Diffusion in a fluid is usually accompanied by fluid motion, as we will discussin Section 13.4. However, in this section, we consider the situation where the medium that the species is diffusing through is *stationary*; for example, the diffusion of a gas through a solid. One purpose of this analysis is to illustrate the similarities that exist between the equations that govern conduction problems and those that govern mass diffusion problems. Many of the solution techniques discussed in Chapters 2 through 6 for conduction heat transfer can be applied directly to mass transfer problems of this type.

Consider the transient diffusion problem shown in Figure 13.5. A semi-infinite, stationary medium initially has a uniform mass concentration of species 1, $\rho_{1,ini}$, when the surface (at $x = 0$) is exposed to a fluid that contains species 1. Therefore, the surface of the medium experiences a step change in mass concentration of species 1, from $\rho_{1,ini}$ to $\rho_{1,s}$. There are no chemical reactions occurring in the medium.

The governing differential equation for the diffusion of species 1 through the medium can be derived using the same techniques that were used throughout this text for heat transfer problems; the difference is that here we will balance the mass of species 1 rather than energy. A differential control volume is defined, as shown in Figure 13.5, and a mass balance on species 1 within the control volume leads to:

$$\dot{m}_{1,x} = \dot{m}_{1,x+dx} + \frac{\partial m_1}{\partial t}, \tag{13.30}$$

where $\dot{m}_1$ is the mass flow rate of species 1, and m_1 is the mass of species 1 within the control volume. Expanding the $x+dx$ term in the usual manner leads to:

$$\dot{m}_{1,x} = \dot{m}_{1,x} + \frac{\partial \dot{m}_{1,x}}{\partial x}dx + \frac{\partial m_1}{\partial t}. \tag{13.31}$$

The only mechanism for the flow of species 1 is due to diffusion since the medium is stationary. Therefore, the mass flow rate of species 1 is given by Fick's Law, Eq. (13.20):

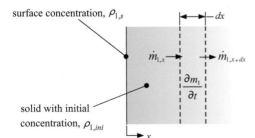

Figure 13.5 Transient diffusion of species 1 through a stationary semi-infinite solid.

$$\dot{m}_1 = - A_c\, D_{1,m}\, \frac{\partial \rho_1}{\partial x},$$
(13.32)

where A_c is the cross-sectional area for diffusion and $D_{1,m}$ is the diffusion coefficient of species 1 through the medium. The rate of change of the mass of species 1 in the differential control volume can be expressed as:

$$\frac{\partial m_1}{\partial t} = A_c\, dx \frac{\partial \rho_1}{\partial t}.$$
(13.33)

Substituting Eqs. (13.32) and (13.33) into the mass balance, Eq. (13.31), results in

$$0 = \frac{\partial}{\partial x}\left[- A_c\, D_{1,m} \frac{\partial \rho_1}{\partial x}\right] dx + A_c\, dx \frac{\partial \rho_1}{\partial t}.$$
(13.34)

Assuming that the diffusion coefficient and cross-sectional area are constant, Eq. (13.34) can be rearranged to yield the governing differential equation for transient 1-D diffusion through a stationary medium:

$$D_{1,m} \frac{\partial^2 \rho_1}{\partial x^2} = \frac{\partial \rho_1}{\partial t}.$$
(13.35)

The boundary conditions for Eq. (13.35) are related to the concentration at the surface

$$\rho_{1,x=0} = \rho_{1,s},$$
(13.36)

the initial concentration in the solid

$$\rho_{1,t=0} = \rho_{1,ini},$$
(13.37)

and the concentration far from the surface

$$\rho_{1,x\to\infty} = \rho_{1,ini}.$$
(13.38)

It is important to recognize the similarity in the governing relations describing heat and mass transfer. Equation (13.35) is analogous to the transient 1-D conduction equation that governs the temperature within a semi-infinite body, derived in Section 6.2.1:

$$\alpha \frac{\partial^2 T}{\partial x^2} = \frac{\partial T}{\partial t}.$$
(13.39)

If temperature (T) in Eq. (13.39) is replaced with mass concentration of species 1 (ρ_1), and the thermal diffusivity (α) in Eq. (13.39) is replaced with the diffusion coefficient ($D_{1,m}$), then Eq. (13.39) is identical to Eq. (13.35). The solutions for the transient conduction problems presented in Chapter 6 can be directly employed for mass transfer problems, assuming that the boundary conditions are also analogous. Section 6.2 presents several solutions for semi-infinite and bounded transient conduction problems and all of these solutions can be applied to diffusive mass transfer problems in a stationary medium. Further, the diffusion of mass is characterized by a mass diffusion penetration depth, δ_d, just as the diffusion of energy and momentum are characterized by thermal and momentum penetration depths, δ_t and δ_m, respectively. The mass diffusion penetration depth grows approximately according to:

$$\delta_d = 2\sqrt{D_{1,m}\, t}.$$
(13.40)

Example 13.4

A thin plate of borosilicate glass has been placed in a helium environment for a long time so that it has reached a uniform concentration of helium across its thickness, $\rho_{He,ini} = 0.002\ \text{kg/m}^3$. It is removed from the helium environment and placed in an oven filled with nitrogen gas for $t = 7$ days. The half-thickness of the

Continued

Example 13.4 (cont.)

plate is $L = 1$ mm and the temperature of the oven is $T = 100°C$. Estimate the concentration of helium at the center of the plate at the end of the process.

Known Values

Figure 1 illustrates a sketch of the problem with the known values indicated.

borosilicate glass at $T = 100°C$

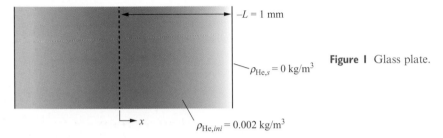

Figure 1 Glass plate.

Assumptions

- The concentration of helium at the surface after being placed in the oven is $\rho_{He,s} = 0$.
- The plate is thin enough that the concentration distribution is 1-D.
- The one-term solution for transient conduction in a bounded plane wall presented in Section 6.2.3 can be used. This implies that the mass transfer equivalent of the Fourier number is sufficiently large, greater than about 0.2.

Analysis and Solution

The diffusion coefficient for helium in borosilicate glass is obtained using Eq. (13.24) with $D_o = 1.94 \times 10^{-9}$ m²/s, $E_a = 2.34 \times 10^4$ kJ/kmol:

$$D = D_o \exp\left(-\frac{E_a}{R_{univ}\,T}\right)$$

$$= 1.94 \times 10^{-9}\,\frac{m^2}{s}\, \exp\left(-\frac{2.34 \times 10^4\,kJ}{kmol}\left|\frac{kmol\,K}{8.314\,kJ}\right|\frac{1}{373.1\,K}\right) = 1.029 \times 10^{-12}\,\frac{m^2}{s}. \tag{1}$$

The solution to the diffusion of energy (conduction) in a bounded plane wall is presented in Section 6.2.3 and culminates in the approximate, one-term solution for the dimensionless temperature difference:

$$\tilde{\theta}(\tilde{x}, Fo) \approx \left[\frac{2 \sin(\zeta_1)}{\zeta_1 + \cos(\zeta_1)\sin(\zeta_1)}\right] \exp\left(-\zeta_1^2\,Fo\right) \cos(\zeta_1\,\tilde{x}), \tag{2}$$

where $\tilde{\theta}$, $\tilde{x}$, and Fo are defined according to:

$$\tilde{x} = \frac{x}{L} \tag{3}$$

$$\tilde{\theta} = \frac{T - T_\infty}{T_{ini} - T_\infty} \tag{4}$$

$$Fo = \frac{t\,\alpha}{L^2}. \tag{5}$$

Example 13.4 (cont.)

The parameter ζ_1 is the first eigenvalue for the problem, which depends on the Biot number. The case where the surface temperature or, in this case, concentration is specified corresponds to an infinitely large heat transfer coefficient or $Bi \to \infty$ which leads to $\zeta_1 = \pi/2$ according to Table 6.2.

The conduction problem is analogous to a mass diffusion problem with the substitution of ρ_{He} for T and D for α:

$$\tilde{\theta} = \frac{\rho_{He} - \rho_{He,s}}{\rho_{He,ini} - \rho_{He,s}} \tag{6}$$

$$Fo = \frac{t\,D}{L^2} = \frac{7\text{ day}}{} \left|\frac{1.029 \times 10^{-12}\text{ m}^2}{\text{s}}\right| \frac{1}{\left|(0.001)^2\text{ m}^2\right|} \left|\frac{24\text{ hr}}{\text{day}}\right| \frac{60\text{ min}}{\text{hr}} \left|\frac{60\text{ s}}{\text{min}}\right| = 0.6221. \tag{7}$$

Because the Fourier number is larger than 0.2 the single term approximation is appropriate. Equation (2) is used to evaluate the dimensionless concentration at $x = 0$ (corresponding to the center of the plate):

$$\tilde{\theta} \approx \left[\frac{2\sin(\zeta_1)}{\zeta_1 + \cos(\zeta_1)\sin(\zeta_1)}\right] \exp\left(-\zeta_1^2\,Fo\right)\cos(\zeta_1\,\tilde{x})$$

$$= \left[\frac{2\sin\left(\frac{\pi}{2}\right)}{\frac{\pi}{2} + \cos\left(\frac{\pi}{2}\right)\sin\left(\frac{\pi}{2}\right)}\right] \exp\left[-\left(\frac{\pi}{2}\right)^2(0.6221)\right]\cos(0) = 0.2744, \tag{8}$$

which provides the concentration by using Eq. (6):

$$\rho_{He} = \rho_{He,s} + \tilde{\theta}\left(\rho_{He,ini} - \rho_{He,s}\right) = 0 + 0.2744\left(0.002\,\frac{\text{kg}}{\text{m}^3} - 0\right) = \boxed{0.000549\,\frac{\text{kg}}{\text{m}^3}}. \tag{9}$$

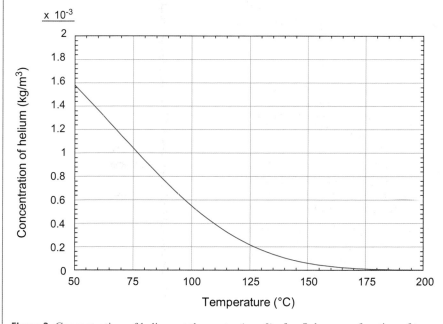

Figure 2 Concentration of helium at the center ($x = 0$) after 7 days as a function of temperature.

Continued

Example 13.4 (cont.)

Discussion

The problem shows how energy and mass diffusion are analogous mechanisms and therefore solutions obtained for conduction problems can be used to analyze mass diffusion problems. Note that the diffusion process is accelerated by the increased temperature because the diffusion coefficient is a strong function of temperature. Figure 2 illustrates the concentration at the center of the plate after 7 days as a function of temperature and shows that increasing the temperature beyond about 180°C would essentially eliminate any helium in the glass.

13.4 Diffusion of a Species in a Fluid

Section 13.3 considers the diffusion of a species through a stationary solid. The diffusion of a species through a fluid is more complex because of the bulk flow of fluid that may result from the diffusion process. Even if the fluid itself is otherwise stationary, the diffusive mass transfer process itself may cause a bulk fluid motion. This phenomenon is referred to as **Stefan Flow**.

13.4.1 Diffusive and Advective Mass Transfer

Fick's Law predicts the mass transfer of species i due to *diffusion only*. If there is any bulk motion associated with the medium that species i is diffusing through, then there is also a **bulk mass transfer** (sometimes called an **advective mass transfer**) that is associated with the bulk velocity. Both mass transfers must be considered when a gas or liquid is being analyzed.

Diffusion is driven by differences or gradients in concentration. In an ideal gas mixture, the concentration differences are associated with a spatial variation in mass density, which complicates the analysis. At a constant temperature and pressure, however, the molar density, c (kmol/m^3), of an ideal gas mixture is constant. For this reason, it is more convenient to present the mass flux of a species in a gas mixture on a molar basis, i.e., as a molar flux. The molar flux of species i can then be converted to a mass flux by multiplying it by the molar mass of species i.

If species i is diffusing in the x-direction through a fluid medium (designated by the subscript m) then the total molar flux of species i ($\dot{n}_i''$) is the sum of the rates of transfer due to diffusion (obtained by Fick's Law) and due to the bulk motion of the fluid:

$$\dot{n}_i'' = \underbrace{-D_{i,m}\frac{\partial c_i}{\partial x}}_{\text{diffusive flux}} + \underbrace{u\,c_i}_{\substack{\text{advective}\\\text{flux}}}, \tag{13.41}$$

where u is the total velocity of the mixture in the x-direction and c_i is the concentration of species i. The product of the total velocity of the mixture in the x-direction and the molar density of the mixture is the total molar flux

$$\dot{n}'' = u\,c, \tag{13.42}$$

where c is the molar density of the mixture. The total molar flux is the sum of the molar fluxes for each component:

$$\dot{n}'' = \sum_{j=1}^{N_s} \dot{n}_j''. \tag{13.43}$$

Substituting Eq. (13.43) into Eq. (13.42) and solving for u provides:

$$u = \frac{1}{c}\sum_{j=1}^{N_s} \dot{n}_j''. \tag{13.44}$$

Substituting Eq. (13.44) into Eq. (13.41) leads to:

$$\dot{n}_i'' = -D_{i,m}\frac{\partial c_i}{\partial x} + \frac{c_i}{c}\sum_{j=1}^{N_s}\dot{n}_j''. \tag{13.45}$$

Substituting Eq. (13.12) into Eq. (13.45) provides:

$$\dot{n}_i'' = -c\,D_{i,m}\frac{\partial y_i}{\partial x} + y_i\sum_{j=1}^{N_s}\dot{n}_j''. \tag{13.46}$$

13.4.2 Evaporation through a Layer of Gas

Figure 13.6 illustrates a steady-state mass diffusion process involving two chemical species. Species 1 is evaporating from the surface of pure liquid at the bottom of a tube and diffusing through the gas above the liquid. The gas is a binary mixture consisting of species 1 vapor and species 2.

The mole fraction of species 1 in the gas at the top of the tube is fixed and equal to $y_{1,\infty}$. The mole fraction of species 1 in the gas immediately above the liquid is equal to its larger, saturation value, $y_{1,sat}$, as discussed in Section 13.2.3. Therefore, the mass transfer of species 1 is driven from the liquid surface to the top of the container by the resulting concentration gradient. If the process continues for a sufficiently long time then all of the liquid will finally evaporate if it is not continuously replenished. It is evident that there must be a net flow of species 1 in the positive x-direction. However, it is less evident that species 2 must be diffusing in the *opposite* direction due to an opposing gradient in its mole fraction. This diffusion process must actually induce a bulk flow of the fluid in the column of gas; this bulk flow is referred to as **Stefan Flow**.

A differential control volume is defined as shown in Figure 13.6 and a balance on species 2 is carried out on a molar basis. Steady-state conditions can occur if the mole fraction of species 1 at the top of the tube and at $x = 0$ are constant; this requires that the level of liquid 1 be maintained constant by admitting liquid to the tube at the same rate as it evaporates. At steady state, the molar flux is constant and therefore

$$\dot{n}_{2,x}''\,A_c = \dot{n}_{2,x+dx}''\,A_c, \tag{13.47}$$

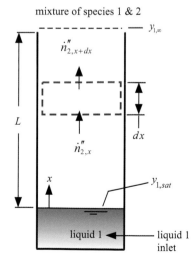

Figure 13.6 Mass transfer through a gas.

where A_c is the cross-sectional area of the tube. Dividing Eq. (13.47) through by A_c and expanding the $x + dx$ term leads to:

$$\dot{n}''_{2,x} = \dot{n}''_{2,x} + \frac{d\dot{n}''_2}{dx} dx, \tag{13.48}$$

which can be simplified to:

$$\frac{d\dot{n}''_2}{dx} = 0. \tag{13.49}$$

Equation (13.49) requires that the rate of flow of species 2 must be constant everywhere since there is no storage of the species in the tube and also no sources or sinks. Integrating Eq. (13.49) one time leads to:

$$\dot{n}''_2 = C_1, \tag{13.50}$$

where C_1 is a constant of integration. A similar differential balance carried out on species 1 leads to:

$$\dot{n}''_1 = C_2, \tag{13.51}$$

which indicates that the net mass flux of species 1 must also be constant for all values of x.

The solubility of gas species 2 in the liquid is assumed to be zero (i.e., the liquid is pure species 1). Therefore, the liquid surface effectively acts as a barrier to the flux of species 2 and the boundary condition at the surface of the liquid is

$$\dot{n}''_{2\,x=0} = 0. \tag{13.52}$$

Substituting Eq. (13.52) into Eq. (13.50) leads to:

$$\dot{n}''_2 = 0. \tag{13.53}$$

Equation (13.53) indicates that the net molar flux of species 2 must be zero everywhere along the length of the tube. The net molar flux of species 1 is given by Eq. (13.46):

$$\dot{n}''_1 = -c\,D_{1,2}\frac{dy_1}{dx} + y_1\left(\dot{n}''_1 + \dot{n}''_2\right). \tag{13.54}$$

Note that at constant temperature and pressure, the molar concentration of the mixture, c, is constant both in time and in position. Substituting Eq. (13.53) into Eq. (13.54) leads to:

$$\dot{n}''_1 = -c\,D_{1,2}\frac{dy_1}{dx} + y_1\,\dot{n}''_1, \tag{13.55}$$

which can be rearranged to give

$$\dot{n}''_1(1-y_1) = -c\,D_{1,2}\frac{dy_1}{dx}. \tag{13.56}$$

Equation (13.56) can be separated and integrated:

$$-\frac{\dot{n}''_1}{c\,D_{1,2}}\int_{x=0}^{x=L} dx = \int_{y_{1,sat}}^{y_{1,\infty}} \frac{dy_1}{(1-y_1)}. \tag{13.57}$$

Notice that $\dot{n}''_1$ was removed from the integral on the left side of Eq. (13.57) because it is not a function of x, as shown by Eq. (13.51). An integration variable, w, is defined as

$$w = 1 - y_1 \tag{13.58}$$

and substituted into Eq. (13.57):

$$-\frac{\dot{n}''_1}{c\,D_{1,2}}\int_{x=0}^{x=L} dx = -\int_{1-y_{1,sat}}^{1-y_{1,\infty}} \frac{dw}{w}. \tag{13.59}$$

Carrying out the integration in Eq. (13.59) leads to:

$$-\frac{\dot{n}_1'' L}{c\, D_{1,2}} = -\ln\left[\frac{(1 - y_{1,\infty})}{(1 - y_{1,sat})}\right]. \tag{13.60}$$

Solving for the molar flux of species 1:

$$\dot{n}_1'' = \frac{c\, D_{1,2}}{L}\ln\left[\frac{(1 - y_{1,\infty})}{(1 - y_{1,sat})}\right] = \frac{\dot{m}_1''}{MW_1}. \tag{13.61}$$

Equation (13.61) is called **Stefan's Law**. Note that $\dot{n}_1''$ will be positive when $y_{1,sat} > y_{1,\infty}$ and negative when $y_{1,sat} < y_{1,\infty}$.

When the mole fraction of species 1 is small (e.g., less than about 0.1), then the natural log terms in Eq. (13.61) can be represented by the approximation

$$\ln(1 - y) \approx -y \quad \text{for small } y. \tag{13.62}$$

With this approximation, Eq. (13.61) can be written as

$$\dot{n}_1'' \approx c\, D_{1,2}\frac{(y_{1,sat} - y_{1,\infty})}{L} \quad \text{for small } y_1. \tag{13.63}$$

Notice the similarity between Eq. (13.63) and the thermal resistance formula associated with conduction through a plane wall. At low values of concentration (i.e., small mole fraction), Eq. (13.63) indicates that the steady-state transfer of mass through a fluid is driven by a mole fraction difference and resisted (on a per unit area basis) by the quantity $L/(c\, D_{1,2})$. At higher values of concentration, the analogy is disrupted by the bulk flow that is induced by the mass transfer process and Eq. (13.61) must be used.

Example 13.5

Perhaps the easiest and least costly way to measure the concentration of an air pollutant is with a **diffusion tube**. A typical diffusion tube has a length $L = 71$ mm and internal diameter $D = 11$ mm. A chemical trap that either absorbs or reacts with the pollutant is placed at one end of the tube while the other end of the tube is open to the atmosphere. The tubes are usually mounted with the open end facing down so that rain and dirt are not collected in the tube. After exposure to the atmosphere for a known period of time, the tubes are sealed and sent to a laboratory where the amount of pollutant in the trap is measured. Based on this information, the average level of pollutant in the atmosphere can be determined. In a particular case, a diffusion tube that has been exposed to the environment for $t_{exp} = 30$ days has accumulated a total of $m_{NO_2} = 2$ µg of nitrogen dioxide. The average ambient temperature during the month was $T = 18°C$. The mole fraction of nitrogen dioxide at the surface of the trap is very small. Estimate the average mass concentration of nitrogen dioxide in the environment during the measurement period ($\rho_{NO_2,\infty}$ in µg/m³).

Known Values

Figure 1 illustrates a sketch of the problem with the known values indicated.

Continued

Example 13.5 (cont.)

$t_{exp} = 30$ day

$T = 18°C$

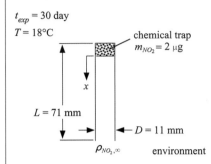

chemical trap
$m_{NO_2} = 2$ μg

$L = 71$ mm

x

$D = 11$ mm

$\rho_{NO_2, \infty}$ environment

Figure I Diffusion tube.

Assumption

- The mole fraction of nitrogen dioxide at the surface ($x = 0$) of the trap is $y_{NO_2, s} = 0$.

Analysis

The average mass flux of nitrogen dioxide from the atmosphere to the sample trap during the measurement period is:

$$\dot{m}_{NO_2} = \frac{m_{NO_2}}{t_{exp}}. \tag{1}$$

The associated average mass flux in the x-direction is

$$\dot{m}''_{NO_2} = -\frac{\dot{m}_{NO_2}}{A_c}, \tag{2}$$

where A_c is the cross-sectional area of the diffusion tube:

$$A_c = \pi \frac{D^2}{4}. \tag{3}$$

The negative sign is included in Eq. (2) because the mass flow rate is from the atmosphere to the trap (i.e., in the negative x-direction according to the sign convention used to derive Stefan's Law). Stefan's Law for this problem, Eq. (13.61) is

$$\dot{m}''_{NO_2} = \frac{c\, MW_{NO_2}\, D_{NO_2, air}}{L} \ln \left[1 - y_{NO_2, \infty}\right], \tag{4}$$

where the molar density of the mixture (c) is taken to be the molar density of air at atmospheric pressure and the average temperature:

$$c = \frac{p_{atm}}{R_{univ}\, T}. \tag{5}$$

Equation (4) is solved for the average mole fraction of nitrogen dioxide in the atmosphere:

$$y_{NO_2, \infty} = 1 - \exp \left(\frac{L\, \dot{m}''_{NO_2}}{c\, MW_{NO_2} D_{NO_2, air}}\right). \tag{6}$$

The molar concentration of nitrogen dioxide in the atmosphere is therefore:

$$c_{NO_2, \infty} = c\, y_{NO_2, \infty}. \tag{7}$$

Example 13.5 (cont.)

The mass concentration is the product of the molar concentration and the molar mass of NO_2:

$$\rho_{NO_2,\infty} = c_{NO_2,\infty} \, MW_{NO_2}. \tag{8}$$

Solution

The inputs together with Eqs. (1) through (8) are entered in EES. The diffusion coefficient of nitrogen dioxide in air is estimated using the **D_12_gas** function in EES.

```
$UnitSystem SI Mass J K Pa
L = 71 [mm]*Convert(mm,m)                       "length of diffusion tube"
D = 11 [mm]*Convert(mm,m)                        "diameter of diffusion tube"
m_NO2=2 [microgram]*Convert(microgram,kg)        "mass collected"
T=ConvertTemp(C,K,18 [C])                        "average temperature"
time_exp=30 [day]*Convert(day,s)                 "exposure time"

m_dot_NO2=m_NO2/time_exp                          "average mass flow rate to the trap"
m``_dot_NO2= -m_dot_NO2/Ac                        "average mass flux in x-direction"
Ac=pi*D^2/4                                       "cross-sectional area"
D_NO2_air=D_12_gas('NO2','Air',T,Po#)            "diffusion coefficient"
c=Po#/(R#*T)                                      "molar density of air"
MW_NO2=MolarMass(NO2)                             "molar mass of NO2"
y_NO2_infinity=1-Exp(L*m``_dot_NO2/(c*MW_NO2*D_NO2_air))
                                                  "average mole fraction in atmosphere"
c_NO2_infinity=y_NO2_infinity*c                   "molar concentration of NO2 in atmosphere"
rho_NO2_infinity=c_NO2_infinity*MW_NO2           "average concentration in atmosphere"
```

Solving leads to $\boxed{\rho_{NO_2,\infty} = 4.033 \times 10^{-8} \text{ kg/m}^3 \ (40.33 \ \mu\text{g/m}^3)}$.

Discussion

The computed value of concentration can be compared to target values of 20 $\mu\text{g/m}^3$ for ecosystem protection and an upper daily average regulatory limit of 100 $\mu\text{g/m}^3$ for human health.

13.5 Momentum, Energy, and Mass Transfer Analogies

The mass transfer rate of species i from a surface to a free stream is driven by a gradient in its composition. Figure 13.7 shows the steady flow of a gas (e.g., air) over the surface of liquid i (e.g., water) that is evaporating and thus providing a source for the mass transfer of the vapor component of species i into the free stream. The flow is steady and at a sufficiently low velocity that the boundary layer is laminar. Further, the shear is small enough that there is a minimal effect on the velocity of the liquid at the interface.

The concentration of species i in the free stream is $c_{i,\infty}$ and the concentration of species i at the surface of the liquid is $c_{i,sat}$, its saturation value. A mass diffusion boundary layer thickness (δ_d) can be defined as the distance from the liquid surface over which 99 percent of the composition change occurs; note that this is analogous to the definitions of the thermal and momentum boundary layers that are discussed in Section 7.1. The difference in the concentration at the surface and the concentration in the free stream drives a diffusive mass transfer that causes species i to transfer in the y-direction, into the gas stream. As the flow proceeds downstream, there is

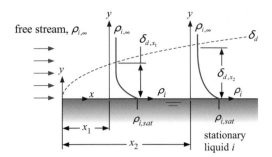

Figure 13.7 Concentration profiles of species i for laminar flow over a stationary, evaporating liquid.

additional time for diffusion to occur so that the mass diffusion penetration depth will increase, as illustrated by the concentration profiles sketched at positions x_1 and x_2 in Figure 13.7. The dashed line indicates the thickness of the diffusion boundary layer as a function of position, x.

Figure 13.7 should look familiar because it is essentially the same illustration that is used in Chapter 7 to discuss the growth of a thermal boundary layer in a laminar external flow over a flat plate. In Section 7.1.2, it is shown that the thickness of the thermal boundary layer for laminar flow is given approximately by

$$\delta_t \approx 2\sqrt{\alpha \frac{x}{u_\infty}}, \tag{13.64}$$

where α is the thermal diffusivity of the fluid. The transport of momentum by molecular diffusion is analogous to the transport of energy; a momentum boundary layer will develop according to

$$\delta_m \approx 2\sqrt{\upsilon \frac{x}{u_\infty}}, \tag{13.65}$$

where υ is the kinematic viscosity of the fluid. In exactly the same manner, the transport of mass by molecular diffusion of species i will be characterized by a mass diffusion boundary layer thickness given approximately by

$$\delta_d \approx 2\sqrt{D_{i,m} \frac{x}{u_\infty}}. \tag{13.66}$$

Note that α, υ, and $D_{i,m}$ all have units of m²/s and, according to Eqs. (13.64) through (13.66), are analogous fluid transport properties. The ratio of the kinematic viscosity to the thermal diffusivity is defined as the Prandtl number:

$$Pr = \frac{\upsilon}{\alpha}. \tag{13.67}$$

The Prandtl number provides a measure of the relative ability of the fluid to transport momentum and energy. The analogous ratio of the kinematic viscosity to the diffusion coefficient is the **Schmidt number** (Sc), which provides a measure of the relative ability of the fluid to diffusively transport momentum and mass of species i:

$$Sc = \frac{\upsilon}{D_{i,m}} = \frac{\mu}{\rho D_{i,m}}. \tag{13.68}$$

The Schmidt number can be used in convective mass transfer problems in the same manner that the Prandtl number was used in convective heat transfer problems. The **Lewis number** (Le) is the ratio of the thermal diffusivity to the diffusion coefficient and provides a measure of the relative ability of a fluid to transport energy and mass under hydrodynamically similar conditions:

$$Le = \frac{\alpha}{D_{i,m}} = \frac{k}{\rho c D_{i,m}}. \tag{13.69}$$

Note that the Prandtl, Schmidt, and Lewis numbers are related according to

$$Le = \frac{Sc}{Pr}. \tag{13.70}$$

A **mass transfer coefficient** (h_d) can be defined for a convective mass transfer problem based on the rate of mass transfer and the concentration difference driving the process:

$$\dot{m}_i'' = h_d \left(\rho_{i,sat} - \rho_{i,\infty} \right) = - D_{i,m} \frac{\partial \rho_i}{\partial y} \bigg|_{y=0}. \tag{13.71}$$

The mass transfer coefficient defined by Eq. (13.71) has units of length/time (m/s). Notice that the mass transfer coefficient is defined in a manner that is analogous to Newton's Law of Cooling for a convective heat transfer problem:

$$\dot{q}'' = h \left(T_s - T_\infty \right) = - k \frac{\partial T}{\partial y} \bigg|_{y=0}. \tag{13.72}$$

Chapter 8 provides correlations for the Nusselt number that can be used to determine the heat transfer coefficient for external flow problems:

$$Nu = \frac{h \, L_{char}}{k}, \tag{13.73}$$

where L_{char} is a characteristic dimension that depends on the geometry of the problem (e.g., for a flat plate, L_{char} is the distance from the leading edge of the plate, x). The analogous dimensionless parameter in a mass transfer problem is referred to as the **Sherwood number** (Sh) and is used to correlate the solutions to mass transfer problems:

$$Sh = \frac{h_d \, L_{char}}{D_{i,m}}. \tag{13.74}$$

The Sherwood number is the dimensionless mass transfer coefficient just as the Nusselt number is the dimensionless heat transfer coefficient.

In Section 13.3, we found that the governing differential equations for energy and mass transport in a stationary medium are the same if temperature is replaced with concentration and thermal diffusivity is replaced with the diffusion coefficient. This similarity in the energy and mass transport problems allow us to use correlations that have been developed for convective heat transfer to solve convective mass transfer problems. The Prandtl number and Nusselt number in the convective heat transfer correlations should be replaced by the Schmidt number and the Sherwood number, respectively, in order to apply them to convective mass transfer situations. In some cases, the situation is reversed. Convection heat transfer relations can be determined by measurements of the mass transfer rate.

The average Sherwood number is defined based on the average mass transfer coefficient over the surface of interest ($\overline{h}_d$):

$$\overline{Sh} = \frac{\overline{h}_d \, L_{char}}{D_{i,m}}. \tag{13.75}$$

When the boundary conditions are analogous, the correlations provided in Chapters 8 and 9 for the average Nusselt number expressed in terms of the Reynolds and Prandtl numbers, can be also used to estimate the average Sherwood number in terms of the Reynolds and Schmidt numbers:

$$\text{if } \overline{Nu} = \text{a function of } (Re, Pr) \text{ then } \overline{Sh} \approx \text{the same function of } (Re, Sc). \tag{13.76}$$

The analogy between heat and mass transfer allows many mass transfer problems to be solved using correlations that were developed for heat transfer problems. However, there are some limitations to this method. Although the equations describing heat and mass transfer may be mathematically identical, their boundary conditions often differ. For example, heat transfer convection correlations are developed assuming zero velocity at the solid–fluid interface. When mass transfer occurs at the interface, the velocity at the surface will not be zero due to the presence of Stefan Flow, as discussed in Section 13.4. Even for a fluid with no external forces acting on it, there will be a velocity in the y-direction that is induced at the surface due to the diffusive mass transfer. Consequently, the heat transfer correlations can only be applied to convective mass transfer problems that are characterized by mass fluxes where this velocity is low. One method for correcting, approximately, for the impact of the surface velocity on the mass transfer coefficient is discussed in Lienhard (2005). The mass transfer coefficient computed using the heat/mass transfer analogy embodied by Eq. (13.76) is multiplied by a "blowing factor", BF:

$$BF = \frac{\ln(1+B)}{B},\qquad(13.77)$$

where the parameter B is related to the mass fraction of species i:

$$B = \frac{mf_{i,\infty} - mf_{i,s}}{mf_{i,s} - 1},\qquad(13.78)$$

where $mf_{i,\infty}$ and $mf_{i,s}$ are the mass fractions of species i in the free stream and at the surface, respectively. If both $mf_{i,\infty}$ and $mf_{i,s}$ are small (as is typically the case), then B approaches 0 and the blowing factor predicted by Eq. (13.77) approaches unity. Notice that the form of the blowing factor is similar to the form of Stefan's Law derived in Section 13.4.2 and given in Eq. (13.61). Indeed, the blowing factor is meant to correct a zero-velocity solution for the presence of velocities induced by diffusive mass transfer.

The method described in this section uses the analogy between heat and mass transfer in order to employ convective heat transfer correlations to solve mass transfer problems. These relations can be used to provide order-of-magnitude estimates of mass transfer rates and to indicate trends. Although limited, these capabilities are still of value.

Example 13.6

A puddle of water is spilled onto an outdoor table. The height of the puddle is $H = 2$ mm and the diameter is $D = 4$ cm. The wind is blowing over the puddle with velocity $u_\infty = 3.2$ mph. Both the water on the table and the surrounding air are at $T = 20°C$. The relative humidity is 50 percent. Estimate the rate of evaporation of water from the puddle.

Known Values

Figure 1 illustrates a sketch of the problem with the known values indicated.

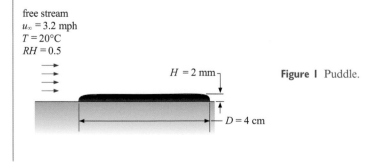

free stream
$u_\infty = 3.2$ mph
$T = 20°C$
$RH = 0.5$

$H = 2$ mm

$D = 4$ cm

Figure 1 Puddle.

Example 13.6 (cont.)

Assumptions

- The mass fraction of water is small enough that any bulk velocity induced by diffusional effects can be neglected.
- The mass transfer from the puddle can be modeled using correlations for external flow over a flat plate.
- The effect of surface tension is negligible.
- Assume that the temperature of the water remains at 20°C.

Analysis

The partial pressure of the water vapor at the interface between the puddle and the air is the saturation pressure at the temperature of the water:

$$p_{w,s} = p_{sat} \left(\text{Water}, T \right). \tag{1}$$

The partial pressure of the water vapor in the free stream is the product of the relative humidity and the saturation pressure:

$$p_{w,\infty} = RH \, p_{sat} \left(\text{Water}, T \right). \tag{2}$$

The concentrations of water vapor on the gas side of the interface and in the free stream are evaluated:

$$\rho_{w,s} = \rho \left(\text{Water}, T, \text{sat. vapor} \right) \tag{3}$$

$$\rho_{w,\infty} = \rho \left(\text{Water}, p_{w,\infty}, T \right). \tag{4}$$

The density of the air is computed at the total pressure and temperature:

$$\rho = \rho \left(\text{Air}, p_o, T \right), \tag{5}$$

where p_o is atmospheric pressure. The viscosity of the air is computed at the given temperature:

$$\mu = \mu \left(\text{Air}, T \right). \tag{6}$$

The Reynolds number associated with the flow is computed according to

$$Re = \frac{\rho \, \mu_\infty \, D}{\mu}, \tag{7}$$

where the characteristic length is taken to be the diameter of the puddle. The diffusion coefficient for water vapor in air is computed using Eq. (13.22):

$$D_{a,w} = 1.97 \times 10^{-5} \left[\frac{\text{m}^2}{\text{s}} \right] \left(\frac{T}{256 \, [\text{K}]} \right)^{1.685}. \tag{8}$$

The Schmidt number is computed using Eq. (13.68):

$$Sc = \frac{\mu}{\rho \, D_{a,w}}. \tag{9}$$

The correlation for a flat plate is used to predict the average Sherwood number associated with mass transfer:

$$\overline{Sh} = \overline{Sh} \left(Re, Sc \right). \tag{10}$$

The average mass transfer coefficient is computed using the definition of the average Sherwood number:

$$\overline{h}_d = \frac{\overline{Sh} \, D_{a,w}}{D}. \tag{11}$$

Continued

Example 13.6 (cont.)

The rate of evaporation is obtained from Eq. (13.71):

$$\dot{m}_{evap} = h_d \, \pi \left(\frac{D}{2}\right)^2 (\rho_{w,s} - \rho_{w,\infty}). \tag{12}$$

Solution

The inputs together with Eqs. (1) through (12) are entered in EES. The properties required by Eqs. (1) through (6) are obtained using the built-in property functions in EES. The correlation for a flat plate associated with Eq. (10) is implemented using the External_Flow_Plate_ND procedure.

```
$UnitSystem SI Mass J K Pa
T=ConvertTemp(C,K,20 [C])                          "ambient & surface temperature"
RH=0.5 [-]                                          "relative humidity"
H=2 [mm]*Convert(mm,m)                              "height of puddle"
u_infinity=3.2 [mph]*Convert(mph,m/s)              "wind velocity"
D=4 [cm]*Convert(cm,m)                              "puddle diameter"
p_sat=P_sat(Water,T=T)                             "saturation pressure"
p_w_s=p_sat                                        "partial pressure of water at surface"
p_w_infinity=p_sat*RH                              "partial pressure of water in free stream"
rho_w_s=Density(Water,x=1 [-],T=T)                 "concentration of water at surface"
rho_w_infinity=Density(Water,P=p_w_infinity,T=T)   "concentration of water in free stream"
rho= Density (Air,T=T,P=Po#)                       "density of air"
mu=Viscosity(Air,T=T)                              "viscosity of air"
Re=rho*D*u_infinity/mu                             "Reynolds number"
D_a_w=1.97e-5 [m^2/s]*(T/256 [K])^1.685            "diffusion coefficient"
Sc=mu/(rho*D_a_w)                                  "Schmidt number"
Call External_Flow_Plate_ND(Re, Sc: Sh_bar, )      "use correlations to estimate Sh"
h_d_bar=D_a_w*Sh_bar/D                             "mass transfer coefficient"
m_dot_evap=h_d_bar*pi*(D/2)^2*(rho_w_s-rho_w_infinity)  "rate of water evaporation"
```

Solving leads to $\boxed{\dot{m}_{evap} = 2.283 \times 10^{-7} \, \text{kg/s} \, (0.0002283 \, \text{g/s})}$.

Discussion

The temperature of the water was assumed to remain at 20°C in this example, but due to the simultaneous heat transfer that occurs when water evaporates the temperature would drop to the wet-bulb temperature, which is about 14°C, as explained in Section 13.6.

Equation (12) is appropriate provided that the diffusion velocity is low. This assumption can be checked by computing the blowing factor using Eqs. (13.77) and (13.78). The mass fractions of water vapor at the surface and in the free stream are computed according to:

$$mf_{w,s} = \frac{\rho_{w,s}}{\rho} \tag{13}$$

$$mf_{w,\infty} = \frac{\rho_{w,\infty}}{\rho}. \tag{14}$$

Example 13.6 (cont.)

The blowing factor is computed according to:

$$B = \frac{mf_{i,\infty} - mf_{i,sat}}{mf_{i,sat} - 1} \tag{15}$$

$$BF = \frac{\ln(1+B)}{B}. \tag{16}$$

Equations (13) through (16) are added to the EES model.

mf_w_s=rho_w_s/rho	"mass fraction at surface"
mf_w_infinity=rho_w_infinity/rho	"mass fraction in free stream"
B=(mf_w_infinity-mf_w_s)/(mf_w_s-1)	
BF=ln(1+B)/B	"blowing factor"

Solving provides $BF = 0.9964$, which suggests that the diffusional velocity is not important for this problem. The rate of evaporation can be used to estimate how long it will take for the puddle to evaporate. The mass of water in the puddle is

$$m_{puddle} = \pi \left(\frac{D}{2}\right)^2 H \rho_l, \tag{17}$$

where ρ_l is the density of liquid water. Assuming that the rate of evaporation remains constant, the time to evaporate is:

$$t_{evap} = \frac{m_{puddle}}{\dot{m}_{evap}}. \tag{18}$$

rho_l=Density(Water,T=T,P=Po#)	"density of water"
m_puddle=pi*(D/2)^2*H*rho_l	"mass of puddle"
time_evap=m_puddle/m_dot_evap	"approximate time to evaporate"

Solving provides $t_{evap} = 10,987$ s (3.05 hr).

13.6 Simultaneous Heat and Mass Transfer

The First Law of Thermodynamics requires that energy be provided to evaporate a liquid because the specific internal energy of a vapor is higher than that of a liquid. In many situations this energy comes from the liquid that is evaporating and therefore the evaporation process will result in a reduction in the temperature of the remaining liquid. The combined effect of heat and mass transfer can result in significant temperature changes and increased energy transfer rates at a wet surface in devices such as cooling towers, evaporative condensers, and wet-surface evaporators. Simultaneous heat and mass transfer problems can be considered using the convection heat transfer correlations that are provided in Chapters 8 through 10 together with the heat and mass transfer analogy discussed in Section 13.5.

Example 13.7

The amount of water (i.e., the humidity) in an air–water vapor mixture can be measured using a device called a **psychrometer**. A simple psychrometer consists of two ordinary thermometers. The bulb surface of

Continued

Example 13.7 (cont.)

one of the thermometers (the dry-bulb thermometer) is dry while the other thermometer bulb (the wet-bulb thermometer) is wetted by covering it with a wick made of cotton or another material that has been soaked in water. The thermometer bulbs are then exposed to a high-velocity flow of ambient air, either by swinging the thermometers or by blowing air over them with a small fan. The dry thermometer measures what is called the **dry-bulb temperature**, which is just the temperature of the air. The surface temperature of the wet-bulb thermometer will decrease due to the evaporation that occurs. The temperature reduction will continue until the energy required by evaporation is exactly provided by convection heat transfer from the surrounding air. The resulting temperature is referred to as the **wet-bulb temperature**. The difference between the wet- and dry-bulb temperatures is related to the humidity of the air.

Estimate the wet-bulb temperature that will be measured by a thermometer with diameter $D = 3$ mm that is exposed to a flow of air with velocity $u_\infty = 5$ m/s when the air temperature is $T_\infty = 27°C$ and its relative humidity is $RH = 0.5$.

Known Values

Figure 1 illustrates a sketch of the problem with the known values indicated.

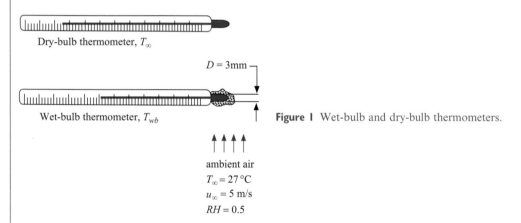

Figure 1 Wet-bulb and dry-bulb thermometers.

Assumptions

- The mass fraction of water is small enough that the diffusional velocity can be neglected.
- The thermometer can be modeled as cross-flow over a cylinder.
- The wet-bulb thermometer has reached steady state.

Analysis

The concentration of water vapor at the liquid–gas interface on the wet-bulb thermometer is obtained from the density of saturated water vapor at the wet-bulb temperature:

$$\rho_{w,s} = \rho \left(\text{Water}, T_{wb}, \text{sat.vapor} \right). \tag{1}$$

The partial pressure of the water vapor in the free stream is the product of the relative humidity and the saturation pressure at the dry bulb temperature:

$$p_{w,\infty} = RH \, p_{sat} \left(\text{Water}, T_\infty \right). \tag{2}$$

Example 13.7 (cont.)

The concentration of water vapor in the free stream is evaluated using the partial pressure and dry-bulb temperature:

$$\rho_{w,\infty} = \rho\left(\text{Water}, p_{w,\infty}, T_\infty\right). \tag{3}$$

The properties of the air used in the correlations are evaluated at the film temperature:

$$T_{film} = \frac{T_{wb} + T_\infty}{2}. \tag{4}$$

The density, viscosity, conductivity, and Prandtl number of the air are computed at the total pressure and film temperature:

$$\rho = \rho\left(\text{Air}, p_o, T_{film}\right) \tag{5}$$

$$\mu = \mu\left(\text{Air}, T_{film}\right) \tag{6}$$

$$k = k\left(\text{Air}, T_{film}\right) \tag{7}$$

$$Pr = Pr\left(\text{Air}, T_{film}\right). \tag{8}$$

The Reynolds number associated with the flow is computed according to:

$$Re = \frac{\rho\, u_\infty\, D}{\mu}. \tag{9}$$

The diffusion coefficient for water vapor in air is computed using Eq. (13.22):

$$D_{a,w} = 1.97 \times 10^{-5} \left[\frac{\text{m}^2}{\text{s}}\right]\left(\frac{T}{256\,[\text{K}]}\right)^{1.685}. \tag{10}$$

The Schmidt number is computed using Eq. (13.68):

$$Sc = \frac{\mu}{\rho D_{a,w}}. \tag{11}$$

The correlation for cross-flow over a cylinder is used to predict the average Nusselt number associated with convective heat transfer:

$$\overline{Nu} = \overline{Nu}\,(Re, Pr). \tag{12}$$

The average heat transfer coefficient is computed according to:

$$\bar{h} = \frac{\overline{Nu}\,k}{D}. \tag{13}$$

The correlation for cross-flow over a cylinder is also used to predict the average Sherwood number associated with mass transfer:

$$\overline{Sh} = \overline{Sh}\,(Re, Sc). \tag{14}$$

The average mass transfer coefficient is computed using the definition of the average Sherwood number:

$$\bar{h}_d = \frac{\overline{Sh}\,D_{a,w}}{D}. \tag{15}$$

The rate of convective heat transfer to the wet-bulb thermometer per unit area is:

$$\dot{q}''_{conv} = \bar{h}\,(T_\infty - T_{wb}). \tag{16}$$

Continued

Example 13.7 (cont.)

The rate of evaporation per unit area from the wet-bulb thermometer is obtained from Eq. (13.71):

$$\dot{m}''_{evap} = \bar{h}_d \left(\rho_{w,s} - \rho_{w,\infty} \right). \tag{17}$$

The enthalpy of vaporization of water vapor evaluated at the wet-bulb temperature:

$$\Delta i_{vap} = \Delta i_{vap} \left(\text{Water}, T_{wb} \right). \tag{18}$$

An energy balance on the wet-bulb thermometer requires that the rate of convection balance the rate of enthalpy change associated with the evaporating liquid:

$$\dot{q}''_{conv} = \dot{m}''_{evap} \, \Delta i_{vap}. \tag{19}$$

Solution

Equations (1) through (19) are 19 equations in the 19 unknowns T_{wb}, $\rho_{w,s}$, $p_{w,\infty}$, $\rho_{w,\infty}$, T_{film}, ρ, μ, k, Pr, Re, $D_{a,w}$, Sc, $\overline{Nu}$, $\bar{h}$, $\overline{Sh}$, $\bar{h}_d$, $\dot{q}''_{conv}$, $\dot{m}''_{evap}$, and Δi_{vap}. These equations are entered in EES and solved. Note that in order to help with convergence, initially Eqs. (1) through (18) are entered with an *assumed* value of T_{wb} and then the guess values are updated. Finally, the assumed value of wet-bulb temperature is removed and Eq. (19) is entered.

```
$UnitSystem SI Mass J K Pa
D = 3 [mm]*Convert(mm,m)                               "diameter of thermometer"
u_infinity=5 [m/s]                                     "air velocity"
T_infinity=ConvertTemp(C,K,27 [C])                     "dry-bulb temperature"
RH=0.5 [-]                                             "relative humidity"

//T_wb=290 [K]                                         "assumed wet-bulb temperature"
rho_w_s=Density(Water,T=T_wb,x=1)                      "concentration of water vapor at surface"
p_w_infinity=RH*P_sat(Water,T=T_infinity)             "partial pressure in free stream"
rho_w_infinity=Density(Water,T=T_infinity,P=p_w_infinity) "concentration of water vapor in free stream"
T_film=(T_infinity+T_wb)/2                             "film temperature"
rho=Density(Air,T=T_film,P=Po#)                        "density of air"
mu=Viscosity(Air,T=T_film)                             "viscosity of air"
k=Conductivity(Air,T=T_film)                           "conductivity of air"
Pr=Prandtl(Air,T=T_film)                               "Prandtl number of air"
Re=rho*D*u_infinity/mu                                 "Reynolds number"
D_a_w=1.97e-5 [m^2/s]*(T_film/256 [K])^1.685          "diffusion coefficient"
Sc=mu/(rho*D_a_w)                                      "Schmidt number"
CALL External_Flow_Cylinder_ND(Re,Pr: Nusselt_bar,)  "correlations to get Nusselt"
h_bar=Nusselt_bar*k/D                                  "average heat transfer coefficient"
CALL External_Flow_Cylinder_ND(Re, Sc: Sh_bar, )     "correlations to get Sh"
h_d_bar=Sh_bar*D_a_w/D                                 "average mass transfer coefficient"
q``_dot_conv=h_bar*(T_infinity-T_wb)                   "convective flux"
m``_dot_evap=h_d_bar*(rho_w_s-rho_w_infinity)          "convective mass flux"
DELTAi_vap=Enthalpy_Vaporization(Water,T=T_wb)         "latent heat of vaporization"
q``_dot_conv=DELTAi_vap*m``_dot_evap                   "energy balance"
```

Solving leads to $\boxed{T_{wb} = 292.2 \text{ K } (19.06°\text{C})}$. The wet-bulb temperature has been reduced by nearly 8°C relative to the dry-bulb temperature due to the evaporation effect. This difference can be used to determine the relative humidity when the psychrometer is used as a measurement tool.

Example 13.7 (cont.)

Discussion

The difference between the wet-bulb and dry-bulb temperatures is a strong function of the relative humidity. Figure 2 illustrates the wet-bulb temperature depression ($T_\infty - T_{wb}$) as a function of relative humidity and shows why this device is so useful for measuring humidity.

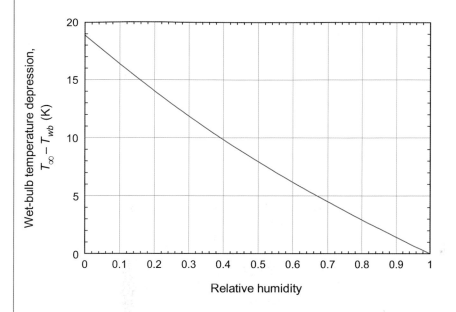

Figure 2 Wet-bulb temperature depression as a function of the relative humidity.

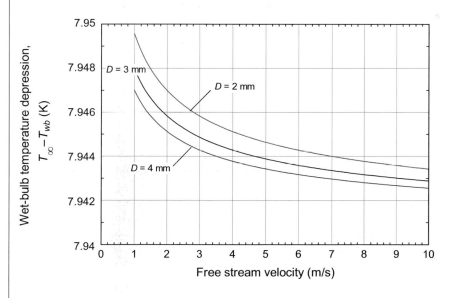

Figure 3 Wet-bulb temperature depression as a function of the air velocity for different values of the diameter.

Continued

Example 13.7 (cont.)

On the other hand, the difference between the dry-bulb and wet-bulb temperatures is a weak function of the other parameters in the problem, including free stream velocity and dry-bulb bulb diameter. Figure 3 illustrates the wet-bulb temperature depression as a function of free stream velocity for different values of bulb diameter. Notice the y-axis range is quite small indicating that varying any of these parameters has a weak effect on the result. The heat and mass transfer analogy causes any change in convection to be nearly exactly cancelled by a corresponding change in evaporation rate leaving the temperature difference unaffected.

The wet bulb is sufficiently close to the adiabatic saturation temperature from Thermodynamics that they are often used interchangeably.

13.7 Conclusions and Learning Objectives

This chapter provides an introduction to diffusive mass transfer. Mass transfer in a mixture is driven by diffusion due to gradients in the concentration of one of the species. These concentrations can be expressed either on a mass or a molar basis. The diffusion mass transfer rate equation used for this type of problem is referred to as Fick's Law which provides the definition of a diffusion coefficient. Methods for estimating the diffusion coefficient for gas, liquids, and solids are presented as well as some typical boundary conditions that are encountered for mass transfer problems.

Fick's Law is used to examine some mass transfer problems. It is shown that when the medium is *completely stationary* (i.e., diffusion in a solid) the governing differential equation reduces to the same equations that were solved for transient conduction problems and therefore, with proper substitutions, the solutions obtained in Chapter 6 can be applied. However, when the medium can flow (e.g., gas or liquid) then some bulk motion will be generated by the diffusion even if the medium would otherwise be stationary. This phenomenon is referred to as Stefan Flow.

Finally, mass transfer in the presence of externally driven bulk velocity (convective mass transfer) is considered. The heat/mass transfer analogy allows the convective heat transfer correlations provided in Chapter 8 to be applied, approximately, to these situations by replacing the Prandtl number with the Schmidt number and the Nusselt number with the Sherwood number. In this way, it is possible to estimate the rate of mass transfer using commonly accepted heat transfer correlations. In many situations both heat and mass transfer together play a role and the energy associated with the evaporation of a fluid can be quite important to the performance of a device.

Some specific concepts that you should understand are listed below.
- The diffusion of the mass of one species in a mixture is driven by gradients in its concentration.
- The concentration of a species can be expressed on either a mass or molar basis.
- The use of the Ideal Gas Law to compute concentrations of low-density gas mixtures.
- Fick's Law and the significance of the diffusion coefficient.
- Methods for estimating the diffusion coefficient of various mixtures.
- Common techniques for estimating the concentration at interfaces.
- The governing differential equations that are used to analyze transient diffusion through a solid and how these are similar to transient conduction problems.
- The method of using conduction solutions to solve diffusion problems in the limit that the medium is stationary.
- The process of diffusion of a species in a fluid, including the ability to derive the governing differential equation.
- The mechanism that causes Stefan Flow and the derivation of Stefan's Law.
- The significance of the Schmidt and Lewis numbers and their similarity to the Prandtl number.
- The heat/mass transfer analogy and its use to estimate mass transfer coefficients for convective mass transfer problems.
- The ability to deal with problems where both mass and heat transfer are occurring.

References

Bolz, R. E. and G. L. Tuve, *Handbook of Tables for Applied Engineering Science*, Second Edition, CRC Press (1976).

Lienhard, J. H, IV and J. H. Lienhard V, *A Heat Transfer Textbook*, Third Edition, Phlogiston Press, Cambridge, MA (2005).

Mills, A. F., *Mass Transfer*, Second Edition, Prentice Hall, Upper Saddle River (2001).

Perry, J. H., C. H. Chilton, and S. D. Kirkpatrick, *Chemical Engineers' Handbook*, Fourth Edition, McGraw-Hill, New York (1963).

Poling, B. E., J. M. Prausnitz, and J. O' Connell, *The Properties of Gases and Liquids*, Fifth Edition, McGraw-Hill, New York (2000).

Sander, R., "Compilation of Henry's Law constants (version 4.0) for water as solvent," *Atmos. Chem. Phys.*, Vol. **15**, pp. 4399–4981 (2015). www.atmos-chem-phys.net/15/4399/2015/

Problems

Concentration Relationships

13.1 An ideal gas mixture at 350 K and 300 kPa has the following volumetric analysis: 65 percent N_2, 20 percent O_2, and 15 percent CO_2.
  (a) Determine the molar mass of the mixture.
  (b) Determine the mass fraction of each gas.
  (c) Determine the mass concentration of each gas.
  (d) Determine the molar concentration of each gas.

13.2 Dry air is a gas mixture approximately consisting of 78.1 percent nitrogen, 21.0 percent oxygen, and 0.9 percent argon on a molar basis.
  (a) What are the mass fractions of each of these gases?
  (b) What is the molar mass of this mixture?
  (c) The air is humidified such that the mass fraction of the water is 0.01. Determine the mole fraction of water.

13.3 A volumetric analysis of a gas mixture shows that it consists of CO_2 ($y_{CO_2} = 0.036$), H_2O ($y_{H_2O} = 0.072$), oxygen ($y_{O_2} = 0.131$), argon ($y_{Ar} = 0.008$) with the remainder being nitrogen. Determine:
  (a) the molar mass of this gas mixture, and
  (b) the mass fraction of each gas.

13.4 A mixture is formed mixing $m_m = 0.25$ kg of methane (with molar mass $MW_m = 16$ kg/kmol), $m_e = 0.15$ kg of ethane ($MW_e = 30$ kg/kmol) and $m_n = 0.1$ kg of nitrogen ($MW_n = 28$ kg/kmol). The mixture is placed in a container that is maintained at $T = 25°C$ and $p = 5$ bar. At these conditions, the mixture behaves in accordance with the Ideal Gas Law. Determine:

  (a) the volume of the mixture,
  (b) the equivalent molecular weight of the mixture,
  (c) the density of the mixture on a mass basis,
  (d) the density of the mixture on a molar basis,
  (e) the mass fractions of each species,
  (f) the mole fractions of each species,
  (g) the mass concentration of each species, and
  (h) the molar concentration of each species.

13.5 The composition of mixtures of air and water vapor are often reported in terms of the humidity ratio. The humidity ratio, ω, is defined as the mass of water vapor per mass of dry air. The humidity ratio is related to, but not exactly the same as the mass fraction. In a particular case, the humidity ratio is $\omega = 0.0078$ at temperature $T = 30°C$ and pressure $p = 101.3$ kPa. Determine:
  (a) the mass fraction of the water vapor,
  (b) the mole fraction of the water vapor,
  (c) the mass concentration of the water vapor,
  (d) the molar concentration of the water vapor, and
  (e) the maximum possible value for the mole fraction of the water vapor at equilibrium.

The Diffusion Coefficient and Boundary Conditions

13.6 Plot the mole fraction of oxygen near the air–liquid interface for liquid water as a function of temperature from 0°C to 40°C. These results should indicate why trout prefer cold water.

13.7 Ammonia gas is maintained at a partial pressure of 20 kPa over a solution of water and ammonia. The liquid is maintained at a constant temperature and diffusion is allowed to occur until equilibrium conditions are obtained. Prepare a plot of the mole fraction of ammonia at the liquid–gas interface as a function of temperature.

13.8 Prepare a plot of the binary diffusion coefficient of hydrogen in air and carbon dioxide in air as a function of temperature for temperatures between 200 K and 500 K at 1 atm and at 2 atm. Based on your plots, indicate how the binary diffusion coefficient varies with molecular weight, temperature, and pressure. Offer an explanation for the behavior you observe in the plots.

13.9 A 12 oz aluminum soda can contains 350 ml of liquid, which can be assumed to have the properties of water, and 5 ml of gas space above the liquid. The can is pressurized to 135 kPa with carbon dioxide gas and cooled to 5°C.
(a) What is the mole fraction of water vapor in the gas phase?
(b) Determine the mass of carbon dioxide dissolved in the water, assuming that the concentration of carbon dioxide is uniform and equal to value at the interface at 5°C.
(c) The can is allowed to come to the ambient temperature of 27°C. The can expands a bit and so the pressure of the carbon dioxide gas in the can remains at 135 kPa. Determine the mass of carbon dioxide dissolved in the water at this temperature.
(d) The can is opened and eventually the partial pressure of the carbon in the air comes to 0.000045 atm at 27°C at 1 atm total pressure. What is the equilibrium mass of carbon dioxide dissolved in the liquid at this condition?

13.10 The air-conditioning load for a building can be broken into latent and sensible contributions. The latent component represents the energy that must be expended to remove the water vapor from the building. Water vapor enters by infiltration as outdoor air leaks inside and by diffusion through the walls and ceiling. The building in question is rectangular with outer dimensions of 40 ft by 60 ft with 8 ft ceilings. The infiltration rate is estimated at 0.65 air changes per hour. The diffusion coefficient for water through 3/8 inch gypsum board (without a vapor barrier) is approximately 4.5×10^{-5} ft^2/s at atmospheric pressure.

Estimate and compare the rates of moisture transfer by infiltration and diffusion on a day in which the outdoor conditions are 95°F and 45 percent relative humidity and indoor conditions are 75°F and 40 percent relative humidity. Is the contribution by diffusion significant? If not, then why are people concerned with water vapor diffusion in a building?

Transient Diffusion through a Solid

13.11 Pure helium gas is contained at 25°C and 104 kPa within a Pyrex cylinder that is 0.10 m in height and 0.025 m in inside diameter. The walls are 2 mm thick. The air surrounding the cylinder is also at 25°C and it contains no measureable amount of helium. The concentration of helium in Pyrex at 25°C at 100 kPa is known to be 0.008 cm^3 (measured at standard temperature and pressure) per cm^3 of glass; this is the concentration boundary condition on the inside surface of the cylinder.
(a) Determine the rate at which helium leaks from this vessel.
(b) Estimate the time required for the pressure to be reduced to 100 kPa.

13.12 Natural gas (methane) is transported at 25°C and 100 bar over long distances through 1.2 m diameter pipelines at a velocity of 10 m/s. The pipeline is made of steel with a wall thickness of 2.0 cm. It has been suggested that hydrogen gas could be transported in these same pipelines. However, hydrogen is a small molecule that diffuses through most materials. The diffusion coefficient for hydrogen in steel is about 7.9×10^{-9} m^2/s at 25°C.
(a) Calculate the power transported by methane (assuming it will be combusted) through the pipeline. The lower heating value of methane is 5.002×10^7 J/kg.
(b) Estimate the velocity required to provide the same power if hydrogen rather than methane is transported through the pipeline at the same temperature and pressure. The lower heating value of hydrogen is 1.200×10^8 J/kg.
(c) Compare the pumping power required to transport the natural gas and hydrogen a distance of 100 km.
(d) Estimate the rate of hydrogen loss from a 100 km pipeline. Do you believe this loss is significant?

13.13 A balloon made of a synthetic rubber is inflated with helium to a pressure of $p_{ini} = 130$ kPa at which point its diameter is $D_{ini} = 0.12$ m. The mass of the balloon material is $m_{bal} = 0.53$ g and its thickness is $\delta = 0.085$ mm. The balloon is released in a room that is maintained at $T = 25°C$ filled with air ($y_{N_2} = 79$ percent nitrogen and $y_{O_2} = 21$ percent oxygen) at $p_{atm} = 100$ kPa. Over a period of time, helium diffuses out of the balloon and oxygen and nitrogen diffuse in. The pressure in the balloon above atmospheric pressure is linearly proportional to the balloon volume. The diffusion coefficients for helium, oxygen and nitrogen through this synthetic rubber are $D_{He,rubber} = 60 \times 10^{-8}$, $D_{O_2,rubber} = 16 \times 10^{-8}$, and $D_{N_2,rubber} = 15 \times 10^{-8}$ cm²/s, respectively.

(a) Prepare a numerical model of the balloon deflation process. Plot the volume and pressure within the balloon as a function of time. Plot the mole fraction of helium, oxygen, and nitrogen in the balloon as a function of time.

(b) At what time does the balloon lose its buoyancy?

Diffusion of a Species in a Fluid

13.14 One way to measure the binary diffusion coefficient of two gases is to use an apparatus as shown in the figure. Two glass spheres are connected with a tube having length L and inner diameter D. A stopcock is provided in the tube. The stopcock is initially closed and sphere 1 is filled with pure gas A and sphere 2 is filled with pure gas B. The pressures in spheres 1 and 2 are the same and the entire apparatus is isothermal at temperature T. The experiment is started when the stopcock is opened which allows gas A to diffuse from sphere A into sphere B. (Gas B must diffuse into sphere A at the same molar rate.)

(a) Derive an expression for the molar rate of diffusion of gas A at an arbitrary time when the mole fractions of gas A in spheres 1 and 2 are $y_{A,1}$ and $y_{A,2}$.

(b) Assume that the rate of diffusion is slow relative the time required to change the concentration in the spheres. Using the result derived in (a) and an unsteady-state molar balance on gas A in sphere 1, show that $2 \ln\left(\frac{2y_{A,1}-1}{1}\right) = -\frac{A D_{AB}}{V_1 L} t$

where $y_{A,1}$ is the mole fraction of gas A in sphere 1 at time t,

A is the cross-sectional area of the tube,

L is the length of the tube,

D_{AB} is the binary diffusion coefficient,

V_1 is the volume of sphere 1, and

t is time.

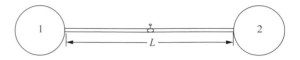

13.15 The experimental apparatus described in Problem 13.14 is used to measure the binary gas diffusion coefficient for CH_4 in air at 25°C and 101.3 kPa. The volume of both spheres is 1 liter. The tube connecting the spheres has an inner diameter of 1 cm and a length of 15 cm. Sphere 1 is initially filled with pure CH_4 while sphere 2 initially contains pure air. The mole fraction of CH_4 in sphere 1 is measured each hour for a 48 hour period. A plot of the measured data is prepared and a curve fit that describes the plot is:

$$y_{CH_4} = 1.0038 - 0.0091 \cdot hour + 0.000057 \cdot hour^2$$

where *hour* is the time in hours.

(a) Use the experimental data to estimate the binary diffusion coefficient for CH_4 in air.

(b) Compare the estimated diffusion coefficient with the value provided by the D_12_gas function in EES.

13.16 One method to measure the binary diffusion coefficient is to use a Stefan tube. In this case, 5.0 g of methanol liquid is placed in a 1 inch inner diameter tube that is 12 in long in a nitrogen environment at 25°C and 1 atm. The methanol evaporates from the liquid surface and diffuses into the nitrogen. After exactly 5 days, the mass of methanol is measured and it is found that 2.4 g remain in the tube. Based on this information, estimate the binary diffusion coefficient for methanol in nitrogen at 25°C and compare it with the value provided in the EES Diffusion Library.

13.17 A cylindrical tank having an internal diameter of $D_t = 0.42$ m and an internal height of $H_t = 1.4$ m was originally used to store hot water, but is no longer used for that purpose. A custodian cut the two pipes that were used to charge and discharge the tank leaving $L = 0.12$ m of pipe sticking up from the top of the tank. The internal diameter of the pipes is $D_p = 2.2$ cm. The water remaining in the tank is at room temperature, $T = 25°C$. The building in which the tank is located is at atmospheric pressure and it is maintained at an average relative humidity of 40 percent.

(a) What is the rate of moisture transport to the building from the tank?

(b) How much time is required for the level in the tank to drop by 1.0 m?

13.18 A desiccant is a material that absorbs water vapor. Its ability to adsorb water is represented in terms of an isotherm which relates the moisture content (W, defined as the ratio of the mass of water absorbed to the mass of dry desiccant) to the relative humidity (RH, defined as the ratio of the partial pressure of water vapor in air to the saturation vapor pressure of water at the air temperature). The desiccant under investigation has an isotherm that can be approximately represented as $W = 0.046\ RH$. The desiccant initially has a moisture content of 0.012 at its surface and it is at 25°C. It is placed at the bottom of a glass tube that is 54 mm high and 24 mm in diameter. Air above the tube is at atmospheric pressure, 25°C and 68 percent relative humidity. Determine the mass flux of water vapor to the surface of the desiccant at these conditions.

13.19 The vapor pressure of water over a concentrated sulfuric acid solution is near zero. As a result, it can be used to measure the humidity ratio in air. In a particular case, a concentrated solution of sulfuric acid is weighed and then placed at the bottom of a tube that is 4 cm diameter and 15 cm long above the acid solution. The tube that is open to atmospheric air. Exactly 24 hours later, the solution is weighed again and found to have increased in mass by 0.25 g. The pressure and temperature are 101.3 kPa and 25°C. Determine:

(a) the average mole fraction of water in the air, and

(b) the corresponding humidity ratio.

13.20 A bottle of household ammonia contains a liquid mixture of 6 percent (by weight) ammonia

in water at 25°C. The gas space above the liquid is a mixture of ammonia, water vapor and air.

(a) Determine the mole fraction of ammonia in the liquid solution.

(b) The gas space above the liquid in the bottle consists of ammonia vapor, water vapor, and air. The Henry constant for ammonia in water solution at 25°C is 0.9255 atm. Assuming the liquid solution is sufficiently dilute for Henry's Law to apply, determine the mole fraction of ammonia in the gas space.

(c) Assume that the partial pressure of water vapor in the gas space is the saturation pressure of water. Determine the mole fraction of water vapor in the gas space.

(d) The cap is left off of the bottle and both ammonia and water vapor diffuse into the environment. The diffusion coefficients of ammonia vapor in air and water vapor in atmospheric air are both about 0.000021 m^2/s at 25°C. Assuming that the diffusion rates of water and ammonia are governed by Stefan's Law and independent of each other, estimate the ratio of the rate of diffusion of ammonia to that of water from the bottle into the atmospheric air at 25°C, 50 percent relative humidity surrounding the bottle.

13.21 A janitor is about to clean a large window at one end of a corridor with an ammonia-water solution. The corridor is 2.5 m high, 2 m wide, and 3 m in length. The conditions in the corridor are 25°C and 101 kPa. The concentration of the ammonia that evaporates from the window is estimated to be 100 ppm. Many humans can detect ammonia by smell at levels of 1 ppm. Estimate the time required for a person standing at the other end of the corridor to detect the ammonia after the janitor starts to wash the window.

Heat Mass Transfer Analogy

13.22 An outdoor swimming pool measuring 24 m by 8 m is filled with water at 25°C. The air outdoors is also at 25°C with a wet-bulb temperature of 19°C. The wind speed over the pool surface is 6 m/s. Estimate the rate of energy loss from the water due to mass transfer.

13.23 A spherical raindrop is falling through air at $T_\infty = 20°C$, atmospheric pressure, and a relative humidity of $RH = 0.5$. The diameter of the sphere is $D_{ini} = 1$ mm. You may assume that the sphere is always at its terminal velocity (i.e., the velocity at which the drag and gravitational forces are balanced) and at the temperature where evaporation and convection are balanced. Assume that the droplet remains spherical. Plot the diameter of the droplet as a function of time.

13.24 A square slab of dry ice (solid carbon dioxide) that is $th = 1$ inch thick and $W = 9$ inches on each side is placed on an insulated surface in a large room filled with air at $T_\infty = 75°F$ and $p = 1$ atm. Dry ice has a density of $\rho_s = 93.6$ lb$_m$/ft^3 and it sublimes with a vapor pressure of $p_v = 1$ atm. During this phase change, the dry ice remains at $T_s = -109.4°F$. Estimate the time required for the dry ice to disappear.

13.25 Naphthalene is an aromatic hydrocarbon with a molecular weight of $MW = 128.2$ kg/kgmol that sublimes at a reasonable rate a room temperature. Naphthalene was commonly used for moth balls, but is now considered to be a carcinogen. At $T = 25°C$, solid naphthalene has a density of $\rho = 1.16$ g/cm^3 and a vapor pressure of $p_v = 0.082$ mm Hg. An engineer has recognized that heat and mass transfer are analogous processes and plans to estimate the heat transfer coefficient for an unusual geometry by measuring how much mass of naphthalene is sublimed. A review of the literature indicates that the Schmidt number for naphthalene is $Sc = 2.5$. To test accuracy of the heat/mass transfer analogy, the engineer first measures the mass of naphthalene that sublimes from a sphere of $D = 2.5$ cm diameter when exposed to a stream of pure air at temperature $T = 25°C$, pressure $p = 101.3$ kPa, and velocity $u_\infty = 10$ m/s. The test is run for $t_{test} = 2$ hr and during this time the mass of the naphthalene sphere is reduced by $\Delta m = 250$ mg. Determine the error relative to accepted correlations for this geometry.

13.26 Data for naphthalene at $25°C$ are provided in Problem 13.25. Determine the time required for 90 percent of the mass in a 1.0 cm sphere of naphthalene to sublime into an air stream at $25°C$ and 100 kPa that is flowing at 5 m/s.

13.27 It is necessary to determine the mass transfer coefficient from a long square bar is covered with a thin film of hexane liquid. Each side of the bar is $W = 2$ cm and the bar is positioned in an air flow in a diamond configuration as shown in the figure. The air temperature is 20°C, but the velocity is not known. However, a previous heat transfer test is available that has found the electric power needed to maintain surface of the bar at 34°C in a 20°C air flow is 100 W per m of length at the same air flow rate.

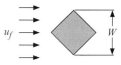

(a) Determine the convection coefficient based on the heat transfer test.
(b) Determine the diffusion coefficient for saturated hexane vapor at 20°C into 20°C air.
(c) Estimate the mass transfer rate of hexane per unit length of the bar.

Simultaneous Heat and Mass Transfer

13.28 A small lake has an average diameter of 0.5 km. The water at the surface of the lake is 16°C. Air at 25°C and 42 percent relative humidity is blowing over the lake at 3 m/s.
(a) Estimate the rate at which water moves from the lake to the atmosphere.
(b) Estimate the rate of heat transfer from the lake surface if mass transfer is neglected.
(c) Determine the rate of heat transfer from the lake when mass transfer is considered.

13.29 A runner is moving at a pace of 8 minute per mile outdoors where the conditions are 84°F and 52 percent relative humidity. As an approximation, the runner can be considered to be a cylinder having a diameter of 0.75 ft and a height of 5.6 ft. The temperature of the runner's skin is at 90°F and he is perspiring profusely.
(a) Calculate the rate of heat transfer from the runner by convection (not including mass transfer).
(b) Calculate the rate of heat transfer from the runner considering mass transfer. Assume that the runner's skin is entirely wetted by the perspiration and the perspiration has the properties of water.

13.30 A cup having a 3 inch diameter is filled to the top with 92°C water. Atmospheric air at 25°C and 40 percent relative humidity is blown across the surface of the cup at 0.4 m/s.

(a) Estimate the rate of heat transfer from the water by radiation, assuming an emissivity of 1.

(b) Estimate the rate of heat transfer from the water by convection.

(c) Estimate the rate of energy transfer from the water by evaporation due to mass transfer.

13.31 In low-humidity locations, water can be cooled by placing it in a porous container. A small amount of water diffuses through the container walls and wets the outside surface. The water on the surface subsequently evaporates and some of the energy need to supply the enthalpy change of evaporation is provided from the water remaining the container, which lowers its temperature.

(a) Calculate the steady-state temperature of the container wall and water in the container if the ambient conditions are 100 kPa, 35°C, and 20 percent relative humidity.

(b) Is it possible for the water in the container to freeze when the outdoor temperature is 35°C? If so, at what relative humidity will this occur? If not, what is the lowest temperature that the water could possibly get to?

13.32 The wick of a wet-bulb thermometer is dipped in liquid ethanol and exposed to a stream of dry air at 27°C and 101.3 kPa flowing at 5 m/s. Determine the temperature that the thermometer will display.

13.33 You have asked to join the team of engineers responsible for the design of an air-washer. Your part of this project is to prepare an analysis that will determine the diameter, velocity, and temperature of droplets as they fall in an upward flowing air stream. You are considering a single water droplet with an initial diameter of 1.5 mm and initial temperature of 45°C that is released into a 25°C, 35 percent relative humidity, 100 kPa air stream that is flowing upward at 30 m/s.

Plot the diameter, velocity, and temperature of the droplet as a function of time. Assume that the droplet remains spherical and that it can be considered to have a uniform temperature at any time.

13.34 One type of household humidifier operates by expelling water droplets into the air. The droplets have an average diameter of 10 μm. After leaving the dehumidifier, the droplets "float" around the room and evaporate. In a particular case, the room is maintained at 25°C, 100 kPa, and 25 percent relative humidity. You may assume that the droplet is at the temperature where evaporation and convection are balanced.

(a) Plot the mass of the droplet as a function of time and determine the time required for the droplets to completely evaporate.

(b) The humidifier requires a work input to form the droplets. The work input is related to the change in area of the water as it is transformed from one large "drop" to many smaller droplets. Calculate the energy required to distribute 1 kg of droplets with this vaporizer and compare it to the energy needed to vaporize 1 kg of water at 25°C. Comment on whether you believe that this humidifier saves energy compared to traditional vaporization process based on boiling water.

Projects

13.35 In order to detect chemical threats that are being smuggled into the country within a shipping container, the government is working on a system that samples the air inside the container on the dock as it is being unloaded. The chance of detecting the chemical threat is strongly dependent upon its concentration distribution at the time that the container is sampled. Therefore, you have been asked to prepare a simple model of the migration of the threat species from its release point within a passage formed by the space between two adjacent boxes. The problem is not a simple diffusion problem because the threat chemical is adsorbed onto the walls of the passage. The situation is simplified as 1-D diffusion through a duct. One end of the duct is exposed to a constant concentration of the threat chemical that is equal to its saturation concentration, $c_{sat} = 0.026$ kg/m^3. The duct is filled with clean air and the walls of the duct are clean (i.e., at time $t = 0$ there is no threat chemicals either in the air in the duct or on the walls of the duct). The hydraulic diameter of the duct is $D_h = 10$ cm. The length of the duct is infinite. The diffusion coefficient for the threat chemical in air is $D = 2.2 \times 10^{-5}$ m^2/s. The mass of threat chemical per unit area adsorbed on the wall of the container (M_w'') is related to the concentration of the chemical in the air (c) according to:

$$\frac{M_w''}{M_{w,m}''} = \frac{A \dfrac{c}{c_{sat}}}{\left(1 - \dfrac{c}{c_{sat}}\right)\left[1 + (A - 1)\dfrac{c}{c_{sat}}\right]},$$

where $M_{w,m}'' = 4 \times 10^{-4}$ kg/m^2 is the mass per unit area associated with a single monolayer and $A = 20$ is a dimensionless constant. The total time available for diffusion between

loading the container and unloading is $t_{transit} = 14$ days. Because the length of the duct is so much larger than the hydraulic diameter of the duct, it is reasonable to assume that the concentration distribution is 1-D. Further, because the concentration of the threat chemical is so small, it is reasonable to neglect any bulk velocity induced by the diffusion process; that is, only mass transfer by diffusion is considered.

(a) Prepare a 1-D transient model of the diffusion process using the ode45 solver in MATLAB.

(b) Plot the concentration distribution within the passage at various times.

(c) Plot the concentration distribution within the passage at $t = t_{transit}$ and overlay on this plot the zero-adsorption solution to show how adsorption has retarded the migration of the threat chemicals within the container.

14 | Radiation

From a thermodynamic perspective, thermal energy can be transferred across a boundary (i.e., heat transfer can occur) by only two mechanisms: conduction and radiation. Conduction is the process in which energy exchange occurs due to the interactions of molecular (or smaller) scale energy carriers within a material. The conduction process is intuitive; it is easy to imagine energy carriers having a higher level of energy (represented by their temperature) colliding with neighboring particles and thereby transferring some of their energy to them. Radiation is a very different heat transfer process because energy is transferred *without* the involvement of any molecular interactions. Radiation energy exchange is related to electromagnetic waves and therefore can occur over long distances through a complete vacuum. For example, the energy that our planet receives from the Sun is a result of radiation exchange. This chapter presents an introduction to radiation heat transfer with a focus on providing methods for solving radiation problems.

14.1 Introduction

14.1.1 Electromagnetic Radiation

All substances emit energy in the form of **electromagnetic radiation** as a result of molecular and atomic activity; molecular electronic, vibrational or rotational transitions result in the emission of energy in the form of radiation. Radiation that is emitted by atoms or molecules that lie inside of an object is mostly absorbed by the matter surrounding it. However, matter that lies on or near the surface of an object can emit radiation that travels through space before hitting another surface where it might be absorbed, typically by matter very close to the surface. As a result, most radiation heat transfer is considered to be a surface-to-surface phenomenon. The radiation heat transfer considered in this chapter is assumed to be from one surface to another.

The characteristics and amount of radiation emitted by a substance are dependent on its temperature as well as other properties of the surface. Radiation is exchanged between two surfaces even when they are at the same temperature. However, in this case the *net* radiation exchange is zero; that is, the rate at which each surface is emitting radiation is equal to the rate at which it is absorbing the incident radiation that was emitted by the other surface. In this text, we will examine radiation from an engineering perspective with an eye towards solving radiation heat transfer problems involving several surfaces at various temperatures.

14.1.2 The Electromagnetic Spectrum

Electromagnetic radiation propagates through a vacuum at the speed of light, $c = 299{,}792$ km/s. The propagation speed is the product of the **wavelength** of the radiation (λ) and its **frequency** (v):

$$c = \lambda v. \tag{14.1}$$

When classified according to its wavelength, **thermal radiation** (i.e., the radiation that is emitted by virtue of an object's temperature) is only a subset of electromagnetic radiation. The electromagnetic radiation **spectrum** is shown in Figure 14.1. At one extreme, the spectrum includes very short wavelength radiation such as gamma rays. According to Eq. (14.1), these waves must have high frequency and therefore high energy. At the other extreme, the spectrum includes long wavelength, low frequency radiation such as radio waves. **Visible light** is the portion of the electromagnetic spectrum that our eyes can detect, between 0.38 μm and 0.78 μm (representing colors ranging from violet to red). Our eyes are most sensitive to the green radiation that occurs at about the center of this range, 0.55 μm.

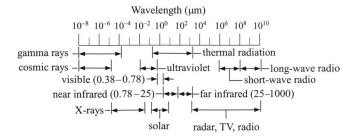

Figure 14.1 Electromagnetic spectrum (reproduced from Duffie and Beckman, 2013).

Approximately 6.5 percent of the energy emitted by the Sun is **ultraviolet** (UV) radiation, which lies at wavelengths below violet light (0.38 μm); our atmosphere absorbs most of this UV radiation. UV radiation is sub-classified into three bands: UV-C (0.20 μm to 0.28 μm), UV-B (0.28 μm to 0.32 μm), and UV-A (0.32 μm to 0.38 μm). UV-A radiation has the lowest energy of the three bands and a large percentage of this radiation is transmitted through our atmosphere. This is the type of radiation that is responsible for a sun tan. Overexposure to UV-A without UV blocking sun tan lotion can result in skin damage. However, UV-B radiation is of greater concern since radiation in this wavelength band can damage the DNA of plants and animals and cause skin cancer in humans. UV-B is strongly absorbed by the ozone (O_3) that resides in the stratosphere, approximately 25 miles above ground level. For this reason, the thinning of the ozone layer that has been caused by reactions that are catalyzed by chlorofluorocarbon (CFC) refrigerants has raised concern and resulted in an international ban on CFC refrigerant production, an initiative that began in 1996. UV-C radiation is more energetic and potentially more dangerous than UV-B radiation. However, UV-C radiation is strongly absorbed by gases in our atmosphere and therefore essentially none of this radiation reaches ground level. Exposure to UV-C radiation is a concern for astronauts carrying out extra-vehicular activities.

Thermal radiation (i.e., radiation generated by a surface due to its temperature) is the portion of the electromagnetic spectrum between the wavelengths of approximately 0.2 μm and 1000 μm. The other portions of the spectrum are largely generated by nonthermal processes. For example, gamma rays are produced by radioactive disintegration and radio waves are artificially produced by electrical oscillations. The radiation problems considered in this chapter will involve only thermal radiation.

14.2 Emission of Radiation by a Blackbody

14.2.1 Introduction

The net radiation heat transfer from a surface is a consequence of the difference between the amount of radiation that is emitted by the surface and the amount of radiation that is absorbed by the surface. Therefore, in order to understand radiation heat transfer, it is necessary to understand both of these processes. The absorption of radiation by a surface is complicated as it depends not only on the properties of the surface, but also on how the surface is oriented relative to all other surfaces involved in the problem and the amount of radiation that they emit. Therefore, it is easier to start by considering the amount of radiation that a surface emits because that depends only on the surface and its own temperature and properties.

Thermal radiation is emitted by a surface due to its temperature. The magnitude of the radiation that is emitted by a surface at a given temperature may be a complicated function of wavelength (i.e., the radiation is distributed **spectrally**) and direction (i.e., the radiation is distributed **directionally**). A **diffuse** surface emits radiation uniformly in all directions. There is an upper limit on the rate that radiation can be emitted by any surface at a specified temperature defined by the **blackbody**. The behavior of any real surface is characterized by comparison with the limiting case of a blackbody.

14.2.2 Blackbody Emission

A blackbody has the following characteristics.

1. It absorbs all of the radiation that is incident upon it (regardless of the direction or wavelength of the incident radiation).
2. It emits radiation uniformly in all directions (i.e., it is a diffuse emitter).
3. It emits the maximum possible amount of radiation at a given temperature and wavelength.

Planck's Law

The radiation that is emitted by a blackbody is referred to as the **blackbody spectral emissive power**, $E_{b,\lambda}$. The blackbody spectral emissive power is a function of temperature and wavelength, as shown in Figure 14.2. A blackbody at a specific temperature will emit thermal radiation over a large range of wavelengths. As the temperature of the blackbody increases, the spectral emissive power increases at all wavelengths and a larger fraction of the emitted power occurs at lower wavelengths. The amount of thermal radiation emitted by a blackbody is an extremely strong function of temperature. (Note that the emissive power axis in Figure 14.2 is logarithmic.) We can often neglect radiation when dealing with heat transfer problems that occur near room temperature but radiation may dominate problems at high temperatures. Even at room temperature, radiation can be significant in comparison with the heat transfer rate resulting from free convection.

The surface temperature of the Sun is approximately 5780 K. Figure 14.2 shows that much of the radiation emitted by the Sun is in the visible region (0.38 μm to 0.78 μm); it is not a coincidence that our eyes have evolved to make use of the radiation emitted at this temperature. Also notice that as the temperature of an object is increased (starting at room temperature), the peak emission moves through invisible, infrared bands and then enters the visible band at wavelengths that we perceive as red. This behavior explains why objects that are heated first appear to be red and eventually become "white" hot when there is emission at all of the wavelengths in the visible region.

There is a peak emissive power in the blackbody spectrum that occurs at wavelength λ_{max}. The wavelength at maximum emissive power is related to temperature according to:

$$\lambda_{max}\, T = 2897.8 \text{ μm-K}. \tag{14.2}$$

Equation (14.2) is referred to as **Wien's Law**, named for Wilhelm Wien who reported this observation in 1893 based on experimental data.

The units of the blackbody spectral emissive power in Figure 14.2 are power/area/wavelength (W/m²-μm) and the quantity should be interpreted as the power emitted per unit area per unit of wavelength. The area under any constant temperature curve in Figure 14.2 is the **total blackbody emissive power**, E_b, which has units of power/area (W/m²) and is the rate that radiation is emitted per unit area. The blackbody emissive power was shown

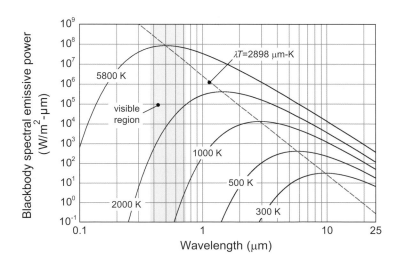

Figure 14.2 Blackbody spectral emissive power, $E_{b,\lambda}$, as a function of wavelength for various values of temperature.

experimentally by Stefan in 1879 and theoretically by Boltzmann in 1884 to be proportional to the fourth power of absolute temperature. The proportionality constant, σ, is called the **Stefan–Boltzmann constant**:

$$E_b = \int_0^\infty E_{b,\lambda} \, d\lambda = \sigma T^4 \quad \text{where} \quad \sigma = 5.67 \times 10^{-8} \frac{\text{W}}{\text{m}^2\text{-K}^4}. \tag{14.3}$$

In 1901, Max Planck published an empirical equation that fit the experimental data at both long and short wavelengths. His equation, called **Planck's Law**, is

$$E_{b,\lambda} = \frac{C_1}{\lambda^5 \left[\exp\left(\dfrac{C_2}{\lambda T}\right) - 1 \right]}, \tag{14.4}$$

where $C_1 = 3.742 \times 10^8$ W-μm^4/m^2 and $C_2 = 14{,}388$ μm-K. Setting the derivative of $E_{b,\lambda}$, Eq. (14.4), with respect to wavelength to zero yields Wien's Law, Eq. (14.2). Integrating Eq. (14.4) over all wavelengths leads to Eq. (14.3).

Blackbody Emission in Specified Wavelength Bands

The blackbody spectral emissive power, Eq. (14.4), can be integrated over all possible wavelengths in order to arrive at the blackbody emissive power, σT^4. This is equivalent to the total area under the blackbody spectral emissive power curve, as shown by the cross-hatched area in Figure 14.3:

$$E_b = \int_{\lambda=0}^\infty \frac{C_1}{\lambda^5 \left[\exp\left(\dfrac{C_2}{\lambda T}\right) - 1 \right]} \, d\lambda = \sigma T^4. \tag{14.5}$$

However, Eq. (14.4) cannot be integrated in closed form between arbitrary wavelength limits. For example, there is no analytical solution to the integral:

$$E_{b,0-\lambda_1} = \int_0^{\lambda_1} \frac{C_1}{\lambda^5 \left[\exp\left(\dfrac{C_2}{\lambda T}\right) - 1 \right]} \, d\lambda, \tag{14.6}$$

where $E_{b,0-\lambda_1}$ is the amount of radiation per unit area emitted by a blackbody at wavelengths less than λ_1. The integration in Eq. (14.6) corresponds to the area under the blackbody emissive power curve for $0 < \lambda < \lambda_1$, as shown by the darker area under the left side of the curve in Figure 14.3.

The fraction of the total emissive power that is emitted in the wavelength band from 0 to λ_1 ($F_{0-\lambda_1}$, sometimes referred to as the **external fractional function**) is defined as the ratio of the integral given by Eq. (14.6) to the

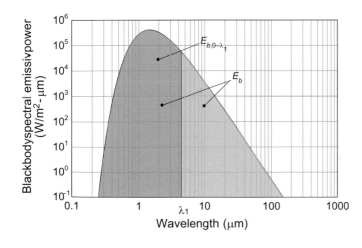

Figure 14.3 Blackbody spectral emissive power curve for 2000 K showing the areas corresponding to E_b and $E_{b,0-\lambda_1}$.

Table 14.1 Fraction of total blackbody emission that occurs between wavelengths 0 and λ_1 as a function of $\lambda_1 T$.

$\lambda_1 T$ (μm-K)	$F_{0-\lambda_1}$	$\lambda_1 T$ (μm-K)	$F_{0-\lambda_1}$
0	0	9000	0.8900
500	1.3×10^{-9}	10,000	0.9142
1000	3.207×10^{-4}	12,000	0.9451
2000	0.0667	14,000	0.9628
3000	0.2732	16,000	0.9738
4000	0.4809	18,000	0.9808
5000	0.6337	20,000	0.9855
6000	0.7378	30,000	0.9953
7000	0.8081	40,000	0.9978
8000	0.8562	∞	1

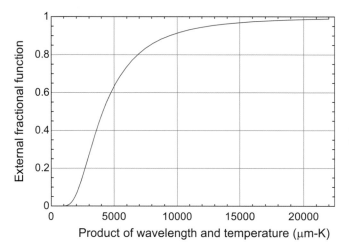

Figure 14.4 Fraction of the total blackbody emission that occurs between wavelengths 0 and λ_1 as a function of $\lambda_1 T$.

integral over all wavelengths, Eq. (14.5) (or equivalently as the ratio of the darker to the total areas shown in Figure 14.3):

$$F_{0-\lambda_1} = \frac{E_{b,0-\lambda_1}}{E_b} = \int_0^{\lambda_1} \frac{C_1}{\sigma T^4 \lambda^5 \left[\exp\left(\frac{C_2}{\lambda T}\right) - 1 \right]} \, d\lambda. \tag{14.7}$$

The integral in Eq. (14.7) appears to be a function of both λ and T. However, if the integration variable is transformed from λ to the product of λ and T, then the integral can be written as:

$$F_{0-\lambda_1} = \int_0^{\lambda_1 T} \frac{C_1}{\sigma (\lambda T)^5 \left[\exp\left(\frac{C_2}{\lambda T}\right) - 1 \right]} \, d(\lambda T). \tag{14.8}$$

Equation (14.8) has only one integration variable, the product λT, and therefore the quantity $F_{0-\lambda_1}$ is only a function of $\lambda_1 T$. Figure 14.4 shows $F_{0-\lambda_1}$ as a function of $\lambda_1 T$ and this information is tabulated in Table 14.1.

The blackbody emissive power that is within a range of wavelengths from λ_1 and λ_2 can be found using the difference between two values of the external fractional function, as indicated by Eq. (14.9):

$$
F_{\lambda_1-\lambda_2} = \int_{\lambda_1}^{\lambda_2} \frac{C_1}{\sigma T^4 \lambda^5 \left[\exp\left(\frac{C_2}{\lambda T} \right) - 1 \right]} \, d\lambda
$$

$$
= \int_0^{\lambda_2} \frac{C_1}{\sigma T^4 \lambda^5 \left[\exp\left(\frac{C_2}{\lambda T} \right) - 1 \right]} \, d\lambda - \int_0^{\lambda_1} \frac{C_1}{\sigma T^4 \lambda^5 \left[\exp\left(\frac{C_2}{\lambda T} \right) - 1 \right]} \, d\lambda \tag{14.9}
$$

$$
= F_{0-\lambda_2} - F_{0-\lambda_1}.
$$

The value of the external fractional function, $F_{\lambda_1-\lambda_2}$, can also be obtained using the **Blackbody** function that is provided in EES.

Example 14.1

A plant can be considered to be a solar collector that converts solar energy to chemical energy via photosynthesis. Consider a typical plant that can utilize solar energy only between the wavelengths λ_{low} = 400 nm and λ_{high} = 700 nm. The plant is able to convert approximately η_t = 7 percent of the energy in this band into chemical energy. Estimate the total efficiency of a plant relative to its ability to convert all of the solar radiation that falls on it to chemical energy.

Known Values

Figure 1 illustrates a sketch of the problem with the known values indicated.

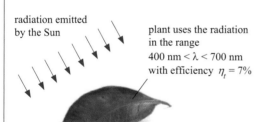

radiation emitted by the Sun

plant uses the radiation in the range
400 nm < λ < 700 nm
with efficiency η_t = 7%

Figure 1 Plant.

Assumption

- Radiation from the Sun can be approximated as being emitted from a blackbody at T_{sun} = 5780 K.

Analysis and Solution

The fraction of the solar energy that lies in the range of the spectrum that is useful to plants can be obtained using the external fractional function information from either Figure 14.4 or Table 14.1. The external fractional functions of interest are evaluated at

$$
\lambda_{low} T_{sun} = (0.4 \ \mu m)(5800 \ K) = 2320 \ \mu m \ K \tag{1}
$$

Continued

Example 14.1 (cont.)

$$\lambda_{high} T_{sun} = (0.7 \ \mu m)(5800 \ K) = 4060 \ \mu m \ K, \tag{2}$$

which, according to Figure 14.4, approximately leads to

$$F_{0-\lambda_{low}} \approx 0.11 \tag{3}$$

$$F_{0-\lambda_{high}} \approx 0.49. \tag{4}$$

The fraction of the solar spectrum that lies between λ_{low} and λ_{high} is therefore:

$$fs = F_{0-\lambda_{high}} - F_{0-\lambda_{low}} = 0.49 - 0.11 = 0.38. \tag{5}$$

So 38 percent of the solar energy is useful to plants and this energy is used at an efficiency of 7 percent. Therefore, the total efficiency of the plant is

$$\eta_{plant} = fs \, \eta_{plant} = (0.38)(0.07) = \boxed{0.027 \ (2.7 \ \text{percent})}. \tag{6}$$

Discussion

The Blackbody function in EES requires the temperature and two wavelengths and returns the fraction of the total energy emitted by a blackbody at that temperature between those two wavelengths, $F_{\lambda_1-\lambda_2}$ from Eq. (14.9). The EES code below calculates the plant efficiency more precisely as it avoids reading information approximately from Figure 14.4.

```
$UnitSystem SI Mass J K Pa
lambda_low=0.4 [micron]                        "low wavelength"
lambda_high=0.7 [micron]                        "high wavelength"
T=5800 [K]                                      "blackbody temperature of sun"
eta_t=0.07 [-]                                  "efficiency of use of 400 nm to 700 nm radiation"
fs=BlackBody(5800 [K],lambda_low,lambda_high)   "fraction of energy between 400 nm and 700 nm"
eta_plant=fs*eta_t                              "total efficiency of a plant as a solar collector"
```

Solving provides $\eta_{plant} = 0.0257$ (2.57 percent).

14.3 Radiation Exchange between Black Surfaces

14.3.1 Introduction

The net rate of radiation heat transfer from a surface is the difference between the rate of radiation that is emitted by the surface and the rate at which the radiation that is incident on the surface is absorbed. The emission of radiation from a blackbody was discussed in Section 14.2. It is now necessary to consider the rate that radiation is absorbed by a surface to complete the problem. In this section, we will examine radiation heat transfer between *black* surfaces (i.e., surfaces that behave as a blackbody, discussed in Section 14.2.2). Black surfaces provide a particularly simple place to start, because they absorb *all* incident radiation; that is, none of the incident radiation is reflected or transmitted. As a result, it is only necessary to understand how much radiation is incident on the surface in order to complete the radiation problem.

The amount of incident radiation is determined by considering the radiation emitted by other surfaces and their geometric orientation with respect to the surface of interest. The geometric orientation of surfaces relative to each other is captured in a parameter that is referred to as the **view factor**. View factors are dimensionless ratios that characterize the degree to which two surfaces "see" one another and therefore how efficiently they exchange radiation.

14.3.2 View Factors

Diffuse surfaces emit radiation uniformly in all directions. In the limit that the surfaces involved in a radiation problem are all diffuse emitters, the fraction of the radiation emitted by one surface that hits another depends only on the relative geometric orientation of the two surfaces, and not on the characteristics of the surfaces themselves or their temperatures. The geometric orientation is captured by the view factor (sometimes referred to as the shape factor, angle factor, or configuration factor). The view factor, $F_{i,j}$, is defined as the fraction of the total radiation that leaves surface i that goes directly to surface j:

$$F_{i,j} = \frac{\text{radiation leaving surface } i \text{ that goes directly to surface } j}{\text{total radiation leaving surface } i}. \tag{14.10}$$

The words "goes directly" in the definition of the view factor exclude the possibility of radiation emitted by surface i reflecting off of other surfaces before finally reaching surface j; black surfaces do not reflect radiation in any case, but this possibility does exist for the nonblack surfaces that are considered in Section 14.4. Also, the word "leaving" in the definition of the view factor recognizes radiation will be reflected *and* emitted by nonblack surfaces. For the black surfaces considered in this section, only radiation emitted by the surface will be "leaving" the surface, but again this definition makes view factors useful even for nonblack surface problems.

Inspection

Based on its definition, the view factor between any two surfaces must lie between 0 (for two surfaces that do not see each other at all, e.g., $F_{3,4} = 0$ in Figure 14.5) to unity (for two surfaces that see only one another, e.g., $F_{1,2} = 1$ in Figure 14.5). In some cases, the view factor between two surfaces can be identified by inspection (for example, any two of the surfaces shown in Figure 14.5). However, in most cases the view factor is not this obvious.

The View Factor Integral

The general formula that may be used to determine the view factors for two surfaces i and j is the double integral:

$$A_i F_{i,j} = A_j F_{j,i} = \int_{A_j} \int_{A_i} \frac{\cos(\theta_i) \cos(\theta_j)}{\pi \, r^2} \, dA_i \, dA_j, \tag{14.11}$$

where, as illustrated in Figure 14.6, r is the distance between differential areas dA_i and dA_j on the two surfaces, θ_i is the angle between the normal to dA_i and a vector connecting areas dA_i and dA_j, and θ_j is the angle between the normal to dA_j and the same vector.

The actual setup and evaluation of Eq. (14.11) is usually difficult and often requires numerical integration. Fortunately, view factors for many situations have already been determined. The use of these view factor solutions is facilitated by the view factor rules and relationships that are presented in this section.

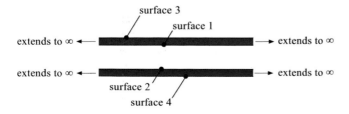

Figure 14.5 Four surfaces, numbered 1 through 4.

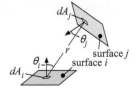

Figure 14.6 Two surfaces exchanging radiation.

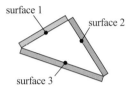

Figure 14.7 Three surfaces in an enclosure.

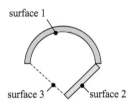

Figure 14.8 A three-surface enclosure including a nonflat surface (1) and an imaginary surface (3).

The Enclosure Rule

All of the radiation emitted by a surface that is part of an enclosure (a set of closed surfaces) must strike some surface within the enclosure. For example, consider the enclosure shown in Figure 14.7 where the surfaces extend indefinitely into and out of the paper. All of the radiation that is emitted by surface 1 must hit one of the surfaces in the enclosure (i.e., surface 1, 2, or 3); therefore

$$F_{1,1} + F_{1,2} + F_{1,3} = 1. \tag{14.12}$$

Similarly, all of the radiation emitted by surface 2 must hit surfaces 1, 2, or 3:

$$F_{2,1} + F_{2,2} + F_{2,3} = 1. \tag{14.13}$$

In general, if surface i is part of an N surface enclosure then:

$$\sum_{j=1}^{N} F_{i,j} = 1. \tag{14.14}$$

Equation (14.14) is the **enclosure rule** and it can be written for each of the N surfaces in the enclosure.

It is important to note that the view factor between a surface and itself ($F_{i,i}$) is not necessarily zero. For the flat surfaces shown in Figure 14.7, it is clear that $F_{1,1}$, $F_{2,2}$, and $F_{3,3}$ *will* all be zero because no part of these surfaces can see another part of the same surface. However, many surfaces are not flat. For example, some of the radiation leaving surface 1 in Figure 14.8 will also hit surface 1 and therefore $F_{1,1}$ will not be zero and must be included in the enclosure rule.

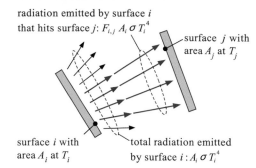

radiation emitted by surface i
that hits surface j: $F_{i,j} A_i \sigma T_i^4$

surface j with
area A_j at T_j

Figure 14.9 Two surfaces, i and j, exchanging radiation.

surface i with
area A_i at T_i

total radiation emitted
by surface i: $A_i \sigma T_i^4$

It is sometimes useful to define an imaginary surface where none exists in order to invoke the enclosure rule for a set of surfaces that would not otherwise form an enclosure. The imaginary surface characterizes radiation that passes through a region of space. For example, in Figure 14.8 surface 3 is defined in order to include all of the radiation that passes through the opening between the edges of surfaces 1 and 2.

Reciprocity

Figure 14.9 shows two black surfaces, i and j. Surface i has area A_i and is at temperature T_i, while surface j has area A_j and is at temperature T_j. The total radiation that is emitted by surface i is the product of its area and the blackbody emissive power of surface i:

$$\text{radiation emitted by surface } i = A_i \sigma\, T_i^4. \tag{14.15}$$

According to the definition of the view factor, the radiation that is emitted by surface i and directly hits surface j is:

$$\text{radiation emitted by surface } i \text{ that hits surface } j = F_{i,j} A_i\, \sigma\, T_i^4. \tag{14.16}$$

Similarly, the radiation that is emitted by surface j and directly hits surface i is:

$$\text{radiation emitted by surface } j \text{ that hits surface } i = F_{j,i} A_j\, \sigma\, T_j^4. \tag{14.17}$$

The net radiation heat transfer from surface i to surface j, $\dot{q}_{i \text{ to } j}$, is the difference between Eq. (14.16) and Eq. (14.17):

$$\dot{q}_{i \text{ to } j} = F_{i,j} A_i\, \sigma\, T_i^4 - F_{j,i} A_j \sigma\, T_j^4. \tag{14.18}$$

In the limit that surface i and surface j are at the same temperature (i.e., $T_i = T_j = T$), the Second Law of Thermodynamics requires that $\dot{q}_{i \text{ to } j} = 0$, and therefore Eq. (14.18) reduces to:

$$F_{i,j} A_i\, \sigma\, T^4 - F_{j,i} A_j\, \sigma\, T^4 = 0. \tag{14.19}$$

In order for Eq. (14.19) to be true, it is necessary that

$$A_i\, F_{i,j} = A_j\, F_{j,i}, \tag{14.20}$$

which is the view factor rule known as **reciprocity**. Equation (14.20) was derived in the limit that the temperatures of surfaces 1 and 2 are equal. However, the areas and view factors in Eq. (14.20) are geometric quantities that do not depend on temperature; therefore, the relationship must be true whether or not the temperatures of the surface are equal.

Other View Factor Relationships

There are some additional helpful view factor relationships. Any two surfaces (j and k) can be combined; then, according to the definition of the view factor:

$$F_{i,jk} = F_{i,j} + F_{i,k}. \tag{14.21}$$

Using reciprocity together with Eq. (14.21) results in

$$\left(A_i + A_j\right)F_{ij,k} = A_i F_{i,k} + A_j F_{j,k}. \tag{14.22}$$

The property of symmetry can also sometimes be used to help determine view factors. For example, consider a sphere, surface 1, placed between two infinite parallel plates, surfaces 2 and 3, as shown in Figure 14.10. Since the sphere cannot see itself, Eq. (14.14) requires that $F_{1,2} + F_{1,3} = 1$. However, by symmetry (i.e., the fact that surface 1 "sees" surfaces 2 and 3 equally), $F_{1,2} = F_{1,3}$. Therefore, $F_{1,2} = F_{1,3} = 0.50$.

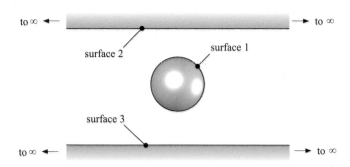

Figure 14.10 Spherical surface between two infinite parallel plates.

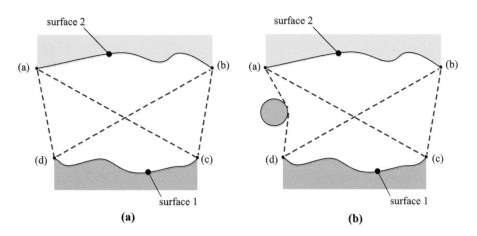

Figure 14.11 Two-dimensional surfaces (a) without and (b) with an obstruction.

The Crossed and Uncrossed Strings Method

A useful method for calculating view factors in 2-D geometries is the **crossed and uncrossed strings method**, described in Hottel and Sarofim (1967) and McAdams (1954). The process is illustrated in Figure 14.11(a) and involves defining four "strings" that are drawn from the corners of each surface to both corners of the other surface. In Figure 14.11(a) notice the four dotted lines corresponding to two "crossed strings" (L_{ac} and L_{bd}) and two "uncrossed" strings (L_{ad} and L_{bc}).

The crossed and uncrossed strings method provides the view factors between surfaces 1 and 2 in terms of the length of the crossed and uncrossed strings:

$$A_1 F_{1,2} = A_2 F_{2,1} = W \frac{\sum L_{\text{crossed}} - \sum L_{\text{uncrossed}}}{2} = W \frac{(L_{ac} + L_{bd}) - (L_{ad} - L_{bc})}{2}, \tag{14.23}$$

where W is the width of the surfaces into the page; note that W will also appear in the areas in Eq. (14.23) and therefore cancel. The crossed and uncrossed strings method is applicable even if there is an obstruction between the surfaces, as indicated in Figure 14.11(b). In this case, the string L_{ad} must "wrap" around the obstruction, which affects its length. The method is only valid for 2-D geometries.

Example 14.2

Consider two flat plates that are parallel to one another, as shown in Figure 1. The plates are very long in the direction into the page.

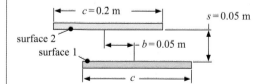

Figure 1 Two long plates.

Surface 1 is the top of the lower plate while surface 2 is the bottom of the upper plate. The plates are the same width, $c = 0.2$ m, and the distance between the plates is $s = 0.05$ m. The center of the plates are offset by a distance $b = 0.05$ m. Determine the view factor from surface 1 to surface 2.

Known Values

Figure 1 illustrates a sketch of the problem with the known values indicated.

Assumptions

- The plates are sufficiently long that this can be modeled as a 2-D problem.
- Both plates are diffuse surfaces.

Analysis and Solution

The crossed and uncrossed strings method is used to determine the view factor. The four lengths that must be determined are L_{ad}, L_{bc}, L_{ac}, and L_{bd}, as shown in Figure 2.

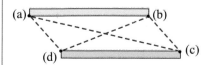

Figure 2 Crossed and uncrossed strings.

The length of the uncrossed strings, L_{ad} and L_{bc}, are identical:

$$L_{ad} = L_{bc} = \sqrt{b^2 + s^2} = \sqrt{(0.05\,\text{m})^2 + (0.05\,\text{m})^2} = 0.0707\,\text{m}. \tag{1}$$

Continued

Example 14.2 (cont.)

The lengths of the crossed strings, L_{ac} and L_{bd}, are computed:

$$L_{ac} = \sqrt{(c+b)^2 + s^2} = \sqrt{(0.2\,\text{m} + 0.05\,\text{m})^2 + (0.05\,\text{m})^2} = 0.255\,\text{m} \tag{2}$$

$$L_{bd} = \sqrt{(c-b)^2 + s^2} = \sqrt{(0.2\,\text{m} - 0.05\,\text{m})^2 + (0.05\,\text{m})^2} = 0.158\,\text{m}. \tag{3}$$

Equation (14.23) is used to compute the view factor, where $A_1 = W\,c$ and W is the width of the plates into the page:

$$F_{1,2} = W \frac{\sum L_{\text{crossed}} - \sum L_{\text{uncrossed}}}{2A_1} = \frac{(L_{ac} + L_{bd}) - (L_{ad} + L_{bc})}{2\,c}$$

$$= \frac{(0.255 + 0.158) - (0.0707 + 0.0707)}{2(0.2)} = \boxed{0.6791}. \tag{4}$$

Discussion

The view factor indicates that about 68 percent of the radiation emitted by surface 1 will directly hit surface 2, which is reasonable given their close proximity. One way of checking view factors is to look at their behavior in well-defined limits. For example, Figure 3 illustrates the view factor $F_{1,2}$ as a function of the plate-to-plate spacing, s. When s approaches zero the view factor limits to 0.75 because 75 percent of surface 1 is directly adjacent to surface 2. As s becomes large the view factor asymptotically approaches 0.

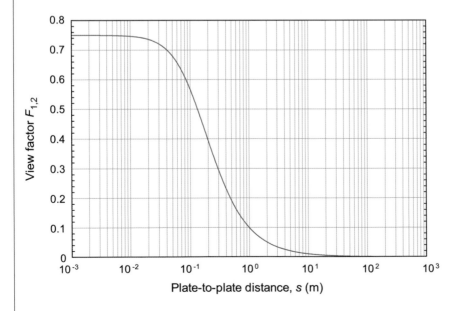

Figure 3 View factor as a function of plate-to-plate distance.

The View Factor Libraries

The double integral that provides the view factor between two arbitrary surfaces, Eq. (14.11), is difficult to evaluate even for relatively simple geometries. Fortunately, view factors have already been determined for many common situations. For example, Siegel and Howell (2002) provide a summary of view factor formulae and a larger collection of view factors in either graphical or analytical form. This information is provided on a web site compiled by Howell (www.me.utexas.edu/~howell/). A few of these view factor formulae are summarized in Table 14.2.

A comprehensive set of view factor relations have been implemented as EES library functions; these are accessed from the Function Information Window by selecting Radiation View Factors from the drop down menu of heat exchanger functions. The view factors are further classified into 2-D, 3-D, or differential view factors. The 2-D view factors are appropriate when one dimension is much longer than the other dimensions in the problem so that this dimension can be assumed to be infinite. The view factors for these situations can be derived from the crossed and uncrossed string method presented in the previous section. The 3-D view factors relate surfaces that have finite size in all dimensions. An index for the view factors available in EES is at http://fchartsoftware.com/ees/heat_transfer_library/shape_factors/hs1.htm.

14.3.3 Blackbody Radiation Calculations

Figure 14.12 shows two black surfaces, i and j that are exchanging radiation; surface i has area A_i and is at temperature T_i, while surface j has area A_j and is at temperature T_j. The equation that describes this energy exchange is derived following the steps previously used in Section 14.3.2 in order to derive the reciprocity relation, but without making the assumption that the two surfaces have the same temperature.

The total rate of radiation emitted by surface i is the product of the area and the blackbody emissive power:

$$\text{radiation emitted by surface } i = A_i \, \sigma \, T_i^4. \tag{14.24}$$

According to the definition of the view factor, the radiation that is emitted by surface i and incident on surface j is

$$\text{radiation emitted by surface } i \text{ that hits surface } j = F_{i,j} \, A_i \, \sigma \, T_i^4. \tag{14.25}$$

Because surface j is black, all of the radiation that is incident on it must be absorbed. Similarly, the radiation that is emitted by surface j and hits surface i where it is absorbed is

$$\text{radiation emitted by surface } j \text{ that hits surface } i = F_{j,i} \, A_j \, \sigma \, T_j^4. \tag{14.26}$$

The net rate of radiation exchange from surface i to surface j, $\dot{q}_{i \text{ to } j}$, is the difference between Eq. (14.25) and Eq. (14.26):

$$\dot{q}_{i \text{ to } j} = F_{i,j} \, A_i \, \sigma \, T_i^4 - F_{j,i} \, A_j \sigma \, T_j^4. \tag{14.27}$$

Reciprocity, Eq. (14.20), requires that

$$F_{i,j} \, A_i = F_{j,i} \, A_j, \tag{14.28}$$

so that Eq. (14.27) can be simplified to

$$\dot{q}_{i \text{ to } j} = A_i \, F_{i,j} \, \sigma \left(T_i^4 - T_j^4 \right) = A_i F_{i,j} \left(E_{b,i} - E_{b,j} \right). \tag{14.29}$$

The Space Resistance

Equation (14.29) can be rearranged to have the same form as a resistance equation. The net rate of radiation heat transfer between two black surfaces i and j ($\dot{q}_{i \text{ to } j}$) is driven by a difference in their blackbody emissive

Table 14.2 View factors (adapted from www.me.utexas.edu/~howell/).

Parallel plates ($L \gg W$)

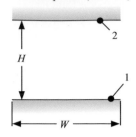

$$F_{1,2} = \sqrt{1 + \left(\frac{H}{W}\right)^2} - \frac{H}{W}$$

Plates joined at an angle ($L \gg W$)

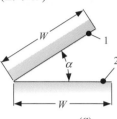

$$F_{1,2} = 1 - \sin\left(\frac{\alpha}{2}\right)$$

Plate to cylinder ($L \gg r, b_1$)

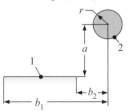

$$F_{1,2} = \frac{r}{(b_1 - b_2)}\left[\tan^{-1}\left(\frac{b_1}{a}\right) - \tan^{-1}\left(\frac{b_2}{a}\right)\right]$$

Semicircle to itself with concentric cylinder ($L \gg r_1$)

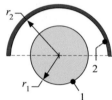

$$F_{2,2} = 1 - \frac{2}{\pi}\left[\sqrt{1 - \left(\frac{r_1}{r_2}\right)^2} + \frac{r_1}{r_2}\sin^{-1}\left(\frac{r_1}{r_2}\right)\right]$$

Sphere to coaxial disk

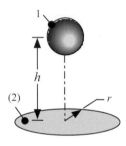

$$F_{1,2} = \frac{1}{2}\left[1 - \frac{1}{\sqrt{1 + \left(\frac{r}{h}\right)^2}}\right]$$

Cylinder to cylinder ($L \gg r$)

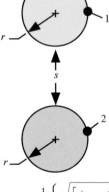

$$F_{1,2} = \frac{1}{\pi}\left\{\sqrt{\left[\left(1 + \frac{s}{2r}\right)^2 - 1\right]} + \sin^{-1}\left[\left(1 + \frac{s}{2r}\right)^{-1}\right] - \left(1 + \frac{s}{2r}\right)\right\}$$

Coaxial disks

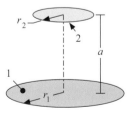

$$F_{1,2} = \frac{1}{2}\left[S - \sqrt{S^2 - 4\left(\frac{r_2}{r_1}\right)^2}\right]$$

$$S = 1 + \frac{1 + \left(\frac{r_2}{a}\right)^2}{\left(\frac{r_1}{a}\right)^2}$$

Sphere to a cylinder

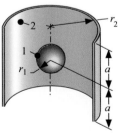

$$F_{1,2} = \frac{1}{\sqrt{1 + \left(\frac{r_2}{a}\right)^2}}$$

Base of cylinder to internal sides

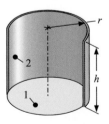

$$F_{1,2} = \frac{h}{r}\left(\sqrt{1 + \left(\frac{h}{2r}\right)^2} - \frac{h}{2r}\right)$$

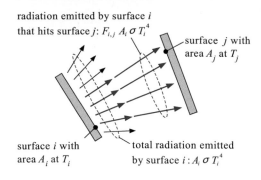

radiation emitted by surface i
that hits surface j: $F_{i,j} A_i \sigma T_i^4$

surface j with
area A_j at T_j

surface i with
area A_i at T_i

total radiation emitted
by surface i: $A_i \sigma T_i^4$

Figure 14.12 Two black surfaces, i and j, at different
temperatures, T_i and T_j, exchanging radiation.

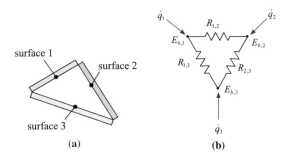

surface 1

surface 2

surface 3

(a)

(b)

Figure 14.13 (a) Three-surface enclosure and (b)
corresponding resistance network.

powers, $E_{b,i} - E_{b,j}$, and the resistance to the radiation heat transfer is the inverse of the product of the area and
view factor:

$$\dot{q}_{i \text{ to } j} = \frac{(E_{b,i} - E_{b,j})}{R_{i,j}} \text{ where } R_{i,j} = \frac{1}{A_i F_{i,j}} = \frac{1}{A_j F_{j,i}}. \tag{14.30}$$

The resistance $R_{i,j}$ in Eq. (14.30) is sometimes referred to as the surface-to-surface resistance, the geometrical
resistance, or the **space resistance**. The space resistance tends to increase as either the area of the surface or the
view factor between the surfaces is reduced; this makes sense, as reducing the area or view factor will reduce the
ease with which two surfaces can interact radiatively.

For radiation problems involving only a few black surfaces it is convenient to draw a resistance network
where each node represents the emissive power of an isothermal surface and these nodes are connected by
space resistances that are calculated using Eq. (14.30). For example, the three-surface enclosure shown in
Figure 14.13(a) can be represented by the resistance network shown in Figure 14.13(b). The blackbody emissive
power of each node is related to the temperature of the surface while the heat transfer rate entering the circuit at
each node represents the net radiation heat transfer *from* the surface *to* its surroundings.

Example 14.3

A three-surface enclosure is formed by the internal surfaces of a cylinder (top, bottom, and side). The
radius of the cylinder is $R = 0.1$ m and the height is $H = 0.25$ m. The bottom surface is steadily heated at a
rate $\dot{q}_1 = 100$ W, the top surface is cooled so that its temperature is held at $T_3 = 300$ K, and the side surface is

Continued

Example 14.3 (cont.)

insulated. Determine the temperature of the heated surface and the temperature of the side, assuming that all surfaces are blackbodies.

Known Values

Figure 1 illustrates a sketch of the problem with the known values indicated.

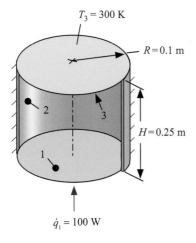

$T_3 = 300$ K

$R = 0.1$ m

$H = 0.25$ m

$\dot{q}_1 = 100$ W

Figure 1 Three-surface enclosure.

Assumptions

- Radiation is the only heat transfer mechanism.
- All three surfaces are isothermal.

Analysis and Solution

The three-surface problem is represented using the resistance network shown in Figure 2.

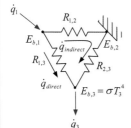

$\dot{q}_1$

$R_{1,2}$

$E_{b,1}$

$\dot{q}_{indirect}$

$E_{b,2}$

$R_{1,3}$

$R_{2,3}$

$\dot{q}_{direct}$

$E_{b,3} = \sigma T_3^4$

$\dot{q}_3$

Figure 2 Resistance network.

The areas of the three surfaces are computed:

$$A_1 = A_3 = \pi R^2 = \pi (0.1 \text{ m})^2 = 0.0314 \text{ m}^2 \tag{1}$$

Example 14.3 (cont.)

$$A_2 = 2 \pi R H = 2 \pi (0.1 \text{ m})(0.25 \text{ m}) = 0.157 \text{ m}^2. \tag{2}$$

The view factor between surfaces 1 and 2 and between surfaces 3 and 2 are computed using the relationship given in Table 14.2:

$$F_{1,2} = F_{3,2} = \frac{H}{R}\left(\sqrt{1 + \left(\frac{H}{2R}\right)^2} - \frac{H}{2R}\right) = \frac{0.25}{0.1}\left(\sqrt{1 + \left(\frac{0.25}{2(0.1)}\right)^2} - \frac{0.25}{2(0.1)}\right) = 0.877. \tag{3}$$

The view factor between surface 1 and surface 3 is obtained using the enclosure rule (note that $F_{1,1} = 0$):

$$F_{1,3} = 1 - F_{1,2} = 1 - 0.877 = 0.123. \tag{4}$$

The space resistances shown in Figure 2 can be computed:

$$R_{1,2} = \frac{1}{A_1 F_{1,2}} = \frac{1}{0.0314 \text{ m}^2}\bigg|\frac{}{0.877} = 36.3 \text{ m}^{-2} \tag{5}$$

$$R_{2,3} = \frac{1}{A_3 F_{3,2}} = \frac{1}{0.0314 \text{ m}^2}\bigg|\frac{}{0.877} = 36.3 \text{ m}^{-2} \tag{6}$$

$$R_{1,3} = \frac{1}{A_1 F_{1,3}} = \frac{1}{0.0314 \text{ m}^2}\bigg|\frac{}{0.123} = 258.7 \text{ m}^{-2}. \tag{7}$$

The boundary conditions associated with the network are shown in Figure 2. The temperature of surface 3 is known, which allows its blackbody emissive power, $E_{b,3}$, to be computed. In a radiation network, the blackbody emissive power corresponds to the potential of the surface:

$$E_{b,3} = \sigma T_3^4 = \frac{5.67 \times 10^{-8} \text{ W}}{\text{m}^2 \text{ K}^4}\bigg|\frac{300^4 \text{ K}^4}{} = 459.3 \frac{\text{W}}{\text{m}^2}. \tag{8}$$

The heat transfer entering at node 1 in Figure 2 must leave at node 3 since surface 2 is adiabatic. Therefore, with these boundary conditions, the network places $R_{1,3}$ in parallel with the series combination of $R_{1,2}$ and $R_{2,3}$. The total equivalent resistance of the network between nodes 1 and 3 is given by

$$R_{total} = \left[\frac{1}{R_{1,3}} + \frac{1}{(R_{1,2} + R_{2,3})}\right]^{-1} = \left[\frac{1}{258.7} + \frac{1}{(36.3 + 36.3)}\right]^{-1} = 56.7 \text{ m}^{-2}. \tag{9}$$

The blackbody emissive power of the bottom (surface 1) can be computed according to

$$E_{b,1} = E_{b,3} + \dot{q}_1 R_{total} = 459.3 \frac{\text{W}}{\text{m}^2} + \frac{100 \text{ W}}{}\bigg|\frac{56.7}{\text{m}^2} = 6128 \frac{\text{W}}{\text{m}^2}. \tag{10}$$

The temperature of surface 1 can be computed from its blackbody emissive power:

$$T_1 = \left(\frac{E_{b,1}}{\sigma}\right)^{1/4} = \left(\frac{6128 \text{ W}}{\text{m}^2}\bigg|\frac{\text{m}^2 \text{ K}^4}{5.67 \times 10^{-8} \text{ W}}\right)^{1/4} = \boxed{573.4 \text{ K}}. \tag{11}$$

The heat transfer from the bottom surface (node 1 in Figure 2) can be transferred to surface 3 either directly or indirectly (from the bottom surface to the side to the top surface). The heat transfer that passes through the indirect path ($\dot{q}_{indirect}$ in Figure 2) is calculated according to:

$$\dot{q}_{indirect} = \frac{E_{b,1} - E_{b,3}}{R_{1,2} + R_{2,3}} = \frac{(6128 - 459.3)\text{W}}{\text{m}^2}\bigg|\frac{\text{m}^2}{(36.3 + 36.3)} = 78.1 \text{ W}. \tag{12}$$

The blackbody emissive power of the sides (surface 2) can be computed according to:

$$E_{b,2} = E_{b,1} - \dot{q}_{indirect} R_{1,2} = 6128 \frac{\text{W}}{\text{m}^2} - \frac{78.1 \text{ W}}{}\bigg|\frac{36.3}{\text{m}^2} = 3294 \frac{\text{W}}{\text{m}^2}. \tag{13}$$

Continued

Example 14.3 (cont.)

The temperature of surface 2 can be computed according to:

$$T_2 = \left(\frac{E_{b,2}}{\sigma}\right)^{1/4} = \left(\frac{3294 \text{ W}}{\text{m}^2}\left|\frac{\text{m}^2 \text{ K}^4}{5.67 \times 10^{-8} \text{ W}}\right.\right)^{1/4} = \boxed{490.9 \text{ K}}. \tag{14}$$

Discussion

The relative size of the radiation resistances, like the thermal resistances discussed in Chapter 2, allow us to understand what radiation paths are most important. For example, the radiation resistances between the side and the top ($R_{2,3}$) and between the side and the bottom ($R_{1,2}$) are both much smaller than the radiation resistance between the top and bottom directly ($R_{1,3}$). As a result, very little radiation travels directly between the top and bottom ($\dot{q}_{direct}$ in Figure 2) while most radiation travels from the bottom to the side and then from the side to the top ($\dot{q}_{indirect}$). Resistance networks are useful for radiation problems having a small number of surfaces. Other methods, shown in the following sections, are more convenient for radiation problems having more than three surfaces.

Example 14.4

The Earth interacts radiatively with two surfaces, the Sun and outer space. The effective surface temperature of the Sun (surface 1) is approximately $T_1 = 5780$ K and the diameter of the Sun is $D_1 = 1.390 \times 10^9$ m. The effective temperature of space (surface 3) is approximately $T_3 = 2.7$ K. The diameter of the Earth is $D_2 = 1.276 \times 10^7$ m. The distance between the Earth and the Sun varies throughout the year but is, on average, $R = 1.497 \times 10^{11}$ m. Estimate the average temperature of the Earth using the concepts discussed in this section.

Known Values

Figure 1 illustrates a sketch of the problem with the known values indicated.

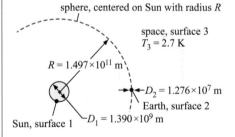

Figure 1 The Earth and its relationship with the Sun (not to scale).

Assumptions

- The Earth can be considered an adiabatic surface.
- All three surfaces are black.
- Steady-state conditions exist.

Analysis and Solution

There are three surfaces involved in the problem; therefore, the resistance network that represents the situation must include three nodes, as shown in Figure 2.

Example 14.4 (cont.)

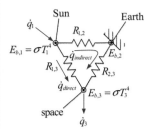

Figure 2 Resistance network.

The nodes represent the blackbody emissive powers of each surface and the resistances between the nodes represent the degree to which the surfaces interact. We will assume that the Earth is adiabatic; that is, all of the energy received from the Sun is radiated to space, resulting in a steady-state temperature for the Earth. The boundary conditions associated with the resistance network include the specified temperature of the Sun (surface 1) and space (surface 3); their blackbody emissive powers are

$$E_{b,1} = \sigma T_1^4 = \frac{5.67 \times 10^{-8} \text{ W}}{\text{m}^2 \text{ K}^4} \left| \frac{5800^4 \text{ K}^4}{} = 6.417 \times 10^7 \frac{\text{W}}{\text{m}^2} \right.$$ (1)

$$E_{b,3} = \sigma T_3^4 = \frac{5.67 \times 10^{-8} \text{ W}}{\text{m}^2 \text{ K}^4} \left| \frac{2.7^4 \text{ K}^4}{} = 3.013 \times 10^{-6} \frac{\text{W}}{\text{m}^2} \right. .$$ (2)

The surface areas of the Sun and the Earth are

$$A_1 = \pi D_1^2 = \pi \left(1.390 \times 10^9\right)^2 \text{ m}^2 = 6.070 \times 10^{18} \text{ m}^2$$ (3)

$$A_2 = \pi D_2^2 = \pi \left(1.276 \times 10^7\right)^2 \text{ m}^2 = 5.115 \times 10^{14} \text{ m}^2.$$ (4)

The view factor between the Sun and space is essentially unity ($F_{1,3} \approx 1$; an observer on the surface of the Sun would see much more of space than anything else). Therefore, the Sun-to-space space resistance is

$$R_{1,3} = \frac{1}{A_1 F_{1,3}} = \frac{1}{6.070 \times 10^{18} \text{ m}^2} \left| \frac{}{1} = 1.647 \times 10^{-19} \text{ m}^{-2}. \right.$$ (5)

The view factor between the Earth and space is also nearly unity ($F_{2,3} \approx 1$, because the fraction of the total view of space from Earth that is blocked by the Sun is very small).

Therefore, the Earth-to-space space resistance ($R_{2,3}$) is

$$R_{2,3} = \frac{1}{A_2 F_{2,3}} = \frac{1}{5.115 \times 10^{14} \text{ m}^2} \left| \frac{}{1} = 1.955 \times 10^{-15} \text{ m}^{-2}. \right.$$ (6)

The view factor between the Sun and the Earth ($F_{1,2}$) is nearly zero. However, clearly $F_{1,2}$ is not exactly zero or the Earth would be a very cold place. The value of $F_{1,2}$ can be determined by imagining a sphere of radius R that is centered at the Sun (see Figure 1); the view factor is the ratio of the projected area occupied by the Earth on this sphere to the surface area of the sphere itself:

$$F_{1,2} = \frac{\pi \dfrac{D_2^2}{4}}{4 \pi R^2} = \frac{D_2^2}{16 R^2} = \frac{\left(1.276 \times 10^7\right)^2}{16 \left(1.497 \times 10^{11}\right)^2} = 4.541 \times 10^{-10}.$$ (7)

This result indicates that about 5 out of every 10 billion photons emitted by the Sun ultimately strike the surface of the Earth. The Sun-to-Earth resistance ($R_{1,2}$) is

Continued

Example 14.4 (cont.)

$$R_{1,2} = \frac{1}{A_1\,F_{1,2}} = \frac{1}{6.070 \times 10^{18}\ \mathrm{m}^2}\left|\frac{1}{4.541 \times 10^{-10}}\right. = 3.628 \times 10^{-10}\ \mathrm{m}^{-2}. \tag{8}$$

Because the Earth–Sun system is assumed to be at steady state, the rate at which energy is transferred from the Sun to the Earth and then re-radiated to space is

$$\dot{q}_{indirect} = \frac{(E_{b,1} - E_{b,3})}{R_{1,2} + R_{2,3}}$$

$$= \frac{\left(6.417 \times 10^7 - 3.013 \times 10^{-6}\right)\ \mathrm{W}}{\mathrm{m}^2}\left|\frac{\mathrm{m}^2}{\left(3.628 \times 10^{-10} + 1.955 \times 10^{-15}\right)}\right. = 1.769 \times 10^{17}\ \mathrm{W}. \tag{9}$$

The blackbody emissive power of the Earth ($E_{b,2}$) is:

$$E_{b,2} = E_{b,1} - \dot{q}_{indirect}R_{1,2} = 6.417 \times 10^7\ \frac{\mathrm{W}}{\mathrm{m}^2} - \frac{1.769 \times 10^{17}\ \mathrm{W}}{}\left|\frac{3.628 \times 10^{-10}}{\mathrm{m}^2}\right. = 345.8\ \frac{\mathrm{W}}{\mathrm{m}^2}. \tag{10}$$

The temperature of the Earth (T_2) is related to its blackbody emissive power:

$$T_2 = \left(\frac{E_{b,2}}{\sigma}\right)^{1/4} = \left(345.8\ \frac{\mathrm{W}}{\mathrm{m}^2}\left|\frac{\mathrm{m}^2\ \mathrm{K}^4}{5.67 \times 10^{-8}\ \mathrm{W}}\right.\right)^{1/4} = \boxed{279.4\ \mathrm{K}\ (43.33°\mathrm{F})}. \tag{11}$$

Discussion

These calculations predict that the temperature of the Earth is only 279.4 K (6.3°C or 43.3°F). This result leads into a discussion of several important issues, most notably the **greenhouse effect** and global warming (or climate change). The calculated Earth temperature is reasonable, but low relative to the observed average temperature of the Earth. The average temperature of the Earth varies somewhat from year to year but is approximately 57.2°F, which is about 14°F higher than the predicted temperature from Eq. (11). One of the major reasons for the discrepancy between the observed and calculated temperature of the Earth is the greenhouse effect that is caused by absorption of the radiation emitted by the Earth in the atmosphere. The Sun-to-Earth heat transfer that passes through the atmosphere through resistance $R_{1,2}$ and the Earth-to-space heat transfer that passed through resistance $R_{2,3}$ are, according to our analysis, identical as we have assumed that there is no source of energy on the Earth itself and the Earth is not increasing or decreasing in temperature at a significant rate. However, the radiation associated with these two heat transfers is spectrally distributed very differently. The energy that is incident on the Earth is emitted by an object at high temperature (the Sun) and therefore, according to Planck's Law (discussed in Section 14.2), it is concentrated at short wavelengths. The energy leaving the Earth is emitted by an object at much lower temperature (the Earth) and therefore this radiation is spread out over a range of longer wavelengths. Figure 3 shows the spectral distribution of the total power emitted by the Earth (assuming that it is at 280 K) and received from the Sun (assuming that it is at 5780 K). Note that the integral of these spectral distributions over all wavelengths is the same (1.74×10^{17} W), even though the logarithmic x-scale makes the areas look different.

Example 14.4 (cont.)

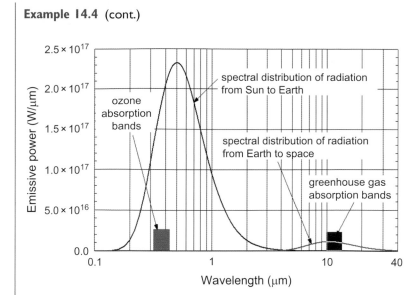

Figure 3 Spectrum of the blackbody radiation received from the Sun and emitted by the Earth.

There are strong absorption bands associated with the carbon dioxide and water vapor in the atmosphere. These absorption bands lie at long wavelengths and therefore tend to preferentially "trap" the energy emitted by the Earth so that some of it does not make it through the atmosphere to space. This selective absorption of energy in specific wavelength bands is referred to as the greenhouse effect and it is one reason that the temperature of the Earth is higher than the value that we calculated. Global warming (or climate change) is primarily related to the gradual buildup of carbon dioxide in the atmosphere due to the combustion of fossil fuels. The carbon dioxide released by combustion tends to increase the greenhouse gas absorption bands shown in Figure 3 and therefore the average temperature of the Earth must rise in order to re-establish thermal equilibrium.

N-Surface Solutions

The representation of a blackbody radiation problem with a resistance network, as in the previous two examples, can be a useful method of visualizing the problem. However, for even a relatively modest number of surfaces, this technique is not very practical. Typically, each surface interacts with all of the other surfaces involved in a radiation problem and therefore a network involving more than three or four surfaces becomes more confusing than it is helpful. However, it is possible to systematically solve a blackbody radiation problem that involves an arbitrary number of surfaces provided that the view factors and areas are known and a boundary condition can be defined for each surface.

The net rate of radiation exchange from surface i to all of the N surfaces in a problem is obtained by summing Eq. (14.29) over all of the surfaces involved:

$$\dot{q}_i = A_i \sigma \sum_{j=1}^{N} F_{i,j} \left(T_i^4 - T_j^4 \right) = A_i \sum_{j=1}^{N} F_{i,j} \left(E_{b,i} - E_{b,j} \right) \text{ for } i = 1 \ldots N. \tag{14.31}$$

Equation (14.31), written for each of the N surfaces, provides N equations in $2N$ unknowns: the blackbody emissive power (or temperature) and the net heat transfer associated with each of the N surfaces. A complete set of boundary conditions will include a specification of either the temperature or net heat transfer rate (or a relationship between these quantities) for each of the surfaces, providing N additional equations and therefore a completely specified problem.

Example 14.5

Figure 1 illustrates a 2-D radiation problem involving four surfaces. Surface 1 is a heater plate with width $W = 0.2$ m that is maintained at $T_H = 1000$ K. The surface extends infinitely into the page. One side of the heater plate radiatively interacts with a pipe that has radius $R_p = 0.05$ m and temperature $T_p = 400$ K. The distance between the center of the pipe and the heater is $s = 0.06$ m. The pipe lies at the center of a semicircular shield with radius $R_s = 0.1$ m. The surface of the shield that is towards the pipe participates in the radiation problem; the other side of the shield (facing the surroundings) does not. The surroundings are at $T_{sur} = 300$ K.

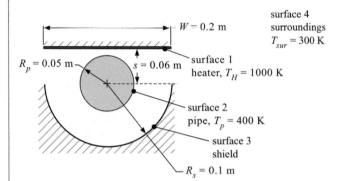

Figure 1 Heater plate used to heat a pipe.

Determine the efficiency of the heater; the efficiency is defined here as the ratio of the heat transfer rate to the pipe to the radiation heat transfer rate from the heater.

Known Values

Figure 1 illustrates a sketch of the problem with the known values indicated.

Assumptions

- All surfaces are black.
- The problem is 2-D and steady state.
- The only heat transfer mechanism is radiation.

Analysis

There are four surfaces involved in the problem; therefore, a resistance network is likely to be too complex to be useful. Instead, we will evaluate the areas and view factors that characterize the problem and employ Eq. (14.31) together with a complete set of boundary conditions to solve the problem.
The areas of surfaces 1 to 3 are computed:

$$A_1 = W\,L \tag{1}$$

$$A_2 = 2\,\pi\,R_p\,L \tag{2}$$

$$A_3 = \pi\,R_s\,L. \tag{3}$$

The area of the surroundings, A_4, can be set to some arbitrary, large value.

Example 14.5 (cont.)

A complete set of view factors is needed. The view factor from the pipe to the heater plate ($F_{2,1}$) can be obtained using the View Factor Library in EES (function F2D_6). The view factor from the pipe to itself is $F_{2,2} = 0$. The view factor from the pipe to the shield is $F_{2,3} = 0.5$, by inspection. The view factor from the pipe to the surroundings is obtained using the enclosure rule:

$$F_{2,4} = 1 - F_{2,1} - F_{2,2} - F_{2,3}. \tag{4}$$

The view factor from the heater plate to itself is $F_{1,1} = 0$. The view factor from the heater plate to the pipe can be obtained from reciprocity:

$$F_{1,2} = \frac{A_2 \, F_{2,1}}{A_1}. \tag{5}$$

The view factor from the heater plate to the shield, $F_{1,3}$, is obtained by subtracting the view factor from the plate to the pipe from the view factor from the plate to the shield that would exist without the pipe, obtained from the View Factor Library (function F2D_10). The view factor from the plate to the surroundings is obtained using the enclosure rule:

$$F_{1,4} = 1 - F_{1,1} - F_{1,2} - F_{1,3}. \tag{6}$$

The view factor between the shield and the plate ($F_{3,1}$) and the shield and the pipe ($F_{3,2}$) are obtained by reciprocity:

$$F_{3,1} = \frac{A_1 \, F_{1,3}}{A_3} \tag{7}$$

$$F_{3,2} = \frac{A_2 \, F_{2,3}}{A_3}. \tag{8}$$

The view factor between the shield and itself, $F_{3,3}$, is obtained from the View Factor Library (function F2D_8). The view factor between the shield and the surroundings is obtained using the enclosure rule:

$$F_{3,4} = 1 - F_{3,1} - F_{3,2} - F_{3,3}. \tag{9}$$

The view factors from the surroundings to each of the other surfaces are obtained using reciprocity:

$$F_{4,1} = \frac{A_1 \, F_{1,4}}{A_4} \tag{10}$$

$$F_{4,2} = \frac{A_2 \, F_{2,4}}{A_4} \tag{11}$$

$$F_{4,3} = \frac{A_3 \, F_{3,4}}{A_4}. \tag{12}$$

The view factor from the surroundings to itself, $F_{4,4}$, is obtained from the enclosure rule:

$$F_{4,4} = 1 - F_{4,1} - F_{4,2} - F_{4,3}. \tag{13}$$

Equation (14.31) is written for each of the four surfaces:

$$\dot{q}_i = A_i \, \sigma \sum_{j=1}^{4} F_{i,j} \left(T_i^4 - T_j^4 \right) \text{ for } i = 1 \dots 4. \tag{14}$$

The four boundary conditions for the problems are:

$$T_1 = T_H \tag{15}$$

$$T_2 = T_p \tag{16}$$

$$\dot{q}_3 = 0 \tag{17}$$

Continued

Example 14.5 (cont.)

$$T_4 = T_{sur}. \tag{18}$$

Solution

Once the areas and view factors are computed, Eqs. (14) through (18) represent eight equations in the eight unknowns T_i and $\dot{q}_i$ for $i = 1\ldots4$. The equations derived in the previous section are entered in EES and solved. The efficiency is computed according to:

$$\eta = \frac{-\dot{q}_2}{\dot{q}_1}, \tag{19}$$

where the negative is used because $\dot{q}_2$ calculated using Eq. (14) refers to the heat transfer rate *from* the pipe.

```
$UnitSystem SI Mass J K Pa Radian
T_H=1000 [K]                          "heater plate temperature"
T_p=400 [K]                           "pipe temperature"
T_sur=300 [K]                         "surrounding temperature"
W=0.2 [m]                             "width of heater"
R_p=0.05 [m]                          "radius of pipe"
R_s=0.1 [m]                           "radius of shield"
s=0.06 [m]                            "spacing from center of pipe to heater"
L=1 [m]                               "per unit length"

"areas"
A[1]=W*L                              "heater area"
A[2]=2*pi*R_p*L                       "pipe area"
A[3]=pi*R_s*L                         "shield area"
A[4]=9e9 [m^2]                        "surroundings area"

"view factors"
F[2,1]=F2D_6(s,W/2)
F[2,2]=0
F[2,3]=0.5
F[2,4]=1-F[2,1]-F[2,2]-F[2,3]

F[1,1]=0
F[1,2]=A[2]*F[2,1]/A[1]
F[1,3]=F2D_10(s,W,2*R_s)-F[1,2]
F[1,4]=1-F[1,1]-F[1,2]-F[1,3]

F[3,1]=A[1]*F[1,3]/A[3]
F[3,2]=A[2]*F[2,3]/A[3]
F[3,3]=F2D_8(R_p,R_s)
F[3,4]=1-F[3,1]-F[3,2]-F[3,3]

F[4,1]=A[1]*F[1,4]/A[4]
F[4,2]=A[2]*F[2,4]/A[4]
```

Example 14.5 (cont.)

```
F[4,3]=A[3]*F[3,4]/A[4]
F[4,4]=1-F[4,1]-F[4,2]-F[4,3]

"net heat transfer from each surface"
Duplicate i=1,4
     q_dot[i]=A[i]*sigma#*sum(F[i,j] *(T[i]^4-T[j]^4),j=1,4)
End

"boundary conditions"
T[1]=T_H
T[2]=T_p
q_dot[3]=0
T[4]=T_sur

"efficiency"
eta= -q_dot[2]/q_dot[1]
```

Solving provides $\boxed{\eta = 0.6972 \ (67.72 \text{ percent})}$.

Discussion

Figure 2 shows the Arrays Window that results from the solution. Notice that the solution results in a prediction of the heat transfer rate for any surface where the temperature is set (i.e., surfaces 1, 2, and 4) and a temperature for any surface where the heat transfer rate is set (i.e., surface 3). A positive heat transfer rate indicates that radiation is being transferred from the surface; for example, $\dot{q}_1 = 10,593$ W indicates that there is 10.593 kW of radiation from the heater plate. A negative heat transfer rate indicates that radiation is being transferred to the surface; for example, $\dot{q}_2 = -7385$ W indicates that the pipe is receiving 7.385 kW of radiation. The shield is specified as being adiabatic and the solution predicts that the temperature that the shield must be at $T_3 = 686.1$ K in order to experience no net radiation heat transfer.

Main							
Sort	A_i	$F_{i,1}$	$F_{i,2}$	$F_{i,3}$	$F_{i,4}$	T_i	$\dot{q}_i$
	[m²]	[-]	[-]	[-]	[-]	[K]	[W]
[1]	0.2	0	0.5152	0.2288	0.256	1000	10593
[2]	0.3142	0.328	0	0.5	0.172	400	-7385
[3]	0.3142	0.1457	0.5	0.282	0.07231	686.1	0
[4]	9.000E+09	5.688E-12	6.005E-12	2.524E-12	1	300	-3208

Figure 2 Arrays Window from EES Solution.

It is easy to change the boundary conditions and obtain a new solution. For example, if the temperature of the shield, surface 3, were set to 500 K (rather than using an adiabatic boundary condition) then it is only necessary to change one line in the code to achieve a new solution.

```
//q_dot[3]=0
T[3]=500 [K]
```

Continued

Example 14.5 (cont.)

Solving will lead to $\dot{q}_3 = -2034$ W indicating that surface 3 is receiving 2.034 kW of radiation if its temperature is 500 K. In the absence of external cooling, the temperature of surface 3 will increase with time given this heat input and eventually the shield will reach its equilibrium temperature of 686.1 K.

14.4 Radiation Characteristics of Real Surfaces

14.4.1 Introduction

Radiation heat transfer from a surface is a consequence of the difference between the rate of radiation that is emitted by a surface and the rate at which radiation is absorbed by the surface. The emission of radiation by a blackbody was investigated in Section 14.2 and in Section 14.3 radiation exchange between black surfaces was discussed. The calculations for radiation heat transfer associated with blackbodies are simplified by the fact that blackbodies emit the maximum possible amount of radiation, given by Planck's Law, and also absorb all incident radiation. In this section, real surfaces are considered. Real surfaces emit less radiation than a blackbody and also do not absorb all of the incident radiation; instead, some radiation may be reflected or transmitted. The surface properties that characterize the emission, absorption, reflection, and transmission of radiation from a real surface are emissivity (ε), absorptivity (α), reflectivity (ρ), and transmissivity (τ). For a given surface, these properties are generally functions of temperature, wavelength and direction.

14.4.2 Emission from Real Surfaces

Spectral, Directional Emissivity
The radiation emitted by a real surface is a function of its temperature, wavelength, and direction (relative to the surface normal). Direction is most conveniently defined in **spherical coordinates**, which are based on the location that the radiation intercepts a hemisphere that is placed above and centered on the differential surface area of interest, as shown in Figure 14.14.

The spherical coordinate θ is the angle relative to the normal vector. Therefore $\theta = 0$ indicates radiation emitted normal to the surface while $\theta = \pi/2$ radian (90°) indicates radiation emitted parallel to the surface. The intensity of the radiation that is emitted may be a strong function of θ. The spherical coordinate ϕ is the angle that defines the rotation about the normal; the intensity of emitted radiation is not typically a strong function of ϕ.

The **spectral, directional emissivity** ($\varepsilon_{\lambda,\theta,\phi}$) of a surface is defined as the ratio of the intensity of the radiation emitted by the surface ($I_{\lambda,\theta,\phi}$) to the intensity of radiation emitted by a blackbody at the same wavelength ($I_{b,\lambda}$).

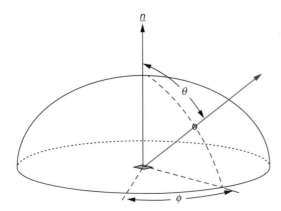

Figure 14.14 Spherical coordinates used to define direction.

Intensity is defined as the rate of radiation emitted per unit wavelength per unit area and per unit of solid angle (i.e., per unit of differential surface on the hemisphere in Figure 14.14, defined by $d\theta$ and $d\phi$). Notice that the spectral, directional emissivity is a function of wavelength and direction, thus the subscripts λ, θ, and ϕ:

$$\varepsilon_{\lambda,\theta,\phi} = \frac{I_{\lambda,\theta,\phi}}{I_{b,\lambda}}. \tag{14.32}$$

Hemispherical Emissivity

The spectral emissive power of a surface, E_λ, is the intensity of radiation integrated over all directions (i.e., from $0 < \theta < \pi/2$ and from $0 < \phi < 2\pi$) at a particular wavelength. The **hemispherical emissivity** is defined as the ratio of the spectral emissive power of a surface to the blackbody spectral emissive power at the same wavelength, $E_{b,\lambda}$. The hemispherical emissivity of a surface (ε_λ) is only a function of wavelength at a specified temperature:

$$\varepsilon_\lambda = \frac{E_\lambda}{E_{b,\lambda}}. \tag{14.33}$$

Total Hemispherical Emissivity

The **total hemispherical emissivity** is *spectrally* averaged (i.e., integrated over all wavelengths) as well as being *directionally* averaged. The total hemispherical emissivity of a surface is defined as the ratio of the total emissive power (E) to the total emissive power of a blackbody (E_b):

$$\varepsilon = \frac{E}{E_b}. \tag{14.34}$$

The total emissive power is obtained by integrating the spectral emissive power of the surface over all wavelengths and the blackbody emissive power is given by Eq. (14.5). With these substitutions, Eq. (14.34) becomes

$$\varepsilon = \frac{\int_0^\infty E_\lambda \, d\lambda}{\sigma T^4}. \tag{14.35}$$

Substituting Eq. (14.33) into Eq. (14.35) leads to:

$$\varepsilon = \frac{\int_0^\infty \varepsilon_\lambda E_{b,\lambda} d\lambda}{\sigma T^4}. \tag{14.36}$$

Given information about the hemispherical emissivity of a surface as a function of wavelength, it is possible to use Eq. (14.36) to determine the total hemispherical emissivity of that surface. Total hemispherical emissivity data for a few surfaces are included in EES as part of the Heat Transfer Library. Select Heat Transfer & Fluid Flow from the Function Information dialog in the Options menu. Select Total Hemispherical Emissivity from the dropdown menu in order to examine the surfaces accessible by the Emissivity_TotHemi function. The values returned by this function or obtained from a reference should be used with some caution as the emissivity of a surface may vary substantially depending on how it is handled.

Example 14.6

The surface of a solar collector has been engineered such that it has a high value of hemispherical emissivity ($\varepsilon_{\lambda,1} = 0.92$) at low wavelengths (below $\lambda_c = 6$ μm) and a low value of hemispherical emissivity ($\varepsilon_{\lambda,2} = 0.2$) at high wavelengths (above λ_c). If the temperature of the surface is $T = 350$ K then determine the total hemispherical emissivity of the surface.

Continued

Example 14.6 (cont.)

Known Values

Figure 1 illustrates a sketch of the problem with the known values indicated.

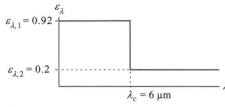

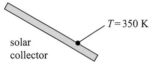

Figure 1 Solar collector with a selective surface.

Assumption

• The emissivity is constant within the two wavelength bands defined in the problem statement.

Analysis and Solution

The total hemispherical emissivity is computed using Eq. (14.36):

$$\varepsilon = \frac{\int_0^\infty \varepsilon_\lambda E_{b,\lambda} d\lambda}{\sigma T^4}. \tag{1}$$

The integral is broken into two parts corresponding to the two wavelength bands:

$$\varepsilon = \frac{\int_0^{\lambda_c} \varepsilon_{\lambda,1} E_{b,\lambda} d\lambda}{\sigma T^4} + \frac{\int_{\lambda_c}^\infty \varepsilon_{\lambda,2} E_{b,\lambda} d\lambda}{\sigma T^4}. \tag{2}$$

Because the hemispherical emissivity is constant within each band it can be removed from the integrand:

$$\varepsilon = \varepsilon_{\lambda,1} \underbrace{\frac{\int_0^{\lambda_c} E_{b,\lambda} d\lambda}{\sigma T^4}}_{F_{0-\lambda_c}} + \varepsilon_{\lambda,2} \underbrace{\frac{\int_{\lambda_c}^\infty E_{b,\lambda} d\lambda}{\sigma T^4}}_{1-F_{0-\lambda_c}}. \tag{3}$$

Equation (3) can be written in terms of the external fractional function discussed in Section 14.2.2:

$$\varepsilon = \varepsilon_{\lambda,1} F_{0-\lambda_c} + \varepsilon_{\lambda,2}(1 - F_{0-\lambda_c}). \tag{4}$$

The product of λ_c and the temperature T is

$$\lambda_c T = (6 \ \mu\text{m}) \ (350 \ \text{K}) = 2100 \ \mu\text{m K}. \tag{5}$$

Example 14.6 (cont.)

The value of $F_{0-\lambda_c}$ is approximately 0.08 according to Table 14.1. Therefore, the total hemispherical emissivity of the surface is

$$\varepsilon = 0.92\,(0.08) + 0.2(1 - 0.08) = \boxed{0.258}. \tag{6}$$

Discussion

The type of surface analyzed here is referred to as a **semi-gray surface** because the hemispherical emissivity is assumed to be constant within different wavelength bands. In Section 14.4.3 we will find that the emissivity is related to the absorptivity. Therefore, the selective surface analyzed here is ideal for a solar collector because the emissivity (and therefore absorptivity) is high at short wavelengths (which is where the Sun emits most of its energy) and low at longer wavelengths (which is where the collector itself emits most its energy). The collector is able to absorb a large amount of the incident solar energy but does not transfer a lot of energy due to radiation to its surroundings.

Notice that the total hemispherical emissivity depends on the temperature of the surface. As the temperature of the collector increases, we should expect that the total hemispherical emissivity will increase because the collector surface will tend to emit more of its energy at shorter wavelengths where the hemispherical emissivity is higher. Figure 2 illustrates the total hemispherical emissivity as a function of temperature.

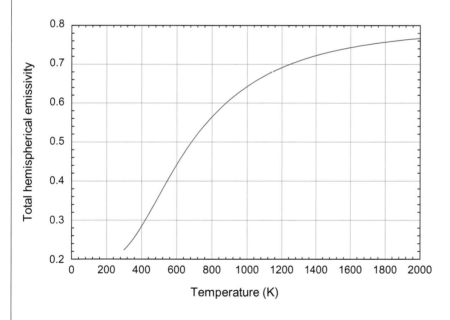

Figure 2 Total hemispherical emissivity as a function of temperature.

The Diffuse Surface Approximation

The diffuse surface approximation assumes that a surface has an emissivity that is independent of direction (θ and ϕ). Therefore, the emissivity for a diffuse surface at a particular temperature is only a function of wavelength.

The Diffuse Gray Surface Approximation

The diffuse gray surface approximation assumes that a surface has an emissivity that is independent of direction (θ and ϕ) and wavelength (λ). The diffuse gray surface approximation is often made in order to simplify radiation exchange calculations. Radiation exchange between diffuse gray surfaces is discussed in Section 14.5.

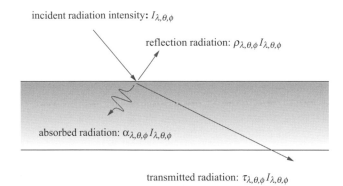

incident radiation intensity: $I_{\lambda,\theta,\phi}$

reflection radiation: $\rho_{\lambda,\theta,\phi} I_{\lambda,\theta,\phi}$

absorbed radiation: $\alpha_{\lambda,\theta,\phi} I_{\lambda,\theta,\phi}$

transmitted radiation: $\tau_{\lambda,\theta,\phi} I_{\lambda,\theta,\phi}$

Figure 14.15 Intensity of radiation striking a surface may be reflected, absorbed, or transmitted.

14.4.3 Reflectivity, Absorptivity, and Transmissivity

Section 14.4.2 discussed how a real surface emits radiation as a function of direction and wavelength. In order to complete a radiation heat transfer problem, we must also consider how a real surface will deal with incident radiation. In general, incident radiation that strikes a surface can be reflected, absorbed, or transmitted, as shown in Figure 14.15.

The ratio of the reflected intensity to the incident intensity at a given wavelength and direction is called the **spectral, directional reflectivity**, $\rho_{\lambda,\theta,\phi}$. The value of reflectivity must be between 0 and 1. Similar definitions are used for the **spectral, directional absorptivity**, $\alpha_{\lambda,\theta,\phi}$, and **spectral, directional transmissivity**, $\tau_{\lambda,\theta,\phi}$. At any wavelength, λ, and direction θ, ϕ, an energy balance requires that

$$\rho_{\lambda,\theta,\phi} + \alpha_{\lambda,\theta,\phi} + \tau_{\lambda,\theta,\phi} = 1. \tag{14.37}$$

Opaque surfaces are defined as those that transmit no radiation and so $\tau_{\lambda,\theta,\phi} = 0$. Black surfaces are by definition opaque surfaces that absorb all radiation regardless of wavelength and direction. Therefore $\alpha_{\lambda,\theta,\phi} = 1.0$ for a black surface, and Eq. (14.37) indicates that no radiation is reflected from a black surface ($\rho_{\lambda,\theta,\phi} = \tau_{\lambda,\theta,\phi} = 0$ for a black surface).

Most objects that we come into contact with in our day-to-day lives are at a temperature that is too low for them to emit any substantial amount of radiation in the visible spectrum. Therefore, what our eyes perceive is mostly the visible radiation from other sources (e.g., the Sun or a light bulb) that is being reflected by these objects. The term blackbody arises because a surface that does not reflect any visible radiation will appear black to our eyes (unless it is so hot that it emits visible radiation). This terminology can be misleading, however, since the visible range is a small section of the electromagnetic spectrum. Snow, for example, reflects less than 2 percent of the infrared radiation that it receives; therefore, snow closely approximates a blackbody in the infrared wavelength band even though it clearly does not *look* black to our eyes. Many white paints behave in a similar manner. A true blackbody does not exist in nature since some radiation is always reflected from a surface; however, a few substances (e.g., carbon black) absorb nearly all incident radiation within the visible and near infrared range. Perhaps the closest approximation to a blackbody is a cavity with a pinhole opening. Radiation that enters the cavity through the pinhole may be reflected from the cavity walls many times before being absorbed; however, it has a very small chance of exiting through the pinhole. As a result, essentially all of the radiation entering the pinhole will ultimately be absorbed and none reflected.

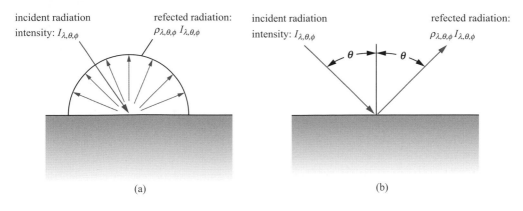

Figure 14.16 Reflection from (a) a diffuse surface and (b) a specular surface.

Diffuse and Specular Surfaces

The directional dependence of the surface properties ($\rho_{\lambda,\theta,\phi}$, $\tau_{\lambda,\theta,\phi}$, and $\alpha_{\lambda,\theta,\phi}$) is complicated; however, the complication is reduced by classifying surfaces as either **diffuse** or **specular**. The reflected radiation from a diffuse surface is assumed to be distributed uniformly over all directions regardless of the direction of the incident radiation, as illustrated in Figure 14.16(a). At the other extreme, radiation that strikes a specular surface at a particular angle is reflected at the same angle, as indicated in Figure 14.16(b). The behavior of real surfaces is somewhere between these two extremes, being partially diffuse and partially specular. However, many materials tend to be nearly diffuse or nearly specular. For example, a highly polished metal surface or a mirror tends to exhibit specular behavior, at least within the narrow visible wavelength band that is detectable by our eyes. This is why it is possible to see a clear reflection in a mirror. Roughened surfaces or surfaces with an oxidized coating tend to reflect radiation diffusely. Only radiation exchange involving diffuse surfaces are considered in this text. A detailed discussion of radiation exchange between surfaces that are specular or partially specular and partially diffuse is provided by Siegel and Howell (2002).

Hemispherical and Total Hemispherical Reflectivity, Absorptivity, and Transmissivity

The hemispherical values of the reflectivity, absorptivity, and transmissivity are the ratios of the total rate of incident radiation at a particular wavelength that is reflected, absorbed, and transmitted, respectively, to the total rate of radiation that is incident on the surface at that wavelength. These hemispherical values are integrated over all directions and are therefore only a function of wavelength at a specified temperature (ρ_λ, α_λ, and τ_λ). The total hemispherical values of reflectivity, absorptivity, and transmissivity (ρ, α, and τ) are integrated overall all directions and wavelengths.

Kirchhoff's Law

Kirchhoff's Law relates the spectral, directional absorptivity of a surface to its spectral, directional emissivity:

$$\varepsilon_{\lambda,\theta,\phi} = \alpha_{\lambda,\theta,\phi}. \tag{14.38}$$

Kirchhoff's Law is formally derived by considering an object at thermal equilibrium in an enclosure and recognizing that the rate at which the object absorbs and emits radiation must be the same.

The Diffuse Surface Approximation

A diffuse surface is defined as a surface with an emissivity that is independent of direction (θ and ϕ). According to Kirchhoff's Law, Eq. (14.38), the absorptivity is equal to the emissivity; therefore, the absorptivity of a diffuse surface must also be independent of direction. Further, if the diffuse surface is opaque then Eq. (14.37) indicates that the reflectivity is given by

$$\rho_{\lambda,\theta,\phi} = 1 - \alpha_{\lambda,\theta,\phi}. \qquad (14.39)$$

Therefore, the reflectivity of an opaque, diffuse surface must also be independent of direction.

The Diffuse Gray Surface Approximation

A diffuse gray surface is defined as a surface that has an emissivity that is independent of direction (θ and ϕ) and wavelength (λ). According to Kirchhoff's Law, the absorptivity of a diffuse gray surface must also be independent of direction and wavelength. An opaque, diffuse gray surface will have reflectivity given by

$$\rho = 1 - \alpha. \qquad (14.40)$$

Example 14.7

In Example 14.6 the total hemispherical emissivity of a solar collector that employs a selective surface was calculated. The surface has been engineered such that it has a high value of hemispherical emissivity ($\varepsilon_{\lambda,1} = 0.92$) at short wavelengths (below $\lambda_c = 6$ μm) and a low value of hemispherical emissivity ($\varepsilon_{\lambda,2} = 0.2$) at long wavelengths (above λ_c). Because the temperature of the surface is relatively low, $T = 350$ K, the total hemispherical emissivity of the surface was calculated to be relatively low, $\varepsilon = 0.258$; this low value is attractive because it suggests that little energy will be transferred from the surface due to radiation emitted by the surface. However, in order to be a good solar collector it is also necessary that the surface be highly absorptive to solar radiation and Kirchhoff's Law indicates that the emissivity and absorptivity are related. Fortunately, the Sun emits radiation at relatively short wavelengths where the emissivity and therefore the absorptivity is high. Calculate the total hemispherical absorptivity of the surface relative to radiation emitted by the Sun.

Known Values

Figure 1 illustrates a sketch of the problem with the known values indicated.

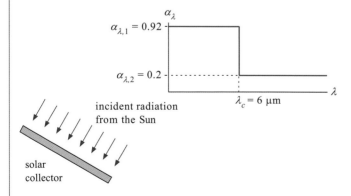

Figure 1 Solar collector with a selective surface.

Assumptions

- The emissivity is constant within the two wavelength bands defined in the problem statement.
- The collector is diffuse.
- The Sun emits radiation as if it were a blackbody at $T_{sun} = 5800$ K.

Example 14.7 (cont.)

Analysis and Solution

According to Kirchhoff's Law, Eq. (14.38), the spectral, directional absorptivity must be equal to the spectral, directional emissivity. In this problem the surface is assumed to be diffuse and therefore there is no directional dependence:

$$\alpha_\lambda = \varepsilon_\lambda. \tag{1}$$

The total hemispherical absorptivity is computed by integrating the hemispherical absorptivity over all wavelengths of the *incident* radiation. For this problem, the incident radiation is assumed to be emitted by a blackbody at the temperature of the surface of the Sun:

$$\alpha = \frac{\int_0^\infty \alpha_\lambda E_{b,\lambda, T = T_{sun}} \, d\lambda}{\sigma T_{sun}^4}. \tag{2}$$

The integral is broken into two parts corresponding to the two wavelength bands:

$$\alpha = \frac{\int_0^{\lambda_c} \alpha_{\lambda, 1} E_{b,\lambda, T = T_{sun}} \, d\lambda}{\sigma T_{sun}^4} + \frac{\int_{\lambda_c}^\infty \alpha_{\lambda, 2} E_{b,\lambda, T = T_{sun}} \, d\lambda}{\sigma T_{sun}^4}. \tag{3}$$

Because the absorptivity is constant within each band it can be removed from the integrand:

$$\alpha = \alpha_{\lambda, 1} \underbrace{\frac{\int_0^{\lambda_c} E_{b,\lambda, T = T_{sun}} \, d\lambda}{\sigma T_{sun}^4}}_{F_{0-\lambda_c}} + \alpha_{\lambda, 2} \underbrace{\frac{\int_{\lambda_c}^\infty E_{b,\lambda, T = T_{sun}} \, d\lambda}{\sigma T_{sun}^4}}_{1 - F_{0-\lambda_c}}. \tag{4}$$

Equation (3) can be written in terms of the external fractional function discussed in Section 14.2.2, evaluated using the temperature of the Sun:

$$\alpha = \alpha_{\lambda, 1} F_{0-\lambda_c} + \alpha_{\lambda, 2} \left(1 - F_{0-\lambda_c}\right). \tag{5}$$

The product of λ_c and the temperature of the Sun, T_{sun}, is

$$\lambda_c T_{sun} = (6 \ \mu m) \, (5800 \ K) = 34,800 \ \mu m \cdot K. \tag{6}$$

The value of $F_{0-\lambda_c}$ is approximately 0.997 according to Table 14.1. Therefore, the total hemispherical absorptivity of the surface is:

$$\alpha = 0.92 \, (0.997) + 0.2 \, (1 - 0.997) = \boxed{0.918}. \tag{7}$$

Discussion

The selective surface analyzed here is ideal for a solar collector because the absorptivity relative to solar radiation is 91.8 percent, whereas the emissivity of the surface is quite small. Notice that the total hemispherical absorptivity must be evaluated by considering the wavelength distribution of the radiation that is *incident* on the surface whereas the total hemispherical emissivity is evaluated by considering the temperature of the *surface itself*. As a result, even though Kirchhoff's Law requires that $\varepsilon_{\lambda, \theta, \phi} = \alpha_{\lambda, \theta, \phi}$, it is not generally necessary that $\varepsilon = \alpha$. However, for the diffuse, gray surface radiation problems considered in the next section Kirchhoff's Law does require that $\varepsilon = \alpha$ since neither value can vary with either direction or wavelength.

14.5 Diffuse Gray Surface Radiation Exchange

14.5.1 Introduction

A blackbody absorbs all of the radiation that hits it, regardless of wavelength or angle. None of the incident radiation is reflected or transmitted. A surface that exhibits this behavior has an absorptivity of $\alpha = 1$ and therefore a blackbody is a perfect absorber of radiation. A blackbody is also a perfect emitter of radiation because, according to Kirchhoff's Law, a blackbody must have an emissivity of $\varepsilon = 1$ regardless of wavelength. Radiation emitted by a blackbody is diffuse, i.e., the intensity of the radiation is independent of direction.

No real surface exhibits the properties of a blackbody, although some materials approach this behavior within particular wavelength bands. Real surfaces have emissivity values that are lower than 1 and their emissivity may vary with wavelength and direction, as described in Section 14.4.2. However, an average value of emissivity, the total hemispherical emissivity, can be defined by convolution of the hemispherical emissivity and the wavelength dependence of emitted radiation using Eq. (14.36).

The concept of a diffuse, gray surface was introduced in Section 14.4.3. A diffuse gray surface is another idealized surface, one that has a constant emissivity at all wavelengths and emits radiation uniformly in all directions. While no real surface exhibits these characteristics completely, the results of a diffuse gray surface radiation exchange calculation are more accurate than the blackbody radiation calculations presented in Section 14.3 and are often sufficient for engineering calculations. This section presents methods for calculating radiation exchange between diffuse gray surfaces that are opaque.

14.5.2 Radiosity

Consider a diffuse gray surface i that is receiving **irradiation** from its surroundings, G_i, as shown in Figure 14.17. The source of the irradiation is likely a combination of radiation emitted and reflected from other surfaces near surface i (or perhaps even from surface i itself). The radiation leaving surface i includes the portion of the irradiation that is reflected from surface i ($\rho_i G_i$) as well as the radiation emitted by surface i ($\varepsilon_i E_{b,i}$). The sum of the reflected and emitted radiation per unit area is called **radiosity**, J_i. The radiosity is the rate of radiation that is leaving surface i per unit area. Note that the radiosity associated with a surface may be a particularly complex function of wavelength since the spectral distribution of the radiation emitted from the surface may differ substantially from the spectral distribution of radiation reflected from the surface. However, this complexity does not factor into gray surface calculations, since it is assumed that none of the surface characteristics depend on wavelength.

The radiosity is the sum of the reflected and emitted radiation in Figure 14.17:

$$J_i = \rho_i G_i + \varepsilon_i E_{b,i}, \tag{14.41}$$

where ε_i is the emissivity of the surface and ρ_i is the reflectivity of the surface. For an opaque, diffuse gray surface, the reflectivity and absorptivity are related according to:

$$\rho_i = 1 - \alpha_i. \tag{14.42}$$

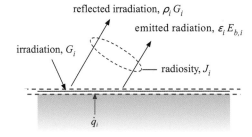

Figure 14.17 Definition of radiosity for surface i.

According to Kirchhoff's Law, the absorptivity and emissivity for a diffuse gray surface must be equal and therefore:

$$\rho_i = 1 - \varepsilon_i. \tag{14.43}$$

Substituting Eq. (14.43) into Eq. (14.41) leads to:

$$J_i = \underbrace{(1 - \varepsilon_i)\, G_i}_{\text{reflected}} + \underbrace{\varepsilon_i\, E_{b,i}}_{\text{emitted}}. \tag{14.44}$$

The first term in Eq. (14.44) is the reflected irradiation and the second term is the emitted radiation. Note that if ε_i approaches 1 (i.e., the surface is black) then the radiosity will only consist of the radiation that is emitted by the surface (the blackbody emissive power) because all incident radiation is absorbed. At the other extreme, if ε_i approaches 0 (i.e., the surface is a perfect reflector) then the radiosity will only consist of the irradiation, diffusely reflected from the surface with no additional emitted power.

The net rate of radiation heat transfer provided to the surface i from an external source is equal to $\dot{q}_i$, in Figure 14.17. According to an energy balance, $\dot{q}_i$ can be calculated according to:

$$\underbrace{\dot{q}_i}_{\substack{\text{external energy transfer} \\ \text{to the surface, or} \\ \text{net radiation heat transfer} \\ \text{from the surface}}} = \underbrace{A_i J_i}_{\text{radiosity}} - \underbrace{A_i G_i}_{\text{irradiation}}. \tag{14.45}$$

Note that if the surface is adiabatic (i.e., no energy is provided to the surface so $\dot{q}_i = 0$), then Eq. (14.45) requires that the radiosity must be equal to the irradiation. Also note that $\dot{q}_i$ can be thought of as the net radiation heat transfer from the surface to the surroundings.

Rearranging Eq. (14.44) in order to solve for the irradiation, G_i, leads to:

$$G_i = \frac{J_i - \varepsilon_i E_{b,i}}{1 - \varepsilon_i}. \tag{14.46}$$

Substituting Eq. (14.46) into Eq. (14.45) results in

$$\dot{q}_i = A_i \left(J_i - \frac{J_i - \varepsilon_i E_{b,i}}{1 - \varepsilon_i} \right) = A_i \left(J_i \frac{1 - \varepsilon_i}{1 - \varepsilon_i} - \frac{J_i - \varepsilon_i E_{b,i}}{1 - \varepsilon_i} \right) = A_i \left(\frac{-\varepsilon_i J_i + \varepsilon_i E_{b,i}}{1 - \varepsilon_i} \right), \tag{14.47}$$

which can be rearranged:

$$\dot{q}_i = \left(\frac{\varepsilon_i A_i}{1 - \varepsilon_i} \right) (E_{b,i} - J_i). \tag{14.48}$$

Equation (14.48) is important because it allows us to compute the radiosity leaving a surface from its surface characteristics (A_i and ε_i), its temperature (T_i, which is related to $E_{b,i}$), and the net radiation heat transfer from the surface ($\dot{q}_i$). The equivalent to Eq. (14.48) was not required for blackbody radiation exchange problems because the radiosity leaving a blackbody was equal to the blackbody emissive power.

14.5.3 Diffuse Gray Surface Radiation Calculations

Equation (14.48) can be applied to any gray surface that is involved in radiation exchange. Notice that Eq. (14.48) can be written in the form of a resistance equation, as shown in Figure 14.18. The driving force for radiation heat transfer is the difference between the surface's blackbody emissive power and its radiosity ($E_{b,i} - J_i$) and the resistance is the quantity $(1 - \varepsilon_i)/(\varepsilon_i A_i)$. The resistance between the surface's blackbody emissive power and its radiosity is called the **surface resistance** for surface i ($R_{s,i}$) and should not be confused with the space resistance between surface i and another surface j ($R_{i,j}$), discussed in Section 14.3.3:

$$R_{s,i} = \frac{1 - \varepsilon_i}{A_i \varepsilon_i}. \tag{14.49}$$

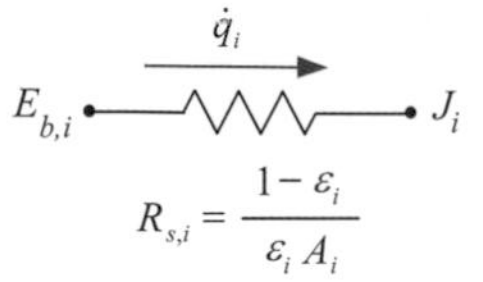

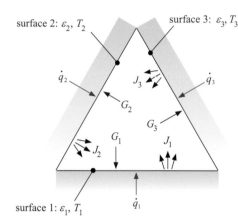

$$E_{b,i} \;\text{—}\!\!\!\bigwedge\!\!\!\text{—}\; J_i$$

$$R_{s,i} = \dfrac{1-\varepsilon_i}{\varepsilon_i A_i}$$

Figure 14.18 Definition of a surface resistance.

Figure 14.19 Gray surface enclosure consisting of three isothermal surfaces.

The surface resistance and the space resistance both have units m^{-2} (in the SI system). Note that if the surface is black (i.e., if $\varepsilon_i = 1$), then the surface resistance limits to 0 and the resistance "disappears"; this explains why it was not necessary to consider surface resistances when doing the blackbody radiation exchange problems in Section 14.3. Also, if the surface is a perfect reflector (i.e., if $\rho_i = 1$ so that $\varepsilon_i = 0$), then the surface resistance becomes infinitely large. In this limit, the surface does not communicate radiatively with its environment; all incident radiation is reflected. It is often desirable to isolate a surface from radiation, such as in a cryogenic experiment placed in a vacuum vessel or an apparatus that is launched in space. The surfaces of such devices are often made as "shiny" as possible in order to closely emulate a perfect reflector and therefore maximize the value of the surface resistance in Eq. (14.49). Multiple layers of reflective material (called radiation shields or multi-layer insulation, MLI) can provide very high levels of thermal isolation for radiation.

The radiosity from a surface plays the same role that the blackbody emissive power played in blackbody radiation exchange problems. The radiosity is the amount of radiation per area that is leaving the surface. The radiosity interacts with other surfaces in the vicinity by becoming irradiation, i.e., radiation that strikes other surfaces. The radiation exchange between surfaces can therefore be represented using the space resistances that were introduced in Section 14.3.3; however, these space resistances extend between nodes that represent the radiosity of each surface rather than their blackbody emissive powers.

Resistance Network

Based on this discussion, radiation exchange between diffuse gray surfaces is only one step more complex than radiation exchange between blackbodies. An extra resistance must be added to the resistance network that is used to model the problem; this extra resistance is the surface resistance from Eq. (14.48), which relates a surface's radiosity to its blackbody emissive power.

Consider the enclosure shown in Figure 14.19, which includes three surfaces. The temperature of each surface is uniform (at T_1, T_2, and T_3) and the rate at which energy is radiated from each of these surfaces is $\dot{q}_1, \dot{q}_2,$ and $\dot{q}_3$. If each of the surfaces is black ($\varepsilon_1 = \varepsilon_2 = \varepsilon_3 = 1.0$), then the resistance network that represents this enclosure

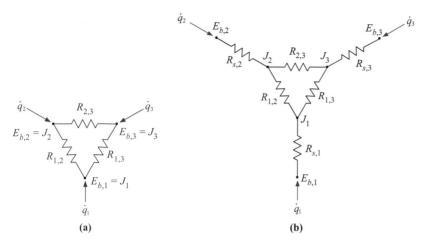

Figure 14.20 Resistance network that represents the enclosure shown in Figure 14.19 if (a) the surfaces are black and (b) the surfaces are diffuse and gray.

is shown in Figure 14.20(a); the blackbody emissive power of each surface interacts directly via space resistances. If each surface is gray and diffuse, then the resistance network that represents the enclosure is shown in Figure 14.20(b); the blackbody emissive power of each surface is related to its radiosity by a surface resistance and the radiosities of each surface interact via space resistances.

Example 14.8

A spherical object (surface 1) is placed at the center of a cubical enclosure. The diameter of the object is $D_s = 0.3$ m. The width of each side of the enclosure is $W = 0.75$ m. The surface of the object is unpolished stainless steel; the surface can be considered diffuse and gray with an emissivity of $\varepsilon_s = 0.30$. The top surface of the enclosure (surface 2) is black and maintained at $T_C = 300$ K. The remaining five sides (together referred to as surface 3) are adiabatic and have emissivity $\varepsilon_e = 0.25$. The spherical object is being heated at a rate $\dot{q}_s = 120$ W. Determine the steady-state surface temperature of the sphere.

Known Values

Figure 1 illustrates a sketch of the problem with the known values indicated.

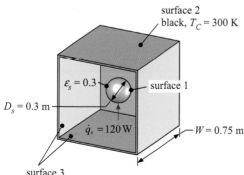

surface 2
black, $T_C = 300$ K

$\varepsilon_s = 0.3$

surface 1

$D_s = 0.3$ m

$\dot{q}_s = 120$ W

$W = 0.75$ m

surface 3
five remaining sides are insulated with $\varepsilon_e = 0.25$

Figure 1 Sphere at the center of a cubical enclosure.

Continued

Example 14.8 (cont.)

Assumptions

- Radiation is the only heat transfer mechanism.
- All surfaces are isothermal.
- Steady state.

Analysis and Solution

This three surface problem is represented using the resistance network shown in Figure 2. Notice that there is no surface resistance necessary for surface 2 because it is black and therefore $R_{s,2}= 0$, indicating that the blackbody emissive power and radiosity are the same for this surface.

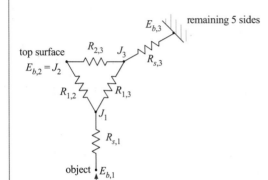

Figure 2 Resistance network.

The areas of the three surfaces are computed:

$$A_1 = 4\pi\left(\frac{D_s}{2}\right)^2 = 4\pi\left(\frac{0.3\ \text{m}}{2}\right)^2 = 0.2827\ \text{m}^2 \tag{1}$$

$$A_2 = W^2 = (0.75\ \text{m})^2 = 0.5625\ \text{m}^2 \tag{2}$$

$$A_3 = 5W^2 = 5(0.75\ \text{m})^2 = 2.813\ \text{m}^2. \tag{3}$$

The view factors from surface 1 to 2 and from surface 1 to 3 are obtained by inspection:

$$F_{1,2} = \frac{1}{6} \tag{4}$$

$$F_{1,3} = \frac{5}{6}. \tag{5}$$

The view factor from surface 2 to surface 1 is obtained from reciprocity:

$$F_{2,1} = \frac{A_1}{A_2}F_{1,2} = \frac{(0.2827)}{(0.5625)}\frac{1}{6} = 0.08378. \tag{6}$$

The view factor from surface 2 to surface 3 is obtained from the enclosure rule (recognizing that the view factor from surface 2 to itself is zero):

$$F_{2,3} = 1 - F_{2,1} = 1 - 0.08378 = 0.9162. \tag{7}$$

Example 14.8 (cont.)

The three space resistances in Figure 2 are computed:

$$R_{1,3} = \frac{1}{A_1 F_{1,3}} = \frac{1}{0.2827 \text{ m}^2} \Big|\frac{6}{5} = 4.244 \text{ m}^{-2} \tag{8}$$

$$R_{2,3} = \frac{1}{A_2 F_{2,3}} = \frac{1}{0.5625 \text{ m}^2} \Big|_{0.9162} = 1.940 \text{ m}^{-2} \tag{9}$$

$$R_{1,2} = \frac{1}{A_1 F_{1,2}} = \frac{1}{0.2827 \text{ m}^2} \Big|\frac{6}{1} = 21.22 \text{ m}^{-2}. \tag{10}$$

The surface resistances shown in Figure 2 are computed:

$$R_{s,1} = \frac{(1 - \varepsilon_s)}{A_1 \varepsilon_s} = \frac{(1 - 0.3)}{0.2827 \text{ m}^2} \Big|_{0.3} = 8.252 \text{ m}^{-2} \tag{11}$$

$$R_{s,3} = \frac{(1 - \varepsilon_e)}{A_3 \varepsilon_e} = \frac{(1 - 0.25)}{2.813 \text{ m}^2} \Big|_{0.25} = 1.067 \text{ m}^{-2}. \tag{12}$$

The blackbody emissive power of surface 2 is computed:

$$E_{b,2} = \sigma T_C^4 = \frac{5.67 \times 10^{-8} \text{ W}}{\text{m}^2 \text{ K}^4} \Big| \frac{300^4 \text{ K}^4}{} = 459.3 \frac{\text{W}}{\text{m}^2}. \tag{13}$$

Because of the boundary conditions associated with the problem, all of the energy entering at surface 1 must leave at surface 2. Therefore, the equivalent resistance between these two points can be computed according to:

$$R_{eq} = R_{s,1} + \left[\frac{1}{R_{1,2}} + \frac{1}{(R_{1,3} + R_{2,3})} \right]^{-1} = 8.252 + \left[\frac{1}{21.22} + \frac{1}{(4.244 + 1.940)} \right]^{-1} = 13.04 \text{ m}^{-2}. \tag{14}$$

The blackbody emissive power of surface 1 can be computed according to:

$$E_{b,1} = E_{b,2} + \dot{q}_s R_{eq} = 459.3 \frac{\text{W}}{\text{m}^2} + \frac{120 \text{ W}}{} \Big| \frac{13.04}{\text{m}^2} = 2024 \frac{\text{W}}{\text{m}^2}. \tag{15}$$

The temperature of surface 1 can be computed:

$$T_1 = \left(\frac{E_{b,1}}{\sigma} \right)^{1/4} = \left(\frac{2024 \text{ W}}{\text{m}^2} \Big| \frac{\text{m}^2 \text{ K}^4}{5.67 \times 10^{-8} \text{ W}} \right)^{1/4} = \boxed{434.7 \text{ K}}. \tag{16}$$

Discussion

Notice that the surface resistance associated with surface 3, $R_{s,3}$, played no part in the solution; the variable $R_{s,3}$ did not show up in any of Eqs. (14) through (16) because surface 3 is adiabatic and therefore no heat is transferred through the surface resistance $R_{s,3}$ in Figure 2. Such a surface is referred to as a **re-radiating surface**. The emissivity of a re-radiating surface does not matter; there is no flow of heat through the resistance and therefore $J_3 = E_{b,3}$ regardless of the emissivity of surface 3.

N-Surface Solutions

The representation of a diffuse, gray surface radiation problem using a resistance network can be a useful method of visualizing the problem. However, just as with the blackbody problems, if even a relatively modest number of surfaces are involved then the resistance diagram becomes hopelessly complicated. Each surface interacts with all (or most) of the other surfaces and therefore a network involving more than three or four surfaces becomes more confusing than useful.

It is possible to systematically solve a diffuse, gray surface radiation problem that involves an arbitrary number of surfaces, provided that the view factors and areas are known and a boundary condition can be defined for each surface. The system of equations that is required can be understood by examining Figure 14.20 (b). The net rate of radiation exchange from any surface i is obtained from

$$\dot{q}_i = \frac{\varepsilon_i A_i (E_{b,i} - J_i)}{(1 - \varepsilon_i)} \text{ for } i = 1 \ldots N. \tag{14.50}$$

Also, an energy balance written for each of the radiosity nodes leads to

$$\dot{q}_i = A_i \sum_{j=1}^{N} F_{i,j} (J_i - J_j) \text{ for } i = 1 \ldots N. \tag{14.51}$$

When written for each of the N surfaces, Eqs. (14.50) and (14.51) provide $2N$ equations in $3N$ unknowns corresponding to the blackbody emissive power ($E_{b,i}$), the net heat transfer rate ($\dot{q}_i$), and the radiosity (J_i) for each of the $i = 1 \ldots N$ surfaces. A complete set of boundary conditions will include a specification of either the blackbody emissive power (temperature) or the net heat transfer rate for each surface (or a relationship between these quantities). The boundary conditions provide N additional equations and therefore a completely specified problem.

Example 14.9

Example 14.5 examined the 2-D radiation problem involving four surfaces that is shown again below in Figure 1. Surface 1 is a heater plate with width $W = 0.2$ m that is maintained at $T_H = 1000$ K. The surface extends indefinitely into the page. One side of the heater plate radiatively interacts with a pipe that has radius $R_p = 0.05$ m and temperature $T_p = 400$ K. The distance between the center of the pipe and the heater is $s = 0.06$ m. The pipe lies at the center of a semicircular shield with radius $R_s = 0.1$ m. The surface of the shield that is towards the pipe participates in the radiation problem; the other side of the shield (facing the surroundings) is insulated and does not radiate. The surroundings are at $T_{sur} = 300$ K. In Example 14.5, all of these surfaces were assumed to be black. Let's solve the problem again, this time assuming that the heater plate has emissivity $\varepsilon_H = 0.85$, the pipe has emissivity $\varepsilon_p = 0.65$, and the shield has emissivity $\varepsilon_s = 0.25$.

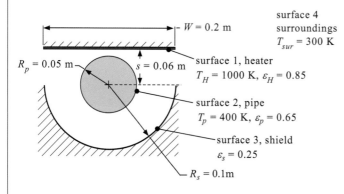

Figure 1 Heater plate used to heat a pipe.

Determine the efficiency of the heater (the ratio of the heat transfer rate to the pipe to the radiation heat transfer rate from the heater).

Known Values

Figure 1 illustrates a sketch of the problem with the known values indicated.

Example 14.9 (cont.)

Assumptions

- All surfaces are diffuse and gray.
- The problem is 2-D and steady state.
- The only heat transfer mechanism is radiation.

Analysis

There are four surfaces involved in the problem; therefore, a resistance network is likely to be too complex to be useful. The areas (A_i for $i = 1..4$) and view factors ($F_{i,j}$ for $i = 1\ldots4$ and $j = 1\ldots4$) that characterize the problem were previously determined in Example 14.5 and are unchanged by the emissivities so they can be used here. The emissivity of each surface must be assigned:

$$\varepsilon_1 = \varepsilon_H \tag{1}$$
$$\varepsilon_2 = \varepsilon_p \tag{2}$$
$$\varepsilon_3 = \varepsilon_s \tag{3}$$
$$\varepsilon_4 = 1. \tag{4}$$

Note that the surroundings are assumed to act as a black surface, absorbing all radiation and reflecting none. The emissivity of the surroundings (surface 4) is therefore set to 1. In fact, it is better to set ε_4 to a value that is close to, but not quite equal to 1 in order to avoid dividing by zero when Eq. (14.50) is written for surface 4.

Equations (14.50) and (14.51) are written for each surface:

$$\dot{q}_i = \frac{\varepsilon_i A_i (E_{b,i} - J_i)}{(1 - \varepsilon_i)} \text{ for } i = 1\ldots4 \tag{5}$$

$$\dot{q}_i = A_i \sum_{j=1}^{4} F_{i,j}(J_i - J_j) \text{ for } i = 1\ldots4. \tag{6}$$

Equations (5) and (6) are written for all four surfaces and represent eight total equations. The boundary conditions for the problem remain the same as Example 14.5:

$$T_1 = T_H \tag{7}$$
$$T_2 = T_p \tag{8}$$
$$\dot{q}_3 = 0 \tag{9}$$
$$T_4 = T_{sur}. \tag{10}$$

The blackbody emissive power and temperature of each surface are related according to:

$$E_{b,i} = \sigma T_i^4 \text{ for } i = 1\ldots4. \tag{11}$$

The efficiency of the heater is computed according to:

$$\eta = \frac{-\dot{q}_2}{\dot{q}_1}. \tag{12}$$

Solution

Given the areas and view factors from Example 14.5, Eqs. (1) through (12) represent 21 equations in the 21 unknowns ε_i, T_i, $E_{b,i}$, J_i, and $\dot{q}_i$ for $i = 1\ldots4$ as well as η. The equations are entered in EES and solved.

```
$UnitSystem SI Mass J K Pa Radian
T_H=1000 [K]                      "heater plate temperature"
T_p=400 [K]                       "pipe temperature"
```

Continued

Example 14.9 (cont.)

```
T_sur=300 [K]                          "surrounding temperature"
W=0.2 [m]                              "width of heater"
R_p=0.05 [m]                           "radius of pipe"
R_s=0.1 [m]                            "radius of shield"
s=0.06 [m]                             "spacing from center of pipe to heater"
L=1 [m]                                "per unit length"
e_H = 0.85 [-]                         "emissivity of the heater"
e_p = 0.65 [-]                         "emissivity of the pipe"
e_s = 0.25 [-]                         "emissivity of the shield"

"areas"
A[1]=W*L                               "heater area"
A[2]=2*pi*R_p*L                        "pipe area"
A[3]=pi*R_s*L                          "shield area"
A[4]=9999 [m^2]                        "surroundings area"

"view factors"
F[2,1]=F2D_6(s,W/2)
F[2,2]=0
F[2,3]=0.5
F[2,4]=1-F[2,1]-F[2,2]-F[2,3]

F[1,1]=0
F[1,2]=A[2]*F[2,1]/A[1]
F[1,3]=F2D_10(s,W,2*R_s)-F[1,2]
F[1,4]=1-F[1,1]-F[1,2]-F[1,3]

F[3,1]=A[1]*F[1,3]/A[3]
F[3,2]=A[2]*F[2,3]/A[3]
F[3,3]=F2D_8(R_p,R_s)
F[3,4]=1-F[3,1]-F[3,2]-F[3,3]

F[4,1]=A[1]*F[1,4]/A[4]
F[4,2]=A[2]*F[2,4]/A[4]
F[4,3]=A[3]*F[3,4]/A[4]
F[4,4]=1-F[4,1]-F[4,2]-F[4,3]

"emissivities"
e[1]=e_H
e[2]=e_p
e[3]=e_s
e[4]=0.999 [-]

"surface heat transfer rates"
Duplicate i=1,4
    q_dot[i]=e[i]*A[i]*(E_b[i]-J[i])/(1-e[i])
End
```

Example 14.9 (cont.)

"radiosity energy balances"
Duplicate i=1,4
    q_dot[i]= A[i]***Sum**(F[i,j]*(J[i]-J[j]),j=1,4)
End

"boundary conditions"
T[1]=T_H
T[2]=T_p
q_dot[3]=0
T[4]=T_sur

Duplicate i=1,4
    E_b[i]=sigma#*T[i]^4
End

"efficiency"
eta= -q_dot[2]/q_dot[1]

Solving provides $\boxed{\eta = 0.5845 \ (58.45 \ \text{percent})}$.

Discussion

The efficiency computed in this example is significantly less than the value that was obtained in Example 14.5 in which all of the surfaces were assumed to be black (58.45 percent vs. 69.72 percent). This result is a consequence of the additional surface resistances that have been placed between the heater and the pipe making the heat transfer path more resistive to radiation.

14.6 Conclusions and Learning Objectives

This chapter provides an introduction to radiation heat transfer, focusing on the tools that are needed to solve a multi-surface heat transfer problem. Radiation is related to energy transfer due to electromagnetic waves that travel between surfaces and these waves are distributed spectrally (i.e., over a range of wavelengths). The emission of radiation by a blackbody is governed by Planck's Law and the blackbody model provides a convenient limiting behavior for all surfaces. The radiation exchange between black surfaces requires that we also understand how radiation moves between a set of interacting surfaces due to their geometry and orientation; this aspect of the problem is considered using view factors. View factors can be obtained in a number of ways including libraries of solutions that have been developed by researchers. Using view factors, it is possible to solve black surface radiation exchange problems using a resistance analogy. For larger numbers of surfaces the equations that must be solved are derived and presented in this chapter.

Real surfaces are more complex than black surfaces because they emit less radiation; the emissivity is the ratio of the amount of radiation emitted by a real surface to the amount emitted by a black surface. The emissivity can be a function of direction and wavelength. Real surfaces may also reflect or transmit some part of the incident radiation (irradiation), in addition to absorbing it. The absorptivity, reflectivity, and transmissivity are all surface characteristics used to indicate how real surfaces deal with irradiation. In this text we confine ourselves to surfaces that are opaque (transmit nothing), diffuse (have surface characteristics independent of direction), and gray (have surface characteristics that are independent of wavelength). The radiosity method provides a useful technique for dealing with radiation heat transfer problems involving these diffuse gray surfaces.

Some specific concepts that you should understand are listed below.
- The electromagnetic spectrum and where thermal radiation lies on the spectrum.
- The behavior of the radiation emitted by a blackbody and Planck's Law.
- How to use the external fractional function to determine the amount of radiation emitted over wavelength bands.
- Methods for determining view factors and the view factor rules.
- How to calculate space resistances and solve blackbody radiation exchange problems.
- Radiation emission by real surfaces and the meaning of emissivity.
- The difference between emissivity, hemispherical emissivity, and total hemispherical emissivity.
- The meaning of the diffuse and the gray surface assumptions.
- The meaning of absorptivity, reflectivity, and transmissivity.
- The meaning of irradiation and radiosity.
- How to calculate surface resistances and deal with diffuse gray surface problems.
- The meaning of a re-radiating surface.

References

Duffie, J. A. and W. A. Beckman, *Solar Engineering of Thermal Processes*, Fourth Edition, Wiley (2013).
Hottel, H. C. and A. F. Sarofim, *Radiative Transfer*, McGraw-Hill, New York (1967).
McAdams, W. H., *Heat Transmission*, Third Edition, McGraw-Hill, New York (1954).
Siegel, R. and J. R. Howell, *Thermal Radiation Heat Transfer*, Fourth Edition, Taylor and Francis (2002).

Problems

Blackbody Radiation and the Electromagnetic Spectrum

14.1 A new stove top uses a halogen lamp that is placed under a glass surface as the heat source for each burner. The advantages of this design are that the lamp delivers instant heat and responds very quickly to changes in the temperature setting. The heating element of the lamp is a circular disk that is insulated on its back and has a diameter of $D_l = 2.4$ cm. The design specification for the stove top requires that it be capable of heating $V_w = 2$ liters of water from $T_{ini} = 25°C$ to boiling in less than $t_b = 8$ minutes at a location that is at sea level. Assume that the heating element radiates as a blackbody and that all of the radiation emitted by the heating element is absorbed by the pot containing the water. Ignore convection for this problem.

(a) If the efficiency of the burner is 100 percent (i.e., all electrical power provided to the halogen heating element is transferred to the water) then what is the minimum required electrical power input to the unit in order to meet the design specification?

(b) Estimate the temperature of the heating element required.

(c) Will the radiant energy from this stove top unit be visible? What is the fraction of the radiation emitted by the element that is visible?

14.2 The solar constant, G_{sc}, is the defined as energy from the Sun per unit time that would be received on a unit area of surface perpendicular to the direction of the propagation of the radiation at the mean Earth–Sun distance if there were no atmosphere surrounding Earth. We know that the diameter of the Sun is approximately $D_{sun} = 1.39 \times 10^9$ m and that the surface of the Sun is at an equivalent temperature of approximately $T_{sun} = 5780$ K. The diameter of the Earth is $D_{earth} = 1.276 \times 10^7$ m and the mean Earth–Sun distance is $R = 1.497 \times 10^{11}$ m.

(a) Estimate the value of the solar constant.

(b) In 2003, the amounts of primary energy consumed in the world as a result of combustion of coal, natural gas, and oil was 140×10^9 GJ/yr, 95×10^9 GJ/yr, and 190×10^9 GJ/yr, respectively. Compare the amount of energy that is radiated to Earth from the Sun to the annual energy consumed by combustion of these fossil fuels.

(c) The First Law of Thermodynamics indicates that energy cannot be destroyed. If your answer to part (b) indicated that more energy strikes the planet in year than we use then explain why we are experiencing an energy shortage.

14.3 Radiation that passes through the atmosphere surrounding our planet is absorbed to an extent that depends on its wavelength due to the presence of gases such as water vapor, oxygen, carbon dioxide, and methane. However, there is a relatively large range of wavelengths between 8 μm and 13 μm for which there is relatively little absorption in the atmosphere and, thus, the transmittance of the atmosphere is high. This wavelength band is called the *atmospheric window*. Infrared detectors on satellites measure the relative amount of infrared radiation emitted from the ground in this wavelength band in order to provide an indication of the ground temperature.

(a) What fraction of the radiation from the Sun is in the atmospheric window? The Sun can be approximated as a blackbody source at 5780 K.

(b) Prepare a plot of the fraction of the thermal radiation emitted between 8 μm and 13 μm to the total radiation emitted by the ground for temperatures between $-10°C$ and $30°C$.

(c) Based on your answers to (a) and (b), indicate whether radiation in the atmospheric window can provide a clear indication of surface temperature to satellite detectors.

14.4 A black surface is at an absolute temperature T_1 and emits radiation at a rate $E_{b,1}$. If the absolute temperature is doubled (to $T_2 = 2\,T_1$) then what

is the ratio of $E_{b,2}$ (the amount of radiation that the plate emits at T_2) to $E_{b,1}$?

14.5 A tungsten filament is placed in an evacuated glass enclosure. Assume that the glass enclosure is completely transparent for wavelengths less than 3 μm but completely opaque for wavelengths greater than 3 μm. The filament is heated to 2550 K.

(a) What fraction of the radiant energy emitted by the filament is in the visible region (0.38 μm to 0.78 μm)?

(b) What fraction of the radiant energy emitted by the filament cannot be transmitted through the enclosure?

14.6 The temperature of the surface of the Sun is approximately 5800 K. Assuming that the Sun can be considered to be a blackbody, determine:

(a) the fraction of the radiation emitted by the Sun that is in the UV band between 0.20 μm and 0.38 μm,

(b) the fraction of the radiation emitted by the Sun that is in the visible band between 0.38 μm and 0.78 μm, and

(c) the fraction of the radiation emitted by the Sun that is in nonvisible infrared bands between 0.78 μm and 1000 μm.

14.7 The incident solar flux on the Earth at the Earth's mean orbital radius (i.e., the mean distance from the Earth to the Sun) is approximately equal to $\dot{q}''_{s-e} = 1350$ W/m². The Earth's mean orbital radius is $R_{o,e} = 149.6$ million km.

(a) Based on these facts, what is the rate of energy generated by the Sun?

(b) Using the answer from (a), determine the incident solar flux on the planet Mercury. The mean orbital radius of Mercury is $R_{o,m} = 57.9$ million km. Note that you do not need to know the diameter of Mercury in order to answer this question.

(c) Estimate the equilibrium temperature on Mercury based on an energy balance on the planet (assume that Mercury behaves as a blackbody and has no atmosphere that would cause a greenhouse effect). You do not need to know the diameter of Mercury to solve this problem.

14.8 A quartz crystal has a transmittance of 0.86 for radiation between 0.25 μm and 0.78 μm. Outside of this wavelength band, the crystal is essentially opaque. Prepare a plot of the fraction of incident radiation that is transmitted through the crystal from a black emission source at temperature T for 1000 K $< T <$ 5280 K.

14.9 Photovoltaic cells convert a portion of the radiation that is incident on their surface into electrical power. The efficiency of the cells is defined as the ratio of the electrical power produced to the incident radiation. The efficiency of solar cells is dependent upon the wavelength distribution of the incident radiation. An explanation for this behavior was originally provided by Einstein and initiated the discovery of quantum theory. Radiation can be considered to consist of flux of photons. The energy per photon (e) is: $e = hc/\lambda$, where h is Planck's constant, c is the speed of light, and λ is the wavelength of the radiation. The number of photons per unit area and time is the ratio of the spectral emissive power, $E_{b,\lambda}$ to the energy of a single photon, e. When radiation strikes a material, it may dislodge electrons. However, the electrons are held in place by forces that must be overcome. Only those photons that have energy above a material-specific limit, called the bandgap energy limit (i.e., photons with wavelengths lower than $\lambda_{bandgap}$) are able to dislodge an electron. In addition, photons having energy above the bandgap limit are still only able to dislodge one electron per photon; therefore, only a fraction of their energy, equal to $\lambda/\lambda_{bandgap}$, is useful for providing electrical current. Assuming that there are no imperfections in the material that would prevent dislodging of an electron and that none of the dislodged electrons recombine (i.e., a quantum efficiency of 1), the efficiency of a photovoltaic cell can be expressed as:

$$\eta = \frac{\displaystyle\int_0^{\lambda_{bandgap}} \frac{\lambda}{\lambda_{bandgap}} E_{b,\lambda}\, d\lambda}{\displaystyle\int_0^{\infty} E_{b,\lambda}\, d\lambda}.$$

(a) Calculate the maximum efficiency of a silicon solar cell that has a bandgap wavelength of $\lambda_{bandgap}$ = 1.12 μm and is irradiated by solar energy having an equivalent blackbody temperature of 5780 K.

(b) Calculate the maximum efficiency of a silicon solar cell that has a bandgap wavelength of $\lambda_{bandgap}$ = 1.12 μm that is irradiated by incandescent light produced by a black tungsten filament at 2700 K.

(c) Repeat part (a) for a gallium arsenide cell that has a bandgap wavelength of $\lambda_{bandgap}$ = 0.73 μm, corresponding to a bandgap energy of 1.7 eV.

(d) Plot the efficiency versus bandgap wavelength for solar irradiation. What bandgap wavelength provides the highest efficiency?

14.10 A novel hybrid solar lighting system has been proposed in which concentrated solar radiation is collected and then filtered so that only radiation in the visible range (0.38 μm to 0.78 μm) is transferred to luminaires in the building by a fiber-optic bundle. The unwanted heating of the building caused by lighting can be reduced in this manner. The nonvisible energy at wavelengths greater than 0.78 μm can be used to produce electricity with thermal photovoltaic cells. Solar radiation can be approximated as radiation from a blackbody at 5780 K. See Problem 14.9 for a discussion of a model of the efficiency of a photovoltaic cell.

(a) Determine the maximum theoretical efficiency of silicon photovoltaic cells ($\lambda_{bandgap}$ = 1.12 μm) if they are illuminated with solar radiation that has been filtered so that only wavelengths greater than 0.78 μm are available.

(b) Determine the bandgap wavelength ($\lambda_{bandgap}$) that maximizes the efficiency for this application.

14.11 Light is "visually evaluated radiant energy", i.e., radiant energy that your eyes are sensitive to (just like *sound* is pressure waves that your ears are sensitive to). Because light is both radiation and an observer-derived quantity, two different systems of terms and units are used to describe it: radiometric (related to its fundamental electromagnetic character) and photometric (related to the visual sensation of light). The radiant power ($\dot{q}$) is the total radiation emitted from a source and is a radiometric quantity (with units W). The radiant energy emitted by a blackbody at a certain temperature is the product of the blackbody emissive power (E_b, which is the integration of blackbody spectral emissive power over all wavelengths) and the surface area of the object (A):

$$\dot{q} = A E_b = A \int_{\lambda=0}^{\infty} E_{\lambda,b} \, d\lambda = A \, \sigma \, T^4.$$

On the other hand, luminous power (F) is the "light" emitted from a source and is a photometric quantity (with units of lumen, which are abbreviated lm). The radiant and luminous powers are related by

$$F = A \, K_m \int_0^{\infty} E_{\lambda,b} \, (\lambda) \, V \, (\lambda) \, d\lambda,$$

where K_m is a constant (683 lm/W photopic) and $V(\lambda)$ is the relative spectral luminous efficiency curve. Notice that without the constant K_m, the luminous power is just the radiant power filtered by the function $V(\lambda)$ and has units of W; the constant K_m can be interpreted as converting W to lumen, the photometric unit of light. The filtering function $V(\lambda)$ is derived based on the sensitivity of the human eye to different wavelengths (in much the same way that sound meters use a scale based on the sensitivity of your ear in order to define the acoustic unit, decibel or dB). The function $V(\lambda)$ is defined as the ratio of the sensitivity of the human eye to radiation at a particular wavelength to the sensitivity of your eye at 0.555 μm; 0.555 μm is selected because your eye is most sensitive to this wavelength (which corresponds to green). An approximate equation for $V(\lambda)$ is: $V(\lambda) = \exp\left[-285.4 \, (\lambda - 0.555)^2\right]$, where λ is the wavelength in micron (μm). The luminous efficiency of a light source (η_l) is defined as the number of lumens produced per watt of radiant power:

$$\eta_l = \frac{F}{\dot{q}} = \frac{K_m \int_0^{\infty} E_{b,\lambda} \, (\lambda) \, V \, (\lambda) \, d\lambda}{\sigma \, T^4}.$$

The conversion factor from W to lumen, K_m, is defined so that the luminous efficiency of sunlight is 100 lm/W; most other, artificial light sources will be less than this value. The most commonly used filament in an incandescent light bulb is tungsten; tungsten will melt around 3650 K. An incandescent light bulb with a tungsten filament is typically operated at 2770 K in order to extend the life of the bulb. Determine the luminous efficiency of an incandescent light bulb with a tungsten filament.

View Factors

14.12 A room in a building has a floor area of 4 m × 3 m and a ceiling height of 3 m. There is a single door to the room that is 1 m wide.

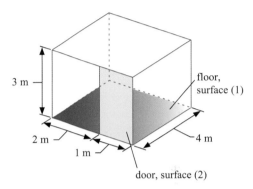

Determine the view factor between the door (surface 2) and the floor (surface 1).

14.13 The figure shows two plates that are parallel and aligned at their center lines.

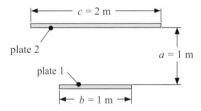

Plate 1 has width $b = 1$ m and plate 2 has width $c = 2$ m. The plates are separated by an amount $a = 1$ m. Assume that the plates extend a long way into the page so that the situation is 2-D.

(a) Determine the view factor $F_{1,2}$ using the crossed and uncrossed strings method.

(b) Check your result from (a) against the value obtained using the EES function F2D_10.

14.14 Find the view factor $F_{1,2}$ for the geometry shown in the figure in the following two ways and compare the results.

(a) Use the view factor function F3D_2 in EES (you will need to call the function more than once).

(b) Use the differential view factor relation FDiff_4 and do the necessary integration.

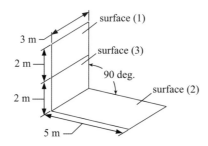

14.15 A homeowner has installed a skylight in a room that measures 6 m x 4 m with a 2.5 m ceiling height, as shown in the figure. The skylight is located in the center of the ceiling and it is square, 2 m on each side. A desk is to be located in a corner of the room. The surface of the desk is 0.9 m high and the desk surface is 0.5 m × 1 m in area. The skylight has a diffusing glass so that the visible light that enters the skylight should be uniformly distributed.

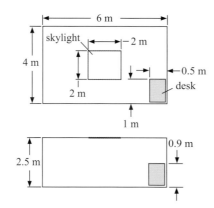

Determine the fraction of the light emanating from the skylight that will directly illuminate the desktop. Does it matter which wall the desk is positioned against (i.e., if you turned the desk 90° would the result be different)?

Blackbody Radiation Exchange

14.16 A spherical blackbody surface having a diameter of 0.1 m is located in blackbody surroundings that are at 25°C. For sphere surface temperatures between 100°C and 1000°C, prepare plots of the following quantities:

(a) the net radiation heat transfer rate between the sphere and its surroundings,

(b) the wavelength at which emission from the sphere is at a maximum, and

(c) the fraction of the energy emitted by the surface between wavelengths of 4 μm and 20 μm.

14.17 A bare-bead thermocouple of 0.1 inch diameter is used to measure the steady-state temperature of a very hot flowing gas in a duct with cooled walls. Would a smaller bead size give a more accurate reading? You may neglect conduction along the thermocouple wires.

14.18 A braze furnace is made of a top surface (surface 1) that is heated to 500 K. The heated surface transfers heat radiatively to the side (surface 2) at 400 K and bottom (surface 3) at 300 K. Assume that all surfaces are black and that the furnace is very long (into the page).

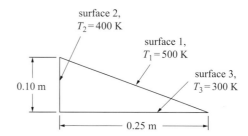

(a) How much radiation is emitted by the top surface (1) per unit length of the furnace (W/m)?

(b) How much radiation is emitted by the side surface (2) per unit length of the furnace (W/m)?

(c) How much of the radiation emitted by the side surface hits the top surface (W/m)?

(d) How much radiation is emitted by the bottom surface (3) per unit length of the furnace (W/m)?

(e) How much of the radiation emitted by the bottom surface hits the top surface (W/m)?

(f) How much heat must be supplied to the top surface in order to keep it at 500 K (W/m)?

14.19 The figure illustrates a concept for rejecting heat from a spacecraft. Liquid water is pumped through a bank of pipes with outer diameter $D_o = 1.0$ cm. The pipes run parallel to one another; they are spaced apart a fixed distance,

$s = 2.0$ cm, and are positioned a distance, $w = 4.0$ cm, from the outside wall of the space. Assume for this problem that the pipes are $L = 1.0$ m long and there are $N = 100$ pipes.

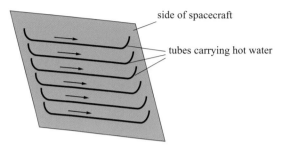

side of spacecraft

tubes carrying hot water

The second figure illustrates a cutaway view of the heat rejection system, shown from the side. Assume that the system can be treated using a 2-D model – that is, neglect the end effects of the pipes (where the water enters and exits). Also neglect the fact that the outer tubes in the array do not have pipes on both sides. The pipe surface temperature is 50°C and the pipe surface is black as is the outer surface of the spacecraft. The outer surface of the spacecraft is at 20°C. Assume that the spacecraft shades the pipes from the Sun or any other radiating body (i.e., the pipes are exposed to deep space with an effective temperature of $T_{space} = 4$ K). Use a resistance network to solve this problem that includes three surfaces: the surface of the pipes (p), the wall (w), and space (s).

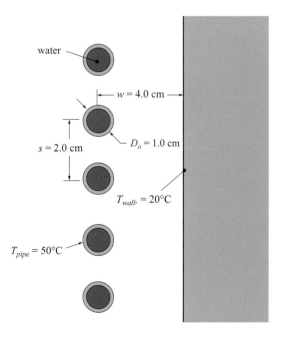

water

$w = 4.0$ cm

$s = 2.0$ cm

$D_o = 1.0$ cm

$T_{wall} = 20°C$

$T_{pipe} = 50°C$

(a) Calculate the value of each of the resistors in your network.
(b) What is the total rate of heat transferred from the pipes to space (W)?
(c) What is the total rate of heat transferred from the pipes to the wall (W)?
(d) What is the net rate of heat transferred from the pipes (W)?
(e) What is the heat transfer rate from the wall to space (W)?

14.20 A cylindrical thermos bottle is made of a two stainless steel cylinders having the dimensions shown in the figure. Except for a few low-conductivity struts, the space between the cylinders is filled with air at atmospheric pressure. The wall thickness of the metal is $th = 0.7$ mm. The thermos is filled with coffee at $T_H = 95°C$ and surrounded by air at $T_C = 25°C$. Assume that the inner wall of the thermos is at T_H and the outer wall is at T_C. Neglect heat loss from the bottom or top.

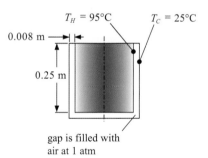

$T_H = 95°C$ $T_C = 25°C$

0.008 m

0.25 m

gap is filled with air at 1 atm

(a) Estimate the rate of heat loss by radiation assuming that the stainless steel surfaces are black.
(b) Is the radiative heat loss significant compared to the convective heat loss due to natural convection in the gap?

14.21 A 50 mm diameter hole is drilled through a 25 mm thick plate of steel.
(a) Determine the view factor between the inside surface of the hole and the surroundings above and below the plate.
(b) Determine the rate of radiation heat transfer between the inside surface of the steel when it is at 350°C. Assume that the surface radiates as a blackbody and that the surroundings are at 25°C.

14.22 A transfer line is used to provide liquid nitrogen to an experiment, as shown in the figure.

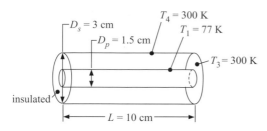

The pipe is $L = 10$ cm long with outer diameter $D_p = 1.5$ cm. The surface of the pipe is $T_1 = 77$ K. The space between the pipe and the outer shield is evacuated. The diameter of the outer shield is $D_s = 3$ cm and the temperature of the shield is $T_4 = 300$ K. One end is insulated and the other end has temperature $T_3 = 100$ K. All surfaces are black. Determine the rate of heat transfer to the pipe.

14.23 A rectangular building warehouse has dimensions of 50 m × 30 m with a ceiling height of 10 m. The floor of this building is heated. On a cold day, the inside surface temperature of the walls are found to be 16°C, the ceiling surface is 12°C, and the heated floor is at a temperature of 32°C. Estimate the radiant heat transfer from the floor to the walls and the ceiling assuming that all surfaces are black. What fraction of the heat transfer is radiated to the ceiling?

14.24 A furnace wall has a 4 cm hole in the insulated wall for visual access. The wall is 8 cm wide. The temperature inside the furnace is 1900°C and it is 25°C outside of the furnace. Assuming that the insulation acts as a black surface at a uniform temperature, estimate the radiative heat transfer through the hole.

14.25 A cubical enclosure has a height, width and length of 0.12 m. The top side of the enclosure is maintained at 100°C while the bottom is maintained at 0°C. The remaining four surfaces are perfectly insulated. Assuming all surfaces radiate as blackbodies, determine the heat transfer rate between the 100°C and 0°C surfaces and the temperatures of all surfaces.

14.26 The bottom surface of the cylindrical cavity shown in the figure is heated to $T_{bottom} = 750°C$

while the top surface is maintained at $T_{top} = 100°C$. The sides of the cavity are insulated externally and isothermal (i.e., the sides are made of a conductive material and therefore come to a single temperature).

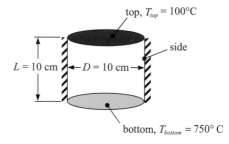

The diameter of the cylinder is $D = 10$ cm and its length is $L = 10$ cm. Assume that the cylinder is evacuated so that the only mechanism for heat transfer is radiation. All surfaces are black ($\varepsilon = 1.0$).

(a) Calculate the net rate of heat transfer from the bottom to the top surface. How much of this energy is radiated directly from the bottom surface to the top and how much is transferred indirectly (from the bottom to the sides to the top)?
(b) What is the temperature of the sides?
(c) If the sides were not insulated but rather also cooled to $T_{side} = 100°C$ then what would be the total heat transfer rate from the bottom surface?

14.27 A homeowner inadvertently left a spray can near the barbeque grill. The spray can is $H = 8$ inch high with a diameter of $D = 2.25$ inch. The side of the barbeque grill is $H = 8$ inch high and $W = 18$ inch wide. The spray can is located with its center aligned with the center of the grill wall and it is $a = 6$ inch from the wall, as shown in the figure. Assume the can to be insulated on its top and bottom.

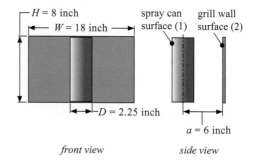

front view   *side view*

(a) What is the view factor between the spray can, surface 1, and the grill wall, surface 2? What is the view factor between the spray can and the surroundings, surface 3?

(b) Assuming both surfaces to be black, what is the net rate of radiation heat transfer to the spray can when the grill wall is at $T_2 = 350°F$, the spray can exterior is $T_1 = 75°F$, and the surroundings are at $T_3 = 75°F$?

(c) If the grill wall is at $T_2 = 350°F$ and the surroundings are at $T_3 = 75°F$ then what is the equilibrium temperature of the spray can if radiation is the only heat transfer mechanism?

14.28 Two long pipes, each having an outside diameter of 10 cm are oriented parallel to each other with the centers of the pipe spaced a distance of 30 cm. The surface of one pipe is maintained at 224°C as a result of hot oil flowing within the pipe. The surface of the other pipe is maintained at 12°C, as a result of water flow. Consider both pipes to be black surfaces. The surroundings are at 25°C.

(a) Determine the total rate of radiation heat transfer from the hot pipe per unit length.

(b) Determine the total rate of radiation heat transfer to the cold pipe per unit length.

14.29 A cylindrical heating element has a length of $L = 15$ cm and a diameter of $D_1 = 0.5$ cm. It is surrounded by a cylindrical enclosure having a diameter of $D_2 = 2$ cm that is open at the ends. The surface of the enclosure is at $T_2 = 240°C$ and the electrical dissipation of the heating element is $\dot{q}_h = 70$ W. The surroundings are at $T_3 = 20°C$.

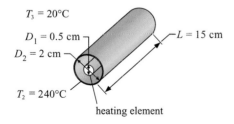

$T_3 = 20°C$

$D_1 = 0.5$ cm

$D_2 = 2$ cm

$T_2 = 240°C$

$L = 15$ cm

heating element

(a) Determine the temperature of the heating element assuming that radiation is the only mechanism for heat transfer. The ends of the heating element may be assumed to be insulated.

(b) How much error would result if the geometry were approximated to be 2-D?

14.30 Two parallel plates, each 1 m × 2 m are spaced a distance d apart. One plate is maintained at 25°C whereas the other is at 324°C. Assuming the plates are black, plot the net rate of heat transfer between the plates as a function of d for values between 0 and 10 m.

14.31 A rectangular enclosure has dimensions 3 m, 3 m, and 9 m. One of the two 3 m × 3 m sides is maintained at 500 K. One of the four 3 m × 9 m surfaces is maintained at 300 K. All other surfaces are insulated. Determine the temperatures of all surfaces in the enclosure, assuming that they are black and interact only by radiation.

14.32 The figure illustrates a furnace in a power plant. Just inside the furnace wall is a series of regularly spaced pipes containing water that is being heated by combustion flames at $T_f = 870°C$. Because of the high temperatures involved, you may assume that all of the heat transfer within the furnace occurs by radiation.

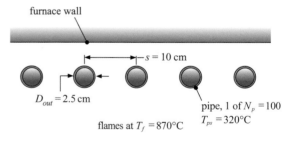

furnace wall

$s = 10$ cm

$D_{out} = 2.5$ cm

pipe, 1 of $N_p = 100$

$T_{ps} = 320°C$

flames at $T_f = 870°C$

The pipes are spaced $s = 10$ cm (center-to-center distance) apart and traverse the entire outer extent of the furnace; there are a total of $N_p = 100$ pipes, although only five are shown in the figure. The pipes have outer diameter $D_{out} = 2.5$ cm. The surface temperature of the pipes is $T_{ps} = 320°C$. All surfaces can be assumed to be black. For the purposes of this problem you may treat the flame as a black surface that extends parallel to the furnace wall on the other side of the pipes. End effects can be neglected.

(a) If the furnace wall is perfectly insulated, estimate the total rate of heat transfer to the water (i.e., to the pipe surface) per unit length. What is the wall temperature in this limit?

(b) Assume that the furnace wall is not insulated but rather it is cooled to a temperature, $T_w = 800$ K. Compute the net heat transfer rate to the furnace wall (this is the heat that would have to be conducted through the wall and transferred to the ambient air outside of the furnace in order to maintain the temperature of the wall at T_w).

(c) Use your model from (b) to prepare a plot that shows the rate of heat transferred to the wall as a function of the wall temperature; does your plot make sense? Is it consistent with your answer from parts (a) and (b)?

(d) The wall has an area-specific resistance (that includes both conduction through the wall and convection from its outer surface) of $R''_w = 0.2$ K-m^2/W that characterizes the heat transfer from the wall's internal surface to the ambient air outside of the furnace at $T_{amb} = 20°$C. Overlay a plot of the heat transferred through the wall as a function of the wall temperature on your result from part (c); there should be a location where the two lines intersect so that the heat transfer into the wall predicted using your radiation solution matches the heat transfer through the wall based on the wall resistance.

(e) What is the wall temperature and furnace efficiency (defined as the ratio of the heat transfer to the water to the heat transfer from the flame)?

14.33 A solar thermal collector consists of a 3 ft × 6 ft plate of blackened metal, insulated on its bottom side. A 3 ft × 6 ft sheet of glass is placed 1 inch above the metal plate. Owing to absorption of solar radiation, the plate is at a temperature of 84°C. The glass is opaque to thermal radiation and it can be considered to act as a black surface. The edges of the collector are insulated. The surroundings above the glass cover are at 19°C.

(a) Determine the temperature of the glass, neglecting convection. Assume that the glass is isothermal.

(b) Determine the rate of radiative heat transfer from the metal plate.

(c) Compare the result from part (a) with the rate of radiative heat transfer from the plate that would result if there were no glass cover.

Properties of Real Surfaces

14.34 A solar collector is used to absorb energy from the Sun. The solar collector losses are based in part on how much energy it emits to its surroundings. The figure illustrates the variation of emissivity with wavelength for semi-gray surfaces that are referred to as surfaces A, B, and C.

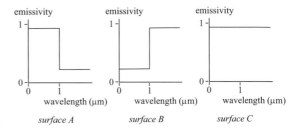

Which of the three surfaces shown in the figure is a better choice for your solar collector surface? Justify your answer.

14.35 Three surfaces each with the same area have spectral emissivity shown in the figure. The surfaces are each laid horizontally on the ground at noon on a clear day and exposed to sunlight. Most solar radiation is at wavelengths that are less than about 3 μm while most energy emitted by a surface near room temperature is at wavelengths greater than 3 μm.

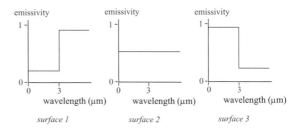

Select the statement below that is true about the temperature of the three surfaces.

(a) There is insufficient information to determine the relative temperature of the three surfaces.

(b) All three surfaces will reach the same temperature ($T_1 = T_2 = T_3$).

(c) Surface 1 will be cooler than surface 2 which will be cooler than surface 3 ($T_1 < T_2 < T_3$).

(d) Surface 1 will be warmer than surface 2 which will be warmer than surface 3 ($T_1 > T_2 > T_3$).

(e) Both surface 1 and 3 will be hotter than surface 2 ($T_1 > T_2$ and $T_3 > T_2$).

14.36 Consider a small body at temperature T_b that is completely enclosed by a large black surface at temperature T_s. Assume that there is a vacuum between the body and surface. Assume that the temperature of the surface is less than that of the small body ($T_s < T_b$). The emissivity of the small body is 0.10.

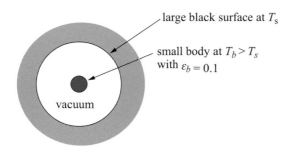

Circle the correct answer from those below. The net energy transfer is:

(a) from the surface to the body,

(b) from the body to the surface,

(c) zero, or

(d) we cannot determine this unless we know the specific values of the two temperatures.

14.37 A surface has wavelength-dependent properties as listed in the table. The surface is maintained at 500 K.

Wavelength range (μm)	α_λ
0–0.6	0.8
0.6–2.6	0.25
2.6–100	0.10

(a) Determine the total hemispherical absorptance of this surface for solar radiation.

(b) Determine the total hemispherical emissivity of this 500 K surface.

14.38 The figure shows the hemispherical emissivity of tungsten as a function of wavelength for various temperatures; these data are provided in an EES Lookup table (Tungsten_spectral. lkt in the Userlib/Heat Transfer/Radiation Properties). The emissivity of tungsten is of particular engineering interest because tungsten is used as a filament for incandescent light bulbs.

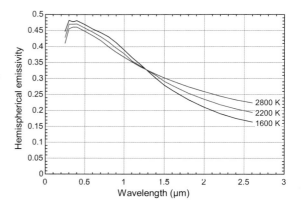

Calculate the total hemispherical emissivity for tungsten at 2800 K.

14.39 Calculate and plot the total reflectance of polished aluminum at 697 K for radiation emitted from sources between 300 K and 6000 K. The spectral emittance of polished aluminum is provided in the EES Radiation Properties folder as the Lookup Table Aluminum-Spectral.lkt.

14.40 A 10,000 sq. ft. office building requires approximately $\dot{q}_v'' = 1.0$ W/ft^2 of visible radiant energy for lighting; this is energy emitted between the wavelengths $\lambda_{v,low} = 0.38$ μm and $\lambda_{v,high} = 0.78$ μm. The efficiency of a lighting system (η_v) can be calculated as the ratio of the visible radiant energy that is emitted to the total energy emitted.

(a) Compute the efficiency of a light source that consists of a blackbody at $T = 2800$ K.

(b) Plot the efficiency of a blackbody lighting system as a function of temperature.

There are two costs associated with providing the lighting that is required by the office. The electricity required to heat the blackbody to its temperature and the electricity that is required to run the cooling system that must remove the energy provided by the light source (note that both the visible and the invisible radiation are deposited as thermal energy in the building). Assume that the building cooling system has an average coefficient of performance of $COP = 3.0$ and the building is occupied for 5 days per week and 12 hours per day. Assume that the cost of electricity is $0.12/kW-hr.

(c) What is the total cost associated with providing lighting to the office building for one year? How much of this cost is direct (that is, associated with buying electricity to run the light bulbs) versus indirect (that is, associated with running air conditioning equipment in order to remove the energy dumped into the building by the light bulbs). Assume that you are using a light bulb that is a black body with a temperature of 2800 K.

An advanced light bulb has been developed that is not a black body but rather has an emissivity that is a function of wavelength. The temperature of the advanced light bulb remains 2800 K, but the filament can be modeled as being semi-gray; the emissivity is equal to $\varepsilon_{low} = 0.80$ for wavelengths from 0 to $\lambda_c = 1.0$ µm and $\varepsilon_{high} = 0.25$ for wavelengths above 1.0 µm.

(d) What is the efficiency of the new light bulb?
(e) What is the yearly savings in electricity that can be realized by replacing your old light bulbs (the blackbody at 2800 K) with the advanced light bulbs?

Diffuse Gray Surface Radiation Exchange

14.41 A sphere having a surface area of 0.375 m^2 has an emissivity of 0.35 and a uniform temperature of 675 K. The sphere is completely surrounded by a 300 K surface having a surface area of 3.345 m^2 and an emissivity of 0.75. Determine the net heat transfer rate between these surfaces.

14.42 A small object is located in a large room. The object is maintained at a constant temperature of 626 K and it can be considered to be a gray surface. The room is maintained at 300 K and it too can be considered to be a gray surface.

(a) What will happen to the heat transfer rate if the emissivity of the object is increased?
(b) What will happen to the heat transfer rate if the emissivity of the room walls is increased?
(c) Provide an explanation for your answers to (a) and (b).

14.43 A solar panel on a space station has an efficiency that varies with temperature according to $\eta = 0.25[1 - 0.002(T - 298)]$, where T is the temperature in K. The efficiency is the fraction of the solar energy incident on its surface that is transformed into electricity. One side of the panel is effectively insulated while the other radiates to space (2 K) with a thermal emissivity of 0.48. Prepare a plot of the efficiency as a function of the solar irradiation for values between 200 and 1250 W/m^2. Assume all of the irradiation is absorbed by the panel.

14.44 A power transistor is designed to operate at a maximum temperature of 480°F in 140°F air. The transistor has a surface area of 1 in^2 and a reflectance of 0.15. The convection coefficient between the air and the transistor is 1.0 Btu/hr-ft^2-R.

(a) Determine the maximum power this transistor can dissipate.
(b) In an attempt to increase the dissipation rate, the transistor is modified to include a fin which increases its effective surface area to 2 inch^2. However, the view factor between the transistor/fin and its surroundings is reduced to 0.8. What is the maximum power than can be dissipated with this design, assuming that the reflectance and convection coefficient are unchanged?

14.45 A thermocouple having a surface area of 0.3 inch^2 is located long the centerline of a long duct of diameter 3 inches. The duct walls are maintained at 200°F. The thermocouple indicates a temperature of 310°F. The heat transfer coefficient between the thermocouple and the air is 31 Btu/hr-ft^2-R. The emissivity of the duct wall is 0.9 and that of the thermocouple is 0.60. Using this information estimate the actual temperature of the air flowing in this duct.

14.46 A kitchen oven is cubic with each dimension being 2 ft. The oven walls have an emissivity of 0.75 and are maintained at 400°F. A roast having a surface area of 0.5 ft² and an emissivity of 0.5 is placed in the center of the oven. The roast is initially at a uniform temperature of 50°F.

(a) Determine the initial rate of radiation heat transfer to the roast.

(b) Determine the percent change in the initial rate of radiation heat transfer if the oven walls are coated with a reflective surface such that their emissivity is reduced from 0.75 to 0.1. Indicate if this change will reduce the cooking time for the roast.

14.47 It has been proposed that considerable energy for air conditioning could be saved if the roofs of buildings were made of reflective material. The purpose of this problem is to investigate this claim. Consider a building having a flat roof that is $W = 12$ m on each side with the interior air maintained at a uniform temperature of $T_{in} = 25°C$. The heat transfer coefficient between the inside of the roof and the internal air is $\bar{h}_{in} = 5$ W/m²-K. Fiberglass insulation is provided between the ceiling of the building the roof surface; the thickness of the fiberglass is $th_{ins} = 15$ cm. It is $T_\infty = 35°C$ outdoors with solar radiation at $G = 800$ W/m² and a wind speed of $u_\infty = 5$ m/s. Assume that the absorptivity for solar radiation is equal to the emissivity.

(a) Compare the rate of heat gain per unit area for roof surfaces having emissivity values of $\varepsilon = 0.1$ and $\varepsilon = 0.9$.

(b) Plot the rate of heat gain per unit area for $\varepsilon = 0.1$ and $\varepsilon = 0.9$ as a function of the fiberglass insulation thickness from th_{ins} ranging from 2 cm to 20 cm.

(c) What is your conclusion regarding the merit of using a reflective roof surface?

14.48 A spherical object, 15 cm in diameter, is placed 96 cm from a wall that extends a long distance in every direction from the sphere. The sphere is maintained at 480 K. The wall is adiabatic. The surroundings are at 290 K.

(a) Assume the sphere is a blackbody. Plot the temperature of the wall and the net radiative heat transfer to the surroundings (surface 3) as a function of the emissivity of the wall ranging from 0.01 to 0.99.

(b) Assume that the wall has an emissivity of 0.5. Plot the temperature of the wall and the radiative heat transfer rate to the surroundings as a function of the emissivity of the sphere ranging from 0.1 to 0.99.

14.49 A cylindrical heating element has length $L = 15$ cm and diameter $D_1 = 0.5$ cm. The emissivity of the heating element is $\varepsilon_1 = 0.22$. It is surrounded by a cylindrical enclosure having a diameter $D_2 = 2$ cm that is open at the ends. The emissivity of the enclosure is $\varepsilon_2 = 0.9$. The heater and enclosure surroundings are at $T_3 = 23°C$. The electrical dissipation of the heating element is $\dot{q}_{he} = 70$ W. Assume that radiation is the only heat transfer mechanism and the ends of the heating element are insulated.

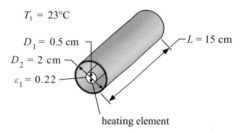

Calculate and plot the temperature of the heating element and the enclosure as a function of the emissivity of the enclosure.

14.50 A cylinder is placed at the center of a square enclosure.

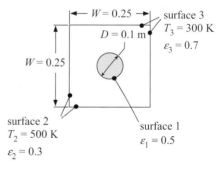

The cylinder is not being heated or cooled and it is at steady state. The surface of the cylinder is surface 1 and the diameter of the cylinder is $D = 0.1$ m. The emissivity of the surface of the cylinder is $\varepsilon_1 = 0.5$. The enclosure is square and $W = 0.25$ m on each side. Two of the sides of the enclosure are surface 2 and are maintained at $T_2 = 500$ K. The emissivity of surface 2 is

$\varepsilon_2 = 0.3$. The other two sides of the enclosure are surface 3 and are maintained at $T_3 = 300$ K. The emissivity of surface 3 is $\varepsilon_3 = 0.7$. The problem is two-dimensional and the view factor between surface 2 and itself and surface 3 and itself are $F_{2,2} = F_{3,3} = 0.25$. What is the temperature of the cylinder?

14.51 The figure illustrates (in cross-section) a spherical cryogenic experiment that is placed at the center of a cubical enclosure (a box).

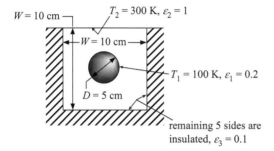

$W = 10$ cm
$T_2 = 300$ K, $\varepsilon_2 = 1$
$W = 10$ cm
$T_1 = 100$ K, $\varepsilon_1 = 0.2$
$D = 5$ cm
remaining 5 sides are insulated, $\varepsilon_3 = 0.1$

The spherical experiment (surface 1) has diameter $D = 5.0$ cm and emissivity $\varepsilon_1 = 0.2$. The temperature of the experiment is maintained at $T_1 = 100$ K. Each face of the cubical enclosure is $W = 10$ cm $\times$ $W = 10$ cm. The top surface of the enclosure (surface 2) is maintained at $T_2 = 300$ K and is black. The other five sides (surface 3) are insulated externally and have emissivity $\varepsilon_3 = 0.1$.

(a) What is the view factor between the experiment and the top surface ($F_{1,2}$)?
(b) What is the view factor between the experiment and the five insulated sides ($F_{1,3}$)?
(c) What is the view factor between the top surface and the experiment ($F_{2,1}$)?
(d) Draw and clearly label a resistance network that represents the radiation heat transfer problem.
(e) Calculate the values of all of the resistances in your diagram from (d). You may assume that the view factor between the insulated sides and the top ($F_{3,2}$) is 0.70.
(f) What is the net rate of heat transfer to the experiment?
(g) How would your answer to (f) change if the emissivity of the experiment were reduced (would the heat transfer to the experiment increase, decrease, or stay the same)?
(h) How would your answer to (f) change if the emissivity of the insulated sides were reduced

(would the heat transfer to the experiment increase, decrease, or stay the same)?

14.52 The figure illustrates a 2-D radiation problem. A thin plate is being heated radiatively by a semicircular heater. The bottom side of the plate is insulated, as is the outer surface of the heater. Neglect any heat transfer mechanism other than radiation for this problem.

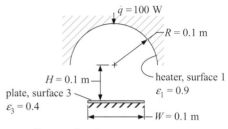

$\dot{q} = 100$ W
$R = 0.1$ m
$H = 0.1$ m
plate, surface 3
$\varepsilon_3 = 0.4$
heater, surface 1
$\varepsilon_1 = 0.9$
$W = 0.1$ m
surroundings, surface 2
$T_2 = 300$ K

The width of the plate is $W = 0.1$ m. The distance between the plate and the center of the heater is $H = 0.1$ m. The radius of the heater is $R = 0.1$ m. The heater is centered over the plate, as shown. The problem is 2-D because both the heater and plate extend indefinitely into the page; we will do the problem on a per unit length basis, $L = 1$ m. There are three surfaces involved in the problem: (1) the inner surface of the heater, (2) the surroundings, and (3) the top of the plate. The emissivity of the plate is $\varepsilon_3 = 0.4$ and the emissivity of the heater is $\varepsilon_1 = 0.9$. The heater is provided with an external heat transfer (to the heater surface) of 100 W (per unit length). The plate is externally adiabatic (it is not being heated or cooled externally). The surroundings are at $T_2 = 300$ K. The view factor from the heater to itself is $F_{1,1} = 0.36$.

(a) Determine the view factor between the plate and the heater, $F_{3,1}$, using the Crossed and Uncrossed Strings Method.

For the remainder of this problem, assume that your answer to part (a) was $F_{3,1} = 0.75$ (this may or may not be the correct answer).

(b) Draw a resistance network that represents this radiation problem. Clearly label the resistances, the nodes, and indicate the boundary conditions.

(c) Determine the values of each of the resistances in your network from (b).

(d) Calculate the temperature of the heater and the temperature of the plate.

(e) If the emissivity of the heater, ε_1, is decreased, what is the effect on the surface temperature of the heater? Justify your answer.

(f) If the emissivity of the plate, ε_3, is decreased, what is the effect on the surface temperature of the plate? Justify your answer.

14.53 A pipe carrying cryogenic fuel for a spacecraft is shown in the figure. The pipe has diameter, $D_1 = 5.0$ cm and surface temperature $T_1 = 110$ K. The emissivity of the pipe surface is $\varepsilon_1 = 0.2$. This pipe is placed $s = 15$ cm away from another pipe with diameter $D_2 = 10$ cm, surface temperature $T_2 = 320$ K, and emissivity $\varepsilon_2 = 0.5$ that carries hot gas. These pipes run parallel to one another and are exposed to space, which can be approximated as having a temperature of $T_3 = 4$ K.

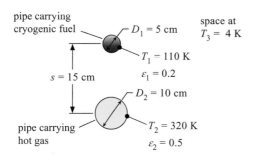

(a) Draw a resistance network to represent this problem. Clearly label what each node and resistor represents.

(b) Calculate the values for each of the resistances in your network (on a per unit length basis, $L = 1$ m).

(c) Calculate the net radiation heat transfer per unit length to the pipe carrying the cryogenic fluid.

14.54 A spherical temperature sensor is installed in a combustion chamber in order to measure the temperature of the flame. The diameter of the sensor is $D_1 = 0.01$ m and the emissivity of the sensor is $\varepsilon_1 = 0.5$. The temperature of the flame is $T_2 = 1150$ K and the view factor from the

sensor to the flame is $F_{1,2} = 0.85$. The sensor has two sources of error. First, there is a conduction heat transfer from the sensor to its leads of $\dot{q}_{lead} = 0.5$ W. Second, in addition to experiencing radiation with the flame, the sensor also experiences radiation with the surroundings at $T_3 = 300$ K. Both the flame and the surroundings can be assumed to be black. Neglect convection from the sensor.

(a) Determine the temperature of the sensor and therefore the error in the flame temperature measurement.

(b) Is the temperature measurement error primarily attributed to conduction to the leads or radiation exchange with the surroundings? Justify your answer with a calculation.

14.55 The figure illustrates an ice skating rink that can be modeled as a large rectangular building with a floor space that is $L = 50$ m $\times$ $W = 20$ m and a ceiling that $H = 3.0$ m high.

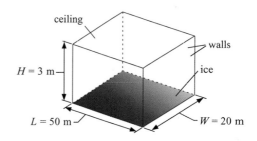

The air in the rink is maintained at $T_\infty = 60°$F with a relative humidity of $RH = 65$ percent. The heat transfer coefficient between the air and the ceiling, ice surface, and walls is $h = 6.5$ W/m²-K. The ceiling is insulated from the external air, as shown in the figure. The walls (all four of them) are black at $T_{wall} = 50°$F. The ice is maintained at $T_{ice} = -12°$F and can also be considered black. This problem deals with the selection of the ceiling surface; specifically, is it better to make the ceiling have a high or low emissivity (ε_c)? The issue is condensation; if the ceiling surface temperature is too low then water will condense out of the air onto the ceiling and drip onto the ice surface, ruining it. Therefore, it is necessary to keep the ceiling surface above the dew point temperature of the air. This is a multi-mode problem because the ceiling experiences both convection (with the internal air) and radiation (with the walls and the ice). We will solve this problem by separately considering the radiation and convection aspects of the problem and then finally force these two components to be consistent.

(a) Draw a resistance network that represents the radiation heat transfer between the ceiling, ice, and the walls. Clearly label what each node and resistance represents.

(b) Calculate the values of each of your resistances from (a). Assume that the emissivity of the ceiling is $\varepsilon_c = 0.5$ for this question.

(c) If the temperature of the ceiling is $T_c = 40°F$ then what is the net rate of radiation heat transfer from the ceiling?

(d) If the temperature of the ceiling is $T_c = 40°F$ then what is the net rate of convection heat transfer to the ceiling?

(e) Adjust the value of T_c so that your answers from (c) and (d) match; that is, determine the value of T_c where the rate of convection to the ceiling matches the rate of radiation from the ceiling.

(f) What is the dew point temperature of the air in the ice rink? You may find the function **DewPoint** for the fluid **AirH2O** (an ideal gas mixture of air and water vapor) to be useful for this calculation.

(g) Prepare a plot of the ceiling surface temperature and the dew point temperature as a function of the emissivity of the ceiling. Note that the dew point temperature is not affected by the emissivity (it should be a horizontal line); what range of emissivity will prevent condensation? In general, should architects provide ice rink ceilings with high or low emissivity?

(h) Prepare a plot showing the refrigeration load on the ice rink (i.e., the total rate of

heat transfer to the ice) as a function of the emissivity of the ceiling.

14.56 The figure illustrates a pipe that is being radiatively heated by two plates.

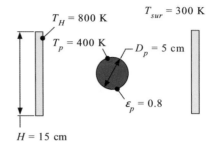

The pipe is centered between the two plates. The pipe has diameter $D = 5$ cm and the surface of the pipe has emissivity $\varepsilon_p = 0.8$. The temperature of the pipe surface is $T_p = 400$ K. The heater plates are black and have surface temperature $T_H = 800$ K. The height of the plates is $H = 15$ cm. The surroundings are at $T_{sur} = 300$ K. You may assume that this is a 2-D problem. The view factor from the pipe to the heater plates (both of them) is $F_{p,H} = 0.5$.

(a) Determine the rate of heat transfer to the pipe per unit length.

(b) The pipe contains a fluid with conductivity $k = 50$ W/m-K. The flow is laminar and fully developed. Estimate the mean temperature of the fluid in the pipe.

14.57 A homeowner has inadvertently left a spray can of lighter fluid near his barbeque grill, as shown in the figure. The can is $L = 7$ cm high with a diameter of $D = 3$ cm. The emissivity of the surface of the can is $\varepsilon_{sc} = 0.4$. The side of the barbeque grill is black and has a temperature $T_g = 450$ K. The view factor between the can and the grill is $F_{sc,g} = 0.42$. The can also experiences a solar heat flux of $G''_s = 920$ W/m². You may assume that the solar flux is uniformly distributed over the can surface. The can experiences convection and radiation with its surroundings at $T_{sur} = 300$ K. The heat transfer coefficient for convection is $\bar{h} = 5$ W/m²-K. Assume the can to be insulated on its top and bottom so that only its sides experience heat transfer.

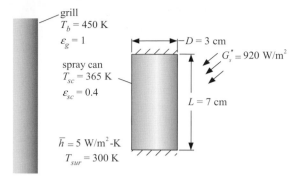

grill
$T_b = 450$ K
$\varepsilon_g = 1$

$D = 3$ cm

$G_s'' = 920$ W/m^2

spray can
$T_{sc} = 365$ K
$\varepsilon_{sc} = 0.4$

$L = 7$ cm

$\bar{h} = 5$ W/m^2-K
$T_{sur} = 300$ K

At a particular instant of time, the temperature of the spray can is $T_{sc} = 365$ K.

(a) Draw a resistance network that can be used to model this problem. Clearly label each of the resistances in your network and show clearly where the solar flux appears.

(b) Calculate the rate of radiation heat transfer to the spray can.

(c) Is the temperature of the spray can increasing or decreasing? Justify your answer with a calculation.

(d) Eventually, the spray can will achieve a steady-state temperature. Would reducing the emissivity of the spray increase, decrease, or have no effect on this steady-state temperature? You do not need to compute the steady-state temperature – you only need to explain how it will be affected by ε_{sc} and clearly justify your answer.

14.58 A square metal conductor is cooled radiatively by an adjacent tube, as shown in the figure.

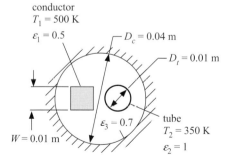

conductor
$T_1 = 500$ K
$\varepsilon_1 = 0.5$

$D_c = 0.04$ m
$D_t = 0.01$ m

$\varepsilon_3 = 0.7$

tube
$T_2 = 350$ K
$\varepsilon_2 = 1$

$W = 0.01$ m

The tube and the conductor run parallel to one another and are contained within a round conduit. For this problem, the surfaces will be labeled according to: 1 = conductor, 2 = tube, 3 = conduit. The temperature of the conductor surface is $T_1 = 500$ K and the conductor has a square cross-section with width $W = 0.01$ m. The emissivity of the surface of the conductor is $\varepsilon_1 = 0.5$. The diameter of cooled tube is $D_t = 0.01$ m. The surface of the tube is maintained at $T_2 = 350$ K by a flow of gas through the tube. The tube surface is black ($\varepsilon_2 = 1.0$). The diameter of the conduit is $D_c = 0.04$ m and the emissivity of the conduit surface is $\varepsilon_3 = 0.7$. The external surface of the conduit is insulated. The conduit is evacuated so that only radiation heat transfer occurs. The view factor between the conductor and the tube is $F_{1,2} = 0.2$ and the view factor from the conduit to itself is $F_{3,3} = 0.7$.

(a) Draw a resistance network that represents the problem. Clearly label each resistance and each node.

(b) Determine the view factor between the conduit and the tube, $F_{3,2}$.

For the remainder of this problem, assume that $F_{3,2} = 0.1$; note that this may or may not be the correct answer to part (b).

(c) Determine the rate of radiation heat transfer from the conductor per unit length (into the page).

(d) What is the temperature of the conduit?

(e) If the emissivity of the conductor surface (ε_1) were reduced, what effect would this have on the rate of radiation heat transfer from the conductor? Explain your answer.

(f) If the emissivity of the conduit surface (ε_3) were reduced, what effect would this have on the rate of radiation heat transfer from the conductor? Explain your answer.

The gas flowing in the tube has properties $Pr = 0.7$, $k = 0.02$ W/m-K, $\mu = 1 \times 10^{-5}$ Pa-s, and $\rho = 10$ kg/m^3. The mass flow rate of gas is $\dot{m} = 0.002$ kg/s. Assume that the pipe wall is thin and conductive and the internal surface of the pipe is smooth. Further, assume that the flow is fully developed.

(g) Determine the mean temperature of the gas in the flowing in the tube.

14.59 A thermocouple has a diameter $D_{tc} = 0.02$ m. The thermocouple is made of a material with density $\rho_{tc} = 8000$ kg/m^3 and specific heat capacity $c_{tc} = 450$ J/kg-K. The temperature of the thermocouple (you may assume that the thermocouple is at a uniform temperature) is

T_{tc} = 320 K and the emissivity of the thermo-couple's surface is ε_{tc} = 0.50. The thermocouple is placed between two very large (assume infinite in all directions) black plates. One plate is at T_1 = 300 K and the other is at T_2 = 500 K. The thermocouple is also exposed to a flow of air at T_a = 300 K. The heat transfer coefficient between the air and thermocouple is $\bar{h}$ = 50 W/m²-K.

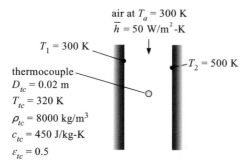

air at T_a = 300 K
$\bar{h}$ = 50 W/m²-K

T_1 = 300 K

T_2 = 500 K

thermocouple
D_{tc} = 0.02 m
T_{tc} = 320 K
ρ_{tc} = 8000 kg/m³
c_{tc} = 450 J/kg-K
ε_{tc} = 0.5

(a) What is the rate of convective heat transfer from the thermocouple?
(b) What is the net rate of radiative heat transfer to the thermocouple?
(c) What is the rate of temperature change of the thermocouple?
(d) If you want the thermocouple to accurately measure the temperature of the air (and therefore be unaffected by radiation), would you try to increase or decrease its emissivity? Justify your answer.

14.60 The Earth radiates to space, which has an effective temperature of about 4 K. However, the Earth is surrounded by an atmosphere consisting of gases that absorb radiation in specific wavelength bands. For this reason, the equivalent blackbody temperature of the sky is greater than 4 K but generally lower than the ambient temperature by 5°C to 30°C, depending on the extent of cloud cover and amount of moisture in the air. The largest difference between the ambient and equivalent blackbody sky temperature occurs during nights in which there is no cloud cover and low humidity. An important multi-mode heat transfer problem is related to determining the night-time temperature at which there is a danger that citrus fruit will freeze. Consider the following situation. During a clear calm night, an orange with diameter D = 6.5 cm experiences radiation heat

transfer with the sky and the ground as well as convection to the ambient air. The ground temperature is approximately T_{ground} = 10°C, regardless of the ambient temperature and is constant during the night. The equivalent blackbody temperature of the sky, T_{sky}, is ΔT_{sky} = 15°C lower than the ambient temperature, T_∞. The emissivity of the ground is ε_{ground} = 0.8 and the sky can be considered to be black. The emissivity of the orange is ε_{orange} = 0.5. Estimate the ambient temperature, T_∞, at which the orange will freeze; assume that the orange achieves a steady-state condition. Oranges consist of mostly water and therefore they freeze at about 0°C.

14.61 A spherical object is suspended between two walls, as shown in the figure.

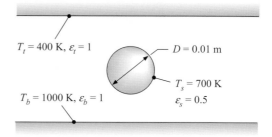

T_t = 400 K, ε_t = 1

D = 0.01 m

T_s = 700 K
ε_s = 0.5

T_b = 1000 K, ε_b = 1

The two walls are infinitely large (they extend to infinity in each direction) and black. The upper wall has a temperature of T_t = 400 K and the bottom wall has a temperature of T_b = 1000 K. The sphere has diameter D = 0.01 m and emissivity ε_s = 0.5. The sphere experiences only radiation (no convection) and has mass m = 0.005 kg and specific heat capacity c = 800 J/kg-K. At one point in time, the temperature of the sphere is T_s = 700 K.
(a) Draw a resistance network that represents this problem.
(b) Determine the net radiation heat transfer from the object when it has a temperature of T_s = 700 K.
(c) Determine the time rate of change of the object's temperature when it has a temperature of T_s = 700 K.
(d) Determine the steady-state temperature of the object (i.e., determine the value of T_s such that the time rate of change of temperature is zero).

(e) If the emissivity of the object (ε_s) was reduced, how would your answers from parts (c) and (d) change? Justify your answer.

14.62 Two spheres of the same diameter, $D = 0.1$ m, are placed close to each other in a vacuum environment. Sphere 1 is black and is heated such that its temperature is $T_1 = 700$ K. Sphere 2 has emissivity $\varepsilon_2 = 0.5$ and is electrically heated at a rate $\dot{q}_2 = 10$ W. The surroundings are also black and are at $T_3 = 300$ K. The view factor from sphere 1 to sphere 2 is $F_{1,2} = 0.07$. Determine the temperature of sphere 2.

14.63 Consider the radiation problem shown in the figure.

surroundings at $T_4 = 300$ K

sphere - surface 2
$T_2 = 200$ K, $\varepsilon_2 = 1$

sphere - surface 1
$T_1 = 400$ K, $\varepsilon_1 = 1$

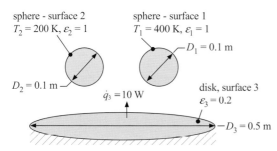

$D_1 = 0.1$ m

$D_2 = 0.1$ m

$\dot{q}_3 = 10$ W

disk, surface 3
$\varepsilon_3 = 0.2$

$D_3 = 0.5$ m

Two spheres are suspended above a disk. The spheres each have the same diameter ($D_1 = D_2 = 0.1$ m) and both are black ($\varepsilon_1 = \varepsilon_2 = 1$). The disk has diameter $D_3 = 0.5$ m and emissivity $\varepsilon_3 = 0.2$. The disk is heated such that the net radiation heat transfer from the disk is $\dot{q}_3 = 10$ W. The surroundings are at $T_4 = 300$ K. The view factor from sphere 1 to sphere 2 is $F_{1,2} = 0.2$. The view factor from sphere 1 to the disk and from sphere 2 to the disk are both the same: $F_{1,3} = F_{2,3} = 0.2$.
(a) Determine the view factor from the disk to the surroundings, $F_{3,4}$.
(b) Draw a resistance network that represents the radiation interactions associated with the disk (surface 3). Clearly show on your diagram what the boundary conditions are associated with each of the nodes.
(c) Determine the values of each of the resistances in your network from (b).
(d) Determine the temperature of the disk.

14.64 The figure illustrates a radiation problem.

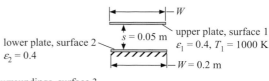

W

lower plate, surface 2
$\varepsilon_2 = 0.4$

$s = 0.05$ m

upper plate, surface 1
$\varepsilon_1 = 0.4$, $T_1 = 1000$ K

$W = 0.2$ m

surroundings, surface 3
$T_3 = 300$ K

An upper plate is heated to $T_1 = 1000$ K and has width $W = 0.2$ m. A bottom plate is insulated on its back side but exchanges radiation with the upper plate and the surroundings. The distance between the plates is $s = 0.05$ m. The bottom plate is centered with respect to the upper plate and has the same width, W. The problem is 2-D because both the heater and plate extend infinitely into the page; we will do the problem on a per unit length basis, $L = 1$ m. There are three surfaces involved in the problem: (1) the bottom surface of the upper plate, (2) the top surface of the lower plate, and (3) the surroundings. The surroundings are at $T_3 = 300$ K. The emissivity of both plates are the same, $\varepsilon_1 = \varepsilon_2 = 0.4$. The bottom plate is externally adiabatic (it is not being heated or cooled externally). Neglect any heat transfer mechanism other than radiation for this problem.
(a) Draw a resistance network that represents this radiation problem. Clearly label the resistances, the nodes, and indicate the boundary conditions.
(b) Determine the values of each of the resistances in your network from (a).
(c) Calculate the temperature of the lower plate.
(d) If the emissivity of the upper plate, ε_1, is decreased, what is the effect on the rate of heat transfer to the upper plate? Justify your answer.
(e) If the emissivity of the lower plate, ε_2, is decreased, what is the effect on the surface temperature of the lower plate? Justify your answer.

14.65 A spherical metal part with diameter $D = 4$ cm is placed in an industrial oven. The surface of the part has emissivity $\varepsilon = 0.5$ and experiences radiation with the walls of the oven and a heating element in the oven. Both the walls and the heating element can be modeled as black. The view factor between the part and the heating element is $F_{p,h} = 0.2$. The walls are at temperature $T_w = 50°C$ and the heating element is at temperature $T_h = 800°C$. At a certain time during the heating process the temperature of the surface of the part is $T_p = 200°C$.

(a) Determine the rate of radiation heat transfer to the part.

The part also experiences natural convection with the air in the oven at $T_\infty = 70°C$.

(b) Determine the rate of natural convection heat transfer from the part to the air.

The part is made of a material with density $\rho = 8000 \, kg/m^3$, $k = 20 \, W/m\text{-}K$, and $c = 200 \, J/kg\text{-}K$.

(c) Is a lumped capacitance approximation valid for this object? Justify your answer.

(d) Assume your answer from part (c) is yes (this may or may not be the correct answer). Determine the rate at which the temperature of the object is changing.

14.66 Two pipes run parallel to each other and parallel to a wall, as shown in the figure. Both pipes have diameter $D = 0.03$ m. We will do this problem on a per unit length basis, $L = 1$ m. The surface of pipe 1 is held at $T_1 = 250$ K and the pipe has an emissivity of $\varepsilon = 0.3$. The surface of pipe 2 is held at $T_2 = 400$ K

and pipe 2 is black. The wall has temperature $T_3 = 320$ K and is also black. The wall is very large relative to the pipes – you may assume that it extends infinitely in both directions. The surroundings are at $T_\infty = 300$ K. Pipe 1 experiences solar irradiation at a rate of $\dot{q}_s = 75$ W.

(a) Draw a resistance network that represents this radiation problem. Be sure to show where the solar irradiation appears in the network.

(b) The view factor from pipe 1 to pipe 2 is $F_{1,2} = 0.05$. Determine the view factor from pipe 1 to the wall, $F_{1,3}$.

(c) Determine the net rate of radiation heat transfer to pipe 1.

Pipe 1 experiences a very light breeze at $u_\infty = 2$ m/s and T_∞. Assume that the properties of air are $k = 0.026 \, W/m\text{-}K$, $\mu = 1.9 \times 10^{-5}$ Pa-s, $\rho = 1.1 \, kg/m^3$, $c = 1005 \, J/kg\text{-}K$, $\alpha = 2.5 \times 10^{-5} \, m^2/s$, $v = 1.7 \times 10^{-5} \, m^2/s$, $Pr = 0.7$, and $\beta = 0.0036 \, K^{-1}$.

(d) Determine a mixed convection heat transfer coefficient between the surface of pipe 1 and the surrounding air that accounts for both forced and natural convection effects.

(e) If the local heat transfer coefficient between the fluid that is flowing within pipe 1 and the internal surface of the pipe is $h_{in} = 86$ W/m²-K then what is the mean temperature of the fluid in the pipe? Assume that the pipe wall is negligibly thin.

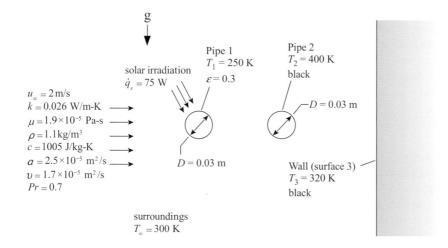

Figure for Problem 14.66.

14.67 A cylindrical heating element is used to heat a flow of water to an appliance. Typically, the element is exposed to water and therefore it is well cooled. However, you have been asked to assess the fire hazard associated with a scenario in which the appliance is suddenly drained (i.e., the water is removed) but the heat to the heating element is not deactivated. You want to determine the maximum temperature that the element will reach under this condition.

The heating element and passage wall are shown in the figure. The length of the element is $L = 9.0$ cm and its diameter is $D_e = 0.5$ cm. The element is concentric to a passage wall with diameter $D_o = 2.0$ cm. The emissivity of the element is $\varepsilon_e = 0.5$ and the emissivity of the passage wall is $\varepsilon_o = 0.9$. The surroundings are at $T_{sur} = 25°C$. The worst-case situation occurs if the outer passage wall is assumed to be insulated externally (i.e., there is no conduction or convection from the passage). The heating element dissipates $\dot{q}_e = 60$ W.

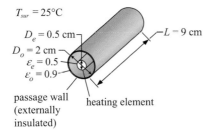

(a) What is the temperature of the element? Assume that radiation is the only important heat transfer for this problem. Note that your problem should include three surfaces (the element, the passage, and the surroundings); that is, you should not neglect the radiation exchange between the element and passage and the surroundings. However, you may assume that the ends of the element are adiabatic.

(b) What is the temperature of the passage wall?

(c) Other calculations have shown that the passage wall will not reach temperatures greater than 80°C because it is thermally communicating with surroundings. If the passage wall is maintained at $T_o = 80°C$ then what is the maximum temperature that the heating element will reach?

14.68 The figure illustrates a pipe that runs parallel to a plate.

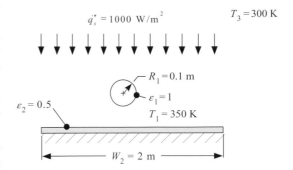

The problem is 2-D. The surface of the plate has an emissivity $\varepsilon_2 = 0.5$ and the width of the plate is $W_2 = 2$ m. The back side of the plate is insulated. The pipe has outer radius $R_1 = 0.1$ m and contains fluid that is being heated. The pipe surface is black ($\varepsilon_1 = 1$) and the temperature of the pipe surface is $T_1 = 350$ K. Both the plate and the pipe are exposed to a solar flux of $\dot{q}_s'' = 1000$ W/m^2 in a direction that is normal to the plate surface. The surroundings are at $T_3 = 300$ K. Model this problem using three surfaces: (1) = pipe, (2) = plate, and (3) = surroundings. The view factor between the pipe and the plate is $F_{1,2} = 0.45$. All surfaces are diffuse and gray.

(a) Determine the view factor between the pipe and the surroundings, $F_{1,3}$, and the view factor between the plate and the surroundings, $F_{2,3}$.

(b) Draw a resistance network that represents this radiation problem. Be sure to indicate the node that represents each surface as well as the heat transfer associated with the solar flux on the pipe, the heat transfer associated with the solar flux on the plate, and the heat transfer to the fluid in the pipe.

(c) Use your resistance network from (b) to determine the temperature of the plate.

(d) Determine the rate of heat transfer to the fluid per unit length.

(e) How do you think that changing the emissivity of the plate will affect your answer from (d)?

14.69 A satellite orbits the Earth at a height (above the Earth) of $H_{orbit} = 3.5 \times 10^5$ m. The satellite can be modeled as a sphere of diameter

D_{sat} = 1 m. The emissivity of the satellite surface is ε_{sat} = 0.5. There is an internal dissipation of energy within the satellite of $\dot{q}_{sat}$ = 100 W. The diameter of the Earth is D_{earth} = 1.29×10^7 m, the temperature of the Earth is T_{earth} = 300 K, and the surface of the Earth can be considered to be black. The temperature of space is T_{space} = 3 K. Depending on the position in the orbit, the solar flux ranges from G = 0 W/m^2 (on the night-side of Earth) to G = 1350 W/m^2 (on the day-side).

(a) Develop a model that can predict the temperature of the satellite as a function of the solar flux. What is the temperature of the satellite on the night-side of the Earth? What is the temperature on the day-side?

(b) Plot the temperature of the satellite as a function of G for various values of the satellite surface emissivity. Is there a value of the emissivity that keeps the satellite temperature within the acceptable range of 280 K < T_{sat} < 400 K during its entire orbit?

14.70 Three metal plates, each W = 40 cm by L = 60 cm, are parallel and centered as shown in the figure. Each of the plates have an emissivity of ε = 0.15. The top and bottom plates (surfaces 1 and 3) are separated by a vertical distance of H = 50 cm. The bottom and middle plates (surfaces 1 and 2) are separated by a vertical distance a. The temperature of the bottom plate is maintained at T_1 = 584°C. The plates radiatively interact with the surroundings at T_4 = 25°C. The underside of the bottom plate is insulated.

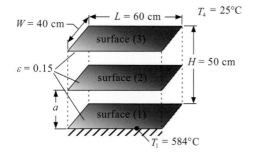

Calculate and plot the temperature of upper plate and the net rate of radiative heat transfer from the lower plate as a function of a for 1 cm < a < 49 cm.

14.71 A long rectangular duct has dimensions H = 1 m by W = 2 m, as shown in the figure. The bottom of the duct is maintained at T_1 = 1200 K, the top is maintained at T_2 = 1000 K and the sides are both at T_3 = 1100 K. A small spherical thermocouple is placed within the duct at a location that is a = 0.25 m from the bottom and b = 1 m from the side. The walls of the duct have an emissivity of ε_d. The thermocouple emissivity is ε_{tc} = 0.5. Assume radiation to be the only heat transfer mechanism.

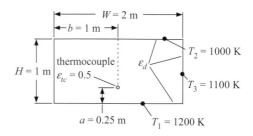

Plot the temperature that the thermocouple will measure as a function of the emissivity of the duct walls, ε_d.

14.72 A satellite orbits the Earth at a height (above the surface of the Earth) of H_{orbit} = 3.5×10^5 m. The diameter of the Earth is D_{earth} = 1.29×10^7 m and the temperature of the Earth is T_{earth} = 300 K. The distance between the Earth and the Sun is approximately R = 1.497×10^{11} m. The Sun has diameter D_{sun} = 1.39×10^9 m and the surface temperature of the Sun is approximately T_{sun} = 5780 K. The Earth and the Sun can be considered black. The satellite is spherical with diameter D_{sat} = 1 m. The emissivity of the satellite surface is ε_{sat} = 0.5. There is $\dot{q}_{sat}$ = 100 W of power dissipation within the satellite that must be rejected from its surface through radiation.

(a) Estimate the steady-state temperature of the satellite when it is on the day-side of the Earth.

(b) Estimate the steady-state temperature of the satellite when it is on the night-side of the Earth.

(c) Plot the steady-state day-side and night-side temperature as a function of the satellite emissivity. Explain the shape of your plot.

(d) Overlay on your plot from (c) the steady-state day-side and night-side temperature as a function of the satellite emissivity if the satellite power dissipation were $\dot{q}_{sat} = 0$ W. Explain the shape of your plot.

Projects

14.73 A braze oven is made by heating the bottom surface of a cylindrical enclosure as shown in the figure.

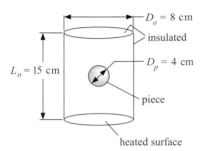

heated surface

The sides and top of the enclosure are insulated. The piece to be heated is suspended at the center of the oven. All surfaces are black. The diameter of the piece is $D_p = 4$ cm and the piece is made of material with density $\rho_p = 7500$ kg/m^3 and specific heat capacity $c_p = 510$ J/kg-K. The diameter of the oven is $D_o = 8$ cm and the length of the oven is $L_o = 15$ cm. The temperature of the heater and the piece is initially $T_{ini} = 300$ K. The heater temperature is ramped linearly from T_{ini} to $T_{max} = 1000$ K over $t_{ramp} = 10$ min.
  (a) Develop a numerical model using the Integral command in EES that can predict the temperature of the piece as a function of time for 30 minutes after the heater is activated.
  (b) Plot the temperature of the piece and the heater as a function of time.

14.74 You are working on an advanced detector for biological agents; the first step in the process is to ablate (i.e., vaporize) individual particles in an air stream so that their constituent molecules can be identified through mass spectrometry. There are various methods available for providing the energy to the particle required for ablation; for example, using multiple pulses of a high-power laser. You are analyzing a less-expensive technique for vaporization

that utilizes radiation energy. A very-high-temperature element is located at the bottom of a cylinder, as shown in the figure.

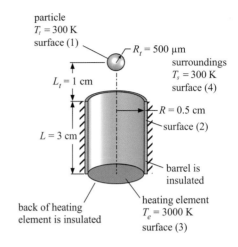

The length of the cylinder which is the "barrel" of the heat source is $L = 3.0$ cm and the radius of the cylinder and the heating element is $R = 0.5$ cm. The heating element is maintained at a very high temperature, $T_e = 3000$ K. The back side of the heating element and the external surfaces of the barrel of the heat source are insulated. The particle that is being ablated may be modeled as a sphere with radius $R_t = 500$ μm and is located $L_t = 1.0$ cm from the mouth of the barrel and is on the centerline of the barrel. The particle is at $T_t = 300$ K and the surroundings are at $T_s = 300$ K. All surfaces are black. For this problem, the particle is surface (1), the cylindrical barrel is surface (2), the disk-shaped heating element is surface (3), and the surroundings is surface (4).
  (a) Determine the areas of all surfaces and the view factors between each surface. This should result in an array A and matrix F that are both completely filled.
  (b) Determine the net radiation heat transferred to the target.
  (c) What is the efficiency of the ablation system (i.e., what is the ratio of the energy delivered to the particle to the energy required by the element)?
  (d) The particle has density $\rho_t = 7000$ kg/m^3 and specific heat capacity $c_t = 300$ J/kg-K. Use your radiation model as the basis of a

transient, lumped capacitance numerical model of the particle that can predict the temperature of the particle as a function of time. Assume that the particle is initially at $T_{t,in}$ = 300 K. Use the Integral function in EES and prepare a plot showing the particle temperature as a function of time.

14.75 Consider two parallel plates that are separated by a distance of H = 0.5 m. The plates are each L = 1 m by W = 2 m. The lower plate (surface 1) is maintained at T_1 = 400 K and has emissivity ε_1 = 0.4. The surroundings (surface 2) are at T_2 = 4 K. The upper plate has a temperature profile that varies linearly in the x-direction from T_C = 500 K at one edge (x = 0) to T_H = 1000 K at the other edge (x = W). The temperature is uniform in the y-direction. The emissivity of the upper plate is ε_3 = 0.6. This problem can be solved numerically by discretizing the upper plate into N equal-area segments, each at a constant temperature. Assume that the upper surface of the upper plate and the lower surface of the lower plate are both insulated.

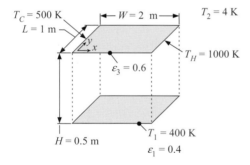

(a) Calculate the total energy that must be provided to the upper plate.
(b) Plot the total energy provided to the upper plate as a function of N for N = 1 to 10. From your results, how many segments do you believe are needed to represent the effect of the temperature distribution in the upper plate?

14.76 A photovoltaic panel having dimensions of 1 m × 2 m is oriented directly towards the Sun (i.e., south) at a 45° angle. The panel is exposed to solar radiation at 720 W/m². The efficiency of the panel, defined as the electrical power produced divided by the incident solar radiation, is

11.2 percent. The back side of the photovoltaic panel is well insulated. The emissivity of the photovoltaic material is estimated to be 0.90. The ambient and ground temperature during the test is 22°C and there is no measurable wind. The sky is clear and the equivalent temperature of the sky for radiation is 7°C. Estimate the steady-state surface temperature of the photovoltaic panel assuming that all of the radiation that strikes the panel is absorbed. What fraction of the thermal energy transfer to the air is due to radiation?

14.77 Droplets of molten metal are injected in a vacuum environment, as shown in the figure.

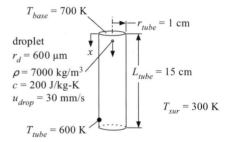

The droplets are injected at T_{ini} = 700 K into a tube that is maintained at T_{tube} = 600 K. The base of the tube at x = 0 is kept at T_{base} = 700 K. The droplets have radius r_d = 600 μm and have density ρ = 7000 kg/m³ and heat capacity c = 200 J/kg-K. The tube radius is r_{tube} = 1 cm and the tube length is L_{tube} = 15 cm. The surroundings are at T_{sur} = 300 K. The droplet velocity is constant and equal to u_{drop} = 30 mm/s. All surfaces are black and the droplet travels along the center line of the tube.

(a) Develop an EES function that can provide the view factor between the droplet and the tube as it moves from x = r_d to x = 2 L_{tube}. Plot the view factor between the droplet and the tube as a function of x. Explain the shape of the curve.
(b) Given an arbitrary value of the time relative to the start of the injection (t) and the droplet temperature (T_d), develop an EES model that can provide the rate of change of the droplet temperature.
(c) Use the EES Integral command to determine the temperature as a function of

position of the droplet. Plot the temperature as a function of position for $r_d < x < 2\,L_{tube}$.

(d) Overlay on your plot from (c) the solution for various values of T_{tube} ranging from 300 K to 700 K.

14.78 One method of storing hydrogen (for example, in your future hydrogen-fueled car) is as a liquid. Unfortunately, the temperature of hydrogen must be reduced to 20.3 K before it liquefies (at 1 atm pressure). It is difficult to provide sufficient insulation to prevent excessive boil-off of the liquid hydrogen due to parasitic heat leak from ambient. Typically, the liquid hydrogen is surrounded by a vacuum space in order to eliminate convection and conduction. However, radiation may still cause a substantial rate of boil-off.

It is not possible to completely eliminate heat leak to the hydrogen and it is very difficult to provide an on-board cryogenic refrigeration system capable of providing refrigeration at 20 K; such a cryocooler would need to be multi-stage and very expensive and complicated. On the other hand, a relatively simple and inexpensive, single-stage cryocooler can be used to provide refrigeration at 80 K or above. Clearly, such a cryocooler cannot re-liquefy any boil-off. However, it is possible to use such an 80 K cryocooler to actively cool a shroud that provides a guard against radiation to the hydrogen tank, as shown in the figure.

In this problem you will analyze the benefit of such an actively cooled shroud. Assume that the surroundings driving heat leak to the system are at 300 K and the surface associated with these surrounding is black. Further, assume that there are five layers of radiation shields that are very thin, shiny ($\varepsilon_s = 0.1$) sheets of aluminized mylar placed between the surroundings and the active shroud. The active shroud surface can be assumed to be isothermal at T_{shroud} – assume that the internal and external surfaces of the shroud are at the same temperature and are black. The shroud is cooled so that its temperature T_{shroud} can be controlled (within some limits). There are five additional radiation shields between the shroud and the surface of the hydrogen tank. Assume that the hydrogen tank's surface is black.

Additionally, assume that the surface area of each surface (each radiation shield, the shroud, etc.) is 0.35 m^2 (which is nominally consistent with a spherical, 5 gallon tank) and the view factor between each adjacent surface is 1.0. Assume that you have a cryocooler available to you that is capable of reaching 80 K with no applied load; the load temperature increases by 10 K for every 1 W of refrigeration applied to the cryocooler.

(a) Prepare a plot that shows the rate that heat must be removed from the shroud as a function of the shroud temperature. On

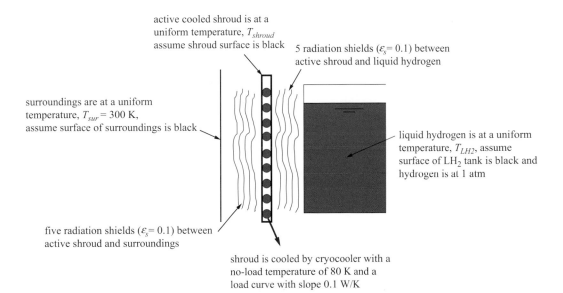

active cooled shroud is at a uniform temperature, T_{shroud}
assume shroud surface is black

5 radiation shields ($\varepsilon_s = 0.1$) between active shroud and liquid hydrogen

surroundings are at a uniform temperature, $T_{sur} = 300$ K, assume surface of surroundings is black

liquid hydrogen is at a uniform temperature, T_{LH2}, assume surface of LH$_2$ tank is black and hydrogen is at 1 atm

five radiation shields ($\varepsilon_s = 0.1$) between active shroud and surroundings

shroud is cooled by cryocooler with a no-load temperature of 80 K and a load curve with slope 0.1 W/K

Figure for problem 14.78.

the same plot, show the refrigeration capacity of the cryocooler as a function of temperature. Indicate on the plot the temperature of the shroud when it is actively cooled and the temperature of the shroud when it is not cooled.

(b) Prepare a plot showing the rate of hydrogen boil-off (in units of gallons of liquid/day) as a function of the shroud temperature. Using the results of (a), indicate the rate of boil-off for an actively cooled and uncooled shroud.

14.79 An infrared detector requires that a semiconductor such as mercury cadmium telluride be cooled to a low operating temperature. Research has pushed the operating temperature up substantially so that the cooling requirements on the semiconductor are not so strenuous. The figure illustrates a disk-shaped detector that is cooled to $T_d = 230$ K and placed within an optical cavity.

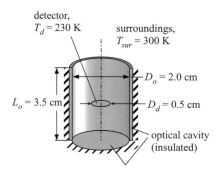

detector,
$T_d = 230$ K

surroundings,
$T_{sur} = 300$ K

$D_o = 2.0$ cm

$L_o = 3.5$ cm

$D_d = 0.5$ cm

optical cavity
(insulated)

The diameter of the detector is $D_d = 0.5$ cm and the detector is thin. The cavity has an inner diameter of $D_o = 2.0$ cm and a length of $L_o = 3.5$ cm. The detector is mounted at the midpoint of the cavity. One end of the cavity is closed and the other is open to the surroundings. The cavity surface (the sides and bottom) are black, adiabatic, and isothermal. The detector is also black and isothermal (but not adiabatic – it is being cooled by a cryocooler). The surroundings are at $T_{sur} = 300$ K. There are three surfaces in this problem: the surroundings (*sur*, model this surface as an imaginary surface that forms the top of the detector), the optical cavity (*o*, this includes both the sides and bottom of the detector), and the detector (*d*, the top and bottom of the disk).

(a) Draw a resistance network that represents this problem.
(b) Determine the values of each of the resistances in your network from (a).
(c) Determine the rate of cooling required by the detector in order to maintain its temperature at $T_d = 230$ K.
(d) Prepare a plot showing the cooling power required by the detector as a function of the detector temperature for 220 K $< T_d <$ 290 K.

It is difficult to integrate a cryocooler with the detector directly. Therefore, you are looking at the option of cooling the surfaces of the optical cavity in order to indirectly cool the detector. The detector itself is therefore adiabatic (i.e., the radiation incident on the detector from the surroundings is balanced by the radiation from the detector to the optical cavity) while the optical cavity is not.

(e) Determine the temperature that the optical cavity surfaces must be cooled to in order to maintain the detector at $T_d = 230$ K.
(f) What is the cooling required at the surface of the optical cavity in order to maintain the detector at $T_d = 230$ K?
(g) Overlay on your plot from (d) the cooling power required as a function of the detector temperature for this technique in which the optical cavity is cooled rather than the detector itself. You will probably need to use a log-scale in order for this plot to be very informative.

14.80 The figure illustrates a set of three reactor beds that are heated radiantly by three heating elements.

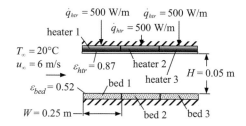

$\dot{q}_{htr} = 500$ W/m $\dot{q}_{htr} = 500$ W/m

$\dot{q}_{htr} = 500$ W/m

heater 1

$T_\infty = 20°C$
$u_\infty = 6$ m/s $\varepsilon_{htr} = 0.87$ heater 2

$H = 0.05$ m

heater 3

$\varepsilon_{bed} = 0.52$ bed 1

$W = 0.25$ m   bed 2   bed 3

The reactants are provided as a flow over the beds. The temperature of the reactant flow is $T_\infty = 20°C$ and the free stream velocity is $u_\infty = 6$ m/s; you may assume that the properties of the reactant flow are consistent with those of air at atmospheric pressure. Each of the heaters and beds are $W = 0.25$ m wide and very long (the problem is 2-D). The heaters and beds are separated by $H = 0.05$ m. The beds are insulated on their back sides but transfer heat to the free stream by convection. The surface of the beds has emissivity, $\varepsilon_{bed} = 0.52$. The heaters are each provided with $\dot{q}_{htr} = 500$ W/m; there is a piece of glass that protects the heaters from the reactants and prevents convective heat loss from the heaters. The upper surfaces of the heaters are insulated. You may assume that the three heaters and three beds are all isothermal (i.e., they are each at a unique but uniform temperature). The surface of the heaters has emissivity, $\varepsilon_{htr} = 0.87$. The surroundings are at $T_{sur} = 20°C$.

(a) Determine the temperature of each of the beds.
(b) What is the efficiency of the heating system?
(c) Determine the heater power that should be applied to each of the three heaters in order to keep each of the three beds at $T_{bed} = 65°C$.

14.81 The figure illustrates two identical metal plates that are inserted into the sides of a radiant oven.

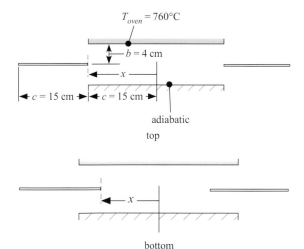

The plates are very thin and have very high conductivity; therefore, they can be assumed to be at a uniform temperature. The top wall of the oven is heated to $T_{oven} = 760°C$ and the bottom edge is adiabatic. The half-width of the oven is $c = 15$ cm and the plates are centered in the gap with a distance from the plate to the oven wall of $b = 4$ cm. The width of each plate is the same as the half-width of the oven. The

plates begin withdrawn – their edges are aligned with the edge of the oven (i.e., $x = c$ as shown in the top figure). The plates are very slowly inserted into the oven (i.e., x is reduced to zero – note that both plates move towards the center together), as shown in the bottom figure. Because the insertion process is slow, the plates may always be assumed to be at steady state. The problem is 2-D (the plates and oven extend a long way into the page). The temperature of the surroundings (and the temperature of any oven walls outside of the oven itself) is $T_{sur} = 20°C$. Assume that the only heat transfer mechanism for this problem is radiation. In your analysis, refer to the top oven wall as surface 1, the adiabatic oven wall as surface 2, the plates as surface 3, and the surroundings as surface 4.

(a) Determine the temperature of the plate and the temperature of the adiabatic wall when $x = c/2$ (i.e., the insertion process is 50 percent complete). Be sure to print out the arrays table containing the view factors between each surface.
(b) Plot the temperature of the plates and the temperature of the adiabatic wall as a function of x. Explain the shape of your plot.
(c) Plot the temperature of the plates as a function of x for various values of b. Explain the shape of your plot.

14.82 This problem considers a (fictitious) power generation system for a spacecraft orbiting the planet Mercury. The surface of Mercury can reach 700 K and therefore you are considering the possibility of collecting radiation emitted from Mercury in order to operate a heat engine. The details of the collector are shown schematically in the figure.

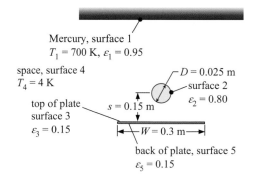

The collector geometry consists of a pipe and a backing plate; this geometry is 2-D, so the problem will be solved on a per unit length basis, $L = 1$ m, into the page. The diameter of the pipe is $D = 0.025$ m. The pipe surface (surface 2) is maintained at a constant temperature (T_2) and has emissivity $\varepsilon_2 = 0.8$. Energy that is transferred to the pipe is provided to the power generation system. The pipe is oriented so that it is parallel to the surface of the planet (surface 1) which is at $T_1 = 700$ K and has an emissivity of $\varepsilon_1 = 0.95$. You may assume that the surface of the planet extends infinitely in all directions. There is a back plate positioned $s = 0.15$ m away from the centerline of the collector pipe. The back plate is $W = 0.30$ m wide and is centered with respect to the pipe. The top surface of the back plate (the surface oriented towards the collector pipe, surface 3) has emissivity $\varepsilon_3 = 0.15$. The bottom surface of the back plate (the surface oriented towards space, surface 5) also has emissivity $\varepsilon_5 = 0.15$. The collector and back plate are surrounded by outer space, which has an effective temperature $T_4 = 4$ K; assume that the collector is shielded from the Sun. Assume that the back plate is isothermal.

(a) Prepare a plot showing the net rate of radiation heat transfer to the collector from Mercury as a function of the collector temperature, T_2.

The energy transferred to the collector pipe is provided to the hot end of a heat engine that operates between T_2 and $T_{radiator}$, where T_2 is the collector temperature and $T_{radiator}$ is the temperature of a radiator panel that is used to reject heat, as shown in the figure. The heat engine has a Second Law efficiency $\eta_2 = 0.30$; that is, the heat engine produces 30 percent of the power that a reversible heat engine would produce operating between the same temperature limits (T_2 and $T_{radiator}$). The heat engine radiator rejects heat to space; assume that the radiator panel has an emissivity $\varepsilon_{radiator} = 0.90$ and a surface area $A_{radiator} = 10$ m^2. Also, assume that the radiator only sees space at $T_4 = 4$ K.

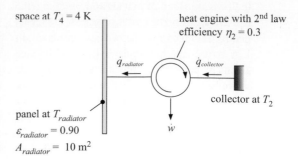

(b) Prepare a plot showing the amount of power generated by the heat engine ($\dot{w}$) and the radiator temperature as a function of the collector temperature, T_2.

14.83 The figure illustrates a 2-D model of the cavity receiver in a solar energy power plant.

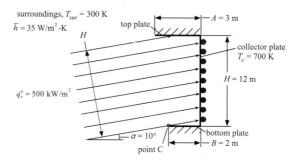

The receiver consists of a collector plate with height $H = 12$ m that is cooled by the flow of a molten salt through tubes mounted on its back side. You may assume that the collector plate temperature is uniform and equal to $T_c = 700$ K. The bottom plate has width $B = 2$ m and the top plate has width $A = 3$ m. Each of these surfaces are black and both convect and radiate to surroundings at $T_{sur} = 300$ K. The heat transfer coefficient is $\bar{h} = 35$ W/m^2-K. The cavity receiver is subjected to a radiation heat flux directed from a system of mirrors. You may model the heat flux as a collimated beam of radiation of width H and strength $\dot{q}_s'' = 500$ kW/m^2. Depending on the time of day, the beam of radiation may not be perfectly aligned with the collector plate. The beam is focused so that its edge always passes through the point labeled point C in the figure, but it can have an angle α that ranges from 0° (perfectly aligned)

to 20° (misaligned so that some of the radiation falls on the top plate rather than the collector plate). Assume a nominal value of $\alpha = 10°$.

Assume that the problem is 2-D (the plates extend a long way into the page). In your analysis, refer to the collector plate as surface 1, the bottom plate as surface 2, the top plate as surface 3, and the surroundings as surface 4. The bottom plate experiences only radiation and convection. The top plate experiences radiation and convection as well as some incident heat transfer from the beam of radiation (when $\alpha > 0$). The collector plate experiences radiation, convection, incident heat transfer from the beam of radiation, and heat transfer to the molten salt. The purpose of the collector is to transfer the energy associated with the beam of radiation to the molten salt. Therefore, the collector efficiency is defined as the ratio of the rate of heat transfer to the molten salt to the radiation heat transfer that is carried by the beam.

(a) Determine the view factors between each surface. Print out a 4 × 4 matrix containing these view factors.

(b) Determine the temperature of the top and bottom plate, the rate of heat transfer to the molten salt, and the collector efficiency.

(c) Plot the collector efficiency as a function of the collector plate temperature for various values of the heat transfer coefficient, ranging from 20 W/m²-K $< \bar{h} <$ 80 W/m²-K.

(d) Plot the collector efficiency as a function of the angle α.

14.84 The figure illustrates an air-cooled heating system.

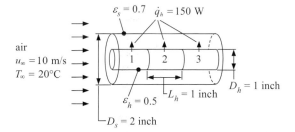

The heater consists of three rods in a row, each rod is isothermal (although the heaters may have different temperatures) and each heater

rod experiences ohmic heating at a rate of $\dot{q}_h = 150$ W. The heater rods have diameter $D_h = 1$ inch and length $L_h = 1$ inch. The ends of the rod may be considered to be adiabatic. The surface of the heater rods have emissivity $\varepsilon_h = 0.5$. There is a thin metallic shield that surrounds the rods. The shield can be assumed to be isothermal and is centered over the heaters. The length of the shield is $3L_h$ and the diameter is $D_s = 2$ inch. The surface of shield (inside and outside) has emissivity $\varepsilon_s = 0.7$. Air at $T_\infty = 20°$C is blown over the rods and the shield with velocity $u_\infty = 10$ m/s. The heaters and the shield radiate to surroundings at $T_{sur} = T_\infty$.

(a) What are the temperature of each of the heaters and the shield if convection is neglected?

(b) What are the temperature of each of the heaters and the shield if convection is included? You may model the convection on the heaters and shield as external flow over a flat plate.

(c) If the heaters cannot be independently controlled (i.e., the ohmic heating applied to each heater must be the same) and the heater temperature cannot go above 850°C then what is the maximum total ohmic heating that can be achieved (i.e., the sum of the heating applied to all three heaters)?

(d) Plot the maximum total ohmic heating as a function of the air velocity.

(e) Overlay on your plot from (d) the maximum ohmic heating that can be achieved if the heaters can be independently controlled so that the ohmic dissipation in each heater rod can be different.

14.85 The figure illustrates an annular heater used in an appliance to heat a flow of gas. The outer shell has an inner diameter of $D_{out} = 1.5$ cm and experiences a uniform heat flux due to electrical heating, $\dot{q}'' = 3500$ W/m². The emissivity of the inner surface of the outer shell is $\varepsilon_{out} = 0.85$. The outer shell is conductive and therefore achieves a uniform temperature. The external surface of the outer shell is insulated. The center rod in the annulus has an outer diameter of $D_{in} = 0.8$ cm and an emissivity of $\varepsilon_{in} = 0.75$. The length of the heater is $L = 5$ cm. The surroundings are at $T_{sur} = 20°$C. Normally,

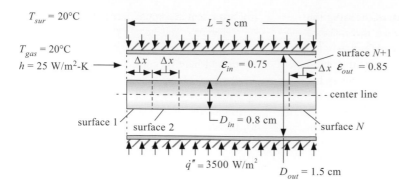

$T_{sur} = 20°C$

$T_{gas} = 20°C$
$h = 25$ W/m²-K

$L = 5$ cm

surface $N+1$

$\varepsilon_{in} = 0.75$

$\varepsilon_{out} = 0.85$

center line

surface 1

surface 2

$D_{in} = 0.8$ cm

surface N

$\dot{q}'' = 3500$ W/m²

$D_{out} = 1.5$ cm

Figure for Problem 14.85.

gas flows through the annular space. The bulk temperature of the gas is $T_{gas} = 20°C$ and the heat transfer coefficient between the gas and the surface of the duct (both the inner and outer surfaces) is $h = 25$ W/m²-K. You may assume that the bulk temperature of the gas does not change substantially as it passes through the duct and that the gas is transparent to radiation.

Because the inner cylinder is thin and non-conductive it will not achieve a uniform temperature. Instead, you should split the inner cylinder into N segments, each of equal length. Each segment interacts radiatively with the shell and the surroundings and convectively with the gas. You may neglect axial conduction along the cylinder. In your analysis, surfaces 1 through N should be the cylinder segments, surface $N+1$ should be the inner surface of the outer shell, and surface $N+2$ should be the surroundings. Solve the problem using EES.

(a) Setup the area vector and view factor matrix. Your program should be capable of dealing with an arbitrary number of segments (N) but you should print out your solution for $N = 5$.

(b) What is the temperature of the outer shell? Plot the temperature of the cylinder as a function of position (x – measured from the inlet to the duct) for a few values of N.

(c) Plot the efficiency of the heater (i.e., the ratio of the heat transferred to the fluid to the heat transferred from the heater) as a function of the length of the heater.

(d) One failure scenario that must be investigated is that the flow of gas is shut off (i.e., h approaches zero) but the heater

power remains on. What is the temperature of the outer surface in a failure such as this?

14.86 The figure illustrates a temperature sensor that is being used to measure the temperature of a flow of air in a duct.

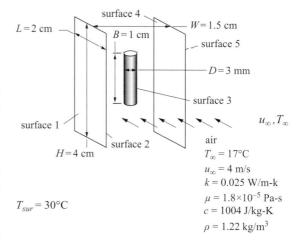

$L = 2$ cm

$B = 1$ cm

surface 4

$W = 1.5$ cm

surface 5

$D = 3$ mm

surface 3

surface 1

surface 2

$H = 4$ cm

air
$T_\infty = 17°C$
$u_\infty = 4$ m/s
$k = 0.025$ W/m-k
$\mu = 1.8 \times 10^{-5}$ Pa-s
$c = 1004$ J/kg-K
$\rho = 1.22$ kg/m³

u_∞, T_∞

$T_{sur} = 30°C$

The sensor is mounted in an environment where the surroundings are at a different temperature than the air stream. Therefore, the sensor is guarded by radiation shields on either side. The air flow over the sensor and the shields is at $T_\infty = 17°C$, $u_\infty = 4$ m/s and has properties $k = 0.025$ W/m-K, $\mu = 1.8 \times 10^{-5}$ Pa-s, $\rho = 1.22$ kg/m³, and $c = 1004$ J/kg-K. The surroundings are at $T_{sur} = 30°C$. The radiation shields are thin plates aligned with the direction of the flow. The length of the plates in the flow direction is $L = 2$ cm, the height of the plates is $H = 4$ cm, and the spacing between the plates is $W = 1.5$ cm. The temperature sensor is a cylinder that is located at the vertical and horizontal center of the plates and is centered between the plates. The diameter of the sensor is $D = 3$ mm and the length is $B = 1$ cm. The emissivity of the outer surfaces of both shields (surfaces 1 and 5 facing away from the sensor as shown in the figure) are $\varepsilon_{shield,out} = 0.25$. The emissivity of the inner surfaces of both shields (surfaces 2 and 4 facing towards the sensor as shown in the figure) is $\varepsilon_{shield,in} = 0.12$. The emissivity of the temperature sensor is $\varepsilon_{sensor} = 0.7$. You may model the flow over the plates as external flow over a flat plate and the flow over the cylinder as external flow over a cylinder. You may neglect radiation from the ends of the cylinder. Use the surface numbering scheme shown in the figure to solve the problem.

(a) Print out your emissivity vector, area vector, and view factor matrix.

(b) Determine the temperature of the shields and the temperature of the sensor. What is the measurement error (i.e., the difference between the temperature of the sensor and T_∞)?

(c) Plot the temperature sensor error as a function of the spacing between the plates, W. Explain the shape of this plot.

(d) Your plot from part (c) should show that the error is reduced as W is reduced. However, you cannot reduce the spacing to the point where the boundary layer growing on the shield interferes with the boundary layer growing on the sensor. Estimate the minimum allowable shield spacing.

(e) Plot the temperature sensor error as a function of the emissivity of the sensor. Explain the shape of your plot.

(f) Plot the temperature sensor error as a function of the free stream velocity, u_∞.

(g) Plot the temperature sensor air as a function of the emissivity of the inner surface of the shields, $\varepsilon_{shield,in}$. Explain the shape of your plot.

(h) In some installations it is possible that solar flux may be applied to the external surface of the left shield (i.e., surface 1). The solar flux may be as high as $G_s = 1000$ W/m^2. You are concerned that this solar flux may cause the temperature sensor error to increase. Modify your model so that it includes an additional irradiation on surface 1 related to the solar flux and plot the temperature sensor error as a function of G_s.

Appendix A Thermophysical Properties of Solids

This appendix provides the density, specific heat, thermal conductivity, and thermal diffusivity for selected solid substances at 25°C. The data were abstracted from the property database in EES, which provides data for these (and other) properties over a range of temperatures as well as data for many additional substances. References for the sources of these data can be found in the EES program. Separate tables are provided for metals (Table A.1) and nonmetallic substances (Table A.2).

Table A.1 Thermophysical properties of metals at 25°C.

Name	ρ (kg/m^3)	c (J/kg-K)	k (W/m-K)	$\alpha \times 10^6$ (m^2/s)
Aluminum	2699	900.9	235	96.6
Beryllium	1850	1814.4	201	60.0
Bismuth	9778	122.0	7.93	6.65
Boron	2460	1098.3	27.1	10.0
Brass	8528	378.7	79.3	24.5
Bronze	8802	384.5	79.9	23.6
Cadmium	8646	230.9	96.8	48.5
Carbon steel	7849	462.0	65.6	18.1
Chromium	7189	449.0	93.8	29.1
Cobalt	8898	425.3	100	26.5
Copper	8958	389.3	396	114
Germanium	5350	322.3	60.5	35.1
Gold	19288	129.2	80.5	32.3
Inconel 718	8192	449.9	9.77	2.65
Indium	7310	232.5	82.1	48.3
Invar	8055	530.2	13.8	3.23
Iridium	22558	130.0	147	50.1
Iron	7873	441.4	80.5	23.2
Lead	11335	128.9	35.3	24.2
Magnesium	1737	1022.3	156	87.9
Molybdenum	10279	250.5	138	53.6
Nichrome	8400	420.0	12.0	3.40
Nickel	8901	442.9	91.0	23.1
Niobium	8569	264.7	53.7	23.7
Palladium	12021	244.6	72.0	24.5
Platinum	21447	132.9	73.0	25.6
Rhenium	21018	135.8	48.0	16.8
Rhodium	12450	242.6	150	49.7
Silicon	2329	709.1	150	90.6
Silver	10487	234.8	429	174
Stainless steel 304	7997	475.6	14.9	3.91
Tantalum	16688	139.9	57.5	24.6

Table A.1 (cont.)

Name	ρ (kg/m^3)	c (J/kg-K)	k (W/m-K)	$\alpha \times 10^6$ (m^2/s)
Thorium	11697	117.9	54.0	39.2
Tin	7362	226.8	68.3	40.9
Titanium	4505	520.9	21.9	9.35
Tungsten	19249	131.8	174	68.7
Uranium	19096	115.9	27.9	12.6
Vanadium	5999	482.2	31.0	10.7
Zinc	7137	388.7	121	43.7
Zirconium	6519	277.7	22.7	12.6

Table A.2 Thermophysical properties of nonmetallic materials at 25°C.

Name	ρ (kg/m^3)	c (J/kg-K)	k (W/m-K)	$\alpha \times 10^6$ (m^2/s)
Asphalt	2115	920	0.062	0.0319
Bakelite	1297	1465	1.4	0.737
Brick	1920	835	0.72	0.449
Cement mortar	1860	780	0.72	0.496
Clay	1460	880	1.3	1.01
Coal (anthracite)	1350	1260	0.26	0.153
Concrete mix	2300	880	1.4	0.692
Cotton	80	1256	0.04	0.398
Fiberglass batts	28		0.038	
Fused silica	2220	731.8	1.376	0.847
G10	1950	992.7	0.6049	0.312
Glass (Pyrex)	2225	2139	1.153	0.242
Glass (soda lime)	2530	880	0.937	0.421
Naphthalene	1162	1344	0.4411	0.282
Paper	930	1400	0.05	0.0384
Paraffin wax	818	2926	0.24	0.100
Plywood	545	1215	0.12	0.181
Polystyrene	49.95	1181	0.02488	0.000422
Polystyrene (bead)	16		0.03975	
Polyurethane	48.96	1530	0.03538	0.472
Rock (granite)	2630	775	2.79	1.37
Rock (limestone)	2320	810	2.15	1.14
Rock (marble)	2680	830	2.8	1.26
Rubber	1190	1680	0.013	0.00650
Sand	1515	800	0.027	0.0223
Sheathing	290	1300	0.055	0.146
Silicon carbide	3210	1017	134.1	41.1
Soil	2050	1840	0.52	0.138
Teflon	2192	1028	0.2727	0.121

(cont.)

Table A.2 (cont.)

Name	ρ (kg/m^3)	c (J/kg-K)	k (W/m-K)	$\alpha \times 10^6$ (m^2/s)
Vermiculite	122		0.06763	
Wood (aspen)	447.9	1652	0.1194	0.161
Wood (balsa)	140	1652	0.04977	0.215
Wood (fir)	436.9	1652	0.1145	0.159
Wood (oak)	738.9	1652	0.1792	0.147
Wood (pine)	447.9	1652	0.1194	0.161

Appendix B Thermophysical Properties of Liquids

This appendix provides the density, specific heat, thermal conductivity, viscosity and Prandtl number for liquids and brine mixtures. The data were abstracted from the property database in EES, which provides data for these properties for a range of temperatures as well as data for additional substances. References for the sources of these data can be found in the EES program. Separate tables are provided for liquids (Table B.1) and brine mixtures (Table B.2). Note that the thermophysical properties of saturated fluids can be found in Appendix D.

Table B.1 Thermophysical properties of selected liquids at 25°C.				
Name	ρ (kg/m^3)	c (J/kg-K)	k (W/m-K)	$\mu \times 10^3$ (Pa-s)	Pr
Aniline	18,118	2072	0.1678	4.411	54.45
Benzyl alcohol	1040	2008	0.1599	5.855	73.54
Benzyl chloride	1097	1440	0.1356	1.142	12.13
Bromobenzene	1487	963	0.1101	1.093	9.560
Carbon tetrachloride	1584	850	0.1016	0.9056	7.576
Chlorobenzene	1101	1330	0.1267	0.7548	7.923
Chloroform	1480	955	0.1152	0.5420	4.493
Cycloheptane	806	1907	0.1243	1.446	22.18
Diesel#2	847.2	1934	0.1377	3.716	52.20
Diethylamine	701	2313	0.1348	0.3203	5.495
Diethylene glycol	1116	2310	0.2212	30.21	315.5
Dowtherm A	1056	1587	0.1379	4.316	49.66
Dowtherm G	1043	1563	0.1259	11.43	141.9
Dowtherm J	867	1846	0.1321	8.533	119.2
Engine oil 10W	885.2	1901	0.1442	133.0	1753
Engine oil 20W	885.2	1901	0.1442	287.1	3786
Ethyl acetate	894.5	1986	0.1415	0.4280	6.009
Ethylene glycol	1110	2392	0.2567	18.63	173.6
Ethylene diamine	893.1	2873	0.2526	1.923	21.88
Fluorinert FC40	1855	1053	0.0653	3.928	63.36
Fluorinert FC70	1936	1053	0.0698	26.95	406.8
Fluorobenzene	1019	1523	0.1259	0.5627	6.803
Formic Acid	1214	2160	0.2560	1.615	13.63
Iodobenzene	1672	775	0.1351	1.656	9.495
Isobutanol	796.9	2457	0.1335	3.512	64.63
Isopentanol	806.8	2387	0.1334	5.464	97.7878
Mercury	13,533	139	8.5169	1.532	0.02508
n-Butanol	805	2405	0.1490	2.844	45.90
n-Hexanol	815.6	2367	0.1488	4.813	76.57
n-Octanol	823	2340	0.1524	8.037	123.4
n-Pentanol	810.6	2378	0.1482	3.726	59.77
n-Propanol	799.1	2410	0.1519	2.125	33.72

(cont.)

Table B.I (cont.)

Name	ρ (kg/m^3)	c (J/kg-K)	k (W/m-K)	$\mu \times 10^3$ (Pa-s)	Pr
n-Propyl amine	711.9	2777	0.1730	0.3887	6.242
Pyridine	978.6	1680	0.1600	0.8814	9.255
Tetrahydrofuran	880	1720	0.1200	0.4660	6.679
Therminol 59	971.5	1695	0.1210	6.004	84.05
Therminol 66	1004	1575	0.1174	92.35	1239
Therminol VP1	1060	1559	0.1358	3.771	43.28
Triethylamine	723.2	2175	0.1178	0.3528	6.515

Table B.2 Thermophysical properties of brine mixtures at 25°C. The percentage provided in the name is the percent of the substance on a weight basis in a mixture with water.

Name	ρ (kg/m^3)	c (J/kg-K)	k (W/m-K)	$\mu \times 10^3$ (Pa-s)	Pr
Ethyl alcohol (5%)	987.9	4225	0.5696	1.088	8.068
Ethyl alcohol (10%)	980.4	4285	0.5348	1.319	10.57
Ethyl alcohol (20%)	966.6	4321	0.4709	1.821	16.71
Ethyl alcohol (30%)	950.9	4227	0.4140	2.210	22.56
Ethyl alcohol (40%)	931.6	4056	0.3633	2.388	26.65
Ethyl alcohol (50%)	909.6	3861	0.3184	2.395	29.04
Ethylene glycol (5%)	1003	4115	0.5835	0.9939	7.009
Ethylene glycol (10%)	1009	4049	0.5596	1.121	8.111
Ethylene glycol (20%)	1022	3905	0.5132	1.450	11.03
Ethylene glycol (30%)	1036	3733	0.4695	1.875	14.91
Ethylene glycol (40%)	1049	3539	0.4291	2.424	19.99
Ethylene glycol (50%)	1062	3338	0.3922	3.156	26.86
Lithium chloride (5%)	1026	3916	0.5933	1.061	7.001
Lithium chloride (10%)	1054	3680	0.5811	1.263	7.999
Lithium chloride (20%)	1114	3287	0.5592	1.898	11.16
Sodium chloride (5%)	1032	3933	0.6018	0.9658	6.311
Sodium chloride (10%)	1069	3729	0.5967	1.065	6.658
Sodium chloride (20%)	1145	3415	0.5858	1.369	7.980
Ammonia (5%)	975.5	4186	0.5695	0.9735	7.154
Ammonia (10%)	955.8	4192	0.5392	1.001	7.780
Ammonia (20%)	920.3	4207	0.4876	1.113	9.601
Ammonia (30%)	888.6	4234	0.4554	1.144	10.64
Propylene glycol (5%)	1001	4129	0.5785	1.064	7.594
Propylene glycol (10%)	1005	4082	0.5504	1.252	9.286
Propylene glycol (20%)	1013	3987	0.4973	1.739	13.94
Propylene glycol (30%)	1021	3870	0.4485	2.480	21.41
Propylene glycol (40%)	1029	3723	0.4035	3.580	33.04
Propylene glycol (50%)	1036	3549	0.3619	5.120	50.21

Appendix C Thermophysical Properties of Gases

This appendix provides thermophysical data for selected gases at atmospheric pressure. The data were abstracted from the property database in EES, which provides data for these properties for a range of temperatures as well as data for additional substances. References for the sources of these data can be found in the EES program. The data provided are the molar mass (MW), density (ρ), constant pressure specific heat (c_p), thermal conductivity (k), dynamic viscosity (μ), Prandtl number (Pr), thermal diffusivity (α), and kinematic viscosity (v) as a function of temperature. Note that the volumetric coefficient of thermal expansion (β) can be estimated as the inverse of the absolute temperature for these gases at atmospheric pressure. Also note that:

$$R = \frac{R_{univ}}{MW} \quad \text{and} \quad c_v = c_p - R \quad \text{where} \quad R_{univ} = 8314 \text{ J/kmol-K}.$$

Table C.1 Thermophysical properties of air ($MW = 28.965$) at atmospheric pressure.

T (K)	ρ (kg/m^3)	c_p (J/kg-K)	$k \times 10^3$ (W/m-K)	$\mu \times 10^6$ (Pa-s)	Pr	$\alpha \times 10^6$ (m^2/s)	$v \times 10^6$ (m^2/s)
100	3.605	1041	9.471	7.107	0.7808	2.525	1.971
200	1.769	1007	18.50	13.33	0.7257	10.39	7.538
300	1.177	1007	26.38	18.54	0.7072	22.27	15.75
400	0.8821	1014	33.45	23.06	0.6991	37.39	26.14
500	0.7056	1030	39.94	27.09	0.6986	54.96	38.39
750	0.4704	1087	54.53	35.80	0.7137	106.6	76.10
1000	0.3528	1141	67.68	43.28	0.7298	168.1	122.7
1500	0.2352	1211	91.78	56.33	0.7434	322.1	239.5
2000	0.1764	1250	114.5	68.07	0.7435	519.0	385.8

Table C.2 Thermophysical properties of ammonia ($MW = 17.031$) at atmospheric pressure.

T (K)	ρ (kg/m^3)	c_p (J/kg-K)	$k \times 10^3$ (W/m-K)	$\mu \times 10^6$ (Pa-s)	Pr	$\alpha \times 10^6$ (m^2/s)	$v \times 10^6$ (m^2/s)
250	0.8492	2242	21.47	8.393	0.8764	11.28	9.884
300	0.6990	2165	25.10	10.16	0.8762	16.59	14.54
350	0.5964	2205	30.47	12.04	0.8715	23.16	20.19
400	0.5207	2283	37.22	13.97	0.8572	31.30	26.83
450	0.4622	2378	44.93	15.92	0.8428	40.87	34.44
500	0.4157	2483	53.12	17.86	0.8349	51.47	42.97
550	0.3777	2593	61.22	19.79	0.8379	62.51	52.38
600	0.3462	2706	68.61	21.68	0.8552	73.24	62.63
650	0.3195	2821	74.56	23.55	0.8910	82.73	73.72

Table C.3 Thermophysical properties of carbon dioxide (*MW* = 44.010) at atmospheric pressure.

T (K)	ρ (kg/m³)	c_p (J/kg-K)	$k \times 10^3$ (W/m-K)	$\mu \times 10^6$ (Pa-s)	Pr	$\alpha \times 10^6$ (m²/s)	$\nu \times 10^6$ (m²/s)
250	2.165	805.2	12.94	12.56	0.7819	7.422	5.804
300	1.797	852.6	16.77	15.02	0.7635	10.95	8.361
350	1.537	899.1	20.92	17.40	0.7480	15.14	11.32
400	1.343	941.8	25.15	19.70	0.7374	19.88	14.66
450	1.193	980.4	29.38	21.90	0.7309	25.11	18.35
500	1.073	1015	33.54	24.02	0.7272	30.77	22.37
550	0.9756	1047	37.61	26.05	0.7253	36.81	26.70
600	0.8941	1076	41.59	28.00	0.7244	43.22	31.31
650	0.8252	1103	45.49	29.87	0.7242	49.99	36.20

Table C.4 Thermophysical properties of helium (*MW* = 4.003) at atmospheric pressure

T (K)	ρ (kg/m³)	c_p (J/kg-K)	$k \times 10^3$ (W/m-K)	$\mu \times 10^6$ (Pa-s)	Pr	$\alpha \times 10^6$ (m²/s)	$\nu \times 10^6$ (m²/s)
100	0.4871	5194	73.71	9.778	0.6890	29.14	20.08
200	0.2437	5193	118.0	15.14	0.6665	93.22	62.13
300	0.1625	5193	156.0	19.93	0.6635	184.8	122.6
400	0.1219	5193	190.4	24.29	0.6627	300.7	199.3
500	0.09753	5193	222.3	28.36	0.6626	438.9	290.8
600	0.08128	5193	252.4	32.22	0.6628	598.0	396.3
700	0.06967	5193	281.1	35.89	0.6632	776.8	515.2
800	0.06096	5193	308.5	39.43	0.6637	974.5	646.8
900	0.05419	5193	335.0	42.85	0.6642	1190	790.7

Table C.5 Thermophysical properties of hydrogen (*MW* = 2.016) at atmospheric pressure.

T (K)	ρ (kg/m³)	c_p (J/kg-K)	$k \times 10^3$ (W/m-K)	$\mu \times 10^6$ (Pa-s)	Pr	$\alpha \times 10^6$ (m²/s)	$\nu \times 10^6$ (m²/s)
100	0.2457	11230	67.29	4.039	0.6739	24.39	16.44
200	0.1228	13540	124.6	6.776	0.7362	74.99	55.20
300	0.08184	14310	177.9	9.093	0.7315	151.9	111.1
400	0.06139	14480	227.1	11.07	0.7058	255.5	180.3
500	0.04912	14510	272.1	12.79	0.6818	381.8	260.3
600	0.04093	14550	313.1	14.32	0.6657	525.7	350.0
700	0.03509	14610	349.9	15.77	0.6586	682.5	449.4
800	0.03070	14710	382.6	17.20	0.6612	847.3	560.2
900	0.02729	14840	411.2	18.70	0.6746	1016	685.2

Table C.6 Thermophysical properties of oxygen (MW = 31.999) at atmospheric pressure.

T (K)	ρ (kg/m^3)	c_p (J/kg-K)	$k \times 10^3$ (W/m-K)	$\mu \times 10^6$ (Pa-s)	Pr	$\alpha \times 10^6$ (m^2/s)	$v \times 10^6$ (m^2/s)
100	3.995	935.4	9.087	7.712	0.7939	2.432	1.931
200	1.956	914.3	18.24	14.72	0.7378	10.20	7.525
300	1.301	919.3	26.49	20.65	0.7168	22.15	15.88
400	0.9749	940.6	34.03	25.84	0.7143	37.11	26.50
500	0.7798	970.5	41.05	30.49	0.7208	54.24	39.10
600	0.6498	1001	47.66	34.73	0.7293	73.29	53.44
700	0.5569	1028	53.97	38.65	0.7362	94.27	69.40
800	0.4873	1050	60.02	42.33	0.7408	117.3	86.86
900	0.4332	1069	65.87	45.81	0.7431	142.3	105.7

Table C.7 Thermophysical properties of nitrogen (MW = 28.013) at atmospheric pressure.

T (K)	ρ (kg/m^3)	c_p (J/kg-K)	$k \times 10^3$ (W/m-K)	$\mu \times 10^6$ (Pa-s)	Pr	$\alpha \times 10^6$ (m^2/s)	$v \times 10^6$ (m^2/s)
100	3.483	1072	9.382	6.959	0.7950	2.513	100
200	1.711	1044	18.28	12.91	0.7370	10.24	200
300	1.138	1041	25.97	17.89	0.7174	21.91	300
400	0.8532	1045	32.81	22.21	0.7074	36.80	400
500	0.6825	1056	39.04	26.06	0.7052	54.15	500
600	0.5687	1075	44.84	29.58	0.7092	73.33	600
700	0.4875	1098	50.31	32.83	0.7167	93.98	700
800	0.4266	1122	55.51	35.89	0.7255	116.0	800
900	0.3792	1146	60.52	38.78	0.7342	139.3	900

Table C.8 Thermophysical properties of water vapor (MW = 18.015) at atmospheric pressure.

T (K)	ρ (kg/m^3)	c_p (J/kg-K)	$k \times 10^3$ (W/m-K)	$\mu \times 10^6$ (Pa-s)	Pr	$\alpha \times 10^6$ (m^2/s)	$v \times 10^6$ (m^2/s)
375	0.5944	2072	25.22	12.34	1.013	20.48	20.76
400	0.5549	2009	27.02	13.28	0.9878	24.23	23.94
425	0.5209	1984	29.02	14.25	0.9749	28.07	27.37
450	0.4910	1976	31.17	15.25	0.9663	32.13	31.05
475	0.4646	1976	33.46	16.25	0.9597	36.45	34.98
500	0.4409	1982	35.86	17.27	0.9542	41.05	39.17
525	0.4196	1990	38.37	18.30	0.9492	45.94	43.60
550	0.4003	2001	40.96	19.33	0.9445	51.12	48.28
575	0.3827	2014	43.63	20.37	0.9400	56.61	53.21
600	0.3667	2027	46.37	21.41	0.9358	62.39	58.38
625	0.3519	2041	49.18	22.45	0.9317	68.47	63.79
650	0.3383	2056	52.05	23.49	0.9277	74.85	69.44
675	0.3257	2071	54.98	24.53	0.9239	81.51	75.31
700	0.3140	2087	57.96	25.56	0.9203	88.47	81.42

Appendix D Thermophysical Properties of Saturated Liquids

This appendix provides thermophysical data for saturated liquids and vapors. Table D.1 provides vapor pressure (P_{sat}), density (ρ_l), specific heat (c_l), thermal conductivity (k_l), dynamic viscosity (μ_l), Prandtl number (Pr_l), enthalpy of vaporization (Δi_{vap}), surface tension (σ), and volumetric coefficient of thermal expansion (β_l) for saturated liquid water, ammonia, nitrogen, R134a, and propane for a range of temperatures. Table D.2 provides vapor pressure (P_{sat}), density (ρ_v), specific heat ($c_{p,v}$), thermal conductivity (k_v), dynamic viscosity (μ_v), Prandtl number (Pr_v), and volumetric coefficient of thermal expansion (β_v) for the same fluids in saturated vapor form. The data were abstracted from the property database in EES, which provides data for these properties for a range of temperatures as well as data for additional substances. References for the sources of these data can be found in the EES program.

Table D.1 Thermophysical properties of saturated liquids.

Water (MW = 18.02)

T (K)	P_{sat} (kPa)	ρ_l (kg/m^3)	$c_{p,l}$ (J/kg-K)	k_l (W/m-K)	$\mu_l \times 10^3$ (Pa-s)	Pr_l	Δi_{vap} (kJ/kg)	$\sigma \times 10^3$ (N/m)	$\beta_l \times 10^3$ (K^{-1})
273.15	0.6112	999.8	4220	0.5563	1.792	13.59	2501	75.65	−0.06817
283.15	1.228	999.7	4196	0.5798	1.306	9.451	2477	74.22	0.08768
293.15	2.339	998.2	4184	0.5994	1.002	6.993	2454	72.73	0.2066
300	3.537	996.5	4181	0.6111	0.8538	5.841	2437	71.68	0.2747
310	6.231	993.3	4180	0.6264	0.6935	4.628	2414	70.10	0.3608
320	10.55	989.4	4181	0.6396	0.5770	3.772	2390	68.47	0.4359
330	17.21	984.7	4184	0.6510	0.4895	3.146	2365	66.78	0.5033
340	27.19	979.5	4188	0.6608	0.4220	2.675	2341	65.04	0.5653
350	41.68	973.7	4195	0.6691	0.3688	2.312	2316	63.24	0.6236
360	62.19	967.4	4202	0.6759	0.3261	2.027	2291	61.40	0.6794
380	128.9	953.3	4224	0.6856	0.2627	1.619	2238	57.58	0.7873
400	245.8	937.5	4256	0.6902	0.2186	1.348	2183	53.57	0.895
420	437.3	919.9	4299	0.6902	0.1867	1.163	2124	49.41	1.008
440	733.7	900.7	4357	0.6860	0.1628	1.034	2059	45.09	1.132
460	1171	879.6	4433	0.6779	0.1443	0.9438	1989	40.65	1.273
480	1790	856.6	4533	0.6660	0.1296	0.8823	1913	36.10	1.439
500	2639	831.4	4663	0.6506	0.1177	0.8433	1827	31.47	1.644
520	3769	803.6	4837	0.6317	0.1076	0.8237	1731	26.78	1.908
540	5237	772.8	5075	0.6093	0.09882	0.8232	1623	22.07	2.263
560	7106	738.0	5421	0.5830	0.09089	0.8450	1498	17.39	2.778
580	9448	697.8	5964	0.5525	0.08338	0.9001	1352	12.79	3.599
600	12340	649.5	6948	0.5162	0.07579	1.020	1173	8.370	5.133

Table D.1 (cont.)

Ammonia (MW = 17.03)

T (K)	P_{sat} (kPa)	ρ_l (kg/m³)	$c_{p,l}$ (J/kg-K)	k_l (W/m-K)	$\mu_l \times 10^3$ (Pa-s)	Pr_l	Δi_{vap} (kJ/kg)	$\sigma \times 10^3$ (N/m)	$\beta_l \times 10^3$ (K⁻¹)
225	45.49	699.7	4372	0.7154	0.3186	1.947	1411	37.53	1.692
250	164.9	669.1	4499	0.6320	0.2230	1.587	1339	31.74	1.915
275	460.1	636.1	4626	0.5538	0.1669	1.394	1256	25.86	2.184
300	1062	600.1	4798	0.4804	0.1294	1.293	1158	20.06	2.574
325	2133	559.6	5093	0.4109	0.1020	1.264	1041	14.48	3.226
350	3864	512.2	5676	0.3437	0.08037	1.327	895.1	9.229	4.51
400	10300	344.1	23097	0.1870	0.04173	5.153	347.3	0.5505	49.62

Nitrogen (MW = 28.01)

T (K)	P_{sat} (kPa)	ρ_l (kg/m³)	$c_{p,l}$ (J/kg-K)	k_l (W/m-K)	$\mu_l \times 10^3$ (Pa-s)	Pr_l	Δi_{vap} (kJ/kg)	$\sigma \times 10^3$ (N/m)	$\beta_l \times 10^3$ (K⁻¹)
65	17.41	859.6	2003	0.1695	0.2821	3.334	213.6	11.76	4.831
70	38.55	838.5	2014	0.1595	0.2202	2.782	208.1	10.58	5.129
75	76.04	816.7	2031	0.1495	0.1767	2.402	202.1	9.416	5.981
80	136.9	793.9	2056	0.1395	0.1451	2.138	195.7	8.284	5.907
90	360.5	745.0	2141	0.1198	0.1028	1.837	180.5	6.113	7.113
100	778.3	689.4	2318	0.1001	0.07576	1.754	161.0	4.086	9.269
110	1466	621.5	2743	0.0804	0.05599	1.910	134.3	2.244	14.21
120	2511	523.4	4503	0.0610	0.03843	2.837	92.05	0.6774	35.87

R134a (MW = 102.0)

T (K)	P_{sat} (kPa)	ρ_l (kg/m³)	$c_{p,l}$ (J/kg-K)	k_l (W/m-K)	$\mu_l \times 10^3$ (Pa-s)	Pr_l	Δi_{vap} (kJ/kg)	$\sigma \times 10^3$ (N/m)	$\beta_l \times 10^3$ (K⁻¹)
240	72.53	1398	1267	0.1077	0.4186	4.925	221.5	16.25	2.113
260	177.0	1337	1308	0.1001	0.3153	4.121	208.2	13.30	2.361
280	373.0	1272	1361	0.0916	0.2435	3.615	193.3	10.47	2.726
300	703.2	1200	1432	0.0823	0.1900	3.305	176.1	7.786	3.303
320	1217	1117	1543	0.0721	0.1476	3.158	155.5	5.272	4.326
340	1973	1015	1751	0.0607	0.1118	3.228	129.3	2.977	6.566
360	3043	870.3	2435	0.0470	0.07822	4.051	90.91	0.9989	15.36

Propane (MW = 44.1)

T (K)	P_{sat} (kPa)	ρ_l (kg/m³)	$c_{p,l}$ (J/kg-K)	k_l (W/m-K)	$\mu_l \times 10^3$ (Pa-s)	Pr_l	Δi_{vap} (kJ/kg)	$\sigma \times 10^3$ (N/m)	$\beta_l \times 10^3$ (K⁻¹)
240	147.9	570.9	2294	0.1273	0.1838	3.314	416.1	14.23	2.104
260	310.7	546.2	2415	0.1155	0.1478	3.090	392.3	11.63	2.391
280	581.8	519.2	2564	0.1043	0.1195	2.936	364.7	9.104	2.803
300	997.8	489.1	2757	0.0940	0.09651	2.829	332.0	6.692	3.435
320	1599	454.6	3039	0.0846	0.07715	2.772	291.6	4.420	4.551
340	2431	411.8	3587	0.0756	0.05984	2.840	238.6	2.345	7.234
360	3556	346.0	5979	0.0667	0.04180	3.748	154.4	0.5896	22.22

Table D.2 Thermophysical properties of saturated vapors.

Water ($MW = 18.02$)

T (K)	P_{sat} (kPa)	ρ_v (kg/m³)	$c_{p,v}$ (J/kg-K)	$k_v \times 10^3$ (W/m-K)	$\mu_v \times 10^6$ (Pa-s)	Pr_v	$\beta_v \times 10^3$ (K⁻¹)
273.15	0.6112	0.004851	1884	16.76	9.216	1.036	3.681
283.15	1.228	0.009406	1895	17.41	9.461	1.029	3.559
293.15	2.339	0.01731	1906	18.09	9.727	1.025	3.447
300	3.537	0.02559	1914	18.56	9.920	1.023	3.376
310	6.231	0.04366	1927	19.28	10.21	1.021	3.281
320	10.55	0.07166	1942	20.02	10.52	1.020	3.195
330	17.21	0.1136	1959	20.80	10.83	1.020	3.119
340	27.19	0.1744	1979	21.61	11.16	1.022	3.052
350	41.68	0.2603	2003	22.46	11.49	1.025	2.994
360	62.19	0.3785	2033	23.35	11.82	1.029	2.948
380	128.9	0.7482	2110	25.30	12.50	1.043	2.886
400	245.8	1.369	2218	27.50	13.19	1.064	2.873
420	437.3	2.352	2366	30.04	13.88	1.094	2.914
440	733.7	3.833	2560	32.97	14.57	1.131	3.014
460	1171	5.983	2801	36.38	15.26	1.175	3.181
480	1790	9.014	3098	40.37	15.95	1.224	3.428
500	2639	13.20	3463	45.06	16.65	1.280	3.778
520	3769	18.90	3926	50.64	17.38	1.347	4.274
540	5237	26.63	4540	57.38	18.15	1.436	4.993
560	7106	37.15	5410	65.77	19.01	1.564	6.091
580	9448	51.74	6761	76.71	20.02	1.765	7.905
600	12340	72.84	9181	92.12	21.35	2.128	11.35

Ammonia ($MW = 17.03$)

T (K)	P_{sat} (kPa)	ρ_v (kg/m³)	$c_{p,v}$ (J/kg-K)	$k_v \times 10^3$ (W/m-K)	$\mu_v \times 10^6$ (Pa-s)	Pr_v	$\beta_v \times 10^3$ (K⁻¹)
225	45.49	0.4209	2189	20.30	7.625	0.8220	4.800
250	164.9	1.404	2391	21.56	8.355	0.9268	4.729
275	460.1	3.691	2708	23.54	9.113	1.048	4.956
300	1062	8.253	3177	26.41	9.894	1.190	5.561
325	2133	16.57	3887	30.49	10.74	1.369	6.774
350	3864	31.32	5122	36.56	11.79	1.652	9.325
400	10300	130.8	34462	74.69	18.51	8.539	85.40

Nitrogen ($MW = 28.01$)

T (K)	P_{sat} (kPa)	ρ_v (kg/m³)	$c_{p,v}$ (J/kg-K)	$k_v \times 10^3$ (W/m-K)	$\mu_v \times 10^6$ (Pa-s)	Pr_v	$\beta_v \times 10^3$ (K⁻¹)
65	17.41	0.9132	1063	5.816	4.512	0.8251	15.96
70	38.55	1.896	1082	6.352	4.883	0.8316	15.25
75	76.04	3.54	1108	6.906	5.262	0.8442	14.83
80	136.9	6.089	1145	7.488	5.652	0.8642	14.72
90	360.5	15.08	1266	8.789	6.482	0.9333	15.53
100	778.3	31.96	1503	10.44	7.429	1.070	18.42
110	1466	62.58	2062	12.80	8.626	1.389	26.46
120	2511	125.1	4631	16.99	10.62	2.895	64.57

Table D.2 (cont.)

R134a ($MW = 102.0$)

240	72.53	3.839	770.6	8.813	9.603	0.8397	4.797
260	177.0	8.912	841.8	10.81	10.39	0.8092	4.960
280	373.0	18.24	929.8	12.77	11.19	0.8151	5.423
300	703.2	34.22	1044	14.75	12.05	0.8531	6.338
320	1217	60.76	1211	16.85	13.05	0.9387	8.133
340	1973	105.9	1525	19.28	14.47	1.145	12.30
360	3043	194.1	2622	22.75	17.33	1.997	28.88

Propane ($MW = 44.1$)

240	147.9	3.439	1519	12.39	6.733	0.8255	4.964
260	310.7	6.909	1670	14.34	7.355	0.8562	5.163
280	581.8	12.63	1855	16.57	8.036	0.8997	5.655
300	997.8	21.65	2098	19.24	8.815	0.9612	6.625
320	1599	35.78	2464	22.64	9.770	1.063	8.589
340	2431	58.96	3200	27.30	11.11	1.302	13.47
360	3556	105.4	6585	34.94	13.70	2.581	38.78

Appendix E Engineering Equation Solver (EES) Software

E.1 Introduction to EES

EES (pronounced "ease") is an acronym for Engineering Equation Solver. The basic function provided by EES is the numerical solution of nonlinear algebraic and differential equations. EES is an *equation-solver*, rather than a *programming language*, since it can automatically solve sets of implicit, nonlinear equations. EES also provides many other capabilities including unit checking of equations, parametric studies, optimization, uncertainty analyses, and high-quality plotting tools. EES provides array variables that can be used to carry out the finite-difference calculations that are common in heat transfer problems. In addition, EES provides high-accuracy thermodynamic and transport property functions for many fluids and solid materials that can be integrated with the equations. The combination of all of these capabilities together with an extensive library of heat transfer functions, discussed throughout this text, makes EES a powerful tool for solving engineering heat transfer problems.

In order to become familiar with the basic features of EES it is recommended that you access the video tutorials listed below. These (and other) video tutorials can also be accessed by selecting YouTube Tutorials from the Help menu in EES. You will be linked to a web page where all of the currently available tutorials are accessible (http://fchartsoftware.com/ees/youtube.php).

- Entering and Solving Equations: The Basics (8:49)
- Setting and Checking Units (9:25)
- Units of Temperature and Temperature Differences (5:34)
- Suggested Method for using Units in EES & Alternate Units (5:21)
- Creating and Solving Parametric Tables (7:04)
- Generating a Plot (5:48)
- The Unit System (6:46)
- Real Fluid Properties (9:38)
- The Ideal Gas Model (5:37)
- Printing (3:24)

E.2 Using EES Effectively

In the subsequent section, EES will be used to solve a heat transfer problem in order to demonstrate its capabilities and reinforce the basic use of EES that you learned in the preceding video tutorials. The solution steps will attempt to show not only the basics of using the software, but also some higher level concepts associated with using EES *effectively*. EES has the ability to do something that most software packages cannot do; it can solve equations regardless of the order that they are entered. This is a powerful feature. However, in our experience the ability to enter equations in arbitrary order can also lead to misuse and confusion and so it is worth establishing firmly a best-practice methodology for using EES. We will attempt to adhere rigorously to this practice within the book and suggest that you do the same.

Most of the variables involved in heat transfer problems represent physical quantities rather than pure numbers. Therefore, our variables have *units* in addition to values. The ability to work with problems and measurements that are expressed in a variety of units is a basic engineering skill. Used correctly, EES can essentially guarantee that you have not made any unit conversion errors. Furthermore, EES can check for unit consistency among your equations, which provides an additional check on your problem solution.

The first step in solving your problem should be to set the unit system. Although EES can work with any set of SI or English units, we recommend that you set the units to the basic SI unit system (Pa, N, m, kg, J, W, s, etc.). This is accomplished by selecting Unit System from the Options menu and selecting SI and the Kelvin, Pa, J, as shown in Figure E.1.

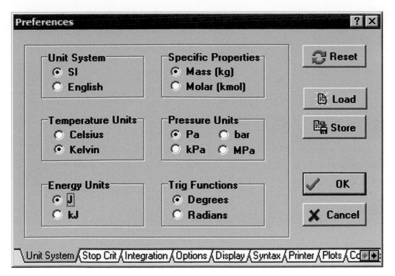

Figure E.1 Selecting the basic SI unit system.

Alternatively, the unit system can be specified using the $UnitSystem directive placed at the top of the Equations Window.

$UnitSystem SI Mass J K Pa Radian

The known information (the values provided by the problem statement) should be entered and immediately converted to these base SI units. It is often necessary to convert from units such as inch, torr, atm, tons (of mass or of cooling) etc. to base SI units. All of the variables used in the problem statement and solution should have the appropriate base SI units; EES allows the units of each variable to be explicitly set, as discussed in the subsequent example. The advantage of this technique is that the base SI units are self-consistent and therefore it is not necessary to do any unit conversions as you enter the equations required to solve the problem. Once the problem has been solved, then you can convert the solution to whatever units are most appropriate (or use the alternate unit feature to see the results in a dual set of units).

EES allows equations to be entered in any order; however, it is a good idea to enter the equations in a logical order. When possible, equations should be entered in a *sequential* manner so that an intermediate solution can be obtained after each equation is entered. This is similar to the way you would solve the problem by hand or if you were using a conventional programming language. Each new equation should use only information that was stated or obtained from a previous solution. There are several reasons for entering equations in this manner. First, it allows your program to be debugged as it is written. It is tempting to enter all of the equations involved in a problem into EES before trying to solve the problem. When a solution cannot be obtained, it is a frustrating process to debug the *entire* program. It is better to enter a *single* equation and see if it solves. When problems are encountered using this approach, they can immediately be isolated to the last equation that was entered. This is good engineering practice; when something goes wrong with an experiment or device, an experienced engineer will immediately try to isolate the problem by testing each sub-system of the device.

There is another, more subtle advantage to using this sequential approach. EES solves a set of nonlinear equations using an iterative approach; the approach starts from an initial point that is characterized by a set of "guess" values for each variable and then iteratively improves the solution. The solution process is easier if the initial guess values for each variable are close to the final solution. By obtaining a series of intermediate solutions it is possible to always have a good set of guess values available. This process will greatly increase the likelihood that EES provides the correct solution to the equations at the point where it is no longer possible for them to be entered sequentially.

Finally, it is imperative that you separate the analysis step of the problem in which the equations that must be solved are derived from the solution step where they are solved. EES is extremely useful in the solution step, but it cannot help you at all in the analysis step. Carefully derive all of the equations that describe the problem, verify that they are complete, and only then move to EES to help solve them.

E.3 Example Problem

In this section, a heat transfer problem is worked from start to finish in order to demonstrate some of the capabilities of the EES program. EES is particularly appropriate for this problem since the solution requires iterations that would be tedious to do by hand.

A $D = 0.5$ mm diameter platinum wire that is $L = 1.0$ cm long is used as a hot wire anemometer, a device to measure the velocity of air. The free stream air temperature (T_∞) is 20°C and the air pressure is $p = 100$ kPa. The electrical resistance of the platinum wire (R_e) depends on its temperature (T) according to:

$$R_e = \frac{0.30 \left[\text{ohm}/\text{K}^{0.5}\right]}{\sqrt{T - 270 \, [\text{K}]}}, \tag{E.1}$$

where R is in (ohm) and T is in (K). The current (I_c) that is provided to the hot wire is adjusted in order to maintain the wire temperature at a fixed temperature, $T = 60°C$. The electrical dissipation associated with the current is convected to the free stream. The rate of heat transfer associated with convection is:

$$\dot{q}_{conv} = \bar{h} A_s (T - T_\infty), \tag{E.2}$$

where A_s is the surface area of the wire exposed to the air and $\bar{h}$ is the heat transfer coefficient. The heat transfer coefficient is estimated according to:

$$\bar{h} = 0.51 \left(\frac{k}{D}\right) \left(\frac{\rho \, D u_\infty}{\mu}\right)^{0.5}, \tag{E.3}$$

where k, ρ, and μ are the thermal conductivity, density, and viscosity of the air (evaluated at the average of T and T_∞), D is the diameter of the wire, and u_∞ is the free stream velocity. Because the heat transfer coefficient depends on the air velocity, the amount of current and therefore the voltage measured across the wire can be related to the free stream velocity. Calculate and plot the voltage measured across the wire (V) for a range of free stream velocity (u_∞) values between 20 m/s and 100 m/s.

Known Values

The known values are shown in the sketch in Figure E.2.

Assumptions

- Steady-state conditions exist.
- The temperature of the wire does not vary axially or radially.

air
$T_\infty = 20°C$
$p = 1$ atm
u_∞

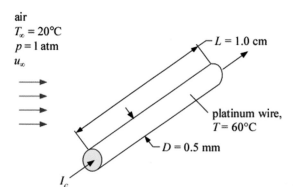

$L = 1.0$ cm

platinum wire,
$T = 60°C$

$D = 0.5$ mm

I_c

Figure E.2 A hot wire anemometer used to measure air velocity.

Analysis

The steady-state energy balance on the wire requires that the generation due to ohmic dissipation ($\dot{g}$) must equal the heat transfer from the surface due to convection ($\dot{q}_{conv}$):

$$\dot{g} = \dot{q}_{conv}, \tag{E.4}$$

where $\dot{g}$ is the rate of thermal energy generation by ohmic dissipation and $\dot{q}_{conv}$ is the convective heat transfer rate from the wire surface. The energy generation rate can be calculated according to:

$$\dot{g} = I_c^2 R_e. \tag{E.5}$$

The resistance of the platinum wire is a known function of temperature, given by Eq. (E.1). The product of the current and the resistance is the voltage:

$$V = I_c R_e. \tag{E.6}$$

The surface area of the wire is computed according to:

$$A_s = \pi D L. \tag{E.7}$$

The properties k, ρ, and μ are the thermal conductivity, density, and viscosity of air; all of these are evaluated at the film temperature, which is the average of T_∞ and T:

$$T_f = \frac{T + T_\infty}{2}, \tag{E.8}$$

$$k = k(T_f), \tag{E.9}$$

$$\rho = \rho(T_f, p), \tag{E.10}$$

$$\mu = \mu(T_f). \tag{E.11}$$

Equations (E.1) through (E.11) are 11 equations in the 11 unknowns R_e, $\dot{q}_{conv}$, $\bar{h}$, $\dot{g}$, I_c, V, A_s, k, μ, ρ, and T_f. Therefore, we can move from the analysis step to the solution step at this point. Notice that the process of deriving these 11 equations did not involve EES and it is independent of the technique used to obtain a solution.

Solution

Start EES or select the New command from the File menu if you have already been using the program. A blank Equations Window will appear. Since this problem will require the use of the built-in properties for air, it is necessary to specify the unit system that will be used to obtain property information with the Unit System command in the Options menu (see Figure E.1) or by using the $UnitSystem directive. Note that temperature will be expressed in K, pressure will be expressed in Pa, and energy units will be expressed in J with these choices.

```
$UnitSystem SI Mass J K Pa Radian
```

The first section of your EES program should contain the known information and accomplish the process of converting this information to the base SI unit system. Your code can be organized into sections that are identified with comments, as shown below:

```
"Appendix E: Heat Transfer Example"
"Input Information"
```

Because the text entered in the Equation Window is enclosed by quotes, it is not processed as equations entered in EES. Rather, these comments are used to make the code more readable. It is good engineering practice to

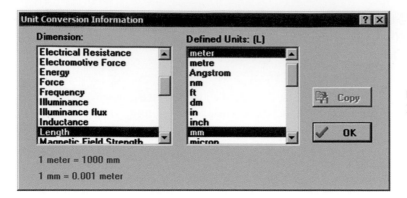

Figure E.3 Unit Conversion Information dialog.

provide comments with almost every line of code that you write so that you can, at a glance, recall what the purpose of the code is. The other method for entering comments is to use curly brackets, {}.

The known information is entered, starting with the dimensions of the wire. The diameter of the wire is stated as 0.5 mm, which is not in a base SI unit; therefore, the diameter must be converted from mm to m. This can be accomplished by dividing by 1000; however, you may not remember that mm and m are related by a factor of 1000. Certainly, the conversions between other units are not as easy to remember. Fortunately, EES has the built-in function **Convert** which provides the conversion constant that is required to convert between any two dimensionally consistent units. The protocol for calling the **Convert** function is:

Convert('From','To')

where 'From' and 'To' are unit designations. For example, to get the conversion factor required to convert *from* mm *to* m would require the function **Convert**(mm,m). To view the units that are programmed in EES, select Unit Conversion Info from the Options menu as shown in Figure E.3; the units are separated by dimension (the left scroll box).

The wire dimensions are entered and converted according to:

D=0.5 [mm]***Convert**(mm,m)	"diameter of wire"
L=1.0 [cm]***Convert**(cm,m)	"length of wire"

Notice that the diameter is entered as 0.5 [mm] (the square brackets indicate that the constant 0.5 has units mm) which is converted from mm to m using the **Convert** function. The constant returned by the **Convert** function has units of m/mm in this case. Thus, the variable D has units m. The units of the variables D and L must be explicitly set by the user. There are a number of ways to set the units. One way is to select Variable Info from the Options menu. The Variable Information Window that is accessed includes a list of all variables involved in the problem; the units for each variable can be entered in the column Units, as shown in Figure E.4.

Select Check Units from the Calculate menu and EES will indicate that no unit problems are detected because the units of the constants, the variables, and the conversion factors are all consistent. To see this more clearly, select Formatted Equations from the Windows menu; the units of the constants and conversion appropriately cancel as shown in Figure E.5. The Formatted Equations Window provides the equations in a more readable format. Further, these equations can be copied and pasted into a Word document in order to facilitate report writing.

The remaining variables in the problem statement are entered and converted to base SI units:

T_infinity=**converttemp**(C,K,20 [C])	"free stream temperature"
T=**converttemp**(C,K,60 [C])	"surface temperature"
u_infinity=20 [m/s]	"free stream velocity"
p=100 [kPa]***convert**(kPa,Pa)	"free stream pressure"

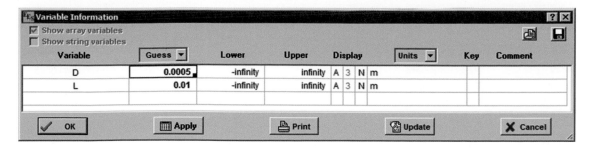

Figure E.4 Variable Information dialog.

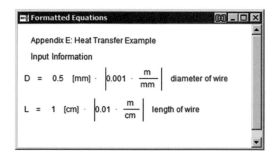

Figure E.5 Formatted Equation Window

Notice that the process of converting the free stream temperature from °C to K does not involve a simple multiplication and therefore a special function (ConvertTemp) is required. The calling protocol for ConvertTemp is:

ConvertTemp("From","To",value)

where "From" and "To" are the temperature scales to be converted from and converted to, respectively, and value is the value of the temperature to be converted.

At this point, we are ready to begin to solve the problem, keeping in mind the strategy that the equations should be entered sequentially so that solutions can be obtained periodically. This approach provides the best chance to debug the program efficiently.

The temperature of the wire is used to compute the resistance of the wire using Eq. (E.1):

R_e=0.3 [ohm/K^0.5]*Sqrt(T-270[K])   "wire resistance"

The units for the variable that has just been introduced (i.e., R_e) should be set in the Variable Information Window (or the Solutions Window). Note that it is possible at this point to check the units and obtain a solution. In short, the validity of the solution to this point can be examined. If problems are found, then they can be addressed quickly because there is only a single line of code that could be the source of any errors.

The surface area of the wire is computed using Eq. (E.7).

A_s=pi#*D*L   "surface area"

Note that the variable pi# is a built-in constant that contains the value of π. (EES also recognizes pi, without the hashtag.) Symbols ending with the hashtag indicate a built-in constant; a list of such constants can be obtained by selecting Constants from the Options menu, as shown in Figure E.6.

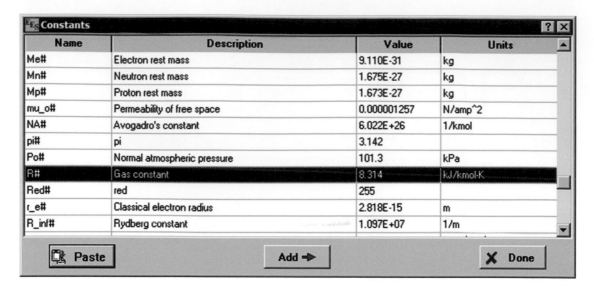

Figure E.6 Constants dialog.

The required properties of air must be calculated at the film temperature, defined by Eq. (E.8).

T_f=(T+T_infinity)/2	"film temperature"

The properties of air (and many other substances) are built into EES. To access these properties, select Function Info from the Options menu and then select Thermophysical properties (Figure E.7).

EES provides property information for several types of substances including real fluids, ideal gas fluids, air/water mixtures, etc. Here we will model air as an ideal gas and therefore select the radio button Ideal gases, as shown in the figure. The left scroll box includes the various properties that can be obtained, and the right scroll box includes the substances that are available. Scroll to the property function **Conductivity** in the left scroll box and the substance **Air** in the right scroll box. Select the Paste button and the function will be pasted into the Equations Window, with a little modification the equation becomes:

k=**Conductivity**(***Air***,T=T_f)	"conductivity of air"

The function **Conductivity** requires two calling parameters; the first indicates the substance (Air) and the second parameter (T = T_f) actually includes two pieces of information. The quantity on the left side of the equal sign tells EES that the independent variable used to specify conductivity will be temperature (T) while the right side of the equal sign tells EES the value of the independent variable (equal to the variable T_f in the problem).

The units for the conductivity returned by the function **Conductivity** will be consistent with the unit system that you set (in Figure E.1). Therefore, you should set the units of the variable k to W/m-K. Along the same line, the temperature that you supply to the function will be interpreted as having units K because this is consistent with the unit system specified. The viscosity and density of air are calculated using the same process:

mu=**Viscosity**(***Air***,***T***=T_f)	"viscosity of air"
rho=**Density**(***Air***,***T***=T_f,***P***=P)	"density of air"
Pr=**Prandtl**(Air,T=T_f)	"Prandtl number of air"

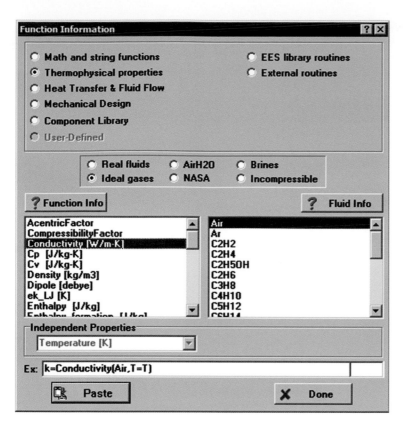

Figure E.7 Function Information Window showing the Fluid property functions.

Note that density depends on both temperature and pressure. These four equations represent Eqs. (E.9) through (E.11) in the Analysis section.

The substance Air in EES corresponds to the ideal gas model of air. Many other, more complex and nonideal substances are included in EES; for example, Air_ha corresponds to a more accurate model for the properties of air (including its saturation properties). For nonideal substances, two independent properties must be used to specify any dependent property. For example, to obtain the conductivity of air using the substance Air_ha would require:

k=**Conductivity**(*Air_ha*, T=T_f, P=P) "conductivity of air using the more accurate Air_ha substance"

The conductivity could also be specified with a different set of independent properties (for example enthalpy and pressure rather than temperature and pressure). In this case, the first argument would change from T= to h=, as shown below

k=**Conductivity**(*Air_ha*, h=100000 [J/kg], P=P)
"conductivity of air using the more accurate Air_ha substance based on enthalpy and pressure"

For this problem the ideal gas model of air is sufficient, and therefore the correct equation remains:

k=**Conductivity**(*Air*,T=T_f) "conductivity of air"

It is now possible to compute the heat transfer coefficient using Eq. (E.3):

h_bar=0.51*(k/D)*(rho*u_infinity*D/mu)^0.5 "heat transfer coefficient"

Note that because all of the variables are in their base SI units, the units of the heat transfer coefficient that is calculated should be W/m²-K; it is a good idea to set the units for the variable h_bar. EES will check the unit consistency of the equations that have been entered when the Solve or Check Units command in the Calculate menu is issued. The rate of convection heat transfer can be determined from Eq. (E.2) and the rate of electrical dissipation from Eq. (E.4):

```
q_dot_conv=h_bar*A_s*(T-T_infinity)          "convection heat transfer rate"
g_dot = q_dot_conv                           "energy balance"
```

Finally, Eq. (E.5) can be used to determine the required current and Eq. (E.6) the voltage.

```
g_dot=I_c^2*R_e                              "generation in the wire"
V=I_c*R_e                                    "voltage"
```

The complete code is shown in the Equations Window in Figure E.8.

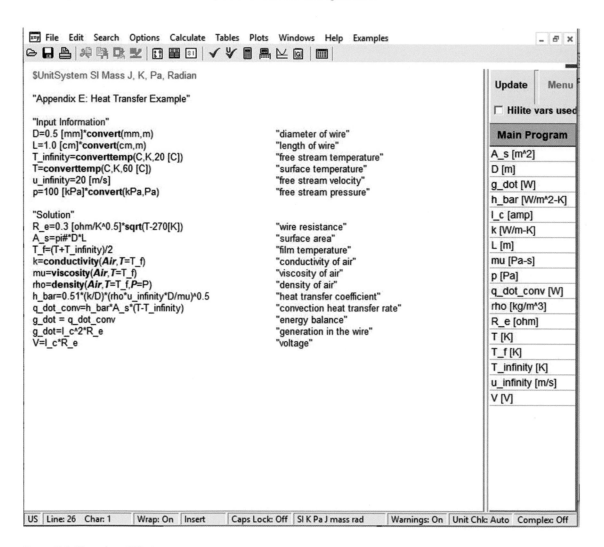

Figure E.8 Equations Window.

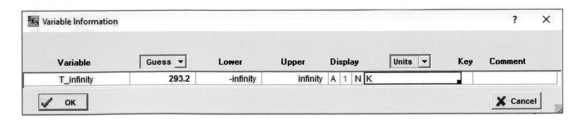

Figure E.9 Variable Information Window for T_infinity.

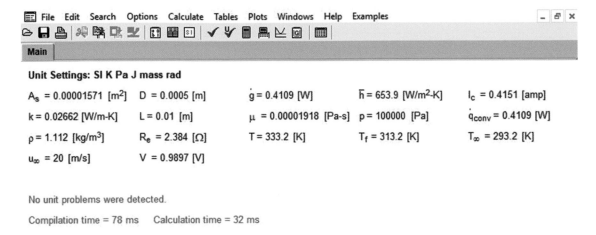

Figure E.10 Solution Window.

The variables used in the equations are shown on the right side of the Equations Window in the Variable Information palette. Clicking on a variable name brings up an abbreviated version of the Variable Information dialog for just the selected variable. For example, clicking the left button on T_infinity will cause the dialog in Figure E.9 to appear, in which the units, guess value, lower and upper bounds and other information can be viewed or changed. Clicking the right mouse button on a variable name in the palette will locate that variable in the Equations Window.

The Solution Window is shown in Figure E.10 and indicates that the voltage measured by the hot wire when the free stream velocity is 20 m/s will be 1.458 V; note that the results shown here are for the US/UK system, in which a decimal point (.) is used as the decimal separator. The display will use a comma (,) as the decimal separator if your computer is set for EU defaults. The US/EU choice can be made within EES using the $Keyboard directive.

Units of variables can also be changed in the Solution Window by right-clicking on the variable. For example, right-clicking on the variable T_infinity will cause the dialog shown in Figure E.11 to appear. The display format and units of the selected variable can be viewed or changed in this dialog. Note that it also possible to add a second unit specification, e.g., C, which will result in the variable being displayed in both sets of units in the Solutions Window.

One of the most useful features of EES is its ability to carry out parametric studies. In this problem, for example, we were asked to determine the voltage (V) as a function of air velocity, (u_∞) for values ranging between 20 and 100 m/s. The solution that is shown in Figure E.10 is for $u_\infty = 20$ m/s. A series of calculations at other velocities can be obtained automatically and plotted in EES.

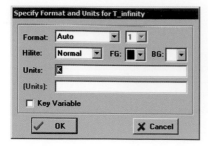

Figure E.11 Specify Format and Units dialog.

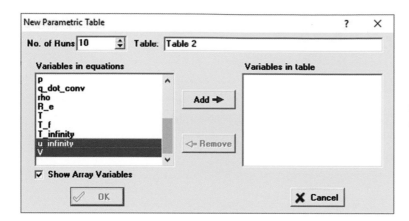

Figure E.12 New Parametric Table dialog.

Select the New Parametric Table command in the Tables menu. A dialog will be displayed listing the variables appearing in the Equations Window (Figure E.12). In this case, we will construct a table containing the variables u_infinity and V. Click on these variables in the list on the left. Click the Add button to move the selected variables to the list on the right. As a short cut, you can double-click on a variable name in the list on the left to directly move it to the list on the right. Note that you can set the number of runs (rows) in the Parametric Table and give the table a name. These changes can be made now or at a later time. Click the OK button to create the table. The Parametric Table works something like a spreadsheet. You can type numbers directly into the cells; numbers that you enter are shown in black (with the default color scheme) and when you solve the table, these entries produce the same effect as if you set the corresponding variable to that value with an equation in the Equations Window. We will be entering values for u_∞ into the table. For that reason, it will be necessary to remove the equation that sets the value of u_∞ from the Equations Window (otherwise you will over-constrain the problem by defining the variable twice). There are several ways to remove the variable but the simplest is to enclose the equation u_infinity=20 [m/s] in comment brackets { }; this can be done by highlighting the equation, right-clicking and selecting Comment from the popup menu. Now it is necessary to enter values of u_infinity in the Parametric Table. You could simply type values into the table cells, but it is easier to have EES enter the values automatically. Automatic entry is initiated by clicking on the triangular icon at the upper right of the header cell for the u_infinity column in the table. This action will bring up the Alter values dialog. Enter 20 for the first value and 100 for the last value (Figure E.13). EES will automatically fill in the values for all rows in the Parametric Table when the dialog is dismissed by clicking the OK button. At this point, the Parametric Table should appear as shown in Figure E.14.

Select Solve Table from the Calculate menu or press F3. The Solve Table dialog will appear (Figure E.15) allowing you to choose the runs for which calculations will be done. Note that the Update Guess Values control is selected in Figure E.15; this control causes the solution for the last run (or row) to be used as the guess values

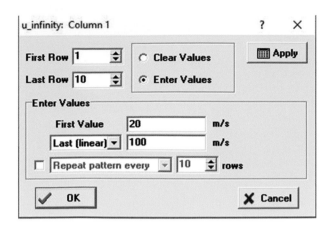

Figure E.13 Alter Values dialog.

Table 1

1..10	u_∞ [m/s]	V [V]
Run 1	20	
Run 2	28.89	
Run 3	37.78	
Run 4	46.67	
Run 5	55.56	
Run 6	64.44	
Run 7	73.33	
Run 8	82.22	
Run 9	91.11	
Run 10	100	

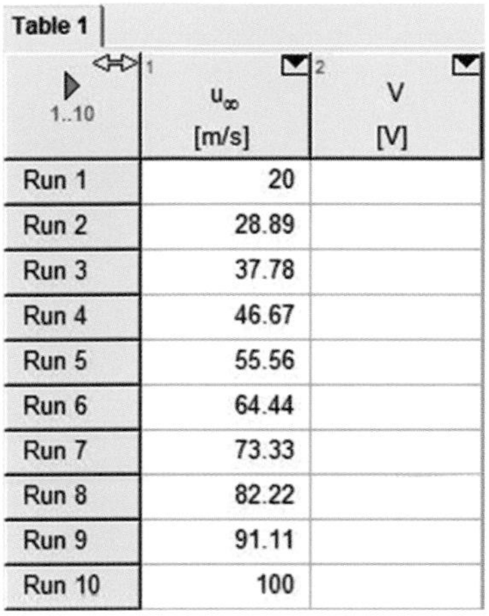

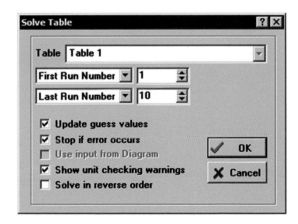

Figure E.14 Parametric Table, before solving.

Figure E.15 Solve Table dialog.

for the subsequent run. Click the OK button. A status window will be displayed, indicating the progress of the solution. When the calculations are completed, the values calculated for V and T are entered into the table (Figure E.16). The values calculated by EES will be displayed in blue, bold or italic type depending on the setting made in the Screen Display tab of the Preferences dialog in the Options menu.

1..10	u_∞ [m/s]	V [V]
Run 1	20	0.9897
Run 2	28.89	1.085
Run 3	37.78	1.16
Run 4	46.67	1.223
Run 5	55.56	1.278
Run 6	64.44	1.326
Run 7	73.33	1.37
Run 8	82.22	1.409
Run 9	91.11	1.446
Run 10	100	1.48

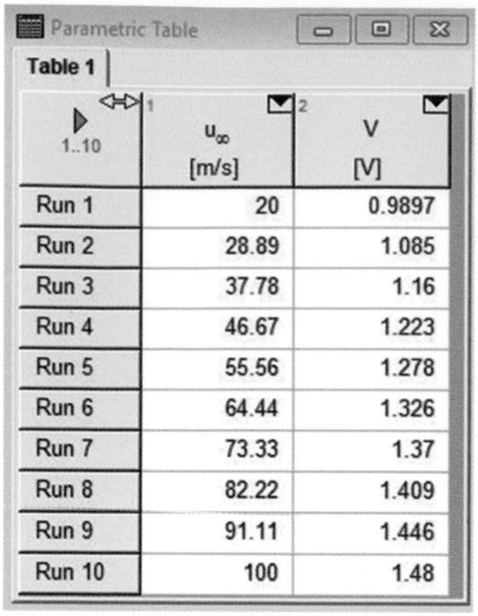

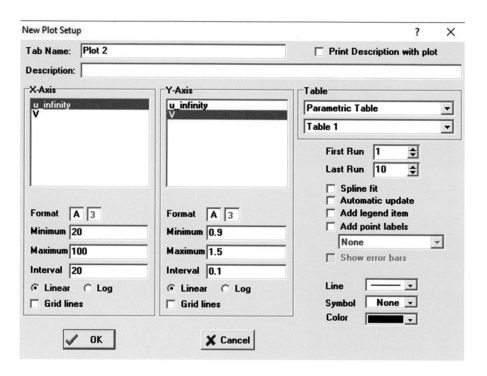

Figure E.16 Parametric Table, after solving.

Figure E.17 New Plot Setup dialog.

The relationship between variables such as V and u_infinity is apparent by examining Figure E.16; however, it can be seen more clearly with a plot. Select New Plot Window from the Plot menu. The New Plot Setup dialog shown in Figure E.17 will appear. Choose u_infinity to be the x-axis by clicking on it in the X-Axis list and click on V in the Y-Axis list. You may wish to adjust the scale limits or add grid lines. The spline fit option will draw a smooth line through the data points, which is useful when there are only a few points being plotted. When you click the OK button, the plot will be constructed and the plot window will appear as shown in Figure E.18. Once created, there are a variety of ways in which the appearance of the plot can be changed. Double-click the mouse in the plot rectangle or on the plot axis to see some of these options.

This problem was relatively simple, but it did illustrate most of the basic features of EES that you'll need to know to get started. More features are introduced throughout the book.

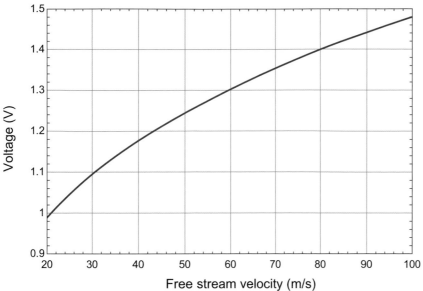

Figure E.18 Plot Window.

Appendix F Finite Element Heat Transfer (FEHT) Software

FEHT (pronounced "feet") is an acronym for Finite Element Heat Transfer. The basic function provided by FEHT is the numerical solution of 2-D conduction problems using the finite element technique. FEHT is intuitive to use and therefore it can be learned very quickly by students and researchers. This tutorial allows you to quickly familiarize yourself with some of the basic features of FEHT.

A demonstration version of FEHT can be downloaded from http://fchartsoftware.com/feht/. The demonstration version allows up to 1000 nodes and it can be used to solve most of the problems in this text. The full version allows up to 20,000 nodes. Contact F-Chart Software (info@fchart.com) for information on academic site licenses.

F.1 Introduction

The process of solving a problem using FEHT has three parts: (1) problem specification, (2) setting up the mesh and carrying out the calculations, and (3) examining the results. The problem specification includes setting up the geometry associated with the problem in a simple drawing environment and then setting the material properties and boundary conditions.

In order to obtain a solution it is necessary to specify a mesh of triangular elements. The FEHT software has an auto mesh option, which will automatically discretize the computational domain for most geometries. Alternatively, the user can manually specify an initial, coarse mesh. In either case, the mesh can be subsequently refined (i.e., more and more elements can be added) in order to get a more accurate solution. Once the problem is specified and a mesh is setup then the calculations are initiated in order to compute the temperature at each node by solving a set of algebraic equations similar to those described in Sections 4.3 and 6.4 for steady-state and transient finite difference solutions, respectively.

A variety of output capabilities are provided by FEHT in order to examine the solution. For steady-state problems, the temperature at each node may be shown using one of several types of contour plots. Temperature gradients can be superimposed on the contour plots using arrows that indicate both the direction and magnitude in order to show the flow of heat through the system.

The remainder of this appendix illustrates the use of FEHT in the context of the problem shown in Figure F.1. A bracket is connected to a hot oven wall. The face where the bracket is bolted to the oven has temperature $T_H = 600°C$ while the opposite face experiences convection to a fluid at temperature $T_w = 20°C$ with a convection coefficient of 24 W/m²-K. All other surfaces are adiabatic. The material has conductivity $k = 5.7$ W/m-K. The bracket depth is 15 cm. The geometry is shown in cross-section in Figure F.2. The ends of the bracket are assumed to be adiabatic so that the problem can be modeled as being 2-D with temperature gradients only in the x- and y-directions, as shown in Figure F.2.

F.2 Getting Started

FEHT is a 32-bit application that will run on all 32- and 64-bit Microsoft Windows Operating systems. Run the SETUP_FEHT.exe program to install the program on your computer system. The installation program will create a menu item in the Programs section of the Start menu. Context-sensitive help is available at any point in FEHT after pressing the F1 key.

Commands are distributed among pull-down menus; a brief summary of their functions follows.
- The **File** menu provides commands for loading and saving work files, printing, and copying information to and from the clipboard.

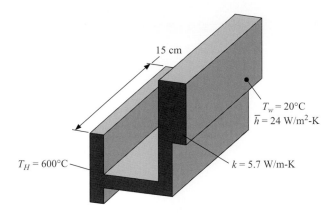

Figure F.1 A bracket bolted to an oven wall.

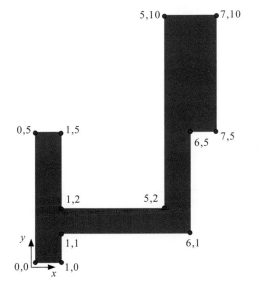

Figure F.2 Cross-section of bracket with the location of various points (all shown in cm).

- The **Subject** menu allows the problem discipline to be selected from the areas of heat transfer, electrical current, electrostatics, scalar magnetostatics, bio-heat transfer, potential flow, and porous media flow.
- The **Setup** menu commands allow specification of the unit system, size, scale, coordinate system (Cartesian or cylindrical), and problem type (steady-state or transient).
- The **Draw** menu contains the commands to outline a material, manually or automatically construct element lines, delete or group selected items, reposition nodes, and add text.
- The **Display** menu contains a variety of commands that affect the screen display including a zoom command that can be used to enlarge a selected part of the screen.
- The **Specify** menu allows material properties, internal generation, boundary conditions, and initial conditions to be specified.
- The **Run** menu contains commands to check a problem definition and to initiate or continue the calculations.
- The **View** menu provides the means to make any of the FEHT windows active.
- The **Examples** menu provides convenient access to a number of example problems. Examples can also be opened with the **Open** command in the **File** menu.
- The **Help** menu provides access to the on-line help.

Online Help for any command in the pull-down menus can be obtained by first pressing and holding the mouse button down with the cursor positioned on the menu item and then pressing the F1 key. Clicking the Contents button will present the Help index. Clicking on an underlined word will provide help relating to that subject.

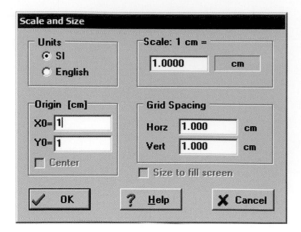

Figure F.3 Scale and Size dialog.

F.3 Problem Specification

By default, FEHT is configured for steady-state heat transfer problems in Cartesian coordinates. These characteristics apply to the practice problem shown in Figure F.1 and do not need to be changed. It is usually best to set the unit system, scale, and grid spacing at the start of a problem, although they can be changed at any time. Pull down the Setup menu and select the **Scale and Size** command, which will bring up a dialog (Figure F.3) in which the scale attributes can be entered.

To change the unit system, click on the appropriate radio button; as usual, we will select SI units. A reasonable scale for this problem is to have 1 cm on the screen represent 1 cm of the bracket. The unit for length is, by default, centimeters but it can be set to mm, cm, m, or km by clicking in the units box to the right of the scale value. (In the English system, the length unit can be inches, feet, yards, or miles.) Note that double-clicking within any edit box causes the characters to be highlighted allowing you to replace the highlighted field with the entered character. The origin can be set by entering the values for the variables X0 and Y0 which designates the location of the origin for the coordinate system on the screen *in screen coordinates* (i.e., relative to the bottom left corner of the screen). The default values, X0 = 0.0, Y0 = 0.0 correspond to the origin being placed at the very lower left of the screen. In Figure F.3 the origin is placed 1 cm vertically and horizontally from the lower left corner, which makes it more accessible. (The Center and Set Display Size command in the Display menu can be applied later to change the origin and scale.) Gridlines make the drawing easier to prepare and the grid spacing can be specified using the dialog in the lower right of the window. Click the OK button or press the Enter key to set these scale attributes.

The first step is to sketch material outlines. It is easier to prepare a scale drawing with a coordinate grid. Select Show Grid from the Display menu (if it is not already selected). Select Outline (free form) from the Draw menu. There is also an Outline (circular) draw command to enter circular shapes which must eventually be represented by short straight line segments. Note that the x and y coordinates of the cursor position are shown in the small window at the upper left of the screen, just below the menu bar. Move the mouse to locate the cursor approximately at position $x = 0$ cm, $y = 0$ cm. (Note that it is not necessary or even possible to have the cursor positioned exactly at the desired position. Just place the cursor close to the desired position; the exact node coordinates will be entered subsequently.)

Click the mouse to fix a node at the corner. The first node is shown as a small closed circle. Now, position the cursor approximately at $x = 0$ cm, $y = 5$ cm and click the mouse. You can hold the shift key down to provide a drawing aid for horizontal, vertical, or 45° lines. Click on the remaining corners at $x = 1$ cm, $y = 5$ cm; $x = 1$ cm, $y = 2$ cm; etc. Finally, click on the first corner ($x = 0$ cm, $y = 0$ cm). The outline must begin and end on the same node without crossing any existing lines. At this point, the outline will flash, indicating that the outlining process is completed and the material within the flashing boundary is selected. The outline number and name are shown in the center information window at the top of the screen below the menu bar. The area

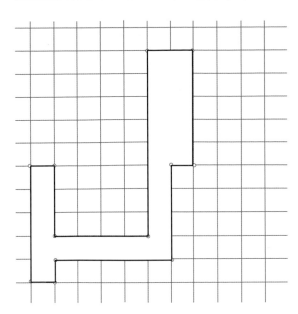

Figure F.4 Approximate outline of the bracket.

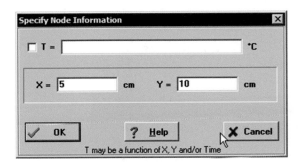

Figure F.5 Entering node coordinates in the Specify Node Information dialog.

enclosed by the outline is shown in the right information window. The screen should appear approximately as shown in Figure F.4. Next, right-click on each of the nodes (or left-click and select Node Information from the Specify menu) in turn and enter the exact coordinates for the node as shown in Figure F.5. The final outline is shown in Figure F.6. Note that you can now center and scale your outline to fill the screen if you wish by selecting Center and Set Display Size from the Display menu.

The material properties must be specified. A material (i.e., region within an outline) can be selected by clicking the mouse anywhere within its outline, which should cause it to flash. Select Material Properties from the Specify menu in order to obtain the property dialog box shown in Figure F.7. There are a number of default materials programmed in the database that can be selected. The material associated with the bracket was not specified, but the value of its conductivity was. Therefore, select "not specified" from the list and enter the given conductivity, as shown in Figure F.7. The values of density and Specific heat are not needed because this problem is steady state. Arbitrary values (1.0) have been entered. If a transient analysis is desired then the correct values for these properties will be required. Material properties may also be entered as a function of temperature and position. You can change the name of the material to "Bracket" if you wish.

Finally, it is necessary to set the boundary conditions. The boundary at $x = 0$ cm has a specified temperature. Move the cursor to a point near the center of this line and click the button. The line should now be flashing. Once a boundary has been selected (i.e., is flashing), the Boundary Conditions menu item in the Specify menu becomes accessible. Select this menu item to bring up the Boundary Conditions dialog (Figure F.8). Enter the specified temperature (600°C) and click the OK button.

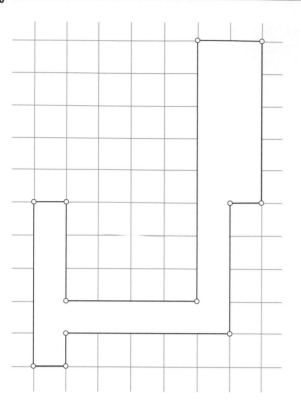

Figure F.6 Exact outline of the bracket.

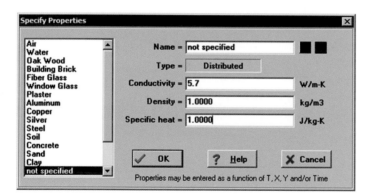

Figure F.7 Specify Properties dialog.

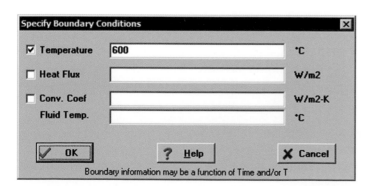

Figure F.8 Specify Boundary Conditions dialog.

The boundary line is now shown in bold to indicate that the boundary condition has been specified. Continue to specify the boundary conditions of each boundary; note that all of the adiabatic boundaries can be set at once clicking on each line sequentially. Set the boundary at $x = 7$ cm to a convective boundary with a convection coefficient of 24 W/m²-K and a fluid temperature of 20°C.

F.4 Meshing and Calculation

To complete the problem definition, it is necessary to discretize or *mesh* the material into triangular elements. The triangular elements can either be entered manually or automatically. In either case, it is easier to view the element lines when the material pattern is hidden; the pattern can be removed by selecting Hide Patterns from the Display menu. The grid lines are also no longer needed; select Hide Grid from the Display menu.

To take advantage of the auto mesh option, select Auto Mesh from the Draw menu. The result for the bracket is shown in Figure F.9. Notice that the mesh that was automatically generated has some elements with very large aspect ratios, resulting from the geometry of this material. The automatically generated mesh will work, but it is easy to generate a better mesh by creating a few element lines manually before applying Auto Mesh. Select Undo Auto Mesh from the Draw menu and then select Element Lines from the Draw menu. Move the cursor to the lower left node and click the button. Now, click at another location where either a node exists or you would like to place one. If no node exists then a node will be created and an element line will be drawn between the two nodes. If a node does exist at the selected location then an element line will be added. Repeat this process of adding element lines and nodes so that the screen appears as shown in Figure F.10 (left). Clicking in the area surrounding the drawing or pressing the Esc key cancels the Element Lines command. Next select Auto Mesh to complete the meshing into triangular elements, as shown in Figure F.10 (right).

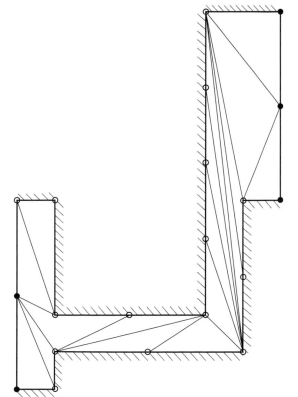

Figure F.9 Result of auto mesh.

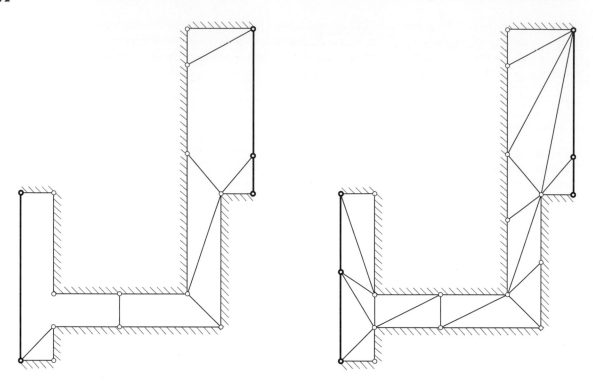

Figure F.10 Display after (left) manually creating element lines and (right) applying Auto Mesh.

You have considerable freedom in selecting your mesh and your mesh does not need to match Figure F.10. Any combination of manually or automatically generated element lines that result in a triangular mesh is acceptable. The following rules apply to manual element line construction.

- The first end of the line must be on an existing line or node. A new node will form at this point if one is not already there.
- Element lines cannot cross existing lines.

It is not necessary to construct a very fine mesh since the Reduce Mesh command in the Draw menu can refine a mesh once it has been constructed. Selecting the Reduce Mesh command results in the mesh shown in Figure F.11; this command can be reapplied until the upper limit of nodes is reached.

Select the Calculate command from the Run menu to initiate the calculations. FEHT will first check the problem definition to ensure that the distributed materials are properly discretized and all properties and boundary conditions are specified. Any errors detected will be listed in the information window at the upper right of the screen, just below the menu bar. This example problem is assumed to be steady-state. Had this been a transient problem (specified by selecting Transient from the Setup menu), a dialog box would have appeared in which the start, stop, and step times would be entered. If no errors are found, a dialog will appear indicating that the calculations are in progress and when the calculations have been completed, the dialog box will display the elapsed time and other information (Figure F.12).

F.5 Output

Click the Continue button. A number of output display windows in the View menu will now be accessible. The temperatures at each node can be displayed by selecting Temperatures from the View menu (Figure F.13). Hold the left mouse button down to display the temperature at the cursor in the status bar at the upper left of the screen.

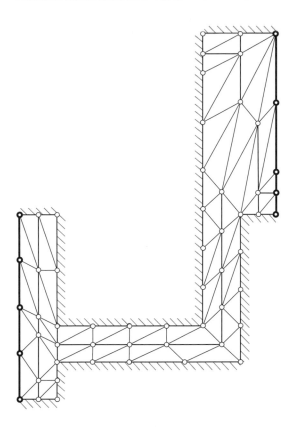

Figure F.11 Display after applying the Reduce Mesh command.

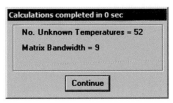

Figure F.12 Calculations completed message.

Temperature can also be displayed in the form of a contour plot by selecting Temperature Contours which will bring up the dialog shown in Figure F.14. Three types of contour plots are available: a banded plot showing gradations of hot to cold in sections; a continuous color plot showing temperatures as colors ranging either from red to blue, shades of gray, or full spectrum; and a contour plot of lines of constant temperatures. The minimum and maximum temperature values in the contour plot can be entered manually or FEHT will automatically find the limits if you click in the User/Auto box at the upper left. Click the OK button or press the enter key to show the contour plot (Figure F.15). The legend is shown in the status bar at the top of the display. The temperature at the cursor position will be displayed at the upper left of the screen below the menu bar when the mouse button is depressed within the contours. The heat flow can be visualized by plotting the temperature gradients as arrows with direction and magnitude. Select Gradients from the View menu and then select Superimpose on contours to obtain the plot shown in Figure F.16.

It is possible to determine the total heat flow through the bracket. Select Heat Flows from the View menu. The screen will be redrawn with the nodes hidden. Click on any line segment of the bracket outline and an arrow will appear indicating the direction of heat flow. The magnitude of the heat flow is shown in the information window at the top of the screen below the menu bar. Clicking on adjacent lines forming the inside boundary in a

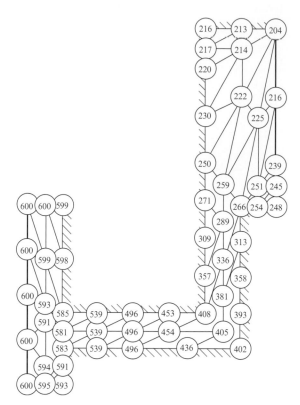

Figure F.13 Temperatures at each node.

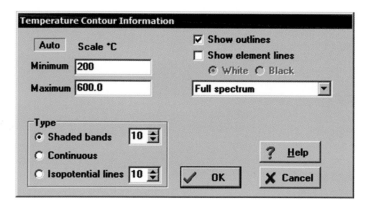

Figure F.14 Temperature Contour Information dialog.

clockwise or counterclockwise manner will allow the heat flows to be summed. Alternatively you can select multiple lines by pressing the mouse outside of the material and dragging it while holding the mouse down to create a selection rectangle. All lines that lie completely within the selection rectangle will be selected. After selecting all of the lines, you may wish to group them using the Group command in the Draw menu. Clicking on any one of the lines in a group selects all lines in that group. The total rate of heat flow for the bracket can be determined by selecting the right boundary as shown in Figure F.17. The heat flow rate is approximately 246 W per meter of depth into the screen. Taking into account that the bracket dimension into the screen is 15 cm allows the heat flow to be calculated as 36.9 W.

Select Input from the View menu to return to the drawing window. At this point, you may wish to explore. Try using smaller triangular elements by selecting the Reduce Mesh Size command in the Draw

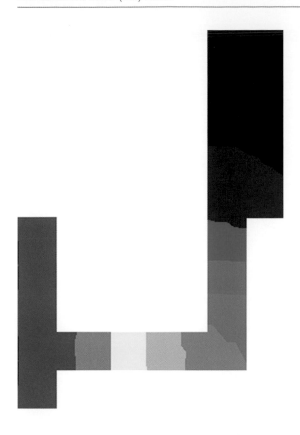

Figure F.15 Contour plot.

Figure F.16 Temperature gradients superimposed on temperature contour plot.

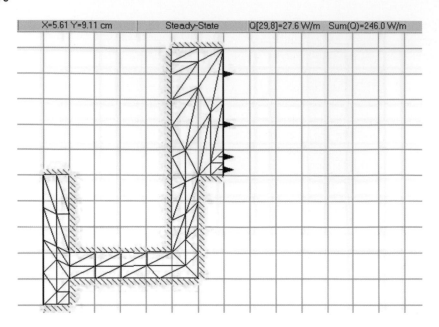

Figure F.17 Heat flow rate from the bracket due to convection.

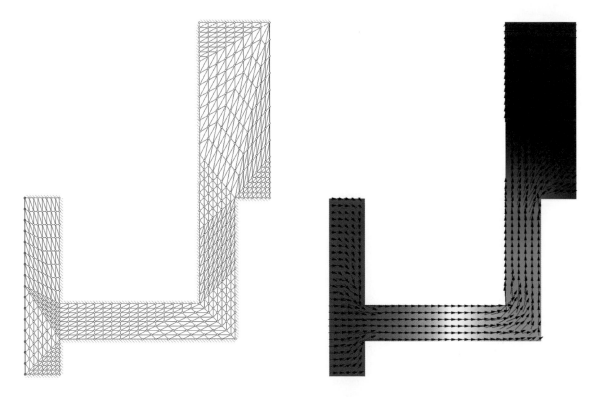

Figure F.18 Bracket problem with a reduced mesh and the resulting temperature contours.

menu. As the mesh is refined the solution should approach some limit indicating that the numerical model has converged. For example, Figure F.18 shows the solution with a reduced mesh. The convective heat flow from the right border for this mesh is 235.9 W/m or 35.38 W, which is not much different from the result obtained for the coarse mesh.

This problem was a steady-state 2-D problem. A transient analysis could be done by selecting Transient in the Setup menu. It would be necessary to set the density and specific heat of the bracket material and the initial temperatures using the Properties and Initial Temperatures commands in the Specify menu.

FEHT offers many other options. For example, it is possible to analyze a material with a hole. To do this, draw the outline of the material with the Outline (free form) command. Identify the location of the center of the hole by moving the cursor and observing the coordinates in the status bar at the top of the screen. Then draw the hole with the Outline (circular) command. Finally select the hole (so that its outline is flashing) and select the Subtract command from the Draw menu. Don't be concerned if you make a mistake because you can always undo your changes using the Undo command in the Draw menu.

Appendix G MATLAB Software

MATLAB is a sophisticated software package and we are only going to touch upon a few of its capabilities within this book. MATLAB is essentially a programming language; one of the reasons that it is so powerful is that its basic data structure is the array. Therefore, you can solve problems that involve large vectors and matrices intuitively and manipulate the results easily. MATLAB is used in this book almost exclusively for carrying out numerical simulations.

Most of the examples in this book that use MATLAB were completed using Version R2014a; however, these examples could be solved with little or no modifications using earlier or later versions of MATLAB. Additional information concerning MATLAB can be obtained from the distributor.

G.1 Getting Started

When you open MATLAB, you will be in the Command Window or working environment, as shown in Figure G.1. This is the set of tools and facilities that you work with as the MATLAB user or programmer. The working environment includes facilities for managing the variables in your workspace and importing and exporting data. There are also tools for developing, managing, debugging, and profiling M-files, which are

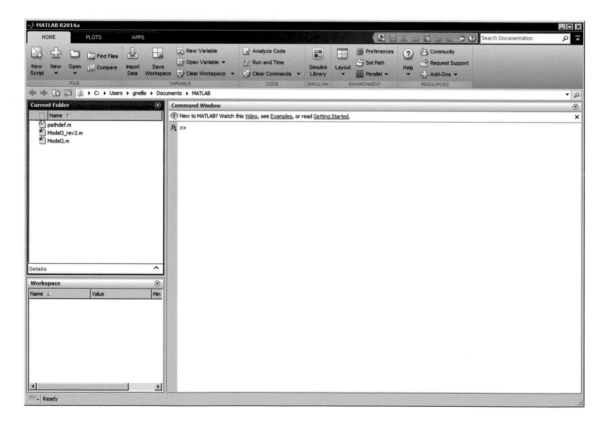

Figure G.1 MATLAB working environment.

MATLAB programs or applications. Notice the window that shows both the files in the current directory (the Current Folder tab) and the variables currently in the workspace (the Workspace tab). You can change the directory that you are working in from within the Current Directory tab.

You can enter variables and manipulate them, call functions, etc. in the working environment (the Command Window). The commands are entered at the command-line prompt (>>). The best way for you to get started with MATLAB is to learn how to enter and manipulate matrices as these are the fundamental data type used by MATLAB. In MATLAB, a matrix is a rectangular array of numbers; special meaning is sometimes attached to 1×1 matrices (i.e., scalars) and to matrices with only one row or column (i.e., vectors). The operations in MATLAB are designed to be as natural as possible. Where other programming languages work with numbers one at a time, MATLAB allows you to work with entire matrices. Therefore, it is much easier to work with large sets of data (such as the temperatures predicted by a numerical simulation) in MATLAB.

You can enter matrices into MATLAB in several different ways, including:
- manually, as an explicit list of elements,
- as data loaded from an external file,
- generated using functions built into MATLAB, or
- generated with your own functions (i.e., your own M-files).

To enter matrices manually, you have to follow a few basic conventions:
- separate the elements of a row with blanks or commas,
- use a semicolon (;) to indicate the end of each row, and
- surround the entire list of elements with square brackets, [ ].

For example, to enter the 3×3 matrix $\underline{\underline{A}}$, shown below:

$$\underline{\underline{A}} = \begin{bmatrix} 2 & 3 & 1 \\ 1 & 5 & 1 \\ 7 & 1 & 2 \end{bmatrix} \tag{G.1}$$

it would be necessary to type the following into the MATLAB environment.

```
>> A = [2 3 1; 1 5 1; 7 1 2]
```

Note that MATLAB will echo the results of carrying out each command as it is entered; in this case, MATLAB will display the matrix that was created after you press the Enter key.

```
A =
     2   3   1
     1   5   1
     7   1   2
```

The echo feature can be turned off by using a semicolon after your command. For example, typing the following would not be echoed in the working environment.

```
>> A = [2 3 1; 1 5 1; 7 1 2];
```

If you want to see what the MATLAB variable A is, just type A and press the Enter key.

```
>> A
A =
     2   3   1
     1   5   1
     7   1   2
```

The 3×1 vector $\underline{b}$, defined as:

$$\underline{b} = \begin{bmatrix} 1 \\ 2 \\ 5 \end{bmatrix} \tag{G.2}$$

may be entered as follows.

```
>> b = [1; 2; 5]
b =

     1
     2
     5
```

G.2 Working with Matrices

Working with matrices in MATLAB is easy. For example, the solution to the matrix equation:

$$\begin{bmatrix} 2 & 3 & 1 \\ 1 & 5 & 1 \\ 7 & 1 & 2 \end{bmatrix} \begin{bmatrix} x_1 \\ x_2 \\ x_3 \end{bmatrix} = \begin{bmatrix} 1 \\ 2 \\ 5 \end{bmatrix} \text{ or } \underline{\underline{A}}\,\underline{X} = \underline{b} \tag{G.3}$$

is obtained simply by typing the following.

```
>> X=A\b
X =

    11.0000
     6.0000
   -39.0000
```

You can read more about the backslash operation (\) by typing the following.

```
>> help slash
Matrix division.
  \    Backslash or left division.
       A\B is the matrix division of A into B, which is roughly the
       same as INV(A)*B , except it is computed in a different way.
       If A is an N-by-N matrix and B is a column vector. . ..
```

MATLAB has an extensive help system that can be accessed either by typing the function you are interested in, or by selecting MATLAB Help from the Help menu. To obtain help relative to the use of the exponential function, type the following.

```
>> help exp
exp    Exponential.
       exp(X) is the exponential of the elements of X, e to the X.
       For complex Z=X+i*Y, exp(Z) = exp(X)*(COS(Y)+i*SIN(Y)).

       See also expm1, log, log10, expm, expint.

       Overloaded methods:
           zpk/exp
           tf/exp
           codistributed/exp
```

```
        gpuArray/exp
        sym/exp

Reference page in Help browser
        doc exp
```

The details of the help message may differ depending on what version of MATLAB you are using; however, the specific capabilities of MATLAB that we will be using should be compatible with essentially any version. Note that, unlike EES, variable and function names in MATLAB *are* case-sensitive.

Typically, we will be assigning or addressing the elements in a vector or array using subscripts. The element in row i and column j of matrix A is denoted by A(i,j). For example, A(2,1) is the number in the second row and first column; for matrix A in Eq. (G-1), A(2,1) is 1 and A(3,1) is 7. If you want to change the value of A(2,1) from 1 to 8, simply re-assign it.

```
>> A(2,1)=8;
>> A
A =
     2   3   1
     8   5   1
     7   1   2
```

It is possible to extract or manipulate entire rows or columns of a matrix. To obtain the second column of A, use the : symbol which indicates all entries; therefore, A(:,2) would indicate all rows in column 2.

```
>> A(:,2)
ans =
     3
     5
     1
```

The mathematical operators +, -, *, /, etc., have their typical meanings when applied to scalar arguments. However, if these (and any other) operators are applied to a matrix then the operator will refer to the matrix form of the operation. For example, $A\,b$ is the matrix multiplication of the matrix A by the vector b (you may need to review your linear algebra to remember what this means).

```
>> A*b
ans =
    13
    23
    19
```

The order matters for matrix multiplication; that is $b\,A$ will not be equal to $A\,b$, in fact, $b\,A$ is not even possible, as you cannot multiply a 3 × 1 matrix with a 3 × 3 matrix.

```
>> b*A
??? Error using ==> mtimes
Inner matrix dimensions must agree.
```

You may want to avoid carrying out matrix operations and, instead, operate on each element in a matrix individually. This can be accomplished by placing a period (.) in front of the operator. For example, the operation A*A leads to the matrix multiplication of A by itself.

```
>> A*A
ans =
    35   22    7
    63   50   15
    36   28   12
```

However, the operation A .*A leads to each element in A being squared.

```
>> A.*A
ans =
     4    9    1
    64   25    1
    49    1    4
```

The results are quite different.

G.3 Using M-Files

MATLAB is both a programming language and an interactive computational environment. The files that contain code in the MATLAB language are called M-files; the easiest way of developing a numerical solution is as an M-file because it can be saved, manipulated, and debugged without having to be re-typed in each time. You create M-files using a text editor and then access them from the working environment as you would any other MATLAB function or command.

There are two kinds of M-files.

- Scripts, which do not accept input arguments or return output arguments. Scripts operate on data contained in the workspace and are exactly equivalent to typing commands directly into the workspace.
- Functions, which can accept input arguments and return output arguments. Functions do not operate on the data in the workspace and can only access variables that are local to the function.

The easiest way to create an M-file is to select New>Script or New>Function from the File section of the Home tab; this will open an M-file editor where you can enter the code and save the file for later use. You can save your M-files in any directory. However, in order to run the file from the MATLAB environment you need to make sure that the file is located in a directory that is in MATLAB's search path. To place a new directory in the search path, type the command pathtool and then click the Add Folder. . . button in the Set Path dialog that pops up (Figure G.2) and navigate to the directory that you want to add.

When you run a script, MATLAB simply executes the commands that are contained in the M-file. Scripts can operate on existing data in the workspace, or they can create new data on which to operate. Although scripts do not formally return output arguments, any variables that they create remain in the workspace and can therefore be used in subsequent computations. In addition, scripts can produce graphical output using functions like plot. Functions are M-files that can accept input arguments and return output arguments. The name of the M-file and of the function should be the same. Functions operate on variables within their own workspace which is separate from the workspace at the MATLAB command prompt. Both types of M-files will be needed to generate numerical solutions to heat transfer problems.

Open a new M-File script and save it as practice.m in a directory that is contained in MATLAB's search path. Comments can be used to clarify the code contained in M-files. Comments are indicated by the % symbol; anything following the % symbol is not executed. In your practice.m script, specify the radius and calculate the area of a disk.

```
r=2;                    %radius of disk
A=pi*r^2;               %area of disk
```

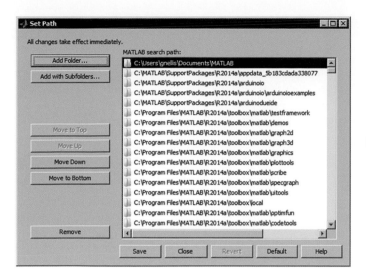

Figure G.2 Set Path dialog.

Save the file and execute it from the main workspace; nothing appears to happen, but if you type r or A then you can see that these variables have been assigned in the script.

```
>> practice
>> A
A =
     12.5664
>> r
r =
     2
```

One of the nice things about using MATLAB is that it is easy to debug as you go along; for example, if you were unsure as to the value of *r* in the script then simply remove the semicolon.

```
r=2                              %radius of disk
A=pi*r^2;                        %area of disk
```

Then MATLAB will echo the value to the workspace as it executes the script.

```
>> practice
r =
     2
```

Scripts can be turned into functions. The header of the function has the following format.

```
function[output argument(s)]=function name(input argument(s))
```

The keyword function tells MATLAB that you are creating a function rather than a script. In MATLAB (unlike EES) more than one output argument can be returned from the function; these go within the square brackets. For example, our script practice.m can be turned into a function that takes r as an input and returns A as an output.

```
function[A]=practice(r)
A=pi*r^2;      %area of disk
```

You can call the function practice from any other function or the main workspace.

```
>> A=practice(1)
A =
      3.1416
```

It is important to understand that functions have their own workspace and only communicate with the calling workspace via the input/output protocol. For example, we could try to re-assign the input argument r within the function.

```
function[A]=practice(r)
A=pi*r^2;      %area of disk
r=5;           %attempt to re-assign r
```

Calling the function practice will not affect the value of the variable r in the workspace.

```
>> r=1;
>> A=practice(1);
>> r
r =
      1
```

G.4 MATLAB Programming Constructs

MATLAB allows the use of programming constructs that should be familiar to anyone who has used a formal programming language like FORTRAN, Basic, etc. However, you may not be familiar with these other languages and therefore this section introduces a few of the most commonly used logic statements in MATLAB.

The if/else statements are very common and useful. These logic statements are used to conditionally execute a statement or set of statements based on an expression. For example, we could rewrite the practice function so that it returns the area of a disk if a radius greater than 5 is entered but otherwise it returns the value 999.

```
function[A]=practice(r)
if(r<5)
   A=pi*r^2;   %area of disk
else
   A=999;
end
```

Note that the statements within the if and else statements are indented; this is good form and makes it easy to understand the structure of the code. Execute the function practice from the workspace and check to ensure that it is working.

```
>> practice(1)
ans =
       3.1416
>> practice(6)
ans =
     999
```

Notice that the if statement is not followed by then as it must be in most other languages. Also, the relational operators used by MATLAB should be familiar with the possible exception of equality, which is indicated by = = rather than =. To see the list of relational operators (and other things) type help relop in the workspace. Finally, don't forget that all of these programming constructs must be terminated with an end keyword.

The for loop in MATLAB is analogous to the Duplicate statement in EES; it executes a statement (or several statements) a specific number of times. The for loop is important when preparing a numerical solution in heat transfer. The for keyword is followed by an integer variable and a range. Consider the following code, for example.

```
for i=1:10
         statement(s)
end
```

This would execute the statements sandwiched between for and end keywords 10 times; each time the variable i would take on a different value. We can make the function practice return a 10×1 vector in which the first element is the input argument r and each subsequent element is twice the previous one.

```
function[A]=practice(r)

A(1,1)=r;
for i=2:10
A(i,1)=A(i-1,1)*2;
end
```

You can call the function practice from the command-line prompt.

```
>> practice(r)
ans =
       1
       2
       4
       8
      16
      32
      64
     128
     256
     512
```

By nesting for loops you can generate or operate on matrices rather than vectors.

Finally, the while statement repeats a set of statements until some break condition is met; often, the while statement is used to terminate an iteration process within a numerical method based on achieving convergence. Here we can rewrite the practice function so that it stops adding values to the vector when they get larger than 4000.

```
i=1;
A(i,1)=r;
while(A(i,1)<4000)
    i=i+1;
    A(i,1)=A(i-1,1)*2;
end
```

There are many other capabilities of MATLAB and we will introduce several additional aspects of the software in the context of various problems; however, this should be sufficient to get you started.

Appendix H Answers to Selected Problems

Chapter 1

1.1 10.2 kW

1.2 43 W/m²-K

1.3 10 W

1.4 22.1 W/m²-K

1.5 336°F

1.6 (a) 75.2 W/m²-K

1.7 (a) 83.4°C

1.8 (a) 5.55°C

1.9 (a) 216 W/m²

1.10 167°F

1.11 211.4 W/m²

1.12 (a) 516.2 Pa

1.13 87.9 percent

1.14 (c) 98°F

1.15 (a) 6584 MJ

1.16 (a) 262.1 W

1.17 (a) Radiation heat flux = 4560 W/m²

1.19 (a) 1.574 hr

1.20 (a) 684.6 W/m²-K

1.21 94.8 W/m²

1.22 –0.429 K/s

1.23 (a) 142.2 W

1.24 (a) 14917 W

1.25 (a) 283.6°C

Chapter 2

2.4 Heat flux in x-direction = 1000 W/m²

2.6 (g) 1333 W

2.7 (g) 0.06667 W

2.10 (e) 18316 W/m²

2.11 (h) 186.6 W/m²

2.13 (h) 19.51 W

2.14 0.0352 W/m-K

2.15 68083 W/m²

2.16 5.6 cm

2.17 353.9 W

2.18 0.00549 m

2.19 (d) 128.3 W

2.22 (b) 363.3 K

2.23 20.0 W/m²-K

2.24 (a) 0.653 m²-K/W

2.26 (a) 315.6 Btu/hr-ft²

2.27 (a) 270.8 W

2.28 (a) 5.758 W/m

2.29 2.858 W

2.30 4188 W

2.31 304.4 K

2.32 (c) 548.4 W

2.33 (b) 70.3 W

2.35 (a) 900 K

2.36 (a) 0.393 W/m

2.37 (c) 1.91 W

2.38 36 percent increase

2.39 1849 W

2.40 (c) 81.7 W/m²

2.41 (a) 325.4 K

2.42 (d) 268.2 K

2.43 (a) 69 W/m²

2.44 (b) 410 K

2.45 (c) 83.2 percent

2.46 (b) 82 W

2.47 (b) 132.1 W

2.48 (b) 50 W/m-K

2.49 18.66 hr-ft²-R/Btu

2.50 (c) 625.8 W

2.51 (a) 17.9°C

2.52 (a) 4.216 hr-ft²-R/Btu

2.53 (b) 583.1 K and 566.4 K

2.54 (b) 666.7 W

2.55 (b) 70.7°C

2.56 (a) 13.27°C

2.57 (a) 1676 W

2.58 (c) 0.94 cm/hr

2.65 (c) 53°C

2.66 (b) 233.3 W

2.67 (a) 9696 W

2.68 (a) 179.3°C

2.70 (b) 85.5 percent

2.73 (a) 172.6°C

2.77 (a) 750 W/m

2.78 (b) 79.8 percent

2.82 (a) 3.96E6 W/m³

2.83 (c) 1.5617E6 W/m³

2.84 (a) 322.3 K (120.6°F)

2.85 (b) 204.1 W

2.87 (b) 21.3 W/m²

2.92 (b) 233.3 W

2.99 541.6 K

2.100 52.59 W

2.101 (e) 3.615 m

2.103 (c) 255.1°C at insulated face

2.106 419.0 K (outside) and 442.7 K (inside)

2.107 0.131 W

2.108 (b) 452.2 K

2.118 (a) 39.1 W

Chapter 3

3.2 The 1-D extended surface approximation is valid.

3.3 The 1-D extended surface approximation is valid.

3.4 The 1-D extended surface approximation is not valid.

3.9 The 1-D extended surface approximation is valid.

3.10 The problem should be analyzed as a 2-D conduction problem.

3.13 (c) 68 W

3.14 (a) 211.3°C and 3.986 W

3.17 (a) 8.35 W/m^2-K

3.18 (b) 304 K

3.22 (b) 2.62 W

3.23 (a) 15.1 W/m^2-K

3.27 (a) The 1-D extended surface approximation is valid.

3.28 (a) The 1-D extended surface approximation is valid.

3.29 (f) 22 W/m^2

3.30 (b) 2.43 W

3.33 (b) 2.526 W

3.35 (b) 0.924

3.36 432 W

3.37 (a) 93 percent and 260 W/m

3.38 (a) 7.5 percent increase

3.39 (a) 443 W

3.40 (a) 17.4 kW

3.42 (c) Total resistance =0.0393 K/W

3.43 (a) 0.0080 K/W

3.46 (b) Fin efficiency= 0.477

3.50 (c) 0.656

3.51 (a) 2.55 W

3.52 No danger of the camper burning his/her hand.

3.54 40 percent

3.55 (c) 0.9 mm

3.56 (f) 0.0009 m

3.58 (c) 0.75 cm

3.59 (e) 1.56 W

3.60 (d) 5.2 kg

3.61 (c) 3.19 W and 57.5 percent

3.62 (g) 4.34 K

Chapter 4

4.1 (b) $\left.\frac{\partial T}{\partial y}\right|_{y=0} = -\frac{q''}{k}$

4.3 $r\frac{\partial^2 T}{\partial x^2} + \frac{\partial}{\partial r}\left[r\frac{\partial T}{\partial r}\right] = 0$

4.5 $\frac{\partial^2 T}{\partial x^2} + \frac{\partial^2 T}{\partial y^2} = 0$

4.7 $\frac{\partial^2 T}{\partial x^2} + \frac{\partial^2 T}{\partial y^2} - \frac{2\bar{h}}{k\,th}(T - T_\infty) = 0$

4.9 $S \approx 685$ m

4.10 (a) 137.3 W

4.11 (a) 9.49 W/m

4.12 12.95 kW

4.13 (a) 3075 W for 100 ft

4.14 (a) 0.422 W

4.16 (b) 134.9°C

4.17 (b) 21.7 lb$_m$/hr

4.19 (a) $S = \dfrac{2\pi L}{\ln\left(\dfrac{r_{out}}{r_{in}}\right)}$

4.20 (a) 1872 W

4.23 (a) *keff, x* = 23.83 W/m-K and *keff, y* = 4.15 W/m-K

4.29 (c) 12 rpm

4.31 $\left(\frac{\Delta x}{2} + \frac{\Delta y}{2}\right)\cdot\bar{h}\cdot\left(T_{inf} - T_5\right) + \frac{\Delta x}{2}\cdot k\cdot(T_1 \quad T_5) +$ $\Delta x\cdot k\cdot(T_9 - T_5) + \Delta y\cdot k\cdot(T_6 - T_5) = 0$

4.37 (b) 1032 W/m

4.38 (b) $7.98/season

4.40 151.8 W; 126.4 W without studs

4.41 36.6 W and 94.3 percent

4.42 55.4 W/m

4.43 (a) 0.643 R-ft2-hr/Btu (0.113 K-m2/W)

4.44 (b) 611.4 W/m

4.46 $\approx$65 W/m

4.47 322 W/m

4.48 (a) 730 W

4.49 496.5 W/m^2

4.50 132.4 W/m^2-K

Chapter 5

5.1 No

5.2 (b) 1.12 s

5.3 (b) 2217 s

5.4 (a) 2.1 W/m-K (*x* and *y*) and 5.9 W/m-K (*z*)

5.5 (c) 887.2 s

5.8 1.36 hr

5.9 24.3 W/m^2-K

5.10 (b) 11.6 s

5.11 (b) 16 s

5.12 32.5 s

5.13 200 W/m²-K

5.14 78.4 Btu/hr-ft²-R

5.15 500.8 s

5.16 (a) 152.1 m

5.17 12160 s

5.18 118.5 Btu/hr-ft²-R

5.19 (b) 150 s

5.20 149.6 < time < 322.9 s

5.22 (b) 120°C

5.23 (b) 260 s

5.25 (a) 10.6 cm

5.28 (b) 0.72 s

5.30 (b) 121.4 s

5.33 (a) 120°C

5.34 (a) 485.8°C

5.35 (a) 3.3 hr

5.36 (a) 22.4°C

5.38 (b) 613 s

5.42 (b) 0.6 s

5.43 (c) 468 s

5.47 3834 s

5.51 (c) 12.5 W/m²-K

5.57 (a) 50 K/W

5.58 (a) 21.2 W

5.59 (b) 47 μs

5.60 (a) 769.6 K (source) and 353.6 K (target)

5.61 (a) 0.516 W

Chapter 6

6.1 0.346 m

6.2 (a) 0.25 s

6.6 (a) 120°C

6.7 (b) 0.21°C

6.8 (a) 2.84 s

6.9 (a) 0.25 s

6.10 (c) 0.32 s

6.11 (b) 23 s

6.12 (a) 25 s

6.13 (a) 1000 s

6.14 (b) 87°C

6.15 (a) 6.1 s

6.16 A significant reduction is expected.

6.17 (a) 0.06 s

6.18 (b) ≈23,200 s

6.19 $\frac{\alpha}{r}\frac{\partial}{\partial r}\left(r\frac{\partial T}{\partial r}\right) = \frac{2\bar{h}}{\rho c\, th}(T_\infty - T) + \frac{\partial T}{\partial t}$

6.20 (a) 87°C (surface) and 99°C (center)

6.22 $\frac{\partial^2 T}{\partial x^2} - \frac{\bar{h}\, per}{k\, A_c}(T - T_\infty) = \frac{1}{\alpha}\frac{\partial T}{\partial t}$

6.23 $\frac{\partial^2 T}{\partial x^2} + \frac{g'''}{k}\exp(-\gamma x) = \frac{1}{\alpha}\frac{\partial T}{\partial t}$

6.28 (c) 1986 kJ

6.29 (a) 3.54 × 10⁵ m²/s

6.30 (b) 0.85 m

6.31 (b) 25.6 s

6.32 (a) 98.8°C

6.33 (a) 18,000 s

6.34 (a) 346 s

6.35 (a) 0.105 m

6.36 (a) 2579 s

6.37 (a) 625.8 s

6.38 (a) 48.3 s

6.39 (b) 8258 s

6.40 (a) 0.75 W/m-K

6.41 (a) 0.2355 m

6.42 (b) 2171 J

6.43 (a) 13.5 W/m²-K

6.44 (b) 0.06 s (diffusive) and 0.12 s (lumped)

6.45 (a) −7.5°C

6.46 20°C

6.47 (c) $\frac{dT_1}{dt} = \frac{8k}{\rho c L^2}(T_2 - T_1) + \frac{4h_g}{D\rho c}(T_g - T_1)$

6.48 (c) $k\cdot A\cdot\frac{T_{i-1}-T_i}{\Delta x} - k\cdot A\cdot\frac{T_i - T_{i+1}}{\Delta x}\,\rho\cdot c\cdot A\cdot\Delta x\cdot\frac{dT_i}{dt}$

6.52 (a) $\frac{d^2 T}{dx^2} + \frac{aT}{k} = -\frac{g'''}{k} + \frac{aT_o}{k}$

6.54 1.30 hr

6.56 (a) 465 s

6.57 (c) 640.6 W

6.59 (a) 106,666 W/m³

6.62 (d) 149.6 s (lumped) and 8.9 s (diffusive)

6.63 (c) 36 s (diffusive) and 24 s (lumped)

6.66 11,860 s (convection oven) and 8650 (oil bath)

6.69 (b) 1104 s

6.70 (b) 3.08 MJ/m

6.71 (d) $3548

6.72 (b) 0.15 m

6.75 (a) 1.9 mm

6.76 (a) 0.000075 s

Chapter 7

7.2 B will have a higher rate of heat transfer.

7.3 1.4

7.4 (a) 3817

7.6 (a) 12.91 W/m²-K

7.7 (a) 227.9 W/m²-K

7.8 (a) 0.0019

7.13 (a) 85.4 W

7.14 (c) 0°C, 44.7°C and 63.2°C

7.15 (c) 0.272 m

7.17 (a) 1 m
7.18 (a) 0.54 mm (momentum) and 0.31 mm (thermal)
7.19 (b) 1890 W/m^2-K
7.20 (c) 0.007 m
7.23 (b) 370 W/m^2-K
7.24 0.0033
7.25 $\frac{dp}{dx} = \mu \frac{d^2u}{dy^2}$
7.27 246 m/s
7.29 27.6 N/m
7.30 0.01 m/s
7.31 (c) 1090 N/m
7.32 125 N
7.34 (d) 456.5 W/m^2-K

Chapter 8

8.1 (a) 4.45 m/s
8.2 (a) 317 W
8.3 (a) 335.7 K and 8.1 W
8.4 (a) 47.3°C, 88.1°C, and 58.1 W
8.5 (a) 0.0183 m
8.6 (b) 177.7 W/m^2-K
8.7 (a) 3.077 m
8.8 (b) 0.6465 N
8.9 (b) 8.575 W/m^2-K (air) and 6600 W/m^2-K (water)
8.10 (a) 40.1°C
8.11 (a) 0.000297 N and 1008 W
8.12 (a) 47 W
8.13(a) $162.8 for unit #5
8.14 (a) 17.35 W/m^2-K
8.16 (a) 19.89 W/m^2-K
8.17 (a) 0.291 K/W
8.18 (a) 254.7 W/m^2-K and 0.394 N
8.20 (a) 50 m/s
8.22 323.5 W/m^2-K
8.23 143.8 W
8.25 (a) 0.52 m
8.26 2
8.29 (a) –0.52 K/s
8.29 (a) 112.4°F
8.30 (d) 7.0 s (external) and 3.0 s (internal)
8.31 (a) 39.8 W/m^2-K

Chapter 9

9.9 (b) 285.4 K
9.10 (a) 255.1 kPa
9.11 (a) 3714 W/m^2-K
9.12 (a) 70.14 W

9.13 (b) 60.9 W
9.14 (a) 47.57 W
9.15 (a) 0.34 liter/min
9.16 (a) 16.1 kg/s, 162676 Pa, and 50 percent
9.17 (a) 213 kPa
9.18 (b) 12290 W
9.20 (a) 275.8 kPa and 46.1°C
9.21 (a) Decrease.
9.22 (a) 4.1 liter/min and 0.062 kg/s
9.23 (a) 3097 W
9.25 61.9°C and 110 W
9.26 80.2°C
9.27 (a) 317.2 K
9.28 344.6 K
9.29 (b) 73.47°F
9.31 6894 Pa, 136 C and 11,831 W
9.32 (a) 22.4 ft
9.33 (a) 274.1 kWE
9.34 (a) 307.1 K
9.35 (a) 3389 W/m^2-K
9.36 (a) 5.5e5 Pa
9.37 (a) 4744 Pa
9.38 (a) 8.34 W/m^2-K
9.39 (a) 286.1 K
9.40 (a) 61.5 W/m^2-K
9.41 (a) 318.2 K
9.42 (b) 0.7 m (hydrodynamic and thermal)
9.43 (c) 150 W/m^2-K
9.44 (a) 20480 W/m^2-K
9.45 (a) 35.4 kW
9.46 (a) 3162 W/m^2-K
9.47 (c) 89.4°F
9.48 (a) 601 K
9.50 (a) 9049 W/m^2-K
9.51 (e) $2313

Chapter 10

10.2 (a) 0.000279 1/K (liquid) and 0.003333 1/K (vapor)
10.3 (a) 5.287e6
10.5 6979 W and 138.4°C
10.6 69.1 W
10.7 (a) 138 W
10.8 (a) 38.96 W
10.10 (a) 2790 W
10.11 (a) 78.2°C
10.12 (c) 380.8 W
10.13 (a) 6.192 W
10.14 26.93 W (radiation is important)
10.15 4.42 W/m^2-K

10.16 Horizontally
10.17 (a) 9.499 W/m^2-K
10.19 1
10.20 b) 3.934 W/m^2-K
10.21 (a) 59°F
10.25 (a) 179 W
10.27 (a) 120.7°C
10.28 (a) 137.7°C
10.29 (a) 8.6 W/m^2-K
10.30 (a) 6.94 W/m^2-K
10.32 (a) 86.6°C
10.33 (a) 94.83 W

Chapter 11

11.1 (a) 25380 W
11.3 10 K
11.5 283,914 W/m^2 and 43.7 K
11.6 16.4 min
11.7 (b) 2974 W/m^2-K
11.9 283.5 K (water) and 273.4 K (refrigerant)
11.10 (a) 21.96 kW
11.11 (a) 196.3 W
11.12 (a) 1.98 kg/s per tube
11.13 (a) 191 kW
11.15 0.358 kg/s
11.17 (b) Horizontal downward facing.
11.19 7594 W
11.20 (a) 317.8 K, 9776 Pa
11.22 0.005484 kg/s
11.24 (a) 21745 W/m^2-K
11.25 0.00256 kg/s
11.26 (a) 54.69 W/m^2-K
11.27 (a) 226.6 W/m
11.29 (a) 0.00010 kg/s
11.30 (c) 0.00341 kg/s
11.31 22.4 W/m

Chapter 12

12.1 41 W/K
12.2 (a) 5041 ft^2
12.3 (a) 1450 W/m^2-K
12.4 (a) 2.518 W/K
12.5 (b) 2116 W/K
12.6 (a) 2008 W/K
12.7 (b) 8 percent
12.8 (a) 45°C
12.10 (a) 92.97 Pa
12.11 (a) 209.2 Pa

12.13 (a) 19.91 K
12.14 568.7 Btu/hr-R
12.15 73.56 Btu/min
12.16 (d) 11.14 K
12.17 8.74°C
12.18 2.9 Btu/min-°F at 1 gpm
12.19 (a) 560.1 W/K
12.20 (b) 157.8 kW
12.21 (a) 0.53
12.22 (a) 0.449
12.23 (a) 6.898 ft
12.24 (e) 20.7 W/K
12.25 (a) NTU = 2, 315 W
12.26 (a) 0.929 kg/s
12.27 (a) 46.4 m
12.28 (a) 61.5 W/m^2-K
12.29 (a) 282.2 kg/hr
12.30 (b) 0.5
12.31 (a) 0.475
12.32 (b) 0.2332
12.33 (a) 130.54 kPa
12.34 (a) 0.4718
12.36 (a) 1.505e6 W
12.37 (a) 64.48 K
12.38 (b) 4.48 W and 54.8°C
12.39 (a) 67.9°C and 187.8 W
12.40 (a) 4650 W
12.41 (a) 549 W/m^2-K
12.42 (a) 9.06°C
12.43 (a) 40.2 W, 73.2°C, and 29.8 kW
12.45 (a) 92.9°C
12.46 (a) 0.702
12.49 (a) 204 W/m^2-K
12.50 (a) 0.003537 kg/s and 2.581 m^2
12.52 (a) 0.22
12.53 (b) 0.793
12.54 (b) 10 kW/K
12.55 (b) 57.6°C
12.56 (b) 0.522
12.57 (a) 2130 W/K
12.58 (a) 0.904 for the recuperator

Chapter 13

13.1 (a) 31.21 kg/kmol
13.2 (c) 0.000555
13.3 (a) 28.49 kg/kmol
13.4 (d) 0.202 kmol/m^3
13.5 (b) 0.0124
13.9 (a) 0.006463

13.10 0.00149 kg/s (diffusion) and 0.000898 kg/s
     (infiltration)

13.11 (b) 1260 day

13.12 (b) 41.3 m/s

13.13 (b) 3.5 hr

13.15 (a) 0.0002087 m^2/s

13.16 (a) 0.0000155 m^2/s

13.17 (a) 2.26e-9 kg/s

13.18 (a) 0.000004956 k/s-m^2

13.19 (b) 0.0142

13.20 (a) 0.06325

13.21 31800 s

13.22 58.739 kW

13.24 2.69 hr

13.25 4.4 percent

13.26 40 hr

13.27 (a) 89.29 W/m^2-K

13.28 (a) 6.901 kg/s

13.29 (b) 735 W

13.30 (a) 2.554 W

13.31 (b) 12.5°C

13.32 4213°C

Chapter 14

14.1 (a) 1305 W

14.2 (b) 13,000

14.3 (a) 0.00104

14.4 16

14.5 (a) 6.5 percent

14.6 (a) 0.100

14.7 (a) 3.8e26 W

14.9 (a) 43.9 percent

14.10 (a) 40 percent

14.11 13.35 lm/W

14.12 0.2056

14.13 0.6847

14.14 (a) 0.103

14.15 0.0031

14.16 (a) 20.46 W

14.18 (a) 954.2 W

14.19 (b) 856.7 W

14.20 (a) 34.2 W

14.21 (a) 0.618

14.22 2.083 W

14.23 0.66

14.24 838.5 W

14.25 6.771 W

14.26 (a) 281 W

14.27 0.172 and 0.828

14.28 (a) 1918 W

14.29 (a) 876.9 K

14.32 (a) 505 kW and 1040 K

14.33 (a) 0.7743

14.37 (a) 0.4536

14.38 0.308

14.40 (a) 0.097

14.41 1446 W

14.44 (a) 2.62 W

14.45 424.6 K

14.46 (b) 8.2 percent longer

14.47 a) 579 W (low emissivity) and 1730 W (high
    emissivity)

14.50 393.3 K

14.51 (a) 1/6

14.52 (a) 0.685

14.53 (c) 0.7575 W

14.54 1094 K (56 K error)

14.55 (c) 39.2 kW

14.56 (a) 1306 W

14.57 (b) 3.052 W

14.58 (b) 0.0454

14.59 (b) 0.88 W

14.60 275.1 K

14.61 (b) 2.439 W

14.62 434.3 K

14.63 (a) 0.936

14.64 (b) 7.5 m^{-2}

14.65 (a) 31.9 W

14.66 (b) 0.45

14.67 (a) 1146 K

14.68 (a) 0.86

14.69 (a) 266.1 K (night) and 388.9 K (day)

14.72 (a) 315.5 K

14.74 (b) 0.41 W

14.75 (a) 23.17 kW

14.76 70.8°C

14.79 (c) 10.8 mW

14.80 (b) 73 percent

14.81 (a) 936.7 K (surface 2) and 830.9 K (surface 3)

14.84 (a) 1221 K, 1229 K, 1221 K, and 859.7 K for
    surfaces 1,4

14.85 (b) 141.1°C

14.86 (a) 1306 W

Index